Atomic Weights of the Elements (Based on Carbon-12)

Name	Symbol	Atomic Number	Atomic Weight[a]	Name	Symbol	Atomic Number	Atomic Weight
Actinium	Ac	89	(227)	Mercury	Hg	80	200.6
Aluminum	Al	13	26.98	Molybdenum	Mo	42	95.94
Americium	Am	95	(243)	Meitnerium[b]	Mt	109	(266)
Antimony	Sb	51	121.8	Neodymium	Nd	60	144.2
Argon	Ar	18	39.95	Neon	Ne	10	20.18
Arsenic	As	33	74.92	Neptunium	Np	93	(237)
Astatine	At	85	(210)	Nickle	Ni	28	58.69
Barium	Ba	56	137.3	Niobium	Nb	41	92.91
Berkelium	Bk	97	(247)	Nitrogen	N	7	14.01
Beryllium	Be	4	9.012	Nobelium[b]	No	102	(259)
Blismuth	Bi	83	209.0	Osmium	Os	76	190.2
Bohrium[b]	Bh	107	(262)	Oxygen	O	8	16.00
Boron	B	5	10.81	Palladium	Pd	46	106.4
Bromine	Br	35	79.90	Phosphorus	P	15	30.97
Cadmium	Cd	48	112.4	Platinum	Pt	78	195.1
Calcium	Ca	20	40.08	Plutonium	Pu	94	(244)
Californium	Cf	98	(251)	Polonium	Po	84	(209)
Carbon	C	6	12.01	Potassium	K	19	39.10
Cerium	Ce	58	140.1	Praseodymium	Pr	59	140.9
Cesium	Cs	55	132.9	Promethium	Pm	61	(145)
Chlorine	Cl	17	35.45	Protactinium	Pa	91	(231)
Chromium	Cr	24	52.00	Radium	Ra	88	(226)
Cobalt	Co	27	58.93	Radon	Rn	86	(222)
Copper	Cu	29	63.55	Rhenium	Re	75	186.2
Curium	Cm	96	(247)	Rhodium	Rh	45	102.9
Dubnium[b]	Db	104	(261)	Rubidium	Rb	37	85.47
Dysprosium	Dy	66	162.5	Ruthenium	Ru	44	101.1
Einsteinium	Es	99	(252)	Rutherfordium[b]	Rf	106	(263)
Erbium	Er	68	167.3	Samarium	Sm	62	150.4
Europium	Eu	63	152.0	Scandium	Sc	21	44.96
Fermium	Fm	100	(257)	Selenium	Se	34	78.96
Fluorine	F	9	19.00	Silicon	Si	14	28.09
Francium	Fr	87	(223)	Silver	Ag	47	107.9
Gadolinium	Gd	64	157.3	Sodium	Na	11	22.99
Gallium	Ga	31	69.72	Strontium	Sr	38	87.62
Germanium	Ge	32	72.59	Sulfur	S	16	32.06
Gold	Au	79	197.0	Tantalum	Ta	73	180.9
Hafnium	Hf	72	178.5	Technetium	Tc	43	(98)
Hahnium[b]	Hn	108	(265)	Tellurium	Te	52	127.6
Helium	He	2	4.003	Terbium	Tb	65	158.9
Holmium	Ho	67	164.9	Thallium	Tl	81	204.4
Hydrogen	H	1	1.008	Thorium	Th	90	232.0
Indium	In	49	114.8	Thulium	Tm	69	168.9
Iodine	I	53	126.9	Tin	Sn	50	118.7
Iridium	Ir	77	192.2	Titanium	Ti	22	47.88
Iron	Fe	26	55.85	Tungsten	W	74	183.9
Joliotium[b]	Jl	105	(262)	Uranium	U	92	238.0
Krypton	Kr	36	83.80	Vanadium	V	23	50.94
Lanthanum	La	57	138.9	Xenon	Xe	54	131.3
Lawrencium[b]	Lr	103	(260)	Ytterbium	Yb	70	173.0
Lead	Pb	82	207.2	Yttrium	Y	39	88.91
Lithium	Li	3	6.941	Zinc	Zn	30	65.38
Lutetium	Lu	71	175.0	Zirconium	Zr	40	91.22
Magnesium	Mg	12	24.31	—[c]	—	110	(269)
Manganese	Mn	25	54.94	—[c]	—	111	(272)
Mendelevium[b]	Md	101	(258)				

[a] The atomic weights in parentheses are the mass numbers of the longest lived isotopes.
[b] Recommended element names by International Union of Pure & Applied Commission (IUPAC) on Nomenclature of Inorganic Chemistry.
[c] Element 110 was first produced Nov. 9, 1994 and Element 111 was first produced Dec. 8, 1994 both in Darmstadt, Germany.

Sixth Edition

Chemistry

An Introduction to General, Organic, and Biological Chemistry

Karen C. Timberlake
Los Angeles Valley College

HarperCollins*CollegePublishers*

This book is dedicated

to all my family who support my work and share my hopes and dreams, to my friend Susan who taught me to persevere in the joy of life, and to you, the student, and the realization of your endeavors.

The whole art of teaching is only the art of awakening the natural curiosity of young minds.—ANATOLE FRANCE

One must learn by doing the thing; though you think you know it, you have no certainty until you try.—SOPHOCLES

Discovery consists of seeing what everybody has seen and thinking what nobody has thought.—ALBERT SZENT-GYORGI

Executive Editor: Doug Humphrey
Development Editor: Cathleen E. Petree
Project Editor: Ginny Guerrant
Design Administration: Jess Schaal
Text and Cover Design: Ellen Pettengell
Front Cover Photo: Bruce Iverson PHOTOMICROGRAPHY, Portsmouth, NH
Art Development: Kelly Mountain
Photo Researcher: Karen Koblik
Production Administrator: Randee Wire
Compositor: Interactive Composition Corporation
Printer and Binder: R. R. Donnelley & Sons Company
Cover Printer: Phoenix Color Corporation

For permission to use copyrighted material, grateful acknowledgment is made to the copyright holders on page 705, which is hereby made part of this copyright page.

Front and back cover photo: Photomicrograph of sulphur crystals

Chemistry: An Introduction to General, Organic, and Biological Chemistry, Sixth Edition.

Library of Congress Cataloging-in-Publication Data

Timberlake, Karen C.
Chemistry : an introduction to general, organic, and biological chemistry / Karen C. Timberlake. -- 6th ed.
p. cm.
Includes index.
ISBN 0-673-99054-0
1. Chemistry. I. Title
QD31.2.T55 1996
540--dc20 95-17
CIP

95 96 97 98 99 9 8 7 6 5 4 3 2 1

A Visual Guide to the Book

Each chapter opens with an overview of the material called **Looking Ahead.** The **Learning Goals** show what concepts are covered in each text section.

Questions to Think About show the connections between chemistry and everyday life.

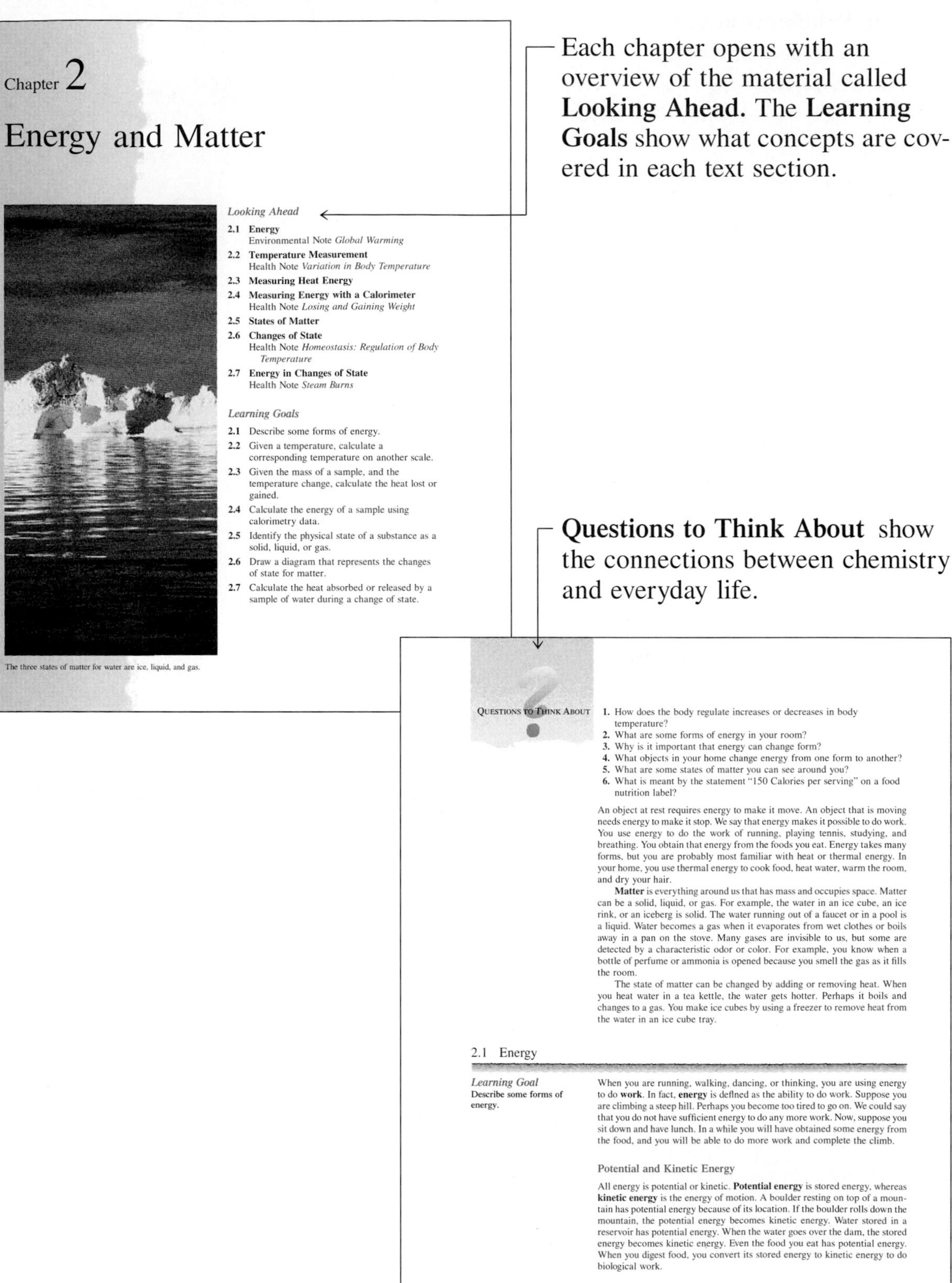

Chapter 2

Energy and Matter

Looking Ahead

2.1 Energy
Environmental Note *Global Warming*
2.2 Temperature Measurement
Health Note *Variation in Body Temperature*
2.3 Measuring Heat Energy
2.4 Measuring Energy with a Calorimeter
Health Note *Losing and Gaining Weight*
2.5 States of Matter
2.6 Changes of State
Health Note *Homeostasis: Regulation of Body Temperature*
2.7 Energy in Changes of State
Health Note *Steam Burns*

Learning Goals

2.1 Describe some forms of energy.
2.2 Given a temperature, calculate a corresponding temperature on another scale.
2.3 Given the mass of a sample, and the temperature change, calculate the heat lost or gained.
2.4 Calculate the energy of a sample using calorimetry data.
2.5 Identify the physical state of a substance as a solid, liquid, or gas.
2.6 Draw a diagram that represents the changes of state for matter.
2.7 Calculate the heat absorbed or released by a sample of water during a change of state.

The three states of matter for water are ice, liquid, and gas.

QUESTIONS TO THINK ABOUT

1. How does the body regulate increases or decreases in body temperature?
2. What are some forms of energy in your room?
3. Why is it important that energy can change form?
4. What objects in your home change energy from one form to another?
5. What are some states of matter you can see around you?
6. What is meant by the statement "150 Calories per serving" on a food nutrition label?

An object at rest requires energy to make it move. An object that is moving needs energy to make it stop. We say that energy makes it possible to do work. You use energy to do the work of running, playing tennis, studying, and breathing. You obtain that energy from the foods you eat. Energy takes many forms, but you are probably most familiar with heat or thermal energy. In your home, you use thermal energy to cook food, heat water, warm the room, and dry your hair.

Matter is everything around us that has mass and occupies space. Matter can be a solid, liquid, or gas. For example, the water in an ice cube, an ice rink, or an iceberg is solid. The water running out of a faucet or in a pool is a liquid. Water becomes a gas when it evaporates from wet clothes or boils away in a pan on the stove. Many gases are invisible to us, but some are detected by a characteristic odor or color. For example, you know when a bottle of perfume or ammonia is opened because you smell the gas as it fills the room.

The state of matter can be changed by adding or removing heat. When you heat water in a tea kettle, the water gets hotter. Perhaps it boils and changes to a gas. You make ice cubes by using a freezer to remove heat from the water in an ice cube tray.

2.1 Energy

Learning Goal
Describe some forms of energy.

When you are running, walking, dancing, or thinking, you are using energy to do **work**. In fact, **energy** is defined as the ability to do work. Suppose you are climbing a steep hill. Perhaps you become too tired to go on. We could say that you do not have sufficient energy to do any more work. Now, suppose you sit down and have lunch. In a while you will have obtained some energy from the food, and you will be able to do more work and complete the climb.

Potential and Kinetic Energy

All energy is potential or kinetic. **Potential energy** is stored energy, whereas **kinetic energy** is the energy of motion. A boulder resting on top of a mountain has potential energy because of its location. If the boulder rolls down the mountain, the potential energy becomes kinetic energy. Water stored in a reservoir has potential energy. When the water goes over the dam, the stored energy becomes kinetic energy. Even the food you eat has potential energy. When you digest food, you convert its stored energy to kinetic energy to do biological work.

39

Sample Problems appear throughout the book to reinforce chemical concepts and help students develop problem-solving skills and critical thinking. **Study Checks** below the problems challenge students to immediately apply the ideas they just learned. **Answers to Study Checks** appear at the end of the chapter.

1.5 Problem Solving Using Conversion Factors 21

Sample Problem 1.9
Problem Solving Using Conversion Factors

On a recent bicycle trip, Maria averaged 35 miles per day. How many days did it take her to cover 175 miles?

Solution

Step 1 Identify the given quantity in the question.

Given: 175 miles

Step 2 Write a unit plan.

Unit plan: miles → days Unit for answer

Step 3 Identify relationships needed and write conversion factor(s).

Relationship (from Problem): 35 miles = 1 day

Conversion Factors: $\frac{1\text{ day}}{35\text{ miles}}$ and $\frac{35\text{ miles}}{1\text{ day}}$

Step 4 Set up the problem starting with the given, cancel units, and carry out calculations. The answer is given with two significant figures.

Problem Setup:

Unit for answer goes here

$175\ \cancel{\text{miles}} \times \frac{1\text{ day}}{35\ \cancel{\text{miles}}} = 5.0\text{ days}$

3 SFs 2 SFs 2 SFs

(175 ÷ 35 = 5.0)

Suppose the problem were set up in the following way. If the units do not cancel, you know there is an error in the setup.

$175\text{ miles} \times \frac{35\text{ miles}}{\text{day}} = 6215\ \textbf{miles}^2\textbf{/day}$ Wrong units!

Study Check
A recipe for shark fin soup calls for 3.0 quarts of chicken broth. If 1 quart contains 4 cups, how many cups of broth are needed?

Problem Solving Using Metric Factors

The metric conversion factors from Table 1.6 are used to change from one metric unit to another, as seen in the following Sample Problem.

134 Chapter 4 Compounds and Their Bonds

Steps for Writing Electron-Dot Structures

Step 1 Determine the total number of valence electrons in all of the atoms.

Step 2 Place single bonds between each set of atoms. Each single bond uses two of the available valence electrons.

Step 3 Subtract the electrons used to bond the atoms, and arrange the remaining valence electrons to give each atom an octet. Hydrogen needs only a single bond, or two electrons.

Step 4 If octets of all the atoms cannot be completed using the remaining electrons, rearrange some of the electrons so that another pair or two are shared between two of the atoms as a double or triple bond. Check that all of the atoms have octets.

Sample Problem 4.16
Drawing Electron-Dot Structures Having Multiple Bonds

Draw the electron-dot structure of the compound CO_2. (Carbon is the central atom.)

O C O

Solution
Oxygen has six valence electrons, and carbon has four. Each oxygen atom needs two electrons, and the carbon atom needs four electrons.

Step 1 Using group numbers, calculate the total number of valence electrons available:

O C O

$6\,e^- + 4\,e^- + 6\,e^- = 16$ valence electrons

Step 2 Connect the atoms by single bonds:

O:C:O

Uses $2\,e^-$ Uses $2\,e^-$

Step 3 Arrange the remaining valence electrons to satisfy octets:

:O:C:O:

Not octets

Step 4 Because octets cannot be completed using the remaining 12 electrons, double bonds are needed instead of single bonds between atoms. Rearrange the electrons, placing them between atoms to form double bonds to give octets to all the atoms.

Rearranging electron dots Octets

:O: C :O: → :O::C::O:

Study Check
Determine the number of valence electrons and arrange them in the electron-dot structure of the HCN molecule.

Art and text complement the text discussion by giving both a written description and a visual interpretation of concepts like calculating nutritional content or balancing chemical equations.

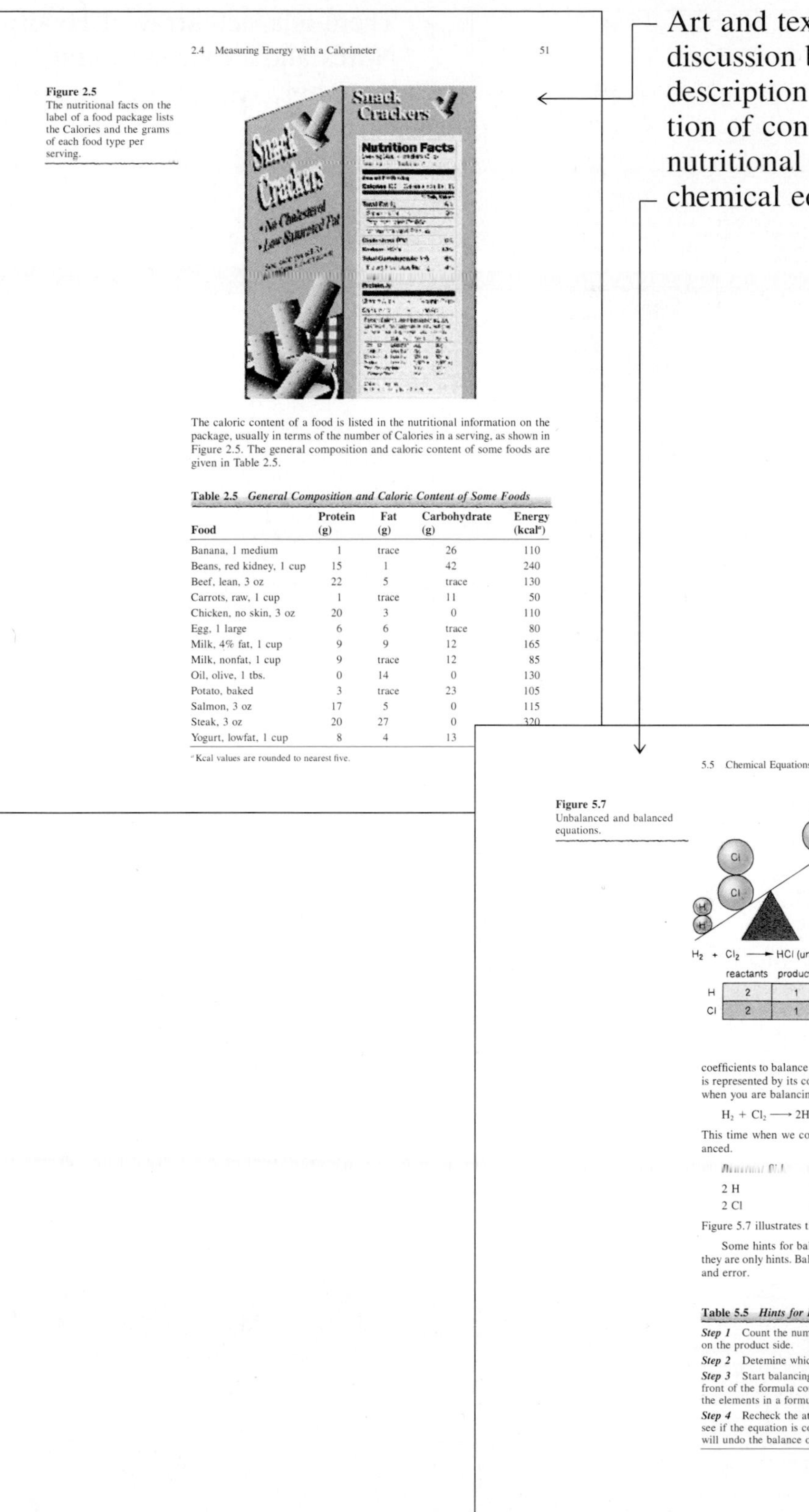

2.4 Measuring Energy with a Calorimeter 51

Figure 2.5
The nutritional facts on the label of a food package lists the Calories and the grams of each food type per serving.

The caloric content of a food is listed in the nutritional information on the package, usually in terms of the number of Calories in a serving, as shown in Figure 2.5. The general composition and caloric content of some foods are given in Table 2.5.

Table 2.5 ***General Composition and Caloric Content of Some Foods***

Food	Protein (g)	Fat (g)	Carbohydrate (g)	Energy (kcal[a])
Banana, 1 medium	1	trace	26	110
Beans, red kidney, 1 cup	15	1	42	240
Beef, lean, 3 oz	22	5	trace	130
Carrots, raw, 1 cup	1	trace	11	50
Chicken, no skin, 3 oz	20	3	0	110
Egg, 1 large	6	6	trace	80
Milk, 4% fat, 1 cup	9	9	12	165
Milk, nonfat, 1 cup	9	trace	12	85
Oil, olive, 1 tbs.	0	14	0	130
Potato, baked	3	trace	23	105
Salmon, 3 oz	17	5	0	115
Steak, 3 oz	20	27	0	320
Yogurt, lowfat, 1 cup	8	4	13	

[a] Kcal values are rounded to nearest five.

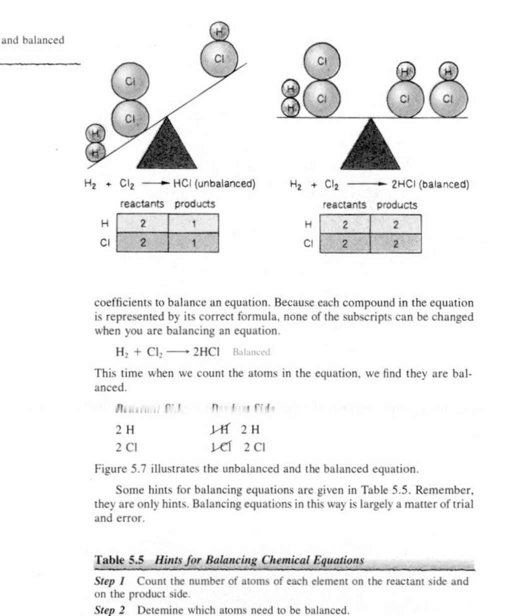

5.5 Chemical Equations 163

Figure 5.7
Unbalanced and balanced equations.

coefficients to balance an equation. Because each compound in the equation is represented by its correct formula, none of the subscripts can be changed when you are balancing an equation.

$H_2 + Cl_2 \longrightarrow 2HCl$ Balanced

This time when we count the atoms in the equation, we find they are balanced.

2 H	~~1 H~~ 2 H
2 Cl	~~1 Cl~~ 2 Cl

Figure 5.7 illustrates the unbalanced and the balanced equation.

Some hints for balancing equations are given in Table 5.5. Remember, they are only hints. Balancing equations in this way is largely a matter of trial and error.

Table 5.5 ***Hints for Balancing Chemical Equations***

Step 1 Count the number of atoms of each element on the reactant side and on the product side.

Step 2 Detemine which atoms need to be balanced.

Step 3 Start balancing, one element at a time, by placing coefficients in front of the formula containing that element. A typical starting place is with the elements in a formula with subscripts.

Step 4 Recheck the atoms on the reactant side and on the product side to see if the equation is completely balanced. Sometimes, balancing one element will undo the balance of another. If this happens, repeat the process.

There is a rich array of **Health Notes** and **Environmental Notes** that apply chemical topics to real issues.

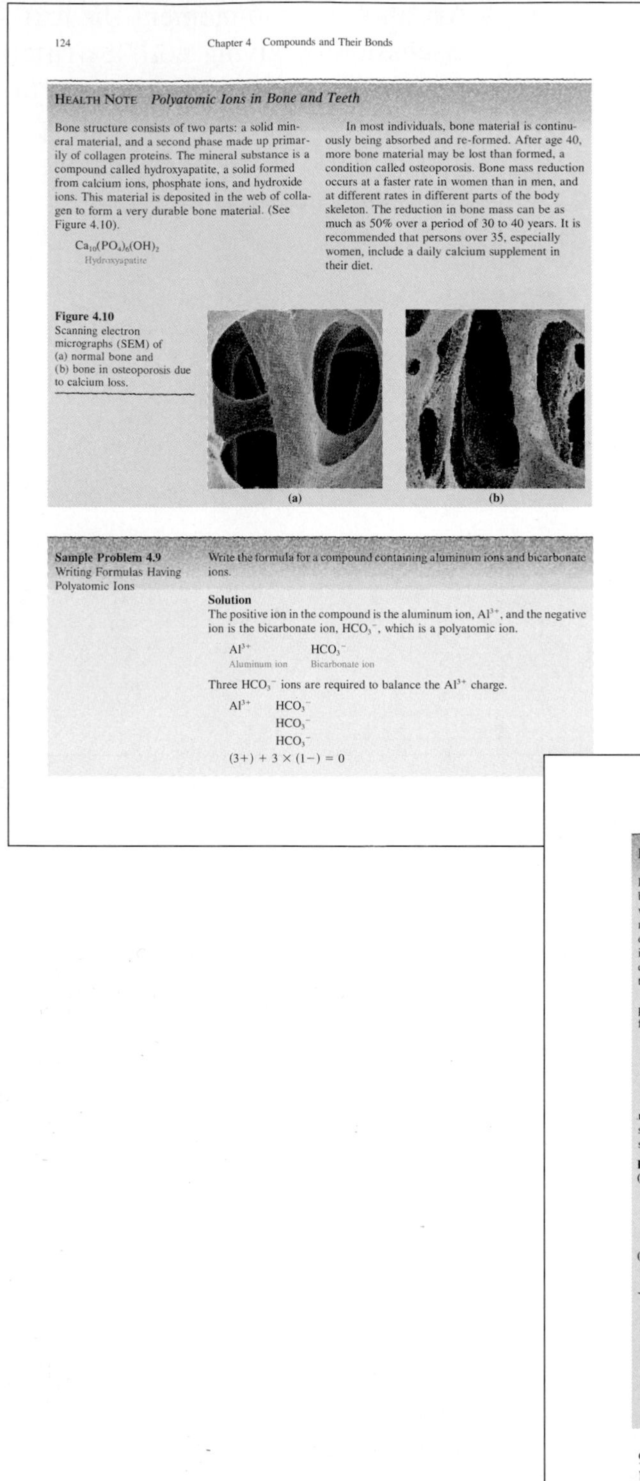

124 Chapter 4 Compounds and Their Bonds

HEALTH NOTE *Polyatomic Ions in Bone and Teeth*

Bone structure consists of two parts: a solid mineral material, and a second phase made up primarily of collagen proteins. The mineral substance is a compound called hydroxyapatite, a solid formed from calcium ions, phosphate ions, and hydroxide ions. This material is deposited in the web of collagen to form a very durable bone material. (See Figure 4.10).

$Ca_{10}(PO_4)_6(OH)_2$
Hydroxyapatite

In most individuals, bone material is continuously being absorbed and re-formed. After age 40, more bone material may be lost than formed, a condition called osteoporosis. Bone mass reduction occurs at a faster rate in women than in men, and at different rates in different parts of the body skeleton. The reduction in bone mass can be as much as 50% over a period of 30 to 40 years. It is recommended that persons over 35, especially women, include a daily calcium supplement in their diet.

Figure 4.10
Scanning electron micrographs (SEM) of (a) normal bone and (b) bone in osteoporosis due to calcium loss.

(a) (b)

Sample Problem 4.9
Writing Formulas Having Polyatomic Ions

Write the formula for a compound containing aluminum ions and bicarbonate ions.

Solution
The positive ion in the compound is the aluminum ion, Al^{3+}, and the negative ion is the bicarbonate ion, HCO_3^-, which is a polyatomic ion.

Al^{3+} HCO_3^-
Aluminum ion Bicarbonate ion

Three HCO_3^- ions are required to balance the Al^{3+} charge.

Al^{3+} HCO_3^-
HCO_3^-
HCO_3^-
$(3+) + 3 \times (1-) = 0$

9.8 Buffers 309

ENVIRONMENTAL NOTE *Acid Rain*

Rain typically has a pH of 6.2. It is slightly acidic because the carbon dioxide in the air combines with water to form carbonic acid. However, in many parts of the world, rain has become considerably more acidic, with pH values as low as 3 being reported. One cause of acid rain is the sulfur dioxide (SO_2) gas produced when coal that contains sulfur is burned.

In the air, the SO_2 gas reacts with oxygen to produce SO_3, which then combines with water to form sulfuric acid, H_2SO_4, a strong acid.

$$S + O_2 \longrightarrow SO_2$$
$$2SO_2 + O_2 \longrightarrow 2SO_3$$
$$SO_3 + H_2O \longrightarrow H_2SO_4$$

In parts of the United States, acid rain has made lakes so acidic they are no longer able to support fish and plant life. Limestone ($CaCO_3$) is sometimes added to these lakes to neutralize the acid. In Eastern Europe, acid rain has brought about an environmental disaster. Nearly 40% of the forests in Poland have been severely damaged, and some parts of the land are so acidic that crops will not grow. Throughout Europe, monuments made of marble (a form of $CaCO_3$) are deteriorating as acid rain dissolves the marble. See Figure 9.10.

$$2H^+ + CaCO_3 \longrightarrow Ca^{2+} + H_2O + CO_2$$

Efforts to slow or stop the damaging effects of acid rain include the reduction of sulfur emissions. This will require installation of expensive equipment in coal-burning plants to absorb more of the SO_2 gases before they are emitted. In some outdated plants, this may be impossible, and they will need to be closed. It is a difficult problem for engineers and scientists, but one that must be solved.

Figure 9.10
(a) Detail of the marble that is part of the Washington Square Arch in Washington Square Park completed on July 17, 1935.
(b) The destructive effect of acid rain on the same marble, June, 1994.

(a) (b)

9.8 Buffers

Learning Goal
Describe the role of buffers in maintaining the pH of a solution.

When an acid or base is added to pure water, the pH changes drastically. However, if a solution is buffered, there is little change in pH when small amounts of acid or base are added. A **buffer solution** resists a change in pH when small amounts of acid or base are added. For example, blood is a buffer

Chapter Summaries emphasize the main ideas in each section of the chapter. Pertinent chapters include summaries of reactions.

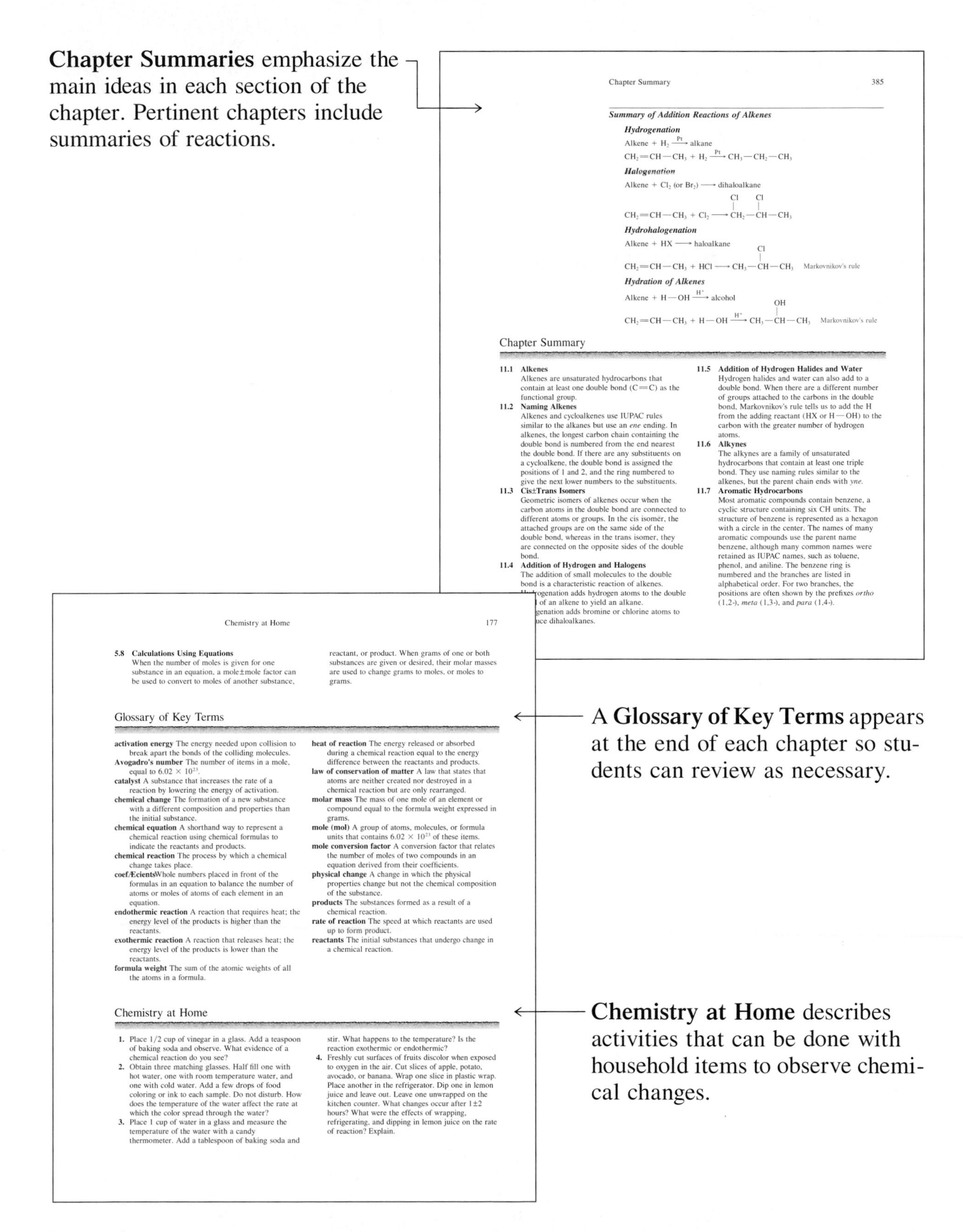

Chapter Summary 385

Summary of Addition Reactions of Alkenes

Hydrogenation

Alkene + $H_2 \xrightarrow{Pt}$ alkane

$CH_2{=}CH{-}CH_3 + H_2 \xrightarrow{Pt} CH_3{-}CH_2{-}CH_3$

Halogenation

Alkene + Cl_2 (or Br_2) $\longrightarrow$ dihaloalkane

$CH_2{=}CH{-}CH_3 + Cl_2 \longrightarrow CH_2(Cl){-}CH(Cl){-}CH_3$

Hydrohalogenation

Alkene + HX $\longrightarrow$ haloalkane

$CH_2{=}CH{-}CH_3 + HCl \longrightarrow CH_3{-}CH(Cl){-}CH_3$ Markovnikov's rule

Hydration of Alkenes

Alkene + $H{-}OH \xrightarrow{H^+}$ alcohol

$CH_2{=}CH{-}CH_3 + H{-}OH \xrightarrow{H^+} CH_3{-}CH(OH){-}CH_3$ Markovnikov's rule

Chapter Summary

11.1 Alkenes
Alkenes are unsaturated hydrocarbons that contain at least one double bond (C=C) as the functional group.

11.2 Naming Alkenes
Alkenes and cycloalkenes use IUPAC rules similar to the alkanes but use an *ene* ending. In alkenes, the longest carbon chain containing the double bond is numbered from the end nearest the double bond. If there are any substituents on a cycloalkene, the double bond is assigned the positions of 1 and 2, and the ring numbered to give the next lower numbers to the substituents.

11.3 Cis±Trans Isomers
Geometric isomers of alkenes occur when the carbon atoms in the double bond are connected to different atoms or groups. In the cis isomer, the attached groups are on the same side of the double bond, whereas in the trans isomer, they are connected on the opposite sides of the double bond.

11.4 Addition of Hydrogen and Halogens
The addition of small molecules to the double bond is a characteristic reaction of alkenes. ...rogenation adds hydrogen atoms to the double ...l of an alkene to yield an alkane. ...genation adds bromine or chlorine atoms to ...uce dihaloalkanes.

11.5 Addition of Hydrogen Halides and Water
Hydrogen halides and water can also add to a double bond. When there are a different number of groups attached to the carbons in the double bond, Markovnikov's rule tells us to add the H from the adding reactant (HX or H—OH) to the carbon with the greater number of hydrogen atoms.

11.6 Alkynes
The alkynes are a family of unsaturated hydrocarbons that contain at least one triple bond. They use naming rules similar to the alkenes, but the parent chain ends with *yne*.

11.7 Aromatic Hydrocarbons
Most aromatic compounds contain benzene, a cyclic structure containing six CH units. The structure of benzene is represented as a hexagon with a circle in the center. The names of many aromatic compounds use the parent name benzene, although many common names were retained as IUPAC names, such as toluene, phenol, and aniline. The benzene ring is numbered and the branches are listed in alphabetical order. For two branches, the positions are often shown by the prefixes *ortho* (1,2-), *meta* (1,3-), and *para* (1,4-).

Chemistry at Home 177

5.8 Calculations Using Equations
When the number of moles is given for one substance in an equation, a mole±mole factor can be used to convert to moles of another substance, reactant, or product. When grams of one or both substances are given or desired, their molar masses are used to change grams to moles, or moles to grams.

Glossary of Key Terms

activation energy The energy needed upon collision to break apart the bonds of the colliding molecules.
Avogadro's number The number of items in a mole, equal to 6.02×10^{23}.
catalyst A substance that increases the rate of a reaction by lowering the energy of activation.
chemical change The formation of a new substance with a different composition and properties than the initial substance.
chemical equation A shorthand way to represent a chemical reaction using chemical formulas to indicate the reactants and products.
chemical reaction The process by which a chemical change takes place.
coefÆcientsWhole numbers placed in front of the formulas in an equation to balance the number of atoms or moles of atoms of each element in an equation.
endothermic reaction A reaction that requires heat; the energy level of the products is higher than the reactants.
exothermic reaction A reaction that releases heat; the energy level of the products is lower than the reactants.
formula weight The sum of the atomic weights of all the atoms in a formula.
heat of reaction The energy released or absorbed during a chemical reaction equal to the energy difference between the reactants and products.
law of conservation of matter A law that states that atoms are neither created nor destroyed in a chemical reaction but are only rearranged.
molar mass The mass of one mole of an element or compound equal to the formula weight expressed in grams.
mole (mol) A group of atoms, molecules, or formula units that contains 6.02×10^{23} of these items.
mole conversion factor A conversion factor that relates the number of moles of two compounds in an equation derived from their coefficients.
physical change A change in which the physical properties change but not the chemical composition of the substance.
products The substances formed as a result of a chemical reaction.
rate of reaction The speed at which reactants are used up to form product.
reactants The initial substances that undergo change in a chemical reaction.

Chemistry at Home

1. Place 1/2 cup of vinegar in a glass. Add a teaspoon of baking soda and observe. What evidence of a chemical reaction do you see?
2. Obtain three matching glasses. Half fill one with hot water, one with room temperature water, and one with cold water. Add a few drops of food coloring or ink to each sample. Do not disturb. How does the temperature of the water affect the rate at which the color spread through the water?
3. Place 1 cup of water in a glass and measure the temperature of the water with a candy thermometer. Add a tablespoon of baking soda and stir. What happens to the temperature? Is the reaction exothermic or endothermic?
4. Freshly cut surfaces of fruits discolor when exposed to oxygen in the air. Cut slices of apple, potato, avocado, or banana. Wrap one slice in plastic wrap. Place another in the refrigerator. Dip one in lemon juice and leave out. Leave one unwrapped on the kitchen counter. What changes occur after 1±2 hours? What were the effects of wrapping, refrigerating, and dipping in lemon juice on the rate of reaction? Explain.

A **Glossary of Key Terms** appears at the end of each chapter so students can review as necessary.

Chemistry at Home describes activities that can be done with household items to observe chemical changes.

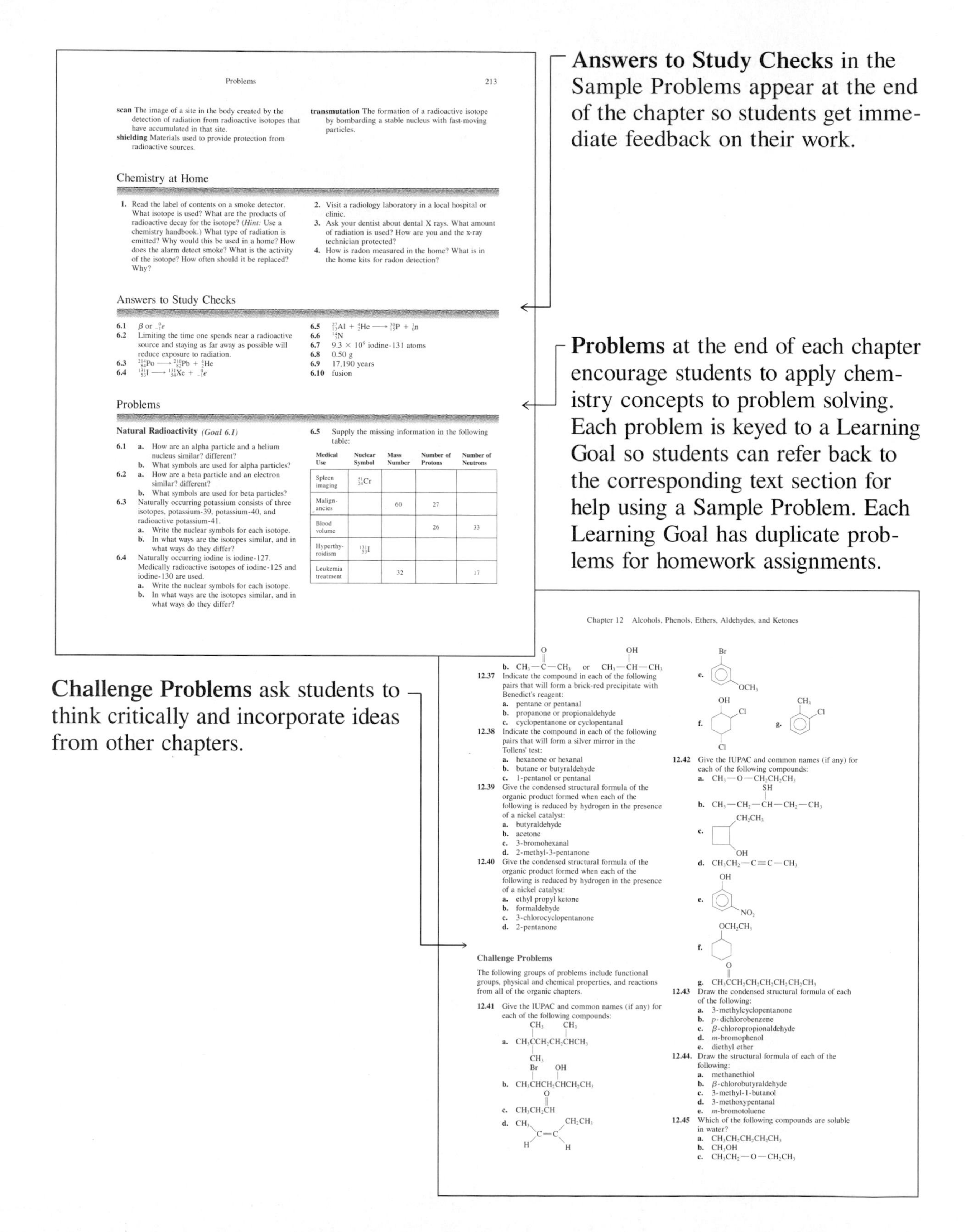

Answers to Study Checks in the Sample Problems appear at the end of the chapter so students get immediate feedback on their work.

Problems at the end of each chapter encourage students to apply chemistry concepts to problem solving. Each problem is keyed to a Learning Goal so students can refer back to the corresponding text section for help using a Sample Problem. Each Learning Goal has duplicate problems for homework assignments.

Challenge Problems ask students to think critically and incorporate ideas from other chapters.

Problems 213

scan The image of a site in the body created by the detection of radiation from radioactive isotopes that have accumulated in that site.

shielding Materials used to provide protection from radioactive sources.

transmutation The formation of a radioactive isotope by bombarding a stable nucleus with fast-moving particles.

Chemistry at Home

1. Read the label of contents on a smoke detector. What isotope is used? What are the products of radioactive decay for the isotope? (*Hint:* Use a chemistry handbook.) What type of radiation is emitted? Why would this be used in a home? How does the alarm detect smoke? What is the activity of the isotope? How often should it be replaced? Why?
2. Visit a radiology laboratory in a local hospital or clinic.
3. Ask your dentist about dental X rays. What amount of radiation is used? How are you and the x-ray technician protected?
4. How is radon measured in the home? What is in the home kits for radon detection?

Answers to Study Checks

6.1 β or $^{0}_{-1}e$
6.2 Limiting the time one spends near a radioactive source and staying as far away as possible will reduce exposure to radiation.
6.3 $^{214}_{84}Po \longrightarrow ^{210}_{82}Pb + ^{4}_{2}He$
6.4 $^{131}_{53}I \longrightarrow ^{131}_{54}Xe + ^{0}_{-1}e$
6.5 $^{27}_{13}Al + ^{4}_{2}He \longrightarrow ^{30}_{15}P + ^{1}_{0}n$
6.6 $^{14}_{7}N$
6.7 9.3×10^{9} iodine-131 atoms
6.8 0.50 g
6.9 17,190 years
6.10 fusion

Problems

Natural Radioactivity *(Goal 6.1)*

6.1 a. How are an alpha particle and a helium nucleus similar? different?
b. What symbols are used for alpha particles?
6.2 a. How are a beta particle and an electron similar? different?
b. What symbols are used for beta particles?
6.3 Naturally occurring potassium consists of three isotopes, potassium-39, potassium-40, and radioactive potassium-41.
a. Write the nuclear symbols for each isotope.
b. In what ways are the isotopes similar, and in what ways do they differ?
6.4 Naturally occurring iodine is iodine-127. Medically radioactive isotopes of iodine-125 and iodine-130 are used.
a. Write the nuclear symbols for each isotope.
b. In what ways are the isotopes similar, and in what ways do they differ?
6.5 Supply the missing information in the following table:

Medical Use	Nuclear Symbol	Mass Number	Number of Protons	Number of Neutrons
Spleen imaging	$^{51}_{24}Cr$			
Malignancies		60	27	
Blood volume			26	33
Hyperthyroidism	$^{131}_{53}I$			
Leukemia treatment		32		17

Chapter 12 Alcohols, Phenols, Ethers, Aldehydes, and Ketones

b. $CH_3-C(=O)-CH_3$ or $CH_3-CH(OH)-CH_3$
12.37 Indicate the compound in each of the following pairs that will form a brick-red precipitate with Benedict's reagent:
a. pentane or pentanal
b. propanone or propionaldehyde
c. cyclopentanone or cyclopentanal
12.38 Indicate the compound in each of the following pairs that will form a silver mirror in the Tollens' test:
a. hexanone or hexanal
b. butane or butyraldehyde
c. 1-pentanol or pentanal
12.39 Give the condensed structural formula of the organic product formed when each of the following is reduced by hydrogen in the presence of a nickel catalyst:
a. butyraldehyde
b. acetone
c. 3-bromohexanal
d. 2-methyl-3-pentanone
12.40 Give the condensed structural formula of the organic product formed when each of the following is reduced by hydrogen in the presence of a nickel catalyst:
a. ethyl propyl ketone
b. formaldehyde
c. 3-chlorocyclopentanone
d. 2-pentanone

Challenge Problems

The following groups of problems include functional groups, physical and chemical properties, and reactions from all of the organic chapters.

12.41 Give the IUPAC and common names (if any) for each of the following compounds:
a. $CH_3C(CH_3)_2CH_2CH(CH_3)CH_3$
b. $CH_3CH(Br)CH_2CH(OH)CH_2CH_3$
c. $CH_3CH_2CH(=O)$
d. $CH_3(H)C=C(H)CH_2CH_3$
e. Br, OCH_3 (benzene ring)
f. OH, Cl, Cl (cyclohexane ring)
g. CH_3, Cl (benzene ring)
12.42 Give the IUPAC and common names (if any) for each of the following compounds:
a. $CH_3-O-CH_2CH_2CH_3$
b. $CH_3-CH_2-CH(SH)-CH_2-CH_3$
c. CH_2CH_3, OH (cyclobutane ring)
d. $CH_3CH_2-C\equiv C-CH_3$
e. OH, NO_2 (benzene ring)
f. OCH_2CH_3 (cyclohexane ring)
g. $CH_3C(=O)CH_2CH_2CH_2CH_2CH_2CH_3$
12.43 Draw the condensed structural formula of each of the following:
a. 3-methylcyclopentanone
b. *p*-dichlorobenzene
c. β-chloropropionaldehyde
d. *m*-bromophenol
e. diethyl ether
12.44. Draw the structural formula of each of the following:
a. methanethiol
b. β-chlorobutyraldehyde
c. 3-methyl-1-butanol
d. 3-methoxypentanal
e. *m*-bromotoluene
12.45 Which of the following compounds are soluble in water?
a. $CH_3CH_2CH_2CH_2CH_3$
b. CH_3OH
c. $CH_3CH_2-O-CH_2CH_3$

Brief Contents

Chapter

Appendices

Contents

3 Atoms and Elements 72

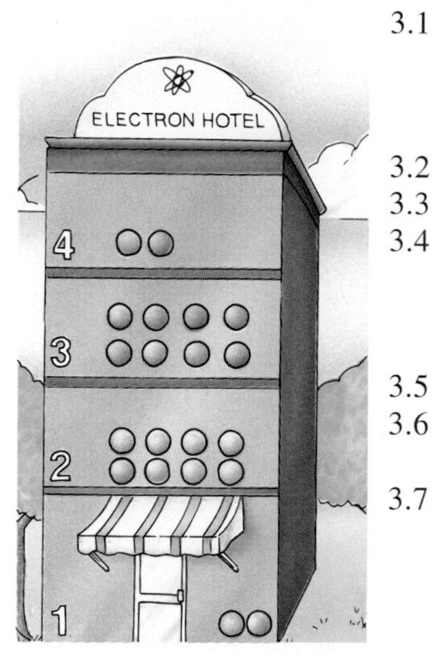

4 Compounds and Their Bonds 107

5 Chemical Quantities and Reactions 148

6 Nuclear Radiation 183

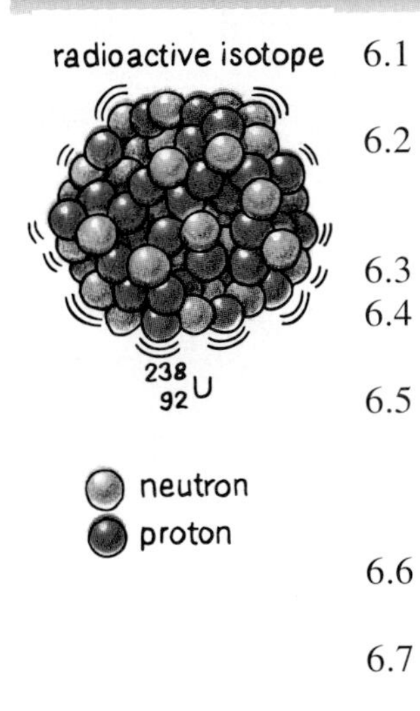

7 Gases 217

8 Solutions 250

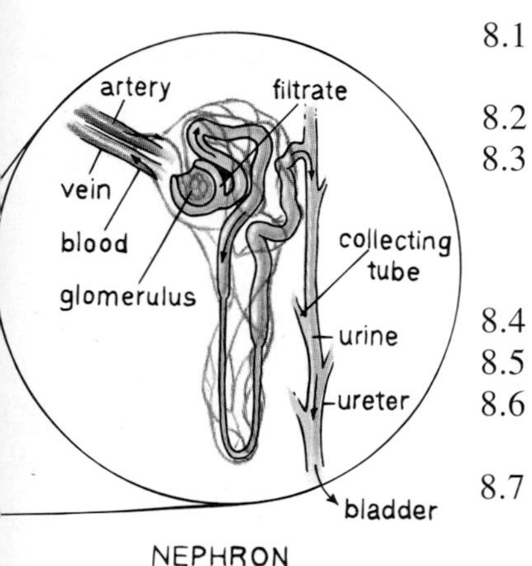

11 Alkenes, Alkynes, and Aromatic Hydrocarbons 359

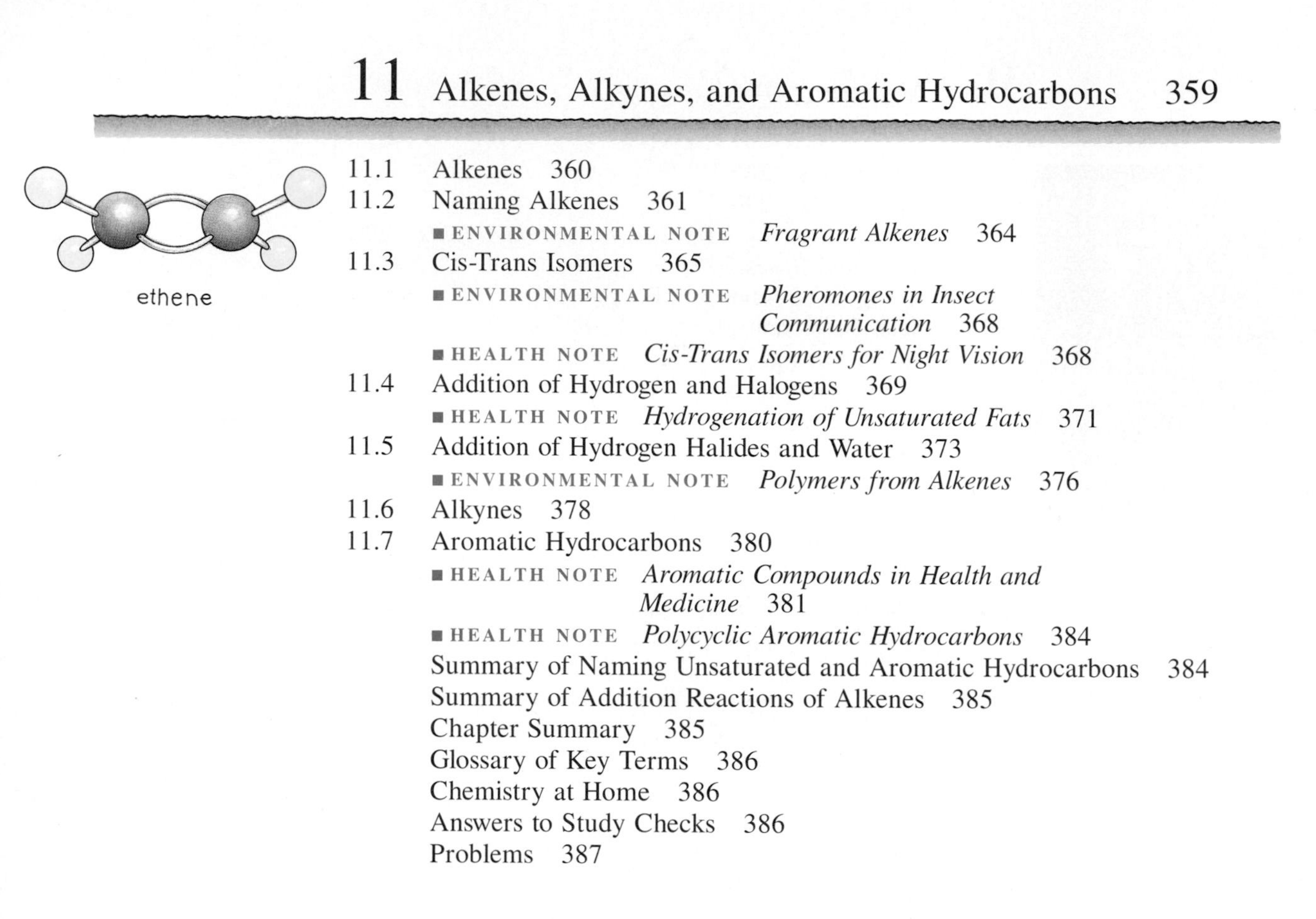

12 Alcohols, Phenols, Ethers, Aldehydes, and Ketones 391

13 Carboxylic Acids, Esters, Amines, and Amides 431

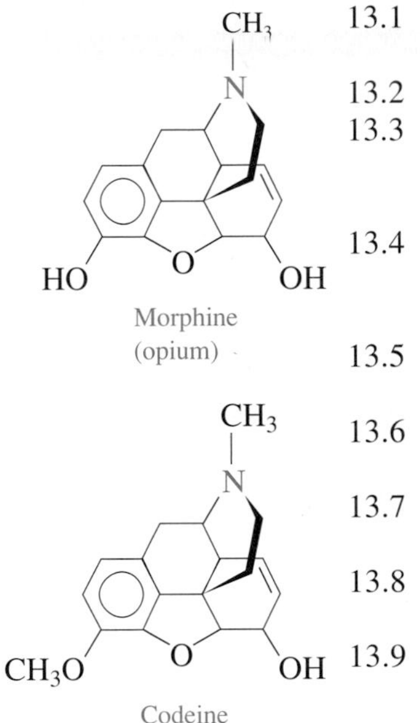

Morphine (opium)

Codeine

14 Carbodydrates 474

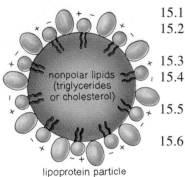

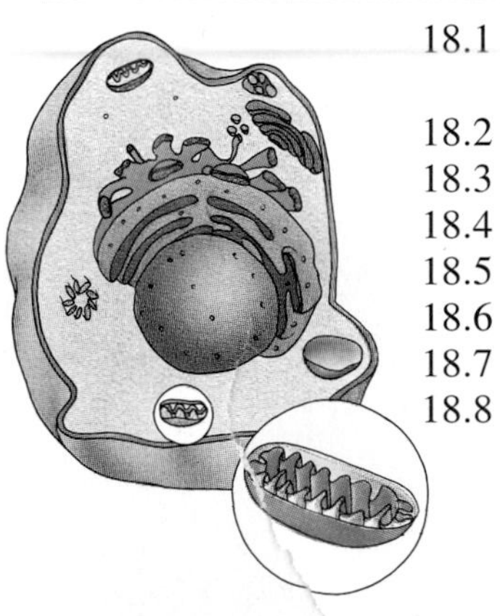

Appendices

Preface

Welcome to the Sixth Edition of *Chemistry: An Introduction to General, Organic, and Biological Chemistry*. It is my hope that the reshaping of this text over six editions has resulted in a text that makes teaching and learning chemistry an enthusiastic and positive experience for both the instructor and student. It remains my goal to assist students in their development of critical thinking, understanding of scientific concepts, and problem-solving techniques. These skills will be needed to make decisions about major issues in our lives concerning our environment, medicine, and health.

This textbook is written for students with no previous background in chemistry who are preparing for health-related careers or wish to take a preliminary course in chemistry. Over the years I have found that chemistry is less formidable when concepts are associated with applications to the health and environmental sciences. Connections made between chemistry and the real world help students recognize how chemistry impacts their lives and future careers. I have found discussions of applications of chemistry valuable in increasing student interest, motivation, concentration, and performance in class.

New to This Edition

In response to the needs of students and suggestions of instructors and reviewers, many improvements have been made in this Sixth Edition. Each chapter begins with a new preview section that contains *Learning Goals, Questions To Think About,* and *Looking Ahead.* This preview section gives students an overview of the content of the chapter and ways that the chemical concepts impact their lives. Throughout each chapter, a large number of sample problems, all with complete solutions and study checks, challenge the student to participate actively in problem solving. In each chapter, the *Health Notes* and *Environmental Notes* connect the chemical principles to real-life issues of health, medicine, and the environment.

New in this edition are the *Chapter Summaries* to review and emphasize the major topics in each chapter section. Also new in this edition is *Chemistry At Home* to challenge and excite students about the chemistry that takes place in the common materials around them. All of the problem sets have been rewritten with duplicate exercises to give more problems that engage the student in problem solving with a greater range of difficulty. In this Sixth Edition, the answers to the odd-number chapter problems are given in an appendix to give instructors the option of quiz, group, or homework assignments. Also new are the *Challenge Problems* that require students to use critical thinking that is more complex and related to concepts in other chapters.

In this text, more difficult topics that may be time consuming are placed at the end of chapters. Depending on the time limitations, these topics can be covered or omitted without affecting the flow of learning from one chapter to another.

Features of the Learning Program

This textbook has been designed with many instructional features to enhance your students' learning.

1. **Looking Ahead:** Each chapter begins with an outline of topics, Health Notes, and Environmental Notes.
2. **Learning Goals:** Each chapter contains a set of learning goals that specify the concepts to be covered. This gives students an overview of the material in each chapter and prepares them for what they are expected to accomplish.
3. **Questions to Think About:** These are intended to stimulate students to think about ways that chemistry impacts their daily lives. The preview questions enhance learning by relating students experiences with the chemical concepts and topics to be presented in the chapter. The preview questions stimulate critical thinking early, rather than waiting until the end of the chapter.
4. **Sample Problems:** Sample problems with complete solutions throughout the text reinforce the concepts in each chapter. This format continually involves students with the material through active participation in the patterns of problem solving. A completely worked-out solution and explanation in each sample provides a model of the problem solving and calculations that students must do to answer the question successfully.
5. **Study Checks and Answers to Study Checks:** With every sample problem, there is a *Study Check* that challenges students to apply the information and methods in a similar type of problem, but without the solution. Students can interact with the subject matter by immediately applying the problem solving in the sample problem. Immediate feedback is available by looking in the *Answers to Study Checks* at the end of the chapter.
6. **Health Notes and Environmental Notes:** In every chapter, boxed notes relate and apply chemical principles to real issues and concerns that students may have about health, medicine, and the environment. Clinical examples interest and motivate students interested in careers in the health sciences. All of the Health Notes and Environmental Notes are listed in the table of contents and include discussions about global warming, radon in our homes, loss and gain of weight, ozone depletion, acid rain, cholesterol, anabolic steroids, pheromones, hyperbaric chambers, kidney stones, alcohol, recycling of plastics,

alkaloids, blood types, genetic diseases, ketone bodies, recombinant DNA, viruses, AIDS, and cancer. Many of the notes are updated with current information.

7. **Chapter Summaries:** At the end of each chapter, a chapter summary reviews major topics and emphasizes the main ideas in each section.
8. **Glossary of Key Terms:** A glossary of key terms at the end of each chapter defines terms in the chapter so that you can review new vocabulary.
9. **Chemistry at Home:** These projects can be done at home to apply what students have learned or can also be used for extra credit work. The extension of chemistry into real life can stimulate thinking and increase your students' interest in science.
10. **End-of-Chapter Problems:** There are a large number of problems to work at the end of each chapter. Problems are keyed to the section heads and learning goals, allowing instructors to select appropriate problems for any section. A full range of problems are now structured into duplicate sets, with each problem followed by a similar type of problem (duplicate problem). Students may compare their answers to the odd-number problems with the answers given in the appendix. The duplicate problems can be used by instructors for classroom quizzes, homework, group problem solving, or other problem-solving exercises where answers are not available. Each problem set concludes with a group of *Challenge Problems* that engage students in solving problems with a greater range of difficulty. These problems are more complex mathematically, require more critical thinking and material from other chapters, and have real-life relationships.
11. **Appendices:** Several appendices provide additional reference materials for your students' learning needs. These appendices review common conversion factors, scientific notations, using percent in calculations, and some basic operations with the calculator including changing signs and recognizing scientific notation. The answers to the odd-number problems are also in the appendices.
12. **Art Program:** Throughout the text full color is used to highlight major ideas. Photographs and new graphics are used to increase understanding and enhance learning. The wider margins include figures, diagrams, tables, and photographs that illustrate the concepts.

Content Changes in Each Chapter

In Chapter 1, the section on conversion factors now precedes significant figures. The section on metric and SI prefixes has been combined with equalities. Chapter 2 begins with the topics of energy and its measurement, food energy, states of matter, and calculations of heat energy. New material includes equilibrium of solid–liquid and liquid–vapor states.

In Chapter 3, the discussion on subshells and orbitals appears at the end of the chapter, making it optional. In Chapter 4, the cations of transition elements are discussed along with the ions of representative elements. The section on naming ionic compounds now includes the naming of compounds containing polyatomic ions. Bond polarity and bond types now appears at the end of the chapter.

In Chapter 5, the topic of chemical equations is now covered in one section, which is followed by a new section that discusses energy changes, heats, and rates of reactions. New in this chapter is the effect of catalysts on the energy of activation (moved from enzymes), energy diagrams for exothermic and endothermic reactions, and a new Health Note on hot packs and cold packs. The material on moles, molar mass, and equations has been rewritten. Writing conversion factors for the mole relationships in equations is carefully explained and then utilized in calculating quantities in reactions.

In Chapter 6, the material on producing radioactive isotopes now follows nuclear equations and the biological effect of ionizing radiation is now a Health Note. A new Health Note combines maximum permissible dose, radiation sickness, and lethal dose. The table on average annual radiation per person is updated to include radon, and global fallout is deleted.

In Chapter 7, Gases, new graphics illustrate the gas relationships more clearly and a new Health Note on blood pressure has been added. All of the discussions of gas laws have been rewritten to clarify concepts. The Environmental Note on ozone depletion was moved to Chapter 12 to be included with the discussion of chlorofluorohydrocarbons.

Chapter 8 begins with a discussion on hydrogen bonding in water and leads into solution formation. Factors affecting rates of solution now include nature of solute and solvent, saturation, temperature effect on solubility of solids and gases in liquids, and solubility rules (moved from Chapter 9 in the last edition).

Chapter 9 begins with a rewritten section on electrolytes and equivalents. Acids and bases are now defined by both the Arrhenius and Brønsted–Lowry concepts and the term *hydronium ion* is used more consistently. The strengths of acids and bases is now a separate section and the neutralization of acids and bases now precedes the ionization of water. The reverse reaction in weak electrolytes, weak acids, and bases introduces the idea of equilibrium between molecules and ions. The concentrations of ions in water has been simplified to 1×10^{-7} M. The relationship between H_3O^+ and OH^- is now more conceptual (increase/decrease) and less calculator work is required. The discussion of pH includes an explanation of the mathematical meaning of pH and logs. Instructions have been added to calculate pH using a calculator. Buffers are now defined as solutions that resist changes in pH when small amounts of acid or base are added. Acid–base titration is now the last section in the chapter.

Chapter 10 introduces organic chemistry by discussing the hydrocarbons and the components in crude oil. The comparison of organic and inorganic compounds is now between butane and sodium chloride, compounds of similar masses. Structural isomers now follow complete structural formulas. The sections on nomenclature have been rewritten for clarity. New figures contain more three-dimensional drawings of the ball-and-stick models. A new Health Note on the cycloalkanes and haloalkanes currently used as anesthetics, an area of interest to allied health majors, is added to the section on cycloalkanes. The Environmental Note on haloalkanes now includes a photo of spraying with pesticides and the note on ozone depletion has been moved into this section.

Chapter 11 has a more descriptive title. The nature of the alkene family and the names of alkenes are now two separate sections. The rules for naming alkenes now appear in a series of steps to allow a clearer explanation. The sections on naming and on cis–trans isomers stand alone and can be omitted by instructors who do not include those topics. The addition reactions are now in two sections, one for the addition of X_2 reagents, and the other for the addition of HX and HOH. This section also includes a clearer explanation of Markovnikov's rule. The polymerization of alkenes has been rewritten and now includes notes and a photo on biopolymers. The last section in the chapter on benzene has been reworked. A new Health Note on aromatic compounds in health and medicine has been added. In Chapter 12, phenols are now introduced with alcohols in the first section. The Health Note on alcohols now includes levels of ethanol intoxication, behavior, and dangers. A discussion on thiols is included in the alcohol section, but the discussion on disulfides has been moved to Chapter 16 (on proteins). The properties of solubility of alcohols and ionization of phenol have been retained, but lists of boiling points have been deleted. The reactions of alcohols, dehydration, and oxidation, are now in the same section for easy reference. The note on biological oxidation now includes the Breathalyzer test for drunk drivers. In Chapter 13, the naming of acids is explained more carefully. A new Health Note on α-hydroxy acids and their current use in skin care products has been added. Another new note discusses plastics and recycling identification codes. The naming of amines and amides has been reworked and the note on amines in health and medicine has been expanded. There is a new Health Note on heterocyclic amines and their physiological activity. A new note was added on the Alar controversy, in which hormones were sprayed on apples.

The biomolecules section begins with a rewritten Chapter 14 on carbohydrates. A new explanation of chirality of sugars emphasizes nonsuperimposable mirror images, introduces the term *enantiomers,* and deletes the use of symmetry. A Health Note on chiral compounds is expanded to include more examples such as thalidomide and L-ibuprofen as well as the development of *chiral technology*. After describing the open-chain structures of the monosaccharides, hemiacetal and hemiketal formation is used to introduce their cyclic structures. In the structures of disaccharides, the α-1,4-glycosidic bond is now drawn as a single line. New Health Notes on the sweetness of sugars and the role of carbohydrates in blood typing and new figures for fiber and polysaccharides found in nature have been added. The number of tests to identify carbohydrates has been reduced. Projects for Chemistry at Home include photosynthesis, hydrolysis of starch, solubility of carbohydrates, and tests for starch with iodine.

The classes of lipids now introduce Chapter 15 to give an overview followed by some ways a fatty acid can be written. Monounsaturated fatty acids are now defined separately from polyunsaturated fatty acids, and palmitoleic acid has been added to the list of fatty acids. A new figure depicting the fit of saturated compared with unsaturated fatty acids has been added as well as a figure comparing the percentage of saturated, monounsaturated, and polyuunsaturated fatty acids in some typical fats and oils. The section on phospholipids and glycolipids now includes a description of cell membranes. Fat-soluble vitamins, steroid hormones, and anabolic steroids are now included in this chapter.

Chapter 16 combines two chapters from the last edition on proteins and enzymes. Amino acid structures are described and the zwitterion form used more extensively throughout the chapter. The discussion on peptide bonds has been rewritten and hydrolysis added. The sections on protein structure have been rewritten and the discussion on denaturation shortened. With the earlier discussion of catalysts in Chapter 5, the discussion of enzymes describes enzyme action and inhibition, and now includes a discussion of the induced-fit model. The old section on digestion enzymes has been moved to Chapter 18. Water-soluble vitamins are discussed as coenzymes in this chapter.

In Chapter 17, the components of nucleic acids have been combined into one figure. The section on replication has been rewritten to emphasize the role of complementary base pairing. The process of protein synthesis is followed by a discussion on genetic mutations. The last section reviews the cellular regulation of protein synthesis. A Health Note on viruses has been expanded to include the HIV virus.

In Chapter 18, metabolism and energy production describe ATP, catabolic reactions, and anabolic reactions. It has been completely rewritten beginning with digestion of foods (moved from enzymes) to the production of ATP from the oxidation products of the food molecules. The process of energy production now starts with glycolysis and carbohydrate degradation, acetyl coenzyme A, citric acid cycle, and the electron transport chain. The anion form of the carboxylic acids is used consistently and throughout the discussion, the types of oxidation and their coenzymes are emphasized. The discussion of the electron transport chain has been rewritten to emphasize the transfer of electrons and includes the chemiosmotic theory for energy production. Finally, the total ATP is calculated for the complete oxidation of glucose. Fatty acid synthesis is used as an example of an anabolic pathway, and the connection of amino acids to the citric acid intermediates and urea is shown.

The Supplement Package to Accompany the Text

For the Student

Laboratory Manual

The early experiments in the ***Laboratory Manual*** introduce students to basic laboratory skills. Students carry out laboratory investigations, develop the skills of manipulating laboratory equipment, gather and report data, solve problems, calculate, and draw conclusions. In this edition, as in the past, there is an emphasis on safety in the laboratory. Hazardous chemicals and procedures considered dangerous have been omitted. For each experiment, there is a report page and a set of questions that relate the laboratory to the corresponding information in the text. Some questions require essay-type answers to promote writing skills in science.

Study Guide

The ***Study Guide*** reviews the basic concepts, provides learning drills, and gives a practice examination, all with answers, for each chapter. All sections are keyed to the learning goals in the text so students may cross-reference working tools. Students can grade their own practice exams and check to determine if they have mastered the material. In this way, they can identify areas of difficulty and review the material again.

Tutorial Software

Chemistry Study Pak, prepared by Tom Hall, is an interactive software package designed to reinforce the concepts presented in the text and help students prepare for exams. Tutorials and Practice Tests are provided for each chapter. Test questions may be generated randomly, if desired. Immediate feedback and explanations are provided. Students needing additional help are directed to the appropriate parts of the text for further review. ***Chemistry Study Pak*** is available for computers running Windows 3.1 or higher.

For the Instructor

Instructor's Manual

The ***Instructor's Manual*** gives chapter overviews, suggestions for lecture demonstrations, worked-out solutions, and answers to the even-number problems in the textbook. Experiments that relate to each chapter are described, and the materials required for 10 students are listed. Instructions for preparing special solutions are included, and the laboratory skills to be demonstrated for the students are described. A complete set of sample student laboratory reports from the laboratory manual completes the Instructor's Manual.

Test Bank

The printed ***Test Bank*** contains more than 500 multiple-choice, true-false, and matching questions and their answers. The Test Bank available on Test-Master software disks for IBM PC and Macintosh allows instructors to scramble questions, add new questions, and select questions based on level of difficulty.

Transparencies

Instructors who adopt the text will receive a set of 100 full-color transparency acetates providing figures and illustrations from the text.

Safety Video

An American Chemical Society videotape on Safety in the Chemistry Laboratory is available to qualified instructors who adopt the text. Contact your local HarperCollins representative.

Video Demonstrating Alternative Teaching Methods in Chemistry

Karen Timberlake's video, Promoting Student Success in Chemistry, shows actual classroom activities that are student-centered, including problem solving using cooperative learning in a large lecture setting, peer presentations, and study teams in discussion hours. This video is available to qualified instructors who adopt the text. Contact your local HarperCollins representative.

Acknowledgments

I wish to thank my husband, Bill, for his enthusiasm, invaluable assistance in writing and reviewing manuscripts, and late dinners out. Thanks to my son, John, for his cooperation and help in the preparation of this edition of the text.

I am most grateful to the following instructors who reviewed portions of the manuscript, provided support, and gave helpful criticism and suggestions.

Mary Lee Abkemeier, La Guardia Community College
Edith Bartley, Tarrant County Jr. College/South Campus
Thomas P. Carey, Berkshire Community College
Raymond Chamberlain, City College of San Francisco
John Ferrara, Cuyahoga Community College/East Campus
Kevin Gratton, Johnson County Community College
Ann A. Hicks, Thomas More College
Terri Jakuboski, Lewis and Clark Community College
Mani Jayaswal, Florida Community College
Craig R. Johnson, Carlow College
Sharon Kapica, County College of Morris
Edward King, South Plains College
Victor N. Kingery, Garland County Community College
Joanne Kirvaitis, Moraine Valley College
Bernard Koser, Westchester Community College
William Lehman, Johnson County Community College
Irving Lillien, Miami Dade Community College/South Campus
David Macaulay, William Rainey Harper College
Voya Moon, Western Iowa Tech College
Deborah Nycz, Broward Community College/Central Campus
Thomas Nycz, Broward Community College/North Campus

John A. Paparelli, San Antonio College
Phil Reedy, Metropolitan Community College
Diane Riebeth, Riverside Community College
David Saltzman, Santa Fe Community College
Dennis Sardella, Boston College
Somnath Sengupta, Quincy College
S. B. Sharma, Columbus College
Ursula Simonis, San Francisco State University
Darold Skerritt, Hartnell College
Robert Smith, Skyline College
Roy Stein, University of Toledo Community & Tech College
Mary Vennos, Essex Community College
Dave Waggoner, Monroe County Community College
David Ziegler, Hannibal-LaGrange College
W. C. Zipperer, Armstrong College

I am grateful to the staff at HarperCollins Glenview for their encouragement and professional work during the preparation and achievement of this project: Doug Humphrey, Kathy Richmond, Cathleen Petree, Karen Koblik, Kelly Mountain, Kathy Jordan, Ginny Guerrant, and Karen Capel. Writing a chemistry text is an ongoing process as students, teachers, and problem-solving techniques, and helping students to develop their reasoning powers. With this aim, I have revised ***Chemistry***. I look forward to your use of this text. I welcome any suggestions, questions, criticism, or overall comments on this revision. My e-mail address is khemist@aol.com or write to me at the address below.

Karen C. Timberlake
Los Angeles Valley College
Van Nuys, CA 91401

To the Student

Here you are in chemistry, perhaps because you need a science course or just because you want to find out something about chemistry. Maybe you want to be a nurse, respiratory therapist, dietician, or other professional in the health sciences. If so, as you progress through this text, you will discover that chemistry is indeed exciting to learn and that it has an important relationship to the world around you. Every chapter in this Sixth Edition includes many applications of chemistry to health, medicine, and the environment. Your interest in the sciences will help you learn chemistry, and by learning chemistry you will gain a deeper understanding of physiology, medical care, and major issues of today, including pollution, global warming, the ozone layer, acid rain, nuclear energy, recombinant DNA, and AIDS.

I have designed this text with you in mind. To aid your learning process, each chapter begins with a set of learning goals that tell you what to expect in the chapter and what you need to accomplish. Each chapter begins with **Looking Ahead** and **Questions To Think About** to help you recall experiences that involve chemistry in your life and set the stage for the chemical concepts discussed throughout the chapter.

As you progress through each chapter, take time to consider the learning goal for each section. Read the material in that section and master the goal by doing the **Sample Problems** and **Study Checks.** The **Answers to the Study Checks** are at the end of the chapter. If you have difficulty with a sample problem or its study check, review that part of the unit again. For further self-testing, work the end-of-chapter **Problems** that are keyed to each section. You can check your answers to the odd-number problems by referring to the appendix in the back of the text. It is not necessary to study a chapter all the way through at one time. Instead, you may wish to cover only a few of the goals each time you study. Also, you can prepare for lectures by reading ahead in the text.

To review your knowledge of the important ideas in a chapter, read over the glossary at the end of the chapter. Study the tables and figures, which emphasize important concepts.

The study of chemistry involves some hard work, but I hope that you will find the effort rewarding when you see and understand the role of chemistry in many related fields. If you would like to share your feelings about chemistry or comments about this text, I would appreciate hearing from you.

Karen C. Timberlake
Los Angeles Valley College
Van Nuys, CA 91401
e-mail: khemist@aol.com

Chapter 1

Measurements

The metric system is used for measurement by scientists and health professionals all over the world.

Looking Ahead

Learning Goals

1.1 Write the names and abbreviations for the metric units used in measurements of length, volume, and mass.

1.2 Use the numerical values of prefixes to write a metric equality.

1.3 Write a conversion factor for two units that describe the same quantity.

1.4 Report answers to calculations using the correct number of significant figures.

1.5 Use a conversion factor to change from one unit to another.

1.6 Calculate the density or specific gravity of a substance, and use the density or specific gravity to calculate the mass or volume of a substance.

Questions to Think About

1. In Canada, a road sign gives a speed limit as 80 km/hr. Would you be exceeding the speed limit of 55 mi/hr if you were in the United States?
2. At a grocery store in Mexico, tomatoes are priced by the kilogram, not by the pound. Why?
3. In winter, why does ice form on the surface of a lake and not sink to the bottom?

What kinds of measurements did you make today? Perhaps you checked your weight by stepping on a scale this morning. Perhaps you measured out 2 cups of flour and a cup of milk to make your pancake batter. You may have filled the gas tank of your car with 10 gallons of gasoline and driven 12 miles to school. In each case, you were measuring mass (weight), volume, or distance. Try to recall some of the measurements you made today.

Men and women in the health sciences use measurement every day to evaluate the health of a patient. Temperatures are taken, weights and heights are recorded, and samples of blood and urine are collected for laboratory testing. Medications are given in dosages that must be measured accurately. In the dental office, hygienists measure solutions used in fluoride treatments and in the preparation of dental materials.

1.1 Units of Measurement

Learning Goal
Write the names and abbreviations for the metric units used in measurements of length, volume, and mass.

Suppose that during a recent medical examination the nurse recorded your mass (weight) as 70.0 kilograms (70.0 kg), your height as 1.78 meters (1.78 m), and your temperature as 37.0 degrees Celsius (37.0°C). The system of measurement used in these clinical evaluations is the *metric system.* Perhaps you would be more familiar with these measurements if they were stated in the American system. Then your weight would be 154 pounds (154 lb), your height 5 feet 10 inches (5 ft 10 in.), and your temperature 98.6 degrees Fahrenheit (98.6°F).

The **metric system** is used by scientists and health professionals throughout the world. It is also the common measuring system in all but a few countries in the world. In 1960, a modification of the metric system called the *International System of Units,* Système International **(SI),** was adopted to provide additional uniformity. In this text, we will use metric units and introduce some of the SI units that are in use today.

Length

The **meter (m)** is used to measure length in both the metric system and SI. It is 39.4 inches (in.), which makes a meter slightly longer than a yard (yd). A smaller unit of length, the **centimeter (cm),** is about as wide as your little finger. For comparison, there are 2.54 cm in 1 in.

1 m = 39.4 in.

2.54 cm = 1 in.

Figure 1.1 compares metric and American units for length.

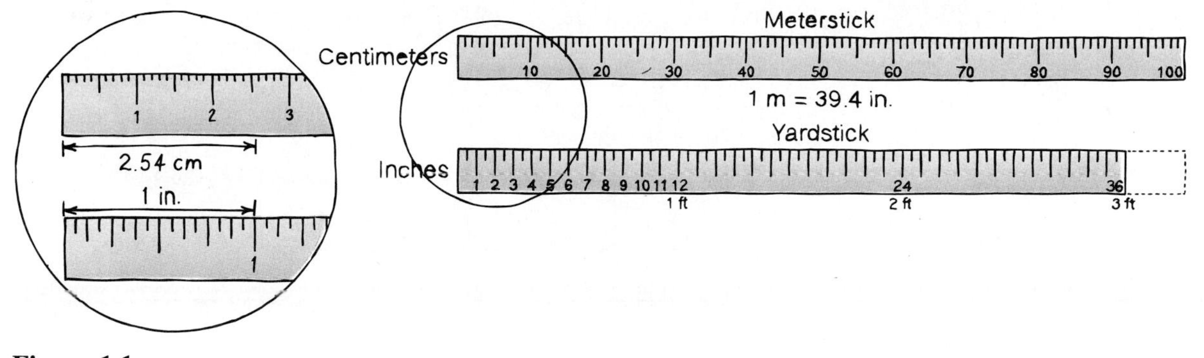

Figure 1.1
Comparison of the metric and American units for measuring length.

Volume

Volume is the amount of space occupied by a substance. The metric unit *liter* is commonly used to measure volume. A **liter (L)** is slightly larger than the quart (qt), which is used in the American system. The **milliliter (mL)** is more convenient for measuring smaller volumes of fluids in hospitals and laboratories.

1 L = 1.06 qt
946 mL = 1 qt

A comparison of metric and American units for volume is shown in Figure 1.2.

Figure 1.2
Comparison of the metric and American units for measuring volume.

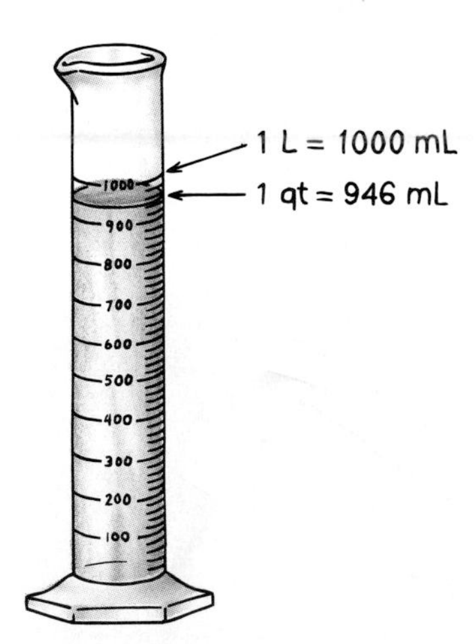

Mass

The **mass** of an object is a measure of the quantity of material it contains. Everything has mass, including rocks, water, people, and dogs. In the metric system, the unit for mass is the **gram (g).** The SI unit of mass, the **kilogram (kg),** is used for larger masses such as body weight. It takes 2.20 lb to make 1 kg, and 454 g is needed to equal 1 pound.

1 kg = 2.20 lb
454 g = 1 lb

Figure 1.3 illustrates the relationship between kilogram and pounds. A summary of some metric and SI units is given in Table 1.1.

Table 1.1 ***Units of Measurement***

Measurement	Metric	SI
Length	Meter (m)	Meter (m)
Volume	Liter (L)	Cubic meter (m^3)
Mass	Gram (g)	Kilogram (kg)
Time	Second (s)	Second (s)
Temperature	Celsius (°C)	Kelvin (K)

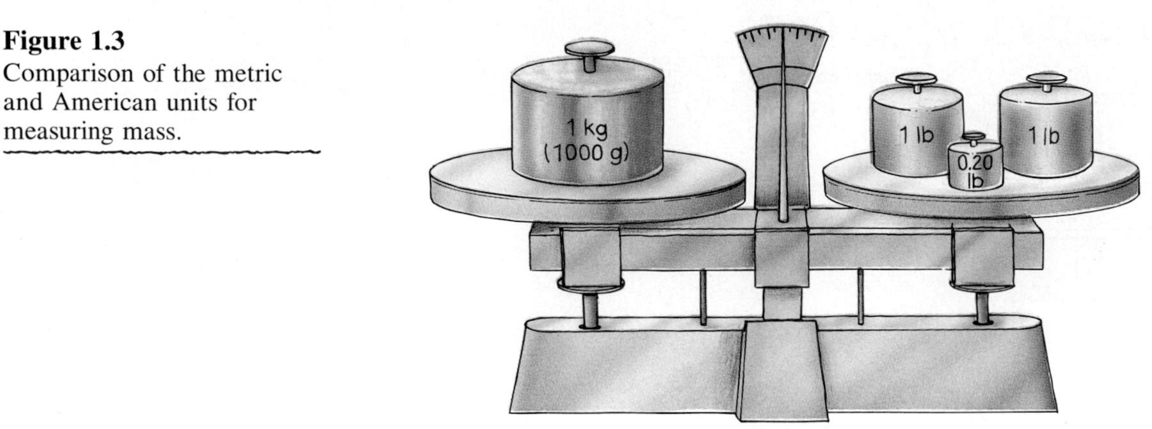

Figure 1.3
Comparison of the metric and American units for measuring mass.

Figure 1.4
The mass of a sample is given as a digital readout on an electronic balance.

You may be more familiar with the term weight than with mass. Weight is a measure of the gravitational pull on an object. On Earth, an astronaut with a mass of 75.0 kg has a weight of 165 lb. On the moon where the gravitational pull is one-sixth that of Earth, the astronaut has a weight of 27.5 lb. However, the mass of the astronaut is the same as on Earth, 75.0 kg. Scientists measure mass rather than weight because mass does not vary.

In a chemistry laboratory, a balance is used to measure the mass of a substance. When we "weigh" a substance, we are actually comparing its mass with a standard object with a known mass. An example of measuring the mass of a substance on a balance is shown in Figure 1.4.

Sample Problem 1.1
Units of Measurement

Complete the following table:

Type of Measurement	*Metric Unit*	*Abbreviation*
length	________	________
________	liter	________
________	________	g

Solution

length	meter	m
volume	liter	L
mass	gram	g

Study Check
What type of measurement uses the unit centimeter?

1.2 Metric Prefixes and Equalities

Learning Goal

Use the numerical values of prefixes to write a metric equality.

The U.S. Food and Drug Administration has determined the daily values (DV) of nutrients for adults and children age 4 or older. Some of these recommended daily values are listed in Table 1.2.

Table 1.2 ***Daily Values for Selected Nutrients***

Nutrient	Amount Recommended	Nutrient	Amount Recommended
Protein	44 grams	Iron	18 milligrams
Vitamin C	60 milligrams	Iodine	150 micrograms
Vitamin B_{12}	6 micrograms	Sodium	2400 milligrams
Calcium	1 gram	Zinc	15 milligrams

The special feature of the metric system of units is that a **prefix** can be attached to any physical quantity to increase or decrease its size by some factor of 10. For example, in the daily values, prefixes *milli* and *micro* are used to make the smaller units, milligram (mg) and microgram (μg). Table 1.3 lists some of the metric prefixes, their symbols, and their decimal values. These prefix values may also be expressed in scientific notation using powers of 10, a topic reviewed in the Appendix A.

The relationship of a unit to its base unit can be expressed by replacing the prefix with its numerical value. For example, when the prefix *kilo* in *kilometer* is replaced with its value of 1000, we find that a kilometer is equal to 1000 meters. Other examples follow.

1 **kilo**meter (1 km) = **1000** meters (1000 m)
1 **kilo**liter (1 kL) = **1000** liters (1000 L)
1 **kilo**gram (1 kg) = **1000** grams (1000 g)

Table 1.3 ***Metric and SI Prefixes***[a]

Prefix	Symbol	Meaning	Numerical Value	Scientific Notation
Prefixes that Increase the Size of the Unit				
mega	M	one million	1,000,000	10^6
kilo	k	one thousand	1,000	10^3
hecto	h	one hundred	100	10^2
deka	da	ten times	10	10^1
Prefixes that Decrease the Size of the Unit				
deci	d	one-tenth	0.1	10^{-1}
centi	c	one-hundredth	0.01	10^{-2}
milli	m	one-thousandth	0.001	10^{-3}
micro	μ	one-millionth	0.000001	10^{-6}
nano	n	one-billionth	0.000000001	10^{-9}

[a]Prefixes in boldface are used most often.

Sample Problem 1.2
Prefixes

Fill in the blanks with the correct numerical value:

a. kilogram = _____ grams
b. millisecond = _____ second
c. deciliter = _____ liter

Solution

a. The numerical value of *kilo* is 1000; 1000 grams.
b. The numerical value of *milli* is 0.001; 0.001 second.
c. The numerical value of *deci* is 0.1; 0.1 liter.

Study Check

Write the correct prefix in the blanks:

a. 1,000,000 seconds = _____ second
b. 0.01 meter = _____ meter

Length Equalities

An ophthalmologist may measure the diameter of the retina of the eye in centimeters (cm), whereas a surgeon may need to know the length of a nerve in millimeters (mm). When the prefix *centi* is used with the unit *meter,* it indicates the unit *centimeter,* a length that is one-hundredth of a meter (0.01 m). A *millimeter* measures a length of 0.001 m. There are 1000 mm in a meter.

If we compare the lengths of a millimeter and a centimeter, we find that 1 mm is 0.1 cm; there are 10 mm in 1 cm. These comparisons are examples of **equalities,** which show the relationship between two units that measure the same quantity. For example, in the equality 1 m = 100 cm, each quantity describes the same length but in a different unit. Note that each quantity in the equality expression has both a number and a unit.

An Equality

First Quantity			*Second Quantity*	
1	m	=	100	cm
↑	↑		↑	↑
Number	+ unit		Number	+ unit

Some Length Equalities

1 m = 100 cm

1 m = 1000 mm

1 cm = 10 mm

Some metric units for length are compared in Figure 1.5, and examples are shown in Figure 1.6.

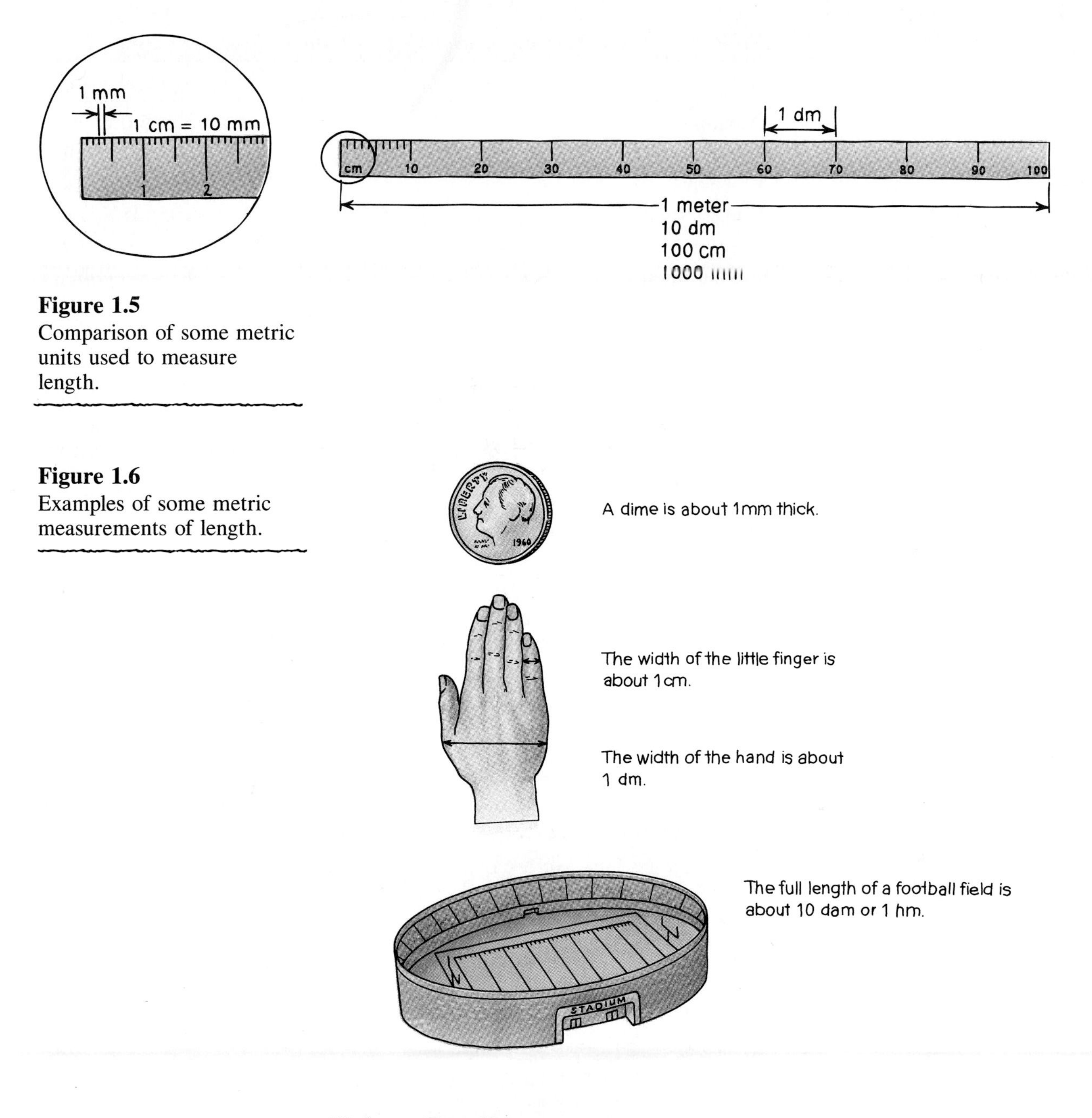

Figure 1.5
Comparison of some metric units used to measure length.

Figure 1.6
Examples of some metric measurements of length.

Volume Equalities

Volumes of 1 L or smaller are common in the health sciences. When a liter is divided into 10 equal portions, each portion is a deciliter (dL). There are 10 dL in 1 L. Laboratory results for blood work are often reported in deciliters. Notice the values listed in Table 1.4 for some substances in the blood.

When a liter is divided into a thousand parts, each of the smaller volumes is called a milliliter. In a 1-L bottle of physiological saline, there are 1000 mL

Table 1.4 *Some Typical Laboratory Test Values*

Substance in Blood	Typical Range
Albumin	3.5–5.0 g/dL
Ammonia	20–150 μg/dL
Calcium	8.5–10.5 mg/dL
Cholesterol	105–250 mg/dL
Iron (male)	80–160 μg/dL
Protein (total)	6.0–8.0 g/dL

of solution. Bottles of intravenous (IV) liquids typically contain solution volumes of 500 mL or 250 mL. Small amounts of liquids measured in milliliters or microliters (μL) may be added to the IV solution or given by injection.

Some Volume Equalities

1 L = 10 dL

1 L = 1000 mL

1 dL = 100 mL

Some typical equipment for measuring the volume of liquids in the hospital or in the chemistry laboratory is shown in Figure 1.7.

The **cubic centimeter** (**cm^3** or **cc**) is the volume of a cube whose dimensions are 1 cm on each side. A cubic centimeter has the same volume as a milliliter, and the units are often used interchangeably.

1 cm^3 = 1 cc = 1 mL

When you see *1 cm,* you are reading about length; when you see *1 cc* or *1 cm^3* or *1 mL* , you are reading about volume. A comparison of units of volume is illustrated in Figure 1.8.

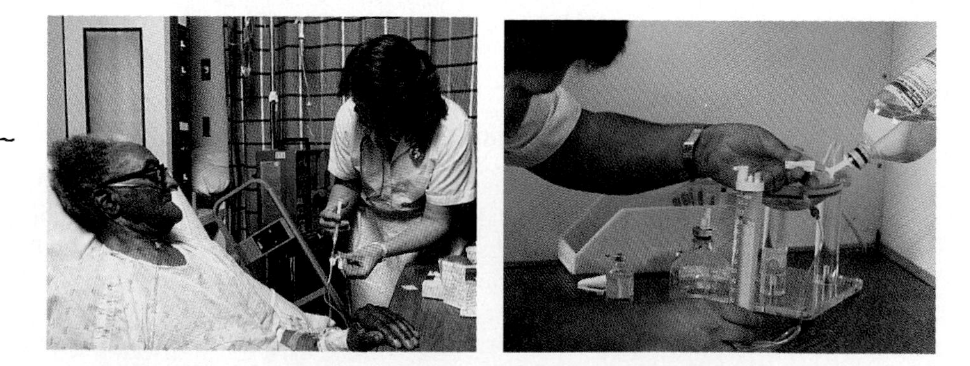

Figure 1.7
A nurse and a pharmacist measure specific volumes of medications.

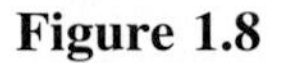

Figure 1.8
A cube measuring 10 cm on each side has a volume of 1000 cm^3, or 1 L; a cube measuring 1 cm on each side has a volume of 1 cm^3 or 1 mL.

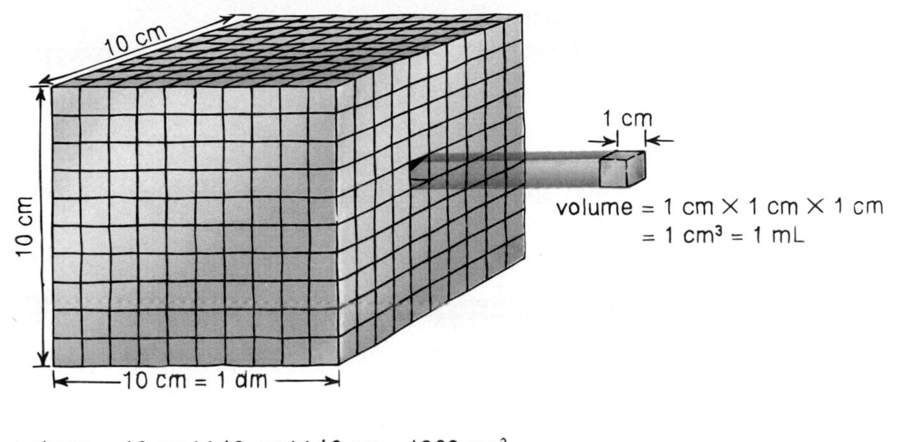

volume = 10 cm × 10 cm × 10 cm = 1000 cm^3
= 1000 mL
= 1 L

Mass Equalities

When you get a physical examination, your mass is recorded in kilograms, whereas the results of your laboratory tests are reported in grams, milligrams (mg), or micrograms (μg). A kilogram is equal to 1000 g. One gram represents the same mass as 1000 mg, and one mg equals 1000 μg.

Some Mass Equalities

1 kg = 1000 g
1 g = 1000 mg
1 mg = 1000 μg

Sample Problem 1.3
Writing Metric Relationships

1. Identify the largest unit in each of the following pairs:
   a. centimeter or kilometer
   b. L or dL
   c. mg or μg
2. Complete the following list of metric equalities:
   a. 1 L = __________ dL **b.** 1 km = __________ m
   c. 1 m = __________ cm **d.** 1 cm^3 = __________ mL

Solution

1. a. kilometer **b.** L **c.** mg
2. a. 10 dL **b.** 1000 m **c.** 100 cm **d.** 1 mL

Study Check
Complete the following equalities:
a. 1 kg = __________ g **b.** 1 mL = __________ L

1.3 Conversion Factors

Learning Goal

Write a conversion factor for two units that describe the same quantity.

Many problems in chemistry and the health sciences require a change of units. You make changes in units every day. For example, suppose you spent 2.0 hours (hr) on your homework, and your instructor asked you how many minutes that was. You would answer 120 minutes (min). You knew how to change from hours to minutes because you knew an equality (1 hr = 60 min) that related the two units. To do the problem, the equality is written in the form of a fraction called a **conversion factor.** One of the quantities is the numerator, and the other is the denominator. Be sure to include the units when you write the conversion factors. Two factors are always possible from any equality.

Two Conversion Factors for the Equality 1 hr = 60 min

$$\frac{\text{Numerator}}{\text{Denominator}} \begin{matrix}\rightarrow \\ \rightarrow\end{matrix} \frac{60\text{ min}}{1\text{ hr}} \quad \text{and} \quad \frac{1\text{ hr}}{60\text{ min}}$$

These factors are read as "60 minutes per 1 hour," and "1 hour per 60 minutes." The term *per* means "divide." Some common relationships and their corresponding conversion factors are given in Table 1.5. It is important that the equality you select to construct a conversion factor is a true relationship.

Metric–Metric Conversion Factors

We can write metric conversion factors for the metric relationships we have studied. For example, from the equality for meters and centimeters, we can write the following factors:

Metric Equality	*Conversion Factors*
1 m = 100 cm	$\frac{100\text{ cm}}{1\text{ m}}$ and $\frac{1\text{ m}}{100\text{ cm}}$

Table 1.5 ***Some Common Equalities and Their Corresponding Conversion Factors***

Equality	Conversion Factors		
1 yard = 3 feet	$\frac{1\text{ yd}}{3\text{ ft}}$	and	$\frac{3\text{ ft}}{1\text{ yd}}$
1 dollar = 100 cents	$\frac{1\text{ dollar}}{100\text{ cents}}$	and	$\frac{100\text{ cents}}{1\text{ dollar}}$
1 hour = 60 minutes	$\frac{1\text{ hr}}{60\text{ min}}$	and	$\frac{60\text{ min}}{1\text{ hr}}$
1 gallon = 4 quarts	$\frac{1\text{ gal}}{4\text{ qt}}$	and	$\frac{4\text{ qt}}{1\text{ gal}}$

Table 1.6 ***Some Useful Metric Conversion Factors***

Metric Relationship (Equality)	Conversion Factors		
Length			
1 m = 1000 mm	$\frac{1\text{ m}}{1000\text{ mm}}$	and	$\frac{1000\text{ mm}}{1\text{ m}}$
1 cm = 10 mm	$\frac{1\text{ cm}}{10\text{ mm}}$	and	$\frac{10\text{ mm}}{1\text{ cm}}$
Volume			
1 L = 1000 mL	$\frac{1\text{ L}}{1000\text{ mL}}$	and	$\frac{1000\text{ mL}}{1\text{ L}}$
1 dL = 100 mL	$\frac{1\text{ dL}}{100\text{ mL}}$	and	$\frac{100\text{ mL}}{1\text{ dL}}$
Mass			
1 kg = 1000 g	$\frac{1\text{ kg}}{1000\text{ g}}$	and	$\frac{1000\text{ g}}{1\text{ kg}}$
1 g = 1000 mg	$\frac{1\text{ g}}{1000\text{ mg}}$	and	$\frac{1000\text{ mg}}{1\text{ g}}$

Both are proper conversion factors for the relationship; one is just the inverse of the other. The usefulness of conversion factors is enhanced by the fact that we can turn a conversion factor over and use its inverse. Table 1.6 lists some conversion factors from the metric relationships we have discussed.

Metric–American Conversion Factors

Suppose you need to convert from pounds, a unit in the American system, to kilograms in the metric (or SI) system. A relationship you could use is

$$1\text{ kg} = 2.20\text{ lb}$$

The corresponding conversion factors would be

$$\frac{2.20\text{ lb}}{1\text{ kg}} \quad \text{and} \quad \frac{1\text{ kg}}{2.20\text{ lb}}$$

Table 1.7 lists some useful metric–American relationships, and Figure 1.9 illustrates the different units used to describe the contents of some packaged foods. A summary of equalities and conversion factors can be found in Appendix B.

Table 1.7 *Some Metric–American Relationships and Conversion Factors*

Relationship	Conversion Factors		
Length			
2.54 cm = 1 in.	$\frac{2.54\ \text{cm}}{1\ \text{in.}}$	and	$\frac{1\ \text{in.}}{2.54\ \text{cm}}$
1 m = 39.4 in.	$\frac{1\ \text{m}}{39.4\ \text{in.}}$	and	$\frac{39.4\ \text{in.}}{1\ \text{m}}$
Volume			
946 mL = 1 qt	$\frac{946\ \text{mL}}{1\ \text{qt}}$	and	$\frac{1\ \text{qt}}{946\ \text{mL}}$
1 L = 1.06 qt	$\frac{1\ \text{L}}{1.06\ \text{qt}}$	and	$\frac{1.06\ \text{qt}}{1\ \text{L}}$
Mass			
454 g = 1 lb	$\frac{454\ \text{g}}{1\ \text{lb}}$	and	$\frac{1\ \text{lb}}{454\ \text{g}}$
1 kg = 2.20 lb	$\frac{1\ \text{kg}}{2.20\ \text{lb}}$	and	$\frac{2.20\ \text{lb}}{1\ \text{kg}}$

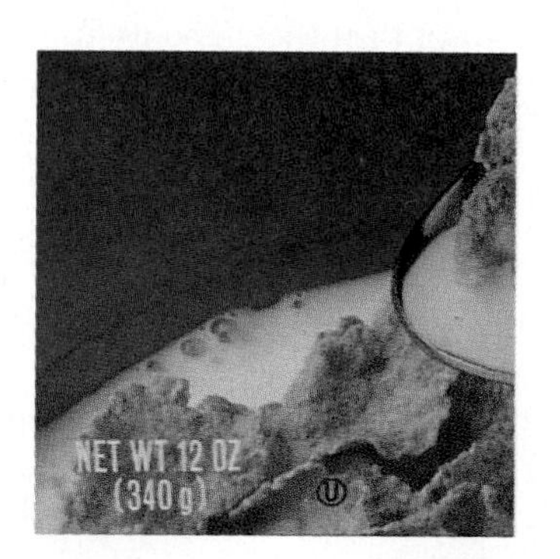

Figure 1.9
Metric and American units describe the contents of some packaged foods.

Sample Problem 1.4
Writing Conversion Factors for Equalities

Write conversion factors for the following equalities:
a. There are 1000 mg in 1 g.
b. One day has 24 hours.
c. There are 12 eggs in 1 dozen eggs.

Solution

Equality	*Conversion Factors*		
a. 1 g = 1000 mg	$\frac{1\text{g}}{1000\text{ mg}}$	and	$\frac{1000\text{ mg}}{1\text{ g}}$
b. 1 day = 24 hr	$\frac{1\text{ day}}{24\text{ hr}}$	and	$\frac{24\text{ hr}}{1\text{ day}}$
c. 12 eggs = 1 dozen eggs	$\frac{12\text{ eggs}}{1\text{ dozen eggs}}$	and	$\frac{1\text{ dozen eggs}}{12\text{ eggs}}$

Study Check
Write the equality and conversion factors for a relationship between inches and centimeters.

1.4 Significant Figures

Learning Goal
Report answers to calculations using the correct number of significant figures.

In science, we measure many things including the length of a bacterium, the volume of a liquid, and the amount of cholesterol in a blood sample. The numbers used to report our measurements are called measured numbers. Often, these measured numbers are used in calculations. It is therefore necessary to know how to use measured numbers in calculations and how to report final answers properly.

Significant Figures

In a clinical exercise, several students measured the height of a patient and had the following results:

172.7 cm   172.6 cm   172.4 cm   172.8 cm

Although the first three figures (172) in all four measurements are identical, there are differences in the value of the last digit. Such variations occur when we reach the limit of measurement provided by our measuring device. The last digit in a measured number is obtained by estimation. That means that measurements of this patient's height by different observers can have some variation or uncertainty in the last digit. The numbers that are reported in a measurement including the last estimated digit are called **significant figures.** All of the measured heights of the patient have four significant figures.

When different measuring tools are used, the reported measurements can have different numbers of significant figures. As shown in Figure 1.10, there are no divisions on cylinder A between 30 and 40 mL. A volume of 35 mL

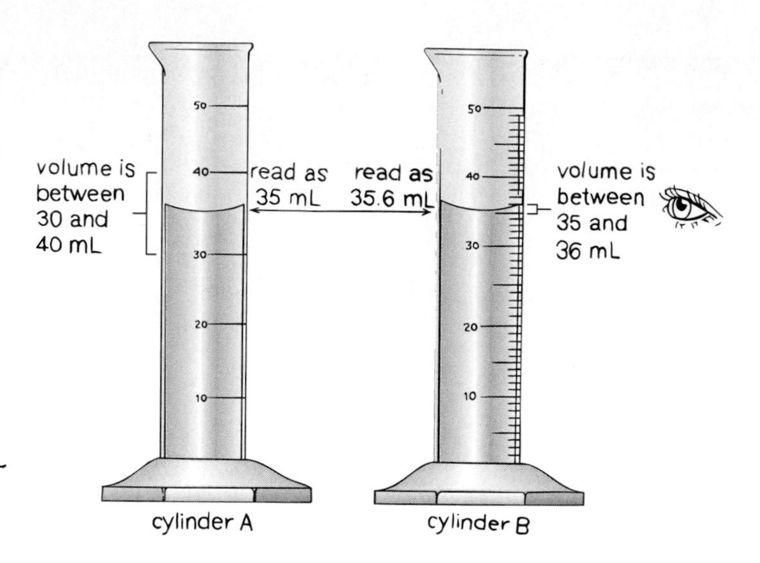

Figure 1.10
The number of units that can be used to report a volume of liquid differs with the markings on the graduated cylinder used. Using cylinder A, a measurement of 35 mL, which has two significant figures, can be reported. Using cylinder B, a measurement of 35.6 mL, which has three significant figures, can be reported.

is recorded by estimating where the mL divisions would be. However, on cylinder B, the 1-mL divisions are marked. On that cylinder, the volume can be reported as 35.6 by estimating the tenths (0.1) mL divisions between the 35 and 36 mL.

Zeros in Measured Numbers

When a zero is a part of a measured number, it counts as a significant figure. However, when a zero is used as a placeholder, it is not significant. Placeholder zeros are zeros that occur at the front of a decimal number or at the end of very large numbers without decimal points. Table 1.8 gives the rules for counting significant figures.

Exact Numbers

An *exact number* is a number used to count something; it is not measured. If you buy eight doughnuts, you have exactly eight doughnuts to eat. There is no uncertainty here because you counted the number of doughnuts. However, measured numbers are never exact. The last figure in any measured number is an estimation. The number of significant figures in a measured number is determined by the accuracy of the measuring device.

Exact numbers are also used in equalities between units of the same system of measurement. In the American system, 1 ft is defined as equal to 12 in. In the metric system, 1 m is defined as 100 cm. These relationships are not obtained by measurement; there is no uncertainty associated with them. Therefore, numbers in a definition are exact and the rules for significant figures do not apply. Some examples of exact numbers are listed in Table 1.9.

Table 1.8 ***Rules for Counting Significant Figures***

	Examples of Measured Numbers	Number of Significant Figures
1. A *significant number* is:		
a. a nonzero digit	122.35 cm	5
	4.5 g	2
b. a zero between nonzero digits	205 mm	3
	5.082 kg	4
c. a zero at the end of decimal number	5.0 cm	2
	2.10 kg	3
2. A *nonsignificant number* is:		
a. a zero in front of a decimal number	0.055 m	2
	0.0004 lb	1
b. a zero at the end of a large number without a decimal point (placeholder zero)	84,000 m	2
	1,065,000 kg	4

Table 1.9 ***Examples of Some Exact Numbers***

	Defined Equalities	
Counted Numbers	**American System**	**Metric System**
Eight doughnuts	1 ft = 12 in.	1 L = 1000 mL
Two baseballs	1 qt = 4 cups	1 m = 100 cm
Five capsules	1 lb = 16 ounces	1 kg = 1000 g

Sample Problem 1.5
Significant Figures

Identify each of the following numbers as measured or exact; give the number of significant figures in each of the measured numbers.

a. 42.2 g **b.** 3 eggs **c.** 0.0005 nm
d. 450,000 km **e.** 9 planets

Solution

a. measured; three **b.** exact **c.** measured; one
d. measured; two **e.** exact

Study Check

State the number of significant figures in each of the following measured numbers:

a. 0.00035 g **b.** 2000 m **c.** 2.0045 L

Rounding Off Calculator Results

In mathematical calculations, the degree of uncertainty in the measurements is indicated by the number of significant figures in the final answer. For example, dividing a mass of 7.2 g by a volume of 3.8 mL, gives a calculator answer of 1.894736842 g/mL (some calculators will give only 1.8947368). However, because each of the measurements has *two significant figures,* the number shown on the calculator is rounded off to give a final answer with two significant figures.

$$\frac{7.2\text{ g}}{3.8\text{ mL}} = 1.894736842 = 1.9\text{ g/mL}$$

Calculator display

Final answer, rounded off to two significant figures

Rules for Rounding Off

1. If the first digit to be dropped is **4 or less,** it and all following digits are simply dropped from the number.
2. If the first digit to be dropped is **5 or greater,** the last retained digit of the number is increased by 1.

		Three Significant Figures	*Two Significant Figures*
Example 1:	8.4234 rounds off to	8.42	8.4
Example 2:	14.780 rounds off to	14.8	15

Adding Significant Zeros

Sometimes, a calculator result is a small whole number. Then, you may add significant zeros to obtain an answer with the correct number of significant figures (SFs). For example, suppose the calculator display is 4, but you need three significant numbers in your answer. Adding two significant zeros after the 4 gives a final answer, 4.00.

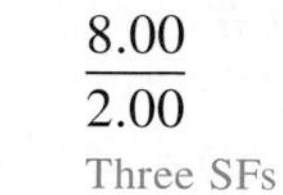

$$\frac{8.00}{2.00} = 4 \rightarrow 4.00$$

Three SFs

Calculator display

Final answer, two zeros added to give three SFs

Sample Problem 1.6
Rounding Off

Round off each of the following numbers to three significant figures:

a. 35.7823 m **b.** 0.002627 L
c. 3826.8 g **d.** 1.2836 kg

Solution

a. 35.8 m **b.** 0.00263 L
c. 3830 g **d.** 1.28 kg

Study Check

Round off each of the numbers in the above exercise to two significant figures.

Significant Figures in Multiplication and Division

Using a calculator will permit you to solve problems much faster than you can without one. However, calculators cannot think for you. It is up to you to enter the numbers correctly, press the right keys, and adjust the calculator result to give a proper answer. In multiplication and division, the calculator's answer is adjusted to have the same number of digits as the measurement with the *fewest* significant figures (here abbreviated SFs). See Appendix C to review the use of a calculator.

In multiplication and division, the final answer has the same number of digits as the measurement with the *fewest* significant figures.

Example 1

Multiply the following measured numbers: 24.65 × 0.67.

24.65	×	0.67	=	16.5155	→ 17
Four SFs		Two SFs		Calculator display	Final answer, rounded to two SFs

The measurement 0.67 has the least number of significant figures, two. Therefore, the calculator answer is rounded off to two significant figures.

Example 2

Solve the following:

$$\frac{2.85 \times 67.4}{4.39}$$

To do this problem on a calculator, enter the number and then press the operation key. In this case, we might press the keys in the following order:

2.85	×	67.4	÷	4.39	=	43.756264	→ 43.8
Three SFs		Three SFs		Three SFs		Calculator display	Final answer, rounded off to three SFs

All of the measurements in this problem have three significant figures. Therefore, the calculator result is rounded off to give an answer 43.8 that has three significant figures.

Sample Problem 1.7
Significant Figures in Measured Numbers

Perform the following calculations of measured numbers. Give the answers with the correct number of significant figures.

a. 56.8×0.37 **b.** $\frac{71.4}{11}$

c. $\frac{(2.075)(0.585)}{(8.42)(0.0045)}$ **d.** $\frac{25.0}{5.00}$

Solution

a. Calculator answer: 21.016 $\longrightarrow$ 21
b. Calculator answer: 6.4909091 $\longrightarrow$ 6.5
c. Calculator answer: 32.036817 $\longrightarrow$ 32
d. Calculator answer: 5 $\longrightarrow$ 5.00 (Must add significant zeros)

Study Check

Solve:

a. 45.26×0.01088
b. $2.6 \div 324$
c. $\frac{4.0 \times 8.00}{16}$

Significant Figures in Addition and Subtraction

In addition or subtraction, the final answer is reported to the same number of decimal places as the least accurately measured number in the problem.

In addition and subtraction, the final answer has the same number of decimal places as the least accurate measurement.

Example 3

Add:

	2.045	Thousandths
+	34.1	Tenths; less accurate measurement
	36.145	Calculator display
	36.1	Answer, rounded off to tenths place.

Example 4

Subtract:

	255	Ones; less accurate measurement
−	175.65	Hundredths
	79.35	Calculator display
	79	Answer, rounded off to ones place

Sample Problem 1.8
Significant Figures In Addition and Subtraction

Perform the following calculations and give the answer with the correct number of significant figures:

a. 27.8 cm + 0.235 cm
b. 104.45 mL + 0.838 mL + 46 mL
c. 153.247 g − 14.82 g

Solution

a. 28.0 cm **b.** 151 mL **c.** 138.43 g

Study Check

Solve:

a. 82.45 mg + 1.245 mg + 0.00056 mg
b. 4.259 L − 3.8 L

1.5 Problem Solving Using Conversion Factors

Learning Goal
Use a conversion factor to change from one unit to another.

The process of problem solving in chemistry is much the same as the process you use when you work out everyday problems, such as changing dollars to cents or hours to minutes. To illustrate this process, let's take a look at an everyday situation.

Suppose you go to the store to buy some apples. The sign at the fruit stand states that 1 lb of apples costs 48 cents as shown in Figure 1.11. When you weigh the apples, you have 2.0 lb. How much will you pay, in cents, for the

Figure 1.11
Converting units.

apples? A close look at the way we think about this problem might show the following:

Step 1 The quantity 2.0 lb, is the given or stated quantity.

Given: 2.0 lb apples

Step 2 It is helpful to decide on a unit plan. For this problem, our unit plan is to change from pounds to cents.

Unit Plan: lb → cents

Given — Unit for answer

Step 3 The relationship stated in the problem as 48 cents per pound of apples can be written in the form of conversion factors.

Relationship: 1 lb apples = 48 cents

Conversion Factors: $\frac{\text{1 lb apples}}{\text{48 cents}}$ and $\frac{\text{48 cents}}{\text{1 lb apples}}$

Step 4 Now we can write the setup for the problem using our unit plan (step 2) and the conversion factors (step 3). First, write down the given quantity, 2.0 lb. Then multiply it by the conversion factor that has the unit lb in the denominator, because that will cancel out the given unit of lb in the numerator. The unit of cents in the numerator (top number) gives the correct unit for the answer.

Problem Setup:

Unit for answer goes here

$$2.0\ \cancel{\text{lb}} \times \frac{\text{48 cents}}{1\ \cancel{\text{lb}}} = \text{96 cents}$$

Given (stated unit) — Conversion factor (cancels out given) — Answer (desired unit)

Take a look at the way the units cancel. The unit that you want in the answer is the one that remains after all the other units have canceled out. This is a helpful way to check a problem.

$$\frac{\text{Numerator} \rightarrow}{\text{Denominator} \rightarrow} \cancel{\text{lb}} \times \frac{\text{cents}}{\cancel{\text{lb}}} = \text{cents}$$ Answer unit

Do the calculation on your calculator to give the following:

2.0 × 48 = 96

The number 96 combined with the desired unit cents gives the final answer of 96 cents. With few exceptions, answers to numerical problems must contain a number and a unit.

96 cents Answer

Sample Problem 1.9
Problem Solving Using Conversion Factors

On a recent bicycle trip, Maria averaged 35 miles per day. How many days did it take her to cover 175 miles?

Solution

Step 1 Identify the given quantity in the question.

Given: 175 miles

Step 2 Write a unit plan.

Unit plan: miles → days Unit for answer

Step 3 Identify relationships needed and write conversion factor(s).

Relationship (from Problem): 35 miles = 1 day

Conversion Factors: $\frac{1\text{ day}}{35\text{ miles}}$ and $\frac{35\text{ miles}}{1\text{ day}}$

Step 4 Set up the problem starting with the given, cancel units, and carry out calculations. The answer is given with two significant figures.

Problem Setup:

Unit for answer goes here

$$175\ \cancel{\text{miles}} \times \frac{1\text{ day}}{35\ \cancel{\text{miles}}} = 5.0\text{ days}$$

3 SFs 2 SFs 2 SFs

(175 ÷ 35 = 5.0)

Suppose the problem were set up in the following way. If the units do not cancel, you know there is an error in the setup.

$$175\text{ miles} \times \frac{35\text{ miles}}{\text{day}} = 6215\ \textbf{miles}^2\textbf{/day}$$ Wrong units!

Study Check
A recipe for shark fin soup calls for 3.0 quarts of chicken broth. If 1 quart contains 4 cups, how many cups of broth are needed?

Problem Solving Using Metric Factors

The metric conversion factors from Table 1.6 are used to change from one metric unit to another, as seen in the following Sample Problem.

Sample Problem 1.10
Problem Solving Using Metric Factors

The Daily Value (DV) for sodium is 2400 mg. How many grams of sodium is that?

Solution

Step 1 **Given:** 2400 mg

Step 2 **Unit Plan:** mg $\rightarrow$ g Unit for answer

Step 3 **Metric Equality:** 1 g = 1000 mg

Conversion Factors: $\frac{1\text{g}}{1000 \text{ mg}}$ and $\frac{1000 \text{ mg}}{1 \text{ g}}$

Step 4 **Problem Setup:**

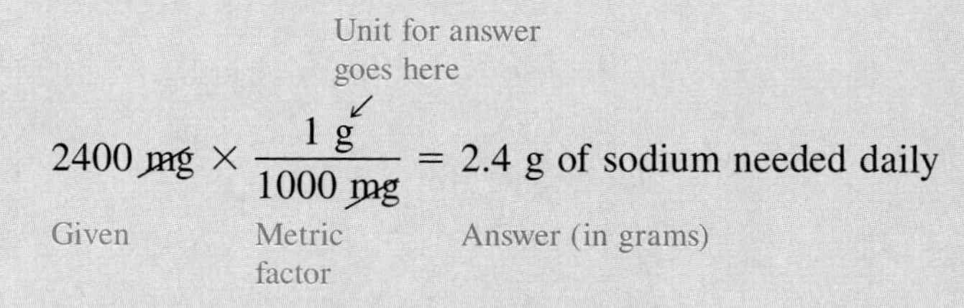

Unit for answer goes here
2400 mg × 1 g / 1000 mg = 2.4 g of sodium needed daily
Given
Metric factor
Answer (in grams)

Study Check
A can containing 473 mL of frozen orange juice is diluted with 1415 mL of water. How many liters of orange juice were prepared?

Problem Solving with Metric–American Factors

Conversion factors shown in Table 1.7 can be used to link the metric (or SI) and American systems of measurement in solving problems. Figure 1.12 shows the quantities of some common items in both metric and American units.

Sample Problem 1.11
Problem Solving Using Metric–American Factors

The length of a newborn infant is 19.0 in. What is the length of the baby, in centimeters?

Solution

Step 1 **Given:** 19.0 in.

Step 2 **Unit Plan:** in. $\rightarrow$ cm

Step 3 **Metric–American Relationship:** 1 in. = 2.54 cm

Conversion Factors: $\frac{1 \text{ in.}}{2.54 \text{ cm}}$ and $\frac{2.54 \text{ cm}}{1 \text{ in.}}$

Step 4 **Problem Setup:**

$$19.0\ \not{\text{in.}} \times \frac{2.54\ \text{cm}}{1\ \not{\text{in.}}} = 48.3\ \text{cm}$$

Study Check

A student traveling in Mexico buys 1.5 kg of grapes. How many pounds of grapes did she purchase?

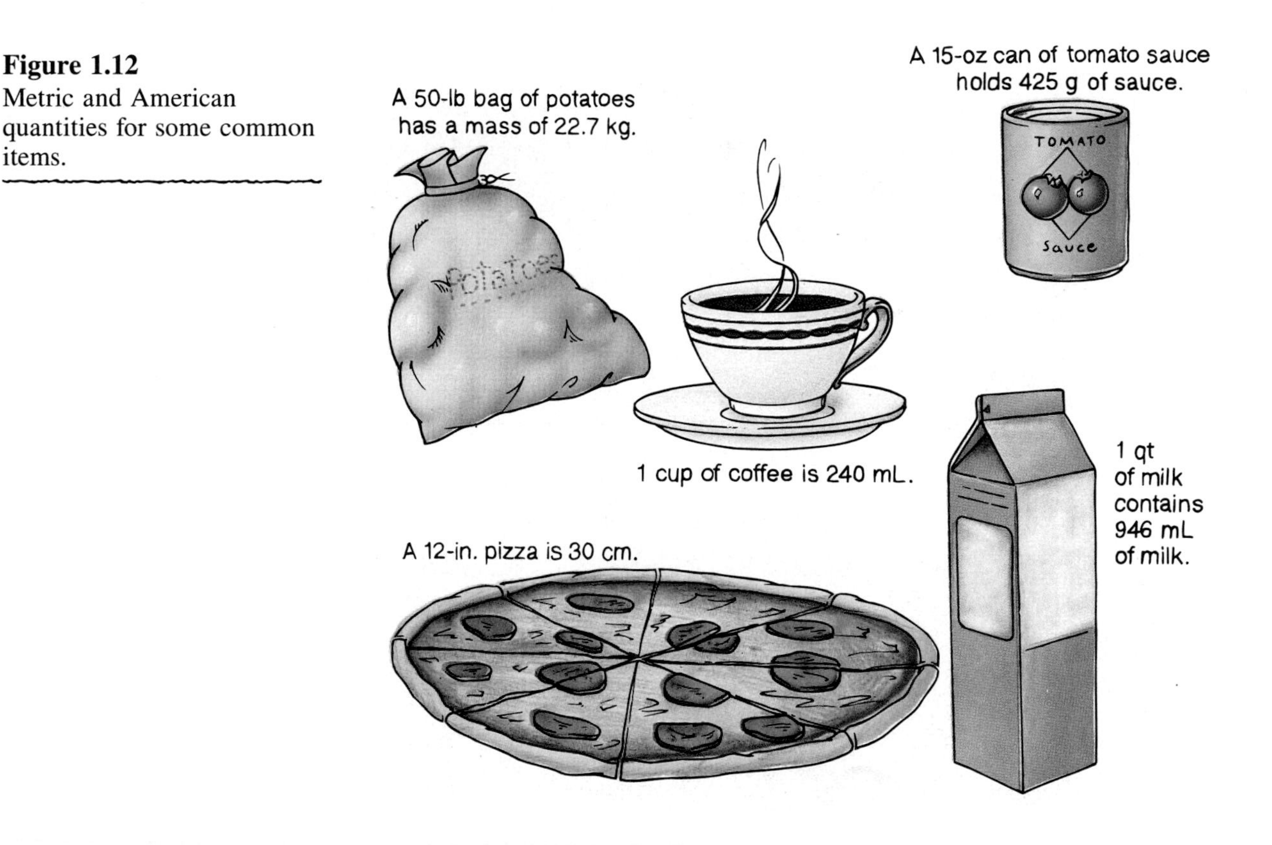
A 15-oz can of tomato sauce holds 425 g of sauce.
TOMATO
Sauce
A 50-lb bag of potatoes has a mass of 22.7 kg.
Potatoes
1 cup of coffee is 240 mL.
1 qt of milk contains 946 mL of milk.
A 12-in. pizza is 30 cm.

Figure 1.12
Metric and American quantities for some common items.

Using Two or More Conversion Factors in Sequence

In many problems you will need to use two or more steps in your unit plan. Then two or more conversion factors will be required. These can be constructed from the equalities you have learned or those which are stated in the problem. In setting up the problem, one factor follows the other. Each factor is arranged to cancel the preceding unit until you obtain the desired unit.

Sample Problem 1.12
Problem Solving Using Two Factors

A recipe for salsa requires 3.0 cups of tomato sauce. If only metric measures are available, how many milliliters of tomato sauce are needed? (There are 4 cups in 1 quart.)

Solution

You may not know a relationship between cups and milliliters. However, you do know how to change cups to quarts, and quarts to milliliters.

Step 1 **Given:** 3.0 cups

Step 2 **Unit Plan:** cups → quarts → milliliters   Unit for answer

Step 3 **Relationships and Conversion Factors:**

1. 1 qt = 4 cups $\frac{1\text{ qt}}{4\text{ cups}}$ and $\frac{4\text{ cups}}{1\text{ qt}}$

2. 1 qt = 946 mL $\frac{1\text{ qt}}{946\text{ mL}}$ and $\frac{946\text{ mL}}{1\text{ qt}}$

Step 4 **Problem Setup:** Use factor 1 to convert from cups to quarts:

$$3.0\ \cancel{\text{cups}} \times \frac{1\text{ qt}}{4\ \cancel{\text{cups}}} \times \textbf{?}$$

Then use factor 2 to convert from quarts to milliliters:

cups → qt → mL

Unit for answer ↙

$$\cancel{\text{cups}} \times \frac{\cancel{\text{qt}}}{\cancel{\text{cups}}} \times \frac{\text{mL}}{\cancel{\text{qt}}} = \text{mL}$$

In the setup, each factor is arranged to cancel all units except the unit for the answer. The complete setup appears as follows:

$$3.0\ \cancel{\text{cups}} \times \frac{1\ \cancel{\text{qt}}}{4\ \cancel{\text{cups}}} \times \frac{946\text{ mL}}{1\ \cancel{\text{qt}}} = 710\text{ mL}$$

Given quantity | American factor | Metric–American factor | Answer (in milliliters)

The calculations are done in sequence on a calculator. Pay attention here to the use of significant figures. Recall that relationships within a measurement system are exact, and most relationships between the American and metric system are approximate.

3.0	÷	4	×	946	=	710
Two SFs		Exact		Three SFs		Two SFs

Study Check

One medium bran muffin contains 4.2 g of fiber. How many ounces (oz) of fiber are obtained by eating three medium bran muffins, if 1 lb = 16 oz? (*Hint:* Number of muffins → g of fiber → lb → oz)

Table 1.10 ***Steps in Problem Solving***

Step 1 Identify the given quantity amount and unit.

Step 2 Write a unit plan to help you think about changing units from the given to the answer unit. Be sure you can supply a conversion factor for each change.

Step 3 Determine the equalities and corresponding conversion factors you will need to change from one unit to another.

Step 4 Set up the problem according to your unit plan. Arrange each conversion factor to cancel the preceding unit. Check that the units cancel to give the unit of the answer. Carry out the calculations and give a final answer with the correct number of significant figures and unit.

Using a sequence of two or more conversion factors is a very efficient way to set up and solve problems especially if you are using a calculator. Once you have the problem set up, the calculations can be done without writing out the intermediate values. This process is worth practicing until you understand unit cancellation and the mathematical calculations. A summary of steps in problem solving is given in Table 1.10.

Clinical Calculations Using Conversion Factors

Conversion factors are also useful for calculating medications. For example, if an antibiotic is available in 5-mg tablets, the dosage can be written as a conversion fraction, 5 mg/1 tablet. In many hospitals, the apothecary unit of grains (gr) is still in use; there are 65 mg in 1 gr. When you do a medication problem, you often start with a doctor's order that contains the quantity to give the patient. The medication dosage is used as a conversion factor.

Sample Problem 1.13
Clinical Calculations with Factors

Dr. Alvarez orders 0.050 g of a medication for your patient. If each tablet contains 10 mg of medication, how many tablets are needed?

Solution

Step 1 **Given:** 0.050 g

Step 2 **Unit Plan:** g → mg → tablets   Unit for answer

Step 3 **Relationships and Conversion Factors:**

1. 1 g = 1000 mg $\frac{1\text{ g}}{1000\text{ mg}}$ and $\frac{1000\text{ mg}}{1\text{ g}}$

2. 1 tablet = 10 mg of medication

$\frac{1\text{ tablet}}{10\text{ mg}}$ and $\frac{10\text{ mg}}{1\text{ tablet}}$

Step 4 **Problem setup:** g → mg → tablets

$$\underset{\text{Given}}{0.050\ \cancel{g}} \times \underset{\text{Metric factor}}{\frac{1000\ \cancel{mg}}{1\ \cancel{g}}} \times \underset{\text{Dosage factor}}{\frac{1\ \text{tablet}}{10\ \cancel{mg}}} = \underset{\text{Answer (in tablets)}}{5\ \text{tablets}}$$

(Unit for answer: tablet)

Study Check

An aspirin tablet contains 5.0 grains (gr) of aspirin. How many milligrams (mg) of aspirin are in two aspirin tablets, if 1 gr = 65 mg?

1.6 Density and Specific Gravity

Learning Goal

Calculate the density or specific gravity of a substance, and use the density or specific gravity to calculate the mass or volume of a substance.

We can measure the mass of a substance on a balance, and we can determine its volume. However, the separate measurements do not tell us how tightly packed the substance might be or whether its mass is spread out over a large volume or a small one. If we compare the mass of a substance to its volume, we obtain its density. **Density** is defined as the ratio of the mass of the substance to the volume of the substance at a given temperature.

$$\text{Density} = \frac{\text{mass of substance}}{\text{volume of substance}}$$

In the metric system, the densities of solids and liquids are usually expressed as grams per cubic centimeter (g/cm^3) or grams per milliliter (g/mL). The density of gases is usually stated as grams per liter (g/L). Table 1.11 gives the densities of some common substances.

Table 1.11 ***Density of Some Common Substances***

Solids (at 25°C)	Density (g/mL)	Liquids (at 25°C)	Density (g/mL)	Gases (at 0°C)	Density (g/L)
Cork	0.26	Gasoline	0.66	Hydrogen	0.090
Wood (maple)	0.75	Ethyl alcohol	0.79	Helium	0.179
Ice	0.92	Olive oil	0.92	Methane	0.714
Sugar	1.59	Water (at 4°C)	1.000	Neon	0.90
Bone	1.80	Plasma (blood)	1.03	Nitrogen	1.25
Aluminum	2.70	Urine	1.003–1.030	Air (dry)	1.29
Cement	3.00	Milk	1.04	Oxygen	1.45
Diamond	3.52	Mercury	13.6	Carbon dioxide	1.96
Silver	10.5				
Lead	11.3				
Gold	19.3				

Sample Problem 1.14
Calculating Density

A 50.0-mL sample of buttermilk has a mass of 56.0 g. What is the density of the buttermilk?

Solution

To calculate density, substitute the mass (g) and volume (mL) of the buttermilk into the expression for density.

$$\text{Density} = \frac{\text{mass}}{\text{volume}} = \frac{56.0\text{ g}}{50.0\text{ mL}} = \frac{1.12\text{ g}}{1\text{ mL}} = 1.12\text{ g/mL}$$ Two units in this answer

Study Check

A copper sample has a mass of 44.65 g and a volume of 5.0 cm^3. What is the density of copper, expressed as grams per cubic centimeter (g/cm^3)?

Density of Solids

The density of a solid is calculated from its mass and volume. When a solid is completely submerged, it displaces a volume of water that is *equal to its own volume*. In Figure 1.13, the water level rises from 25 mL to 35 mL. This means that 10 mL of water is displaced and that the volume of the object is m10 mL.

Figure 1.13
Determining the density of a solid by using volume displacement.

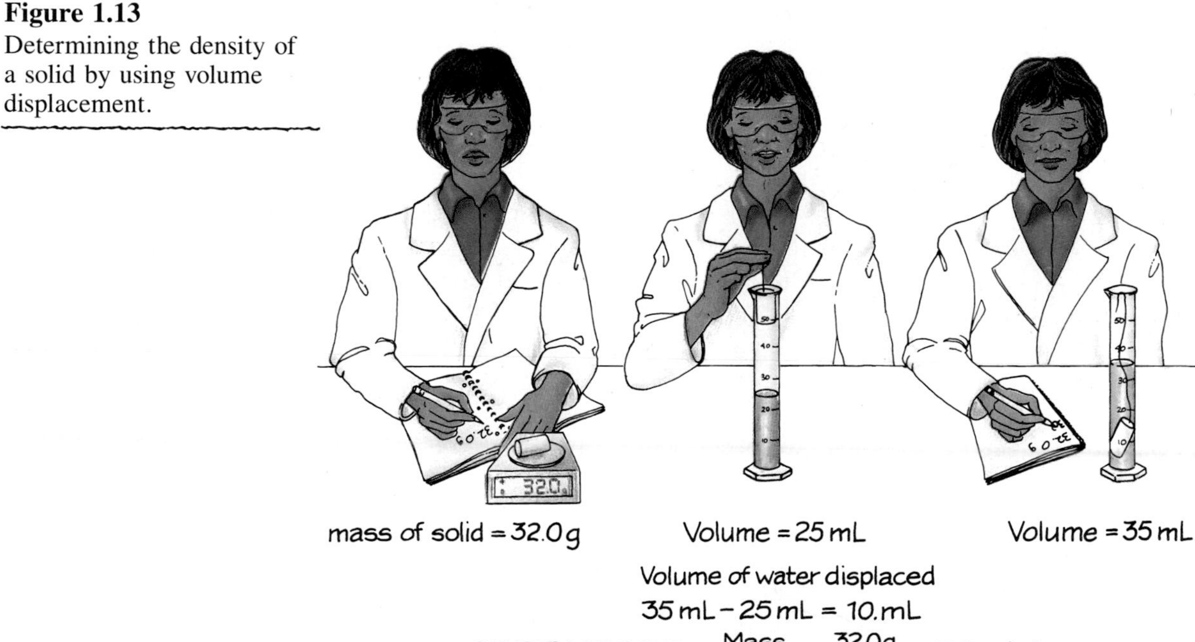

Sample Problem 1.15
Using Volume Displacement to Calculate Density

A lead weight used in the belt of a scuba diver has a mass of 226 g. When the weight is carefully placed in a graduated cylinder containing 200.0 mL of water, the water level rises to 220.0 mL. What is the density of the lead weight?

Solution

Both the mass and volume of the lead weight are needed to calculate its density. Its mass, 226 g, is given. The volume of the lead weight is equal to the volume of water displaced, which is calculated as follows:

$$\text{Volume displaced} \quad 220.0\text{ mL} - 200.0\text{ mL} = 20.0\text{ mL} \quad \text{Addition to first decimal place}$$

$$\text{Volume of the lead weight} = 20.0\text{ mL}$$

$$\text{Density of lead} = \frac{\text{mass of lead}}{\text{volume of lead}} = \frac{226\text{ g (Three SFs)}}{20.0\text{ mL (Three SFs)}} = 11.3\text{ g/mL} \quad \text{Three SFs in answer}$$

In the expression for density, be sure to use the volume of water the object displaces and *not* the original volume of water.

Study Check

A total of 0.50 lb of glass marbles is added to 425 mL of water. The water level rises to a volume of 528 mL. What is the density of the glass marbles?

Why do Objects Sink or Float?

Whether an object sinks or floats is determined by the relative densities of the object and the substance it is immersed in. Take a look at Figure 1.14, which shows samples of lead and cork in water. Because lead is more dense than water, it sinks. However, the cork floats, because it is less dense than water.

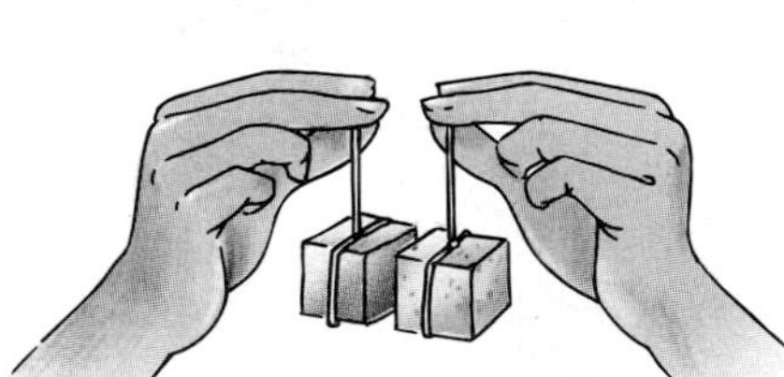

Figure 1.14 Differences in density determine whether an object will sink or float.

Sample Problem 1.16
Sink or Float

From their densities, determine whether each of the following substances will sink or float when placed in sea water, which has a density of 1.025 g/mL.

Substance	*Density*
Gasoline	0.66 g/mL
Asphalt	1.2 g/mL
Cardboard	0.69 g/cm^3

Solution

The gasoline and cardboard will float because they are less dense than sea water. Asphalt will sink because it is more dense than water.

Study Check

Using Table 1.11, explain why helium-filled balloons float in air at 25°C.

Problem Solving Using Density as a Conversion Factor

Because density is a comparison of two units, it can be used as a conversion factor. For example, if the mass and the density of a sample are known, the volume of the sample can be calculated.

Sample Problem 1.17
Problem Solving Using Density

The cast on Malcolm's leg has a density of 2.32 g/cm^3. What is the volume of the cast if it weighs 4.25 lb?

Solution

Step 1 **Given:** 4.25 lb

Step 2 **Unit Plan:** lb → g → cm^3 Unit for answer

Step 3 **Relationships and Conversion Factors:**

$$\text{Metric–American:}\quad 1\text{ lb} = 454\text{ g};\ \frac{454\text{ g}}{1\text{ lb}}\quad\text{and}\quad\frac{1\text{ lb}}{454\text{ g}}$$

$$\text{Density:}\quad 2.32\text{ g} = 1\text{ cm}^3;\ \frac{2.32\text{ g}}{1\text{ cm}^3}\quad\text{and}\quad\frac{1\text{ cm}^3}{2.32\text{ g}}$$

Step 4 **Problem Setup:**

$$4.25\ \cancel{\text{lb}} \times \frac{454\ \cancel{\text{g}}}{1\ \cancel{\text{lb}}} \times \frac{1\text{ cm}^3}{2.32\ \cancel{\text{g}}} = 832\text{ cm}^3$$

Given; Metric–American Factor; Density factor; Answer to three SFs

Study Check

Milk has a density of 1.04 g/mL. What is the mass of milk in a 1-qt container, if 1 qt = 946 mL?

HEALTH NOTE *Determination of Percentage of Body Fat*

Body mass is made up of protoplasm, extracellular fluid, bone, and adipose tissue. One way to determine the amount of adipose tissue is to measure the whole-body density. After the on-land mass of the body is determined, the underwater body mass is obtained by submerging the person in water. (See Figure 1.15.) Because water helps support the body by giving it buoyancy, the underwater body mass is less. A higher percentage of body fat will make a person more buoyant, causing the underwater mass to be even lower. This occurs because fat has a lower density than the rest of the body.

The difference between the on-land mass and underwater mass, known as the buoyant force, is used to determine the body volume. Several adjustments such as subtracting the volume of air trapped in the lungs and the intestine are made. Then the mass and volume of the person are used to calculate body density. For example, suppose a 70.0-kg person has a body volume of 66.7 L. The body density is calculated as

$$\frac{\text{Body mass}}{\text{Body volume}} = \frac{70.0 \text{ kg}}{66.7 \text{ L}}$$
$$= 1.05 \text{ kg/L} \quad \text{or} \quad 1.05 \text{ g/mL}$$

When the body density is determined, it is compared with a chart that correlates the percentage of adipose tissue with body density. A person with a body density of 1.05 g/mL has 21% body fat, according to such a chart. This procedure is used by athletes to determine exercise and diet programs.

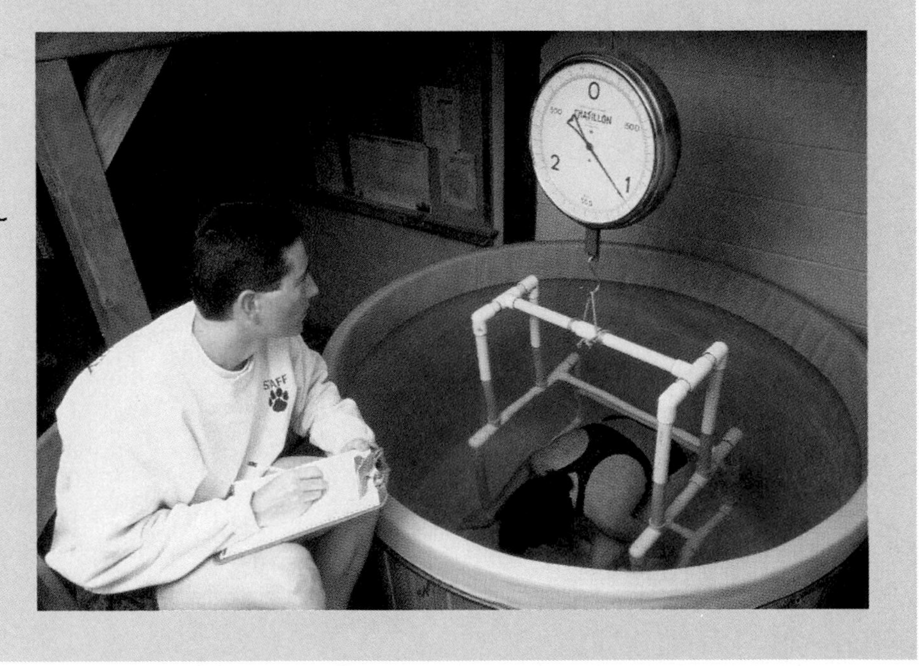

Figure 1.15
A person is submerged in a water tank to determine underwater body mass and to calculate percentage of body fat.

Specific Gravity

Specific gravity (sp gr) is a ratio between the density of a substance and the density of water. Specific gravity is calculated by dividing the density of a sample by the density of water. In this text, we will use the metric value of 1.00 g/mL for the density of water. A substance with a specific gravity of 1.00

Figure 1.16
The specific gravity of a beer is measured with a hydrometer.

has the same density as water. A substance with a specific gravity of 3.00 is three times as dense as water; whereas a substance with a specific gravity of 0.50 is just one-half as dense as water.

$$\text{Specific gravity} = \frac{\text{density of sample}}{\text{density of water}}$$

In the calculations for specific gravity, the units of density must match. Then, all units cancel to leave only a number. *Specific gravity is one of the few unitless values you will encounter in chemistry.*

An instrument called a hydrometer is often used to measure the specific gravity of fluids such as battery fluid or a sample of urine. In Figure 1.16, a hydrometer is used to measure the specific gravity of a fluid.

Sample Problem 1.18
Specific Gravity

What is the specific gravity of coconut oil that has a density of 0.925 g/mL?

Solution

$$\text{sp gr oil} = \frac{\text{density of oil}}{\text{density of water}} = \frac{0.925\ \cancel{\text{g/ml}}}{1.00\ \cancel{\text{g/ml}}} = 0.925 \quad \text{No units}$$

Study Check
What is the specific gravity of ice if 35.0 g of ice has a volume of 38.2 mL?

Health Note *Specific Gravity of Urine*

The specific gravity of urine is often determined as part of a laboratory evaluation of the health of an individual. The specific gravity of urine is normally in the range 1.003–1.030. This is somewhat greater than the specific gravity of water because compounds such as urea are dissolved in water in the kidneys to form urine.

If the specific gravity of a person's urine is too low or too high, a physician might suspect a problem with the kidneys. For example, if a urine sample shows a specific gravity of 1.001, significantly lower than normal, malfunctioning of the kidneys is a possibility.

Sample Problem 1.19
Problem Solving with Specific Gravity

John took 2.0 teaspoons (tsp) of cough syrup (sp gr 1.20) for a persistent cough. If there are 5.0 mL in 1 tsp, what was the mass (in grams) of the cough syrup?

Solution

Step 1 **Given:** 2.0 tsp

Step 2 **Unit Plan:** tsp → mL → g Unit for answer

Step 3 **Relationships and Conversion Factors:**

$$1 \text{ tsp} = 5.0 \text{ mL}; \quad \frac{1 \text{ tsp}}{5.0 \text{ mL}} \quad \text{and} \quad \frac{5.0 \text{ mL}}{1 \text{ tsp}}$$

For problem solving, it is convenient to convert the specific gravity of a sample to its density.

Density of sample = sp gr of sample × density of water

$$1.20 \times \frac{1.00 \text{ g}}{1 \text{ mL}} = 1.20 \text{ g/mL}$$

Relationships: 1 mL = 1.20 g

$$\textbf{Conversion Factors:} \quad \frac{1 \text{ mL}}{1.20 \text{ g}} \quad \text{and} \quad \frac{1.20 \text{ g}}{1 \text{ mL}}$$

Step 4 **Problem Setup:**

$$2.0 \cancel{\text{tsp}} \times \frac{5.0 \cancel{\text{mL}}}{1 \cancel{\text{tsp}}} \times \frac{1.20 \text{ g}}{1 \cancel{\text{mL}}} = 12 \text{ g of syrup}$$

Study Check

An ebony carving has a mass of 275 g. If ebony has a specific gravity of 1.33, what is the volume of the carving?

Chapter Summary

1.1 Units of Measurements
In science, physical quantities are described in units of the metric or International System (SI). Some important units are meter (m) for length, liter (L) for volume, and gram (g) and kilogram (kg) for mass.

1.2 Metric Prefixes and Equalities
Prefixes placed in front of a unit change the size of the unit by factors of 10. Prefixes such as *centi, milli,* and *micro* provide smaller units; prefixes such as *kilo* provide large units. An equality can be written for two metric units that measure the same quantity of length, volume, or mass. Examples of metric equalities are: 1 m = 100 cm; 1 L = 1000 mL; 1 kg = 1000 g.

1.3 Conversion Factors
Conversion factors are used to express a relationship in the form of a fraction. Two factors can be written for any relationship in the metric or American system.

1.4 Significant Figures
Significant figures are the numbers reported in a measurement including the last estimated digit. Zeros in front of a decimal number or at the end of a large number are not significant. When doing calculations, answers must be limited to the number of significant figures used in the problem.

1.5 Problem Solving Using Conversion Factors
Factors are useful when changing a quantity expressed in one unit to one expressed in another

unit. A useful approach to problem solving is to identify the given quantity and its unit, write a unit plan, select the appropriate conversion factors and place after the given. The factors in a problem setup are arranged to cancel the preceding unit and provide the desired unit. The final answer is reported with the correct number of significant figures.

1.6 Density and Specific Gravity
The density of a substance is a ratio of its mass to its volume, usually g/mL or g/cm^3. The units of density can be used as a factor to convert between the mass and volume of a substance. Specific gravity (sp gr) is a unitless relationship of the density of a substance divided by the density of water, 1.00 g/mL.

Glossary of Key Terms

centimeter (cm) A unit of length in the metric system; there are 2.54 cm in 1 in.

conversion factor A ratio in which the numerator and denominator are quantities from an equality or given relationship. For example, the conversion factors for the relationship 1 kg = 2.20 lb are written as the following:

$$\frac{2.20 \text{ lb}}{1 \text{ kg}} \quad \text{and} \quad \frac{1 \text{ kg}}{2.20 \text{ lb}}$$

cubic centimeter (cm^3, cc) The volume of a cube that has 1-cm sides, equal to 1 mL.

density The relationship of the mass of an object to its volume expressed as grams per cubic centimeter (g/cm^3), grams per milliliter (g/mL), or grams per liter (g/L).

gram (g) The metric unit used in measurements of mass.

equality A relationship between two units that measure the same quantity.

exact number A number obtained by counting or definition.

kilogram (kg) A metric mass of 1000 g, equal to 2.20 lb. The kilogram is the SI standard unit of mass.

liter (L) The metric unit for volume that is slightly larger than a quart.

mass A measure of the quantity of material in an object.

meter (m) The metric unit for length that is slightly longer than a yard. The meter is the SI standard unit of length.

metric system A system of measurement used by scientists and in most countries of the world.

milliliter (mL) A metric unit of volume equal to one-thousandth of a L (0.001 L).

prefix The part of the name of a metric unit that precedes the base unit and specifies the size of the measurement. All prefixes are related on a decimal scale.

SI units An International System of units that modifies the metric system.

significant figures The numbers recorded in a measurement.

specific gravity (sp gr) A unitless relationship between the density of a substance and the density of water:

$$\text{sp gr} = \frac{\text{density of sample}}{\text{density of water}}$$

volume The amount of space occupied by a substance.

Chemistry at Home

1. Read the labels on some food products in your refrigerator or on kitchen shelves. Write out some of the relationships you see. How can you use them to derive a metric–American conversion factor?
2. Calculate your height and weight in metric (SI) units.
3. Fill a bucket or sink with water. Place cans of diet and nondiet soft drinks in the water. What happens? Using information on the label how might you account for your observations?
4. Design an experiment to determine the most dense in each of the following:
   a. water and vegetable oil
   b. water and ice
   c. rubbing alcohol and ice
   d. vegetable oil, water, and ice

Answers to Study Checks

1.1 length
1.2 **a.** megasecond **b.** centimeter
1.3 **a.** 1000 **b.** 0.001
1.4 Equality: 1 in. = 2.54 cm

Conversion Factors: $\frac{1 \text{ in.}}{2.54 \text{ cm}}$ and $\frac{2.54 \text{ cm}}{1 \text{ in.}}$

1.5 **a.** two **b.** one **c.** five
1.6 **a.** 36 m **b.** 0.0026 L **c.** 3800 g **d.** 1.3 kg
1.7 **a.** 0.4924 **b.** 0.0080 or 8.0×10^{-3} **c.** 2.0
1.8 **a.** 83.70 mg **b.** 0.5 L
1.9 12 cups
1.10 1.888 L
1.11 3.3 lb
1.12 0.44 oz
1.13 650 mg
1.14 8.9 g/cm^3
1.15 2.2 g/mL
1.16 Air has a density of 1.29 g/L. Because helium has a density of 0.179 g/L, a helium-filled balloon is less dense than air and floats in air.
1.17 984 g of milk
1.18 0.916
1.19 207 mL

Problems

Answers to the odd-numbered problems are found in Appendix E of this book.

Units of Measurement *(Goal 1.1)*

1.1 What are the base units used for measuring length in the metric system and the American system?

1.2 What are the base units used for measuring volume and mass in the metric system and the American system?

1.3 What is the name of the unit and the type of measurement (mass, volume, or length) indicated for each of the following quantities?
a. 4.8 m **b.** 325 g
c. 1.5 mL **d.** 480 m

1.4 What is the name of the unit and the type of measurement (mass, volume or length) indicated for each of the following quantities?
a. 0.8 L **b.** 3.6 m **c.** 4 kg **d.** 35 g

Metric Prefixes and Equalities *(Goal 1.2)*

1.5 How does the prefix *kilo* affect the gram unit in *kilogram*?

1.6 How does the prefix *centi* affect the meter unit in *centimeter*?

1.7 Write the abbreviation for each of the following units:
a. milligram **b.** deciliter **c.** kilometer
d. kilogram **e.** microliter

1.8 Write the complete name for each the following units:
a. cm **b.** kg **c.** dL
d. mm **e.** μg

1.9 Write the numerical values for each of the following prefixes:
a. centi **b.** kilo **c.** milli
d. deci **e.** mega

1.10 Write the complete name (prefix + unit) for each of the following numerical values:
a. 0.10 g **b.** 10 g **c.** 1000 g
d. 1/100 g **e.** 0.001 g

1.11 Complete the following metric relationships:
a. 1 m = ____ cm **b.** 1 km = ____ m
c. 1 mm = ____ m **d.** 1 L = ____ mL

1.12 Complete the following metric relationships:
a. 1 L = ____ dL **b.** 1 dL = ____ L
c. 1 g = ____ kg **d.** 1 g = ____ mg

1.13 For each of the following pairs, which is the larger unit?
a. milligram or kilogram
b. centiliter or microliter
c. cm or km

1.14 For each of the following pairs, which is the smaller unit?
a. mg or g
b. centimeter or millimeter
c. mL or μm

Conversion Factors *(Goal 1.3)*

1.15 Why does an equality such as 1 m = 100 cm always give two conversion factors?

1.16 What is the equality expressed by the conversion factor $\frac{1000\text{ g}}{1\text{ kg}}$?

1.17 Write a numerical relationship and conversion factors for each of the following statements:
- **a.** There are 3 ft in 1 yd.
- **b.** One mile is 5280 feet.
- **c.** One minute is 60 seconds.
- **d.** A car goes 27 mi on 1 gallon of gas.

1.18 Write a numerical relationship and conversion factors for each of the following statements:
- **a.** There are 4 quarts in 1 gallon.
- **b.** At the store, oranges are $1.29 per lb.
- **c.** There are 7 days in 1 week.
- **d.** One dollar has four quarters.

1.19 Write the numerical relationship and conversion factors for the following metric units:
- **a.** centimeters and meters
- **b.** milligrams and grams
- **c.** liters and milliliters
- **d.** deciliters and milliliters

1.20 Write the numerical relationship and corresponding conversion factors for the following metric–American units:
- **a.** centimeters and inches
- **b.** pounds and kilograms
- **c.** pounds and grams
- **d.** quarts and milliliters

Significant Figures *(Goal 1.4)*

1.21 When is it important to count significant figures in a number?

1.22 What are exact numbers?

1.23 Are the numbers, in each of the following statements, measured numbers or exact numbers?
- **a.** A patient weighs 155 lb.
- **b.** The basket holds 8 apples.
- **c.** In the metric system, 1 kg is equal to 1000 g.
- **d.** The distance from Denver, Colorado, to Houston, Texas, is 1720 km.

1.24 Are the numbers, in each of the following statements, measured numbers or exact numbers?
- **a.** There are 31 students in the laboratory.
- **b.** The oldest known flower lived 120,000,000 years ago.
- **c.** The largest gem ever found, an aquamarine, had a mass of 104 kg.
- **d.** A laboratory test shows a blood cholesterol level of 184 mg/dL.

1.25 When are zeros in a measured number counted as significant figures? When are they not significant?

1.26 How are the number of significant figures determined for answers of multiplication or division? Of addition and subtraction?

1.27 How many significant figures are in each of the following measured quantities?
- **a.** 11.005 g **b.** 0.00032 m
- **c.** 36,000,000 km **d.** 185.34 kg

1.28 How many significant figures are in each of the following measured quantities?
- **a.** 20.60 mL **b.** 1036.48 g
- **c.** 4.00 m **d.** 5.005 kL

1.29 Round off each of the following numbers to three significant figures.
- **a.** 1.854 **b.** 184.2038
- **c.** 0.004738265 **d.** 8807
- **e.** 1.832149

1.30 Round off each of the numbers in problem 1.29 to two significant figures.

1.31 For the following multiplication and division problems, give answers with the correct number of significant figures:
- **a.** 45.7×0.034 **b.** 0.00278×5
- **c.** $\frac{34.56}{1.25}$ **d.** $\frac{(0.2465)(25)}{1.78}$

1.32 For the following multiplication and division problems, give answers with the correct number of significant figures:
- **a.** 400×185 (1) **b.** $\frac{2.40}{(4)(125)}$ (1)
- **c.** $0.825 \times 3.6 \times 5.1$ 3 2 (2) **d.** $\frac{3.5 \times 0.261}{8.24 \times 20.0}$ (2)

1.33 For the following addition and subtraction problems, give answers rounded off to the correct number of significant figures:
- **a.** 45.48 cm + 8.057 cm
- **b.** 23.45 g + 104.1 g + 0.025 g
- **c.** 145.675 mL − 24.2 mL
- **d.** 1.08 L − 0.585 L

1.34 For the following addition and subtraction problems, give answers rounded off to the correct number of significant figures:
- **a.** 5.08 g + 25.1 g
- **b.** 85.66 cm + 104.10 cm + 0.025 cm
- **c.** 24.568 mL − 14.25 mL
- **d.** 0.2654 L − 0.2585 L

Problem Solving Using Conversion Factors *(Goal 1.5)*

1.35 How would you explain conversion factors to a friend?

1.36 What advice would you give to your friend for solving problems?

1.37 Use American conversion factors to solve the following problems:
- **a.** How many yards are in 24 ft? (1 yd = 3 ft)
- **b.** How many seconds are in 15 min?
- **c.** You need 3.5 qt of oil for your car. How many gallons is that? (1 gal = 4 qt)

1.38 Use American conversion factors to solve the following problems:
- **a.** A necklace contains 2.6 oz of gold. How many pounds is that, if 1 lb = 16 oz?
- **b.** You ran a total of 2.4 miles today. How far is that in feet? (1 mi = 5280 ft)
- **c.** One game at the arcade requires one quarter. You have \$3.50 in your pocket. How many games can you play?

1.39 Use metric conversion factors (Table 1.6 or Appendix B) to solve the following problems:
- **a.** A student's height is 175 centimeters. How tall is the student, in meters?
- **b.** A cooler has a volume of 5500 mL. What is the capacity of the cooler, in liters?
- **c.** A hummingbird has a mass of 0.055 kg. What is the mass of the hummingbird, in grams?

1.40 Use metric conversion factors (Table 1.6 or Appendix B) to solve the following problems:
- **a.** The Daily Value of phosphorus is 800 mg. How many grams of phosphorus are recommended?
- **b.** A glass of orange juice contains 0.85 dL of juice. How many milliliters of orange juice is that?
- **c.** A package of chocolate instant pudding contains 2840 mg of sodium. How many grams of sodium is that?

1.41 Solve the following problems using one or more conversion factors (see Table 1.7 or Appendix B for metric–American equalities):
- **a.** A container holds 0.750 qt. How many milliliters of lemonade will it hold?
- **b.** What is the mass in kilograms of a person who weighs 165 lb?
- **c.** The femur, or thigh bone, is the longest bone in the body. In a 6-ft-tall person, the femur might be 19.5 in. long. What is the length of that person's femur, in millimeters?
- **d.** A person on a diet has been losing weight at the rate of 3.5 lb per week. If the person has been on the diet for 6 weeks, how many kilograms were lost?

1.42 Solve the following problems using one or more conversion factors (see Table 1.7 or Appendix B for metric–American equalities):
- **a.** You need 4.0 ounces of a steroid ointment. If there are 16 oz in 1 lb, how many grams of ointment does the pharmacist need to prepare?
- **b.** During surgery, a patient receives 5.0 pints of plasma. How many milliliters of plasma were given? (1 qt = 4 pt)
- **c.** A piece of plastic tubing measuring 560 mm in length is needed for an intravenous setup. How many feet of tubing are required?
- **d.** Zippy the snail moves at the rate of 2.0 inches per hour. How many centimeters does Zippy travel in 4.0 hours?

1.43 Using conversion factors, solve the following clinical problems:
- **a.** You have used 250 L of distilled water for a dialysis patient. How many gallons of water is that? (*Hint:* L → qt → gal)
- **b.** A patient needs 0.024 g of a sulfa drug. There are 8-mg tablets in stock. How many tablets should be given?
- **c.** The daily dose of ampicillin for the treatment of an ear infection is 115 mg/kg of body weight. What is the daily dose for a 34-lb toddler?

1.44 Using conversion factors, solve the following clinical problems:
- **a.** The physician has ordered 1.0 g of tetracycline to be given every 6 hours to a patient. If your stock on hand is 500-mg tablets, how many will you need for 1 day's treatment?
- **b.** An intramuscular medication is given at 5.00 mg/kg of body weight. If you give 425 mg of medication to a patient, what is the patient's weight in pounds?
- **c.** A physician has ordered 325 mg of atropine, intramuscularly. If atropine is available as 0.50 g/mL, how many milliliters would you need to give?

Density and Specific Gravity *(Goal 1.6)*

1.45 What is the density (g/mL) of each of the samples described?
- **a.** 20.0 mL of a salt solution that has a mass of 24.0 g.
- **b.** A solid object with a mass of 1.65 lb and a volume of 170 mL.
- **c.** A gem that has a mass of 45.0 g. When placed in a graduated cylinder containing 20.0 mL of water, the water level rises to 34.5 mL.

1.46 What is the density of each of the samples described?
- **a.** A medication, if the contents of a syringe filled to 3.00 mL has a mass of 3.85 g?
- **b.** The fluid in a car battery, if it has a volume of 125 mL and a mass of 155 g?
- **c.** A 5.00-mL urine sample from a patient suffering from symptoms resembling those of diabetes mellitus. The mass of the urine sample is 5.025 g.

1.47 Use density values to solve the following problems:

a. What is the mass, in grams, of 150 mL of a liquid with a density of 1.4 g/mL?
b. What is the mass of a glucose solution that fills a 500-mL intravenous bottle if the density of the glucose solution is 1.15 g/mL?
c. Kari, a sculptor, has prepared a mold for casting a bronze figure. The figure has a volume of 225 mL. If bronze has a density of 7.8 g/mL, how many ounces of bronze does Kari need to melt in the preparation of the bronze figure?

1.48 Use density values to solve the following problems:
a. A graduated cylinder contains 18.0 mL of water. What is the new water level after 35.6 g of silver metal with a density of 10.5 g/mL is submerged in the water?
b. A thermometer containing 8.3 g of mercury has broken. If mercury has a density of 13.6 g/mL, what volume spilled?
c. A fish tank holds 35 gallons of water. Using the density of 1.0 g/mL for water, determine the number of pounds of water in the fish tank.

1.49 Solve the following specific gravity problems:
a. A urine sample has a density of 1.030 g/mL. What is the specific gravity of the sample?
b. A liquid has a volume of 40.0 mL and a mass of 45.0 g. What is the specific gravity of the liquid?
c. The specific gravity of an oil is 0.85. What is its density?

1.50 Solve the following specific gravity problems:
a. A 5% glucose solution has a specific gravity of 1.02. What is the mass of 500 mL of glucose solution?
b. A bottle containing 325 g of cleaning solution is used for carpets. If the cleaning solution has a specific gravity of 0.850, what volume of solution was used?
c. Butter has a specific gravity of 0.86. What is the mass, in grams, of 0.250 L of butter?

Challenge Problems

1.51 A fish company delivers 22 kg of salmon, 5.5 kg of crab, and 3.48 kg of oysters to your seafood restaurant. The prices are:

Salmon	\$3.75/lb
Crab	6.25/lb
Oysters	5.70/lb

a. What is the total mass, in kilograms, of the seafood?
b. What was the total cost in dollars?

1.52 In France, grapes are 14 francs per kilogram. What is the cost of grapes in dollars per pound, if the exchange rate is 5.5 francs per dollar?

1.53 Bill's recipe for onion soup calls for 2.0 lb of thinly sliced onions. If an onion has an average mass of 100 g, how many onions does Bill need?

1.54 Celeste's diet restricts her intake of protein to 24 g per day. If she eats an 8.0-oz burger that is 15.0% protein, has she exceeded her protein limit for the day? How many ounces of a burger would be allowed for Celeste? (See Appendix D for calculations using percentage.)

1.55 Some of the nutrition information listed on a box of crackers includes:

Serving size	1/2 ounce (6 crackers)
Fat	4 grams
Sodium	140 milligrams

a. If the box has a net weight (contents only) of 8 oz, about how many crackers are in the box?
b. If you ate 10 crackers, how many ounces of fat are you consuming?
c. How many grams of sodium are used to prepare crackers for 50 boxes?

1.56 A dialysis unit requires 75,000 mL of distilled water. How many gallons of water are needed? (1 gal = 4 qt)

1.57 A letter has arrived for you that reads as follows:

Dear Cousin:

How are you? I hear you are taking chemistry now. Maybe you can help me. I have been working in our mine and recently found a brilliant stone that looks like a diamond, but I am not sure. Do you know how I could determine its density? I read in a mineralogy magazine that diamond has a density of 3.51 g/cm^3. Please describe how I can determine the density of this stone, including the tools I need to buy, the measurements I need to make, and the necessary calculations. I hope you are enjoying your chemistry class.

Sincerely, Cousin Emma

Write a letter to Emma using complete sentences.

1.58 A graduated cylinder contains 155 mL of water. A 15.0-g piece of iron (Density = 7.86 g/cm^3) and a 20.0-g piece of lead (Density = 11.3 g/cm^3) are added. What is the new water level in the cylinder?

1.59 How many cubic centimeters (cm^3) of olive oil have the same mass as 1.00 L of gasoline?

1.60 Ethyl alcohol has a specific gravity of 0.79. What is the volume, in quarts, of 1.50 kg of alcohol?

Chapter 2

Energy and Matter

The three states of matter for water are ice, liquid, and gas.

Looking Ahead

Learning Goals

2.1 Describe some forms of energy.

2.2 Given a temperature, calculate a corresponding temperature on another scale.

2.3 Given the mass of a sample, and the temperature change, calculate the heat lost or gained.

2.4 Calculate the energy of a sample using calorimetry data.

2.5 Identify the physical state of a substance as a solid, liquid, or gas.

2.6 Draw a diagram that represents the changes of state for matter.

2.7 Calculate the heat absorbed or released by a sample of water during a change of state.

Questions to Think About

1. How does the body regulate increases or decreases in body temperature?
2. What are some forms of energy in your room?
3. Why is it important that energy can change form?
4. What objects in your home change energy from one form to another?
5. What are some states of matter you can see around you?
6. What is meant by the statement "150 Calories per serving" on a food nutrition label?

An object at rest requires energy to make it move. An object that is moving needs energy to make it stop. We say that energy makes it possible to do work. You use energy to do the work of running, playing tennis, studying, and breathing. You obtain that energy from the foods you eat. Energy takes many forms, but you are probably most familiar with heat or thermal energy. In your home, you use thermal energy to cook food, heat water, warm the room, and dry your hair.

Matter is everything around us that has mass and occupies space. Matter can be a solid, liquid, or gas. For example, the water in an ice cube, an ice rink, or an iceberg is solid. The water running out of a faucet or in a pool is a liquid. Water becomes a gas when it evaporates from wet clothes or boils away in a pan on the stove. Many gases are invisible to us, but some are detected by a characteristic odor or color. For example, you know when a bottle of perfume or ammonia is opened because you smell the gas as it fills the room.

The state of matter can be changed by adding or removing heat. When you heat water in a tea kettle, the water gets hotter. Perhaps it boils and changes to a gas. You make ice cubes by using a freezer to remove heat from the water in an ice cube tray.

2.1 Energy

Learning Goal

Describe some forms of energy.

When you are running, walking, dancing, or thinking, you are using energy to do **work**. In fact, **energy** is defined as the ability to do work. Suppose you are climbing a steep hill. Perhaps you become too tired to go on. We could say that you do not have sufficient energy to do any more work. Now, suppose you sit down and have lunch. In a while you will have obtained some energy from the food, and you will be able to do more work and complete the climb.

Potential and Kinetic Energy

All energy is potential or kinetic. **Potential energy** is stored energy, whereas **kinetic energy** is the energy of motion. A boulder resting on top of a mountain has potential energy because of its location. If the boulder rolls down the mountain, the potential energy becomes kinetic energy. Water stored in a reservoir has potential energy. When the water goes over the dam, the stored energy becomes kinetic energy. Even the food you eat has potential energy. When you digest food, you convert its stored energy to kinetic energy to do biological work.

Sample Problem 2.1
Forms of Energy

Identify the energy in each of the following as potential or kinetic:
a. gasoline **b.** a waterfall **c.** a candy bar

Solution
a. potential energy (stored) **b.** kinetic energy (Motion)
c. potential energy (stored)

Study Check
Would the energy in a stretched rubber band be kinetic or potential?

Forms of Energy

Solar energy from the sun is stored as chemical energy during the formation of trees, plants, and fuels. When wood is burned, chemical energy is released as heat or thermal energy, and light, which is radiant energy. **Heat** is associated with the motion of particles in a substance. A frozen pizza feels cold because the particles in the pizza have little heat, or thermal energy, and are moving very slowly. As heat is added, the motions of the particles increase, and the pizza becomes warm. Eventually the particles have enough thermal energy to make the pizza hot and ready to eat.

Figure 2.1
Examples of electrical energy converted into light energy, mechanical energy,

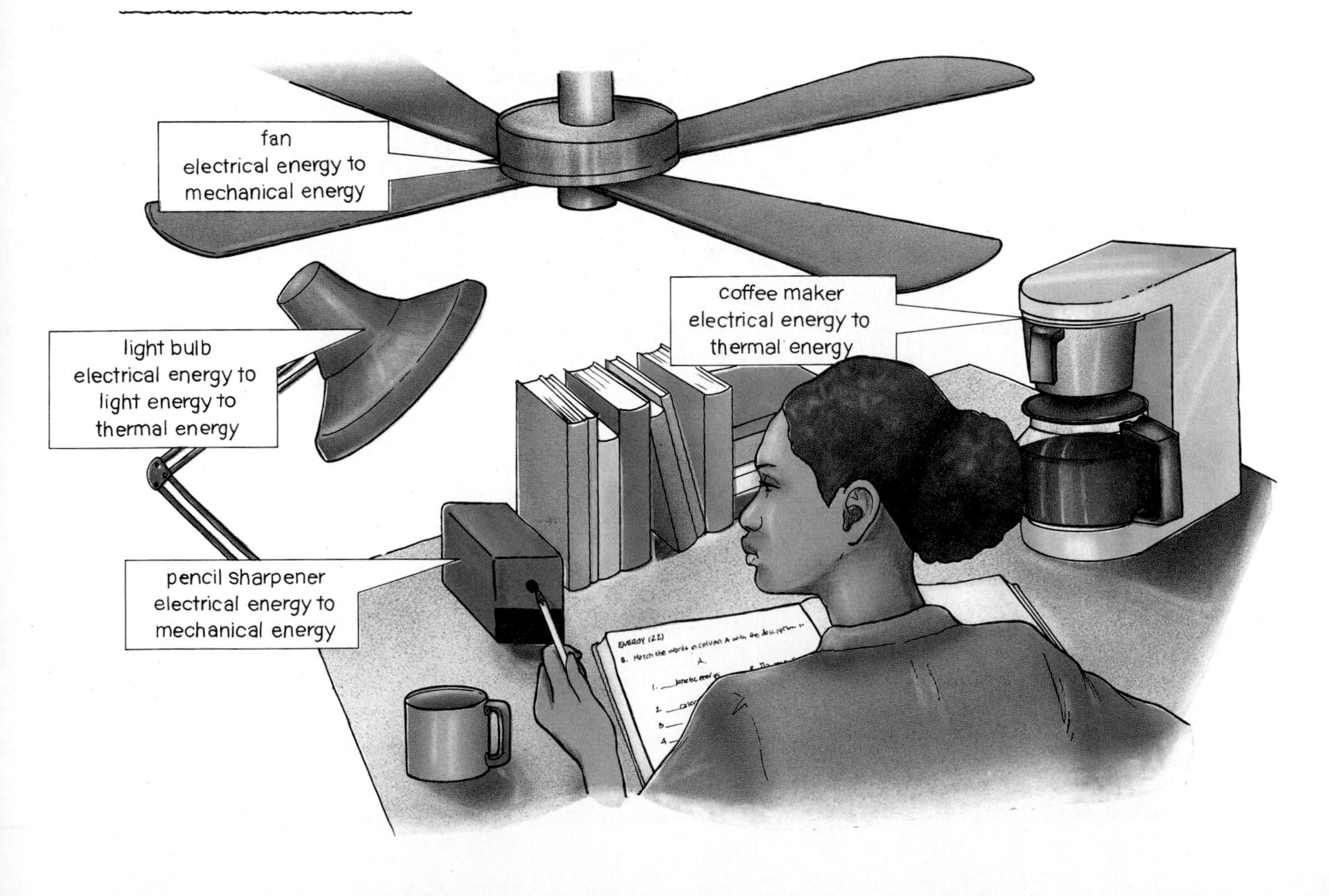

In an electrical power plant, thermal energy from the burning of fossil fuels such as natural gas or coal is used to produce steam, which turns the turbines that produce electrical energy. In your home, electrical energy is converted to radiant energy when you switch on a light bulb, to mechanical energy when you use a mixer or a washing machine, and to thermal energy when you use a hair dryer or a toaster (see Figure 2.1). In your body, chemical energy from the food you eat is converted to mechanical energy when you use muscles to ride a bicycle, run a marathon, or mow the lawn.

Environmental Note *Global Warming*

The amount of carbon dioxide (CO_2) gas in our atmosphere is on the increase as we burn more gasoline, coal, and natural gas. The oceans and forests normally absorb carbon dioxide, but they cannot keep up with the continued increase. The cutting of trees in the rain forests (deforestation) reduces the amount of carbon dioxide removed from the atmosphere. Many of the trees are also burned as land is cleared. It has been estimated that deforestation may account for 15–30% of the carbon dioxide that goes into the atmosphere each year.

Some scientists think that the carbon dioxide in the atmosphere may be acting like the glass in a greenhouse. Sunlight passes through to warm the Earth's surface, but the heat it produces is trapped by the carbon dioxide as shown in Figure 2.2. It is not yet clear how severe the effects of global warming might be. Some scientists estimate that by around the year 2030, the atmospheric level of carbon dioxide could double and cause the temperature of the Earth's atmosphere to rise by 2–5°C. If that should happen, it could have a profound impact on the Earth's climate. For example, an increase in the melting of snow and ice could raise the ocean levels by as much as 2 m, which is enough to flood many cities located on the ocean shorelines.

Worldwide efforts already are being made to reduce fossil fuel use and to slow or stop deforestation. It will require cooperation throughout the world to avoid the bleak future that some scientists predict should the global warming continue unchecked.

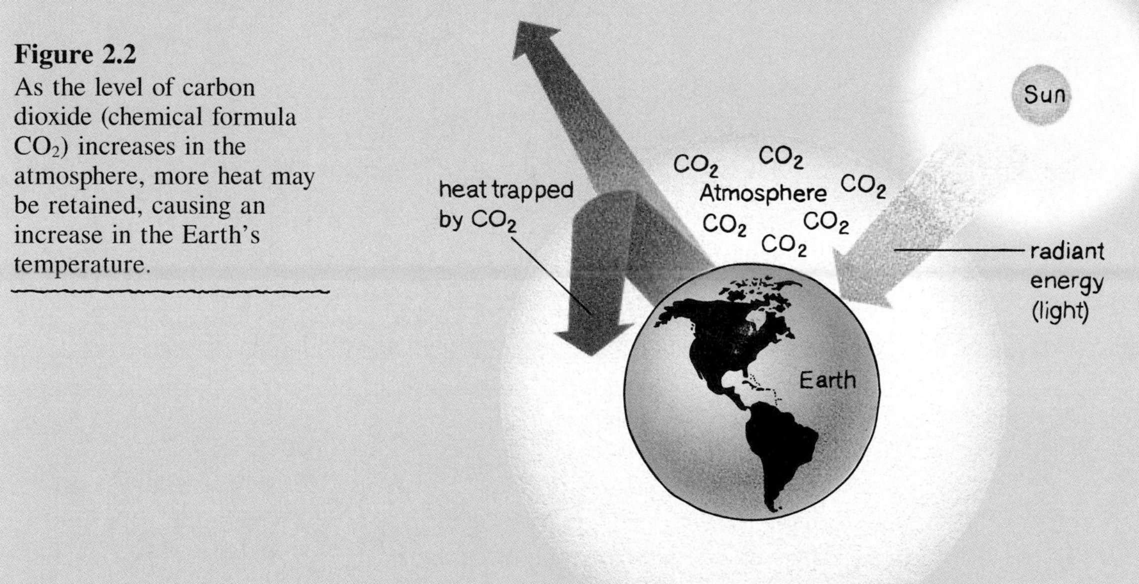

Figure 2.2
As the level of carbon dioxide (chemical formula CO_2) increases in the atmosphere, more heat may be retained, causing an increase in the Earth's temperature.

2.2 Temperature Measurement

Learning Goal

Given a temperature, calculate a corresponding temperature on another scale.

When we take the **temperature** of a substance, we are measuring the intensity of heat. Temperature tells us the direction that heat will flow between two substances. Heat always flows from a substance with a higher temperature to a cooler one until the temperatures of both substances are the same. When you drink hot coffee or touch a hot stove, heat flows to your mouth or hand, which is at a lower temperature. When you touch an ice cube, it feels cold because heat flows from your hand to the colder ice cube.

We use thermometers to determine how "hot" or "cold" a substance is. When a thermometer is placed in a substance, the heat flow causes the liquid in the thermometer, alcohol, or mercury to rise or fall. When it stops, the thermometer and the substance have reached the same temperature.

Celsius and Fahrenheit Temperatures

Temperatures in science, and in most of the world, are measured and reported in *Celsius (°C)* units. In the United States, everyday temperatures are commonly reported in *Fahrenheit (°F)* units. A typical room temperature of 22°C would be the same as 72°F. A normal body temperature of 37°C is 98.6°F.

On the Celsius and Fahrenheit scales, the temperatures of melting ice and boiling water are used as reference points. On the **Celsius scale,** the freezing point of water is 0°C, and the boiling point is 100°C. On a **Fahrenheit scale,** water freezes at 32°F and boils at 212°F. On each scale, the temperature difference between freezing and boiling is divided into smaller units, or degrees. On the Celsius scale, there are 100 units between the freezing and boiling of water compared with 180 units on the Fahrenheit scale. That makes a Celsius unit almost twice the size of a Fahrenheit degree 1°C = 1.8°F (See Figure 2.3.)

$$\text{180 Fahrenheit units} = \text{100 Celsius units}$$

$$\frac{\text{180 Fahrenheit units}}{\text{100 Celsius units}} \quad \text{or} \quad \frac{\text{1.8 Fahrenheit units}}{\text{1 Celsius unit}}$$

In chemistry, a laboratory, or the hospital, you will measure temperatures in Celsius units. If you do need to convert to Fahrenheit, multiply the Celsius temperature by the above conversion factor to change to Fahrenheit units, and add 32 degrees. The 32 you add adjusts the zero point on the Celsius scale of 0°C to 32°F on the Fahrenheit scale.

$$°F = \underbrace{\frac{1.8°F}{1°C}}_{\text{Changes °C to °F}}(°C) + \underbrace{32°}_{\text{Adjusts zero points}} \longrightarrow °F = 1.8\,(°C) + 32°$$

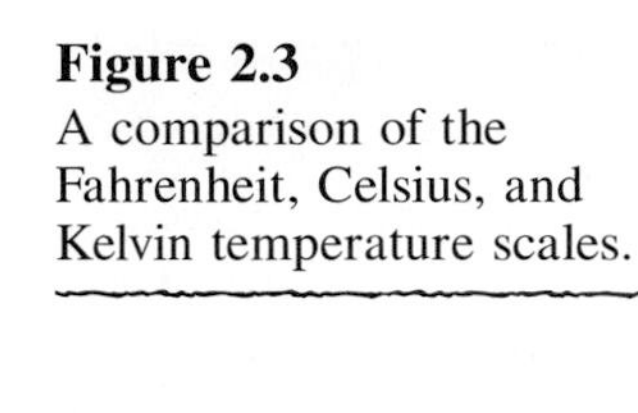

Figure 2.3
A comparison of the Fahrenheit, Celsius, and Kelvin temperature scales.

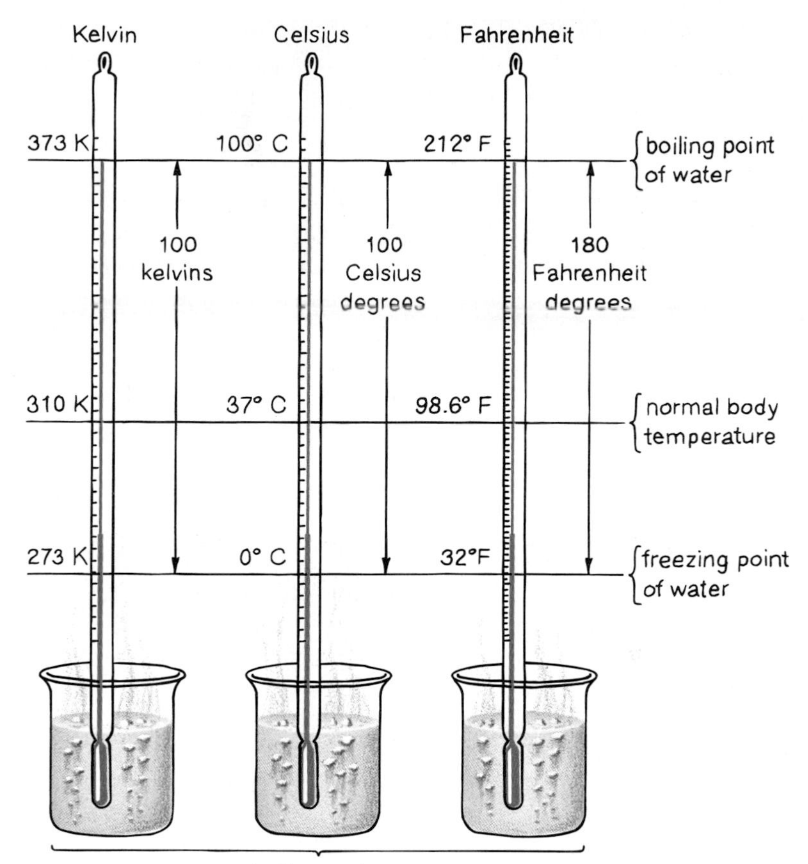

Sample Problem 2.2
Temperature Conversion

While traveling in China, James discovers that his temperature is 38.2°C. What is his temperature on a Fahrenheit thermometer?

Solution
Using the temperature conversion equation, we write:

$$°F = 1.8(°C) + 32°$$

In the temperature expression, *the values of 1.8 and 32 are exact numbers.* Therefore, the answer is reported to the same decimal place as the initial temperature.

$$\begin{aligned} °F &= 1.8(38.2) + 32 \\ &= 68.8 + 32 \quad \text{Exact} \\ &= 100.8°F \quad \text{Answer to tenth's place} \end{aligned}$$

Study Check
When making ice cream, Azita uses rock salt to chill the mixture. If the temperature drops to −11°C, what is it in °F?

Sample Problem 2.3
Converting to Celsius Temperature

Donyelle is going to cook a turkey at 325°F. If she uses an oven thermometer with Celsius units, at what temperature should she set the oven?

Solution

To convert from Fahrenheit to Celsius, we rearrange the temperature equation.

$$°F = 1.8\,(°C) + 32$$

$$°F - 32 = 1.8\,(°C)$$

$$\frac{(°F - 32)}{1.8} = °C$$

Entering the temperature of 325°F into the equation gives:

$$°C = \frac{(325 - 32)}{1.8} = \frac{\overset{\text{Three SFs}}{293}}{\underset{\text{Exact}}{1.8}} = 163°C$$

On your calculator, first subtract (325 − 32), and then divide by 1.8.

Study Check

Your patient, Mr. Lee, has a temperature of 103.6°F. What is his temperature on a Celsius thermometer?

HEALTH NOTE *Variation in Body Temperature*

Normal body temperature is considered to be 37.0°C, although it varies throughout the day and from person to person. Oral temperatures of 36.1°C are common in the morning and climb to a high of 37.2°C between 6 P.M. AND 10 P.M. Elevations of temperature above 37.2°C for a person at bed rest are usually an indication of disease. Individuals involved in prolonged exercise may also experience elevated temperatures. Body temperatures of marathon runners can range from 39°C to 41°C as heat production during exercise exceeds the body's ability to lose heat.

Changes of more than 3.5°C from the normal body temperature begin to interfere with bodily functions. Temperatures above 41°C can lead to convulsions, particularly in children, and cause permanent brain damage. Heatstroke (hyperpyrexia) occurs above 41.1°C. Sweat production stops, and the skin becomes hot and dry. The pulse rate is elevated, and respiration becomes weak and rapid. The person can become lethargic and lapse into a coma. Damage to internal organs is a major concern, and treatment, which must be immediate, may include immersing the person in an ice-water bath.

In hypothermia, body temperature can drop as low as 28.5°C. The person may appear cold and pale and have an irregular heartbeat. Unconsciousness can occur if the body temperature drops below 26.7°C. Respiration becomes slow and shallow, and oxygenation of the tissues decreases. Treatment involves providing oxygen and increasing blood volume with glucose and saline fluids. Internal temperature may be restored by injecting warm fluids (37.0°C) into the peritoneal cavity.

Table 2.1 *A Comparison of Temperatures*

Example	Fahrenheit (°F)	Celsius (°C)	Kelvin (K)
Sun	9937	5503	5776
A hot oven	450	232	505
A desert	120	49	322
A high fever	104	40	313
Room temperature	72	22	295
Water freezes	32	0	273
A northern winter	−76	−60	213
Helium boils	−452	−269	4
Absolute zero	−459	−273	0

Kelvin Temperature Scale

Scientists tell us that the coldest temperature possible is −273°C (more accurately, −273.16°C). On the **Kelvin scale,** this temperature, called *absolute zero,* has the value of 0 Kelvin (0 K). Units on the Kelvin scale are called kelvins (K); no degree symbol is used. Because there are no lower temperatures, the Kelvin scale has no negative numbers. Between the freezing and boiling points of water, there are 100 kelvins, which makes a kelvin equal in size to a Celsius unit.

$$1\ \text{K} = 1°\text{C}$$

To calculate a Kelvin temperature, add 273 to the Celsius temperature:

$$\text{K} = °\text{C} + 273$$

Table 2.1 gives a comparison of some temperatures on the three scales.

Sample Problem 2.4
Converting Temperature to Kelvins

What is a normal body temperature of 37°C on the Kelvin scale?

Solution
To convert from Celsius temperature to a Kelvin temperature, we use the equation:

$$\text{K} = °\text{C} + 273$$
$$\text{K} = 37 + 273$$
$$= 310\ \text{K}$$

Study Check
You are cooking a pizza at 375°F. What is that temperature in kelvins? (*Hint:* Convert to Celsius temperature first.)

Sample Problem 2.5
Converting from Kelvin to Celsius Temperature

In a cryogenic laboratory, Miriam freezes a sample using liquid nitrogen at 77 K. What is the temperature in °C?

Solution
To find the Celsius temperature, we rearrange the Kelvin equation:

$$K = °C + 273$$
$$K - 273 = °C$$

Using our given temperature of 77 K, we calculate the Celsius temperature for the freezing of the sample.

$$°C = 77\ K - 273 = -196°C$$

Study Check
On the planet Mercury, the average night temperature is 13 K, and the average day temperature is 683 K. What are these temperatures in Celsius degrees?

2.3 Measuring Heat Energy

Learning Goal
Given the mass of a sample, and the temperature change, calculate the heat lost or gained.

In chemistry and the health sciences, heat is commonly measured in calories, from Latin *caloric* meaning "heat." A **calorie (cal)** is the amount of heat needed to raise the temperature of 1 g of water by 1°C. A **kilocalorie (kcal),** which is also the nutritional "big" *Calorie,* is used for larger amounts of heat. The **joule (J),** pronounced "jewel," is the SI unit of energy. One calorie is the same amount of energy as 4.18 joules. One kilojoule is 1000 joules.

$$1\ kcal = 1000\ calories$$
$$1\ cal = 4.18\ J$$
$$1\ kJ = 1000\ J$$

Specific Heat

All substances absorb energy, but some need only a small amount of heat to become hot and others need more. The amount of heat that raises the temperature of 1 g of a substance by 1°C is known as its **specific heat.**

$$\text{Specific heat} = \frac{\text{calories}}{1\ g \times 1°C}$$

On a hot day at the beach, the ocean water is cool to the touch, the air is warm; but the sand is so hot, it can burn your feet. Sand has a lower specific heat than water. Therefore, sand reaches a higher temperature than water when it absorbs the same amount of heat. Because water has a high specific heat, it

Table 2.2
Specific Heats of Some Substances

Substance	Specific Heat (cal/g °C)
Water	1.00
Alcohol	0.58
Wood	0.42
Aluminum	0.22
Sand	0.19
Iron	0.106
Copper	0.092
Silver	0.057
Gold	0.031

functions in the body to absorb large amounts of heat without causing fluctuations in body temperature.

When you cook, you may use aluminum or copper pans. If you compare specific heats, you find that the amount of heat that raises 1 g of water 1°C increases the temperature of 1 g of aluminum by about 5°C, and 1 g of copper by 10°C. Materials such as aluminum and copper can heat quickly to cook the food. Table 2.2 lists the specific heats of a variety of materials.

Calculations Using Specific Heat

The heat gained or lost by a substance is calculated by multiplying the mass of the substance, the temperature change, and its specific heat. The temperature change often noted as ΔT is the difference in the temperatures of the sample before and after heat is added or removed. (The Greek letter Delta, symbol Δ, means "change in"; read ΔT as "delta T.")

$$\text{Heat gain or loss} = \text{mass} \times \underset{\text{change } (\Delta T)}{\text{temperature}} \times \text{specific heat}$$

$$\text{Calories} = \text{g} \times {}^\circ\text{C} \times \frac{\text{cal}}{\text{g } {}^\circ\text{C}}$$

$$\text{Joules} = \text{g} \times {}^\circ\text{C} \times \frac{\text{J}}{\text{g } {}^\circ\text{C}}$$

Sample Problem 2.6
Calculating Heat Energy

How many calories are used to warm 15 g of water from 12°C to 36°C?

Solution
The temperature change (ΔT) is the difference between the final and initial temperatures.

$$\Delta T = 36^\circ\text{C} - 12^\circ\text{C} = 24^\circ\text{C}$$

The heat is calculated by multiplying the mass of the water, the temperature change, and the specific heat of water (1.00 cal/g °C).

$$\underset{\text{Mass}}{15\ \cancel{\text{g}}} \times \underset{\Delta T}{24^\circ\cancel{\text{C}}} \times \underset{\text{Specific heat}}{\frac{1.00\ \text{cal}}{\cancel{\text{g}}\ ^\circ\cancel{\text{C}}}} = \underset{\text{Heat energy}}{360\ \text{cal}}$$

Study Check
How many calories are needed to heat 30.0 g of copper from 75°C to 125°C if the specific heat of copper is 0.092 cal/g °C?

Sample Problem 2.7
Calculating Heat Loss in Joules

A 225-g sample of hot tea cools from 95.0°C to 23.0°C. What is the energy loss in joules assuming that tea has the same specific heat as water (4.18 J/g °C)?

Solution

$$\Delta T = 95.0°C - 23.0°C = 72.0°C$$

The number of joules lost during cooling is calculated as follows:

$$\underset{\text{Mass}}{225\ \cancel{g}} \times \underset{\Delta T}{72.0\cancel{°C}} \times \underset{\text{Specific heat}}{\frac{4.18\ J}{\cancel{g}\ \cancel{°C}}} = \underset{\text{Heat released}}{67{,}700\ J} \quad \text{or} \quad 6.77 \times 10^4\ J$$

Study Check
How much heat in joules is released when 15 g of gold cools from 215°C to 35°C? (See Table 2.2.)

2.4 Measuring Energy with a Calorimeter

Learning Goal
Calculate the energy of a sample using calorimetry data.

When you are watching your food intake, the Calories you are counting are actually kilocalories. In the field of nutrition, it is common to use the **Calorie, Cal** (with an uppercase C) to mean 1000 cal, or 1 kcal. Nutritional values may also be given in kilojoules (kJ).

Nutritional Calories

1 Cal = 1 kcal
1 kcal = 1000 cal
1 kcal = 4.18 kJ

1 cal = 4.18 J

The number of Calories in a food is determined by using an apparatus called a **calorimeter,** shown in Figure 2.4. A sample is placed in a steel container within the calorimeter and water is added to fill a surrounding chamber. When the food burns (combustion), the heat causes an increase in the temperature of the water. By calculating the kilocalories (or Calories) absorbed by the water, we can determine the caloric content of the food as shown in Sample Problem 2.8. The caloric content of many foods, including the breakfast foods given in Table 2.3, are found this way.

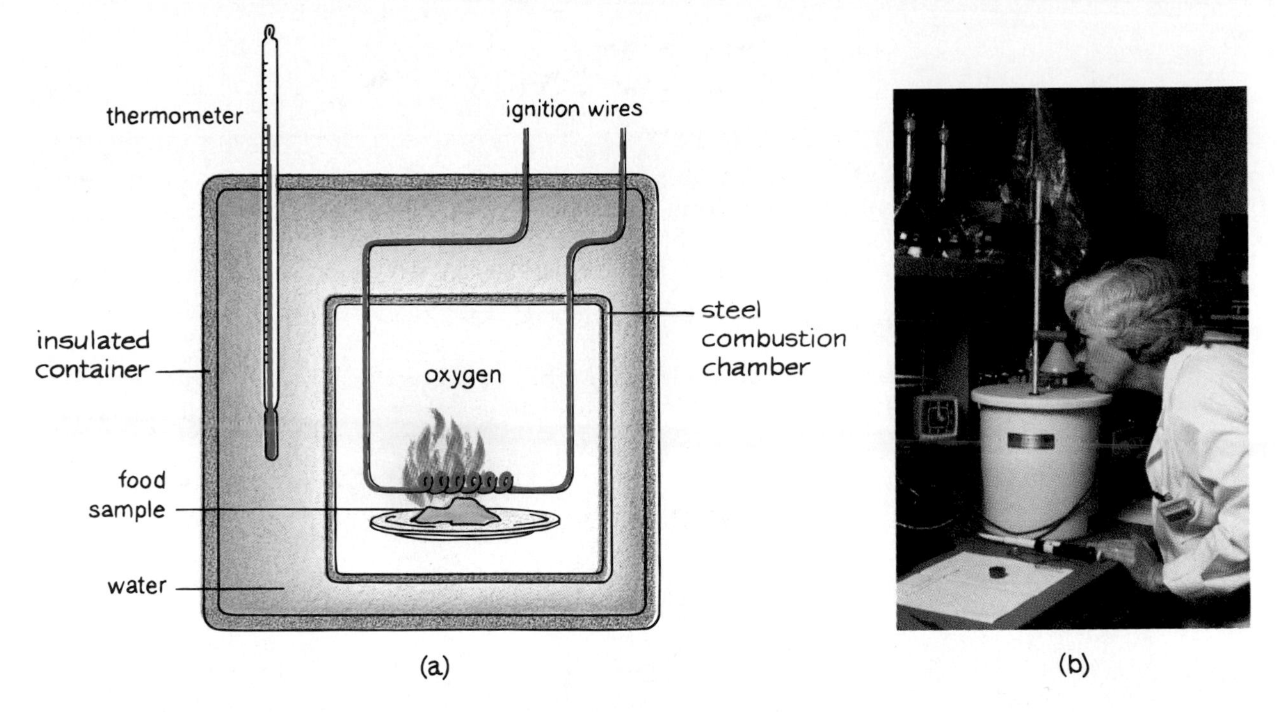

Figure 2.4
(a) A calorimeter. The energy released by the combustion of a food sample is determined by measuring the mass of the food sample, the mass of water in the calorimeter, and the change in temperature of the water.
(b) The temperature change is observed during the combustion of a food sample in a calorimeter.

Table 2.3 ***Energy Values of Some Breakfast Foods***

Item	Quantity	Energy (kcal)
Breads and Cereals		
Bagel	1 medium, plain	135
Corn flakes	1 cup	100
Doughnut	1	125
Muffin	1, 3-in. diameter	120
Oatmeal	1 cup	130
Pancake	2, 4-in. diameter	120
Toast	1 slice, wheat	65
Fruit		
Banana	1 medium	110
Cantaloupe	1/2	80
Grapefruit	1/2, pink	40
Fruit Juice		
Apple	1 cup	120
Orange	1 cup	110
Tomato	1 cup	35
Meats, Eggs, and Milk		
Bacon	2 slices	180
Eggs	2 scrambled with milk and oil	220
Milk	1 cup, 4%	165
	1 cup, 2%	130
	1 cup, nonfat	85

Sample Problem 2.8
Calculating Food Energy

A slice of whole-wheat bread is placed in a calorimeter. The mass of the water is 1000 g at an initial temperature of 22°C. The heat given off during the combustion of the bread raises the temperature of the surrounding water to 87°C. What is the energy in Calories (kilocalories) for the slice of bread?

Solution

Step 1 **Calculate heat absorbed by water**
The heat absorbed by the water in the calorimeter is calculated as follows:

$$\underset{\text{Mass of water}}{1000\ \cancel{g}\ \text{water}} \times \underset{\Delta T}{65\cancel{^\circ C}} \times \underset{\text{Specific heat}}{\frac{1\ \text{cal}}{\cancel{g}\ \cancel{^\circ C}}} = 65{,}000\ \text{cal}$$

To convert to kilocalories, we use the equality:

$$1\ \text{kcal} = 1000\ \text{cal}$$

$$65{,}000\ \cancel{\text{cal}} \times \frac{1\ \text{kcal}}{1000\ \cancel{\text{cal}}} = 65\ \text{kcal} = 65\ \text{Cal}$$

Step 2 **Assign Calories to food sample**
Since the combustion of the bread provided the heat to warm the water, we can say that a slice of whole-wheat bread contains 65 kcal. In nutritional Calories, it is the same as 65 Cal.

Energy of 1 slice of whole-wheat bread = 65 kcal or 65 Cal

Study Check
When 0.20 oz of a chocolate candy bar undergoes combustion in a calorimeter, the temperature of the surrounding 500.0 g of water rises from 18°C to 76°C. How many kilocalories (Calories) are available in 1.0 oz of chocolate?

Caloric Food Values

The **caloric values** are the energy in kilocalories per gram of the three types of food: carbohydrates, fats, and proteins. These values determined by calorimetry are listed in Table 2.4. If the composition of a food is known in terms of the mass of each food type, the caloric content, the total number of Calories, can be calculated.

$$\text{Kilocalories} = \underset{\text{Mass of carbohydrate, fat, or protein}}{\cancel{g}} \times \underset{\text{Caloric value}}{\frac{\text{kcal}}{\cancel{g}}}$$

Table 2.4 ***Caloric Values for the Three Food Types***

Food Type	Carbohydrate	Fat (lipid)	Protein
Caloric value	$\frac{4\ \text{kcal}}{1\ \text{g}}$	$\frac{9\ \text{kcal}}{1\ \text{g}}$	$\frac{4\ \text{kcal}}{1\ \text{g}}$

Figure 2.5
The nutritional facts on the label of a food package lists the Calories and the grams of each food type per serving.

The caloric content of a food is listed in the nutritional information on the package, usually in terms of the number of Calories in a serving, as shown in Figure 2.5. The general composition and caloric content of some foods are given in Table 2.5.

Table 2.5 ***General Composition and Caloric Content of Some Foods***

Food	Protein (g)	Fat (g)	Carbohydrate (g)	Energy (kcal[a])
Banana, 1 medium	1	trace	26	110
Beans, red kidney, 1 cup	15	1	42	240
Beef, lean, 3 oz	22	5	trace	130
Carrots, raw, 1 cup	1	trace	11	50
Chicken, no skin, 3 oz	20	3	0	110
Egg, 1 large	6	6	trace	80
Milk, 4% fat, 1 cup	9	9	12	165
Milk, nonfat, 1 cup	9	trace	12	85
Oil, olive, 1 tbs.	0	14	0	130
Potato, baked	3	trace	23	105
Salmon, 3 oz	17	5	0	115
Steak, 3 oz	20	27	0	320
Yogurt, lowfat, 1 cup	8	4	13	120

[a] Kcal values are rounded to nearest five.

Sample Problem 2.9
Calculating Caloric Values

A 2.3-g sample of butter, a fat, is placed in a calorimeter with 600. g of water at an initial temperature of 17°C. After combustion of the butter, the water has a temperature of 52°C. What is the caloric value (kcal/g) of a fat?

Solution

The heat absorbed by the water is calculated as follows:

$$600.\ \not{g} \times 35^\circ\not{C} \times \frac{1\ \text{cal}}{\not{g}\ ^\circ\not{C}} = 21{,}000\ \text{cal}$$

Mass of water × ΔT × Specific heat

$$21{,}000\ \not{\text{cal}} \times \frac{1\ \text{kcal}}{1000\ \not{\text{cal}}} = 21\ \text{kcal}$$

Because the fat provided the 21 kcal of heat, the caloric value of the fat is calculated as follows:

$$\frac{21\ \text{kcal}}{2.3\ \text{g fat}} = 9.1\ \text{kcal/g fat}$$

Study Check

A 5.4-g sample of glucose, a carbohydrate, is placed in a calorimeter. The water in the container has a mass of 1000. g, and an initial temperature of 24°C. After all the glucose is burned, the water temperature is 46°C. What is the caloric value for a carbohydrate?

Sample Problem 2.10
Caloric Content for a Food

How many kilocalories are in a piece of chocolate cake that contains 5 g of protein, 35 g of carbohydrate, and 10 g of fat?

Solution

Using the caloric values for carbohydrate, fat, and protein (Table 2.4), we can calculate the total number of kcal:

Carbohydrate: $35\ \not{g} \times \frac{4\ \text{kcal}}{1\ \not{g}} = 140\ \text{kcal}$

Fat: $10\ \not{g} \times \frac{9\ \text{kcal}}{1\ \not{g}} = 90\ \text{kcal}$

Protein: $5\ \not{g} \times \frac{4\ \text{kcal}}{1\ \not{g}} = 20\ \text{kcal}$

Total caloric content $= 250\ \text{kcal}$

Study Check

A 1-oz (28 g) serving of oat bran hot cereal with half a cup of whole milk contains 10 g of protein, 22 g of carbohydrate, and 7 g of fat. If you eat two servings of the oat bran for breakfast, how many kilocalories will you obtain?

HEALTH NOTE *Losing and Gaining Weight*

The number of kilocalories needed in your daily diet depends on your age and physical activity. Some general levels of energy needs are given in Table 2.6.

Table 2.6 ***Typical Energy Requirements***

Age (yr)	Weight (lb)	Mass (kg)	Energy (kcal)
Young Adult			
Female (13–19)	115	52	2400
Male (13–19)	130	59	2980
Adult			
Female	121	55	2200
Male	143	65	3000

When food intake exceeds energy output, a person's body weight increases. Food intake is usually regulated by the hunger center in the hypothalamus, located in the brain. The regulation of food intake is normally proportional to the nutrient stores in the body. If these nutrient stores are low, you feel hungry; if they are high, you do not feel like eating.

Weight reduction occurs when food intake is less than energy output. Many diet products contain cellulose, which has no nutritive value but provides bulk and makes you feel full. Some diet drugs depress the hunger center and must be used with caution because they excite the nervous system and can elevate blood pressure. Because muscular exercise is an important way to expend energy, an increase in daily exercise aids weight loss. Table 2.7 lists some activities and the amount of energy they require.

Table 2.7 ***Energy Expended by a 70-kg (155-lb) Person***

Activity	Energy Expended (kcal/hr)
Sleeping	60
Sitting	100
Walking	200
Swimming	500
Running	550

2.5 States of Matter

Learning Goal

Identify the physical state of a substance as a solid, liquid, or gas.

Matter is anything that occupies space and has mass. This book, the food you eat, the water you drink, your cat or dog, and the air you breathe are just a few examples of matter. Suppose I asked you to look into a mirror and describe what you see. You might tell me the color of your hair, eyes, and skin, or the length and texture of your hair. Perhaps you have freckles or dimples. These are descriptions of **physical properties** or characteristics, properties of matter that can be observed or measured without affecting the identity of a substance. Substances are also described by their chemical properties, which we will discuss in Chapter 5. The physical properties of a substance include its shape, physical state, color, and its melting or boiling point. A sample of copper shown in Figure 2.6 has the physical properties listed in Table 2.8.

Figure 2.6
Cake molds made of copper metal.

Table 2.8
Some Physical Properties of Copper

Color	Reddish-orange
Odor	Odorless
Melting point	1083°C
Boiling point	2567°C
State at 25°C	Solid
Luster	Very shiny
Conduction of electricity	Excellent
Conduction of heat	Excellent

Figure 2.7
The solid states of sugar crystals, quartz, and amethyst, a purple form of quartz.

Figure 2.8
A liquid with a volume of 100 mL takes the shape of its container.

Figure 2.9
A gas takes the shape and volume of its container.

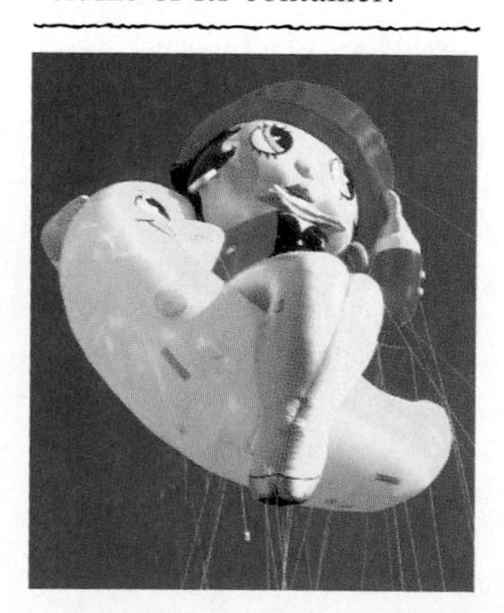

On Earth, matter exists in one of three *physical states*: solid, liquid, or gas. Because all matter is made up of tiny particles, each state is determined by the interaction between its particles. In a **solid,** strong attractive forces hold the particles so close together that they touch. They are arranged in such a rigid pattern that they can only vibrate slowly in their fixed positions. This gives a solid its own shape and own volume. For many solids, this rigid structure produces a crystal as seen in sugar, quartz, and amethyst (see Figure 2.7).

The most common liquid is water. In a **liquid,** the particles have enough energy to move freely in random directions. They are still close to each other with sufficient attractions to maintain a definite volume, but there is no rigid structure. Thus, when oil or vinegar is poured frcm one container to another, the liquid maintains its own volume but takes the shape of the new container (see Figure 2.8).

The air you breathe is made of gases, mostly nitrogen and oxygen. In a **gas,** the particles move so fast that they no longer attract each other. As a result, they can move far apart to put great distances between gas particles. This behavior allows gases to expand to fill their container or to compress if the container size is reduced. Gases have no definite shape or volume of their own; they take the shape and volume of their container, as shown in Figure 2.9. Table 2.9 compares some of the properties of the three states of matter.

Table 2.9 ***Some Properties of Solids, Liquids, and Gases***

Property	Solid	Liquid	Gas
Shape	Has its own shape	Takes the shape of the container	Takes the shape of its container
Volume	Has its own volume	Has its own volume	Fills the volume of the container
Arrangement of particles	Fixed, very close	Random, close	Random, far apart
Interaction between particles	Very strong	Strong	Essentially none
Movement of particles	Very slow	Moderate	Very fast
Examples	Ice Salt Iron	Water Oil Vinegar	Water vapor Helium Air

Sample Problem 2.11
States of Matter

Identify each of the following as a solid, a liquid, or a gas:

a. oxygen in the air
b. a substance with a fixed arrangement of particles
c. safflower oil
d. a substance that has great distances between its particles.

Solution

a. gas **b.** solid **c.** liquid **d.** gas

Study Check

What state of water has its own volume but takes the shape of its container?

2.6 Changes of State

Learning Goal
Draw a diagram that represents the changes of state for matter.

When you cut paper into strips, mince a clove of garlic, or roll clay into thin coils, you have made a physical change in a substance. When matter undergoes a **physical change,** its shape, size, appearance, or state may change but not its identity. Cutting paper or reshaping a piece of clay changes its shape or size but not the type of substance. Dissolving salt in water changes the appearance of the salt, but it can be reformed by heating the mixture and

Table 2.10 ***Examples of Some Physical Changes***

Change of shape	Hammering gold into gold leaf Drawing copper into thin wire
Change of size	Cutting cloth to make a shirt Grinding ginger root into fine particles for spice jars
Change of state	Steam fogging the mirror Making ice cubes in a freezer
Change of appearance	Dissolving sugar in tea

Figure 2.10
(a) Evaporation occurs at the surface of a liquid;
(b) evaporation and condensation in equilibrium in a closed container;
(c) boiling occurs as bubbles of gas form throughout the liquid.

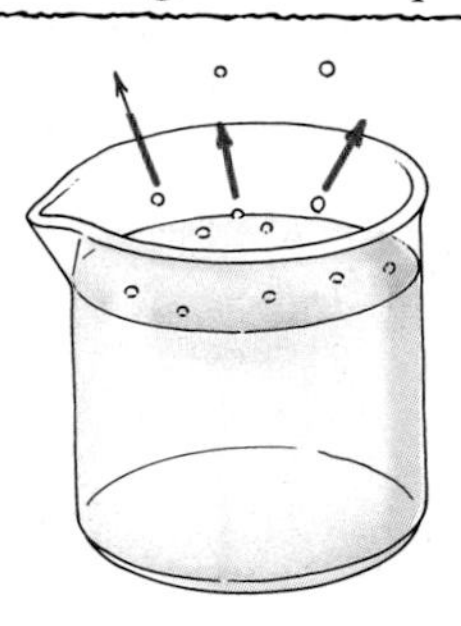
(a)

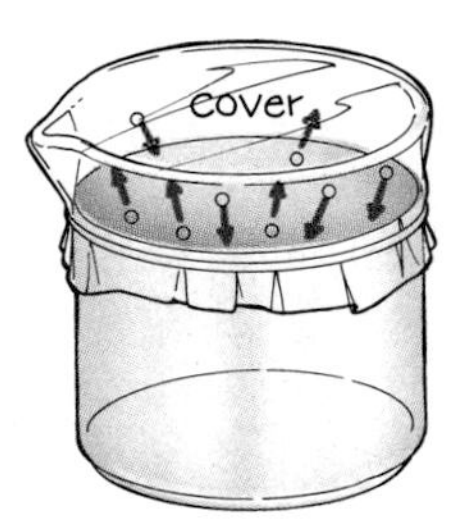

(b)

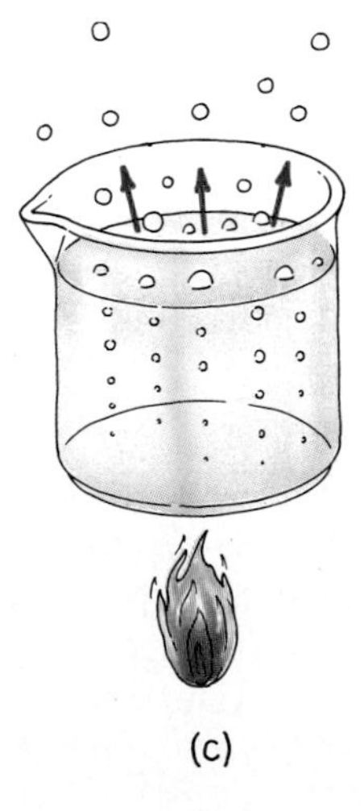
(c)

evaporating the water. Thus, in a physical change, no new substances are produced.

Matter undergoes a **change of state** when a change in temperature converts it to another state. When an ice cube melts in a drink, water boils in a teapot, or fog forms on windows on a cold morning, a change of state has occurred. Table 2.10 gives examples of some physical changes.

Melting and Freezing

When the temperature of a solid rises, its particles become so lively that they can overcome the forces holding them together. At a temperature called the **melting point (mp),** the particles separate from the solid and move about in random patterns. The substance is **melting,** changing from a solid to a liquid.

If the temperature is lowered, the reverse process takes place. Kinetic energy is lost, the particles slow down, and attractive forces pull the particles close together. The substance is **freezing.** A liquid changes to a solid at the **freezing point (fp),** which is the same temperature as the melting point. Freezing occurs when heat is removed, and melting occurs when heat is added. Every substance has its own freezing (melting) point: water freezes at 0°C; ice melts at 0°C; gold freezes (melts) at 1064°C; nitrogen freezes (melts) at −210°C.

Suppose we have a glass containing ice and water. At warm temperatures, the ice melts, forming more liquid. At cold temperatures, heat is lost and the liquid freezes. However, at the melting (freezing) point of 0°C, ice melts at the same rate that water freezes. No more ice melts as long as the water is at a temperature of 0°C. The reversible processes of melting and freezing are in balance, shown by writing double arrows in opposite directions. We say that the ice–liquid system has reached dynamic equilibrium.

$$\text{Solid} \underset{\text{Freezing}}{\overset{\text{Melting}}{\rightleftharpoons}} \text{liquid}$$

Evaporation and Condensation

Water in a mud hole disappears, unwrapped food dries out, and clothes hung on a line dry. Water **evaporates** when the faster moving water molecules have enough energy to escape from the surface of the liquid. (See Figure 2.10.) The loss of the "hot" water molecules removes heat and leaves the liquid cooler.

As the water vapor particles cool, they lose kinetic energy and slow down. In **condensation,** the water molecules form liquid again as attractive forces pull them together, a process that is the reverse of evaporation. You may have noticed that condensation forms on the mirror when you take a shower. Because heat is lost in condensation, it is a warming process. That is why, when a rain storm is approaching, we notice a warming of the air as water molecules in the air condense to rain.

HEALTH NOTE *Homeostasis: Regulation of Body Temperature*

In a physiological system of equilibrium called homeostasis, changes in our environment are balanced by changes in our bodies. It is crucial to our survival that we balance heat gain with heat loss. If we do not lose enough heat, our body temperature rises. At high temperatures, the body can no longer regulate our metabolic reactions. If we lose too much heat, body temperature drops. At low temperatures, essential functions proceed too slowly.

The skin plays an important role in the maintenance of body temperature. When the outside temperature rises, receptors in the skin send signals to the brain. The temperature-regulating part of the brain stimulates the sweat glands to produce perspiration. As perspiration evaporates from the skin, heat is removed and the body temperature is lowered.

In cold temperatures, epinephrine is released, causing an increase in metabolic rate, which increases the production of heat. Receptors on the skin signal the brain to contract the blood vessels. Less blood flows through the skin, and heat is conserved. The production of perspiration stops to lessen the heat lost by evaporation. A summary of the body's temperature regulation is illustrated in Figure 2.11.

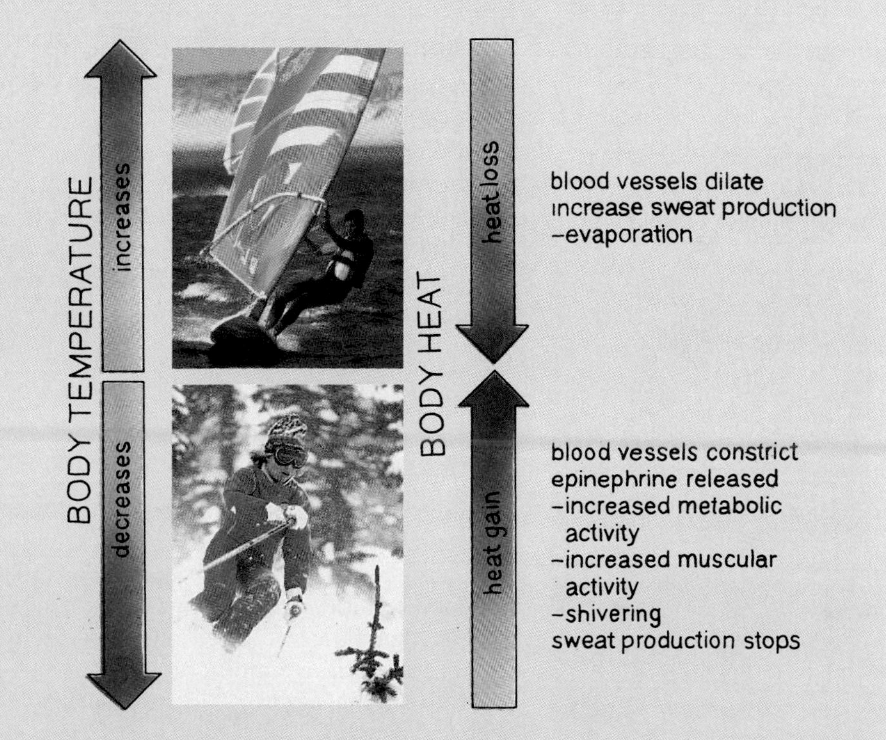

Figure 2.11
In homeostasis, a physiological equilibrium maintains a constant internal body temperature.

Liquid–Gas Equilibrium

With enough time, the water in an open container will all evaporate. However, if a tight-fitting cover is placed on the container, the water level will go down just a small amount. At first, the water molecules evaporate from the surface. Then, some of the vapor molecules begin to condense and return to liquid. Eventually, the number of evaporating molecules is equal to the number condensing. The reverse processes of evaporation and condensation have equalized. As a result of the liquid–gas equilibrium, the water level does not go any lower.

$$\text{Liquid} \underset{\text{Condensation}}{\overset{\text{Evaporation}}{\rightleftharpoons}} \text{gas (vapor)}$$

Boiling

As the heat added to a liquid increases, more and more particles can evaporate. At the **boiling point,** the particles throughout the liquid have the energy needed to change into a gas. The **boiling** of the liquid occurs as gas bubbles form throughout the liquid, rise to the surface, and escape. (See Figure 2.10c.)

Sublimation

In a process called **sublimation,** the particles on the surface of a solid absorb enough heat to change directly to a gas. For example, dry ice, which is solid carbon dioxide, sublimes at −78°C. It was called "dry" ice because it did not go through a liquid state to become a gas. You may have noticed that mothballs seem to disappear at room temperature. They have sublimed, a process we can detect by the odor of their vapor. In very cold areas, snow does not melt but sublimes directly to vapor. In a frost-free refrigerator, ice on the walls of the freezer sublimes when warm air is circulated through the compartment.

Figure 2.12
Freeze-dried foods are produced by sublimation.

$$\text{Solid} \underset{\text{Crystallization}}{\overset{\text{Sublimation}}{\rightleftharpoons}} \text{gas}$$

Freeze-dried foods prepared by sublimation are convenient for long storage, and for camping and hiking. (See Figure 2.12.) A food that has been frozen is placed in a vacuum chamber where it dries as the ice sublimes. The dried food retains all of its nutritional value and needs only water to be edible. A food that is freeze-dried does not need refrigeration because bacteria cannot grow without moisture. The reverse process of sublimation is called **crystallization.** Figure 2.13 summarizes our discussion of the various changes of state.

Figure 2.13
A summary of the changes of state.

Sample Problem 2.12
Identifying States and Changes of States

Give the state or change of state described in the following:

a. particles on the surface of a liquid escaping to form vapor
b. a liquid changing to a solid
c. gas bubbles forming throughout a liquid

Solution

a. evaporation **b.** freezing **c.** boiling

Study Check

What is the phase change that occurs in condensation?

Sample Problem 2.13
Physical Properties and Changes

Classify each of the following as a physical property or physical change:

a. cutting your nails **b.** aluminum foil being shiny
c. grinding peppercorns into flakes **d.** a melting ice cube

Solution

a. physical change **b.** physical property
c. physical change **d.** physical change

Study Check

Red tomatoes are chopped up to make salsa. Which part of this statement is a physical property, and which is a physical change?

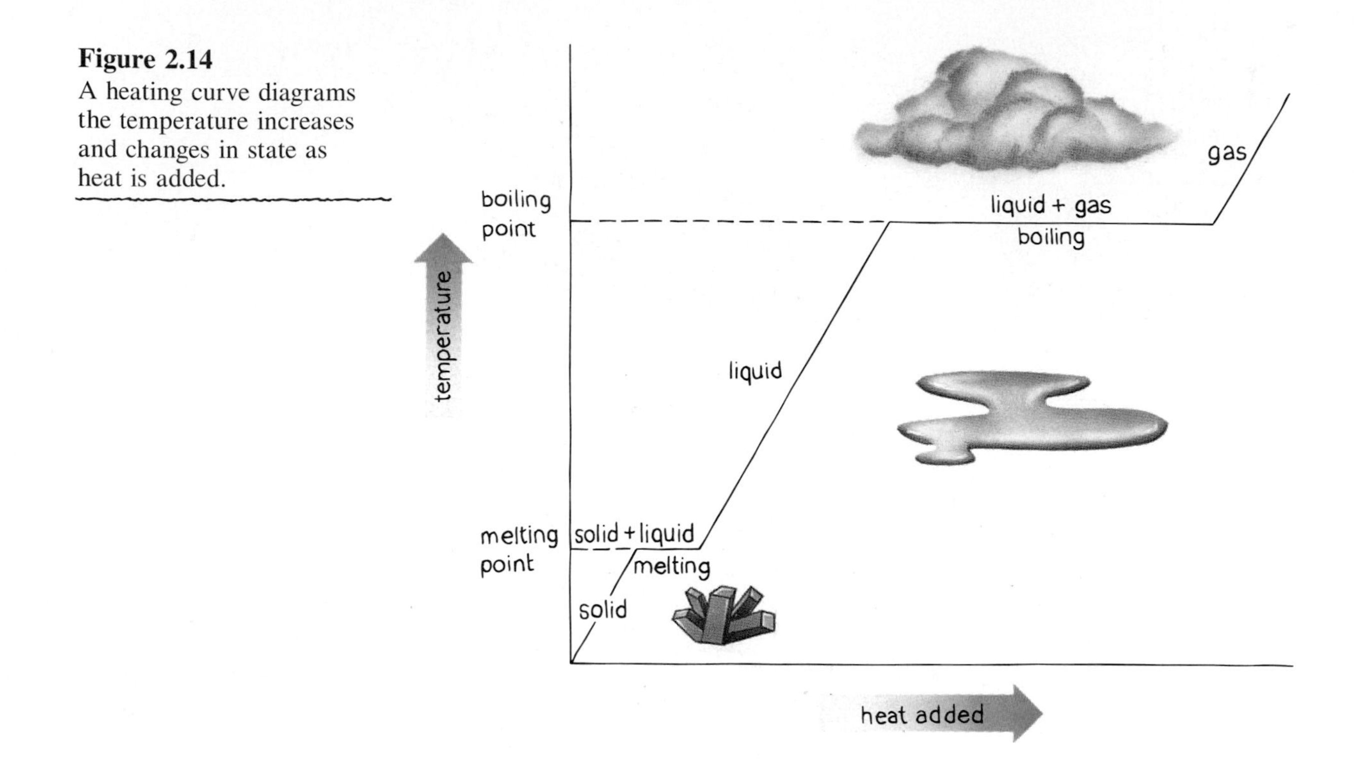

Figure 2.14
A heating curve diagrams the temperature increases and changes in state as heat is added.

Diagrams for Changes of State

When heat is added to a solid, it begins to melt, forming a liquid. As more heat is added, the liquid warms until it boils and forms a gas. Each of these changes of state can be illustrated visually as a diagram called a **heating curve.** (See Figure 2.14.) In a heating curve, the temperature is shown on the vertical axis. The progress of the reaction is shown on the horizontal axis. Each of the segments represents a warming step or a change of state during the heating of the solid.

Steps on a Heating Curve

The first diagonal line indicates a warming of the solid as the temperature rises. When the melting temperature is reached, solid begins to change to a liquid. There is no change in temperature as the solid changes to liquid. This is shown as a flat line, or plateau, at the melting point on the heating curve. During the melting process, all of the heat is used to break apart the particles of the solid while temperature remains constant.

After all the particles are in the liquid state, the temperature begins to rise again as shown by the next diagonal line. As more heat is added, the liquid warms and the particles move about more vigorously until the boiling point is reached. Now the liquid begins to boil indicated as another flat line on the

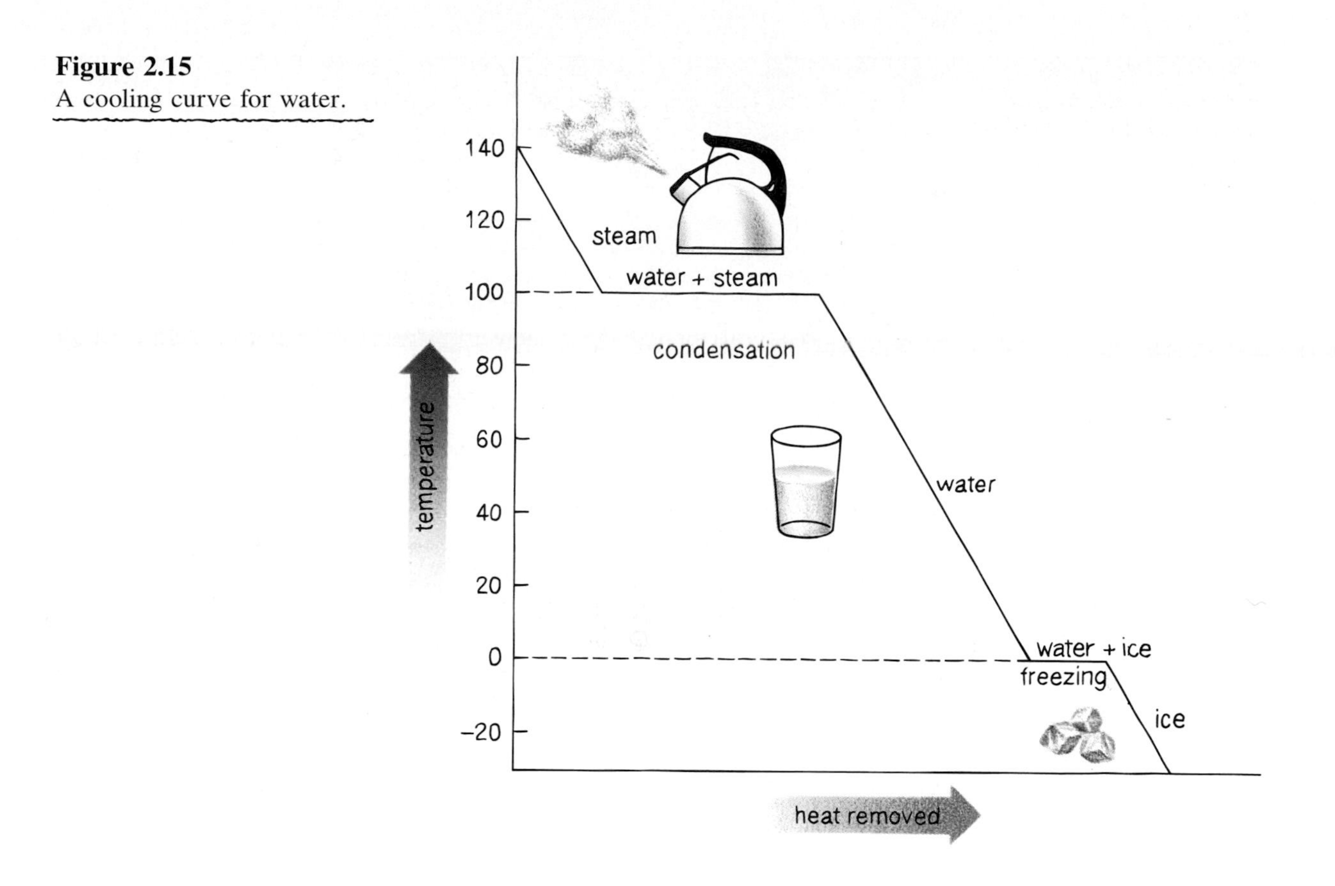

Figure 2.15
A cooling curve for water.

diagram at the boiling point. Again there is no temperature change (it is constant) as long as liquid is forming gas. After all the liquid has changed to gas, additional heat will cause the temperature to rise again.

Steps on a Cooling Curve

When heat is removed from a substance such as water vapor, the gas cools and condenses. A **cooling curve** is a diagram of this cooling process. (See Figure 2.15.) On the cooling curve, the temperature is plotted on the vertical axis and the reaction progress on the horizontal axis. If a gas is cooled, heat is lost and the temperature drops. At the condensation point (same as the boiling point), the gas begins to condense, forming liquid. This process is indicated by a flat line (plateau) on the cooling curve at the condensation point.

After all of the gas has changed into liquid, the particles within the liquid cool as indicated by the downward sloping line that shows the temperature decrease. At the freezing point, the particles in the liquid slow so much that solid begins to form. A flat line at the freezing point indicates the change of state from liquid to solid (freezing). Once all of the substance is frozen, more heat can be lost, which lowers the temperature below the freezing point.

Sample Problem 2.14
A Heating Curve for Water

Imagine that you have an ice cube at −10°C in a pan. As you heat the ice cube in the pan, describe the heating curve if you heat it to 110°C.

Solution

As it warms, the temperature of the ice cube rises, from −10°C to 0°C, shown as a diagonal line on the heating curve. At the melting point of 0°C, the ice cube begins to melt. To show this on the heating curve, a flat line is drawn at 0°C. Once it is all melted, the liquid warms from 0°C to 100°C, a temperature change that is shown as a diagonal line. The water begins to boil, a change of state shown as a flat line at 100°C. After all the liquid vaporizes, the temperature rises again, shown by the diagonal line from 100°C to 110°C.

Study Check

Silver has a melting point of 962°C and a boiling point of 2212°C. In what state or change of state is silver at the following temperatures?

a. 450°C **b.** 962°C **c.** 1500°C

2.7 Energy in Changes of State

Learning Goal

Calculate the heat absorbed or released by a sample of water during a change of state.

During melting, energy called the **heat of fusion** is needed to separate the particles of a solid. For example, 80. calories (two significant figures) of heat are needed to melt 1 g of ice.

Heat of Fusion for Water

$$\frac{80.\ \text{cal}}{1\ \text{g ice}}$$

The heat of fusion (80. cal/g) is also the heat that must be removed to freeze 1 g of water. Water is sometimes sprayed in fruit orchards during very cold weather. If the air temperature drops to 0°C, the water begins to freeze. As heat is removed, the air is warmed to protect the fruit.

To determine the heat needed to melt a sample of ice, multiply the mass of the ice by its heat of fusion. There is no temperature change in the calculation because temperature remains constant as long as the ice is melting. Remember that all changes of state appeared as *flat lines* on our heating curve

Calculating Heat to Melt (or Freeze) Water

$$\text{Heat} = \text{mass} \times \text{heat of fusion}$$
$$\text{cal} = \text{g} \times 80\ \text{cal/g}$$

Sample Problem 2.15
Heat of Fusion

Ice cubes at 0°C with a mass of 75 g are added to your soft drink.

a. How much heat (calories) will be absorbed to completely melt the ice at 0°C?

b. What happens to the temperature of your drink? Why?

Solution

a. The heat in calories to melt the ice is calculated as follows:

$$\underset{\text{Mass}}{75\,\not{g}} \times \underset{\text{Heat of fusion}}{\frac{80.\text{ cal}}{1\,\not{g}}} = 6000\text{ cal}$$

b. The drink will be colder because heat from the drink is providing the energy to melt the ice.

Study Check

In a freezer, 120 g of water at 0°C are placed in an ice cube tray.

a. How many joules of heat must be removed to form ice cubes, if 1 cal = 4.18 J?

b. How is the heat removed to form ice?

Heat of Vaporization

The energy needed to vaporize 1 g of liquid to gas is the **heat of vaporization.** For water, 540 calories, a great amount of heat, are needed to convert 1 g of water to vapor.

Heat of Vaporization for Water

$$\frac{540\text{ cal}}{1\text{ g water (Liquid)}}$$

When 1 g of water condenses, the heat of vaporization, 540 calories, is the amount of energy that must be removed. Therefore, 540 cal/g is also the *heat of condensation* of water. To calculate the heat to vaporize (or condense) a sample of water, multiply the mass of the sample by the heat of vaporization. As before, no temperature change is used in the heat calculation for a change of state.

Calculating Heat to Vaporize (or Condense) Water

$$\underset{\text{cal}}{\text{Heat}} \underset{=}{=} \underset{\text{g}}{\text{mass}} \times \underset{\times\ 540\text{ cal/g}}{\text{heat of vaporization}}$$

Sample Problem 2.16
Calculating Heat for Vaporization

In a sauna, 150 g of water are converted to steam at 100°C. How many kilocalories of heat are needed?

Solution
The heat in calories to vaporize 150 g of water at 100°C is calculated by multiplying its mass by the heat of vaporization of water.

$$150\ \cancel{g} \times \frac{540\ \text{cal}}{1\ \cancel{g}} = 81{,}000\ \text{cal}$$

To convert calories to kilocalories, we use the equality:

$$1\ \text{kcal} = 1000\ \text{cal}$$

$$81{,}000\ \cancel{\text{cal}} \times \frac{1\ \text{kcal}}{1000\ \cancel{\text{cal}}} = 81\ \text{kcal}$$

Study Check
When steam from a pan of boiling water reaches the window, it condenses. How much heat in kilojoules (kJ) is released when 25 g of steam condenses at 100°C? (1 cal = 4.18 J)

Combining Energy Calculations

Up to now, we have calculated one step in a heating or cooling curve. A summary of energy calculations is listed in Table 2.11. However, many problems require a combination of steps along the curve that include a temperature change as well as a change of state. Then the heat is calculated for each step separately and added together to find the total energy.

Table 2.11 *A Summary of Energy Calculations*

Number of State(s)	Change in State	Change in Temperature	Calculation of Heat Absorbed or Released
One (solid, liquid, or gas)	None	Increases or decreases	Heat = mass × temperature change × specific heat
Two	Solid to liquid; liquid to solid	None 80	Heat = mass × heat of fusion
	Liquid to gas; gas to liquid	None 540	Heat = mass × heat of vaporization

Sample Problem 2.17
Combined Heat Calculations

Calculate the calories needed to convert 40.0 g of water at 25°C to steam at 100°C. You will need to use the specific heat of liquid water, 1.00 cal/g °C, and the heat of vaporization for water, 540 cal/g.

Solution

If we sketched a heating curve for water, we would find that two steps are needed for the calculation of heat. One step is the heat needed to warm the liquid from 25°C to 100°C, and the second step is the heat for the vaporization of the water to gas.

Step 1 Heat to warm the water (liquid) from 25°C to 100°C:

$$\underset{\text{Mass}}{40.0\ \cancel{g}} \times \underset{\Delta T}{75\cancel{^\circ C}} \times \underset{\text{Specific heat of water}}{\frac{1.00\text{ cal}}{\cancel{g}\ \cancel{^\circ C}}} = 3000\text{ cal}$$

Step 2 Heat to vaporize the water at 100°C:

$$\underset{\text{Mass}}{40.0\ \cancel{g}} \times \underset{\text{Heat of vaporization of water}}{\frac{540\text{ cal}}{1\ \cancel{g}}} = 21{,}600\text{ cal}$$

We calculate the total heat by adding step 1 and step 2:

1. Heating water	3,000 cal
2. Changing liquid to steam	21,600 cal
Total heat needed:	24,600 cal

Study Check

How many calories are released when 25 g of steam condenses at 100°C, cools to 0°C, and freezes? (*Hint:* The solution will require three energy steps.)

Health Note *Steam Burns*

Hot water at 100°C will cause burns and damage to the skin. However, getting steam on the skin is even more dangerous. Let us consider 100 g of hot water at 100°C. If this water falls on a person's skin, the temperature of the water will drop to body temperature, 37°C. The heat released during cooling burns the skin. The amount of heat can be calculated from the temperature change of 63°C:

$$100\ \cancel{g} \times 63\cancel{^\circ C} \times \frac{1\text{ cal}}{\cancel{g}\ \cancel{^\circ C}} = 6300\text{ cal heat}$$

For comparison, we can calculate the amount of heat released when 100 g of steam at 100°C hits the skin. First, the steam condenses to water (liquid) at 100°C:

$$100\ \cancel{g} \times \frac{540\text{ cal}}{1\ \cancel{g}} = 54{,}000\text{ cal heat released}$$

Now the temperature of the 100 g of water drops from 100°C to 37°C, releasing still more heat—6300 cal, as we saw earlier. Now we can calculate the total amount of heat released from the condensation of the steam as follows:

Condensation (100°C)	=	54,000 cal
Cooling (100°C to 37°C)	=	6,300 cal
Heat released:	=	60,300 cal

The amount of heat released from steam is almost 10 times greater than the amount from hot water.

Chapter Summary

2.1 Energy
Energy is the ability of an object to do work. Potential energy is stored energy; kinetic energy is the energy of motion.

2.2 Temperature Measurement
In science, temperature is measured in Celsius units, °C, or kelvins, K. In the United States, the Fahrenheit scale, °F, is still in use.

2.3 Measuring Heat Energy
Specific heat is the amount of energy required to raise the temperature of 1 g of a substance by 1°C. The heat lost or gained by a substance is determined by multiplying its mass (g), the temperature change (ΔT), and its specific heat (cal/g °C).

2.4 Measuring Heat with a Calorimeter
The nutritional Calorie is the same energy as 1 kcal or 1000 calories. The energy of a sample can be determined by the heat absorbed by water in a calorimeter in which the sample has undergone combustion. The caloric content of a food is the sum of calories from carbohydrate, fat, and protein. It may be calculated by using their number of grams in a food and the caloric values of 4 kcal/g for carbohydrate and protein, 9 kcal/g for fat.

2.5 States of Matter
Matter is anything that has mass and occupies space. Physical properties such as shape, state, or color can be observed without affecting the identity of a substance. The three states of matter are solid, liquid, and gas.

2.6 Changes of State
A substance undergoes a physical change when the shape, size, or state of a substance changes, but not the substance itself. A change of state involves an increase or decrease of attraction between the particles of a substance when it melts, freezes, evaporates, boils, condenses, or sublimes. During a change of state, the temperature remains constant. A heating or cooling curve illustrates the changes in temperature and states as heat is added to or removed from a substance.

2.7 Energy in Changes of State
The heat of fusion is the amount of heat needed to convert 1 g of solid to liquid at the melting (freezing) point. For water, 80. cal must be added to melt 1 g of ice, or removed to freeze 1 g of water. The heat of vaporization is the amount of heat needed to convert 1 g of liquid to vapor at the boiling point. For water, 540 cal must be added to vaporize 1 g of liquid water, or removed to condense 1 g of steam. The heat given off or gained by a substance undergoing temperature changes as well as changes of state is the sum of two or more energy calculations.

Glossary of Key Terms

absolute zero The lowest possible temperature of matter.

boiling The formation of bubbles of gas throughout a liquid.

boiling point (bp) The temperature at which a substance exists as a liquid and gas; liquid changes to gas (boils), and gas changes to liquid (condenses).

caloric value The kilocalories obtained per gram of the food types, carbohydrate, fat, and protein.

calorie (cal) The amount of heat energy that raises the temperature of 1 g of water 1°C.

Calorie (Cal) The dietary unit of energy, equal to 1000 cal, or 1 kcal.

calorimeter An instrument used to measure the amount of heat released by a sample that undergoes combustion.

Celsius scale A temperature scale on which the freezing point of water is 0°C and the boiling point is 100°C.

change of state The transformation of one state of matter to another; for example, solid to liquid, liquid to solid, liquid to gas.

condensation The change of state of a gas to a liquid.

cooling curve A diagram that illustrates the temperature change and changes of states for a substance as heat is removed.

crystallization The change of a gas directly to a solid; the reverse of sublimation.

energy The ability to do work.

evaporation The formation of a gas (vapor) by the escape of high-energy particles from the surface of a liquid.

Fahrenheit scale A temperature scale on which ice melts at 32°C and water boils at 212°C.

freezing A change of state from liquid to solid.
freezing point (fp) The temperature at which the solid and liquid forms of a substance are in equilibrium; liquid changes to a solid (freezes), solid changes to liquid (melts).
gas A state of matter characterized by no definite shape or volume. Particles in a gas move rapidly and exhibit little or no attraction to each other.
heat The energy associated with the motion of particles in a substance.
heat of fusion The energy required to melt 1 g of a substance. For water, 80. cal are needed to melt 1 g of ice; 80. cal are released when 1 g of water freezes.
heat of vaporization The energy required to vaporize 1 g of a substance. For water, 540 calories are needed to vaporize 1 g of liquid: 1 g of steam gives off 540 cal when it condenses.
heating curve A diagram that shows that temperature changes and changes of state of a substance as it is heated.
joule The SI unit of heat energy; 4.18 J = 1 cal.
Kelvin scale An SI temperature scale that assigns zero to the lowest possible temperature (absolute zero). One kelvin (K) also represents a unit of temperature on the Kelvin scale: K = °C + 273.
kinetic energy A type of energy that is required for actively doing work; energy of motion.
kilocalorie An amount of heat energy equal to 1000 calories.
liquid A state of matter that takes the shape of its container but has its own volume. A liquid has moderate attractions between particles.
matter Anything that has mass and occupies space.
melting The conversion of a solid to a liquid.
melting point (mp) The temperature at which a solid becomes a liquid (melts). It is the same temperature as the freezing point.
physical change The change in the state, size, shape, or appearance of a substance without any change in the identity of the substance.
physical property A description of a substance such as its color, shape, or melting or boiling point.
physical state The states of matter: solid, liquid, and gas.
potential energy An inactive type of energy that is stored for use in the future.
solid A state of matter that has its own shape and volume, with little motion—only vibrations—and strong attractions between particles.
specific heat A quantity of heat that changes the temperature of 1 g of a substance by 1°C.
sublimation The change of state in which a solid is transformed directly to a gas without forming a liquid.
temperature An indication of the direction of heat flow, high to low.

Chemistry at Home

1. In your home or dorm room, find something that is colder than you, the same temperature as you, and warmer than you. Why do they feel hotter or colder to you?
2. Heat a spoon in a pan of hot water. Using tongs, transfer the spoon to a bowl of cool water. If you have a candy thermometer, check the temperature of the cool water first. After a few minutes, check the temperature of water in the bowl. How has it changed? Is the spoon still hot? Why? What happened to the energy of the spoon?
3. Place some nickels in a plastic bag and some ice cubes in another bag. Put both in your freezer. In a few hours, or the next day, remove the bags. In 10 minutes, pick up the samples, one in each hand. How do their temperatures compare? Are they the same? Why or why not? Recheck in 20 minutes.
4. Obtain two identical containers, such as bowls or pie tins, and two thermometers. Place water in one container and an equal amount of sand in the other. Set the containers in the sun and place a thermometer in each to the same depth. Record the temperature initially and every 10 minutes after that for a total of 60 minutes. Which sample had the higher final temperature? Why does one substance not heat up as quickly as the other?
5. Place some water in a teakettle. Hold a small mirror a few inches from the spout. What happens? Heat the water in a teakettle. As it warms, hold the mirror a few inches from the spout. What happens? When the water boils, *very carefully and quickly* hold the mirror near the spout. *Keep your fingers away! Steam burns!* In the old western movies, a mirror was used to see if somebody was alive. Why?
6. Seal some slices of a fruit or vegetable such as celery or carrot in a plastic bag and place the same number in an open container. Place both in a frost-free freezer. Every few days, check the food. How have they changed? Why?

7. Obtain three Styrofoam coffee cups. Fill one about two-thirds full of very warm tap water and a second with ice water. If you have a candy thermometer, determine the temperature of each. Otherwise, use your finger to test each quickly. Pour about one-half of each into a third coffee cup and stir. How does its temperature compare with the original water sample? Why did the temperature change?

Answers to Study Checks

2.1 potential energy
2.2 12°F
2.3 39.8°C
2.4 464 K
2.5 night, −260°C; day, 410°C
2.6 140 cal
2.7 350 J
2.8 150 kcal; 150 Cal
2.9 4.1 kcal/g
2.10 380 kcal
2.11 liquid water
2.12 gas to liquid
2.13 The color red is a physical property; the change of size is a physical change.
2.14 **a.** solid **b.** melting (freezing) **c.** liquid
2.15 **a.** 40,000 J or 4.0×10^4 J
b. Heat will flow from the water to the freezer compartment that is colder.
2.16 56 kJ
2.17 18,000 cal

Problems

Energy *(Goal 2.1)*

2.1 Discuss the changes in the potential and kinetic energy of a roller coaster ride as the roller coaster climbs up a hill and goes down the other side.

2.2 Discuss the changes in the potential and kinetic energy of a ski jumper taking the elevator to the top of the jump, and going down the ramp.

2.3 Indicate whether each statement describes potential or kinetic energy:
a. water at the top of a waterfall
b. water falling in a waterfall
c. the energy in a lump of coal
d. a skier at the top of a hill

2.4 Indicate whether each statement describes potential or kinetic energy:
a. the energy in your food
b. a tightly wound spring
c. an earthquake
d. a car speeding down the freeway

2.5 What are the forms of energy as they change in the following examples?
a. using a hair dryer
b. using an electric fan
c. burning gasoline in a car engine
d. sunlight falling on a solar water heater

2.6 What are the forms of energy as they change in the following examples?
a. turning on a light switch
b. cooking food on a gas stove
c. using an iron
d. burning a log in a fireplace

Temperature Measurement *(Goal 2.2)*

2.7 Your friend who is visiting from France just took her temperature. When she reads 99.8, she becomes concerned that she is quite ill. How would you explain the temperature to your friend?

2.8 You have a friend who is using a recipe for flan from a Mexican cookbook. You notice that he set your oven temperature at 175°F. What would you advise him to do?

2.9 Solve the following temperature conversions:
a. 37.0°C = ___ °F **b.** 65.3°F = ___ °C
c. −27°C = ___ K **d.** 62°C = ___ K
e. 110°F = ___ °C **f.** 72°F = ___ K

2.10 Solve the following temperature conversions:
a. 25°C = ___ °F **b.** 155°C = ___ °F
c. −25°F = ___ °C **d.** 224 K = ___ °C
e. 545 K = ___ °C **f.** 875 K = ___ °F

2.11 **a.** A patient with heat stroke has a temperature of 106°F. What does this read on a Celsius thermometer?
b. Because high fevers can cause convulsions in children, the doctor wants to be called if the child's temperature goes over 40°C. Should the doctor be called if a child has a temperature of 103°F?

2.12 **a.** Hot compresses are being prepared for the patient in room 32B. The water is heated to 145°F. What is the temperature of the hot water in °C?
b. During extreme hypothermia, a young woman's temperature dropped to 20.6°C. What was her temperature on the Fahrenheit scale?

Measuring Heat Energy *(Goal 2.3)*

2.13 Why are cooking pans made of aluminum, iron, or copper?

2.14 What information is needed to calculate the amount of heat energy required to heat a substance?

2.15 What is the amount of heat energy needed in each of the following?
a. the number of calories to heat 25 g of water from 15°C to 25°C
b. the number of calories to heat 10.0 g of copper from 25°C to 275°C if the specific heat of copper is 0.092 cal/g °C
c. the number of joules to heat 150 g of water from 0°C to 75°C
d. the number of kilocalories to heat 350 g of water in a kettle from 25°C to 100°C

2.16 What is the amount of heat energy given off by each of the following?
a. the number of calories given off when 85 g of water cools from 45°C to 25°C
b. the number of calories given off when 0.50 kg of water cools from 80°C to 60°C
c. the number of kilocalories given off when 250 g of sand cools from 455°C to 225°C, if the specific heat of sand is 0.19 cal/g °C
d. the number joules given off when 25 g of water cools from 85°C to 0°C

2.17 Silver has a specific heat of 0.057 cal/g °C. How many calories are needed to raise the temperature of 10.0 g of silver from 15°C to 237°C?

2.18 An electric power plant releases heat into a nearby stream. The temperature of 5.0 kg of stream water increases from 22°C to 28°C. How many kilocalories of heat were absorbed by the water?

Measuring Heat with a Calorimeter *(Goal 2.4)*

2.19 The combustion of the following foods has been determined by calorimetry. Use the data given below to calculate the number of kilocalories for each food:

Food	*Water in Calori-meter (g)*	*Temperature (°C)*	
		Initial	*Final*
a. celery (1 stalk)	500	25	35
b. waffle (7-in. diameter)	5000	20	62
c. popcorn (1 cup, no oil)	1000	25	50

2.20 A 0.50-g sample of octane, a component in gasoline, is burned in a calorimeter. If the water in the calorimeter has a mass of 1200 g and an initial temperature of 22°C, what is the amount of heat in kilojoules per gram produced by the octane if the final temperature is 26°C?

2.21 Use the caloric values for carbohydrates (4 kcal/g), fats (9 kcal/g), and proteins (4 kcal/g) to complete the following table:

Food	*Carbohydrate (g)*	*Fat (g)*	*Protein (g)*	*Energy (kcal)*
a. orange, 1	16	0	1	_____
b. apple, 1	_____	0	0	72

2.22 Use the caloric values for carbohydrates (4 kcal/g), fats (9 kcal/g), and proteins (4 kcal/g) to complete the following table:

Food	*Carbo-hydrate (g)*	*Fat (g)*	*Protein (g)*	*Energy (kcal)*
a. danish pastry, 1	30	15	5	_____
b. avocado, 1	13	_____	5	405

2.23 One cup of clam chowder contains 9 g of protein, 12 g of fat, and 16 g of carbohydrate. How many kilocalories are in the clam chowder?

2.24 A high-protein diet contains 70 g of carbohydrate, 150 g of protein, and 5 g of fat. How many kilocalories does this diet provide?

States of Matter *(Goal 2.5)*

2.25 Indicate whether each of the following describes a gas, a liquid, or a solid:
a. This substance has no definite volume nor shape.

b. The particles in this substance are not attracted to each other.

c. The particles of this substance are held in a definite structure.

2.26 Indicate whether each of the following describes a gas, a liquid, or a solid:

a. The substance has a definite volume but takes the shape of the container.

b. The particles of this substance are very far apart.

c. This substance occupies the entire volume of the container.

Changes of State *(Goal 2.6)*

2.27 What is the change of state in each of the following? Identify as melting, freezing, evaporation, boiling, sublimation, or condensation.

a. Particles at the surface of a liquid escape as a gas.

b. The substance changes from a liquid to a solid.

c. Bubbles of gas form throughout a liquid.

d. A solid changes to a gas, but no liquid forms.

2.28 What is the change of state in each of the following? Identify as melting, freezing, evaporation, boiling, sublimation, or condensation.

a. The water vapor in the clouds changes to rain.

b. The solid structure of a substance breaks down as a liquid forms.

c. Snow on the ground turns to liquid water.

d. Coffee is freeze-dried.

2.29 Classify the following as a physical property or a physical change:

a. An iceberg is solid water.

b. Water on the street freezes during a cold wintry night.

c. A wood log is chopped to make kindling.

d. Butter is melted to put on popcorn.

2.30 Classify the following as a physical property or a physical change:

a. Apple chips are made by freeze-drying apple slices.

b. Dough is used to make different pasta shapes.

c. The boiling point of nitrogen is 77 K.

d. Salt is dissolved in water to make a throat.

2.31. a. How does perspiration during heavy exercise cool the body?

b. Why do clothes dry more quickly on a hot, summer day than on a cold, winter day?

c. Why do wet clothes dry when hung on a clothes line but stay wet in a plastic bag?

2.32 a. For sports injuries, a spray such as ethyl chloride may be used to numb an area of the skin. Explain how a substance like ethyl chloride that evaporates quickly can numb the skin.

b. Why does water in a wide, flat, shallow dish evaporate more quickly than the same amount of water in a tall, narrow glass?

c. Which will dry out faster, a sandwich on a plate or in plastic wrap?

2.33 Draw a heating curve for ice that is heated from −20°C to 140°C. Indicate the segment of the graph that corresponds to each of the following:

a. solid **b.** melting point **c.** liquid
d. boiling point **e.** gas

2.34 Draw a cooling curve for steam that cools from 110°C to −10°C? Indicate the segment of the graph that corresponds to each of the following:

a. solid **b.** freezing point **c.** liquid
d. condensation point (boiling point)
e. gas

Energy in Changes of State *(Goal 2.7)*

2.35 How many calories are needed in each of the following problems? Use the heat of fusion or the heat of vaporization.

a. to melt 115 g of ice

b. to melt 8.0 kg of ice

c. to vaporize 10.0 g of water

d. to vaporize 50.0 g of water

2.36 How many calories are removed in each of the following problems? Use the heat of fusion or the heat of vaporization.

a. to freeze 10.0 g of water

b. to freeze 250 g of water

c. to condense 40.0 g of steam

d. to condense 5.0 kg of steam

2.37 How much heat energy in calories is absorbed or removed in each of the following?

a. warming 20.0 g of water at 15°C to 70°C (one step)

b. melting 50.0 g of ice at 0°C, and warming the liquid to 65°C (two steps)

c. condensing 100 g of steam at 100°C to liquid, and cooling the liquid to 0°C (two steps)

d. melting 80.0 g of ice at 0°C, warming the liquid to 100°C, and vaporizing it at 100°C (three steps)

2.38 How much heat energy in calories is absorbed or removed in each of the following?

a. condensing 125 g of steam at 100°C, and cooling the liquid to 15°C (two steps)

b. melting a 5.0-kg block of ice at 0°C, and warming the liquid to 15°C (two steps)
c. condensing 250 g of steam at 100°C, cooling the liquid, and freezing it at 0°C (three steps)
d. warming 150 mL of water (density = 1.0 g/mL) from 10°C to 100°C, and vaporizing it at 100°C (two steps)

2.39 A bag of ice containing 220 g of ice at 0°C was placed on a burn on a patient's hand. When the ice bag was removed, the ice had melted, and the liquid water had a temperature of 21°C. How many kilocalories of heat were absorbed by the ice? How many kilojoules is that?

2.40 A 115-g sample of steam at 100°C escapes from a volcano. It condenses, cools, and falls as snow at 0°C. How many kilocalories of heat are released? How many kilojoules is that?

Challenge Problems

2.41 If you used the 2,000 Calories you expend in energy in 1 day to heat 50,000 g of water at 20°C, what would be its new temperature?

2.42 A 0.50-g sample of vegetable oil is placed in a calorimeter. During combustion, 18.9 kJ were given off. What is the caloric value (kcal/g) of the oil? (1 cal = 4.18 J)

2.43 A typical diet in the United States provides 15% of the calories from protein, 45% from carbohydrates, and 40% from fats. Calculate the total grams of protein, carbohydrate, and fat to be included each day in diets having the following caloric requirements:
a. 1000 kcal **b.** 1800 kcal
c. 2600 kcal

2.44 You have just eaten a quarter-pound cheeseburger, french fries, and chocolate shake. With the following nutritional information, determine how many hours you will need to run to "burn off" the kilocalories in your meal. See Table 2.8 (assume you are a 70-kg person).

Item	*Protein* (*g*)	*Fat* (*g*)	*Carbohydrate* (*g*)
quarter-pound cheeseburger	31	29	34
french fries	3	11	26
chocolate shake	11	9	60

2.45 If you want to lose 1 pound of "fat," which is 15% water, how many kilocalories do you need to expend? How many kilojoules is that? (1 lb = 454 g)

2.46 The melting point of benzene is 5°C and its boiling point is 80°C. Sketch a heating curve for benzene from 0°C to 100°C.
a. What is the state of benzene at 15°C?
b. What happens on the curve at 5°C?
c. What is the state of benzene at 60°C?
d. What is the state of benzene at 90°C?
e. At what temperature will both liquid and gas be present?

2.47 Water is sprayed on the ground of an orchard when temperatures are near freezing.
a. How does water protect the fruit from freezing?
b. How many kilocalories of heat are released if 5.0 kg of water at 15°C is sprayed on the ground cools and freezes at 0°C?

2.48 When 1 g of gasoline burns, 11,500 calories of energy are given off. If the density of gasoline is 0.74 g/mL, how many kilocalories of energy are obtained from 1 gallon of gasoline? (1 gal = 4 qt; 1 qt = 946 mL)

2.49 An ice cube tray holds 1.2 cups of water. If the water has a temperature of 25°C, how many joules of heat must be removed to cool and freeze the water at 0°C? (1 qt = 4 cups; 1 qt = 946 mL; density of water = 1.00 g/mL; 1 cal = 4.18 J)

2.50 At what temperature are the Fahrenheit and Celsius scales identical?

2.51 A 3.0-kg block of lead is taken from a furnace at 300°C and placed on a large block of ice at 0°C. The specific heat of lead is 0.028 cal/g °C. If all the heat given up by the lead is used to melt ice, how much ice is melted after the temperature of the lead drops to 0°C?

Chapter 3

Atoms and Elements

Fireworks display vibrant colors that result when electrons in atoms of various elements are excited by heat. Atoms of strontium, sodium, and copper produce the colors red, yellow, and green.

Looking Ahead

Learning Goals

3.1 Given the name of an element, write its correct symbol; from the symbol, write the correct name.

3.2 Describe the electrical charge, mass (amu), and location in an atom for a proton, neutron, and electron.

3.3 Given the atomic number and mass number of an atom, state the number of protons, neutrons, and electrons.

3.4 Use the periodic table to identify the group and the period of an element, and whether it is a metal or nonmetal.

3.5 Given the name or symbol of one of the first 20 elements in the periodic table, write the electron arrangement.

3.6 Use the electron arrangement of an element to state its group number and to explain periodic law.

3.7 Write the electron configuration using subshell notation.

Questions to Think About

1. What elements can you find in your home?
2. On a dry day, why does your hair fly away after you brush it?
3. Where are atoms found?
4. What particles are inside an atom?

All of the matter that surrounds us is composed of primary substances called *elements.* There are different kinds of elements. The elements calcium and phosphorus build your teeth and bones. The hemoglobin that carries oxygen in your blood contains the element iron. The elements carbon, hydrogen, oxygen, and nitrogen derived from the digestion of food are used by the cells of your body to build proteins. Today, there are 111 different elements. Of these, 88 occur naturally and are found in different combinations, providing the great number of compounds that make up our world. The rest of the elements have been produced artificially and are not found in nature.

You have probably seen the element aluminum. Imagine that you are tearing a piece of aluminum foil into smaller and smaller pieces. Now imagine that you have a piece so small that you can no longer break it down further. Then you would have an atom of aluminum, the smallest particle of an element that still retains the characteristics of that element.

3.1 Elements and Symbols

Learning Goal
Given the name of an element, write its correct symbol; from the symbol, write the correct name.

Elements are primary substances from which all other things are built. They cannot be broken down into simpler substances. Many of the elements were named for planets, mythological figures, minerals, colors, geographic locations, and famous people. Some sources of the names of elements are listed in Table 3.1.

Chemical Symbols

Chemical symbols are one- or two-letter abbreviations for the names of the elements. Only the first letter of an element's symbol is capitalized; a second letter, if there is one, is lower case. That way, we know when a different

Table 3.1 *Some Elements and Their Names*

Element	Source of Name
Uranium	The planet Uranus
Titanium	Titans (mythology)
Chlorine	*Chloros,* "greenish yellow" (Greek)
Iodine	*Ioeides,* "violet" (Greek)
Magnesium	Magnesia, a mineral
Californium	California
Curium	Marie and Pierre Curie

Table 3.2 Names and Symbols of Some Common Elements

Name[a]	Symbol	Name[a]	Symbol
Aluminum	Al	Lead (*plumbum*)	Pb
Argon	Ar	Lithium	Li
Barium	Ba	Magnesium	Mg
Boron	B	Mercury (*hydrargyrum*)	Hg
Bromine	Br	Neon	Ne
Cadmium	Cd	Nickel	Ni
Calcium	Ca	Nitrogen	N
Carbon	C	Oxygen	O
Chlorine	Cl	Phosphorus	P
Cobalt	Co	Potassium (*kalium*)	K
Copper (*cuprum*)	Cu	Silicon	Si
Fluorine	F	Silver (*argentum*)	Ag
Gold (*aurum*)	Au	Sodium (*natrium*)	Na
Helium	He	Strontium	Sr
Hydrogen	H	Sulfur	S
Iodine	I	Tin (*stannum*)	Sn
Iron (*ferrum*)	Fe	Zinc	Zn

[a] Names given in parentheses are ancient Latin or Greek words from which the symbols are derived.

element is indicated. If both letters are capitalized, it represents the symbols of two different elements. For example, the element nickel has the symbol Ni. However, the two capital letters, NI, specify two elements, nitrogen (N) and iodine (I).

One-Letter Symbols		*Two-Letter Symbols*	
C	carbon	Co	cobalt
S	sulfur	Si	silicon
N	nitrogen	Ne	neon
I	iodine	Ni	nickel

Although most of the symbols use letters from the current names, some are derived from their ancient Latin or Greek names. For example, Na, the symbol for sodium, comes from the Latin word *natrium.* The symbol for iron, Fe, is derived from the Latin name *ferrum.* Table 3.2 lists the names and symbols of some common elements. Learning their names and symbols will greatly help your learning of chemistry. A complete list of all the elements and their symbols appears on the inside front cover of this text. Figure 3.1 shows examples of some common elements.

Health Note *Latin Names for Elements in Clinical Usage*

In medicine, the Latin names for sodium and potassium are often used. The condition in which there is too much sodium in the body is called hypernatremia, and a low sodium level is called hyponatremia. In the case of potassium, both the modern name and the Latin name are used. For example, a high potassium level may be called hyperpotassemia or hyperkalemia; a below-normal potassium level may be called hypopotassemia or hypokalemia.

Figure 3.1
Some common elements and objects made of them: carbon (C) as graphite in a pencil, and in a diamond; copper (Cu) in a penny; nickel (Ni) in a nickel; aluminum (Al) in aluminum foil.

Sample Problem 3.1
Writing Chemical Symbols

What are the chemical symbols for the following elements?

a. carbon **b.** nitrogen **c.** chlorine **d.** copper

Solution

a. C **b.** N **c.** Cl **d.** Cu

Study Check

What are the chemical symbols for the elements silicon, sulfur, and silver?

Sample Problem 3.2
Naming Chemical Elements

Give the name of the element that corresponds to each of the following chemical symbols:

a. Zn **b.** K **c.** H **d.** Fe

Solution

a. zinc **b.** potassium **c.** hydrogen **d.** iron

Study Check

What are the names of the elements with chemical symbols Mg, Al, and F?

Health Note *Elements Essential to Health*

Several elements are essential for the well-being and survival of the human body. Some examples and the amounts present in a 60-kg person are listed in Table 3.3.

Table 3.3 ***Elements Essential to Health***

Element	Symbol	Amount in a 60-kg Person	Where Found
Oxygen	O	39 kg	Water, carbohydrates, fats, proteins
Carbon	C	11 kg	Carbohydrates, fats, proteins
Hydrogen	H	6 kg	Water, carbohydrates, fats, proteins
Nitrogen	N	2 kg	Proteins, DNA, RNA
Calcium	Ca	1 kg	Bones, teeth
Phosphorus	P	0.6 kg	Bones, teeth, DNA, RNA
Potassium	K	0.2 kg	Inside cells (important in conduction of nerve impulses)
Sulfur	S	0.2 kg	Some amino acids
Sodium	Na	0.1 kg	Body fluids (important in nerve conduction and fluid balance)
Magnesium	Mg	0.1 kg	Bone (important in enzyme function)
Chlorine	Cl	0.1 kg	Outside cells (major electrolyte)

3.2 The Atom

Learning Goal
Describe the electrical charge, mass (amu), and location in an atom for a proton, a neutron, and an electron.

Atoms are the smallest particles of an element that retain the characteristics of that element. Billions of atoms are packed together to build you and all the matter around you. The paper in this book contains atoms of carbon, hydrogen, and oxygen. The ink on this paper, even the dot over the letter *i*, contains huge numbers of atoms. There are as many atoms in that dot as there are seconds in 10 billion years.

The concept of the atom is relatively recent. Although the Greek philosophers in 500 B.C. reasoned that matter must contain minute particles they called *atomos,* the idea of atoms did not become a scientific theory until 1808. Then John Dalton (1766–1844) developed an atomic theory that proposed that atoms were responsible for the combinations of elements found in compounds.

Atomic Theory

1. All matter is made up of tiny particles called atoms.
2. All atoms of a given element are similar to one another and different from atoms of other elements.
3. Atoms of two or more different elements combine to form compounds. A particular compound is always made up of the same kinds of atoms and always has the same number of each kind of atom.

4. A chemical reaction involves the rearrangement, separation, or combination of atoms. Atoms are never created or destroyed during a chemical reaction.

Atoms are the building blocks of everything we see around us; yet we cannot see an atom or even a billion atoms with the naked eye. However, when billions and billions of atoms are packed together, the characteristics of each atom are added to those of the next until we can see the characteristics we associate with the element. For example, a small piece of the shiny, copper-colored element we call copper consists of many, many copper atoms. Through a special kind of microscope called a tunneling microscope, we can now see images of individual atoms, such as the atoms of carbon in graphite shown in Figure 3.2.

Figure 3.2
Graphite, a form of carbon, magnified millions of times by a scanning tunneling microscope. This instrument generates an image of the atomic structure. The round yellow objects are atoms.

Subatomic Particles

By the early part of the twentieth century, growing evidence indicated that the atom was not a solid sphere, as Dalton had imagined. New experiments showed that atoms were composed of even smaller bits of matter called **subatomic particles.** Much of the chemistry of an element depends upon the subatomic particles that are the building blocks of the atoms.

There are three subatomic particles of interest to us, the proton, neutron, and electron. Two of these carry electrical charges. The **proton** has a positive charge (+), and the **electron** carries a negative charge (−). The **neutron** has no electrical charge; it is neutral.

Like charges repel; they push away from each other. When you brush your hair on a dry day, electrical charges that are alike build up on the brush and in your hair; as a result your hair flies away from the brush. Opposite or unlike charges attract. The crackle of clothes taken from the clothes dryer indicates the presence of electrical charges. The clinginess of the clothing is due to the attraction of opposite, unlike charges, as shown in Figure 3.3.

Figure 3.3
Attraction and repulsion of electrical charges.

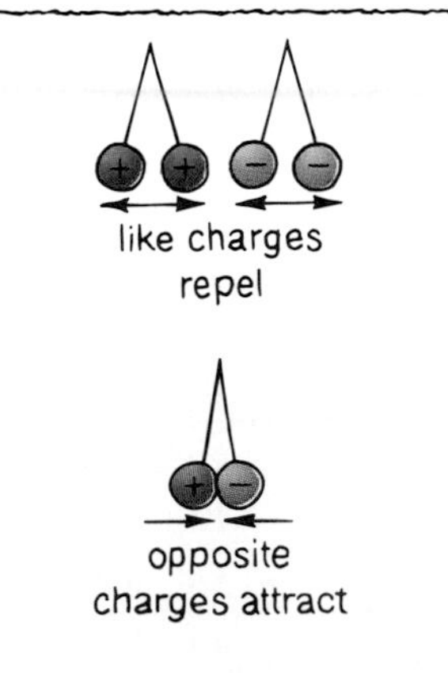

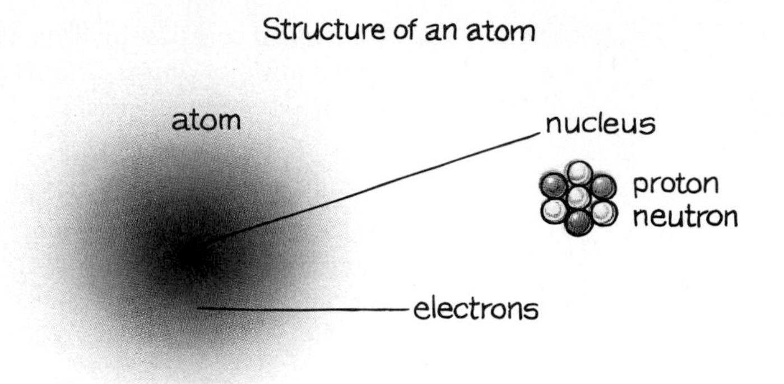

Figure 3.4
Arrangement of the subatomic particles in an atom. The protons and neutrons are located in the nucleus at the center of the atom; the electrons are located outside the nucleus.

Nucleus of the Atom

The protons and neutrons in an atom are tightly packed in a tiny space at the center of the atom called the **nucleus.** The rest of the atom is extremely large, but it consists mostly of empty space and the electrons. If we imagine a football stadium, the nucleus would be about the size of a golf ball in the center. The rest of the large volume would be mostly empty space occupied only by the fast-moving electrons. (See Figure 3.4.)

Mass

All of the subatomic particles are extremely small compared with the things you see around you. One proton has a mass of 1.7×10^{-24} g, and the neutron is about the same. The mass of the electron is even less. Because the mass of a subatomic particle is so minute, chemists use a smaller unit of mass called an **atomic mass unit (amu).** An amu is defined as one-twelfth of the mass of a carbon-12 atom, a standard with which the mass of every other atom is compared. In biology, the atomic mass unit may be called a dalton in honor of John Dalton. On the amu scale, the proton and neutron each have a mass of about 1 amu. Because the electron is so light, its mass is usually ignored in atomic mass calculations. Table 3.4 summarizes some information about the subatomic particles in an atom.

Table 3.4 ***Subatomic Particles in the Atom***

Subatomic Particle	Symbol	Electrical Charge	Approximate Mass	Location in Atom
Proton	p or p^+	1+	1 amu	Inside nucleus
Neutron	n or n^0	None	1 amu	Inside nucleus
Electron	e^-	1−	0	Outside nucleus

Sample Problem 3.3
Identifying Subatomic Particles

Complete the following table for subatomic particles:

Name	*Symbol*	*Mass (amu)*	*Charge*	*Location in Atom*
Electron	________	________	________	________
________	________	1	0	________

Solution

Electron	e^-	0	1−	Outside nucleus
Neutron	n^0	1	0	Nucleus

Study Check
Give the symbol, mass, electrical charge, and location of a proton in the atom.

3.3 Atomic Number and Mass Number

Learning Goal
Given the atomic number and the mass number of an atom, state the number of protons, neutrons, and electrons.

All of the atoms of the same element always have the same number of protons. This feature distinguishes atoms of one element from atoms of all the other elements. An **atomic number,** which is equal to the number of protons in the nucleus of an atom, is used to identify each element.

Atomic number = number of protons in an atom

On the inside front cover of this text is a list of all the elements, their chemical symbols, and their atomic numbers. Next to it is a periodic table, which gives all of the elements in order of increasing atomic number. In the periodic table, the atomic number is the whole number that appears above the symbol for each element, as illustrated in Figure 3.5. For example, a hydrogen atom, with atomic number 1, has 1 proton; a helium atom, with atomic number 2, has 2 protons; an atom of oxygen, with atomic number 8, has 8 protons; and gold, with atomic number 79, has 79 protons.

Atoms are Neutral

An atom is electrically neutral. That means that the number of protons in an atom is equal to the number of electrons. This electrical balance gives an atom an overall charge of zero. Thus, in every atom, the atomic number also gives the number of electrons. Table 3.5 lists the atomic number and corresponding number of protons and electrons for some atoms.

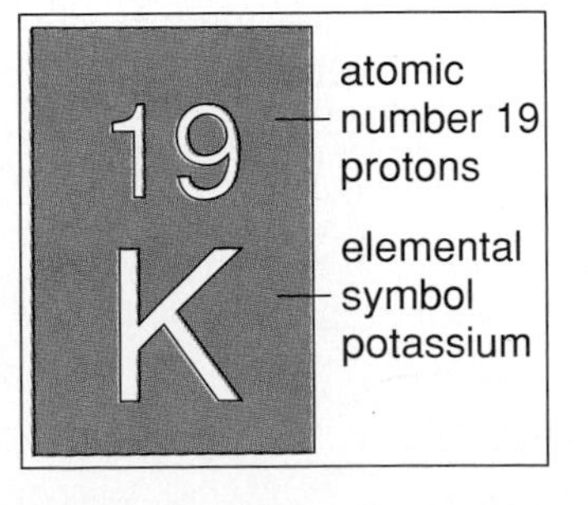

Figure 3.5
Atomic numbers appear above the symbols of the elements listed in the periodic table. The atomic number indicates the number of protons in an atom of that element.

Table 3.5 *Number of Protons and Electrons in Some Atoms*

Element	Symbol	Number	Number of Protons	Number of Electrons	Overall Charge
Hydrogen	H	1	1	1	0
Helium	He	2	2	2	0
Oxygen	O	8	8	8	0
Sodium	Na	11	11	11	0
Iron	Fe	26	26	26	0
Gold	Au	79	79	79	0

Sample Problem 3.4
Using Atomic Number to Find the Number of Protons and Electrons

Using the periodic table, state the atomic number, number of protons, and number of electrons for an atom of each of the following elements:
a. nitrogen **b.** magnesium **c.** bromine

Solution
a. atomic number 7; 7 protons and 7 electrons
b. atomic number 12; 12 protons and 12 electrons
c. atomic number 35; 35 protons and 35 electrons

Study Check
Consider an atom that has 26 electrons.
a. How many protons are in its nucleus?
b. What is its atomic number?
c. What is its name, and what is its symbol?

Mass Number

The protons and neutrons in an atom provide essentially all of its mass. The **mass number** of an atom is equal to the sum of the number of protons and the number of neutrons in the nucleus. Because we are counting the subatomic particles in the nucleus, the mass number is always a whole number.

Mass number = total number of protons and neutrons in one atom of an element

For example, an atom of potassium with 19 protons and 20 neutrons has a mass number of 39. By knowing the atomic number and the number of neutrons, you can calculate the mass number of an atom.

Element:	K	O	Al	Fe
Protons:	19	8	13	26
Neutrons:	20	8	14	30
Mass number (total):	39	16	27	56

Table 3.6 illustrates the relationship between atomic number, mass number, and the number of protons, neutrons, and electrons in some atoms of different elements.

Table 3.6 ***Composition of Some Atoms of Different Elements***

Element	Symbol	Atomic Number	Mass Number	Number of Protons	Number of Neutrons	Number of Electrons
Hydrogen	H	1	1	1	0	1
Nitrogen	N	7	14	7	7	7
Phosphorus	P	15	31	15	16	15
Chlorine	Cl	17	37	17	20	17
Iron	Fe	26	56	26	30	26

Sample Problem 3.5
Calculating Mass Number

Calculate the mass number of an atom by using the information given:

a. 5 protons and 6 neutrons
b. 18 protons and 22 neutrons
c. atomic number 48, and 64 neutrons

Solution

a. mass number = 5 + 6 = 11
b. mass number = 18 + 22 = 40
c. mass number = 48 + 64 = 112

Study Check
What is the mass number of a silver atom that has 60 neutrons?

Sample Problem 3.6
Calculating Numbers of Protons and Neutrons

For an atom of phosphorus that has a mass number of 31, determine the following:

a. the number of protons
b. the number of neutrons
c. the number of electrons

Solution

a. On the periodic table, the atomic number of phosphorus is 15. A phosphorus atom has 15 protons.
b. The number of neutrons in this atom is found by subtracting the atomic number from the mass number. The number of neutrons is 16.

Mass number	−	atomic number	=	number of neutrons
31	−	15	=	16

c. Because an atom is neutral, there is an electrical balance of protons and electrons. Because the number of electrons is equal to the number of protons, the phosphorus atom has 15 electrons.

Study Check

How many neutrons are in the nucleus of a bromine atom that has a mass number of 80?

Isotopes

We have seen that all atoms of the same element have the same number of protons and electrons. However, the atoms are not completely identical because they can have different numbers of neutrons. **Isotopes** are atoms of the same element that have different numbers of neutrons. For example, all atoms of the element magnesium (Mg) have 12 protons. However, some magnesium atoms have 12 neutrons, others have 13 neutrons, and still others have 14 neutrons. The differences in numbers of neutrons for these magnesium atoms cause their mass numbers to be different but not their chemical behavior. The three isotopes of magnesium have the same atomic number but different mass numbers.

To distinguish between the different isotopes of an element, we can write an **isotope symbol** that indicates the mass number and the atomic number of the atom.

Symbol for an Isotope of Magnesium

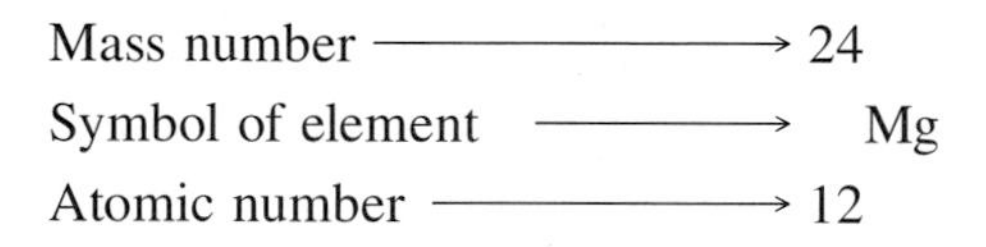

Magnesium has three naturally occurring isotopes, shown in Table 3.7. Table 3.8 lists nuclear symbols for the isotopes of some selected elements.

Table 3.7 ***Isotopes of Magnesium***

	12e⁻ 12p⁺ 12n°	12e⁻ 12p⁺ 13n°	12e⁻ 12p⁺ 14n°
Nuclear symbol	$^{24}_{12}Mg$	$^{25}_{12}Mg$	$^{26}_{12}Mg$
Number of protons	12	12	12
Number of electrons	12	12	12
Mass number	24	25	26
Number of neutrons	12	13	14

Table 3.8 ***Isotopes of Some Common Elements***

Element	Isotope Symbols			
Lithium	$^{6}_{3}Li$	$^{7}_{3}Li$		
Carbon	$^{12}_{6}C$	$^{13}_{6}C$	$^{14}_{6}C$	
Oxygen	$^{16}_{8}O$	$^{17}_{8}O$	$^{18}_{8}O$	
Sulfur	$^{32}_{16}S$	$^{33}_{16}S$	$^{34}_{16}S$	$^{36}_{16}S$
Chlorine	$^{35}_{17}Cl$	$^{37}_{17}Cl$		
Copper	$^{63}_{29}Cu$	$^{65}_{29}Cu$		

Sample Problem 3.7
Identifying Protons and Neutrons in Isotopes

State the number of protons and neutrons in the following isotopes of **neon** (Ne):
a. $^{20}_{10}Ne$ **b.** $^{21}_{10}Ne$ **c.** $^{22}_{10}Ne$

Solution
The atomic number of Ne is 10; each isotope has 10 protons. The number of neutrons in each isotope is found by subtracting the atomic number (10) from each mass number.
a. 10 protons; 10 neutrons (20 − 10)
b. 10 protons; 11 neutrons (21 − 10)
c. 10 protons; 12 neutrons (22 − 10)

Study Check
Write a symbol for the following isotopes:
a. a nitrogen atom with 8 neutrons
b. an atom with 20 protons and 22 neutrons
c. an atom with mass number 27 and 14 neutrons

Atomic Weight

In laboratory work, a scientist generally uses samples that contain many atoms of an element. Among those atoms are all of the various isotopes with their different masses. To obtain a convenient mass to work with, chemists use the mass of an "average atom" of each element. This average atom has an **atomic weight,** which is the weighted average mass of all of the naturally occurring isotopes of that element.

To calculate an atomic weight, the percent abundance of each isotope must be known. For example, in chlorine (Cl), 75.5 percent of the atoms have a mass number of 35; the other 24.5 percent have a mass number of 37. Using these values and the mass number of each isotope (which is quite close to each isotope's mass in amu), we can calculate the atomic weight for the element.

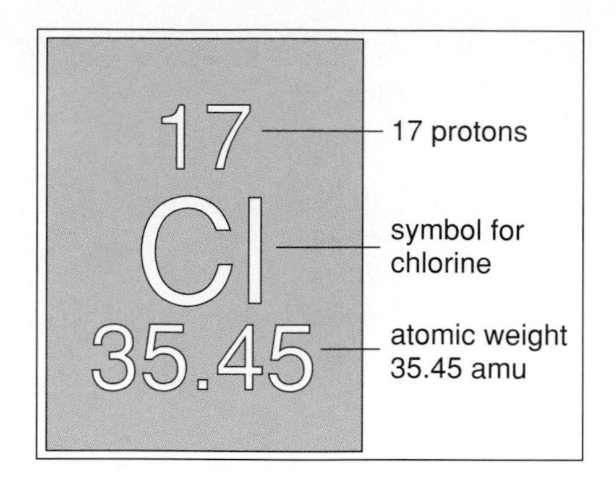

Figure 3.6
Information that is given for chlorine in the periodic table.

Determination of the Atomic Weight of Chlorine

Isotope	*Percent of Sample*		*Mass Number*		*Contribution to Average Atom*
$^{35}_{17}Cl$	$\frac{75.5}{100}$	×	35	=	26.4 amu
$^{37}_{17}Cl$	$\frac{24.5}{100}$	×	37	=	9.1 amu
Atomic weight Cl				=	35.5 amu

This calculation illustrates the way in which isotopes determine the overall atomic weight for an element. Most elements consist of several isotopes, and this is one reason atomic weights are seldom whole numbers.

On the periodic table, the atomic weight is given below the symbol of each element; the mass number is not given. The element chlorine is shown in Figure 3.6 as it appears in the periodic table.

Sample Problem 3.8
Identifying Atomic Weight

Using the periodic table, give the atomic weight to one decimal place (tenth's place) for each of the following elements.

a. hydrogen **b.** iron **c.** sulfur **d.** potassium

Solution

a. 1.0 amu **b.** 55.9 amu **c.** 32.1 amu **d.** 39.1 amu

Study Check

What is the name of the element with atomic weight 65.38 amu?

3.4 The Periodic Table

Learning Goal

Use the periodic table to identify the group and the period of an element, and whether it is a metal or a nonmetal.

By the late 1800s, scientists began to find similarities in the behavior of the elements that were known at the time. A Russian chemist, Dmitri Mendeleev, suggested that the elements could be arranged in groups that showed similar chemical and physical properties. Today, we have the modern **periodic table,** which arranges the elements by increasing atomic number in such a way that similar properties repeat at periodic intervals.

Periods and Groups

A single horizontal row in the periodic table is called a **period.** Each row is counted from the top of the table down as Period 1, Period 2, and so on. The first period contains only the elements hydrogen (H) and helium (He). The second period, which is the second row of elements, contains lithium (Li), beryllium (Be), boron (B), carbon (C), nitrogen (N), oxygen (O), fluorine (F), and neon (Ne). The third period begins with sodium (Na) and ends at argon (Ar). There are seven periods in the periodic table.

Each column is a **group** of elements that have similar physical and chemical properties. The groups are identified by numbers that go across the top of the periodic table. For many years the periodic table was divided into the following two sections: the A groups containing the **representative elements** (1A–8A) and the **transition elements** (B), as shown in Figure 3.7. Recently, international chemistry groups agreed to number the columns (groups) 1 to 18 from left to right. Both numbering systems appear on the periodic chart inside the cover of this text. For our discussions in this text, we will use the traditional A and B group designations.

Sample Problem 3.9
Groups and Periods in the Periodic Table

State whether each set represents elements in a group, a period, or neither:
a. F, Cl, Br, I **b.** Na, Al, P **c.** K, Al, O

Solution

a. The elements F, Cl, Br, and I are part of a group of elements; they all appear in the vertical column 7A.
b. The elements Na, Al, and P all appear in the third row or third period in the periodic table.
c. Neither. The elements K, Al, and O are not part of the same group and they do not belong to the same period.

Study Check

a. What elements are found in Period 2?
b. What elements are found in Group 2A?

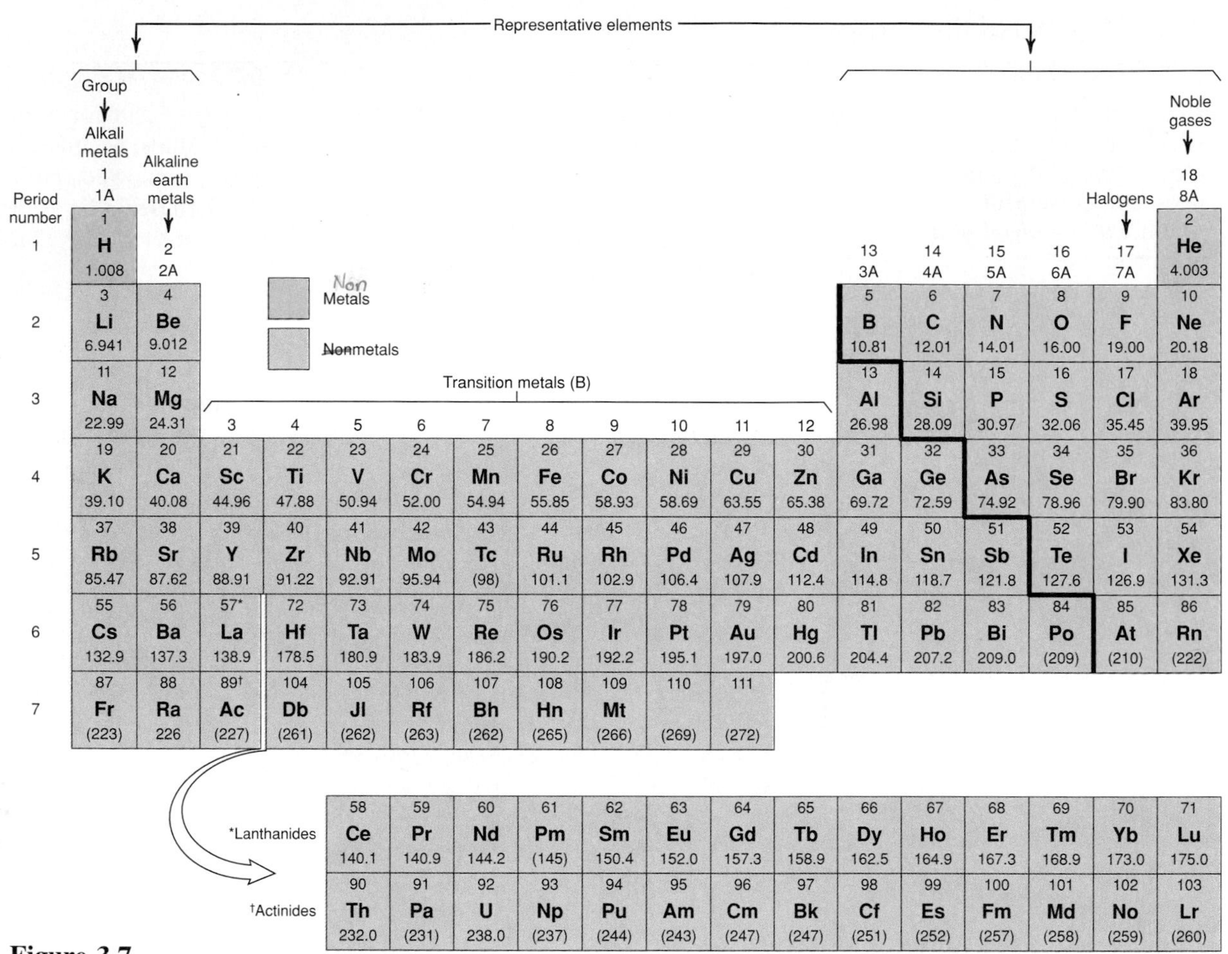

Figure 3.7
Groups and periods in the periodic table.

Sample Problem 3.10
Stating Group and Period Number

Identify the period number and the group number for the following elements:
a. calcium **b.** tin

Solution

The period is found by counting down the horizontal rows of the elements on the periodic table, and the group number is found at the top of the vertical column that contains that element.

a. Calcium (Ca) is in Period 4 and Group 2A.
b. Tin (Sn) is in Period 5 and Group 4A.

Study Check

Give the symbols of the elements that are represented by the following period and group numbers:

a. Period 3, Group 5A
b. Period 6, Group 8A

Lithium (Li)

Sodium (Na)

Potassium (K)

Figure 3.8
Lithium (Li), sodium (Na), and potassium (K), some alkali metals from Group 1A.

Classification of Groups

Several groups in the periodic table have special names, as Figure 3.7 shows. Group 1A elements, lithium (Li), sodium (Na), potassium (K), rubidium (Rb), cesium (Cs), and francium (Fr), are part of a family of elements known as the **alkali metals.** The elements within this group exhibit similar properties (see Figure 3.8). The alkali metals are soft, shiny metals that are good conductors of heat and electricity and have relatively low melting points. They react vigorously with water and form white products when they combine with oxygen.

Group 2A elements, beryllium (Be), magnesium (Mg), calcium (Ca), strontium (Sr), barium (Ba), and radium (Ra), are called the **alkaline earth metals.** They are also shiny metals like those in Group 1A, but they are not quite so reactive.

The **halogens** are found on the right side of the periodic table in Group 7A. They include the elements fluorine (F), chlorine (Cl), bromine (Br), and iodine (I), as shown in Figure 3.9. On the periodic table, each halogen is listed as a single symbol. However, halogens exist as combinations of two atoms joined to form diatomic molecules. For example, a sample of chlorine consists of Cl_2 molecules. The other halogens exist as F_2, Br_2, and I_2 molecules.

Group 8A contains the **noble gases,** helium (He), neon (Ne), argon (Ar), krypton (Kr), xenon (Xe), and radon (Rn). They are quite unreactive and are seldom found in combination with other elements.

Figure 3.9
Chlorine (Cl_2), bromine (Br_2), and iodine (I_2), some halogens from Group 7A.

Chlorine (Cl_2)

Bromine (Br_2)

Iodine (I_2)

HEALTH NOTE ***Calcium and Strontium***

Calcium (Ca) and strontium (Sr) are two elements of the alkaline earth group, Group 2A. The chemical behavior of strontium is so similar to that of calcium that when strontium is ingested, it replaces some of the calcium in the bones and teeth. This similarity in behavior caused great concern among scientists during nuclear testing, because radioactive strontium is a product of certain nuclear reactions. If the radioactive strontium were to drift to cattle grazing lands, it could become a part of cow's milk and eventually find its way to the bones of young children. Once it is there, the effects of the radioactivity are detrimental to proper growth and development. This is what happened in Chernobyl in 1986.

Metals and Nonmetals

Another feature of the periodic table is the heavy zigzag line that separates the elements into *metals* and the *nonmetals.* The metals are those elements on the left of the line except for hydrogen, and the nonmetals are the elements on the right.

In general, most **metals** are shiny solids. They can be shaped into wires (ductile) or hammered into a flat shape (malleable). Metals are often good conductors of heat and electricity. They usually melt at higher temperatures than nonmetals. All of the metals are solids at room temperature, except for mercury (Hg), which is a liquid. Some typical metals are sodium (Na), magnesium (Mg), copper (Cu), gold (Au), silver (Ag), iron (Fe), and tin (Sn).

Nonmetals are not very shiny, malleable, or ductile, and they are often poor conductors of heat and electricity. They typically have low melting points and low densities. You may have heard of nonmetals such as hydrogen (H), carbon (C), nitrogen (N), oxygen (O), chlorine (Cl), and sulfur (S). Table 3.9 compares some characteristics of silver, a metal, with those of sulfur, a nonmetal.

Table 3.9 ***Some Characteristics of Silver, a Metal, and Sulfur, a Nonmetal***

Silver	Sulfur
A shiny metal	A dull, yellow nonmetal
Extremely ductile	Brittle
Can be hammered into sheets (malleable)	Shatters when hammered
Good conductor of heat and electricity	Poor conductor, good insulator
Used in coins, jewelry, tableware	Used in gunpowder, rubber, fungicides
Density 10.5 g/mL	Density 3.1 g/mL
Melting point 962°C	Melting point 113°C

Sample Problem 3.11
Classifying Elements as Metals and Nonmetals

Using a periodic table, classify the following elements as metals or nonmetals:

a. Na **b.** Si **c.** Cl **d.** Cu

Solution

a. Metal; sodium is located to the left of the heavy zigzag line that separates metals and nonmetals.
b. Nonmetal; silicon is on the right side of the zigzag line.
c. Nonmetal; chlorine is on the right side of the zigzag line.
d. Metal; copper is located to the left of the zigzag line.

Study Check

Which of the following elements, nickel or nitrogen, would be a gas and a poor conductor of heat?

HEALTH NOTE *Some Important Trace Elements in the Body*

Some metals and nonmetals known as trace elements are essential to the proper functioning of the body. Although they are required in very small amounts, their absence can disrupt major biological processes and cause illness. The trace elements listed in Table 3.10 are present in the body combined with other elements.

Table 3.10 *Some Important Trace Elements in the Body*

Element	Adult DV[a]	Biological Function	Deficiency Symptoms	Dietary Sources
Iron (Fe)	10 mg (males) 18 mg (females)	Formation of hemoglobin; enzymes	Dry skin, spoon nails, decreased hemoglobin count, anemia	Beef, kidneys, liver, egg yolk, oysters, spinach, beans, apricots, raisins, whole-wheat bread
Copper (Cu)	2.0–5.0 mg	Necessary in many enzyme systems; growth; aids formation of red blood cells and collagen	Uncommon; anemia, decreased white cell count; bone demineralization	Nuts, organ meats, whole-wheat grains, shellfish, eggs, poultry, leafy green vegetables
Zinc (Zn)	15 mg	Amino acid metabolism; enzyme systems; energy production; collagen	Retarded growth and bone formation; skin inflammation; loss of taste and smell; poor healing	Wheat germ, shellfish, milk, lima beans, fish, eggs, whole grains, turkey (dark meat), cheddar cheese

Table 3.10 (*cont.*)

Element	Adult DV[a]	Biological Function	Deficiency Symptoms	Dietary Sources
Manganese (Mn)	2.5–5.0 mg	Necessary for some enzyme systems; collagen formation; bone formation; central nervous system; fat and carbohydrate metabolism; blood clotting	Abnormal skeletal growth; impairment of central nervous system	Cereals, peas, beans, lettuce, wheat bran, meat, poultry, fish
Iodine (I)	150 μg	Necessary for activity of thyroid gland	Hypothyroidism; goiter; cretinism	Iodized table salt, seafood
Fluorine (F)	1.5–4.0 mg	Necessary for solid teeth formation and retention of calcium in bones with aging	Dental cavities	Tea, fish, milk, eggs, water in some areas, supplementary drops, toothpaste

[a] DV, daily value.

3.5 Electron Arrangement in the Atom

Learning Goal

Given the name or symbol of one of the first 20 elements in the periodic table, write the electron arrangement.

Table 3.11
Capacity of Some Electron Shells

Electron Shell	Maximum Number of Electrons
1	2
2	8
3	18
4	32

The chemical behavior of an element is primarily determined by the way the electrons are arranged about the nucleus. Every electron has a specific amount of energy. Electrons of similar energy are grouped in an energy level called a **shell.** The shells closest to the nucleus contain electrons with the lowest energies, whereas shells farther away from the nucleus contain electrons with higher energies.

Each electron shell can hold a different number of electrons. As shown in Table 3.11, shell 1, the lowest energy level, can hold up to 2 electrons; shell 2 can hold up to 8 electrons; shell 3 can take 18 electrons; and shell 4 has room for 32 electrons.

There are additional energy levels or shells in an atom, but they are beyond our consideration in this text. In the atoms of the elements known today, electrons can occupy energy levels as high as shell 7.

Electron Arrangements for the First Twenty Elements

The **electron arrangement** of an atom gives the number of electrons in each shell. We might imagine electron shells as floors in a hotel. The ground floor fills first, then the second floor, and so on. In Figure 3.10, two electrons are placed in shell 1, the lowest energy level. The next eight electrons go into shell 2. Both shell 1 and shell 2 are now filled. Shell 3 initially takes eight electrons; it then stops filling for a while, even though it is capable of holding more electrons. This break in filling is due to an overlapping in the energy values of shell 3 and shell 4. At this point, shell 4 takes the next two electrons.

The electron arrangements for the first 20 elements can be written by placing electrons in shells beginning with the lowest energy. The single elec-

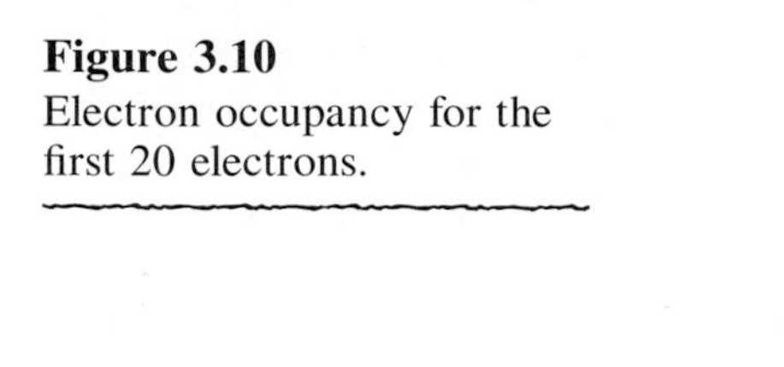

Figure 3.10
Electron occupancy for the first 20 electrons.

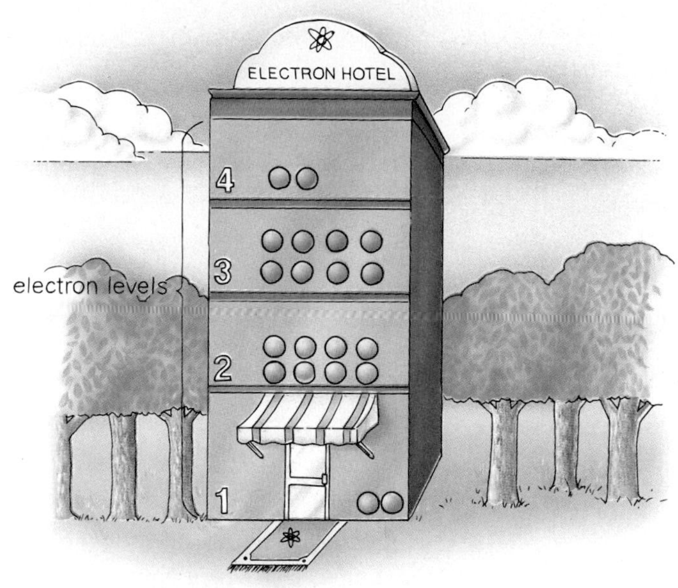

tron of hydrogen and the two electrons of helium can be placed in shell 1. When we wish to draw a simple diagram of the electron arrangement, we indicate the nucleus of the atom and draw curved lines to represent each of the occupied shells. In the illustrations that follow, one isotope has been chosen for each element shown. The electron configurations for hydrogen and helium would appear as follows:

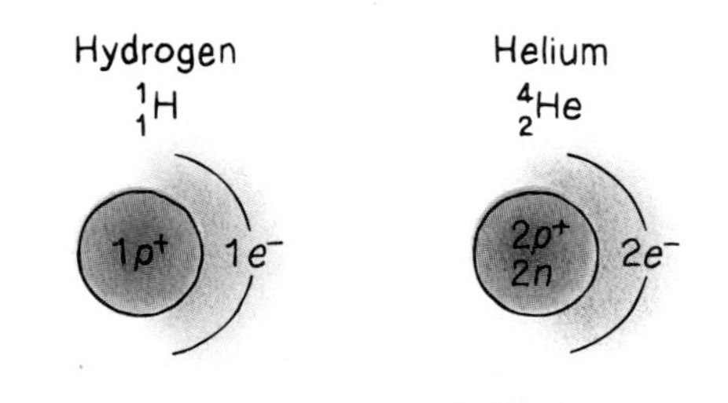

The elements of the second period (lithium, Li, to neon, Ne) have enough electrons to fill the first shell and begin filling the second shell. For example, lithium has 3 electrons. Two of those electrons complete shell 1. The remaining electron goes into the second shell. As we go across the period, more and more electrons enter the second shell. For example, an atom of carbon, with a total of 6 electrons, fills shell 1 with 2 electrons; and 4 remaining electrons enter the second shell. The last element in Period 2 is neon. The 10 electrons in an atom of neon completely fill the first and second shells.

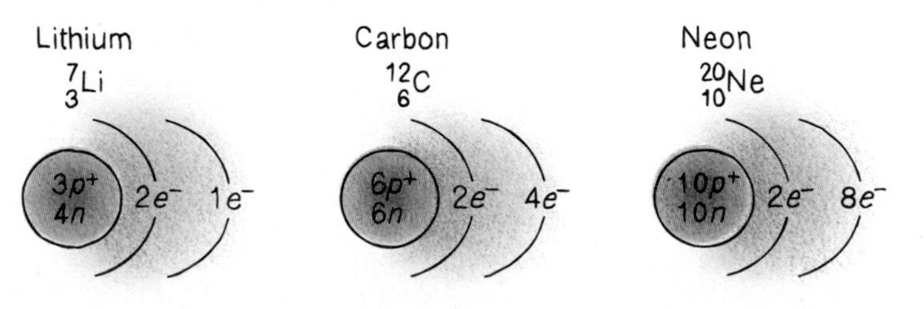

In an atom of sodium, atomic number 11, the first and second electron shells are filled and the last electron enters the third shell. The rest of the elements in the third period continue to add to the third shell. For example, a sulfur (s) atom with 16 electrons has 2 electrons in the first shell, 8 electrons in the second shell, and 6 electrons in the third shell. At the end of the period, we find that argon has 8 electrons in the third shell.

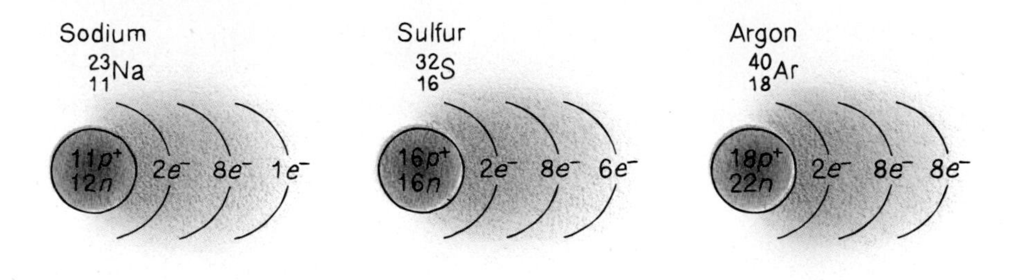

At this point, the third shell with 8 electrons stops filling for a while. The remaining electrons in potassium and calcium actually begin to fill the fourth shell. The third shell will continue filling in elements with higher atomic numbers. The electron arrangements for the first 20 elements are summarized in Table 3.12. Although there are many more elements in the periodic table, we will not consider the electron arrangements of elements beyond atomic number 20.

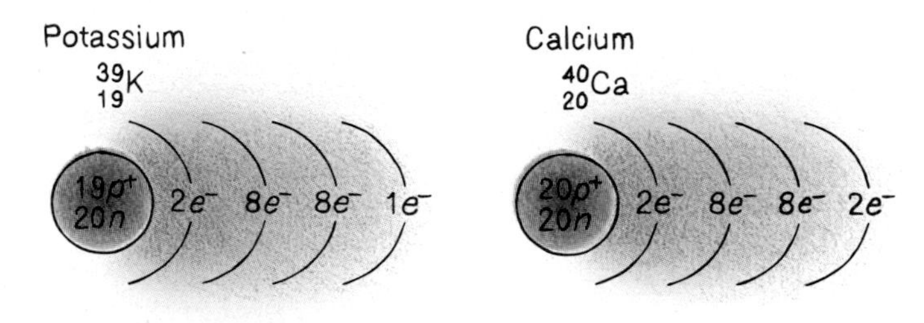

Table 3.12 ***Electron Arrangements for the First Twenty Elements***

Element	Symbol	Atomic Number	Number of Electrons in Shell 1	2	3	4	Element	Symbol	Atomic Number	Number of Electrons in Shell 1	2	3	4
Hydrogen	H	1	1				Sodium	Na	11	2	8	1	
Helium	He	2	2				Magnesium	Mg	12	2	8	2	
Lithium	Li	3	2	1			Aluminum	Al	13	2	8	3	
Beryllium	Be	4	2	2			Silicon	Si	14	2	8	4	
Boron	B	5	2	3			Phosphorus	P	15	2	8	5	
Carbon	C	6	2	4			Sulfur	S	16	2	8	6	
Nitrogen	N	7	2	5			Chlorine	Cl	17	2	8	7	
Oxygen	O	8	2	6			Argon	Ar	18	2	8	8	
Fluorine	F	9	2	7			Potassium	K	19	2	8	8	1
Neon	Ne	10	2	8			Calcium	Ca	20	2	8	8	2

Sample Problem 3.12
Writing Electron Arrangements

Write the electron arrangement for each of the following:
a. oxygen **b.** chlorine

Solution

a. Oxygen has an atomic number of 8. Therefore, there are 8 electrons in the electron arrangement:

$2\ e^-$ $6\ e^-$

b. An atom of chlorine has 17 protons and 17 electrons. The electrons are arranged as follows:

$2\ e^-$ $8\ e^-$ $7\ e^-$

Study Check

What element has the following electron arrangement?

$2\ e^-$ $8\ e^-$ $2\ e^-$

Energy-Level Changes

Whenever possible, electrons occupy the lowest energy levels. However, if a source of energy such as heat or light is available, an electron can absorb a certain amount of energy and jump to a higher energy level. Electrons in this higher energy state, however, are unstable. They drop back to one of the lower, more stable energy levels, releasing some of their energy, as illustrated in Figure 3.11. Neon lights are an example of how electrons that gain energy can emit energy as light (see Figure 3.12).

The amount of energy released when electrons drop to lower energy levels can be large or small. High-energy emissions include x-rays and gamma rays. Low-energy emissions include infrared rays (heat), radio waves, and microwaves. Figure 3.13 illustrates some forms of energy emitted when electrons drop to lower energy levels in atoms.

Only the energies in the visible region are detected by our eyes. For example, when sunlight passes through a prism or a raindrop, it is separated into the colors known as the visible spectrum, which we see in rainbows.

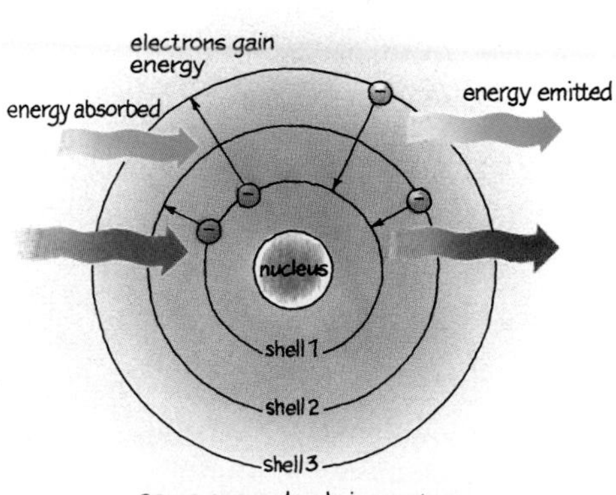

Figure 3.11
Electrons gain energy and jump to higher energy levels. When they drop back to lower energy levels, energy is emitted.

Figure 3.12
When an electric current flows through a gas-filled tube, electrons jump to higher energy levels and fall again, emitting light of different colors.

Figure 3.13
The waves of the electromagnetic spectrum show a pattern of increasing wavelengths and decreasing energy.

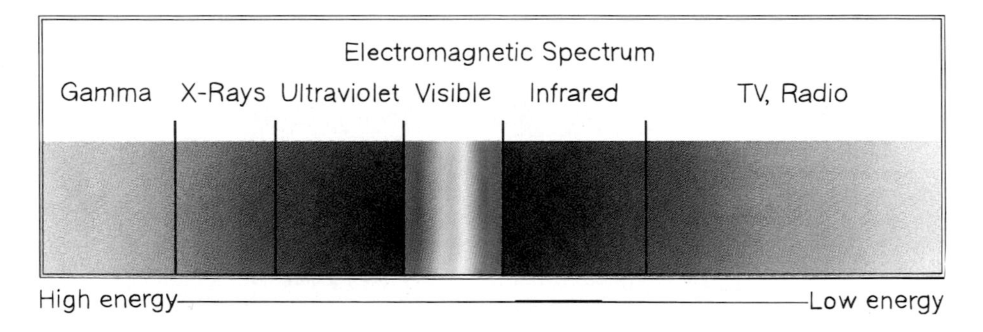

3.6 Periodic Law

Learning Goal
Use the electron arrangement of an element to state its group number and to explain periodic law.

We have seen that the groups in the periodic table contain elements having similar properties. This repetition of properties with increasing atomic number is known as the **periodic law.** If we now observe the electron arrangements for the elements in a group, we find another similarity. All the elements in a group (also called family) of elements have the same number of electrons in their outermost shells.

For example, the elements lithium, sodium, and potassium are part of Group 1A. All of these elements have one electron in their outer shells, as seen in Table 3.13. The outer shell is the shell with the highest energy that contains one or more electrons. Thus, the similarity of properties among elements in a group can now be attributed to their having the same number of electrons in each of their outermost shells.

Table 3.13 *A Comparison of Electron Arrangements of Some Group 1A Elements*

Element	Atomic Number	Number of Electrons in Shell 1	2	3	4	
Lithium	3	2	1			One electron in each outer shell
Sodium	11	2	8	1		
Potassium	19	2	8	8	1	

Group Number

The **group numbers** 1A–8A appear at the top of the periodic table. Each group number is equal to the number of electrons in the outer shell of the elements in that column. All elements in Group 1A have 1 electron in their outer shells, elements in Group 2A have 2 electrons in their outer shells, elements in Group 3A have 3 electrons in their outer shells, and so on. We are most interested in the number of electrons in the outer shells because these electrons have the greatest effect on the way an atom forms compounds. Table 3.14 shows the electron arrangement by group for the first 20 elements.

Table 3.14 *Electron Arrangements, by Group, for the First Twenty Elements*

		Number of Electrons in Shell			
Group Number	**Element**	**1**	**2**	**3**	**4**
1A	Hydrogen	1			
	Lithium	2	1		
	Sodium	2	8	1	
	Potassium	2	8	8	1
2A	Beryllium	2	2		
	Magnesium	2	8	2	
	Calcium	2	8	8	2
3A	Boron	2	3		
	Aluminum	2	8	3	
4A	Carbon	2	4		
	Silicon	2	8	4	
5A	Nitrogen	2	5		
	Phosphorus	2	8	5	
6A	Oxygen	2	6		
	Sulfur	2	8	6	
7A	Fluorine	2	7		
	Chlorine	2	8	7	
8A	Helium	2			
	Neon	2	8		
	Argon	2	8	8	

Sample Problem 3.13
Using Group Numbers

Using the periodic table, write the group number and the number of electrons in the outer electron level of the following elements:
a. sodium **b.** sulfur **c.** aluminum

Solution

a. Sodium is in Group 1A; sodium has 1 electron in the outer electron level.
b. Sulfur is in Group 6A; sulfur has 6 electrons in the outer electron level.
c. Aluminum is in Group 3A; aluminum has 3 electrons in the outer electron level.

Study Check

What is the group number, total number of electrons, and name of an atom that has 5 electrons in the third electron level?

ENVIRONMENTAL NOTE *Biological Reactions to Sunlight*

Our everyday life depends on sunlight, but exposure to sunlight can have damaging effects on living cells, and too much exposure can even cause their death. The list of damaging effects of sunlight includes sunburn; wrinkling; premature aging of the skin; changes in the DNA of the cells, which can lead to skin cancers and melanomas; inflammation of the eyes; and, perhaps, cataracts. Some drugs, like the acne medications Accutane and Retin-A, as well as antibiotics, diuretics, sulfonamides, and estrogens, make the skin extremely photosensitive and can cause undesirable changes in its reaction to sunlight. Using a sunscreen is now recommended by doctors to prevent the adverse effects of sun exposure.

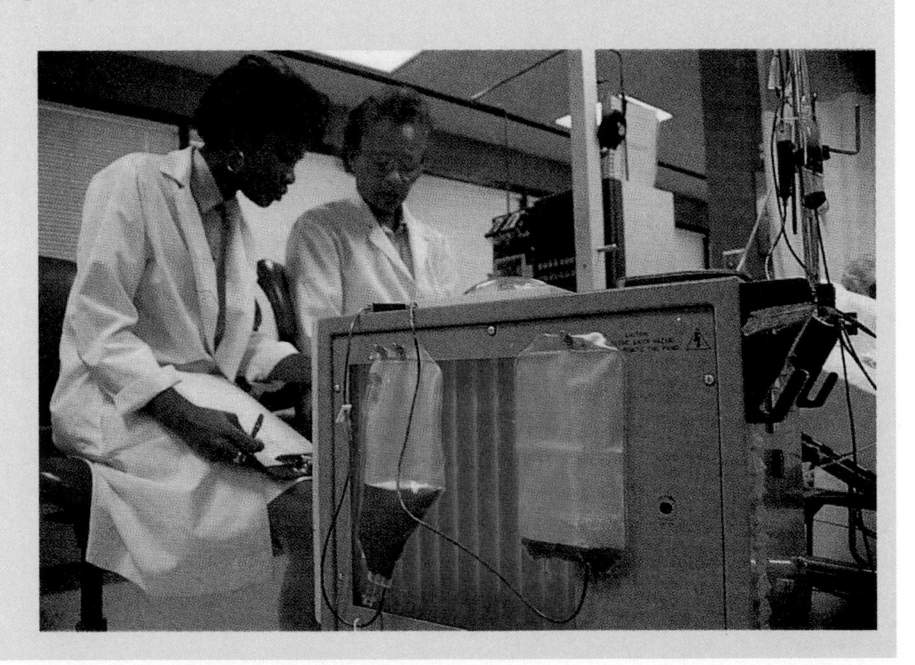

Figure 3.14
In cutaneous T-cell lymphoma, an abnormal increase in T cells causes painful ulceration of the skin. The skin is treated by photopheresis, in which the patient receives a photosensitive chemical, and then blood is removed from the body and exposed to ultraviolet light. The blood is returned to the patient, and the treated T cells stimulate the immune system to respond to the cancer cells.

High-energy radiation is the most damaging biologically. Most of the radiation in this range is absorbed in the epidermis of the skin. The degree to which radiation is absorbed depends on the thickness of the epidermis, the hydration of the skin, the amount of coloring pigments and proteins of the skin, and the arrangement of the blood vessels. In light-skinned people, 85–90% of the radiation is absorbed by the epidermis, with the rest reaching the dermis layer. In dark-skinned people, 90–95% of the radiation is absorbed by the epidermis, with a smaller percentage reaching the dermis.

However, medicine does take advantage of the beneficial effect of sunlight. Phototherapy can be used to treat certain skin conditions, including psoriasis, eczema, and dermatitis. In the treatment of psoriasis, for example, oral drugs are given to make the skin more photosensitive; exposure to ultraviolet (UV) radiation follows. Low-energy radiation is used to break down bilirubin in neonatal jaundice. Sunlight is also a factor in stimulating the immune system. (See Figure 3.14.)

3.7 Subshells and Orbitals

Learning Goal

Write the electron configuration using subshell notation.

The electrons in the electron shells or main energy levels can be described in yet more detail. Within each shell, the electrons with identical energy are grouped as **subshells.** The different types of subshells are identified by the letters *s*, *p*, *d*, and *f*. The *s* subshell is lowest in energy, followed by the *p* subshell, then the *d* subshell, and finally the highest energy subshell, the *f* subshell.

Order of Increasing Energy for Subshells

$s \longrightarrow p \longrightarrow d \longrightarrow f$

Lowest energy — Highest energy

The number of subshells in each shell is equal to the numerical value of that shell. As shown in Figure 3.15, shell 1 has only one subshell, 1*s*. Shell 2 has two subshells, 2*s* and 2*p*. The 2*s* subshell is lower in energy than the 2*p*.

Figure 3.15
The energy of the electron subshells increases as their distance from the nucleus increases. Electrons fill lowest available subshells first.

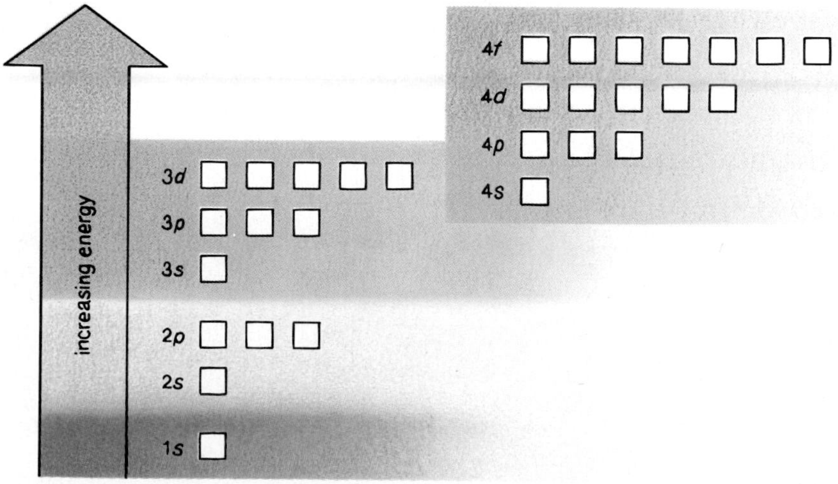

Table 3.15 ***Number of Electrons Within the Subshells***

Shell	Subshell	Number of Electrons	Shell Capacity
1	1*s*	2	2
2	2*s*	2	8
	2*p*	6	
3	3*s*	2	18
	3*p*	6	
	3*d*	10	
4	4*s*	2	32
	4*p*	6	
	4*d*	10	
	4*f*	14	

Shell 3 has three subshells, 3*s*, 3*p*, and 3*d*. The fourth shell (shell 4) consists of four subshells, 4*s*, 4*p*, 4*d*, and 4*f*. Note that some of the subshells in the third and fourth shells are so close in energy that they overlap. The 4*s* subshell is lower in energy than the 3*d* subshell. In electron arrangements with subshells, the 4*s* is filled before the 3*d*.

Number of Electrons in Subshells

Each type of subshell holds a specific number of electrons. Any *s* subshell holds just 2 electrons. A *p* subshell can hold up to 6 electrons, a *d* subshell takes up to 10 electrons, and an *f* subshell holds a maximum of 14 electrons. The number of electrons in the subshells adds up to give the number of electrons in each electron shell, as Table 3.15 shows.

Sample Problem 3.14
Subshells

Describe the third electron shell in terms of the following:

a. maximum number of electrons
b. number and designation of subshells
c. number of electrons in each subshell

Solution

a. The maximum number of electrons for the third shell is 18.
b. The third shell has three subshells: 3*s*, 3*p*, and 3*d*.
c. The 3*s* subshell can accommodate 2 electrons, the 3*p* subshell can hold up to 6 electrons, and the 3*d* subshell will take up to 10 electrons.

Study Check

State the number of electrons that would fill the following:

a. the second shell
b. the 3*p* subshell
c. the 5*s* subshell

Orbitals

An **orbital** is described as a region in space around the nucleus in which an electron of a certain energy is most likely to be found. An orbital can hold only one or two electrons.

There is a special shape for orbitals in each type of subshell. An *s* orbital is spherical, with the nucleus at the center. You might think of a 1*s* orbital as analogous to a golf ball. That means that the electrons of the 1*s* subshell are most likely to be in a spherical space. The 2*s* orbital might be like a baseball, which is bigger than a golf ball but has the same shape. A 3*s* orbital is spherical too, but perhaps it is like a basketball, fitting over the other two. Each of the *s* orbitals has a maximum of two electrons, and there is just one *s* orbital for every *s* subshell.

A *p* subshell consists of three *p* orbitals, which have shapes like dumbbells. Recall that a *p* subshell can hold six electrons. If each *p* orbital holds just two electrons, then three *p* orbitals are needed to build the *p* subshell. The three dumbbell-shaped *p* orbitals are arranged in three different directions (*x*, *y*, *z* axes) around the nucleus. As shown in Figure 3.16, all *p* orbitals have the same shape; only their size increases as the energy level increases.

Electron Arrangements Using Subshells

In an electron arrangement with subshells, the number of electrons in each subshell is shown as a superscript. For example, the electron arrangement for a fluorine atom with nine electrons shows two electrons in the 1*s* subshell, two electrons in the 2*s* subshell, and five electrons in the 2*p* subshell. The shorthand method of writing this electron arrangement for fluorine is as follows:

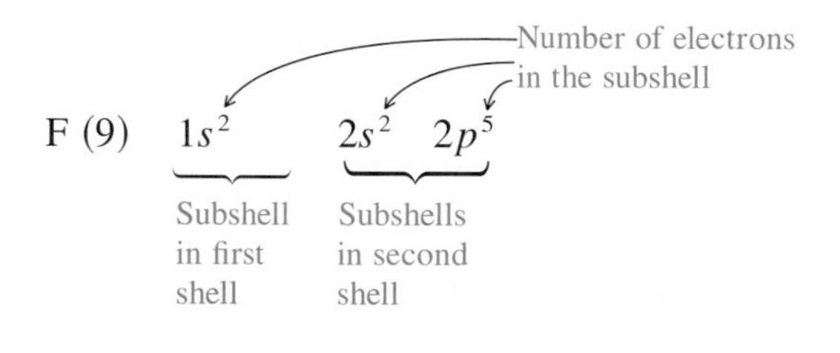

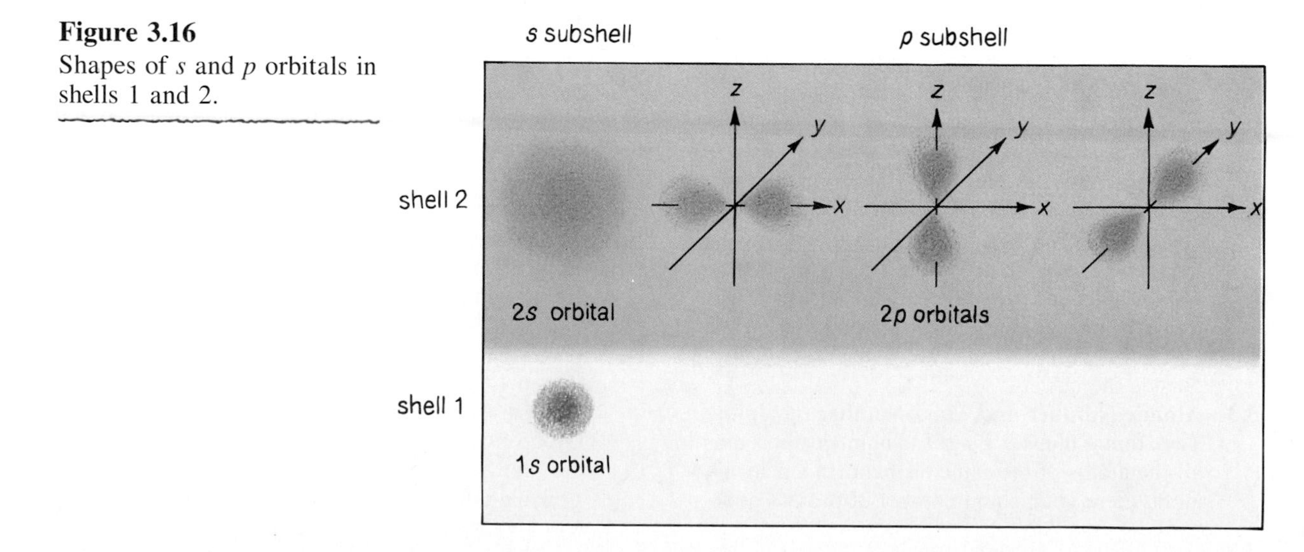

Figure 3.16
Shapes of *s* and *p* orbitals in shells 1 and 2.

Examples of electron arrangements for atoms of oxygen, sodium, phosphorus, chlorine, and calcium follow.

Element (Atomic Number)	*Electron Arrangement with Subshells*
O (8)	$1s^2 2s^2 2p^4$
Na (11)	$1s^2 2s^2 2p^6 3s^1$
P (15)	$1s^2 2s^2 2p^6 3s^2 3p^3$
Cl (17)	$1s^2 2s^2 2p^6 3s^2 3p^5$
Ca (20)	$1s^2 2s^2 2p^6 3s^2 3p^6 4s^2$

Sample Problem 3.15
Writing Electron Arrangements with Subshells

Write the electron arrangement with subshells for argon.

Solution

An atom of argon has 18 electrons. The order of filling the subshells begins with the $1s$, followed by the $2s$ and $2p$. This completes the first and second shells, with a total of 10 electrons.

$$1s^2\ 2s^2\ 2p^6$$

There are still 8 electrons to account for. The next subshells to fill are the $3s$ and $3p$, giving us a total of 18 electrons and the correct electron arrangement for argon.

$$1s^2\ 2s^2\ 2p^6\ 3s^2\ 3p^6$$

Study Check

Write the electron arrangement with subshells for sulfur.

Chapter Summary

3.1 Elements and Symbols
Elements are the primary substances of matter. Chemical symbols are one- or two-letter abbreviations of the names of the elements.

3.2 The Atom
An atom is the smallest particle that retains the characteristics of an element. Atoms are composed of three subatomic particles. Protons have a positive charge (+), electrons carry a negative charge (−), and neutrons are electrically neutral. The protons and neutrons each with a mass of about 1 amu are found in the tiny, dense nucleus. Electrons with a much smaller mass are located outside the nucleus.

3.3 Atomic Number and Mass Number
The atomic number gives the number of protons in all the atoms of the same element. In a neutral atom, there is an equal number of protons and electrons. The mass number is the total number of protons and neutrons in an atom. Atoms that have the same number of protons but different numbers of neutrons are called isotopes. The atomic weight of an element is the weighted average mass of all the isotopes in a naturally occurring sample of that element.

3.4 The Periodic Table
The periodic table is an arrangement of the elements by increasing atomic number. A vertical column on the periodic table containing elements with similar properties is called a group. A horizontal row is called a period. Elements in Group 1A are called the alkali metals: Group 2A, alkaline earth metals; Group 7A, the halogens; and Group 8A, the noble gases. Together, the A groups are known as the representative elements. On the periodic table, elements known as metals are

located on the left of the heavy zigzag line, and nonmetals are to the right of the heavy zigzag line.

3.5 Electron Arrangement in the Atom
Every electron has a specific amount of energy. In an atom, the electrons of similar energy are grouped in specific energy levels or shells. The first level nearest the nucleus can hold 2 electrons, the second level can hold 8 electrons, the third level will take up to 18 electrons. The electron arrangement is written by placing the number of electrons in that atom in order from the lowest energy levels and filling to higher levels.

3.6 Periodic Law
The periodic law states that physical and chemical properties of elements will be repeated as the atomic number increases. The similarity of behavior for the elements in a group is related to having the same number of electrons in their outermost shells. The group number for an element gives the number of electrons in its outermost shell.

3.7 Subshells and Orbitals
Within the electrons shells, each group of electrons with identical energies are called subshells. The first four subshells are identified by the letters, *s*, *p*, *d*, and *f*. In each subshell, the electrons occupy regions in space called orbitals. Orbitals may contain one or two electrons. The electron arrangement using subshells describes the order of filling beginning with $1s^22s^22p^63s^23p^64s^2$.

Glossary of Key Terms

alkali metals Elements of Group 1A except hydrogen that are soft, shiny metals with one outer shell electron.

alkaline earth metals Group 2A elements having two electrons in their outer shells.

atom The smallest particle of an element that retains the characteristics of the element.

atomic mass unit (amu) A small mass unit used to describe the mass of very small particles such as atoms and subatomic particles; 1 amu is equal to one-twelfth the mass of a carbon-12 atom.

atomic number A number that is equal to the number of protons in an atom.

atomic weight The weighted average mass of all the naturally occurring isotopes of an element.

chemical symbol An abbreviation that represents that name of an element.

electron A negatively charged subatomic particle having a very small mass that is usually ignored in calculations; its symbol is e^-.

electron arrangement An organization of electrons within the atom by increasing energy shells.

element A primary substance that cannot be separated into any simpler substances.

group A vertical column in the periodic table that contains elements having similar physical and chemical properties.

group number A number that appears at the top of each vertical column (group) in the periodic table and indicates the number of electrons in the outermost shell.

halogen Group 7A elements of fluorine, chlorine, bromine, and iodine.

isotope An atom that differs only in mass number from another atom of the same element. Isotopes have the same atomic number (number of protons) but different numbers of neutrons.

isotope symbol An abbreviation used to indicate the mass number and atomic number of an isotope.

mass number The total number of neutrons and protons in the nucleus of an atom.

metal An element that is shiny, malleable, and a good conductor of heat and electricity. The metals are located to the left of the zigzag line in the periodic table.

neutron A neutral subatomic particle having a mass of 1 amu and found in the nucleus of an atom; its symbol is n or n^0.

noble gas An element in Group 8A of the periodic table, generally unreactive and seldom found in combination with other elements.

nonmetal An element with little or no luster that is a poor conductor of heat and electricity. The nonmetals are located to the right of the zigzag line in the periodic table.

nucleus The compact, very dense center of an atom, containing the protons and neutrons of the atom.

orbital A region in space in which an electron is most likely to be found.

period A horizontal row of elements in the periodic table.

periodic law The repetition of similar chemical and physical properties with increasing atomic number due to the reappearance of the same number of electrons in the outermost shells of atoms of those elements.

periodic table An arrangement of elements by inc3reasing atomic number such that elements having similar chemical behavior are grouped in vertical columns.
proton A positively charged subatomic particle having a mass of 1 amu and found in the nucleus of an atom; its symbol is p or p^+.
representative element An element found in Groups 1A through 8A of the periodic table.
shell An energy level containing electrons of similar energies.
subatomic particle A particle within an atom; protons, neutrons, and electrons are subatomic particles.
subshell A group of electrons having identical energies.
transition element An element located between Groups 2A and 3A on the periodic table.

Chemistry at Home

1. Using a reference such as the *Handbook of Chemistry and Physics,* determine the elements and some of their properties that are associated with the following characters in mythology: Prometheus, Tantalos, Mercury, Neptune, Pluto, Niobe, Pallas, Thor, Titans, and Vanadis.
2. Tear up a piece of paper into small bits. Brush your hair several times and place the brush just above the bits of paper. Use your knowledge of electrical charges to give an explanation for your observations. Try the same experiment with a comb.
3. Obtain a box of toothpicks and some miniature marshmallows or gum drops. If they are different colors, use one color for protons, one for neutrons, and one for electrons. If the same, use color or ink pens to mark dots of three different colors on the marshmallows. Use pieces of toothpicks to build a nucleus of an atom. Then use a whole toothpick to add the electrons. Make models of atoms helium (2 neutrons), lithium (4 neutrons), carbon (6 neutrons), nitrogen (7 neutrons). Give the element and symbol for each. How would you make an isotope of the carbon atom or nitrogen atom?

Answers to Study Checks

3.1 Si, S, and Ag
3.2 magnesium, aluminum, and fluorine
3.3 A proton, symbol p or p^+, has a mass of 1 amu, carries a charge of 1+, and is found in the nucleus of an atom.
3.4 **a.** 26 **b.** 26 **c.** iron, Fe
3.5 Because the atomic number of silver is 47, it has 47 protons. The mass number is 107, which is the sum of 47 protons and 60 neutrons.
3.6 45 neutrons
3.7 **a.** $^{15}_{7}N$ **b.** $^{42}_{20}Ca$ **c.** $^{27}_{13}Al$
3.8 zinc
3.9 **a.** Period 2: Li, Be, B, C, N, O, F, Ne
b. Group 2A: Be, Mg, Ca, Sr, Ba, Ra
3.10 **a.** P **b.** Rn
3.11 nitrogen
3.12 magnesium
3.13 5A; 15; phosphorus
3.14 **a.** 8 **b.** 6 **c.** 2
3.15 $1s^22s^22p^63s^23p^4$

Problems

Elements and Symbols *(Goal 3.1)*

3.1 Write the symbols for the following elements:
a. copper **b.** silicon
c. potassium **d.** nitrogen
e. iron **f.** barium
g. lead **h.** strontium

3.2 Write the symbols for the following elements:
a. oxygen **b.** lithium
c. sulfur **d.** aluminum
e. hydrogen **f.** neon
g. tin **h.** gold

3.3 Write the name of the element for each symbol.
a. C **b.** Cl **c.** I **d.** Hg
e. F **f.** Ar **g.** Zn **h.** Ni

3.4 Write the name of the element for each symbol.
a. He **b.** P **c.** Na **d.** Mg
e. Ca **f.** Br **g.** Cd **h.** Si

The Atom *(Goal 3.2)*

3.5 Is a proton, neutron, or electron described by the following?
- **a.** has the smallest mass
- **b.** carries a positive charge
- **c.** is found outside the nucleus
- **d.** is electrically neutral

3.6 Is a proton, neutron, or electron described by the following?
- **a.** has a mass about the same as a proton's
- **b.** is found in the nucleus
- **c.** is found in the largest part of the atom
- **d.** carries a negative charge

Atomic Number and Mass Number *(Goal 3.3)*

3.7 Would you use atomic number, mass number, or both to obtain the following?
- **a.** number of protons in an atom
- **b.** number of neutrons in an atom
- **c.** number of particles in the nucleus
- **d.** number of electrons in an atom

3.8 What do you know about the subatomic particles from the following?
- **a.** atomic number
- **b.** mass number
- **c.** mass number − atomic number
- **d.** mass number + atomic number

3.9 Write the names and symbols of the elements with the following atomic numbers:
a. 3 **b.** 9 **c.** 20 **d.** 30
e. 10 **f.** 14 **g.** 53 **h.** 8

3.10 Write the names and symbols of the elements with the following atomic numbers:
a. 1 **b.** 11 **c.** 19 **d.** 26
e. 35 **f.** 47 **g.** 15 **h.** 2

3.11 How many protons and electrons are there in a neutral atom of the following?
a. magnesium **b.** zinc
c. iodine **d.** potassium

3.12 How many protons and electrons are there in a neutral atom of the following?
a. carbon **b.** fluorine
c. calcium **d.** sulfur

3.13 Complete the following table for neutral atoms.

Name of Element	Symbol	Atomic Number	Mass Number	Number of Protons	Number of Neutrons	Number of Electrons
	Al		27			
		12			12	
Potassium					20	
				16	15	
			56			26

3.14 Complete the following table for neutral atoms.

Name	Symbol	Atomic Number	Mass Number	Number of Protons	Number of Neutrons	Number of Electrons
	N		15			
Calcium			42			
				38	50	
		14			16	
		56	138			

3.15 What are the number of protons, neutrons and electrons in the following isotopes?
a. $^{27}_{13}Al$ **b.** $^{52}_{24}Cr$ **c.** $^{34}_{16}S$ **d.** $^{56}_{26}Fe$

3.16 What are the number of protons, neutrons and electrons in the following isotopes?
a. $^{2}_{1}H$ **b.** $^{14}_{7}N$ **c.** $^{26}_{14}Si$ **d.** $^{70}_{30}Zn$

3.17 Write the symbol for isotopes with the following:

Example: $^{23}_{11}Na$

a. 15 protons and 16 neutrons
b. 35 protons and 45 neutrons
c. 13 electrons and 14 neutrons
d. a chlorine atom with 18 neutrons

3.18 Write the symbol for isotopes with the following:

Example: $^{23}_{11}Na$

a. an oxygen atom with 10 neutrons
b. 4 protons and 5 neutrons
c. 26 electrons and 30 neutrons
d. a mass number of 24 and 13 neutrons

3.19 There are four isotopes of sulfur with mass numbers 32, 33, 34, and 36.
a. Write the isotope symbol for each of these atoms.
b. How are these isotopes alike?
c. How are they different?
d. Why is the atomic weight of sulfur not a whole number?

3.20 There are four isotopes of strontium with mass numbers 84, 86, 87, 88.
a. Write the isotope symbol for each of these atoms.
b. How are these isotopes alike?
c. How are they different?
d. Why is the atomic weight of strontium not a whole number?

The Periodic Table *(Goal 3.4)*

3.21 Give the group or period number described by each of the following statements:
a. contains the elements C, N, and O
b. begins with helium
c. the alkali metals
d. ends with neon

3.22 Give the group or period number described by each of the following statements:
a. contains Na, K, and Rb
b. begins with atomic number 3
c. the noble gases
d. contains F, Cl, Br, and I

3.23 Classify the following as an alkali metal, alkaline earth metal, transition element, halogen, or noble gas:
a. Ca **b.** Fe **c.** Xe
d. Na **e.** Cl

3.24 Classify the following as an alkali metal, alkaline earth metal, transition element, halogen, or noble gas:
a. Ne **b.** Mg **c.** Fe
d. Br **e.** Ba

3.25 Give the symbol of the element described by the following:
a. Group 4A, Period 2
b. a noble gas in Period 1
c. an alkali metal in Period 3
d. Group 2A, Period 4
e. Group 3A, Period 3

3.26 Give the symbol of the element described by the following:
a. an alkaline earth metal in Period 2
b. Group 5A, Period 3
c. a noble gas in Period 4
d. a halogen in Period 5
e. Group 4A, Period 4

3.27 Is each of the following a metal or nonmetal?
a. calcium
b. sulfur
c. is shiny
d. does not conduct heat
e. in Group 7A

3.28 Is each of the following a metal or nonmetal?
a. in Group 2A
b. good conductor of electricity
c. chlorine
d. iron
e. is not shiny

3.29 Are the following a metal or nonmetal?
a. phosphorus **b.** magnesium
c. silver **d.** fluorine
e. sulfur **f.** silicon
g. nitrogen **h.** aluminum

3.30 The following are trace elements that have been found to be crucial to the biochemical and physiological processes in the body. Indicate whether each is a metal or nonmetal.
a. zinc **b.** cobalt
c. manganese (Mn) **d.** iodine
e. copper **f.** selenium (Se)
g. nickel **h.** iron

Electron Arrangement in the Atom *(Goal 3.5)*

3.31 Write the electron arrangement for each of the following elements:

Example: sodium 2, 8, 1

a. carbon **b.** argon
c. sulfur **d.** potassium
e. an atom with 13 protons and 14 neutrons
f. nitrogen

3.32 Write the electron arrangements for each of the following atoms:

Example: sodium 2, 8, 1

a. phosphorus **b.** neon **c.** $^{18}_{8}O$
d. an atom with atomic number 18
e. aluminum **f.** silicon

3.33 Identify the element that has the following electron arrangements:

Energy Level:	*1*	*2*	*3*
a.	$2e^-$	$1e^-$	
b.	$2e^-$	$8e^-$	$2e^-$
c.	$1e^-$		
d.	$2e^-$	$8e^-$	$7e^-$
e.	$2e^-$	$6e^-$	

3.34 Identify the element that has the following electron arrangements:

Energy Level:	*1*	*2*	*3*
a.	$2e^-$	$5e^-$	
b.	$2e^-$	$8e^-$	$6e^-$
c.	$2e^-$	$4e^-$	
d.	$2e^-$	$8e^-$	$8e^-$
e.	$2e^-$	$8e^-$	$3e^-$

3.35 **a.** Electrons can jump to higher energy levels when they _____ a specific amount of energy.
b. When electrons drop to lower energy levels, they _____ a certain amount of energy.

3.36 **a.** Comparing x rays, microwaves, and radio waves as shown in Figure 3.13, which has the highest radiation energy?
b. Why are we protected from x rays but not radiowaves?

Periodic Law *(Goal 3.6)*

3.37 The elements boron and aluminum are in the same group on the periodic table.
a. Write the electron arrangements for B and Al.
b. How many electrons are in the outer energy level of each atom?
c. What is their group number?

3.38 The elements fluorine and chlorine are in the same group on the periodic table.
a. Write the electron arrangements for F and Cl.
b. How many electrons are in each of their outer energy levels?
c. What is their group number?

3.39 What is the number of electrons in the outer energy level and the group number for each of the following elements?

Example: fluorine $7e^-$; Group 7A

a. magnesium **b.** chlorine
c. oxygen **d.** nitrogen

3.40 What is the number of electrons in the outer energy level and the group number for each of the following elements?

Example: fluorine $7e^-$; Group 7A

a. lithium **b.** silicon
c. neon **d.** argon

3.41 Would you expect Mg, Ca, and Sr to display similar physical and chemical behavior? Why?

3.42 Name two elements that would exhibit physical and chemical behavior similar to chlorine.

Subshells and Orbitals *(Goal 3.7)*

3.43 What would be the maximum number of electrons in the following?
a. $2p$ orbital **b.** $2p$ subshell
c. shell 2 **d.** $3s$ orbital

3.44 What would be the maximum number of electrons in the following?
a. $1s$ subshell **b.** $3s$ subshell
c. shell 3 **d.** $4s$ orbital

3.45 Write the electron arrangement using subshell notation ($1s^2 2s^2$) for the following elements:
a. magnesium
b. phosphorus
c. argon
d. sulfur
e. chlorine

3.46 Write the electron arrangement using subshell notation ($1s^2 2s^2$) for the following elements:
a. potassium
b. sodium
c. beryllium
d. nitrogen
e. carbon

3.47 Identify the element that has the following electron arrangement:
a. $1s^1$
b. $1s^2 2s^2 2p^3$
c. $1s^2 2s^2 2p^6 3s^1$
d. $1s^2 2s^2 2p^6$

3.48 Identify the element that has the following electron arrangement:
a. $1s^2 2s^2 2p^2$
b. $1s^2 2s^1$
c. $1s^2 2s^2 2p^6 3s^2 3p^4$
d. $1s^2 2s^2 2p^6 3s^2 3p^6 4s^2$

3.49 Distinguish between:
  a. atoms and isotopes
  b. atomic number and mass number

3.50 Distinguish between:
  a. chemical symbol and isotope symbol
  b. atomic number and atomic weight

Challenge Problems

3.51 The following statements are *false*. Reword the underlined word or phrase to make a *true* statement.
  a. The proton is a neutral particle.
  b. The electrons are found in the nucleus.
  c. The nucleus is the largest part of the atom.
  d. The neutron has a negative charge.
  e. Most of the mass of an atom is due to its electrons.

3.52 State the number of protons and neutrons in the following atoms:
  a. ^{2}H **b.** ^{37}Cl
  c. ^{106}Cd **d.** ^{209}Bi

3.53 The most abundant isotope of iron is ^{56}Fe.
  a. How many protons, neutrons, and electrons are in this isotope?
  b. What is the symbol of a radioactive isotope of iron with 25 neutrons?
  c. What is the symbol of a radioactive isotope with the same mass number, 51, and 27 neutrons?

3.54 Cadmium, atomic number 79, consists of eight naturally occurring isotopes. Do you expect any of the isotopes to have the atomic weight given on the periodic table for cadmium? Explain.

3.55 A sample of copper has two naturally occurring isotopes. ^{63}Cu has a mass of 62.93 with an abundance of 69.1%, and ^{65}Cu has a mass of 64.93 and an abundance of 30.9%. What is the atomic weight of copper?

3.56 A lead atom has a mass of 3.4×10^{-22} g. How many lead atoms are in a cube of lead that has a volume of 2.00 cm^3, if the density of lead is 11.3 g/cm^3?

3.57 If the diameter of a sodium atom is 3.14×10^{-8} cm, how many sodium atoms would fit along a line 1 inch long.

3.58 Consider the following atoms in which the chemical symbol of the element is represented by X.

$^{16}_{8}X$ $^{16}_{9}X$ $^{18}_{10}X$ $^{17}_{8}X$ $^{18}_{8}X$

  a. What atoms have the same number of protons? How many?
  b. Which atoms are isotopes? Of what element?
  c. Which atoms have the same mass number?
  d. What atoms have the same number of neutrons?

Chapter 4

Compounds and Their Bonds

The elements silicon and oxygen combine to form a compound we know as quartz, which is seen in crystals and some semi-precious stones.

Looking Ahead

Learning Goals

4.1 Using the periodic table, write the electron-dot structures for the first 20 elements.

4.2 Illustrate the octet rule using the electron-dot structures of the noble gases.

4.3 Write the formulas of the simple ions for metals and nonmetals.

4.4 Using charge balance, write the correct formula for an ionic compound.

4.5 Write a formula of a compound containing a polyatomic ion.

4.6 Given the formula of an ionic compound, write the correct name.

4.7 Diagram the electron-dot structure for a covalent molecule.

4.8 Given the formula of a covalent compound, write its correct name; given the name of a covalent compound, write its formula.

4.9 Using electronegativity values, classify a bond as covalent, polar covalent, or ionic.

QUESTIONS TO THINK ABOUT

1. How does an element differ from a compound?
2. What are some compounds listed on a vitamin bottle?
3. What are ions?

Most of the things you see and use every day are made of *compounds,* in which atoms of one element are combined with atoms of other elements. Although there are 111 elements known today, there are millions of different compounds because of the many different ways in which atoms may combine. In ionic compounds such as table salt (sodium chloride, NaCl), electrons are transferred from one atom to another to form positive and negative ions. Strong attractive forces called ionic bonds hold the ions together. In covalent compounds such as water (H_2O), electrons are shared by the atoms, an attraction called a covalent bond.

The substances necessary for life include ionic and covalent compounds. For example, the human body is about 60–65% water, a covalent compound made from the elements hydrogen and oxygen. Some other compounds necessary for life are carbohydrates, fats, and proteins. All are made of the elements carbon, hydrogen, and oxygen; proteins contain nitrogen and sulfur also. These compounds, obtained from your diet, are needed to build and repair cells, provide energy, and direct metabolic processes. Ionic compounds such as sodium chloride and potassium chloride (KCl) are needed for the proper functioning of the heart muscle and nerve conduction.

4.1 Valence Electrons

Learning Goal

Using the periodic table, write the electron-dot structures for the first 20 elements.

For the atoms of the representative elements in the periodic table, the electrons in the outer shell play an important role in determining their chemical properties. These influential electrons, called **valence electrons,** are located in the valence shell, which is the highest or outermost energy level of an atom. For example, all of the elements in Group 1A, such as potassium, have one valence electron. Recall that in an atom of potassium, the electrons are arranged in the first four energy levels as 2, 8, 8, 1. Because the fourth shell is the outermost shell for potassium, the electron in the fourth shell is a valence electron. In an electron-dot structure, this valence electron is placed next to the symbol for the element, as shown in Figure 4.1.

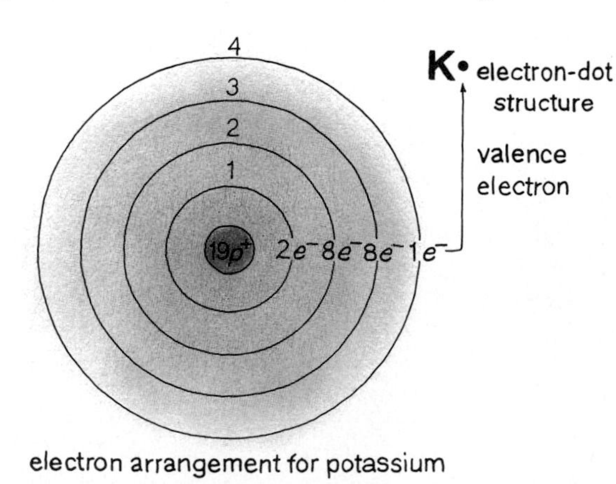

Figure 4.1
The electron arrangement for potassium and its electron-dot structure.

The elements in Group 2A have two valence electrons; similarly, the elements in Group 3A have three valence electrons, elements in Group 5A have five valence electrons, and elements in Group 7A have seven valence electrons. If we compare the electron arrangement and the group number of an element, we find that the number of valence electrons in an atom is equal to the group number for that element. Some examples of the relationship between electron arrangement, group number, and number of valence electrons are shown in Table 4.1.

Table 4.1 ***Electron Arrangement***

Element	Atomic Number	Group Number	Number of Electrons in Energy Level 1	2	3	4	Number of Valence Electrons
Li	3	1A	2	[1]			1
Ca	20	2A	2	8	8	[2]	2
Al	13	3A	2	8	[3]		3
C	6	4A	2	[4]			4
N	7	5A	2	[5]			5
S	16	6A	2	8	[6]		6
Cl	17	7A	2	8	[7]		7

Sample Problem 4.1
Counting Valence Electrons

Using the electron arrangement, state the number of valence electrons in atoms of each of the following elements:
a. O **b.** Na

Solution

a. The element oxygen has six valence electrons, as shown in its electron arrangement.

2 [6]

b. The element sodium has one valence electron, because there is one electron in its highest energy level, shell 3.

2 8 [1]

Study Check

How many electrons are in the outer shell of sulfur?

Electron-Dot Structures

An **electron-dot structure** is a convenient way to represent the valence shell. Valence electrons are shown as dots placed on the sides, top, or bottom of the symbol for the element. It does not matter on which of the four sides you place the dots. However, one to four valence electrons are usually arranged as single dots. When there are more than four electrons, the electrons begin to pair up. Any of the following would be an acceptable electron-dot structure for magnesium, which has two valence electrons:

Possible Electron-Dot Structures for the Two Valence Electrons in Magnesium

Ṁg· Mg̣ ·Ṁg ·Mg· Mg̣·

Electron-dot structures for the first 20 elements are given in Table 4.2. Note that although helium has just two valence electrons, it appears in Group 8A with the rest of the noble gases because shell 1, its outermost shell, is filled.

Table 4.2 ***Electron-Dot Structures for the First Twenty Elements***

	Group Number							
	1A	**2A**	**3A**	**4A**	**5A**	**6A**	**7A**	**8A**
Number of Valence Electrons	1 e^-	2 e^-	3 e^-	4 e^-	5 e^-	6 e^-	7 e^-	8 e^-
	H·							He:
	Li·	Ḃe·	·Ḃ·	·Ċ·	·N̈·	·Ö:	·F̈:	:N̈e:
	Na·	Ṁg·	·Ȧl·	·Ṡi·	·Ṗ·	·S̈:	·C̈l:	:Är:
	K·	Ċa·						

Sample Problem 4.2
Writing Electron-Dot Structures

Write the electron-dot structure for each of the following elements:
a. chlorine **b.** aluminum **c.** argon

Solution

a. Because the group number for chlorine is 7A, we can state that chlorine has seven valence electrons.

·C̈l:

b. Aluminum, in Group 3A, has three valence electrons.

·Ȧl·

c. Argon, in Group 8A, has eight valence electrons.

:Är:

Study Check
What is the electron-dot structure for the alkaline-earth metal in Period 4?

4.2 The Octet Rule

Learning Goal

Illustrate the octet rule using the electron-dot structures of the noble gases.

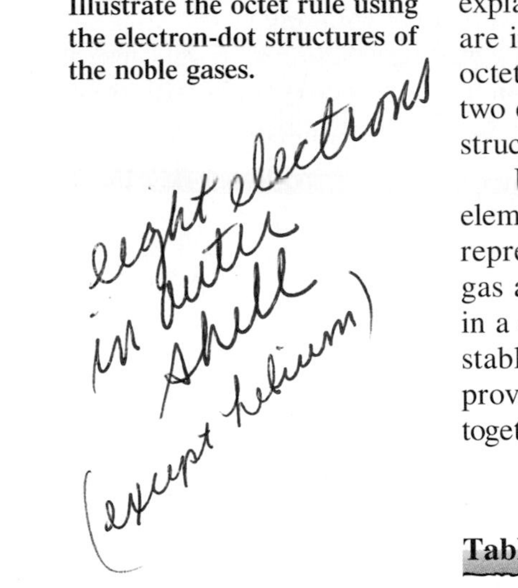

The noble gases do not combine readily with other elements. One explanation for their lack of reactivity is that their eight valence electrons are in a particularly stable arrangement. All of the noble gases have an octet of eight valence electrons, except for helium, which is stable with two electrons in its first shell. The electron arrangements and electron-dot structures for some noble gases are shown in Table 4.3.

Except for the noble gases, most elements tend to combine with other elements to form a *compound.* In most of the **compounds** formed by representative elements (Groups 1A–7A), each atom has acquired a noble gas arrangement. The **octet rule** indicates that the atoms of the elements in a compound lose, gain, or share valence electrons in order to produce a stable, noble gas arrangement of eight electrons. Thus, the octet rule provides a basic key to understanding the ways in which atoms bond together in compounds.

Table 4.3 ***Electron Arrangement for Some Noble Gases***

	Number of Electrons in Shell				
Noble Gas	**1**	**2**	**3**	**4**	**Electron-Dot Structure**
He	2				He:
Ne	2	8			:N̤̈e:
Ar	2	8	8		:Ä̤r:
Kr	2	8	18	8	:K̤̈r:

Sample Problem 4.3
Octet Rule

Write the electron-dot structures for neon (Ne), sodium (Na), and fluorine (F). Which atoms have octets and which do not? Which atoms will most likely form compounds? Why?

Solution

:N̤̈e: Na· ·F̤̈:

Neon, a noble gas, already has an octet of eight valence electrons and is not likely to form compounds. Sodium, however, has just one valence electron, and fluorine has seven. They would be expected to form compounds so that each of them will have an octet of valence electrons.

Study Check

Which of the following elements have octets?

a. carbon **b.** calcium **c.** krypton (Kr)

HEALTH NOTE *Some Uses for Noble Gases*

Noble gases may be used when it is necessary to have a substance that is unreactive. Scuba divers normally use a pressurized mixture of nitrogen and oxygen gases for breathing under water. However, when the air mixture is used at depths where pressure is high, the nitrogen gas is absorbed into the blood, where it can cause mental disorientation. To avoid this problem, a breathing mixture of oxygen and helium may be substituted. The diver still obtains the necessary oxygen, but the unreactive helium that dissolves in the blood does not cause mental disorientation. However, its lower density does change the vibrations of the vocal cords, and the diver will sound like Donald Duck.

Helium is also used to fill blimps and balloons. When dirigibles were first designed, they were filled with hydrogen, a very light gas. However, when they came in contact with any type of spark or heating source, they exploded violently because of the extreme reactivity of hydrogen gas with oxygen present in the air. Today blimps are filled with unreactive helium gas, which presents no danger of explosion. (See Figure 4.2.)

Lighting tubes are generally filled with a noble gas such as neon or argon. While the electrically heated filaments that produce the light get very hot, the surrounding noble gases do not react with the hot filament. If heated in air, the elements that constitute the filament would soon burn up.

Figure 4.2
The Goodyear blimp filled with helium gas is ready for takeoff.

4.3 Ions

Learning Goal
Write the formulas of the simple ions for metals and nonmetals.

Positive Ions

In ionic compounds, metals lose their valence electrons in order to attain a noble gas arrangement. By losing electrons, they form **ions,** which are atoms that have a different number of electrons than protons. For example, when a sodium atom loses its outer electron, the remaining electrons have a noble gas arrangement, as shown in Figure 4.3.

By losing an electron, sodium now has 10 electrons instead of 11. Because there are still 11 protons in its nucleus, the atom is no longer neutral. It has become a sodium ion and has an electrical charge, called an **ionic charge,** of 1+. In the symbol for the ion, the ionic charge is written in the upper right-hand corner, as in Na^+.

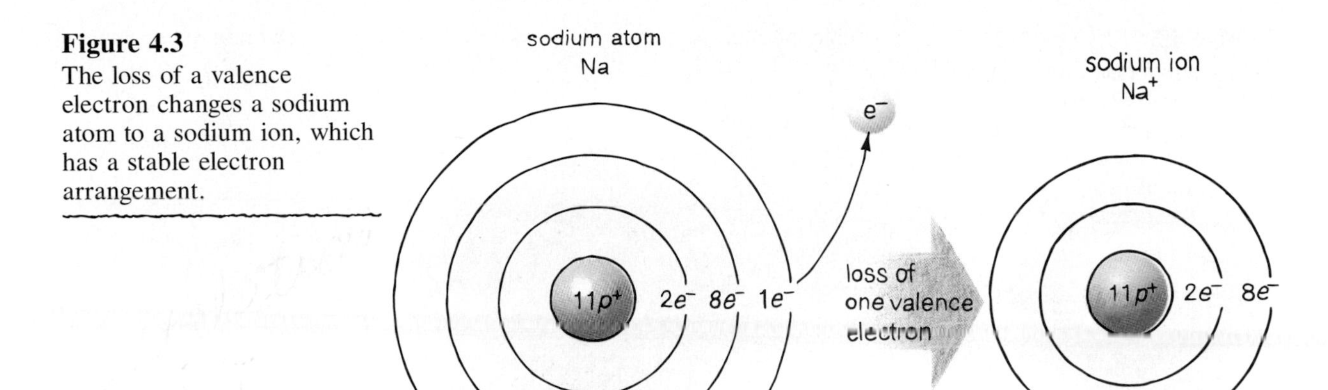

Figure 4.3
The loss of a valence electron changes a sodium atom to a sodium ion, which has a stable electron arrangement.

Comparing the Charges in the Sodium Atom and the Sodium Ion

	Na Atom	*Na^+ Ion*
Charge from electrons	11−	10−
Charge from protons	11+	11+
Overall ionic charge	0	1+

In general, metals in ionic compounds have lost their valence electrons to form positively charged ions. Positive ions are also called **cations** (pronounced *cat'-ions*). Another metal, magnesium, in Group 2A, attains a noble gas arrangement by losing two valence electrons to form a positive ion with a 2+ ionic charge (See Figure 4.4.)

Comparing the Charges in the Magnesium Atom and the Magnesium Ion

	Mg Atom	*Mg^{2+} Ion*
Charge from electrons	12−	10−
Charge from protons	12+	12+
Overall ionic charge	0	2+

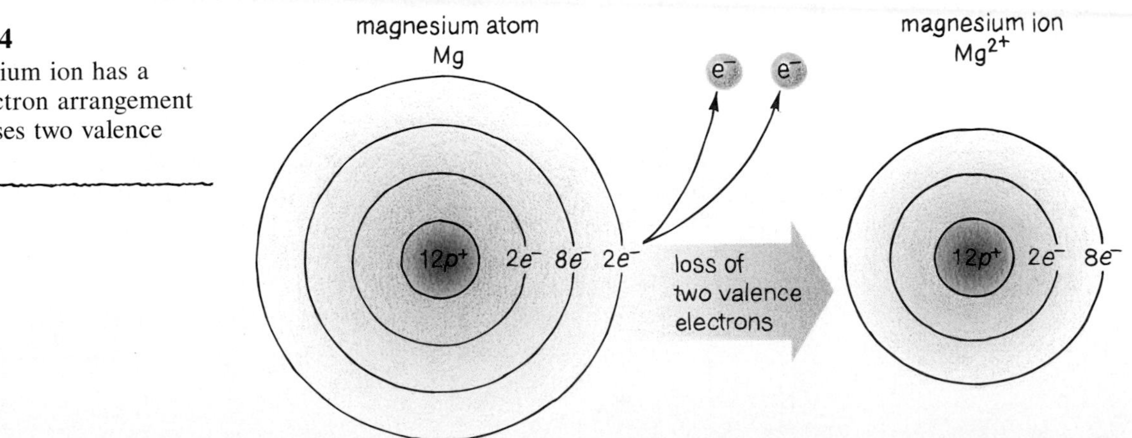

Figure 4.4
A magnesium ion has a stable electron arrangement after it loses two valence electrons.

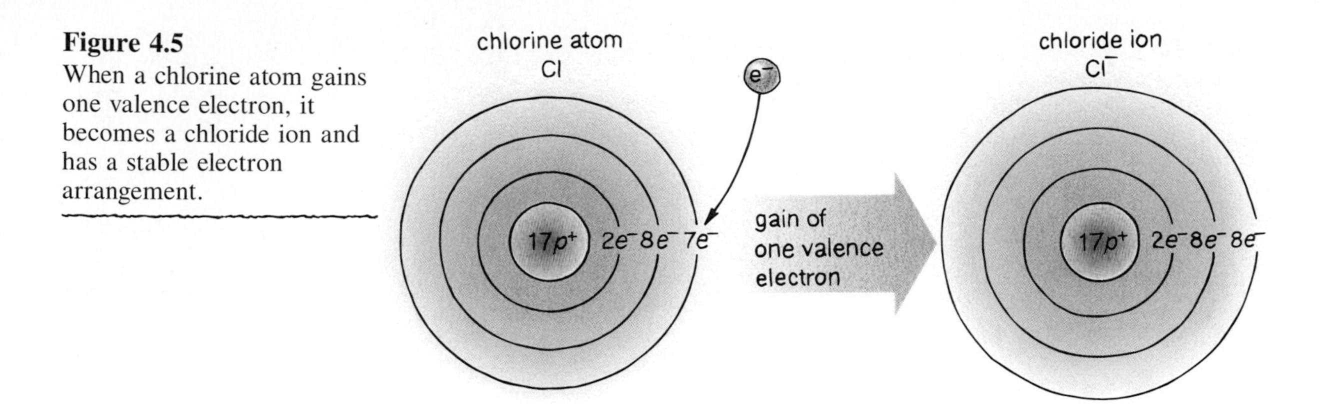

Figure 4.5
When a chlorine atom gains one valence electron, it becomes a chloride ion and has a stable electron arrangement.

Negative Ions

A nonmetal forms a negative ion when it gains one or more valence electrons in the valence shell. For example, when an atom of chlorine with seven valence electrons obtains one more electron, it completes a noble gas arrangement, as shown in Figure 4.5. The resulting particle is a chloride ion having a negative ionic charge (Cl^-). Ions with negative ionic charges are called **anions** (pronounced *an'-ions*).

Comparing the Charges in the Chlorine Atom and the Chloride Ion

	Cl Atom	*Cl^- Ion*
Charge from electrons	17−	18−
Charge from protons	17+	17+
Overall ionic charge	0	1−

Sample Problem 4.4
Calculating Ionic Charge

Write the symbol and the ionic charge for each of the following ions:

a. a nitrogen ion that has 7 protons and 10 electrons
b. a calcium ion that has 20 protons and 18 electrons
c. a lithium ion that has 3 protons and 2 electrons

Solution

a. $7\ p^+$ and $10\ e^- = N^{3-}$
b. $20\ p^+$ and $18\ e^- = Ca^{2+}$
c. $3\ p^+$ and $2\ e^- = Li^+$

Study Check
How many protons and electrons are in a bromide ion, Br^-?

Ionic Charges from Group A Numbers

Group numbers can be used to determine the ionic charges of the ions in that group. We have seen that metals lose electrons to form positive ions. The elements in Groups 1A, 2A, and 3A lose the same number of electrons (1, 2, and 3, respectively) to produce ions with the same ionic charge. Group 1A metals form ions with 1+ charges, Group 2A metals form ions with 2+ charges, and Group 3A metals form ions with 3+ charges.

When the nonmetals from Groups 5A, 6A, and 7A form ions, they acquire negative charges. Group 5A nonmetals form ions with 3− charges, Group 6A nonmetals form ions with 2− charges, and Group 7A nonmetals form ions with 1− charges. The elements of Group 4A are not discussed here because they do not typically form simple ions. Table 4.4 lists the ionic charges for ions within groups of representative elements.

Table 4.4 ***Ionic Charges for Representative Elements***

Group Number	Number of Electrons	Electron Change to Give an Octet	Ionic Charge	Examples
Metals				
1A	1	Lose 1	1+	Li^+, Na^+, K^+
2A	2	Lose 2	2+	Mg^{2+}, Ca^{2+}
3A	3	Lose 3	3+	Al^{3+}
Nonmetals				
5A	5	Gain 3	3−	N^{3-}, P^{3-}
6A	6	Gain 2	2−	O^{2-}, S^{2-}
7A	7	Gain 1	1−	F^-, Cl^-, Br^-, I^-

Sample Problem 4.5
Writing Ions

Consider the elements aluminum and oxygen.

a. Identify each as a metal or a nonmetal.
b. State the number of valence electrons for each.
c. State the number of electrons that must be lost or gained for each to acquire an octet.
d. Write the symbol of each resulting ion, including its ionic charge.

Solution

	Aluminum	*Oxygen*
a.	metal	nonmetal
b.	3 valence electrons	6 valence electrons
c.	loses 3 e^-	gains 2 e^-
d.	Al^{3+}	O^{2-}

Study Check
What are the symbols for the ions formed by potassium and sulfur?

Metals that Form Two Positive Ions

We have seen that the metals in Group 1A and Group 2A form only one type of positive ion. Therefore, we are able to predict their ionic charges from the periodic table. For example, sodium always forms Na^{+}, and magnesium always forms Mg^{2+}. Aluminum, in Group 3A, always forms the same ion, Al^{3+}. The other metals in the periodic table, such as the transition metals, also form positive ions. However, we are not able to predict their ionic charges because they can form more than one type of positive ion.

For example, iron, a transition metal, forms Fe^{2+} and Fe^{3+}. In some compounds, iron loses two electrons, producing the Fe^{2+} ion. In other compounds, iron loses three electrons and forms the Fe^{3+} ion. (The reasons for these different ionic charges are more complex than the octet rule and will not be considered in this text.)

When two different ions are possible for the same element, a naming system is needed that will differentiate between the ions. Using the name *iron ion,* for example, does not specify whether the Fe^{2+} or the Fe^{3+} ion is the positive ion. Therefore, for metals that form two or more positive ions, a Roman numeral placed after the name of the metal in parentheses indicates the charge for that particular ion.

Ionic charge shown as superscript:	Fe^{2+}	Fe^{3+}
Ionic charge indicated by Roman numeral:	Iron(II)	Iron(III)

An older system names Fe^{2+} as the *ferrous ion* and Fe^{3+} as the *ferric ion.* The ending *ous* is added to the root of the Latin name, in this case *ferrum,* to identify the ion with the lower ionic charge; the *ic* ending is used to indicate the ion with the higher ionic charge.

Ferrous	indicates the Fe^{2+} ion (the lower charge)
Ferric	indicates the Fe^{3+} ion (the higher charge)

Table 4.5 lists the common ions of some metals that produce more than one positive ion. Figure 4.6 shows how some of the ions are positioned in the periodic table.

Table 4.5 ***Some Metals That Form Two Positive Ions***

Element	Possible Ions	Systematic Name	Older Name
Iron	Fe^{2+}	Iron(II)	Ferrous
	Fe^{3+}	Iron(III)	Ferric
Copper	Cu^{+}	Copper(I)	Cuprous
	Cu^{2+}	Copper(II)	Cupric
Gold	Au^{+}	Gold(I)	Aurous
	Au^{3+}	Gold(III)	Auric
Tin	Sn^{2+}	Tin(II)	Stannous
	Sn^{4+}	Tin(IV)	Stannic
Lead	Pb^{2+}	Lead(II)	Plumbous
	Pb^{4+}	Lead(IV)	Plumbic

Figure 4.6
Location on the periodic table of some ions formed by metals and nonmetals.

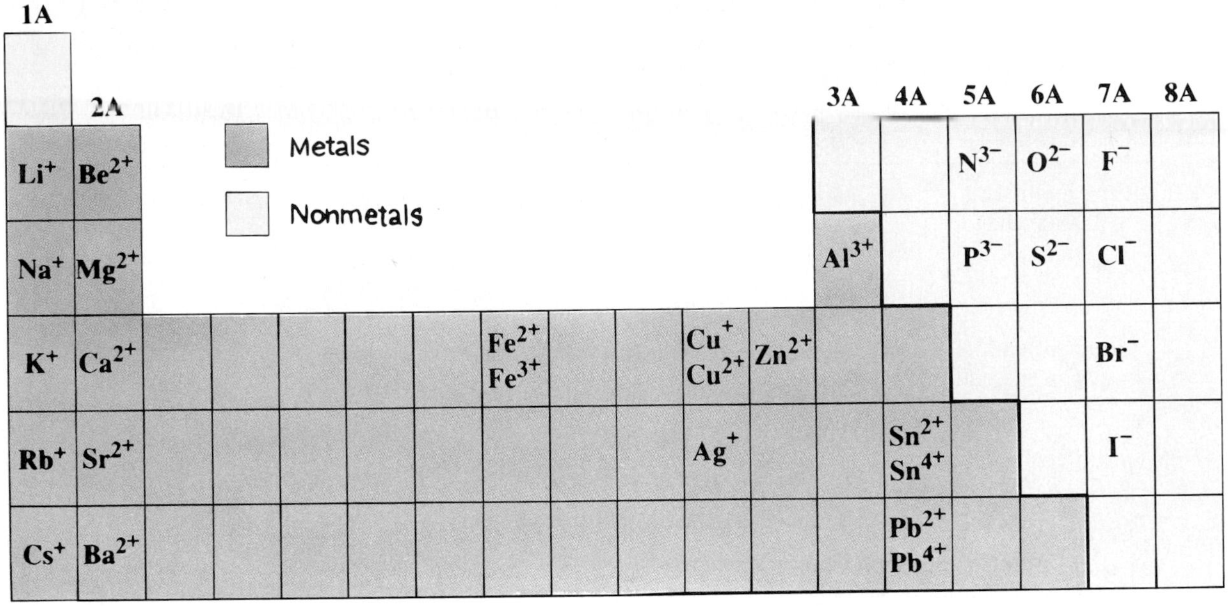

1A
2A
3A 4A 5A 6A 7A 8A

Metals
Nonmetals

Li^+ Be^{2+} N^{3-} O^{2-} F^-
Na^+ Mg^{2+} Al^{3+} P^{3-} S^{2-} Cl^-
K^+ Ca^{2+} Fe^{2+} Fe^{3+} Cu^+ Cu^{2+} Zn^{2+} Br^-
Rb^+ Sr^{2+} Ag^+ Sn^{2+} Sn^{4+} I^-
Cs^+ Ba^{2+} Pb^{2+} Pb^{4+}

Typically, the transition metals form at least two positive ions. However, zinc and silver form only one type of positive ion. Their ionic charges are fixed, so their elemental names are sufficient. The zinc ion is always Zn^{2+}, and the silver ion is Ag^+.

Zinc	Zn^{2+}
Silver	Ag^+

Sample Problem 4.6
Writing Simple Ions

Write the symbol of the ion for the following:
a. fluorine **b.** iron(II) **c.** zinc

Solution
a. F^- **b.** Fe^{2+} **c.** Zn^{2+}

Study Check
What are the symbols of the ions of copper?

HEALTH NOTE *Some Important Ions in the Body*

There are a number of ions in body fluids that have important physiological and metabolic functions. Some of them are listed in Table 4.6.

Table 4.6 *Ions in the Body*

Ion	Occurrence	Function	Source	Result of Too Little	Result of Too Much
Na^+	Principal cation outside the cell	Regulation and control of body fluids	Salt, seafood, meat	Hyponatremia, anxiety, diarrhea, circulatory failure, decrease in body fluid	Hypernatremia, little urine, thirst, edema
K^+	Principal cation inside the cell	Regulation of body fluids and cellular functions	Bananas, orange juice, skim milk, prunes, meat	Hypokalemia (hypopotassemia), lethargy, muscle weakness, failure of neurological impulses	Hyperkalemia (hyperpotassemia), irritability, nausea, little urine, cardiac arrest
Ca^{2+}	Cation outside the cell; 90% of calcium in the body in bone as $Ca_3(PO_4)_2$ or $CaCO_3$	Major cation of bone; muscle smoothant	Milk, cheese, butter, meat, some vegetables	Hypocalcemia, tingling fingertips, muscle cramps, tetany	Hypercalcemia, relaxed muscles, kidney stones, deep bone pain, nausea
Mg^{2+}	Cation outside the cell; 70% of magnesium in the body in bone structure	Essential for certain enzymes, muscles, and nerve control	Widely distributed (part of chlorophyll of all green plants), nuts, grains	Disorientation, hypertension, tremors, slow pulse	Drowsiness
Cl^-	Principal anion outside the cell	Gastric juice, regulation of body fluids	Salt, seafood, meat	Same as for Na^+	Same as for Na^+

4.4 Ionic Compounds

Learning Goal

Using charge balance, write the correct formula for an ionic compound.

Ionic compounds consist of positive and negative ions. The ions are held together by strong electrical attractions between the opposite charges, called **ionic bonds.** Consider the ionic compound sodium chloride, NaCl. Its **formula,** NaCl, indicates that it consists of Na^+ ions and Cl^- ions. Figure 4.7

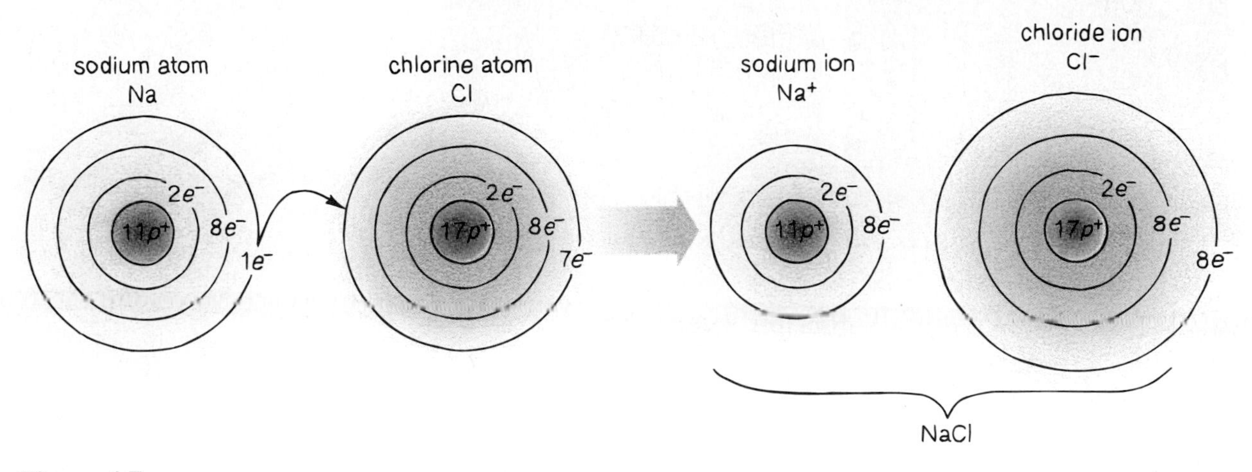

Figure 4.7
In sodium chloride, NaCl, the valence electron of sodium is transferred to complete the outer shell of a chlorine atom, forming an Na^+ cation and a Cl^- anion. The resulting ionic compound is held together by ionic bonds between the positive and negative ions.

illustrates how NaCl is formed through the transfer of an electron from the metal sodium to the nonmetal chlorine.

Sodium atom	Chlorine atom		Sodium ion	Chloride ion		Sodium chloride an ionic compound
Na·	:Cl: (with 7 dots)	⟶	Na^+	$[:\ddot{Cl}:]^-$	=	NaCl

Charge Balance in Ionic Compounds

The formula of an ionic compound indicates the number and kinds of ions that make up the ionic compound. The sum of the ionic charges in the formula is always zero. For example, the NaCl formula indicates that there is one sodium ion, Na^+, for every chloride ion, Cl^-, in the compound. Note that the ionic charges of the ions do not appear in the formula of the ionic compound.

Charge Balance for Ions in the Ionic Compound NaCl

Na^+	+	Cl^-	=	NaCl
1+	+	1−	=	0
Total positive charge		Total negative charge		Overall charge of formula

The physical and chemical properties of an ionic compound such as NaCl are very different from those of the original elements. For example, the original elements of NaCl were sodium, a soft, shiny metal, and chlorine, a yellow-green poisonous gas. Yet, as positive and negative ions, they form table salt, NaCl, a white, crystalline substance that is common in our diet. Strong ionic bonds resulting from the electrical attraction of Na^+ ions and Cl^- ions are responsible for the high melting point of NaCl, 800°C. To account for these properties, scientists diagram the solid form of NaCl as an alternating pattern of Na^+ and Cl^- ions packed together in a lattice structure, as shown in Figure 4.8.

Figure 4.8
(a) Crystals of the ionic compound sodium chloride.
(b) Crystals of NaCl under magnification.
(c) A diagram of the arrangement of Na^+ and Cl^- packed together in a NaCl crystal.

Subscripts in Formulas

Consider another ionic compound, one containing ions of magnesium and chlorine. To achieve noble gas configuration, a magnesium atom loses its two valence electrons, and two chlorine atoms each gain one electron. When more than one ion of an element is required to complete the electron transfer, that number is shown below the line as a subscript following the symbol of that element. For the ionic compound formed between magnesium and chlorine, $MgCl_2$, the 2 below the line is the subscript indicating that there are two chlorine atoms in the formula.

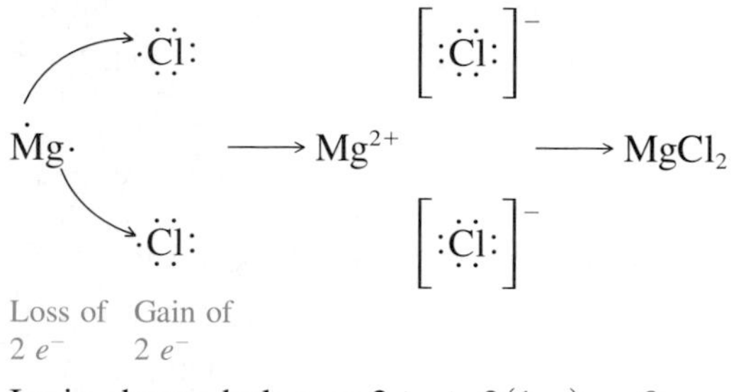

Ionic charge balance: $2+ \; + \; 2(1-) = 0$

Sample Problem 4.7
Diagramming an Ionic Compound

Diagram the formation of the ionic compound aluminum fluoride, AlF_3.

Solution
In their electron-dot structures, aluminum has three valence electrons and fluorine has seven. The aluminum loses its three valence electrons, and each fluorine atom gains an electron, to give ions with noble gas arrangements in the ionic compound AlF_3.

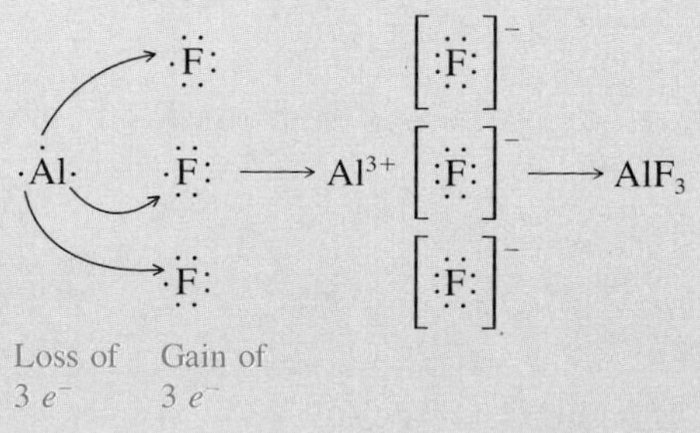

Ionic charge balance: $3+ \; + \; 3(1-) = 0$

Study Check
Diagram the formation of lithium sulfide, Li_2S.

Writing Ionic Formulas from Ionic Charges

We have seen that the formula of an ionic compound represents the number of positive ions and negative ions in that compound that give an overall charge of zero. Thus, we can now write a formula directly from the ionic charges of the positive and negative ions. Suppose we wish to write the formula of the ionic compound containing Na^+ and S^{2-} ions. To balance the ionic charge of the S^{2-} ion, we will need to place two Na^+ ions in the formula. This gives the formula Na_2S, which has an overall charge of zero.

Na^+				
Na^+	+	S^{2-}	=	Na_2S
Positive charge		Negative charge		Net charge of formula
$2 \times (1+)$	+	$2-$	=	0

Sample Problem 4.8
Writing Formulas From Ionic Charges

Use ionic charge balance to write the formula for the ionic compound containing Na^+ and N^{3-}.

Solution
Determine the number of each ion needed for charge balance. The charge for nitrogen (3−) is balanced by three Na^+ ions. Writing the positive ion first gives the formula Na_3N.

$$\begin{matrix} Na^+ & & \\ Na^+ & N^{3-} \longrightarrow & Na_3N \\ Na^+ & & \end{matrix}$$

Charge balance: $3(1+) + 3- = 0$

Study Check
Use ionic charges to determine the formula of the compound that would form when calcium and chlorine react.

4.5 Polyatomic Ions

Learning Goal
Write a formula of a compound containing a polyatomic ion.

There are ionic compounds that contain three or more elements. Such compounds contain a **polyatomic ion,** which is a group of atoms that has an overall electrical charge. Most polyatomic ions consist of a nonmetal such as phosphorus, sulfur, carbon, or nitrogen covalently bonded to one or more oxygen atoms. These oxygen-containing polyatomic ions have negative charges of 1−, 2−, or 3− because one, two, or three electrons were added to the atoms in the group to complete their octets. Only one of the common polyatomic ions, NH_4^+, is positively charged.

Naming Polyatomic Ions

The name of the most common form of the oxygen-containing polyatomic ion ends in *ate*. The *ite* ending is used for the name of a related ion that has one fewer oxygen atom. Recognizing these endings will help you identify polyatomic ions in the name of a compound. The hydroxide ion (OH^-) and cyanide ion (CN^-) are exceptions to this naming pattern. There is no easy way to learn polyatomic ions. You will need to memorize the number of oxygen atoms and the charge associated with each ion, as shown in Table 4.7. By learning the formulas and the names of the ions shown in the boxes, you can derive the related ions. For example, by learning that sulfate is written SO_4^{2-}, we can write the formula of sulfite, which has one fewer oxygen atom, as SO_3^{2-}. Or the formula of hydrogen carbonate can be written by placing a hydrogen in front of the formula for carbonate (CO_3^{2-}) and decreasing the charge from 2− to 1− to give HCO_3^-.

Writing Formulas for Compounds Containing Polyatomic Ions

No polyatomic ion exists by itself. Like any ion, a polyatomic ion must be associated with ions of opposite charge. The bonding between polyatomic ions and other ions is one of electrical attraction and is thus ionic. For example, the compound sodium sulfate consists of sodium ions (Na^+) and

Table 4.7 ***Names and Formulas of Some Common Polyatomic Ions***

Nonmetal	Formula of Ion[a]	Name of Ion
Hydrogen	OH^-	Hydroxide
Nitrogen	NH_4^+	Ammonium
	$\boxed{NO_3^-}$	Nitrate
	NO_2^-	Nitrite
Chlorine	$\boxed{ClO_3^-}$	Chlorate
	ClO_2^-	Chlorite
Carbon	$\boxed{CO_3^{2-}}$	Carbonate
	HCO_3^-	Hydrogen carbonate (or bicarbonate)
	CN^-	Cyanide
Sulfur	$\boxed{SO_4^{2-}}$	Sulfate
	HSO_4^-	Hydrogen sulfate (or bisulfate)
	SO_3^{2-}	Sulfite
	HSO_3^-	Hydrogen sulfite (or bisulfite)
Phosphorus	$\boxed{PO_4^{3-}}$	Phosphate
	HPO_4^{2-}	Hydrogen phosphate
	$H_2PO_4^-$	Dihydrogen phosphate
	PO_3^{3-}	Phosphite

[a]Boxed formulas are the most common polyatomic ion for that element.

Figure 4.9
Diagram of the formation of the ionic compound sodium sulfate, which contains a polyatomic ion.

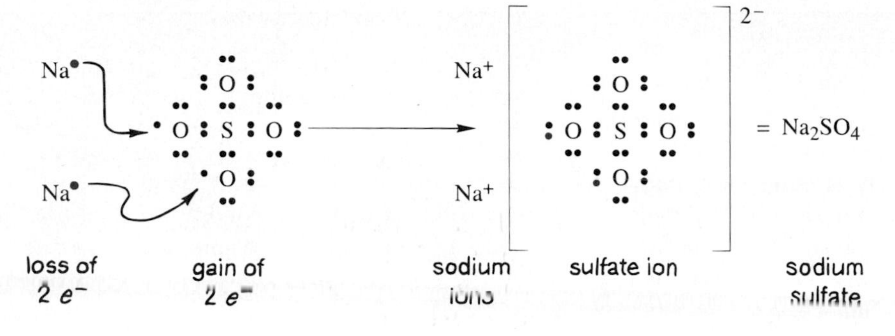

sulfate ions (SO_4^{2-}) held together by ionic bonds. Figure 4.9 diagrams the formation of Na_2SO_4.

Writing correct formulas for compounds containing polyatomic ions follows the same rules of charge balance that we used for writing the formulas of ionic compounds. The total negative and positive charges must equal zero. For example, consider the formula for a compound containing calcium ions and carbonate ions. The *ate* ending of carbonate indicates that it is a polyatomic ion. The ions are written as

Ca^{2+}	CO_3^{2-}
Calcium ion	Carbonate ion

Ionic charge: $(2+) + (2-) = 0$

Because one ion of each balances the charge, the formula can be written as

$CaCO_3$
Calcium carbonate

When more than one polyatomic ion is needed for charge balance, parentheses are used to enclose the formula of the ion. A subscript is written outside the closing parenthesis. Consider the formula for magnesium nitrate. The ions in this compound are the magnesium ion and the nitrate ion, a polyatomic ion.

Mg^{2+}	NO_3^-
Magnesium ion	Nitrate ion

To balance the positive charge of 2+, two nitrate ions are needed. The formula, including the parentheses around the nitrate ion, is as follows:

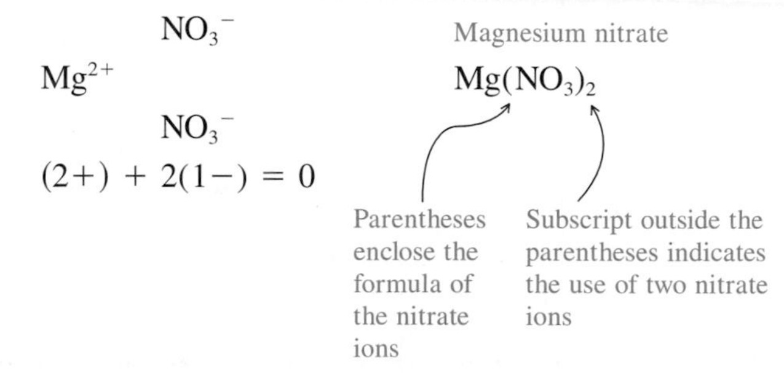

HEALTH NOTE *Polyatomic Ions in Bone and Teeth*

Bone structure consists of two parts: a solid mineral material, and a second phase made up primarily of collagen proteins. The mineral substance is a compound called hydroxyapatite, a solid formed from calcium ions, phosphate ions, and hydroxide ions. This material is deposited in the web of collagen to form a very durable bone material. (See Figure 4.10).

$Ca_{10}(PO_4)_6(OH)_2$
Hydroxyapatite

In most individuals, bone material is continuously being absorbed and re-formed. After age 40, more bone material may be lost than formed, a condition called osteoporosis. Bone mass reduction occurs at a faster rate in women than in men, and at different rates in different parts of the body skeleton. The reduction in bone mass can be as much as 50% over a period of 30 to 40 years. It is recommended that persons over 35, especially women, include a daily calcium supplement in their diet.

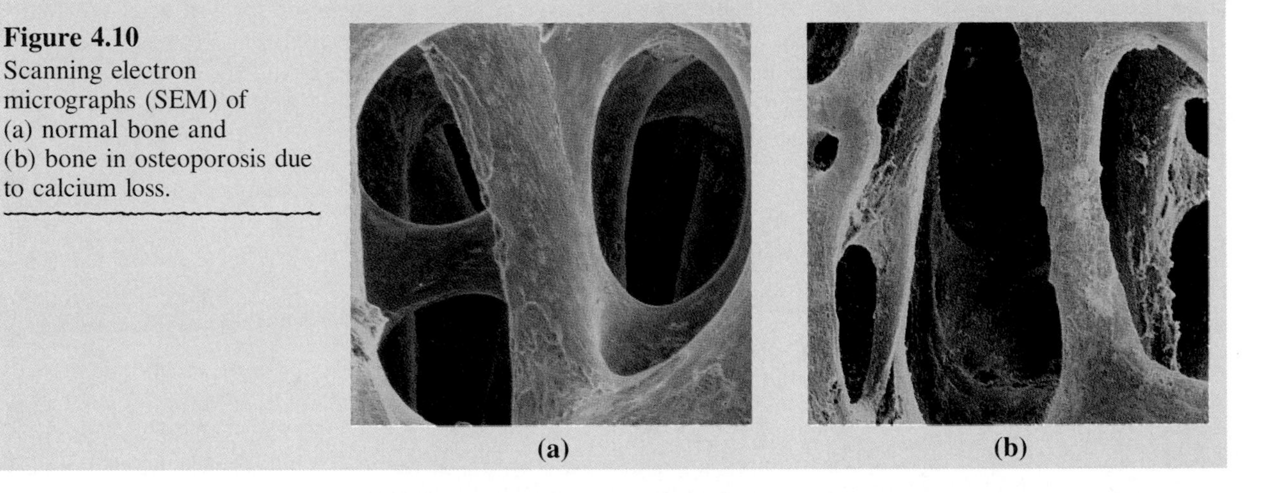

Figure 4.10 Scanning electron micrographs (SEM) of (a) normal bone and (b) bone in osteoporosis due to calcium loss.

Sample Problem 4.9
Writing Formulas Having Polyatomic Ions

Write the formula for a compound containing aluminum ions and bicarbonate ions.

Solution
The positive ion in the compound is the aluminum ion, Al^{3+}, and the negative ion is the bicarbonate ion, HCO_3^-, which is a polyatomic ion.

Al^{3+}	HCO_3^-
Aluminum ion	Bicarbonate ion

Three HCO_3^- ions are required to balance the Al^{3+} charge.

$$Al^{3+} \quad HCO_3^- \quad HCO_3^- \quad HCO_3^-$$

$$(3+) + 3 \times (1-) = 0$$

The formula for the compound is written by enclosing the formula of the bicarbonate ion, HCO_3^-, in parentheses and writing the subscript 3 just outside the last parenthesis:

$Al(HCO_3)_3$

Study Check

Write the formula for a compound containing ammonium ions and phosphate ions.

4.6 Naming Ionic Compounds

Learning Goal

Given the formula of an ionic compound, write the correct name.

In Section 4.4, we saw that an ionic compound consists of the positive ion of a metal and the negative ion of a nonmetal. The metal ion is named by its elemental name. The nonmetal ion is named by replacing the ending of its elemental name by *ide.* Table 4.8 lists the names of some important metal and nonmetal ions.

Table 4.8 ***Formulas and Names of Some Common Ions***

Group Number	Formula of Ion	Name of Ion	Group Number	Formula of Ion	Name of Ion
Metals			***Nonmetals***		
1A	Li^+	Lithium	5A	N^{3-}	Nitride
	Na^+	Sodium		P^{3-}	Phosphide
	K^+	Potassium			
2A	Mg^{2+}	Magnesium	6A	O^{2-}	Oxide
	Ca^{2+}	Calcium		S^{2-}	Sulfide
	Ba^{2+}	Barium			
3A	Al^{3+}	Aluminum	7A	F^-	Fluoride
				Cl^-	Chloride
				Br^-	Bromide
				I^-	Iodide
Transition Metals					
	Fe^{2+}	Iron(II)			
	Fe^{3+}	Iron(III)			
	Cu^+	Copper(I)			
	Cu^{2+}	Copper(II)			
	Ag^+	Silver			
	Zn^{2+}	Zinc			

Sample Problem 4.10
Naming Ions

Write the name of each of the following ions:
a. Al^{3+} **b.** S^{2-} **c.** Fe^{3+}

Solution

a. Aluminum ion; a metal ion uses the elemental name.
b. Sulfide ion: a nonmetal changes the ending of the name of the element to *ide.*
c. Iron(III) ion; a transition metal ion uses the name of the element and shows the ionic charge.

Study Check

Write the formula of each of the following ions:
a. nitride ion **b.** calcium ion

Naming Ionic Compounds Containing Two Elements

In the name of an ionic compound made up of two elements, the positive metal ion is named first, and the negative nonmetallic ion is named next. Subscripts are never mentioned; they are understood as a result of the charge balance of the ions in the compound.

Formula of ionic compound:	$MgCl_2$	
Ions:	Mg^{2+}	Cl^-
	Metal	Nonmetal
Name of ionic compound:	Magnesium chloride	

Table 4.9 gives formulas and names of some ionic compounds.

Table 4.9 ***Formulas and Names of Some Ionic Compounds***

	Ion		
Formula	**Metal**	**Nonmetal**	**Name**
MgS	Mg^{2+}	S^{2-}	Magnesium sulfide
K_2O	K^+	O^{2-}	Potassium oxide
$CaCl_2$	Ca^{2+}	Cl^-	Calcium chloride
Na_3N	Na^+	N^{3-}	Sodium nitride
Al_2O_3	Al^{3+}	O^{2-}	Aluminum oxide

Sample Problem 4.11
Naming Ionic Compounds

Write the name of each of the following ionic compounds:
a. Na_2O **b.** Al_2S_3

Solution

Compound	*Ions and Names*		*Name of Compound*
a. Na_2O	Na^+	O^{2-}	
	Sodium	Oxide	Sodium oxide
b. Al_2S_3	Al^{3+}	S^{2-}	
	Aluminum	Sulfide	Aluminum sulfide

Study Check
Name the compound Ca_3P_2.

Naming Compounds that Include Transition Metals

As we saw earlier, when a metal forms more than one type of positive ion, the ionic charge of that metal in a particular formula must be included as a Roman numeral when the metal is named. The selection of the correct Roman numeral depends upon the calculation of the ionic charge of the metal in the formula of a specific ionic compound. For example, we know that in the formula $CuCl_2$ the positive charge of the copper ion must balance the negative charge of two chloride ions. Because we know that chloride ions each have a 1− charge, there must be a total negative charge of 2−. Balancing the 2− by the positive charge gives a charge of 2+ for Cu (a Cu^{2+} ion):

$CuCl_2$

$$\begin{array}{lcll} \text{Cu charge} & + & Cl^- \text{ charge} & = 0 \\ (?) & + & 2(1-) & = 0 \\ (2+) & + & 2- & = 0 \end{array}$$

Because copper forms ions with two different charges, Cu^+ and Cu^{2+}, we need to use the Roman numeral system and place II after *copper* when naming the compound:

Copper(II) chloride

Table 4.10 ***Some Ionic Compounds of Metals that Form Two Kinds of Positive Ions***

Compound	Systematic Name	Older Name
$FeCl_2$	Iron(II) chloride	Ferrous chloride
$FeCl_3$	Iron(III) chloride	Ferric chloride
Cu_2S	Copper(I) sulfide	Cuprous sulfide
$CuCl_2$	Copper(II) chloride	Cupric chloride
$SnCl_2$	Tin(II) chloride	Stannous chloride
$PbBr_4$	Lead(IV) bromide	Plumbic bromide

Table 4.10 lists names of some ionic compounds of metals that form more than one type of positive ion. A summary of how ionic compounds are named follows.

Rules for Naming Ionic Compounds

1. For compounds in which the metal forms only one type of positive ion (Groups 1A and 2A; Al, Zn, Ag):
   a. Name the metal by its element name.
   b. Name the nonmetal by its element name, but change the ending of its name to *ide.*
2. For compounds in which the metal forms more than one type of positive ion:
   a. Assign the known ionic charge to the negative ion.
   b. Multiply the charge of the negative ion by its subscript to find the total negative charge.
   c. State the total positive charge required to balance the negative charge.
   d. Assign the positive charge to the metal ion. If there are two or more positive ions, first divide the total positive charge by the number of metal ions.
   e. Name the compound using a Roman numeral after the name of the metal ion.

Sample Problem 4.12
Naming Ionic Compounds

Write the name for each of the following ionic compounds:
a. FeO **b.** $AlBr_3$ **c.** Cu_2S

Solution

a. Because iron forms Fe^{2+} and Fe^{3+}, it is necessary to determine the ionic charge of Fe in FeO:

Fe charge	+ O^{2-} charge	= 0
?	+ (2−)	= 0
(2+)	+ (2−)	= 0

The charge on Fe is therefore 2+. So we place the Roman numeral II in parentheses after *iron:*

Iron(II) oxide

b. Because aluminum is in Group 3A and forms only Al^{3+}, the name of the metal is sufficient. A Roman numeral is not needed.

$AlBr_3$ Aluminum bromide

c. As a transition metal, copper forms more than one positive ion. Balancing the charge in Cu_2S, we determine the ionic charge of the copper ion:

2(Cu charge) + S^{2-} charge = 0
2(?) + (2−) = 0
2(1+) + (2−) = 0

The charge on Cu is therefore 1+. So we place the Roman numeral I in parentheses after copper:

Copper(I) sulfide

Although the total positive charge is 2+, it is divided between two copper ions; therefore, each copper ion in the formula is indicated as Cu^+.

Study Check
Write the name of the compound whose formula is $AuCl_3$.

Writing Formulas from the Name of an Ionic Compound

The formula of an ionic compound can be written from its name because the first term in the name describes the metal ion and the second term specifies the nonmetal ion. Subscripts are added as needed to balance the charge. The steps for writing a formula from the name of an ionic compound follow.

Steps for Writing a Formula from the Name of an Ionic Compound

Step 1 Write the positive ion from the first term in the name, and the negative ion from the second term.

Step 2 Balance the ionic charges to give an overall charge of zero.

Step 3 Write the formula, placing the symbol for the metal first and the symbol for the nonmetal second. Use appropriate subscripts for charge balance.

Sample Problem 4.13
Writing Formulas for Ionic Compounds

Write the correct formula for iron(III) chloride.

Solution

Step 1 Write the names of the metal ion and the nonmetal ion. Because the Roman numeral (III) indicates the ionic charge of the particular iron ion, we can then write Fe^{3+} for the positive ion. The chloride ion has an ionic charge of 1−.

Iron(III)	chloride
Fe^{3+}	Cl^-

Step 2 Balance the charges.

Ions needed to balance charge:	Fe^{3+}	Cl^-
		Cl^-
		Cl^-
Total charge:	$(3+) + 3(1-) = 0$	

Step 3 Write the formula of the compound, metal first, using subscripts when needed.

$FeCl_3$

Study Check
What is the correct formula of the ionic compound aluminum sulfide?

Naming Compounds Containing Polyatomic Ions

When naming ionic compounds containing polyatomic ions, we write the positive ion, usually a metal, first, and then we write the name of the polyatomic ion. It is important that you learn to recognize the polyatomic ion in the formula and name it correctly. As with other ionic compounds, no prefixes are used.

Na_2SO_4	$FePO_4$	$Al_2(CO_3)_3$
$Na_2\ \boxed{SO_4}$	$Fe\ \boxed{PO_4}$	$Al_2(\boxed{CO_3})_3$
Sodium sulfate	Iron(III) phosphate	Aluminum carbonate

Table 4.11 lists the formulas and names of some ionic compounds that include polyatomic ions, and also gives their uses in medicine and industry.

Table 4.11 ***Some Compounds that Contain Polyatomic Ions***

Formula	Name	Use
$BaSO_4$	Barium sulfate	Radiopaque medium
$CaCO_3$	Calcium carbonate	Antacid, calcium supplement
$Ca_3(PO_4)_2$	Calcium phosphate	Calcium replenisher
$CaSO_3$	Calcium sulfite	Preservative in cider and fruit juices
$CaSO_4$	Calcium sulfate	Plaster casts
$AgNO_3$	Silver nitrate	Topical anti-infective
$NaHCO_3$	Sodium bicarbonate	Antacid
$Zn_3(PO_4)_2$	Zinc phosphate	Dental cements
$FePO_4$	Iron(III) phosphate	Food and bread enrichment
K_2CO_3	Potassium carbonate	Alkalizer, diuretic
$Al_2(SO_4)_3$	Aluminum sulfate	Antiperspirant, anti-infective
$AlPO_4$	Aluminum phosphate	Antacid
$MgSO_4$	Magnesium sulfate	Cathartic, Epsom salts

Sample Problem 4.14
Naming Compounds Containing Polyatomic Ions

Name the following ionic compounds:
a. $CaSO_4$ **b.** $Cu(NO_2)_2$

Solution

	Compound Formula	*Ions Present*	*Name of Compound*
a.	$CaSO_4$	Ca^{2+}, calcium SO_4^{2-}, sulfate	calcium sulfate
b.	$Cu(NO_2)_2$	Cu^{2+}, copper(II) NO_2^-, nitrite One less oxygen than nitrate, NO_3^-	copper(II) nitrite

Study Check
What is the name of $Ca_3(PO_4)_2$?

4.7 Covalent Bonds

Learning Goal
Diagram the electron-dot structure for a covalent molecule.

When atoms of two nonmetals combine, covalent compounds result. The atoms, which are similar or even identical, are held together by sharing electrons. This type of bond is called a **covalent bond.**

The simplest covalent molecule is hydrogen, H_2. A hydrogen atom attains a noble gas arrangement when its first shell is filled with two electrons. To accomplish this, two hydrogen atoms share their single electrons and form a covalent bond:

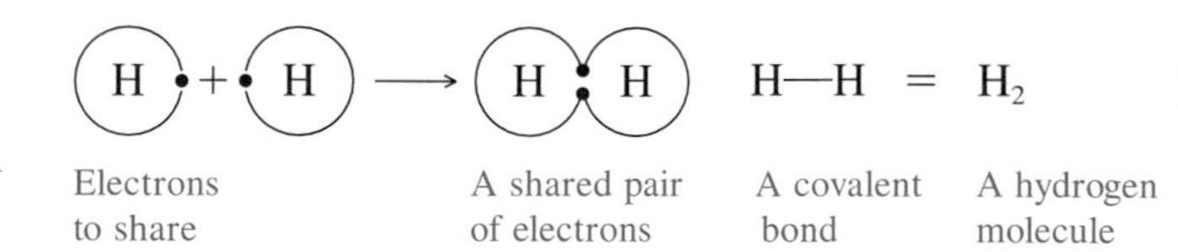

The covalent bond representing a shared pair of electrons can also be written as a line between the two atoms. Atoms that are bonded through covalent bonds produce a **molecule.**

In most covalent molecules, each of the atoms acquires an octet of eight electrons. For example, a fluorine molecule, F_2, consists of two fluorine atoms. Each atom needs one more valence electron for a noble gas arrangement. An octet for each is achieved when the fluorine atoms share a pair of electrons in their outer shells.

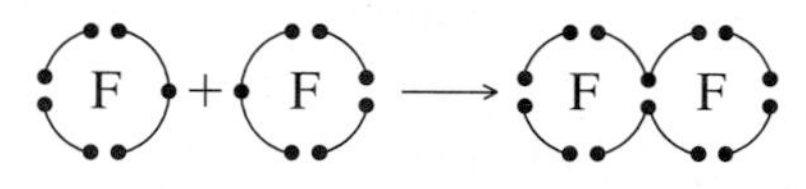

or F_2

Electrons to share

A shared pair of electrons

A fluorine molecule

Table 4.12 ***Elements that Exist as Diatomic, Covalent Molecules***

Element	Diatomic Molecule	Name	Element	Diatomic Molecule	Name
H	H_2	Hydrogen	F	F_2	Fluorine
N	N_2	Nitrogen	Cl	Cl_2	Chlorine
O	O_2	Oxygen	Br	Br_2	Bromine
			I	I_2	Iodine

Hydrogen (H_2) and fluorine (F_2) are examples of several nonmetal elements whose natural state is diatomic; that is, they consist of molecules of two atoms. The elements that exist as diatomic molecules are listed in Table 4.12.

The number of electrons that an atom shares and the number of covalent bonds it forms are equal to the number of electrons needed to acquire a noble gas arrangement, usually an octet. For example, carbon has four valence electrons. Because carbon needs to acquire four more electrons for an octet, it forms four covalent bonds by sharing its four valence electrons. Table 4.13 relates the group numbers of some elements to the number of covalent bonds the elements typically form.

Table 4.13 ***Covalent Bonds Required by Various Nonmetals***

Group Number	Nonmetal	Number of Valence Electrons	Number of Electrons Shared to Complete an Octet	Number of Covalent Bonds
1A	H	1	1	1
4A	C, Si	4	4	4
5A	N, P	5	3	3
6A	O, S	6	2	2
7A	F, Cl, Br, I	7	1	1

Figure 4.11
Diagram of a carbon atom sharing electrons with four atoms of hydrogen to form four covalent bonds. The carbon atom is stable with eight electrons in the outer shell, and each hydrogen is stable with a filled outer shell of two electrons.

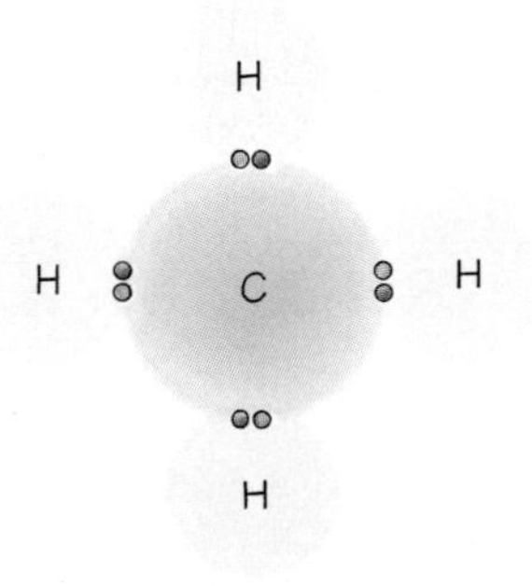

Consider a compound of carbon and hydrogen. To attain an octet, carbon shares four electrons, and hydrogen shares one electron. In a molecule, a carbon atom can form four covalent bonds with four hydrogen atoms. (See Figure 4.11.) The electron-dot structure for the molecule is written with the carbon atom in the center and the hydrogen atoms on the sides. Each valence electron in carbon is paired with a hydrogen electron.

Table 4.14 gives the electron-dot structures for several covalent molecules.

Table 4.14 *Electron-Dot Structures for Some Covalent Compounds*

H_2	CH_4	NH_3	SCl_2	Cl_2
Structures Using Electron Dots Only				
H:H	H H:C̤̈:H H	H:N̤̈:H H	:S̤̈:C̤̈l: :C̤l:	:C̤̈l:C̤̈l:
Structures Using Bonds and Electron Dots				
H—H	H | H—C—H | H	H—N̈—H | H	:S̈—C̤̈l: | :C̤l:	:C̤̈l—C̤̈l:

Sample Problem 4.15
Writing Electron-Dot Structures for Covalent Compounds

Draw an electron-dot structure for water, H_2O.

Solution
Oxygen, which has six valence electrons, shares two electrons to form two covalent bonds. Two atoms of hydrogen, each having one valence electron, will form two covalent bonds, one for each atom:

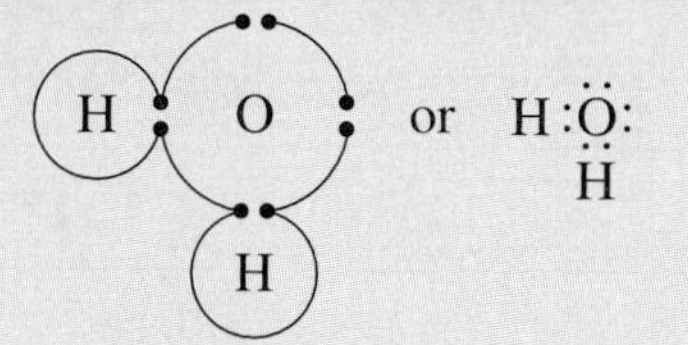

Electron-dot structure for H_2O

Study Check
Write the electron-dot structure for NH_3.

Multiple Covalent Bonds

Up to now, we have looked at covalent bonding in molecules having only single bonds. In many covalent compounds, atoms share two or three pairs of electrons to complete their electron octets. A **double bond** is the sharing of two pairs of electrons as shown in Sample Problem 4.16, and in a **triple bond,** three pairs of electrons are shared.

Double and triple bonds are formed when single covalent bonds fail to complete the octets of all the atoms in the molecule. For example, in the electron-dot structure for the covalent compound N_2, an octet is achieved when each nitrogen atom shares three electrons. Thus, three covalent bonds, or a triple bond, will form.

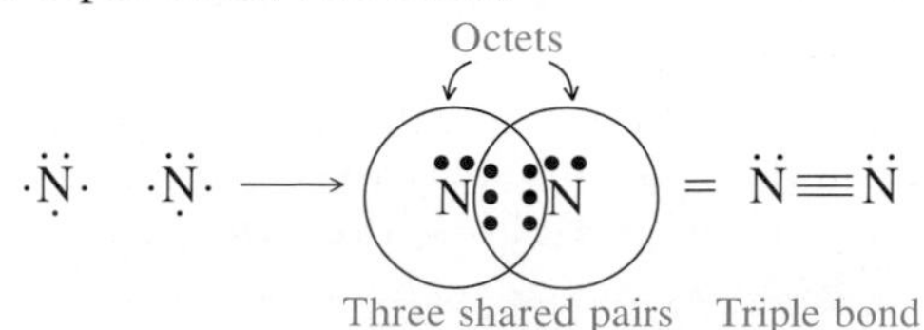

We can use the following steps to write electron-dot structures:

Steps for Writing Electron-Dot Structures

Step 1 Determine the total number of valence electrons in all of the atoms.

Step 2 Place single bonds between each set of atoms. Each single bond uses two of the available valence electrons.

Step 3 Subtract the electrons used to bond the atoms, and arrange the remaining valence electrons to give each atom an octet. Hydrogen needs only a single bond, or two electrons.

Step 4 If octets of all the atoms cannot be completed using the remaining electrons, rearrange some of the electrons so that another pair or two are shared between two of the atoms as a double or triple bond. Check that all of the atoms have octets.

Sample Problem 4.16
Drawing Electron-Dot Structures Having Multiple Bonds

Draw the electron-dot structure of the compound CO_2. (Carbon is the central atom.)

O C O

Solution
Oxygen has six valence electrons, and carbon has four. Each oxygen atom needs two electrons, and the carbon atom needs four electrons.

Step 1 Using group numbers, calculate the total number of valence electrons available:

$$\begin{matrix} O & & C & & O & \\ 6\,e^- & + & 4\,e^- & + & 6\,e^- & = 16 \text{ valence electrons} \end{matrix}$$

Step 2 Connect the atoms by single bonds:

O:C:O

Uses 2 e^- Uses 2 e^-

Step 3 Arrange the remaining valence electrons to satisfy octets:

:Ö:C̤̈:Ö:

Not octets

Step 4 Because octets cannot be completed using the remaining 12 electrons, double bonds are needed instead of single bonds between atoms. Rearrange the electrons, placing them between atoms to form double bonds to give octets to all the atoms.

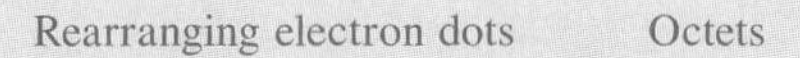

Study Check
Determine the number of valence electrons and arrange them in the electron-dot structure of the HCN molecule.

4.8 Naming Covalent Compounds

Learning Goal

Given the formula of a covalent compound, write its correct name; given the name of a covalent compound, write its formula.

In the name of a covalent compound, the first nonmetal in the formula is named by its elemental name; the second nonmetal is named by its elemental name with the ending changed to *ide*. All subscripts indicating two or more atoms of an element are expressed as prefixes and attached in front of each name. Table 4.15 lists some prefixes used in naming covalent compounds.

Unlike the names of ionic compounds, the names of covalent compounds need prefixes because several different compounds can be formed from the same two nonmetals. For example, carbon and oxygen can form two different compounds, carbon monoxide, CO, and carbon dioxide, CO_2. Nitrogen and oxygen also form several different covalent molecules. We could not distinguish between the following compounds by using the name *nitrogen oxide*. Therefore, prefixes must be used.

Table 4.15 ***Prefixes Used in Naming Covalent Compounds***

Number of Atoms	Prefix
1	Mono
2	Di
3	Tri
4	Tetra
5	Penta
6	Hexa
7	Hepta
8	Octa

Some Covalent Compounds Formed by Nitrogen and Oxygen

NO	Nitrogen monoxide
N_2O	Dinitrogen monoxide
N_2O_3	Dinitrogen trioxide
N_2O_4	Dinitrogen tetroxide
N_2O_5	Dinitrogen pentoxide

In the name of a covalent compound, the prefix *mono* is understood for the first element and is usually omitted. When the vowels *o* and *o* or *a* and *o* appear together, the first vowel is omitted. Table 4.16 lists the formulas, names, and commercial uses of some other covalent compounds.

Table 4.16 ***Some Common Covalent Compounds***

Formula	Name	Commercial Uses
CS_2	Carbon disulfide	Manufacture of rayon
CO_2	Carbon dioxide	Carbonation of beverages; fire extinguishers, propellant in aerosols, dry ice
SiO_2	Silicon dioxide	Manufacture of glass
NCl_3	Nitrogen trichloride	Bleaching of flour in some countries (prohibited in U.S.)
SO_2	Sulfur dioxide	Preserving fruits, vegetables; disinfectant in breweries, bleaching textiles
SO_3	Sulfur trioxide	Manufacture of explosives
SF_6	Sulfur hexafluoride	Electrical circuits
ClO_2	Chlorine dioxide	Bleaching pulp (for making paper), flour, leather
ClF_3	Chlorine trifluoride	Rocket propellant

Sample Problem 4.17
Naming Covalent Compounds

Name each of the following covalent compounds:
a. NCl_3 **b.** P_4O_6

Solution

a. In NCl_3, the first nonmetal is nitrogen. We could use the term *mononitrogen,* but the prefix *mono* is usually dropped from the name of the first element. The second nonmetal is chlorine, which is named *chloride.* The subscript indicating three chlorine atoms is shown as the prefix *tri* in front of the name of the second element, trichloride.

NCl_3
Nitrogen trichloride

b. In P_4O_6, the first nonmetal is phosphorus. Because there are four P atoms, it is named *tetraphosphorus.* The second nonmetal, consisting of six oxygen atoms, is named *hexoxide.* (When the vowels *a* and *o* appear together, as would happen in *hexa* plus *oxide,* the ending of the prefix is dropped.) Putting the names for the two nonmetals together, we have

P_4O_6
Tetraphosphorus hexoxide

Study Check
Using high temperatures and pressures, scientists have recently combined some of the noble gases with other elements. One of the compounds produced was XeF_6. How would you name this compound?

Sample Problem 4.18
Writing Formulas From Names of Covalent Compounds

Write the formulas of the following covalent compounds:
a. sulfur dichloride **b.** dinitrogen pentoxide

Solution

a. The first nonmetal is sulfur. Since there is no prefix given, we can assume that there is one atom of sulfur. The second nonmetal in the formula is chlorine. The prefix *di* indicates that there are two atoms of chlorine, which means that the subscript 2 appears with chlorine in the formula.

Sulfur dichloride
SCl_2

b. The first nonmetal in the name is nitrogen. The prefix *di* means that there are two atoms of nitrogen and therefore the subscript 2 is needed. The second nonmetal is oxygen. The prefix *pent(a)* means "five"—there are five atoms of oxygen so the subscript 5 is needed.

Dinitrogen pentoxide
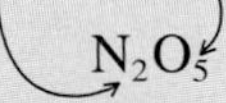

Study Check
What is the formula of iodine pentafluoride?

Summary of Naming Compounds

Throughout this chapter we have examined strategies for naming ionic and covalent compounds. Now we can summarize the rules. In general, compounds having two elements are named by stating the first element, followed by the second element with an *ide* ending. If the first element is a metal, the compound is ionic: if the first element is a nonmetal, the compound is covalent. For ionic compounds, it is necessary to determine whether the metal can form more than one type of positive ion; if so, a Roman numeral following the name of the metal indicates the particular ionic charge. In naming covalent compounds having two elements, prefixes are necessary to indicate the number of atoms of each nonmetal as shown in that particular formula. Ionic compounds having three or more elements include some type of polyatomic ion. They are named by ionic rules, but have an *ate* or *ite* ending when the polyatomic ion has a negative charge. Table 4.17 summarizes some naming rules for elements, ionic compounds, and covalent compounds.

Table 4.17 ***Rules for Naming Elements, Ionic Compounds, and Covalent Compounds***

Type	Formula Feature	Naming Procedure
Element	Symbol of element; may be diatomic. *Examples*: Ca Cl_2	Use name of element. *Examples:* Calcium Chlorine
Ionic compound (two elements)	Symbol of metal followed by symbol of nonmetal; subscripts used for charge balance. *Examples:* Na_2O Fe_2S_3	Use element name for metal; Roman numeral required if two or more positive ions possible. For nonmetal use element name with *ide* ending. *Examples:* Sodium oxide Iron(III) sulfide
Ionic compound (three elements)	Usually symbol of metal followed by a polyatomic ion composed of nonmetals; parentheses may enclose polyatomic ion for charge balance. *Examples:* $Mg(NO_3)_2$ $CuSO_4$ $(NH_4)_2CO_3$	Use element name for metal, with Roman numeral if needed, followed by name of polyatomic ion. *Examples:* Magnesium nitrate Copper(II) sulfate Ammonium carbonate
Covalent compound (two elements)	Symbols of two nonmetals; subscripts show number of atoms in a molecule. *Examples:* N_2O_3 CCl_4	Place prefixes before each element name if there are two or more atoms in the formula; add *ide* ending to second element. *Examples:* Dinitrogen trioxide Carbon tetrachloride

4.9 Bond Polarity

Learning Goal

Using electronegativity values, classify a bond as covalent, polar covalent, or ionic.

In covalent bonds, as we have seen, electrons may be shared between atoms that are identical, as in Cl_2, or between atoms that are different, as in HCl. When electrons are shared between identical nonmetal atoms, they are shared equally. However, when they are shared between atoms of different nonmetals, they are shared unequally because one of the atoms in the covalent bond has a stronger attraction for the pair of electrons than the other. The ability of an atom to attract the shared electrons is called its electronegativity, and it is indicated by a number called an **electronegativity value.** The element with the greater electronegativity value pulls the shared electron closer to its nucleus.

In the table of electronegativities shown in Figure 4.12, fluorine, the most electronegative element, is assigned a value of 4.0. By contrast, the metals in Groups 1A and 2A have relatively low electronegativities. As we read from left to right across any period of the periodic table, we see that the electronegativity values increase, which means that the atoms of those elements to the right have a greater attraction for electrons.

The **polarity** of a bond is a result of the difference in the electronegativity values of the two atoms in a bond. When the atoms are identical in a covalent bond, they have the same attraction for the shared pairs of electrons and the same electronegativity values. Therefore, the electronegativity difference is zero.

When electrons are shared between nonmetal atoms having different electronegativity values, they are pulled closer to the atom with the higher

Figure 4.12
Electronegativity values for representative elements in the periodic table.

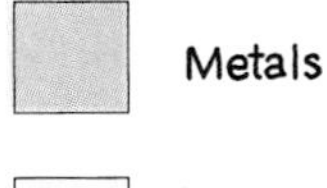

1A	2A	3A	4A	5A	6A	7A
H 2.1						
Li 1.0	Be 1.5	B 2.0	C 2.5	N 3.0	O 3.5	F 4.0
Na 0.9	Mg 1.2	Al 1.5	Si 1.8	P 2.1	S 2.5	Cl 3.0
K 0.8	Ca 1.0	Ga 1.6	Ge 1.8	As 2.0	Se 2.4	Br 2.8
Rb 0.8	Sr 1.0	In 1.7	Sn 1.8	Sb 1.9	Te 2.1	I 2.5
Cs 0.7	Ba 0.9					

Figure 4.13
Equal and unequal sharing of electrons in the covalent bond of H_2 and the polar bond of HCl.

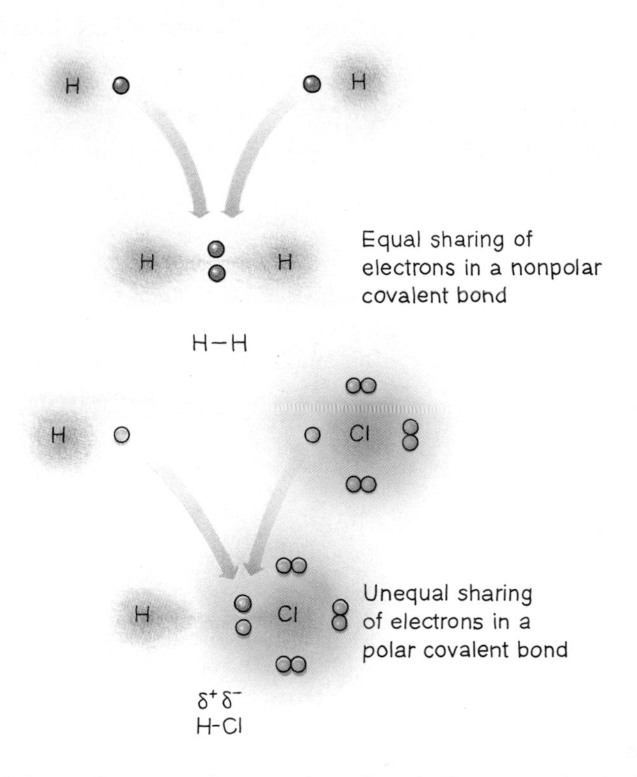

electronegativity to form a **polar covalent bond.** For example, in HCl the shared pair of electrons is pulled closer to the more electronegative chlorine atom, as shown in Figure 4.13.

In a polar bond, the more electronegative atom may be shown with a partial negative charge, indicated by the lowercase Greek letter delta with a negative sign, δ^-. The atom with the lower electronegativity is farther from the electron pair and is shown with a partial positive charge, indicated by the Greek letter delta with a positive sign, δ^+.

Covalent bonds with electronegativity differences up to 1.6 are considered to be polar. When the difference in electronegativity is greater than 1.6, the differences in electron attraction are so great that electrons are transferred from one atom to the other. Ions form and the bonding is ionic. Table 4.18 gives some examples of how bond type can be predicted from electronegativity differences.

Table 4.18 ***Predicting Bond Type From Electronegativity Differences***

Molecule		Type of Electron Sharing	Electronegativity Difference[a]	Bond Type
H_2	H—H	Shared equally	2.1 − 2.1 = 0	Covalent
Cl_2	Cl—Cl	Shared equally	3.0 − 3.0 = 0	Covalent
HI	$\overset{\delta+}{H}$—$\overset{\delta-}{I}$	Shared unequally	2.5 − 2.1 = 0.4	Polar Covalent
HCl	$\overset{\delta+}{H}$—$\overset{\delta-}{Cl}$	Shared unequally	3.0 − 2.1 = 0.9	Polar Covalent
NaCl	$Na^+\ Cl^-$	Electron transfer	3.0 − 0.9 = 2.1	Ionic
MgO	$Mg^{2+}\ O^{2-}$	Electron transfer	3.5 − 1.2 = 2.3	Ionic

[a]Values are taken from Figure 4.12.

Sample Problem 4.19
Identifying Bond Type

Indicate whether bonds between the following atoms would be covalent, polar covalent, or ionic. If polar covalent, indicate the partial charges with δ^+ and δ^- notation.

a. carbon and chlorine **b.** fluorine and fluorine

Solution

a. The bond between the two nonmentals C and Cl has an electronegativity difference of 0.5. The bond is polar covalent. Because the chlorine atom has the greater electronegativity (3.0), the Cl is designated as partially negative, δ^-, in the polar bond.

$$\overset{\delta^+}{\mathrm{C}}-\overset{\delta^-}{\mathrm{Cl}}$$

b. The bond F—F occurs between identical atoms. It is a covalent bond, as indicated by an electronegativity difference of zero (0).

Study Check

Determine if the bonds between the following would be covalent, polar covalent, or ionic. If polar covalent, show the partial charges with δ^+ and δ^-.

a. hydrogen and hydrogen **b.** hydrogen and oxygen

A Review of Bonding

As we have seen, the types of bonds that hold atoms together range from the covalent bonds, in which electrons are shared equally, all the way to the ionic bond, in which electrons are completely transferred from one atom to another. This variation in bonding is continuous; there is no definite point where

Figure 4.14
A summary of electronegativity difference and bond type.

Typical Elements	Two Nonmetals (identical)	Two Nonmetals (different)	Metal and Nonmetal
Electron Bonding	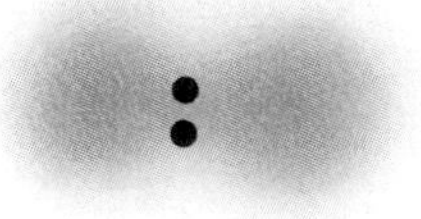 : shared equally	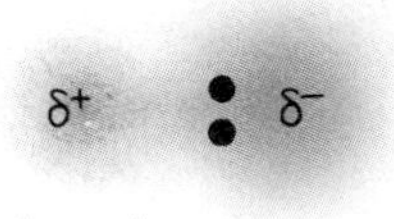 δ^+ : δ^- shared unequally	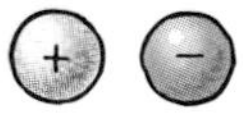 (+) (−)
Electronegativity Difference	0	1.6 or less	over 1.6
Bond Type	nonpolar covalent	polar covalent	ionic
Examples	H_2 Cl_2 Br_2	H_2O CCl_4 HBr	Li_2O NaCl $BaCl_2$

Table 4.19 ***Examples of Types of Bonding***

Elements to Bond	Electron-Dot Structure	Type of Elements	Type of Bonding	Molecule or Ionic Unit	Formula
F and F	:F· :F·	Two nonmetals (the same element)	Covalent	F—F	F_2
P and Cl	·P· :Cl·	Two nonmetals (different elements)	Polar covalent	Cl—P—Cl with Cl bonded to P	PCl_3
Na and F	Na· ·F:	Metal, nonmetal	Ionic	Na^+ F^-	NaF
Ca and O	Ca· :O·	Metal, nonmetal	Ionic	Ca^{2+} O^{2-}	CaO

one type of bond stops and the next starts. However, for purposes of discussion, we can state some general rules (summarized in Figure 4.14) that may be used to predict the type of bond formed between two elements. Examples of the different bonding types are given in Table 4.19.

Rules for Predicting Bond Type

1. A bond between two atoms of the same element, or between atoms having the same electronegativity value is a covalent bond; we refer to it just as covalent, for simplicity.
2. A covalent bond between atoms of two different nonmetals having different electronegativities is polar covalent.
3. A bond between a metal and a nonmetal is usually ionic.

Sample Problem 4.20
Predicting Bond Polarity

Predict the type of bond between each of the following pairs of elements:
a. Ca and Cl **b.** P and S **c.** Br and Br

Solution

a. A bond between calcium, a metal, and chlorine, a nonmetal, would be ionic.
b. A bond between phosphorus and sulfur, two different nonmetals, is polar covalent.
c. A bond between atoms of the same element would be covalent.

Study Check

Predict the type of bond for the following:
a. S—O **b.** Na—O **c.** O—O

Chapter Summary

4.1 Valence Electrons
The electrons located in the outermost energy level of an atom play a major role in determining its chemical properties. An electron-dot structure represents the valence electrons in an atom.

4.2 The Octet Rule
The nonreactivity of the noble gases is associated with the stable arrangement of eight electrons, an octet, in their valence shells; helium needs two electrons for stability. Atoms of elements other than the noble gases achieve stability by losing, gaining, or sharing their valence electrons with other atoms in the formation of compounds.

4.3 Ions
Metals of the representative elements form an octet by losing their valence electrons to form positively charged cations; Group 1A, 1+, Group 2A, 2+, and Group 3A, 3+. When they react with metals, nonmetals gain electrons to form octets in their valence shells. As anions, they have negative charges; Group 5A, 3−, Group 6A, 2−, Group 7A, 1−. Typically transition metals form cations with two or more ionic charges. The charge is given as a Roman numeral in the name such as iron (II) and iron (III) for the cations of iron with 2+ and 3+ ionic charges.

4.4 Ionic Compounds
The total positive and negative ionic charge must balance in the formula of an ionic compound. Charge balance in a formula is achieved by using subscripts after each symbol so that overall charge is zero.

4.5 Polyatomic Ions
A group of nonmetal atoms that carries an electrical charge, for example, the carbonate ion has the formula CO_3^{2-}. Most polyatomic ions have names that end with *ate* or *ite*.

4.6 Naming Ionic Compounds
In naming ionic compounds, the positive ion is given first, followed by the name of the negative ion. Ionic compounds containing two elements end with *ide*, and those with three elements (polyatomic ions) end with *ate* or *ite*. When the metal forms more than one positive ion, its ionic charge is determined from the total negative charge in the formula.

4.7 Covalent Bonds
In a covalent bond, electrons are shared by atoms of nonmetals. By sharing, each of the atoms achieves a noble gas arrangement. In a covalent bond, electrons are shared equally by atoms, usually of the same element. In a polar covalent bond, electrons are unequally shared because they are attracted by the more electronegative atom. In some covalent compounds, double or triple bonds are needed to provide an octet.

4.8 Naming Covalent Compounds
Two nonmetals can form two or more different covalent compounds. In their names, prefixes are used to indicate the subscript in the formula. The ending of the second nonmetal is changed to *ide*. The formula of a covalent compound is written using the symbol of the nonmetals in the name followed by subscripts given by the prefixes.

4.9 Bond Polarity
Electronegativity values indicate the ability of atoms to attract electrons. In general, the electronegativity of a metal is low; a nonmetal is high. In a covalent bond, the difference in electronegativities is zero; in a polar covalent bond, the difference can be up to 1.6. Greater than 1.6, the bond is considered to be ionic.

Glossary of Key Terms

anion A negatively charged ion such as Cl^-, O^{2-}, or SO_4^{2-}.

cation A positively charged ion such as Na^+, Mg^{2+}, or Al^{3+}.

compound A combination of atoms in which noble gas arrangements are attained through electron transfer or electron sharing.

covalent bond A sharing of valence electrons by atoms.

double bond A sharing of two pairs of electrons by two atoms.

electron-dot structure The representation of an atom that shows each valence electron as a dot above, below, or on the sides of the symbol of the element.

electronegativity value A number that indicates the relative ability of an element to attract electrons.

formula The group of symbols that represent the atoms or ions of a unit of a compound.

ion An atom or group of atoms having an electrical charge because of a loss or gain of electrons.

ionic bond The attraction between oppositely charged ions that results from the transfer of electrons.

ionic charge The difference between the number of protons (positive) and the number of electrons (negative), written in the upper right corner of the symbol for the element.

ionic compound A compound of positive and negative ions held together by ionic bonds.

molecule The smallest unit of two or more atoms held by covalent bonds.

octet rule Elements in Groups 1A–7A react with other elements by forming ionic or covalent bonds that produce a noble gas arrangement, usually eight electrons in the outer shell.

polar covalent bond A covalent bond in which the electrons are shared unequally, resulting in a partially negative end (δ^-) and a partially positive end (δ^+).

polarity A measure of the unequal sharing of electrons, indicated by the difference in electronegativity values.

polyatomic ion A group of covalently bonded nonmetal atoms that has an overall electrical charge.

triple bond A sharing of three pairs of electrons by two atoms.

valence electrons The electrons present in the outermost shell of an atom, which are largely responsible for the chemical behavior of the element.

Chemistry at Home

1. Read the labels on some products such as toothpaste, shampoo, and deodorant. What are some of the compounds listed? What are some chemicals listed on the labels of food packages in your kitchen? Use a Merck Index or other reference book to obtain their formulas.
2. A simple gargle is made of salt and warm water. What are the chemical formulas of the gargle mixture? Prepare a gargle by placing 2–3 teaspoons of salt in a glass of warm water. How do you know there is salt in the solution? Pour some of the gargle into a shallow bowl. Let the bowl sit on the kitchen counter for a few days. What happens? What part of the gargle is left?

Answers to Study Checks

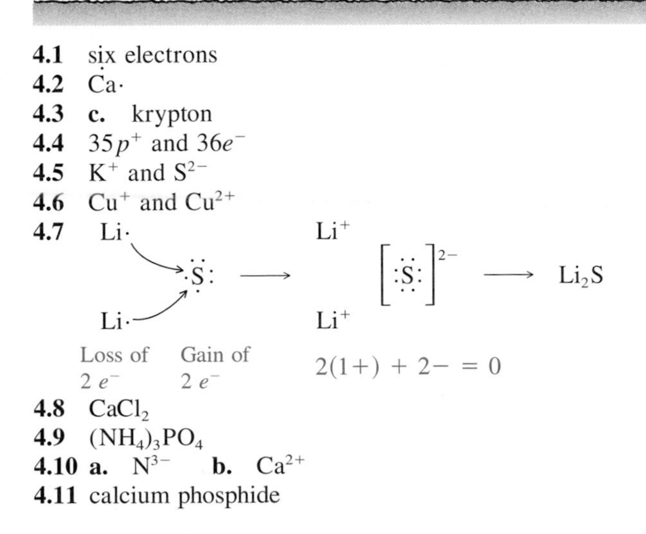

4.1 six electrons
4.2 Ca·
4.3 **c.** krypton
4.4 $35p^+$ and $36e^-$
4.5 K^+ and S^{2-}
4.6 Cu^+ and Cu^{2+}
4.7

4.8 $CaCl_2$
4.9 $(NH_4)_3PO_4$
4.10 **a.** N^{3-} **b.** Ca^{2+}
4.11 calcium phosphide
4.12 gold(III) chloride
4.13 Al_2S_3
4.14 calcium phosphate
4.15 H : N : H with H below N (overlapping circles around H, N, H, H) or H:N̈:H with H below
4.16 Number of valence electrons: $1 + 4 + 5 = 10e^-$
A triple bond gives octets to C and N. H:C:::N: or H—C≡N
4.17 xenon hexafluoride
4.18 IF_5
4.19 **a.** covalent **b.** polar covalent $\overset{\delta^+}{H}—\overset{\delta^-}{O}$
4.20 **a.** polar covalent **b.** ionic **c.** nonpolar

Problems

Valence Electrons *(Goal 4.1)*

4.1 Write the electron arrangement and the number of valence electrons in an atom of each the following elements:
a. N **b.** O **c.** Ar
d. K **e.** S

4.2 Write the electron arrangement and the number of valence electrons in an atom of each the following elements:
a. Na **b.** Al **c.** Cl
d. Mg **e.** P

4.3 Write the group number and electron-dot structure for each element:
a. sulfur **b.** nitrogen **c.** calcium
d. sodium **e.** potassium

4.4 Write the group number and electron-dot structure for each element:
a. carbon **b.** oxygen **c.** fluorine
d. lithium **e.** chlorine

The Octet Rule *(Goal 4.2)*

4.5 Write the electron-dot arrangement for each of the following elements. Indicate which elements would form compounds, and which would be stable as written.
a. neon **b.** oxygen
c. lithium **d.** argon

4.6 Write the electron-dot arrangement for each of the following elements. Indicate which elements would form compounds, and which would be stable as written.
a. magnesium **b.** nitrogen
c. helium **d.** potassium

Ions *(Goal 4.3)*

4.7 Write the symbols of the ions with the following number of protons and electrons:
a. 3 protons, 2 electrons
b. 9 protons, 10 electrons
c. 12 protons, 10 electrons
d. 26 protons, 23 electrons
e. 30 protons, 28 electrons

4.8 How many protons and electrons are in the following ions?
a. O^{2-} **b.** K^+ **c.** Br^- **d.** S^{2-} **e.** Sr^{2+}

4.9 State the number of electrons lost or gained when the following elements form ions:
a. Mg **b.** P **c.** Group 7A
d. Na **e.** Al

4.10 State the number of electrons lost or gained when the following elements form ions:
a. O **b.** Group 2A **c.** F
d. Li **e.** N

4.11 Write the symbol for the ion of each of the following:
a. chloride **b.** potassium
c. oxide **d.** aluminum

4.12 Write the symbol for the ion of each of the following:
a. fluoride **b.** calcium
c. sodium **d.** lithium

4.13 What is the name of each of the following ions:
a. K^+ **b.** S^{2-} **c.** Ca^{2+} **d.** N^{3-}

4.14 What is the name of each of the following ions:
a. Mg^{2+} **b.** Ba^{2+} **c.** I^- **d.** Cl^-

4.15 Write the names of the ions of the following transition metals (include the Roman numeral when necessary):
a. Fe^{2+} **b.** Cu^{2+} **c.** Zn^{2+} **d.** Pb^{4+}

4.16 Write the names of the following ions (include the Roman numeral when necessary):
a. Ag^+ **b.** Cu^+ **c.** Fe^{3+} **d.** Sn^{2+}

Ionic Compounds *(Goal 4.4)*

4.17 Using electron-dot structures, diagram the formation of the following ionic compounds:
a. KCl **b.** $MgCl_2$ **c.** Na_3N

4.18 Using electron-dot structures, diagram the formation of the following ionic compounds:
a. MgS **b.** $AlCl_3$ **c.** Li_2O

4.19 Write the correct ionic formula for compounds formed between the following ions:
a. Na^+ and O^{2-}
b. Al^{3+} and Br^-
c. Ba^{2+} and Cl^-
d. Zn^{2+} and Cl^-
e. Al^{3+} and S^{2-}

4.20 Write the correct ionic formula for compounds formed between the following ions:
a. Fe^{3+} and Cl^-
b. Cu^{2+} and S^{2-}
c. Li^+ and S^{2-}
d. Cu^+ and O^{2-}
e. K^+ and I^-

4.21 Write the correct formula for ionic compounds formed by the following metals and nonmetals:
a. sodium and sulfur
b. potassium and nitrogen
c. aluminum and iodine
d. lithium and oxygen

4.22 Write the correct formula for ionic compounds formed by the following metals and nonmetals:
a. calcium and chlorine
b. barium and bromine
c. sodium and phosphorus
d. magnesium and oxygen

Polyatomic Ions *(Goal 4.5)*

4.23 Write the formulas including the charge for the following polyatomic ions:
a. bicarbonate **b.** ammonium
c. phosphate **d.** hydrogen sulfate

4.24 Write the formulas including the charge for the following polyatomic ions:
a. nitrite **b.** sulfite
c. hydroxide **d.** phosphite

4.25 Name the following polyatomic ions:
a. SO_4^{2-} **b.** CO_3^{2-}
c. PO_4^{3-} **d.** NO_3^-

4.26 Name the following polyatomic ions:
a. OH^- **b.** SO_3^{2-}
c. HCO_3^- **d.** NO_2^-

4.27 Complete the following table with the formula of the compound:

	OH^-	NO_2^-	CO_3^{2-}	HSO_4^-	PO_4^{3-}
Li^+					
Cu^{2+}					
Ba^{2+}		$Ba(NO_2)_2$			

4.28 Complete the following table with the formula of the compound:

	OH^-	NO_3^-	HCO_3^-	SO_4^{2-}	PO_4^{3-}
NH_4^+					
Al^{3+}				$Al_2(SO_4)_3$	
Pb^{4+}					

Naming Ionic Compounds *(Goal 4.6)*

4.29 Write names for the following ionic compounds:
a. Al_2O_3 **b.** $CaCl_2$ **c** Na_2O
d. Mg_3N_2 **e.** KI

4.30 Write names of the following ionic compounds:
a. $FeCl_2$ **b.** K_3P **c.** $SnCl_4$
d. LiBr **e.** MgO

4.31 Indicate the ionic charge of the metal ion in each of the following ionic compounds:
a. $SnCl_2$ **b.** FeO
c. Cu_2S **d.** CuS

4.32 Indicate the ionic charge of the metal ion in each of the following ionic compounds:
a. $AuCl_3$ **b.** Fe_2O_3
c. PbI_4 **d.** $SnCl_2$

4.33 Write names for the following ionic compounds:
a. $FeCl_2$ **b.** CuO
c. Fe_2S_3 **d.** AlP

4.34 Write names for the following ionic compounds:
a. Ag_3P **b.** Cu_2S
c. SnO_2 **d.** $AuCl_3$

4.35 Write formulas for the following ionic compounds:
a. magnesium chloride
b. sodium sulfide
c. copper(I) oxide
d. zinc phosphide
e. gold(III) nitride

4.36 Write formulas for the following ionic compounds:
a. iron(III) oxide
b. barium fluoride
c. tin(IV) chloride
d. silver sulfide
e. copper(II) chloride

4.37 Circle the polyatomic ion in each of the following formulas and give the name of the compound:
a. Na_2CO_3 **b.** NH_4Cl **c.** Li_3PO_4
d. $Cu(NO_2)_2$ **e.** $FeSO_3$

4.38 Circle the polyatomic ion in each of the following formulas and give the name of the compound:
a. KOH **b.** $NaNO_3$ **c.** $CuCO_3$
d. $NaHCO_3$ **e.** $BaSO_4$

4.39 Write the correct formula for the following compounds:
a. barium hydroxide
b. sodium sulfate
c. iron(II) nitrate
d. zinc phosphate
e. iron(III) carbonate

4.40 Write the correct formula for the following compounds:
a. aluminum chloride
b. ammonium oxide
c. magnesium bicarbonate
d. sodium nitrite
e. copper(I) sulfate

Covalent Bonds *(Goal 4.7)*

4.41 Write the electron-dot structure for the following covalent molecules:
a. Br_2 **b.** H_2 **c.** HF **d.** OF_2

4.42 Write the electron-dot structure for the following covalent molecules:
a. NCl_3 **b.** CCl_4 **c.** H_2 **d.** SiF_4

4.43 The following covalent compounds have double or triple bonds. Write their electron-dot structures. (The order of atoms is indicated in parentheses.)

a. SO_2 (O S O)

b. CH_2O
  O
(H C H)

c. SO_3
  O
(O S O)

4.44 The following covalent compounds have double or triple bonds. Write their electron-dot structures. (The order of atoms is indicated in parentheses.)

a. N_2 (N N)

b. N_2Cl_2 (Cl N N Cl)

c. HNO_2 (H O N O)

Naming Covalent Compounds *(Goal 4.8)*

4.45 Name the following covalent compounds:

a. H_2S **b.** CBr_4 **c.** SCl_2
d. HF **e.** NI_3

4.46 Name the following covalent compounds:

a. CS_2 **b.** P_2O_5 **c.** Cl_2O
d. PCl_3 **e.** N_2O_4

4.47 Write the formulas of the following covalent compounds:

a. carbon tetrachloride
b. carbon monoxide
c. phosphorus trichloride
d. dinitrogen tetroxide

4.48 Write the formulas of the following covalent compounds:

a. oxygen difluoride
b. hydrogen monochloride
c. iodine pentafluoride
d. dinitrogen monoxide

4.49 Name the compounds that are found in the following sources:

a. $Al_2(SO_4)_3$ antiperspirant
b. $CaCO_3$ antacid
c. N_2O "laughing gas," inhaled anesthetic
d. Na_3PO_4 cathartic
e. $(NH_4)_2SO_4$ fertilizer

4.50 Name the compounds that are found in the following sources:

a. N_2 freezing food
b. $Mg_3(PO_4)_2$ antacid
c. Fe_2O_3 pigment in paint, paper, ceramics, glass
d. $MgSO_4$ Epsom salts
e. Cu_2O fungicide

Bond Polarity *(Goal 4.9)*

4.51 Place the symbols for partially positive (δ^+) and partially negative (δ^-) above the appropriate atoms in the following covalent bonds:

a. H—F **b.** C—Cl
c. N—O **d.** N—F

4.52 Place the symbols for partially positive (δ^+) and partially negative (δ^-) above the appropriate atoms in the following covalent bonds:

a. O—H **b.** O—S
c. P—Cl **d.** S—Cl

4.53 Identify the bonding between the following pairs of elements as covalent, polar covalent, or ionic. If no bond forms, write **none.**

a. Cl and Cl
b. Ne and O
c. F and F
d. Mg and F
e. S and Cl

4.54 Identify the bonding between the following pairs of elements as covalent, polar covalent, or ionic. If no bond forms, write **none.**

a. Na and Cl
b. O and F
c. K and S
d. H and H
e. N and H

Challenge Problems

4.55 What are the elements in Period 3 represented by the following where X is the symbol?

a. $\cdot\underset{\cdot}{\ddot{X}}\cdot$ **b.** $X\cdot$ **c.** $\cdot\dot{X}\cdot$ **d.** $\cdot\underset{\cdot}{\dot{X}}\cdot$

4.56 Write the electron arrangement of the following:

a. Ar **b.** Na^+ **c.** S^{2-}
d. Cl^- **e.** Ca^{2+}

4.57 Write the electron arrangement of the following:

a. N^{3-} **b.** Ne **c.** F^-
d. Al^{3+} **e.** Li^+

4.58 What noble gas has the same arrangement as the following ions?

a. Al^{3+} **b.** Cl^- **c.** Ca^{2+}
d. Na^+ **e.** N^{3-}

4.59 Consider an ion with the symbol Z^{2+}.

a. What is the group number of the element?
b. What is the electron-dot structure of the element?
c. If Z is a representative element in Period 2, what is the element?

4.60 Consider the following electron-dot structures for elements X and Y:

X·   ·Ÿ· (Y with two dots above, one dot on each side, and two dots below)

a. What are the group numbers of X and Y?
b. Will a compound of X and Y be ionic or covalent?
c. What ions would be formed by X and Y?
d. What would be the formula of a compound of X and Y?
e. What would be the formula of a compound of X and chlorine?
f. What would be the formula of a compound of Y and chlorine?

4.61 One of the ions of tin is tin(IV).
a. What is the symbol for this ion?
b. How many protons and electrons are in the ion?
c. What is the formula of tin(IV) oxide?
d. What is the formula of tin(IV) phosphate?

4.62 As shown in the health note, the mineral portion of bone is composed of hydroxyapatite also called durapatite, $Ca_{10}(PO_4)_6(OH)_2$. It is also named as a sequence of the ions that compose this mineral. How would it be named in this way?

Chapter 5

Chemical Quantities and Reactions

When exposed to air and water, iron reacts with oxygen to form rust.

Looking Ahead

Learning Goals

5.1 Given a chemical formula, determine the formula weight.

5.2 Given the chemical formula of a substance, calculate the molar mass.

5.3 Given the number of moles of a substance, calculate the mass in grams; given the mass, calculate the number of moles.

5.4 Classify a change in matter as chemical or physical.

5.5 Write a balanced equation for a chemical reaction from the formulas of the reactants and products.

5.6 Describe the energy in exothermic and endothermic reactions; list factors that change the rate of a reaction.

5.7 From a balanced equation, write conversion factors for any of the mole relationships.

5.8 Given the quantity of a reactant or product in a balanced equation, calculate the quantity of another substance in the reaction.

QUESTIONS TO THINK ABOUT

1. What does a fever do to physiological or metabolic processes such as the rate of breathing, pulse rate, and so on?
2. Why can a person who falls through thin ice survive in the cold water without oxygen for several minutes?
3. If a carburetor is tuned to increase the oxygen in the engine, how would it change the rate of the combustion of gasoline?
4. To speed up the healing of burns, patients are placed in chambers containing a high concentration of oxygen. Why does this help?
5. When fatty acids in the skin react with water and oxygen, they are broken down to smaller molecules that have strong odors. During exercise, body odor is more prevalent. Why would this be so?

In Chapter 4, we saw that a chemical formula gives the number of atoms of each element in a compound. In this chapter, we will describe how those atoms can be rearranged through chemical change to form different substances. When we know the formulas of the substances participating in a chemical change, we can write an equation to represent their chemical reaction.

When we know the equation for a reaction, we can measure the correct amounts of the reactants and calculate the amounts of product. We do much the same thing at home when we use a recipe to make cookies or add the right amount of water to make soup. At the automotive repair shop, a mechanic adjusts the carburetor or fuel-injection system of an engine to allow the correct amounts of fuel and oxygen from the air to mix together so the engine will run properly. In medicine, the dosages of medications are calculated to give an amount that is appropriate for a person's body mass.

5.1 Formula Weight

Learning Goal
Given a chemical formula, determine the formula weight.

In Chapter 4, we learned that the symbols and subscripts in a formula tell us the number of atoms of each type of element. For example, the formula for aspirin, $C_9H_8O_4$, indicates that there are nine carbon atoms, eight hydrogen atoms, and four oxygen atoms in each molecule.

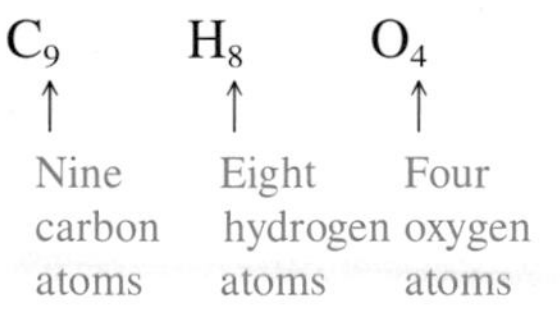

Sample Problem 5.1
Counting the Atoms in a Formula

State the number and kinds of atoms indicated by each of the following formulas:

a. caffeine, $C_8H_{10}N_4O_2$ **b.** $Ca_3(PO_4)_2$

Solution

a. $C_8H_{10}N_4O_2$: 8 carbon atoms, 10 hydrogen atoms, 4 nitrogen atoms, and 2 oxygen atoms
b. $Ca_3(PO_4)_2$: 3 calcium atoms, 2 phosphorus atoms, and 8 oxygen atoms

Study Check
A molecule of the antibiotic streptomycin contains 21 carbon atoms, 39 hydrogen atoms, 7 nitrogen atoms, and 12 oxygen atoms. What is the formula for streptomycin?

To determine the **formula weight** of a compound, multiply the atomic weight of each element by its subscript and add the results. For example, the formula weight (44.0 amu) for carbon dioxide, CO_2, is obtained by adding the atomic weights of one carbon atom and two oxygen atoms. In this text, we will take the atomic weight to one decimal place (tenths place).

Number of Atoms in the Formula		*Atomic Weight*		*Total Weight for Each Element*
1 ~~carbon atom~~	×	$\frac{12.0 \text{ amu}}{1 \text{ carbon atom}}$	=	12.0 amu
+2 ~~oxygen atoms~~	×	$\frac{16.0 \text{ amu}}{1 \text{ oxygen atom}}$	=	32.0 amu
		Formula weight of CO_2	=	44.0 amu

Sample Problem 5.2
Calculating Formula Weight

Calculate the formula weight of $CaCO_3$, a compound used as an antacid and as a means of replenishing calcium in the diet.

Solution
In the formula $CaCO_3$, the subscript 3 indicates that there are three atoms of oxygen. The symbols for calcium and carbon do not show subscripts, which means that there is only one atom of calcium and one atom of carbon. The atomic weight of each element is obtained from the periodic table, and the formula weight is calculated as follows:

Number and Type of Atoms in the Formula		*Atomic Weight*		*Total Weight for Each Element*
1 ~~atom Ca~~	×	$\frac{40.1 \text{ amu}}{1 \text{ atom Ca}}$	=	40.1 amu
+1 ~~atom C~~	×	$\frac{12.0 \text{ amu}}{1 \text{ atom C}}$	=	12.0 amu
+3 ~~atoms O~~	×	$\frac{16.0 \text{ amu}}{1 \text{ atom O}}$	=	48.0 amu
		Formula weight $CaCO_3$	=	100.1 amu

Study Check
Calculate the formula weight of Aspartame, $C_{14}H_{18}N_2O_5$, an artificial sweetener.

Sample Problem 5.3
Calculating Formula Weight with Polyatomic Ions

The compound aluminum nitrate, $Al(NO_3)_3$, is an ingredient in antiperspirants. What is its formula weight?

Solution

When a formula contains a polyatomic ion, we must consider the subscript outside of the parentheses. In the formula for the antiperspirant $Al(NO_3)_3$, we have one atom of aluminum, three atoms of nitrogen (3×1), and nine atoms of oxygen (3×3).

Number and Type of Atoms in the Formula		*Atomic Weight*		*Total Weight for Each Element*
1 atom Al	$\times$	$\frac{27.0 \text{ amu}}{1 \text{ atom Al}}$	=	27.0 amu
3 atoms N	$\times$	$\frac{14.0 \text{ amu}}{1 \text{ atom N}}$	=	42.0 amu
9 atoms O	$\times$	$\frac{16.0 \text{ amu}}{1 \text{ atom O}}$	=	144 amu
		Formula weight $Al(NO_3)_3$	=	213 amu

Study Check

Calcium bisulfite, $Ca(HSO_3)_2$, is used in papermaking and as a disinfectant to wash brewery casks. Calculate the formula weight of calcium bisulfite.

5.2 The Mole

Learning Goal

Given the chemical formula of a substance, calculate the molar mass.

At the store, you buy eggs by the dozen. In an office, pencils are ordered by the gross and paper by the ream. In a restaurant, soda is ordered by the case. In each of these examples, the terms *dozen, gross, ream, and case* count the number of items present. For example, when you buy a dozen donuts, you know you will get 12 donuts in the box. (See Table 5.1 and Figure 5.1.) In chemistry, particles such as atoms and molecules are counted by the **mole, (mol),** a unit that contains 6.02×10^{23} items. This very large number, called

Table 5.1 *Number of Items in Some Typical Collections*

Collection	Number of Items
1 trio of singers	3 singers
1 dozen eggs	12 eggs
1 case of soda	24 cans of soda
1 gross of pencils	144 pencils
1 ream of paper	500 sheets of paper
1 mol of atoms	6.02×10^{23} atoms
1 mol of molecules	6.02×10^{23} molecules

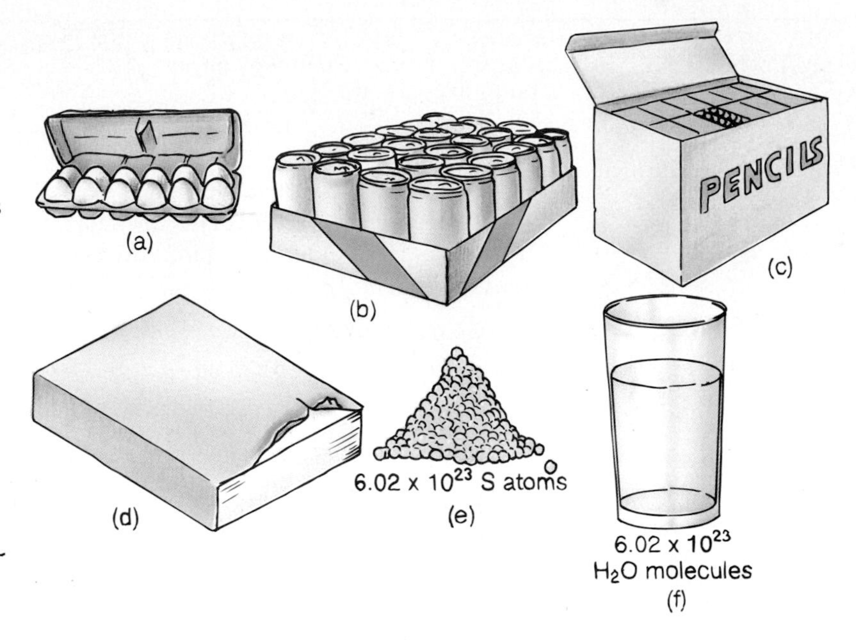

Figure 5.1
Some common collections and the number of items in them.
(a) A dozen eggs is 12 eggs.
(b) A case of soda contains 24 cans of soda.
(c) A gross of pencils is 144 pencils.
(d) A ream of paper contains 500 sheets.
(e) A mole of sulfur contains 6.02×10^{23} sulfur atoms.
(f) A mole of water contains 6.02×10^{23} water molecules.

Avogadro's number, after Amedeo Avogadro, an eighteenth-century Italian physicist, looks like this when written out:

Avogadro's Number

$$602{,}000{,}000{,}000{,}000{,}000{,}000{,}000 = 6.02 \times 10^{23}$$

One mole of any element always has Avogadro's number of atoms. For example, 1 mol of carbon contains 6.02×10^{23} carbon atoms; 1 mol of aluminum contains 6.02×10^{23} aluminum atoms; 1 mol of sulfur contains 6.02×10^{23} sulfur atoms.

$$1 \text{ mol of an element} = 6.02 \times 10^{23} \text{ atoms of that element}$$

One mole of a compound also contains Avogadro's number of molecules or formula units. (Molecules are the particles of covalent compounds; formula units are the group of ions given by the formula of an ionic compound). For example, 1 mol of CO_2, a covalent compound, contains 6.02×10^{23} molecules of CO_2. One mole of NaCl, an ionic compound, contains 6.02×10^{23} formula units of NaCl (Na^+, Cl^-). Table 5.2 gives some examples of the number of particles in some 1-mol quantities.

Molar Mass

You will never weigh a single atom or molecule. They are much too small to weigh, even on the most accurate balance. Instead, you will work in a laboratory with samples that can be weighed in grams, on a balance. If you weigh the formula weight of an element or compound in grams, you would have

Table 5.2 ***Number of Particles in One-Mole Samples***

Substance	Number and Type of Particles
1 mol carbon	6.02×10^{23} carbon atoms
1 mol aluminum	6.02×10^{23} aluminum atoms
1 mol sulfur	6.02×10^{23} sulfur atoms
1 mol water (H_2O)	6.02×10^{23} H_2O molecules
1 mol NaCl	6.02×10^{23} NaCl formula units
1 mol sucrose ($C_{12}H_{22}O_{11}$)	6.02×10^{23} sucrose molecules
1 mol vitamin C ($C_6H_8O_6$)	6.02×10^{23} vitamin C molecules

1 mol of that substance. This mass is known as the **molar mass,** which is the number of grams numerically equal to the formula weight. For example, the atomic weight of carbon is 12.0, and its molar mass is 12.0 grams. In one molar mass (12.0 g) of carbon, there are 6.02×10^{23} carbon atoms.

Atomic mass Unit

$$1 \text{ atom C} = 12.0 \text{ amu} \quad \text{Atomic mass of carbon}$$
$$1 \text{ mole C} = 12.0 \text{ g C}$$
$$\text{Molar mass of C} = \frac{12.0 \text{ g}}{1 \text{ mol}}$$

Figure 5.2
One mole samples starting from front to back: sodium chloride (NaCl 58.5 g), copper (II) sulfate pentahydrate ($CuSO_4 \cdot 5H_2O$ 249.7 g), aluminum (Al 27.0 g), and sulfur (S 32.1 g).

Earlier, we saw that a molecule of carbon dioxide (CO_2) had a formula weight of 44.0 amu. Now we can say that 1 mol of carbon dioxide molecules has a mass of 44.0 g and contains 6.02×10^{23} carbon dioxide molecules. Using the periodic table, we can calculate the molar mass of any substance.

$$1 \text{ molecule } CO_2 = 44.0 \text{ amu}$$
$$1 \text{ mole } CO_2 = 44.0 \text{ g}$$
$$\text{Molar mass } CO_2 = \frac{44.0 \text{ g}}{1 \text{ mol}}$$

Figure 5.2 shows some 1-mol qualities of substances. Table 5.3 lists the molar mass for several 1-mol samples.

Table 5.3 ***The Molar Mass and Number of Particles in One-Mole Quantities***

Substance	Molar Mass	Number of Particles in One Mole
Carbon (C)	12.0 g	6.02×10^{23} C atoms
Sodium (Na)	23.0 g	6.02×10^{23} Na atoms
Iron (Fe)	55.9 g	6.02×10^{23} Fe atoms
NaF (preventative for dental caries)	42.0 g	6.02×10^{23} NaF formula units
$CaCO_3$ (antacid)	100.1 g	6.02×10^{23} $CaCO_3$ formula units
$C_6H_{12}O_6$ (glucose)	180.0 g	6.02×10^{23} glucose molecules
$C_8H_{10}N_4O_2$ (caffeine)	194.0 g	6.02×10^{23} caffeine molecules

Sample Problem 5.4
Finding the Molar Mass

Calculate the molar mass of salicylic acid, $C_7H_6O_3$.

Solution

Using the atomic weights on the periodic table (rounded to one decimal place), the molar mass is calculated as follows:

$$7 \text{ moles C} \times \frac{12.0 \text{ g}}{1 \text{ mol C}} = 84.0 \text{ g}$$

$$+6 \text{ moles H} \times \frac{1.0 \text{ g}}{1 \text{ mol H}} = 6.0 \text{ g}$$

$$+3 \text{ moles O} \times \frac{16.0 \text{ g}}{1 \text{ mol O}} = 48.0 \text{ g}$$

$$\text{Molar mass of salicylic acid} = 138.0 \text{ g}$$

Study Check

Calculate the mass of 1 mol of carmine, $C_{22}H_{20}O_{13}$, a red pigment used in paints, inks, and stains in microbiology.

5.3 Calculations Using Molar Mass

Learning Goal

Given the number of moles of a substance, calculate the mass in grams; given the mass, calculate the number of moles.

The molar mass of an element or a compound is one of the most useful numerical values in chemistry. Molar mass is used to change from moles of a substance to grams, or from grams to moles. To do these calculations, we use the molar mass as a conversion factor. For example, 1 mol of magnesium has a mass of 24.3 g. To express molar mass as an equality, we can write

$$1 \text{ mol Mg} = 24.3 \text{ g Mg}$$

From this equality, two conversion factors can be written.

$$\frac{24.3 \text{ g Mg}}{1 \text{ mol Mg}} \quad \text{and} \quad \frac{1 \text{ mol Mg}}{24.3 \text{ g Mg}}$$

Conversion factors are written for compounds in the same way. For example, the molar mass of the compound H_2O is 18.0 g.

$$1 \text{ mol } H_2O = 18.0 \text{ g } H_2O$$

The conversion factors from the molar mass of H_2O are written as

$$\frac{18.0 \text{ g } H_2O}{1 \text{ mol } H_2O} \quad \text{and} \quad \frac{1 \text{ mol } H_2O}{18.0 \text{ g } H_2O}$$

We can now change from moles to grams, or grams to moles, using the conversion factors derived from the molar mass. (Remember to calculate the molar mass of the substance first.) Using the unit-factor method discussed in Chapter 1, we can now solve the following sample problems.

Sample Problem 5.5
Converting Moles of an Element to Grams

Silver metal is used in the manufacture of tableware, mirrors, jewelry, and dental alloys. If the design for a piece of jewelry requires 0.750 mol of silver, how many grams of silver would be needed?

Solution
Using the atomic weight, we write the molar mass of silver as an equality.

$$1 \text{ mol Ag} = 107.9 \text{ g}$$

From the equality, we can write conversion factors that relate grams and moles.

$$\frac{107.9 \text{ g Ag}}{1 \text{ mol Ag}} \quad \text{and} \quad \frac{1 \text{ mol Ag}}{107.9 \text{ g Ag}}$$

The problem is set up by multiplying the given (0.750 mol Ag) by the conversion factor that cancels "mol Ag." For this problem, we use the first conversion factor.

$$\underset{\text{Given}}{0.750 \cancel{\text{mol Ag}}} \times \underset{\text{Molar mass conversion factor}}{\frac{107.9 \text{ g Ag}}{1 \cancel{\text{mol Ag}}}} = \underset{\text{Answer (rounded to three significant figures)}}{80.9 \text{ g Ag}}$$

Study Check
Calculate the number of grams of gold (Au) present in 0.124 mol of gold.

Sample Problem 5.6
Converting Moles of a Compound to Mass

Salicylic acid has a formula of $C_7H_6O_3$. How many grams of salicylic acid are in 2.25 mol?

Solution
In an earlier problem, we calculated the molar mass of salicylic acid as 138.0 g. The molar mass gives the equality

$$1 \text{ mol } C_7H_6O_3 = 138.0 \text{ g } C_7H_6O_3$$

From the molar mass, we can write two conversion factors:

$$\frac{138.0 \text{ g salicylic acid}}{1 \text{ mol salicylic acid}} \quad \text{and} \quad \frac{1 \text{ mol salicylic acid}}{138.0 \text{ g salicylic acid}}$$

In the problem setup, the given, 2.25 mol of salicylic acid is multiplied by the first factor to cancel the unit "mol salicylic acid."

$$\underset{\text{Given}}{2.25 \cancel{\text{mol salicylic acid}}} \times \underset{\text{Molar mass conversion factor}}{\frac{138.0 \text{ g salicylic acid}}{1 \cancel{\text{mol salicylic acid}}}} = \underset{\text{Answer (rounded to 3 SFs)}}{311 \text{ g salicylic acid}}$$

Study Check
Calculate the mass (grams) of 0.65 mol of ammonium carbonate, $(NH_4)_2CO_3$, a substance found in baking powders.

Sample Problem 5.7
Converting the Mass of an Element to Moles

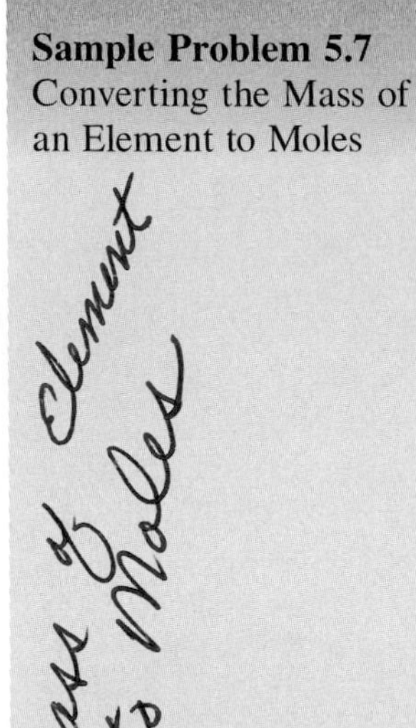

The total bone mass of a typical 70.0-kg person contains 115 g of phosphorus. How many moles of phosphorus are in the bone mass?

Solution
Phosphorus has a molar mass of 31.0 g.

$$1 \text{ mol P} = 31.0 \text{ g P}$$

The two conversion factors for this molar mass are

$$\frac{31.0 \text{ g P}}{1 \text{ mol P}} \quad \text{and} \quad \frac{1 \text{ mol P}}{31.0 \text{ g P}}$$

In the problem setup, the given, 115 g P, is multiplied by the conversion factor that cancels "g P."

$$115 \cancel{\text{g P}} \times \frac{1 \text{ mol P}}{31.0 \cancel{\text{g P}}} = 3.71 \text{ mol P}$$

Given — Molar mass conversion factor — Answer (rounded to 3 SFs)

Study Check
How many moles of iron are contained in 255 g of iron (Fe) metal?

Sample Problem 5.8
Converting the Mass of a Compound to Moles

Iodized salt provides sufficient iodine in a person's diet to prevent goiter, a condition that occurs in the thyroid when iodine levels are too low. Iodized salt is prepared by adding 0.010% potassium iodide (KI) to table salt. If 0.0737 g KI are added to each box of salt (737 g), how many moles of KI are required in the preparation of 250 boxes of salt?

Solution
The molar mass of KI is

$$1 \text{ mol KI} = 166.0 \text{ g KI}$$

Written in the form of conversion factors, this molar mass gives

$$\frac{166.0 \text{ g KI}}{1 \text{ mol KI}} \quad \text{and} \quad \frac{1 \text{ mol KI}}{166.0 \text{ g KI}}$$

The problem is set up with the given, 250 boxes of salt, multiplied by a factor of 0.0737 g KI per box. The final conversion factor cancels "g KI."

$$250 \cancel{\text{boxes of salt}} \times \frac{0.0737 \cancel{\text{g KI}}}{1 \cancel{\text{box of salt}}} \times \frac{1 \text{ mol KI}}{166.0 \cancel{\text{g KI}}} = 0.111 \text{ mol KI}$$

Given (exact) — Conversion factor — Molar mass conversion factor — Answer (rounded to 3 SFs)

Study Check
One gelcap of an antacid contains 311 mg $CaCO_3$ and 232 mg $MgCO_3$. In a recommended dosage of two gelcaps, how many moles each of $CaCO_3$ and $MgCO_3$ are present?

5.4 Chemical Changes

Learning Goal
Classify a change in matter as a chemical or a physical change.

Every day, chemical changes are taking place around us. You may have seen a nail or other iron object turn to rust when the object was exposed to oxygen and water in air. Perhaps you have cleaned a silver dish or a piece of jewelry because the metal had changed from a shiny, silver color to a dull, gray-black color as it was exposed to sulfur. You may have dropped an antacid tablet in a glass of water and noticed fizzing and bubbling as a gas was released. In a car, gasoline reacts with oxygen and burns producing carbon dioxide (CO_2), water (H_2O), and the energy that makes it move.

When a chemical change occurs, new substances form that have different characteristics. Iron (Fe) reacts with oxygen (O_2) and silver (Ag) reacts with sulfur (S) to produce the new substances, rust (Fe_2O_3) and tarnish (Ag_2S), that have different properties than the reactants. The bubbling of some antacid tablets in water is the result of a gas, carbon dioxide (CO_2), that is formed from the sodium hydrogen carbonate ($NaHCO_3$) contained in the tablet. In each example, a **chemical change** occurs because atoms of the initial substances form new combinations that have new identities; a **chemical reaction** has taken place.

In Chapter 2, we learned that there are no new substances formed in a **physical change.** Let us look at some everyday examples of chemical and physical changes in Table 5.4 and Figure 5.3.

Table 5.4 ***Comparison of Some Chemical and Physical Changes***

Chemical Changes	Physical Changes
Rusting nail	Melting ice
Bleaching a stain	Boiling water
Burning a log	Sawing a log in half
Tarnishing silver	Tearing paper
Fermenting grapes	Breaking a glass
Souring of milk	Pouring milk

Figure 5.3
Examples of chemical change include a rusty nail, tarnish on a silver fork, and the fizzing (bubbling) of an antacid tablet in water.

Sample Problem 5.9
Classifying Chemical and Physical Change

Classify each of the following changes as physical or chemical:

a. water freezing into an icicle **b.** burning a match
c. breaking a chocolate bar **d.** digesting a chocolate bar

Solution

a. Physical. Freezing water involves only a change from liquid water to ice. No change has occurred in the substance water.
b. Chemical. Burning a match causes the formation of new substances that were not present before striking the match.
c. Physical. Breaking a chocolate bar does not affect its composition.
d. Chemical. The digestion of the chocolate bar converts its components into new substances.

Study Check

Classify the following changes as physical or chemical:

a. chopping a carrot
b. developing a Polaroid picture
c. inflating a balloon

5.5 Chemical Equations

Learning Goal
Write a balanced equation for a chemical reaction from the formulas of the reactants and products.

Whenever you put together a bicycle, build a model airplane, prepare a new recipe, mix a medication, or clean a patient's teeth, you follow a set of directions. These directions tell you what materials to use, how much you need, and the results you will obtain. In chemistry, a **chemical equation** tells us the materials we need and products that will form in a chemical reaction.

As an analogy, we might imagine that we are going to make pancakes. From the recipe, we see that we need milk, eggs, and pancake mix. These ingredients are called **reactants** by the chemist, and the pancakes we make are the new substances that form called the **products.** We write the recipe in the form of an equation placing the ingredients on the left of an arrow and the products on the right:

Writing an Equation

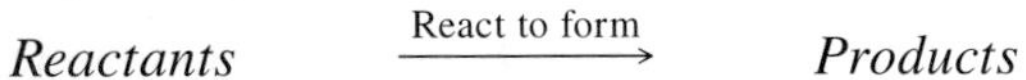

Milk + eggs + pancake mix $\longrightarrow$ pancakes

However, our recipe equation does not give us enough information. We need to know how much of each reactant to use. The amount is shown by a number called a **coefficient** placed in front of the name of each reactant and product. Our equation for making pancakes is now complete and can be written as the following:

1 cup milk + 3 eggs + 2 cups pancake mix $\longrightarrow$ 16 pancakes

Figure 5.4 illustrates the reactants and the products of our cooking experiment.

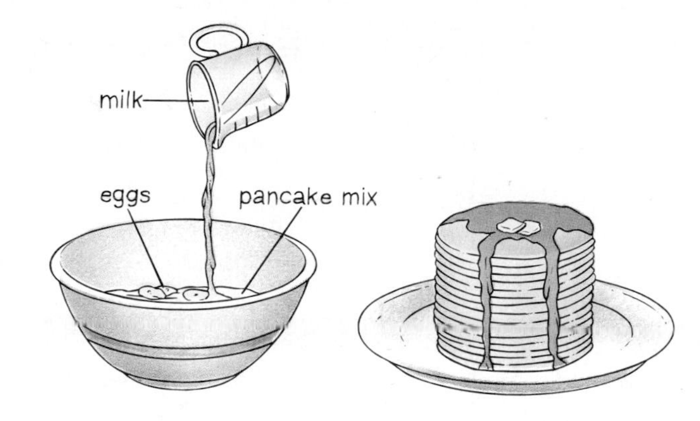

Figure 5.4
In preparing pancakes, the proper amounts of milk, eggs, and pancake mix are mixed together. After the ingredients undergo a reaction, the pancakes are ready to eat.

In a chemical equation, the chemical formulas are written for the reactants and the products. For example, the gas methane is used in many homes for cooking and heating as seen in Figure 5.5. When methane (CH_4) reacts with oxygen, the reaction produces carbon dioxide, water, and heat. We can look at this reaction at the atomic level.

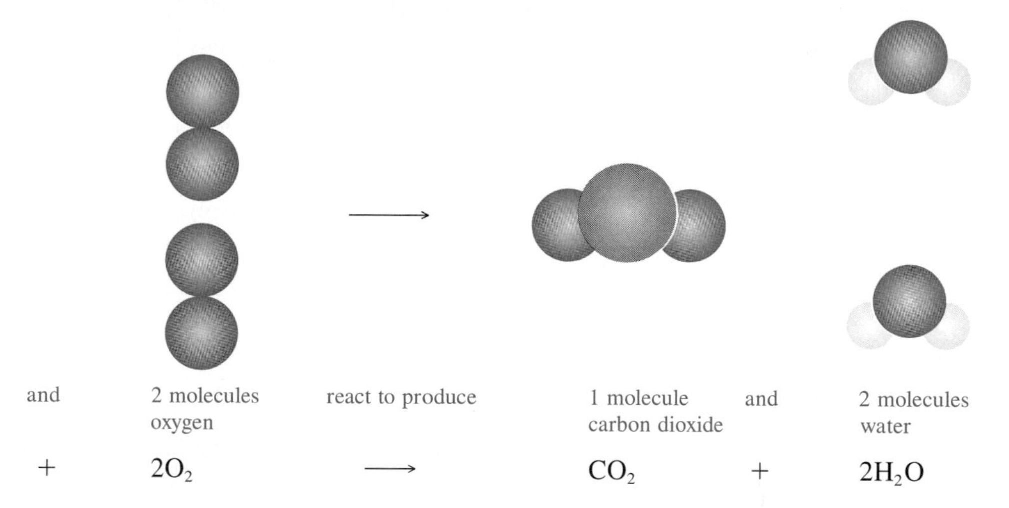

1 molecule methane	and	2 molecules oxygen	react to produce	1 molecule carbon dioxide	and	2 molecules water
CH_4	+	$2O_2$	$\longrightarrow$	CO_2	+	$2H_2O$

Figure 5.5
In a gas burner, methane (CH_4) reacts with oxygen to produce carbon dioxide (CO_2) and water (H_2O).

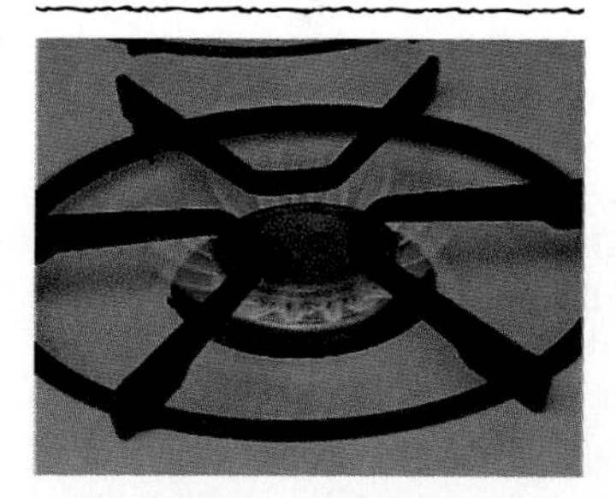

In the equation, the numbers that appear in front of the formulas for oxygen (O_2) and water (H_2O) are the coefficients. They tell us that one molecule of methane reacts with two molecules of oxygen to produce one molecule of carbon dioxide and two molecules of water.

Physical States of Reactants and Products

In an equation, letters may be used that give the physical state (solid, liquid, or gas) of the reactants and products. If a substance is dissolved in water, it is called an aqueous solution. The following list gives some of these abbreviations.

Abbreviation	*Meaning*	*Example*	*Meaning*
(*s*)	Solid	Fe(*s*)	Iron is in the solid state
(*l*)	Liquid	$H_2O(l)$	Water is a liquid
(*g*)	Gas	$Cl_2(g)$	Chlorine is a gas
(*aq*)	Dissolved in water	NaCl(aq)	Sodium chloride is dissolved in water (aqueous solution)

The abbreviations in the following equation tell us that solid calcium hydride reacts with water, a liquid, and produces an aqueous solution of calcium hydroxide and hydrogen gas.

$$CaH_2(s) + 2H_2O(l) \longrightarrow Ca(OH)_2(aq) + 2H_2(g)$$

Sample Problem 5.10
Writing a Chemical Equation

1. State the meaning in words for the following chemical equations:

$$\underset{\text{Ammonia}}{4NH_3} + 3O_2 \longrightarrow 2N_2 + 6H_2O$$

2. Write an equation including the physical states that represents the following reaction: Two molecules of sulfur dioxide gas react with one molecule of oxygen gas to produce two molecules of sulfur trioxide gas.

Solution

1. Four molecules of ammonia and three molecules of oxygen react to form two molecules of nitrogen and six molecules of water.

2. $2SO_2(g) + O_2(g) \longrightarrow 2SO_3(g)$

Study Check

Three molecules of oxygen react with one molecule of carbon disulfide to form one molecule of carbon dioxide and two molecules of sulfur dioxide. If all the substances are gases, write an equation for the reaction including their physical states.

Atoms Are Conserved in a Chemical Reaction

Atoms cannot be created or destroyed during a chemical reaction. This statement known as the **law of conservation of matter** means that the atoms of the reactants are just rearranged to form the products of the reaction. No new atoms enter the reaction, nor do any of the original atoms disappear. Therefore, a balanced chemical equation shows the same number of atoms for each kind of element on both sides of the equation.

Consider the reaction in which hydrogen reacts with oxygen to form water. In this reaction, two molecules of hydrogen (H_2) react with one molecule of oxygen (O_2) to form two molecules of water. The four hydrogen atoms shown in the reactants are balanced by four hydrogen atoms in the water product; two oxygen atoms in the reactants are balanced by two oxygen atoms in the water product.

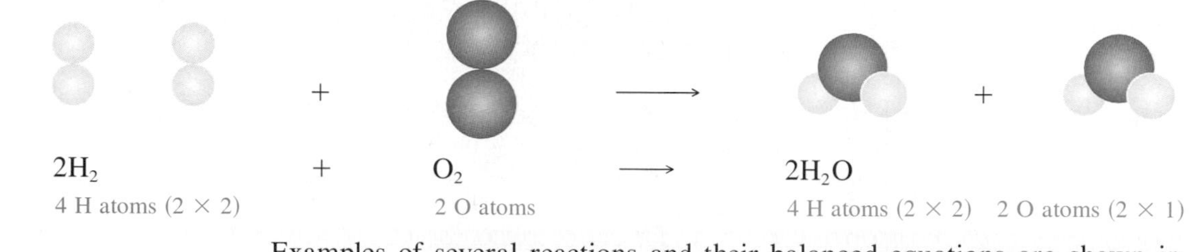

Examples of several reactions and their balanced equations are shown in Figure 5.6.

Figure 5.6
Some chemical reactions. The balanced equations for these reactions are
(a) $Zn(s) + CuSO_4(aq) \longrightarrow Cu(s) + ZnSO_4(aq)$
(b) $Na_2SO_4(aq) + BaCl_2(aq) \longrightarrow BaSO_4(s) + 2NaCl(aq)$
(c) $2Mg(s) + O_2(g) \longrightarrow 2MgO(s)$
(d) $Zn(s) + 2HCl(aq) \longrightarrow ZnCl_2(aq) + H_2(g)$

(a)

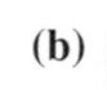

(b)

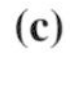

(c)

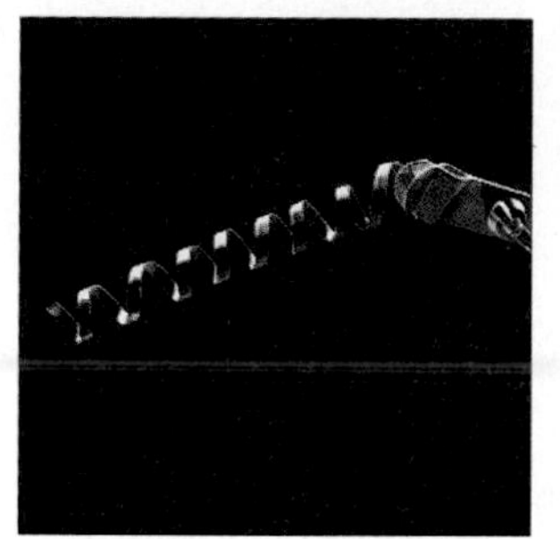

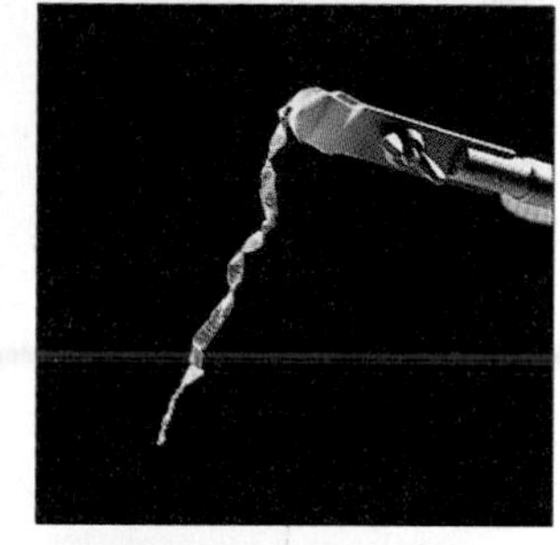

(d)

Sample Problem 5.11
Counting Atoms in a Balanced Equation

State the number of atoms of each element in the reactants and products for the following balanced equations:

a. $2Fe + 3Cl_2 \longrightarrow 2FeCl_3$ **b.** $C_4H_8 + 6O_2 \longrightarrow 4CO_2 + 4H_2O$

Solution

The number of atoms can be calculated by multiplying the subscripts in a formula by the coefficient. When there is no subscript, the subscript 1 is understood.

Coefficients

a. $2Fe + 3Cl_2 \longrightarrow 2FeCl_3$

Subscripts

Reactant Side	*Product Side*
2 Fe	2 Fe
6 Cl (3 × 2)	6 Cl (2 × 3)

b. $C_4H_8 + 6O_2 \longrightarrow 4CO_2 + 4H_2O$

Reactant Side	*Product Side*
4 C	4 C
8 H	8 H (4 × 2)
12 O (6 × 2)	12 O (4 × 2) + (4)

Study Check

One of the components of acid rain is nitric acid (HNO_3), produced when nitrogen dioxide in smog reacts with the water and oxygen in the air. State the number and kind of atoms on both sides of the equation for the reaction of nitorgen dioxide in air.

$$4NO_2 + 2H_2O + O_2 \longrightarrow 4HNO_3$$

Nitric acid

Balancing a Chemical Equation

When we know the reactants and products of a chemical reaction, we can write their formulas in an equation:

$$H_2 + Cl_2 \longrightarrow HCl$$

When we count the atoms of hydrogen and chlorine on the reactant side with the number of atoms on the product side, we find that there are more hydrogen and chlorine atoms on the reactant side. The equation is *not balanced.*

Reactant Side	*Product Side*
2 H	1 H
2 Cl	1 Cl

To balance the number of hydrogen and chlorine atoms in the product, the coefficient 2 must be written in front of the HCl formula. Use only

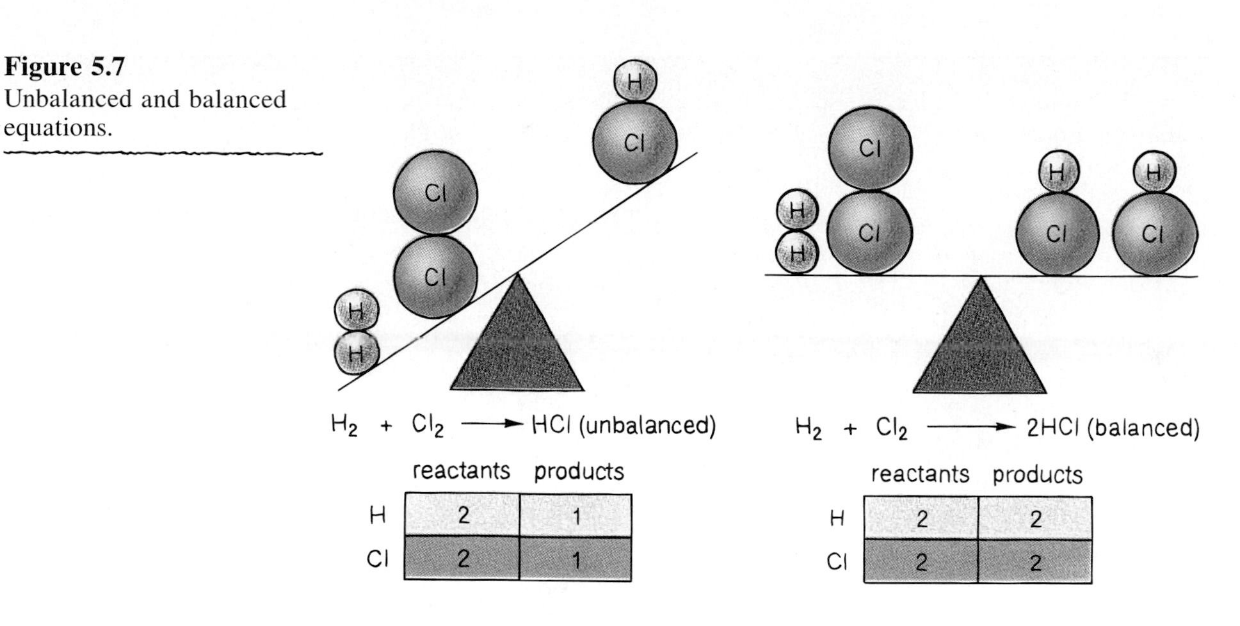

Figure 5.7
Unbalanced and balanced equations.

coefficients to balance an equation. Because each compound in the equation is represented by its correct formula, none of the subscripts can be changed when you are balancing an equation.

$H_2 + Cl_2 \longrightarrow 2HCl$ Balanced

This time when we count the atoms in the equation, we find they are balanced.

Reactant Side	*Product Side*
2 H	~~1 H~~ 2 H
2 Cl	~~1 Cl~~ 2 Cl

Figure 5.7 illustrates the unbalanced and the balanced equation.

Some hints for balancing equations are given in Table 5.5. Remember, they are only hints. Balancing equations in this way is largely a matter of trial and error.

Table 5.5 ***Hints for Balancing Chemical Equations***

Step 1 Count the number of atoms of each element on the reactant side and on the product side.

Step 2 Detemine which atoms need to be balanced.

Step 3 Start balancing, one element at a time, by placing coefficients in front of the formula containing that element. A typical starting place is with the elements in a formula with subscripts.

Step 4 Recheck the atoms on the reactant side and on the product side to see if the equation is completely balanced. Sometimes, balancing one element will undo the balance of another. If this happens, repeat the process.

Sample Problem 5.12
Balancing Equations

Use coefficients to balance the following equations:

a. $S_8 + O_2 \longrightarrow SO_2$ **b.** $Al + Cl_2 \longrightarrow AlCl_3$

Solution

a. We start by checking the number of atoms on the reactant and product sides.

Reactant Side	*Product Side*	
8 S	1 S	Not balanced
2 O	2 O	

To balance the S, we place a coefficient of 8 in front of SO_2.

$S_8 + O_2 \longrightarrow 8SO_2$ Balances S

Rechecking, we find that we need more oxygen on the reactant side.

Reactant Side	*Product Side*	
8 S	~~1 S~~ 8 S	Balanced
2 O	~~2 O~~ 16 O	Not balanced

The oxygen is balanced by placing a coefficient of 8 in front of O_2.

$S_8 + 8O_2 \longrightarrow 8SO_2$ Balances O

Rechecking the atom balance, we now have equal numbers of each type of atom. The equation is balanced.

Reactant Side	*Product Side*
8 S	~~1 S~~ 8 S
~~2 O~~ 16 O	~~2 O~~ 16 O

b. We start by checking the number of each atom on the reactant and product sides.

Reactant Side	*Product Side*	
1 Al	1 Al	
2 Cl	3 Cl	Not balanced

To balance the Cl, we place a coefficient of 3 in front of Cl_2 and a coefficient of 2 in front of $AlCl_3$, to give a total of 6 Cl on both sides.

$Al + 3Cl_2 \longrightarrow 2AlCl_3$ Balances Cl

Rechecking the balance of atoms, we find that Al becomes unbalanced. Al is balanced by placing a coefficient of 2 in front of Al.

$2Al + 3Cl_2 \longrightarrow 2AlCl_3$ Balances Al and Cl

All the atoms in the equation are now balanced as shown by the equal numbers of each atom on the reactant side and on the product side.

Reactant Side	*Product Side*
~~1 Al~~ 2 Al	~~1 Al~~ 2 Al
~~2 Cl~~ 6 Cl	~~3 Cl~~ 6 Cl

Study Check
Balance the following equation:

$$C + SO_2 \longrightarrow CS_2 + CO$$

HEALTH NOTE *Smog and Health Concerns*

There are two types of smog. One, photochemical smog, requires sunlight to initiate reactions that produce pollutants such as nitrogen oxides and ozone. The other type of smog, industrial or London smog, occurs in areas where coal containing high amounts of sulfur is burned and sulfur dioxide is given off.

Photochemical smog is most prevalent in cities where people are dependent on cars for transportation. On a typical day in Los Angeles, for example, nitrogen monoxide (NO) emissions from car exhausts increases as traffic increases on the roads. The nitrogen monoxide is formed when N_2 and O_2 react at high temperatures in car and truck engines.

$$N_2 + O_2 \xrightarrow{\text{Heat}} 2NO$$

In the air, NO reacts with oxygen in the air to produce NO_2, a reddish brown gas that is irritating to the eyes and damaging to the respiratory tract. See Figure 5.8.

$$2NO + O_2 \longrightarrow 2NO_2$$

When NO_2 is exposed to sunlight, it is converted into NO and oxygen atoms.

$$NO_2 \xrightarrow{\text{Sunlight}} NO + \underset{\text{Oxygen atoms}}{O}$$

Oxygen atoms are so reactive, they react with the oxygen molecules in the atmosphere and ozone forms.

$$O + O_2 \longrightarrow \underset{\text{Ozone}}{O_3}$$

In the upper atmosphere (the stratosphere), ozone is beneficial because it protects us from harmful ultraviolet radiation that comes from the sun. However, in the lower atmosphere, ozone irritates the eyes and respiratory tract, where it causes coughing, decreased lung function, and fatigue. It also causes deterioration of fabrics, cracks rubber, and damages trees and crops.

Industrial smog is prevalent in areas where coal with a high sulfur content is burned to produce electricity. During combustion, the sulfur is

Figure 5.8
Nitrogen dioxide, a reddish brown gas that irritates the respiratory tract, is one of the pollutants in photochemical smog.

converted to sulfur dioxide:

$$S + O_2 \longrightarrow SO_2$$

The SO_2 is damaging to plants, suppressing growth, and it is corrosive to metals such as steel. SO_2 is also damaging to humans and can cause lung impairment and respiratory difficulties. The SO_2 in the air reacts with more oxygen to form SO_3, which combines with water to form sulfuric acid, an extremely corrosive acid:

$$2SO_2 + O_2 \longrightarrow 2SO_3$$

$$SO_3 + H_2O \longrightarrow \underset{\text{Sulfuric acid}}{H_2SO_4}$$

The presence of sulfuric acid in rivers and lakes causes an increase in the acidity of the water, reducing the ability of animals and plants to survive.

5.6 Energy in Chemical Reactions

Learning Goal

Describe the energy in exothermic and endothermic reactions; list factors that change the rate of a reaction.

For a chemical reaction to take place, the molecules of the reactants must come in contact with each other. If there are no collisions between the molecules (or atoms), no reaction is possible. When the molecules collide, bonds between atoms are broken and new bonds can form. The energy needed to break apart those bonds is called the **activation energy.** If the energy of a collision is less than the activation energy, the molecules bounce apart without reacting.

The concept of activation energy is analogous to climbing over a hill. To reach a destination on the other side, we must expend energy to climb to the top of the hill. Once we are at the top, we can easily roll down to the other side. The energy needed to get us from our starting point to the top of the hill would be our activation energy.

Exothermic and Endothermic Reactions

As the bonds of the products form, energy is returned to the reaction system. However, that energy return may be greater or less than the activation energy used. The **heat of reaction** is the energy difference between the reactants and the products. In **exothermic reactions,** the energy of the products is lower than that of the reactants and heat is given off. In **endothermic reactions,** the energy of the products is higher than the energy of the reactions and heat is absorbed. (See Figure 5.9.)

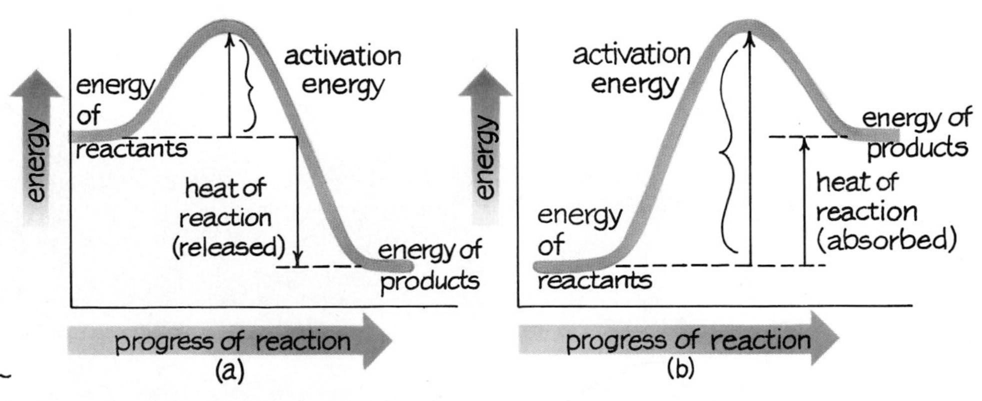

Figure 5.9
(a) In an exothermic reaction, the energy of the products is lower than the energy of the reactants; energy is released.
(b) In an endothermic reaction, the energy of the products is higher than the energy of the reactants; energy is absorbed.

For example, the combustion of methane (CH_4) is an exothermic reaction. Heat (213 kcal per mole) is released along with the products CO_2 and H_2O.

Exothermic Reaction $\underset{\text{Methane}}{CH_4} + 2O_2 \longrightarrow CO_2 + 2H_2O + \underset{\text{Heat released}}{213 \text{ kcal of heat}}$

The formation of hydrogen iodide is an endothermic reaction. In the overall energy change, heat is absorbed when the hydrogen and iodine undergo reaction.

Endothermic Reaction $H_2 + I_2 + \underset{\text{Heat absorbed}}{12 \text{ kcal of heat}} \longrightarrow 2HI$

Reaction	*Energy Change*	*Heat in the Equation*
Exothermic	Heat released	Product side
Endothermic	Heat absorbed	Reactant side

Sample Problem 5.13
Energy of Reaction

The gases from the air, N_2 and O_2, react in the hot engine of a car to give one of the smog-forming pollutants called nitrogen monoxide (NO). The reaction requires 43 kcal of energy. Write a balanced equation for the reaction including the energy.

Solution
First, we write the unbalanced equation in terms of the reactants and products.

$$N_2 + O_2 \longrightarrow NO$$

Then we balance the equation by placing the coefficient 2 in front of NO.

$$N_2 + O_2 \longrightarrow 2NO$$

Because energy is required, the reaction is endothermic, which means that the energy is part of the reactants.

$$N_2 + O_2 + 43 \text{ kcal} \longrightarrow 2NO$$

Study Check
When ethanol (C_2H_5OH) reacts with oxygen (O_2), the products are carbon dioxide, water, and 326 kcal of heat. Write a balanced equation including the heat of reaction.

Sample Problem 5.14
Exothermic and Endothermic Reactions

In the reaction between solid carbon and oxygen gas, the energy of the product carbon dioxide (CO_2) is 94 kcal lower than the energy of the reactants.

a. Is the reaction exothermic or endothermic?
b. What is the equation for the reaction?

Solution

a. exothermic **b.** $C + O_2 \longrightarrow CO_2 + 94\text{ kcal}$

Study Check

A reaction is endothermic. Is the energy of the products higher or lower than the energy of the reactants?

HEALTH NOTE *Hot Packs and Cold Packs*

In the hospital, at a first-aid station, or at an athletic event, a *cold pack* may be used to reduce swelling from an injury, remove heat from inflammation, or decrease capillary size to lessen the effect of hemorrhaging. Inside the plastic container of a cold pack, there is a compartment containing solid ammonium nitrate (NH_4NO_3) that is separated from a compartment containing water as illustrated in Figure 5.10. The pack is activated when it is hit or squeezed hard enough to break the walls between the compartments and cause the ammonium nitrate to mix with the water (shown as H_2O over the reaction arrow). In an endothermic process, each gram of NH_4NO_3 that dissolves absorbs 79 cal of heat from the water. The temperature drops and the pack becomes cold and ready to use:

Endothermic Reaction in a Cold Pack

$$NH_4NO_3(s) + \text{heat} \xrightarrow{H_2O} NH_4NO_3(aq)$$

Hot packs are used to relax muscles, lessen aches and cramps, and increase circulation by expanding capillary size. Constructed in the same way as cold packs, a hot pack may contain the salt $CaCl_2$. The dissolving of the salt in water is exothermic releasing 160 cal per gram of salt. The temperature rises and the pack gets hot and ready to use:

Exothermic Reaction in a Hot Pack

$$CaCl_2\ (s) \xrightarrow{H_2O} CaCl_2(aq) + \text{heat}$$

Figure 5.10
In a cold pack, an endothermic reaction lowers the temperature of the pack. In a hot pack, an exothermic reaction raises the temperature.

Rate of Reaction

The **rate** (or speed) **of a reaction** is measured by the amount of reactant used up, or the amount of product formed, in a certain period of time. Reactions with low activation energy go faster than reactions with high activation energies. For a particular reaction, the rate of reaction is affected by the temperature, the amounts of reactants in the container and by catalysts.

At higher temperatures, the increase in kinetic energy makes the reactants move faster and collide more often, and it provides collisions with the energy of activation. Reactions go faster at higher temperatures. For every approximately 10°C increase in temperature, most reaction rates double. If we want food to cook faster, we raise the temperature. When body temperature rises, there is an increase in the pulse rate, rate of breathing, and metabolic rate. On the other hand, we slow down a reaction by lowering the temperature. We refrigerate perishable foods to make them last longer. In some cardiac surgeries, body temperature is lowered to 28°C so the heart can be stopped when less oxygen is required by the brain. This is also the reason that some people have survived submersion in icy lakes for long periods of time.

The rate of a reaction also increases when more reacting molecules are added. Then there are more collisions between the reactants and the reaction goes faster. For example, a patient having difficulty breathing may be given a breathing mixture with a higher oxygen content than the atmosphere. The increase in the number of oxygen molecules in the lungs increases the rate at which oxygen combines with hemoglobin. The increased rate of oxygenation of the blood means that the patient can breathe more easily.

$$\underset{\text{Hemoglobin}}{Hb} + \underset{\text{Oxygen}}{O_2} \longrightarrow \underset{\text{Oxyhemoglobin}}{HbO_2}$$

Another way to speed up a reaction is to lower the energy of activation. This can be done by adding a **catalyst,** which speeds up the reaction but is not itself changed or used up. (See Figure 5.11.) Suppose in our earlier trip, we

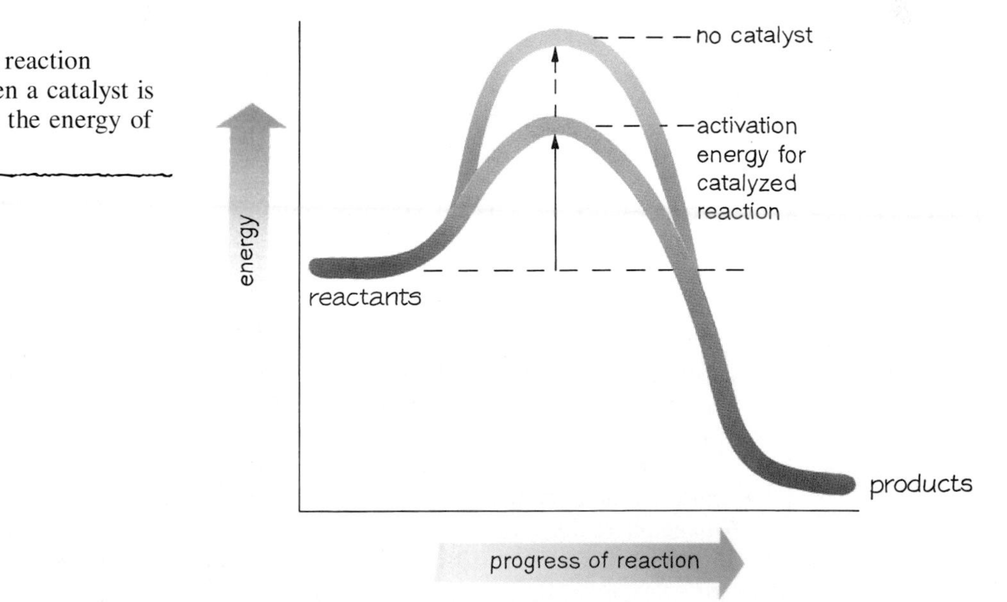

Figure 5.11
The rate of a reaction increases when a catalyst is used to lower the energy of activation.

found a tunnel through the hill. If we don't have to climb to the top, we don't need as much energy to get to the other side. A catalyst acts by providing an alternate pathway with a lower energy requirement. That makes it possible for more collisions to form product successfully. Catalysts have found many uses in industry. In the production of margarine, the reaction of hydrogen with vegetable oils is normally very slow. However, when finely divided platinum is added, the reaction occurs rapidly. In the body, biocatalysts called enzymes make almost all metabolic reactions go at the fast rates necessary for proper cellular activity.

A summary of the factors affecting reaction rate is given in Table 5.6.

Table 5.6 ***Factors that Increase Reaction Rate***

Factor	Reason
More reactants	More collisions
Higher temperature	More collisions, more collisions with energy of activation
Adding a catalyst	Lowers energy of activation

Sample Problem 5.15
Factors that Affect the Rate of Reaction

Indicate whether the following changes will increase, decrease, or have no effect upon the rate of reaction:

a. increase in temperature
b. decrease in the number of reactants
c. adding a catalyst

Solution

a. increase **b.** decrease **c.** increase

Study Check

How does the lowering of temperature affect the rate of reaction?

5.7 Mole Relationships in Chemical Equations

Learning Goal
From a balanced equation, write conversion factors for any of the mole relationships.

In the first part of this chapter, we saw that equations were balanced in terms of the numbers of atoms and molecules in a chemical reaction. However, when we do experiments in the laboratory, or prepare medications in the hospital, or follow a recipe for pancakes, we work with samples measured in grams that contain billions of atoms and molecules.

The law of conservation of matter tells us that the total amount of matter in the reactants must be equal to that of the products. Let us go back to our pancake recipe and see what this means. If we weighed all of the ingredients, we would find that their total mass was equal to the mass of the pancakes.

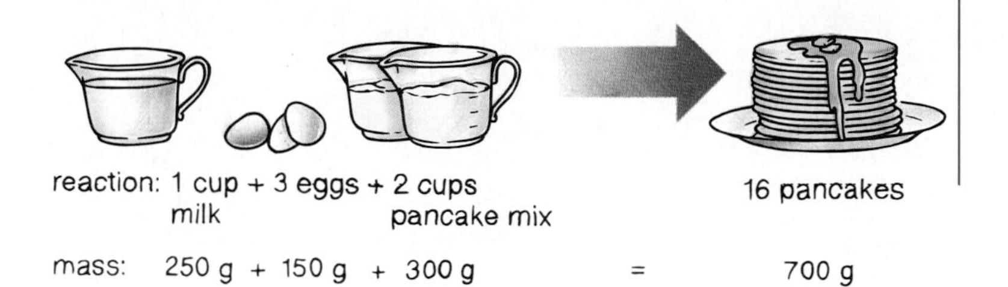

Figure 5.12
Mass is conserved in a recipe. The pancakes produced are equal in mass to the ingredients.

After those substances undergo a chemical reaction, their mass becomes part of the mass of the pancakes. The conservation of mass is illustrated in Figure 5.12.

1 cup milk	+ 3 eggs	+ 2 cups pancake mix	$\longrightarrow$	16 pancakes
250 g	+ 150 g	+ 300 g	=	700 g

Conservation of mass:

Mass of the reactants	= mass of the products
700 g	= 700 g

The total mass of the reactants and products in a chemical equation can be compared in a similar way. Consider the reaction in which carbon and oxygen react to give carbon dioxide.

$$C + O_2 \longrightarrow CO_2$$

Earlier we interpreted this equation as one carbon atom and one oxygen molecule react to form one carbon dioxide molecule. Now that we have learned the mole concept, we can use the coefficients to describe the number of moles of the reactants and products in the reaction. Let us look at the ways we can interpret a chemical equation:

	C	+	O_2	$\longrightarrow$	CO_2
Particles:	1 atom C		1 molecule O_2	form	1 molecule CO_2
	100 atoms C		100 molecules O_2	form	100 molecules CO_2
	6.02×10^{23} atoms C		6.02×10^{23} molecules O_2	form	6.02×10^{23} molecules CO_2
Moles:	1 mol C		1 mol O_2	form	1 mol CO_2

Conservation of Mass in an Equation

If there are the same number of atoms of each element in the reactants and products, then the total mass of the reactants and the products must also be equal. (See Figure 5.13.) Such a result is expected according to the law of conservation of matter. Using another equation, we convert the mole relationships to mass using the molar masses.

	$3H_2$	$+ N_2$	$\longrightarrow$	$2NH_3$
Moles:	3 mol H_2	+ 1 mol N_2	form	2 mol NH_3
Molar mass:	2.0 g/mol	28.0 g/mol		17.0 g/mol
Moles:	$\times$ 3 mol	$\times$ 1 mol		$\times$ 2 mol
Mass:	6.0 g H_2	28.0 g N_2		34.0 g NH_3
Total mass:		34.0 g reactants	=	34.0 g products

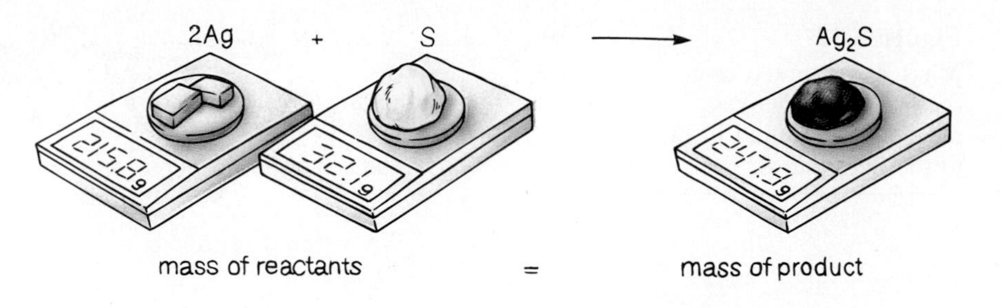

Figure 5.13
In the reaction of silver and sulfur to form silver sulfide (Ag_2S), mass is conserved: the mass of the reactants on the laboratory balance is equal to the mass of the product.

Sample Problem 5.16
Conservation of Mass

Demonstrate the law of conservation of mass by calculating the total mass of the reactants and the total mass of the products in the following equation:

$$CH_4 + 2O_2 \longrightarrow CO_2 + 2H_2O$$

Solution
Multiplying each molar mass by the number of moles (coefficients) in the equation gives the mass of the reactants and products.

	Reactants			*Products*	
	CH_4 +	$2O_2$	$\longrightarrow$	CO_2 +	$2H_2O$
Moles:	1 mole CH_4	2 mol O_2		1 mol CO_2	2 mol H_2O
Molar mass:	× 16.0 g/mol	× 32.0 g/mol		× 44.0 g/mol	× 18.0 g/mol
Mass:	16.0 g	+ 64.0 g		44.0 g	36.0 g
Total mass:	80.0 g reactants		=	80.0 g products	

Study Check
Show that mass is conserved in the following equation:

$$4NH_3 + 3O_2 \longrightarrow 2N_2 + 6H_2O$$

Mole Conversion Factors from Balanced Equations

Because an equation's coefficients indicate the number of moles, we can use the coefficients to write conversion factors for the mole relationship between any two substances in the equation. For example, we read the following equation as "4 mol Al and 3 mol O_2 react to form 2 mol Al_2O_3."

$$4Al + 3O_2 \longrightarrow 2Al_2O_3$$

By considering any two substances at a time, we can state the following mole–mole relationships:

4 mol Al react with 3 mol O_2
4 mol Al produce 2 mol Al_2O_3
3 mol O_2 produce 2 mol Al_2O_3

From these relationships, six ***mole–mole conversion factors*** can be written.

$$\text{Al and O}_2: \quad \frac{4 \text{ mol Al}}{3 \text{ mol O}_2} \quad \text{and} \quad \frac{3 \text{ mol O}_2}{4 \text{ mol Al}}$$

$$\text{Al and Al}_2\text{O}_3: \quad \frac{4 \text{ mol Al}}{2 \text{ mol Al}_2\text{O}_3} \quad \text{and} \quad \frac{2 \text{ mol Al}_2\text{O}_3}{4 \text{ mol Al}}$$

$$\text{O}_2 \text{ and Al}_2\text{O}_3: \quad \frac{3 \text{ mol O}_2}{2 \text{ mol Al}_2\text{O}_3} \quad \text{and} \quad \frac{2 \text{ mol Al}_2\text{O}_3}{3 \text{ mol O}_2}$$

Sample Problem 5.17
Writing Mole Conversion Factors

Consider the following balanced equation:

$$4\text{Na} + \text{O}_2 \longrightarrow 2\text{Na}_2\text{O}$$

Write the mole conversion factors for
a. Na and O_2 **b.** Na and Na_2O

Solution

a. $\dfrac{4 \text{ mol Na}}{1 \text{ mol O}_2}$ and $\dfrac{1 \text{ mol O}_2}{4 \text{ mol Na}}$

b. $\dfrac{4 \text{ mol Na}}{2 \text{ mol Na}_2\text{O}}$ and $\dfrac{2 \text{ mol Na}_2\text{O}}{4 \text{ mol Na}}$

Study Check
For the above equation, what are the mole conversion factors for O_2 and Na_2O?

5.8 Calculations Using Chemical Equations

Learning Goal
Given the quantity of a reactant or product in a balanced equation, calculate the quantity of another substance in the reaction.

Whether you are preparing a recipe, adjusting a carburetor for the proper mixture of fuel and air, or preparing medicines in a pharmaceutical laboratory, you need to know the proper amounts of reactants to use and how much product the reaction will produce. Using mole conversion factors, we can use the moles of one substance (A) to determine the moles of any other substance (B) in the equation.

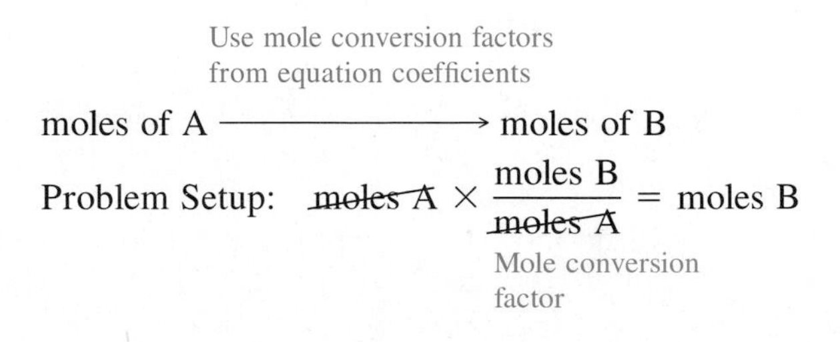

Sample Problem 5.18
Using Mole Conversion Factors

At high temperatures, iron(III) oxide reacts with carbon to produce iron and carbon monoxide.

$$Fe_2O_3 + 3C \longrightarrow 2Fe + 3CO$$

a. How many moles of C are required to react with 2.5 mol Fe_2O_3?
b. How many moles of Fe can be produced from 4.0 mol C?

Solution

a. In this problem, we need to convert the number of moles of Fe_2O_3 to the corresponding number of moles of C using a mole conversion factor.

$$\text{moles of } Fe_2O_3 \xrightarrow{\text{Mole conversion factor}} \text{moles of C}$$

The molar relationship from the equation coefficients gives the following possible conversion factors:

$$\frac{1 \text{ mol } Fe_2O_3}{3 \text{ mol C}} \quad \text{and} \quad \frac{3 \text{ mol C}}{1 \text{ mol } Fe_2O_3}$$

The problem is solved by multiplying the given, 2.5 mol Fe_2O_3, by the mole factor that cancels "moles Fe_2O_3."

$$\underset{\text{Given}}{2.5 \cancel{\text{mol } Fe_2O_3}} \times \underset{\text{Mole conversion factor}}{\frac{3 \text{ mol C}}{1 \cancel{\text{mol } Fe_2O_3}}} = \underset{\text{Answer}}{7.5 \text{ mol C}}$$

b. In this problem, we convert from "moles C" to "moles Fe" using their mole conversion factor.

$$\text{moles of C} \xrightarrow{\text{Mole conversion factor}} \text{moles of Fe}$$

The equation coefficients of Fe and C give two possible conversion factors.

$$\frac{2 \text{ mol Fe}}{3 \text{ mol C}} \quad \text{and} \quad \frac{3 \text{ mol C}}{2 \text{ mol Fe}}$$

The problem is set up by multiplying the given, 4.0 mol C, by the mole factor that cancels "moles C."

$$\underset{\text{Given}}{4.0 \cancel{\text{mol C}}} \times \underset{\text{Mole conversion factor}}{\frac{2 \text{ mol Fe}}{3 \cancel{\text{mol C}}}} = \underset{\text{Answer}}{2.7 \text{ mol Fe}}$$

Study Check

If the above reaction produces 0.600 mol CO, how many moles each of Fe_2O_3 and C reacted?

Calculations with Moles and Mass in Chemical Reactions

Many problems require us to determine the mass of a reactant or product. To do the calculation, we need to use the molar masses of the substances and mole factors from the equation coefficients in the following steps:

Steps in Calculating Moles and Grams of Substances in a Balanced Chemical Equation

Step 1 If the substance (A) is given in grams, change to moles (A) using the molar mass of A.

grams of A ⟶ moles of A

Step 2 Convert moles of the given substance (A) to moles of the desired substance (B) using a mole conversion factor derived from the equation coefficients.

moles of A ⟶ moles of B

Step 3 If necessary, change the moles (B) of the desired substance to grams using the molar mass (B).

moles of B ⟶ grams of B

This process of converting grams to moles of one substance and moles to grams of a related substance is the key in calculating the grams of any substance in a reaction.

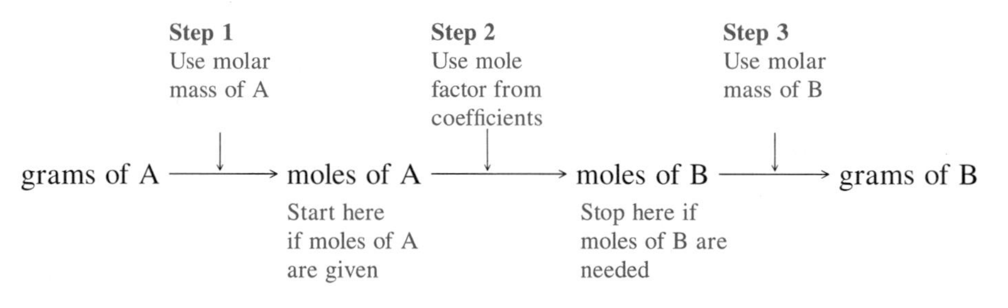

Sample Problem 5.19
Calculating the Mass of a Reactant or Product

Propane gas, C_3H_8, a fuel for camp stoves and some specially equipped automobiles, reacts with oxygen to produce carbon dioxide and water. The equation for this reaction is

$$C_3H_8(g) + 5O_2(g) \longrightarrow 3CO_2(g) + 4H_2O(g)$$

a. How many grams of C_3H_8 are required to react with 10.0 mol of O_2?
b. How many grams of H_2O are produced when 66.0 g of C_3H_8 react?

Solution

a. To solve this problem, we use two conversion factors: the mole factor that converts moles O_2 to moles C_3H_8, and the molar mass of C_3H_8 to change moles C_3H_8 to grams C_3H_8.

Moles O_2 → Mole C_3H_8 → Grams C_3H_8

$$10.0\ \cancel{\text{mol } O_2} \times \frac{1\ \cancel{\text{mol } C_3H_8}}{5\ \cancel{\text{mol } O_2}} \times \frac{44.0\text{ g } C_3H_8}{1\ \cancel{\text{mol } C_3H_8}} = 88.0\ C_3H_8$$

Given | Mole factor from coefficients | Molar mass C_3H_8

b. To solve this problem, we use three conversion factors. First, we change grams C_3H_8 to moles C_3H_8. Then, moles C_3H_8 are converted to moles of H_2O, and finally, moles H_2O are changed to grams H_2O.

$$66.0\ \cancel{\text{g } C_3H_8} \times \frac{1\ \cancel{\text{mol } C_3H_8}}{44.0\ \cancel{\text{g } C_3H_8}} \times \frac{4\ \cancel{\text{mol } H_2O}}{1\ \cancel{\text{mol } C_3H_8}} \times \frac{18.0\text{ g } H_2O}{1\ \cancel{\text{mol } H_2O}} = 108\text{ g } H_2O$$

Given | Molar mass C_3H_8 | Mole factor from coefficients | Molar mass H_2O

Study Check

Using the above equation, how many grams of oxygen reacted if 36.0 g of water are produced?

Chapter Summary

5.1 Formula Weight
The formula weight of a substance is calculated by multiplying the atomic mass of each element in the formula by its subscript and adding the results together.

5.2 The Mole
A mole of any element contains 6.02×10^{23} atoms; a mole of any substance contains 6.02×10^{23} molecules or formula units. The molar mass (g/mol) of any substance is the mass in grams equal numerically to its formula weight.

5.3 Calculations Using Molar Mass
The molar mass is useful as a conversion factor to change a given quantity in grams to moles, or a given number of moles to grams.

5.4 Chemical Changes
A chemical change occurs when the atoms of the initial substances rearrange to form new substances. When new substances form, a chemical reaction has taken place.

5.5 Chemical Equations
A chemical equation shows the formulas and amounts of reactants and products in a chemical reaction. A chemical equation is balanced to show the same number of atoms for each element on both sides of the equation.

5.6 Energy in Chemical Reactions
In a reaction, molecules (or atoms) must collide with energy equal to or greater than the energy of activation. The heat of reaction is the total energy difference between the energy of the reactants and the products. In exothermic reactions, the heat of reaction is the energy released. In endothermic reactions, the heat of reaction is the energy absorbed. The rate of a reaction (the speed at which products form) can be increased by adding more reacting molecules, raising the temperature of the reaction, or by adding a catalyst.

5.7 Mole Relationships in Equations
In a balanced equation, the number of atoms of each element on the reactant and product sides is equal. Mass is conserved: the mass of the reactants is equal to the mass of the products. Coefficients in an equation also describe the moles of reactants and products in the reaction. From the coefficients, mole–mole conversion factors can be written for the reacting relationship between any two substances in the equation.

5.8 Calculations Using Equations
When the number of moles is given for one substance in an equation, a mole–mole factor can be used to convert to moles of another substance, reactant, or product. When grams of one or both substances are given or desired, their molar masses are used to change grams to moles, or moles to grams.

Glossary of Key Terms

activation energy The energy needed upon collision to break apart the bonds of the colliding molecules.
Avogadro's number The number of items in a mole, equal to 6.02×10^{23}.
catalyst A substance that increases the rate of a reaction by lowering the energy of activation.
chemical change The formation of a new substance with a different composition and properties than the initial substance.
chemical equation A shorthand way to represent a chemical reaction using chemical formulas to indicate the reactants and products.
chemical reaction The process by which a chemical change takes place.
coefficients Whole numbers placed in front of the formulas in an equation to balance the number of atoms or moles of atoms of each element in an equation.
endothermic reaction A reaction that requires heat; the energy level of the products is higher than the reactants.
exothermic reaction A reaction that releases heat; the energy level of the products is lower than the reactants.
formula weight The sum of the atomic weights of all the atoms in a formula.
heat of reaction The energy released or absorbed during a chemical reaction equal to the energy difference between the reactants and products.
law of conservation of matter A law that states that atoms are neither created nor destroyed in a chemical reaction but are only rearranged.
molar mass The mass of one mole of an element or compound equal to the formula weight expressed in grams.
mole (mol) A group of atoms, molecules, or formula units that contains 6.02×10^{23} of these items.
mole conversion factor A conversion factor that relates the number of moles of two compounds in an equation derived from their coefficients.
physical change A change in which the physical properties change but not the chemical composition of the substance.
products The substances formed as a result of a chemical reaction.
rate of reaction The speed at which reactants are used up to form product.
reactants The initial substances that undergo change in a chemical reaction.

Chemistry at Home

1. Place 1/2 cup of vinegar in a glass. Add a teaspoon of baking soda and observe. What evidence of a chemical reaction do you see?
2. Obtain three matching glasses. Half fill one with hot water, one with room temperature water, and one with cold water. Add a few drops of food coloring or ink to each sample. Do not disturb. How does the temperature of the water affect the rate at which the color spread through the water?
3. Place 1 cup of water in a glass and measure the temperature of the water with a candy thermometer. Add a tablespoon of baking soda and stir. What happens to the temperature? Is the reaction exothermic or endothermic?
4. Freshly cut surfaces of fruits discolor when exposed to oxygen in the air. Cut slices of apple, potato, avocado, or banana. Wrap one slice in plastic wrap. Place another in the refrigerator. Dip one in lemon juice and leave out. Leave one unwrapped on the kitchen counter. What changes occur after 1–2 hours? What were the effects of wrapping, refrigerating, and dipping in lemon juice on the rate of reaction? Explain.

Answers to Study Checks

5.1 $C_{21}H_{39}N_7O_{12}$
5.2 294 amu
5.3 202.3 amu
5.4 492 g
5.5 24.4 g Au
5.6 62 g
5.7 4.57 mol Fe
5.8 0.00621 mol $CaCO_3$; 0.00550 mol $MgCO_3$
5.9 **a.** physical **b.** chemical **c.** physical
5.10 $3O_2(g) + CS_2(g) \longrightarrow CO_2(g) + 2SO_2(g)$
5.11 4 N atoms, 4 H atoms, 12 O atoms
5.12 $5C + 2SO_2 \longrightarrow CS_2 + 4CO$
5.13 $C_2H_5OH + 3O_2 \longrightarrow 2CO_2 + 3H_2O + 326$ kcal
5.14 higher
5.15 It decreases the number of collisions between reactants, and decreases the number of collisions that have energy of activation, thereby, decreasing the rate of the reaction.
5.16 mass of reactants (164.0 g) equals mass of products (164.0 g)
5.17 $\frac{2 \text{ mol } Na_2O}{1 \text{ mol } O_2}$ and $\frac{1 \text{ mol } O_2}{2 \text{ mol } Na_2O}$
5.18 0.200 mol Fe_2O_3; 0.600 mol C
5.19 80.0 g O_2 reacted

Problems

Formula Weight *(Goal 5.1)*

5.1 State the number of atoms of each element indicated by each of the following formulas:
a. $FeSO_4$ (iron supplement)
b. Al_2O_3 (absorbent and abrasive)
c. $Al(OH)_3$ (gastric antacid)
d. $(NH_4)_2CO_3$ (baking powder)
e. $C_7H_5NO_3S$ (saccharin, a noncaloric sweetener)
f. $C_{22}H_{25}ClN_2O_8$ (tetracycline, an antibiotic)

5.2 State the number of atoms of each element indicated by each of the following formulas:
a. Li_2CO_3 (antidepressant)
b. $Al_2(SO_4)_3$ (antiperspirant)
c. $Mg(OH)_2$ (antacid, laxative)
d. $(NH_4)_2SO_4$ (fertilizer)
e. $C_{14}H_{29}NO_4S$ (penicillin, an antibiotic)
f. $C_{20}H_{24}N_2O_2$ (quinine)

5.3 Calculate the formula weight of each compound in problem 5.1.

5.4 Calculate the formula weight of each compound in problem 5.2.

The Mole *(Goal 5.2)*

5.5 What is a mole?

5.6 What is Avogadro's number?

5.7 **a.** How many carbon atoms are in 1 mol of carbon?
b. How many iron atoms are in 1 mol of iron?
c. How many SO_2 molecules are in 1 mol of SO_2?
d. How many HCl molecules are in 1 mol of HCl?

5.8 **a.** How many silver atoms are in 1 mol of silver?
b. How many helium atoms are in 1 mol of helium?
c. How many molecules are in 1 mol of O_2 molecules?
d. How many formula units are in 1 mol of $NaNO_3$?

5.9 Which is heavier, 1 mol of sodium atoms or 1 mol of iron atoms? Why?

5.10 Which is heavier, 1 mol of O_2 molecules or 1 mol of O_3 molecules? Why?

5.11 What is the mass in grams of 1 mol of each of the following elements:
a. Na **b.** Cl **c.** C
d. Pb **e.** Cu

5.12 What is the mass of 1 mol of each of the following elements:
a. Fe **b.** Mg **c.** Ca
d. H **e.** S

5.13 What is the molar mass (in grams) of each of the following substances:
a. C_3H_6 (cyclopropane, an inhaled anesthetic)
b. $Al(OH)_3$ (an antacid)
c. NH_4NO_3 (a diuretic)
d. $C_{14}H_9Cl_5$ (DDT)
e. $C_{20}H_{18}O_4$ (cyclocumarol, an anticoagulant)

5.14 What is the molar mass (in grams) of each of the following substances:
a. C_2H_6O (ethanol)
b. $NaHCO_3$ (baking powder)
c. NH_4Br (sedative)
d. $C_{29}H_{50}O_2$ (vitamin E)
e. $(NH_4)_2B_4O_7$ (fire retardant)

Calculations Using Molar Mass *(Goal 5.3)*

5.15 Calculate the number of grams in each of the following:
a. 1.50 mol Na **b.** 2.80 mol Ca
c. 0.125 mol Sn **d.** 12.4 mol S
e. 0.0075 mol Cu

5.16 Calculate the number of grams in each of the following:
a. 1.4 mol K **b.** 10.0 mol Al
c. 2.35 mol C **d.** 0.855 mol P
e. 0.0018 mol Ag

5.17 Calculate the number of grams of each of the following:
a. 0.500 mol NH_3 **b.** 1.75 mol Na_2O
c. 0.115 mol $Ca(OH)_2$ **d.** 4.00 mol CO_2
e. 0.015 mol H_3PO_4

5.18 Calculate the number of grams of each of the following:
a. 20.0 mol NaCl **b.** 0.0110 mol $BaSO_4$
c. 3.2 mol $Ca(NO_3)_2$ **d.** 1.55 mol H_2SO_4
e. 0.500 mol $(NH_4)_2SO_4$

5.19 How many grams are needed to have 0.500 mol of each of the following?
a. K **b.** Cl_2 **c.** Na_2CO_3
d. He **e.** H_3PO_4

5.20 How many grams are needed to have 2.25 mol of each of the following?
a. Zn **b.** N_2 **c.** NaBr
d. H_2O **e.** C_6H_{14}

5.21 **a.** The compound $MgSO_4$ is called Epsom salts. How many grams will you need to prepare a bath with 5.0 mol of Epsom salts?
b. In a bottle of soda, there is 0.25 mol CO_2. How many grams of CO_2 are in the bottle?

5.22 **a.** Cyclopropane, C_3H_6, is an anesthetic given by inhalation. How many grams are in a 0.25-mol sample of cyclopropane?
b. The sedative Demerol has the formula $C_{15}H_{22}ClNO_2$. How many grams are present in 0.025 mol of Demerol?

5.23 How many moles are contained in each of the following?
a. 50.0 g Ag **b.** 40.0 g Cu
c. 0.200 g C **d.** 15.0 g NH_3
e. 75.0 g SO_2

5.24 How many moles are contained in each of the following?
a. 50.0 g Fe **b.** 25.0 g Ca
c. 5.00 g S **d.** 40.0 g H_2O
e. 100.0 g O_2

5.25 How many moles are in 25.0 g of each of the following compounds?
a. CO_2 **b.** NaOH **c.** $MgCl_2$
d. C_2H_6O **e.** NH_4Cl

5.26 How many moles are in 4.58 g of each of the following compounds?
a. NH_3 **b.** $Ca(NO_3)_2$ **c.** SO_3
d. H_2SO_4 **e.** $FeCl_3$

5.27 A can of Drano contains 480 g of NaOH. How many moles of NaOH are in the can of Drano?

5.28 A gold nugget weighs 35.0 g. How many moles of gold are in the nugget?

Chemical Changes *(Goal 5.4)*

5.29 Classify each of the following changes as chemical or physical:
a. chewing a gum drop
b. ignition of fuel in the space shuttle
c. drying clothes
d. neutralizing stomach acid with an antacid tablet
e. formation of snowflakes
f. an exploding dynamite stick

5.30 Classify each of the following changes as chemical or physical:
a. fogging the mirror during a shower
b. formation of green leaves in plants
c. breaking a bone
d. mending a broken bone
e. burning a candle
f. picking up the leaves in the yard

Chemical Equations *(Goal 5.5)*

5.31 Interpret the following equations in terms of atoms, molecules, or formula units:
a. $2NO + O_2 \longrightarrow 2NO_2$
b. $2H_2S + 3O_2 \longrightarrow 2SO_2 + 2H_2O$
c. $C_2H_6O + 3O_2 \longrightarrow 2CO_2 + 3H_2O$
Ethanol

5.32 Interpret the following equations in terms of atoms, molecules, or formula units:
a. $2SO_3 \longrightarrow 2SO_2 + O_2$
b. $2Fe_2O_3 + 6C + 3O_2 \longrightarrow 4Fe + 6CO_2$
c. $2Ag_2CO_3 \longrightarrow 4Ag + 2CO_2 + O_2$

5.33 Write an equation for the following reactions using correct formulas for the elements and compounds:
a. Four molecules of ammonia (NH_3) and three molecules of oxygen react to form two molecules nitrogen and six molecules of water.
b. Four iron atoms and three oxygen molecules react to form two formula units of iron(III) oxide.

5.34 Write an equation for the following reactions using correct formulas for the elements and compounds:
- **a.** Two molecules of sulfur dioxide react with oxygen to form two molecules of sulfur trioxide.
- **b.** Two molecules of propanol (C_3H_8O) and nine molecules of oxygen react to form six molecules of carbon dioxide and eight molecules of H_2O.

5.35 State the number of atoms of each element in the reactants and in the products for each of the following equations
- **a.** $2Na + Cl_2 \longrightarrow 2NaCl$
- **b.** $N_2H_4 + 2H_2O_2 \longrightarrow N_2 + 4H_2O$
- **c.** $P_4O_{10} + 6H_2O \longrightarrow 4H_3PO_4$
- **d.** $C_3H_8 + 5O_2 \longrightarrow 3CO_2 + 4H_2O$

5.36 State the number of atoms of each element in the reactants and in the products for each of the following equations:
- **a.** $2N_2 + 3O_2 \longrightarrow 2N_2O_3$
- **b.** $Al_2O_3 + 6HCl \longrightarrow 2AlCl_3 + 3H_2O$
- **c.** $C_5H_{12} + 8O_2 \longrightarrow 5CO_2 + 6H_2O$
- **d.** $6CO_2 + 6H_2O \longrightarrow \underset{\text{Glucose}}{C_6H_{12}O_6} + 6O_2$

5.37 Balance the following equations:
- **a.** $N_2 + O_2 \longrightarrow NO$
- **b.** $HgO \longrightarrow Hg + O_2$
- **c.** $Fe + O_2 \longrightarrow Fe_2O_3$
- **d.** $Na + Cl_2 \longrightarrow NaCl$
- **e.** $Cu_2O + O_2 \longrightarrow CuO$

5.38 Balance the following equations:
- **a.** $Al + Cl_2 \longrightarrow AlCl_3$
- **b.** $P_4 + O_2 \longrightarrow P_4O_{10}$
- **c.** $C_3H_8 + O_2 \longrightarrow CO_2 + H_2O$
- **d.** $Sb_2S_3 + HCl \longrightarrow SbCl_3 + H_2S$
- **e.** $Fe_2O_3 + C \longrightarrow Fe + CO$

5.39 Balance the following equations:
- **a.** $Mg + AgNO_3 \longrightarrow Mg(NO_3)_2 + Ag$
- **b.** $CuCO_3 \longrightarrow CuO + CO_2$
- **c.** $Al + CuSO_4 \longrightarrow Cu + Al_2(SO_4)_3$
- **d.** $Pb(NO_3)_2 + NaCl \longrightarrow PbCl_2 + NaNO_3$
- **e.** $Al + HCl \longrightarrow AlCl_3 + H_2$

5.40 Balance the following equations:
- **a.** $Zn + H_2SO_4 \longrightarrow ZnSO_4 + H_2$
- **b.** $Al + H_2SO_4 \longrightarrow Al_2(SO_4)_3 + H_2$
- **c.** $K_2SO_4 + BaCl_2 \longrightarrow BaSO_4 + KCl$
- **d.** $CaCO_3 \longrightarrow CaO + CO_2$
- **e.** $Al_2(SO_4)_3 + KOH \longrightarrow Al(OH)_3 + K_2SO_4$

Energy in Chemical Reactions *(Goal 5.6)*

5.41
- **a.** Why do chemical reactions require an energy of activation?
- **b.** What is the function of a catalyst?
- **c.** Is the energy of the products higher or lower than the reactants in an exothermic reaction?
- **d.** Draw an energy diagram for an exothermic reaction.

5.42
- **a.** What is measured by the heat of reaction?
- **b.** How does the heat of reaction differ in exothermic and endothermic reactions?
- **c.** Is the energy of the products higher or lower than the reactants in an endothermic reaction?
- **d.** Draw an energy diagram for an endothermic reaction.

5.43 Classify the following as exothermic or endothermic reactions:
- **a.** 125 kcal are released
- **b.** In the energy diagram, the energy level of the products is higher than the reactants.
- **c.** The metabolism of glucose in the body provides energy.

5.44 Classify the following as exothermic or endothermic reactions:
- **a.** In the energy diagram, the energy level of the products is lower than the reactants.
- **b.** In the body, the synthesis of proteins requires energy.
- **c.** 30 kcal are absorbed

5.45 Classify the following as exothermic or endothermic reactions:
- **a.** lighting your Bunsen burner in the laboratory

$$\underset{\text{Methane}}{CH_4(g)} + 2O_2(g) \longrightarrow CO_2(g) + H_2O(g) + 213 \text{ kcal}$$

- **b.** dehydrating limestone

$$Ca(OH)_2 + 15.6 \text{ kcal} \longrightarrow CaO + H_2O$$

- **c.** formation of aluminum oxide and iron from aluminum and iron(III) oxide

$$2Al + Fe_2O_3 \longrightarrow Al_2O_3 + 2Fe + 204 \text{ kcal}$$

5.46 Classify the following as exothermic or endothermic reactions:
- **a.** the combustion of propane

$$C_3H_8 + 5O_2 \longrightarrow 3CO_2 + 4H_2O + 531 \text{ kcal}$$

- **b.** the formation of salt

$$2Na + Cl_2 \longrightarrow 2NaCl + 196 \text{ kcal}$$

- **c.** decomposition of phosphorus pentachloride

$$PCl_5 + 16 \text{ kcal} \longrightarrow PCl_3 + Cl_2$$

5.47
- **a.** What is meant by the *rate of a reaction?*
- **b.** Why does bread grow mold more quickly at room temperature than in the refrigerator?

5.48 **a.** How does a catalyst affect the activation energy?
b. Why is pure oxygen used in respiratory distress?

5.49 How would each of the following change the rate of the reaction?

$$2SO_2 + O_2 \longrightarrow 2SO_3$$

a. adding SO_2
b. raising the temperature
c. adding a catalyst
d. removing some SO_2

5.50 How would each of the following change the rate of the reaction?

$$2NO(g) + 2H_2(g) \longrightarrow N_2(g) + 2H_2O(g)$$

a. adding more NO
b. lowering the temperature
c. removing some H_2
d. adding a catalyst

Mole Relationships in Chemical Equations *(Goal 5.7)*

5.51 Give an interpretation of the following equations in terms of mole relationships:
a. $2SO_2 + O_2 \longrightarrow 2SO_3$
b. $3C + 2SO_2 \longrightarrow CS_2 + 2CO_2$

5.52 Give an interpretation of the following equations in terms of mole relationships:
a. $N_2 + 3\ Cl_2 \longrightarrow 2\ NCl_3$
b. $4HCl + O_2 \longrightarrow 2Cl_2 + 2H_2O$

5.53 Demonstrate the law of conservation of mass by calculating the total mass of the reactants and the total mass of the products in each of the following balanced equations of problem 5.51.

5.54 Demonstrate the law of conservation of mass by calculating the total mass of the reactants and the total mass of the products in each of the following balanced equations of problem 5.52.

5.55 Write all of the mole factors for the equations listed in problem 5.51.

5.56 Write all of the mole factors for the equations listed in problem 5.52.

Calculations Using Equations *(Goal 5.8)*

5.57 Copper metal reacts with sulfur to form copper(I) sulfide by the following equation:

$$2Cu + S \longrightarrow Cu_2S$$

a. How many moles of S are needed to react with 2.0 mol Cu?
b. How many moles of Cu_2S can be produced from 5.0 mol S?
c. How many grams of Cu_2S can be produced when 2.5 mol Cu react?

5.58 Nitrogen gas reacts with hydrogen gas to produce ammonia by the following equation:

$$N_2(g) + 3H_2(g) \longrightarrow 2NH_3(g)$$

a. How many moles of H_2 are needed to react with 1.0 mol N_2?
b. How many grams of H_2 are needed to react with 2.5 mol N_2?
c. How many grams of NH_3 will be produced when 12 g of H_2 react?

5.59 Ammonia and oxygen react to form nitrogen and water.

$$\underset{\text{Ammonia}}{4NH_3} + 3O_2 \longrightarrow 2N_2 + 6H_2O$$

a. How many moles of O_2 are needed to react with 8.0 mol NH_3?
b. How many grams of N_2 will be produced when 6.00 mol NH_3 react?
c. How many grams of O_2 must react to produce 90.0 g H_2O?
d. How many grams of water are formed when 34 g of ammonia undergo reaction?

5.60 Iron(III) oxide reacts with carbon to give iron and carbon monoxide.

$$Fe_2O_3 + 3C \longrightarrow 2Fe + 3CO$$

a. How many moles of C are needed to react with 0.500 mol Fe_2O_3?
b. How many moles of Fe are produced when 36.0 g of C react?
c. How many grams of Fe_2O_3 must react to produce 112 g CO?
d. How many grams of CO are produced from 50.0 g Fe_2O_3?

Challenge Problems

5.61 You have collected 20.0 lb of aluminum cans. How many moles of aluminum do you have?

5.62 The human body contains about 60.0 percent water, H_2O. If a person weighs 70.0 kg, how many moles of water are present in that person's body?

5.63 An ethanol blood level of 400. mg of alcohol per 100 mL of blood can cause coma and possibly be fatal. How many moles of ethanol (C_2H_6O) are present in 100 mL of ethanol?

5.64 At the winery, glucose in grapes undergoes fermentation to produce ethanol and carbon dioxide by the following equation:

$$C_6H_{12}O_6 \longrightarrow 2C_2H_6O + 2CO_2$$

a. How many moles of CO_2 are produced when 50.0 g of glucose undergo fermentation?
b. How many grams of ethanol would be formed from the reaction of 0.240 kg of glucose?

5.65 Gasohol is a fuel containing ethyl alcohol (C_2H_6O) that burns in oxygen (O_2) to give carbon dioxide (CO_2) and water (H_2O).
a. State the reactants and products for this reaction in the form of a balanced equation.
b. How many moles of O_2 are needed to completely react with 4.0 mol ethyl alcohol?
c. If a car produces 88 g of CO_2, how many grams of O_2 are used up in the reaction?
d. How does the rate of reaction change at a higher temperature?

5.66 Balance the following equation:

$$Fe_3O_4 + CO \longrightarrow FeO + CO_2$$

a. How many grams of FeO are produced when 10.0 g of CO react?
b. How does the rate of reaction change if CO is added?
c. How does the rate of reaction change if a catalyst is added?

Chapter 6

Nuclear Radiation

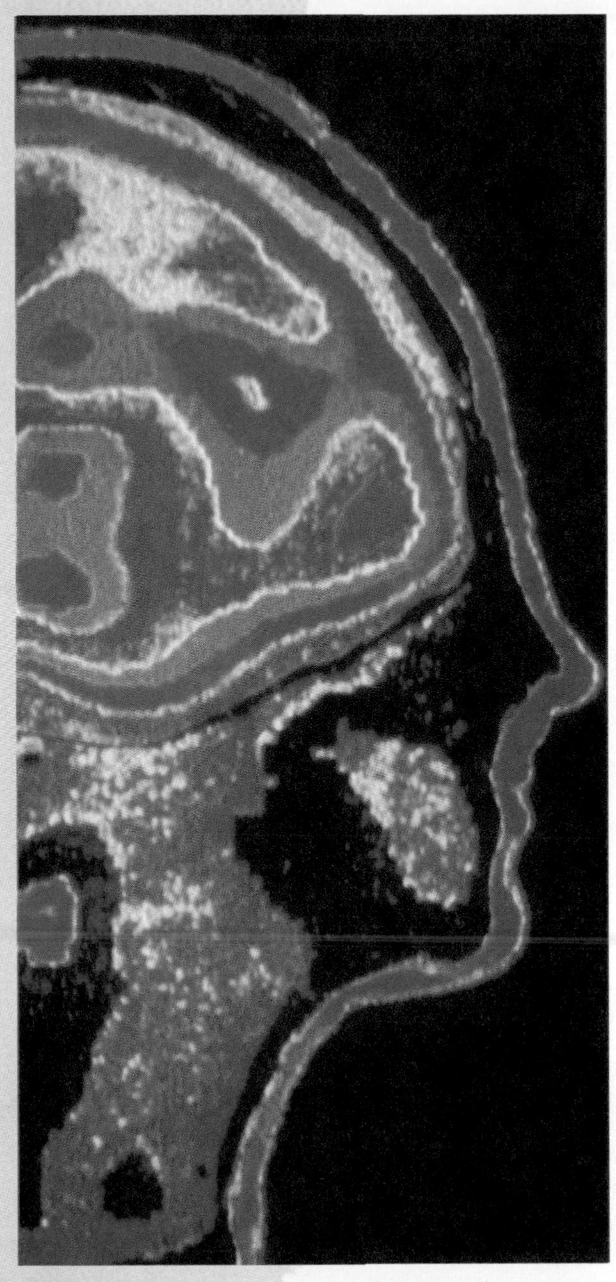

A PET scan is created by a scanning technique using radioisotopes that emit positrons.

Looking Ahead

Learning Goals

6.1 Describe alpha, beta, and gamma radiation.

6.2 Write an equation showing mass numbers and atomic numbers for radioactive decay.

6.3 Write an equation for the formation of a radioactive isotope.

6.4 Describe the detection and measurement of radiation.

6.5 Describe the use of radioisotopes in medicine.

6.6 Given a half-life, calculate the amount of radioisotope remaining after one or more half-lives.

6.7 Describe the processes of nuclear fission and fusion.

Questions to Think About

1. How can you prevent exposure to radiation if you work in a radiology laboratory?
2. How can radiation be both beneficial and harmful?
3. Why does the dental technician cover you with a lead shield when you have dental X rays?
4. What type of radiation makes a smoke alarm work?
5. Why do we have to be careful with nuclear by-products from hospitals and nuclear power plants?

In nuclear medicine, a radiologist uses radioactive isotopes to determine the size and shape of an organ, locate a tumor, or measure the metabolic activity of cells in the thyroid, brain, or liver. Radioactive isotopes are also used to measure the amounts of substances such as drugs and hormones in blood and urine samples.

The amount of radiation received by a patient is carefully monitored, because radiation can cause damage in the cells of the body. Medically, radiation is used to treat cancer and diseases of the blood or bone. The radiation penetrates the abnormal cells, inhibiting their growth. Nearby normal cells are also affected by the radiation, but they have a greater capacity to recover than do the abnormal cells.

6.1 Natural Radioactivity

Learning Goal

Describe alpha, beta, and gamma radiation.

Most naturally occurring isotopes of elements up to atomic number 19 have stable nuclei. Elements with higher atomic numbers (20 to 83) consist of a mixture of isotopes, some of which may have unstable nuclei. When the nucleus of an isotope is unstable, it is *radioactive,* which means that it will spontaneously emit energy to become more stable. This energy, called *radiation,* may take the form of particles such as alpha (α) particles or beta (β) particles or pure energy such as gamma (γ) rays. Elements with atomic numbers of 84 and higher consist only of radioactive isotopes. So many protons and neutrons are crowded together in their nuclei that the strong repulsions between the protons makes those nuclei unstable.

In Chapter 3, we wrote symbols to distinguish among the different isotopes of an element. Recall that an atom's mass number is equal to the sum of the protons and neutrons in the nucleus, and its atomic number is equal to the number of protons. In the symbol for an isotope, the mass number is written in the upper left corner of the symbol, and the atomic number is written in the lower left corner. For example, a radioactive isotope of iodine used in the diagnosis and treatment of thyroid conditions has a mass number of 131 and an atomic number of 53. We can write its symbol as

Mass number (protons and neutrons)
Element
Atomic number (protons)

$$^{131}_{53}\text{I}$$

This isotope is also called iodine-131 or I-131. (Note that radioactive isotopes are named by writing the mass number after the element's name or symbol.) When necessary, we can obtain the atomic number from the periodic table. Table 6.1 compares some stable, nonradioactive isotopes with some radioactive isotopes.

Table 6.1 ***Stable and Radioactive Isotopes of Some Elements***

Magnesium	Iodine	Uranium
Stable Isotopes		
$^{24}_{12}Mg$ Magnesium-24	$^{127}_{53}I$ Iodine-127	None
Radioactive Isotopes		
$^{23}_{12}Mg$ Magnesium-23	$^{125}_{53}I$ Iodine-125	$^{235}_{92}U$ Uranium-235
$^{27}_{12}Mg$ Magnesium-27	$^{131}_{53}I$ Iodine-131	$^{238}_{92}U$ Uranium-238

Types of Radiation

We cannot feel, taste, or smell radiation, but high-energy radiation is capable of creating havoc within the cells of our bodies. Different forms of radiation are emitted from an unstable nucleus when a change takes place among its protons and neutrons. Energy is released, and a new, more stable nucleus is formed. One type of radiation consists of alpha particles. An **alpha particle** contains two protons and two neutrons, which gives it a mass number of 4 and an atomic number of 2. Because it has two protons, an alpha particle has a charge of 2+. That makes it identical to a helium nucleus. In equations, it is written as the Greek letter alpha (α) or as the symbol for helium.

α or $^{4}_{2}He$

Alpha particle

Another type of radiation occurs when a radioisotope emits **beta particles.** A beta particle, which is a high-energy electron, has a charge of 1− and a mass number of 0. It is represented by the Greek letter beta (β) or by the symbol for the electron (e^-). When the symbol e^- is used for a beta particle, the subscript of −1 is used to show its charge.

β or $^{0}_{-1}e$

Beta particle

Beta particles are produced from unstable nuclei when neutrons change into protons. The high-energy electrons did not come from the orbitals of the atoms, and they do not exist in the nucleus until there is a transformation of neutrons within the nucleus.

$$^{1}_{0}n \longrightarrow {}^{1}_{1}H + {}^{0}_{-1}e$$

Neutron in the nucleus	New proton remains in the nucleus	Electron formed and emitted as a beta particle

Gamma rays are high-energy radiation similar to X rays released as an unstable nucleus undergoes a rearrangement to give a more stable, lower energy nucleus. A gamma ray is shown as the Greek letter gamma (γ). Because gamma rays are energy only, there is no mass or charge associated with their symbol.

γ

Gamma ray

Table 6.2 summarizes the types of radiation we will use in nuclear equations.

Table 6.2 ***Some Common Forms of Radiation***

Type of Radiation	Symbol	Mass Number	Atomic Number	Charge
Alpha particle	α, $^{4}_{2}He$	4	2	2+
Beta particle	β, $^{0}_{-1}e$	0	0	1−
Gamma ray, X ray	γ	0	0	0
Proton	$^{1}_{1}H$, $^{1}_{1}p$	1	1	1+
Neutron	$^{1}_{0}n$	1	0	0

Sample Problem 6.1
Writing Formulas for Radiation Particles

Write the nuclear symbol for an alpha particle.

Solution

The alpha particle contains two protons and two neutrons. It has a mass number of 4 and an atomic number of 2.

$$\alpha \quad \text{or} \quad ^{4}_{2}He$$

Study Check

What is the symbol used for beta radiation?

HEALTH NOTE *Biological Effects of Radiation*

When high-energy radiation strikes molecules in its path, electrons may be knocked away. The result of this *ionizing radiation* is the formation of unstable ions or radicals. A radical is a particle that has an unpaired electron. For example, when radiation passes through the human body, it may interact with water molecules, removing electrons and producing H_2O^+ ions or it may produce radicals:

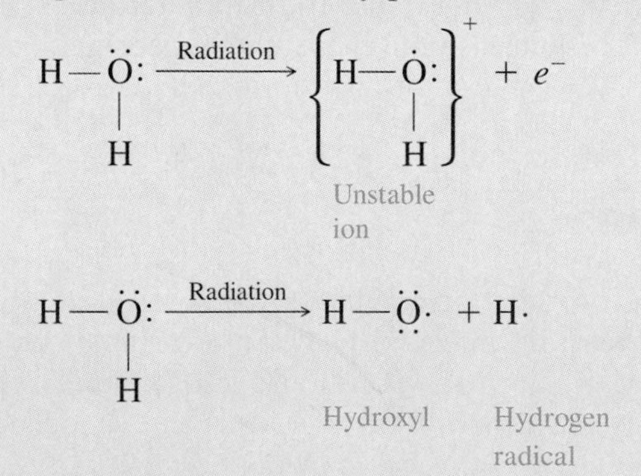

When ionizing radiation strikes the cells of the body, the unstable ions or radicals that form can cause undesirable chemical reactions. The cells in the body most sensitive to radiation are the ones undergoing rapid division—those of the bone marrow, skin, reproductive organs, and intestinal lining, as well as all cells of growing children. Damaged cells may lose their ability to produce necessary materials. For example, if radiation damages cells of the bone marrow, red blood cells may no longer be produced. If sperm cells or ova or the cells of a fetus are damaged, birth defects may result. In contrast, cells of the nerves, muscles, liver, and adult bones are much less sensitive to radiation because they undergo little or no cellular division.

Cancer cells are another example of rapidly dividing cells. Because cancer cells are highly sensitive to radiation, large doses of radiation are used to destroy them. The surrounding normal tissue, dividing at a slower rate, shows a greater resistance to radiation and suffers less damage. In addition, normal tissue is able to repair itself more readily than cancerous tissue. However, this repair is not always complete. Long-range effects of ionizing radiation include a shortened life span, malignant tumors, leukemia, anemia, and genetic mutations.

Radiation Protection

Because many cells in the body are sensitive to radiation, it is important that the radiologist, doctor, and nurse working with radioactive isotopes use proper radiation protection. Proper **shielding** is necessary to prevent exposure. Alpha particles are the heaviest of the radiation particles; they travel only a few centimeters in the air before they collide with air molecules, acquire electrons, and become helium atoms. A piece of paper, clothing, or the skin can be used as protection against alpha particles. Lab coats and gloves will provide sufficient shielding. However, if ingested or inhaled, alpha particles can bring about serious internal damage; because of their mass they cause much ionization in a short distance.

Beta particles have a very small mass and move much faster and farther than alpha particles, traveling as much as several meters through air. They can pass through paper and penetrate as far as 4–5 mm into body tissue. External exposure to beta particles can burn the surface of the skin, but they are stopped before they can reach the internal organs. Heavy clothing such as lab coats and gloves are needed to protect the skin from beta particles. (See Figure 6.1.)

Gamma rays travel great distances through the air and pass through many materials, including body tissues. Only very dense shielding, such as lead or concrete, will stop them. Because they can penetrate so deeply, exposure to gamma rays can be extremely hazardous. Even the syringe used to give an injection of a gamma-emitting radioisotope is placed inside a special lead-glass cover.

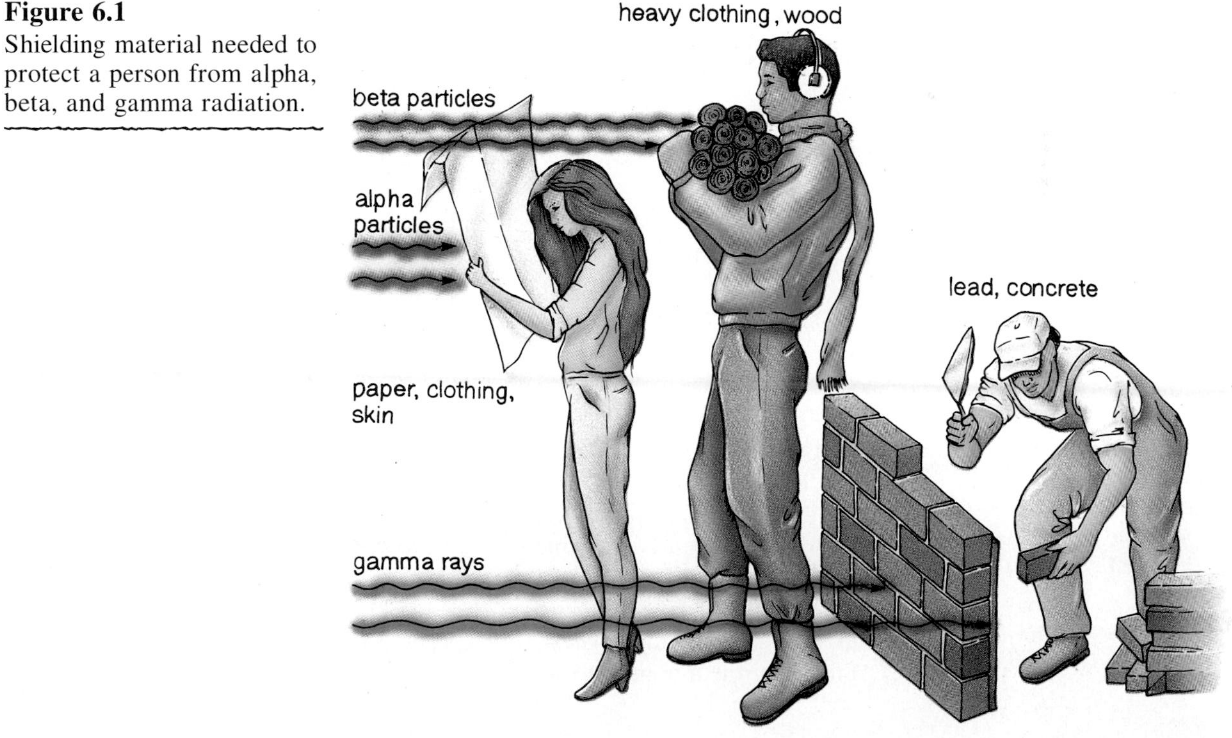

Figure 6.1
Shielding material needed to protect a person from alpha, beta, and gamma radiation.

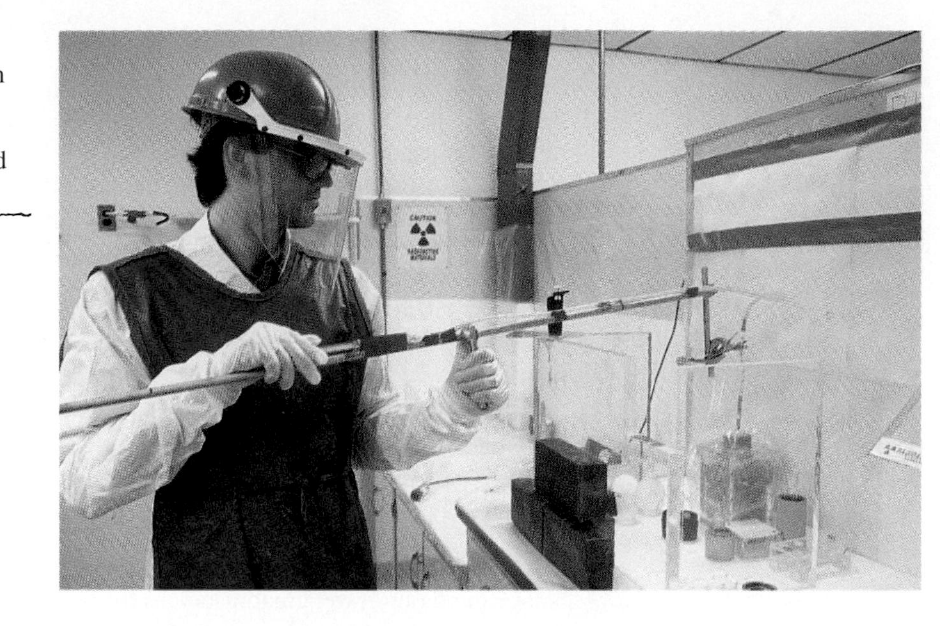

Figure 6.2
A technician in a radiation laboratory wears special protective clothing and stands behind a lead shield when using radioisotopes.

When preparing radioactive materials, the radiologist wears special gloves and works behind leaded windows. Long tongs are used within the work area to pick up vials of radioactive material, keeping them away from the hands and body. (See Figure 6.2.) Table 6.3 summarizes the shielding materials required for the various types of radiation.

Table 6.3 ***Types of Radiation and Shielding Required***

		Distance Particle Travels		
Type	**Symbol**	**Through Air**	**Into Tissue**	**Shielding**
Alpha	α	2–4 cm	0.05 mm	Paper, clothing
Beta	β	200–300 cm	4–5 mm	Heavy clothing, lab coats, gloves
Gamma	γ	500 m	50 cm	Lead, concrete

Try to keep the time you must spend in a radioactive area to a minimum. A certain amount of radiation is emitted every minute. Remaining in a radioactive area twice as long exposes a person to twice as much radiation.

Keep your distance! The greater the distance from the radioactive source, the lower the intensity of radiation received. If you double your distance from the radiation source, the intensity of radiation drops to $(\frac{1}{2})^2$ or one-fourth of its previous value, as Figure 6.3 shows. This is one of the reasons dentists and x-ray technicians leave the room and stand behind a shield or lead-lined wall to take your X rays. They are exposed to radiation every day and must minimize the amount of radiation they receive.

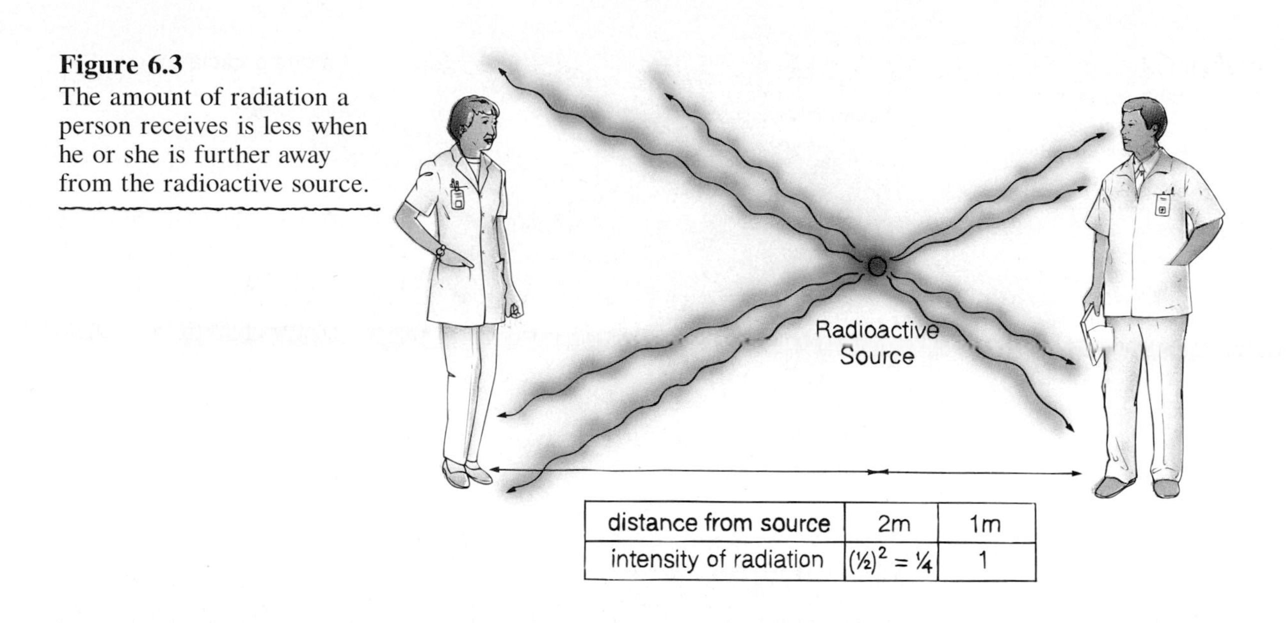

distance from source	2m	1m
intensity of radiation	$(\frac{1}{2})^2 = \frac{1}{4}$	1

Figure 6.3
The amount of radiation a person receives is less when he or she is further away from the radioactive source.

Sample Problem 6.2
Radiation Protection

How does the type of shielding for alpha radiation differ from that used for gamma radiation?

Solution

Alpha radiation is stopped by paper and clothing. However, lead or concrete is needed for protection from gamma radiation.

Study Check

Besides shielding, what other methods help reduce exposure to radiation?

6.2 Nuclear Equations

Learning Goal
Write an equation showing mass numbers and atomic numbers for radioactive decay.

When a nucleus spontaneously breaks down by emitting radiation, the process is called **radioactive decay.** It can be shown as a *nuclear equation* using the symbols for the radioactive nucleus, the new nucleus, and the radiation emitted.

$$\text{Radioactive nucleus} \longrightarrow \text{new nucleus} + \text{radiation } (\alpha, \beta, \gamma)$$

A nuclear equation is balanced when the sum of the mass numbers and the sum of the atomic numbers of the particles and atoms on one side of the equation are equal to their counterparts on the other side.

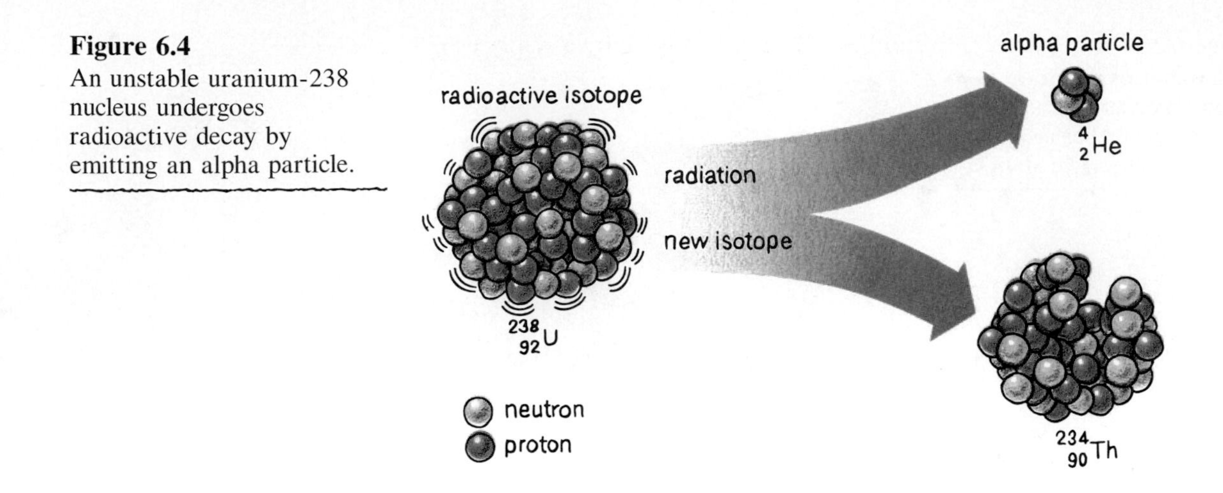

Figure 6.4
An unstable uranium-238 nucleus undergoes radioactive decay by emitting an alpha particle.

Alpha Emitters

Alpha emitters are radioisotopes that decay by emitting alpha particles. For example, uranium-238 decays to thorium-234 by emitting an alpha particle, as shown in Figure 6.4.

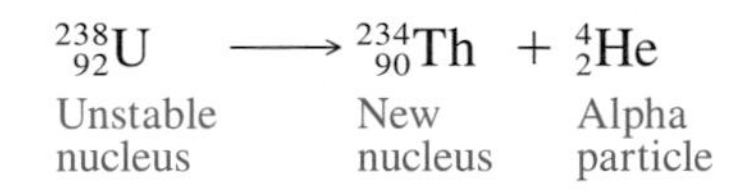

$$^{238}_{92}U \longrightarrow ^{234}_{90}Th + ^{4}_{2}He$$

Unstable nucleus — New nucleus — Alpha particle

The alpha particle emitted contains 2 protons, which gives the new nucleus 2 fewer protons, or 90 protons. That means that the new nucleus has an atomic number of 90 and is therefore thorium (Th). Because the alpha particle has a mass number of 4, the mass number of the thorium isotope is 234, 4 less than that of the original uranium nucleus.

Completing a Nuclear Equation

In another example of radioactive decay, radium-226 emits an alpha particle to form a new isotope whose mass number, atomic number, and identity we must determine. We would first write an incomplete nuclear equation:

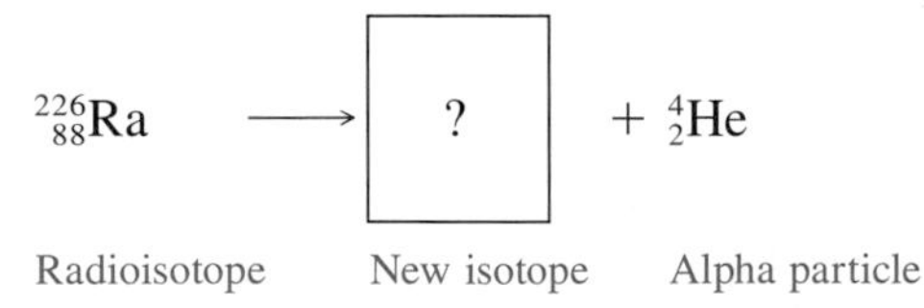

$$^{226}_{88}Ra \longrightarrow \boxed{?} + ^{4}_{2}He$$

Radioisotope — New isotope — Alpha particle

To complete the equation, we must make the mass number of the reactant equal the combined mass numbers of the products. Therefore, the sum of the mass number of the alpha particle and the mass number of the new isotope must equal 226, the mass number of radium. We can calculate the mass number of the new isotope by subtracting:

$226 - 4 = 222$ Mass number of new isotope

The atomic number of radium (88) must equal the sum of the atomic numbers of the alpha particle and the new isotope. Therefore, we obtain the atomic number of the new isotope by the following calculation:

$$88 - 2 = 86 \quad \text{Atomic number of new isotope}$$

In the periodic table, the element that has an atomic number of 86 is radon (Rn). We complete the nuclear equation by writing Rn, the symbol for radon:

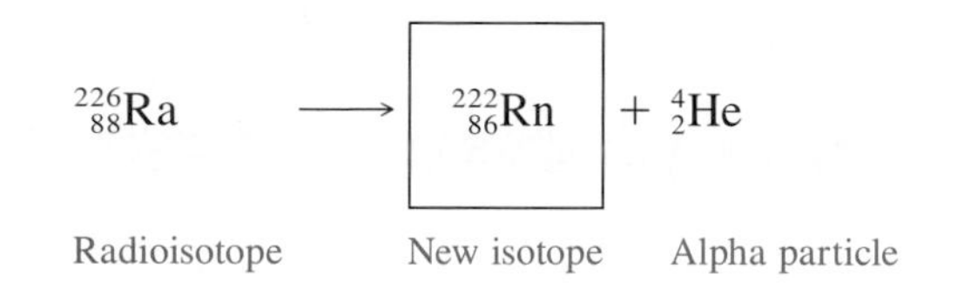

ENVIRONMENTAL NOTE *Radon in Our Homes*

The presence of radon has become a much publicized environmental and health issue because of radiation danger. Radioisotopes that produce radon, such as radium-226 and uranium-238, are naturally present in many types of rocks and soils. Radium-226 emits an alpha particle and is converted into radon gas, which diffuses out of the rocks and soil.

$$^{226}_{88}Ra \longrightarrow ^{222}_{86}Rn + ^{4}_{2}He$$

As uranium-238 decays, it also forms radium-226, which in turn produces radon. Uranium-238 has been found in particularly high levels in an area between Pennsylvania and New England.

Outdoors, radon gas poses little danger, because it dissipates in the air. However, if the source of radon is under a house or building, the gas can enter the house through cracks in the foundation or other openings. Then the radon is inhaled by those living or working there. Inside the lungs, radon emits alpha particles to form polonium-218, which is known to cause cancer when present in the lungs.

$$^{222}_{86}Rn \longrightarrow ^{218}_{84}Po + ^{4}_{2}He$$

Some researchers have estimated that 10% of all lung cancer deaths in the United States are due to radon. The Environmental Protection Agency (EPA) recommends that the maximum level of radon not exceed 4 picocuries (pCi) per liter of air in a home. One (1) picocurie (pCi) is equal to 10^{-12} curies (Ci). In California, 1% of all the houses surveyed exceeded the EPA's recommended maximum radon level.

Sample Problem 6.3
Writing an Equation for Alpha Decay

Smoke detectors, now required in many homes and apartments, contain an alpha emitter such as americium-241. The alpha particles emitted ionize air molecules, producing a constant stream of electrical current. However, when smoke particles enter the detector, they interfere with the formation of ions in the air, and the electric current is interrupted. This causes the alarm to sound and warns the occupants of the danger of fire. Complete the following nuclear equation for the decay of americium-241:

$$^{241}_{95}Am \longrightarrow \boxed{\;?\;} + ^{4}_{2}He$$

Solution

The mass number of the new nucleus is 237, obtained by subtracting the mass number of the alpha particle (4) from the mass number of americium (241).

Mass number of Am	−	mass number of alpha particle	=	mass number of new nucleus
241	−	4	=	237

The atomic number of the new nucleus is obtained by subtracting the atomic number of the alpha particle (2) from the atomic number of americium (95).

Atomic number of Am	−	Atomic number of alpha particle	=	Atomic number of new nucleus
95	−	2	=	93

The element whose atomic number is 93 is neptunium, Np. The completed nuclear equation for the reaction is written as follows:

$$^{241}_{95}\text{Am} \longrightarrow {}^{237}_{93}\text{Np} + {}^{4}_{2}\text{He}$$

Study Check

Write a balanced nuclear equation for the alpha emitter polonium-214.

Beta Emitters

A beta emitter is a radioisotope that decays by emitting beta particles. To form a beta particle, the unstable nucleus converts a neutron into a proton. The newly formed proton adds to the number of protons already in the nucleus and increases the atomic number by 1. However, the mass number of the newly formed nucleus stays the same. For example, carbon-14 decays by emitting a beta particle and forming a nitrogen isotope. (See Figure 6.5.)

In the nuclear equation of a beta emitter, the mass number of the radioisotope and the mass number of the new nucleus are the same, and the

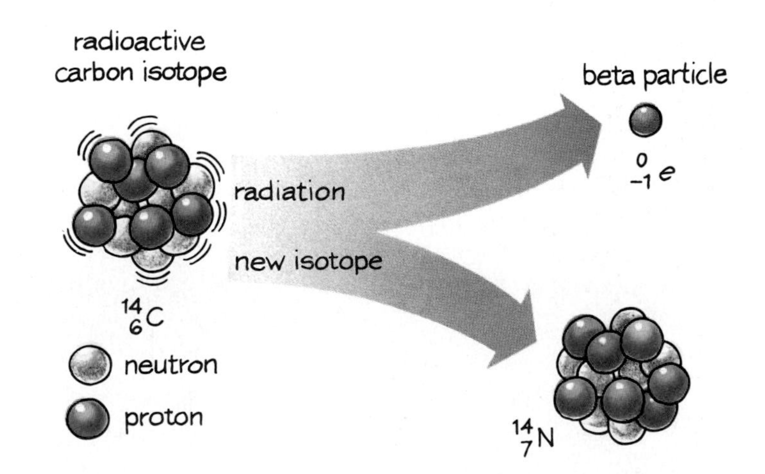

Figure 6.5
A carbon-14 nucleus emits beta radiation when a neutron changes to a proton and electron (beta particle).

atomic number of the new nucleus increases by 1, indicating a change of one element into another. This is the nuclear equation for the beta decay of carbon-14:

Mass number is the same for both nuclei

$$^{14}_{6}C \longrightarrow {}^{14}_{7}N + {}^{0}_{-1}e$$

Atomic number of the new nucleus increases by 1

HEALTH NOTE *Beta Emitters in Medicine*

The radioactive isotopes of several biologically important elements are beta emitters. When a radiologist wants to treat a malignancy within the body, a beta emitter may be used. The short range of penetration into the tissue by beta particles is advantageous for certain conditions. For example, some malignant tumors increase the fluid within the body tissues. A compound containing phosphorus-32, a beta emitter, is injected into the body cavity where the tumor is located. The beta particles travel only a few millimeters through the tissue, so only the malignancy and any tissue within that range are affected. The growth of the tumor is slowed or stopped, and the production of fluid decreases. Phosphorus-32 is also used to treat leukemia, polycythemia vera (excessive production of red blood cells), and lymphomas.

$$^{32}_{15}P \longrightarrow {}^{32}_{16}S + {}^{0}_{-1}e$$

Another beta emitter, iron-59, is used in blood tests to determine the level of iron in the blood, and the rate of production of red blood cells by the bone marrow.

$$^{59}_{26}Fe \longrightarrow {}^{59}_{27}Co + {}^{0}_{-1}e$$

Sample Problem 6.4
Writing an Equation for Beta Decay

Cobalt-60, a radioisotope used in the treatment of cancer, decays by emitting a beta particle. Write the nuclear equation for its decay.

Solution

The radioisotope is a cobalt isotope that has a mass number of 60. Looking at the periodic table, we find that the atomic number of cobalt is 27. The products of the nuclear decay are a beta particle and a new isotope:

$$^{60}_{27}Co \longrightarrow \boxed{\ ?\ } + {}^{0}_{-1}e$$

Because the mass number does not change in beta emission, we can assign a mass number of 60 to the new isotope. The atomic number of the new nucleus will be 1 more than that of cobalt, or 28.

$$\begin{aligned}\text{Atomic number of new isotope} &= \text{Atomic number of Co} + 1\\ 28 &= 27 + 1\end{aligned}$$

Because the element that has an atomic number of 28 is nickel (Ni), we can write the equation for the beta decay of the cobalt radioisotope as

$$^{60}_{27}\text{Co} \longrightarrow \, ^{60}_{28}\text{Ni} + \, ^{0}_{-1}e$$

Study Check
Iodine-131, a beta emitter, is used to check thyroid function and to treat hyperthyroidism. Write its nuclear equation.

Gamma Emitters

There are very few pure gamma emitters, although gamma radiation accompanies most alpha and beta radiation. In radiology, one of the most commonly used gamma emitters is technetium (Tc), a radioactive element that does not occur on Earth. The excited state called metastable technetium may be written as technetium-99m, Tc-99m, or ^{99m}Tc. By emitting energy in the form of gamma rays, the unstable nucleus becomes more stable. Figure 6.6 summarizes the changes in the nucleus for alpha, beta, and gamma radiation.

$$^{99m}_{43}\text{Tc} \longrightarrow \, ^{99}_{43}\text{Tc} + \gamma$$

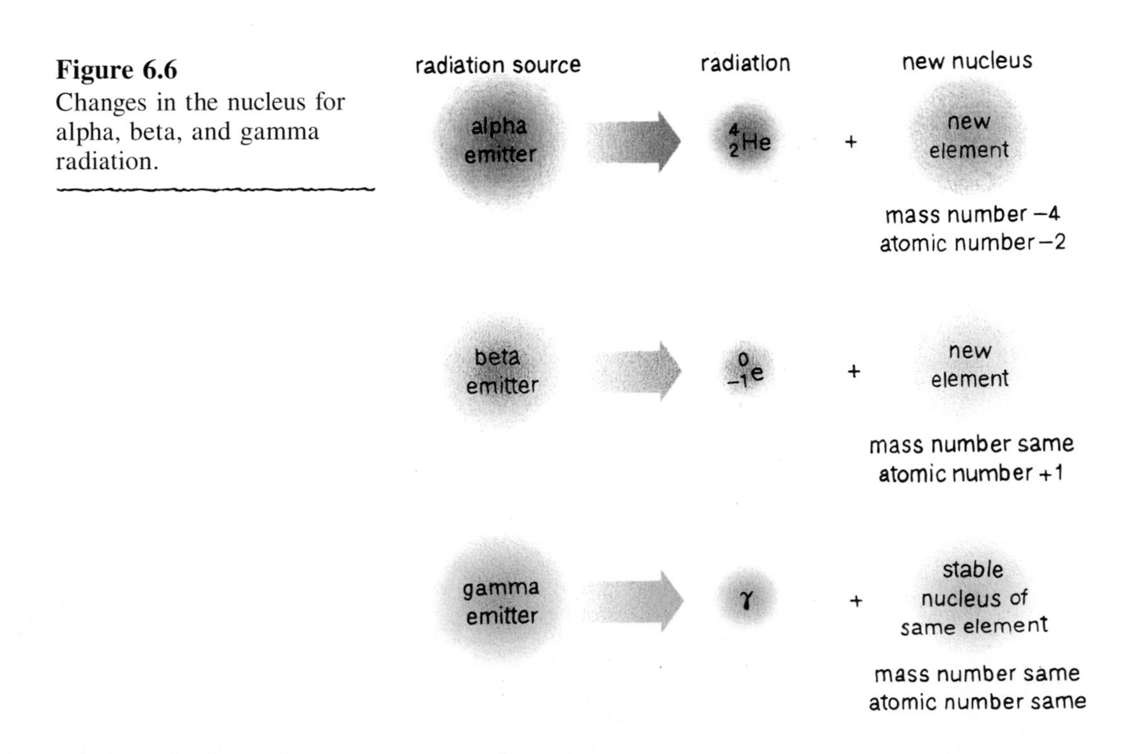

Figure 6.6
Changes in the nucleus for alpha, beta, and gamma radiation.

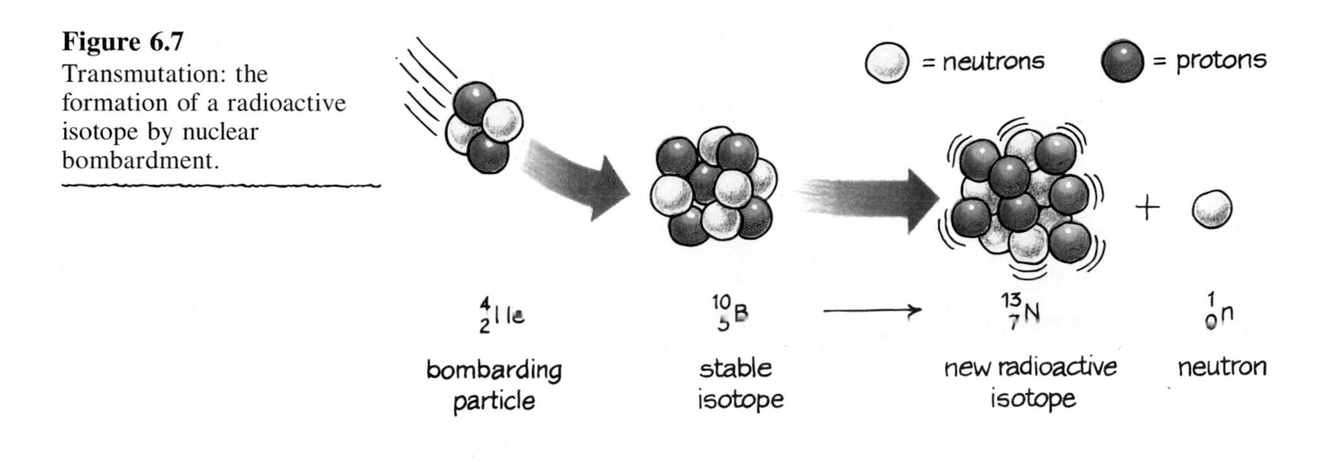

Figure 6.7
Transmutation: the formation of a radioactive isotope by nuclear bombardment.

6.3 Producing Radioactive Isotopes

Learning Goal
Write an equation for the formation of a radioactive isotope.

Today, more than 1500 radioisotopes are produced by converting stable, nonradioactive isotopes into radioactive ones. To do this, a stable atom is bombarded by fast-moving alpha particles, protons, or neutrons. When one of these particles is absorbed by a stable nucleus, the nucleus becomes unstable and the atom is now a radioactive isotope, or **radioisotope.**

When a nonradioactive isotope such as boron-10 is bombarded by an alpha particle, it is converted to nitrogen-13, a radioisotope, as shown in Figure 6.7. In this bombardment reaction, a neutron is emitted.

$$\underset{\text{Bombarding particle}}{{}^{4}_{2}\text{He}} + \underset{\text{Stable isotope}}{{}^{10}_{5}\text{B}} \longrightarrow \underset{\text{Radioactive isotope}}{{}^{13}_{7}\text{N}} + \underset{\text{Neutron}}{{}^{1}_{0}\text{n}}$$

The process of changing one element into another is called **transmutation.**

All of the known elements that have atomic numbers greater than 92 have been produced by bombardment; none of these elements occurs naturally. Most have been produced in only small amounts and exist for such a short time that it is difficult to study their properties. An example is element 105, joliotium, which is produced when californium-249 is bombarded with nitrogen-15.

$${}^{249}_{98}\text{Cf} + {}^{15}_{7}\text{N} \longrightarrow {}^{260}_{105}\text{Jl} + 4\,{}^{1}_{0}\text{n}$$

Technetium-99m is a radioisotope used in nuclear medicine for several diagnostic procedures, including the detection of brain tumors and the examination of the liver and spleen. The source of technetium-99m is molybdenum-99, which is produced in a nuclear reactor by neutron bombardment of molybdenum-98.

$${}^{98}_{42}\text{Mo} + {}^{1}_{0}\text{n} \longrightarrow {}^{99}_{42}\text{Mo}$$

Many radiology laboratories have a small generator containing the radioactive molybdenum-99, which decays to give the technetium-99m radioisotope. (See Figure 6.8.)

$${}^{99}_{42}\text{Mo} \longrightarrow {}^{99m}_{43}\text{Tc} + {}^{0}_{-1}e$$

Figure 6.8
Medical physicist unloading technetium-99m (Tc-99) source from a hospital generator.

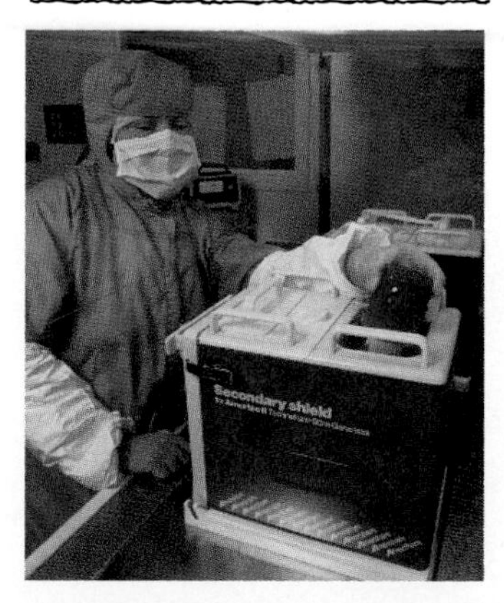

The technetium-99m radioisotope has a half-life of 6 hours and decays by emitting gamma rays. Gamma emission is most desirable for diagnostic work because the gamma rays pass through the body to the detection equipment.

$$^{99m}_{43}\text{Tc} \longrightarrow {}^{99}_{43}\text{Tc} + \gamma$$

Sample Problem 6.5
Writing Equations for Transmutations

Gallium-67 is used in the treatment of lymphomas. It is produced by the bombardment of zinc-66 by a proton. Write the equation for this nuclear bombardment.

Solution
The reactants in this nuclear equation are zinc-66 and a proton. The radioactive isotope formed is gallium-67.

$$^{66}_{30}\text{Zn} + {}^{1}_{1}\text{H} \longrightarrow {}^{67}_{31}\text{Ga}$$

Proton New radioisotope

Study Check
Write the equation for the bombardment of aluminum-27 by an alpha particle to produce the radioactive isotope phosphorus-30 and one neutron.

Sample Problem 6.6
Completing a Nuclear Equation for a Bombardment Reaction

Write the missing symbol in the following bombardment reaction:

$$^{58}_{28}\text{Ni} + {}^{1}_{1}\text{H} \longrightarrow \boxed{\;?\;} + {}^{4}_{2}\text{He}$$

Proton

Solution
The sum of the mass numbers for nickel and hydrogen is 59. Therefore, the mass number of the new isotope must be 59 minus 4, or 55. The sum of the atomic numbers is 29. The atomic number of the new isotope is 29 minus 2, or 27. The element that has atomic number 27 is cobalt (Co).

$$^{58}_{28}\text{Ni} + {}^{1}_{1}\text{H} \longrightarrow \boxed{^{55}_{27}\text{Co}} + {}^{4}_{2}\text{He}$$

Proton New isotope

Study Check
Complete the following bombardment equation:

$$\boxed{\quad} + {}^{4}_{2}\alpha \longrightarrow {}^{17}_{8}\text{O} + {}^{1}_{1}\text{H}$$

6.4 Radiation Detection and Measurement

Learning Goal
Describe the detection and measurement of radiation.

One of the most common instruments for detecting beta and gamma radiation is the Geiger counter, shown in Figure 6.9. It consists of a metal tube filled with a gas such as argon. Wires connect the metal tube to a battery. When radiation enters a window on the end of the tube, it produces ions in the gas; the ions produce an electrical current. Each burst of current is amplified to give a click and a readout on a meter.

$$\text{Ar} + \text{radiation} \longrightarrow \text{Ar}^{+} + e^{-}$$

Curie

When a radiology laboratory obtains a radioisotope, the activity of the sample is measured in terms of the number of disintegrations or nuclear transformations produced by the sample per second. The **curie (Ci)** is the unit used to express nuclear disintegration. One curie is equal to 3.7×10^{10} disintegrations per second (s), the number of atoms that decay in 1.0 g of radium in 1 s.

$$1 \text{ curie (Ci)} = 3.7 \times 10^{10} \text{ disintegrations/s}$$

The curie was named for Marie Curie, a Polish scientist, who along with Pierre Curie discovered the radioactive elements radium and polonium.

Rad

The **rad** (for radiation absorbed dose) is a unit that measures the amount of radiation absorbed by a gram of material such as body tissue. One rad is the absorption of 10^{-2} J of energy per kilogram of tissue. (Recall that joules and calories are units of energy, and that 1 cal = 4.18 J.)

$$1 \text{ rad} = 10^{-2} \text{ J/kg} = 2.4 \times 10^{-3} \text{ cal/kg}$$

Rem

The **rem** (short for radiation equivalent in humans) is a unit that measures the biological damage caused by the various kinds of radiation. The rem considers that the biological effects of alpha, beta, and gamma radiation on tissue

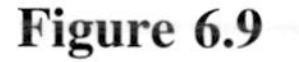

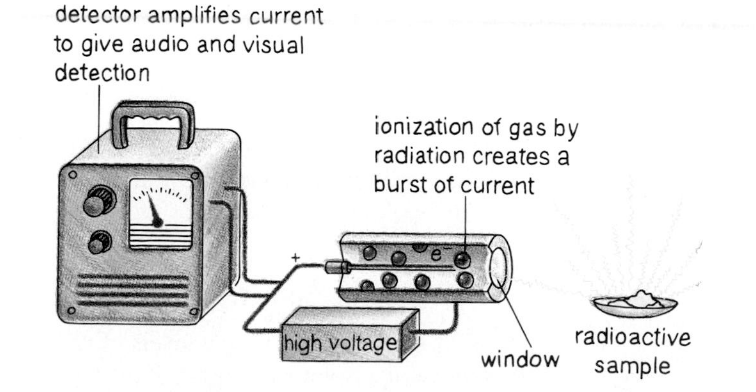
detector amplifies current to give audio and visual detection
ionization of gas by radiation creates a burst of current
e^-
+
high voltage
window
radioactive sample

Figure 6.9
A Geiger counter detects an electrical current produced when radiation ionizes a gas.

are not the same. If the heavy, positively charged alpha particles reach the tissues, they cause more ionization and therefore more damage than do beta particles or gamma rays. To determine the rem dose, the absorbed dose in rads is multiplied by a factor called RBE (for radiation biological effectiveness) that adjusts the biological damage that would result for a particular form of radiation.

Rem	=	rad	×	RBE
Effect of radiation in humans		Dose of radiation absorbed		Biological effect factor

For beta and gamma radiation, and X rays, the RBE is about 1, so the biological damage in rems is about equal to the absorbed radiation (rad). For high-energy protons and neutrons, the RBE factor is about 10, and for alpha particles, it is 20. Thus, the biological damage in rems for alpha particles would be the absorbed dose (rad) multiplied by 20. Table 6.4 summarizes some common units of radiation measurement.

Table 6.4 ***Some Units of Radiation Measurement***

Measurement	Unit	Meaning
Activity	curie (Ci)	3.7×10^{10} disintegrations/s
Absorbed dose	rad	10^{-2} J/kg
Biological damage in humans	rem	rad × RBE

Sample Problem 6.7
Measuring Activity

In a treatment for leukemia, phosphorus-32 which has an activity of 2 millicuries (mCi), is used. If phosphorus-32 is a beta emitter, how many beta particles are emitted in 1 s?

Solution
We can calculate the number of beta particles from a radioisotope's activity. Because 1 Ci is 3.7×10^{10} disintegrations/s, there must be 3.7×10^{10} beta particles produced in a second.

$$2\ \cancel{\text{mCi}} \times \frac{1\ \cancel{\text{Ci}}}{1000\ \cancel{\text{mCi}}} \times \frac{3.7 \times 10^{10}\ \beta\ \text{particles}}{\cancel{\text{s}}\ \cancel{\text{Ci}}} \times 1\ \cancel{\text{s}}$$

$$= 7.4 \times 10^{7}\ \beta\ \text{particles}$$

Study Check
An iodine-131 source has an activity of 0.25 Ci. How many radioactive atoms will disintegrate in 1 min?

Background Radiation

We are all exposed to low levels of radiation every day. Naturally occurring radioactive isotopes are part of the atoms of wood, brick, and concrete in our homes and the buildings where we work and go to school. This radioactivity, called background radiation, is present in the soil, in the food we eat, in the water we drink, and in the air we breathe. For example, one of the naturally occurring isotopes of potassium, potassium-40, is radioactive. It is found in the body because it is always present in any potassium-containing food. Other naturally occurring radioisotopes in air and food are carbon-14, radon-222, strontium-90, and iodine-131.

The atmosphere is another natural source of radiation. We are constantly exposed to radiation (cosmic rays) produced in space by the sun. At higher elevations, the amount of radiation from outer space is greater because there are fewer air molecules to absorb the radiation. People living at high altitudes or flying in an airplane receive more radiation from cosmic rays. For example, a person living in Denver receives about twice the cosmic radiation as a person living in Los Angeles.

You may also receive some radiation from nuclear testing, although such testing has decreased in recent years. If nuclear testing occurs, radioactive isotopes enter the atmosphere and are carried around the world by wind currents. They reach the earth during rainstorms and other weather phenomena. A person living close to a nuclear power plant normally does not receive much additional radiation, perhaps 0.1 millirem (mrem) in 1 year. (One rem equals 1000 mrem.) However, in the accident at the Chernobyl nuclear power plant in 1986, it is estimated that people in a nearby town received as much as 1 rem/hr.

In addition to naturally occurring radiation from construction materials in our homes, we receive radiation from television. In the medical clinic, dental and chest X rays also add to our radiation exposure. Table 6.5 lists some common sources of radiation. The average person in the United States receives about 0.170 rem or 170 mrem of radiation annually.

Table 6.5 ***Average Annual Radiation Received by a Person in the United States***

Source	Dose (mrem)
Natural	
The ground	15
Air, water, food	30
Cosmic rays	40
Wood, concrete, brick	50
Medical	
Chest X ray	50
Dental X ray	20
Upper gastrointestinal tract X ray	200
Other	
Television	2
Air travel	1
Radon[a]	200[a]
Cigarette smoking	35

[a] Varies widely.

HEALTH NOTE *Maximum Permissible Dose*

Any person working with radiation, such as a radiologist, a radiation technician, or a nurse whose patient has received a radioisotope, must wear some type of detection apparatus to measure exposure to radiation. A standard for occupational exposure is the maximum permissible dose (MPD). If this dose is not exceeded, the probability of injury is minimized. The detection apparatus is usually a badge, ring, or pin containing a small piece of photographic film. The window that covers the film absorbs light and beta rays, but gamma rays penetrate the covering and expose the film. The darker the film, the more exposure to gamma radiation the person has had. These badges are checked periodically to prevent exposure to more than the maximum permissible dose. The maximum permissible dose for occupational exposure is 5 rem (5000 mrem) per year.

Radiation Sickness

The larger the dose of radiation received at one time, the greater the effect on the body. Exposure to radiation under 25 rem usually cannot be detected. Whole-body exposure of 100 rem produces a temporary decrease in the number of white blood cells. If the exposure to radiation is 100 rem or higher, the person suffers the symptoms of radiation sickness: nausea, vomiting, fatigue, and a reduction in white cell count. A whole-body dosage greater than 300 rem can lower the white cell count to zero. The patient suffers diarrhea, hair loss, and infection.

Lethal Dose

Exposure to radiation of about 500 rem is expected to cause death in 50% of the people receiving that dose. This amount of radiation is called the lethal dose for one-half the population, or the LD_{50}. The LD_{50} varies for different life-forms, as Table 6.6 shows. Radiation dosages of about 600 rem would be fatal to all humans within a few weeks.

Table 6.6 ***Lethal Doses of Radiation for Some Life-Forms***

Life-Form	LD_{50} (rem)
Insect	100,000
Bacterium	50,000
Rat	800
Human	500
Dog	300

6.5 Medical Applications Using Radioactivity

Learning Goal
Describe the use of radioisotopes in medicine.

Suppose a radiologist wants to determine the condition of an organ in the body. How is this done? In some cases, the patient is given a radioisotope that is known to concentrate in that organ. The cells in the body cannot differentiate between a nonradioactive atom and a radioactive one. All atoms of an element, including any radioactive isotopes, have the same electron arrangement and the same chemistry in the body. The difference is that radioactive atoms can be detected because they emit radiation as they move to the same organs as the nonradioactive atoms of an element. Some radioisotopes used in nuclear medicine are listed in Table 6.7.

After a patient receives a radioisotope, the radiologist determines the level and location of radioactivity emitted by the radioisotope. An apparatus called a scanner is used to produce an image of the organ. The scanner moves slowly across the patient's body above the region where the organ containing

Table 6.7 ***Some Radioisotopes Used in Nuclear Medicine***

Element	Radioisotope	Medical Use
Chromium	^{51}Cr	Spleen imaging, blood volume
Technetium	^{99m}Tc	Brain, lung, liver, spleen, bone, and bone marrow scans
Gallium	^{67}Ga	Treatment of lymphomas
Phosphorus	^{32}P	Treatment of leukemia, polycythemia vera, and lymphomas; detection of brain and breast tumors
Sodium	^{24}Na	Study of vascular disease, extracellular volume, and blood volume determination
Strontium	^{85}Sr	Bone imaging for the diagnosis of bone damage and bone disease
Iodine	^{125}I	Thyroid imaging; study of plasma volume and fat absorption
Iodine	^{131}I	Study of thyroid; treatment of thyroid conditions such as hyperthyroidism

the radioisotope is located. The gamma rays emitted from the radioisotope in the organ can be used to expose a photographic plate, producing a **scan** of the organ. On a scan, an area of decreased or increased radiation can indicate such conditions as a disease of the organ, a tumor, a blood clot, or edema.

A common method of determining thyroid function is the use of radioactive iodine uptake (RAIU). Taken orally, the radioisotope iodine-131 mixes with the iodine already present in the thyroid. Twenty-four hours later, the amount of iodine taken up by the thyroid is determined. A detection tube held up to the area of the thyroid gland detects the radiation coming from the iodine-131 that has located there. (See Figure 6.10.)

Figure 6.10
A scanner is used to detect radiation from radioisotopes in the organs of the chest (left) and thyroid (right).

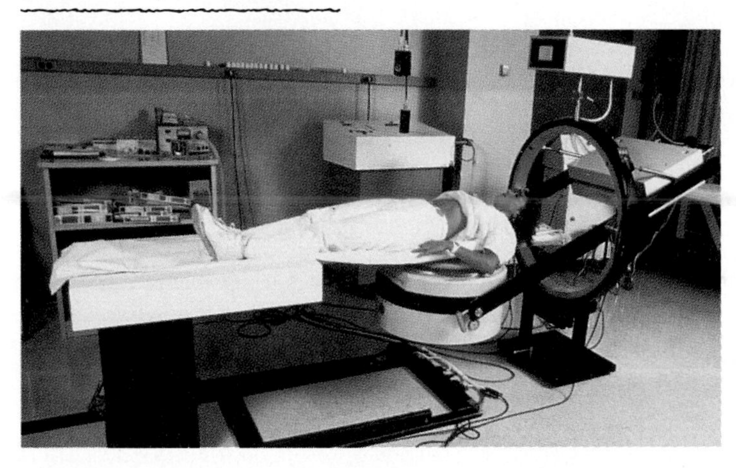

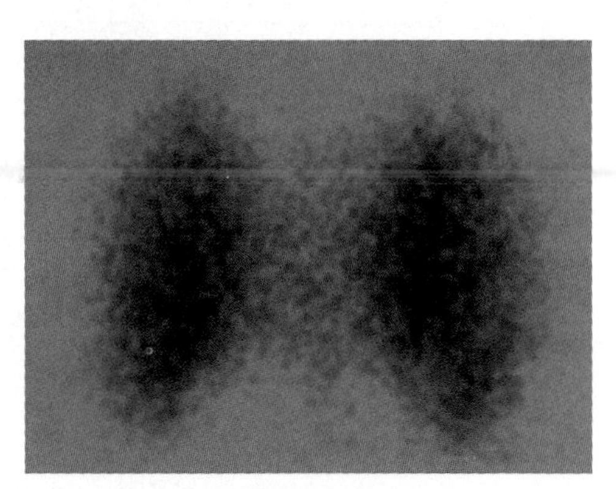

Figure 6.11
Brain scan showing brain with melanoma tumor.

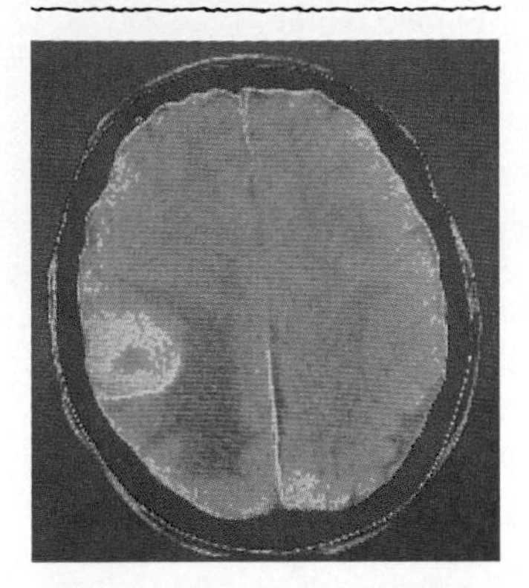

The iodine uptake is directly proportional to the activity of the thyroid. A patient with a hyperactive thyroid will have a higher than normal level of radioactive iodine, whereas a patient with a hypoactive thyroid will record low values.

If the patient has hyperthyroidism, treatment is begun to lower the activity of the thyroid. One treatment involves giving the patient a therapeutic dosage of radioactive iodine, which has a higher radiation count than the diagnostic dose. The radioactive iodine goes to the thyroid where its radiation destroys some of the thyroid cells. The thyroid produces less thyroid hormone, bringing the hyperthyroid condition under control. Normally, radioactive isotopes do not enter brain cells because there is a protective blood-brain barrier. However, if a blood vessel has broken, or if there is a brain tumor drawing on the blood supply, radioisotopic uptake does occur within the brain and appears on a brain scan. (See Figure 6.11.)

HEALTH NOTE *Radiation Doses in Diagnostic and Therapeutic Procedures*

We can compare the levels of radiation exposure commonly used during diagnostic and therapeutic procedures in nuclear medicine. In diagnostic procedures, the radiologist minimizes radiation damage by using only enough radioisotope to evaluate the condition of an organ or tissue. (See Table 6.8.)

The doses used in **radiation therapy** are much greater than those used for diagnostic procedures. For example, a therapeutic dose would be used to destroy the cells in a malignant tumor. Although there will be some damage to surrounding tissue, the healthy cells are more resistant to radiation and can repair themselves. (See Table 6.9.)

Table 6.8 *Radiation Doses Used for Diagnostic Procedures*

Organ	Dose (rem)
Liver	0.3
Thyroid	50.0
Lung	2.0

Table 6.9 *Radiation Doses Used for Therapeutic Procedures*

Condition	Dose (rem)
Lymphoma	4500
Skin cancer	5000–6000
Lung cancer	6000
Brain tumor	6000–7000

Positron Emission Tomography (PET)

Radioisotopes that emit a positron are used in an imaging method called positron emission tomography (PET). A positron is a particle emitted from the nucleus that has the same mass as an electron but has a positive charge.

$$\beta^{+} \quad \text{or} \quad {}^{0}_{+1}e$$

Positron

Carbon-11 is an example of a radioisotope that emits a positron when it decays.

$${}^{11}_{6}\text{C} \longrightarrow {}^{0}_{+1}e + {}^{11}_{5}\text{B}$$

Positron

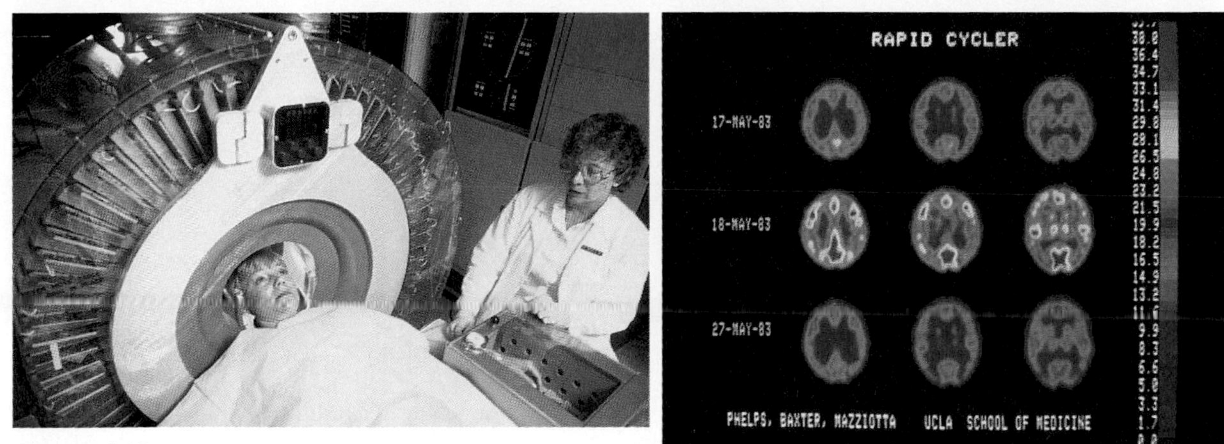

Figure 6.12
A PET scan of the brain.

A positron is produced when a proton changes to a neutron.

$${}^{1}_{1}\text{H} \longrightarrow {}^{0}_{+1}e + {}^{1}_{0}\text{n}$$

The new nucleus retains the same mass number but has a lower atomic number. The positron exists for only a moment before it collides with an electron. Some of a positron's mass is converted into bursts of gamma energy, which are detected by:

$$\underset{\text{Positron}}{{}^{0}_{+1}e} + \underset{\text{Electron in an atom}}{{}^{0}_{-1}e} \longrightarrow \underset{\text{Gamma rays produced}}{2\,\gamma}$$

Medically, positron emitters such as carbon-11, oxygen-15, and nitrogen-13 are used to diagnose conditions involving blood flow, metabolism, and, particularly, the functioning of the brain. Glucose labeled with carbon-11 is used to detect damage in the brain from epilepsy, stroke, and Parkinson's disease. After the radioisotope is injected, the gamma rays from the positrons emitted are detected by computerized equipment to create a three-dimensional image of the organ. (See Figure 6.12.)

HEALTH NOTE *Other Imaging Methods*

Computed Tomography (CT)

Another imaging method used to detect changes within the body is computed tomography (CT). A computer monitors the degree of absorption of 30,000 x-ray beams directed at the brain at successive layers. The differences in absorption based upon the densities of the tissues and fluids in the brain provide a series of images of the brain. This technique is successful in the identification of brain hemorrhages, tumors, and atrophy.

Magnetic Resonance Imaging (MRI)

In another imaging technique called magnetic resonance imaging (MRI), no radiation is used. The atomic nuclei in atoms such as hydrogen emit small amounts of energy when they are excited by a strong magnetic field. Such low-energy changes are harmless to the patient, but they can be detected to create an image on a television monitor. (See Figure 6.13.) Magnetic resonance imaging is used in the detection of multiple sclerosis, abnormalities of the spine and brain, tumors, and birth malformations. In some cases, MRI is replacing the use of X rays and other scanning techniques, including the CT scanner.

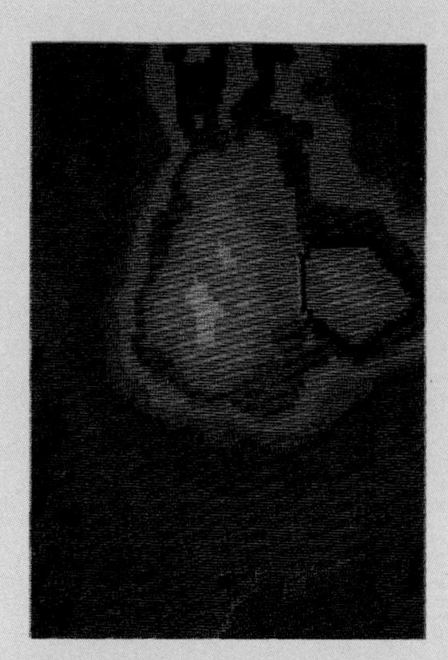

Figure 6.13
An MRI scan of the heart.

6.6 Half-Life of a Radioisotope

Learning Goal

Given a half-life, calculate the amount of radioisotope remaining after one or more half-lives.

The **half-life** of a radioisotope is the time it takes for one-half of a sample to decay. For example, iodine-131, a radioisotope used in diagnosis and treatment of thyroid disorders, has a half-life of 8 days. If we begin with a sample containing 20 grams of iodine-131, there would be 10 g remaining after 8 days. In another 8 days, we would have 5 g of iodine-131 left. Thus, in each half-life of 8 days, one-half of the radioactive sample has disintegrated.

$$20\text{ g }{}^{131}\text{I} \xrightarrow[10\text{ g }{}^{131}\text{I decay}]{8\text{ days}} 10\text{ g }{}^{131}\text{I} \xrightarrow[5\text{ g }{}^{131}\text{I decay}]{8\text{ days}} 5\text{ g }{}^{131}\text{I}$$

This information can be summarized as follows:

Time elapsed	0	8 days	16 days	24 days
Number of half-lives	0	1	2	3
Quantity of iodine-131	20 g	10 g	5 g	2.5 g

A decay curve is a diagram of the decay of a radioactive isotope. Figure 6.14 shows such a curve for the iodine-131 we have discussed.

Naturally occurring isotopes of the elements usually have long half-lives, as shown in Table 6.10. They disintegrate slowly and produce radiation over

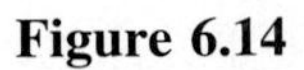

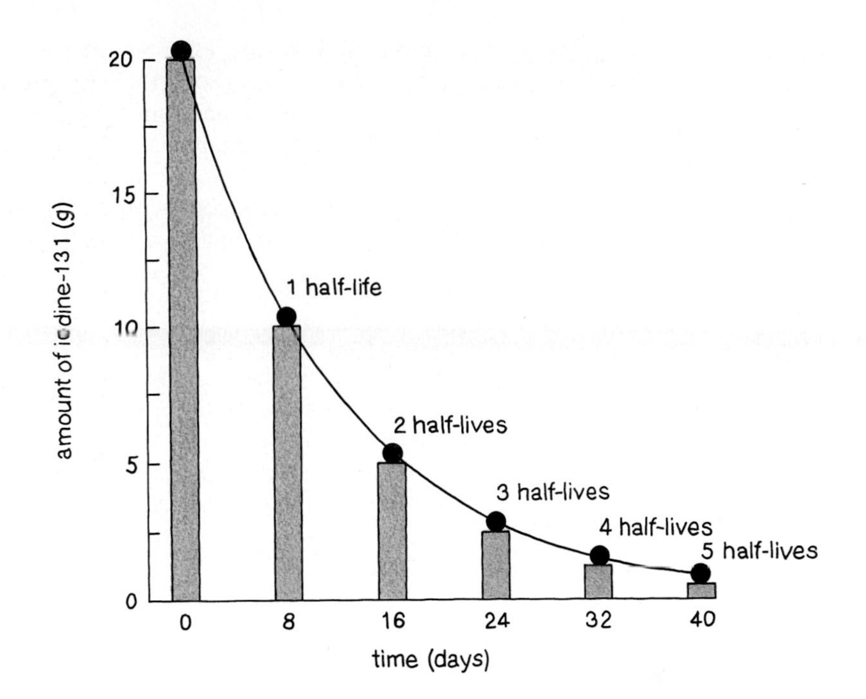

Figure 6.14
Decay curve for iodine-131. One-half of the sample decays with each half-life. Iodine-131 has a half-life of 8 days.

Table 6.10 ***Half-Lives of Some Radioisotopes***

Element	Radioisotope	Half-Life	Type of Radiation
Naturally Occurring Radioisotopes			
Carbon	^{14}C	5730 yr	β
Potassium	^{40}K	1.3×10^9 yr	β, γ
Radium	^{226}Ra	1600 yr	α, γ
Uranium	^{238}U	4.5×10^9 yr	α, γ
Some Medical Radioisotopes			
Carbon	^{11}C	20 min	β^+
Chromium	^{51}Cr	28 days	γ
Iodine	^{131}I	8 days	β, γ
Iodine	^{125}I	60 days	γ
Iron	^{59}Fe	46 days	β, γ
Phosphorus	^{32}P	14 days	β
Oxygen	^{15}O	2 min	β^+
Potassium	^{42}K	12 hr	β, γ
Sodium	^{24}Na	15 hr	β, γ
Strontium	^{85}Sr	64 days	γ
Technetium	^{99m}Tc	6.0 hr	γ

a long period of time, even hundreds or millions of years. In contrast, many of the radioisotopes used in nuclear medicine have much shorter half-lives. They disintegrate rapidly and produce almost all their radiation in a short period of time. For example, technetium-99m emits half of its radiation in the first 6 hr. This means that a small amount of the radioisotope given to a patient is essentially gone within 2 days. The decay products of technetium-99m are totally eliminated by the body.

Sample Problem 6.8
Calculating Quantity of Radioisotopes

Nitrogen-13, which has a half-life of 10 min, is used to image organs in the body. For the diagnostic procedure, the patient receives an injection of a compound containing the radioisotope. Originally, the nitrogen-13 has an activity of 40 microcuries (μCi). If the procedure requires 30 min, what is the remaining activity of the radioisotope?

Solution

$$\text{Number of half-lives} = 30\,\cancel{\text{min}} \times \frac{1\text{ half-life}}{10\,\cancel{\text{min}}}$$

$$= 3$$

The activity of the radioisotope after 3 half-lives is

$$40\ \mu\text{Ci} \xrightarrow{10\text{ min}} 20\ \mu\text{CI} \xrightarrow{10\text{ min}} 10\ \mu\text{Ci} \xrightarrow{10\text{ min}} 5\ \mu\text{Ci}$$

Another way to calculate the activity of radioactive nitrogen-13 left in the sample is to construct a chart to show the number of half-lives, elapsed time, and the amount of radioactive isotope that is left in the sample.

Time elapsed	0	10 min	20 min	30 min
Number of half-lives elapsed	0	1	2	3
Activity of N-13 remaining	40 μCi	20 μCi	10 μCi	5 μCi

Study Check
Iron-59, used in the determination of bone marrow function, has a half-life of 46 days. If the laboratory receives a sample of 8.0 g of iron-59, how many grams are still active after 184 days?

ENVIRONMENTAL NOTE *Dating Ancient Objects*

A technique known as radiological dating is used by geologists, archaeologists, and historians as a way to determine the age of ancient objects. The age of an object derived from plants or animals (such as wood, fiber, natural pigments, bone, and cotton or woolen clothing) is determined by measuring the amount of carbon-14, a naturally occurring radioactive form of carbon. In 1960, Willard Libby received the Nobel Prize for the work he did developing carbon-14 dating techniques during the

1940s. Carbon-14 is produced in the upper atmosphere by the bombardment of $^{14}_{7}N$ by high-energy neutrons from cosmic rays.

$$^{1}_{0}n + ^{14}_{7}N \longrightarrow ^{14}_{6}C + ^{1}_{1}H$$

Neutron from cosmic rays; Nitrogen in atmosphere; Radioactive carbon-14; Proton

The carbon-14 reacts with oxygen to form radioactive carbon dioxide, $^{14}CO_2$. Because carbon dioxide is continuously absorbed by living plants during the process of photosynthesis, some carbon-14 will be taken into the plant. After the plant dies, no more carbon-14 is taken up, and the amount of carbon-14 contained in the plant decreases as it undergoes radioactive decay emitting β particles.

$$^{14}_{6}C \longrightarrow ^{14}_{7}N + ^{0}_{-1}e$$

Scientists use the half-life of carbon-14 (5730 years) to calculate the amount of time that has passed since the plant died, a process called **carbon dating.** The smaller the amount of carbon-14 remaining in the sample, the greater the number of half-lives that have passed. Thus, the approximate age of the sample can be determined. For example, a wooden beam found in an ancient Indian dwelling might have one-half of the carbon-14 found in living plants today. Thus, the dwelling was probably constructed about 5730 years ago, one half-life of carbon-14. (See Figure 6.15.) This technique is useful for dating samples that have ages up to 30,000 years. For older objects, carbon dating is not reliable.

A radiological dating method used for determining the age of rocks is based on the radioisotope uranium-238, which decays through a series of reactions to lead-206. The uranium-238 isotope has a very long half-life, about 4×10^9 (4 billion) years. Measurements of the amounts of uranium-238 and lead-206 enable geologists to determine the age of rock samples. The older rocks will have a higher percentage of lead-206 because more of the uranium-238 has decayed. The age of rocks brought back from the moon by the *Apollo* missions was determined using uranium-238. They were found to be about 4×10^9 years old, approximately the same age calculated for Earth.

Figure 6.15
The age of an ancient boat or wood from a prehistoric village can be determined by carbon-13 dating.

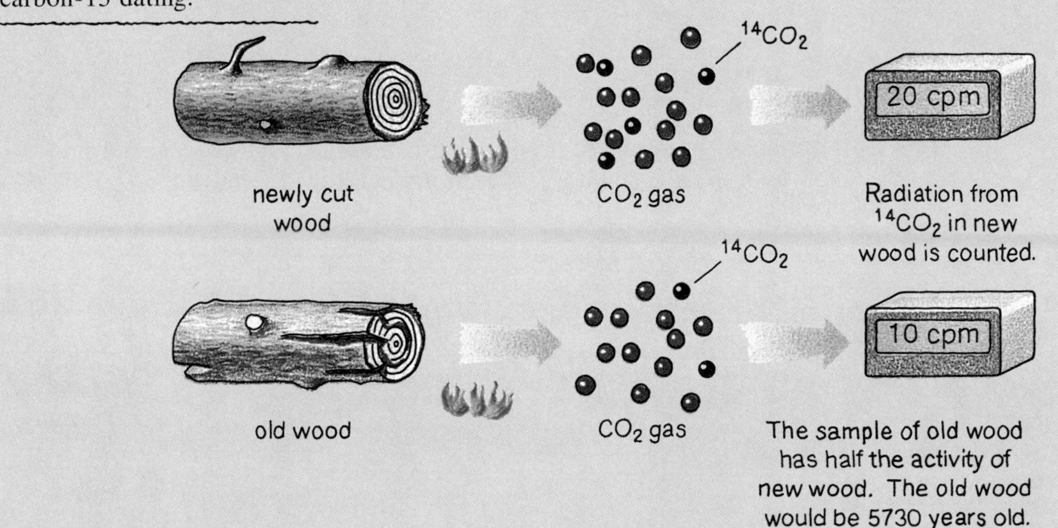
newly cut wood
14CO2
CO2 gas
20 cpm
Radiation from 14CO2 in new wood is counted.
old wood
14CO2
CO2 gas
10 cpm
The sample of old wood has half the activity of new wood. The old wood would be 5730 years old.

Sample Problem 6.9
Dating Using Half-Lives

In Los Angeles, the remains of ancient animals have been unearthed at the La Brea tar pits. Suppose a bone sample from the tar pits is subjected to the carbon-14 dating method. If the sample shows about two half-lives have passed, about when did the animal live?

Solution
We can calculate the age of the bone sample by using the half-life of carbon-14 (5730 years).

$$2 \text{ \sout{half-lives}} \times \frac{5730 \text{ years}}{1 \text{ \sout{half life}}} = 11{,}000 \text{ years}$$

We would estimate that the animal lived about 11,000 years ago, or about 9000 B.C.

Study Check
Suppose that a piece of wood found in a tomb had 1/8 (3 half-lives) of its original C-14 activity. About how many years ago was the wood part of a living tree?

6.7 Nuclear Fission and Fusion

Learning Goal
Describe the processes of nuclear fission and fusion.

Nuclear Fission

In the 1930s, scientists bombarding uranium-235 with neutrons discovered that two medium-weight nuclei were produced along with a great amount of energy. This was the discovery of a new kind of nuclear reaction called nuclear **fission.** The energy generated by nuclear fission, splitting the atom, was called atomic energy. When uranium-235 absorbs a neutron, it breaks apart to form two smaller nuclei, several neutrons, and a great amount of energy. (See Figure 6.16.) A typical equation for nuclear fission is

$$^{235}_{92}\text{U} + {}^{1}_{0}\text{n} \longrightarrow {}^{139}_{56}\text{Ba} + {}^{94}_{36}\text{Kr} + 3\,{}^{1}_{0}\text{n} + \text{Energy}$$

If we could weigh these products with great accuracy, we would find that their total mass is slightly less than the mass of the starting materials. The

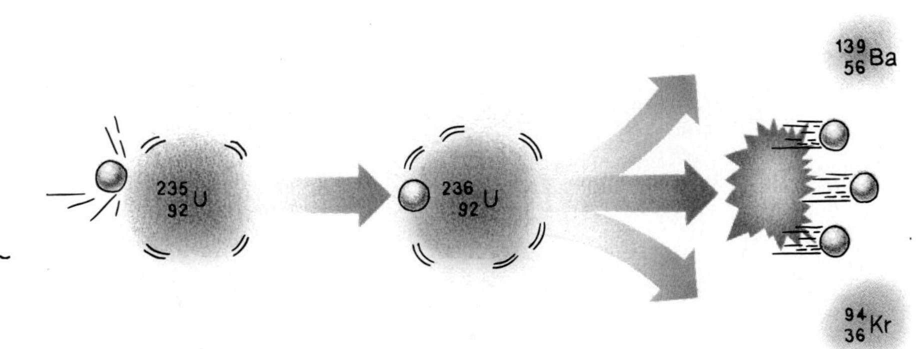

Figure 6.16
Diagram of nuclear fission. After absorbing a fast-moving neutron, a $^{235}_{92}U$ nucleus undergoes fission to produce two smaller nuclei, three neutrons, and a great amount of energy.

missing mass has been converted into energy, consistent with the famous equation derived by Albert Einstein:

$$E = mc^2$$

E is the energy released, m is the mass lost, and c is the speed of light, 3×10^8 m/s. Even though the mass loss is very small, when it is multiplied by the speed of light squared the result is a large value for the energy released. The fission of 1 g of uranium-235 would produce about as much energy as 3 tons of coal.

Chain Reaction

Fission begins when a neutron collides with the nucleus of a uranium atom. The resulting nucleus is unstable and splits into smaller nuclei. This fission process also releases several neutrons and large amounts of gamma radiation and energy. The neutrons emitted have high energies and bombard more uranium-235 nuclei. As fission continues, there is a rapid increase in the number of high-energy neutrons capable of splitting more uranium atoms, a process called a **chain reaction,** shown in Figure 6.17. To sustain a nuclear chain reaction, sufficient quantities of uranium-235 must be brought together to provide a critical mass in which almost all the neutrons immediately collide with more uranium-235 nuclei. So much heat and energy build up that an atomic explosion occurs.

Nuclear Power Plants

In a nuclear power plant, the quantity of uranium-235 is held below a critical mass, so it cannot sustain a chain reaction. The fission reactions are slowed by placing control rods among the samples of uranium. These rods absorb some of the fast-moving neutrons. In this way, less fission occurs, and there is a slower, controlled production of energy. The heat from the controlled

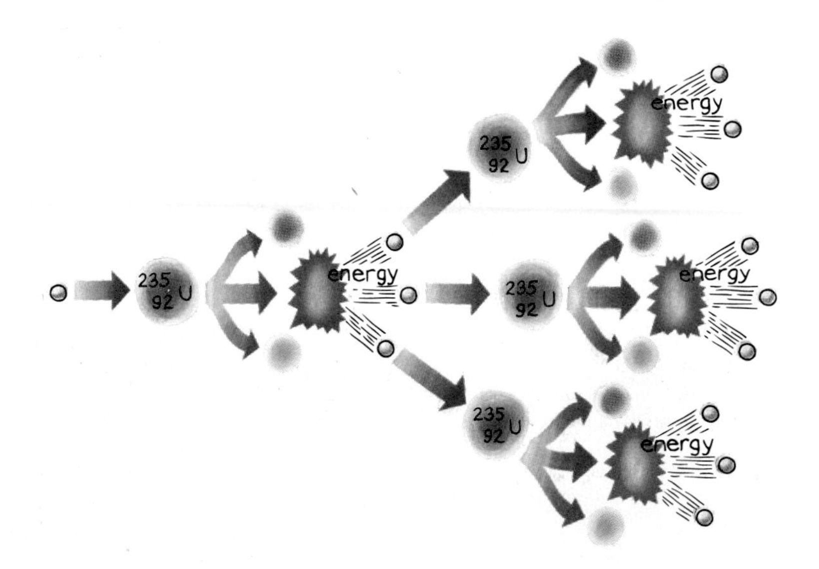

Figure 6.17
Nuclear chain reaction by fission of uranium-235.

fission is used to produce steam. The steam drives a generator, which produces electricity as shown in Figure 6.18. Approximately 10% of the electrical energy produced in the United States is generated in nuclear power plants.

Although nuclear power plants help meet some of our energy needs, there are major problems. One of the most serious problems is the production of radioactive by-products that have very long half-lives. It is essential that these waste products be stored safely for a very long time in a place where they do not contaminate the environment. Early in 1990, the Environmental Protection Agency gave its approval to the storing of radioactive hazardous wastes in chambers 2150 ft underground. It will be a matter of time before we know whether this type of storage is a good idea.

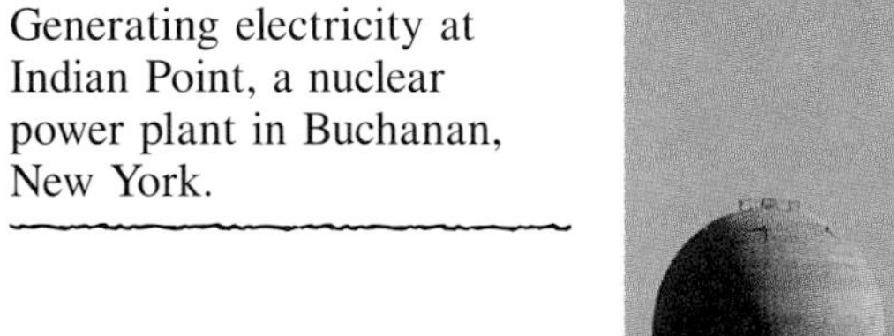

Figure 6.18
Generating electricity at Indian Point, a nuclear power plant in Buchanan, New York.

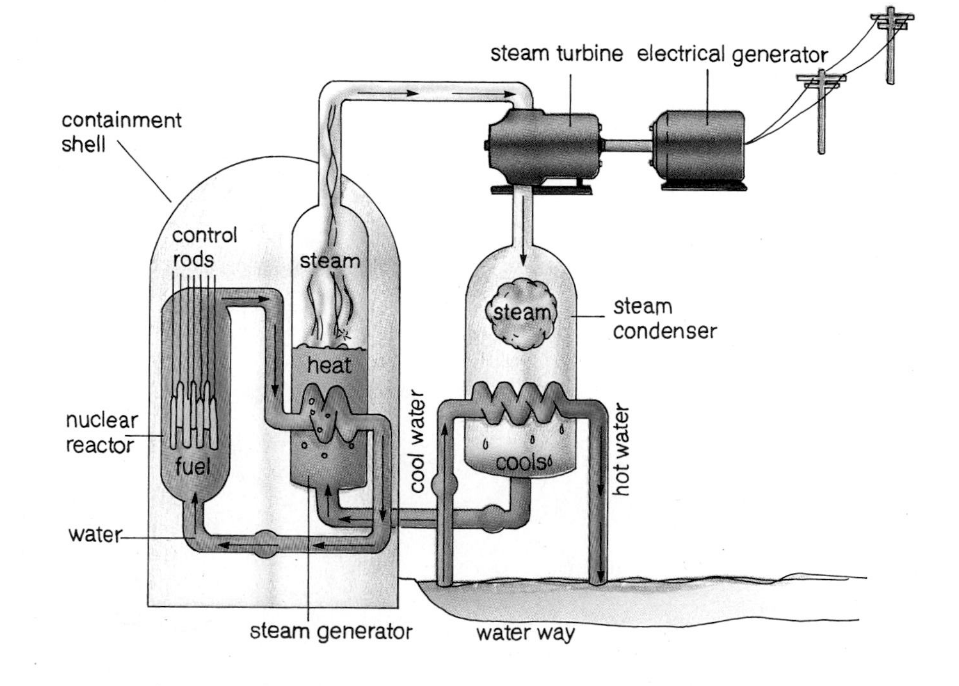

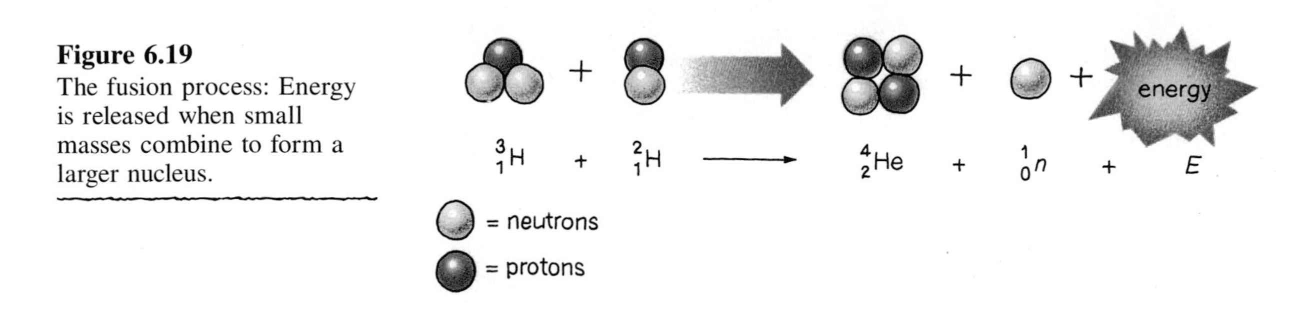

Figure 6.19
The fusion process: Energy is released when small masses combine to form a larger nucleus.

Nuclear Fusion

In **fusion,** two small nuclei, such as from hydrogen, combine to form a larger nucleus. Mass is lost, and a tremendous amount of energy is released, even more than the energy released from nuclear fission. However, a very high temperature (100,000,000°C) is required to overcome the repulsion of the hydrogen nuclei and cause them to undergo fusion. Fusion reactions occur continuously in the sun and other stars, providing us with heat and light. The huge amounts of energy produced by our sun come from the fusion of 6×10^{11} kilograms of hydrogen every second. The fusion reaction shown in Figure 6.19 involves the combination of two isotopes of hydrogen.

$$^3_1\text{H} + {}^2_1\text{H} \longrightarrow {}^4_2\text{He} + {}^1_0\text{n} + \text{energy}$$

The fusion reaction has tremendous potential as a possible source for future energy needs. One of the advantages of fusion as an energy source is that hydrogen is plentiful in the oceans. Although scientists expect some radioactive waste from fusion reactors, the amount is expected to be much less than that from fission, and the waste products should have shorter half-lives. However, fusion is still in the experimental stage because the extremely high temperatures needed have been difficult to reach and even more difficult to maintain. Research groups around the world are attempting to develop the technology needed to make the fusion reaction a reality in our lifetimes.

Sample Problem 6.10
Identifying Fission and Fusion

Classify the following as pertaining to nuclear fission, nuclear fusion, or both:

a. Small nuclei combine to form larger nuclei.
b. Large amounts of energy are released.
c. Very high temperatures are needed for reaction.

Solution

a. fusion **b.** both fusion and fission **c.** fusion

Study Check

Would you expect the following reaction to be an example of a fission or fusion reaction?

$$^2_1\text{H} + {}^1_1\text{H} \longrightarrow {}^3_2\text{He}$$

Chapter Summary

6.1 Natural Radioactivity
Radioactive isotopes have unstable nuclei that break down (decay), spontaneously emitting alpha (α), beta (β), and gamma (γ) radiation. Because radiation can damage the cells in the body, proper protection must be used: shielding, time limitation, and distance.

6.2 Nuclear Equations
A balanced equation is used to represent the changes that take place in the nuclei of the reactants and products. The new isotopes and the type of radiation emitted can be determined from the symbols that show the mass numbers and atomic numbers of the isotopes in the nuclear reaction.

6.3 Producing Radioactive Isotopes
A radioisotope is produced artificially when a nonradioactive isotope is bombarded by a small particle such as a proton, or an alpha or beta particle. Many radioactive isotopes used in nuclear medicine are produced in this way.

6.4 Radiation Detection and Measurement
In a Geiger counter, radiation ionizes gas in a metal tube, which produces an electrical current. The curie (Ci) is a unit of radiation activity that measures the number of nuclear transformations. One curie is 3.7×10^{10} disintegrations in 1 second. The units rad and rem describe the biological effects of radiation on the body.

6.5 Medical Applications Using Radioactivity
In nuclear medicine, radioisotopes are given that go to specific sites in the body. By detecting the radiation they emit, an evaluation can be made about the location and extent of an injury, disease, or tumor, blood flow, or the level of function of a particular organ.

6.6 Half-Life of a Radioisotope
Every radioisotope has its own rate of emitting radiation. The time it takes for one-half of a radioactive sample to decay is called its half-life. For many medical radioisotopes, such as Tc-99m and I-131, half-lives are short; for others, usually naturally occurring ones such as C-14, Ra-226, and U-238, half-lives are extremely long.

6.7 Nuclear Fission and Fusion
In fission, a large nucleus breaks apart into smaller pieces, releasing one or more types of radiation, and a great amount of energy. In fusion, small nuclei combine to form a larger nucleus while great amounts of energy are released.

Glossary of Key Terms

alpha particle A nuclear particle identical to a helium (^4_2He, or α) nucleus (two protons and two neutrons) that is emitted as radiation.

beta particle A particle identical to an electron ($^{0}_{-1}e$, or β) that forms in the nucleus when a neutron changes to a proton and an electron.

carbon dating A technique used to date ancient specimens that contain carbon. The age is determined by the amount of active C-14 that remains in the sample.

chain reaction A fission reaction that will continue once it has been initiated with high-energy neutrons bombarding a heavy nucleus such as U-238.

curie (Ci) A unit of radiation equal to 3.7×10^{10} disintegrations/s.

fission A process in which large nuclei are split into smaller pieces releasing neutrons and large amounts of energy.

fusion A reaction in which large amounts of energy are released when small nuclei combine to form larger nuclei.

gamma ray High-energy radiation (γ) emitted to make a nucleus more stable.

half-life The length of time it takes for one-half of a radioactive sample to decay.

rad Radiation absorbed dose. A measure of an amount of radiation absorbed by the body.

radiation therapy The use of high doses of radiation to destroy harmful tissues in the body.

radioactive decay The process by which a radioactive nucleus breaks down with the release of high-energy radiation.

radioisotope A radioactive atom of an element.

rem Radiation equivalent in humans. A measure of the biological damage by the various kinds of radiation (rad $\times$ radiation biological factor).

scan The image of a site in the body created by the detection of radiation from radioactive isotopes that have accumulated in that site.

shielding Materials used to provide protection from radioactive sources.

transmutation The formation of a radioactive isotope by bombarding a stable nucleus with fast-moving particles.

Chemistry at Home

1. Read the label of contents on a smoke detector. What isotope is used? What are the products of radioactive decay for the isotope? (*Hint:* Use a chemistry handbook.) What type of radiation is emitted? Why would this be used in a home? How does the alarm detect smoke? What is the activity of the isotope? How often should it be replaced? Why?
2. Visit a radiology laboratory in a local hospital or clinic.
3. Ask your dentist about dental X rays. What amount of radiation is used? How are you and the x-ray technician protected?
4. How is radon measured in the home? What is in the home kits for radon detection?

Answers to Study Checks

6.1 β or ${}^{0}_{-1}e$

6.2 Limiting the time one spends near a radioactive source and staying as far away as possible will reduce exposure to radiation.

6.3 ${}^{214}_{84}\text{Po} \longrightarrow {}^{210}_{82}\text{Pb} + {}^{4}_{2}\text{He}$

6.4 ${}^{131}_{53}\text{I} \longrightarrow {}^{131}_{54}\text{Xe} + {}^{0}_{-1}e$

6.5 ${}^{27}_{13}\text{Al} + {}^{4}_{2}\text{He} \longrightarrow {}^{30}_{15}\text{P} + {}^{1}_{0}\text{n}$

6.6 ${}^{14}_{7}\text{N}$

6.7 9.3×10^{9} iodine-131 atoms

6.8 0.50 g

6.9 17,190 years

6.10 fusion

Problems

Natural Radioactivity *(Goal 6.1)*

6.1 **a.** How are an alpha particle and a helium nucleus similar? different?
b. What symbols are used for alpha particles?

6.2 **a.** How are a beta particle and an electron similar? different?
b. What symbols are used for beta particles?

6.3 Naturally occurring potassium consists of three isotopes, potassium-39, potassium-40, and radioactive potassium-41.
a. Write the nuclear symbols for each isotope.
b. In what ways are the isotopes similar, and in what ways do they differ?

6.4 Naturally occurring iodine is iodine-127. Medically radioactive isotopes of iodine-125 and iodine-130 are used.
a. Write the nuclear symbols for each isotope.
b. In what ways are the isotopes similar, and in what ways do they differ?

6.5 Supply the missing information in the following table:

Medical Use	Nuclear Symbol	Mass Number	Number of Protons	Number of Neutrons
Spleen imaging	${}^{51}_{24}\text{Cr}$			
Malign-ancies		60	27	
Blood volume			26	33
Hyperthy-roidism	${}^{131}_{53}\text{I}$			
Leukemia treatment		32		17

6.6 Supply the missing information in the following table:

Medical Use	Nuclear Symbol	Mass Number	Number of Protons	Number of Neutrons
Cancer treatment	$^{60}_{27}Co$			
Brain scan		99	43	
Pancreas scan		75	34	
Bone scan		85		47
Lung function	$^{133}_{54}Xe$			

6.7 Write a symbol for the following:
a. alpha particle **b.** neutron
c. beta particle **d.** nitrogen-15
e. iodine-125

6.8 Write a symbol for the following:
a. proton **b.** gamma ray
c. electron **d.** barium-131
e. cobalt-60

6.9 Identify the symbol for X in each of the following nuclear symbols:
a. $^{0}_{-1}X$ **b.** $^{4}_{2}X$ **c.** $^{1}_{0}X$
d. $^{24}_{11}X$ **e.** $^{14}_{6}X$

6.10 Identify the symbol for X in each of the following nuclear symbols:
a. $^{1}_{1}X$ **b.** $^{32}_{15}X$ **c.** $^{0}_{0}X$
d. $^{59}_{26}X$ **e.** $^{85}_{38}X$

6.11 **a.** Why does beta radiation penetrate further in solid material than alpha radiation?
b. How does ionizing radiation cause damage to cells of the body?
c. Why does the x-ray technician leave the room when you receive an X ray?
d. What is the purpose of wearing gloves when handling radioisotopes?

6.12 **a.** As a nurse in an oncology unit, you sometimes give an injection of a radioisotope. What are three ways you can minimize your exposure to radiation?
b. Why are cancer cells more sensitive to radiation than nerve cells?
c. What is the purpose of placing a lead apron on a patient who is receiving routine dental x-rays?
d. Why are the walls in a radiology office built of thick concrete blocks?

Nuclear Equations *(Goal 6.2)*

6.13 Write a balanced nuclear equation for the alpha decay for each of the following radioactive isotopes:
a. $^{208}_{84}Po$ **b.** $^{232}_{90}Th$ **c.** $^{251}_{102}No$ **d.** $^{220}_{86}Rn$

6.14 Write a balanced nuclear equation for the alpha decay for each of the following radioactive isotopes:
a. $^{243}_{96}Cm$ **b.** $^{252}_{99}Es$ **c.** $^{251}_{98}Cf$ **d.** $^{261}_{107}Bh$

6.15 Write a balanced nuclear equation for the beta decay for each of the following radioactive isotopes:
a. $^{25}_{11}Na$ **b.** $^{20}_{8}O$ **c.** $^{92}_{38}Sr$ **d.** $^{42}_{19}K$

6.16 Write a balanced nuclear equation for the beta decay for each of the following radioactive isotopes:
a. potassium-42 **b.** iron-59
c. iron-60 **d.** barium-141

6.17 Complete each of the following nuclear equations:
a. $^{28}_{13}Al \longrightarrow ? + ^{0}_{-1}e$
b. $? \longrightarrow ^{86}_{36}Kr + ^{1}_{0}n$
c. $^{66}_{29}Cu \longrightarrow ^{66}_{30}Zn + ?$
d. $? \longrightarrow ^{4}_{2}He + ^{234}_{90}Th$

6.18 Complete each of the following nuclear equations:
a. $^{11}_{6}C \longrightarrow ^{7}_{4}Be + ?$
b. $^{35}_{16}S \longrightarrow ? + ^{0}_{-1}e$
c. $? \longrightarrow ^{90}_{39}Y + ^{0}_{-1}e$
d. $^{210}_{83}Bi \longrightarrow ? + ^{4}_{2}He$

Producing Radioactive Isotopes *(Goal 6.3)*

6.19 Complete each of the following bombardment reactions:
a. $^{9}_{4}Be + ^{1}_{0}n \longrightarrow ?$
b. $^{32}_{16}S + ? \longrightarrow ^{32}_{15}P$
c. $? + ^{1}_{0}n \longrightarrow ^{24}_{11}Na + ^{4}_{2}He$

6.20 Complete each of the following bombardment reactions:
a. $^{40}_{18}Ar + ? \longrightarrow ^{43}_{19}K + ^{1}_{1}H$
b. $^{238}_{92}U + ^{1}_{0}n \longrightarrow ?$
c. $? + ^{1}_{0}n \longrightarrow ^{14}_{6}C + ^{1}_{1}H$

Radiation Detection and Measurement *(Goal 6.4)*

6.21 **a.** How does a Geiger counter detect radiation?
b. What radiation unit describes the activity of a sample?
c. What radiation unit describes the radiation dose absorbed by tissue?

6.22 **a.** What is background radiation?
b. What radiation unit describes the biological effect of radiation?
c. What is meant by the terms mCi and mrem?

6.23 **a.** A sample of iodine-131 has an activity of 3.0 Ci. How many disintegrations occur in the iodine-131 sample in 20 sec?
b. The recommended dosage of iodine-131 is 4.20 μCi/kg of body weight. How many microcuries of iodine-131 are needed for a 70.0-kg patient with hyperthyroidism?

6.24 **a.** The dosage of technetium-99m for a lung scan is 20 μCi/kg of body weight. How many millicuries should be given to a 50.0-kg patient? (1 mCi = 1000 μCi)
b. A patient receives 50 mrads in a chest X ray. What would be the absorbed dose in mrems?
c. Suppose a person absorbed 50 mrads of alpha radiation. What would that be in mrems? How does it compare with the mrems in part b?

6.25 Why would an airline pilot be exposed to more background radiation than the airline reservationist who works at the ticket counter?

6.26 In radiation therapy, a patient receives high doses of radiation. What kinds of symptoms of radiation sickness might be exhibited by the patient?

Medical Applications Using Radioactivity *(Goal 6.5)*

6.27 Bone and bony structures consist primarily of $Ca_3(PO_4)_2$. Why are the radioisotopes of calcium-47, phosphorus-32, and strontium-85 used in the diagnosis and treatment of bone lesions and bone tumors?

6.28 A patient with polycythemia vera (excess production of red blood cells) receives radioactive phosphorus-32. Why would this treatment reduce the production of red blood cells in the bone marrow of the patient?

6.29 Treatment with iodine-131 decreases the amount of hormone produced by the thyroid gland. Why?

6.30 A vial contains radioactive iodine-131 with an activity of 2.0 mCi per milliliter. If the thyroid test requires 6.0 mCi in an "atomic cocktail," how many milliliters are used to prepare the test solution?

Half-Life of a Radioisotope *(Goal 6.6)*

6.31 What is meant by the term half-life?

6.32 Why do radioisotopes used for diagnosis in nuclear medicine have short half-lives?

6.33 Technetium-99m is an ideal radioisotope for scanning organs because it has a half-life of 6.0 hr and is a pure gamma emitter. Suppose that 80.0 mg were prepared in the technetium generator this morning. How many milligrams would remain after:
a. one half-life **b.** two half-lives
c. 18 hr **d.** 24 hr

6.34 A sample of sodium-24 with an activity of 12 mCi is used to study the rate of blood flow in the circulatory system. If sodium-24 has a half-life of 15 hr, what is the activity of the sodium after $2\frac{1}{2}$ days?

6.35 Strontium-85, used for bone scans, has a half-life of 64 days. How long will it take for the radiation level of strontium-85 to drop to one-fourth of its original level? To one-eighth?

6.36 Fluorine-18, which has a half-life of 110 min, is used in PET scans. If 100 mg of fluorine-18 is shipped at 8:00 A.M., how many milligrams of the radioisotope are still active if the sample arrives at the radiology laboratory at 1:30 P.M.?

Nuclear Fission and Fusion *(Goal 6.7)*

6.37 What is nuclear fission?

6.38 How does a chain reaction occur in nuclear fission?

6.39 Complete the following fission reaction:
$^{235}_{92}U + ^{1}_{0}n \longrightarrow ^{131}_{50}Sn + ? + 2^{1}_{0}n + \text{energy}$

6.40 In another fission reaction, U-235 bombarded with a neutron produces Sr-94, another small nucleus, and three neutrons. Write the complete equation for the fission reaction.

6.41 Indicate whether each of the following are characteristic of the fission or fusion process or both:
a. Neutrons bombard a nucleus.
b. The nuclear process occurring in the sun.
c. A large nucleus splits into smaller nuclei.
d. Small nuclei combine to form larger nuclei.

6.42 Indicate whether each of the following are characteristic of the fission or fusion process or both:
a. Very high temperatures are required to initiate the reaction.
b. Less radioactive waste results.
c. Hydrogen nuclei are the reactants.
d. Large amounts of energy are released when the nuclear reaction occurs.

Challenge Problems

6.43 Complete each of the following nuclear equations:
a. $^{14}_{7}N + ^{4}_{2}He \longrightarrow ? + ^{1}_{1}H$
b. When two O-16 atoms collide, one of the reaction products is an alpha particle. What is the other product?

6.44 Complete each of the following nuclear equations:

a. $^{235}_{92}U + ^{1}_{0}n \longrightarrow ^{90}_{38}Sr + ? + 3^{1}_{0}n$

b. When californium-249 is bombarded by oxygen-18, a new element, rutherfordium-106, and four neutrons are produced. Write the balanced nuclear equation.

6.45 A nurse was accidentally exposed to potassium-42 while doing some brain scans for possible tumors. The error was not discovered until 36 hours later when the activity of the potassium-42 sample was 2.0 μCi/g. If potassium-42 has a half-life of 12 hr, what was the activity of the sample at the time the nurse was exposed?

6.46 A wooden object from the site of a Mayan temple has a carbon-14 activity of 10 counts per minute compared with a reference piece of wood cut today that has an activity of 40 counts per minute. If the half-life for carbon-14 is 5730 years, what is the age of the ancient wood object?

Chapter 7

Gases

The nitrogen in air breathed under pressure at depths of 100 to 300 feet can be harmful to divers; thus the oxygen in scuba tanks is mixed with helium.

Looking Ahead

Learning Goals

7.1 Describe the kinetic theory of gases.

7.2 Describe the units of measurement used for pressure and change from one unit to another.

7.3 Use the pressure–volume relationship (Boyle's law) to determine the new pressure or volume of a certain amount of gas at a constant temperature.

7.4 Use the temperature–volume relationship (Charles' law) to determine the new temperature or volume of a certain amount of gas at a constant pressure.

7.5 Use the temperature–pressure relationship (Gay-Lussac's law) to determine the new temperature or pressure of a certain amount of gas at a constant volume.

7.6 Use the combined gas law to find the new pressure, volume, or temperature of a gas when changes in two of these properties are given.

7.7 Describe the relationship between the amount of a gas and its volume and use this relationship in calculations.

7.8 Use partial pressures to calculate the total pressure of a mixture of gases.

7.9 Use the ideal gas law to solve for pressure, volume, temperature, or amount of a gas when given values for the other properties.

QUESTIONS TO THINK ABOUT

1. Why do airplanes need to be pressurized?
2. Why does a scuba diver need to exhale air when she ascends to the surface of the water?
3. Why do aerosol cans explode if heated?
4. Why does a bag of chips expand when you take it to a higher altitude?
5. How do respirators (or CPR) help a person obtain oxygen and expel carbon dioxide?

We all live at the bottom of a sea of gases called the atmosphere. The most important of these gases is oxygen, which constitutes about 21% of the atmosphere. Without oxygen life on this planet would be impossible because oxygen is vital to all life processes of plants and animals. Ozone (O_3), formed in the upper atmosphere by the interaction of oxygen with ultraviolet light, absorbs some of the harmful ultraviolet radiation from the sun before it can strike the earth's surface. The other gases in the atmosphere include nitrogen (78% of the atmosphere) and argon, carbon dioxide (CO_2), and water vapor (the chief components of the remaining 1%). Carbon dioxide gas, a product of human cellular metabolism, is used by plants in a process called photosynthesis, which produces the oxygen that is essential to respiration in humans.

The atmosphere has become a dumping ground for other gases, such as methane, chlorofluorohydrocarbons (CFCs), sulfur dioxide, and nitric oxides. The chemical reactions of these gases with sunlight and oxygen in the air are contributing to air pollution, ozone depletion, global warming, and acid rain. Such chemical changes can seriously affect our health and the way all of us live. A knowledge of the behavior of gases and some of the laws that govern gas behavior can help us understand the nature of matter and allow us to make decisions concerning the important environmental and health issues we face today.

7.1 Properties of Gases

Learning Goal

Describe the kinetic theory of gases.

In Chapter 2, we saw that the behavior of gases is quite different from that of liquids and solids. Gas particles are far apart, whereas particles of both liquids and solids are held close together because of strong attractive forces. This means that a gas has no definite shape or volume and will completely fill any container. Because there are great distances between its particles, a gas is less dense than a solid or liquid and can be compressed. A model for the behavior of a gas, called the **kinetic theory,** helps us understand gas behavior.

Kinetic Theory of Gases

1. A gas is composed of very small particles (molecules or atoms).
2. The particles of a gas are very far apart. Thus, a gas is mostly empty space, and we assume that the volume of a container of gas is the same as the volume of the gas.
3. Gas particles move rapidly, colliding with other gas particles and with the walls of the container.
4. Gas particles do not attract or repel one another.
5. The kinetic energy of gas particles is related to the temperature of the gas; particle motion increases when the temperature increases.

The kinetic theory helps explain some of the characteristics of gases. For example, we can quickly smell perfume from a bottle that is opened on the other side of a room, because its particles move rapidly in all directions. They move faster at higher temperatures, and more slowly at lower temperatures. Sometimes tires and gas-filled containers explode when temperatures are too high. From the kinetic theory, we know that gas particles move faster when heated, hit the walls of a container with more force, and cause a buildup of pressure inside a container.

Sample Problem 7.1
Kinetic Theory of Gases

How are the following characteristics considered in the kinetic theory of gases?

a. attraction between gas particles
b. velocity of gas particles

Solution

a. It is assumed that there are no attractions between the particles in a gas.
b. The particles of a gas move rapidly.

Study Check

What effect does increasing the temperature have on the pressure of a gas?

When we study a gas, we describe, measure, and relate four properties: pressure (P), volume (V), temperature (T), and the amount of gas involved, usually expressed as the number of moles (n).

Pressure (P)

The pressure of a gas is the result of a force that is created when gas particles hit the walls of a container. In measuring the pressure of a gas, typical units used are atmosphere (atm) and millimeters of mercury (mm Hg); the latter is also called torr. Other pressure units include inches or centimeters of mercury, and kilopascals.

Volume (V)

Because a gas completely fills its container, the volume of the gas is equal to the volume of the container. The volume of a gas in a 10-mL vial is 10 mL; the volume of oxygen in a 25-L tank is 25 L. The units for the volume of a gas are usually milliliters (mL) or liters (L).

Temperature (T)

All calculations with gases use the Kelvin temperature scale. If a gas were to reach a temperature of absolute zero (0 K), its particles would have no energy or motion. In Chapter 2 you learned that the relationship between degrees Celsius and kelvins is

$$K = {}^\circ C + 273$$

Amount of Gas (n)

When you add air to car or bicycle tires, you increase the quantity of gas and therefore increase the pressure in the tires. In gas laws, the amount of gas in a container is usually stated in moles (n).

A summary of the four properties of a gas is given in Table 7.1.

Table 7.1 ***Properties of a Gas***

Property	Units of Measurement
Pressure (P)	Atmosphere (atm); torr or millimeter of mercury (mm Hg)
Volume (V)	Liter (L); milliliter (mL)
Temperature (T)	Kelvin (K)
Amount (n)	Mole

Sample Problem 7.2
Identifying Gas Properties

Use the word *pressure, volume, temperature,* or *amount* to indicate the property of a gas described in each of the following statements or measurements:

a. 800 torr **b.** 295 K
c. 4.0 L **d.** the space occupied by a gas

Solution

a. pressure **b.** temperature
c. volume **d.** volume

Study Check

When air is added to a bicycle tire, the number of moles of air increases. What property of a gas is described?

7.2 Gas Pressure

Learning Goal

Describe the units of measurement used for pressure and change from one unit to another.

When water boils in a pan covered with a lid, the collisions of the molecules of steam (gas) lift up the lid. The gas molecules exert **pressure,** which is defined as a force acting on a certain area.

$$\text{Pressure } (P) = \frac{\text{force}}{\text{area}}$$

The air that covers the surface of the earth, the atmosphere, contains vast numbers of gas particles. Because the air particles have mass, they are pulled toward the earth by gravity, where they exert an **atmospheric pressure.**

The atmospheric pressure can be measured by using a **barometer,** as shown in Figure 7.1. A long glass tube is closed on one end and filled with

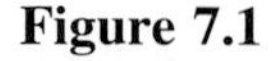

Figure 7.1
A barometer: A column of mercury 760 mm high is supported by the pressure exerted by the gases in the atmosphere at sea level.

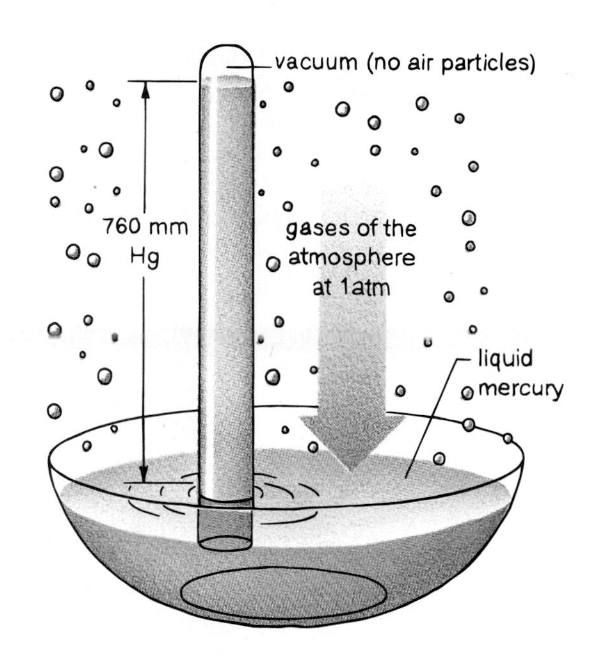

mercury. Then its open end is placed in a dish of mercury. The weight of the mercury in the tube begins to push it out of the tube. However, the pressure of the atmosphere against the mercury in the dish will push the mercury back up the tube.

Eventually, the mercury in the tube reaches a level where its weight is equal to the atmospheric pressure. At a pressure of 1 atmosphere (atm), the mercury column would be 760 mm high. We say that the atmospheric pressure is 760 mm Hg (millimeters of mercury), which is the same as a standard **atmosphere (atm).**

HEALTH NOTE *Measuring Blood Pressure*

The measurement of your blood pressure is one of the important measurements a doctor or nurse makes during a physical examination. Acting like a pump, the heart contracts to create the pressure that pushes blood through the circulatory system. During contraction, the blood pressure called systolic is at its highest. When the heart muscles relax, the blood pressure called diastolic falls. Normal range for systolic pressure is 100–120 mm Hg, and for diastolic pressure, 60–80 mm Hg, usually expressed as a ratio such as 100/80. These values are somewhat higher in older people. When blood pressures are elevated, especially the systolic pressure, such as 140/90, there is a greater risk of stroke, heart attack, or kidney damage. Low blood pressure prevents the brain from receiving adequate oxygen causing dizziness and fainting.

The blood pressures are measured by a sphygmomanometer, an instrument consisting of a stethoscope and an inflatable cuff connected to a tube of mercury called a manometer. (See Figure 7.2.) After the cuff is wrapped around the upper arm, it is pumped up with air until it cuts off the flow of blood through the arm. With the stethoscope over the artery, the air is slowly released from the cuff. When the pressure equals the systolic pressure, blood starts to flow again, and the noise it makes is heard. As air continues to be released, the cuff deflates until no sound in the artery is heard. The second pressure noted is the diastolic pressure, the pressure when the heart is not contracting.

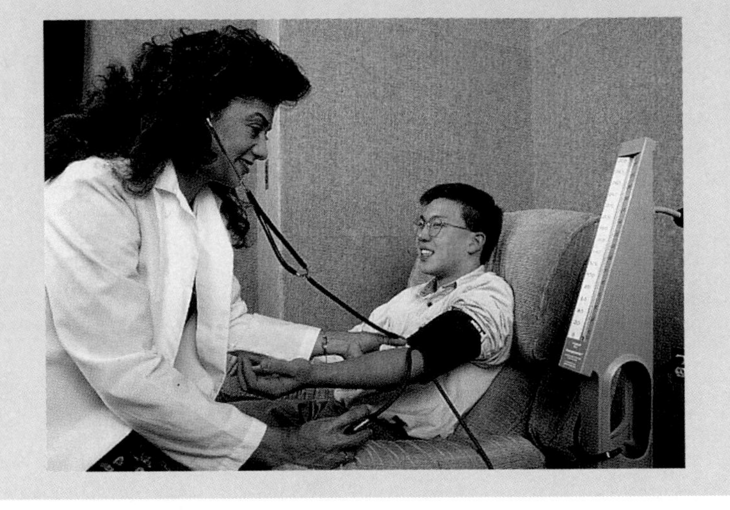

Figure 7.2
A doctor uses a sphygmomanometer to measure blood pressure.

Atmospheric Pressure and Altitude

At sea level, where there are more air particles, the atmospheric pressure is about 1 atm. In Los Angeles, atmospheric pressure can fluctuate between 730 and 760 mm Hg, depending on the weather. At higher altitudes, there are fewer air particles and the atmospheric pressure is less than 1 atm. In Denver (1.6 km), a typical atmospheric pressure might be 630 mm Hg, and if you were climbing Mount Everest (9.3 km) it might get as low as 270 mm Hg.

Deep-sea divers must be concerned about increasing pressures on their ears and lungs when they dive below the surface of the ocean. Because water is denser than air, the pressure on a diver increases rapidly as the diver descends. At a depth of 33 ft below the surface of the ocean, there is an additional atmosphere of pressure exerted by the water on a diver, for a total of 2 atm. At 100 ft down, there is a total of 4 atm on a diver. The air tanks a diver carries continuously adjust the pressure of the breathing mixture to match the increase of pressure on the body.

This pressure is often expressed as 760 **torr,** a pressure unit named to honor Evangelista Torricelli, the inventor of the barometer. Because they are equal, units of torr and mm Hg are used interchangeably.

$$1 \text{ mm Hg} = 1 \text{ torr}$$
$$1 \text{ atm} = 760 \text{ mm Hg} = 760 \text{ torr}$$

In SI units, pressure is measured in pascals (Pa); 1 atm is equal to 101,325 Pa. Because a pascal is a very small unit, it is likely that pressures would be reported in kilopascals. However, the unit is not yet in common use.

$$1 \text{ atm} = 101{,}325 \text{ Pa} = 101.3 \text{ kPa}$$

If you have a barometer in your home, it probably gives pressure in inches of mercury. One atmosphere is also equal to the pressure of a column of mercury that is 29.9 in. high. Weather reports are given in inches of mercury. A lowering of pressure often indicates rain or snow, whereas an increase in pressure (referred to as a high-pressure system) usually brings dry and sunny weather.

The American equivalent of 1 atm is 14.7 pounds per square inch ($lb/in.^2$). This is the measurement you see on a pressure gauge when you check the air pressure in the tires of a car or bicycle. Table 7.2 summarizes the various units used in the measurement of pressure.

Table 7.2 ***Units for Measuring Pressure***

Unit	Abbreviation	Unit Equivalent to 1 atm
Atmosphere	atm	1 atm
Millimeters of Hg	mm Hg	760 mm Hg
Torr	torr	760 torr
Inches of Hg	in. Hg	29.9 in. Hg
Pounds per square inch	$lb/in.^2$ (psi)	14.7 $lb/in.^2$
Pascal	Pa	101,325 Pa

Sample Problem 7.3
Units of Pressure Measurement

A sample of neon gas has a pressure of 0.50 atm. Give the pressure of the neon in:

a. millimeters of Hg **b.** inches of Hg

Solution

a. The equality 1 atm = 760 mm Hg can be written as conversion factors:

$$\frac{760 \text{ mm Hg}}{1 \text{ atm}} \quad \text{or} \quad \frac{1 \text{ atm}}{760 \text{ mm Hg}}$$

Using the appropriate conversion factor, the problem is set up as

$$0.50 \cancel{\text{atm}} \times \frac{760 \text{ mm Hg}}{1 \cancel{\text{atm}}} = 380 \text{ mm Hg}$$

b. One atm is equal to 29.9 in. Hg. Using this equality as a conversion factor in the problem setup, we obtain

$$0.50 \cancel{\text{atm}} \times \frac{29.9 \text{ in. Hg}}{1 \cancel{\text{atm}}} = 15 \text{ in. Hg}$$

Study Check

What is the pressure in atmospheres for a gas that has a pressure of 655 torr?

Table 7.3
Vapor Pressure of Water

Temperature (C°)	Vapor Pressure (mm Hg)
0	5
10	9
20	18
30	32
37	47[a]
40	55
50	93
60	149
70	234
80	355
90	528
100	760

[a]At body temperature.

Vapor Pressure

When liquids evaporate, gas vapor forms at the surface of the liquid and escapes. If the liquid is in an open container, it can all eventually evaporate. In a closed container, the vapor accumulates and creates pressure called **vapor pressure.** Each liquid exerts its own vapor pressure at a given temperature. As temperature increases, more vapor forms, and vapor pressure increases. Table 7.3 lists the vapor pressure of water at various temperatures.

Vapor Pressure and Boiling Point

A liquid reaches its boiling point when its vapor pressure becomes equal to the atmospheric pressure. As boiling occurs, bubbles of the gas form within the liquid and quickly rise to the surface. For example, at an atmospheric pressure of 760 mm Hg, water will boil at 100°C, the temperature at which its vapor pressure reaches 760 mm Hg. (See Figure 7.3.)

At higher altitudes, atmospheric pressures are lower and the boiling point of water is less than 100°C. Earlier, we saw that the typical atmospheric pressure in Denver is 630 mm Hg. This means that water in Denver needs a vapor pressure of 630 mm Hg to boil. Because water has a vapor pressure of 630 mm Hg at 95°C, water boils at 95°C in Denver. Water boils at even lower temperatures at still lower atmospheric pressures, as Table 7.4 shows.

People who live at high altitudes often use pressure cookers to obtain higher temperatures when preparing food. When the pressure exerted is greater than 1 atm, a temperature higher than 100°C is needed before water

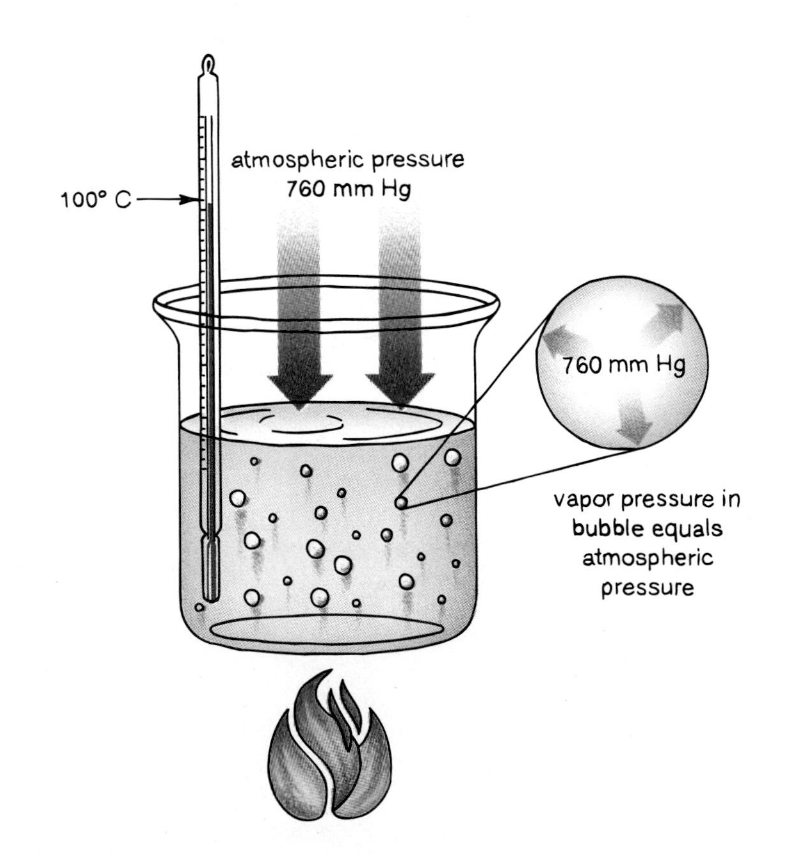

Figure 7.3
Vapor pressure and boiling point for water at 1 atm pressure (760 mm Hg).

Table 7.4 ***Altitude, Atmospheric Pressure, and the Boiling Point of Water***

Location	Altitude (km)	Atmospheric Pressure (mm Hg)	Boiling Point (°C)
Sea level	0	760	100
Los Angeles	0.09	752	99
Las Vegas	0.70	700	98
Denver	1.60	630	95
Mount Whitney	4.00	467	87
Mount Everest	9.30	270	70

Table 7.5
Pressure and the Boiling Point of Water

Pressure (mm Hg)	Boiling Point (°C)
760	100
800	100.4
1075	110
1520 (2 atm)	120
2026	130
7600 (10 atm)	180

will boil. Laboratories and hospitals use devices called autoclaves to sterilize laboratory and surgical equipment. An autoclave, like a pressure cooker, is a closed container that increases the pressure above the water so it will boil at higher temperatures. Table 7.5 shows how the boiling point of water increases as pressure increases.

7.3 Pressure and Volume (Boyle's Law)

Learning Goal
Use the pressure–volume relationship (Boyle's law) to determine the new pressure or volume of a certain amount of gas at a constant temperature.

Imagine that you can see air particles hitting the walls inside a bicycle pump. What happens to the pressure inside the pump as we push down on the handle? As the air is compressed, the air particles are crowded together. In the smaller volume, more collisions occur, and the air pressure increases.

When a change in one property (in this case, volume) causes a change in another property (in this case, pressure), those properties are said to be related to each other. Furthermore, when one change causes a change in the opposite direction, such as an increase in pressure causing a decrease in volume, the properties are said to have an **inverse relationship.** The relationship between the pressure and volume of a gas is known as **Boyle's law.** The law states that the volume (V) of a sample of gas changes inversely with the pressure (P) of the gas as long as there has been no change in the temperature (T) or amount of gas (n) as illustrated in Figure 7.4.

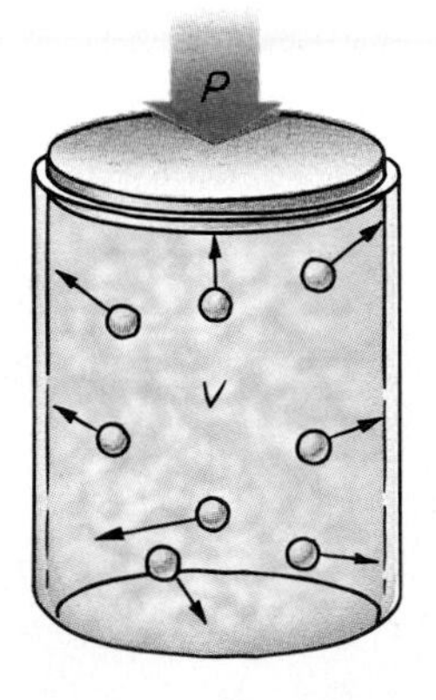

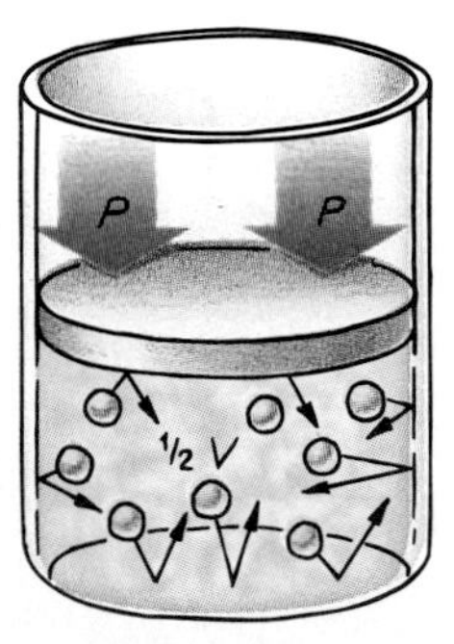

Figure 7.4
When the volume of a gas decreases, its pressure increases as long as there is no change in the temperature or the amount of the gas.

Sample Problem 7.4
Relationship of Pressure and Volume

Complete the information for the pressure and volume of a gas sample at constant temperature.

	Pressure	*Volume*
a.	increases	____
b.	____	increases

Solution

a. Pressure and volume are inversely related. When the pressure of a gas increases, the volume decreases.

b. If the volume of the gas sample increases, the pressure of the gas decreases.

Study Check

If the pressure of a gas decreases, what happens to its volume if no change has occurred in the temperature or in the amount of the gas?

In mathematical terms, the initial pressure (P_i) multiplied by the initial volume (V_i) of a gas gives a constant (C) value.

$$P_iV_i = \text{constant } (C)$$

If we change the volume or pressure of the gas without any change occurring in the temperature or in the amount of the gas, the new pressure and volume will also equal the same constant.

$$P_fV_f = C$$

Because the constant has the same value under both conditions, we can set the initial and final PV values equal to each other.

$$P_iV_i = P_fV_f \quad \text{No change in number of moles and temperature}$$

Sample Problem 7.5
Calculating Pressure When Volume Changes

A sample of hydrogen gas (H_2) has a volume of 4.0 L and a pressure of 1.0 atm. What is the new pressure if the volume is decreased to 2.0 L?

Solution

In calculations with gas laws, it is helpful to organize the data in a table of initial and final conditions.

Initial Conditions	*Final Conditions*	*Change*
$P_i = 1.0$ atm	$P_f = ?$	P must increase
$V_i = 4.0$ L	$V_f = 2.0$ L	V decreases

In this problem, we want to know the final pressure (P_f) for the change in volume.

The problem setup begins with the initial pressure of 1.0 atm followed by a volume factor that is written using the initial and final volumes. There are two possible volume factors; one will decrease pressure, and the other will increase pressure.

Volume Factors

$$\frac{2.0\text{ L}}{4.0\text{ L}} \quad \text{and} \quad \frac{4.0\text{ L}}{2.0\text{ L}}$$

This factor will decrease pressure — This factor will increase pressure

In this problem the volume has decreased. Because pressure and volume are inversely related, the pressure must increase. Therefore, we will use the volume factor that increases pressure.

$$P_f = 1.0\text{ atm} \times \frac{4.0\text{ }\cancel{L}}{2.0\text{ }\cancel{L}} = 2.0\text{ atm}$$

New pressure — Initial pressure — Volume factor that increases pressure

The pressure of the hydrogen gas increases from 1.0 atm to 2.0 atm. This is the change expected, because the volume decreased.

Alternatively, the pressure-volume equation can be solved for the final pressure by dividing both sides by V_f.

$$\frac{P_i V_i}{V_f} = \frac{P_f \cancel{V_f}}{\cancel{V_f}}$$

$$P_f = \frac{P_i V_i}{V_f}$$

Substituting the values listed in the table of information into the equation for P_f gives

$$P_f = \frac{(1.0\text{ atm})(4.0\text{ }\cancel{L})}{(2.0\text{ }\cancel{L})} = 2.0\text{ atm}$$

Study Check

A sample of helium gas has a volume of 250 mL at 800 torr. If the volume is changed to 500 mL, what is the new pressure, assuming no change in temperature or number of moles?

Sample Problem 7.6
Calculating Volume When Pressure Changes

In the hospital respiratory unit, the gauge on a 10.0-L tank of compressed oxygen reads 4500 mm Hg. How many liters of oxygen can be delivered from the tank at a pressure of 750 mm Hg?

Solution

Placing our information in a table gives the following:

Initial Conditions	*Final Conditions*	*Change*
$P_i = 4500$ mm Hg	$P_f = 750$ mm Hg	P decreases
$V_i = 10.0$ L	$V_f = ?$	V must increase

According to Boyle's law, a decrease in the pressure will cause an increase in the volume. The initial volume (V_i) must be multiplied by a pressure factor that gives a greater volume.

$$V_f = V_i \times \frac{P_i}{P_f}$$

$$V_f = 10.0 \text{ L} \times \frac{4500 \text{ torr}}{750 \text{ torr}} = 60. \text{ L}$$

New volume | Initial volume | Pressure factor that increases volume

Study Check

A sample of methane gas (CH_4) has a volume of 125 mL at 0.600 atm pressure and 25°C. How many milliliters will it occupy at a pressure of 1.50 atm and 25°C?

HEALTH NOTE *Pressure-Volume Relationship in Breathing*

The importance of Boyle's law becomes more apparent when you consider the mechanics of breathing. Our lungs are elastic, balloon-like structures contained within an airtight chamber called the thoracic cavity. The diaphragm, a muscle, forms the flexible floor of the cavity.

Inspiration

The process of taking a breath of air begins when the diaphragm flattens, and the rib cage expands, causing an increase in the volume of the thoracic cavity. The elasticity of the lungs allows them to expand when the thoracic cavity expands. According to Boyle's law, the pressure inside the lungs will decrease when their volume increases. This causes the pressure inside the lungs to fall below the pressure of the atmosphere. This difference in pressures produces a **pressure gradient** between the lungs and the atmosphere. In a pressure gradient, molecules flow from an area of greater pressure to an area of lower pressure, a process called **diffusion.** Thus, we inhale as air flows into the lungs *(inspiration)*, until the pressure within the lungs becomes equal to the pressure of the atmosphere.

Expiration

Expiration, or the exhalation phase of breathing, occurs when the diaphragm relaxes and moves back up into the thoracic cavity to its resting position. This reduces the volume of the thoracic cavity, which squeezes the lungs and decreases their volume. Now the pressure in the lungs is greater than the pressure of the atmosphere, so air flows out of the lungs. Thus, breathing is a process in which pressure gradients are continuously created between the lungs and the environment as a result of the changes in the volume and pressure. (See Figure 7.5.)

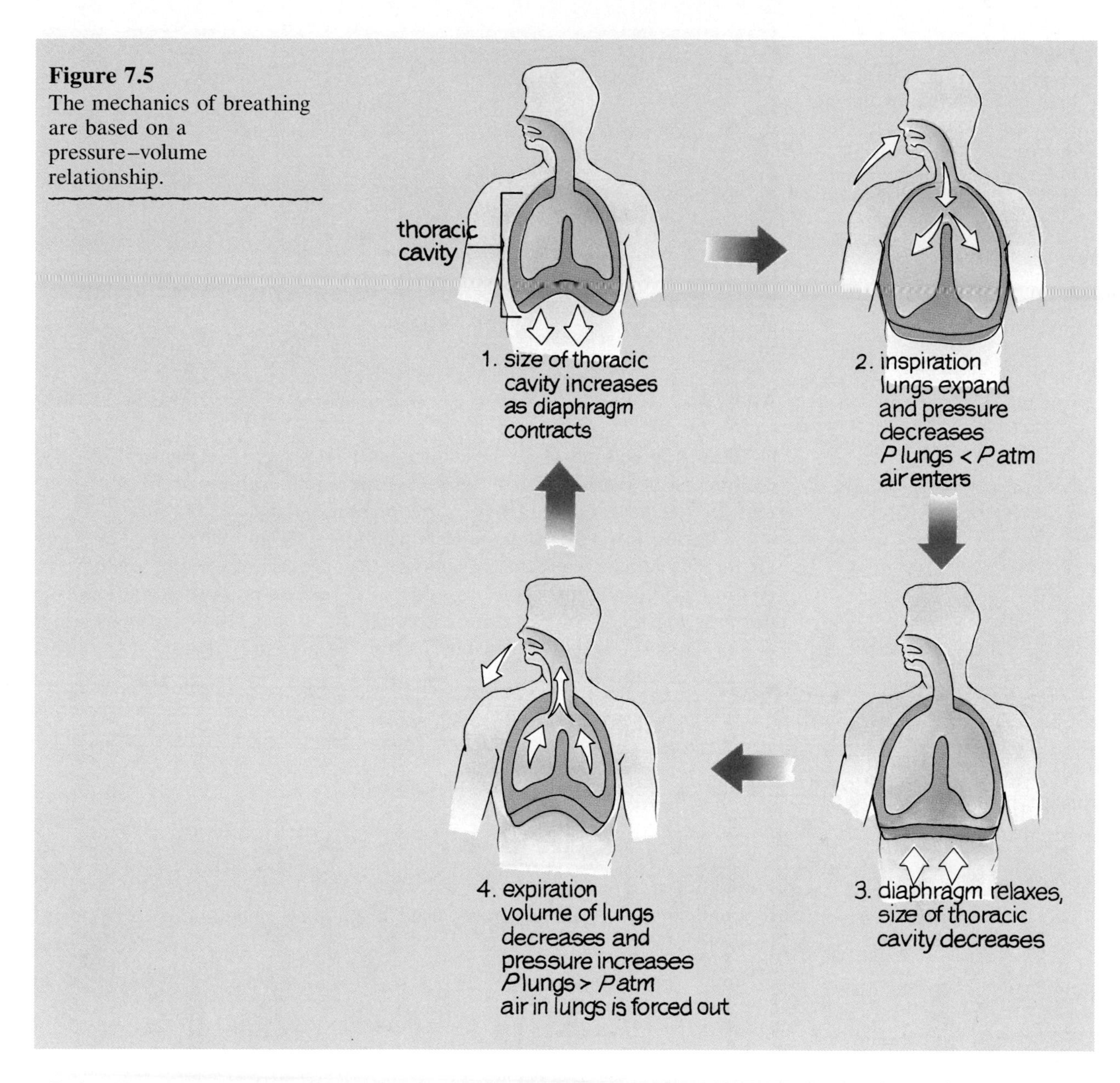
thoracic cavity
1. size of thoracic cavity increases as diaphragm contracts
2. inspiration lungs expand and pressure decreases P lungs < P atm air enters
3. diaphragm relaxes, size of thoracic cavity decreases
4. expiration volume of lungs decreases and pressure increases P lungs > P atm air in lungs is forced out

Figure 7.5
The mechanics of breathing are based on a pressure–volume relationship.

7.4 Temperature and Volume (Charles' Law)

Learning Goal

Use the temperature–volume relationship (Charle's law) to determine the new temperature or volume of a certain amount of gas at a constant pressure.

When preparing a hot-air balloon for a flight, the air in the balloon is heated with a small propane heater. As the air warms, its volume expands. The decrease in density allows the balloon to rise. (See Figure 7.6.)

To study the effect of changing temperature on the volume of a gas, we must not change the pressure or the amount of the gas. Suppose we increase the Kelvin temperature of a gas sample. The kinetic theory shows that the activity (kinetic energy) of the gas will also increase. To keep pressure

Figure 7.6
As the air inside a hot-air balloon is heated, its volume expands and its density decreases. The hot-air balloon rises and floats in the air.

constant, the volume of the container must increase. (See Figure 7.7.) By contrast, if the temperature of the gas is lowered, the volume of the container must be reduced to maintain the same pressure.

Charles' law states that the Kelvin temperature and the volume of a gas are directly related when there is no change in pressure or the amount of gas. A **direct relationship** is one in which the related properties increase or decrease together. For two different conditions, the relationship of volume (V) to temperature (T) is constant as long as pressure (P) and number of moles (n) do not change.

$$\frac{V}{T} = \text{constant } (C)$$

For initial and final conditions, we can write

$$\frac{V_i}{T_i} = \frac{V_f}{T_f} \quad \text{No change in pressure or number of moles}$$

Remember that all temperatures used in gas law calculations must be in kelvins (K).

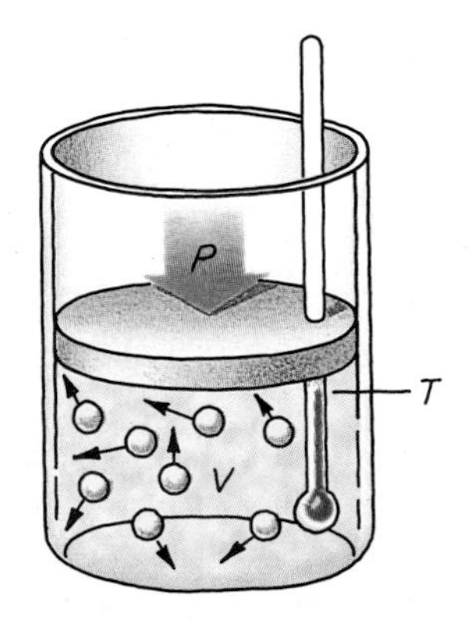

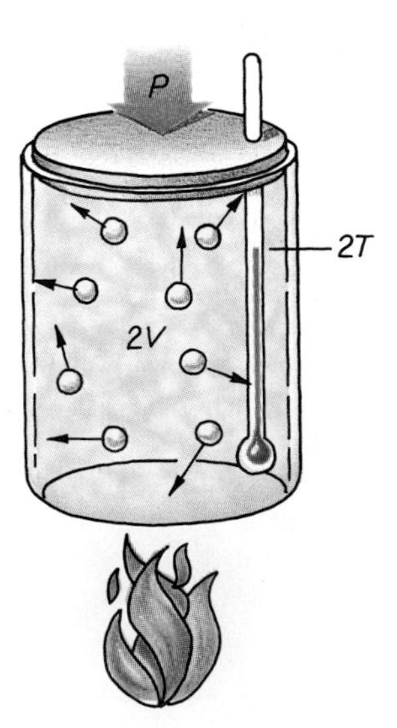

Figure 7.7
The Kelvin temperature of a gas is directly related to the volume of the gas when there is no change in pressure or amount.

Sample Problem 7.7
Relationship of Volume and Temperature

Complete the information for the volume and temperature of a gas when the pressure is kept constant.

	Volume	*Temperature*
a.	____	decreases
b.	increases	____

Solution

a. If the temperature of a gas decreases, its volume also decreases.
b. When the volume of a gas increases, the temperature also increases.

Study Check

If the volume of a sample of gas decreases, how must the temperature change if the pressure is to remain constant?

Sample Problem 7.8
Calculating Volume When Temperature Changes

A sample of neon gas has a volume of 5.0 L and a temperature of 17°C. Find the new volume of the gas after the temperature has been increased to 47°C.

Solution

When the temperatures are given in degrees Celsius, they must be changed to kelvins.

$$T_i = 17°\text{C} + 273 = 290\text{ K}$$
$$T_f = 47°\text{C} + 273 = 320\text{ K}$$

Initial Conditions	*Final Conditions*	*Change*
$T_i = 290$ K	$T_f = 320$ K	T increases
$V_i = 5.0$ L	$V_f = ?$	V must increase

Because an increase in temperature increases the volume, multiply the initial volume by the factor that gives a greater final volume.

$$V_f = V_i \times \text{Temperature factor}$$

$$V_f = 5.0\text{L} \times \frac{320\ \cancel{\text{K}}}{290\ \cancel{\text{K}}} = 5.5\text{ L}$$

New volume — Initial volume — Temperature factor that increases volume

Study Check

A mountain climber inhales 500.0 mL of air at a temperature of −10°C. What volume will the air occupy in the lungs if the climber's body temperature is 37°C?

7.5 Temperature and Pressure (Gay-Lussac's Law)

Learning Goal

Use the temperature–pressure relationship (Gay-Lussac's law) to determine the new temperature or pressure of a certain amount of gas at a constant volume

If we could watch the molecules of a gas as the temperature rises, we would notice that they move faster and hit the sides of the container more often and with greater force. If we keep the volume of the container the same, we would observe an increase in the pressure. A temperature–pressure relationship, also known as **Gay-Lussac'a law,** states that the pressure of a gas is directly related to its Kelvin temperature. This means that an increase in temperature increases the pressure of a gas, and a decrease in temperature decreases the pressure of the gas, as long as the volume and number of moles of the gas stay the same. (See Figure 7.8.)

The ratio of pressure (P) to temperature (T) is the same under all conditions as long as volume (V) and amount of gas (n) do not change.

$$\frac{P_i}{T_i} = \frac{P_f}{T_f}$$

Sample Problem 7.9
Relationship of Temperature and Pressure

Complete the information for the temperature and pressure of a gas when volume is kept constant.

	Temperature	*Pressure*
a.	increases	_____
b.	_____	decreases

Solution

a. If the temperature increases, the pressure of the gas also increases.
b. If the pressure decreases, the temperature of the gas also decreases.

Study Check

If the temperature in a tire decreases, how will pressure change?

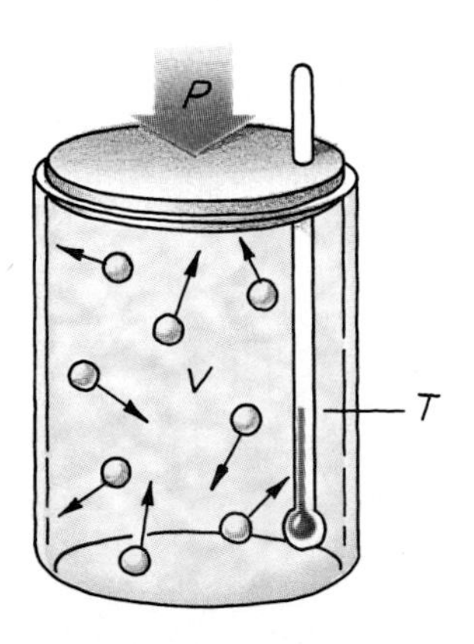

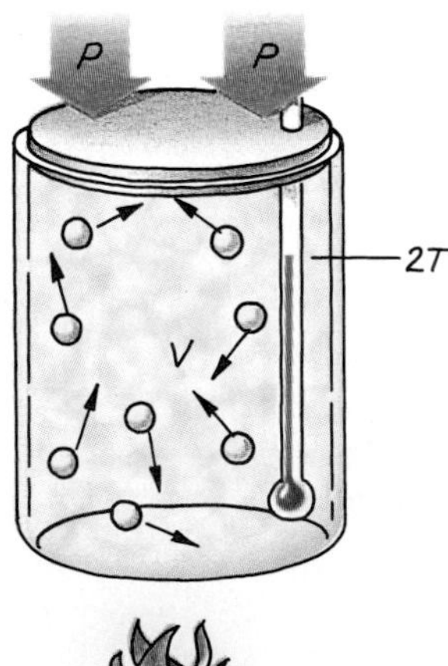

Figure 7.8
The Kelvin temperature of a gas is directly related to the pressure of the gas when there is no change in volume or amount.

Sample Problem 7.10
Calculating Pressure When Temperature Changes

Aerosol cans can be dangerous if they are heated because they can explode. Suppose a can of hair spray with a pressure of 4.0 atm at a room temperature of 27°C is thrown into a fire. If the temperature of the gas inside the aerosol can reaches 402°C, what will be its pressure? The aerosol can may explode if the pressure inside exceeds 8.0 atm. Would you expect the can to explode?

Solution
We must first change the temperatures to kelvins.

$$T_i = 27°C + 273 = 300 \text{ K}$$
$$T_f = 402°C + 273 = 675 \text{ K}$$

Initial Conditions	*Final Conditions*	*Change*
$P_i = 4.0$ atm	$P_f = ?$	P must increase
$T_i = 300$ K	$T_f = 675$ K	T increases

Because the temperature of the gas has increased, the final pressure must increase. In the calculation, we use a factor that gives a greater final pressure.

$$P_f = 4.0 \text{ atm} \times \frac{675 \text{ K}}{300 \text{ K}} = 9.0 \text{ atm}$$

New pressure; Initial pressure; Temperature factor that increases pressure

Because the calculated pressure exceeds 8.0 atm, we might expect the can to explode.

Study Check
In a storage area where the temperature has reached 55°C, the pressure of oxygen gas in a 15.0-L steel cylinder is 965 torr. To what temperature would the gas have to be cooled to reduce the pressure to 850. torr?

7.6 The Combined Gas Law of Pressure, Volume, and Temperature Relationships

Learning Goal
Use the combined gas law to find the new pressure, volume, or temperature of a gas when changes in two of these properties are given.

All of the pressure–volume–temperature relationships for gases that we have studied may be combined into a single relationship called the **combined gas law.** This expression is useful for studying the effect of changes in two of these variables on the third as long as the amount of gas (number of moles) remains constant.

$$\frac{P_i V_i}{T_i} = \frac{P_f V_f}{T_f} \quad \text{No change in amount of gas}$$

By remembering the combined gas law, we can derive the gas laws we have studied by omitting those properties that do not change. Table 7.6 summarizes the pressure–volume–temperature relationships of gases.

Table 7.6 ***Summary of Gas Laws***

Combined Gas Law	Properties Held Constant	Properties that Change	Relationship
$\frac{P_i V_i}{\cancel{T_i}} = \frac{P_f V_f}{\cancel{T_f}}$	n, T	P, V	Boyle's law: $P_i V_i = P_f V_f$
$\frac{\cancel{P_i} V_i}{T_i} = \frac{\cancel{P_f} V_f}{T_f}$	n, P	V, T	Charles' law: $\frac{V_i}{T_i} = \frac{V_f}{T_f}$
$\frac{P_i \cancel{V_i}}{T_i} = \frac{P_f \cancel{V_f}}{T_f}$	n, V	P, T	Gay-Lussac's law: $\frac{P_i}{T_i} = \frac{P_f}{T_f}$

Sample Problem 7.11
Using the Combined Gas Law

A 25.0-mL bubble is released from a diver's air tank at a pressure of 4.00 atm and a temperature of 11°C. What is the volume of the bubble when it reaches the ocean surface, where the pressure is 1.00 atm and the temperature is 18°C?

Solution

We must first change the temperatures to kelvins.

$$T_i = 11°\text{C} + 273 = 284 \text{ K}$$
$$T_f = 18°\text{C} + 273 = 291 \text{ K}$$

Initial Conditions	*Final Conditions*	*Change*
$P_i = 4.00$ atm	$P_f = 1.00$ atm	P decreases
$V_i = 25.0$ mL	$V_f = ?$	
$T_i = 284$ K	$T_f = 291$ K	T increases

To set up the calculation, we will first consider the effect of the pressure change on the volume. Because the pressure change will increase volume (Boyle's law), we multiply the initial volume by a pressure factor that gives a greater final volume.

$$V_f = 25.0 \text{ mL} \times \frac{4.00 \cancel{\text{atm}}}{1.00 \cancel{\text{atm}}} \times ?$$

New volume = Initial volume × Pressure factor that increases volume

Now we will consider the effect of the temperature increase, which would cause an increase in volume (Charles' law). Now we need to multiply by a temperature factor that increases the volume.

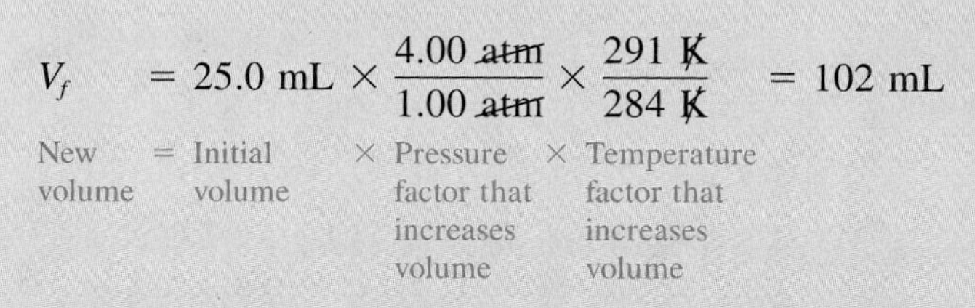

The combined effect of the pressure decrease and temperature increase is an increase in the volume.

Study Check

A weather balloon is filled with 15.0 L of helium at a temperature of 25°C and a pressure of 685 mm Hg. What is the pressure of the helium in the balloon in the upper atmosphere when the temperature is −35°C and the volume becomes 34.0 L?

7.7 Volume and Moles (Avogadro's Law)

Learning Goal

Describe the relationship between the amount of a gas and its volume and use this relationship in calculations.

In our study of the gas laws, we have looked at changes in properties for a specified amount (n) of gas. Now we will consider how the properties of a gas change when there is a change in number of moles or grams. For example, when you blow up a balloon, its volume increases because you add more air molecules. If a basketball gets a hole in it, and some of the air leaks out, its volume decreases. In 1811, Amedeo Avogadro stated that the volume of a gas is directly related to the number of moles of a gas when temperature and pressure are not changed. We refer to this statement as **Avogadro's law.** If the moles of a gas are doubled, then the volume will double as long as we do not change the pressure or the temperature. (See Figure 7.9.) For two conditions, we can write

$$\frac{V_i}{n_i} = \frac{V_f}{n_f}$$

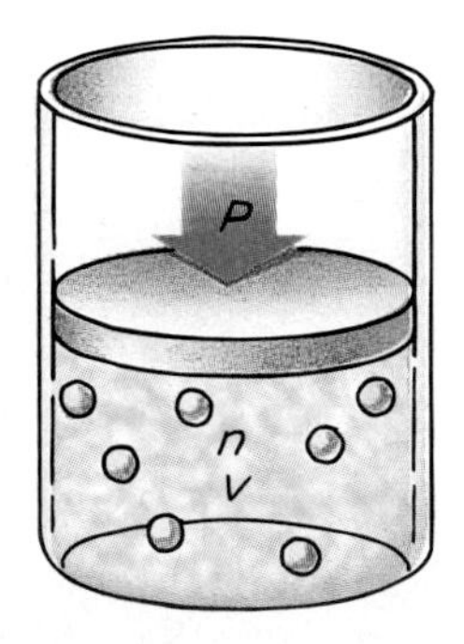

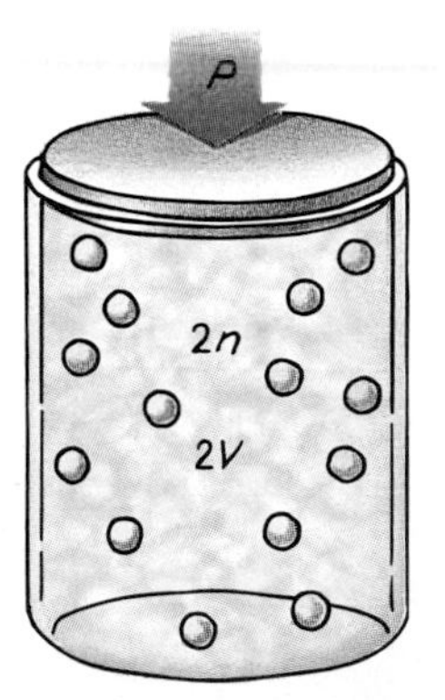

Figure 7.9
If the amount (number of moles) of a gas is doubled, the volume of the gas doubles when there is no change in pressure or temperature.

Sample Problem 7.12
Calculating Volume for a Change in Moles

A balloon with a volume of 220 mL is filled with 2.0 moles of helium. To what volume will the balloon expand if 3.0 moles of helium are added, to give a total of 5.0 moles of helium, and the pressure and temperature do not change?

Solution

A data table for our given information can be set up as follows:

Initial Conditions	*Final Conditions*	*Change*
$V_i = 220$ mL	$V_f = ?$	V must increase
$n_i = 2.0$ mol	$n_f = 5.0$ mol	n increases

Because the amount of gas (number of moles) has increased, the volume of the balloon must increase according to Avogadro's law.

$$V_f = 220 \text{ mL} \times \frac{5.0 \text{ moles}}{2.0 \text{ moles}} = 550 \text{ mL}$$

New volume | Initial volume | Mole factor that increases volume

Study Check

At a certain temperature and pressure, 8.00 g of oxygen has a volume of 5.00 L. What is the volume after 4.00 g of oxygen is added to the balloon?

STP and Molar Volume

Scientists have chosen standard temperature and pressure **(STP)** conditions of 0°C (273 K) and 1 atm (760 mm Hg).

STP Conditions

Standard temperature is 0°C (273 K)

Standard pressure is 1 atm (760 mm Hg)

At STP, 1 mol of any gas occupies a volume of 22.4 L. This value is known as the *molar volume* of a gas. (See Figure 7.10.) Suppose we have

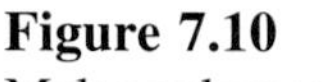

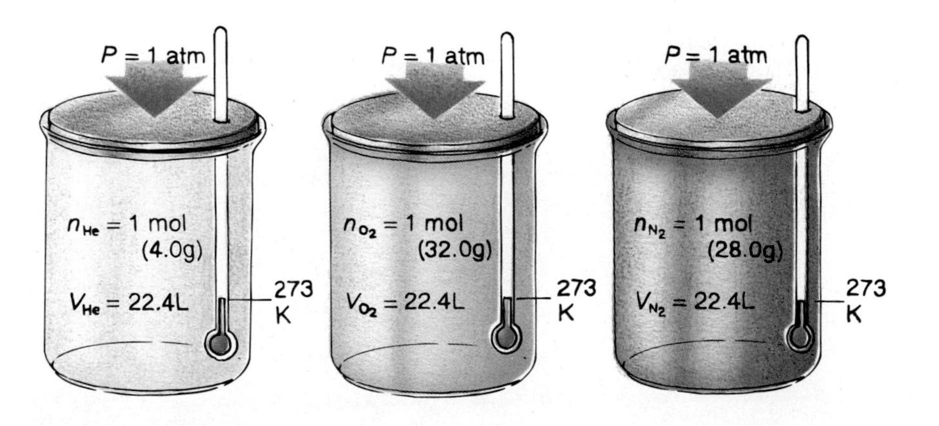

Figure 7.10
Molar volume: One mole of a gas at STP has a volume of 22.4 L.

three containers, one filled with 1 mol of oxygen gas (O_2), another filled with 1 mol of nitrogen gas (N_2), and one filled with 1 mol of helium gas (He). When the gases are at STP conditions (1 atm and 273 K), each has a volume of 22.4 L. Thus, the volume of 1 mol of any gas at STP is 22.4 L, the **molar volume** of a gas.

Molar Volume

The volume of 1 mol of gas at STP = 22.4 L

A long as a gas is at STP conditions (0°C and 1 atm), its molar volume can be used as a conversion factor to convert between the number of moles of gas and its volume.

Molar Volume Conversion Factors

$$\frac{1 \text{ mol gas (STP)}}{22.4 \text{ L}} \quad \text{and} \quad \frac{22.4 \text{ L}}{1 \text{ mol gas (STP)}}$$

Sample Problem 7.13
Calculations Using Molar Volume

What is the volume of 64.0 g O_2 gas at STP?

Solution
The mass of the gas is changed to moles by its molar mass, which is 32.0 g/mol. Because the gas is at STP, we can use the molar volume to convert from moles to liters of gas.

$$64.0 \text{ g } \cancel{O_2} \times \underbrace{\frac{1 \text{ } \cancel{\text{mol } O_2}}{32.0 \text{ g } \cancel{O_2}}}_{\text{Molar mass of } O_2} \times \underbrace{\frac{22.4 \text{ L } O_2}{1 \text{ } \cancel{\text{mol } O_2}}}_{\text{Molar volume of a gas}} = 44.8 \text{ L } O_2$$

Study Check
How many grams of nitrogen (N_2) gas are in 5.6 L of the gas at STP?

7.8 Partial Pressures (Dalton's Law)

Learning Goal
Use partial pressures to calculate the total pressure of a mixture of gases.

Many gas samples are a mixture of gases. For example, the air you breathe is a mixture of mostly oxygen and nitrogen gases. For a gas mixture, we use the gas laws we have studied, because particles of all gases behave in the same way. Therefore, the total pressure of the gases in a mixture is a result of the collisions of the gas particles regardless of what type of gas they are.

In a gas mixture, each gas exerts its **partial pressure**, which is the pressure it would exert if it were the only gas in the container. **Dalton's law**

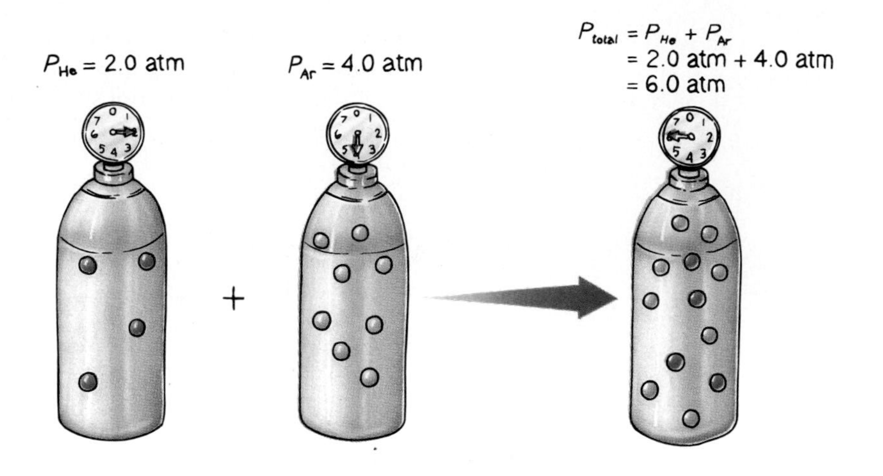

Figure 7.11
The total pressure of a gas mixture, which depends on the total number of gas particles present, is the sum of the partial pressures of the individual gases.

states that the total pressure of a gas mixture is the sum of the partial pressures of the gases in the mixture. (See Figure 7.11.)

$$P_{total} = P_1 + P_2 + P_3 + \cdots$$

Total pressure of a gas mixture = Sum of the partial pressures of the gases in the mixture

Sample Problem 7.14
Calculating the Total Pressure of a Gas Mixture

A 10-L gas tank contains propane (C_3H_8) gas at a pressure of 300 torr. Another 10-L gas tank contains methane (CH_4) gas at a pressure of 500 torr. In preparing a gas fuel mixture, the gases from both tanks are combined in a 10-L container at the same temperature. What is the pressure of the gas mixture?

Solution
Using Dalton's law of partial pressure, we find that the total pressure of the gas mixture is the sum of the partial pressures of the gases in the mixture.

$$\begin{aligned} P_{total} &= P_{propane} + P_{methane} \\ &= 300 \text{ torr} + 500 \text{ torr} \\ &= 800 \text{ torr} \end{aligned}$$

Therefore, when both propane and methane are placed in the same container, the total pressure of the mixture is 800 torr.

Study Check
A gas mixture consists of helium with a partial pressure of 315 mm Hg, nitrogen with a partial pressure of 204 mm Hg, and argon with a partial pressure of 422 mm Hg. What is the total pressure in atmospheres?

Air Is a Gas Mixture

The air you breathe is a mixture of gases. What we call the atmospheric pressure is actually the sum of the partial pressures of the gases in the air. Table 7.7 lists the partial pressures of the gases in air on a typical day.

Table 7.7 ***Composition of Air***

Gas	Pressure (mm Hg)	Percentage
Nitrogen, N_2	594.0	78
Oxygen, O_2	160.0	21
Carbon dioxide, CO_2	0.3	1
Water vapor, H_2O	5.7	
Total air	760.0	100

Sample Problem 7.15
Partial Pressure of a Gas in a Mixture

A mixture of oxygen and helium is prepared for a scuba diver who is going to descend 200 ft below the ocean surface. At that depth, the diver breathes a gas mixture that has a pressure of 7.0 atm. If the partial pressure of the oxygen at that depth is 1.5 atm, what is the partial pressure of the helium?

Solution
From Dalton's law of partial pressures, we know that the total pressure is equal to the sum of the partial pressures:

$$P_{\text{total}} = P_{O_2} + P_{\text{He}}$$

To solve for the partial pressure of helium (P_{He}), we rearrange the expression to give the following:

$$P_{\text{He}} = P_{\text{total}} - P_{O_2}$$
$$P_{\text{He}} = 7.0 \text{ atm} - 1.5 \text{ atm}$$
$$= 5.5 \text{ atm}$$

Thus, in the gas mixture that the diver breathes, the partial pressure of the helium is 5.5 atm.

Study Check
An anesthetic consists of a mixture of cyclopropane gas, C_3H_6, and oxygen gas, O_2. If the mixture has a total pressure of 825 torr, and the partial pressure of the cyclopropane is 73 torr, what is the partial pressure of the oxygen in the anesthetic?

Amount of Gas Dissolved in a Liquid

The amount of gas that dissolves in a liquid depends on the pressure of the gas. **Henry's law** states that the amount of gas that will dissolve in a liquid at a particular temperature is directly related to the pressure of that gas above the liquid. (See Figure 7.12.)

A can of soda contains dissolved carbon dioxide gas. When it is opened, much bubbling occurs. Sometimes the amount of escaping gas is so great that the liquid sprays out of the container. (See Figure 7.13.) When the can was originally filled and sealed, the pressure of the CO_2 gas in it was high, to make sure that sufficient CO_2 dissolved in the liquid. When the can is opened, the CO_2 pressure drops, less CO_2 is soluble, and bubbles of gas rapidly escape from the beverage.

Figure 7.12
Henry's Law: The amount of gas dissolved in a liquid is related to the pressure of that gas above the liquid. When the pressure of the gas increases, the number of gas molecules that enter the liquid increases.

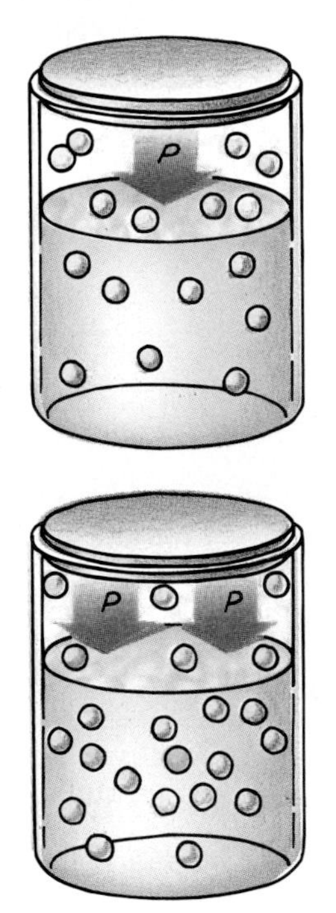

Blood Gases

Our cells continuously use oxygen and produce carbon dioxide. Both gases are transported in the bloodstream between the cells and the lungs. They diffuse in and out of the lungs through the membranes of the alveoli, the tiny air sacs at the ends of the airways in the lungs. An exchange of gases occurs in which oxygen from the air diffuses into the lungs and into the blood, while carbon dioxide produced in the cells is carried to the lungs to be exhaled. In Table 7.8, partial pressures are given for the gases in air that we inhale (inspired air), air in the alveoli, and the air that we exhale (expired air). The partial pressure of water vapor increases within the lungs, because the vapor pressure of water is 47 mm Hg at body temperature.

Oxygen normally has a partial pressure of 100 mm Hg in the alveoli. Because the partial pressure of oxygen in venous blood is 40 mm Hg, oxygen diffuses from the alveoli into the bloodstream. Most of the oxygen combines with hemoglobin, which carries it to the tissues of the body. There the partial pressure of oxygen can be very low, less than 30 mm Hg, causing oxygen to diffuse from the blood into the tissues.

Figure 7.13
When a can of soda is opened, the CO_2 gas in the liquid rapidly escapes.

Table 7.8 ***Partial Pressures of Gases During Breathing***

	Partial Pressure (mm Hg)		
Gas	**Inspired Air**	**Alveolar Air**	**Expired Air**
Nitrogen, N_2	594.0	573	569
Oxygen, O_2	160.0	100	116
Carbon dioxide, CO_2	0.3	40	28
Water vapor, H_2O	5.7	47	47
Total	760.0	760	760

As oxygen is used in the cells of the body during metabolic processes, carbon dioxide is produced, so the partial pressure of CO_2 may be as high as 50 mm Hg or more. Carbon dioxide diffuses from the tissues into the bloodstream and is carried to the lungs. There it diffuses out of the blood, where CO_2 has a partial pressure of 46 mm Hg, into the alveoli, where the CO_2 is at 40 mm Hg and is exhaled. (See Figure 7.14.) Table 7.9 gives the partial pressures of blood gases in the tissues and in oxygenated and deoxygenated blood.

Table 7.9 ***Partial Pressures of Oxygen and Carbon Dioxide in Blood and Tissues***

	Partial Pressure (mm Hg)		
Gas	**Oxygenated Blood**	**Deoxygenated Blood**	**Tissues**
O_2	100	40	30 or less
CO_2	40	46	50 or greater

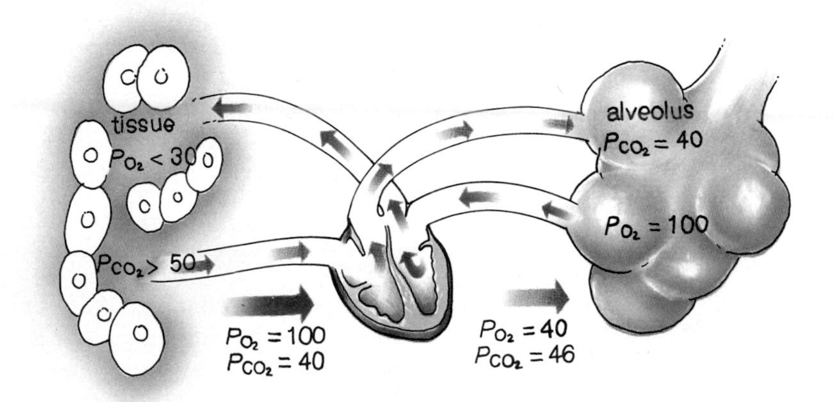

Figure 7.14
The diffusion of oxygen and carbon dioxide across the membranes of the alveoli and tissues during the exchange of blood gases.

HEALTH NOTE *Hyperbaric Chambers*

A burn patient may undergo treatment for burns and infections in a hyperbaric chamber, a device in which pressures can be obtained that are two to three times greater than atmospheric pressure. A greater oxygen pressure increases the level of dissolved oxygen in the blood and tissues, where it fights bacterial infections. The hyperbaric chamber may also be used during surgery, to help counteract carbon monoxide (CO) poisoning, and to treat some cancers. (See Figure 7.15.)

The blood is normally capable of dissolving up to 95% of the oxygen. Thus, if the partial pressure of the oxygen is 2280 mm Hg (3 atm), 95% of that or 2160 mm Hg of oxygen can dissolve in the blood where it saturates the tissues. In the case of carbon monoxide poisoning, this oxygen can replace the carbon monoxide that has attached to the hemoglobin.

A patient undergoing treatment in a hyperbaric chamber must also undergo decompression (reduction of pressure) at a rate that slowly reduces the concentration of dissolved oxygen in the blood. If decompression is too rapid, the oxygen dissolved in the blood may form gas bubbles in the circulatory system.

If divers do not decompress slowly, they suffer a similar condition called the bends. While below the surface of the ocean, divers breathe air at higher pressures. At such high pressures, nitrogen gas will dissolve in their blood. If they ascend to the surface too quickly, the dissolved nitrogen forms bubbles in the blood that can produce life-threatening blood clots. The gas bubbles can also appear in the joints and tissues of the body and be quite painful. A diver suffering from the bends is placed immediately in a decompression chamber where pressure is first increased and then slowly decreased. The dissolved nitrogen can then difffuse through the lungs until atmospheric pressure is reached.

Figure 7.15
A hyperbaric chamber exposes a patient to oxygen at a pressure that is two to three times atmospheric pressure.

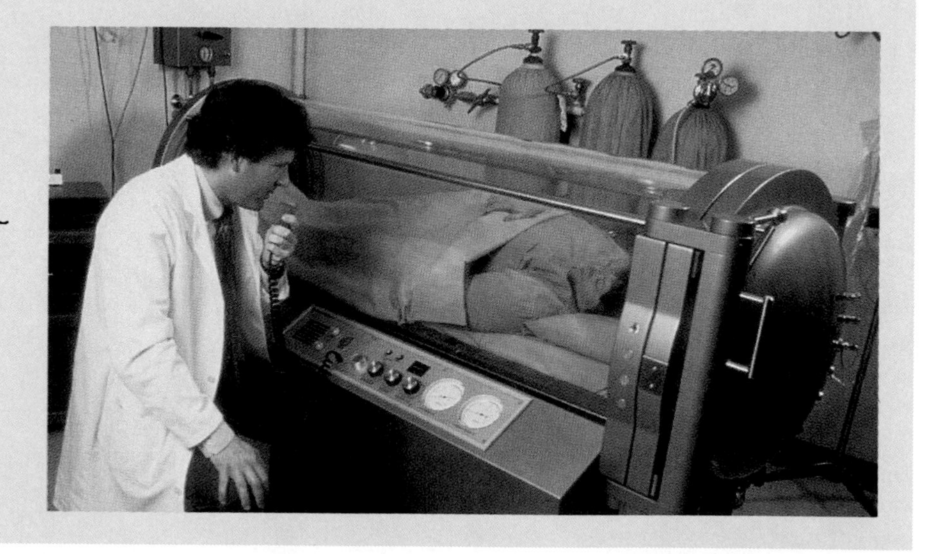

7.9 The Ideal Gas Law

Learning Goal
Use the ideal gas law to solve for pressure, volume, temperature, or amount of a gas when given values for the other properties.

The four properties used in the measurement of a gas—pressure (P), volume (V), temperature (T), and amount of a gas (n)—are combined to form the **ideal gas law** expression.

Ideal Gas Law

$$PV = nRT$$

The ideal gas law is a useful expression when you are given the measurements for any three of the four properties of a gas.

Property of a Gas	*Unit*
Pressure (P)	Atmosphere (atm)
Volume (V)	Liter (L)
Amount (n)	Mole (n)
Temperature (T)	Kelvin (K)

In the ideal gas law, the R value, called the **universal gas constant,** is the combination of the constants from the gas laws relating volume, pressure, temperature, and amount of a gas. The R value can be determined by placing the volume (22.4 L) of 1.00 mol at standard temperature (273 K) and pressure (1.00 atm) in the ideal gas equation and solving the equation for R.

$$\frac{PV}{nT} = \frac{\cancel{n}R\cancel{T}}{\cancel{n}\cancel{T}}$$

$$\frac{PV}{nT} = R$$

Universal gas constant

At STP,

$$\frac{(1.00\ \text{atm})(22.4\ \text{L})}{(1.00\ \text{mol})(273\ \text{K})} = 0.0821\ \frac{\text{L} \cdot \text{atm}}{\text{mol} \cdot \text{K}}$$

So the universal gas constant has a value of $0.0821\ \text{L} \cdot \text{atm/mol} \cdot \text{K}$.

Sample Problem 7.16
Calculations Using the Ideal Gas Law

Nitrous oxide, N_2O, is an anesthetic known as "laughing gas." What is the pressure in atmospheres of 0.35 mole of N_2O gas at 22°C in a 5.00-L tank?

Solution

Organizing the given measurements gives

$P = ?$

$V = 5.00\ \text{L}$

$n = 0.35\ \text{mole}$

$T = 22°\text{C}\ (295\ \text{K})$

To calculate the pressure of the N_2O gas, rearrange the ideal gas law to solve for P:

$$PV = nRT$$

$$\frac{P\cancel{V}}{\cancel{V}} = \frac{nRT}{V}$$

$$P = \frac{nRT}{V}$$

After checking that the measured units match the units of the gas constant R, substitute the values into the expression and solve for pressure. All units cancel except for the pressure in atmospheres.

$$P = \frac{(0.35\ \cancel{mol})(0.0821\ \cancel{L} \cdot atm/\cancel{mol} \cdot \cancel{K})(295\ \cancel{K})}{(5.00\ \cancel{L})} = 1.7\ atm$$

Study Check

Chlorine gas, Cl_2, is used to purify water in swimming pools. How many grams of chlorine are in a 7.00-L tank of gas at a pressure of 760.0 mm Hg and 24°C? (*Hint:* Solve for the number of moles of Cl_2, and change the units of other properties to match the units of the gas constant R.)

Chapter Summary

7.1 Properties of Gases

In a gas, particles are so far apart and moving so fast that they are not attracted to each other. A gas is described by the physical properties of pressure (P), volume (V), temperature (T), and amount in moles (n).

7.2 Gas Pressure

A gas exerts pressure, the force of the gas particles on the surface of a container. Units of gas pressure include torrs, mm Hg, and atmospheres. Vapor pressure is the pressure of the gas that forms when a liquid evaporates. At the boiling point of a liquid, the vapor pressure equals the atmospheric pressure.

7.3 Pressure and Volume (Boyle's Law)

The volume (V) of a gas changes inversely with the pressure (P) of the gas if there is no change in the temperature: $P_i V_i = P_f V_f$. This means that the pressure increases if volume decreases; pressure decreases, if volume increases.

7.4 Temperature and Volume (Charles' Law)

The volume (V) of a gas is directly related to its Kelvin temperature (T) when there is no change in the pressure of the gas:

$$\frac{V_i}{T_i} = \frac{V_f}{T_f}$$

Therefore, if temperature increases, the volume of the gas increases; if temperature decreases, volume decreases.

7.5 Temperature and Pressure (Gay-Lussac's Law)

The pressure (P) of a gas is directly related to its Kelvin temperature (T).

$$\frac{P_i}{T_i} = \frac{P_f}{T_f}$$

This means that an increase in temperature (T) increases the pressure of a gas, or a decrease in temperature decreases the pressure, as long as the volume stays constant.

7.6 The Combined Gas Law of Pressure, Volume, and Temperature Relationships

Gas laws combined into a relationship of pressure (P), volume (V), and temperature (T).

$$\frac{P_i V_i}{T_i} = \frac{P_f V_f}{T_f}$$

This expression is used to determine the effect of changes in two of the variables on the third.

7.7 Volume and Moles (Avogadro's Law)

The volume (V) of a gas is directly related to the number of moles (n) of the gas when the pressure and temperature of the gas do not change.

$$\frac{V_i}{n_i} = \frac{V_f}{n_f}$$

If the moles of gas are increased, the volume must increase; or if the moles of gas are decreased, the volume decreases.

7.8 Partial Pressures (Dalton's Law)

In a mixture of two or more gases, the total pressure is the sum of the partial pressures of the individual gases.

$$P_{total} = P_1 + P_2 + P_3 + \cdots$$

The partial pressure of a gas in a mixture is the pressure it would exert if it were the only gas in the container.

7.9 The Ideal Gas Law

The ideal gas law gives the relationship between the four variables: pressure, volume, moles, and temperature. When measurements of any three variables are given, the fourth can be calculated.

Glossary of Key Terms

atmosphere (atm) The pressure exerted by a column of mercury 760 mm high.

atmospheric pressure The pressure exerted by the atmosphere.

Avogadro's law A law that states that the volume of gas is directly related to the number of moles of gas in the sample when pressure and temperature do not change.

barometer An instrument used to measure atmospheric pressure.

Boyle's law A gas law stating that the pressure of a gas is inversely related to the volume, when temperature and the moles of gas do not change; that is, if volume decreases, pressure increases.

Charles' law A gas law stating that the volume of a gas changes directly with a change in Kelvin temperature, when pressure and the moles of gas do not change.

combined gas law A relationship that combines the gas laws relating pressure, volume, and temperature.

$$\frac{P_i V_i}{T_i} = \frac{P_f V_f}{T_f}$$

Dalton's Law The total pressure exerted by a mixture of gases in a container is the sum of the partial pressures that each gas would exert alone.

diffusion The movement of gas particles from a more concentrated area (high pressure) to a less concentrated area (low pressure).

direct relationship A relationship in which two properties increase or decrease together.

Gay-Lussac's law A gas law stating that the pressure of a gas changes directly with a change in temperature when the number of moles of a gas and its volume are held constant.

Henry's Law The amount of gas dissolved in a liquid at a certain temperature is directly related to the pressure of the gas above the liquid.

ideal gas law A law that combines the four measured properties of a gas in the equation $PV = nRT$.

inverse relationship A relationship in which two properties change in opposite directions.

kinetic theory A model used to explain the behavior of gases.

molar volume A volume of 22.4 L occupied by 1 mol of a gas at STP conditions of 0°C (273 K) and 1 atm.

partial pressure The pressure exerted by a single gas in a gas mixture.

pressure The force exerted by gas particles that hit the walls of a container.

pressure gradient A difference in pressure causing gas particles to move from the area of high pressure to the area of low pressure.

STP Standard conditions of 0°C (273 K) temperature and 1 atm pressure used for the comparison of gases.

torr A unit of pressure equal to 1 mm Hg; 760 torr = 1 atm.

universal gas constant (*R*) A numerical value that relates the properties P, V, n, and T in the ideal gas law, $PV = nRT$.

vapor pressure The pressure exerted by the particles of vapor above a liquid.

Chemistry at Home

1. Pour a small amount of ammonia cleaner in a bowl. How close do you have to be to smell the ammonia? In 5 minutes, return to the bowl of ammonia. How close do you have to get before you can smell the ammonia?
2. Blow up a balloon part way. Place the balloon in the sun, or a warm place. Observe any changes. Then place the balloon in the refrigerator or freezer. What kinds of changes in volume do you observe?
3. Fill a glass one-quarter full of vinegar. Place a small amount of baking soda in the finger of a plastic glove. (The glove should fit tightly around the glass.) Carefully place the glove on the glass and shake the baking soda into the vinegar. What happens to the glove? Explain.
4. Adhere a small candle to the bottom of a shallow bowl. Light the candle and place a glass over the burning candle. What happens? What gas from the air is needed for combustion? Remove the glass and place some water around the candle. Light the candle again and cover it again with a glass. What happens to the candle? What happens to the water? How would you explain your observations?
5. Place small, equal amounts of water in a wide, shallow dish and in a tall, narrow glass. Leave on the counter. In which one does evaporation occur most rapidly? Why?

Answers to Study Checks

7.1 Because increasing the temperature increases the motion of the gas particles, they move faster and hit the walls of the container with more force, increasing the pressure of the gas in the container.
7.2 amount
7.3 0.862 atm
7.4 The volume increases.
7.5 400 torr
7.6 50.0 mL
7.7 Temperature must decrease.
7.8 589 mL
7.9 Pressure will decrease.
7.10 16°C
7.11 241 mm Hg
7.12 7.50 L
7.13 7.0 g N_2
7.14 1.24 atm
7.15 752 torr
7.16 20.4 g Cl_2

Problems

Properties of a Gas *(Goal 7.1)*

7.1 What does the kinetic theory say about the following?
- **a.** Particles of a gas.
- **b.** Motion of gas particles.
- **c.** Distance between gas particles.
- **d.** Attraction between gas particles.
- **e.** Effect of temperature on gases.

7.2 How does the kinetic theory explain the following?
- **a.** Gases move faster at higher temperatures.
- **b.** Gases have very low densities.
- **c.** Gases can be compressed much more than solids.
- **d.** An aerosol can explodes when thrown into a fire.
- **e.** The air in a hot-air balloon is heated to make the balloon rise.

7.3 What are the typical units used to measure the volume, temperature, and quantity of a gas?

7.4 Use the words, *pressure, volume, temperature, or amount* to indicate the property of a gas that is described in each of the following statements or measurements:
- **a.** 350 K
- **b.** 4.00 mol He
- **c.** Determines the kinetic energy of the gas particles.
- **d.** Space occupied by a gas.
- **e.** 10.0 L

Gas Pressure *(Goal 7.2)*

7.5 What units are used to describe the pressure of gas?

7.6 Which of the following statement(s) describes the pressure of a gas?
- **a.** The force of the gas particles on the walls of the container.
- **b.** The number of gas particles in a container.
- **c.** The volume of the container.
- **d.** 3.00 atm
- **e.** 750 torr

7.7 An oxygen tank contains oxygen (O_2) at a pressure of 2.00 atm. What is the pressure in the tank in terms of the following units?
a. torr **b.** $lb/in.^2$ **c.** mm Hg

7.8 On a climb up Mt. Whitney, the atmospheric pressure is 467 mm Hg. What is the pressure in terms of the following units?
a. atm **b.** torr **c.** in. Hg

7.9 Match the terms *vapor pressure, atmospheric pressure, and boiling point* to the following descriptions.
- **a.** The temperature at which bubbles of vapor appear within the liquid.
- **b.** The pressure exerted by a gas above the surface of its liquid.
- **c.** The pressure exerted on the earth by the particles in the air.
- **d.** The temperature at which the vapor pressure of a liquid becomes equal to atmospheric pressure.

7.10 In which pair(s) would boiling occur?

	Atmospheric Pressure	*Vapor Pressure*
a.	760 mm Hg	700 mm Hg
b.	480 torr	480 mm Hg
c.	1.2 atm	912 mm Hg
d.	1020 mm Hg	760 mm Hg
e.	740 torr	1.0 atm

7.11 Give an explanation for the following observations:
a. Water boils at 87°C on the top of Mt Whitney.
b. Food cooks more quickly in a pressure cooker than in an open pan.

7.12 Give an explanation for the following observations:
a. A wet towel dries faster on a windy day.
b. Water used to sterilize surgical equipment is heated to 120°C at 2.0 atm in an autoclave.

Pressure and Volume (Boyle's Law) *(Goal 7.3)*

7.13 Complete the following using the pressure–volume relationship:

	Pressure	*Volume*
a.	decreases	_____
b.	increases	_____

7.14 Complete the following using the pressure–volume relationship:

	Pressure	*Volume*
a.	_____	decrease
b.	_____	increases

7.15 A 10.0-L balloon contains He gas at a pressure of 655 torr. What is the the new pressure of the He gas at each of the following volumes if there is no change in temperature?
a. 20.0 L **b.** 5.00 L
c. 1500 mL **d.** 100.0 mL

7.16 The air in a 5.00-L tank has a pressure of 1.20 atm. What is the new pressure of the air when the air is placed in tanks that have the following volumes if there is no change in temperature?
a. 1.00 L **b.** 2500 mL
c. 750 mL **d.** 20.00 L

7.17 A sample of nitrogen (N_2) gas has a volume of 4.5 L at a pressure of 760 mm Hg. What is the new pressure if the gas sample is compressed to a volume of 1.5 L if there is no change in temperature?

7.18 An emergency tank of oxygen holds 20.0 L of oxygen (O_2) at a pressure of 15.0 atm and 22°C. When the gas is released at 22°C, it provides 300.0 L of oxygen. What is the pressure at which the oxygen is released?

7.19 A sample of nitrogen (N_2) has a volume of 50.0 L at a pressure of 760 mm Hg. What is the volume of the gas at each of the following pressures if there is no change in temperature?
a. 1500 mm Hg **b.** 2.0 atm
c. 0.500 atm **d.** 850 torr

7.20 A sample of methane (CH_4) has a volume of 25 mL at a pressure of 0.80 atm. What is the volume of the gas at each of the following pressures if there is no change in temperature?
a. 0.40 atm **b.** 2.00 atm
c. 2500 mm Hg **d.** 80.0 torr

7.21 Cyclopropane, C_3H_6, is a general anesthetic. A 5.0-L sample has a pressure of 5.0 atm. What is the volume of the anesthetic given to a patient at a pressure of 1.0 atm?

7.22 The volume of air in a person's lungs is 615 mL at a pressure of 760 mm Hg. Inhalation occurs as the pressure in the lungs drops to 752 mm Hg. To what volume did the lungs expand?

7.23 Use the words *inspiration* and *expiration* to describe the part of the breathing cycle that occurs as a result of each of the following:
a. The diaphragm contracts (flattens out).
b. The volume of the lungs decreases.
c. The pressure within the lungs is less than the atmosphere.

7.24 Use the words *inspiration* and *expiration* to describe the part of the breathing cycle that occurs as a result of each of the following:
a. The diaphragm relaxes, moving up into the thoracic cavity.
b. The volume of the lungs expands.
c. The pressure within the lungs is greater than the atmosphere.

Temperature and Volume (Charles' Law) *(Goal 7.4)*

7.25 State the change that occurs for volume or temperature when P and n are not changed.

	Volume	*Temperature*
a.	_____	decreases
b.	increases	_____

7.26 State the change that occurs for volume or temperature when P and n are not changed.

	Volume	*Temperature*
a.	decreases	_____
b.	_____	increases

7.27 A balloon contains 2500 mL of helium gas at 75°C. What is the new volume of the gas when the temperature changes to the following, if n and P are not changed?
a. 55°C **b.** 680 K
c. −25°C **d.** 240 K

7.28 A gas has a volume of 4.00 L at 0°C. What final temperature in degrees Celsius is needed to cause the volume of the gas to change to the following, if n and P are not changed?
a. 10.0 L **b.** 1200 mL
c. 2.50 L **d.** 50.0 mL

Temperature and Pressure (Gay-Lussac's Law) *(Goal 7.5)*

7.29 State the expected change for temperature and pressure according to Gay-Lussac's Law, with no change in n and V.

	Pressure	*Temperature*
a.	_____	increases
b.	_____	decreases

7.30 State the expected change for temperature and pressure according to Gay-Lussac's Law, with no change in n and V.

	Pressure	*Temperature*
a.	decreases	_____
b.	increases	_____

7.31 Solve for the new pressure when each of the following temperature changes occurs, with n and V constant:

a. A gas sample has a pressure of 1200 torr at 155°C. What is the final pressure of the gas after the temperature has dropped to 0°C?

b. An aerosol can has a pressure of 1.40 atm at 12°C. What is the final pressure in the aerosol can if it is used in a room where the temperature is 35°C?

7.32 Solve for the new temperature in degrees Celsius when pressure is changed.

a. A 10.0-L container of helium gas has a pressure of 250 torr at 0°C. To what Celsius temperature does the sample need to be heated to obtain a pressure of 1500 torr?

b. A 500.0-mL sample of air at −10°C is inhaled. What is the new volume when the temperature of the air sample rises to body temperature of 37°C?

The Combined Gas Law of Pressure, Volume, and Temperature Relationships *(Goal 7.6)*

7.33 Write the expression for the combined gas law. What gas laws are combined to make the combined gas law?

7.34 Rearrange the variables in the combined gas law to give an expression for the following:

a. V_f **b.** P_f

7.35 A sample of Cl_2 gas has a volume of 10.0 L at a pressure of 2.00 atm and a temperature of 15°C. What is the pressure of the gas when the volume and temperature of the gas are changed to the following:

a. 5.00 L and 125°C
b. 20.0 L and 0°C
c. 760 mL and 250°C
d. 50.0 L and 350°F

7.36 A sample of argon gas has a volume of 750 mL at a pressure of 1.00 atm and a temperature of 127°C. What is the volume of the gas when the pressure and temperature of the gas are changed to the following:

a. 2.50 atm and 250°C
b. 0.58 atm and 75°C
c. 4.00 atm and −23°C
d. 0.85 atm and 400°C

7.37 A 100.0-mL bubble of hot gases at 225°C and 1.80 atm escapes from an active volcano. What is the new volume of the bubble outside the volcano where the temperature is −25°C and the pressure is 0.80 atm?

7.38 A scuba diver 40 ft below the ocean surface inhales 50.0 mL of compressed air in the scuba tank at a pressure of 3.00 atm and a temperature of 8°C. What is the pressure of air in the lungs if the gas expands to 150.0 mL at a body temperature of 37°C?

Volume and Moles (Avogadro's Law) *(Goal 7.7)*

7.39 Solve the following problems when no changes occur in pressure and temperature:

a. A sample of 4.00 mol of argon gas has a volume of 10.0 L. A small leak causes one-half of the molecules to escape. What is the new volume of the gas?

b. A balloon containing 1.00 mol of helium has a volume of 440 mL. What is the new volume after 2.00 more mol of helium are added to the balloon?

c. A 1500 mL sample of SO_2 contains 6.00 mol SO_2. How many moles of SO_2 are lost when the volume is reduced to 750 mL?

7.40 Solve the following problems when no changes occur in pressure and temperature:

a. A 250-mL sample of neon gas contains 1.0 mol Ne. What is the new volume if 3.0 mol of Ne are added to the container?

b. A 10.0-L sample of helium gas contains 4.00 g He. If 2.00 g of helium are added to the initial sample, what is the new volume of the gas at the same pressure and temperature?

c. A 2500-mL sample of O_2 contains 5.00 mol O_2. If oxygen is released until the volume drops to 1500 mL, how many moles of O_2 were lost?

7.41 Use the molar volume of a gas to solve the following at STP:

a. The number of moles of O_2 in 44.8 L of O_2 gas.

b. The number of moles of CO_2 in 4.00 L of CO_2 gas.

c. The volume (L) of 6.40 g of O_2.
d. The volume (mL) occupied by 50.0 g of neon.

7.42 Use molar volume to solve the following problems at STP:
a. The volume (L) occupied by 2.5 mol N_2.
b. The volume (mL) occupied by 0.420 mol He.
c. The number of grams of neon contained in 11.2 L Ne gas.
d. The number of moles of H_2 in 1600 mL of H_2 gas.

Partial Pressures (Dalton's Law) *(Goal 7.8)*

7.43 In a gas mixture, the partial pressures are nitrogen 425 torr, oxygen 115 torr, and helium 225 torr. What is the total pressure (torr) exerted by the gas mixture?

7.44 In a gas mixture, the partial pressures are argon 415 mm Hg, neon 75 mm Hg, and nitrogen 125 mm Hg. What is the total pressure (atm) exerted by the gas mixture?

7.45 A gas mixture containing oxygen, nitrogen, and helium exerts a total pressure of 925 torr. If the partial pressures are oxygen 425 torr and helium 75 torr, what is the partial pressure (torr) of the nitrogen in the mixture?

7.46 A gas mixture containing oxygen, nitrogen, and neon exerts a total pressure of 1.20 atm. If helium added to the mixture increases the pressure to 1.50 atm, what is the partial pressure (atm) of the helium?

The Ideal Gas Law *(Goal 7.9)*

7.47 What is the pressure, in atmospheres, of 2.0 mol of helium gas in a 10.0-L container at 27°C?

7.48 How many liters are occupied by 5.0 mol of methane gas, CH_4, at a temperature of 0°C and 2.00 atm?

7.49 A steel cylinder of oxygen has a volume of 20.0 L at 22°C, and the oxygen has a pressure of 35 atm. How many mol of oxygen are in the container?

7.50 A cylinder contains 1250 mL of krypton at a temperature of 35°C and a pressure of 475 mm Hg. How many moles of krypton are in the cylinder?

7.51 A gas cylinder contains 64.0 g of oxygen (O_2) at a temperature of 25°C and a pressure of 850 torr. What is the volume in liters of the oxygen gas in the cylinder?

7.52 A sample of ammonia gas, NH_3, is collected in a container that has a volume of 525 mL at a temperature of 15°C and a pressure of 455 torr. How many grams of NH_3 are in the container?

Challenge Problems

7.53 A fire extinguisher has a pressure of 150 lb/in.^{2} at 25°C. What is the pressure in atmospheres if the fire extinguisher is used at a temperature of 75°C?

7.54 A weather balloon has a volume of 750 L when filled with helium at 8°C at a pressure of 380 torr. What is the new volume of the balloon, where the pressure is 0.20 atm and the temperature is −45°C?

7.55 A sample of hydrogen (H_2) gas at 127°C has a pressure of 2.00 atm. At what temperature (°C) will the pressure of the H_2 decrease to 0.25 atm?

7.56 Nitrogen (N_2) is prepared and a sample of 250 mL is collected over water at 30°C and a total pressure of 745 mm Hg.
a. Using the vapor pressure of water in Table 7.3, what is the partial pressure of the nitrogen?
b. What is the volume of the nitrogen at STP?

7.57 A gas mixture with a total pressure of 2400 torr is used by a scuba diver. If the mixture contains 2.0 mol of helium and 6.0 mol of oxygen, what is the partial pressure of each gas in the sample?

7.58 What is the total pressure in mm Hg of a gas mixture containing argon gas at 0.25 atm, helium gas at 350 mm Hg, and nitrogen gas at 360 torr?

7.59 A gas mixture contains oxygen and argon at partial pressures of 0.60 atm and 425 mm Hg. If nitrogen gas added to the sample increases the total pressure to 1250 torr, what is the partial pressure of the nitrogen added?

7.60 A gas mixture contains helium and oxygen at partial pressures of 255 torr and 0.450 atm. What is the total pressure of the mixture after it is placed in a container one-half the volume of the original container?

7.61 Calcium carbonate decomposes with heat to give calcium oxide and carbon dioxide gas.

$$CaCO_3(s) \rightarrow CaO(s) + CO_2(g)$$

If 2.00 mol of $CaCO_3$ react, how many liters of CO_2 gas are produced at STP?

7.62 In the reaction

$$2H_2(g) + O_2(g) \rightarrow 2H_2O(g)$$

a. If 16.00 g O_2 react at 0.800 atm and 127°C, what volume of oxygen was used?
b. How many moles of water are produced in part (a)?
c. If the water is produced at 760 mm Hg and 115°C, what volume of water was collected?

Chapter 8

Solutions

A solution consists of a solute and a solvent that are uniformly mixed.

Looking Ahead

8.1 Water and Solutions
Health Note *Water in the Body*

8.2 Formation of Ionic Solutions

8.3 Factors Affecting Solubility
Health Note *Gout: A Problem of Saturation in Body Fluids*
Health Note *Kidney Stones Are Insoluble Salts*

8.4 Percent Concentration

8.5 Molarity

8.6 Colloids and Suspensions
Health Note *Colloids and Solutions in the Body*

8.7 Osmosis and Dialysis
Health Note *Dialysis by the Kidneys and the Artificial Kidney*

Learning Goals

8.1 Describe hydrogen bonding in water; identify the solute and solvent in a solution.

8.2 Describe the process of dissolving an ionic solute in water.

8.3 Describe the effects of temperature and nature of the solute on its solubility in a liquid.

8.4 Calculate the percent concentration of a solute in a solution; use percent concentration to calculate the amount of solute or solution.

8.5 Calculate the molarity of a solution; use molarity as a conversion factor to calculate the moles of solute or the volume needed to prepare a solution.

8.6 From its properties, identify a mixture as a solution, a colloid, or a suspension.

8.7 Describe the changes in concentration of solute and solvent in the processes of osmosis and dialysis.

Questions to Think About

1. Why does salt preserve foods?
2. How do plants obtain water from the ground?
3. Why is water balance in the body important?
4. Why is a pickle all shriveled up?
5. Why can't you drink sea water?
6. How do your kidneys dialyze toxic substances out of the blood?

Solutions are everywhere around us. Each consists of a mixture of at least one substance dissolved in another. The air we breathe is a solution of oxygen and nitrogen gases. Carbon dioxide gas dissolved in water makes the soda in our carbonated drinks. When we make solutions of coffee or tea, we use hot water to dissolve substances from coffee beans or tea leaves. The ocean is also a solution, consisting of many salts such as sodium chloride dissolved in water. In the hospital, the antiseptic tincture of iodine is a solution of iodine dissolved in alcohol.

Another type of mixture, called a colloid, contains large particles. Mayonnaise, whipped crean, and gelatin are all colloids. In a third type of mixture, called a suspension, the particles are so large they settle out. You may have been given suspensions such as liquid penicillin or calamine lotion, which had to be shaken before use.

In the processes of osmosis and dialysis, we will see how water, essential nutrients, and waste products enter and leave the cells of the body. In osmosis, water flows in and out of the cells of the body. In dialysis, small particles in solution as well as water diffuse through semipermeable membranes.

8.1 Water and Solutions

Learning Goal

Describe hydrogen bonding in water; identify the solute and solvent in a solution.

A solution is a mixture in which one substance called the **solute** is dispersed uniformly in another substance called the **solvent.** Because the solute and the solvent do not react with each other, they can be mixed in varying proportions. A little salt dissolved in water tastes slightly salty. When more salt dissolves, the water tastes very salty. Usually, the solute (in this case, salt) is the substance present in the smaller amount, whereas the solvent (in this case, water) is the larger amount. The salt solution forms when the particles of the solute (salt) become evenly dispersed among the molecules of the solvent (water). (See Figure 8.1.)

Figure 8.1

A solution of copper sulfate ($CuSO_4$) is formed when particles of solute become evenly dispersed among the solvent particles.

Water: An Important Solvent

Water, the most common solvent in nature, is referred to as the universal solvent. In the H_2O molecule, an oxygen atom shares electrons with two hydrogen atoms. Because the oxygen atom is much more electronegative, the O — H bonds are strongly polar. In each of the polar bonds, the oxygen atom has a partially negative (δ^-) charge, and the hydrogen atom has a partially positive (δ^+) charge. As a solvent, we say that water is a *polar solvent.*

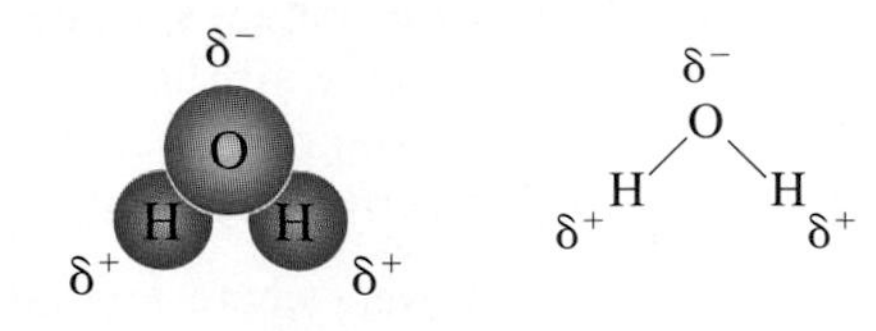

Water, a polar molecule

Hydrogen Bonds in Water

The polarity of water molecules is so great that strong interactions occur *between* the water molecules. The **hydrogen bond** is an attraction between the δ^- of an oxygen atom and the δ^+ of a hydrogen atom in another water molecule. In Figure 8.2, hydrogen bonds are shown as dots between the water molecules. Hydrogen bonds occur between molecules of compounds where hydrogen is bonded to the very electronegative atoms of O, N, and F. Although hydrogen bonds are weaker than covalent or ionic bonds, there are many of them linking molecules together. As a result hydrogen bonding plays an important role in the properties of water, biological compounds such as proteins, and DNA.

Surface Tension

Imagine that you have filled a glass up to the rim with water. Carefully adding a few more drops of water or dropping in some pennies does not cause the water to overflow. Instead the water seems to adhere to itself, forming a dome that rises slightly above the rim of the glass. This effect is the result of the polarity of water. Throughout the liquid in the glass, water molecules are attracted in all directions by surrounding water molecules. However, the water molecules on the surface are pulled like a skin toward the rest of the water in the glass. As a result, the water molecules on the surface become

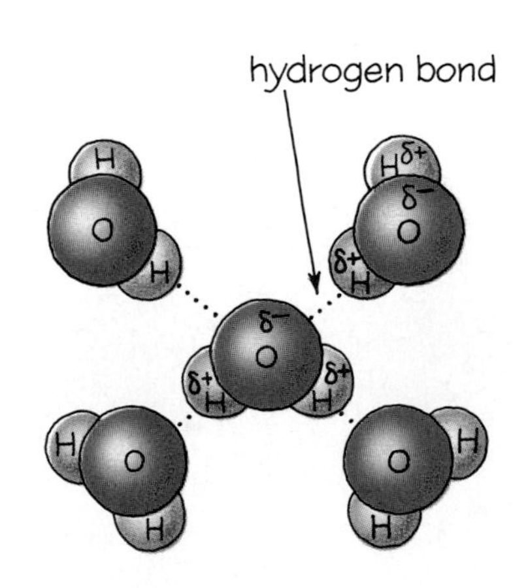

Figure 8.2
Hydrogen bonding links water molecules together.

Figure 8.3
The surface tension of water forms a resting place for this water bug.

more tightly packed, a feature called **surface tension.** Because of surface tension, a needle floats on the top of water, certain water bugs can travel across the surface of a pond or lake, and drops of water are spherical. (See Figure 8.3.)

Types of Solutes and Solvents

Solutes and solvents may be solids, liquids, or gases. The solution that forms has the same physical state as the solvent. When sugar is dissolved in a glass of water, a liquid sugar solution forms. Sugar is the solute, and water is the solvent. The carbonated solutions of soda water and soft drinks are prepared by dissolving CO_2 gas in water. The CO_2 gas is the solute, and water is the solvent. Table 8.1 lists some solutes and solvents and their solutions.

Table 8.1 *Some Examples of Solutions*

Type	Example	Solute	Solvent
Gas Solutions			
Gas in a gas	Air	Oxygen (gas)	Nitrogen (gas)
Liquid Solutions			
Gas in a liquid	Soda water	Carbon dioxide (gas)	Water (liquid)
	Household ammonia	Ammonia (gas)	Water (liquid)
Liquid in a liquid	Vinegar	Acetic acid (liquid)	Water (liquid)
Solid in a liquid	Seawater	Sodium chloride (solid)	Water (liquid)
	Tincture of iodine	Iodine (solid)	Alcohol (liquid)
Solid Solutions			
Liquid in a solid	Dental amalgam	Mercury (liquid)	Silver (solid)
Solid in a solid	Brass	Zinc (solid)	Copper (solid)
	Steel	Carbon (solid)	Iron (solid)

Sample Problem 8.1
Identifying a Solute and a Solvent

Identify the solute and the solvent in each of the following solutions:

a. 1.0 g of sugar dissolved in 100 g of water
b. 50 mL of water mixed with 20 mL of isopropyl alcohol (rubbing alcohol)

Solution

a. Sugar is the smaller quantity that is dissolving. It is the solute; water is the solvent.
b. Because both water and isopropyl alcohol are liquids, the one with the smaller volume, isopropyl alcohol, is the solute. Water is the solvent.

Study Check

A tincture of iodine is prepared with 0.10 g I_2 and 10.0 mL of ethyl alcohol. What is the solute, and what is the solvent?

HEALTH NOTE *Water in the Body*

The average adult contains about 60% water by weight, and the average infant about 75%. About 60% of the body's water is contained within the cells as intracellular fluids; the other 40% makes up extracellular fluids, which include the interstitial fluid in tissues, and the plasma in the blood. These external fluids carry nutrients and waste materials between the cells and the circulatory system.

Every day you lose between 1500 and 3000 mL of water from the kidneys as urine, from the skin as perspiration, from the lungs as you exhale, and from the gastrointestinal tract. Serious dehydration can occur in an adult if there is a 10% net loss in total body fluid, and a 20% loss of fluid can be fatal. An infant suffers severe dehydration with a 5–10% loss in body fluid.

Water lost is continually replaced by the liquids and foods in the diet and from metabolic processes that produce water in the cells of the body. (See Figure 8.4.) Table 8.2 lists the amounts of water contained in some foods.

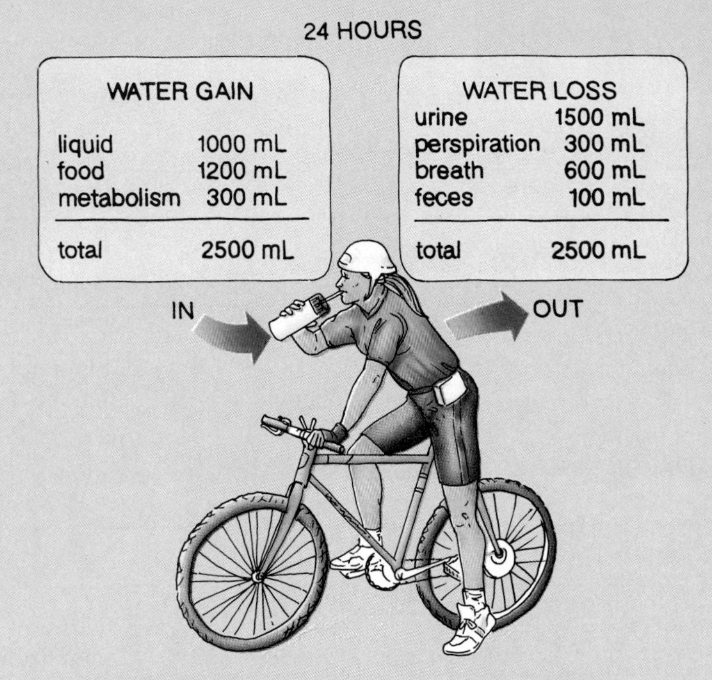

24 HOURS
WATER GAIN
liquid 1000 mL
food 1200 mL
metabolism 300 mL
total 2500 mL
WATER LOSS
urine 1500 mL
perspiration 300 mL
breath 600 mL
feces 100 mL
total 2500 mL
IN
OUT

Figure 8.4
The balance of water loss and gain in the body.

Table 8.2 ***Percentage of Water in Some Foods***

Food	Water (%)	Food	Water (%)
Vegetables		***Meats/Fish***	
Carrot	88	Chicken, cooked	71
Celery	94	Hamburger, broiled	60
Cucumber	96	Salmon	71
Tomato	94	***Grains***	
Fruits		Cake	34
Apple	85	French bread	31
Banana	76	Noodles, cooked	70
Cantaloupe	91	***Milk Products***	
Grapefruit	89	Cottage cheese	78
Orange	86	Milk, whole	87
Strawberry	90	Yogurt	88
Watermelon	93		

8.2 Formation of Ionic Solutions

Learning Goal

Describe the process of dissolving an ionic solute in water.

In solutions of gases, the particles mix easily because they are moving rapidly, are far apart, and are not attracted to or repelled by the other gas particles. However, to form solutions of liquids, or solids dissolved in liquids, there must be an attraction between the solute and the solvent. Then the solute and solvent will separate into individual ions or molecules and form a solution. If no attraction between solute and solvent occurs, the particles of the solute stay together and will not mix with the solvent.

Dissolving Ionic Solutes

The formation of a solution depends on similar polarities of the solute and solvent particles. An ionic compound, such as sodium chloride (NaCl), dissolves in water because water is a polar solvent. As the water molecules bombard the surface of the salt, as shown in Figure 8.5, the Cl^- ions are attracted to the partially positive hydrogen atoms of the water molecules. The Na^+ ions are attracted to the partially negative oxygen atoms. These attractions pull the ions into the water, where they undergo hydration: They become surrounded by several water molecules. A sodium chloride solution has formed.

Figure 8.5
The dissolving of NaCl, an ionic solute, in water. Attracted by polar water molecules, Na^+ and Cl^- move into solution where they are hydrated.

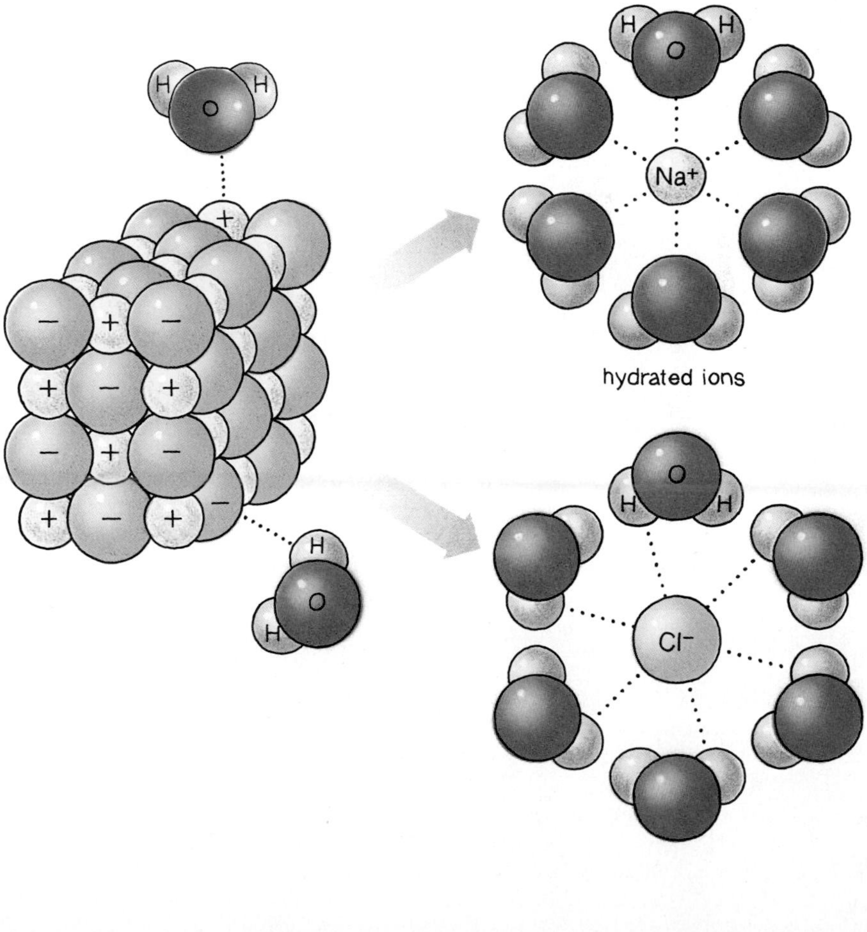

Sample Problem 8.2
Dissolving an Ionic Solute

What happens to the ions in solid KBr when it dissolves in water?

Solution
The ions, K^+ and Br^-, are pulled away from the solid as polar water molecules attract and separate them from the salt. Once dispersed in water, the ions are hydrated.

Study Check
What happens to the ions in solid $CaCl_2$ when it dissolves in water?

Rate of Solution

The time it takes to form a solution depends on how fast the solute leaves the solid and spreads throughout the solvent. One way to form a solution faster is to increase the surface area of the solute. By breaking the solute into smaller particles, more surface area can come in contact with the solvent. The solution then forms more rapidly. It is important to note that we are not changing the solubility of the solute—only the time it takes to form that solution. For example, in a glass of water, the sugar in a spoonful of powdered sugar dissolves faster than the sugar in a sugar cube. The more compact sugar cube has a smaller surface area, so there are relatively fewer molecules on the surface to interact with water. Eventually the final composition of the solutions will be identical.

Another way to increase the rate of solution is by stirring or agitating the solution. If the solute particles are dispersed more quickly, more solvent can interact with the remaining undissolved solute.

A higher temperature also increases the rate of solution by increasing the rate at which the solvent bombards the solute and moves the solute into the solution.

The rate of solution increases when:

1. The solute is crushed.
2. The solution is stirred.
3. The solution is heated.

Sample Problem 8.3
Rate of Solution Formation

You are going to dissolve some salt in a glass of water. Indicate whether each of the following conditions will *increase* or *decrease* the rate of solution for the salt.

a. using hot water **b.** crushing the salt

Solution

a. An increase in the temperature of the water increases the rate of solution.

b. Increasing the surface area of the salt that comes in contact with the water molecules will increase the rate of solution.

Study Check
How will placing a glass of water and some salt in a refrigerator affect the rate of solution formation?

Figure 8.6
The formation of a sugar solution as polar sucrose molecules ($C_{12}H_{22}O_{11}$) are attracted by polar water molecules.

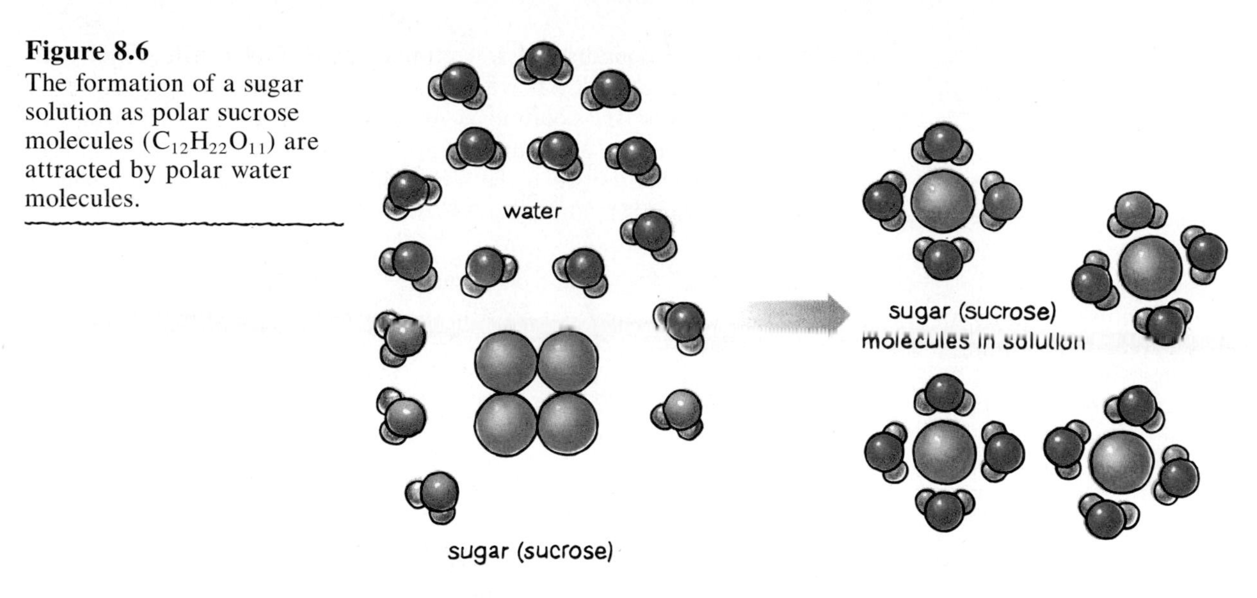

8.3 Factors Affecting Solubility

Learning Goal
Describe the effects of temperature and nature of the solute on its solubility in a liquid.

The term **solubility** is used to describe the amount of a solute that can dissolve in a certain amount of solvent. Many factors such as the type of solute, the type of solvent, and the temperature affect a solute's solubility.

Type of Solute and Solvent

A polar solvent such as water can also act as a solvent for solutes that are polar such as table sugar. In sugar, there are many O—H groups that attract water molecules. In forming a solution, the sugar disperses in water as sugar molecules. (See Figure 8.6.)

Nonpolar compounds such as iodine (I_2), oil and grease do not dissolve in water because they are not attracted by the polar water. Instead, nonpolar solutes form solutions when they dissolve in nonpolar solvents such as pentane (C_5H_{12}). These examples illustrate a rule of thumb, "like dissolves like," which we can use to predict the formation of a solution. Figure 8.7 illustrates some polar and nonpolar solutions.

Figure 8.7
Like dissolves like.
(a) Test tubes contain an upper layer of water, a polar solvent, and a lower layer of CH_2Cl_2, a nonpolar solvent.
(b) The nonpolar solute I_2 dissolves in the nonpolar solvent.
(c) The polar solute $Ni(NO_3)_2$ dissolves in the polar solvent.

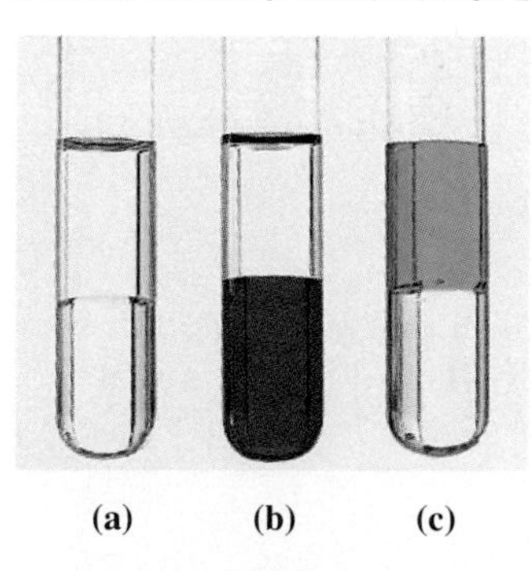

Like Dissolves Like

Polar solutes form solutions with polar solvents;
nonpolar solutes form solutions with nonpolar solvents.

Saturated Solutions

Solubility, expressed in grams of solute in 100 grams of solvent, is the maximum amount of solute that can be dissolved at a certain temperature. If a solute completely dissolves when added to the solvent, the solution does not contain the maximum amount of solute. We call the solution an **unsaturated solution.** When a solution contains all the solute it can dissolve, it is **saturated.** If we try to add more solute, undissolved solute will remain on the bottom of the container. For example, 36 g NaCl will dissolve in 100 grams of water at 20°C. If we add 15 g of NaCl to 100 g of water at 20°C, it all dissolves; the solution is unsaturated. The NaCl solution becomes saturated when we add a total of 36 g NaCl. In the saturated solution, equilibrium is reached between the solute in solution and the undissolved solute. (See Figure 8.8.)

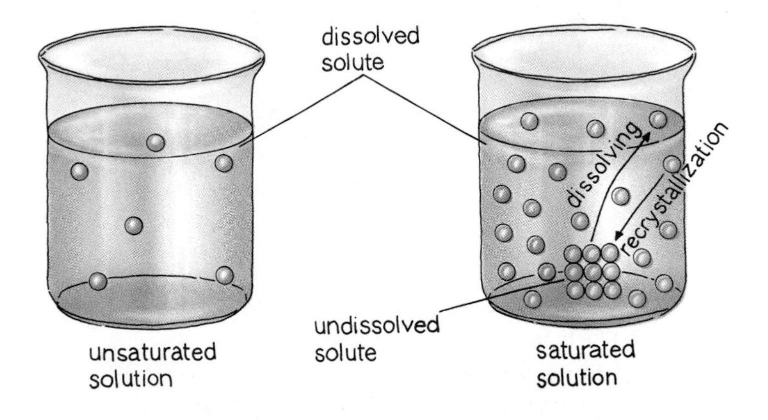

Figure 8.8
In an unsaturated solution, all solute dissolves. In a saturated solution, the maximum amount of solute for a given temperature has dissolved.

Sample Problem 8.4
Predicting Solution Formation

Predict whether KCl, an ionic solute, will form a solution with the following solvents:

a. water **b.** hexane (a nonpolar solvent)

Solution

a. Yes, an ionic compound is most soluble in a polar solvent such as water.

b. No. An ionic compound will not dissolve in a solvent that is not polar.

Study Check

Will oil, a nonpolar solute, dissolve in hexane, a nonpolar solvent?

Sample Problem 8.5
Saturated Solutions

At 70°C, the solubility of KCl is 50 g/100 g of water. In the laboratory, a student mixes 75 g KCl with 100 g of water at a temperature of 70°C.

a. How much of the KCl will dissolve?

b. Is the solution saturated?

c. What is the mass in grams of the undissolved KCl that forms on the bottom of the container?

Solution

a. A total of 50 g of the KCl will dissolve because that is its solubility and, therefore, the maximum amount of KCl allowed in solution at 70°C.

b. Yes, the solution is saturated.

c. The mass of the undissolved KCl is 25 g.

Study Check

At 30°C, the solubility of $CuSO_4$ is 25 g/100 g of water. How many grams of $CuSO_4$ are needed to make a saturated $CuSO_4$ solution with 300 g of water at 30°C?

HEALTH NOTE *Gout: A Problem of Saturation in Body Fluids*

Gout is a disease that affects adults, primarily men, over the age of 40. Attacks of gout occur when the concentration of uric acid in the plasma exceeds its solubility, and undissolved uric acid is deposited in the joints, causing severe pain. At body temperature, 37°C, the solubility of uric acid is 7 mg/100 mL of plasma. Typically, uric acid concentrations range from 2 mg to 6 mg/100 mL of plasma. If the uric acid concentration exceeds 7 mg/100 mL of plasma, solid uric acid crystals can appear in the cartilage, tendons, and soft tissues, as well as in the tissues of the kidneys, where they can cause renal damage.

Foods in the diet that contribute to high levels of uric acid are often rich in protein and include certain meats, sardines, mushrooms, and asparagus. Certain enzyme defects can also cause an overproduction of uric acid. Treatment for gout includes a reduction in protein intake and the restriction of certain foods. Drugs such as colchicine may be used to reduce uric acid production or to increase the excretion of uric acid.

Effect of Temperature

Most solids become more soluble in water as the temperature increases. For example, if you add sugar to ice tea, a layer of undissolved sugar forms on the bottom of the glass when solubility is reached. However, if the tea is heated, more sugar dissolves because the solubility of sugar is greater in hot water. Figure 8.9 illustrates the effect of higher temperatures on the solubility of some solids.

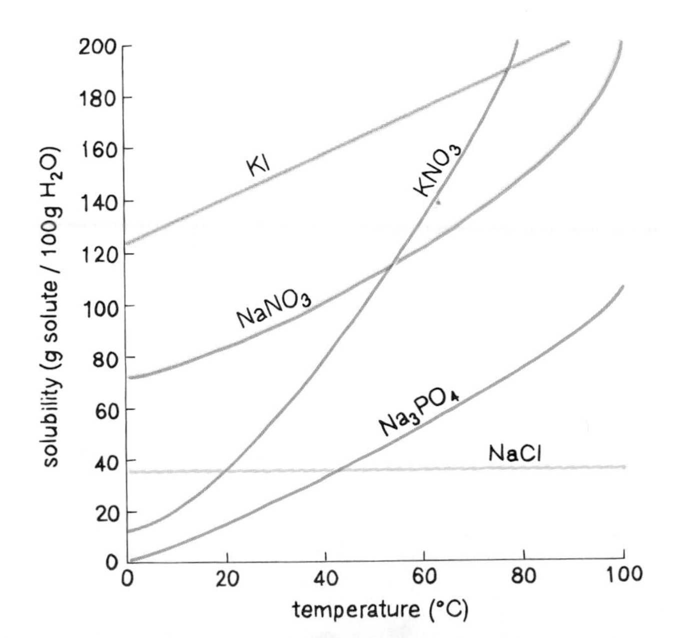

Figure 8.9
Most solids are more soluble in water at higher temperatures.

In contrast, gases are less soluble in water at higher temperature. (See Figure 8.10.) As the temperature increases, increasing numbers of gas molecules escape from the solution. At high temperatures, bottles containing carbonated solutions may burst, as more gas molecules leave the solution and increase the gas pressure inside the bottle. Biologists have found that increased temperatures in rivers and lakes cause the amount of oxygen available to decrease until the warm water can no longer support a biological community. Electrical generating plants are required to have their own ponds of water to use with their cooling towers to lessen the threat of thermal pollution.

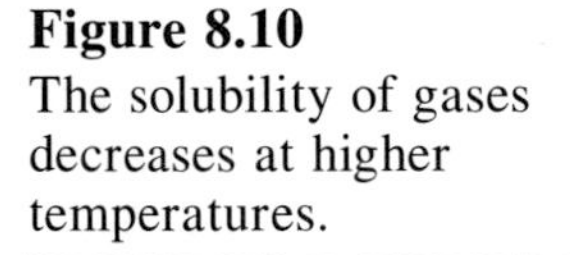

Figure 8.10
The solubility of gases decreases at higher temperatures.

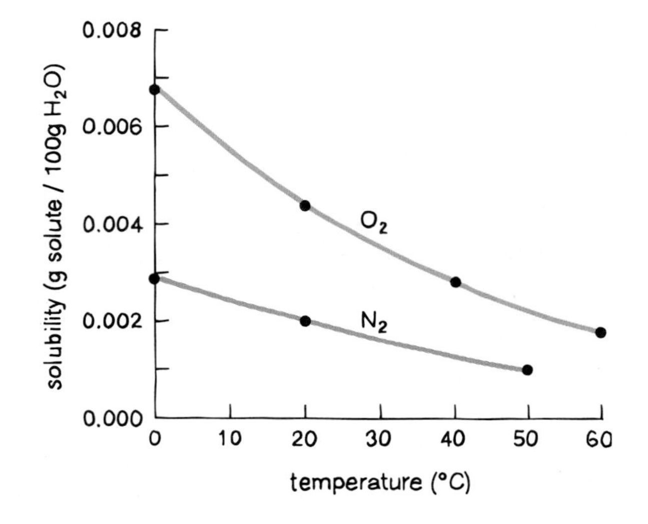

Sample Problem 8.6
Factors Affecting Solubility

Indicate whether the solubility of the solute will increase or decrease in each of the following situations:

a. using 80°C water instead of 25°C water to dissolve sugar
b. increasing the temperature of the water in a lake on the dissolved O_2

Solution

a. An increase in the temperature increases the solubility of the sugar.
b. An increase in the temperature decreases the solubility of O_2 gas.

Study Check

At 10°C, the solubility of KNO_3 is 20 g/100 g H_2O. Would the value of 5 g/100 g H_2O or 65 g/100 g H_2O be the more likely solubility at 40°C? Explain.

Solubility Rules

Up to now, we have considered ionic compounds that dissolve in water; they are *soluble salts.* However, some ionic compounds do not separate into ions in water. They are *insoluble salts* that remain as solids even when in contact

Figure 8.11
Some insoluble salts in water:
(a) yellow lead (II) chromate, $PbCrO_4$ and green nickel (II) hydroxide, Ni $(OH)_2$
(b) yellow cadmium sulfide, CdS and black iron (II) sulfide, (FeS).

(a)

(b)

with water. (See Figure 8.11.) We identify the soluble salts and insoluble salts using the *solubility rules for ionic compounds.*

In medicine, the insoluble salt $BaSO_4$ is used as an opaque substance to enhance X rays of the gastrointestinal tract. (See Figure 8.12.) $BaSO_4$ is so insoluble that it does not dissolve in gastric fluids. Other barium salts cannot be used, because they dissolve in water to release Ba^{2+}, which is poisonous.

Solubility Rules for Ionic Compounds

Always Soluble

1. A salt is soluble in water if it contains at least one of the following ions:

   Li^+ Na^+ K^+ NH_4^+ NO_3^- or $C_2H_3O_2^-$ (Acetate ion)

Mostly soluble

2. Most salts containing Cl^- are soluble, but *not* AgCl, $PbCl_2$, and Hg_2Cl_2.
3. Most salts containing SO_4^{2-} are soluble, but *not* $BaSO_4$, $PbSO_4$, and $CaSO_4$.

Insoluble

4. Salts of other ions (not mentioned in rule 1, 2, or 3) are usually insoluble. They include the following anions.

   OH^- CO_3^{2-} S^{2-} PO_4^{3-}

We use the solubility rules to predict whether a salt (a solid ionic compound) would be expected to be soluble in water. Table 8.3 illustrates the use of these rules.

Figure 8.12
Barium sulfate enhanced X ray of the abdomen showing the large intestine.

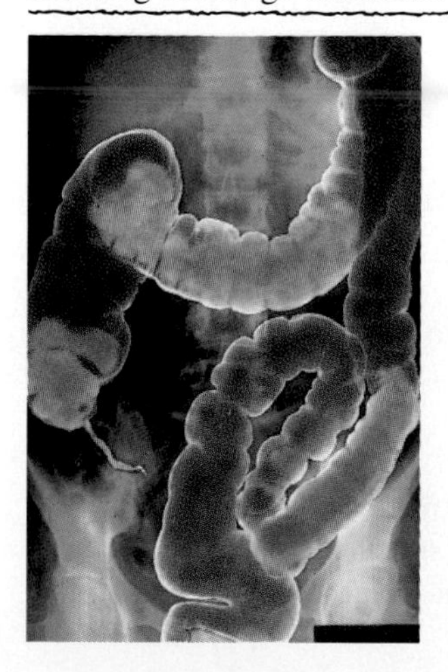

Formation of a Solid

Sometimes, when we mix two solutions, a solid called a *precipitate* forms and separates from the solution. A precipitation reaction occurs when an

Table 8.3 ***Using Solubility Rules***

Ionic Compound	Solubility in Water	Reasoning
K_2S	Soluble	Contains K^+ (rule 1)
$Ca(NO_3)_2$	Soluble	Contains NO_3^- (rule 1)
$PbCl_2$	Insoluble	Is an insoluble chloride (rule 2)
$MgSO_4$	Soluble	Contains SO_4^{2-} (rule 3)
NaOH	Soluble	Contains Na^+ (rule 1)
$AlPO_4$	Insoluble	Contains no soluble ions (rule 4)

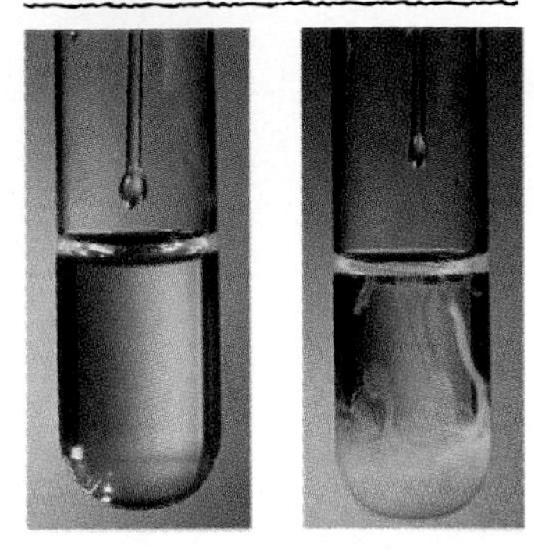

Figure 8.13
A precipitate of AgCl forms when solutions of $AgNO_3$ and NaCl are mixed. The soluble salt, $NaNO_3$, remains in solution.

ion from one solution comes in contact with an ion from the second solution to form crystals of an insoluble salt. For example, if we mix solutions of $AgNO_3(Ag^+, NO_3^-)$ and $NaCl(Na^+, Cl^-)$, a white solid forms. Because AgCl is an insoluble salt, the Ag^+ and Cl^- ion formed a precipitate of AgCl, as shown in Figure 8.13, leaving the Na^+ and NO_3^- ions in solution.

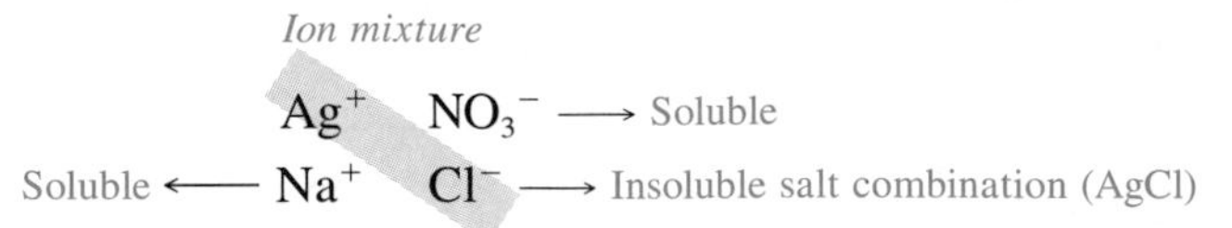

In an *ionic equation,* only the ions that form the insoluble salt (solid) are written.

$$Ag^+ + Cl^- \longrightarrow AgCl(s)$$

In the *molecular equation* for the reaction, the formulas of all the reactants and products are shown.

$$AgNO_3(aq) + NaCl(aq) \longrightarrow AgCl(s) + NaNO_3(aq)$$

HEALTH NOTE *Kidney Stones Are Insoluble Salts*

The kidneys continuously excrete substances that have low solubilities. But if the urine becomes saturated with these substances, they crystallize and form kidney stones. One type, renal calculus, formed in the urinary tract is composed primarily of a calcium phosphate known as hydroxyapatite, $Ca_5(PO_4)_3OH$. The excessive ingestion of mineral salts and a deficiency of water available to form urine can lead to the saturation of the urine with minerals and consequently to the kidney stones. Typically, one stone is formed every 2–3 years, and the tendency to form the stones is strongly familial.

When a kidney stone passes through the urinary tract, it can cause considerable pain and discomfort, necessitating the use of painkillers and surgery. Sometimes lasers can be used to break up large kidney stones in the kidneys. Persons prone to kidney stones are advised to drink six to eight glasses of water every day to prevent saturation levels of minerals in the urine.

Sample Problem 8.7
Soluble Salts

1. Predict whether each of the following salts are soluble in water:
   a. Na_3PO_4 **b.** $CaCO_3$
2. Solutions of $BaCl_2$ and K_2SO_4 are mixed.
   a. What is the insoluble salt that forms?
   b. What ions remain in solution?
   c. What are the ionic and molecular equations for the reaction?

Solution

1. a. Soluble. Any salt containing Na^+ ions is soluble (rule 1).
b. Insoluble. Most salts with CO_3^{2-} are insoluble (rule 4).
2. a. In the mixture, the Ba^{2+} from the $BaCl_2$ solution and the SO_4^{2-} from the K_2SO_4 solution form the insoluble salt $BaSO_4(s)$.

b. The ions in solution are K^+ and Cl^-.
c. The ionic equation for the formation of $BaSO_4$ is

$$Ba^{2+} + SO_4^{2-} \longrightarrow BaSO_4(s)$$

The overall molecular equation is

$$BaCl_2 + K_2SO_4 \longrightarrow BaSO_4(s) + 2KCl$$

Study Check
Would you expect the following salts to be soluble or insoluble in water?
a. $PbCl_2$ **b.** Na_3PO_4 **c.** $FeCO_3$

8.4 Percent Concentration

Learning Goal
Calculate the percent concentration of a solute in a solution; use percent concentration to calculate the amount of solute or solution.

The amount of solute dissolved in a certain amount of solution is called the **concentration** of the solution. Although there are many ways to express a concentration, they all specify a certain amount of solute in a given amount of solution.

$$\text{Concentration of a solution} = \frac{\text{amount of solute}}{\text{amount of solution}}$$

Mass Percent

The *mass percent* of a solution (also called weight percent) is the percent mass of solute in the solution. To calculate the mass (or weight) percent divide the grams of solute by the grams of the solution, and multiply by 100. In the ratio, the units of mass, usually grams, must always match. In this text, we will abbreviate the mass percent of a solute as %(m/m); sometimes you will see %(w/w).

$$\text{mass/mass \%} = \frac{\text{grams of solute}}{\text{grams of solution}} \times 100$$

The mass of the solution is the sum of the grams of solute and the grams of solvent.

$$\text{g solution} = \text{g solute} + \text{g solvent}$$

Suppose we prepared a solution from 10.0 g KCl (solute) and 90.0 g of water (solvent). The mass of the solution is 100.0 g of solution (10.0 g + 90.0 g). The percent concentration by mass of the KCl in the solution is 10.0%.

$$\text{\% KCl (m/m)} = \frac{\text{10.0 g KCl (solute)}}{\text{100.0 g solution}} \times 100 = \text{10.0\% KCl}$$

$$\nearrow \quad \nwarrow$$

$$10.0 \text{ g KCl} + 90.0 \text{ g } H_2O$$

Sample Problem 8.8
Percent Concentration by Mass

If 15 g NaOH are dissolved in 285 g of water, what is the mass percent of NaOH in the solution?

Solution

We find the mass of a solution by adding the mass of the solute and the solvent.

$$15 \text{ g NaOH (solute)} + 285 \text{ g } H_2O \text{ (solvent)} = 300. \text{ g solution}$$

Placing grams of solute and grams of solution in the mass percent expression gives

$$\frac{15 \text{ g NaOH}}{300. \text{ g solution}} \times 100 = 5.0\% \text{ (m/m) NaOH solution}$$

Study Check

What is the mass percent of NaCl in a solution made by dissolving 2.0 g NaCl in 56 g of water?

Volume Percent

Because the volumes of liquids (or gases) can be easily measured, the concentrations of their solutions are often expressed as *volume percent % (v/v)*. The units of volume used in the ratio must be the same, for example, both in milliliters or both in liters.

$$\text{volume/volume \%} = \frac{\text{volume (mL) solute}}{\text{volume (mL) solution}} \times 100$$

↑
Units match

We can also interpret a volume/volume percent as the volume of solute in 100 mL of solution. In the wine industry, a label that reads 12% by volume means 12 mL of alcohol in 100 mL of wine solution.

Mass/Volume Percent

A *mass/volume percent, % (m/v)*, sometimes called a weight/volume percent, % (w/v), is calculated by dividing the grams of the solute by the milliliters (volume) of solution and multiplying by 100. Widely used in hospitals and pharmaceuticals, the preparation of intravenous solutions and medicines involves the mass/volume percent even though the units do not cancel.

$$\text{mass/volume \%} = \frac{\text{grams of solute}}{\text{milliliters of solution}} \times 100$$

The mass/volume percent, % (m/v), indicates the grams of a substance that are contained in exactly 100 mL of a solution. For example, in every 100 mL of a 5% glucose solution, there are 5 g of glucose.

Sample Problem 8.9
Calculating Percent Concentration

A student prepared a solution by dissolving 5 g KI with enough water to give a final volume of 250 mL. (See Figure 8.14.) What is the mass/volume percent of the KI solution?

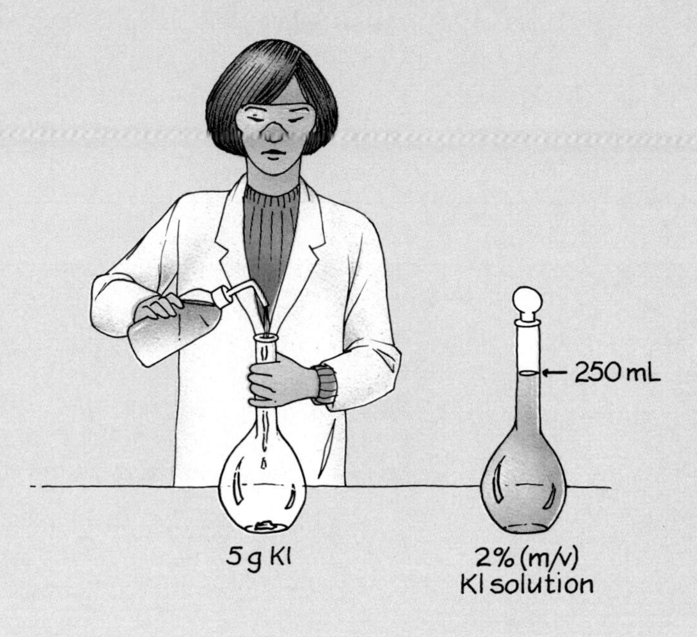

Figure 8.14
A 2% (m/v) KI solution is prepared by placing 5 g KI in a 250-mL flask and filling the flask to the 250-mL mark.

Solution
Placing grams of solute and milliliters of solution in the expression for percent (m/v) concentration of KI gives

$$\text{mass/volume \%} = \frac{\overset{\text{Mass of solute}}{5 \text{ g KI}}}{\underset{\text{Volume of solution}}{250 \text{ mL solution}}} \times 100 = 2\% \text{ (m/v) KI}$$

Study Check
What is the volume percent, % (v/v) of Br_2 in a solution prepared by dissolving 12 mL of bromine (Br_2) in enough carbon tetrachloride to make 250 mL of solution?

Percent Concentrations as Conversion Factors

In the preparation of solutions, we often need to calculate the amount of solute or solution. Then the percent concentration is useful as a conversion factor. Some examples of percent concentrations, their meanings, and possible conversion factors are given in Table 8.4.

Table 8.4 ***Conversion Factors from Percent Concentrations***

Percent Concentration	Meaning	Conversion Factors		
10% (m/m) KCl	There are 10 g of KCl in every 100 g of KCl solution.	$\frac{10\text{ g KCl}}{100\text{ g solution}}$	and	$\frac{100\text{ g solution}}{10\text{ g KCl}}$
5% (m/v) glucose	There are 5 g of glucose in every 100 mL of glucose solution.	$\frac{5\text{ g glucose}}{100\text{ mL solution}}$	and	$\frac{100\text{ mL solution}}{5\text{ g glucose}}$

Sample Problem 8.10
Using Mass/Volume Percent

A laboratory technician must prepare 2.00 L of a 5% (m/v) glucose solution. How many grams of glucose must be measured out?

Solution
Using the 5% (m/v) concentration as a conversion factor, we calculate the grams of glucose to weigh out for the solution as

$$\underset{\text{Given}}{2.00\ \cancel{\text{L}}} \times \underset{\text{Metric factor}}{\frac{1000\ \cancel{\text{mL}}}{1\ \cancel{\text{L}}}} \times \underset{\text{Percent factor}}{\frac{5\text{ g glucose}}{100\ \cancel{\text{mL}}}} = 100\text{ g glucose}$$

Study Check
How many grams of NaCl are needed to prepare 3.0 L of a 1% (m/v) NaCl solution?

Sample Problem 8.11
Using Percent Concentration

A patient requires 50 g of a lipid solution. If the solution on hand is a 10% (m/v) lipid, how many milliliters should be given?

Solution
The percent concentration of the lipid solution can be written as two co version factors.

$$\frac{10\text{ g lipid}}{100\text{ mL of solution}} \quad \text{and} \quad \frac{100\text{ mL of solution}}{10\text{ g lipid}}$$

Using the second factor to cancel grams of lipid (g lipid), the volume needed is calculated as

$$\underset{\text{Solute needed}}{50\ \cancel{\text{g lipid}}} \times \underset{\text{Percent factor}}{\frac{100\text{ mL}}{10\ \cancel{\text{g lipid}}}} = 500\text{ mL lipid solution}$$

The patient must be given 500 mL of the 10% (m/v) lipid solution to obtain 50 g of lipid.

Study Check

An antifreeze mixture is made with ethylene glycol and water. How many liters (L) of a 15% (v/v) antifreeze solution can be prepared from 60.0 mL of ethylene glycol?

8.5 Molarity

Learning Goal

Calculate the molarity of a solution; use molarity as a conversion factor to calculate the moles of solute or the volume needed to prepare a solution.

When the solutes of solutions take part in reactions, chemists are interested in the number of reacting particles. For this purpose, chemists use **molarity (M),** a concentration that states the number of moles of solute in a 1-liter (1 L) volume of a solution. The molarity of a solution can be calculated knowing the moles of solute and the volume of solution.

$$\text{Molarity (M)} = \frac{\text{moles of solute}}{\text{1 liter (1 L) of solution}}$$

For example, if 6.0 mol NaOH were dissolved in enough water to prepare 3.0 L of solution, the resulting NaOH solution has a molarity of 2.0 M. The abbreviation M indicates the units of moles per liter (mol/L).

$$\text{M} = \frac{\text{moles of solute}}{\text{liters of solution}} = \frac{\text{6.0 mol NaOH}}{\text{3.0 L of solution}} = \frac{\text{2.0 mol NaOH}}{\text{1 L}} = \text{2.0 M NaOH}$$

Sample Problem 8.12
Calculating Molarity

What is the molarity (M) of each of the following solutions?

a. 2.0 mol calcium chloride ($CaCl_2$) in 500 mL of solution

b. 60.0 g NaOH in 0.250 L of solution

Solution

a. Because molarity requires liters (L), the volume in liters must be determined.

$$500 \cancel{\text{mL}} \times \frac{1 \text{ L}}{1000 \cancel{\text{mL}}} = 0.5 \text{ L of solution}$$

Placing the moles of solute and the volume in liters into the concentration expression, we calculate the molarity as

$$\text{molarity (M)} = \frac{\text{moles } CaCl_2}{\text{liters of solution}} = \frac{2.0 \text{ mol } CaCl_2}{0.5 \text{ L}} = 4 \text{ M}$$

b. Because molarity requires moles of solute, we must change from grams of NaOH to moles NaOH using the molar mass NaOH (40.0).

$$\underset{\text{Grams of NaOH}}{60.0 \cancel{\text{g NaOH}}} \times \underset{\text{Molar mass}}{\frac{1 \text{ mol NaOH}}{40.0 \cancel{\text{g NaOH}}}} = 1.50 \text{ mol NaOH}$$

Now we use the moles of NaOH and volume of the solution in liters to calculate the molarity of the solution.

$$M = \frac{\overbrace{1.50 \text{ mol NaOH}}^{\text{Moles of solute}}}{\underbrace{0.250 \text{ L}}_{\text{Liters of solution}}} = 6.00 \text{ M solution}$$

Study Check
What is the molarity of a solution that contains 75 g KNO_3 dissolved in 350 mL of solution?

Molarity as a Conversion Factor

When we need to calculate the moles of solute or the volume of solution, the molarity is used as a conversion factor. Examples of conversion factors from molarity are given in Table 8.5.

Table 8.5 *Some Examples of Molar Solutions*

Molarity	Meaning	Conversion Factors
6.0 M HCl	6.0 mol HCl in 1 liter of solution	$\frac{6.0 \text{ mol HCl}}{1 \text{ L}}$ and $\frac{1 \text{ L}}{6.0 \text{ mol HCl}}$
0.20 M NaOH	0.20 mol NaOH in 1 liter of solution	$\frac{0.20 \text{ mol NaOH}}{1 \text{ L}}$ and $\frac{1 \text{ L}}{0.20 \text{ mol NaOH}}$

Sample Problem 8.13
Molarity as a Conversion Factor

How many moles of NaCl are present in 4.0 L of a 2.0 M NaCl solution?

Solution
The molarity (2.0 M) of the solution gives two possible conversion factors.

$$\frac{2.0 \text{ mol NaCl}}{1 \text{ L}} \quad \text{and} \quad \frac{1 \text{ L}}{2.0 \text{ mol NaCl}}$$

Using the factor that cancels volume (L), we can calculate the moles of NaCl.

$$\underbrace{4.0 \text{ L}}_{\text{Given volume}} \times \underbrace{\frac{2.0 \text{ mol NaCl}}{1 \text{ L}}}_{\text{Molarity as a conversion factor}} = 8.0 \text{ mol NaCl}$$

Study Check
How many moles of HCl are present in 750 mL of a 6.0 M HCl solution?

In the preparation of a molar solution, we weigh out the proper number of grams. To do this, we must convert the moles needed in the solution to grams using the molar mass.

Sample Problem 8.14
Using Molarity

How many grams of KCl would you need to weigh out to prepare 0.250 L of a 2.00 M KCl solution?

Solution

The number of moles of KCl is determined by using the given volume of solution and the molarity as a conversion factor.

$$\underset{\text{Given volume}}{0.250\ \cancel{\text{L solution}}} \times \underset{\text{Molarity factor}}{\frac{2.00\ \text{mol KCl}}{1\ \cancel{\text{L solution}}}} = 0.500\ \text{mol KCl}$$

Knowing the moles of solute needed, the number of grams of KCl is calculated using the molar mass of KCl (74.6 g/mol).

$$\underset{\text{Moles KCl}}{0.500\ \cancel{\text{mol KCl}}} \times \underset{\text{Molar mass KCl}}{\frac{74.6\ \text{g KCl}}{1\ \cancel{\text{mol KCl}}}} = 37.3\ \text{g KCl}$$

Study Check

How many grams of baking soda, $NaHCO_3$, are in 325 mL of a 4.50 M $NaHCO_3$ solution?

The volume of a solution can be calculated if the number of moles of solute and the molarity of the solution are given.

Sample Problem 8.15
Calculating Volume of a Molar Solution

What volume, in liters, of 1.5 M HCl solution is needed to provide 6.0 mol of HCl?

Solution

Using the molarity as a conversion factor, we set up the following calculation:

$$\underset{\text{Given}}{6.0\ \cancel{\text{mol HCl}}} \times \underset{\text{Molarity factor}}{\frac{1\ \text{L solution}}{1.5\ \cancel{\text{mol HCl}}}} = 4.0\ \text{L HCl}$$

Study Check

How many milliliters of 2.0 M NaOH solution will provide 20.0 g NaOH? (*Hint:* Convert grams NaOH to moles NaOH first.)

Dilution

A solution can also be prepared by diluting a more concentrated solution. In the process of **dilution,** more of the solvent is added. For example, you might prepare some orange juice by adding three cans of water to the orange juice concentrate. (See Figure 8.15.)

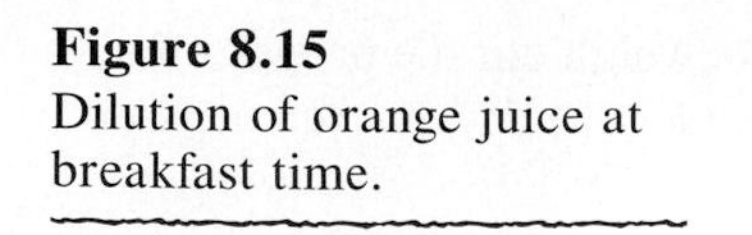

Figure 8.15
Dilution of orange juice at breakfast time.

When more solvent is added, the solution volume increases, causing a decrease in the concentration. However, *the amount of solute (grams or moles) does not change.*
might prepare some orange juice by adding three cans of water to the orange

amount of solute (initial) = amount of solute (diluted)

To calculate the new concentration, we can find the grams (or moles) of solute and divide it by the new, diluted volume.

Sample Problem 8.16
Diluting a Solution

Calculate the new mass/volume percent, % (m/v), of a $CoCl_2$ solution that is prepared by diluting 35 mL of a 10.% (m/v) $CoCl_2$ solution to 100. mL. (See Figure 8.16.)

Figure 8.16
Water is added to 35 mL of a 10% (m/v) $CoCl_2$ to increase the volume to the 100-mL mark. The concentration of the diluted solution is 3.5% (m/v) $CoCl_2$.

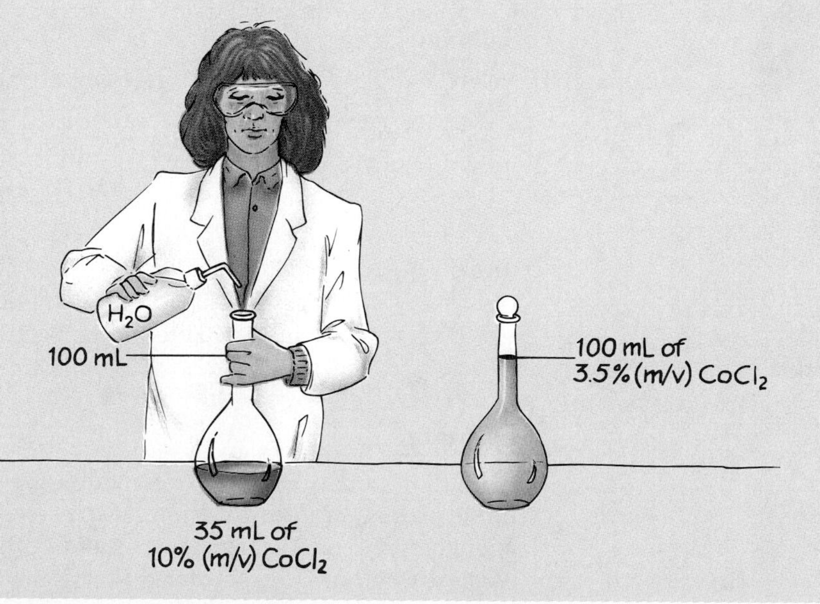

Solution

First, we calculate the number of grams of $CoCl_2$ present in the original solution.

$$35 \text{ mL} \times \frac{10.\text{ g } CoCl_2}{100 \text{ mL}} = 3.5 \text{ g } CoCl_2$$

Second, the percent concentration of the diluted solution is calculated by placing the amount of solute and the new volume of the solution in the expression for mass/volume percent.

$$\frac{3.5 \text{ g } CoCl_2}{100 \text{ mL}} \times 100 = 3.5\% \text{ (m/v) } CoCl_2 \text{ solution}$$

Study Check

What is the new %(m/v) concentration after 75 mL of water are added to 75 mL of a 6% (m/v) vinegar solution?

8.6 Colloids and Suspensions

Learning Goal

From its properties, identify a mixture as a solution, a colloid, or a suspension.

The solute particles in a solution play an important role in determining the properties of that solution. In most of the solutions we discussed so far, the solute is dissolved as small, single particles that are uniformly dispersed throughout the solvent to give a homogeneous solution. The particles are so small that they go through filters and through semipermeable membranes. A semipermeable membrane allows solvent molecules such as water and some very small solute particles to pass through but not large solute molecules. The solute particles do not settle out of solution even after they have been stored for long periods of time. When you observe a true solution, such as salt water, you cannot visually distinguish the solute from the solvent. The solution appears transparent even when a light shines through it because the particles are too small to scatter the light. Although solutions are transparent, they don't have to be colorless.

Colloids

The particles in colloidal dispersions or **colloids,** are much larger than solute particles in a solution. Colloidal particles can be single, large molecules, such as proteins, or groups of molecules or ions. Colloids are homogeneous mixtures that do not separate or settle out. Colloidal particles are small enough to pass through filters but too large to pass through semipermeable membranes.

When a beam of light shines through a colloid, it produces the **Tyndall effect,** in which the large solute particles reflect the light and make the light beam visible as shown in Figure 8.17. The Tyndall effect causes a light beam to be visible in fog because the water droplets in the air reflect the light. When there is smoke or smog in the air, the blues and greens of sunlight are reflected more, so that we observe the oranges and reds of a beautiful sunset.

Figure 8.17
The Tyndall effect: When light is reflected by colloids, it becomes visible.

Types of Colloids

Colloids are classified by the type of substance and the dispersing medium. There are four types of colloids: aerosols, foams, emulsions, and sols. *Aerosols* consist of liquid or solid particles dispersed in a gas. *Foams* are dispersions of gases in liquids or solids. An *emulsion* is a liquid dispersed in another liquid or a solid. For example, when milk is not homogenized, the suspended particles of cream separate. The process of homogenization breaks the cream particles into colloids that remain dispersed within the milk solution. In a *sol,* particles of a solid are dispersed in a liquid or solid to give a more rigid solution. Table 8.6 lists several examples of colloids.

Suspensions

Suspensions are heterogeneous, nonuniform mixtures that are very different from solutions or colloids. The particles of a suspension are so large that they can often be seen with the naked eye. They are trapped by filters and membranes.

The weight of suspended particles causes them to settle out soon after mixing. If you stir muddy water, it mixes but then quickly separates as the suspended particles settle to the bottom and leave clear liquid at the top. You can find suspensions among the medications in a hospital or in your medicine cabinet. These include Kaopectate, calamine lotion, antacid mixtures, and liquid penicillin. It is important to "shake well before using" to suspend all the particles before giving a medication that is a suspension.

Water-treatment plants make use of the properties of suspensions to purify water. When coagulants such as aluminum sulfate or ferric sulfate are added to untreated water, they react with impurities to form large suspended particles called floc. In the water-treatment plant, a system of

Table 8.6 ***Colloids***

Examples	Type	Substance Dispersed	Dispersing Medium
Fog, clouds, sprays	Aerosol	Liquid	Gas
Dust, smoke	Aerosol	Solid	Gas
Shaving cream, whipped cream, soapsuds	Foam	Gas	Liquid
Styrofoam, marshmallows	Foam	Gas	Solid
Mayonnaise, butter, homogenized milk, hand lotions	Emulsion	Liquid	Liquid
Cheese, butter	Emulsion	Liquid	Solid
Blood plasma, paints (latex), gelatin	Sol	Solid	Liquid
Cement, pearls	Sol	Solid	Solid

filters traps the suspended particles, but allows the clean water to go through.

Figure 8.18 illustrates some properties of solutions, colloids, and suspensions, and Table 8.7 compares the different types of solutions.

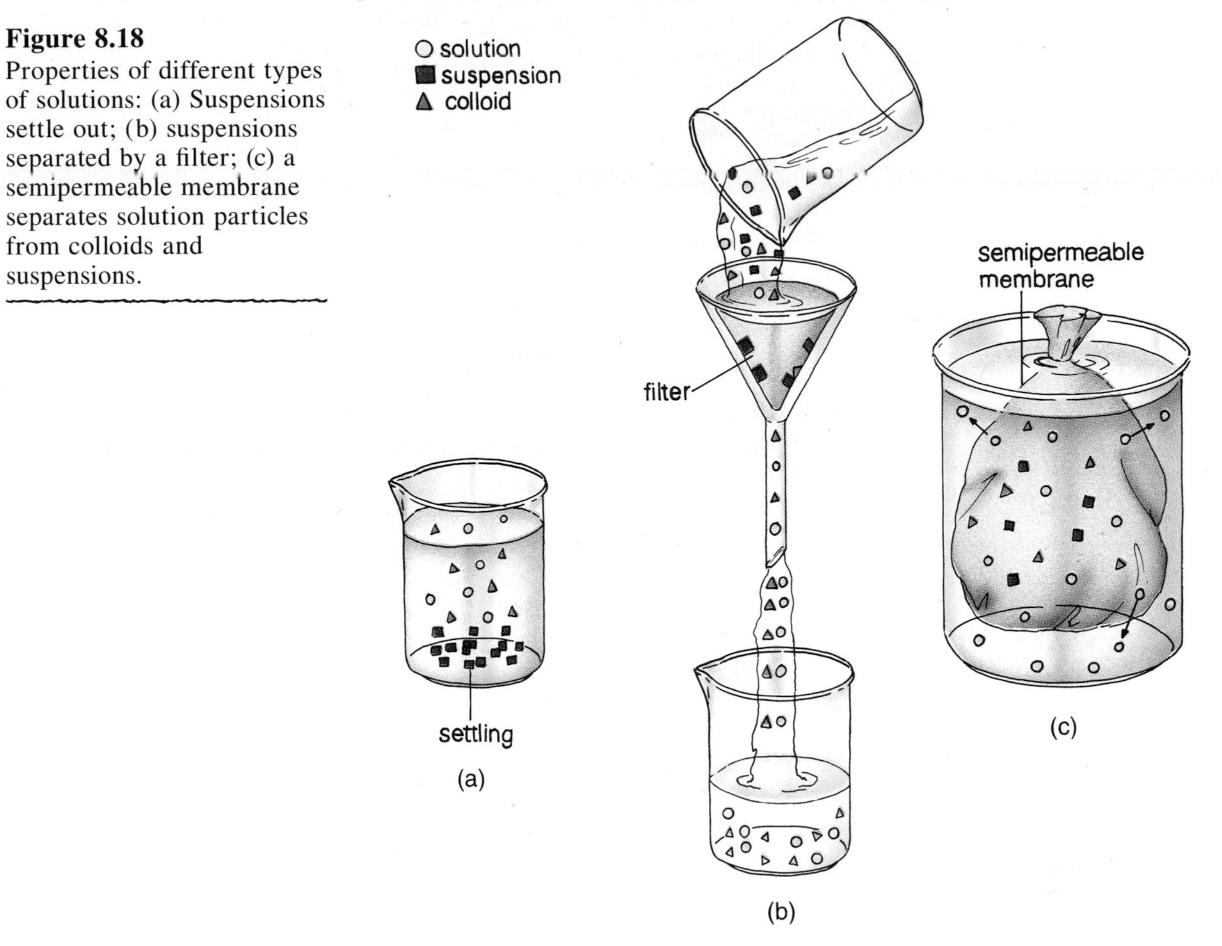

Figure 8.18
Properties of different types of solutions: (a) Suspensions settle out; (b) suspensions separated by a filter; (c) a semipermeable membrane separates solution particles from colloids and suspensions.

Table 8.7 *Comparison of Solutions, Colloids, and Suspensions*

Type of Mixture	Type of Particle	Effect of Light	Settling	Separation
Solution	Small particles such as single atoms, ions, or molecules	Transparent	Particles do not settle	Particles cannot be separated by filters or semipermeable membranes
Colloid	Larger molecules or groups of molecules or ions	Tyndall effect occurs	Particles do not settle	Particles can be separated by semipermeable membranes but not by filters
Suspension	Very large particles that may be visible	Opaque (not transparent)	Particles settle rapidly	Particles can be separated by filters

Sample Problem 8.17
Classifying Types of Mixtures

Classify each of the following as a solution, colloid, or suspension:

a. a homogeneous mixture in which a beam of light is visible
b. a mixture that settles rapidly upon standing
c. a mixture whose solute particles pass through both filters and membranes.

Solution

a. colloid **b.** suspension **c.** solution

Study Check

Enzymes are large, single molecules of protein that catalyze chemical reactions inside the cells of the body. If they cannot pass through the cell membrane, are they solutions or colloids?

HEALTH NOTE *Colloids and Solutions in the Body*

Colloids in the body are separated from solutions by semipermeable membranes. For example, the intestinal lining allows solution particles to pass into the blood and lymph circulatory systems. However, the large colloid molecules from foods are too large to pass through the membrane, and they remain in the intestinal tract. Digestion breaks down large colloidal particles, such as starch and protein, into smaller particles, such as glucose and amino acids, that can pass through the intestinal membrane and enter the circulatory system. Certain foods, such as bran, a fiber, cannot be broken down by human digestive processes, and they move through the intestine intact.

The cell membranes also separate solutions from colloids. For example, large proteins, such as enzymes, are formed inside cells. Because they are colloids, they remain inside the cell. However, many of the nutrients that must be obtained by cells, such as oxygen, amino acids, electrolytes, glucose, and minerals, are solutions that can pass through cellular membranes. Waste products, such as urea and carbon dioxide, also form solutions and pass out of the cell to be excreted.

8.7 Osmosis and Dialysis

Learning Goal
Describe the changes in concentration of solute and solvent in the processes of osmosis and dialysis.

The movement of water into and out of the cells of plants as well as our bodies is an important biological process. In a process called **osmosis,** the solvent water moves through a semipermeable membrane from a solution that has a lower concentration of solute into a solution where the solute concentration is higher. (See Figure 8.19.) This movement, or osmosis, of water happens in a direction that equalizes (or attempts to equalize) the concentrations on both sides of the membrane. For example, when water is separated from sucrose (sugar) solution by a **semipermeable membrane,** water molecules, but not the larger sucrose molecules, move through the membrane to dilute the sucrose solution. As more water flows across the membrane, the level of the sugar solution rises as the level of water on the other side decreases.

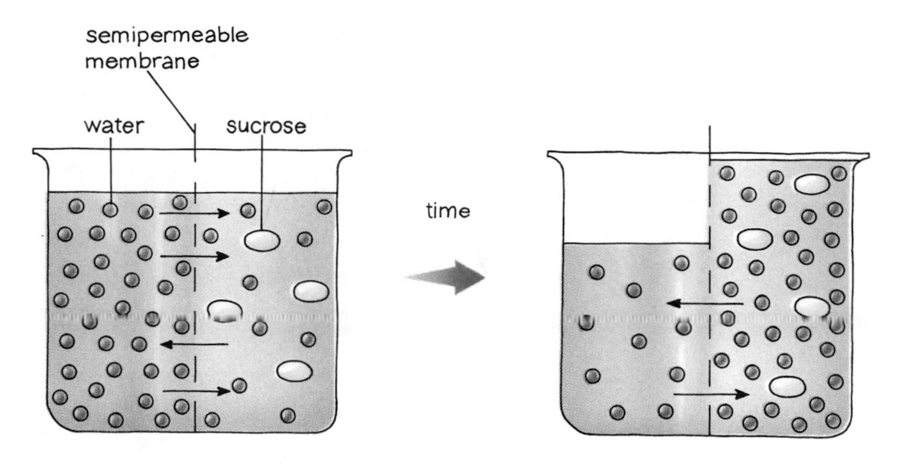

Figure 8.19
In osmosis, water flows through a semipermeable membrane from the solution of lower concentration into the solution of higher concentration.

At some point, the levels of the liquids on each side of the semipermeable membrane stop changing. Because the liquid level of the sugar solution is higher than the solvent side, the weight of the higher level of solution exerts a downward pressure called osmotic pressure. **Osmotic pressure** is the pressure that prevents the flow of water into the more concentrated solution. The water now moves equally between the two compartments as the system comes to equilibrium. The osmotic pressure of a solution depends on the number of particles in the solution. Pure water has no osmotic pressure. The greater the number of particles dissolved, the higher the osmotic pressure.

If a pressure greater than the osmotic pressure is applied to a solution, osmosis is reversed and solvent flows out of the solution and the level of the solvent increases. This process, known as reverse osmosis, is used in desalinization plants in some parts of the world to produce drinking water from sea (salt) water.

Sample Problem 8.18
Osmotic Pressure

A 2% sucrose solution and an 8% sucrose solution are separated by a semipermeable membrane.

a. Which sucrose solution exerts the greater osmotic pressure?
b. In what direction does water flow initially?
c. Which solution will have the higher level of liquid?

Solution

a. The 8% sucrose solution has the higher solute concentration, more solute particles, and the greater osmotic pressure.
b. Initially, water will flow out of the 2% solution into the more concentrated 8% solution.
c. The level of the 8% solution will be higher.

Study Check

If a 10% glucose solution is separated from a 5% glucose solution by a semipermeable membrane, which solution will decrease in volume?

Isotonic Solutions

Because the cell membranes in biological systems are semipermeable, osmosis is an ongoing process. The solutes in the body solutions such as blood, tissue fluids, lymph, and plasma all exert osmotic pressure. Any solution used in the body must be an **isotonic solution,** which exerts the same osmotic pressure as the body fluids. *Iso* means "equal to," and tonic refers to the osmotic pressure of the cell. In the hospital, isotonic solutions or **physiological solutions** include 0.9% NaCl and 5% glucose solution. Although they do not contain the same particles as the body fluids, they exert the same osmotic pressure.

Hypotonic and Hypertonic Solutions

A red blood cell placed in an isotonic solution retains its normal volume because there is an equal flow of water into and out of the cell. (See Figure 8.20a.) However, if a red blood cell is placed in a solution that is not isotonic, the differences in osmotic pressure inside and outside the cell can drastically alter the volume of the cell. When a red blood cell is placed in pure water, a **hypotonic solution** (*hypo* means "lower than"), water flows into the cell by osmosis. (See Figure 8.20b.) The increase in fluid causes the

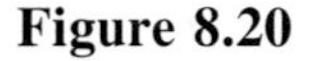

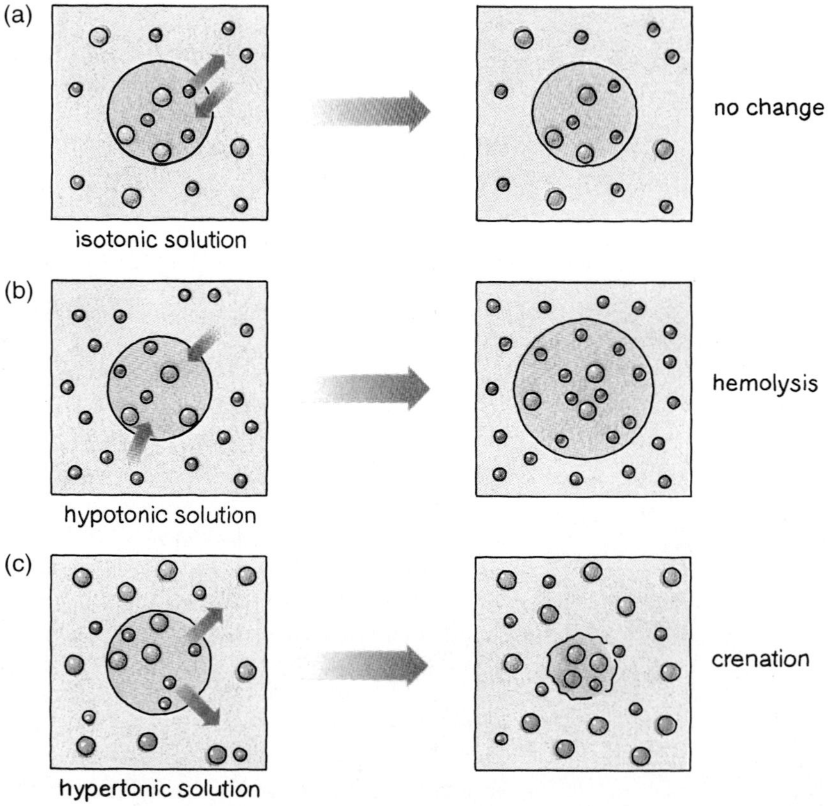

Figure 8.20
(a) In an isotonic solution, a red blood cell retains its normal volume.
(b) In a hypotonic solution, a red blood cell swells (hemolysis) because water flows into the cell.
(c) In a hypertonic solution, water leaves the red blood cell, causing it to shrink (crenation).

cell to swell, and possibly burst, a process called **hemolysis.** A similar process occurs when you place dehydrated food, such as raisins or dried fruit, in water. The water enters the cells and the food becomes plump and smooth.

If a red blood cell is placed in a **hypertonic solution,** which has a higher concentration (*hyper* means "greater than"), water leaves the cells by osmosis. Suppose red blood cells are placed in a 10% NaCl solution. Because the osmotic pressure in the red blood cells is equal to that of a 0.9% NaCl solution, the 10% NaCl solution has a much greater osmotic pressure. As water is lost, the cell shrinks, a process called **crenation.** (See Figure 8.20c.) A similar process occurs when making pickles, in which a hypertonic salt solution causes the cucumbers to shrivel as they lose water.

Sample Problem 8.19
Isotonic, Hypotonic, and Hypertonic Solutions

Describe each of the following solutions as isotonic, hypotonic, or hypertonic. Indicate whether a red blood cell placed in each solution will undergo hemolysis, crenation, or no change.

a. A 5% glucose solution. **b.** A 0.2% NaCl solution.

Solution

a. A 5% glucose solution is isotonic. A red blood cell will not undergo any change.

b. A 0.2% NaCl solution is hypotonic. A red blood cell will undergo hemolysis.

Study Check

What is the effect of a 10% glucose solution on a red blood cell?

Dialysis

Dialysis is a process that is similar to osmosis. In dialysis, a semipermeable membrane, called a dialyzing membrane, permits small solute molecules and ions in solutions as well as water molecules to pass through, but it retains large particles, such as colloids. Dialysis is a way to separate true solution particles from colloids.

Suppose we fill a cellophane bag with a solution of NaCl, glucose, starch, and protein and place it in pure water. Cellophane is a dialyzing membrane, and the sodium ions, chloride ions, and glucose molecules will pass through it into the surrounding water. However, the colloids starch and protein remain inside. Water molecules will flow by osmosis into the colloids within the cellophane bag. Eventually, the total concentrations of sodium ions, chloride ions, and glucose inside and outside the dialysis bag become equal. To remove more NaCl or glucose, the cellophane bag must be placed in a fresh sample of pure water. (See Figure 8.21.)

Figure 8.21
Dialysis, the separation of solutions from colloids in water.

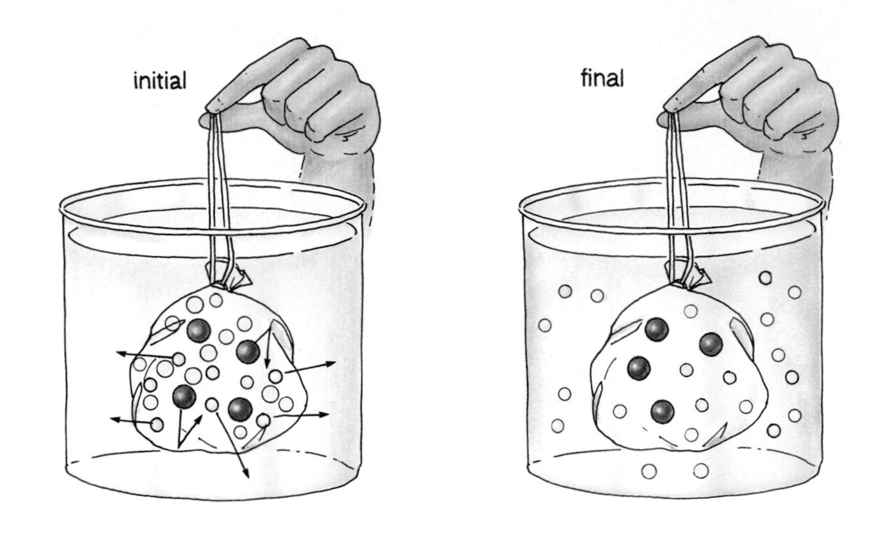

HEALTH NOTE *Dialysis by the Kidneys and the Artificial Kidney*

The fluids of the body undergo dialysis by the membranes of the kidneys, which remove waste materials, excess salts, and water. In an adult, each kidney contains about 2 million nephrons. (See Figure 8.22.) At the top of each nephron, there is a network of arterial capillaries called the glomerulus.

As blood flows into the glomerulus, small particles, such as amino acids, glucose, urea, water, and certain ions, will move through the capillary membranes into the nephron. As this solution moves through the nephron, substances still of value to the body (such as amino acids, glucose, certain ions, and 99% of the water) are reabsorbed. The major waste product, urea, forms urine in the bladder, to be excreted.

Figure 8.22
The nephron, the working unit of the kidney. Water and small particles dialyze out of the blood across the membrane of the glomerulus. Most of the reusable substances, including water, glucose, and amino acids, are reabsorbed from the filtrate, leaving urea and other waste products to be excreted in the urine.

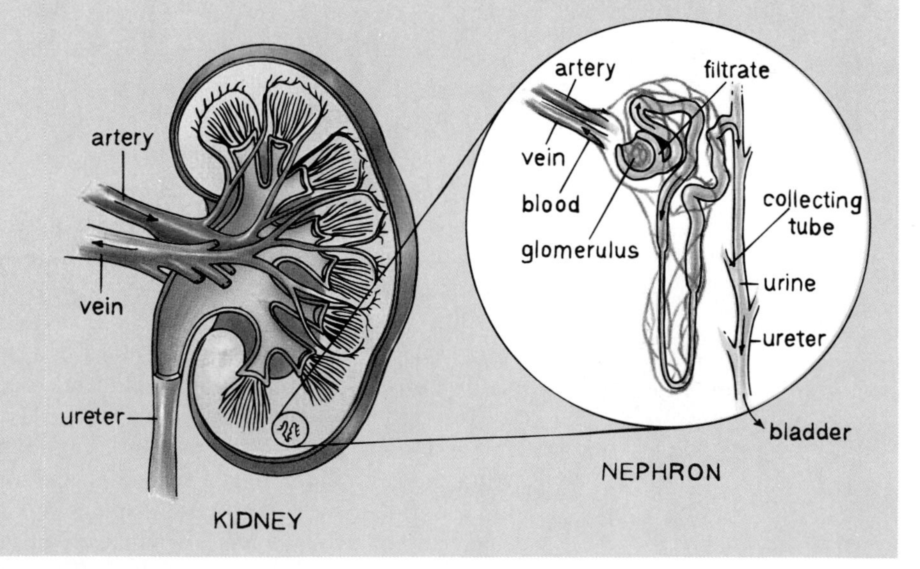

Hemodialysis

If the kidneys fail to dialyze waste products, increased levels of urea can become life-threatening in a relatively short time. A person with kidney failure can use an artificial kidney, which cleanses the blood by **hemodialysis,** as shown in Figure 8.23.

A typical artificial kidney machine contains a large tank filled with about 100 L of distilled water. In the center of this dialyzing bath (dialysate), there is a dialyzing coil or membrane made of cellulose tubing. As the patient's blood flows through the dialyzing coil, the highly concentrated waste products dialyze out of the blood. No blood is lost because the membrane is not permeable to large particles such as red blood cells.

Dialysis patients do not produce much urine. As a result, they retain large amounts of water between dialysis treatments, which produces a strain on the heart. The intake of fluids for a dialysis patient may be restricted to as little as a few teaspoons of water a day. In the dialysis procedure, the pressure of the blood is increased as it circulates through the dialyzing coil so water can be squeezed out of the blood. For some dialysis patients, 2–10 L of water may be removed during one treatment. Dialysis patients have from two to three treatments a week, each treatment requiring about 5–7 hr. Some of the newer treatments require less time. For many patients, dialysis is done at home with a home dialysis unit.

Figure 8.23
Hemodialysis, the dialysis of the blood by an artificial kidney. The initial dialysate consists of water and electrolytes. As blood flows through the dialyzing coil, urea and other waste products dialyze out of the coil and into the dialysate.

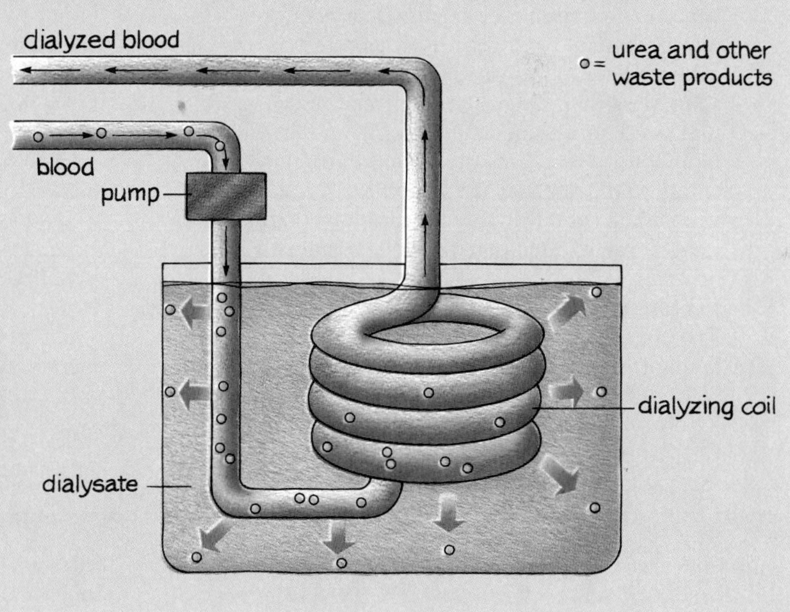

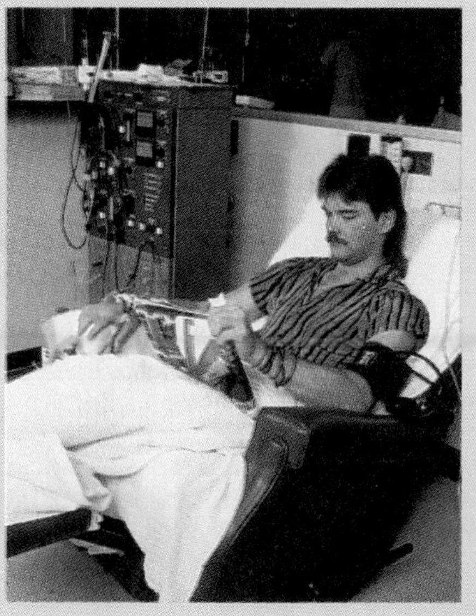

Chapter Summary

8.1 Water and Solutions
The polar O—H bond permits water molecules to hydrogen bond to other water molecules. A solution forms when a solute dissolves in a solvent.

8.2 Formation of Ionic Solutions
An ionic solute dissolves in water, a polar solvent, because the polar water molecules attract and pull the positive and negative ions into solution. The rate of solution formation is increased by crushing the solute, stirring, or heating the solution.

8.3 Factors Affecting Solubility
Solubility describes the amount of a solute that can dissolve in 100 g of solvent. The amount of solute that dissolves depends on the nature of the solute and solvent; "like dissolves like." A solution containing the maximum amount of dissolved solute at a certain temperature is a saturated solution. Any amount of dissolved solute less than the solubility at that temperature makes an unsaturated solution. An increase in temperature increases the solubility of most solids, and decreases the solubility of gases, in water. The solubility rules describe the kinds of ionic combinations that are soluble and insoluble in water.

8.4 Percent Concentration
The concentration of a solute in a solution is the amount of solute dissolved in a certain amount of solution. Mass percent expresses the ratio of the mass of solute to the mass of solution multiplied by 100. Percent concentration can also be expressed as volume/volume and mass/volume ratios. In calculations of grams or milliliters of solute or solution, the percent concentration is used as a conversion factor.

8.5 Molarity
Molarity is the moles of solute per liter of solution. Units of molarity, moles/liter, are used in conversion factors to solve for moles or volume of solution. When a solution is diluted with more solvent, the concentration decreases.

8.6 Colloids and Suspensions
Colloids contain particles that show the Tyndall effect, do not settle out, and pass through filters but not through semipermeable membranes. Suspensions have very large particles that settle out of solution.

8.7 Osmosis and Dialysis
In osmosis, only the solvent (water) passes through a semipermeable membrane from the solution of a lower concentration to a solution of a higher concentration. Isotonic solutions have osmotic pressures equal to that of the body fluids. A red blood cell maintains its volume in an isotonic solution, but swells (hemolysis) in a hypotonic solution, and shrinks (crenates) in a hypertonic solution. In dialysis, water and small solute particles can pass through a dialyzing membrane, while larger particles are retained.

Glossary of Key Terms

colloid A mixture having particles that are moderately large. Colloids scatter light in the Tyndall effect and pass through filters but will not pass through semipermeable membranes.

concentration A measure of the amount of solute that is dissolved in a specified amount of solution.

crenation The shriveling of a cell due to water leaving the cell when the cell is placed in a hypertonic solution.

dialysis A process in which water and small solution particles move through a semipermeable membrane.

dilution The decrease in concentration of a solution by the addition of more solvent.

hemodialysis A mechanical cleansing of the blood by an artificial kidney using the principle of dialysis.

hemolysis A swelling and/or bursting of red blood cells in a hypotonic solution due to an increase in fluid volume.

hydrogen bond The attraction between a partially positive hydrogen atom in one water molecule and the partially negative oxygen atom of another.

hypertonic solution A solution that has a higher osmotic pressure than the red blood cells of the body.

hypotonic solution A solution that has a lower osmotic pressure than the red blood cells of the body.

isotonic solution A solution that has the same osmotic pressure as that of the red blood cells of the body.

mass percent The grams of solute in 100 grams of solution defined as mass/mass.

molarity (M) The number of moles of solute in 1 L of solution.

osmosis The flow of a solvent, usually water, through a semipermeable membrane into a solution of higher concentration.
osmotic pressure The pressure that prevents the flow of water into the more concentrated solution.
physiological solution A solution that exerts the same osmotic pressure as body fluids.
saturated solution A solution containing the maximum amount of solute that can dissolve at a given temperature. Any additional solute will remain undissolved in the container.
semipermeable membrane A membrane that permits the passage of certain substances while blocking or retaining others.
solubility The maximum amount of solute that can dissolve in 100 g of solvent, usually water, at a given temperature.
solute The component in a solution that changes state upon dissolving; if no change in state occurs, it is the component present in smaller quantity.
solution A homogeneous mixture in which the solute is made up of small, single particles (ions or molecules) that can pass through filters and semipermeable membranes.
solvent The substance in which the solute dissolves; usually the component present in greatest amount.
surface tension A characteristic of water in which the hydrogen bonding of water molecules pulls its surface molecules together so tightly that they form a "skin."
suspension A mixture in which the particles are large enough and heavy enough to settle out and/or be retained by both filters and semipermeable membranes.
Tyndall effect The opaque path of a light beam through a colloid as light is reflected by the colloidal particles.
unsaturated solution A solution that contains less solute than its solubility level.

Chemistry at Home

1. Place three to four raisins or dry prunes in water and leave them alone. After 1 hour, how have they changed? Explain.
2. Place a celery stalk in a glass with a layer of food coloring at the bottom. After 1–2 hours, where is the color in the celery stalk? Explain.
3. Place a grape in some salt water. What happens to it after 1 hour? Explain.
4. Using a small flashlight, shine a beam of light into clear glasses containing solutions such as skim milk, water, vinegar, tea, and salt water. Which are solutions and which are colloids? Why?
5. Pour together the following combinations:
   a. oil and water **b.** water and vinegar
   c. salt and water **d.** salt and oil
   In which combinations did a solution form? Why? Why did some combinations not form solutions?
6. Make a solution of salt and water. What happens to the salt crystals? How does it taste? Place the solution in a pan and heat gently. Stop heating when the water is almost gone. What is in the pan? Why?

Answers to Study Checks

8.1 Iodine is the solute, and ethyl alcohol is the solvent.
8.2 The ions Ca^{2+} and Cl^- are pulled away from the solid as polar water molecules attract and separate them from the salt. In solution, they are hydrated.
8.3 The rate of solution formation will decrease at a lower temperature.
8.4 Yes. Both the solute and solvent are nonpolar substances; "like dissolves like."
8.5 75 g $CuSO_4$
8.6 65 g/100 g H_2O because solubility usually increases at higher temperatures.
8.7 **a.** insoluble **b.** soluble **c.** insoluble
8.8 3.4% (m/m) NaCl solution
8.9 4.8% (v/v) Br_2 solution
8.10 30 g
8.11 0.40 L
8.12 2.1 M
8.13 4.5 mol
8.14 123 g
8.15 250 mL
8.16 3% (m/v)
8.17 colloids
8.18 5% glucose
8.19 The red blood cell will shrink (crenate).

Problems

Water and Solutions *(Goal 8.1)*

8.1 **a.** Draw a water molecule and describe its polarity.
b. Show how hydrogen bonding occurs between two water molecules.
c. Why does HF also exhibit hydrogen bonding?
d. Would you expect CH_4 molecules to hydrogen bond? Explain.

8.2 Match the following statements with *hydrogen bonding, surface tension,* or *polar solvent:*
a. The pulling together of water molecules at the surface to form a thick layer that will support the weight of small bugs.
b. Attractive forces between water molecules.
c. Behavior of water in forming solutions with ionic compounds.
d. Formation of beads of water on a waxed surface.

8.3 Identify the solute and the solvent in each solution composed of the following:
a. 10.0 g NaCl and 100.0 g H_2O
b. 50.0 mL ethanol, C_2H_5OH (*l*), and 10.0 mL H_2O
c. 0.20 L oxygen (O_2), 0.80 L nitrogen (N_2)

8.4 Identify the solute and the solvent in each solution composed of the following:
a. 50.0 g silver and 4.0 g mercury
b. 100.0 mL H_2O and 5.0 g sugar
c. 1.0 g I_2 and 50.0 mL alcohol

Formation of Ionic Solutions *(Goal 8.2)*

8.5 Match the following statements with *oxygen atom, hydrogen atom,* or *hydration* to describe the formation of an NaBr solution:
a. The portion of the water molecule attracted to Na^+ ions.
b. The surrounding of Na^+ ions in solution by water molecules.
c. The portion of the water molecule attracted to the Br^- ions.

8.6 Match the following statements with *oxygen atom, hydrogen atom,* or *hydration* to describe the formation of a $CaCl_2$ solution:
a. The portion of the water molecule attracted to the Cl^- ions.
b. The portion of the water molecule attracted to Ca^{2+} ions.
c. The surrounding of Cl^- ions in solution by water molecules.

8.7 Describe the formation of a KI and water solution.

8.8 Describe the formation of an LiBr and water solution.

8.9 Indicate whether each of the following actions will increase or decrease the rate of solution of KI in water:
a. Stirring the mixture.
b. Using large chunks of KI.
c. Placing the beaker of KI and water in ice.
d. Heating the beaker containing the KI and water.

8.10 Indicate whether each of the following actions will increase or decrease the rate of solution of sugar (sucrose) in water:
a. Crushing the sugar into a fine powder.
b. Placing the beaker of sugar and water in ice.
c. Stirring the mixture.
d. Heating the sugar and water.

Factors Affecting Solubility *(Goal 8.3)*

8.11 Water is a polar solvent; CCl_4 is a nonpolar solvent. In which solvent are each of the following more likely to be soluble?
a. KCl, ionic
b. I_2, nonpolar
c. sugar, polar
d. gasoline, nonpolar

8.12 Water is a polar solvent; hexane is a nonpolar solvent. In which solvent are each of the following more likely to be soluble?
a. vegetable oil, nonpolar
b. benzene, nonpolar
c. $LiNO_3$, ionic
d. Na_2SO_4, ionic

8.13 State whether each of the following refers to a saturated or unsaturated solution:
a. A crystal added to a solution does not change in size.
b. A sugar cube completely dissolves when added to a cup of coffee.

8.14 State whether each of the following refers to a saturated or unsaturated solution:
a. A spoonful of salt added to boiling water dissolves.
b. A layer of sugar forms on the bottom of a glass of tea as ice is added.

8.15 Use the table below to determine whether each of the following is a saturated or unsaturated solution:

Solubility (g/100 g H_2O)

T(°C)	KBr	KI
20	65	145
40	80	160
60	90	175
80	100	190
100	110	210

a. 75 g of KBr in 100 g of H_2O at 40°C
b. 140 g of KI in 50 g of H_2O at 40°C
c. 100 g of KBr in 100 g of H_2O at 60°C
d. 100 g of KI in 200 g of H_2O at 20°C

8.16 Use the table in problem 8.15 to determine whether each of the following is a saturated or unsaturated solution:
a. 100 g of KBr in 100 g of H_2O at 100°C
b. 50 g of KI in 50 g H_2O at 60°C
c. 250 g of KI in 100 g H_2O at 100°C
d. 40 g KBr in 200 g H_2O at 80°C

8.17 A solution containing 100 g KBr in 100 g H_2O at 100°C is cooled to 40°C. Use the table in problem 8.15 to answer the following:
a. Is the solution saturated or unsaturated at 100°C?
b. Is the solution saturated or unsaturated at 40°C?
c. How much solid KBr, if any, formed during the cooling process?

8.18 A solution containing 150 g KI in 100 g of water is heated to 80°C and cooled to 20°C. Use the table in problem 8.15 to answer the following:
a. Is the solution saturated or unsaturated at 80°C?
b. Is the solution saturated or unsaturated at 20°C?
c. How much solid KI, if any, formed during the cooling process?

8.19 Explain the following observations:
a. More sugar dissolves in hot tea than in ice tea.
b. Champagne in a warm room goes flat.
c. A warm can of soda has more spray when opened than a cold one.

8.20 Explain the following observations:
a. An open can of soda loses its "fizz" quicker at room temperature than in the refrigerator.
b. Chlorine gas in tap water escapes as the sample warms to room temperature.
c. Less sugar dissolves in iced coffee than in hot coffee.

8.21 Predict whether each of the following ionic compounds is soluble in water:
a. LiCl **b.** $PbCl_2$ **c.** $BaCO_3$
d. K_2O **e.** $Fe(NO_3)_3$

8.22 Predict whether each of the following ionic compounds is soluble in water:
a. PbS **b.** NaI **c.** Na_2S
d. Ag_2O **e.** $CaSO_4$

Percent Concentration *(Goal 8.4)*

8.23 What is the difference between a 5% (m/m) glucose solution and a 5% (m/v) glucose solution?

8.24 What is the difference between a 10% (v/v) methyl alcohol (CH_3OH) solution and a 10% (m/m) methyl alcohol solution?

8.25 Calculate the mass percent of the solute in each of the following solutions:
a. 20.0 g KCl in 150. g of solution
b. 40.0 g $CaCl_2$ in 250. g of solution
c. 50.0 g $NaNO_3$ in 400. g H_2O

8.26 Calculate the mass percent of the solute in each of the following solutions:
a. 8.0 g NaBr in 100. g H_2O
b. 75.0 g KCl in 500.0 g of solution
c. 2.0 g KOH in 20.0 g H_2O

8.27 Calculate the mass/volume percent of the solute in each of the following solutions:
a. 75.0 g Na_2SO_4 in 250 mL of solution
b. 1.5 g Na_2CO_3 in 500. mL of solution
c. 50.0 g of glucose ($C_6H_{12}O_6$) in 0.50 L of solution

8.28 Calculate the mass/volume percent of the solute in each of the following solutions:
a. 2 g sucrose ($C_{12}H_{22}O_{11}$) in 100 mL of solution
b. 0.80 g $MgCl_2$ in 20.0 mL of solution
c. 60.0 g KOH in 150 mL of solution

8.29 Write two conversion factors for the following solution concentrations:
a. 10% (m/m) NaOH
b. 0.5% (m/v) NaCl
c. 15% (v/v) CH_3OH

8.30 Write two conversion factors for the following solution concentrations:
a. 4% (m/v) KCl
b. 1.0% (m/m) K_2CO_3
c. 15% (v/v) isopropyl alcohol solution

8.31 Calculate the amount of solute (g or mL) needed to prepare the following solutions:
a. 50.0 mL of a 5.0% (m/v) KCl solution
b. 225 g of a 10.0% (m/m) K_2CO_3 solution
c. 1250 mL of a 4.0% (m/v) NH_4Cl solution
d. 250 mL of a 10.0% (v/v) acetic acid solution

8.32 Calculate the amount of solute (g or mL) needed to prepare the following solutions:
a. 75 g of a 25% (m/m) NaOH solution
b. 150 mL of a 40.0% (m/v) $LiNO_3$ solution
c. 450 mL of a 2.0% (m/v) KCl solution
d. 100 mL of a 15% (v/v) isopropyl alcohol solution

8.33 A patient receives 100 mL of 20% (m/v) mannitol solution every hour.
a. How many grams of mannitol are given in 1 hour?
b. How many grams of mannitol does the patient receive in 1 day?

8.34 A patient receives 250 mL of a 4% (m/v) amino acid solution twice a day.
a. How many grams of amino acids are in 250 mL of solution?
b. How many grams of amino acids does the patient receive in 1 day?

8.35 Calculate the amount of solution (g or mL) that contains each of the following amounts of solute?
a. 10.0 g HCl from a 1.0% (m/m) HCl solution
b. 5.0 g $LiNO_3$ from a 25% (m/v) $LiNO_3$ solution
c. 40.0 g KOH from a 10.0% (m/v) KOH solution
d. 2.0 mL of acetic acid from a 10.0% (v/v) acetic acid solution

8.36 Calculate the amount of solution (g or mL) that contains each of the following amounts of solute?
a. 50 g of NaCl from a 2.0% (m/v) NaCl solution
b. 4.0 g of NaOH from a 20% (m/v) NaOH solution
c. 20.0 g of KBr from a 8.0% (m/v) solution
d. 40.0 mL of isopropyl alcohol from a 15% (v/v) isopropyl alcohol solution

8.37 A patient needs 100 g of glucose in the next 12 hours. How many liters of a 5% (m/v) glucose solution must be given?

8.38 A patient receives 2.0 g NaCl in 8 hours. How many milliliters of a 1% (w/v) NaCl (saline) were delivered?

Molarity *(Goal 8.5)*

8.39 Calculate the molarity (M) of the following solutions:
a. 4.0 g of KOH in 2.0 L of solution
b. 2.0 mol of glucose in 4.0 L of solution
c. 20.0 g of NaOH in 0.500 L of solution
d. 0.10 mol NaCl in 40.0 mL of solution

8.40 Calculate the molarity (M) of the following solutions:
a. 36.5 g HCl in 1.0 L of solution
b. 0.50 mol glocose in 0.200 L of solution
c. 8.0 g of NaOH in 250 mL of solution
d. 0.300 mol LiBr in 300. mL of solution

8.41 Write two conversion factors for each of the following concentrations:
a. 6.00 M HCl
b. 0.250 M $NaHCO_3$
c. 1.0 M H_2SO_4

8.42 Write two conversion factors for each of the following concentrations:
a. 2.0 M NaOH
b. 0.10 M KBr
c. 3.0 M Li_2CO_3

8.43 Calculate the moles of solute needed to prepare each of the following:
a. 1.0 L of a 3.0 M NaCl solution
b. 0.400 L of a 1.0 M KBr solution
c. 200 mL of a 4.0 M NaCl solution

8.44 Calculate the moles of solute needed to prepare each of the following:
a. 5.0 L of a 2.0 M $CaCl_2$ solution
b. 10.0 L of a 10.0 M NaOH solution
c. 25 mL of a 1.0 M KBr solution

8.45 Calculate the grams of solute needed to prepare each of the following solutions:
a. 1.0 L of a 1.0 M NaOH solution
b. 4.0 L of a 2.0 M KCl solution
c. 500 mL of a 1.0 M NaCl solution

8.46 Calculate the grams of solute needed to prepare each of the following solutions:
a. 2.0 L of a 6.0 M NaOH solution
b. 5.0 L of a 0.10 M $CaCl_2$ solution
c. 1500 mL of a 2.0 M NaCl solution

8.47 What volume in liters provides the following amounts of solute?
a. 2.0 mol NaOH from a 2.0 M NaOH solution
b. 10 mol NaCl from a 1 M NaCl solution
c. 80.0 g NaOH from a 1 M NaOH solution

8.48 What volume in liters provides the following amounts of solute?
a. 0.100 mol KCl from a 4.0 M KCl solution
b. 125 g NaOH from a 1.0 M NaOH solution
c. 5.0 mol HCl from a 6.0 M HCl solution

8.49 Calculate the final concentration of each of the following diluted solutions:
a. 10 mL of 20% (m/v) KCl + 40 mL H_2O
b. 50 mL of 30% (m/v) mannitol + 100 mL H_2O
c. diluting 4.0 L of 6.0 M HCl to 12 L
d. diluting 10.0 mL of a 15% (m/v) NaCl to 50.0 mL

8.50 Calculate the final concentration of each of the following diluted solutions:
- **a.** 200 mL of 10% (m/v) NaCl + 200 mL H_2O
- **b.** adding 150 mL of water to 50 mL of 20% (m/v) KOH
- **c.** diluting 5 mL of 12% (m/v) NaOH to 30 mL
- **d.** diluting 20. mL of 4.0 M $CaCl_2$ to 80. mL

Colloids and Suspensions *(Goal 8.6)*

8.51 Identify the following as characteristic of a solution, colloid, or suspension:
- **a.** A clear mixture that cannot be separated by a membrane.
- **b.** A mixture that appears cloudy when a beam of light is passed through it.
- **c.** A mixture that settles out upon standing.

8.52 Identify the following as characteristic of a solution, colloid, or suspension:
- **a.** Particles of this mixture remain inside a semipermeable membrane.
- **b.** A beam of light is not visible in this solution.
- **c.** The particles of solute in this solution are very large and visible.

8.53 What is an emulsion?

8.54 What is an aerosol?

Osmosis and Dialysis *(Goal 8.7)*

8.55 A 10% (m/v) starch solution is separated from pure water by an osmotic membrane.
- **a.** Which solution has the higher osmotic pressure?
- **b.** In which direction will water flow initially?
- **c.** In which compartment will the volume level rise?

8.56 Two solutions, a 0.1% (m/v) albumin solution and a 2% (m/v) albumin solution are separated by a semipermeable membrane. (Albumins are colloidal proteins.)
- **a.** Which compartment has the higher osmotic pressure?
- **b.** In which direction will water flow initially?
- **c.** In which compartment will the volume level rise?

8.57 Indicate the compartment (A or B) that will increase in volume for each of the following pairs of solutions separated by semipermeable membranes:

	A	*B*
a.	5% (m/v) glucose	10% (m/v) glucose
b.	4% (m/v) albumin	8% (m/v) albumin
c.	0.1% (m/v) NaCl	10% (m/v) NaCl

8.58 Indicate the compartment (A or B) that will increase in volume for each of the following pairs of solutions separated by semipermeable membranes:

	A	*B*
a.	20% (m/v) glucose	10% (m/v) glucose
b.	10% (m/v) albumin	2% (m/v) albumin
c.	0.5% (m/v) NaCl	5% (m/v) NaCl

8.59 Are the following solutions isotonic, hypotonic, or hypertonic compared with a red blood cell?
- **a.** distilled H_2O
- **b.** 1% (m/v) glucose
- **c.** 0.90% (m/v) NaCl
- **d.** 5% (m/v) glucose
- **e.** 10% (m/v) NaCl

8.60 Will a red blood cell undergo crenation, hemolysis, or no change in each of the following solutions?
- **a.** 1% (m/v) glucose
- **b.** 2% (m/v) NaCl
- **c.** 5% (m/v) NaCl
- **d.** 0.1% (m/v) NaCl
- **e.** 10% (m/v) glucose

8.61 Each of the following mixtures is placed in a dialyzing bag and immersed in distilled water. Which substances will be found outside the bag in the distilled water?
- **a.** NaCl solution.
- **b.** Starch, a colloid and alanine, an amino acid solution.
- **c.** NaCl solution and starch, a colloid.
- **d.** Urea solution.

8.62 Each of the following mixtures is placed in a dialyzing bag and immersed in distilled water. Which substances will be found outside the bag in the distilled water?
- **a.** KCl solution and glucose solutions.
- **b.** An albumin solution (colloid).
- **c.** An albumin solution (colloid), KCl solution, and glucose solution.
- **d.** Urea solution and NaCl solution.

Challenge Problems

8.63 A patient receives all her nutrition from fluids given through the vena cava. Every 12 hours, 750 mL of a solution that is 4% (m/v) amino acids from protein and 25% (m/v) glucose is given along with 500 mL of a 10% (m/v) lipid (fat).
- **a.** In 1 day, how many grams of amino acids, glucose, and lipid are given to the patient?
- **b.** How many kilocalories does she obtain in 1 day?

8.64 An 80 proof brandy is 40% (v/v) ethyl alcohol. The "proof" is twice the percent concentration of alcohol in the beverage. How many milliliters of alcohol are present in 750 mL of brandy?

8.65 A solution is prepared with 70.0 g HNO_3 and 130.0 g H_2O. It has a density of 1.21 g/mL.

a. What is the mass percent of the HNO_3 solution?

b. What is the total volume of the solution?

c. What is the mass/volume percent?

d. What is its molarity (M)?

8.66 Why would a dialysis unit (artificial kidney) use isotonic concentrations of NaCl, KCl, $NaHCO_3$, and glucose in the dialysate?

8.67 Why would solutions with high salt content be used to prepare dried flowers?

8.68 A patient on dialysis has a high level of urea, a high level of sodium, and a low level of potassium in the blood. Why is the dialyzing solution prepared with a high level of potassium but no sodium or urea?

8.69 Why is it dangerous to drink sea water even if you are stranded on a desert island?

8.70 In some countries, pure water is obtained from sea water by reverse osmosis. What do you think such a process involves?

Chapter 9

Acids and Bases

Stalactites are downward growing formations of calcium carbonate caused by a solution of Ca^{2+} and HCO_3^- dripping through the ceiling of a limestone cave.

Looking Ahead

Learning Goals

9.1 Identify the components in solutions of electrolytes and nonelectrolytes.

9.2 Calculate the number of equivalents for an electrolyte.

9.3 Describe acids and bases using the Arrhenius and the Brønsted–Lowry concepts.

9.4 Write an equation for the ionization of strong and weak acids.

9.5 Write a balanced equation for the neutralization reaction of an acid and a base.

9.6 Use the ion product of water to calculate the $[H_3O^+]$ and $[OH^-]$ in a solution.

9.7 Calculate pH from $[H_3O^+]$; given the pH, calculate $[H_3O^+]$ and $[OH^-]$ of a solution.

9.8 Describe the role of buffers in maintaining the pH of a solution.

9.9 Calculate the molarity or volume of an acid or base from titration information.

QUESTIONS TO THINK ABOUT

1. What is meant by a label on a shampoo that reads "pH balanced"?
2. Why is it important to check the pH of an aquarium or pool?
3. What are some ingredients found in antacids? What do they do?

Lemons, grapefruits, and vinegar taste sour because they contain acidic solutions. We have acid in our stomach that helps us digest food; we produce lactic acid in our muscles when we exercise. Acid from bacteria turns milk sour to make cottage cheese or yogurt. Bases are solutions that neutralize acids. Sometimes we take antacids such as milk of magnesia to offset the effects of too much acid.

The pH of a solution describes its acidity. The pH of body fluids, including blood and urine, is regulated primarily by the lungs and the kidneys. Major changes in the pH of the body fluids can severely affect biological activities within the cells. Buffers are present to prevent large fluctuations in pH.

In the body fluids, there are electrolytes such as K^+, Na^+, Cl^-, H^+, and HCO_3^-, along with molecules of glucose and urea. Significant changes in their concentrations can indicate illness or injury. Therefore, the measurement of the electrolyte concentration is a valuable diagnostic tool.

9.1 Electrolytes

Learning Goal

Identify the components in solutions of electrolytes and nonelectrolytes.

Solutions containing ions are conductors of electricity. The ions come from substances called **electrolytes.** To test for electrolytes, we can use an apparatus that consists of a pair of electrodes connected by wires to a light bulb. (See Figure 9.1.) The light bulb glows when current flows through it, and that only happens when there are ions moving about the solution.

In water, a **strong electrolyte** exists only as ions in solution. When there are many ions in solution, the light bulb glows brightly. However, in a solution of **weak electrolytes,** there are only a few ions. With a small number of ions in solution, the light bulb glows dimly. When the electrodes are placed in pure water or a sugar solution, the light bulb does not glow at all. Such substances are **nonelectrolytes** because they dissolve in water as covalent molecules, which do not conduct electricity.

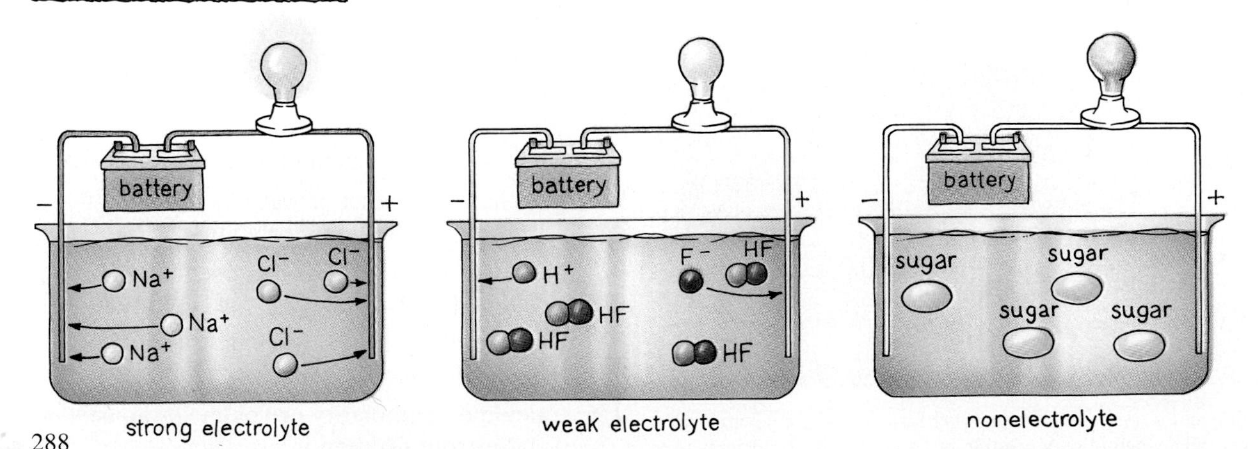

Figure 9.1
In solution a strong electrolyte produces ions and conducts an electrical current; a weak electrolyte produces only a few ions and conducts a weak electrical current. No ions, only molecules, exist in a solution of a nonelectrolyte; there is no electrical current.

Figure 9.2
A solution of NaCl, a strong electrolyte, contains sodium ions (Na^+) and chloride ions (Cl^-).

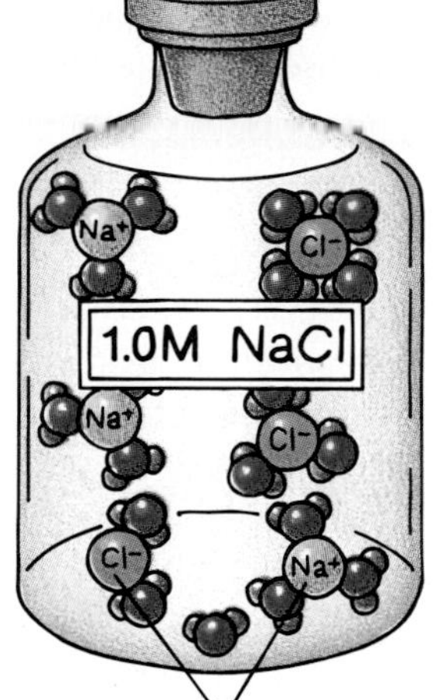

Sodium chloride (NaCl), a strong electrolyte, consists of sodium ions and chloride ions held together by strong ionic bonds. In water, the ions are attracted to water molecules and separate from the solid, a process called dissociation. A solution forms that looks like water, but tastes salty, because it contains dissolved sodium ions and chloride ions. (See Figure 9.2.) We write an equation for the dissociation of NaCl in water to show that solid NaCl separates into ions of $Na^+(aq)$ and $Cl^-(aq)$:

$$NaCl(s) \xrightarrow[\text{100\% dissociation}]{H_2O} Na^+(aq) + Cl^-(aq) \quad \text{Ions in solution}$$

or simplified:

$$NaCl \xrightarrow{H_2O} Na^+ + Cl^-$$

Other soluble ionic compounds dissolve in a similar way. When we write an ionic equation, we balance the electrical charges of the ions. The total positive charge must equal the total negative charge. For example, magnesium chloride, $MgCl_2$, dissolves in water to give one magnesium ion for every two chloride ions in solution.

$$MgCl_2 \xrightarrow{H_2O} Mg^{2+}\ (aq) + 2Cl^-\ (aq)$$

Charge balance for ions in solution

Sample Problem 9.1
Writing Equations for Ionic Compounds

Write the equation for the formation of an aqueous solution of $Mg(NO_3)_2$.

Solution
The formula indicates one magnesium ion and two nitrate ions. The equation for the formation of its solution in water is

$$Mg(NO_3)_2 \xrightarrow{H_2O} Mg^{2+}\ (aq) + 2NO_3^-\ (aq)$$

Study Check
Write an equation for the formation of an aqueous solution of Na_2SO_4.

A weak electrolyte is typically a polar covalent compound that dissolves in water mostly as nondissociated molecules. However, a few of the molecules ionize to give a small number of ions in solution. For example, most HF dissolves in water as molecules. However, a few molecules ionize producing a small number of H^+ and F^- ions. Thus, in water, weak electrolytes are only slightly ionized, which means that most of the dissolved substance is in the molecular form. There are really two reactions taking place in this solution. One is the ionization of HF molecules written as

$$HF \xrightarrow{\text{Ionization}} H^+ + F^-$$

As the number of H^+ and F^- build up in solution, some of the ions recombine to give HF molecules.

$$HF \xleftarrow{\text{Recombination}} H^+ + F^-$$

Eventually the reverse reaction goes as fast as the forward reaction. Then the concentrations of HF, H^+, and F^-, no longer change. An equilibrium exists between the HF molecules and the ions. The forward and reverse reactions are shown in the equation by using a double arrow. (We used a single arrow for strong electrolytes that produced only ions.) The formation of the HF solution (and other weak electrolytes) may be written as follows:

$$HF \rightleftharpoons H^+ + F^-$$

A few ionize

Nonelectrolytes are covalent compounds that dissolve in water as molecules. They produce no ions and their solutions do not conduct electricity. For example, glucose ($C_6H_{12}O_6$) is a nonelectrolyte that dissolves in water to produce a solution of sugar molecules.

$$C_6H_{12}O_6(s) \xrightarrow{H_2O} C_6H_{12}O_6\,(aq)$$

Glucose — Solution of glucose molecules

In an aqueous solution:

A strong electrolyte is completely (100%) ionized.
A weak electrolyte is slightly ionized.
A nonelectrolyte forms no ions, only molecules.

Sample Problem 9.2
Solutions of Electrolytes and Nonelectrolytes

Indicate whether solutions of each of the following contain ions, molecules, or both:

a. Na_2SO_4, a soluble salt **b.** CH_3OH, a nonelectrolyte

Solution

a. A solution of Na_2SO_4 contains the ions of the salt, Na^+ and SO_4^{2-}.
b. A nonelectrolyte such as CH_3OH dissolves in water as molecules.

Study Check

Boric acid, H_3BO_3, is a weak electrolyte. Would you expect a boric acid solution to contain ions, molecules, or both?

9.2 Equivalents

Learning Goal
Calculate the number of equivalents for an electrolyte.

We can describe the amounts of electrolytes, such as Na^+, Cl^-, K^+, and Ca^{2+}, in terms of equivalents. An **equivalent** is the amount of ion that carries 1 mol of electrical charge. For example, 1 mole of Na^+ ions and 1 mol of Cl^- ions are each 1 equivalent because they contain 1 mol of charge. This is the same for ions such as K^+, H^+, OH^-, and HCO_3^-. For an ion with a charge of 2+ or 2−, there are 2 equivalents for each mole. Some examples of ions and equivalents are shown in Table 9.1.

Because equivalents describe the amount of positive or negative charge per mole of ion or compound, we can use the number of equivalents in 1 mol as a conversion factor. For example, Ca^{2+} contains 2 equivalents in 1 mol of ions, which can be written as two possible conversion factors.

$$\frac{2 \text{ equiv } Ca^{2+}}{1 \text{ mol } Ca^{2+}} \quad \text{and} \quad \frac{1 \text{ mol } Ca^{2+}}{2 \text{ equiv } Ca^{2+}}$$

Table 9.1 ***Equivalents of Electrolytes***

Ion	Electrical Charge	Number of Equivalents in 1 mol
1 mol Na^+	1+	1 equiv
1 mol Ca^{2+}	2+	2 equiv
1 mol Fe^{3+}	3+	3 equiv
1 mol Cl^-	1−	1 equiv
1 mol SO_4^{2-}	2−	2 equiv

Sample Problem 9.3
Calculating Equivalents

How many equivalents are in 6.0 g Ca^{2+}?

Solution

The conversion factor of molar mass converts the grams to moles.

$$6.0 \text{ g } Ca^{2+} \times \frac{1 \text{ mol } Ca^{2+}}{40.1 \text{ g } Ca^{2+}} = 0.15 \text{ mol } Ca^{2+}$$

From the positive charge 2+, we know that there are 2 equivalents in 1 mol of calcium ion. Using the factors derived above, we calculate the number of equivalents in our sample.

$$0.15 \text{ mol } Ca^{2+} \times \frac{2 \text{ equiv } Ca^{2+}}{1 \text{ mol } Ca^{2+}} = 0.30 \text{ equiv } Ca^{2+}$$

Study Check

How many grams of Cl^- are present in 2.0 equiv of Cl^-?

Sample Problem 9.4
Electrolyte Concentration

In body fluids, concentrations of electrolytes are often expressed as milliequivalents (meq) per liter. A typical concentration for Na^+ in the blood is 138 meq/L. If 1 equiv = 1000 meq, how many grams of sodium ion are in 1.00 L of blood?

Solution

Using the volume and the electrolyte concentration (in meq/L) we can find the number of milliequivalents in 1.00 L of blood.

$$1.00\ \cancel{L} \times \frac{138\ \cancel{meq}}{1\ \cancel{L}} \times \frac{1\ equiv}{1000\ \cancel{meq}} = 0.138\ equiv\ Na^+$$

We can then convert equivalents to moles (for Na^+ there is 1 equiv per mole), and moles to grams using molar mass.

$$0.138\ \cancel{equiv\ Na^+} \times \frac{1\ \cancel{mol\ Na^+}}{1\ \cancel{equiv\ Na^+}} \times \frac{23.0\ g\ Na^+}{1\ \cancel{mol\ Na^+}} = 3.17\ g\ Na^+$$

Study Check

A Ringer's solution for intravenous replacement contains 155 meq Cl^- per liter of solution. If a patient receives 500. mL of Ringer's solution, how many grams of chloride were given?

HEALTH NOTE *Electrolytes in Body Fluids*

The concentrations of electrolytes present in body fluids and in intravenous fluids given to a patient are often expressed in milliequivalents per liter (meq/L) of solution.

1 equiv = 1000 meq

Table 9.2 gives the concentrations of some typical electrolytes in blood plasma. There is a charge balance, because the total number of positive electrolytes is equal to the total number of negative electrolytes.

Table 9.2 ***Some Typical Concentrations of Electrolytes in Blood Plasma***

Electrolyte	Concentration (meq/L)	Electrolyte	Concentration (meq/L)
Cations (Positive Ions)		***Anions (Negative Ions)***	
Na^+	138	Cl^-	110
K^+	5	HCO_3^-	30
Mg^{2+}	3	HPO_4^{2-}	4
Ca^{2+}	4	Proteins	6
Total cations	150	Total anions	150

The use of a specific intravenous solution depends on the nutritional, electrolyte, and fluid needs of the individual patient. Examples of various types of solutions are given in Table 9.3.

Table 9.3 ***Electrolyte Concentrations in Intravenous Replacement Solutions***

Solution	Electrolytes (meq/L)	Use
Sodium chloride (0.9%)	Na^+ 154, Cl^- 154	Replacement of fluid loss
Potassium chloride with 5% dextrose	K^+ 40, Cl^- 40	Treatment of malnutrition (low potassium levels)
Ringer's solution	Na^+ 147, K^+ 4, Ca^{2+} 4, Cl^- 155	Replacement of fluids and electrolytes lost through dehydration
Maintenance solution with 5% dextrose	Na^+ 40, K^+ 35, Cl^- 40, lactate$^-$ 20, HPO_4^{2-} 15	Maintenance of fluid and electrolyte levels
Replacement solution (extracellular)	Na^+ 140, K^+ 10, Ca^{2+} 5, Mg^{2+} 3, Cl^- 103, acetate$^-$ 47, citrate^{3-} 8	Replacement of electrolytes in extracellular fluids

9.3 Acids and Bases

Learning Goal

Describe acids and bases using the Arrhenius and the Brønsted–Lowry concepts.

The term acid comes from the Latin word *acidus,* which means sour. We are familiar with the sour tastes of vinegar and lemons and other common acids as illustrated in Figure 9.3.

In the 19th century, Arrhenius was the first to describe **acids** as substances that produce hydrogen ions (H^+) when they dissolve in water. For example, hydrogen chloride ionizes in water to give hydrogen ions, H^+, and chloride ions, Cl^-. For polar covalent compounds such as HCl, the formation of ions in water is called an ionization reaction. Using the Arrhenius concept, we write the ionization of HCl in water as

$$HCl(g) \xrightarrow{H_2O} H^+ + Cl^-$$

Polar covalent compound — Ionization in water

The hydrogen ions, H^+, give all Arrhenius acids common characteristics such as a sour taste, changing blue litmus red, and corrosion of some metals.

Figure 9.3
Some common household items that contain weak acids include vinegar, lemons, carbonated beverages, and vitamin C.

Arrhenius Acids:

1. Produce hydrogen ions, H^+, in water.
2. Have a sour taste.
3. Corrode some metals.
4. Turn blue litmus red.
5. Are electrolytes.
6. Neutralize bases.

Naming Acids

Acids are the source of many of the negative ions we named in Chapter 4. Acids that include the *hydro* prefix contain simple anions. Acids that end in *ic acid* contain polyatomic anions with *ate* endings; *ous acids* contain polyatomic anions with *ite* endings.

Acid Prefix/Suffix		*Anion Suffix*	
hydro _______ ic acid		_______ ide	No oxygen
_______ ic acid		_______ ate	Most common form
_______ ous acid		_______ ite	One oxygen less

The names of some common acids and their anions are listed in Table 9.4.

Table 9.4 ***Naming Common Acids***

Acid	Name of Acid	Negative Ion	Name of Anion
HCl	**hydro**chlor**ic acid**	Cl^-	chlor**ide**
HBr	**hydro**brom**ic acid**	Br^-	brom**ide**
HNO_3	nitr**ic acid**	NO_3^-	nitr**ate**
HNO_2	nitr**ous acid**	NO_2^-	nitr**ite**
H_2SO_4	sulfur**ic acid**	SO_4^{2-}	sulf**ate**
H_2SO_3	sulfur**ous acid**	SO_3^{2-}	sulf**ite**
H_2CO_3	carbon**ic acid**	CO_3^{2-}	carbon**ate**
H_3PO_4	phosphor**ic acid**	PO_4^{3-}	phosph**ate**

Bases

Figure 9.4
Some common bases include detergents, antacids, drain openers, and oven and window cleaners.

You may be familiar with some of the bases illustrated in Figure 9.4 such as detergents, antacids, drain openers, and oven and window cleaners. According to Arrhenius, **bases** are ionic compounds that separate into a metal ion and hydroxide ions (OH^-) when they dissolve in water. For example, sodium hydroxide is an Arrhenius base that dissociates in water to give sodium ions, Na^+, and hydroxide ions, OH^-.

$$NaOH \xrightarrow{H_2O} Na^+ + OH^-$$

Ionic compound — Ionization — Hydroxide ion

Most hydroxide bases are from Group 1A and 2A such as KOH, LiOH, and $Ca(OH)_2$. There are other hydroxide bases such as $Al(OH)_3$ and $Fe(OH)_3$, but they are mostly insoluble. The hydroxide ions (OH^-) give Arrhenius bases common characteristics such as a bitter taste and soapy feel.

Arrhenius Bases:

1. Produce hydroxide ions (OH^-) in water.
2. Taste bitter.
3. Feel slippery, soapy.
4. Turn red litmus blue.
5. Are electrolytes.
6. Neutralize acids.

Naming Bases

The typical Arrhenius bases are named as hydroxides.

Bases	*Name*
NaOH	sodium **hydroxide**
KOH	potassium **hydroxide**
$Ca(OH)_2$	calcium **hydroxide**
$Al(OH)_3$	aluminum **hydroxide**

Sample Problem 9.5
Arrhenius Acids and Bases

Identify each of the following as characteristic of an Arrhenius acid or base:
a. is named magnesium hydroxide
b. tastes sour
c. turns blue litmus red

Solution
a. A base contains hydroxide ions.
b. An acid tastes sour.
c. An acid turns blue litmus red.

Study Check
Detergents have a soapy feel. Are they acids or bases?

Brønsted–Lowry Acids and Bases

Early in the 20th century, Brønsted and Lowry expanded the definition of acids and bases to include nonaqueous solutions and bases without hydroxide ions. A **Brønsted–Lowry acid** is a substance that donates a proton (hydrogen ion, H^+) to another substance, and a **Brønsted–Lowry base** is a substance that accepts a proton.

A Brønsted–Lowry acid is a proton (H^+) donor.
A Brønsted–Lowry base is a proton (H^+) acceptor.

According to the Brønsted–Lowry theory, a proton (H^+) does not exist by itself in water. Its attraction to polar water molecules is so strong that the proton bonds to the water molecule and forms a **hydronium ion, H_3O^+**. At times, we use the H^+ symbol for convenience, but understand that it is the H_3O^+ that is actually present in acidic aqueous solutions.

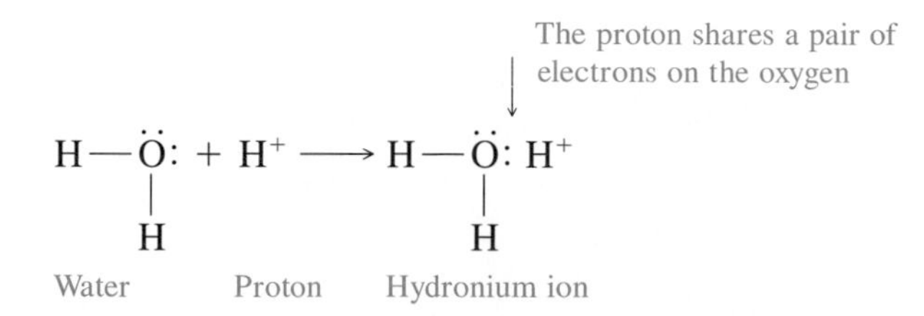

Figure 9.5
Polar covalent hydrogen chloride (HCl) ionizes in water to produce hydrochloric acid, which exists as hydronium ions (H_3O^+) and chloride ions (Cl^-).

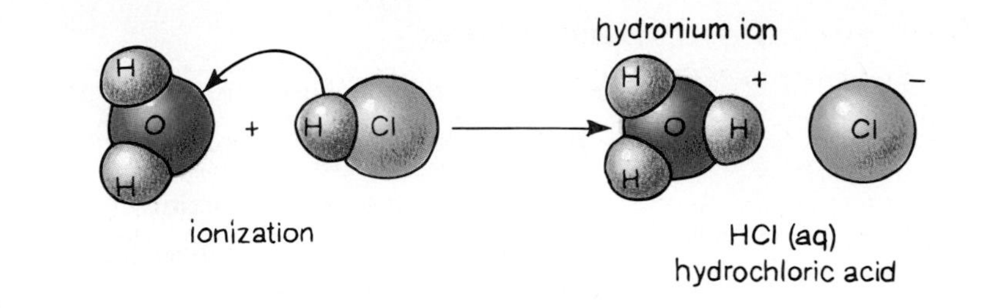

Using the Brønsted–Lowry concept, we can write the formation of a hydrochloric acid solution as a transfer of a proton from hydrogen chloride (HCl) to water. (See Figure 9.5.) By accepting the proton in this reaction, water acts as a base.

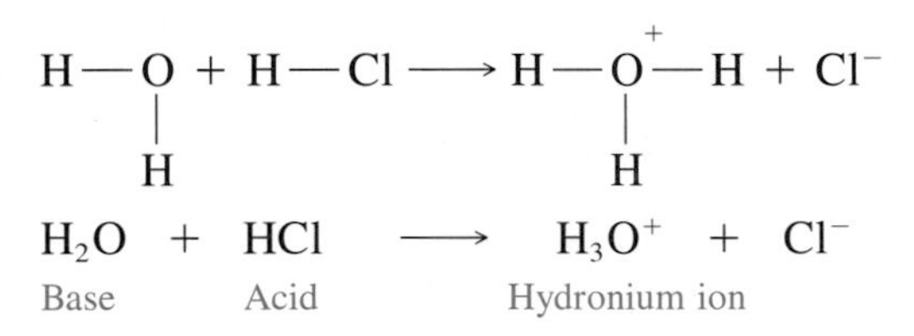

$$\underset{\text{Base}}{H_2O} + \underset{\text{Acid}}{HCl} \longrightarrow \underset{\text{Hydronium ion}}{H_3O^+} + Cl^-$$

With the Brønsted–Lowry definition, we can show that ammonia (NH_3) acts as a base by accepting a proton from water. In this reaction, water acts as an acid.

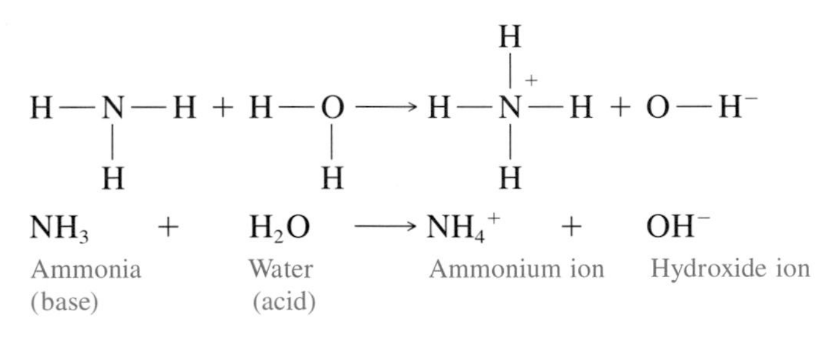

$$\underset{\substack{\text{Ammonia}\\ \text{(base)}}}{NH_3} + \underset{\substack{\text{Water}\\ \text{(acid)}}}{H_2O} \longrightarrow \underset{\text{Ammonium ion}}{NH_4^+} + \underset{\text{Hydroxide ion}}{OH^-}$$

Sample Problem 9.6
Acids and Bases

Write an equation for the ionization of HBr as:

a. an Arrhenius acid. **b.** a Brønsted–Lowry acid.

Solution

a. $HBr \xrightarrow{H_2O} H^+ + Br^-$

b. $HBr + H_2O \longrightarrow H_3O^+ + Br^-$

Study Check

In the following reaction, which reactant acts as an acid, and which is a base?

$$CN^- + H_2O \longrightarrow HCN + OH^-$$

9.4 Strengths of Acids and Bases

Learning Goal

Write an equation for the ionization of strong and weak acids.

Only a few acids, called **strong acids,** ionize completely (100%) to produce ions in solution. Three important strong acids are HCl, HNO_3, and H_2SO_4. We can illustrate their ionization as strong acids by exaggerating the symbols of the ionic products. In the ionization equation for nitric acid, a single arrow is used to indicate complete ionization.

$$HNO_3 + H_2O \longrightarrow \mathbf{H_3O^+ + NO_3^-}$$

Most acids are **weak acids** that do not ionize very much. Even at high concentrations, weak acids contain only a small number of hydronium ions. In fact, you use some weak acids at home, such as citric acid, found in lemon or grapefruit juice, and vinegar, a 5% acetic acid solution. In carbonated soft drinks, CO_2 dissolves in water to form carbonic acid, H_2CO_3, a weak acid. In a weak acid, equilibrium is reached by the time a few dissolved molecules form ions. Using a double arrow, we show the reverse reaction that keeps the concentration of molecules of acetic acid high and that of the ions low.

$$\mathbf{HC_2H_3O_2 + H_2O} \rightleftharpoons H_3O^+ + C_2H_3O_2^-$$

Acetic acid Acetate ion

Figure 9.6 compares the components of a solution of a strong acid (HCl) and a weak acid (HF).

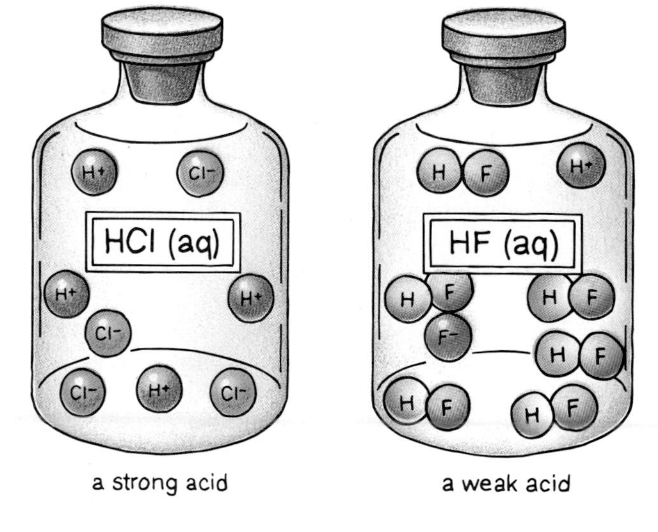

Figure 9.6
A strong acid, HCl, ionizes completely; a small amount of weak acid molecules (HF) ionizes to give a solution of HF molecules and ions.

Sample Problem 9.7
Writing Equations for Ionization of Weak Acids

Write an equation for the ionization of formic acid ($HCHO_2$), a weak acid found in ant stings.

Solution

As a weak acid, formic acid exists in solution mostly as molecules with a few ions. For weak acids, the first hydrogen in the formula is the only hydrogen that ionizes in water.

$$HCHO_2 + H_2O \rightleftharpoons H_3O^+ + CHO_2^-$$

Study Check

Hypochlorous acid, HClO, is a weak acid that forms when chlorine reacts with water in swimming pools and hot tubs. Write an equation for its ionization in water.

Strong and Weak Bases

In water, a **strong base** separates completely into metal ions and hydroxide ions (OH^-). The metal hydroxides of Group 1A as well as $Ba(OH)_2$ in Group 2A are all strong bases.

Sodium hydroxide $NaOH(s) \xrightarrow{H_2O} Na^+ + OH^-$

Potassium hydroxide $KOH(s) \xrightarrow{H_2O} K^+ + OH^-$

Barium hydroxide $Ba(OH)_2(s) \xrightarrow{H_2O} Ba^{2+} + 2OH^-$

Strong bases, such as NaOH (also known as lye), are used in household products to dissolve grease in ovens and to clean drains. Because hydroxide ions cause severe damage to the skin and eyes, the use of such products in the home should be carefully supervised. Most Group 2A hydroxides are not very soluble in water, although they do dissociate completely. One such hydroxide is magnesium hydroxide, $Mg(OH)_2$, an antacid known as milk of magnesia that is ingested to counteract the effects of too much acid in the stomach.

$$Mg(OH)_2 \xrightarrow{H_2O} Mg^{2+} + 2OH^-$$

A **weak base** produces only a few hydroxide ions when it reacts with water. The most common weak base is ammonia (NH_3), found in several household window cleaners. As we saw earlier, ammonia reacts with water by accepting a proton. The resulting hydroxide ion makes the solution basic. The double arrow in the equation indicates an equilibrium in which only a few ammonia molecules have reacted at any given time.

$$\underset{\text{Ammonia}}{NH_3(g)} + \underset{\text{Water}}{H_2O} \rightleftharpoons \underset{\text{Ammonium ion}}{NH_4^+} + \underset{\text{Hydroxide ion}}{OH^-}$$

It is important to know when you are using acids or bases in the laboratory because they can burn the skin severely and damage the eyes. If you spill an acid or a base on your skin or get some in your eyes, be sure to flood the area immediately with water to lessen any damage.

Sample Problem 9.8
Dissociation of a Base in Water

Write an equation for the dissociation of LiOH in water.

Solution

LiOH is a hydroxide of lithium, a Group 1A element. It dissociates in water to give a basic solution of lithium ions (Li^+) and hydroxide ions (OH^-).

$$LiOH \xrightarrow{H_2O} Li^+ + OH^-$$

Study Check

Calcium hydroxide is used in some antacids to counteract excess stomach acid. Write an equation for the formation of a calcium hydroxide solution.

9.5 Acid–Base Neutralization

Learning Goal

Write a balanced equation for the neutralization reaction of an acid and a base.

When solutions of an acid and base are mixed, they lose their acidic and basic properties. In a reaction called **neutralization**, the acid and the base form water and a salt. A **salt** is an ionic compound that contains a metal ion and nonmetal or polyatomic ion. For example, the neutralization reaction of HCl and NaOH is written as

$$\underset{\text{Acid}}{HCl} + \underset{\text{Base}}{NaOH} \longrightarrow \underset{\text{Salt}}{NaCl} + \underset{\text{Water}}{H_2O}$$

To understand the process that occurs in neutralization, we need to write the acid, base, and salt in their ionic forms.

Acid	*Base*	*React to Give*	*Salt*	*Water*
$H^+ + Cl^-$	$Na^+ + OH^-$	$\longrightarrow$	$Na^+ + Cl^-$	H_2O

In solution, HCl exists as H^+ (or H_3O^+) and Cl^-, and NaOH exists as Na^+ and OH^-. In the neutralization, the H^+ reacts with OH^- to form water, leaving the ions Na^+ and Cl^- in the solution. By omitting the unreacted metal and nonmetal ions from the overall equation, we write a net ionic equation for any acid–base *neutralization* as

$$\underset{\text{From an acid}}{H^+} + \underset{\text{From a base}}{OH^-} \longrightarrow \underset{\text{Water forms}}{H_2O}$$

This is the reaction that occurs when a solution of any acid and a solution of a base, strong or weak, are neutralized.

Balancing an Acid–Base Neutralization Equation

In another neutralization reaction, sulfuric acid, H_2SO_4, reacts with sodium hydroxide, NaOH. We can write the unbalanced equation as

$$\underset{\text{Acid}}{H_2SO_4} + \underset{\text{+ base}}{NaOH} \longrightarrow \underset{\text{Salt}}{\text{salt}} + \underset{\text{+ water}}{H_2O}$$

In any neutralization reaction, the amount of H^+ and the OH^- must be equal. Because the acid H_2SO_4 can donate two H^+ ions, it requires two OH^- ions from the base. We therefore place a coefficient of 2 in front of the NaOH. The neutralization of two H^+ ions and two OH^- ions will result in the formation of two molecules of H_2O.

$$H_2SO_4 + 2NaOH \longrightarrow \text{salt} + 2H_2O$$

The ions available for the salt are two Na^+ ions from the base, and the SO_4^{2-} ion from the acid. The formula of the salt, Na_2SO_4, is placed in the equation to give the balanced equation for the neutralization.

$$\underset{\text{Acid}}{H_2SO_4} + \underset{\text{base}}{2NaOH} \longrightarrow \underset{\text{Salt}}{Na_2SO_4} + \underset{\text{water}}{2H_2O}$$

Here are some more examples of balanced neutralization reactions between acids and bases.

Some Acid–Base Neutralization Reactions

Acid		*Base*		*Salt*		*Water*
HNO_3	+	KOH	$\longrightarrow$	KNO_3	+	H_2O
2HCl	+	$Ca(OH)_2$	$\longrightarrow$	$CaCl_2$	+	$2H_2O$
H_3PO_4	+	3NaOH	$\longrightarrow$	Na_3PO_4	+	$3H_2O$

Sample Problem 9.9
Writing a Neutralization Equation

Write and balance an equation for the neutralization of HCl by $Ba(OH)_2$.

Solution

In a neutralization of HCl (acid) and $Ba(OH)_2$(base), the products are a salt and water. The unbalanced reaction is

$$HCl + Ba(OH)_2 \longrightarrow \text{salt} + H_2O$$

A 2 is placed in front of the HCl to provide two H^+ ions to neutralize the two OH^- ions from the base. The product of the neutralization is $2H_2O$.

$$2HCl + Ba(OH)_2 \longrightarrow \text{salt} + 2H_2O$$

The ions that remain in solution are the Ba^{2+} ion from the base and two Cl^- ions from the acid. The formula of the salt is $BaCl_2$.

$$2HCl + Ba(OH)_2 \longrightarrow BaCl_2 + 2H_2O$$

Study Check

Write a balanced equation ion for the neutralization of H_3PO_4 with KOH.

HEALTH NOTE *Antacids*

Antacids are substances used to neutralize excess stomach acid (HCl) and to treat ulcers. Some antacids are mixtures of aluminum hydroxide and magnesium hydroxide. These hydroxides are not very soluble in water, so the levels of available OH^- are not damaging to the intestinal tract. However, aluminum hydroxide has the side effects of producing constipation and binding phosphate in the intestinal tract, which may cause weakness and anorexia. Magnesium hydroxide has a laxative effect. These side effects are less likely when a combination of the antacids is used.

$$Al(OH)_3 + 3HCl \longrightarrow AlCl_3 + 3H_2O$$

$$Mg(OH)_2 + 2HCl \longrightarrow MgCl_2 + 2H_2O$$

Some antacids use calcium carbonate to neutralize excess stomach acid. About 10% of the calcium is absorbed into the bloodstream, where it elevates the levels of serum calcium. Calcium carbonate is not recommended for patients who have peptic ulcers or a tendency to form kidney stones.

$$CaCO_3 + 2HCl \longrightarrow H_2O + CO_2(g) + CaCl_2$$

Still other antacids contain sodium bicarbonate. This type of antacid has a tendency to increase blood pH and elevate sodium levels in the body fluids. It also is not recommended in the treatment of peptic ulcers.

$$NaHCO_3 + HCl \longrightarrow NaCl + CO_2(g) + H_2O$$

The neutralizing substances in some antacid preparations are given in Table 9.5.

Table 9.5 ***Basic Compounds in Some Antacids***

Antacid	Base(s)
Amphojel	$Al(OH)_3$
Milk of magnesia	$Mg(OH)_2$
Mylanta, Maalox, Di-Gel, Gelusil, Riopan	$Mg(OH)_2$, $Al(OH)_3$
Bisodol	$CaCO_3$, $Mg(OH)_2$
Titralac, Tums, Pepto-Bismol	$CaCO_3$
Alka-Seltzer	$NaHCO_3$, $KHCO_3$

9.6 Ion Product of Water

Learning Goal

Use the ion product of water to calculate the $[H_3O^+]$ and $[OH^-]$ in a solution.

Most of the time we think of pure water as a nonelectrolyte that contains only molecules of water. However, careful measurements show that a very small number of water molecules (2 of about 1 billion) have formed ions. When water molecules ionize, a proton (H^+) is transferred from one water molecule to another. In the process, one water molecule acts as an acid and the other as a base. The products of the ionization are an equal number of hydronium ions (H_3O^+) and hydroxide (OH^-) ions. As with other weak acids and weak bases, equilibrium occurs as H_3O^+ and OH^- ions react in the reverse reaction to reform water.

$$\underset{\text{Water}}{H_2O} + \underset{\text{Water}}{H_2O} \rightleftharpoons \underset{\text{Hydronium ion}}{H_3O^+} + \underset{\text{Hydroxide ion}}{OH^-}$$

In pure water, the concentrations of H_3O^+ and OH^- are quite small and equal. The H_3O^+ concentration is 1×10^{-7} moles per liter (mol/L) and the OH^- concentration is also 1×10^{-7} mol/L. Square brackets written around the symbols indicate concentrations in moles per liter (mol/L or M).

Pure Water Concentrations of H_3O^+ and OH^-

$$[H_3O^+] = 1 \times 10^{-7}\ M \qquad [OH^-] = 1 \times 10^{-7}\ M$$

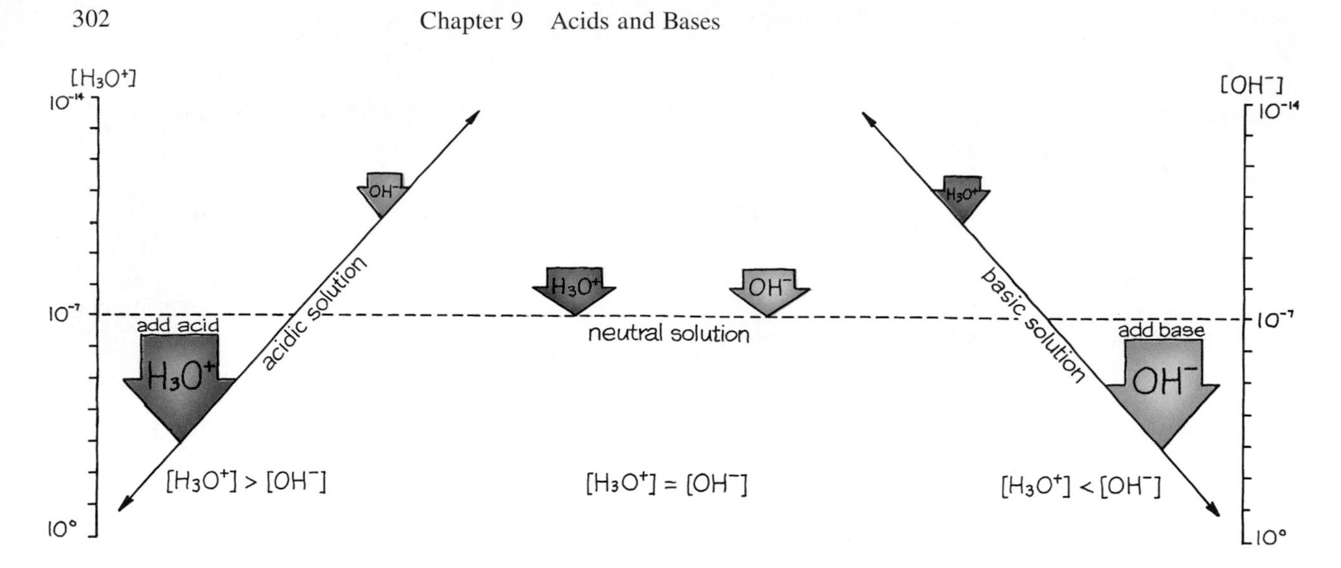

Figure 9.7
In pure water, $[H_3O^+]$ and $[OH^-]$ are equal. In acidic solutions, the $[H_3O^+]$ goes up and the $[OH^-]$ goes down. In basic solutions, the $[OH^-]$ goes up as the $[H_3O^+]$ goes down.

Ion Product for Water

If we multiply the concentrations of the H_3O^+ and OH^- ions in water, we obtain a useful constant K_w, the **ion product of water.** Its numerical value, 1×10^{-14}, is constant for any aqueous solution even if acid or base is added.

Ion Product of Water

$$K_w = [H_3O^+][OH^-]$$
$$= (1 \times 10^{-7})(1 \times 10^{-7})$$
$$= 1 \times 10^{-14}$$

Because the $[H_3O^+]$ and $[OH^-]$ are equal, pure water has no acidic or basic properties; it is **neutral**. In an *acidic solution,* the $[H_3O^+]$ goes up. At the same time, the $[OH^-]$ goes down to keep the $[H_3O^+][OH^-]$ value at 1×10^{-14}. The same idea applies to *basic solutions,* except that a base added to water makes the $[OH^-]$ go up, and the $[H_3O^+]$ go down. Figure 9.7 illustrates the relationship between $[H_3O^+]$ and $[OH^-]$ in neutral, acidic, and basic solution. A comparison of the $[H_3O^+]$ and $[OH^-]$ in a neutral solution and some acidic and basic solutions are given in Table 9.6.

In acids, $[H_3O^+]$ is greater than $[OH^-]$.
In bases, $[OH^-]$ is greater than $[H_3O^+]$.
In neutral solutions, $[H_3O^+]$ equals $[OH^-]$.

Table 9.6 ***Examples of $[H_3O^+]$ and $[OH^-]$ in Neutral, Acidic, and Basic Solutions***

Type of Solution	$[H_3O^+]$	$[OH^-]$	Ion Product, K_w
Neutral: $[H_3O^+] = [OH^-]$	1×10^{-7}	1×10^{-7}	1×10^{-14}
Acidic: $[H_3O^+] > [OH^-]$	1×10^{-2}	1×10^{-12}	1×10^{-14}
	1×10^{-5}	1×10^{-9}	1×10^{-14}
Basic: $[OH^-] > [H_3O^+]$	1×10^{-8}	1×10^{-6}	1×10^{-14}
	1×10^{-11}	1×10^{-3}	1×10^{-14}

If we know the concentration of either the H_3O^+ or OH^- ion, we can calculate the concentration of the other ion by using the ion product expression.

$$[H_3O^+][OH^-] = 1 \times 10^{-14}$$

Dividing the K_w by the $[OH^-]$ gives the $[H_3O^+]$:

$$[H_3O^+] = \frac{1 \times 10^{-14}}{[OH^-]}$$

Dividing the K_w by the $[H_3O^+]$ gives the $[OH^-]$:

$$[OH^-] = \frac{1 \times 10^{-14}}{[H_3O^+]}$$

Sample Problem 9.10
Calculating $[H_3O^+]$ in Solution

A solution of a window cleaner has an $[OH^-]$ of 1×10^{-4} M. What is the $[H_3O^+]$ in the cleaner?

Solution

We can solve for the $[H_3O^+]$ by using the ion product of water.

$$[H_3O^+][OH^-] = 1 \times 10^{-14}$$

$$[H_3O^+] = \frac{1 \times 10^{-14}}{[OH^-]}$$

The value for $[OH^-]$ of 1.0×10^{-4} can be substituted to solve the expression for $[H_3O^+]$.

$$[H_3O^+] = \frac{1 \times 10^{-14}}{1 \times 10^{-4}} = 1 \times 10^{-10}\ \text{M}$$

Study Check

What is the value for $[H_3O^+]$ of orange juice if its $[OH^-]$ is 1×10^{-11} M? Is the solution acidic, basic, or neutral?

Sample Problem 9.11
Calculating $[OH^-]$ in Solution

A vinegar solution has an $[H_3O^+]$ of 2×10^{-3} M. What is the $[OH^-]$ of this solution?

Solution

Rearranging the ion product expression for $[OH^-]$ gives:

$$[H_3O^+][OH^-] = 1 \times 10^{-14}$$

$$[OH^-] = \frac{1 \times 10^{-14}}{[H_3O^+]}$$

Substituting the given value of 2×10^{-3} for the $[H_3O^+]$ completes the calculation.

$$[OH^-] = \frac{1 \times 10^{-14}}{2 \times 10^{-3}}$$

$$= 5 \times 10^{-12}\ \text{M}$$

Study Check
A sample of gastric juice (HCl) has an $[H_3O^+]$ of 5×10^{-2} M. What is the $[OH^-]$ of the sample?

9.7 The pH Scale

Learning Goal
Calculate pH from $[H_3O^+]$; given the pH, calculate $[H_3O^+]$ and $[OH^-]$ of a solution.

In acids and bases, the H_3O^+ concentrations can typically range from 1 M to 1×10^{-14} M. In clinical situations, such values are inconvenient to use, so a simpler way of describing acidity, the **pH** scale, is used. On the pH scale, the pH values range from 0 to 14 as shown in Figure 9.8. The pH values below 7 correspond to acidic solutions, whereas solutions with pH values above 7 are basic. A solution with a pH of 7, the midpoint of the pH scale, would be neutral. Table 9.7 lists the pH values for some common substances.

Neutral solution	pH = 7
Acidic solution	pH below 7
Basic solution	pH above 7

In the laboratory, the pH of a solution is determined by using a pH meter or an indicator that shows specific color changes at certain pH values as seen in Figure 9.9.

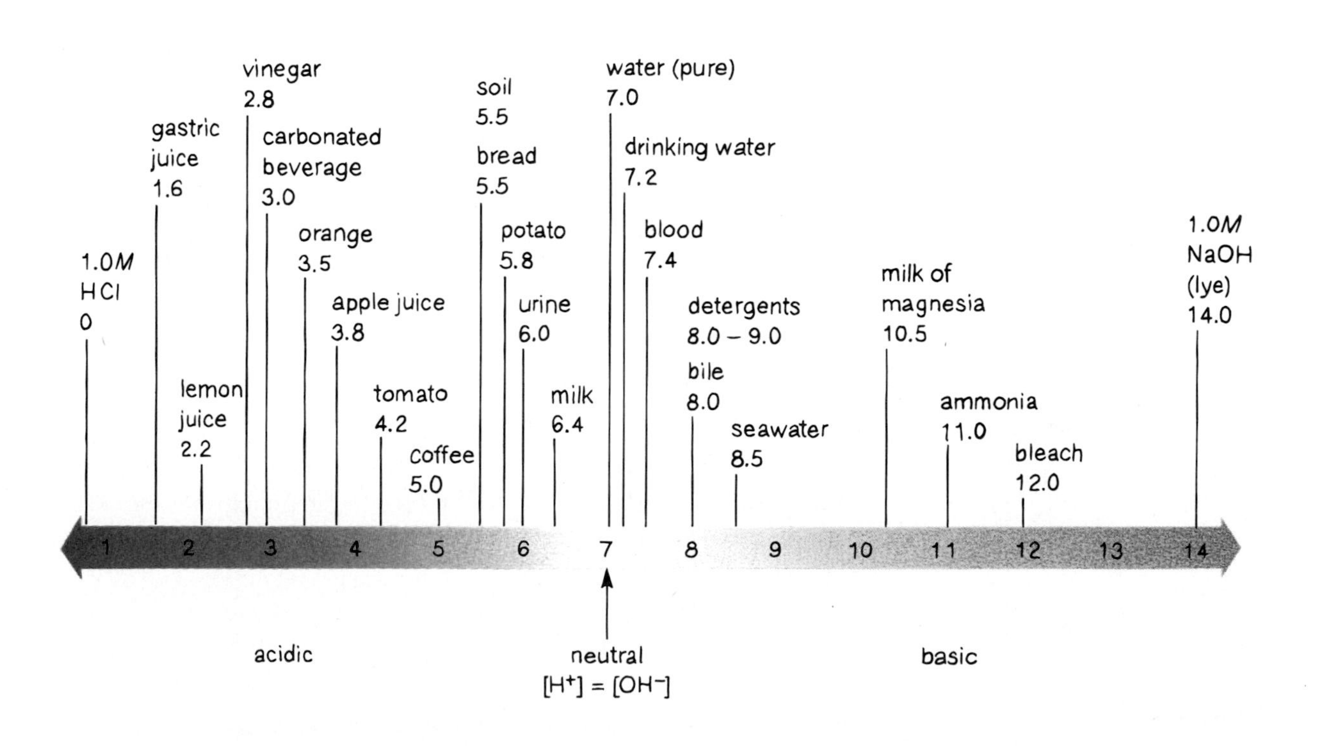

Figure 9.8
On the pH scale values below 7 are acidic, a value of 7 is neutral, and values above 7 are basic.

Table 9.7 ***pH Values of Some Common Substances***

Solution	pH	Solution	pH
1 M HCl	0	Water, pure	7.0
Gastric juice	1.6	Blood	7.4
Lemon juice	2.2	Bile	8.0
Vinegar	2.8	Detergents	8.0–9.0
Carbonated drinks	3.0	Milk of magnesia	10.5
Coffee	5.0	Ammonia solution	11.0
Urine	6.0	Bleach	12.0
Rainwater	6.2	1 M NaOH	14.0

Figure 9.9
The dye in red cabbage juice acts as a pH indicator by turning different colors that correspond to different pH values.

Sample Problem 9.12
Using pH to Identify Acidic and Basic Solutions

Classify the following as acidic, basic, or neutral solutions:
a. tomato juice, pH 4.2
b. saliva, pH 6.6
c. antacid, pH 8.5

Solution
a. Tomato juice is acidic because a pH of 4.2 is below 7 on the pH scale.
b. Saliva is acidic because a pH value of 6.6 is below 7 on the pH scale.
c. The antacid is basic. A pH of 8.5 is above 7 on the pH scale.

Study Check
An NaCl Solution has a pH of 7. Is it acidic, basic, or neutral?

HEALTH NOTE *Stomach Acid, HCl*

When a person sees, smells, thinks about, and/or tastes food, the gastric glands in the stomach begin to secrete an HCl solution that is strongly acidic. In a single day, a person may secrete as much as 2000 mL of gastric juice.

The HCl in the gastric juice activates a digestive enzyme called pepsin that breaks down proteins in food entering the stomach. The secretion of HCl continues until the stomach has a pH of about 2, which is the optimum pH for activating the digestive enzymes without ulcerating the stomach lining. Normally, large quantities of a viscous mucus are secreted within the stomach to protect it from acid and enzyme damage.

When we compare pH values with their corresponding H_3O^+ concentrations, we find that acidic solutions have low pH values and high $[H_3O^+]$. In basic solutions the pH values are high and the $[H_3O^+]$ is low.

Type of Solution	*pH*	*$[H_3O^+]$*
Neutral	7	1×10^{-7} M
Acidic	below 7	more than 1×10^{-7} M
Basic	above 7	less than 1×10^{-7} M

Table 9.8 ***A Comparison of $[H_3O^+]$, $[OH^-]$, and Corresponding pH Values at 25°C***

pH	$[H_3O^+]$	$[OH^-]$	
0	10^0	10^{-14}	
1	10^{-1}	10^{-13}	
2	10^{-2}	10^{-12}	
3	10^{-3}	10^{-11}	**Acidic**
4	10^{-4}	10^{-10}	
5	10^{-5}	10^{-9}	
6	10^{-6}	10^{-8}	
7	10^{-7}	10^{-7}	**Neutral**
8	10^{-8}	10^{-6}	
9	10^{-9}	10^{-5}	
10	10^{-10}	10^{-4}	
11	10^{-11}	10^{-3}	**Basic**
12	10^{-12}	10^{-2}	
13	10^{-13}	10^{-1}	
14	10^{-14}	10^0	

This inverse relationship is a result of the mathematical definition for **pH** as the negative logarithm (log) of the $[H_3O^+]$. The p in pH refers to the negative power of 10 in the hydronium ion concentration.

	Negative value	Power of 10	Hydronium ion concentration
pH =	−	log	$[H_3O^+]$

For example, an acidic solution has a hydronium ion concentration of 1×10^{-6} M. The log of $[H_3O^+]$ is −6 (the power of 10). After we multiply the log by a negative sign −(−6), we obtain a pH value of 6. For $[H_3O^+]$ with coefficients of 1, the pH is obtained simply by dropping the negative sign of the power of 10. Thus, a basic solution with a $[H_3O^+]$ of 1×10^{-11} has a pH of 11. Table 9.8 gives a comparison of some $[H_3O^+]$, $[OH^-]$, and their pH values.

CALCULATOR NOTE *Calculating pH*

When the coefficient in the H_3O^+ concentration is not 1, a calculator with a [log] key, [EXP] or [EE] key, and a change sign, ([+/−]) key is needed to determine pH. Suppose we want to calculate the pH value for a solution with $[H_3O^+]$ of 5×10^{-4} M. We enter the $[H_3O^+]$ using the [EXP] or [EE] key, use the [log] key, and change sign ([+/−]). The steps are shown below.

Calculating pH Using a Typical Calculator

Enter	*Calculator Display*	
[5]	5	
[EXP] (or [EE])	5 00	
[4]	5 04	
[+/−]	5 −04	
[log]	−3.3	Rounded
[+/−]	3.3	This is the pH

The pH of the solution is 3.3. The number of digits after the decimal point in the pH is equal to the number of significant figures in the value of the $[H_3O^+]$. In this case, there is one significant figure in the concentration (5), and one digit after the decimal point (.3) in our answer.

Sample Problem 9.13
Calculating pH

Determine the pH of the following solutions and identify each as acidic, basic, or neutral:

a. $[H_3O^+] = 1 \times 10^{-5}$ M
b. $[H_3O^+] = 6 \times 10^{-8}$ M
c. $[OH^-] = 1 \times 10^{-3}$ M

Solution

a. Because the coefficient in the $[H_3O^+]$ is 1, the pH of the solution is the power of 10 without the negative sign.

$[H_3O^+] = 1 \times 10^{-5}$ pH = 5.0

b. The pH of the $[H_3O^+]$ is determined on a calculator as follows:

6 EXP (EE) 8 +/- log +/- = 7.2

c. Because $[OH^-]$ is given, we have to calculate the $[H_3O^+]$ using the ion product of water.

$$[H_3O^+][OH^-] = 1 \times 10^{-14}$$

$$[H_3O^+] = \frac{1 \times 10^{-14}}{1 \times 10^{-3}}$$

$$= 1 \times 10^{-11}$$

$$\text{pH} = 11.0$$

Study Check

What is the pH of a bleach whose $[OH^-]$ is 1×10^{-4} M? (*Hint:* Calculate the $[H_3O^+]$ first.)

Sometimes, we need to determine the $[H_3O^+]$ from the pH. For whole number pH values, we substitute the pH value as the power of 10 in the concentration.

$$[H_3O^+] = 1 \times 10^{-\text{pH}}$$

Sample Problem 9.14
Calculating $[H_3O^+]$ from pH

Determine $[H_3O^+]$ for solutions having the following pH values:

a. pH = 3.0 **b.** pH = 12.0

Solution

For pH values that are whole numbers, the $[H_3O^+]$ can be written as the exponent of the hydronium ion concentration $[H_3O^+] = 1 \times 10^{-\text{pH}}$.

a. $[H_3O^+] = 1 \times 10^{-3}$ M (pH ↘ exponent) **b.** $[H_3O^+] = 1 \times 10^{-12}$ M (pH ↘ exponent)

Study Check

What is the $[H_3O^+]$ and $[OH^-]$ of beer if it has a pH of 5.0?

ENVIRONMENTAL NOTE *Acid Rain*

Rain typically has a pH of 6.2. It is slightly acidic because the carbon dioxide in the air combines with water to form carbonic acid. However, in many parts of the world, rain has become considerably more acidic, with pH values as low as 3 being reported. One cause of acid rain is the sulfur dioxide (SO_2) gas produced when coal that contains sulfur is burned.

In the air, the SO_2 gas reacts with oxygen to produce SO_3, which then combines with water to form sulfuric acid, H_2SO_4, a strong acid.

$$S + O_2 \longrightarrow SO_2$$
$$2SO_2 + O_2 \longrightarrow 2SO_3$$
$$SO_3 + H_2O \longrightarrow H_2SO_4$$

In parts of the United States, acid rain has made lakes so acidic they are no longer able to support fish and plant life. Limestone ($CaCO_3$) is sometimes added to these lakes to neutralize the acid. In Eastern Europe, acid rain has brought about an environmental disaster. Nearly 40% of the forests in Poland have been severely damaged, and some parts of the land are so acidic that crops will not grow. Throughout Europe and the United States, monuments made of marble (a form of $CaCO_3$) are deteriorating as acid rain dissolves the marble. See Figure 9.10.

$$2H^+ + CaCO_3 \longrightarrow Ca^{2+} + H_2O + CO_2$$

Efforts to slow or stop the damaging effects of acid rain include the reduction of sulfur emissions. This will require installation of expensive equipment in coal-burning plants to absorb more of the SO_2 gases before they are emitted. In some outdated plants, this may be impossible, and they will need to be closed. It is a difficult problem for engineers and scientists, but one that must be solved.

Figure 9.10
(a) Detail of the marble that is part of the Washington Square Arch in Washington Square Park completed on July 17, 1935.
(b) The destructive effect of acid rain on the same marble, June, 1994.

9.8 Buffers

Learning Goal

Describe the role of buffers in maintaining the pH of a solution.

When an acid or base is added to pure water, the pH changes drastically. However, if a solution is buffered, there is little change in pH when small amounts of acid or base are added. A **buffer solution** resists a change in pH when small amounts of acid or base are added. For example, blood is a buffer solution in the body that maintains a pH of about 7.4. If the pH of the blood goes even slightly above or below this value, changes in our uptake of oxygen and our metabolic process can be drastic enough to cause death. Even though we are constantly obtaining acids and bases from foods and biological processes, the buffers in the body so effectively absorb those compounds that blood pH remains unchanged. (See Figure 9.11.)

A **buffer solution** must contain two substances, a weak acid and a salt of that acid or a weak base and a salt of that base. For example, an important buffer in the blood consists of carbonic acid, a weak acid, and its salt, sodium bicarbonate, $NaHCO_3$. The buffering action is the result of the weak acid (or base) in equilibrium with its ions in the solution.

$$\underset{\text{Carbonic acid}}{H_2CO_3} + H_2O \rightleftharpoons H_3O^+ + \underset{\text{Bicarbonate ion}}{HCO_3^-}$$

If we add a salt of the weak acid, $NaHCO_3$, it will dissociate and increase the amount of bicarbonate ion, HCO_3^- in the solution.

$$\underset{\substack{\text{Sodium bicarbonate} \\ \text{(salt of the weak acid)}}}{NaHCO_3} \longrightarrow Na^+ + \underset{\text{More bicarbonate ion}}{HCO_3^-}$$

Although the weak acid itself does not produce very many bicarbonate ions, HCO_3^-, its concentration is higher because of the added salt. Thus, a buffer has high levels of both carbonic acid and bicarbonate ions.

Figure 9.11
Adding an acid or a base to water changes the pH drastically, a buffer resists pH change when an acid or base is added.

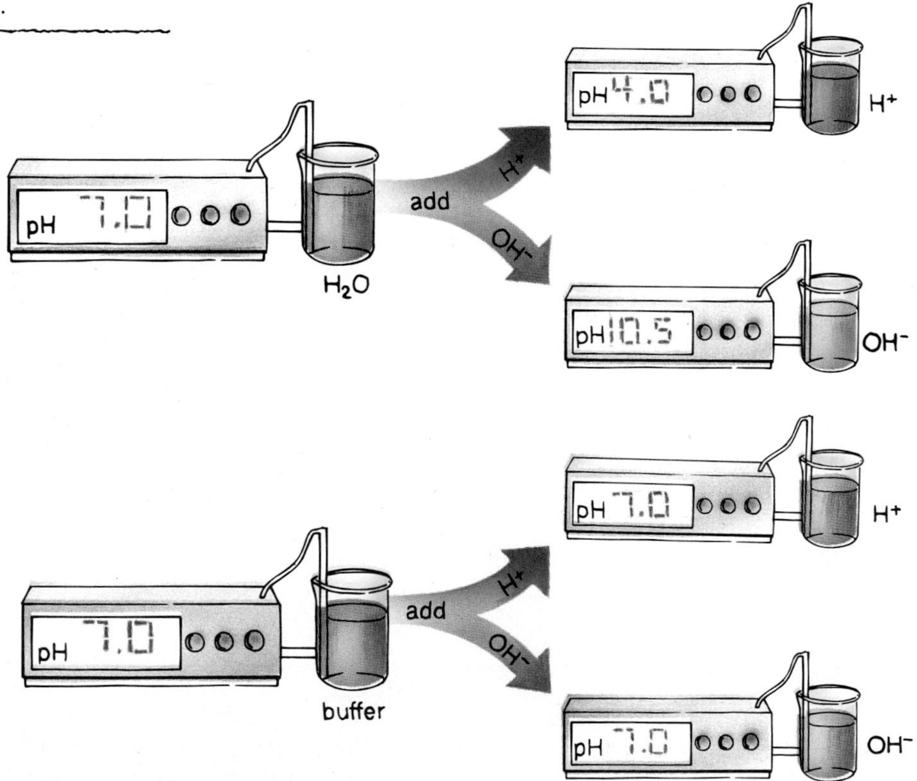

Now let us see how the buffer works to resist pH changes when some H_3O^+ or OH^- is added. Suppose that some OH^- enters the blood. The weak acid available in the blood buffer reacts with this excess OH^- by the following reaction, which neutralizes it:

$$\underset{\text{Carbonic acid}}{H_2CO_3} + \underset{\text{Hydroxide ion}}{OH^-} \longrightarrow H_2O + \underset{\text{Bicarbonate ion}}{HCO_3^-}$$

When excess H_3O^+ ions enter the buffer system, they react with the bicarbonate ion in the reverse direction to reform carbonic acid. This occurs in certain metabolic diseases when the kidneys fail to remove excess hydrogen ions from the blood. The blood becomes more acidic and the pH of the blood falls below 7.35. Then the bicarbonate part of the buffer reacts with the excess H_3O^+ in the blood to neutralize it. The equilibrium between these buffering process is shown in Figure 9.12.

$$\underset{\text{Bicarbonate ion}}{HCO_3^-} + H_3O^+ \longrightarrow \underset{\text{Carbonic acid}}{H_2CO_3} + H_2O$$

In summary, when OH^- ions are added to a buffer solution, they react with the weak acid of the buffer. On the other hand, if additional H_3O^+ ions enter the buffer solution, they are tied up by the negative ions of the weak acid provided by the salt. Both components are necessary for a buffer solution to maintain pH.

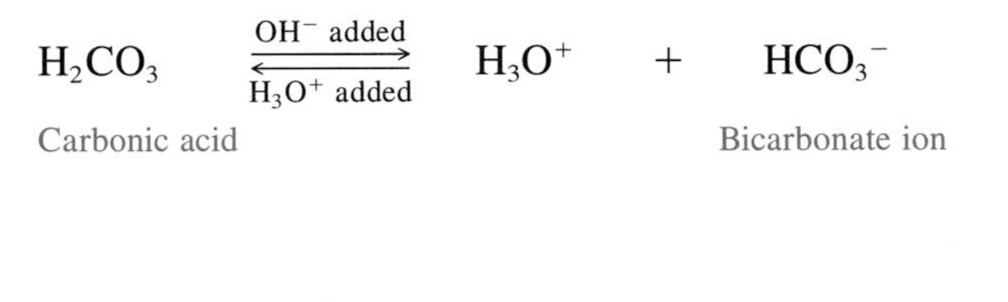

$$\underset{\text{Carbonic acid}}{H_2CO_3} \underset{H_3O^+ \text{ added}}{\overset{OH^- \text{ added}}{\rightleftharpoons}} H_3O^+ + \underset{\text{Bicarbonate ion}}{HCO_3^-}$$

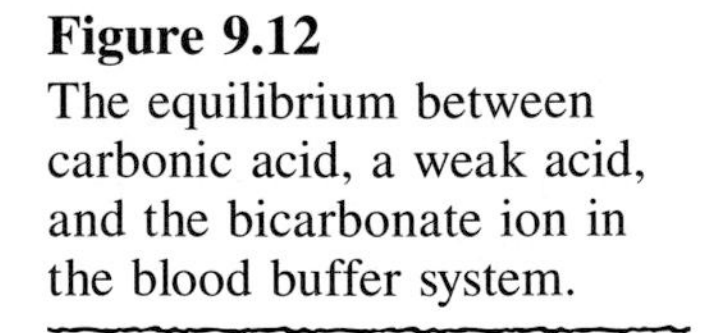

Figure 9.12
The equilibrium between carbonic acid, a weak acid, and the bicarbonate ion in the blood buffer system.

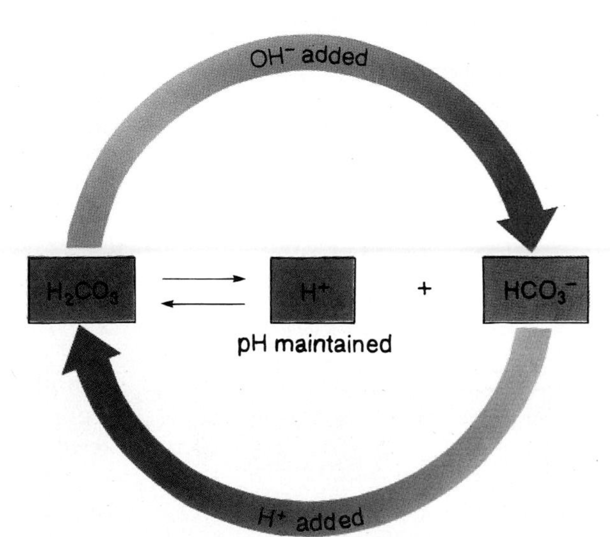

Sample Problem 9.15
Identifying Buffer Solutions

Indicate whether each of the following would make a buffer solution:

a. HCl, a strong acid, and NaCl
b. H_3PO_4, a weak acid
c. HF, a weak acid and NaF

Solution

a. No. A solution of a strong acid and its salt is completely ionized.
b. No. A weak acid is not sufficient for a buffer.
c. Yes. This mixture contains a weak acid and its salt.

Study Check

Will a mixture of NaCl and Na_2CO_3 make a buffer solution? Explain.

HEALTH NOTE *P_{CO_2} and the Carbonate Buffer*

In our cells, carbon dioxide (CO_2) is continually produced as an end product of cellular metabolism. Some CO_2 is carried to the lungs for elimination, and the rest dissolves in the body fluids, forming carbonic acid (H_2CO_3) and bicarbonate ion.

Carbonate Buffer System

$$\underset{\text{Carbon dioxide}}{CO_2} + H_2O \rightleftharpoons \underset{\text{Carbonic acid}}{H_2CO_3} \rightleftharpoons H^+ + \underset{\text{Bicarbonate ion}}{HCO_3^-}$$

The partial pressure of CO_2 gas, P_{CO_2}, determines the concentration of carbonic acid, which in turn affects the pH of the blood. Table 9.9 lists the normal values for the components of the blood buffer in arterial blood.

Table 9.9 ***Normal Values for Blood Buffer in Arterial Blood***

P_{CO_2}	40 mm Hg
H_2CO_3	1.2 meq/L of plasma[a]
HCO_3^-	24 meq/L of plasma
pH	7.4

[a] A milliequivalent (meq) is equal to $\frac{1}{1000}$ of an equivalent (equiv).

A higher level of CO_2 produces more H_2CO_3 in the blood, which produces more H^+, and lowers the pH. This condition is called **acidosis.** When the origin of acidosis is respiratory, it is called respiratory acidosis. Poor respiration or below-normal gas exchange (hypoventilation) causes an increase in the CO_2 in the blood. This can happen in emphysema when CO_2 does not diffuse properly through the lung membrane; it can also happen when the respiratory center in the medulla of the brain is affected by an accident or by depressive drugs. Acidosis caused by conditions that are not respiratory is called metabolic acidosis.

A lowering of the CO_2 level leads to a higher blood pH, a condition called **alkalosis.** Excitement, trauma, or a high temperature may cause a person to breathe rapidly (hyperventilation). If sufficient CO_2 is expelled, its partial pressure in the blood falls below normal. The blood buffer restores the CO_2 level by converting some H_2CO_3 to CO_2 and H_2O. The decrease in H_2CO_3 causes some H^+ to combine with HCO_3^-. The H^+ drops and blood pH rises, a condition called respiratory alkalosis.

Although the amount of CO_2 in the blood depends on the respiratory processes, the kidneys control the amount of H^+ excreted in the urine. The pH of urine varies considerably and can range from 4.5 to 8.2. In severe conditions, even these mechanisms fail, and life-threatening changes in blood pH occur. Table 9.10 lists some of the conditions that lead to changes in the blood pH, and some possible treatments.

Table 9.10 *Acidosis and Alkalosis: Symptoms, Causes, and Treatments*

***Respiratory Acidosis:* CO_2 ↑ pH ↓**	
Symptoms:	Failure to ventilate, suppression of breathing, disorientation, weakness, coma
Causes:	Lung disease blocking gas diffusion (e.g., emphysema, pneumonia, bronchitis, and asthma); depression of respiratory center by drugs, cardiopulmonary arrest, stroke, poliomyelitis, or nervous system disorders
Treatment:	Correction of disorder, infusion of bicarbonate
***Respiratory Alkalosis:* CO_2 ↓ pH ↑**	
Symptoms:	Increased rate and depth of breathing, numbness, light-headedness, tetany
Causes:	Hyperventilation due to anxiety, hysteria, fever, exercise; reaction to drugs such as salicylate, quinine, and antihistamines; conditions causing hypoxia (e.g., pneumonia, pulmonary edema, and heart disease)
Treatment:	Elimination of anxiety-producing state, rebreathing into a paper bag
***Metabolic Acidosis:* H^+ ↑ pH ↓**	
Symptoms:	Increased ventilation, fatigue, confusion
Causes:	Renal disease, including hepatitis and cirrhosis; increased acid production in diabetes mellitus, hyperthyroidism, alcoholism, and starvation; loss of alkali in diarrhea; acid retention in renal failure
Treatment:	Sodium bicarbonate given orally, dialysis for renal failure, insulin treatment for diabetic ketosis
***Metabolic Alkalosis:* H^+ ↓ pH ↑**	
Symptoms:	Depressed breathing, apathy, confusion
Causes:	Vomiting, diseases of the adrenal glands, ingestion of excess alkali
Treatment:	Infusion of saline solution, treatment of underlying diseases

9.9 Acid-Base Titration

Learning Goal

Calculate the molarity or volume of an acid or base from titration information.

Suppose we need to find the molarity of an HCl solution of unknown concentration. We can do this by a laboratory procedure called a **titration** in which we neutralize an acid sample with a known amount of base. In our titration, we first place a measured volume of the acid in a flask, and add a few drops of an **indicator** such as phenolphthalein. In an acidic solution, this indicator is colorless. Then we fill a buret with an NaOH solution of known molarity and carefully add NaOH to the acid in the flask, as shown in Figure 9.13.

In the titration, we neutralize the acid by adding a volume of base that contains a matching number of moles of OH^-. We know that neutralization

Figure 9.13
The titration of an acid. A known volume of an acid is placed in a flask with an indicator and titrated with a measured volume of NaOH to the neutralization point.

has taken place when the phenolphthalein in the solution changes from colorless to pink. The difference in the volume of the NaOH in the buret before and after titration is the volume of NaOH used to reach the neutralization endpoint. From the volume and molarity of the NaOH, we can calculate the number of moles of NaOH and then the concentration of the acid.

Sample Problem 9.16
Titration of an Acid

A 10.0 mL sample of HCl is titrated with 1.0 M NaOH. The equation for the reaction is:

$$NaOH + HCl \longrightarrow NaCl + H_2O$$

If 20.0 mL of base were need to reach the endpoint, what is the molarity of the acid?

Solution

Step 1 **Finding the reacting moles of base**
From the volume and the molarity of the NaOH, we calculate the moles of NaOH used in the titration.

$$20.0\ \cancel{mL} \times \frac{1\ \cancel{L}}{1000\ \cancel{mL}} \times \underbrace{\frac{1.0\ mol\ NaOH}{1\ \cancel{L}}}_{\text{Molarity of NaOH}} = 0.020\ mol\ NaOH$$

Step 2 **Finding the reacting moles of acid**
From the neutralization equation, we know that 1 mol NaOH reacts with 1 mol HCl. Expressing the relationship between HCl and NaOH as conversion factors gives:

$$\frac{1\ mol\ NaOH}{1\ mol\ HCl} \quad \text{and} \quad \frac{1\ mol\ HCl}{1\ mol\ NaOH}$$

The moles of HCl in the sample are calculated using the second factor.

$$0.020\ \cancel{mol\ NaOH} \times \frac{1\ mol\ HCl}{1\ \cancel{mol\ NaOH}} = 0.020\ mol\ HCl$$

Step 3 **Calculating the molarity of the acid**
To calculate the molarity of the acid in the sample, we place the moles of acid and its initial volume in the expression for molarity.

$$\text{Molarity (M) of acid} = \frac{\text{moles of acid}}{\text{volume of acid}} = \frac{0.020\ \text{mol HCl}}{0.0100\ \text{L}} = 2.0\ \text{M HCl}$$

10.0-mL HCl sample expressed in liters

Study Check
In a titration, a 25.0-mL HCl sample required 50.0 mL of a 0.100 M NaOH solution for neutralization. What is the molarity of the HCl sample?

Sample Problem 9.17
Titration of an Acid

A 5.00-mL sample of H_2SO_4 requires 40.0 mL of 2.0 M NaOH to reach the neutralization endpoint. What is the molarity of the acid? The equation for the neutralization is:

$$H_2SO_4 + 2NaOH \longrightarrow Na_2SO_4 + 2H_2O$$

Solution
Using the steps for titration calculations we find the following:

Step 1 From the volume and the molarity of the NaOH, we calculate the moles of NaOH used in the titration.

$$40.0\ \cancel{\text{mL}} \times \frac{1\ \cancel{\text{L}}}{1000\ \cancel{\text{mL}}} \times \frac{2.0\ \text{mol NaOH}}{1\ \cancel{\text{L}}} = 0.080\ \text{mol NaOH}$$

Molarity of NaOH

Step 2 From the neutralization equation, we know that 2 mol NaOH react with 1 mol H_2SO_4. Expressing the relationship as conversion factors gives:

$$\frac{2\ \text{mol NaOH}}{1\ \text{mol}\ H_2SO_4} \quad \text{and} \quad \frac{1\ \text{mol}\ H_2SO_4}{2\ \text{mol NaOH}}$$

The moles of H_2SO_4 in the sample are calculated using the second factor.

$$0.080\ \cancel{\text{mol NaOH}} \times \frac{1\ \text{mol}\ H_2SO_4}{2\ \cancel{\text{mol NaOH}}} = 0.040\ \text{mol}\ H_2SO_4$$

Step 3 To calculate the molarity of the acid in the sample, we place the moles of acid and its initial volume in the expression for molarity.

$$\text{Molarity (M) of acid} = \frac{\text{moles of acid}}{\text{volume of acid}} = \frac{0.040\ \text{mol}\ H_2SO_4}{0.0050\ \text{L}} = 8.0\ \text{M}\ H_2SO_4$$

5.0-mL H_2SO_4 sample expressed in liters

Study Check
What is the volume of 0.100 M NaOH needed to completely neutralize 5.00 mL of 0.200 M H_3PO_4? (*Hint:* Write the balanced equation for neutralization.)

Chapter Summary

9.1 Electrolytes
Electrolytes are conductors of electrical current because they produce ions in aqueous solutions. Strong electrolytes are completely ionized, whereas weak electrolytes are partially ionized.

9.2 Equivalents
An equivalent is the amount of an electrolyte that carries 1 mol of electrical charge (+ or −). Typically, there are 1, 2, or 3 equivalents per mole of a positive or negative ion depending on the charge. The concentrations of electrolytes in body fluids are typically given in milliequivalents (meq) per liter.

9.3 Acids and Bases
In water, an Arrhenius acid produces H^+, and an Arrhenius base produces OH^-. According to the Brønsted–Lowry theory, acids are proton (H^+) donors, and bases are proton acceptors. Protons form hydronium ions, H_3O^+, in water because they bond to polar water molecules.

9.4 Strengths of Acids and Bases
In strong acids, all the H^+ in the acid is donated to H_2O; in a weak acid, only a small percentage of acid molecules produce H_3O^+. Strong bases are hydroxides of Groups 1A and 2A that dissociate completely in water. An important weak base is ammonia, NH_3.

9.5 Acid–Base Neutralization
Neutralization is a reaction of an acid and a base that produces water and a salt. The net ionic equation for any neutralization is

$$H^+ + OH^- \longrightarrow H_2O$$

In a balanced neutralization equation, an equal number of moles of H^+ and OH^- must be reacted.

9.6 Ion Product of Water
In pure water, a few water molecules transfer a proton to other water molecules producing small, but equal, amounts of $[H_3O^+]$ and $[OH^-]$ such that each has a concentration of 1×10^{-7} mol/L. The ion product, K_w, $[H_3O^+][OH^-] = 1 \times 10^{-14}$, applies to all aqueous solutions. In acidic solutions, the $[H_3O^+]$ goes up and the $[OH^-]$ goes down. In basic solutions, the $[OH^-]$ is greater than the $[H_3O^+]$.

9.7 The pH Scale
The pH scale is a range of numbers from 0 to 14 related to the $[H_3O^+]$ of the solution. A neutral solution has a pH of 7. In acidic solutions, the pH is below 7, and in basic solutions the pH is above 7. Mathematically, pH is the negative logarithm of the hydronium ion concentration ($-\log [H_3O^+]$).

9.8 Buffers
A buffer solution resists a change in pH when small amounts of acid or a base are added. A buffer contains either a weak acid and its salt or a weak base and its salt. The weak acid picks up excess OH^-, and the anion of the salt picks up excess H^+. Buffers are important in maintaining the pH of the blood.

9.9 Acid–Base Titration
In a laboratory procedure called a titration, an acid sample is neutralized with a known amount of a base. From the volume and molarity of the base, the concentration of the acid is calculated.

Glossary of Key Terms

acidosis A physiological condition in which the blood pH is lower than 7.35.

alkalosis A physiological condition in which the blood pH is higher than 7.45.

Arrhenius acid and base An acid produces H^+ and a base produces OH^- in aqueous solutions.

Brønsted–Lowry theory of acids and bases An acid is a proton donor, and a base is a proton acceptor.

buffer solution A mixture of a weak acid or a weak base and its salt that resists changes in pH when small amounts of an acid or base are added.

electrolyte A substance that produces ions when dissolved in water.

equivalent The amount of a positive or negative ion that supplies 1 mol of positive (+) or negative (−) electrical charge.

hydronium ion, H_3O^+ The H_3O^+ ion formed by the attraction of a proton (H^+) to an H_2O molecule.

indicator A substance added to a sample for titration that changes color when the acid or base is neutralized.

ion-product constant, K_w The product of the $[H_3O^+]$ and $[OH^-]$ in solution; $K_w = [H_3O^+][OH^-]$.

neutralization A reaction between an acid and a base to form a salt and water.

neutral solution A solution with equal concentrations of $[H_3O^+]$ and $[OH^-]$.
nonelectrolyte A substance that dissolves in water as molecules; its solution will not conduct an electrical current.
pH A measure of the $[H_3O^+]$ in a solution; $pH = -\log[H_3O^+]$.
salt An ionic compound that contains a metal ion and a negative nonmetal ion.
strong acid An acid that completely ionizes in water.
strong base A base that completely ionizes in water.
strong electrolyte An ionic compound, acid, base, or salt that ionizes completely when it dissolves in water. Its solution is a good conductor of electricity.
titration The addition of base to a sample of an acid to determine the concentration of the acid.
weak acid An acid that ionizes only slightly in solution.
weak base A base that ionizes only slightly in solution.
weak electrolyte A substance that produces only a few ions along with many molecules when it dissolves in water. Its solution is a weak conductor of electricity.

Chemistry at Home

1. Prepare an indicator of red cabbage by tearing up several cabbage leaves, covering them with water, and boiling for 5 minutes. Cool and collect the purple indicator solution. Obtain samples of household solutions such as vinegar, lemon juice, other fruit juices. baking soda, antacids, aspirin, window cleaners, soaps, shampoos, and detergents. Add a small amount of cabbage indicator to give an intense color. A pink-orange color indicates a pH range of 1–4, a pink-lavender 5–6, purple 7, green at 8–11, and yellow at 12–13. Classify each of the solutions as acidic (1–6), neutral (7), or basic (8–13).
2. To a solution of baking soda or other base containing the red cabbage indicator, add small amounts of vinegar. How do you know the base is being neutralized? Can you reach a neutral pH? Try other combinations and note the change in indicator color. Is there any difference using a buffered antacid?
3. Read the labels of some "over-the-counter" antacids. What are the formulas of some of the bases used?

Answers to Study Checks

9.1 $Na_2SO_4 \xrightarrow{H_2O} 2Na^+(aq) + SO_4^{2-}(aq)$
9.2 both
9.3 71 g
9.4 2.75 g
9.5 bases
9.6 H_2O is an acid, CN^- is a base.
9.7 $HClO + H_2O \rightleftharpoons H_3O^+ + ClO^-$
9.8 $Ca(OH)_2 \xrightarrow{H_2O} Ca^{2+} + 2OH^-$
9.9 $3KOH + H_3PO_4 \longrightarrow 3H_2O + K_3PO_4$
9.10 $[H_3O^+] = 1 \times 10^{-3}$ M; acidic
9.11 $[OH^-] = 2 \times 10^{-13}$ M
9.12 neutral
9.13 pH = 10.0
9.14 $[H_3O^+] = 1 \times 10^{-5}$ M; $[OH^-] = 1 \times 10^{-9}$ M
9.15 No. The mixture has no weak acid present.
9.16 0.200 M
9.17 30.0 mL

Problems

Electrolytes *(Goal 9.1)*

9.1 KF is a strong electrolyte, and HF is a weak electrolyte. How are they different?

9.2 NaOH is a strong electrolyte, and CH_3OH is a nonelectrolyte. How are they different?

9.3 The following soluble salts are strong electrolytes. Write a balanced equation for their dissociation in water:
a. KCl **b.** $CaCl_2$
c. K_3PO_4 **d.** $Fe(NO_3)_3$

9.4 The following soluble salts are strong electrolytes. Write a balanced equation for their dissociation in water:

a. LiBr **b.** $NaNO_3$
c. $FeCl_3$ **d.** $Mg(NO_3)_2$

9.5 Indicate whether aqueous solutions of the following will contain ions only, molecules only, or molecules and some ions:

a. acetic acid in vinegar ($HC_2H_3O_2$), a weak electrolyte
b. NaBr, a soluble salt
c. fructose, $C_6H_{12}O_6$, a nonelectrolyte

9.6 Indicate whether aqueous solutions of the following will contain ions only, molecules only, or molecules and some ions:

a. Na_2SO_4, a soluble salt
b. ethanol, C_2H_5OH, a nonelectrolyte
c. HCN, hydrocyanic acid, a weak electrolyte

9.7 Write equations for the formation of aqueous solutions of each substance in problem 9.5.

9.8 Write equations for the formation of aqueous solutions of each substance in problem 9.6.

9.9 Indicate the type of electrolyte represented in the following equations:

a. $K_2SO_4 \xrightarrow{H_2O} 2K^+ + SO_4^{2-}$
b. $NH_3 + H_2O \rightleftharpoons NH_4^+ + OH^-$
c. $C_6H_{12}O_6(s) \xrightarrow{H_2O} C_6H_{12}O_6(aq)$

9.10 Indicate the type of electrolyte represented in the following equations:

a. $CH_3OH \xrightarrow{H_2O} CH_3OH(aq)$
b. $MgCl_2 \xrightarrow{H_2O} Mg^{2+} + 2Cl^-$
c. $HClO \underset{}{\overset{H_2O}{\rightleftharpoons}} H^+ + ClO^-$

Equivalents *(Goal 9.2)*

9.11 Indicate the number of equivalents in each of the following:

a. 1 mol K^+ **b.** 2 mol OH^-
c. 1 mol Ca^{2+} **d.** 3 mol CO_3^{2-}

9.12 Indicate the number of equivalents in each of the following:

a. 1 mol Mg^{2+} **b.** 0.5 mol H^+
c. 4 mol Cl^- **d.** 2 mol Fe^{3+}

9.13 Calculate the number of equivalents in each of the following:

a. 25.0 g Cl^- **b.** 15.0 g Fe^{3+}
c. 4.0 g Ca^{2+} **d.** 1.0 g H^+

9.14 Calculate the number of equivalents in each of the following:

a. 10.0 g Na^+ **b.** 8.0 g OH^-
c. 20.0 g Mg^{2+} **d.** 4.0 g Al^{3+}

9.15 If blood plasma typically contains 3 meq/L of Mg^{2+}, how many grams of Mg^{2+} would be in 5 L of plasma?

9.16 If blood plasma typically contains 110 meq/L of Cl^-, how many grams of Cl^- would be in 5 L of plasma?

9.17 A physiological saline solution (0.9% NaCl) contains 154 meq/L each of Na^+ and Cl^-. How many grams each of Na^+ and Cl^- are in 1 L of the saline?

9.18 A solution to replace potassium loss contains 40 meq/L each of K^+ and Cl^-. How many grams each of K^+ and Cl^- are in 1 L of the solution?

9.19 A solution contains 40 meq/L Cl^- and 15 meq/L HPO_4^{2-}. If Na^+ is the only cation in the solution, what is the Na^+ concentration in milliequivalents per liter?

9.20 A sample of Ringer's solution contains the following concentrations (meq/L) of cations: Na^+ 147, K^+ 4, and Ca^{2+} 4. If Cl^- is the only anion in the solution, what is the Cl^- concentration in milliequivalents per liter?

Acids and Bases *(Goal 9.3)*

9.21 Write equations for HBr acting as an Arrhenius acid, and then as a Brønsted–Lowry acid.

9.22 Write equations for HNO_3 acting as an Arrhenius acid, and then as a Brønsted–Lowry acid.

9.23 Indicate whether each of the following statements is characteristic of an acid or base:

a. has a sour taste
b. neutralizes bases
c. produces H_3O^+ ions in water
d. is named potassium hydroxide
e. is a proton (H^+) acceptor

9.24 Indicate whether each of the following statements is characteristic of an acid or base:

a. neutralizes acids
b. produces OH^- in water
c. is a proton (H^+) donor
d. has a soapy feel
e. turns blue litmus red

9.25 Give the name of each of the following acids or bases:

a. HCl **b.** $Ca(OH)_2$
c. LiOH **d.** H_2SO_4

9.26 Give the name of each of the following acids or bases:

a. $Al(OH)_3$ **b.** HBr
c. H_2CO_3 **d.** KOH

Strengths of Acids and Bases *(Goal 9.4)*

9.27 What is the difference between a strong and weak acid?

9.28 What is the difference between a strong and weak base?

9.29 What are three common strong acids?

9.30 What are four common strong bases?

9.31 Write a balanced equation for the ionization of each of the following in water:
a. HI, a strong acid **b.** $Ca(OH)_2$
c. LiOH **d.** HNO_2, a weak acid

9.32 Write a balanced equation for the ionization of each of the following in water:
a. HCN, a weak acid
b. $Mg(OH)_2$
c. NH_3, a weak base
d. HNO_3, a strong acid

Acid–Base Neutralization *(Goal 9.5)*

9.33 Which of the following is a neutralization reaction?
a. $SO_3 + H_2O \longrightarrow H_2SO_4$
b. $HNO_3 + KOH \longrightarrow KNO_3 + H_2O$
c. $2Mg + O_2 \longrightarrow 2MgO$
d. $H_3PO_4 + 3NaOH \longrightarrow Na_3PO_4 + 3H_2O$

9.34 Which of the following is a neutralization reaction?
a. $2HCl + Ca(OH)_2 \longrightarrow CaCl_2 + 2H_2O$
b. $H_2SO_4 + Pb(NO_3)_2 \longrightarrow PbSO_4 + 2HNO_3$
c. $Mg + 2HCl \longrightarrow MgCl_2 + H_2$
d. $HClO_2 + NaOH \longrightarrow NaClO_2 + H_2O$

9.35 Balance each of the following neutralization reactions:
a. $HCl + Mg(OH)_2 \longrightarrow MgCl_2 + H_2O$
b. $H_3PO_4 + LiOH \longrightarrow Li_3PO_4 + H_2O$

9.36 Balance each of the following neutralization reactions:
a. $HNO_3 + Ba(OH)_2 \longrightarrow Ba(NO_3)_2 + H_2O$
b. $H_2SO_4 + Al(OH)_3 \longrightarrow Al_2(SO_4)_3 + H_2O$

9.37 Write a complete balanced equation for the neutralization of each of the following:
a. H_2SO_4 and NaOH
b. HCl and $Fe(OH)_3$
c. H_2CO_3 and $Mg(OH)_2$

9.38 Write a complete balanced equation for the neutralization of each of the following:
a. H_3PO_4 and NaOH
b. HI and LiOH
c. HNO_3 and $Ca(OH)_2$

Ion Product of Water *(Goal 9.6)*

9.39 What are the values for $[H_3O^+]$ and $[OH^-]$ in pure water?

9.40 What is the ion product, K_w, and its value?

9.41 Calculate the $[H_3O^+]$ in each solution if the $[OH^-]$ has the following values:
a. 1×10^{-8} M **b.** 1×10^{-2} M
c. 1×10^{-11} M **d.** 2×10^{-5} M

9.42 Calculate the $[OH^-]$ in each solution if the $[H_3O^+]$ has the following values:
a. 1×10^{-1} M **b.** 1×10^{-7} M
c. 1×10^{-10} M **d.** 5×10^{-4} M

9.43 Indicate whether the following are acidic, basic, or neutral solutions:
a. $[OH^-] = 1 \times 10^{-6}$ M
b. $[H_3O^+] = 1 \times 10^{-10}$ M
c. $[H_3O^+] = 1 \times 10^{-7}$ M
d. $[H_3O^+] = 1 \times 10^{-2}$ M
e. $[OH^-] = 4 \times 10^{-4}$ M
f. $[OH^-] = 8 \times 10^{-12}$ M

9.44 Indicate whether the following are acidic, basic, or neutral solutions:
a. $[OH^-] = 1 \times 10^{-11}$ M
b. $[H_3O^+] = 1 \times 10^{-5}$ M
c. $[OH^-] = 1 \times 10^{-10}$ M
d. $[H_3O^+] = 1 \times 10^{-8}$ M
e. $[OH^-] = 3 \times 10^{-3}$ M
f. $[H_3O^+] = 6 \times 10^{-12}$ M

The pH Scale *(Goal 9.7)*

9.45 State whether each of the following solutions is acidic, basic, or neutral:
a. blood, pH 7.4 **b.** vinegar, pH 2.8
c. drain cleaner, pH 11.2 **d.** coffee, pH 5.5
e. milk, pH 7.0

9.46 State whether each of the following solutions is acidic, basic, or neutral:
a. soda, pH 3.2
b. shampoo, pH 6.0
c. hot tub water, pH 7.8
d. acid rain, pH 6.2
e. laundry detergent, pH 9.5

9.47 Arrange the pH values of the following groups in order from the most acidic to the most basic:
a. 14, 5, 10, 7
b. 8, 10, 1, 13
c. 5.5, 1.6, 2.4, 8.5
d. 3.5, 9.7, 8.8, 11.4

9.48 Arrange the pH values of the following groups in order from the most acidic to the most basic:
  a. 5, 3, 2, 1
  b. 11, 4, 7, 9
  c. 2.4, 1.8, 5.4, 6.3
  d. 11.4, 7.6, 5.2, 10.8

9.49 Calculate the pH of each of the following solutions:
  a. $[H_3O^+] = 1 \times 10^{-4}$ M
  b. $[H_3O^+] = 1 \times 10^{-9}$ M
  c. $[H_3O^+] = 1 \times 10^{-12}$ M
  d. $[OH^-] = 1 \times 10^{-5}$ M
  e. $[OH^-] = 1 \times 10^{-11}$ M

9.50 Calculate the pH of each of the following solutions:
  a. $[H_3O^+] = 1 \times 10^{-1}$ M
  b. $[H_3O^+] = 1 \times 10^{-9}$ M
  c. $[H_3O^+] = 1 \times 10^{-7}$ M
  d. $[OH^-] = 1 \times 10^{-2}$ M
  e. $[OH^-] = 1 \times 10^{-6}$ M

9.51 Use a calculator to determine the pH of each of the following solutions:
  a. $[H_3O^+] = 4 \times 10^{-2}$ M
  b. $[H_3O^+] = 2 \times 10^{-8}$ M
  c. $[OH^-] = 5 \times 10^{-11}$ M

9.52 Use a calculator to determine the pH of each of the following solutions:
  a. $[H_3O^+] = 8 \times 10^{-10}$ M
  b. $[H_3O^+] = 6 \times 10^{-3}$ M
  c. $[OH^-] = 2 \times 10^{-6}$ M

9.53 Determine the $[H_3O^+]$ and $[OH^-]$ for each solution with the following pH values:
  a. pH = 2 **b.** pH = 7
  c. pH = 13 **d.** pH = 5

9.54 Determine the $[H_3O^+]$ and $[OH^-]$ for each solution with the following pH values:
  a. pH = 10 **b.** pH = 4
  c. pH = 1 **d.** pH = 8

9.55 Complete the following table of pH, $[H_3O^+]$, and $[OH^-]$:

$[H_3O^+]$	$[OH^-]$	pH	Acidic, Basic, or Neutral?
	1×10^{-6}		
		2	
1×10^{-5}			
1×10^{-7}			

9.56 Complete the following table for pH, $[H_3O^+]$, and $[OH^-]$:

$[H_3O^+]$	$[OH^-]$	pH	Acidic, Basic, or Neutral?
		10	
	1×10^{-7}		
1×10^{-12}			
	1×10^{-8}		

Buffers *(Goal 9.8)*

9.57 Which of the following represent a buffer system? Explain.
  a. NaOH and NaCl **b.** H_2CO_3 and $NaHCO_3$
  c. HF and KF **d.** KCl and NaCl

9.58 Which of the following represent a buffer system? Explain.
  a. H_3PO_4
  b. $NaHCO_3$
  c. $HC_2H_3O_2$ and $NaC_2H_3O_2$
  d. HCl and NaOH

9.59 Consider the buffer system of hydrofluoric acid, HF, and its salt, NaF.

$$HF \rightleftharpoons H^+ + F^-$$

  a. What is the purpose of the buffer system?
  b. Why is a salt of the acid needed?
  c. How does the buffer react when some H^+ is added?
  d. How does the buffer react when some OH^- is added?

9.60 Consider the buffer system of hydrocyanic acid, HCN, and its salt, NaCN.

$$HCN \rightleftharpoons H^+ + CN^-$$

  a. What is the purpose of the buffer system?
  b. Why is a salt of the acid added?
  c. How does the buffer react when some H^+ is added?
  d. How does the buffer react when some OH^- is added?

Acid–Base Titration *(Goal 9.9)*

9.61 If you need to determine the concentration of a solution of ascorbic acid (vitamin C), how would you proceed?

9.62 If you need to determine the concentration of a sample of vinegar (acetic acid), how would you proceed?

9.63 What is the molarity of the acid in each of the following titrations?

a. A 5.0-mL sample of HCl that is titrated with 22.0 mL of 2.0 M NaOH.

$$HCl + NaOH \longrightarrow NaCl + H_2O$$

b. A 10.0-mL sample of sulfuric acid that is titrated with 15.0 mL of 1.0 M NaOH.

$$H_2SO_4 + 2NaOH \longrightarrow Na_2SO_4 + 2H_2O$$

c. A 10.0-mL sample of phosphoric acid that is titrated with 16.0 mL of 1.0 M NaOH.

$$H_3PO_4 + 3NaOH \longrightarrow Na_3PO_4 + 3H_2O$$

9.64 What is the molarity of the acid in each of the following titrations?

a. A 20.0-mL sample of HCl that is titrated with 8.0 mL of 6.0 M NaOH.

$$HCl + NaOH \longrightarrow NaCl + H_2O$$

b. A 10.0-mL sample of sulfuric acid that is titrated with 25.0 mL of 0.50 M NaOH.

$$H_2SO_4 + 2NaOH \longrightarrow Na_2SO_4 + 2H_2O$$

c. A 30.0-mL sample of phosphoric acid that is titrated with 18.0 mL of 3.0 M NaOH.

$$H_3PO_4 + 3NaOH \longrightarrow Na_3PO_4 + 3H_2O$$

Challenge Problems

9.65 What is the value of $[OH^-]$ in a solution that contains 0.20 g NaOH in 0.25 L of solution?

9.66 What is the value of $[H_3O^+]$ in a solution that contains 1.5 g HNO_3 in 0.50 L of solution?

9.67 An ingredient in some antacids is $Mg(OH)_2$. If the base is not very soluble in water, why is it considered a strong base?

9.68 Acetic acid used to make vinegar is very soluble in water, but it is considered a weak acid. Why?

9.69 How many milliliters of 1.0 M NaOH are needed to titrate 25.0 mL of 2.0 M H_2SO_4?

9.70 If 3.50 g Zn reacts with 50.0 mL HCl, what is the molarity of the HCl solution? What volume of hydrogen is produced at STP?

$$Zn + 2HCl \longrightarrow ZnCl_2 + H_2$$

9.71 What is the pH of each of the following solutions:

a. $[H_3O^+] = 4 \times 10^{-8}$ M

b. $[H_3O^+] = 6 \times 10^{-2}$ M

c. $[OH^-] = 3 \times 10^{-12}$ M

d. $[OH^-] = 0.0001$ M

Chapter 10

Introduction to Organic Chemistry: Alkanes

Crude oil is being pumped below the ocean floor in an off-shore oil well.

Looking Ahead

Learning Goals

10.1 From its properties, classify a compound as organic or inorganic.

10.2 Draw the full structural formula and the condensed structural formula for an alkane.

10.3 Draw the full or condensed structural formulas for the structural isomers of an alkane.

10.4 Use the IUPAC system to name alkanes and draw their condensed structural formulas.

10.5 Give the name for a cycloalkane; draw the condensed structural and geometric formulas.

10.6 Give the name for a haloalkane and draw the condensed structural formula.

10.7 Describe some properties of alkanes.

QUESTIONS TO THINK ABOUT

1. What do you mean when you use the term organic?
2. What do vegetable oil, gasoline, plastics, and sugar have in common?
3. If we are organic, what are our dietary sources of carbon?
4. Why are perfumes soluble in body oils?
5. In oil spills, why does the oil stay on top of the water?
6. Why is it not recommended to use water on an oil fire?
7. What is causing the formation of an ozone hole?

The element carbon has a special role in chemistry because it has the capability to form so many different compounds, many of which are found in biological systems. Gasoline, coal, medicines, shampoos, plastic bottles, and perfumes are all made of organic compounds. The fabrics in your clothes may be naturally occurring organic compounds such as cotton and silk, or they may be synthetic organic compounds such as polyester or nylon. The food we eat is composed of many different organic compounds that supply us with fuel for energy and the carbon atoms needed for building and repairing the cells of our bodies. What we learn about organic molecules will be used to explain the behavior of the very large biomolecules that are important in health and medicine.

Before 1800, scientists used the term *organic* for compounds they thought could only be produced within the cells of living systems. They believed that a "vital force," found only in living organisms, was required to synthesize an organic compound. By the early 1800s, urea, a product of protein metabolism in animals, was prepared from the inorganic compound ammonium cyanate. Since then, millions of organic compounds have been prepared in laboratories and the idea of a vital force has become a part of chemical history.

$$NH_4CNO \xrightarrow{\text{Heat}} H_2N-\overset{\overset{\displaystyle O}{\|}}{C}-NH_2$$

Ammonium cyanate (inorganic) → Urea (organic)

10.1 Organic Compounds

Learning Goal
From its properties, classify a compound as organic or inorganic.

Organic chemistry is the study of compounds containing carbon. The location of carbon in the middle of the periodic table and its low atomic mass makes it ideal as the major element for biological compounds. In organic compounds, carbon is bonded mostly to hydrogen and sometimes to oxygen, nitrogen, sulfur, phosphorus, and halogens. There are a few carbon-containing compounds that are classified as inorganic such as the ions carbonate, bicarbonate, and cyanide as well as the oxides of carbon such as CO_2. In general, **organic compounds** have covalent bonds, low melting and boiling points, burn vigorously, and are soluble in nonpolar solvents. There are a great number of organic compounds because carbon atoms have a unique ability to bond to other carbon atoms to form large molecules.

Until now, most of the chemistry we have studied involved inorganic compounds that were ionic or had polar covalent bonds. In general, these

Table 10.1 ***Properties of Most Organic and Inorganic Compounds***

Organic	**Inorganic**
All are carbon compounds.	Contain metals in salts and oxides.
All have covalent bonds.	All have ionic or polar bonds.
All have low melting points.	All have high melting points.
All have low boiling points.	All have high boiling points.
Most burn in oxygen.	Few burn in oxygen.
Most are soluble in nonpolar solvents.	Most are soluble in polar solvents.
Most are nonelectrolytes.	Many are electrolytes.
All can be large molecules with many atoms.	All are usually small with few atoms.

compounds have high melting and boiling points, do not burn in oxygen, and are soluble in water. Although there are variations among organic and inorganic compounds, it is helpful to compare some of their properties as listed in Table 10.1. Table 10.2 contrasts the properties of butane, an organic compound, and sodium chloride, an inorganic compound.

Table 10.2 ***A Comparison of an Organic and an Inorganic Compound***

Compound	Butane	Sodium chloride
Formula	C_4H_{10}	NaCl
Molar mass	58.1 g/mol	58.5 g/mol
Bonds	Covalent	Ionic
State (room temperature)	Gas	Solid
Melting point	−138°C	801°C
Boiling point	0°C	1413°C
Solubility in water	Low	High
Solubility in a nonpolar solvent	High	Low
Electrolyte	No	Yes
Burns in oxygen	Yes	No

Hydrocarbons

Today chemists know of millions of organic compounds. To study this vast array of structures, the compounds have been organized into *families* according to common features. A list of the families we will discuss is found on the inside back cover. We will study the features (functional groups) of each family and relate them to their chemical behavior.

In the **hydrocarbon** family, all the compounds contain two elements, carbon and hydrogen. The **alkanes** are hydrocarbons that have only single bonds (C—C). They are known as **saturated hydrocarbons** because every carbon atom is bonded to four atoms, the most a carbon atom can have. In Chapter 11, we will study the *unsaturated hydrocarbons,* the alkenes that contain a double bond (C=C), and alkynes with a triple bond (C≡C). We will also look at the aromatic compounds, a group of hydrocarbons that contain benzene rings.

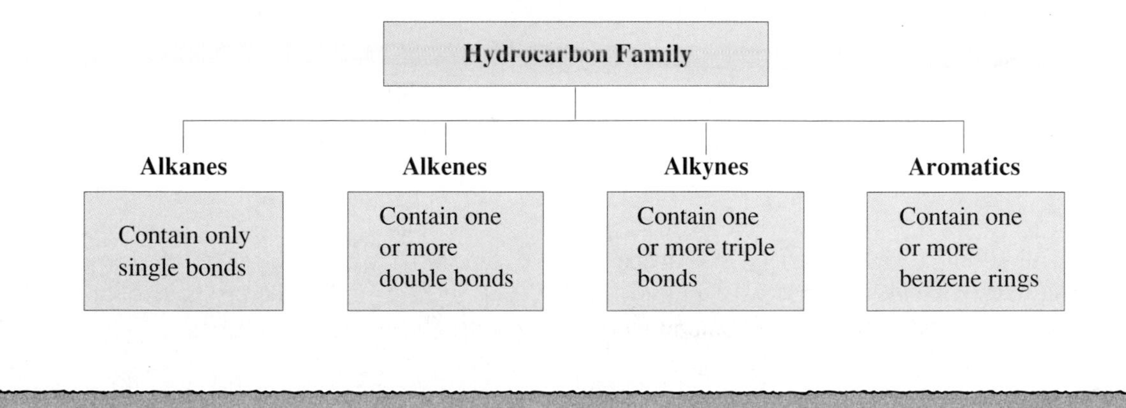
Hydrocarbon Family
Alkanes
Contain only single bonds
Alkenes
Contain one or more double bonds
Alkynes
Contain one or more triple bonds
Aromatics
Contain one or more benzene rings

Sample Problem 10.1
Classifying Properties of Organic Compounds

State whether each of the following properties is typical of an organic or an inorganic compound:

a. soluble in water **b.** low boiling point **c.** burns in air

Solution

a. Inorganic; most inorganic compounds are ionic or polar and soluble in water.
b. Organic; most organic compounds have low boiling points.
c. Organic; organic compounds are often very flammable.

Study Check

Would a compound that dissociates into ions in water most likely be an organic or inorganic compound?

ENVIRONMENTAL NOTE *Hydrocarbons*

Mineral oil is a mixture of liquid hydrocarbons and is used as a laxative and a lubricant. Motor oil is a mixture of high molecular weight liquid hydrocarbons and is used to lubricate the internal compartments of engines. Petrolatum, or Vaseline, is a coloidal system in which liquid hydrocarbons having high boiling points are encapsulated in solid hydrocarbons. It is used in ointments and cosmetics and as a lubricant and a solvent. Applying a hydrocarbon mixture such as Vaseline softens the skin because the water in the skin is retained by the nonpolar compounds. Figure 10.1 shows some examples of mixtures of hydrocarbons.

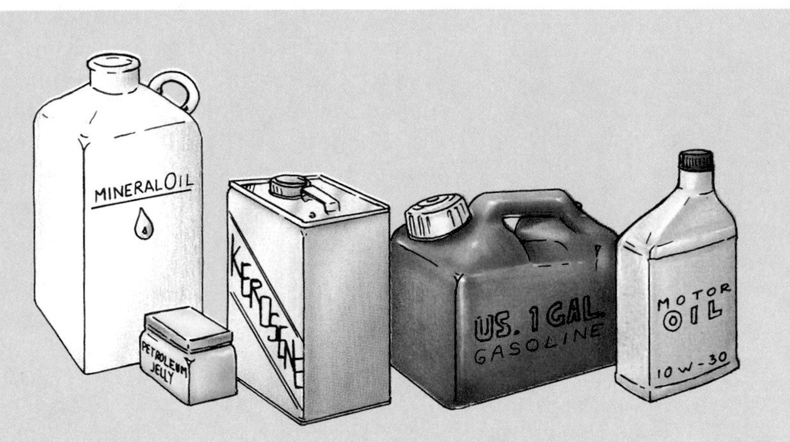

Figure 10.1
Mixtures of hydrocarbons are found in products like these.

Crude oil contains a wide variety of hydrocarbons whose boiling points increase with molar mass. At an oil refinery, the crude oil components are separated by increasing the temperature (See Figure 10.2). Some of the products obtained from crude oil are listed in Table 10.3.

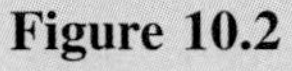
Figure 10.2
At an oil refinery, crude oil is separated into many useful organic products.

Table 10.3 ***Hydrocarbons in Crude Oil***

Product	Number of Carbon Atoms	Temperature Range for Distillation (°C)
Natural gas	1–4	Below 30
Gasoline	5–12	30–200
Kerosene, heating oil	12–15	200–300
Diesel fuel, lubricating oil	15–25	300–400
Asphalt, tar	over 25	Nonvolatile residue

10.2 Alkanes

Learning Goal

Draw the full structural formula and the condensed structural formula for an alkane.

In the simplest hydrocarbon alkane, methane (CH_4), the four valence electrons of carbon are shared with four hydrogen atoms. A hydrogen atom only forms one bond in compounds. In the electron-dot structure, each shared pair of electrons represents a single bond. A **full structural formula** is written when we show all the bonds between all of the atoms.

$$\begin{array}{ccc} & H & \\ H & \!:\!\ddot{\underset{\cdot\cdot}{C}}\!:\! & H \\ & H & \end{array} \qquad \begin{array}{ccccc} & & H & & \\ & & | & & \\ H & — & C & — & H \\ & & | & & \\ & & H & & \end{array} \quad \text{Methane, } CH_4$$

Figure 10.3
The representations of the atoms in a methane (CH_4) molecule.

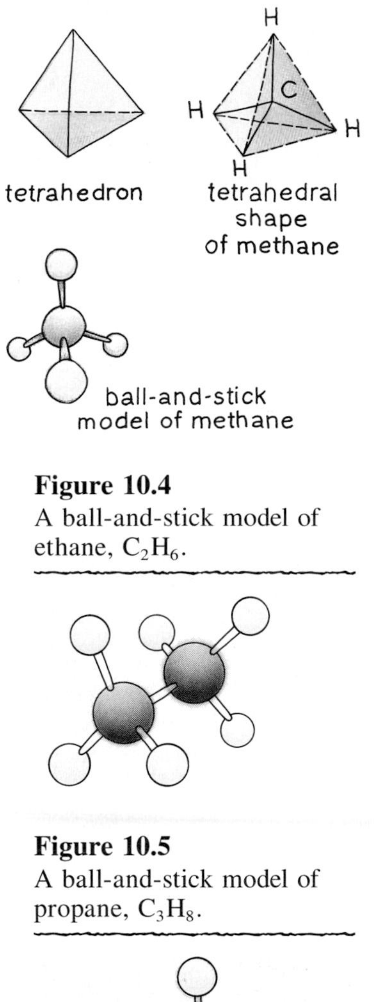

Figure 10.4
A ball-and-stick model of ethane, C_2H_6.

Figure 10.5
A ball-and-stick model of propane, C_3H_8.

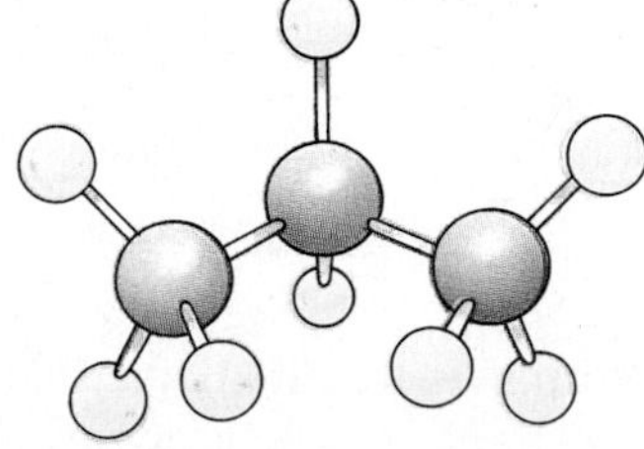

Although our drawing of methane is two-dimensional, methane actually has a three-dimensional shape (shown in Figure 10.3) called a *tetrahedral structure.* The carbon atom is in the center and bonded to four hydrogen atoms, which are as far apart as possible and occupy the corners of the tetrahedron. All carbon atoms in alkanes exhibit the tetrahedral shape, even though we often write the flat formula for convenience.

The alkane with two atoms is ethane, which has the molecular formula C_2H_6. Using single bonds, each carbon is bonded to the other carbon atom and three hydrogen atoms to give a total of four bonds to each carbon atom. A ball-and-stick model of ethane is shown in Figure 10.4.

$$\begin{array}{ccccccc} & & H & & H & & \\ & & | & & | & & \\ H & — & C & — & C & — & H \\ & & | & & | & & \\ & & H & & H & & \end{array} \quad \text{Ethane, } C_2H_6$$

To write the full structural formula of propane, C_3H_8, an alkane with three carbon atoms, we connect the carbon atoms with single bonds.

$$C—C—C$$

Eight hydrogen atoms are attached to complete the four bonds to each carbon atom. A ball-and-stick model of propane is shown in Figure 10.5.

$$\begin{array}{ccccccccc} & & H & & H & & H & & \\ & & | & & | & & | & & \\ H & — & C & — & C & — & C & — & H \\ & & | & & | & & | & & \\ & & H & & H & & H & & \end{array} \quad \text{Propane, } C_3H_8$$

Condensed Structural Formulas

Although full structural formulas are helpful, we will not always draw all of the individual C—H and C—C bonds. In a **condensed structural formula,** we still show the arrangement or structure, but we write each carbon atom and its attached hydrogen atoms as a group. A subscript in each group indicates the number of hydrogen atoms bonded to that carbon atom. The C—H single bonds are understood but not written individually.

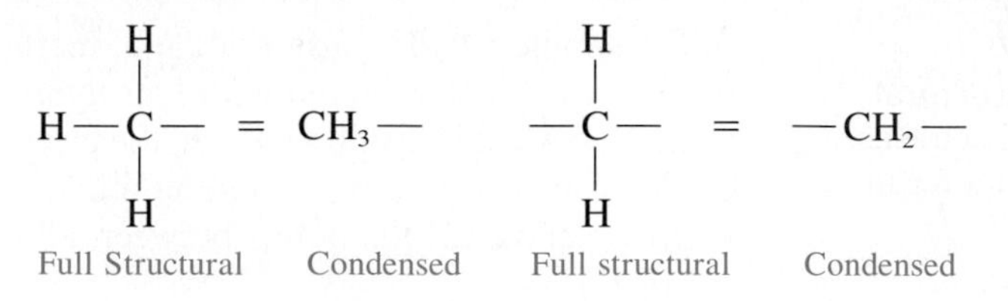

We can simplify further by omitting the line drawn for the carbon–carbon (C—C) bonds. Both the full structural and condensed structural formulas show the arrangement of the carbon atoms in a molecule. The molecular formulas give the total number of each kind of atom, but they indicate nothing about the arrangement of the atoms in the molecule. Here, for example, are the different formulas used to represent propane:

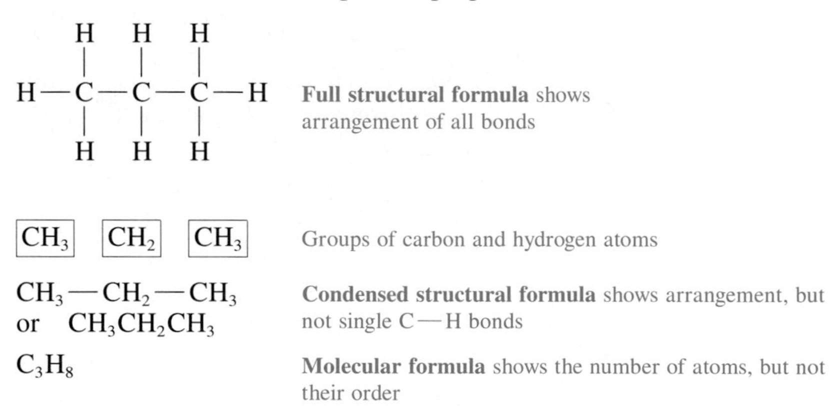

H H H
H—C—C—C—H
H H H
Full structural formula shows arrangement of all bonds
CH_3 CH_2 CH_3
Groups of carbon and hydrogen atoms
CH_3—CH_2—CH_3
or $CH_3CH_2CH_3$
Condensed structural formula shows arrangement, but not single C—H bonds
C_3H_8
Molecular formula shows the number of atoms, but not their order

Sample Problem 10.2
Drawing Structural and Condensed Structural Formulas for Alkanes

A molecule of butane has four carbon atoms in a row. What are its full structural, condensed structural, and molecular formulas?

Solution

In the full structural formula, four carbon atoms are joined to hydrogen atoms using single bonds to give each carbon atom a total of four bonds. In the condensed structural formula, each carbon atom and its attached hydrogen atoms are written as CH_3—, or —CH_2—.

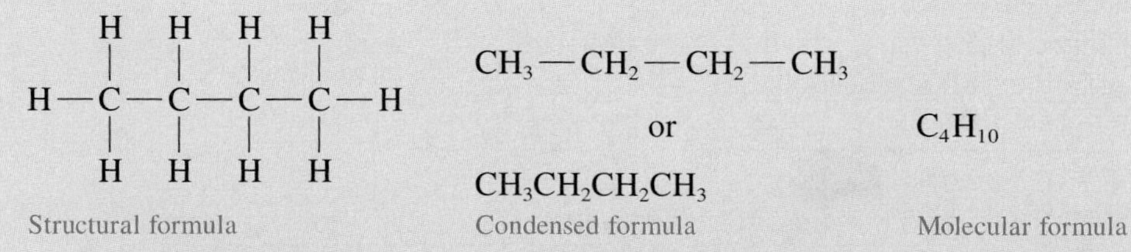

The molecular formula, C_4H_{10}, is the sum of the carbon atoms and hydrogen atoms in the molecule.

Study Check

Write the structural, condensed, and molecular formulas for an alkane that has five carbon atoms in a row.

10.3 Structural Isomers

Learning Goal

Draw the full or condensed structural formulas for the structural isomers of an alkane.

In an **unbranched alkane,** the carbon atoms are connected in a row, one after the other. You could trace a line through the carbon symbols without lifting the pencil off the paper.

Examples of Unbranched Alkanes

$$CH_3—CH_2—CH_3 \qquad CH_3—CH_2—CH_2—CH_3$$

Actually, alkane structures are not really flat or straight molecules. The tetrahedral shape of carbon makes the bonds between carbon atoms appear at angles to each other in a zigzag fashion. However, when we use condensed formulas, we only show the way the carbon atoms are connected. As a result, the formula appears flatter than it really is. Also, one compound can be represented by several diagrams. In Table 10.4, some possible condensed formulas for butane show a chain of four carbon atoms, but it does not need to be drawn in a straight line. However, if you trace the carbon atoms in each formula from one end to the other, you will see that they are all connected in a row; they are all straight-chain alkanes.

When there are four or more carbon atoms in an alkane, the atoms can be arranged so that a **branch,** a side group of carbon atoms also called a *substituent,* is attached to a longer carbon chain. An alkane with at least one carbon branch is called a **branched** (or branched-chain) **alkane.** In the formulas of branched-chain compounds, the branches usually are drawn above or below the carbon chain.

Examples of Some Branched-Chain Alkanes

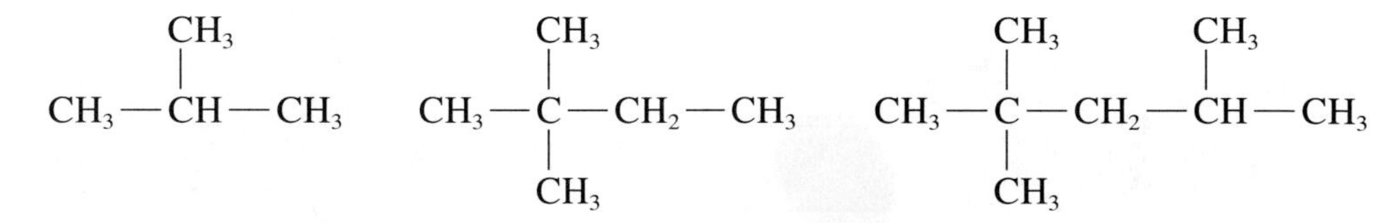

Table 10.4 ***Some Possible Unbranched Formulas for Butane***

$CH_3—CH_2—CH_2—CH_3$

$CH_3—CH_2—CH_2$ [with CH_3 below the last CH_2]

$CH_2—CH_2$ [with CH_3 below each CH_2]

CH_3 / CH_2 / CH_2 / CH_3 [vertical chain]

CH_3 above $CH_2—CH_2$, with CH_3 below the second CH_2

CH_3 / CH_2 / $CH_2—CH_3$

$CH_3—CH_2—CH_2—CH_3$ [zigzag]

Sample Problem 10.3
Formulas for Branched-Chain Alkanes

Draw a condensed structural formula for the following branched-chain alkane:

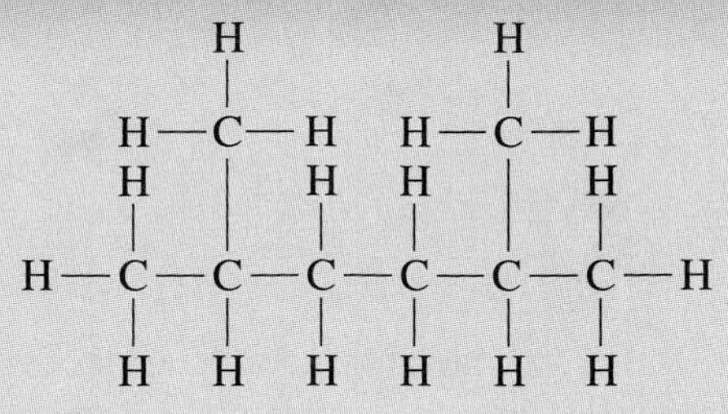

Solution

In the condensed structural formula, each carbon atom and its surrounding hydrogen atoms are shown as a group with a subscript to indicate the number of attached hydrogen atoms.

$$\begin{array}{cccccccccc} & CH_3 & & & & CH_3 & \\ & | & & & & | & \\ CH_3- & CH & -CH_2- & CH_2- & & CH & -CH_3 \end{array} \quad \text{or} \quad \begin{array}{c} \quad CH_3 \qquad\qquad CH_3 \\ \quad | \qquad\qquad\quad | \\ CH_3CHCH_2CH_2CHCH_3 \end{array}$$

Study Check

Write a condensed structural formula for the following compound:

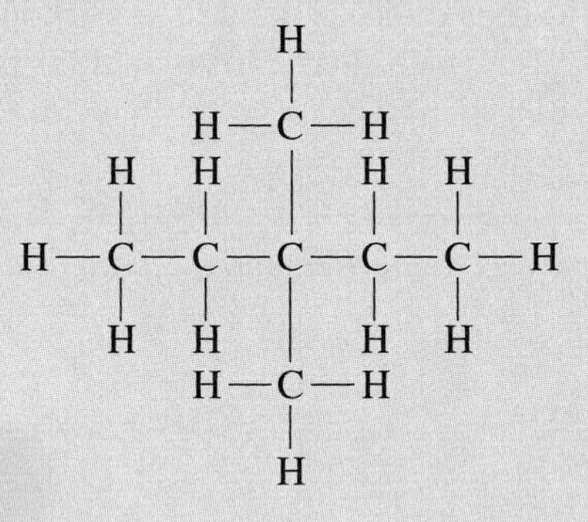

Structural Isomers

Structural isomers are compounds that have the same molecular formula but a different sequence of atoms and therefore different structural formulas. For example, there are two different alkanes with the molecular formula C_4H_{10}. One is the unbranched chain; the other is a branched-chain. The ball-and-stick models of these two isomers are shown in Figure 10.6.

***Condensed Structural Isomers for* C_4H_{10}**

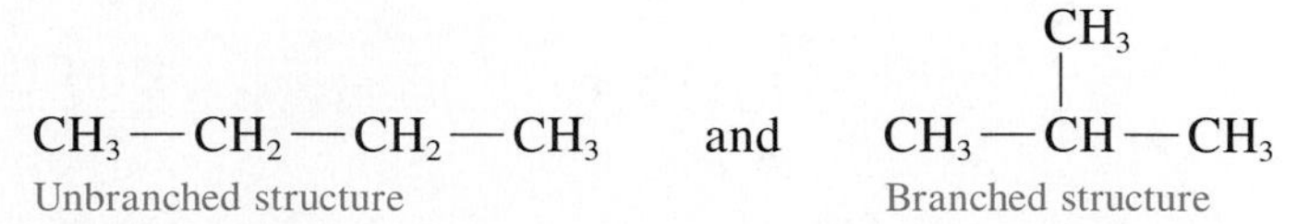

Although these two isomers contain the same type and number of atoms, the atoms are connected in different ways. The different structures of the isomers result in different physical and chemical properties such as boiling point, solubility, and reactivity as shown in Table 10.5.

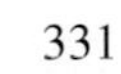

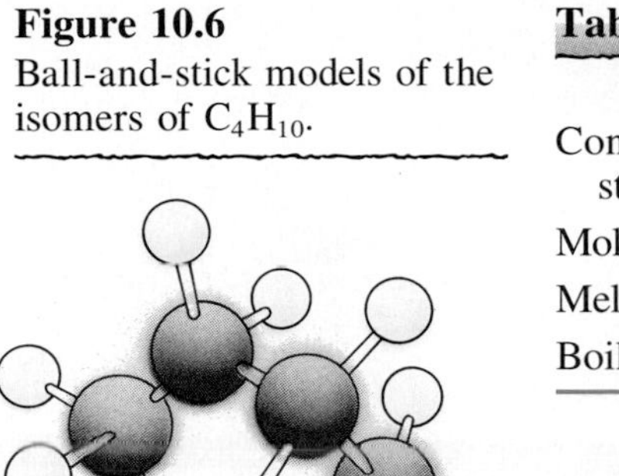

Figure 10.6
Ball-and-stick models of the isomers of C_4H_{10}.

Table 10.5 ***Some Properties of the Structural Isomers of C_4H_{10}***

Condensed structural formulas	$CH_3—CH_2—CH_2—CH_3$	$CH_3—CH(CH_3)—CH_3$
Molar mass	58.1 g/mol	58.1 g/mol
Melting point	−138°C	−159°C
Boiling point	0.5°C	−12°C

Drawing Structural Isomers

There are three structural isomers for the molecular formula C_5H_{12}. Beginning with the straight-chain isomer, we write a carbon chain of five carbon atoms and fill in the appropriate number of carbon atoms:

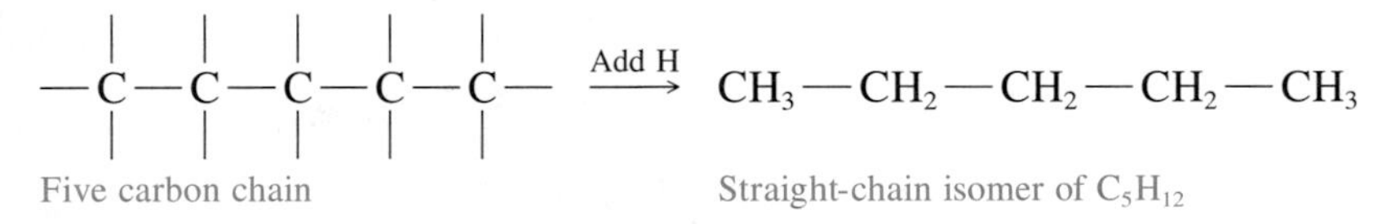

All the other isomers of C_5H_{12} are branched-chain alkanes, but each must still have a total of five carbon atoms. We write a second isomer by writing a shorter chain of four carbon atoms and attaching the fifth carbon as a $CH_3—$ branch. Be sure to attach the $CH_3—$ to a carbon atom inside the chain. If the carbon group is added to an end of the carbon chain, it would repeat the first unbranched isomer.

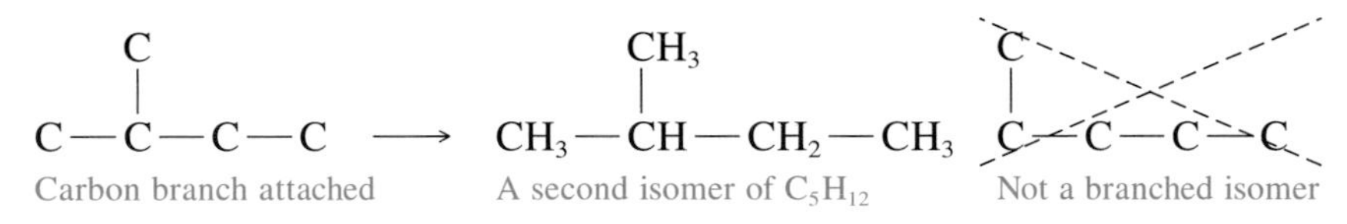

If we move the $CH_3—$ to the other inside carbon in the chain, it gives us the same branched compound, not another isomer.

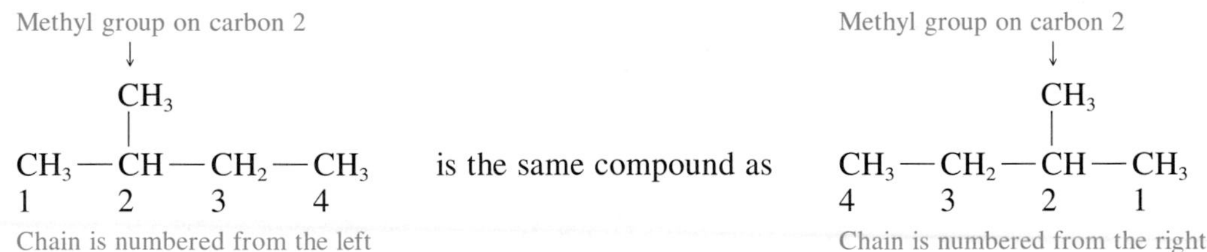

A third isomer is written by drawing a shorter chain of three carbon atoms and attaching two carbon atoms as $CH_3—$ branches to the center carbon. These are the only isomers we can draw with five carbon atoms; any other arrangements will repeat one of the previous structures. The three isomers for C_5H_{12} are illustrated in Figure 10.7.

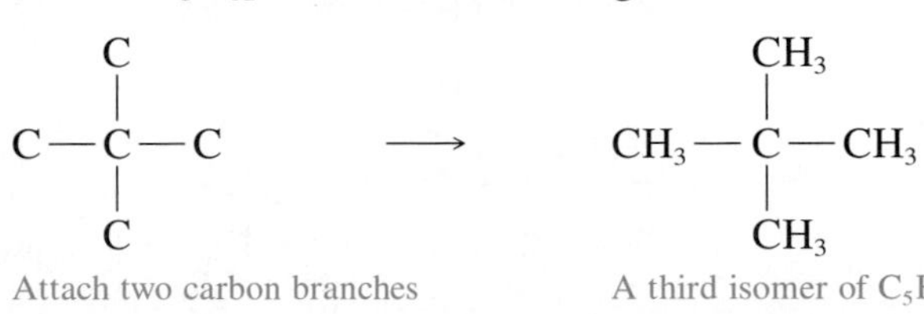

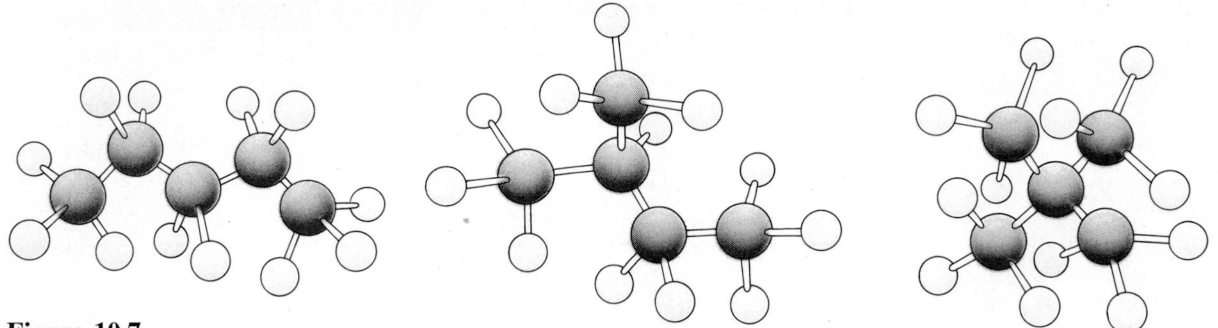

Figure 10.7
Ball-and-stick models of the structural isomers of C_5H_{12}.

Sample Problem 10.4
Drawing Structural Isomers

There are five isomers that have the molecular formula C_6H_{14}. Draw their condensed structural formulas.

Solution

The first isomer is the straight-chain alkane with six carbon atoms.

$$CH_3-CH_2-CH_2-CH_2-CH_2-CH_3$$

In the next two isomers, we write a five-carbon chain and attach a CH_3- to one of the carbon atoms inside the chain. In one isomer, the CH_3- group is attached to carbon 2 in one isomer, and to carbon 3 in the other isomer.

$$\overset{1}{CH_3}-\overset{2}{CH}(CH_3)-\overset{3}{CH_2}-\overset{4}{CH_2}-\overset{5}{CH_3} \qquad \overset{1}{CH_3}-\overset{2}{CH_2}-\overset{3}{CH}(CH_3)-\overset{4}{CH_2}-\overset{5}{CH_3}$$

Two more isomers are drawn using a four-carbon chain and attaching two CH_3- groups to carbon atoms within the chain.

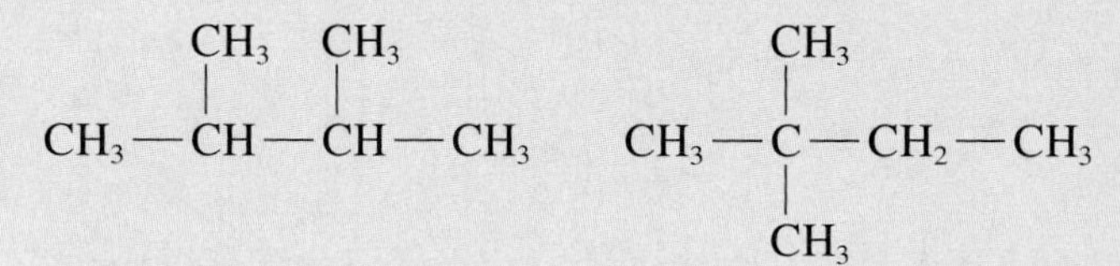

$$CH_3-CH(CH_3)-CH(CH_3)-CH_3 \qquad CH_3-C(CH_3)_2-CH_2-CH_3$$

Any other combination of atoms will repeat one of the above structural isomers.

Study Check

Are the following pairs of compounds isomers or identical structures?

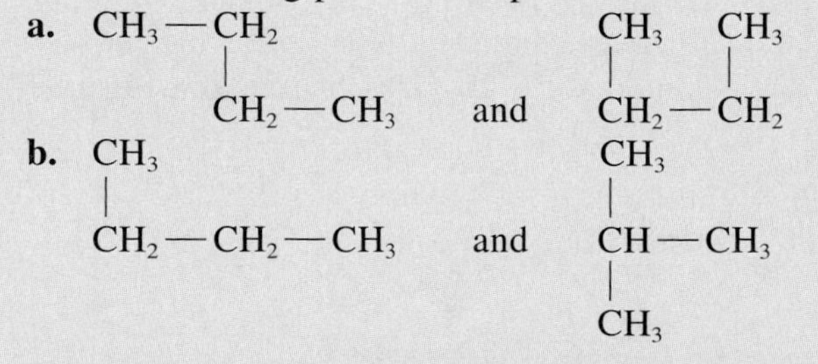

a. CH_3-CH_2 (with CH_2-CH_3 attached below CH_2) and CH_3, CH_3 (attached above) CH_2-CH_2

b. CH_3 (attached above) $CH_2-CH_2-CH_3$ and CH_3 (above) $CH-CH_3$ with CH_3 (below)

10.4 The IUPAC System of Naming

Learning Goal
Use the IUPAC system to name alkanes and draw their condensed structural formulas.

When there were relatively few organic compounds known, names were assigned in a random fashion. Although some of those *common names* are still in use, we now use a naming system known as the *IUPAC rules* (International Union of Pure and Applied Chemistry). Using the IUPAC rules, the stem of a name identifies the number of carbon atoms in the longest chain of a compound, and the suffix describes the family of the compound. The names given in Table 10.6 give the stems for unbranched alkanes with 1 to 10 carbon atoms followed by the suffix *ane*, which indicates the alkane family. It is important to learn these stems because they are used in the names of all the families of organic compounds.

Table 10.6 *IUPAC Names (Stem + ane) for the First 10 Unbranched Alkanes*

Name	Number of Carbon Atoms	Molecular Formula	Condensed Structural Formula
Methane	1	CH_4	CH_4
Ethane	2	C_2H_6	CH_3CH_3
Propane	3	C_3H_8	$CH_3CH_2Ch_3$
Butane	4	C_4H_{10}	$CH_3CH_2CH_2CH_3$
Pentane	5	C_5H_{12}	$CH_3CH_2CH_2CH_2CH_3$
Hexane	6	C_6H_{14}	$CH_3CH_2CH_2CH_2CH_2CH_3$
Heptane	7	C_7H_{16}	$CH_3CH_2CH_2CH_2CH_2CH_2CH_3$
Octane	8	C_8H_{18}	$CH_3CH_2CH_2CH_2CH_2CH_2CH_2CH_3$
Nonane	9	C_9H_{20}	$CH_3CH_2CH_2CH_2CH_2CH_2CH_2CH_2CH_3$
Decane	10	$C_{10}H_{22}$	$CH_3CH_2CH_2CH_2CH_2CH_2CH_2CH_2CH_2CH_3$

Sample Problem 10.5
IUPAC Names

Write the IUPAC names for the following alkanes:

a. $CH_3CH_2CH_3$

b.

$$\begin{array}{lll} CH_3 & & CH_2-CH_3 \\ | & & | \\ CH_2- & CH_2- & CH_2 \end{array}$$

Solution

a. propane (three carbon atoms)

b. hexane (six carbon atoms)

Study Check

Write the condensed structural formula and the name of an unbranched alkane having seven carbon atoms.

Table 10.7 *Some Common Alkyl Substituents*

Alkane	Corresponding Alkyl Group	Condensed Structural Formula	Name
H | H—C—**H** ↙ Remove one hydrogen | H Meth**ane**	H | H—C— | H	CH_3—	Meth**yl**
H H | | H—C—C—**H** ↙ Remove one hydrogen | | H H Eth**ane**	H H | | H—C—C— | | H H	CH_3CH_2—	Eth**yl**
H H H | | | H—C—C—C—H | | | H H H Prop**ane**	Remove end hydrogen H H H | | | H—C—C—C— | | | H H H	$CH_3CH_2CH_2$—	Prop**yl**
	Remove center hydrogen H H | | | H—C—C—C—H | | | H H H	| CH_3CHCH_3	Isoprop**yl**

Alkyl Groups in Branched Alkanes

In the names of branched alkane, the carbon groups attached to a longer carbon chain are named as alkyl groups or alkyl substituents. An **alkyl group** is derived from an alkane by removing one hydrogen. The *ane* ending of the alkane is changed to *yl* to indicate the alkyl group. Alkyl groups do not exist on their own; they must be attached to something such as a carbon chain. Some of the most common alkyl groups are illustrated in Table 10.7. The ball-and-stick models of the alkyl groups are shown in Figure 10.8

IUPAC Names for Branched-Chain Alkanes

We are now ready to name branched-chain alkanes. Suppose you want to name the following branched-chain alkane:

$$\begin{array}{c} \quad CH_3 \\ \quad | \\ CH_3-CH-CH_2-CH_3 \end{array}$$

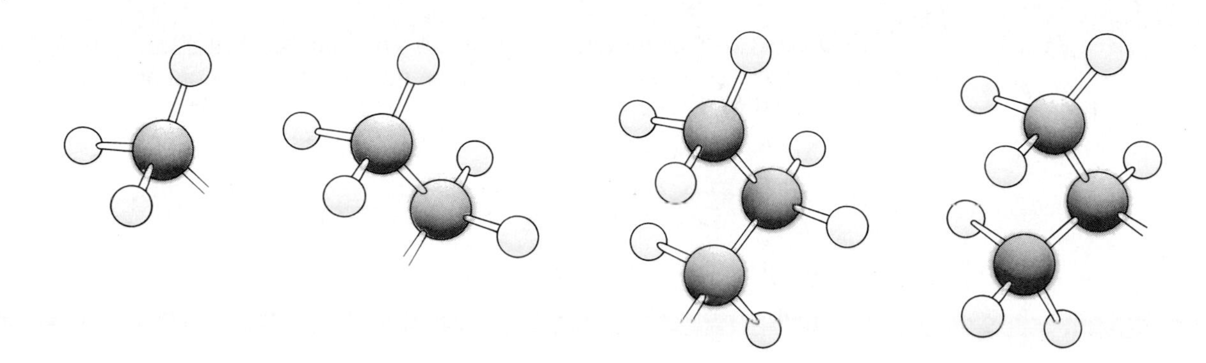

Figure 10.8
Ball-and-stick models of the alkyl groups derived from methane, ethane, and propane.

The following IUPAC rules are used to name this branched-chain alkane:

Step 1 Find the longest chain of carbon atoms and name it as the **parent chain.** In our example, the parent chain contains four carbon atoms and is named *butane.*

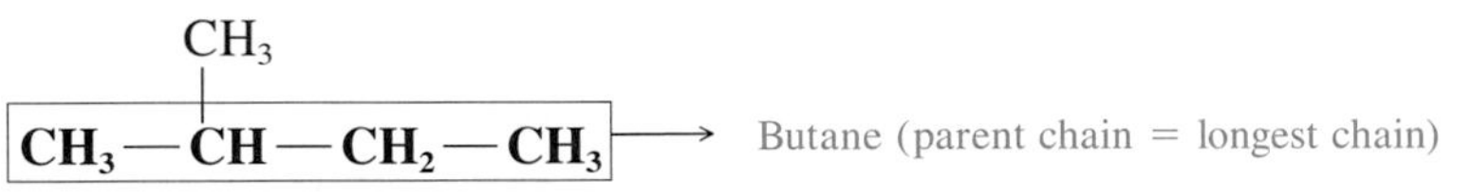

Step 2 Locate and name each substituent attached to the parent chain. In this example, there is a CH_3— attached. Its name, methyl, is placed *in front* of the name of the parent chain.

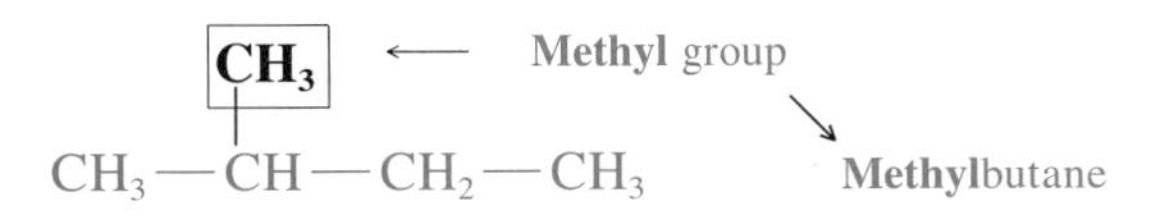

Step 3 Number the parent chain starting from the end of the chain nearest the branch. *If the first alkyl group is nearest the left end of the chain, number from left to right. If it is nearer the right end, number the chain from right to left.*

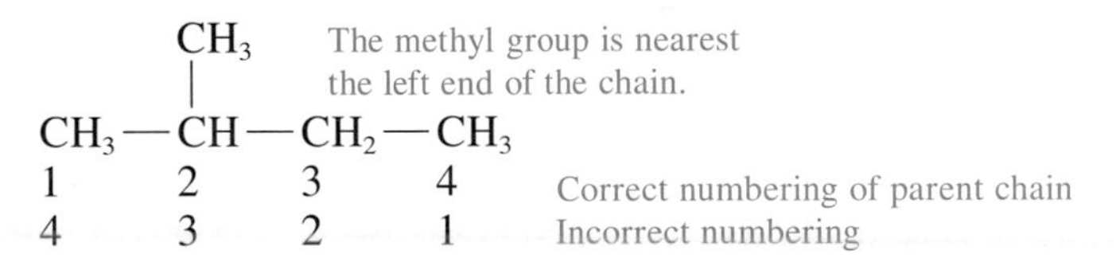

Step 4 Show the location of the substituent by giving it the number of the carbon it is attached to in the parent chain. In our example, the methyl group is attached to the carbon 2 in the chain. A hyphen separates the number from the name of the alkyl group. If the chain were incorrectly numbered in step 2, the number for the alkyl group (3) would not be the lowest possible number and the resulting name would be incorrect.

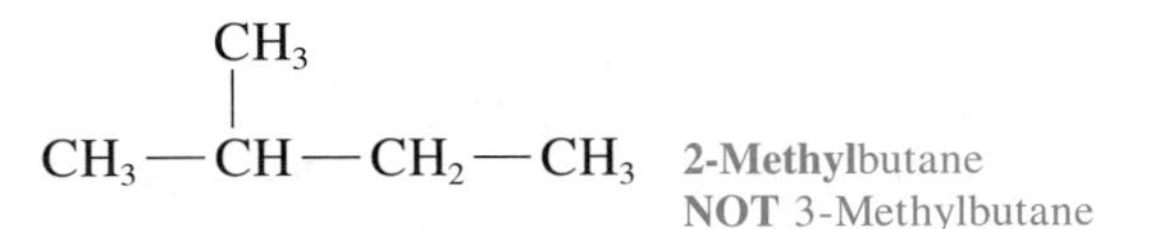

2-Methylbutane
NOT 3-Methylbutane

Step 5 If two or more identical substituents are attached to the parent chain, use a prefix, *di, tri, tetra,* and so on, to indicate the number of times that particular branch occurs.

Number of Identical Substituents	*Prefix*
two	di
three	tri
four	tetra
five	penta

All the identical branches must have numbers, even if they are attached to the same carbon. Number the chain from the end of the carbon chain that gives the lowest combination of numbers, using commas to separate the numbers.

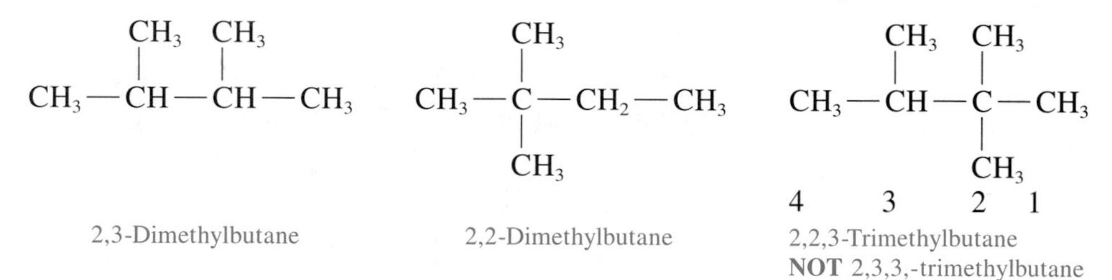

2,3-Dimethylbutane 2,2-Dimethylbutane 2,2,3-Trimethylbutane **NOT** 2,3,3,-trimethylbutane

Once you start numbering a carbon chain with two or more branches, be sure to continue in that same direction to the other end of the chain.

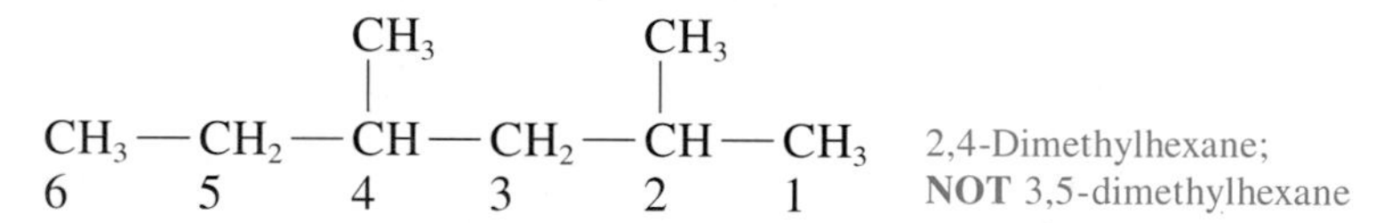

2,4-Dimethylhexane; **NOT** 3,5-dimethylhexane

Step 6 When there are two or more different substituents attached to the carbon chain, their names are listed in alphabetical order with hyphens separating each number and name. Prefixes for identical branches (di, tri, and so on) are not used in deciding alphabetical order.

Examples of Alphabetical Order of Alkyl Substituents

Ethyl, methyl

Triethyl, dimethyl

Ethyl, isopropyl, dimethyl

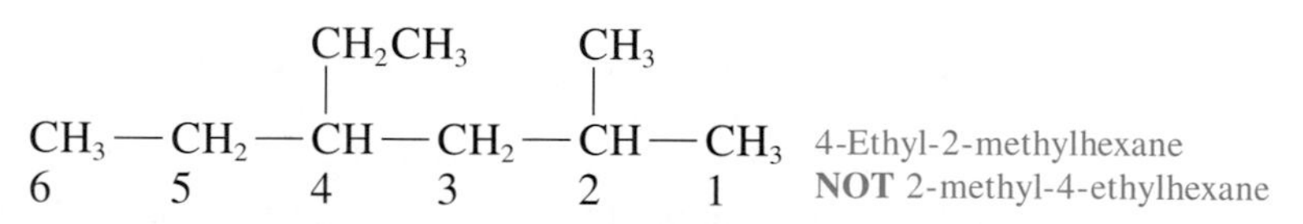

4-Ethyl-2-methylhexane **NOT** 2-methyl-4-ethylhexane

Step 7 If two different branches can be given the same number, assign the lower number to the branch that is first alphabetically.

CH_3 (on C5) CH_2CH_3 (on C3)

$CH_3CH_2CHCH_2CHCH_2CH_3$

1 2 3 4 5 6 7

7 6 5 4 3 2 1 ⟵ Numbering to list ethyl first alphabetically

3-Ethyl-5-methylheptane **NOT** 3-methyl-5-ethylheptane

Be sure that you always find the longest continuous carbon chain. Watch out: The longest chain need not be the most obvious horizontal one. Look at the following structure:

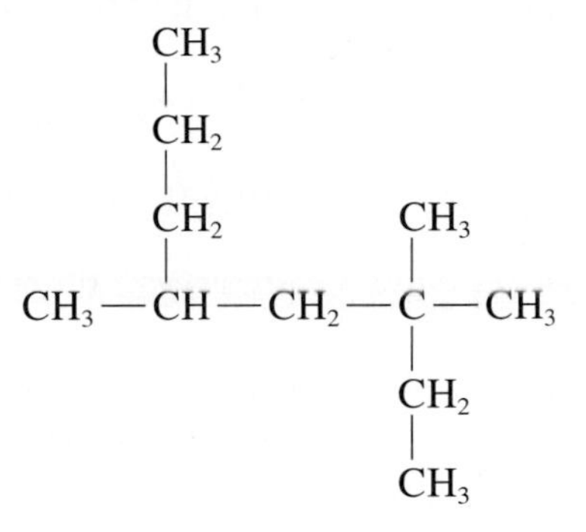

How many carbon atoms are in the parent chain? If you said five, look again. There is a chain of eight carbon atoms that goes around some corners!

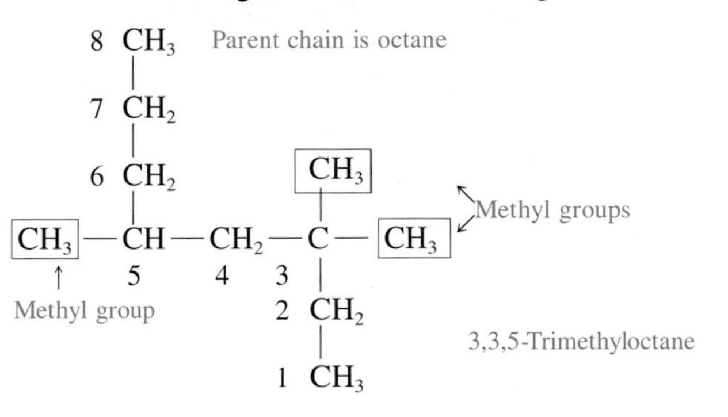

Summary: IUPAC Rules for Naming Branched Alkanes

Step 1 Give an alkane name to the longest chain of carbon atoms.

Step 2 Determine the name(s) of substituents attached to the parent chain.

Step 3 Number the parent chain from the end nearest the first attached group.

Step 4 Give the location of each substituent and its name in front of the alkane name.

Step 5 For identical substituents, attach a prefix of *di, tri, tetra*, and so on to the alkyl name.

Step 6 For two or more different substituents, number and list their names in alphabetical order.

Step 7 If the carbon number can be the same for two different branches, give the lower number to the substituent that comes first alphabetically.

Naming Some Other Groups

During our study of organic compounds, we will encounter compounds that contain other substituents such as -Cl, -NO_2, or -NH_2 attached to the parent

Table 10.8
Names of Some Common Substituents

Substituent	Name
$-NH_2$	Amino
$-Br$	Bromo
$-Cl$	Chloro
$-F$	Fluoro
$-NO_2$	Nitro

chain. The names of some of these are listed in Table 10.8. As we proceed, we will name compounds with these substituents in a way very similar to the rules we used for the branched alkanes. We can practice naming a few of these compounds to show how the IUPAC system is used in the naming of other organic compounds.

$CH_3CH(NO_2)CH_3$ — 2-Nitropropane

$BrCH_2CH_2CH(Br)CH_3$ — 1,3-Dibromobutane

$CH_3CH(NH_2)CH_2CH(CH_3)CH_3$ — 2-Amino-4-methylpentane

Sample Problem 10.6
Writing IUPAC Names for Branched-Chain Alkanes

Name the following compounds using the IUPAC system:

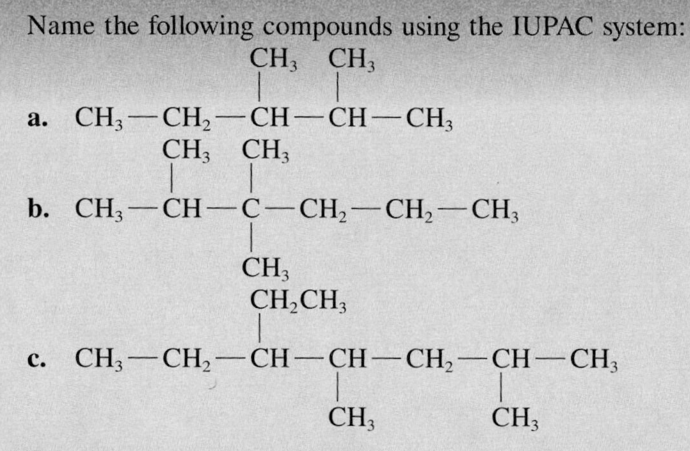

Solution

a. The longest chain of five carbon atoms is named *pentane.* There are two methyl groups on carbons 2 and 3 when the chain is numbered from *right to left.* A prefix of *di* indicates the two methyl branches. The compound is named 2,3-dimethylpentane.

b. The longest chain of six carbon atoms is named *hexane.* There are three methyl groups indicated by the prefix *tri,* one on the second carbon and two on the third carbon. The compound is named 2,3,3-trimethylhexane.

c. The longest chain of seven carbon atoms is named *heptane.* Counting from *right to left,* there are two *(di)* methyl groups on carbons 2 and 4, and one ethyl group on carbon 5. Listing the alkyl groups in alphabetical order gives the name 5-ethyl-2,4-dimethylheptane.

Study Check

Name the following compound using the IUPAC system:

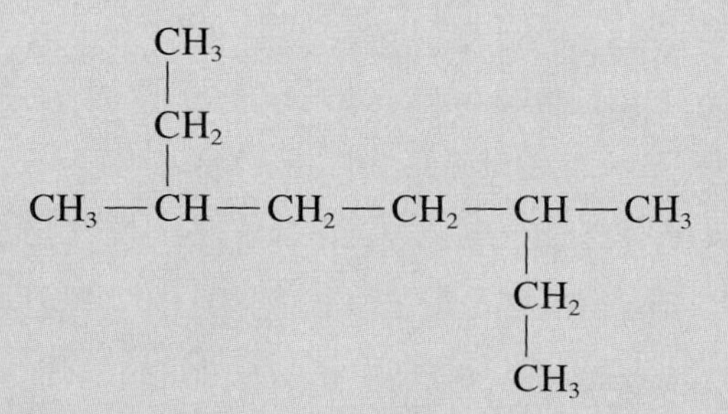

Drawing A Structural Formula

Suppose you are given the name 2,3-dimethylhexane and asked to write the structural formula or condensed structural formula of that compound. The name of the compound tells us the number of carbon atoms in the longest chain, the type of bonding, the name of any substituents and where they are attached to the parent chain. We can break down the name 2,3-dimethylhexane in the following way.

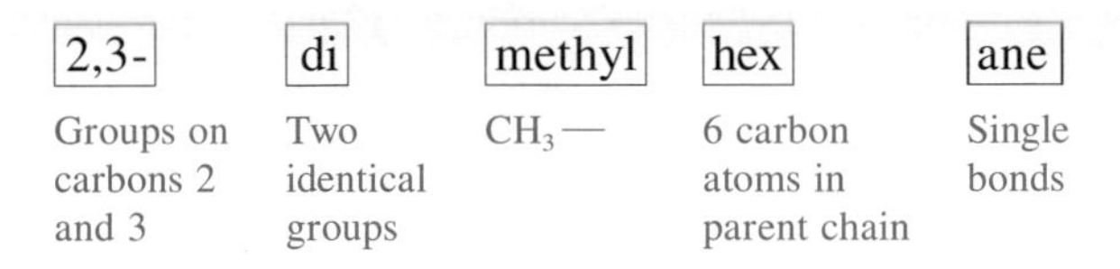

To draw the structural formula for the compound, we proceed as follows:

Step 1 Write the carbon atoms in the parent chain. The name hexane tells us that we need to draw a carbon chain containing six carbon atoms.

C—C—C—C—C—C **Hexane**

Step 2 Number the chain and attach the branches indicated. The numbers 2, 3 in the name indicate that methyl groups, CH_3—, are attached to carbon 2 and carbon 3.

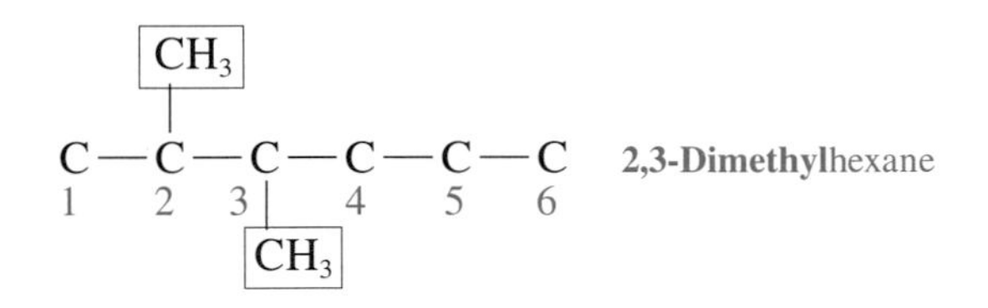

Step 3 Add hydrogen and write a condensed structural formula.

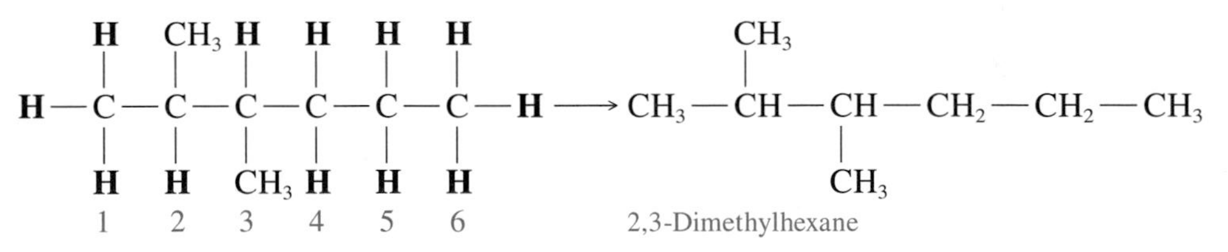

Sample Problem 10.7
Writing a Condensed Structural Formula

Draw the condensed structural formula for 3-ethyl-5-methylheptane.

Solution

Because the parent name is heptane, we start by drawing a chain of seven carbon atoms.

C—C—C—C—C—C—C

Number and attach an ethyl group, $CH_3CH_2—$, to carbon 3, and a methyl group, $CH_3—$, to carbon 5.

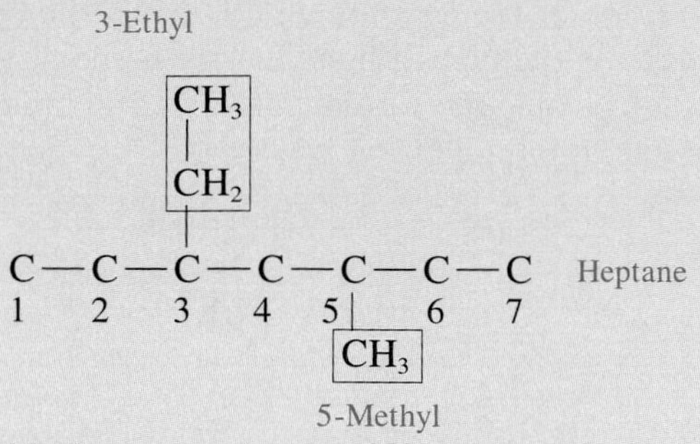

Add hydrogen and write the condensed structural formula.

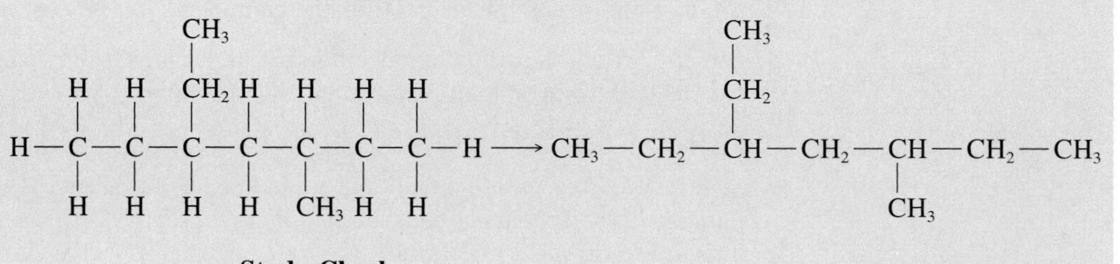

Study Check

Write the condensed structural formula of 2,2,4,4-tetratmethylhexane.

10.5 Cycloalkanes

Learning Goal

Give the name for a cycloalkane; draw the condensed structural and geometric formulas.

Up to now, we have only looked at chains of carbon atoms. However, carbon atoms can also form rings or cyclic structures called **cycloalkanes.** The simplest cycloalkane, cyclopropane, consists of a ring of three carbon atoms. (See Figure 10.9). A cycloalkane has two fewer hydrogen atoms than the corresponding alkane because there are no end carbons. Thus, there are eight hydrogens in the open chain of propane but only six hydrogens in the ring of cyclopropane.

Figure 10.9
A ball-and-stick model of the ring structure of cyclopropane.

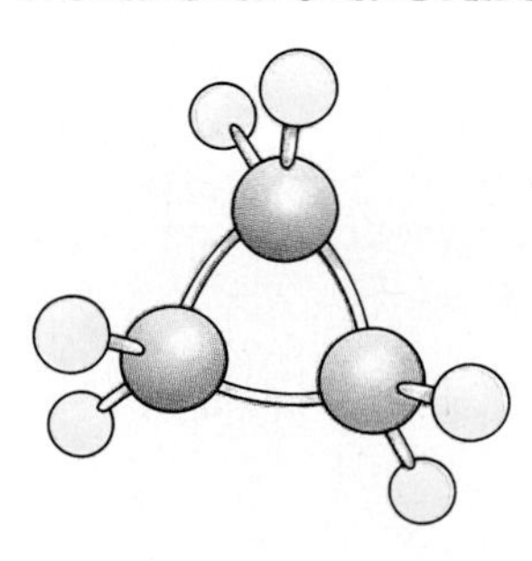

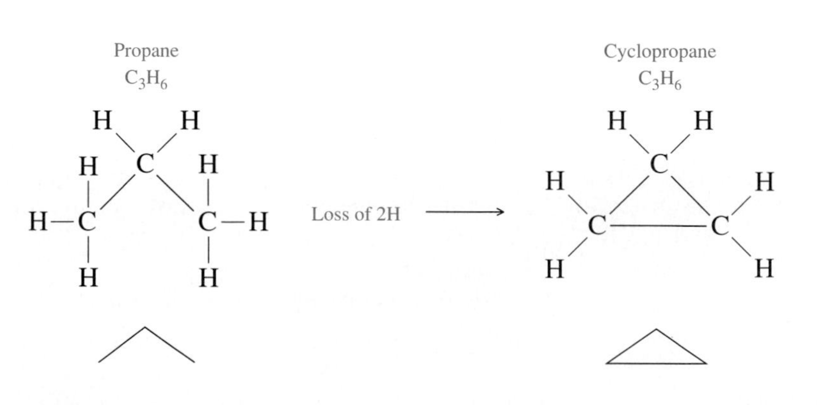

Table 10.9 ***Formulas of Some Common Cycloalkanes***

Condensed Structural Formula	Geometric Formula	Name
CH_2 / H_2C—CH_2		Cyclopropane
H_2C—CH_2 / H_2C—CH_2		Cyclobutane
CH_2 / H_2C CH_2 / H_2C—CH_2		Cyclopentane
CH_2 / H_2C CH_2 / H_2C CH_2 / CH_2		Cyclohexane

The names of cycloalkanes are similar to the names of the straight-chain alkanes with the same number of carbon atoms except that the prefix *cyclo* appears in front of the name. The structures of cycloalkanes are often drawn as simple geometric shapes. No hydrogen atoms are shown, although it is understood that they do exist in the actual compound. For example, cyclopropane can be depicted as a triangle. Each corner of the triangle represents a carbon atom that is attached to two other carbon atoms and two hydrogen atoms. More examples of the geometric shapes for cycloalkanes are given in Table 10.9.

Naming Cycloalkanes with Alkyl Groups

Cycloalkanes with branches are named in a way similar to the alkanes. The prefix *cyclo* is placed in front of the alkane name for the same number of carbon atoms in the ring. The name of an attached alkyl group is placed in front of the cycloalkane name. The ring is not numbered for one side group. If there are two or more side groups, the carbon ring is numbered in the direction that gives the lowest numbers to the branches starting at one of the substituents. You may count clockwise or counterclockwise around the ring, whichever gives the lowest combination of numbers to the substituents.

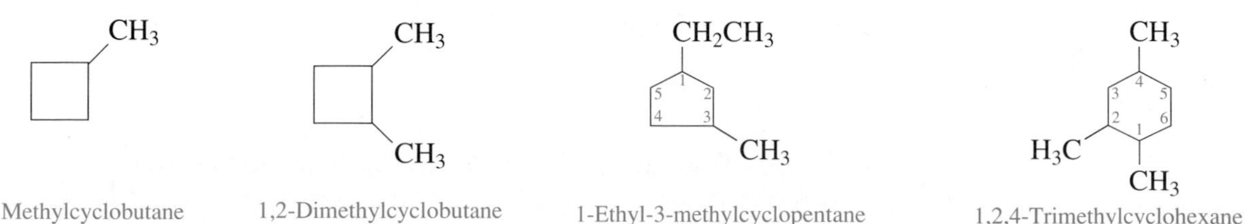

Methylcyclobutane

1,2-Dimethylcyclobutane

1-Ethyl-3-methylcyclopentane

1,2,4-Trimethylcyclohexane
NOT 1,4,5-Trimethylcyclohexane
NOT 1,4,6-Trimethylcyclohexane

Sample Problem 10.8
Naming Cycloalkanes

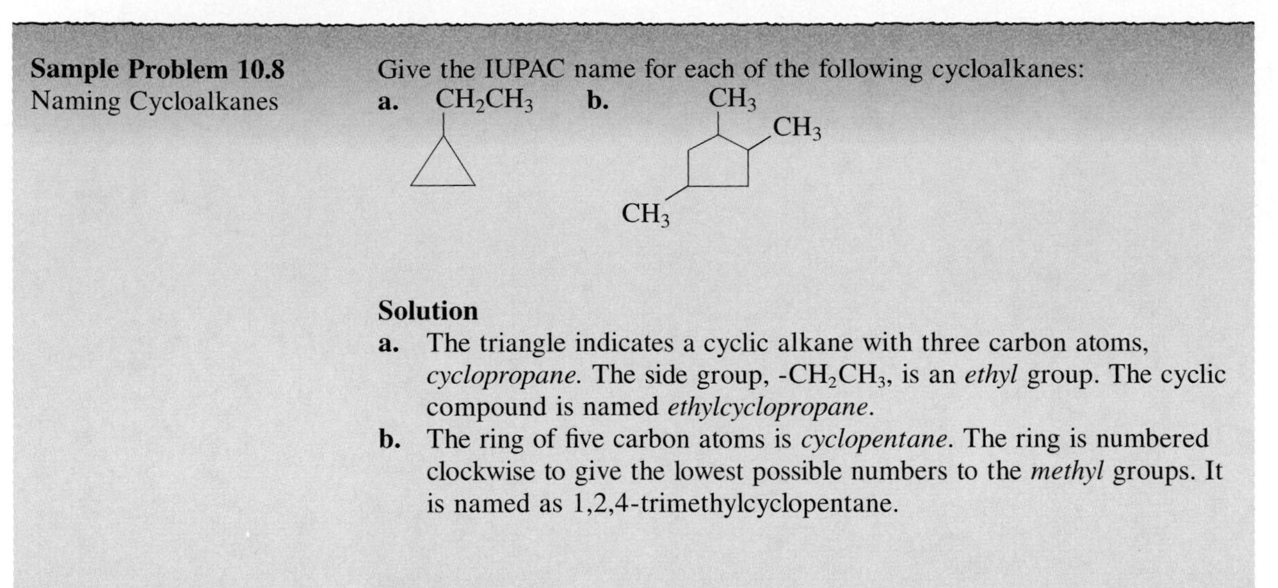

Give the IUPAC name for each of the following cycloalkanes:

a. CH_2CH_3 **b.** CH_3 CH_3 CH_3

Solution

a. The triangle indicates a cyclic alkane with three carbon atoms, *cyclopropane.* The side group, $-CH_2CH_3$, is an *ethyl* group. The cyclic compound is named *ethylcyclopropane.*

b. The ring of five carbon atoms is *cyclopentane*. The ring is numbered clockwise to give the lowest possible numbers to the *methyl* groups. It is named as 1,2,4-trimethylcyclopentane.

Study Check

What is the IUPAC name for the following cyclic alkane?

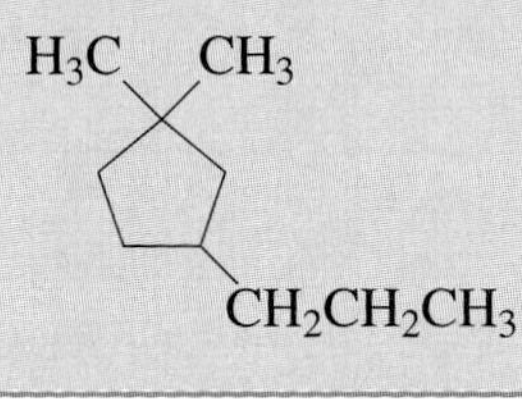

Sample Problem 10.9
Writing Condensed Structural Formulas of Cycloalkanes

Write the structure for 1,4-dimethylcyclohexane.

Solution

The parent name, *cyclohexane,* tells us that the parent chain is a ring of six carbon atoms. The rest of the name tells us there are methyl groups on carbons 1 and 4.

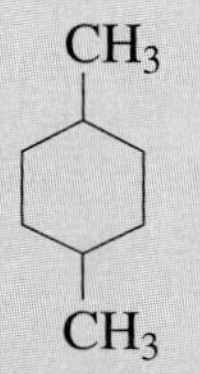

Study Check

What is the structural formula for 1-ethyl-2-methylcyclopentane?

HEALTH NOTE *Cycloalkanes and Haloalkanes as Anesthetics*

General anesthetics are compounds that are inhaled or injected to cause a loss of sensation so that surgery or other procedures can be done without causing pain to the patient. As nonpolar compounds they are soluble in the fats (also nonpolar) of the body including the nerve membranes where they decrease the ability of the nerve cells to conduct the sensation of pain. One of the early anesthetics, carbon tetrachloride, CCl_4, was also used as a cleaning agent to remove oils and grease. However, it was found to be toxic to the liver and its use was discontinued.

In some cases, cyclopropane, C_3H_6, is used as an anesthetic although it has a pungent odor, and forms explosive mixtures with oxygen in air. To prevent such reactions during surgery, helium is added to the inhaled mixture.

One of the most widely used general anesthetics today is Halothane, also called Fluothane. (See Figure 10.10.) It has a pleasant odor, is nonexplosive, has few side effects, undergoes few reactions within the body, and is eliminated quickly.

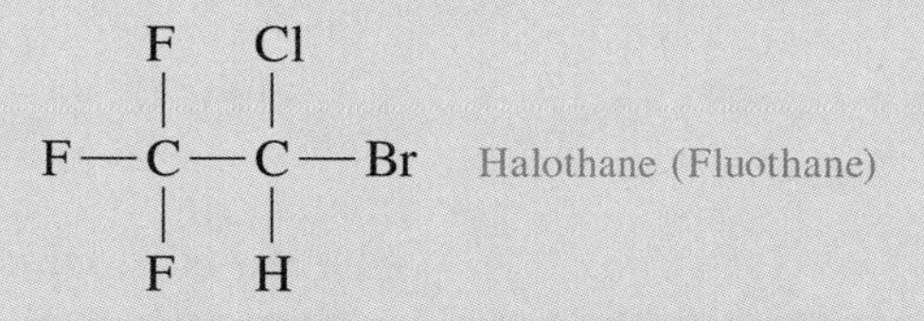

For minor surgeries, a local anesthetic such as chloroethane (ethyl chloride) may be applied to an area of the skin where it evaporates quickly to cool the skin and causes a loss of sensation.

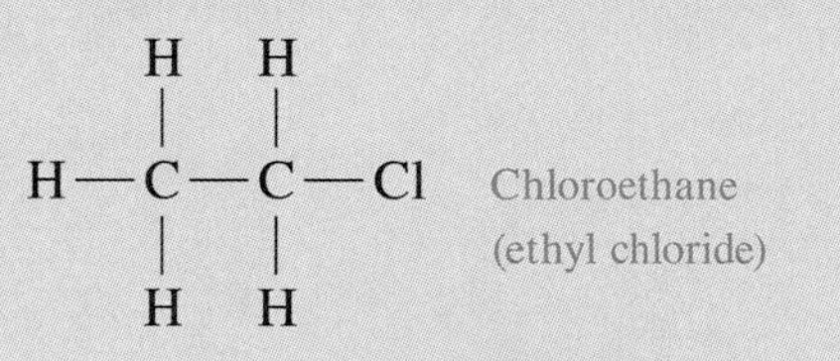

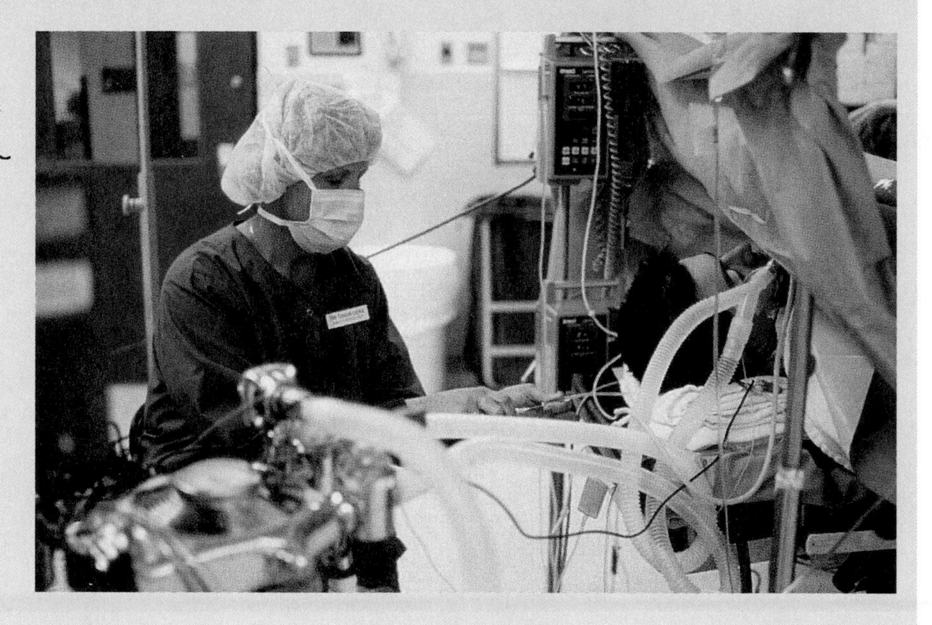

Figure 10.10
Anesthetics used during surgery produce a loss of all sensation.

10.6 Haloalkanes

Learning Goal
Give the name for a haloalkane and draw the condensed structural formula.

In a **haloalkane,** one or more hydrogen atoms of an alkane has been replaced by a halogen atom, F, Cl, Br, or I. In the IUPAC System, the halogen atom is numbered and named in front of the parent alkane using the following prefixes:

F fluoro Cl chloro Br bromo I iodo

Haloalkanes

Characteristic groups:	halogen atom: F, Cl, Br, I
IUPAC:	haloalkane
Common name:	alkyl halide

Many of the common names for simple haloalkanes were given before IUPAC rules were established. Then they are named as *alkyl halides* by naming the alkyl group attached to the halogen atom. (Throughout our discussion of organic chemistry, common names still in use will be placed in parentheses under the IUPAC names.)

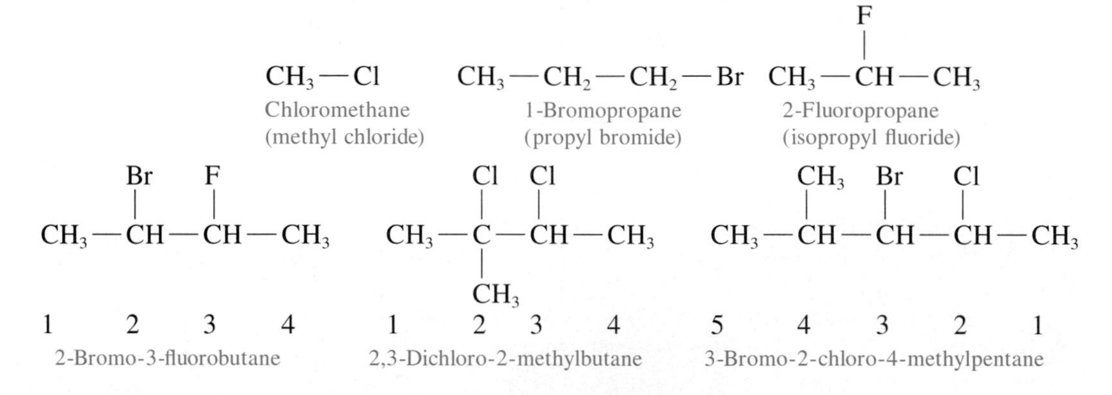

Sample Problem 10.10
Naming Haloalkanes

Freon 11 and Freon 12 are compounds that have been widely used as refrigerants in home and car air conditioners and as propellants for aerosol products. They are also known as CFCs or chlorofluorocarbons, a group of compounds known to destroy the ozone in the upper atmosphere. By 2010, many countries including the United States will stop production of CFCs. What are the IUPAC names for Freon 11 and Freon 12?

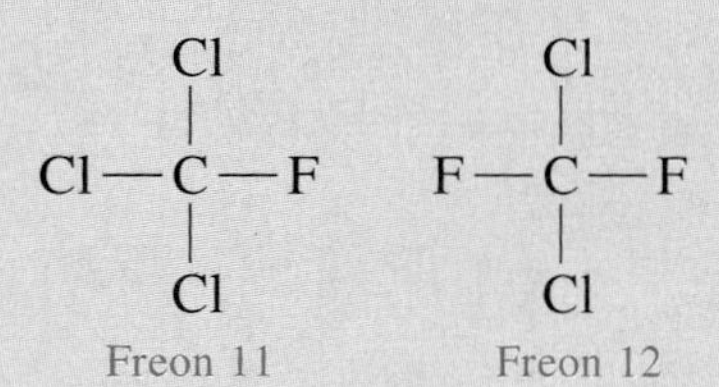

Solution

Freon 11, trichlorofluoromethane; Freon 12, dichlorodifluoromethane.

Study Check

As mentioned in a health note, Halothane or Fluothane is the commercial name of a haloalkane used as an anesthetic. How would it be named using the IUPAC system?

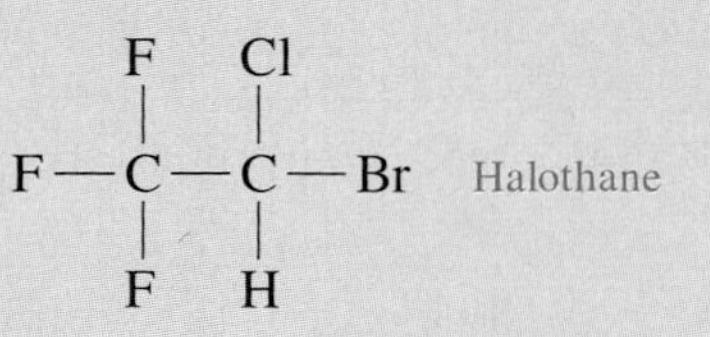

Environmental Note *Chlorinated Cyclic Compounds as Pesticides*

Many insecticides and pesticides are chlorinated cyclic hydrocarbons with commercial names such as aldrin, chlordane, DDT, dieldrin, and lindane. (See Figure 10.11.) They have been used successfully in protecting crops from destruction by insects. In the 1940s, DDT was so effective in controlling the mosquito and tsetse fly that spread malaria, typhus, and sleeping sickness, it was hailed as a miracle.

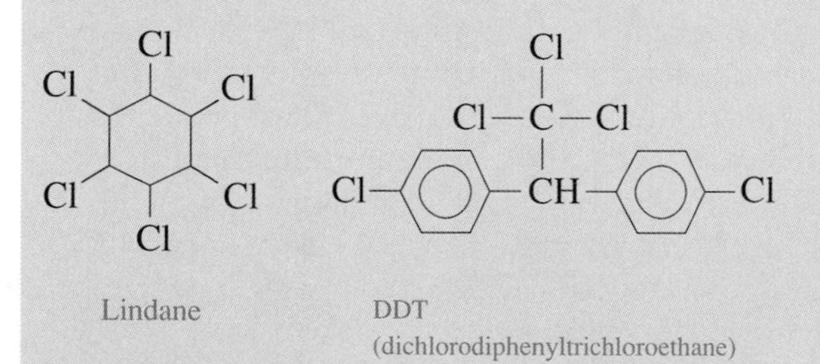

Lindane DDT (dichlorodiphenyltrichloroethane)

Because DDT is a nonpolar compound, it is soluble in nonpolar substances. In lakes and ponds where spraying of DDT occurred, small amounts of the pesticide were ingested and stored in the nonpolar lipids (fats) of microorganisms. Because these microorganisms serve as food for larger organisms and small fish, DDT became more concentrated in the body fat of larger creatures. Predatory birds and animals high in the food chain accumulated the greatest amounts of DDT. Eventually elevated DDT levels affected their reproduction and contributed to the decline and near extinction of some species of hawks, eagles, falcons, and brown pelicans. Many of the chlorinated hydrocarbons, including DDT, have now been banned from use in widespread spraying and the populations of some nearly extinct species have begun to rise again.

A major difficulty with chlorinated hydrocarbons is that they are nonbiodegradable; there are no natural pathways to break them down. They can be stored in human adipose tissue (body fat) and have been found in human breast milk. When they are absorbed by the human body, they act as poisons. Symptoms of pesticide poisoning in humans include dizziness, weakness, kidney damage, and nervous system excitability. Respiratory failure, coma, and death may follow.

Figure 10.11
Snow peas are sprayed with pesticides.

ENVIRONMENTAL NOTE *Ozone Depletion and CFCs*

From 10 to 30 miles above the earth, ultraviolet (UV) radiation produces highly reactive oxygen atoms that react with oxygen, producing ozone (O_3). The ozone in the upper atmosphere absorbs a large portion of the UV rays of the sun and acts as a protective shield for plants and animals on earth. The ultraviolet radiation that does get through can produce sunburn, cataracts, and skin cancers.

Normally, there is a balance between the formation and decomposition of ozone in the atmosphere. However, since the early 1970s scientists have become concerned that certain compounds entering the atmosphere are threatening the stability of the ozone layer. These compounds, called chlorofluorocarbons (CFCs), are the halogenated alkanes that are used as propellants for hair sprays, paints, deodorants, and refrigerants in air conditioners. Although they appear to be unreactive in the lower atmosphere, it has become evident that in the upper atmosphere CFCs decompose in the presence of UV light to produce highly reactive chlorine atoms and other products.

$$\underset{\text{Freon (CFC)}}{CFCl_3} \xrightarrow{\text{Sunlight (UV)}} Cl\cdot + \text{other products}$$

The reactive chlorine atoms break down ozone molecules. It has been estimated that one chlorine atom can destroy as many as 100,000 ozone molecules.

$$2O_3 \xrightarrow{Cl} 3O_2 \quad \text{Ozone destroyed}$$

Since 1985, researchers have observed the thinning of the ozone layer over the South Pole. In some areas as much as 50% of the ozone has been depleted, and at certain times of the year an ozone hole appears. The concern of scientists is that a thinning of the ozone layer over more populated areas will increase the amount of unfiltered UV radiation received by plants and animals, with adverse effects. Already there are reports of an increase in cases of UV retinal damage in people living in the southern regions. (See Figure 10.12.)

By the early 1980s, CFCs were banned by the United States for use in spray containers. By 1987, 24 nations had agreed to cut back on CFC use. Chemical companies are working on substitutes to CFCs that are not damaging to the ozone, but so far they appear to be costly. However, that may be the price for retaining the life-protecting ozone layer.

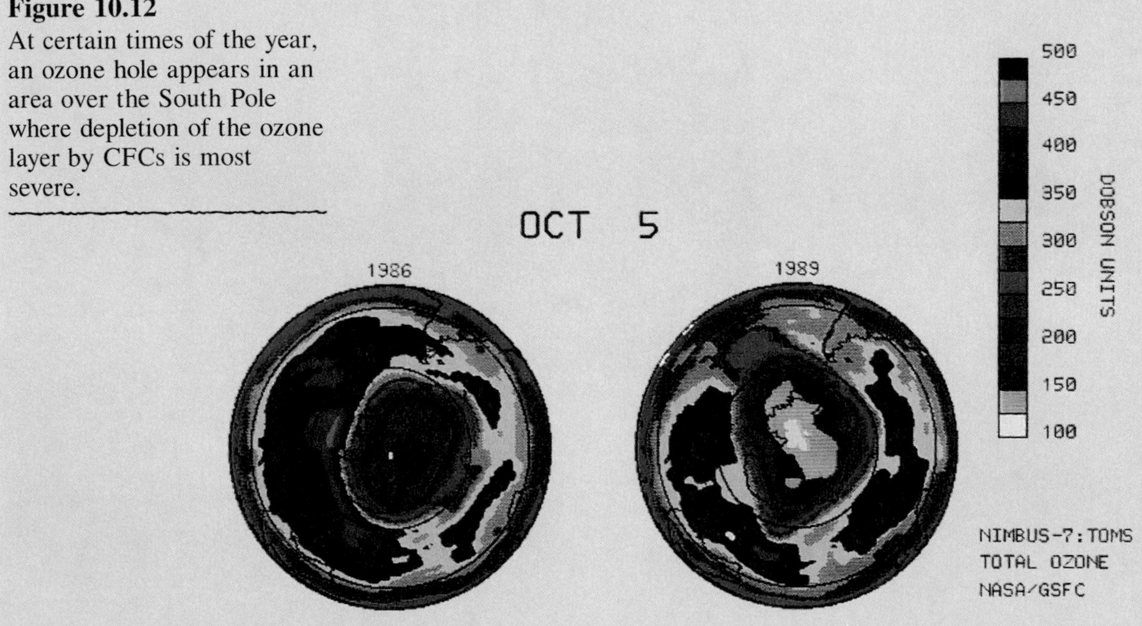

Figure 10.12
At certain times of the year, an ozone hole appears in an area over the South Pole where depletion of the ozone layer by CFCs is most severe.

10.7 Properties of Alkanes

Learning Goal

Describe some properties of alkanes.

The first four alkanes, methane, ethane, propane, and butane, are gases at room temperature and pressure and act as anesthetics if inhaled. They are widely used as heating fuels. Alkanes having 5 to 17 carbon atoms are liquids of room temperature; alkanes having 18 or more carbon atoms are waxy solids.

The bonds of the alkanes, C—C and C—H, are nonpolar, which means that alkanes are nonpolar. Thus, alkanes are soluble in nonpolar solvents such as other alkanes and gasoline. The liquid alkanes float on water because they are less dense and insoluble in water. In oil spills, most of the alkanes remain on the surface, where they spread over a large area.

The alkanes are among the least reactive of any of the organic compounds. The single covalent bonds of alkanes are difficult to break, a feature that makes them chemically inert with most chemical substances. However, alkanes do burn readily in oxygen.

Combustion

The growth of civilization was strongly influenced by the discovery of fire or combustion. Energy produced by burning wood was used to make pottery and glass, extract metals, make weapons, and forge tools. At high temperatures, an organic compound undergoes **combustion** when it reacts rapidly with oxygen to produce carbon dioxide, water and heat. The methane gas we use to cook our foods and heat our homes, the propane in portable heaters and gas barbecues, and the gasoline that powers our cars are all alkanes that undergo combustion. (See Figure 10.13.) In combustion, a great amount of heat is also released. We can write the following equations for the combustion of methane and propane:

Figure 10.13
The propane fuel in the tank undergoes combustion to cook the food on the barbecue.

Combustion

$$\textit{Alkane} + O_2 \longrightarrow CO_2 + H_2O + \textit{heat}$$

Examples of Combustion:

$$CH_4 + 2O_2 \longrightarrow CO_2 + 2H_2O + \text{heat}$$
$$C_3H_8 + 5O_2 \longrightarrow 3CO_2 + 4H_2O + \text{heat}$$

In the cells of the body, glucose acts as a fuel, reacting with the oxygen we take in to produce heat and energy. Although the metabolic combustion of glucose is a slower process that occurs in many small steps, its end products are carbon dioxide, water, and energy—the same as in the combustion reactions we just looked at.

$$C_6H_{12}O_6 + 6O_2 \longrightarrow 6CO_2 + 6H_2O + \text{heat}$$

Sample Problem 10.11
Writing an Equation for Combustion

Gasoline is a mixture of liquid alkanes including pentane, C_5H_{12}. Write and balance the equation for the combustion of pentane in the engine of a car.

Solution
The reactants of this combustion reaction are pentane and oxygen. The products of the combustion are carbon dioxide and water. The unbalanced equation can be written.

$$C_5H_{12} + O_2 \longrightarrow CO_2 + H_2O$$

To write a balanced equation, we match the atoms of C, H, and O in the reactants with those in the products.

$$C_5H_{12} + 8O_2 \longrightarrow 5CO_2 + 6H_2O$$

Study Check
Write a balanced equation for the complete combustion of ethane, C_2H_6

HEALTH NOTE *Incomplete Combustion*

You already know that it is dangerous to use gas appliances in a closed room or to run a car in a closed garage where ventilation is not adequate. A limited supply of oxygen may cause an incomplete combustion of the fuel, which produces carbon monoxide and water:

$$\underset{\text{Limited oxygen supply}}{2CH_4 + 3O_2} \longrightarrow \underset{\text{Carbon monoxide}}{2CO} + 4H_2O + \text{heat}$$

Carbon monoxide is dangerous when it is inhaled because it passes into the blood stream where it attaches to hemoglobin. Normally, the hemoglobin carries oxygen from the lungs to the tissues of the body and brings carbon dioxide (CO_2) back to the lungs, where it is released. However, carbon monoxide (CO) binds so tightly to hemoglobin that it is not easily released. The sites where oxygen would bind to hemoglobin are not available. The hemoglobin is unable to transport oxygen, causing oxygen starvation and death if the victim is not treated immediately.

Halogenation of Alkanes (Substitution)

In the presence of heat or light, an alkane reacts with the halogens bromine or chlorine. A hydrogen atom in a C—H bond is replaced with an atom of bromine or chlorine to give a C—Br or C—Cl bond. A reaction in which one atom replaces another is called a **substitution** reaction; replacement by a halogen is called a **halogenation** reaction. The products of a halogenation are a haloalkane and a hydrogen halide.

Halogenation of Alkanes
Alkane + halogen (Cl_2 or Br_2) $\longrightarrow$ haloalkane and hydrogen halide

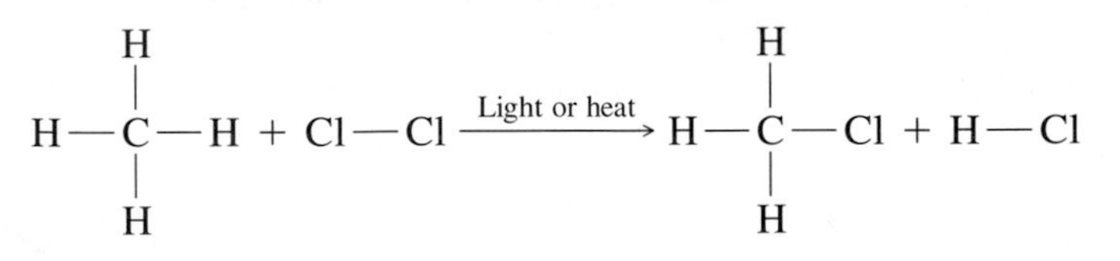

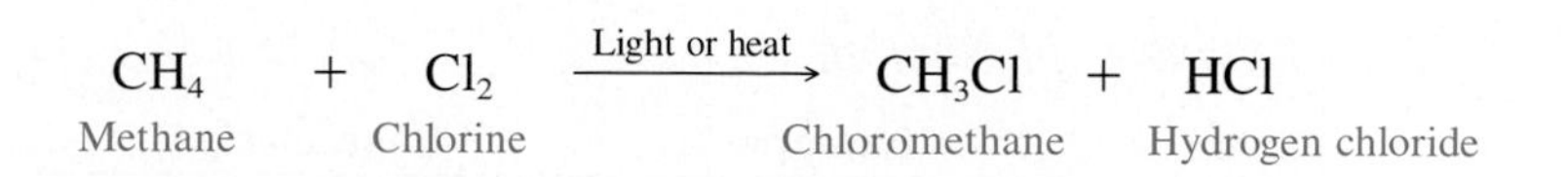

$$CH_4 + Cl_2 \xrightarrow{\text{Light or heat}} CH_3Cl + HCl$$

Methane, Chlorine, Chloromethane, Hydrogen chloride

One of the difficulties with halogenation is that the reaction does not stop with the substitution of one hydrogen atom. Other hydrogen atoms may also be replaced with halogen atoms resulting in a mixture of products as seen below.

$$CH_3Cl + Cl_2 \xrightarrow{\text{Light or heat}} CH_2Cl_2 + HCl$$

Dichloromethane

$$CH_2Cl_2 + Cl_2 \xrightarrow{\text{Light or heat}} CHCl_3 + HCl$$

Trichloromethane (chloroform)

$$CHCl_3 + Cl_2 \xrightarrow{\text{Light or heat}} CCl_4 + HCl$$

Tetrachloromethane (carbon tetrachloride)

When larger alkanes are used, the halogen atoms may substitute for different hydrogen atoms. For example, one-step bromination of propane produces both 1-bromopropane and 2-bromopropane.

$$CH_3CH_2CH_3 + Br_2 \xrightarrow{\text{Light or heat}} CH_3CH_2CH_2Br + CH_3CH(Br)CH_3$$

Propane; 1-Bromopropane (propyl bromide); 2-Bromopropane (isopropyl bromide)

Sample Problem 10.12
Halogenation of Alkanes

Write the condensed structural formula of the product formed when one hydrogen atom is replaced by a halogen atom in the following reactions:

a. $CH_3CH_3 + Br_2 \xrightarrow{\text{Light or heat}}$

b. □ $+ Cl_2 \xrightarrow{\text{Light or heat}}$

Solution

a. $CH_3CH_2Br + HBr$

b. 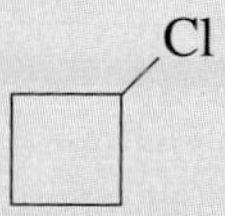 □–Cl $+ HCl$

Study Check

What are the IUPAC names of the monosubstituted products of the chlorination of butane?

Summary of Naming

Type	Example	Characteristic	Structure
Alkane	propane	single C—C, C—H bonds	$CH_3—CH_2—CH_3$
Cycloalkane	cyclopropane	carbon ring	△
Haloalkane	1-bromopropane	halogen atom	$CH_3—CH_2—CH_2—Br$

Summary of Reactions

Combustion

$$\text{Alkane} + O_2 \longrightarrow CO_2 + H_2O + \text{heat}$$

$$CH_4 + 2O_2 \longrightarrow CO_2 + 2H_2O + \text{heat}$$

Halogenation (substitution)

$$\text{Alkane} + \text{halogen} \longrightarrow \text{haloalkane} + \text{hydrogen halide}$$

$$CH_4 + Cl_2 \xrightarrow{\text{Light or heat}} CH_3Cl + HCl$$

Chapter Summary

10.1 Organic Compounds

Organic compounds are compounds of carbon that typically have covalent bonds, low melting and boiling points, burn vigorously, are nonelectrolytes, and are soluble in nonpolar solvents. The hydrocarbon family consists of compounds composed of carbon and hydrogen.

10.2 Alkanes

Alkanes are hydrocarbons that have only single bonds, C—C and C—H. Each carbon in an alkane always has four bonds arranged so that the bonded atoms are in the corners of a tetrahedron. A full structural formula shows a separate line for every bonded atom; a condensed structural formula depicts each carbon atom and its attached hydrogen atoms as groups. A molecular formula shows only the number of each type of atom in the compound but not the arrangement of atoms.

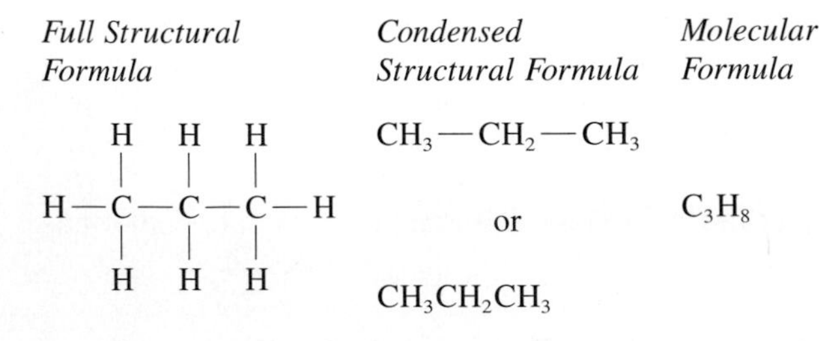

10.3 Structural Isomers

In an unbranched alkane, the carbon atoms are connected one after the other in a chain. A branched-chain alkane contains at least one alkyl branch or substituent. Structural isomers have the same molecular formula but differ in the sequence of atoms in each of their structural formulas.

10.4 The IUPAC System of Naming

The IUPAC system is a set of rules used to name organic compounds in a systematic manner. The name contains a stem that indicates the number of carbon atoms, and an ending that shows the family of the compound. For example, in the name *propane,* the stem *prop* indicates a chain of three carbon atoms and the ending *ane* indicates single bonds (alkane). Branches of carbon atoms attached to a parent chain are named as alkyl groups or alkyl substituents. In a branched alkane, the longest continuous chain is named as the parent alkane. The substituents attached to the chain are numbered and listed alphabetically in front of the name of the parent chain.

10.5 Cycloalkanes

In cycloalkanes, the carbon atoms of alkanes form a cyclic structure. They are named by placing the prefix *cyclo* in front of the name of the alkane with the same number of carbon atoms. Branches are named as alkyl groups and must be numbered if there are two or more on a ring.

10.6 Haloalkanes
When a hydrogen atom in an alkane is replaced by a halogen atom, F, Cl, Br, or I, the compound is a haloalkane. The halogen atoms are named as substituents (fluoro, chloro, bromo, iodo) attached to the alkane chain.

10.7 Properties of Alkanes
The alkanes are nonpolar, less dense than water, and mostly chemically unreactive, except they burn vigorously. Thus, they are widely used as fuels such as natural gas, methane, gasoline, and diesel fuels. In combustion, an alkane at a high temperature reacts rapidly with oxygen to produce carbon dioxide, water, and a great amount of heat. Halogenation, a substitution reaction, takes place in the presence of light or heat when halogen atoms replace one or more hydrogen atoms in an alkane.

Glossary of Key Terms

alkane A hydrocarbon that has only single carbon-to-carbon bonds.

alkyl group A side group derived from an alkane by removing one hydrogen atom. The ending of the alkane name is changed to *yl* to name it as an alkyl branch.

branch A side group of carbon and hydrogen attached to a parent carbon chain or ring.

branched alkane A hydrocarbon containing at least one alkyl substituent attached to the parent chain.

combustion A chemical reaction in which an alkane reacts with oxygen to produce CO_2, H_2O, and heat.

condensed structural formula A type of formula that illustrates the arrangement of the carbon atoms in an organic compound by grouping each carbon atom and its hydrogen atoms.

cycloalkane An alkane that exists as a ring or cyclic structure.

full structural formula A type of formula in which each bond in the compound is shown as a line, C—H or C—C, to give the arrangement of all the atoms in the compound.

haloalkane A type of alkane in which one or more hydrogen atoms has been replaced by a halogen atom.

halogenation A substitution reaction that occurs in the presence of light or heat in which halogen atoms replace one or more hydrogen atoms in an alkane.

hydrocarbon An organic compound consisting of only carbon and hydrogen atoms.

organic compounds Compounds that contain carbon in covalent bonds. Most are nonpolar with low melting and boiling points, insoluble in water, soluble in nonpolar solvents, and flammable.

parent chain The longest continuous chain of carbon atoms in a formula.

saturated hydrocarbon A compound of carbon and hydrogen that has only single carbon–carbon bonds (C—C).

structural isomer An organic compound that has the same molecular formula as another organic compound but a different structure.

substitution The replacement of one atom by another.

unbranched alkane An alkane consisting of carbon atoms that are connected in a continuous chain.

Chemistry at Home

1. Read the label of a bottle or container in the medicine cabinet or the package of a cosmetic. What are the names of some organic compounds contained in the substance?
2. Light a candle. Carefully hold a glass of cool water over the flame. What accumulates on the glass? What are the products of combustion? What observations tell you that a combustion reaction is occuring in a candle flame?
3. Use a wax crayon or some Vaseline to make a pattern on a piece of cotton fabric such as a white T-shirt or sheet. Soak the fabric in a container of a food coloring or dye. Remove the fabric and dry. What happened to the color where you applied the wax or Vaseline? Explain.

Answers to Study Checks

10.1 inorganic

10.2

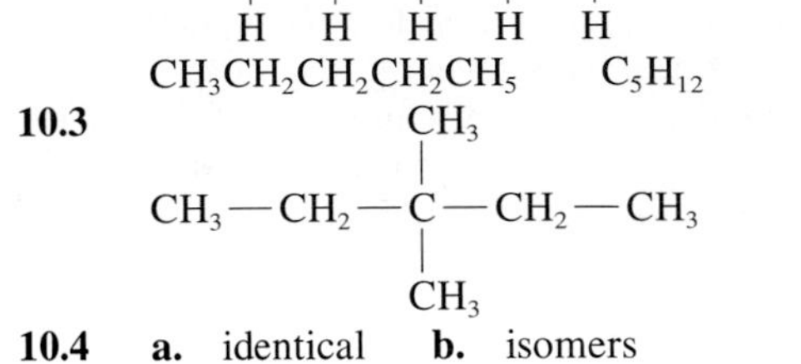

$CH_3CH_2CH_2CH_2CH_3$ C_5H_{12}

10.3

$$CH_3-CH_2-\underset{\displaystyle CH_3}{\overset{\displaystyle CH_3}{\overset{|}{\underset{|}{C}}}}-CH_2-CH_3$$

10.4 **a.** identical **b.** isomers

10.5 $CH_3CH_2CH_2CH_2CH_2CH_2CH_3$, heptane

10.6 3,6-dimethyloctane

10.7

$$CH_3-\underset{\displaystyle CH_3}{\overset{\displaystyle CH_3}{\overset{|}{\underset{|}{C}}}}-CH_2-\underset{\displaystyle CH_3}{\overset{\displaystyle CH_3}{\overset{|}{\underset{|}{C}}}}-CH_2CH_3$$

10.8 1,1-dimethyl-3-propylcyclopentane

10.9 CH_2CH_3 CH_3

10.10 2-bromo-2-chloro-1,1,1-trifluoroethane

10.11 $2C_2H_6 + 7O_2 \longrightarrow 4CO_2 + 6H_2O$

10.12 1-chlorobutane and 2-chlorobutane

Problems

Organic Compounds *(Goal 10.1)*

10.1 How does the historical view of organic chemistry differ from the modern definition of *organic chemistry?*

10.2 Distinguish between the following terms:
a. hydrocarbon and alkane
b. alkane and cycloalkane

10.3 Classify the following compounds as organic or inorganic.
a. $CaCl_2$ **b.** CH_3CH_2Cl
c. C_4H_{10} **d.** $CoCl_3$
e. CH_4

10.4 Classify the following compounds as organic or inorganic.
a. CH_3CH_3 **b.** □
c. Cr_2O_3 **d.** Na_2SO_4
e. $C_{10}H_{22}$

10.5 Match the following physical and chemical properties with the compounds propane, C_3H_8, or potassium chloride, KCl.
a. melts at −190°C
b. burns vigorously
c. melts at 770°C
d. conducts electricity when it dissolves in water
e. is a gas at room temperature

10.6 Match the following physical and chemical properties with the compounds cyclohexane, C_6H_{12}, or calcium nitrate, $Ca(NO_3)_2$.
a. floats on the surface of water (density = 0.78 g/mL)
b. melts above 500°C
c. insoluble in water
d. density of 2.5 g/mL
e. produces ions in water

Alkanes *(Goal 10.2)*

10.7 Give the condensed structural formula for each of the following full structural formulas:

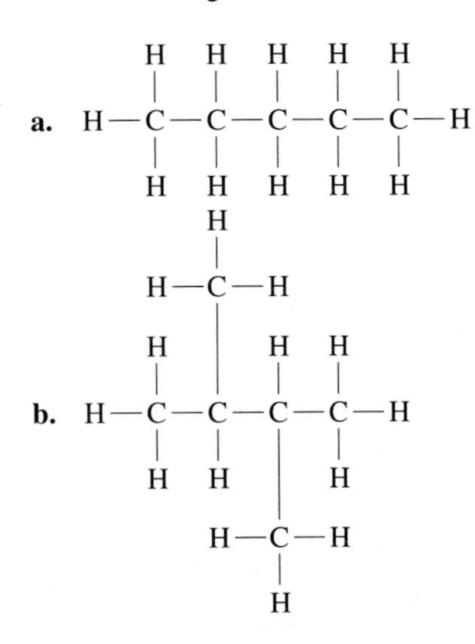

10.8 Give the condensed structural formula for each of the following full structural formulas:

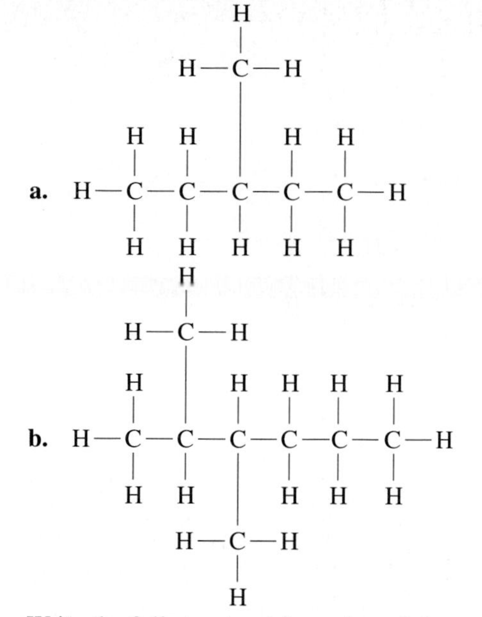

10.9 Write the full structural formula and the condensed structural formula by adding hydrogen atoms to each of the following alkanes:

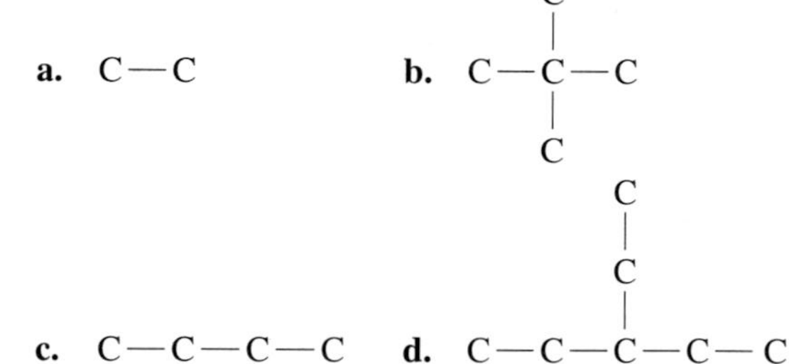

10.10 Write the full structural formula and the condensed structural formula by adding hydrogen atoms to each of the following alkanes:

a. C—C—C

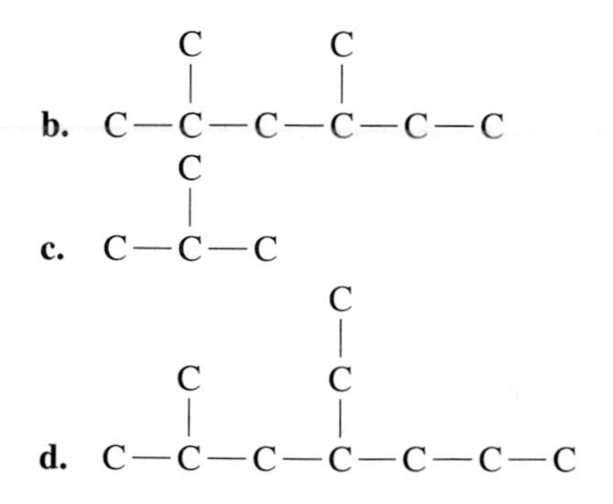

Structural Isomers *(Goal 10.3)*

10.11 What is meant by an unbranched-chain alkane?

10.12 How does an unbranched alkane differ from a branched alkane?

10.13 Classify the following as unbranched or branched alkanes:

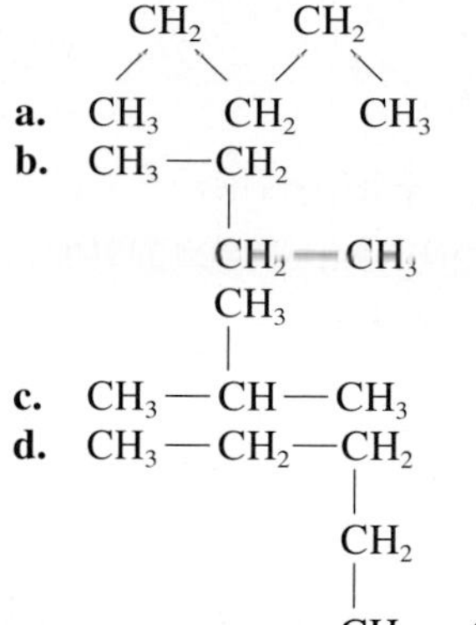

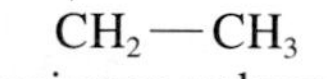

10.14 Classify the following as unbranched- or branched-chain alkanes:

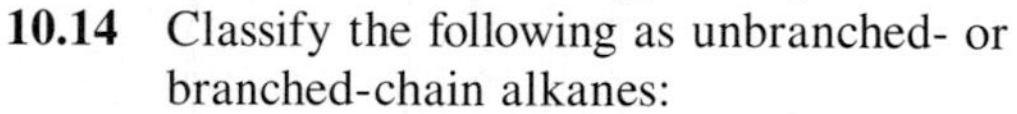

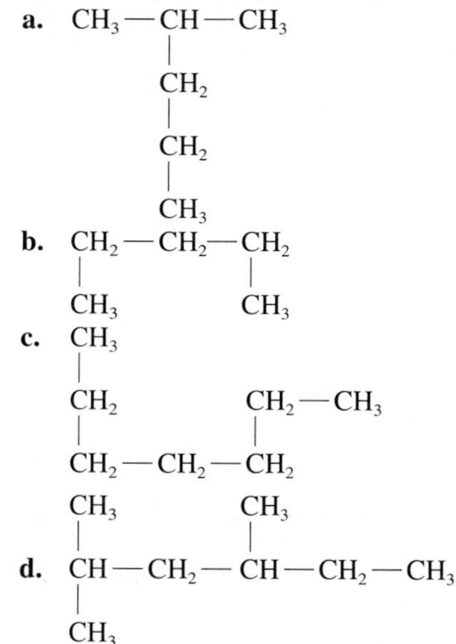

10.15 What are structural isomers?

10.16 How would you determine whether two condensed structural formulas represented structural isomers?

10.17 Which of the following pairs of condensed structural formulas represent structural isomers?

a.
```
CH3—CH2—CH3   and   CH3
                    |
                    CH2—CH3
```

b.
```
CH3                          CH3
|                            |
CH2—CH2—CH3     and     CH3—CH—CH3
```

c.
```
     CH3                       CH3          CH3
     |                         |            |
CH3—CH—CH2—CH3      and        CH2—CH2—CH—CH3
```

d.
```
CH3  CH3          and      CH3—CH2
|    |                         |
CH2—C—CH3                      CH2—CH2—CH2
     |                                 |
     CH3                               CH3
```

10.18 Which of the following pairs of formulas represent structural isomers?

a.
```
     CH3                  CH3
     |                    |
CH3—C—CH3   and   CH3—CH—CH3
     |
     CH3
```

b.
```
     CH3 CH3                         CH3
     |   |                           |
CH3—CH—CH—CH2—CH3       and     CH3—CH—CH2—CH—CH3
                                            |
                                            CH3
```

c.
```
            CH3                     CH3
            |                       |
CH3—CH2—CH—CH3         and     CH3—CH—CH2—CH3
```

d.
```
CH3         CH3                 CH3
|           |                   |
CH2—CH2—CH2         and    CH3—C—CH3
                                |
                                CH3
```

The IUPAC System of Naming *(Goal 10.4)*

10.19 Give the IUPAC name for each of the following alkanes:

a.
```
   H  H
   |  |
H—C—C—H
   |  |
   H  H
```

b.
```
     CH3
     |
CH3—CH—CH2—CH3
```

c.
```
CH3
|
CH2—CH2—CH2
            |
            CH2—CH3
```

d. $CH_3CH_2CH_2CH_2CH_2CH_2CH_2CH_3$

10.20 Give the IUPAC name for each of the following alkanes:

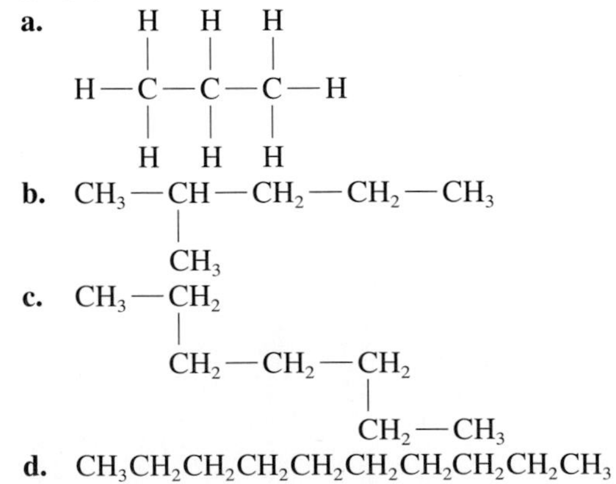

a.
```
   H  H  H
   |  |  |
H—C—C—C—H
   |  |  |
   H  H  H
```

b.
```
CH3—CH—CH2—CH2—CH3
     |
     CH3
```

c.
```
CH3—CH2
     |
     CH2—CH2—CH2
              |
              CH2—CH3
```

d. $CH_3CH_2CH_2CH_2CH_2CH_2CH_2CH_2CH_2CH_3$

10.21 Draw the structure of the following alkyl groups:
a. methyl **b.** propyl

10.22 Draw the structure of the following alkyl groups:
a. ethyl **b.** isopropyl

10.23 Distinguish between the following pairs of terms:
a. methane and methyl group
b. ethane and ethyl group

10.24 Distinguish between the following pairs of terms:
a. methyl group and ethyl group
b. propyl group and isopropyl group

10.25 Give the IUPAC name for each of the following alkanes:

```
       CH3
       |
a. CH3—CH—CH3

                    CH3
                    |
b. CH3—CH2—CH2—CH—CH3

       CH3  CH3
       |    |
c. CH3—CH—CH—CH2—CH3

d. CH3 CH3        CH3
   |   |          |
   CH2—CH—CH2—CH—CH3

e. CH3—CH—CH3
       |
       CH2—CH—CH2—CH2—CH3
           |
           CH3

                   CH3
                   |
f. CH3—CH2—CH—CH—CH3
           |
           CH2
           |
           CH3

                   CH3
                   |
                   CH2
                   |
                   CH2
                   |
g. CH3—CH2—CH2—CH—CH2
                       |
                       CH2
                       |
                       CH3
```

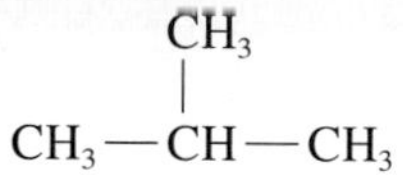

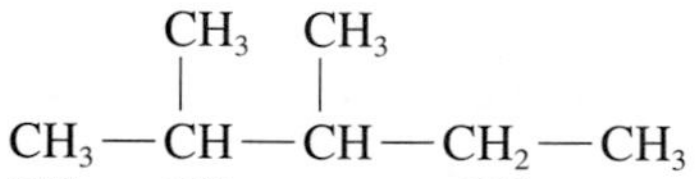

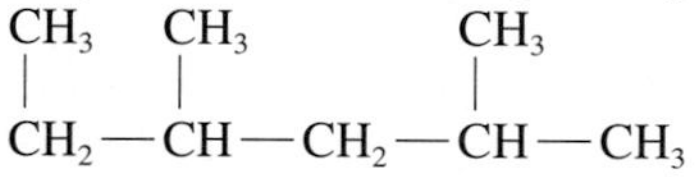

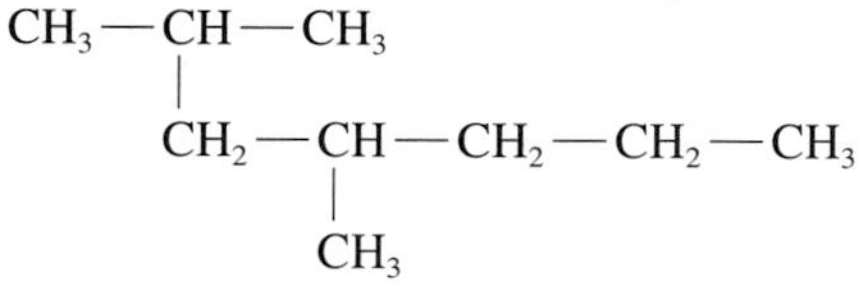

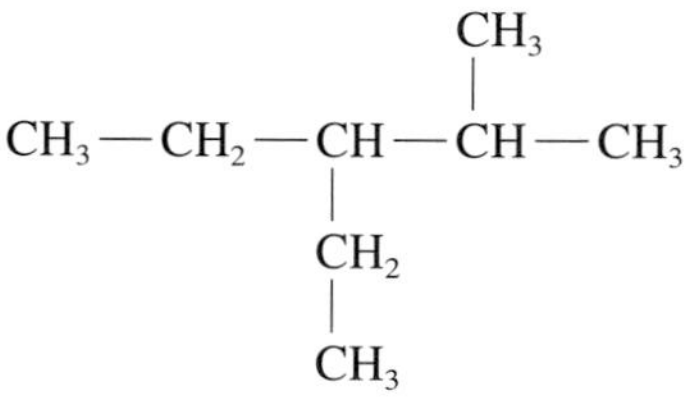

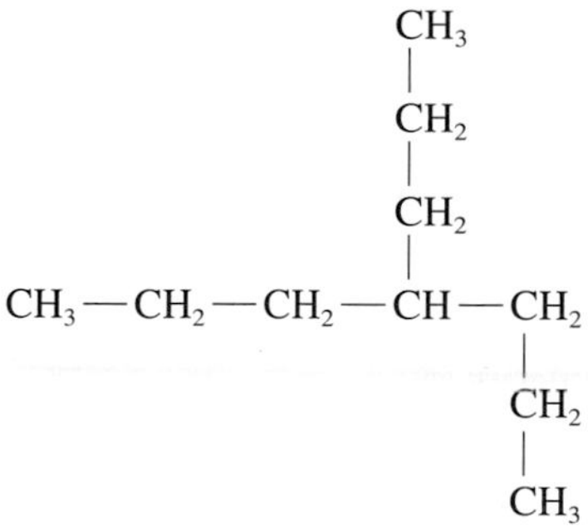

10.26 Give the IUPAC name for each of the following alkanes:

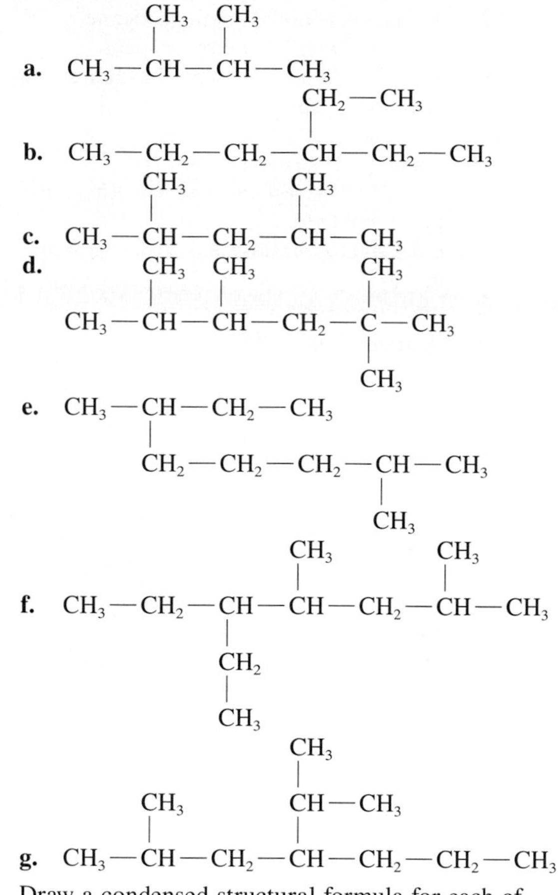

10.27 Draw a condensed structural formula for each of the following alkanes:
a. butane **b.** 2,3-dimethylpentane
c. 2,2,4-trimethyloctane
d. 3,3-dimethylhexane
e. 4-isopropyl-2-methylheptane

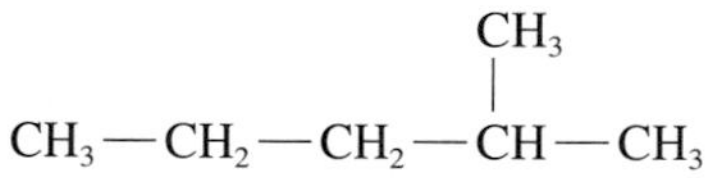

10.28 Draw a condensed structural formula for each of the following alkanes:
a. pentane **b.** 3-ethyl-2-methylpentane
c. 4-propylheptane
d. 2,2,3,5-tetramethylhexane
e. 2,2-dimethyl-4-ethyloctane

10.29 Indicate whether each of the following pairs of alkanes are structural isomers:
a. propane and 2-methylpropane
b. butane and 2-methylpropane
c. 3-methylpentane and 2,2-dimethylpropane
d. hexane and 3,3-dimethylpentane

10.30 Indicate whether each of the following pairs of alkanes are structural isomers:
a. 2-methylpentane and 3-methylpentane
b. 3-methylpentane and 2,2,3-trimethylbutane
c. 3-methylhexane and 3,3-dimethylpentane
d. 3-ethylhexane and 2,2,3-trimethylpentane

Cycloalkanes *(Goal 10.5)*

10.31 How does butane differ from cyclobutane?

10.32 How does 2-methylpentane differ from methylcyclopentane?

10.33 Give the IUPAC name for each of the following cycloalkanes:

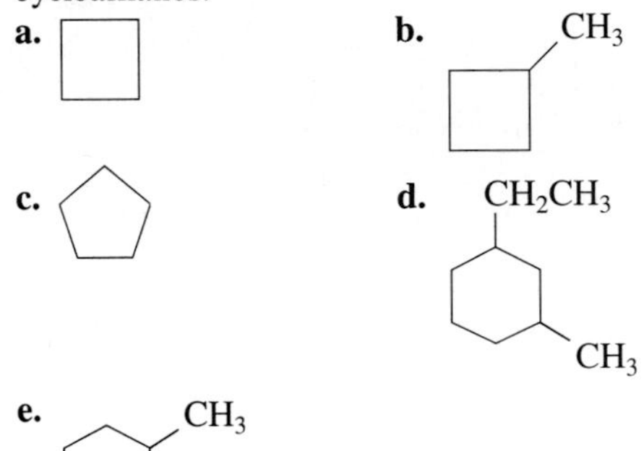

10.34 Give the IUPAC name for each of the following cycloalkanes:

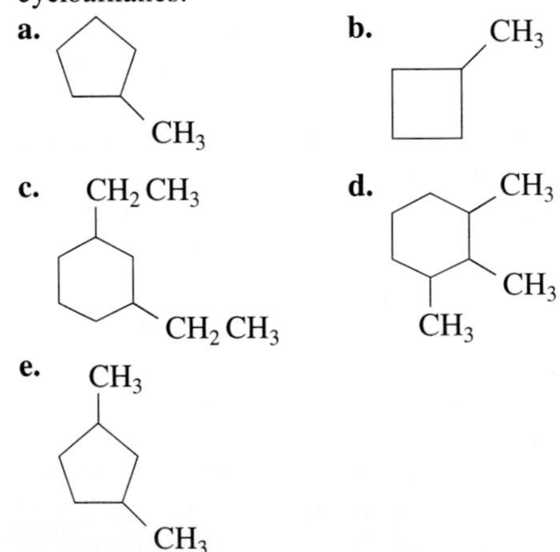

10.35 Draw the structure for each of the following cycloalkanes, using the geometric formula:
a. cyclopropane
b. 1,2-dimethylcyclobutane
c. propylcyclohexane
d. ethylcyclopentane
e. 3-ethyl-1,1-dimethylcyclohexane

10.36 Draw a structure for each of the following cycloalkanes, using the geometric formula:
a. cyclobutane **b.** methylcyclopropane
c. 1,1-dimethylcyclopentane
d. 1-ethyl-4-methylcyclohexane
e. 1,2,4-trimethylcyclopentane

10.37 Are the following pairs of condensed structural formulas isomers, identical, or neither?

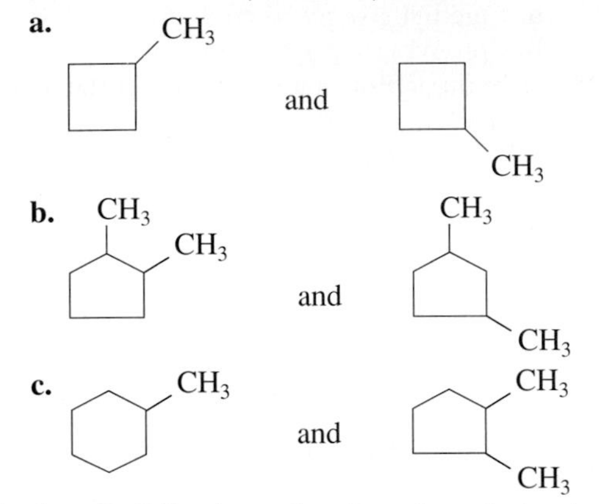

10.38 Are the following pairs of condensed structural formulas isomers, identical, or neither?

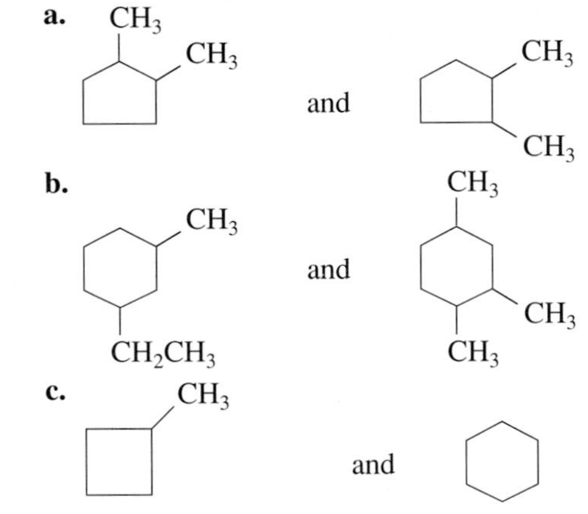

Haloalkanes *(Goal 10.6)*

10.39 Write the IUPAC name for each of the following haloalkanes:

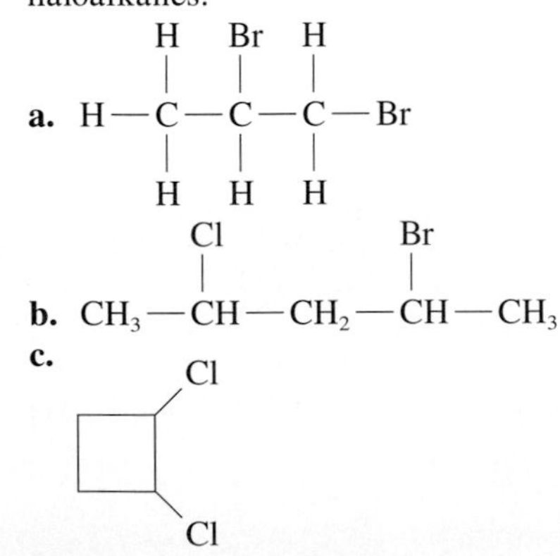

d. $CH_3—C(F)(CH_3)—CH_2—CH_3$

e. $CH_3—C(F)(Br)—CH_2—C(Cl)(Cl)—CH_2—CH_3$

10.40 Write the IUPAC name for each of the following haloalkanes:

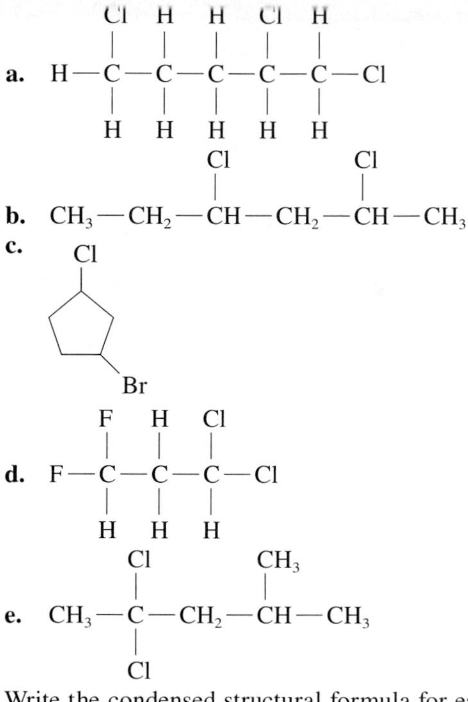

10.41 Write the condensed structural formula for each of the following names:
a. 2-chloropropane
b. 1-bromo-2-chloropropane
c. methyl bromide, a fumigant
d. 1-chloro-1-fluoroethane
e. 1,2-difluorocyclopentane

10.42 Write the condensed structural formula for each of the following names:
a. 1-chlorobutane
b. 1-bromo-2-chlorocyclopropane
c. 3,3-dibromo-2-methylpentane
d. 2,4-dichloro-3-methylhexane
e. 1,1,3,5-tetrachlorocyclohexane

10.43 Chloromethane (methyl chloride) and chloroethane (ethyl chloride) are used as topical anesthetics. When sprayed on the skin, they produce local anesthesia by lowering the temperature in the treated area to freeze the nerve endings. This makes local surgery near the surface of the skin painless. What are the structures of chloromethane and chloroethane?

10.44 The isomers of $C_2H_4Cl_2$ are used as industrial solvents. Write the condensed structural formulas for the two isomers and give their IUPAC names.

10.45 Write the four isomers of C_4H_9Cl and give the IUPAC name for each.

10.46 Write the four isomers of $C_3H_6F_2$ and give the IUPAC name for each.

Properties of Alkanes *(Goal 10.7)*

10.47 Write balanced equations for the complete combustion of each of the following:
a. methane, CH_4
b. propane, C_3H_8
c. cyclobutane, C_4H_8

10.48 Write balanced equations for the complete combustion of each of the following:
a. cyclohexane, C_6H_{12}
b. butane, C_4H_{10}
c. octane, C_8H_{18}

10.49 Heptane, C_7H_{16}, is a straight-chain alkane. It has a density of 0.68 g/mL, melts at −91°C, and boils at 98°C.
a. What is the structural formula of heptane?
b. Is it a solid, liquid, or gas at room temperature?
c. Is it soluble in water?
d. Will it float on the surface of water?
e. Heptane is a component of gasoline. Write the balanced equation for the combustion of heptane.

10.50 Nonane, C_9H_{20}, is a straight-chain alkane. It has a density of 0.79 g/mL, melts at −51°C, and boils at 151°C.
a. What is the condensed structural formula of nonane?
b. Is it a solid, liquid, or gas at room temperature?
c. Is it soluble in water?
d. Will it float on the surface of water?
e. Nonane is a component of gasoline. Write the balanced equation for the complete combustion of nonane.

10.51 Write the condensed structural formulas of the products that have one halogen atom.
a. $CH_3CH_3 + Br_2 \xrightarrow{\text{Light}}$
b. $CH_3CH_2CH_3 + Cl_2 \xrightarrow{\text{Light}}$
c. $CH_4 + Br_2 \xrightarrow{\text{Dark}}$
d. $CH_4 + Cl_2 \xrightarrow{\text{Heat}}$

10.52 Write the condensed structural formulas of the products that have one halogen atom.

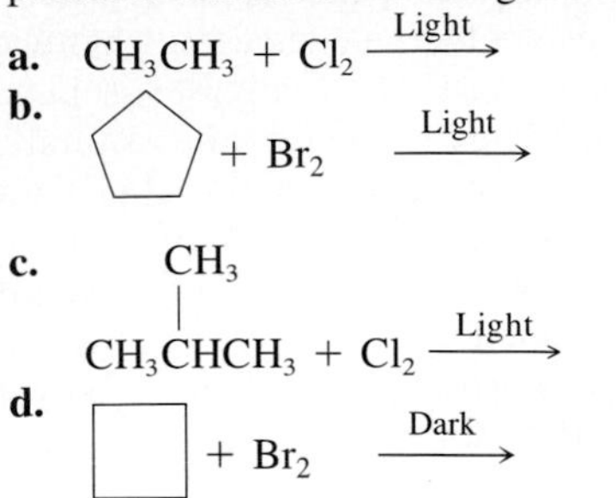

Challenge Problems

10.53 Draw at least six structural isomers of $C_5H_{11}Cl$.

10.54 Some halogenated compounds used as refrigerants have the following names. Draw their structures.

Freon 14	tetrafluoromethane
Freon 114	1,2-dichloro-1,1,2,2-tetrafluoroethane
Freon C318	octafluorocyclobutane

10.55 What is the IUPAC name of lindane, a pesticide?

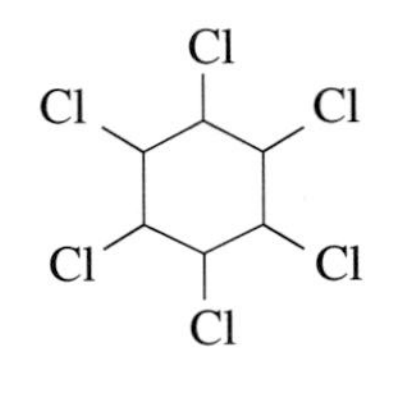

10.56 In an automobile engine, "knocking" occurs when the combustion of gasoline occurs too rapidly. The octane number of gasoline represents the ability of a gasoline mixture to reduce knocking. A sample of gasoline is compared with heptane, rated 0 because it reacts with severe knocking, and 2,2,4-trimethylpentane (isooctane), which has a rating of 100 because of its low knocking. Write the condensed structural formulas, molecular formulas, and equations for complete combustion of these compounds.

10.57 The density of pentane, a component of gasoline, is 0.63 g/ml. The heat of combustion for pentane is 845 kcal per mole (3536 kJ/mol).

a. Write an equation for the complete combustion of pentane.

b. What is the molar mass?

c. How much heat is produced when 1 gallon of pentane is burned (1 gallon = 3.78 liters)?

d. How many liters of CO_2 at STP are produced from the complete combustion of 1 gal of pentane?

10.58 What are the IUPAC names of the monosubstituted products from the bromination of pentane?

10.59 What are the IUPAC names of the disubstituted products from the chlorination of cyclohexane?

Chapter 11

Alkenes, Alkynes, and Aromatic Hydrocarbons

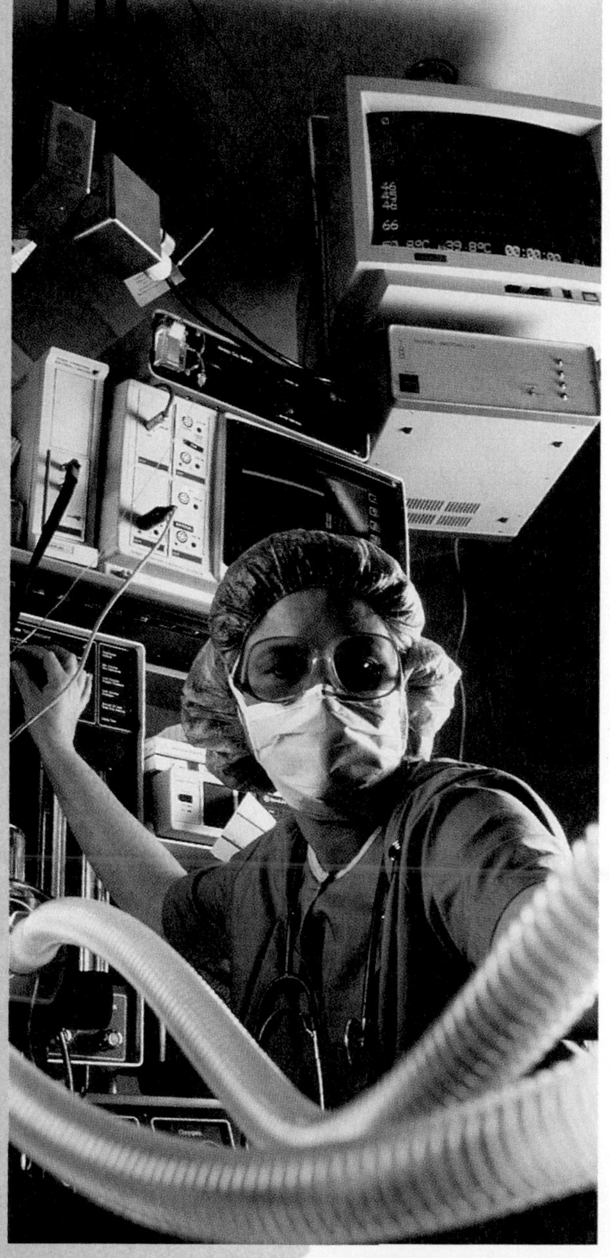

Plastic containers and IV tubing are relatively inert, and will not react with chemicals that are stored in them.

Looking Ahead

Learning Goals

11.1 Describe the carbon–carbon double bond as the functional group of alkenes.

11.2 Name alkenes and cycloalkenes using IUPAC rules and write their structural formulas.

11.3 Identify alkenes that exist as cis–trans isomers, and write their structural formulas and names.

11.4 Write the structural formulas and names for the products of the addition of hydrogen and halogens to alkenes.

11.5 Write the structural formulas and names for the products of the addition of hydrogen halides and water to alkenes, applying Markovnikov's rule when necessary.

11.6 Name alkynes using IUPAC rules and write their condensed structural formulas.

11.7 Describe the bonding in benzene; name aromatic compounds and write their structural formulas.

Questions to Think About

1. What are some polymers you use everyday?
2. A label on a bottle of canola oil lists saturated, polyunsaturated, and monounsaturated fats. What is meant by these terms?
3. A peanut butter label states that it contains hydrogenated vegetable oils. What has happened to the peanut oil?

In the preceding chapter, we looked at the names, structures, and reactions of the alkanes, organic compounds that have only single bonds. In this chapter, we will investigate the unsaturated hydrocarbons, alkenes (double bonds) and the alkynes (triple bonds). We saw that the single bonds in the alkanes were not very reactive. In contrast, unsaturated hydrocarbons are highly reactive because the second and third bonds formed between two carbon atoms are broken more easily than single bonds. As a result, different atoms or groups of atoms can be added to alkenes and alkynes to prepare other kinds of organic compounds.

We will also look at aromatic compounds, a family of hydrocarbons that contains benzene rings. The aromatic compounds of benzene are especially stable because electrons are shared in a six-carbon ring.

11.1 Alkenes

Learning Goal

Describe the carbon–carbon double bond as the functional group of alkenes.

There are millions of organic compounds, and more are synthesized every day. The task of learning organic chemistry seems overwhelming until we discover that each family of compounds has a characteristic structural feature that sets it apart from the other families. This special feature, called a **functional group,** determines the family name and chemical reactivity of all the compounds with that functional group. Thus, when we identify the functional group or groups in an organic compound, even a complex one, we can predict many of its chemical properties.

Alkenes are a family of hydrocarbons that contain a carbon–carbon double bond (C═C) as the functional group. In the simplest alkene, ethene (common name, ethylene), C_2H_4, there is a double bond (two bonds) between two carbon atoms. The double bond forms when each carbon atom shares two valence electrons with the other. (See Figure 11.1.)

Figure 11.1
A ball-and-stick model of ethene, C_2H_4, the simplest alkene.

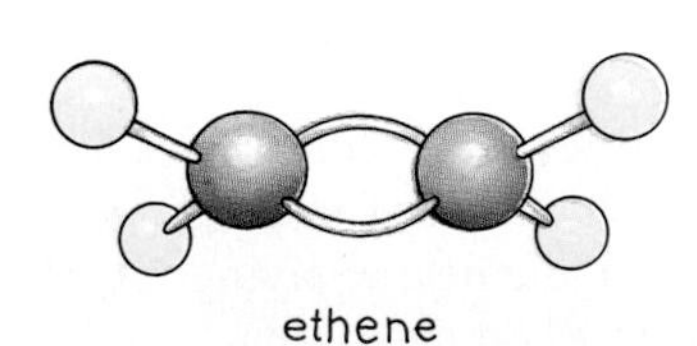

H H
:C::C:
H H

Two valence electrons from each carbon atom form the double bond of alkene

The existence of a double bond makes ethene an unsaturated hydrocarbon because it has two fewer hydrogen atoms compared with the saturated alkane, ethane, C_2H_6. Ethene is a widely used industrial compound and a plant hormone that is used to ripen fruit. The double bond makes an alkene molecule flat (planar): the carbon and hydrogen atoms in the double bond all lie in the same plane.

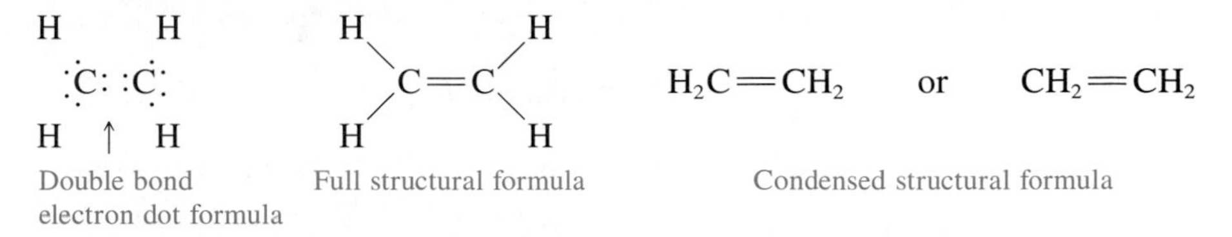

Cycloalkenes are similar to the cycloalkanes, except that there is a double bond between two of the carbon atoms in the ring.

Examples of Condensed Structural Formulas of Cycloalkenes

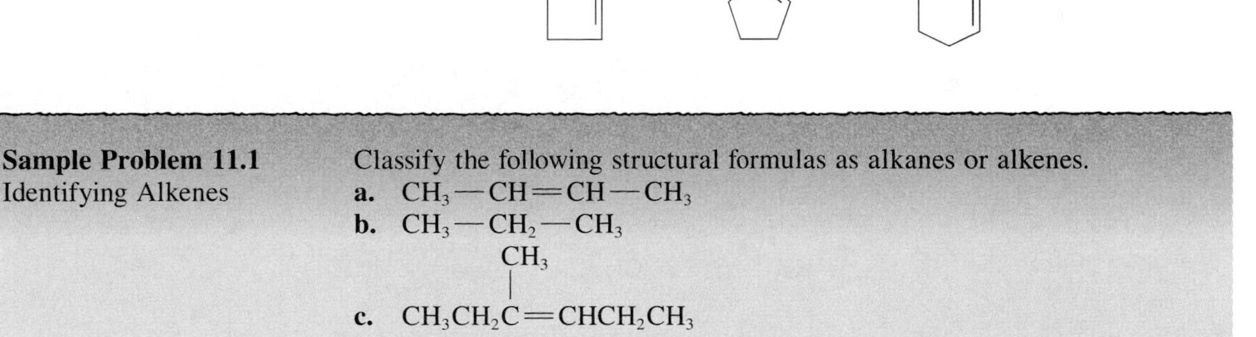

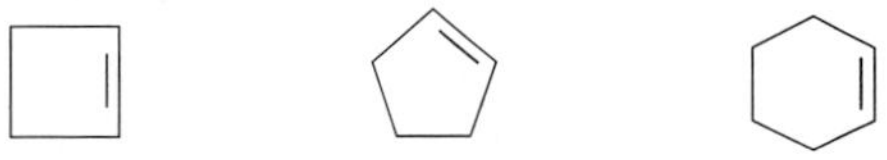

Sample Problem 11.1
Identifying Alkenes

Classify the following structural formulas as alkanes or alkenes.

a. $CH_3{-}CH{=}CH{-}CH_3$

b. $CH_3{-}CH_2{-}CH_3$

c. $CH_3CH_2C(CH_3){=}CHCH_2CH_3$

Solution

The structural formulas with double bonds are alkenes.

a. alkene **b.** alkane **c.** alkene

Study Check

Is the following structure a cycloalkane or cycloalkene?

11.2 Naming Alkenes

Learning Goal

Name alkenes and cycloalkenes using IUPAC rules and write their structural formulas.

The IUPAC rules we learned in Chapter 10 also apply to alkenes with a few new rules added specifically for the double bond. The first two alkenes may also be called by their common names, ethylene and propylene.

$CH_2{=}CH_2$ — Ethene (ethylene)

$CH_3{-}CH{=}CH_2$ — Propene (propylene)

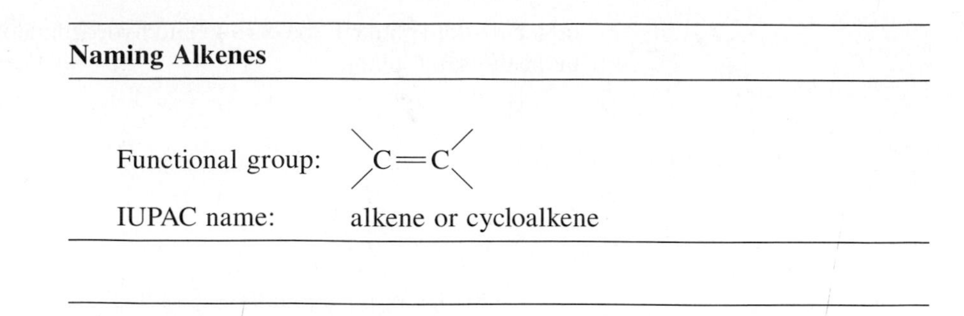

Naming Alkenes

Functional group: $\rangle C{=}C\langle$

IUPAC name: alkene or cycloalkene

IUPAC Rules for Naming Alkenes

Step 1 Select the longest carbon chain *containing the double bond* as the parent chain. Replace the *ane* ending of the corresponding alkane with *ene.*

$CH_3{-}CH_2{-}CH_3$ Propane $\qquad CH_2{=}CH{-}CH_3$ Propene

Step 2 Number the parent chain beginning at the end closest to the double bond. Use the number of the first carbon atom in the double bond to give the location of the double bond. For example, two butene structures can be written for C_4H_8.

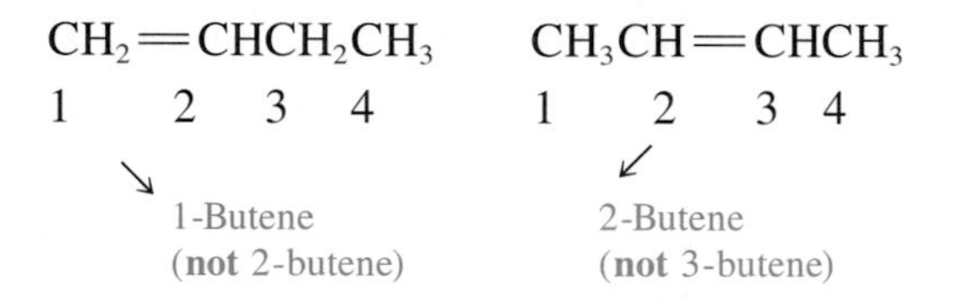

Step 3 Locate by number and name the branches in alphabetical order.

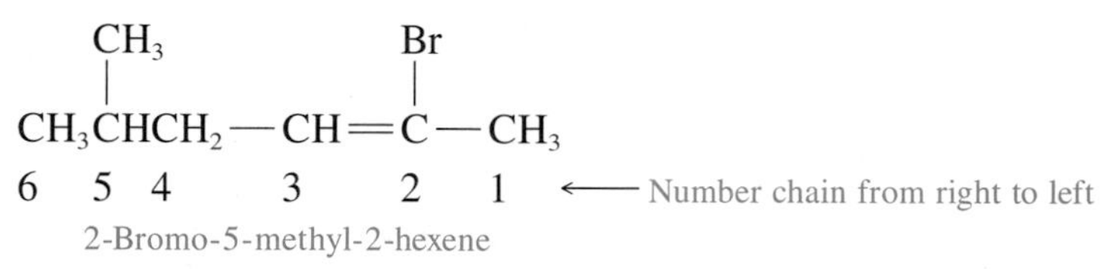

Be sure that you select the longest carbon chain that contains the double bond. Sometimes the longest chain turns a corner.

$$\begin{array}{l} \quad\ \ \ 2 \qquad\ \ 1 \\ \quad\ \ CH_2{-}CH_3 \\ \quad\ \ | \\ CH_3{-}C{=}CH{-}CH_2{-}CH_3 \\ \quad\ \ 3 \quad\ 4 \quad\ \ 5 \qquad 6 \end{array}$$

3-Methyl-3-hexene
(**not** 2-ethyl-2-pentene)

The parent chain has six carbon atoms, not five

Compounds with two double bonds are named as **dienes.**

$CH_2{=}CH{-}CH{=}CH_3$ 1,3-Butadiene

$$\begin{array}{l} \qquad\quad\ \ CH_3 \\ \qquad\quad\ \ | \\ CH_2{=}C{-}CH{=}CH{-}CH_3 \end{array}$$

2-Methyl-1,3-pentadiene

Step 4 If a cyclic compound contains a double bond, name it a **cycloalkene.**

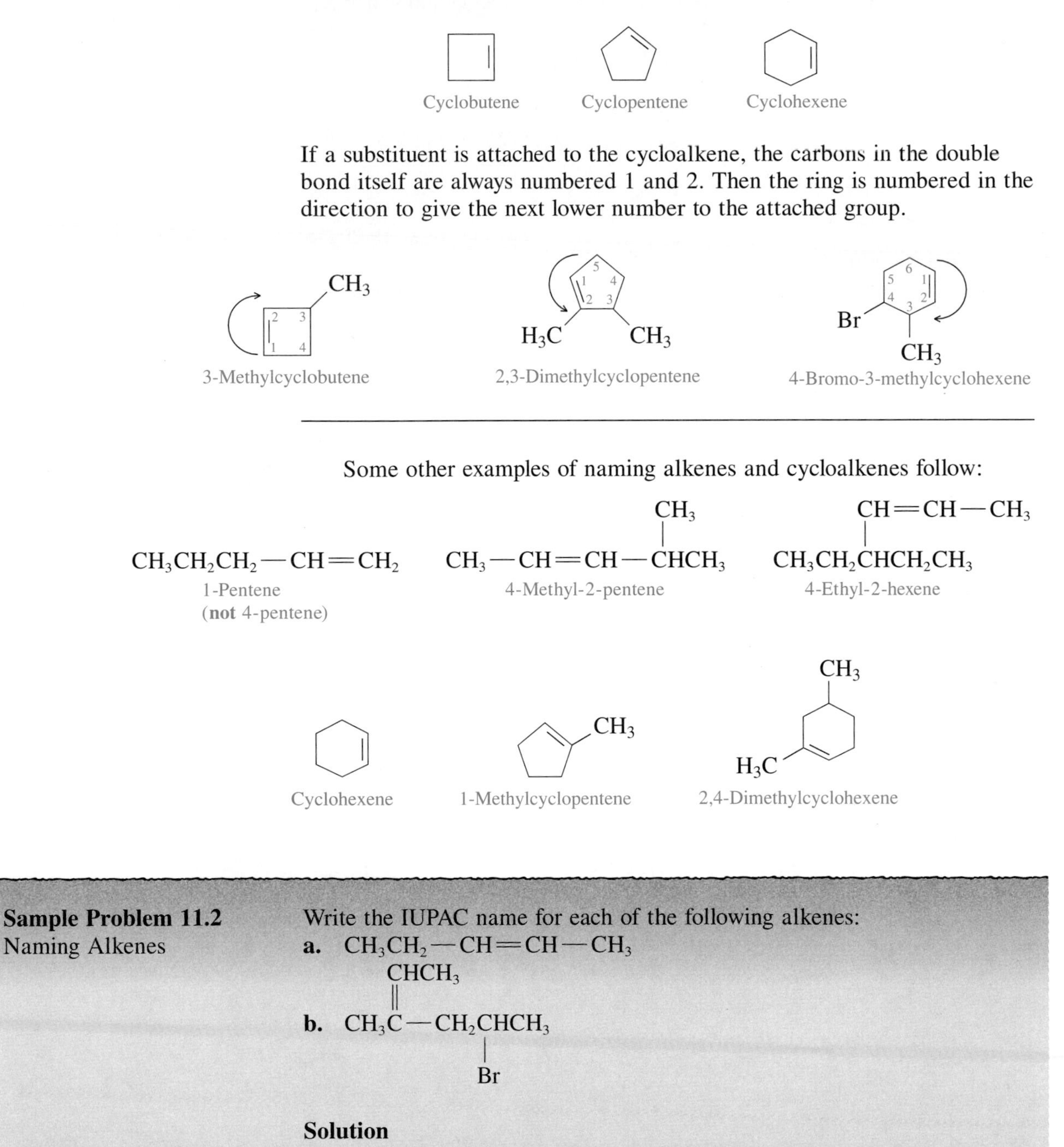

Cyclobutene Cyclopentene Cyclohexene

If a substituent is attached to the cycloalkene, the carbons in the double bond itself are always numbered 1 and 2. Then the ring is numbered in the direction to give the next lower number to the attached group.

3-Methylcyclobutene 2,3-Dimethylcyclopentene 4-Bromo-3-methylcyclohexene

---

Some other examples of naming alkenes and cycloalkenes follow:

$CH_3CH_2CH_2—CH{=}CH_2$
1-Pentene
(**not** 4-pentene)

$CH_3—CH{=}CH—CHCH_3$ (with CH_3 on the CH)
4-Methyl-2-pentene

$CH_3CH_2CHCH_2CH_3$ (with $CH{=}CH—CH_3$ on the CH)
4-Ethyl-2-hexene

Cyclohexene 1-Methylcyclopentene 2,4-Dimethylcyclohexene

Sample Problem 11.2
Naming Alkenes

Write the IUPAC name for each of the following alkenes:

a. $CH_3CH_2—CH{=}CH—CH_3$

b. $CH_3C—CH_2CHCH_3$ (with $=CHCH_3$ on the C and Br on the CH)

Solution

a. The longest chain, pentene, is numbered from right to left to place the double bond on carbon 2. The compound is named 2-pentene.

b. The longest chain *containing the double bond* has six carbons (not five) and is named hexene. The branches in alphabetical order are 5-bromo and 3-methyl. The compound is named 5-bromo-3-methyl-2-hexene.

Study Check

Give the IUPAC name for the following:

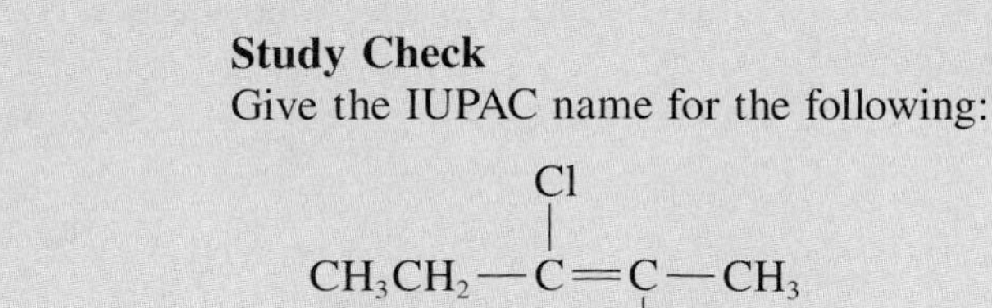

Sample Problem 11.3
Naming Cycloalkenes

Write the structures of all the isomers of chlorocyclopentene and name each.

Solution

The name indicates a cyclic parent chain with five carbon atoms and a double bond. The single chlorine atom can be placed in three different positions on the cyclopentene ring to give three different isomers:

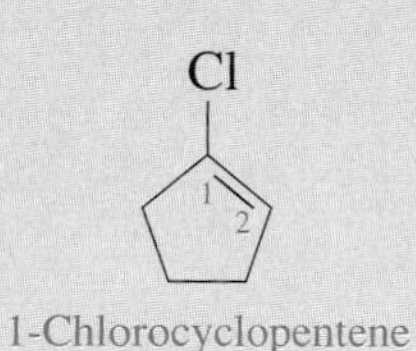

1-Chlorocyclopentene

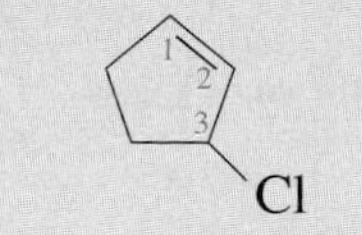

3-Chlorocyclopentene

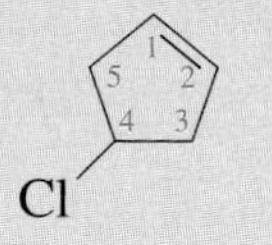

4-Chlorocyclopentene

Study Check

What is the IUPAC name for the following compound?

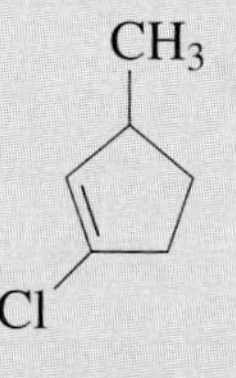

ENVIRONMENTAL NOTE *Fragrant Alkenes*

The odors you associate with cloves and peppermint, or perfumes from roses and lavender, are due to volatile oils that are synthesized by the plants. Unsaturated compounds (known as terpenes) composed of two or more units of isoprene, a five-carbon diene, are responsible for many of these fragrances.

$$CH_2{=}C(CH_3){-}CH{=}CH_2$$

"Isoprene"
2-Methyl-1,3-butadiene

The structures in Figure 11.2 give examples of joined isoprene units.

Figure 11.2
The fragrance of a rose is due to the unsaturated substances in the oils of the flower.

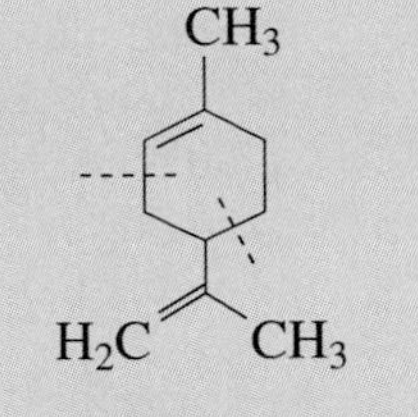

Limonene, found in oils of lemon, orange, and dill

Connected isoprene units

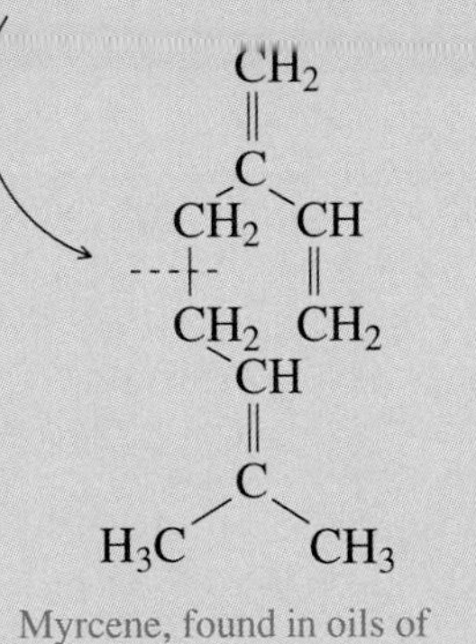

Myrcene, found in oils of bay leaves and verbena

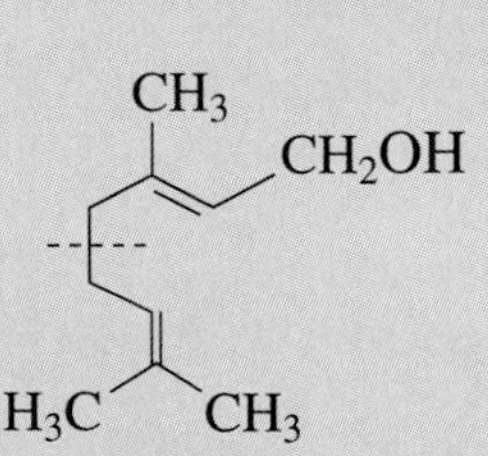

Geraniol, found in oils of rose and citronella

11.3 Cis–Trans Isomers

Learning Goal
Identify alkenes that exist as cis–trans isomers, and write their structural formulas and names.

In an alkene, the carbon atoms in the double bond are not free to rotate. As a result the groups that are attached to the carbons in the double bond remain fixed on one side of the double bond or the other. This gives many alkenes two distinct arrangements in space called **geometric isomers** or **cis–trans isomers.**

The compound 1,2-dichloroethene is an example of an alkene that can exist as two geometric isomers. To differentiate between them, the prefixes *cis* and *trans* are used in front of the alkene name. In the *cis* isomer, the chlorine atoms are on the *same side* of the double bond, while in the trans isomer, the chlorine atoms appear on *opposite sides.* (See Figure 11.3.) The *cis–trans* isomers are actually two different compounds with different physical properties including different boiling points.

Figure 11.3
Ball-and-stick models of the cis and trans isomers of 1,2-dichloroethene.

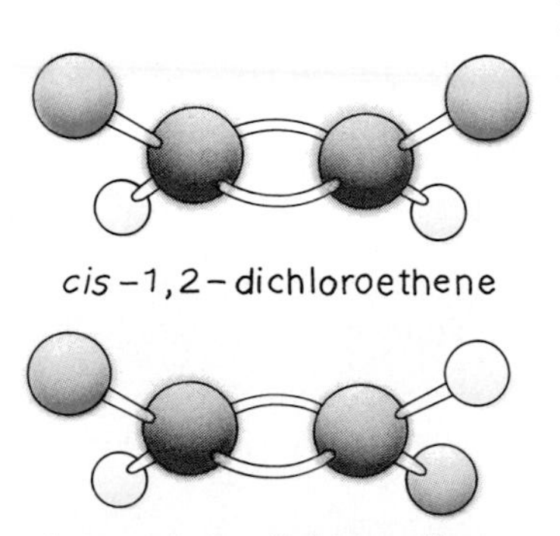

In the cis isomer, chlorine atoms are on the same side of the double bond.

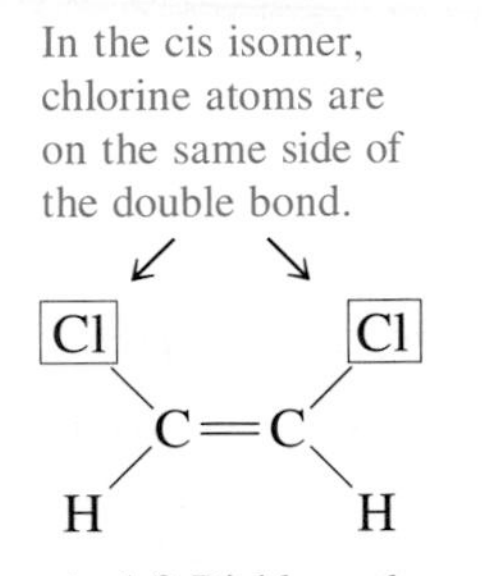

cis-1,2-Dichloroethene (boils at 60°C)

In the trans isomer, the chlorine atoms are on opposite sides of the double bond.

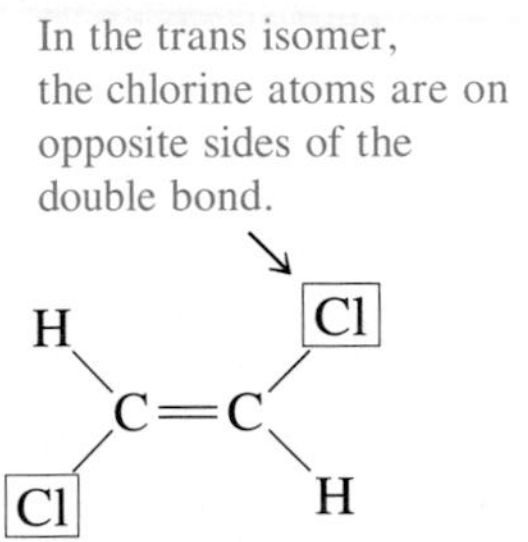

trans-1,2-Dichloroethene (boils at 48°C)

For a molecule to have a cis or trans isomer, the carbons in the double bond must be connected to *two different atoms or groups.* That is, cis–trans isomers are not possible when one of the carbon atoms in the double bond is attached to *identical* atoms or groups. For example, 1,1-dichloro-1-propene has no cis–trans isomers because one of the carbons in the double bond is bonded to two identical Cl atoms.

No cis–trans isomers possible for identical groups on one or both carbon atoms in the double bond

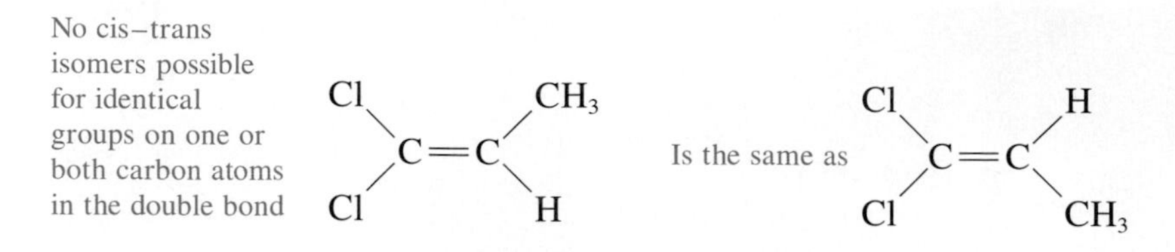

Sample Problem 11.4
Cis–Trans Isomers of Alkenes

Indicate whether each of the following compounds exists as cis–trans isomers. If so, draw and name each isomer.

a. $CH_3CH{=}CBr_2$
b. $Br{-}CH{=}CH{-}Br$

Solution

a. No; one carbon atom in the double bond is attached to two bromine atoms, which are identical atoms.
b. Yes; different atoms are attached to each carbon atom in the double bond.

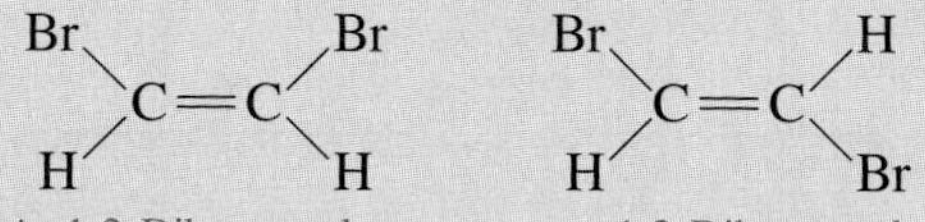

cis-1,2-Dibromoethene *trans*-1,2-Dibromoethene

Study Check
Does the compound with formula $Cl{-}CH{=}CBr_2$ have cis–trans isomers?

Alkene molecules with four or more carbon atoms can also have cis–trans isomers without having other substituents such as chlorine atoms. The simplest alkene that has cis–trans isomers is 2-butene. When the alkyl groups are attached on the same side of the double bond, the compound is *cis*-2-butene.

In *trans*-2-butene, the alkyl groups are on opposite sides of the double bond. (See Figure 11.4.) The cis and trans isomers of 2-butene (as with any pair of cis–trans isomers) are different compounds with different properties such as melting point (mp) and boiling point (bp).

$CH_3{-}CH{=}CH{-}CH_3$

2-Butene

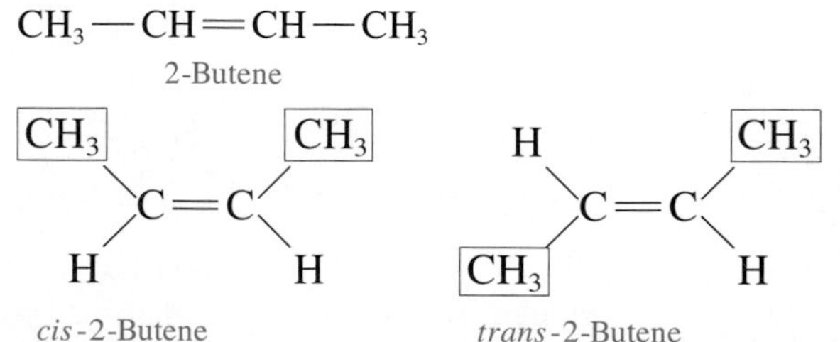

cis-2-Butene
(mp −139°C; bp 3.7°C)

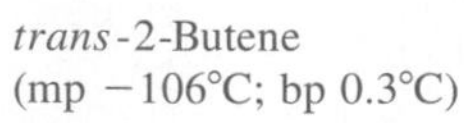

trans-2-Butene
(mp −106°C; bp 0.3°C)

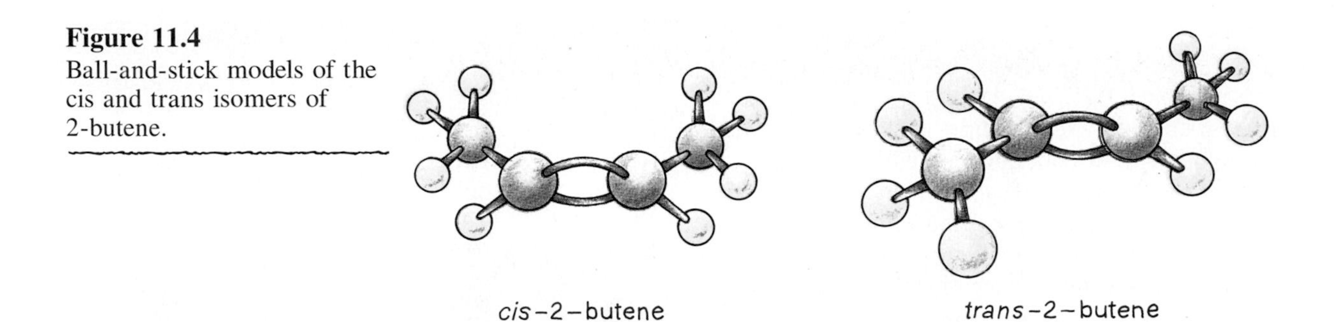

Figure 11.4
Ball-and-stick models of the cis and trans isomers of 2-butene.

The other structural isomers of butene, 1-butene and 2-methylpropene, do not have cis–trans isomers because they have identical groups on at least one of the carbon atoms in the double bond.

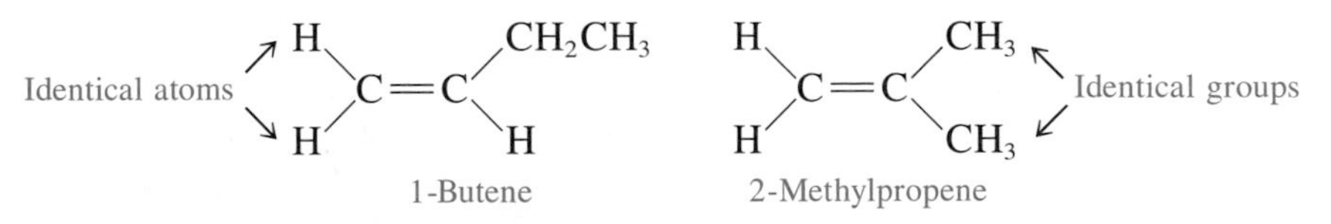

Other examples of cis–trans isomers are given below:

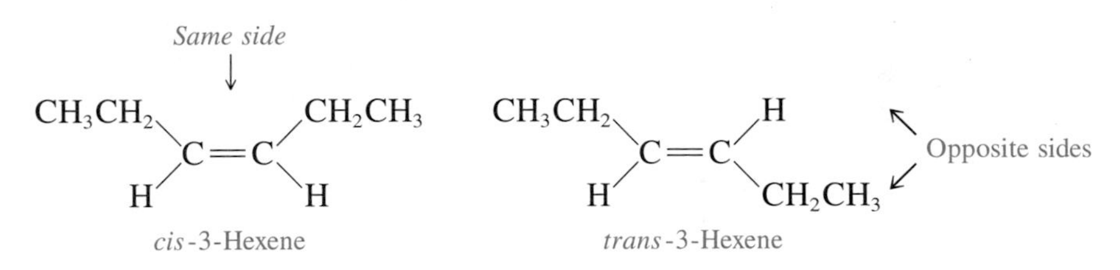

Sample Problem 11.5
Writing and Naming Cis–Trans Isomers

Give the condensed structural formulas and names for the geometric isomers of 2-pentene:

$$CH_3—CH=CH—CH_2CH_3$$

Solution
We can write the cis isomer by placing the alkyl groups on the same side of the double bond, and the trans isomer by placing them on opposite sides.

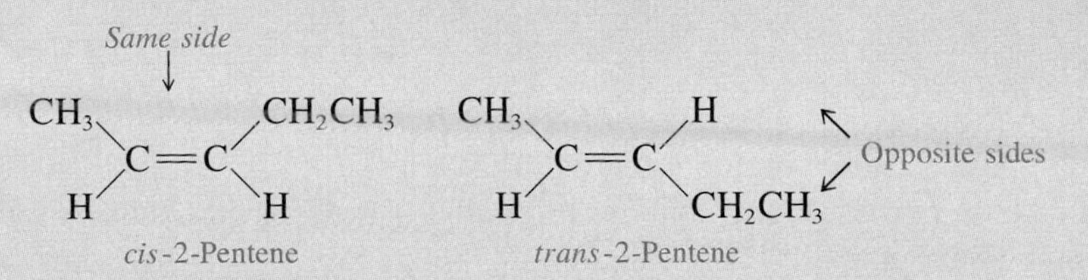

Study Check
Draw the condensed structural formula for *trans*-3-heptene.

ENVIRONMENTAL NOTE *Pheromones in Insect Communication*

Insects emit minute quantities of chemicals called pheromones that communicate with other members of the same species. Some pheromones warn of danger, others call for defense, mark a trail, or are sex attractants. The effectiveness of many of these ten to seventeen carbon atom compounds depends on the cis or trans configuration of the double bond in the molecule. A certain species will respond to one isomer but not the other.

Scientists are interested in synthesizing pheromones to use as nontoxic alternatives to pesticides. For example, when a very small amount of the sex attractant of a certain insect is placed in a trap, the insects are unable to breed successfully. As a result the insects are removed from an agricultural area without affecting a wide range of other organisms. (See Figure 11.5.)

Figure 11.5
A trap in a crab apple tree is baited with codling moth pheromone.

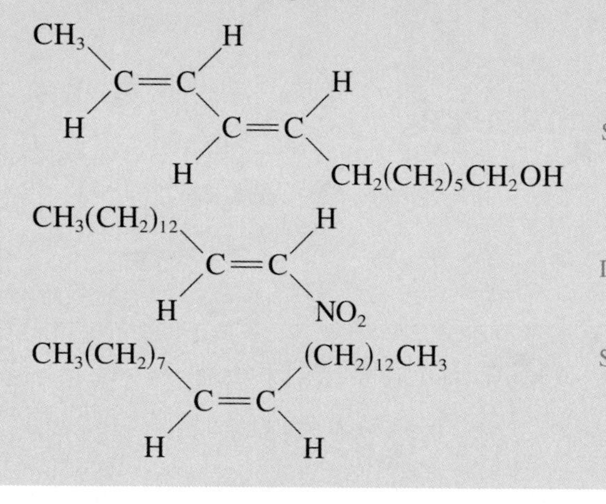

Sex attractant for the codling moth

Defense signal for the termite

Sex attractant for the female housefly

HEALTH NOTE *Cis–Trans Isomers for Night Vision*

The retinas of the eyes consist of two types of cells, rods and cones. The rods on the edge of the retina allow us to see in dim light, and the cones, in the center, produce our vision in bright light. In the rods, there is a substance called rhodopsin (visual purple) that absorbs light. Rhodopsin is composed of *cis*-11-retinal, an unsaturated compound, attached to a protein. When rhodopsin absorbs light, the *cis*-11-retinal isomer is converted to its *trans* isomer, which changes its shape. The *trans* form no longer fits the protein and it separates from the protein. The change from the cis to trans isomer and the separation from the protein generates an electrical signal that the brain perceives as light.

An enzyme (isomerase) converts the trans isomer back to the *cis*-11-retinal isomer and the rhodopsin reforms. If there is a deficiency of rhodopsin in the rods of the retina, night blindness may occur. One common cause is a lack of vitamin A in the diet, because retinal is formed from vitamin A. Without a sufficient quantity of retinal, there is not enough rhodopsin produced to enable us to see adequately in dim light.

Cis-trans isomers of retinal

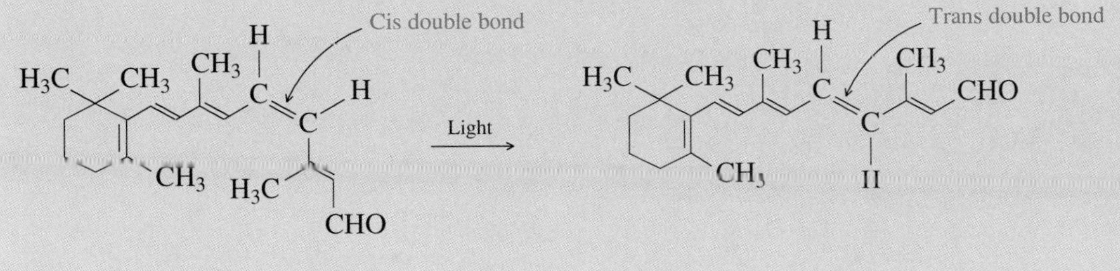

Cis double bond

Light

Trans double bond

11-*cis*-retinal

11-*trans*-retinal

11.4 Addition of Hydrogen and Halogens

Learning Goal

Write the structural formulas and names for the products of the addition of hydrogen and halogens to alkenes.

Earlier we learned that the functional group of a family tells us about the chemical behavior of that family. For alkenes, the most characteristic reaction is the **addition** of atoms or groups of atoms to the carbons of the double bond. Addition occurs because one of the bonds in the double bond is easily broken. We use the symbol A—B to represent a reactant such as H—H, Cl—Cl, H—Cl, or H—OH. The general equation for an addition reaction of an alkene is written as follows:

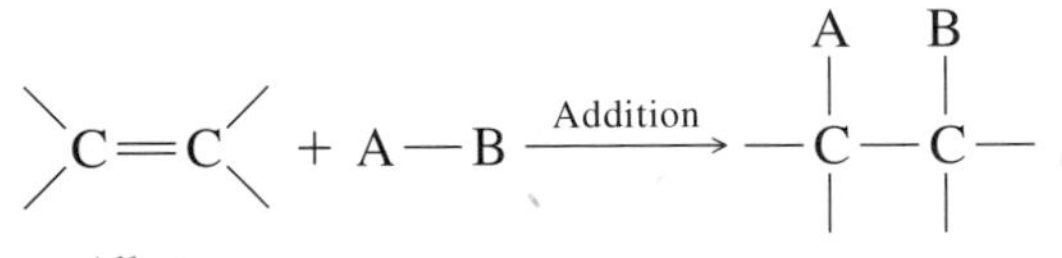

The addition reactions have different names that depend on the type of reactant we add to the alkene.

Reactant Added	Name of Addition Reaction
H_2	Hydrogenation
Cl_2, Br_2	Halogenation
HCl, HBr, HI	Hydrohalogenation
HOH	Hydration

Adding Hydrogen

In the reaction called **hydrogenation,** two atoms of hydrogen are added to a double bond. The hydrogenation of the double bond in an alkene (unsaturated) converts it to a saturated bond of an alkane. A catalyst such as platinum

(Pt) or nickel (Ni) is usually added to speed up the rate of the reaction. The symbol for the catalyst is shown over the arrow in the equation because a catalyst speeds up the reaction but does not appear in the products. The general equation for hydrogenation can be written as follows:

Hydrogenation of Alkenes

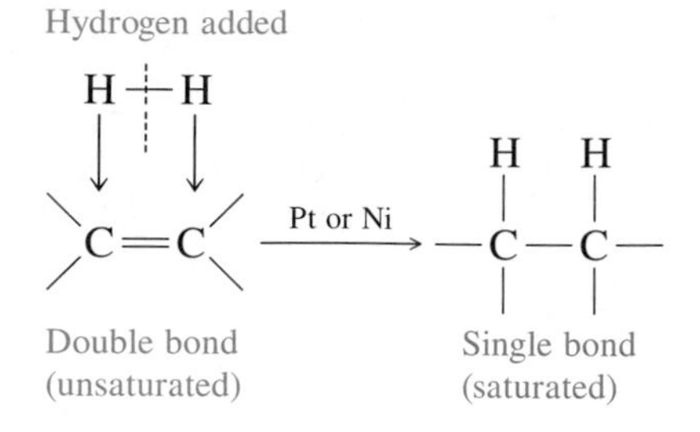

Some examples of the hydrogenation of alkenes follow:

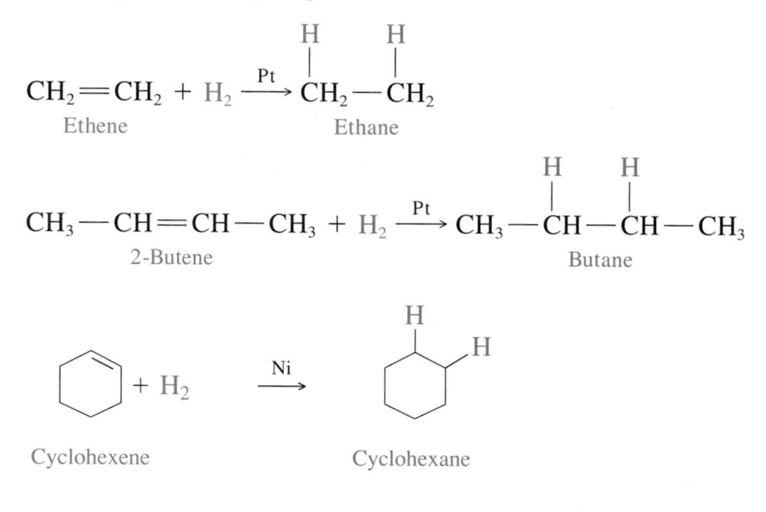

Sample Problem 11.6
Writing Equations for Hydrogenation

Write the structural formula for the product of the following hydrogenation reactions:

a. $CH_3{-}CH{=}CH_2 + H_2 \xrightarrow{Pt}$

b. (cyclopentene) $+ H_2 \xrightarrow{Pt}$

Solution

In an addition reaction, hydrogen adds to the double bond to give an alkane.

a. $CH_3{-}CH_2{-}CH_3$

b.

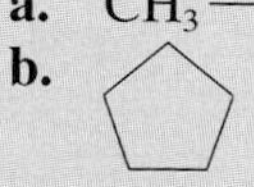

Study Check

Draw the structural formula of the product of the hydrogenation of 2-methyl-1-butene using a platinum catalyst.

HEALTH NOTE *Hydrogenation of Unsaturated Fats*

Vegetable oils such as corn oil or safflower oil are unsaturated fats composed of fatty acids that contain double bonds. The process of hydrogenation is used commercially to convert the double bonds in the unsaturated fats in vegetable oils to saturated fats such as margarine, which are more solid. Adjusting the amount of added hydrogen gives partially hydrogenated fats such as soft margarine, solid margarine in sticks, and shortenings, which are used in cooking. (See Figure 11.6.) For example, oleic acid is a typical unsaturated fatty acid in olive oil and has a cis-double bond at carbon 9. When oleic acid is hydrogenated, it is converted to stearic acid, a saturated fatty acid.

Figure 11.6
The vegetable oils in peanut butter, shortening, and margarine have been hydrogenated to form solid products.

$$CH_3(CH_2)_7-CH{=}CH-(CH_2)_7\overset{\overset{\displaystyle O}{\|}}{C}OH + H_2 \xrightarrow{Pt} CH_3(CH_2)_7-CH_2-CH_2-(CH_2)_7\overset{\overset{\displaystyle O}{\|}}{C}OH$$

Oleic acid (found in olive oil and other unsaturated fats) — Stearic acid (found in saturated fats)

Adding the Halogens Br_2 and Cl_2

Alkenes typically react with chlorine or bromine (the addition of fluorine is explosive and iodine is slow). The reaction, called **halogenation,** occurs readily, without the use of any catalyst, adding the atoms of the halogen to the double bond to yield a dihaloalkane product. In the equation for the general reaction, the symbol X—X or X_2 is used for Br_2 or Cl_2.

Halogenation of Alkenes

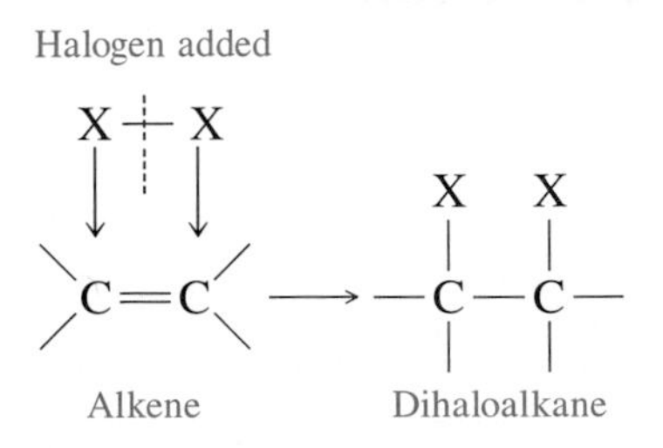

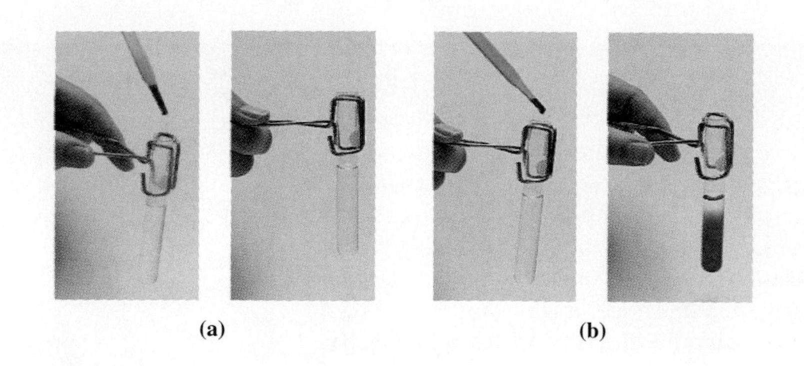

Figure 11.7
(a) When bromine is added to the alkene in the first test tube, the red color is lost immediately as the bromine atoms add to the double bond.
(b) In the second test tube, the red color remains because bromine does not react, or reacts only slowly, with the alkane.

Some examples of adding Cl_2 or Br_2 to alkenes are shown below:

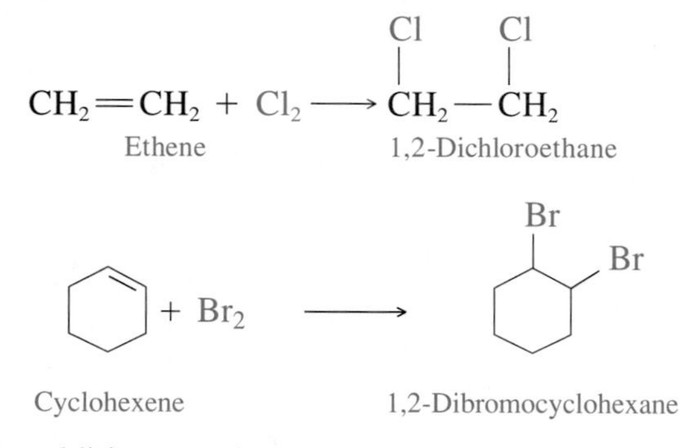

The addition reaction of bromine is sometimes used to test for the presence of double and triple bonds as shown in Figure 11.7.

Sample Problem 11.7
Writing Products of Halogenation

Write the condensed structural formula and name of the product of the following reaction:

$$CH_3{-}\overset{\displaystyle CH_3}{\overset{|}{C}}{=}CH_2 + Br_2 \longrightarrow$$

Solution
The addition of bromine to an alkene places a bromine atom on each of the carbon atoms of the double bond.

$$CH_3{-}\underset{\displaystyle Br}{\underset{|}{\overset{\displaystyle CH_3}{\overset{|}{C}}}}{-}\underset{\displaystyle Br}{\underset{|}{CH_2}} \quad \text{1,2-Dibromo-2-methylpropane}$$

Study Check
What is the name of the product formed when chlorine is added to 1-butene?

11.5 Addition of Hydrogen Halides and Water

Learning Goal
Write the structural formulas and names for the products of the addition of hydrogen halides and water to alkenes, applying Markovnikov's rule when necessary.

In the reaction called **hydrohalogenation,** a hydrogen halide (HCl, HBr, or HI) adds to an alkene to yield a haloalkane. The hydrogen atom bonds to one carbon of the double bond, and the halogen atom adds to the other carbon. The general reaction in which HX represents HCl, HBr, or HI, can be written as follows:

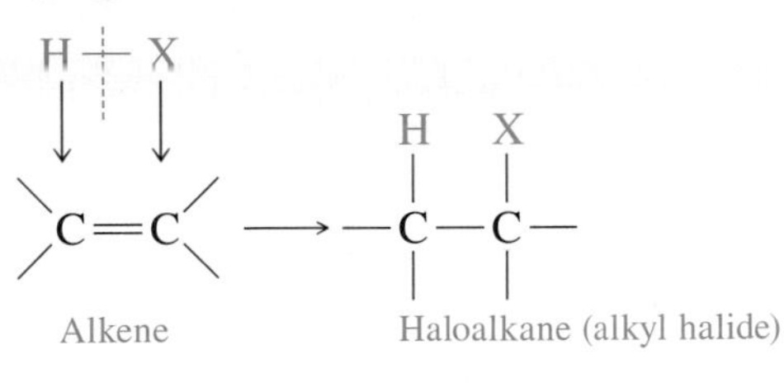

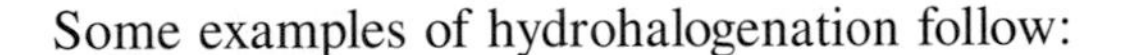

Some examples of hydrohalogenation follow:

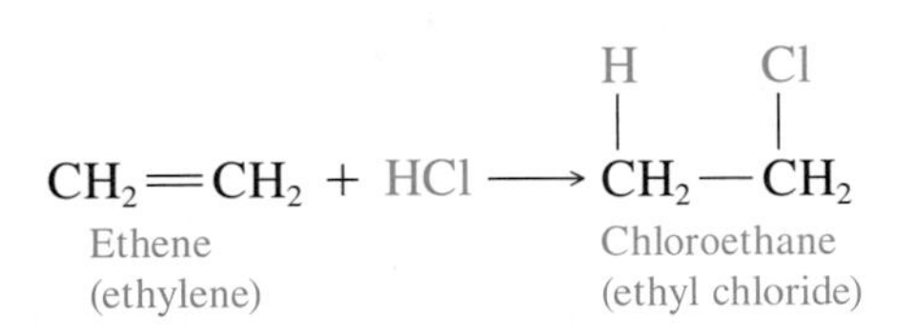

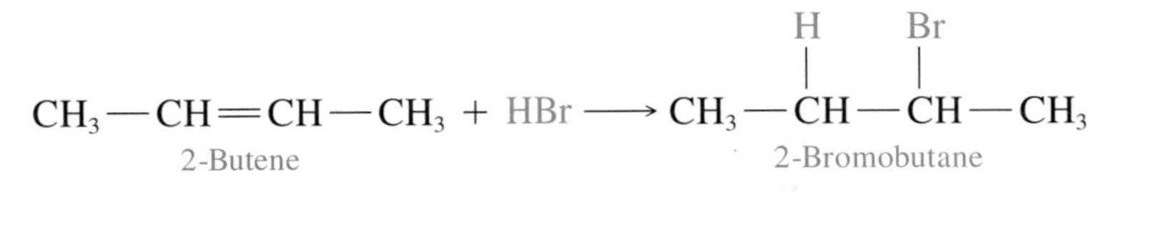

For some alkenes, the addition of H—X yields two products as shown below.

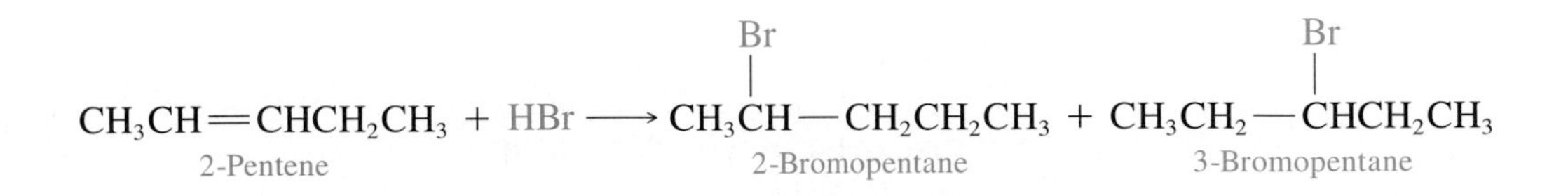

Markovnikov's Rule

When adding HX to some alkenes, you may find you have a choice of where to place the atoms on the double bond. We solve this problem by using **Markovnikov's rule,** which tells us that the hydrogen in HX adds to the carbon that has the greater number of hydrogen atoms, and the halogen adds to the carbon with the fewer hydrogen atoms. Using this rule, the hydrogen is added to the end carbon, and the chlorine to the second carbon in 1-propene.

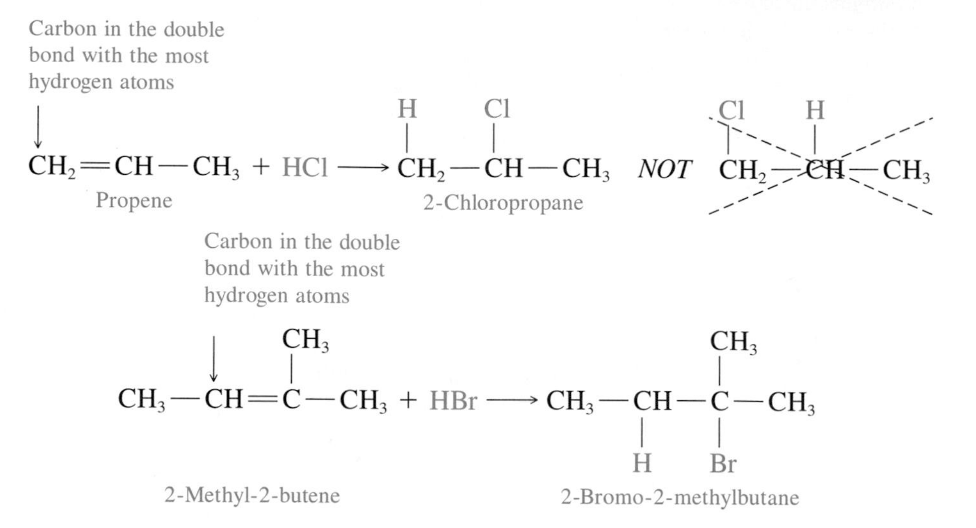

Sample Problem 11.8
Addition Products by Using Markovnikov's Rule

Draw the structural formula for the product of the following reaction:

$$CH_3-\underset{}{\overset{CH_3}{\overset{|}{C}}}=CH_2 + HI \longrightarrow$$

Solution
In the parent chain, carbon 1 of the double bond has more hydrogen atoms than carbon 2. Using Markovnikov's rule, we add the hydrogen atom from HI to carbon 1. The iodine atom adds to carbon 2 in the double bond.

$$CH_3-\underset{\underset{I}{|}}{\overset{\overset{CH_3}{|}}{C}}-CH_3 \quad \text{2-Iodo-2-methylpropane}$$

Study Check
Give the structural formula and name of the product obtained when H_2O is added to 1-methylcyclopentene.

Sample Problem 11.9
Writing a Synthesis by Using an Alkene

What alkene would you start with to prepare the following compound?

$$CH_3-\underset{\displaystyle |}{\overset{\displaystyle Br}{C}}H-CH_2-CH_3$$

Solution

The starting alkene would be 2-butene. The final product is prepared by adding hydrogen bromide.

$$\underset{\text{2-Butene}}{CH_3-CH=CH-CH_3} + HBr \longrightarrow \underset{\text{2-Bromobutane}}{CH_3-\overset{Br}{\overset{|}{C}}H-CH_2-CH_3}$$

Study Check

Write an equation using an alkene and HCl to prepare 2-chloropentane.

Adding Water to Alkenes

Alkenes react with water (HOH) when the reaction is catalyzed by an acid such as H_2SO_4 or H_3PO_4. In this reaction called **hydration,** an H— attaches to one of the carbon atoms in the double bond, and an —OH group to the other carbon. Hydration is used to prepare alcohols, which have an —OH functional group. In the general equation the acid is represented by H^+.

Hydration of Alkenes

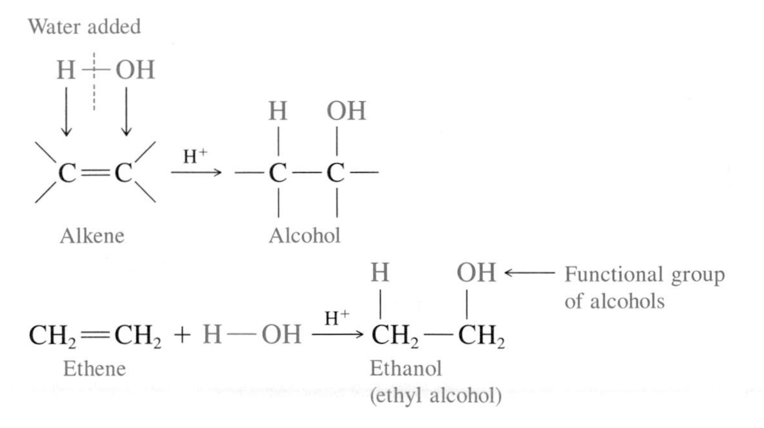

The addition of water to a double bond in which the carbon atoms are attached to different groups follows Markovnikov's rule.

$$\underset{\text{Propene}}{CH_3-CH=CH_2} + H-OH \xrightarrow{H^+} \underset{\text{2-Propanol}}{CH_3-\overset{OH}{\overset{|}{C}}H-\overset{H}{\overset{|}{C}}H_3} \quad \textit{NOT} \quad \cancel{CH_3-CH_2-CH_2-OH}$$

Sample Problem 11.10
Writing Products of Hydration

Write the structural formulas for the products that form in the following hydration reactions:

a. $CH_3CH_2CH_2-CH=CH_2 + HOH \xrightarrow{H^+}$

b. $+ HOH \xrightarrow{H^+}$

Solution

a. Water adds an H— and an —OH to the double bond. We use Markovnikov's rule to add the H— to the CH_2 in the double bond, and the —OH to the CH.

$$CH_3CH_2CH_2-\overset{\overset{\text{OH}}{\downarrow}}{CH}=\overset{\overset{\text{H}}{\downarrow}}{CH_2} \xrightarrow{H^+} CH_3CH_2CH_2-\overset{\overset{\text{OH}}{|}}{CH}-CH_3$$

b. In cyclobutene, the H— adds to one side of the double bond, and the —OH adds to the other side.

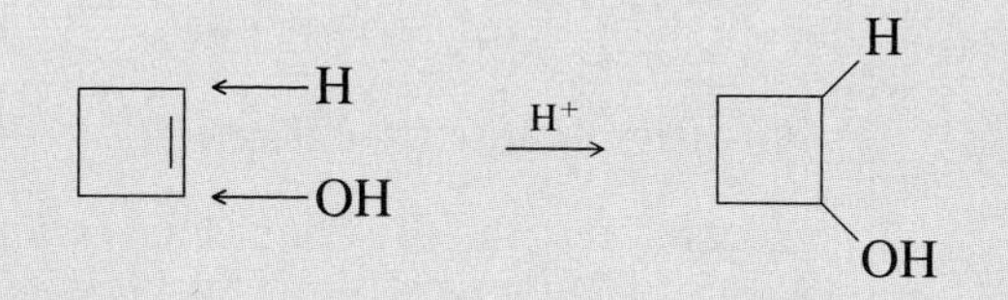

Study Check

Draw the structural formula for the alcohol obtained by the hydration of 2-methyl-2-butene.

Environmental Note *Polymers from Alkenes*

Polymers are long-chain molecules formed from many repeating units of small carbon molecules. Important polymers in nature include cellulose, rubber, silk, wool, starch, proteins, and DNA, the molecule that carries genetic information. Over the past 100 years, the plastics industry has learned to produce synthetic polymers that are used to make many of the materials we use every day such as carpeting, plastic wrap, nonstick pans, nylon, dacron, and other synthetic fibers. In medicine, biopolymers are used to replace diseased or damaged body parts such as hip joints, teeth, heart valves, and blood vessels. (See Figure 11.8.) Most synthetic **polymers** are made by the addition

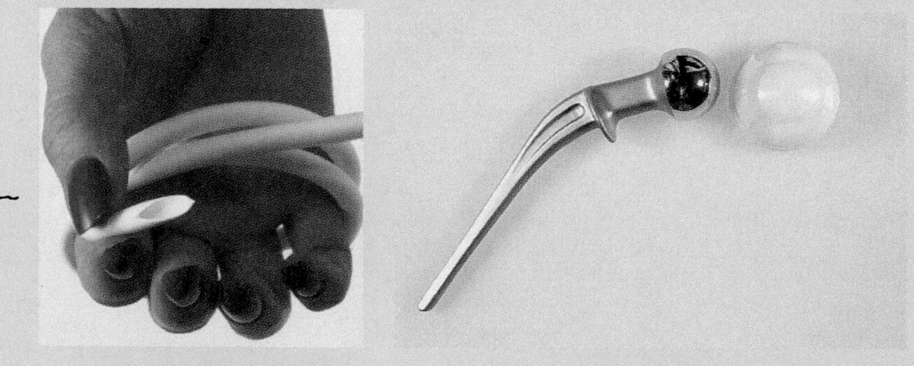

Figure 11.8
Synthetic polymers are finding extensive use in replacing diseased body parts such as an artery or hip joint.

reactions between many small alkenes called monomers at 100°C and high pressures (over 1000 atm). For example, polyethylene, a polymer used in the manufacturing of plastic bottles, film, and plastic dinnerware, is made from molecules of ethylene ($CH_2{=}CH_2$).

The double bonds of the alkene **monomer** units open up to add to the next monomer forming a carbon chain of single (C—C) bonds. This polymerizing process continues until a typical synthetic polymer has from 100 to 1000 of the monomer units in its chain. Other alkenes have been used to make polymers used as a food wrapping, such as Saran Wrap, and Teflon, a nonstick coating for pans and cooking utensils. Another polymer known as polyvinyl chloride, PVC, is used as "vinyl" plastic to make such things as plastic covers, plastic pipes and tubing, body replacement parts, and garden hoses. Table 11.1 lists some alkenes used as monomers in the formation of synthetic polymers and shows examples of each.

The alkane-like nature of these plastic synthetic polymers makes them unreactive. Thus, they do not decompose easily (they are nonbiodegradable) and have become contributors to pollution. Efforts are being made to make them more degradable. It becomes increasingly important to recycle plastic material, rather than add to our growing landfills.

$$\underset{\text{Monomers of ethylene}}{CH_2{=}CH_2 + CH_2{=}CH_2} \xrightarrow{\text{Catalyst}} \underset{\text{Repeating monomer units}}{—CH_2—CH_2—CH_2—CH_2—} + \text{many } CH_2{=}CH_2 \longrightarrow$$

$$\underset{\text{Polyethylene}}{—CH_2—(CH_2—CH_2—)_n—CH_2—} \text{ etc.}$$

$$\underset{\substack{\text{Chloroethene}\\\text{(vinyl chloride)}}}{CH_2{=}\overset{\overset{\large Cl}{|}}{CH} + CH_2{=}\overset{\overset{\large Cl}{|}}{CH}} \longrightarrow \underset{\substack{\text{Polychloroethene}\\\text{(polyvinyl chloride, PVC)}}}{—CH_2—\overset{\overset{\large Cl}{|}}{CH}—CH_2—\overset{\overset{\large Cl}{|}}{CH}—}$$

Table 11.1 ***Some Alkenes and Their Polymers***

Monomer	Polymer	Uses for Polymer
$H_2C{=}CH_2$ Ethene (ethylene)	$—CH_2—CH_2—CH_2—CH_2—$ Polyethylene	Plastic bottles, film, insulation material
$CH_3CH{=}CH_2$ Propene (propylene)	$—CH_2—\overset{\overset{CH_3}{|}}{CH}—CH_2—\overset{\overset{CH_3}{|}}{CH}—$ Polypropylene	Ski and hiking clothing, carpets, artificial joints

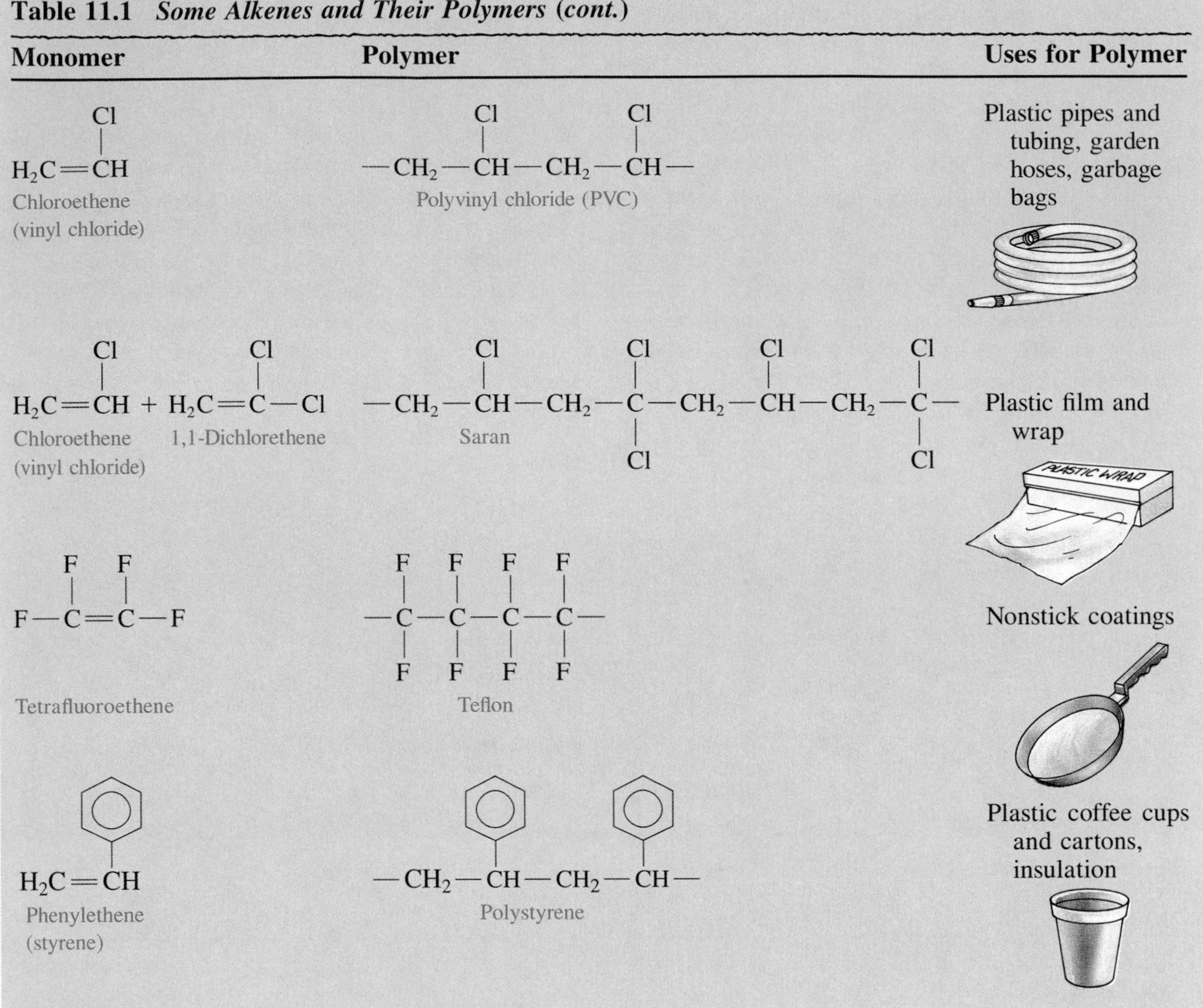

Table 11.1 ***Some Alkenes and Their Polymers (cont.)***

Monomer	Polymer	Uses for Polymer
$H_2C{=}CH(Cl)$ Chloroethene (vinyl chloride)	$-CH_2-CH(Cl)-CH_2-CH(Cl)-$ Polyvinyl chloride (PVC)	Plastic pipes and tubing, garden hoses, garbage bags
$H_2C{=}CH(Cl)$ + $H_2C{=}C(Cl)-Cl$ Chloroethene (vinyl chloride) 1,1-Dichlorethene	$-CH_2-CH(Cl)-CH_2-C(Cl)(Cl)-CH_2-CH(Cl)-CH_2-C(Cl)(Cl)-$ Saran	Plastic film and wrap
$F-C(F){=}C(F)-F$ Tetrafluoroethene	$-CF_2-CF_2-CF_2-CF_2-$ Teflon	Nonstick coatings
$H_2C{=}CH(C_6H_5)$ Phenylethene (styrene)	$-CH_2-CH(C_6H_5)-CH_2-CH(C_6H_5)-$ Polystyrene	Plastic coffee cups and cartons, insulation

11.6 Alkynes

Learning Goal

Name alkynes using IUPAC rules and write their condensed structural formulas.

The **alkynes** are a family of unsaturated hydrocarbons that contain a triple bond. The simplest alkyne is named ethyne in the IUPAC system, but it is commonly known as acetylene. (See Figure 11.9.) A triple bond is found in acetylene, one of the gases used in a welder's torch. (See Figure 11.10.)

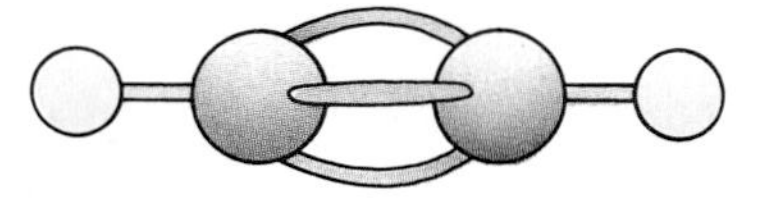

Figure 11.9
A ball-and-stick model of acetylene (IUPAC name *ethyne*).

Figure 11.10
Acetylene is one of the gases used to obtain the high temperatures in a welder's torch.

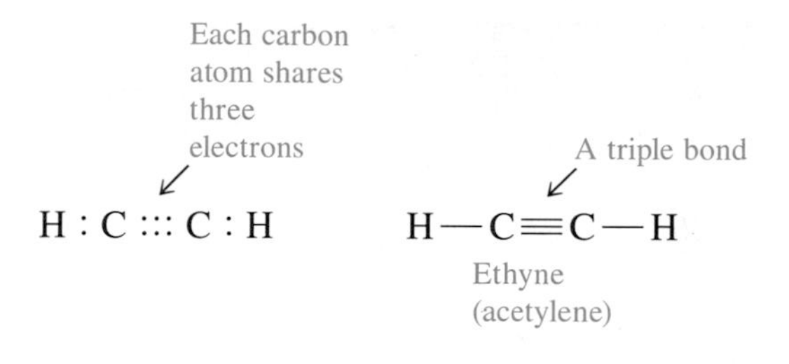

In the IUPAC system, the name of an alkyne ends in *yne*. The alkyne with two carbons is usually called by its common name, acetylene.

$CH \equiv CH$
Ethyne
(acetylene)

Naming Alkynes

Functional group: $—C \equiv C—$
IUPAC name: alkyne

IUPAC Rules for Naming Alkynes

Step 1 Select the longest carbon chain *containing the triple bond* as the parent chain. Replace the ane ending of the corresponding alkane with *yne*.

$CH \equiv C — CH_3$
Propyne

Step 2 Number the parent chain from the end nearest to the triple bond. Use the number of the first carbon atom in the triple bond to give its location. There are no cis–trans isomers for alkynes.

$$\underset{1}{HC}\equiv\underset{2}{C}-\underset{3}{CH_2}\underset{4}{CH_3} \qquad \underset{1}{CH_3}-\underset{2}{C}\equiv\underset{3}{C}-\underset{4}{CH_3}$$

1-Butyne 2-Butyne

Step 3 Locate by number and name the branches in alphabetical order. Some examples of alkynes and their names follow:

$$CH\equiv C-CH_2CH_3 \qquad CH_3-C\equiv C-\overset{\overset{\displaystyle CH_3}{|}}{CH}-CH_3$$

1-Butyne 4-Methyl-2-pentyne (**not** 2-methyl-3-pentyne)

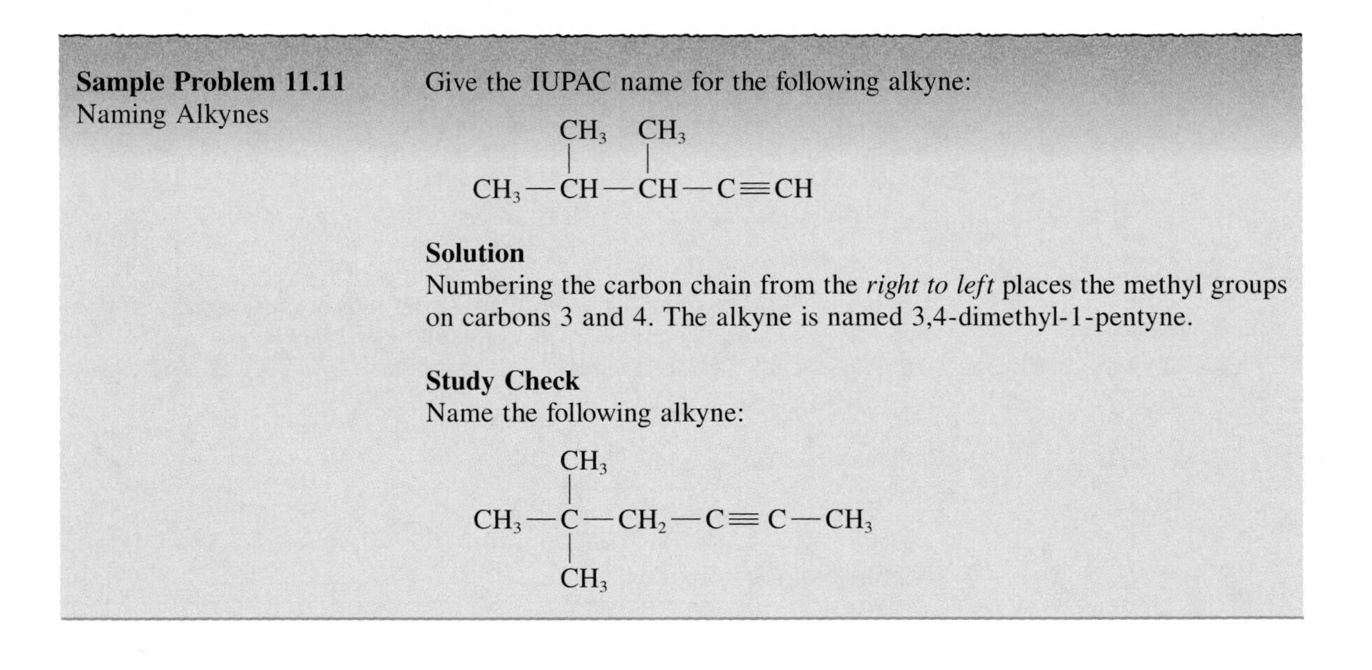

Sample Problem 11.11
Naming Alkynes

Give the IUPAC name for the following alkyne:

$$CH_3-\overset{\overset{\displaystyle CH_3}{|}}{CH}-\overset{\overset{\displaystyle CH_3}{|}}{CH}-C\equiv CH$$

Solution
Numbering the carbon chain from the *right to left* places the methyl groups on carbons 3 and 4. The alkyne is named 3,4-dimethyl-1-pentyne.

Study Check
Name the following alkyne:

$$CH_3-\underset{\underset{\displaystyle CH_3}{|}}{\overset{\overset{\displaystyle CH_3}{|}}{C}}-CH_2-C\equiv C-CH_3$$

11.7 Aromatic Hydrocarbons

Learning Goal
Describe the bonding in benzene; name aromatic compounds, and write their structural formulas.

In 1825, Michael Faraday isolated a hydrocarbon called benzene, which had a molecular formula C_6H_6. Because many compounds containing benzene had fragrant odors, the family of benzene compounds became known as **aromatic compounds.** Benzene consists of six carbons atoms each attached to a hydrogen atom. Although the formula suggests unsaturation, benzene is unusually stable. In reactions, benzene undergoes substitution, and not the addition typical of alkenes. In 1865, August Kekulé dreamed that the CH groups were arranged in a cyclic flat ring with alternating single and double bonds.

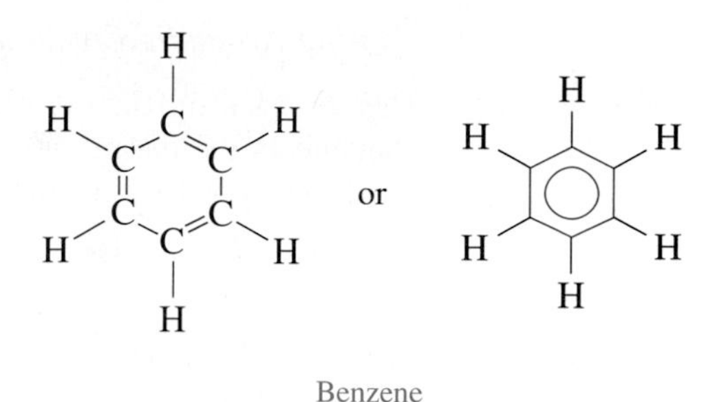

Benzene

However, scientists knew that the bonds in benzene are equal in length and behave more like a blend of a single and double bonds. Kekulé suggested that benzene was better represented as a combination of two structures most often represented by a hexagon with a circle in the center.

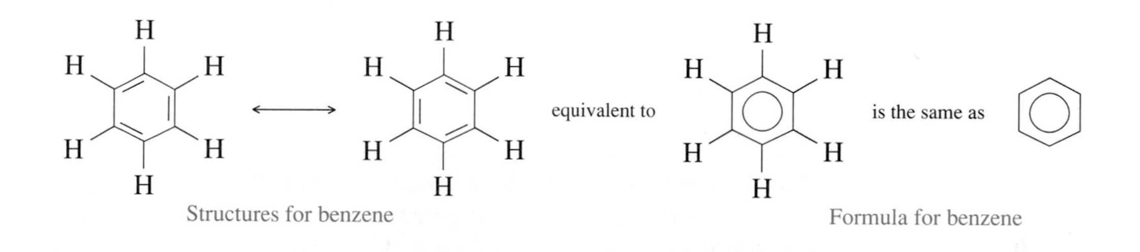

Structures for benzene

Formula for benzene

HEALTH NOTE *Aromatic Compounds in Health and Medicine*

Aromatic compounds are common in nature and in medicine. Toluene is used as a starting reactant to make drugs, dyes, and explosives such as TNT (trinitrotoluene). The benzene ring is found in some amino acids (the building blocks of proteins), in pain relivers such as aspirin, acetaminophen, and ibuprofen, and flavorings such as vanillin.

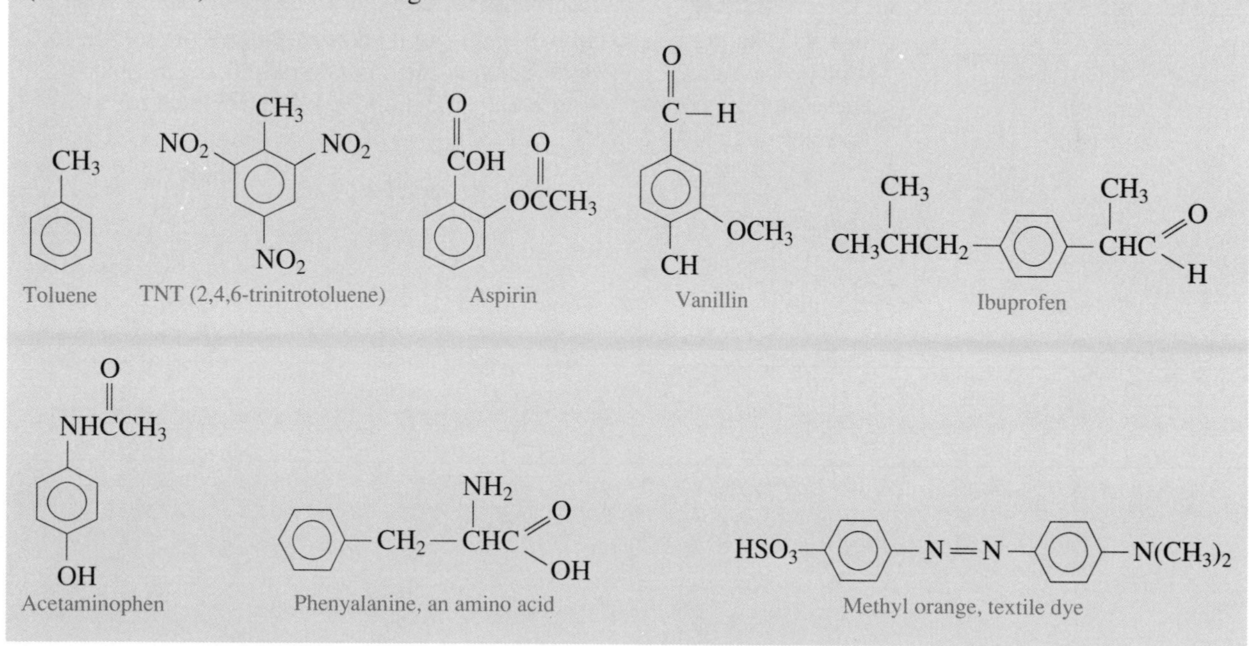

Toluene TNT (2,4,6-trinitrotoluene) Aspirin Vanillin Ibuprofen

Acetaminophen Phenyalanine, an amino acid Methyl orange, textile dye

Naming Aromatic Compounds

Step 1 Identify the benzene ring in a hydrocarbon and assign the name benzene as the parent name. Write the names and locations of any substituents in front of the name *benzene.*

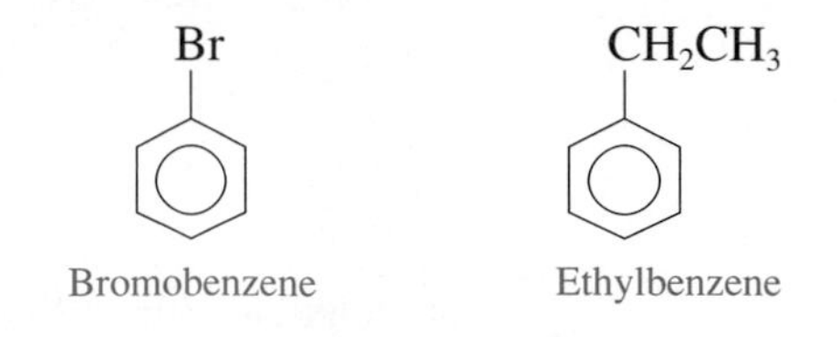

Bromobenzene Ethylbenzene

Step 2 Many of the important monosubstituted aromatic compounds have retained their common names, which are acceptable in the IUPAC system.

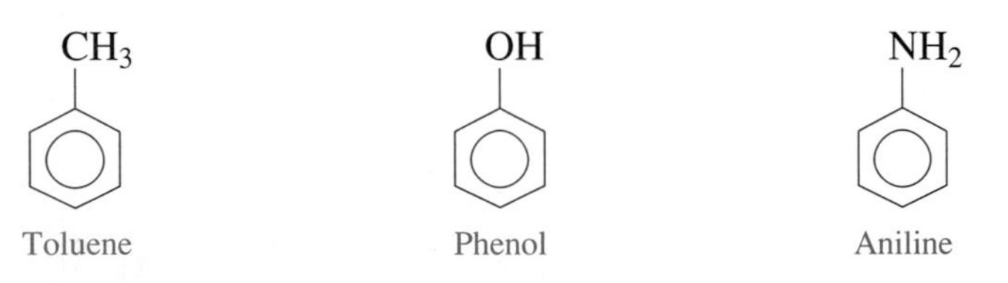

Toluene Phenol Aniline

Step 3 For two or more substituents, number the benzene ring starting with one of the substituents to give the lowest combination of numbers to the substituents.

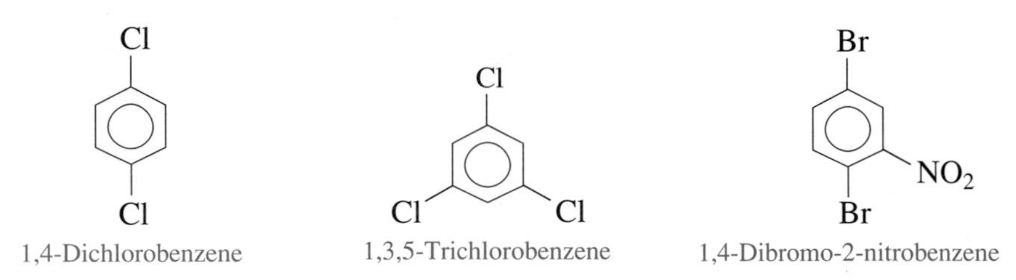

1,4-Dichlorobenzene 1,3,5-Trichlorobenzene 1,4-Dibromo-2-nitrobenzene

Step 4 The prefixes *ortho* (*o*), *meta* (*m*), and *para* (*p*) are often used to indicate the relative position of *two substituents* on the benzene ring. These prefixes correspond to a 1,2-, 1,3-, and a 1,4- placement on a benzene ring.

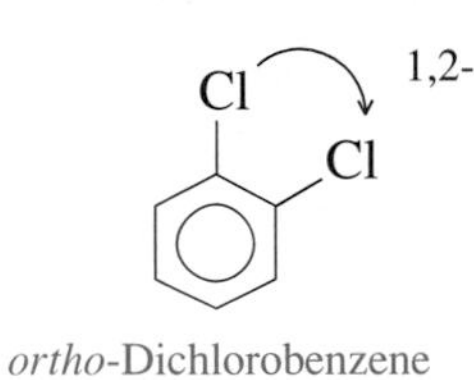

ortho-Dichlorobenzene

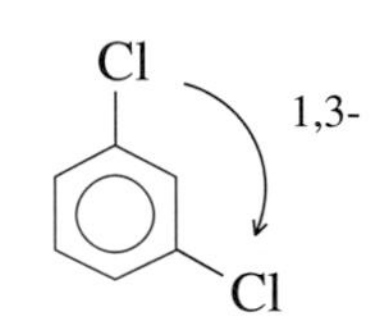

meta-Dichlorobenzene

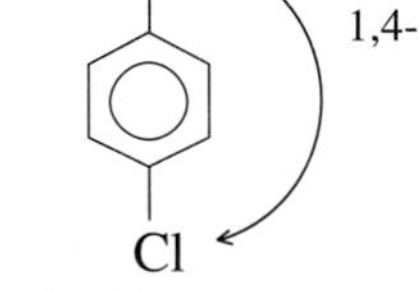

para-Dichlorobenzene

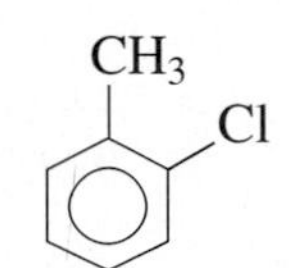

ortho-Chlorotoluene

NH_2

Br

meta-Bromoaniline

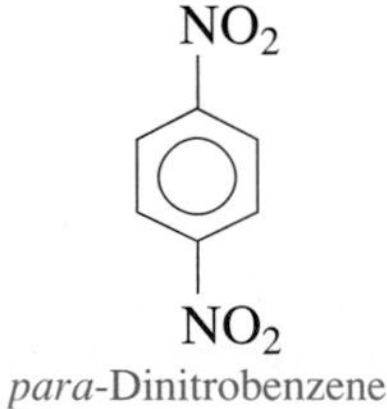

para-Dinitrobenzene

Step 5 When the benzene ring is a substituent on a longer carbon chain, it is named *phenyl:*

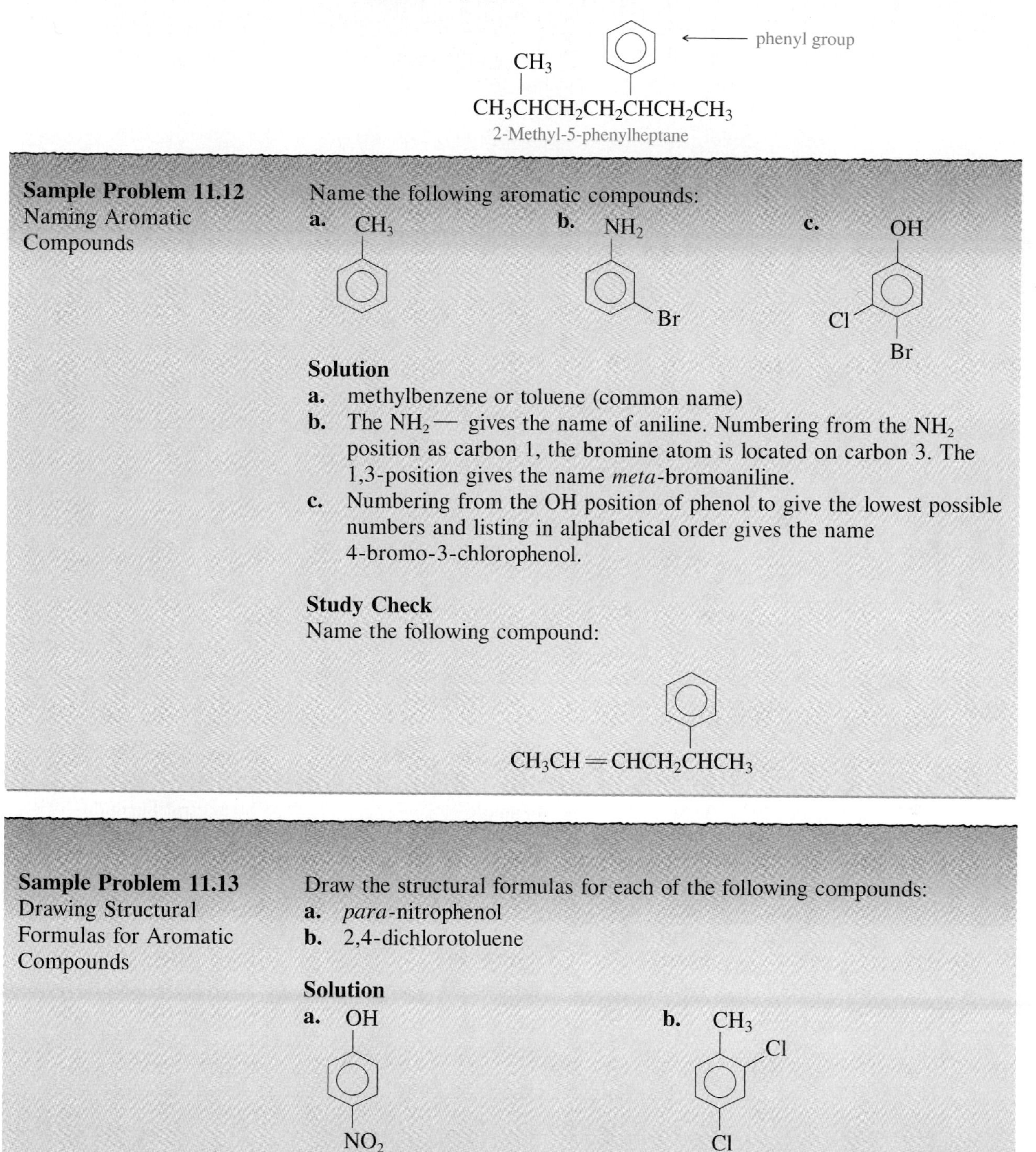

2-Methyl-5-phenylheptane

Sample Problem 11.12
Naming Aromatic Compounds

Name the following aromatic compounds:

a. CH_3 **b.** NH_2, Br **c.** OH, Cl, Br

Solution

a. methylbenzene or toluene (common name)

b. The NH_2— gives the name of aniline. Numbering from the NH_2 position as carbon 1, the bromine atom is located on carbon 3. The 1,3-position gives the name *meta*-bromoaniline.

c. Numbering from the OH position of phenol to give the lowest possible numbers and listing in alphabetical order gives the name 4-bromo-3-chlorophenol.

Study Check

Name the following compound:

$CH_3CH{=}CHCH_2CHCH_3$

Sample Problem 11.13
Drawing Structural Formulas for Aromatic Compounds

Draw the structural formulas for each of the following compounds:

a. *para*-nitrophenol

b. 2,4-dichlorotoluene

Solution

a. OH, NO_2 **b.** CH_3, Cl, Cl

Study Check

Draw the structural formula for *ortho*-bromophenol.

HEALTH NOTE *Polycyclic Aromatic Hydrocarbons*

Large aromatic compounds known as **polycyclic aromatic hydrocarbons** are formed by fusing together two or more benzene rings edge-to-edge. In a fused ring compound, neighboring benzene rings share two or more carbon atoms. Naphthalene with two benzene rings is well known for its use in mothballs. Anthracene with three rings is used in the manufacture of dyes.

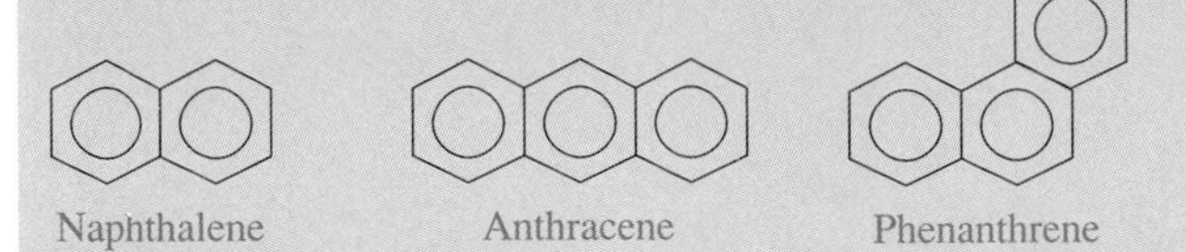

When a polycyclic compound contains phenanthrene, it may act is a carcinogen, a substance known to cause cancer. For example, ben-

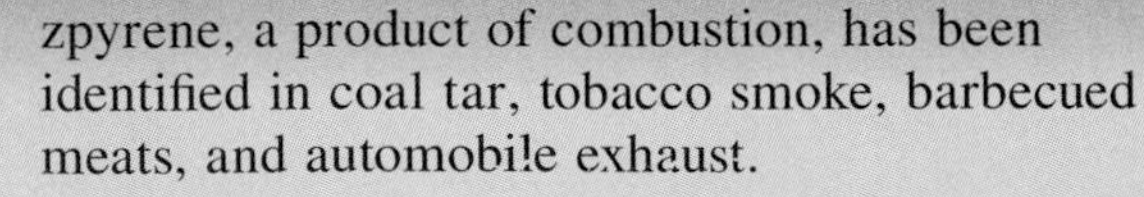
zpyrene, a product of combustion, has been identified in coal tar, tobacco smoke, barbecued meats, and automobile exhaust.

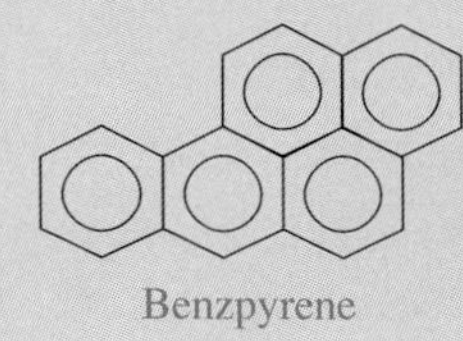

Fused rings containing five or more benzene rings such as benzpyrene are potent cancer-causing chemicals (carcinogens). The molecules interact with the DNA in the cells causing abnormal cell growth and cancer. Increased exposure to carcinogens increases the chance of DNA alterations in the cells. (See Figure 11.11.)

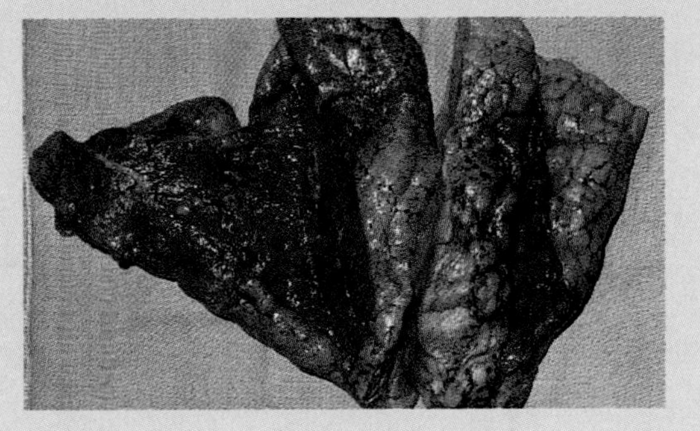

Figure 11.11
Some aromatic compounds in cigarette smoke cause cancer, as seen in the lung tissue of a heavy smoker.

Summary of Naming Unsaturated and Aromatic Hydrocarbons

Type	Example	Structural Formula
Alkene	Propene (propylene)	$CH_3—CH=CH_2$
Cycloalkene	Cyclopropene	(triangle with double bond)
Alkyne	Propyne	$CH_3—C \equiv CH$
Aromatic	Benzene	(benzene ring)
	Methylbenzene or toluene	CH_3 on benzene ring
	1,4-dichlorobenzene or *para*-dichlorobenzene	Cl on benzene ring, Cl at para position

Summary of Addition Reactions of Alkenes

Hydrogenation

$$\text{Alkene} + H_2 \xrightarrow{Pt} \text{alkane}$$

$$CH_2{=}CH{-}CH_3 + H_2 \xrightarrow{Pt} CH_3{-}CH_2{-}CH_3$$

Halogenation

$$\text{Alkene} + Cl_2 \text{ (or } Br_2) \longrightarrow \text{dihaloalkane}$$

$$CH_2{=}CH{-}CH_3 + Cl_2 \longrightarrow \underset{\large Cl}{\overset{}{CH_2}}{-}\underset{\large Cl}{CH}{-}CH_3$$

(Cl on CH_2 and Cl on CH)

Hydrohalogenation

$$\text{Alkene} + HX \longrightarrow \text{haloalkane}$$

$$CH_2{=}CH{-}CH_3 + HCl \longrightarrow CH_3{-}\overset{\displaystyle Cl}{\overset{|}{CH}}{-}CH_3$$ Markovnikov's rule

Hydration of Alkenes

$$\text{Alkene} + H{-}OH \xrightarrow{H^+} \text{alcohol}$$

$$CH_2{=}CH{-}CH_3 + H{-}OH \xrightarrow{H^+} CH_3{-}\overset{\displaystyle OH}{\overset{|}{CH}}{-}CH_3$$ Markovnikov's rule

Chapter Summary

11.1 Alkenes
Alkenes are unsaturated hydrocarbons that contain at least one double bond (C=C) as the functional group.

11.2 Naming Alkenes
Alkenes and cycloalkenes use IUPAC rules similar to the alkanes but use an *ene* ending. In alkenes, the longest carbon chain containing the double bond is numbered from the end nearest the double bond. If there are any substituents on a cycloalkene, the double bond is assigned the positions of 1 and 2, and the ring numbered to give the next lower numbers to the substituents.

11.3 Cis–Trans Isomers
Geometric isomers of alkenes occur when the carbon atoms in the double bond are connected to different atoms or groups. In the cis isomer, the attached groups are on the same side of the double bond, whereas in the trans isomer, they are connected on the opposite sides of the double bond.

11.4 Addition of Hydrogen and Halogens
The addition of small molecules to the double bond is a characteristic reaction of alkenes. Hydrogenation adds hydrogen atoms to the double bond of an alkene to yield an alkane. Halogenation adds bromine or chlorine atoms to produce dihaloalkanes.

11.5 Addition of Hydrogen Halides and Water
Hydrogen halides and water can also add to a double bond. When there are a different number of groups attached to the carbons in the double bond, Markovnikov's rule tells us to add the H from the adding reactant (HX or H—OH) to the carbon with the greater number of hydrogen atoms.

11.6 Alkynes
The alkynes are a family of unsaturated hydrocarbons that contain at least one triple bond. They use naming rules similar to the alkenes, but the parent chain ends with *yne.*

11.7 Aromatic Hydrocarbons
Most aromatic compounds contain benzene, a cyclic structure containing six CH units. The structure of benzene is represented as a hexagon with a circle in the center. The names of many aromatic compounds use the parent name benzene, although many common names were retained as IUPAC names, such as toluene, phenol, and aniline. The benzene ring is numbered and the branches are listed in alphabetical order. For two branches, the positions are often shown by the prefixes *ortho* (1,2-), *meta* (1,3-), and *para* (1,4-).

Glossary of Key Terms

addition A reaction in which atoms or groups of atoms bond to a double bond. Addition reactions include the addition of hydrogen (hydrogenation), halogens (halogenation), hydrogen halides (hydrohalogenation), or water (hydration).

alkene An unsaturated hydrocarbon containing a carbon–carbon double bond.

alkyne An unsaturated hydrocarbon containing a carbon–carbon triple bond.

aromatic compound A compound that contains the ring structure of benzene.

benzene A ring of six carbon atoms each of which is attached to a hydrogen atom, C_6H_6.

cis isomer A geometric isomer in which the groups are connected on the same side of the double bond.

cycloalkene A cyclic hydrocarbon that contains a double bond in the ring.

dienes Compounds with two carbon-carbon double bonds.

functional group An atom or group of atoms that has a strong influence on the overall chemical behavior of an organic compound.

geometric isomers Cis and trans isomers of unsymmetrical alkenes made possible by the lack of rotation around the double bond. In a cis isomer, the groups are attached to the same side of the double bond; in the trans isomer, they appear on opposite sides.

halogenation The addition of Cl_2 or Br_2 to an alkene to form halogen-containing compounds.

hydration An addition reaction in which the components of water, H— and —OH bond to the carbon–carbon double bond to form an alcohol.

hydrogenation The addition of hydrogen (H_2) to the double bond of alkenes to yield alkanes.

hydrohalogenation The addition of a hydrogen halide such as HCl or HBr to a double bond.

Markovnikov's rule When adding HX or HOH to alkenes with different numbers of groups attached to the double bonds, the H adds to the carbon that has the greater number of hydrogen atoms.

monomers Small molecules such as alkenes that are the units that bond to themselves to form a polymer.

polymers Large molecules formed by the combining of many small molecules called monomers.

trans isomer A geometric isomer in which the groups are connected to opposite sides of the double bond in an alkene.

Chemistry at Home

1. Read the labels on bottles of vegetable oils, margarine, and shortenings. What terms deal with the organic chemistry we have studied?
2. Make a list of the items you use or have in your room or home that are made of polymers.
3. Find the marks on the bottom of plastic bottles that give the type of polymer used. Research the environmental resources for a list of the markings and their recycling instructions.

Answers to Study Checks

11.1 cycloalkene

11.2 3-chloro-4-methyl-3-hexene

11.3 1-chloro-3-methylcyclopentene

11.4 No, there are two bromine atoms on one carbon.

11.5

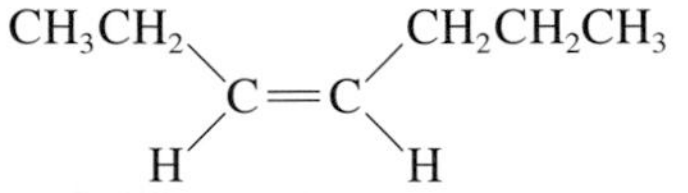

```
CH3CH2          CH2CH2CH3
      \        /
       C == C
      /        \
     H          H
```

11.6

```
   CH3
   |
CH3CHCH2CH3
```

11.7 1,2-dichlorobutane

11.8

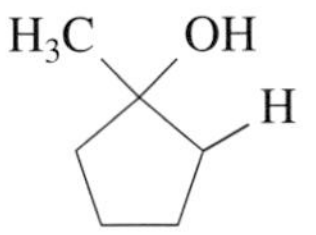

```
H3C   OH
   \ /
    X      H
   / \    /
   cyclopentane ring
```

11.9 $CH_2{=}CH{-}CH_2CH_2CH_3 + HCl \longrightarrow$

```
        Cl
        |
CH3 — CH — CH2CH2CH3
```

11.10 $CH_3-C(CH_3)(OH)-CH_2CH_3$

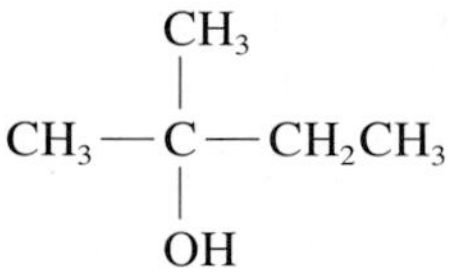

11.11 5,5-dimethyl-2-hexyne

11.12 5-phenyl-2-hexene

11.13 OH, Br (benzene ring with OH and adjacent Br)

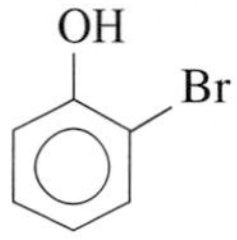

Problems

Alkenes *(Goal 11.1)*

11.1 Compare the structures of cyclohexane and cyclohexene.

11.2 Compare the structures of pentane and 1-pentene.

Naming Alkenes *(Goal 11.2)*

11.3. Give the IUPAC names for each of the following (ignore cis–trans isomers):

a. $CH_2{=}CH_2$

b. 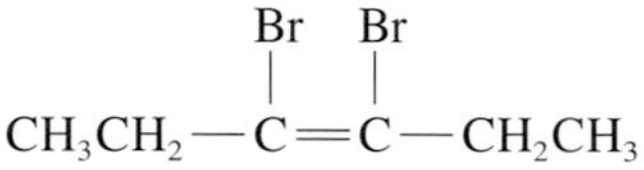 $CH_3CH_2-C(Br){=}C(Br)-CH_2CH_3$

c. $CH_3-CH(Cl)-C(CH_3){=}CH_2$

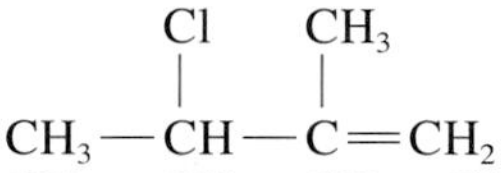

d. $CH_3-CH{=}CH-CH_2-CH_3$

e. 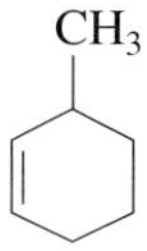 CH_3 (on cyclohexene ring)

11.4 Give the IUPAC names for each of the following (ignore cis–trans isomers):

a. $CH_3-CH{=}CH_2$

b. 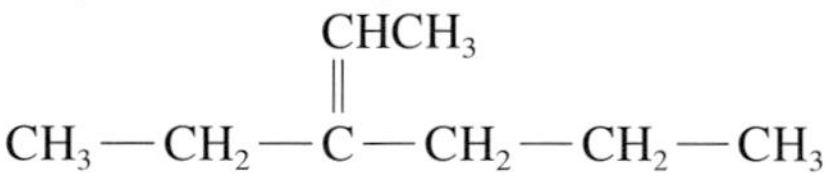 $CH_3-CH_2-C({=}CHCH_3)-CH_2-CH_2-CH_3$

c. H_3C, CH_2CH_3 (on cyclohexene ring)

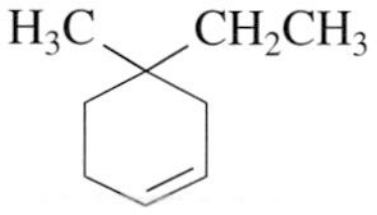

d. 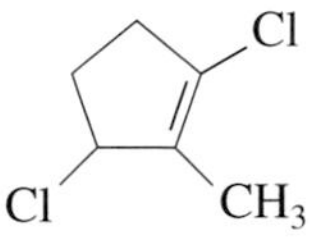 Cl, Cl, CH_3 (on cyclopentene ring)

e. 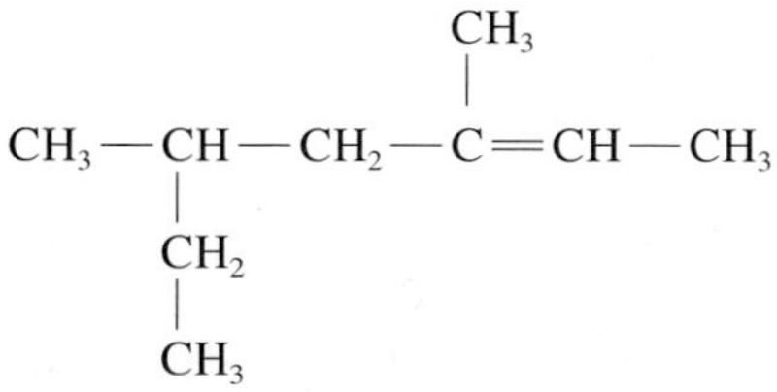 $CH_3-CH(CH_2CH_3)-CH_2-C(CH_3){=}CH-CH_3$

11.5 Draw the condensed structural formulas for each of the following compounds:

a. propene

b. 2-methyl-2-pentene

c. cyclobutene

d. 3-bromocyclopentene

e. 1,2-dichloro-3-methyl-2-heptene

11.6 Draw the condensed structural formulas for each of the following compounds:

a. 3-methyl-1-hexene

b. 1-methylcyclopentene

c. 1,1-dichloro-2-butene

d. 1,3-dimethylcyclobutene

e. 2-bromo-3,4-dimethyl-4-octene

Cis-Trans Isomers *(Goal 11.3)*

11.7 Give the IUPAC name for each of the following cis or trans isomers:

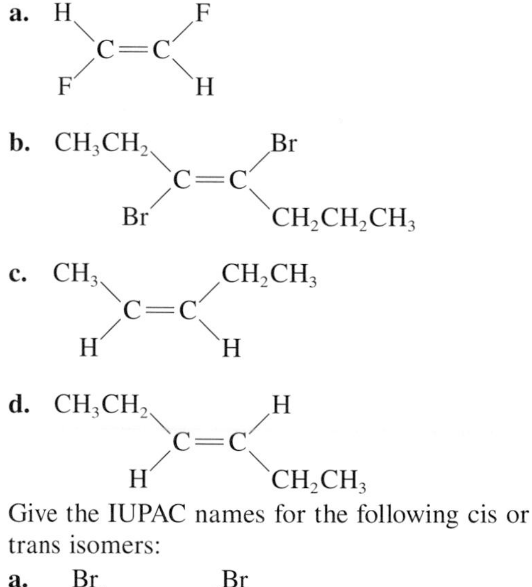

11.8 Give the IUPAC names for the following cis or trans isomers:

a. 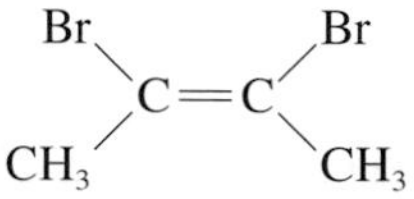 Br, Br, C=C, CH_3, CH_3

b. 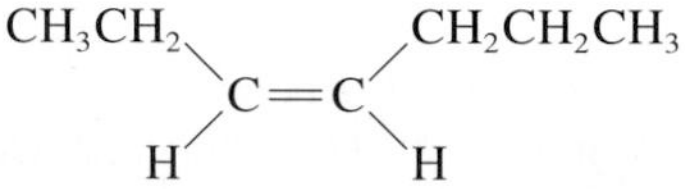 CH_3CH_2, $CH_2CH_2CH_3$, C=C, H, H

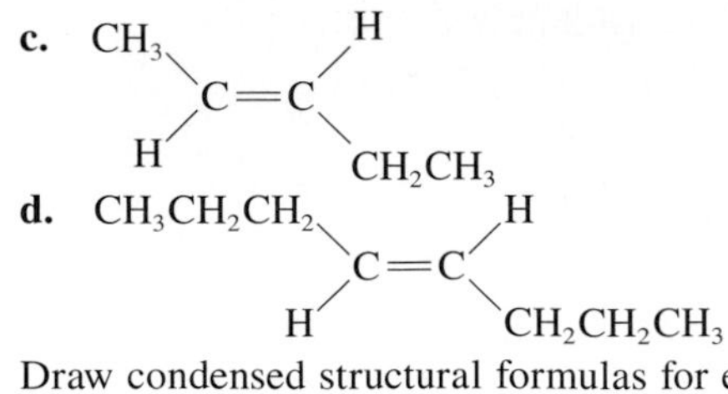

11.9 Draw condensed structural formulas for each of the following compounds:

a. *cis*-1,2-dibromoethene
b. *trans*-2-butene
c. *cis*-1,2-dibromo-1-butene
d. *cis*-3-heptene

11.10 Draw condensed structural formulas for each of the following compounds:

a. *cis*-2-pentene
b. *cis*-1-bromo-2-chloroethene
c. *trans*-2,3-dibromo-2-butene
d. *cis*-3-hexene

Addition of Hydrogen and Halogens *(Goal 11.4)*

11.11 Give the condensed structural formulas and names of the products in each of the following reactions:

a. $CH_3-CH=CH_2 + H_2 \xrightarrow{Pt}$

b. $CH_2=CH-CH(CH_3)-CH_3 + Cl_2 \longrightarrow$ (with CH_3 attached to the third carbon)

c. cyclobutene $+ Br_2 \longrightarrow$

d. cyclopentene $+ H_2 \xrightarrow{Pt}$

e. 2-methyl-2-butene $+ Cl_2 \longrightarrow$

11.12 Give the condensed structural formulas and names of the products in each of the following reactions:

a. $CH_3-CH_2-CH=CH_2 + Br_2 \longrightarrow$

b. cyclohexene $+ H_2 \xrightarrow{Pt}$

c. *cis*-2-butene $+ H_2 \xrightarrow{Pt}$

d. $CH_3-C(CH_3)=CH-CH_2-CH_3 + Cl_2 \longrightarrow$ (with CH_3 attached to the second carbon)

e. 1-methylcyclohexene (ring with CH_3) $+ Br_2 \longrightarrow$

Addition of Hydrogen Halides and Water *(Goal 11.5)*

11.13 Give the condensed structural formulas of the products in each of the following reactions using Markovnikov's rule when necessary:

a. $CH_3-CH=CH-CH_3 + HBr \longrightarrow$

b. cyclopentene $+ HOH \xrightarrow{H^+}$

c. $CH_2=CH-CH_2-CH_3 + HCl \longrightarrow$

d. $CH_3-C(CH_3)=C(CH_3)-CH_3 + HI \longrightarrow$ (with CH_3 attached to each of the double-bonded carbons)

e. $CH_3CH_2-CH=CH-CH_3 + HBr$

11.14 Give the condensed structural formulas of the products in each of the following reactions using Markovnikov's rule when necessary:

a. $CH_3-C(CH_3)=CH-CH_3 + HCl \longrightarrow$ (with CH_3 attached to the second carbon)

b. $CH_3CH_2-CH=CH-CH_2CH_3 + HOH \xrightarrow{H^+}$

c. $CH_3-C(CH_3)=CH_2 + HBr \longrightarrow$ (with CH_3 attached to the second carbon)

d. 1-methylcyclohexene $+ HOH \xrightarrow{H^+}$

e. 1-methylcyclohexene (ring with CH_3) $+ HBr \longrightarrow$

11.15 Write an equation including any catalysts for the following reactions:

a. hydrogenation of 2-methylpropene
b. addition of hydrogen chloride to cyclopentene
c. addition of bromine to 2-pentene
d. hydration of propene

11.16 Write an equation including any catalysts for the following reactions:

a. hydration of 1-methylcyclobutene
b. hydrogenation of 3-hexene
c. addition of hydrogen bromide to 2-methyl-2-butene
d. addition of chlorine to 2,3-dimethyl-2-pentene

Alkynes *(Goal 11.6)*

11.17 Name the following alkynes using the IUPAC rules:

a. $CH\equiv CH$

b. $CH_3C\equiv CH$

c. $CH_3CH_2-C\equiv C-CH_3$

d. $CH_3-CCl_2-CH_2-C\equiv CH$ (with two Cl attached to the second carbon)

e. $CH_3-C\equiv C-CH(CH_3)-CH(CH_3)-CH_3$ (with CH_3 attached to the fourth and fifth carbons)

11.18 Name the following alkynes using the IUPAC rules:

a. $CH{\equiv}C{-}CH_2CH_3$

b. $Cl{-}CH_2{-}C{\equiv}CH$

c. 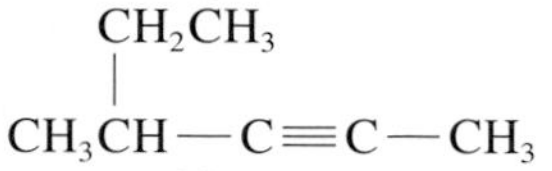 $CH_3CH(CH_2CH_3){-}C{\equiv}C{-}CH_3$ (with CH_2CH_3 on the CH)

d. 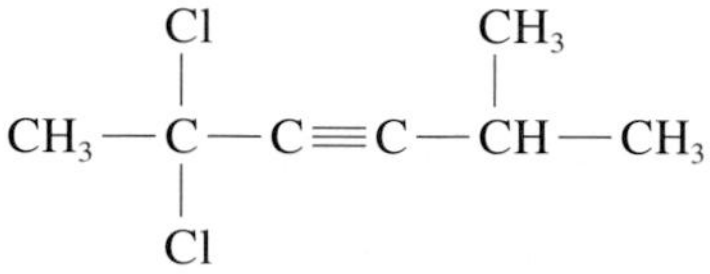 $CH_3{-}CCl_2{-}C{\equiv}C{-}CH(CH_3){-}CH_3$ (Cl above and below the second C; CH_3 on the CH)

e. 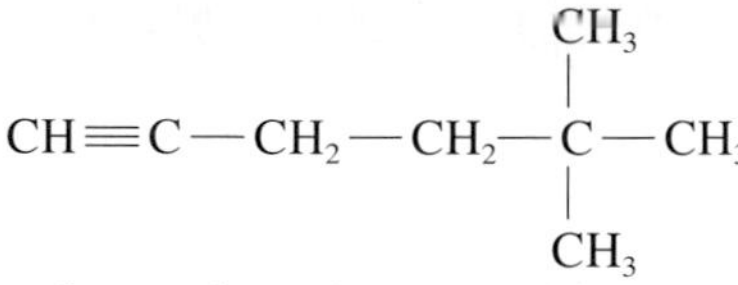 $CH{\equiv}C{-}CH_2{-}CH_2{-}C(CH_3)_2{-}CH_3$ (CH_3 above and below the C)

11.19 Draw the condensed structural formulas for each of the following alkynes:

a. 2-butyne
b. ethyne
c. 1-hexyne
d. 4,4-dimethyl-2-pentyne
e. 1,4-dichloro-2-butyne

11.20 Draw the condensed structural formulas for each of the following alkynes:

a. 3-chloropropyne
b. 3-ethyl-1-pentyne
c. 1-butyne
d. 4-methyl-1-bromo-2-hexyne
e. acetylene

Aromatic Hydrocarbons *(Goal 11.7)*

11.21 Name the following aromatic compounds:

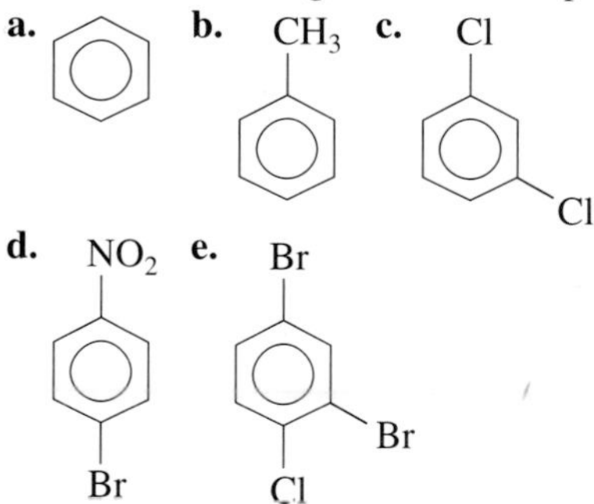

11.22 Name the following aromatic compounds:

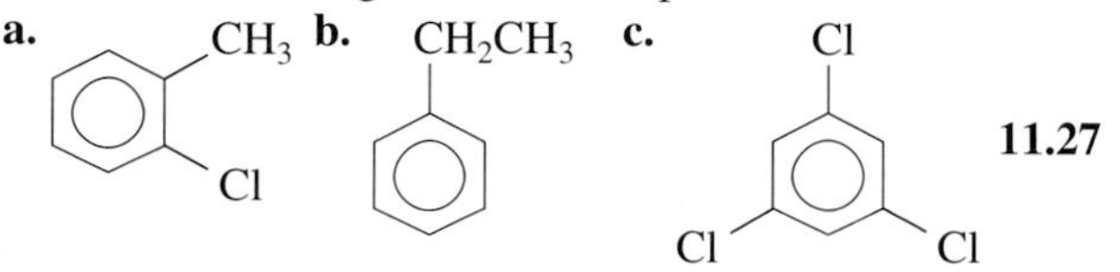

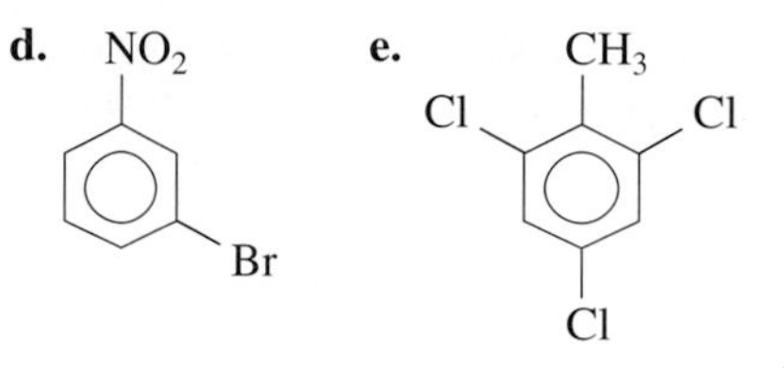

11.23 Draw the condensed structural formulas for each of the following:

a. bromobenzene
b. aniline
c. *p*-dichlorobenzene
d. 2,4-dibromotoluene
e. 1-chloro-2-phenylbutane

11.24 Draw the condensed structural formulas for each of the following:

a. nitrobenzene
b. 1,3,5-trichlorobenzene
c. *meta*-chlorotoluene
d. *o*-ethylnitrobenzene
e. phenylethyne

Challenge Problems

11.25 Give the family for each of the following compounds (alkane, cycloalkane, haloalkane, alkene, cycloalkene, alkyne, or aromatic):

a. $CH_3CH_2CH_2CH{=}CH_2$ **b.**

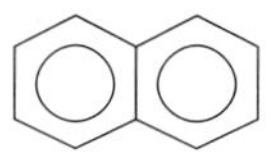

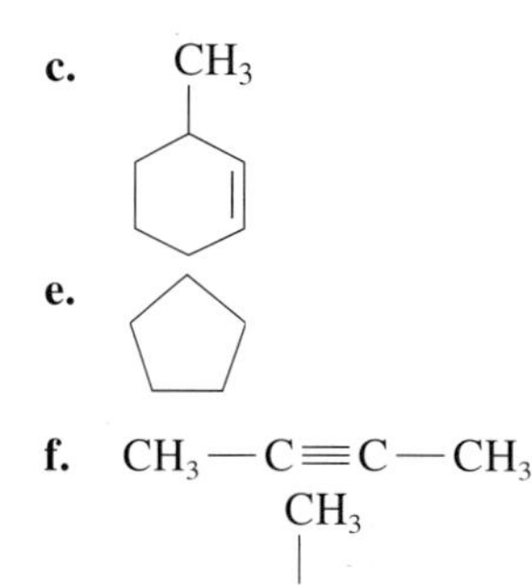

d. CH_3Cl

f. $CH_3{-}C{\equiv}C{-}CH_3$

g. 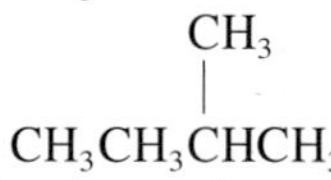 $CH_3CH_2CH(CH_3)CH_3$ (CH_3 on the CH)

11.26 Write the condensed structural formulas of each of the following compounds:

a. 1,2-dimethylcyclopentane
b. *p*-dichlorobenzene
c. 1,3-dibromobutane
d. *o*-chlorotoluene
e. 2,5-dichloro-2-hexene
f. 3-methylcyclobutene
g. *cis*-1,2-dichloro-1-propene
h. 2-butyne

11.27 Draw the structural formulas and give the names for all the isomers of C_4H_8 including cyclic and cis–trans isomers for alkenes.

11.28 What are the structural formulas of the starting materials needed to prepare the following products?

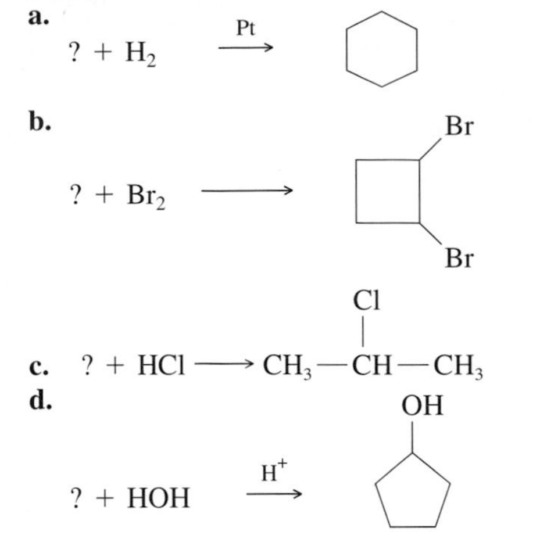

11.29 Three compounds will react with hydrogen (H_2) in the presence of an Ni catalyst to give methylcyclobutane. Draw the condensed structural formulas of these three compounds.

11.30 Write the combustion reaction for the acetylene used in a welder's torch.

11.31 Benzene, C_6H_6, undergoes combustion in the presence of oxygen.

a. Write the balanced equations for the combustion of benzene.

b. How many moles of CO_2 gas are formed by the combustion of 156 g of benzene?

c. How many liters of oxygen are required for the reaction in part (b) at STP?

Chapter 12
Alcohols, Phenols, Ethers, Aldehydes, and Ketones

The process of fermentation which converts glucose to ethanol is one of the earliest known chemical reactions.

Looking Ahead

Learning Goals

12.1 Identify the hydroxyl functional group (—OH) of alcohols and phenols; classify alcohols as primary, secondary, or tertiary.

12.2 Give the IUPAC or common name of an alcohol or phenol; draw the condensed structural formula from the name.

12.3 Describe the behavior of alcohols and phenols in water.

12.4 Write equations for the dehydration and oxidation of an alcohol.

12.5 Give the name of an ether; write the condensed structural formula from the name.

12.6 Give the IUPAC and common names of an aldehyde or ketone; draw the condensed structural formula from the name.

12.7 Describe some physical and chemical properties of aldehydes and ketones.

QUESTIONS TO THINK ABOUT

1. What alcohols are listed on labels of products such as antifreeze, wine, rubbing alcohol, cosmetics, extracts for food flavorings such as oil of almond, and hair spray? What part of the name tells you that it is an alcohol? What is the purpose of the alcohol?
2. What reaction does an alcohol undergo when you have crêpes suzette or banana flambé?
3. What substance is in the *canned heat* used to heat fondue dishes?
4. What is meant by the terms *grain alcohol* and *wood alcohol?*
5. Sailboats use alcohol burners instead of gas burners. What might be the reason?
6. What are some aldehydes or ketones found in products such as paint or fingernail polish remover, shampoos, and paint removers?
7. What is formaldehyde? What are the health hazards of using formaldehyde?

Several families of organic compounds contain the elements oxygen and sulfur in addition to carbon and hydrogen. In alcohols, phenols, and ethers, there is at least one oxygen atom connected by a single bond to a carbon atom. Thiols contain a sulfur atom as —SH bonded to a carbon chain. Aldehydes and ketones contain an oxygen connected to a carbon by a double bond.

The functional group in the alcohol family is called the hydroxyl group (—OH). For centuries, grains, vegetables, and fruits have been used in fermentation processes, which produce the ethyl alcohol in alcoholic beverages. Hydroxyl groups are also important in sugar and starch and in steroids such as cholesterol and estradiol. Menthol is an alcohol used in cough drops, shaving creams, and ointments. Ethyl ether, a compound used for general anesthesia for more than a hundred years, is the best known of the ether family. The simplest aldehyde, formaldehyde, is used for preserving tissue specimens. Naturally occurring aldehydes such as those found in vanilla beans and almonds are used by the food industry as fragrances and flavorings. One of the most widely used ketones is acetone or propanone, which is found in solvents used to remove paint and fingernail polish.

12.1 Alcohols and Phenols

Learning Goal

Identify the hydroxyl functional group (—OH) of alcohols and phenols; classify alcohols as primary, secondary, or tertiary.

In the family of organic compounds known as **alcohols,** a **hydroxyl group (—OH)** has taken the place of a hydrogen atom in an alkane. In a **phenol,** the hydroxyl group is attached to a carbon atom of a benzene ring. We can write the general structures of alcohols and phenols by showing an —OH group connected to an alkyl (R) group or a benzene ring in an aromatic (Ar) compound:

General Formula:

R—OH (Alcohol)   Ar—OH (Phenol)

Examples:

$CH_3—OH$

$CH_3CH_2—OH$

OH (on benzene ring)

Formation of Alcohols

In Chapter 11, we learned that alcohols can be produced by the addition of water (HOH) to an alkene in the presence of an acid (H^+) catalyst. The hydration of an alkene is the method used to prepare alcohols for industrial use. However, ethyl alcohol, the susbstance in alcoholic beverages, is most often prepared by fermentation of sugars.

Hydration of an Alkene Yields an Alcohol

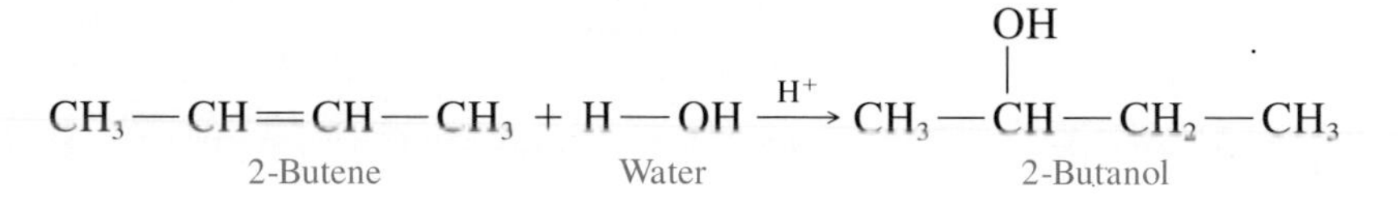

Classification of Alcohols

Alcohols are classified according to the number of alkyl groups attached to the *carbon atom* bonded to the —OH group. In a **primary (1°) alcohol**, there is one alkyl group attached to the carbon bonded to the —OH group. Methanol is considered a primary alcohol. In a **secondary (2°) alcohol,** there are two alkyl groups attached. When there are three alkyl groups, the alcohol is classified as a **tertiary (3°) alcohol.**

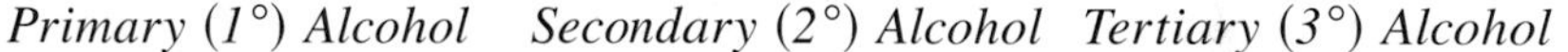

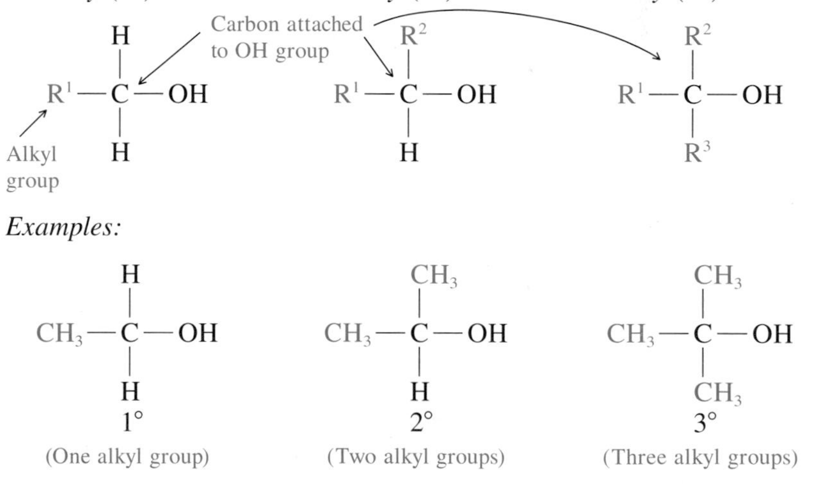

Sample Problem 12.1
Classifying Alcohols

Classify each of the following alcohols as primary, secondary, or tertiary.

a. $CH_3—CH_2—CH_2—OH$

b. $CH_3—CH_2—C(OH)(CH_3)—CH_3$

c. Cyclohexanol (cyclohexane ring bearing OH)

Solution

a. The carbon atom bonded to the —OH group is also attached to one alkyl group. This is a primary (1°) alcohol.

b. The carbon bonded to the —OH is attached to three alkyl groups. This is a tertiary (3°) alcohol.

c. The carbon bonded to the —OH group is bonded to two carbon atoms in the cyclic structure. This is a secondary (2°) alcohol.

Study Check

Classify the following alcohol as 1°, 2°, or 3°:

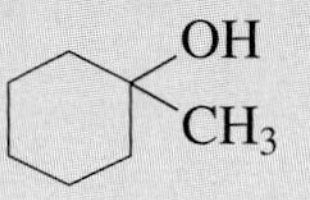

12.2 Names of Alcohols and Phenols

Learning Goal

Give the IUPAC or common name of an alcohol or phenol; draw the condensed structural formula from the name.

Common names, widely used for the smaller alcohols, consist of the name of the alkyl group (R) followed by the family name *alcohol*. To name alcohols using the IUPAC rules, select the parent chain and change the *e* of the corresponding alkane name to *ol*. For example, methane becomes methan*ol* and ethan*e* becomes ethan*ol*. (See Figure 12.1.)

Figure 12.1
Ball-and-stick models of methanol and ethanol.

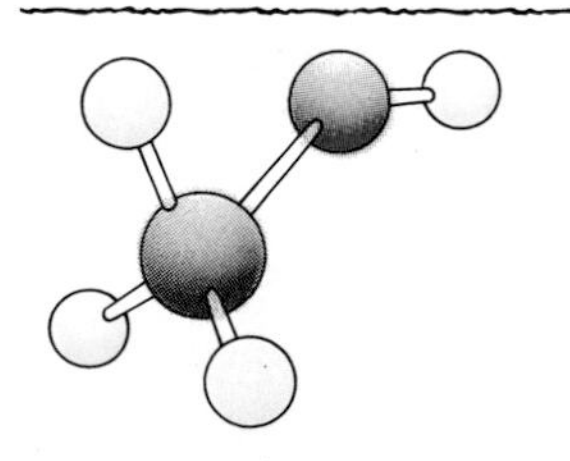

methanol

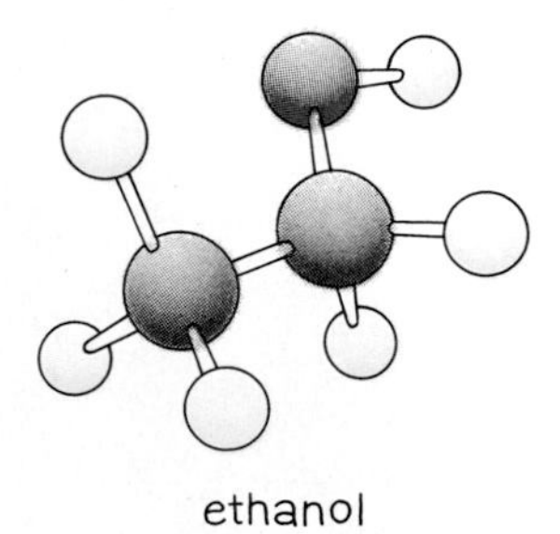

ethanol

Naming Alcohols R—OH

Functional group:	hydroxyl	—OH
IUPAC name:	alkanol	
Common name:	alkyl alcohol	

Some examples of the IUPAC names of alcohols and their common names (in parentheses) follow:

CH_3OH	CH_3CH_2OH	$CH_3CH_2CH_2OH$	$CH_3CH(OH)CH_3$
Methanol (methyl alcohol)	Ethanol (ethyl alcohol)	1-Propanol (propyl alcohol)	2-Propanol (isopropyl alcohol)

Suppose you want to name the following alcohol using the IUPAC rules.

$$CH_3-CH_2-\underset{}{\overset{OH}{\overset{|}{CH}}}-CH_3$$

IUPAC Rules for Naming Alcohols

Step 1 Select the longest carbon chain containing the —OH group as the parent chain and replace the *e* in its name by *ol.*

$$CH_3{-}CH_2{-}\underset{}{\overset{\displaystyle OH}{\overset{|}{CH}}}{-}CH_3$$

Butanol (replace *e* of butane with *ol*)

Step 2 Number the parent chain starting at the end closest to the —OH group.

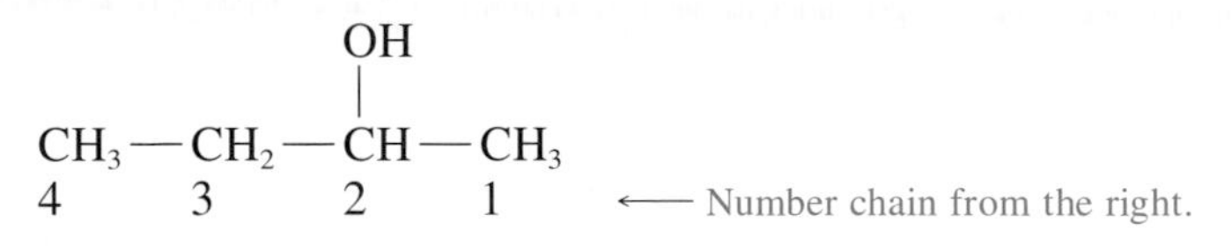

Step 3 Give the number for the position of the hydroxyl group in front of the parent name. In this example, the —OH group is on carbon 2 of the parent chain.

2-Butanol

You can check your answer by reviewing the parts of the name.

2-	but	an	ol
Functional group on second carbon	4 carbon atoms in parent chain	Single bonds between carbons	Alcohol functional group

Step 4 Give the names and numbers of any substituents.

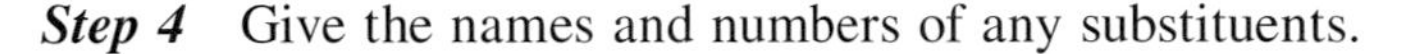

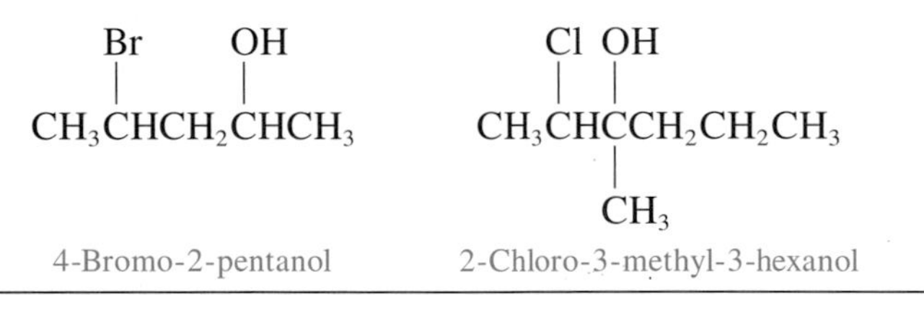

4-Bromo-2-pentanol 2-Chloro-3-methyl-3-hexanol

A cyclic alcohol is named as a *cycloalkanol.* The carbon ring is numbered so that the carbon atom bonded to the —OH group is carbon 1. Any other side groups are shown by numbering from carbon 1. Alcohols containing two —OH groups are named *diols;* three —OH groups, *triols.* Each of the hydroxyl groups is numbered.

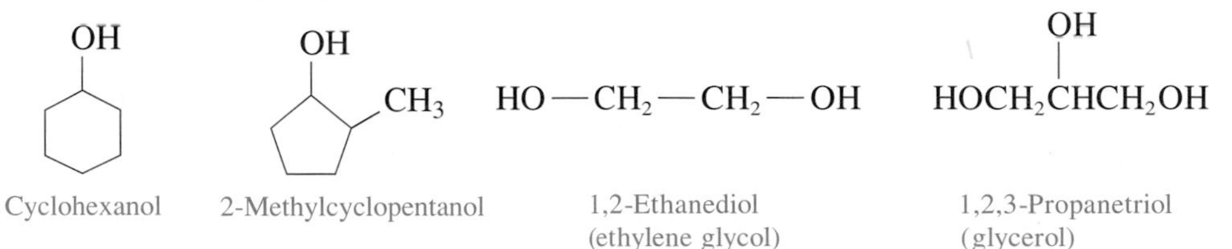

Cyclohexanol 2-Methylcyclopentanol 1,2-Ethanediol (ethylene glycol) 1,2,3-Propanetriol (glycerol)

Naming Phenols

Earlier we named aromatic compounds with —OH groups *phenols.* When there is a second group on the benzene ring, the ring is numbered starting from the carbon 1 bonded to the —OH group, or the prefixes *ortho, meta,* and *para* may be used.

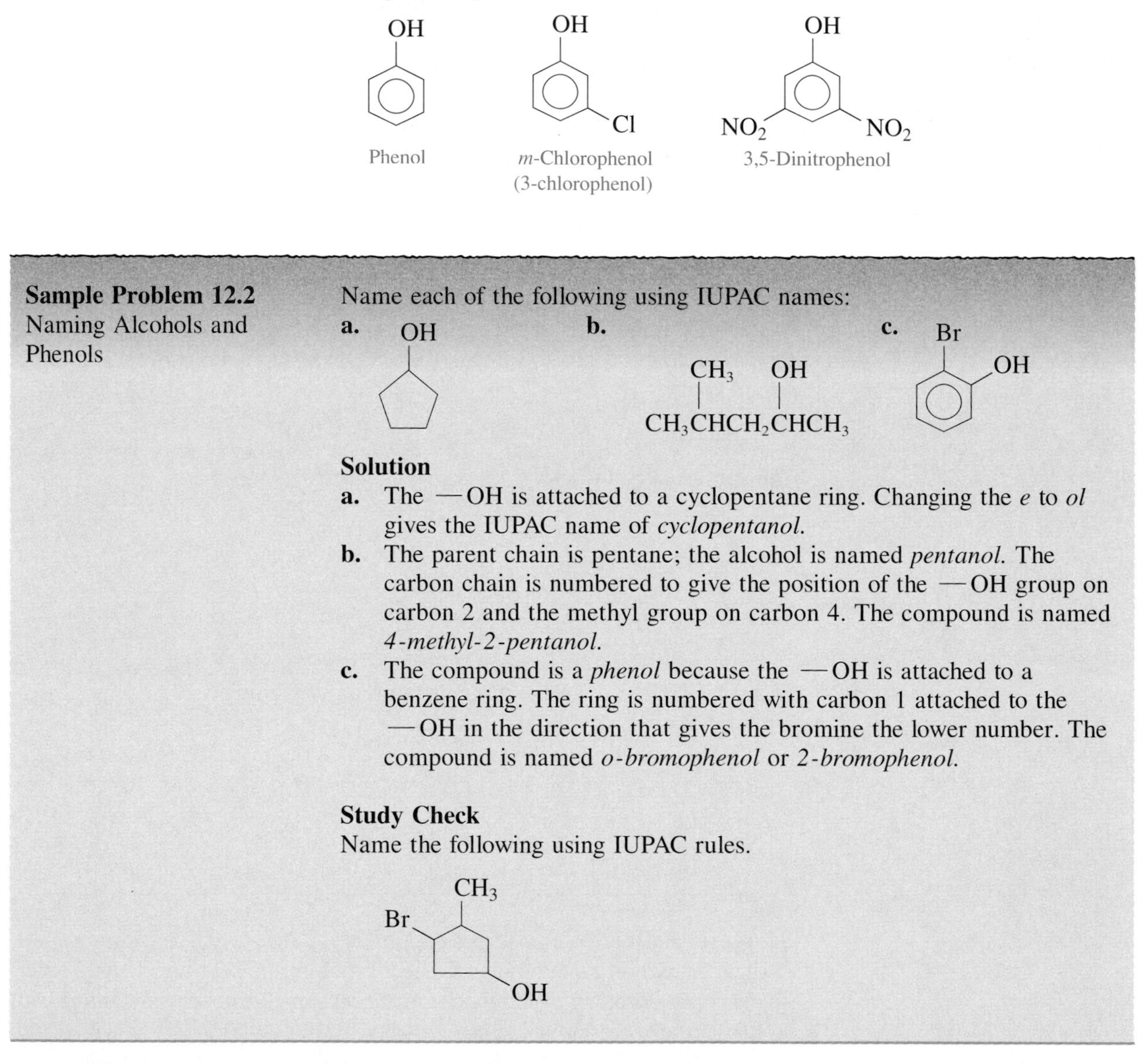

Phenol | *m*-Chlorophenol (3-chlorophenol) | 3,5-Dinitrophenol

Sample Problem 12.2
Naming Alcohols and Phenols

Name each of the following using IUPAC names:

a. (cyclopentane ring with OH)

b.
$$\begin{array}{l} \quad\ \ CH_3 \quad\ \ OH \\ \quad\ \ \ | \qquad\quad\ | \\ CH_3CHCH_2CHCH_3 \end{array}$$

c. (benzene ring with Br and OH on adjacent carbons)

Solution

a. The —OH is attached to a cyclopentane ring. Changing the *e* to *ol* gives the IUPAC name of *cyclopentanol.*

b. The parent chain is pentane; the alcohol is named *pentanol.* The carbon chain is numbered to give the position of the —OH group on carbon 2 and the methyl group on carbon 4. The compound is named *4-methyl-2-pentanol.*

c. The compound is a *phenol* because the —OH is attached to a benzene ring. The ring is numbered with carbon 1 attached to the —OH in the direction that gives the bromine the lower number. The compound is named *o-bromophenol* or *2-bromophenol.*

Study Check

Name the following using IUPAC rules.

(cyclopentane ring with CH_3, Br and OH substituents)

Sample Problem 12.3
Writing Condensed Structural Formulas of Alcohols and Phenols

a. The alcohol 3,5,5-trimethyl-1-hexanol is used as a plasticizer, a substance added to plastics to keep them pliable. Draw its condensed structural formula.

b. Thymol, used as an antiseptic in mouthwashes, has an IUPAC name of 5-methyl-2-isopropylphenol. Draw its condensed structural formula.

Solution

a. The parent chain of six carbon atoms has an —OH group on carbon 1. Methyl groups are bonded to carbon 3 and to carbon 5.

$$\begin{array}{l} \quad CH_3 \;\; CH_3 \\ \quad\;\, | \qquad\; | \\ CH_3CCH_2CHCH_2CH_2OH \\ \quad\;\, | \\ \quad CH_3 \end{array}$$

b. The phenol part of the name tells us to draw a benzene ring with an —OH group on carbon 1. Then the alkyl groups are attached, methyl to carbon 5 and isopropyl to carbon 2.

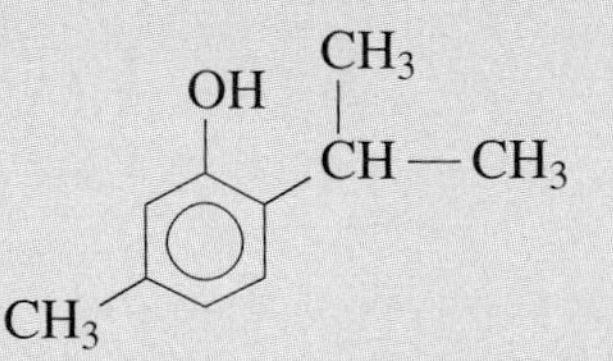

Study Check

Draw the condensed structural formula of 5-chloro-4-methyl-3-heptanol.

HEALTH NOTE *Alcohols in Health and Medicine*

Methanol (CH_3OH), the simplest alcohol, is found in many solvents and paint removers. It is sometimes called *wood alcohol* because methanol is proudced when wood decomposes at high temperatures in the absence of air. Methanol, a colorless and odorous liquid, is extremely poisonous and, if ingested, oxidizes to formaldehyde, a substance that can cause headaches, fatigue, blindness, and death.

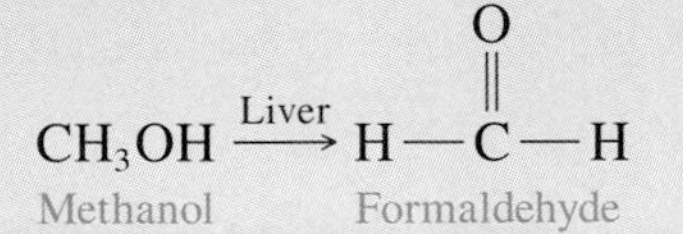

Ethanol (CH_3CH_2OH) or ethyl alcohol is often used as a solvent for perfumes, varnishes, and some medicines, such as tincture of iodine. An antiseptic solution that is 70% by volume ethanol and 30% water is used to prepare an area of skin for injection and to sterilize equipment. It destroys bacteria by coagulating protein.

The term *alcohol* commonly refers to the ethyl alcohol produced by fermentation when yeast is added to sugars in foods such as potatoes, grapes, rice, barley, and rye.

$$\underset{\text{Glucose (a sugar)}}{C_6H_{12}O_6} \xrightarrow{\text{Yeast}} \underset{\text{Ethanol}}{2CH_3CH_2OH} + 2CO_2$$

When ingested in small amounts, ethanol may produce a feeling of euphoria in the body although it is classified as a depressant. In the liver, ethanol oxidizes to acetaldehyde, a substance that impairs mental and physical coordination. If the blood alcohol concentration exceeds 0.4%, coma or death may occur. A dependency on alcohol may lead to hallucinations, liver disease, gastritis, and psychological disturbances.

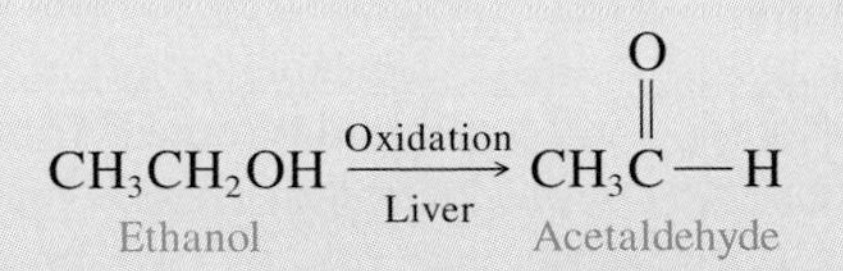

Table 12.1 gives some of the typical behaviors exhibited at various blood alcohol levels.

Table 12.1 ***Typical Behaviors Exhibited by a Person Consuming Alcohol***

Number of Beers (12 oz) or Glasses of Wine (5 oz)	Blood Alcohol Level (%)	Typical Behavior
1	0.025	Slightly dizzy, agreeable talker
2	0.05	Euphoria, loud talking, and laughing
4	0.10	Loss of inhibition, loss of coordination, drowsiness, legally drunk in most states
8	0.20	Intoxicated, quick to anger, exaggerated emotions
12	0.30	Unconscious
16–20	0.40–0.50	Coma and death

Isopropyl alcohol (2-propanol), commonly referred to as *rubbing alcohol*, is used as an astringent because it evaporates rapidly and cools the skin, reducing the size of blood vessels near the surface and decreasing pore size. It is also an antiseptic like ethanol and is used to cleanse the skin before giving an injection or taking a blood sample.

$$CH_3CH(OH)CH_3$$

2-Propanol
(isopropyl alcohol)

Ethylene glycol (1,2-ethanediol) is commonly used as an antifreeze in heating and cooling systems, as a solvent for paints, inks, and plastics, and in the production of synthetic fibers such as Dacron. (See Figure 12.2.) If ingested, it is extremely toxic and causes drowsiness, convulsions, renal damage, coma, and death.

$$HO-CH_2-CH_2-OH$$

1,2-Ethanediol
(ethylene glycol)

Glycerol (glycerin or 1,2,3-propanetriol), a trihydroxy alcohol, is a viscous liquid obtained from oils and fats during the production of soaps. The presence of several polar —OH groups makes it strongly attracted to water, a feature that makes glycerin useful as a skin softener in products such as skin lotions, cosmetics, shaving creams, and liquid soaps. It is also used as an antifreeze and in fluid shock absorbers. Used in the manufacture of dynamite, a mixture of glycerin and a strong oxidizer such as potassium chlorate can be explosive.

$$\begin{array}{l} CH_2-OH \\ | \\ CH-OH \\ | \\ CH_2-OH \end{array}$$

1,2,3-Propanetriol
(glycerol)

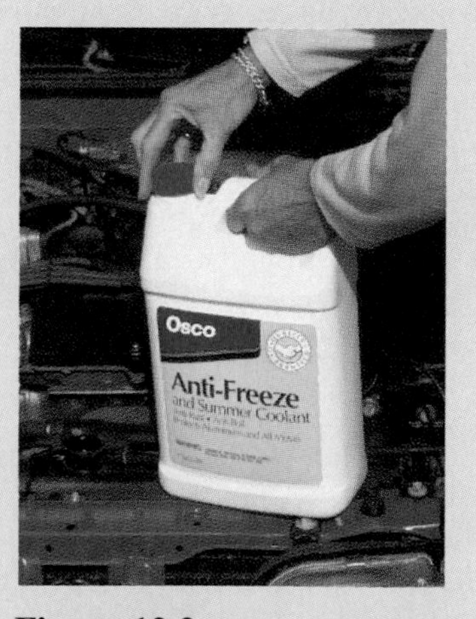

Figure 12.2
The diol ethylene glycol is the major component of many antifreeze fluids.

Menthol is a cyclic alcohol with a peppermint taste and odor that is used in candy, throat lozenges, and nasal inhalers. It causes the mucous membranes to increase their secretions and soothes the respiratory tract.

Cholesterol is the major steroid found in the body, especially in the brain and nerve tissue. It is the starting compound for several hormones, vitamin D, and the bile salts, and it is the principal component of gallstones.

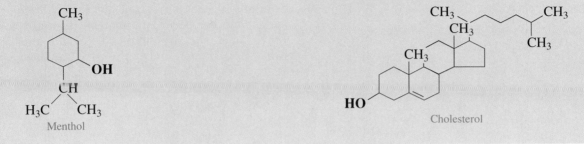

12.3 Properties of Alcohols and Phenols

Learning Goal

Describe the behavior of alcohols and phenols in water.

In Chapters 10 and 11, we learned that hydrocarbons such as alkanes are nonpolar and not soluble in water. However, alcohols with one to four carbon atoms are very soluble in water. The electronegative oxygen atom makes the —OH functional group very polar. As a result, smaller chain alcohols are attracted to polar water molecules because they can form hydrogen bonds. (See Figure 12.3.) In alcohols with five or more carbon atoms, the solubility effect of the nonpolar carbon chain causes them to behave like the corresponding alkanes. For example, methanol with a short carbon chain is highly soluble in water; whereas octanol with a long carbon chain is insoluble. The solubilities of some common alcohols in water are listed in Table 12.2.

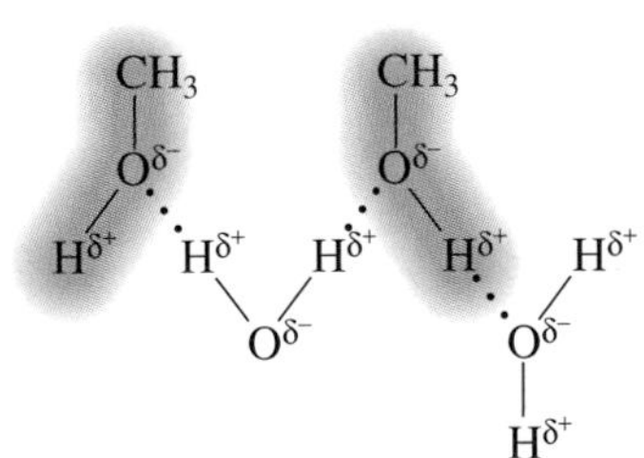

Figure 12.3 Hydrogen bonding occurs between polar hydroxyl groups of methanol molecules and water.

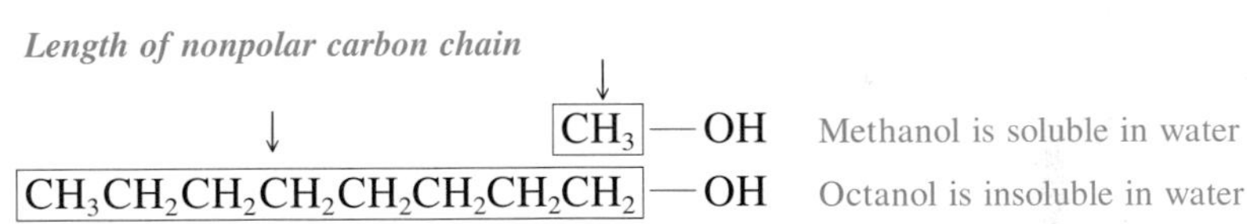

Table 12.2 ***Solubilities of Some Alcohols in Water***

Name	Formula	Water Solubility (g/100 mL H_2O, 20°C)
Methanol	CH_3OH	Any amount
Ethanol	CH_3CH_2OH	Any amount
1-Propanol	$CH_3CH_2CH_2OH$	Any amount
1-Butanol	$CH_3CH_2CH_2CH_2OH$	8.0
1-Pentanol	$CH_3CH_2CH_2CH_2CH_2OH$	2.2
1-Hexanol	$CH_3CH_2CH_2CH_2CH_2CH_2OH$	0.6
1-Heptanol	$CH_3CH_2CH_2CH_2CH_2CH_2CH_2OH$	0.1
1-Octanol	$CH_3CH_2CH_2CH_2CH_2CH_2CH_2CH_2$	Insoluble

HEALTH NOTE *Derivatives of Phenol Are Found in Antiseptics, Essential Oils, and Antioxidants*

Phenol and its derivatives are used as antiseptics in throat lozenges and mouthwashes. In household disinfectant sprays such as Lysol, the active ingredient is *o*-phenylphenol.

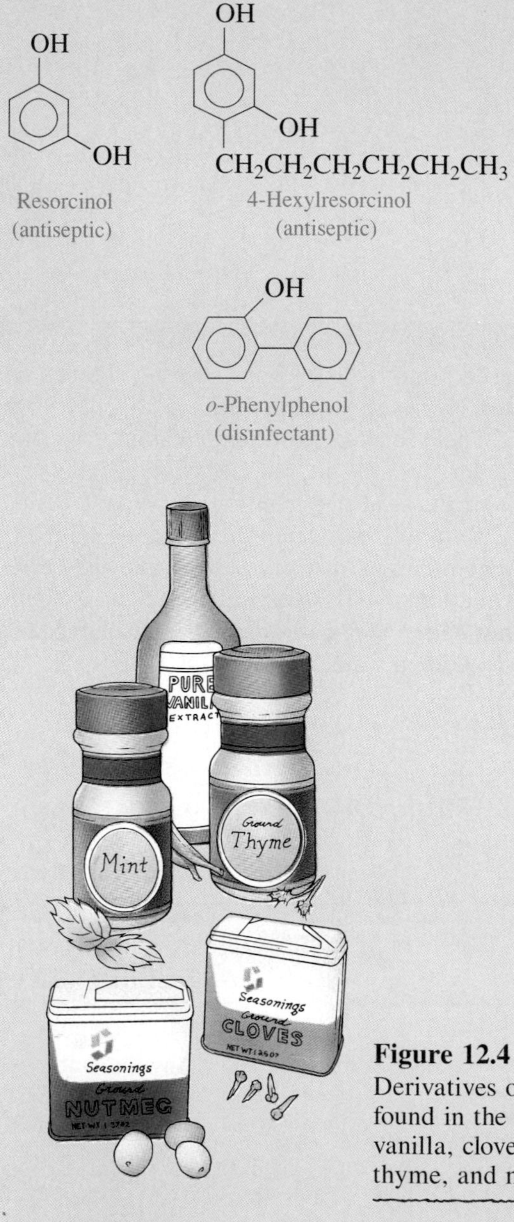

Resorcinol (antiseptic)

4-Hexylresorcinol (antiseptic)

o-Phenylphenol (disinfectant)

Figure 12.4 Derivatives of phenol are found in the essential oils of vanilla, cloves, nutmeg, thyme, and mint.

Several of the essential oils of plants that produce the odor or flavor of the plant are derivatives of phenol. Eugenol is found in cloves, vanillin in the vanilla bean, isoeugenol in nutmeg, and thymol in thyme and mint. Thymol has a pleasant, minty taste and is also used in mouthwashes and by dentists to disinfect a cavity before adding a filling compound. (See Figure 12.4.)

Foods such as cereals and oils spoil when they react with the oxygen in the air. One of the ways to prolong their shelf life is to add a preservative such as BHA or BHT. When a preservative is present, it reacts with the oxygen to slow food spoilage.

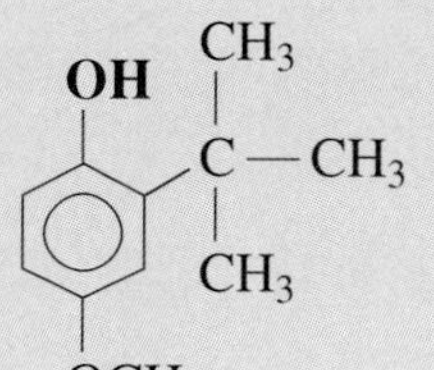

BHA (**b**utylated **h**ydroxy**a**nisole)

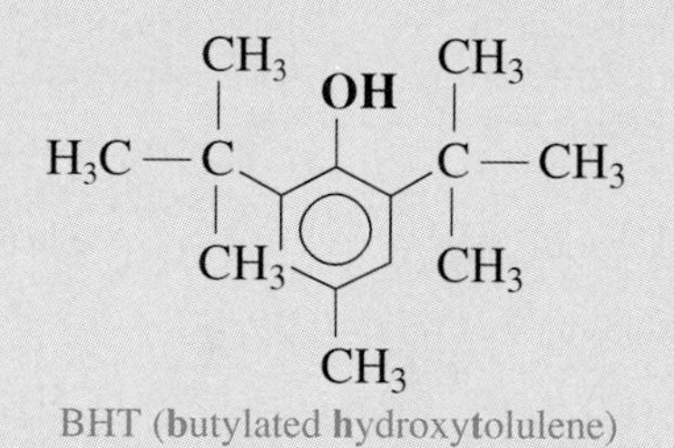

BHT (**b**utylated **h**ydroxy**t**oluene)

Properties of Phenols

The behavior of phenols differs considerably from that of alkyl alcohols. In water, phenol ionizes slightly as a weak acid. In fact, an early name for phenol was *carbolic acid.*

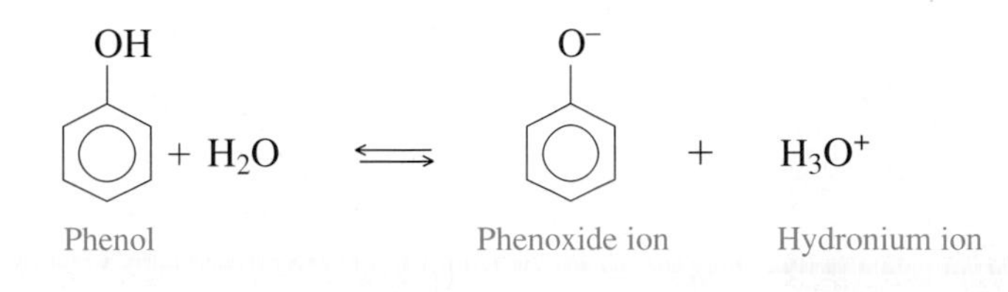

Phenol Phenoxide ion Hydronium ion

A concentrated solution of phenol is very corrosive and highly irritating to the skin, causing severe burns; ingestion can be fatal. Solutions containing phenol should be used with care. Dilute solutions of phenol were previously used in hospitals as antiseptics, but they have generally been replaced.

Thiols

Thiols are a family of sulfur-containing organic compounds similar to alcohols but having an —SH functional group in place of the —OH. In the IUPAC system, thiols are named by adding *thiol* to the name of the longest carbon chain attached to the —SH group. An older term for thiol is mercaptan. The location of the —SH group is indicated by numbering the chain. Some examples of thiols follow:

Naming Thiols R—SH

Functional group:	—SH
IUPAC name:	alkanethiol
Common name:	alkyl mercaptan

CH_3—SH	CH_3CH_2—SH	$CH_3CH(SH)CH_3$
Methanethiol (oysters and cheese)	Ethanethiol (gas detection)	2-Propanethiol

Many short-chain thiols have strong, disagreeable odors. The smell of natural gas is not that of the methane, which is odorless, but of a small amount of ethanethiol that has been added to help detect a gas leak. Methanethiol is the characteristic odor of oysters, cheddar cheese, and "bad breath."

Figure 12.5
Some common substances that have characteristic odors due to thiols.

There are thiols in the spray of a skunk emitted when the skunk senses danger. The characteristic odor of onions is due to 1-propanethiol, which is also a lachrymator, a substance that will make your eyes tear. Garlic contains thiols including 2-propene-1-thiol. (See Figure 12.5.) We can break this name down as follows.

2-	prop	ene	-1-	thiol
Carbon 2 has C=C	3 carbons in chain	Alkene	On carbon 1	—SH group

$$\begin{array}{c} H_3C \quad\quad H \\ \diagdown \quad \diagup \\ C{=}C \\ \diagup \quad \diagdown \\ H \quad\quad CH_2SH \end{array} \qquad CH_3CH_2CH_2{-}SH \qquad CH_2{=}CH{-}CH_2{-}SH$$

trans-2-Butene-1-thiol (in skunk spray) | 1-Propanethiol (in onions) | 2-Propene-1-thiol (in garlic)

Sample Problem 12.4
Thiols

Draw the condensed structural formula of the following:
a. 1-butanethiol **b.** cyclohexanethiol

Solution

a. This compound has a —SH group on the first carbon of a butane chain.

$$CH_3CH_2CH_2CH_2{-}SH$$

b. This compound has a —SH group on a cyclic ring of hexane.

(cyclohexane ring with SH)

Study Check
What is the condensed structural formula of ethanethiol?

12.4 Reactions of Alcohols

Learning Goal
Write equations for the dehydration and oxidation of an alcohol.

In Chapter 10, we learned that hydrocarbons undergo combustion in the presence of oxygen. Alcohols burn with oxygen too. For example, in a restaurant, a banana flambé is prepared by pouring some alcohol on a banana dessert and lighting it. (See Figure 12.6.) The combustion of ethanol proceeds as follows:

$$CH_3CH_2{-}OH + 3O_2 \xrightarrow{\text{Heat}} 2CO_2 + 3H_2O + \text{energy}$$

Figure 12.6
The burning of ethanol produces a flaming dessert.

Dehydration of Alcohols

Earlier we saw that alkenes can add water to yield alcohols. In a reverse reaction, most alcohols can *dehydrate* (lose a water molecule) when they are heated with an acid catalyst such as H_2SO_4. During the **dehydration** of an alcohol, an H— and —OH are removed from *adjacent carbon atoms of the same alcohol* to produce a water molecule. A double bond forms between the same two carbon atoms to produce an alkene product.

Dehydration of an Alcohol (at High Temperatures) Forms an Alkene

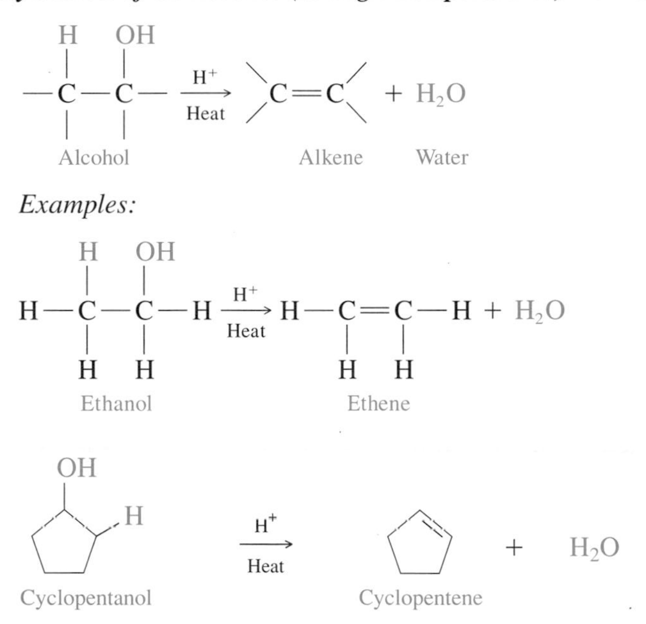

The dehydration of a secondary alcohol can result in the formation of two products. The most prevalent product (major product) is the one that results when the hydrogen (H—) is removed from the carbon atom with the smallest number of hydrogen atoms.

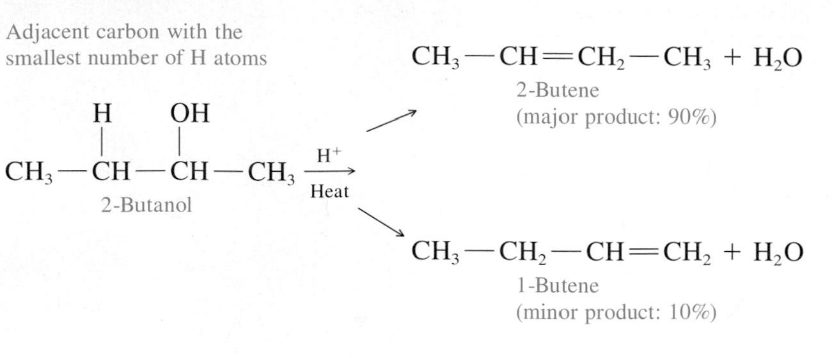

Sample Problem 12.5
Dehydration of Alcohols

Draw the condensed structural formulas for the alkenes produced by the dehydration of the following alcohols:

a. 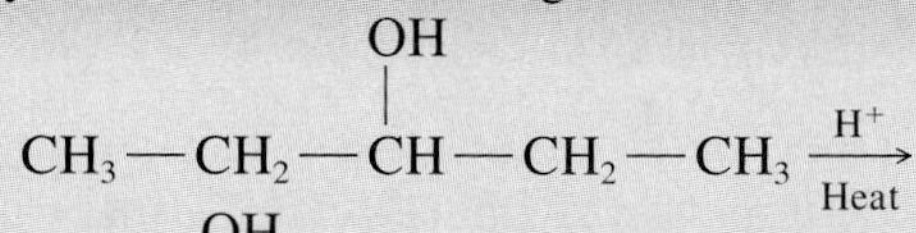 $CH_3—CH_2—CH(OH)—CH_2—CH_3 \xrightarrow[\text{Heat}]{H^+}$

b. 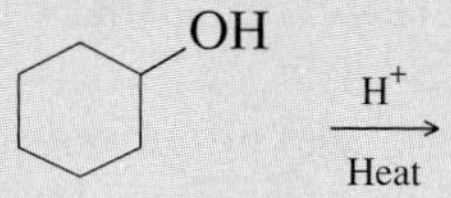 Cyclohexanol (ring with OH) $\xrightarrow[\text{Heat}]{H^+}$

Solution

a. $CH_3—CH_2—CH{=}CH—CH_3 + H_2O$

b. The —OH of this secondary alcohol is removed along with an H from an adjacent carbon. Remember that the hydrogens are not drawn in this type of geometric figure.

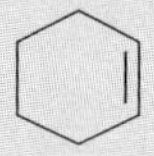

(cyclohexene)

Study Check

What is the name of the alkene produced by the dehydration of cyclopentanol?

Oxidation of Alcohols

In an **oxidation,** hydrogen atoms are lost by a compound or oxygen atoms are added to it. We can recognize the level of oxidation by the number of multiple bonds or the number of bonds to atoms *other* than C and H.

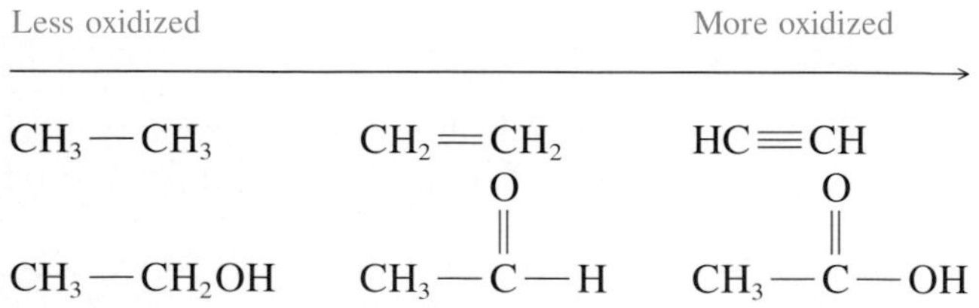

Less oxidized → More oxidized

$CH_3—CH_3$	$CH_2{=}CH_2$	$HC{\equiv}CH$
$CH_3—CH_2OH$	$CH_3—C(=O)—H$	$CH_3—C(=O)—OH$

The oxidation of primary alcohols to aldehydes and secondary alcohols to ketones both occur by the loss of hydrogen. Tertiary alcohols generally do not oxidize. All the oxidation products of alcohols, such as aldehydes, ketones, and carboxylic acids contain the carbon–oxygen double bond $\left(\rangle C{=}O \right)$. The type of product depends on the classification of the alcohol used in the oxidation.

In the body, several important enzyme-catalyzed oxidation reactions take place. In the laboratory, oxidizing agents such as potassium permanganate ($KMnO_4$) or potassium dichromate ($K_2Cr_2O_7$) are used to remove hydrogen or add oxygen. These oxidizing agents are indicated in equations in this text by the symbol [O]. We will now look at the various levels of oxidation that occur for alcohols and some of their products.

Oxidation of Primary Alcohols

In the oxidation of a primary alcohol, one hydrogen is removed from the carbon attached to the —OH, and the other hydrogen is taken from the —OH itself. The **aldehyde** product contains the carbon–oxygen double bond attached to a hydrogen. The other product, water, forms by combining the two "lost" hydrogen atoms with oxygen provided by the oxidizing agent.

Oxidation of Primary Alcohols to Aldehydes

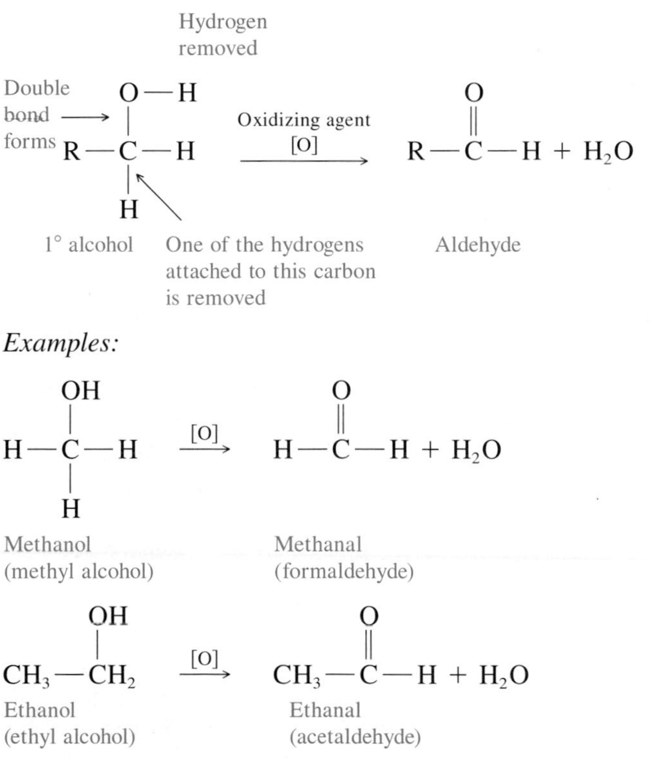

Aldehydes oxidize further, this time by the addition of another oxygen to form a carboxylic acid. This step occurs so readily that it is often difficult to isolate the aldehyde product during oxidation. We will learn more about carboxylic acids in Chapter 13.

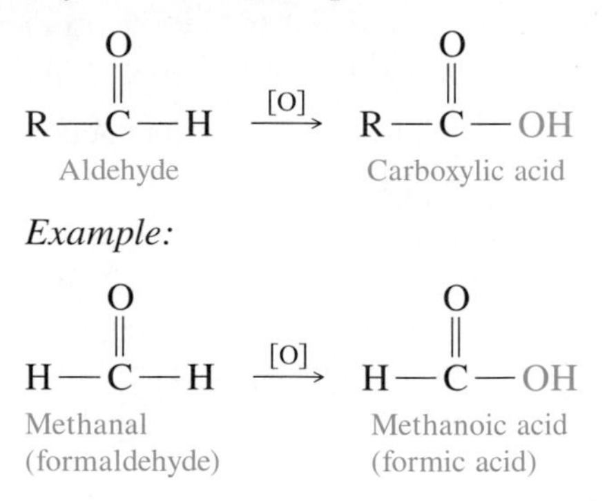

HEALTH NOTE *Biological Oxidation*

Ethylene glycol, the primary alcohol in many antifreeze mixtures, causes kidney malfunction because it is oxidized by enzymes (dehydrogenases) in the liver. Oxalic acid, the oxidized product, contains two carboxylic acid groups that react with calcium ions to form an insoluble calcium salt that impairs the function of the nephron units of the kidney.

$$\underset{\text{Ethylene glycol (primary alcohol)}}{HO{-}CH_2{-}CH_2{-}OH} \xrightarrow{[O]} H{-}\overset{O}{\overset{\|}{C}}{-}\overset{O}{\overset{\|}{C}}{-}H \xrightarrow{[O]} \underset{\text{Oxalic acid}}{HO{-}\overset{O}{\overset{\|}{C}}{-}\overset{O}{\overset{\|}{C}}{-}OH}$$

When the Breathalyzer test is used for suspected drunk drivers, a color change in the oxidizing agent is used to determine the blood alcohol concentration (BAC). The driver exhales a volume of breath into a solution containing the orange Cr^{6+} ion. If there is ethyl alcohol present, it is oxidized, and the Cr^{6+} is reduced to green Cr^{3+}. (See Figure 12.7.)

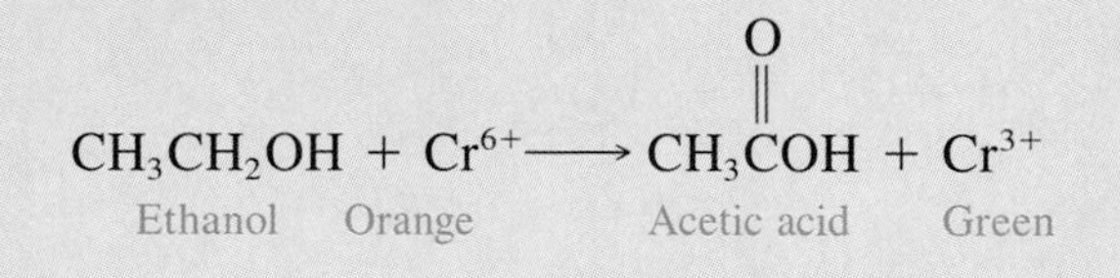

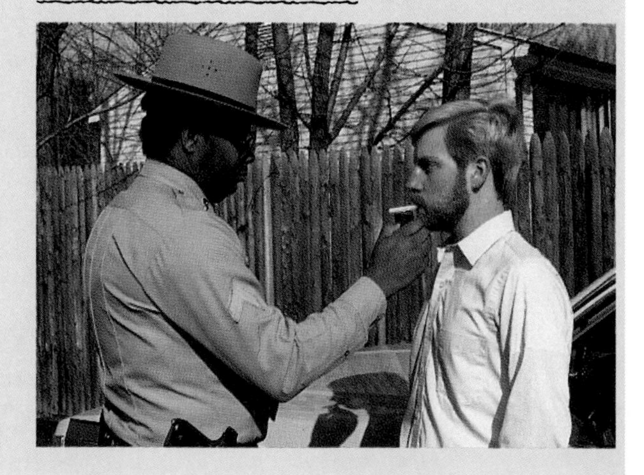

Figure 12.7
The Breathalyzer test determines the blood alcohol level in a possible drunk driver.

HEALTH NOTE *Oxidation of Alcohol in the Body*

In the liver, enzymes oxidize ethanol to acetaldehyde and then to acetic acid. The acetic acid is eventually converted to carbon dioxide and water. Thus, ethanol can be detoxified by the liver. However, its aldehyde and carboxylic acid intermediates cause considerable damage within the cells of the liver.

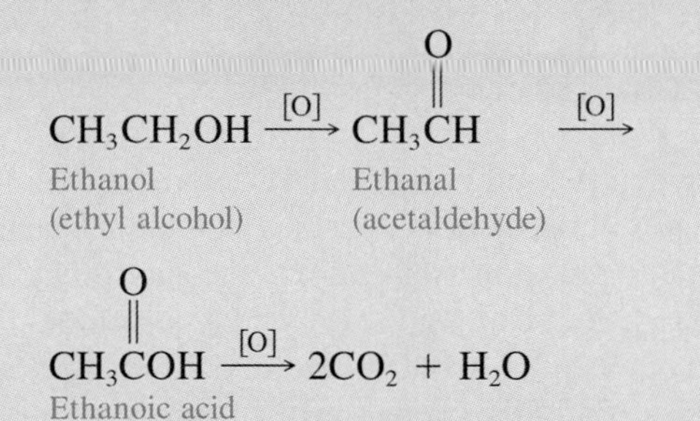

$CH_3COOH \xrightarrow{[O]} 2CO_2 + H_2O$

Ethanoic acid
(acetic acid)

Sometime alcoholics are treated with a drug called Antabuse, which prevents the oxidation of acetaldehyde to acetic acid. This results in an accumulation of acetaldehyde in the blood, which causes nausea, vomiting, and profuse sweating. This severe reaction to alcohol ingestion keeps many people on Antabuse from drinking.

The toxicity of methanol is caused by its oxidation to formaldehyde and formic acid, substances that destroy the retina of the eye. The formic acid, which is not readily eliminated from the body, lowers blood pH so severely that just 30 mL of methanol can lead to a coma and death.

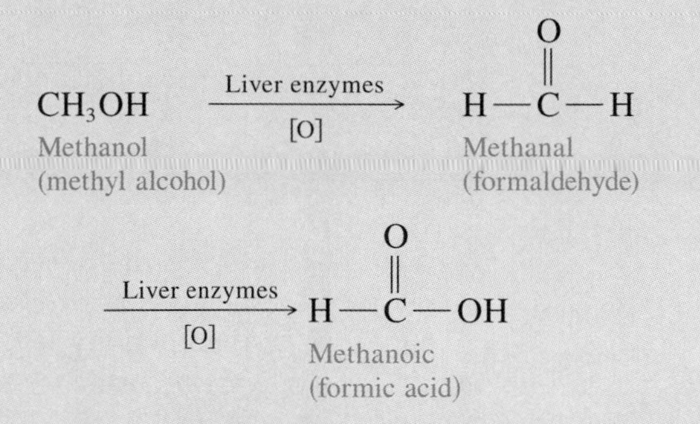

The acidic product of methanol poisoning may be neutralized in the hospital by giving sodium bicarbonate. In some cases, ethanol may be given to the patient. The enzymes in the liver pick up ethanol molecules to oxidize instead of methanol molecules, which gives time for the elimination of methanol via the lungs without the formation of its dangerous oxidation product, formaldehyde.

Oxidation of Secondary Alcohols

The oxidation of the secondary alcohols is similar to that of primary alcohols. The oxidizing agent removes one hydrogen from the —OH and another from the carbon bonded to the —OH group. The result is a ketone that has the carbon–oxygen double bond attached to alkyl groups on both sides. Because there are no hydrogen atoms attached to the double bond in a ketone, no further oxidation is possible.

Oxidation of Secondary Alcohols to Ketones

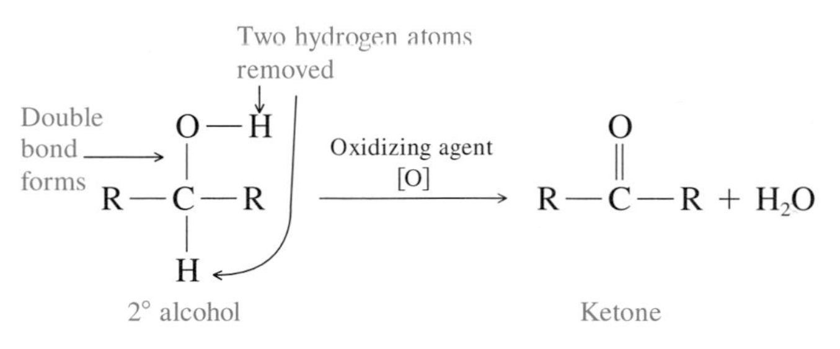

Examples:

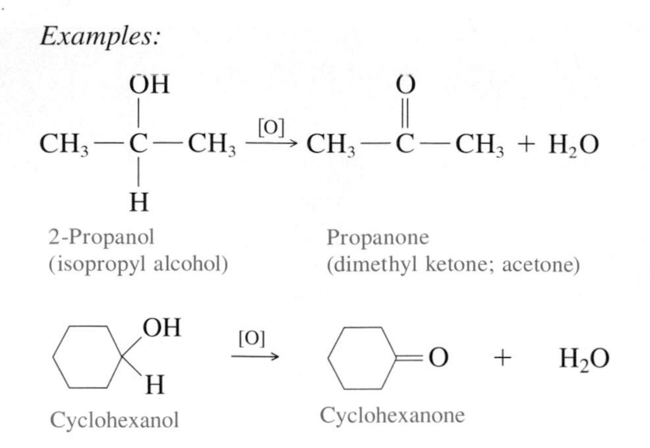

During vigorous exercise, oxygen is used up and lactic acid accumulates in the muscles. Breathing becomes labored and muscles begin to fatigue. When the activity level is decreased, oxygen can convert the lactic acid to pyruvic acid, which breaks down to CO_2, H_2O, and energy. Oxygen enters the muscles and oxidizes the secondary —OH group in lactic acid to a ketone group in pyruvic acid. The muscles in highly trained athletes are capable of taking up greater quantities of oxygen so that vigorous exercise can be maintained for longer periods of time.

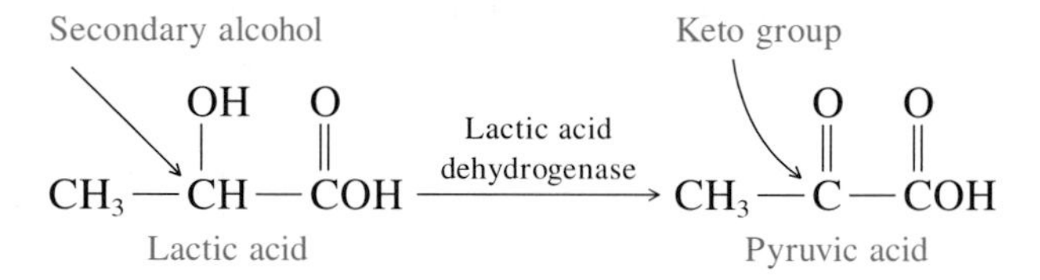

Tertiary alcohols do not oxidize because there are no hydrogen atoms to remove from the alcohol carbon. Therefore, tertiary alcohols are resistant to oxidation.

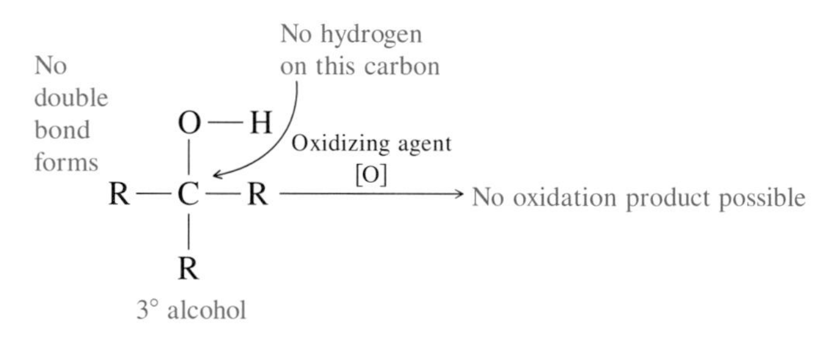

Sample Problem 12.6
Oxidation of Alcohols

Draw the structural formulas of the organic products that form when the following alcohols undergo oxidation.

a. $CH_3CH_2\overset{OH}{\overset{|}{C}}HCH_3 \xrightarrow{[O]}$

b. $CH_3CH_2CH_2OH \xrightarrow{[O]}$

Solution

a. Because the reactant is a secondary alcohol, oxidation produces a ketone and water.

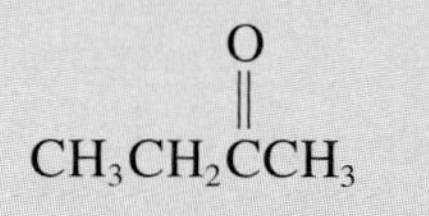

$$CH_3CH_2\overset{\overset{\displaystyle O}{\|}}{C}CH_3$$

b. This primary alcohol can oxidize to an aldehyde, and further oxidize to a carboxylic acid.

$$CH_3CH_2\overset{\overset{\displaystyle O}{\|}}{C}—OH$$

Study Check

When propene is hydrated in the presence of an acid cataylst, the product forms a ketone when it is oxidized. Explain using equations.

12.5 Ethers

Learning Goal

Give the name of an ether; write the condensed structural formula from the name.

An **ether** contains an oxygen atom attached by single bonds to two alkyl or aromatic groups. (See Figure 12.8.) The general formula is $R^1—O—R^2$. The attached groups may be alkyl groups, aromatic, or one of each.

$CH_3—O—CH_3$	$CH_3—O—CH_2CH_3$	$CH_3—O—C_6H_5$
Dimethyl ether	Ethyl methyl ether	Methyl phenyl ether

Figure 12.8
Ball-and-stick model of dimethyl ether.

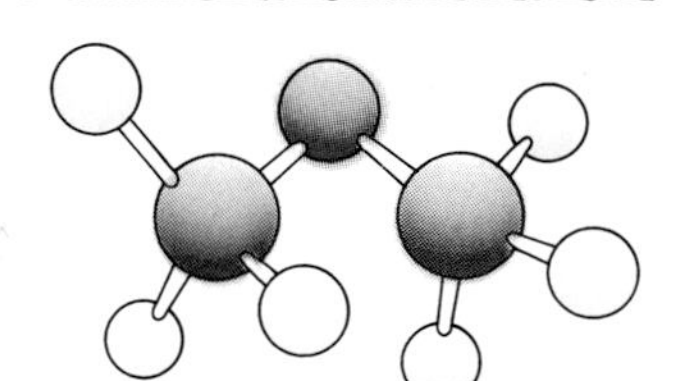

Formation of Ethers

Ethers can be formed from primary alcohols when dehydration occurs at lower temperatures in the presence of an acid catalyst. Then the components of water are removed from two separate molecules: an H— from one alcohol and the —OH from another. When the remaining portions of the two alcohols join, an ether is produced.

Dehydration of Two Alcohols Forms an Ether

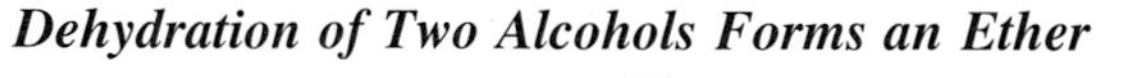

$$R—O\boxed{H + HO}—R \xrightarrow[\text{Heat}]{H^+} R—O—R + H_2O$$

Example:

$$\underset{\text{Methanol}}{CH_3—OH} + \underset{\text{Methanol}}{HO—CH_3} \xrightarrow[\text{Heat}]{H^+} \underset{\text{Dimethyl ether}}{CH_3—O—CH_3} + H_2O$$

Naming Ethers

The simple ethers are generally referred to by their common names. Each of the groups (alkyl or aromatic) attached to the oxygen atom is listed in alphabetical order, followed by the name *ether*. If the groups are identical, the prefix *di* may be used.

$CH_3-O-CH_2-CH_2-CH_3$ Common name: *methyl propyl ether*

Methyl Ether Propyl

Naming Ethers R—O—R

Functional group:	—O—
IUPAC name:	alkoxyalkane
Common name:	dialkyl ether alkyl[1] alkyl[2] ether alkyl phenyl ether

---

IUPAC Rules for Naming Ethers

A complex ether is named according to the IUPAC system, in which the smaller group and the oxygen are named as an *alkoxy* (***alk**yl* + ***oxy**gen*) group attached to the longer alkane chain. Suppose you want to name the above ether by its IUPAC name.

$CH_3-O-CH_2-CH_2-CH_3$

Step 1 Select the longer carbon chain and write its alkane name.

$CH_3-O-\mathbf{CH_2-CH_2-CH_3}$ **Propane**

Step 2 Name the shorter carbon group and oxygen (oxy) as an alkoxy group and number the parent chain beginning at the end nearest to the alkoxy group.

$\mathbf{CH_3-O}-CH_2-CH_2-CH_3$ **Methoxy**propane

Step 3 Give the location of the alkoxy group on the parent chain.

$CH_3-O-\underset{1}{CH_2}-\underset{2}{CH_2}-\underset{3}{CH_3}$ **1-**Methoxypropane

We can check the IUPAC name by breaking it up as follows:

$\boxed{CH_3-O}-\boxed{CH_2-CH_2-CH_3}$

A methoxy — Attached to carbon 1 of propane

Some examples of naming ethers follow:

CH_3-O-CH_3
Methoxymethane
(dimethyl ether)

$CH_3CH_2-O-CH_2CH_3$
Ethoxyethane
(diethyl ether)

$CH_3-\underset{|}{\overset{OCH_3}{CH}}-CH_2-CH_3$
2-Methoxybutane

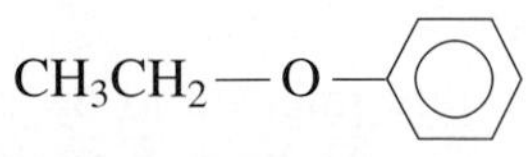

$CH_3CH_2-O-C_6H_5$
Ethoxybenzene
(ethyl phenyl ether)

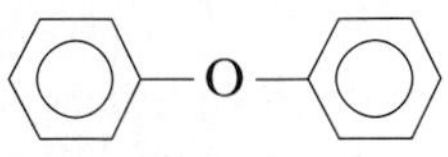

$C_6H_5-O-C_6H_5$
Phenoxybenzene
(diphenyl ether)

Sample Problem 12.7
Ethers

Assign a common name and the IUPAC name to the following ethers:

a. $CH_3CH_2—O—CH_2CH_2CH_3$

b.

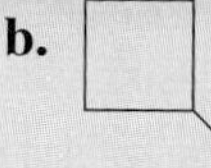

Solution

a. The groups attached to the oxygen are an ethyl group and a propyl group. The common name is *ethyl propyl ether*. Naming the shorter alkyl group and the oxygen as ethoxy gives the IUPAC name of *1-ethoxypropane*.

b. The groups attached to the oxygen are a methyl group and a cyclobutyl group. The common name is *cyclobutyl methyl ether*. In the IUPAC name, the $CH_3O—$ is named methoxy; the cyclobutane is the longer carbon chain, to give the name *methoxycyclobutane*.

Study Check

What is the common name of ethoxybenzene?

Solubility of Ethers

Ethers are less soluble in water than alcohols but more soluble than alkanes. The oxygen atom in ethers makes them slightly polar but not as polar as alcohols. An O—H bond is more polar than an O—R bond. The shorter-chain ethers such as dimethyl ether and diethyl ether are somewhat soluble in water because the polar oxygen in the ether can form a hydrogen bond with water. However, ethers that contain longer carbon chains are not soluble in water.

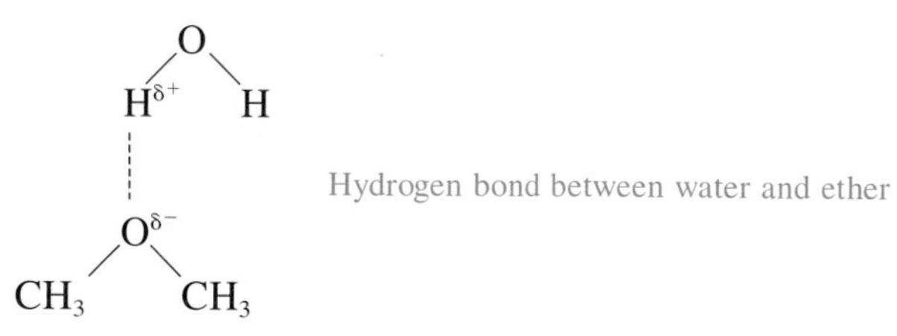

Hydrogen bond between water and ether

Uses of Ethers

The slight polarity of ethers, diethyl ether in particular, makes them very useful as solvents. They are also less reactive than many other organic materials and do not react with the substances they dissolve. However, ether vapors are highly flammable and react explosively with oxygen in the air. The utmost care must be taken when working with ethers. In the process called "freebasing," ether is used as a solvent to extract cocaine from other materials. Sometimes, in such illegal endeavors, a sudden flash fire indicates this use of ether.

HEALTH NOTE *Ethers as Anesthetics*

A general anesthetic causes a loss of all sensation by inhibiting the ability of nerve cells to send signals of pain to the brain. Diethyl ether, or "ether," was used as an anesthetic for more than 100 years. It has anesthetic properties over a wide concentration range and minimal side effects. However, it is extremely volatile at room temperature, and there is always a danger of explosive reaction. For this reason, the use of ethyl ether as an anesthetic was discontinued. Other ethers that have been used as anesthetics are Vinethene, Ethrane (enflurane), and Penthrane (methoxyflurane). (See Figure 12.9.) They are inhalation anesthetics that are less flammable than diethyl ether, but because of side reactions many have been replaced by halothane, which we discussed in Chapter 10.

Figure 12.9
Isoflurane is an inhalation anesthetic used for the induction and maintenance of general anesthesia.

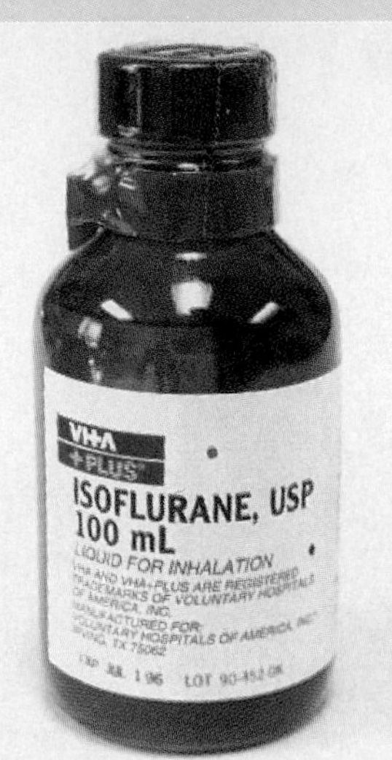

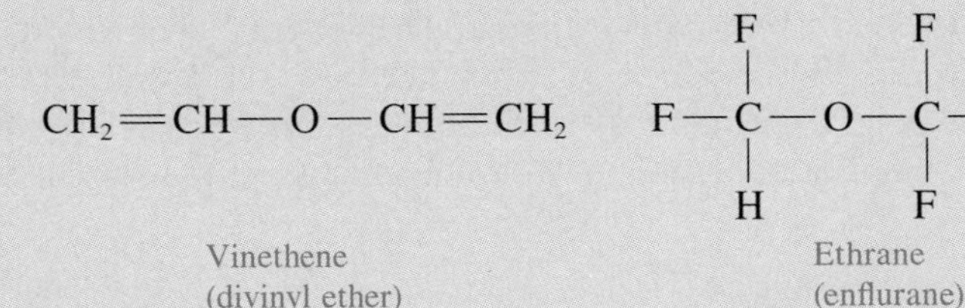

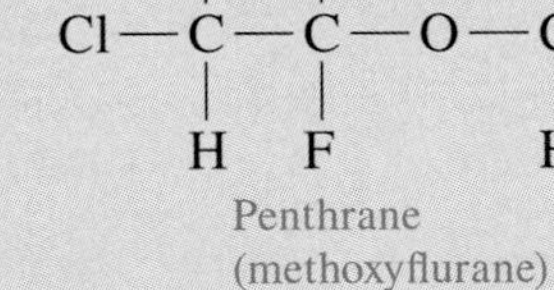

12.6 Aldehydes and Ketones

Learning Goal
Give the IUPAC and common names of an aldehyde or ketone; draw the condensed structural formula from the name.

Many important organic and biochemical compounds contain a **carbonyl group,** $\rangle C{=}O$, which is a carbon atom joined to an oxygen atom by a double bond. Aldehydes and ketones both contain carbonyl groups. In an **aldehyde**, the carbonyl carbon is attached to a hydrogen atom. In a **ketone,** the carbonyl group is bonded to two alkyl or aromatic groups. (See Figure 12.10.) In an aldehyde, the carbonyl group occurs as the first carbon atoms of the carbon chain where it may be abbreviated as $-CHO$. In a ketone, the carbonyl group is found along the carbon chain other than the first carbon atom.

Figure 12.10
Ball-and-stick models of aldehydes and ketones: formaldehyde, acetaldehyde, and acetone.

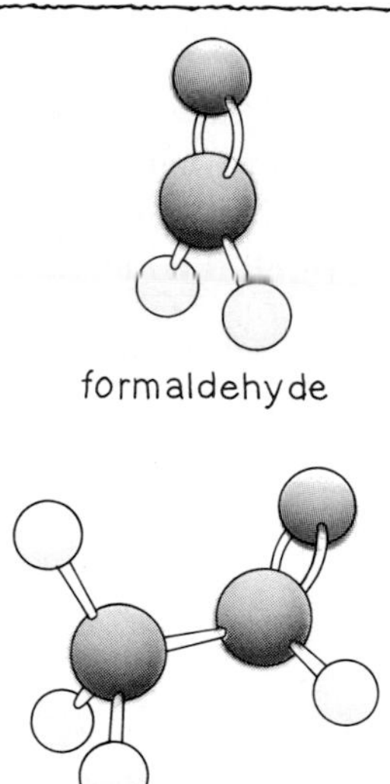

acetone

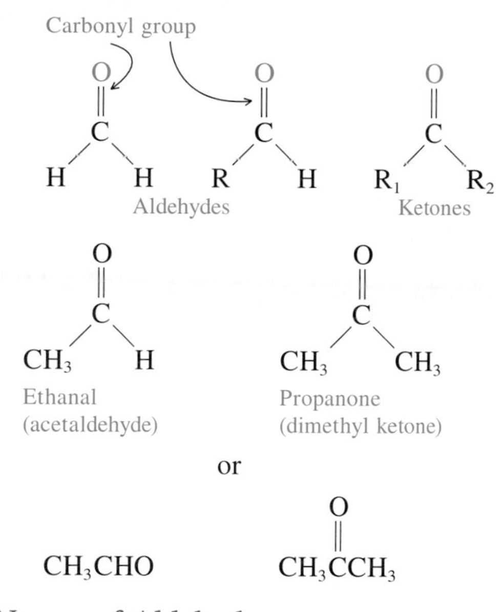

Names of Aldehydes

In the IUPAC system, aldehydes are named by replacing the *e* in the longest carbon chain with *al*. However, common names are widely used for the simple aldehydes. The common names of the first four aldehydes use the prefixes *form, acet, propion,* and *butyr* followed by the word *aldehyde*. For example, methanal is **form**aldehyde, and ethanal is **acet**aldehyde. A 40% solution of formaldehyde called formalin is used as a germicide and as a preservative for tissues. Formaldehyde has also been used in the manufacturing of paper, insulation materials, and cosmetics, including some shampoos. There is now concern that formaldehyde is carcinogenic and should be eliminated from household substances.

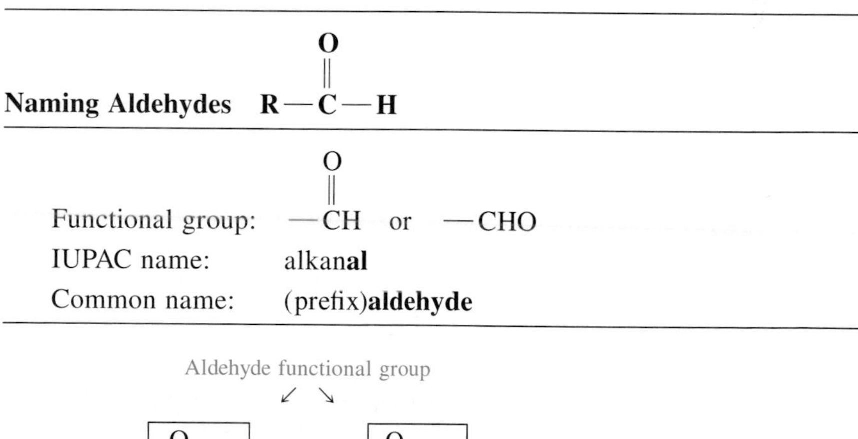

Naming Aldehydes R—C(=O)—H

Functional group:	—CH(=O) or —CHO
IUPAC name:	alkan**al**
Common name:	(prefix)**aldehyde**

Aldehyde functional group

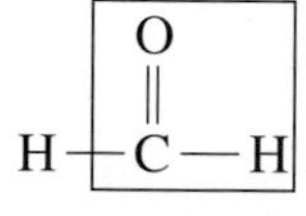

Methanal
(formaldehyde)

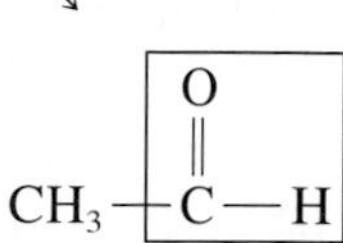

Ethanal
(acetaldehyde)

Some examples of IUPAC and common names for aldehydes follow:

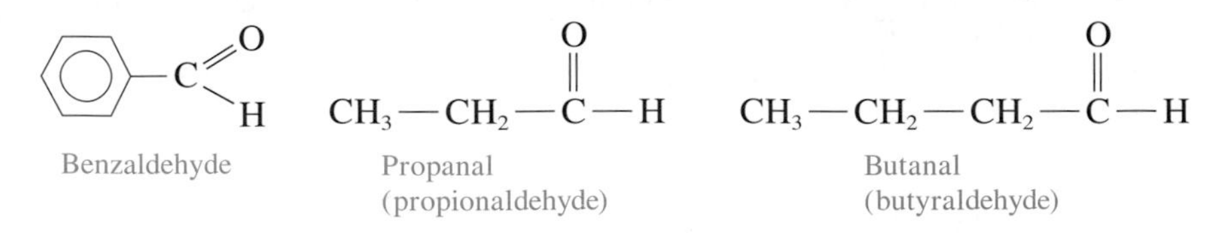

Benzaldehyde

Propanal (propionaldehyde)

Butanal (butyraldehyde)

In an aldehyde, the carbonyl group is always the first carbon in the chain; no number is needed to indicate its position. In the IUPAC system, branches are located by numbering from the carbonyl carbon (carbon 1). In common names, the position of substituents is indicated by the lowercase Greek letters α (alpha), β (beta), and γ (gamma) starting with the carbon next to the carbonyl group.

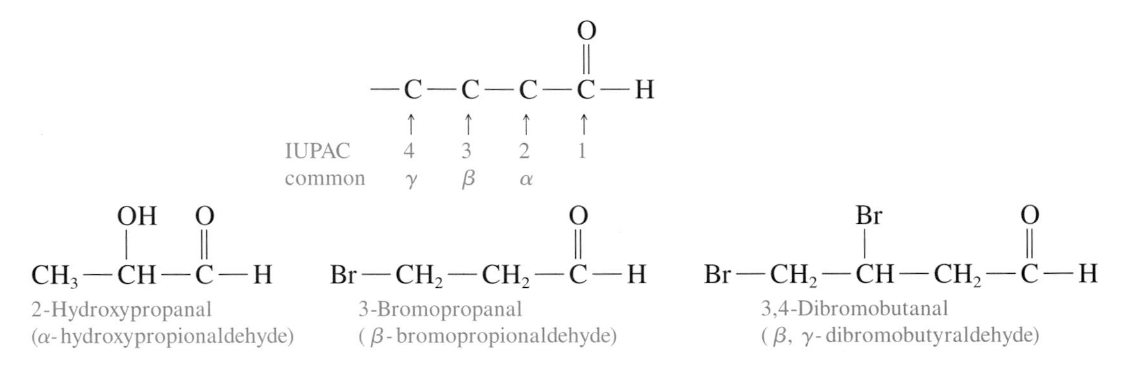

2-Hydroxypropanal (α-hydroxypropionaldehyde)

3-Bromopropanal (β-bromopropionaldehyde)

3,4-Dibromobutanal (β, γ-dibromobutyraldehyde)

Sample Problem 12.8
Names of Aldehydes

Give the IUPAC and common names for the following aldehydes:

a. $Cl-CH_2-\overset{\displaystyle O}{\overset{\|}{C}}-H$

b. $CH_3-\underset{}{\overset{\displaystyle CH_3}{\overset{|}{CH}}}-CH_2-\overset{\displaystyle O}{\overset{\|}{C}}-H$

Solution

a. The longest chain has two carbon atoms with a chlorine atom on the second carbon atom. The IUPAC name is *2-chloroethanal;* the common name is *α-chloroacetaldehyde.*

b. The longest chain has four atoms with a methyl group on the third or the β carbon. The IUPAC name is *3-methylbutanal;* the common name is β-methylbutyraldehyde.

Study Check

Give the IUPAC and common name for the simplest sugar, glyceraldehyde. (As a branch, an —OH is called *hydroxy*).

$HO-CH_2-\overset{\displaystyle OH}{\overset{|}{CH}}-\overset{\displaystyle O}{\overset{\|}{C}}-H$

Names of Ketones

In the IUPAC system, a ketone is named by replacing the *e* with *one* in the name of the longest carbon chain containing the carbonyl group. The chain is numbered starting at the end nearest to the carbonyl group. (Propanone and butanone do not require numbers.) In a cyclic ketone, the carbonyl carbon is designated as carbon 1. In the common names of the simple ketones, which are widely used, the alkyl groups on either side of the carbonyl group are listed in alphabetical order followed by *ketone*.

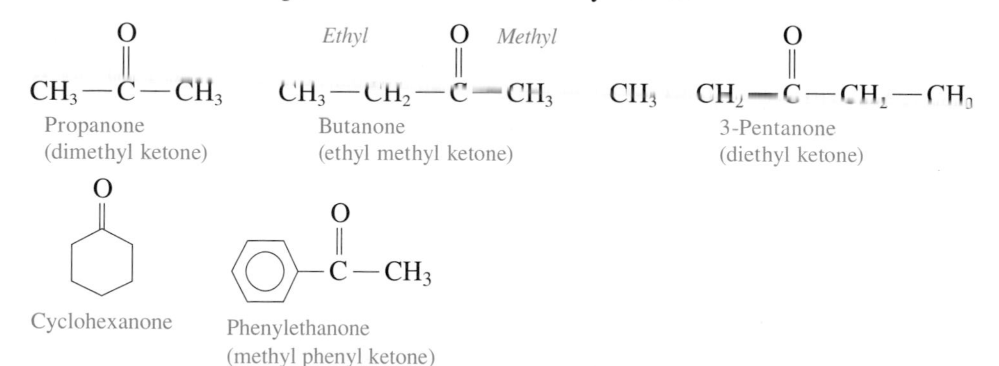

Sample Problem 12.9
Names of Ketones

Give the IUPAC name for the following ketone:

$$CH_3-\underset{}{\overset{CH_3}{\overset{|}{CH}}}-CH_2-\overset{O}{\overset{\|}{C}}-CH_3$$

Solution

The longest chain is five carbon atoms. Counting from the right, the carbonyl group is on carbon 2 and a methyl group is on carbon 4. The IUPAC name is *4-methyl-2-pentanone*.

Study Check

What is the common name of 3-hexanone?

HEALTH NOTE *Aldehydes and Ketones in Health and Medicine*

Several naturally occurring aromatic aldehydes and ketones are responsible for the odor and taste of food flavorings. Benzaldehyde is found in almonds, vanillin in vanilla beans, and cinnamaldehyde in cinnamon.

C_6H_5-CHO

Benzaldehyde
(oil of almonds)

HO, OCH_3, $-\overset{O}{\overset{\|}{C}}-H$

Vanillin
(vanilla bean)

$C_6H_5-CH{=}CH-\overset{O}{\overset{\|}{C}}-H$

Cinnamaldehyde
(oil of cinnamon)

The flavor of butter or margarine is from butanedione, muscone is used to make musk perfumes, and oil of spearmint contains carvone.

$$CH_3-\overset{\overset{\large O}{\|}}{C}-\overset{\overset{\large O}{\|}}{C}-CH_3$$

Butanedione
(butter flavor)

$$\begin{array}{l} \qquad\qquad\qquad\qquad\qquad CH_3 \\ \qquad\qquad\qquad\qquad\qquad\;| \\ CH_2CH_2CH_2CH_2CH_2CHCH_2 \diagdown \\ | \qquad\qquad\qquad\qquad\qquad\qquad\quad C=O \\ CH_2CH_2CH_2CH_2CH_2CH_2CH_2 \diagup \end{array}$$

Muscone
(musk)

CH_3, O, CH_3, C, CH_2

Carvone
(spearmint oil)

In the body, sugars such as glucose and fructose are important sources of energy. The monosaccharides or simple sugars are compounds having an aldehyde or ketone group and several hydroxyl groups.

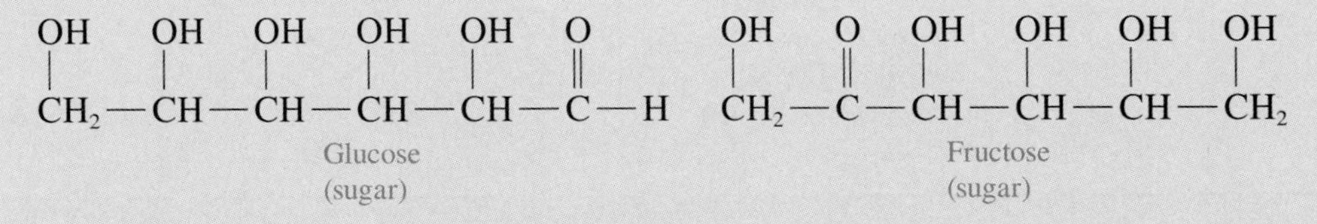

Glucose
(sugar)

Fructose
(sugar)

Important hormones of the steroid family such as cortisone, testosterone, and progesterone contain a carbonyl group.

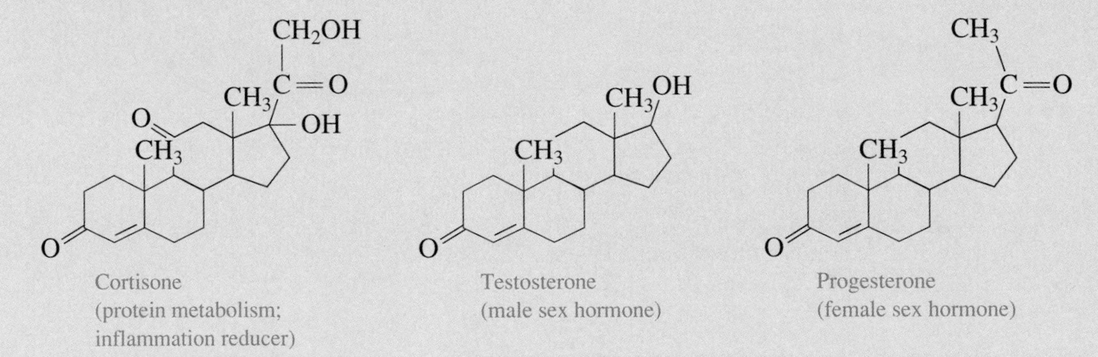

Cortisone
(protein metabolism;
inflammation reducer)

Testosterone
(male sex hormone)

Progesterone
(female sex hormone)

12.7 Properties of Aldehydes and Ketones

Learning Goal
Describe some physical and chemcial properties of aldehydes and ketones.

In a carbonyl group, the highly electronegative oxygen atom attracts the electrons of the double bond. As a result the carbonyl bond is strongly polar having an oxygen that has a partial negative charge and a carbon that is partially positive.

$$\overset{O^{\delta-}}{\underset{\diagup\quad\diagdown}{\overset{\|}{C^{\delta+}}}}$$

Polar carbonyl group

Figure 12.11
Butanone (commercially called methyl ethyl ketone MEK) is used to remove paint.

Solubility of Aldehydes and Ketones in Water

The polarity of the carbonyl bond determines the behavior of the aldehydes and ketones. Compounds with one to four carbons are soluble in water because the oxygen forms hydrogen bonds with water and pulls the small alkane chain into solution.

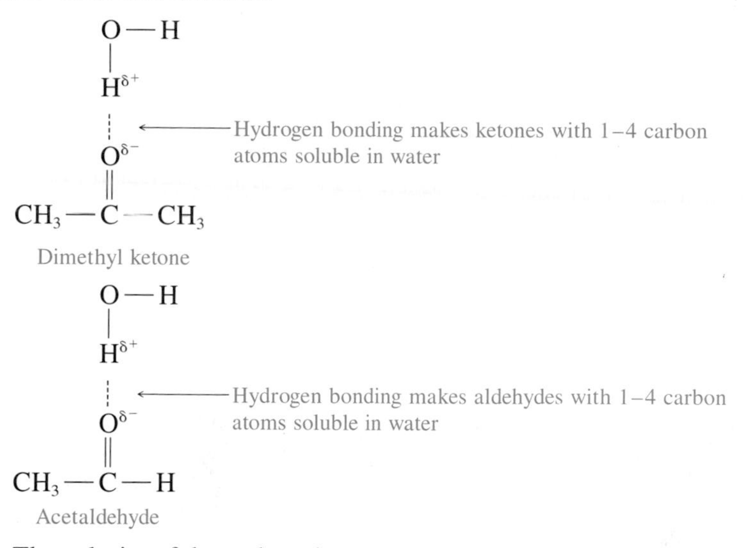

The polarity of the carbonyl group makes ketones good solvents for many organic compounds. (See Figure 12.11.) Propanone (also called acetone or dimethyl ketone) is found in paint removers and fingernail polish remover. In the body, acetone may be produced in uncontrolled diabetes, fasting, and high-protein diets when large amounts of fats are metabolized for energy.

Oxidation of Aldehydes

In Section 12.5, we mentioned that aldehydes were easily oxidized to carboxylic acids, but ketones were not oxidized. The —H attached to the carbonyl group is replaced by a —OH group to give a carboxylic acid.

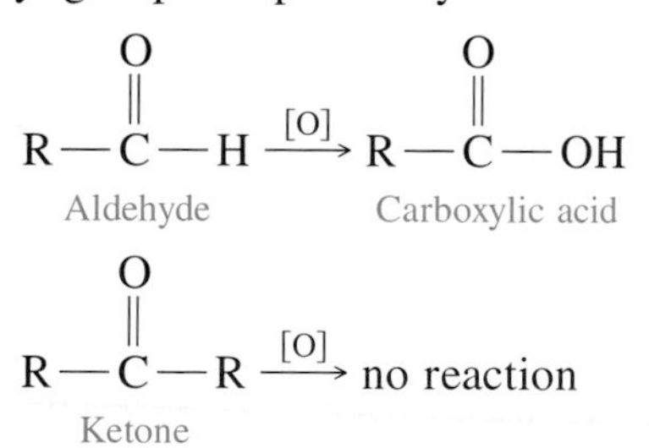

Because aldehydes oxidize easily, mild oxidizing agents can be used to distinguish aldehydes from ketones. In *Tollens' test,* an alkaline solution containing silver ion (Ag^+) is added to an aldehyde. As the aldehyde oxidizes, the silver ions are reduced to metallic silver, which forms a mirror on the inside of the test tube indicating the presence of the aldehyde. Ketones do not react with Tollens' reagent because they cannot be oxidized further.

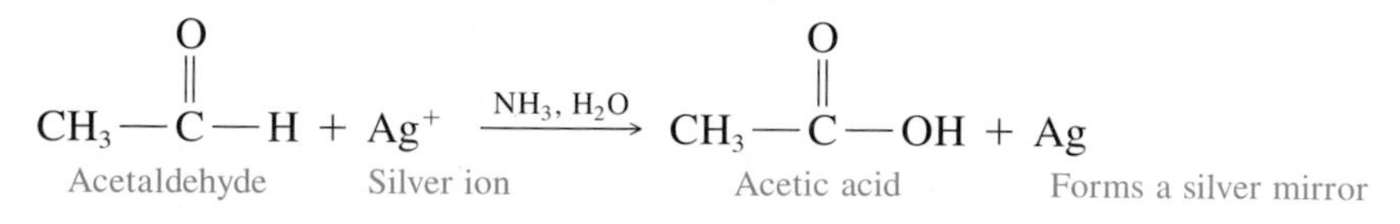

In *Benedict's* and *Fehling's tests*, alkaline solutions of copper(II) ions are added to a sample. An aldehyde in the sample, but not a ketone, will be oxidized. The Cu^{2+} ions are reduced to Cu^{+}, which gives a brick-red precipitate of copper(I) oxide.

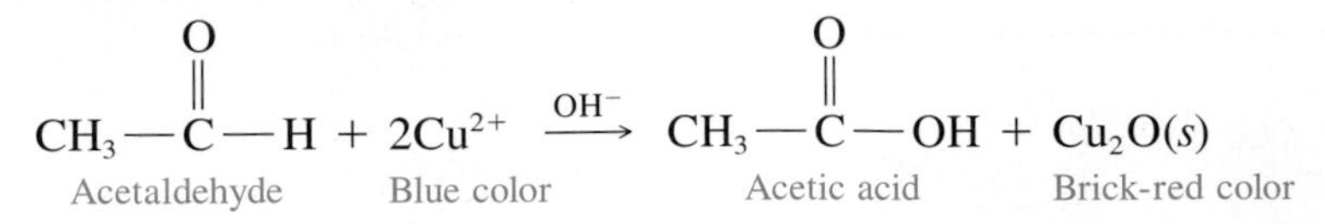

Reduction of Aldehydes and Ketones to Alcohols

Earlier, we saw that alcohols oxidize to aldehydes and ketones. In the reverse reaction, aldehydes and ketones are converted to alcohols by the addition of hydrogen. **Reduction** is the addition of hydrogen to a compound or the loss of oxygen. When hydrogen is added, usually in the presence of a platinum or nickel catalyst, aldehydes are reduced to primary alcohols; ketones are reduced to secondary alcohols.

Reduction

Aldehydes Are Reduced to Primary Alcohols

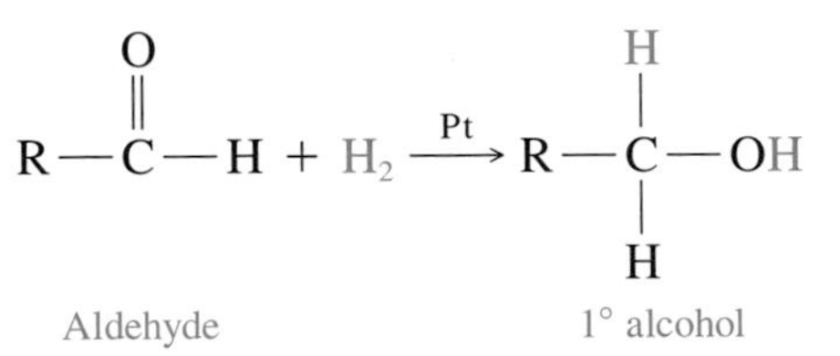

Ketones Are Reduced to Secondary Alcohols

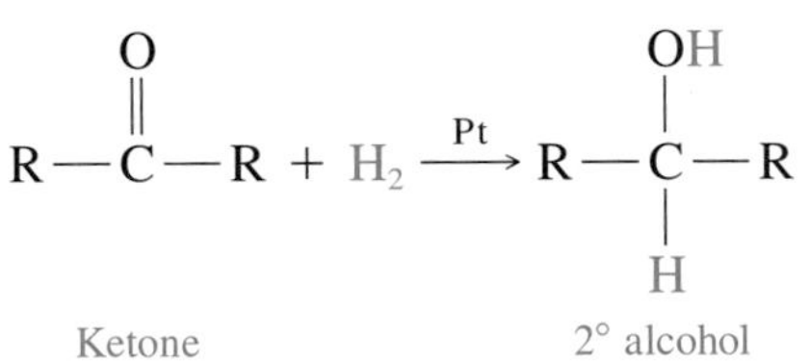

Examples:

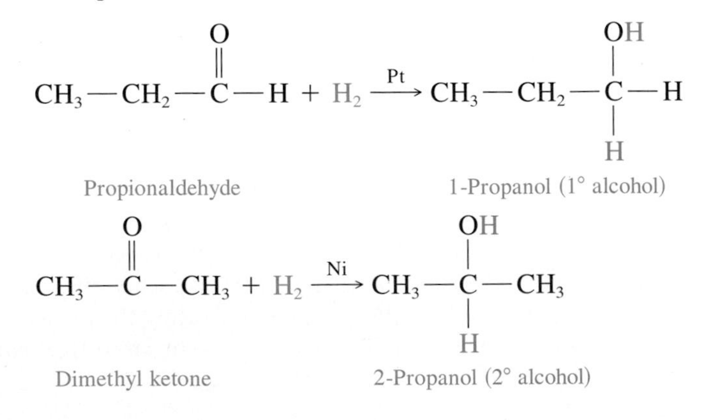

In biological systems, the reduction of aldehydes and ketones is an important process. In an enzyme-catalyzed reaction, hydrogen is provided by a reducing agent that is known as NADH (nicotinamide adenine dinucleotide). In this reaction, NADH is oxidized to NAD^+, which is available to oxidize other biological molecules.

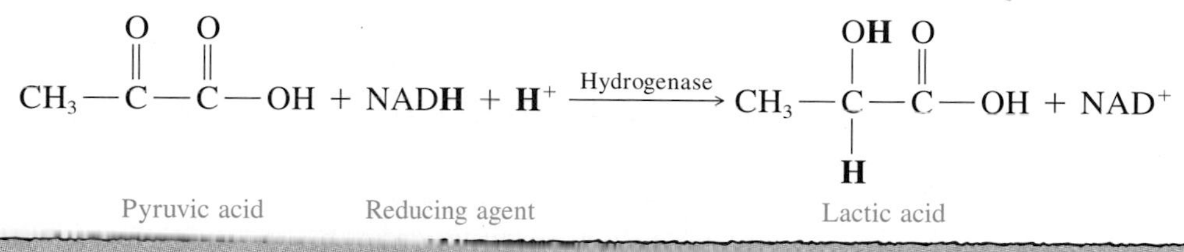

Sample Problem 12.10
Reduction of Carbonyl Groups

Write an equation for the reduction of cyclopentanone in the presence of a nickel catalyst.

Solution

The reacting molecule is a cyclic ketone that has five carbon atoms. Hydrogen atoms will add to the carbon and oxygen in the carbonyl group to form the corresponding secondary alcohol.

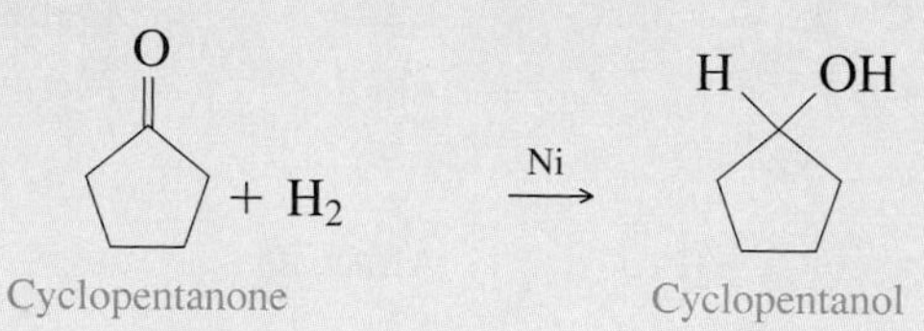

Study Check

What is the name of the product obtained from the hydrogenation of propionaldehyde?

HEALTH NOTE *Glucose Is a Reducing Sugar*

Some of the simple sugars such as glucose contain aldehyde groups that give positive tests with oxidizing agents. A test strip containing Benedict's reagent is used to determine the presence of glucose in the urine of a person who may have diabetes. A blue color indicates that there is no glucose; shades of brown, orange, and brick red indicates increasing glucose levels. The level of glucose present in the urine is found by matching the color produced to a color chart found on the container. (See Figure 12.12.) When a carbohydrate such as glucose is oxidized, the blue Cu^{2+} in

Figure 12.12
A test strip is used to determine the presence and level of glucose in the urine by comparing the color change to a color chart on the container.

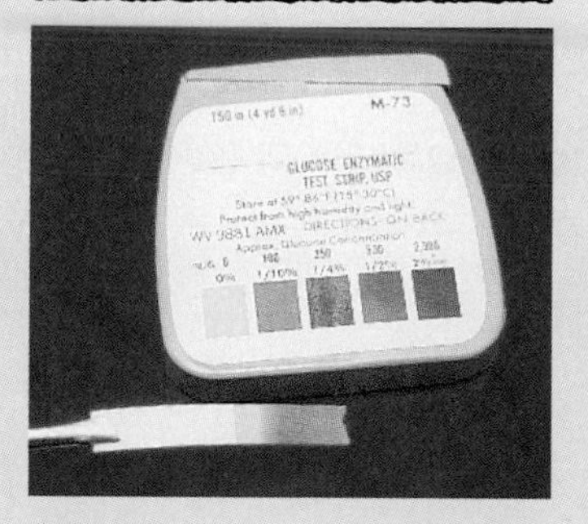

the reagent is reduced to Cu_2O, which has a brick-red color. For this reason, the sugar is called a *reducing sugar*, which refers to the chemical reaction, *not* a loss of weight.

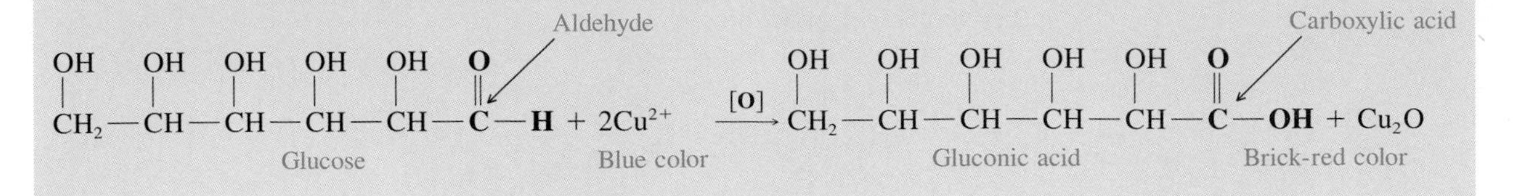

Summary of Naming

Structure	Family	IUPAC Name	Common Name
$CH_3—OH$	Alcohol	Methanol	Methyl alcohol
(benzene ring with OH)	Phenol	Phenol	Phenol
$CH_3—O—CH_3$	Ether	Methoxymethane	Dimethyl ether
$CH_3—SH$	Thiol	Methanethiol	Methyl mercaptan
$H—C(=O)—H$	Aldehyde	Methanal	Formaldehyde
$CH_3—C(=O)—CH_3$	Ketone	Propanone	Dimethyl ketone (acetone)

Summary of Reactions

Dehydration of Alcohols to Form Alkenes

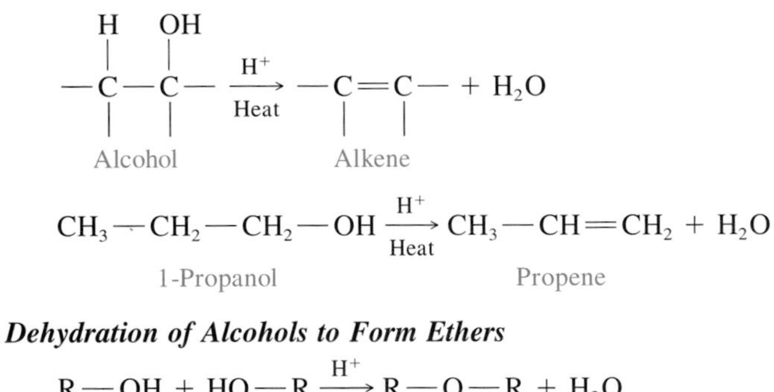

$$—\underset{|}{\overset{H}{\overset{|}{C}}}—\underset{|}{\overset{OH}{\overset{|}{C}}}— \xrightarrow[\text{Heat}]{H^+} —\underset{|}{C}=\underset{|}{C}— + H_2O$$

$$CH_3—CH_2—CH_2—OH \xrightarrow[\text{Heat}]{H^+} CH_3—CH=CH_2 + H_2O$$

Dehydration of Alcohols to Form Ethers

$$R—OH + HO—R \xrightarrow[\text{Heat}]{H^+} R—O—R + H_2O$$

Two alcohols → Ether

$$CH_3—OH + HO—CH_3 \xrightarrow[\text{Heat}]{H^+} CH_3—O—CH_3 + H_2O$$

Methanol → Dimethyl ether

Oxidation of Primary Alcohols to Form Aldehydes and Carboxylic Acids

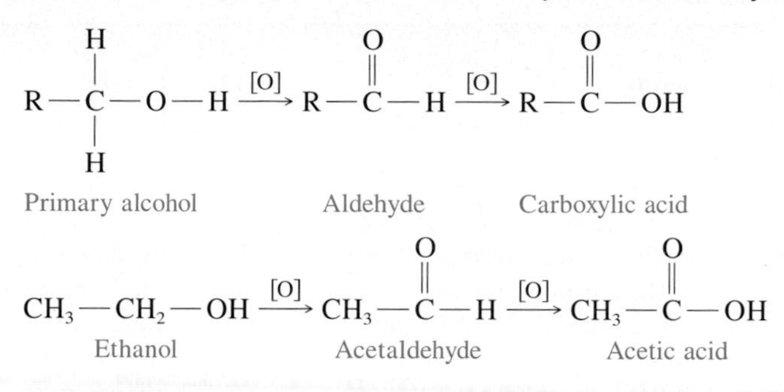

$$R-\underset{\underset{\displaystyle H}{|}}{\overset{\overset{\displaystyle H}{|}}{C}}-O-H \xrightarrow{[O]} R-\overset{\overset{\displaystyle O}{\|}}{C}-H \xrightarrow{[O]} R-\overset{\overset{\displaystyle O}{\|}}{C}-OH$$

Primary alcohol — Aldehyde — Carboxylic acid

$$CH_3-CH_2-OH \xrightarrow{[O]} CH_3-\overset{\overset{\displaystyle O}{\|}}{C}-H \xrightarrow{[O]} CH_3-\overset{\overset{\displaystyle O}{\|}}{C}-OH$$

Ethanol — Acetaldehyde — Acetic acid

Oxidation of Secondary Alcohols to Form Ketones

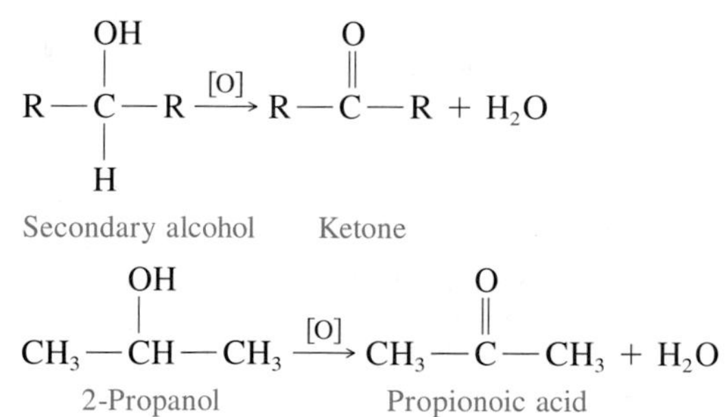

$$R-\underset{\underset{\displaystyle H}{|}}{\overset{\overset{\displaystyle OH}{|}}{C}}-R \xrightarrow{[O]} R-\overset{\overset{\displaystyle O}{\|}}{C}-R + H_2O$$

Secondary alcohol — Ketone

$$CH_3-\overset{\overset{\displaystyle OH}{|}}{CH}-CH_3 \xrightarrow{[O]} CH_3-\overset{\overset{\displaystyle O}{\|}}{C}-CH_3 + H_2O$$

2-Propanol — Propionoic acid

Reduction of Aldehydes to Primary Alcohols

$$R-\overset{\overset{\displaystyle O}{\|}}{C}-H + H_2 \xrightarrow{Ni} R-\underset{\underset{\displaystyle H}{|}}{\overset{\overset{\displaystyle OH}{|}}{C}}-H$$

Aldehyde — Primary alcohol

$$CH_3-\overset{\overset{\displaystyle O}{\|}}{C}-H + H_2 \xrightarrow{Ni} CH_3-CH_2-OH$$

Acetaldehyde — Ethanol

Reduction of Ketones to Secondary Alcohols

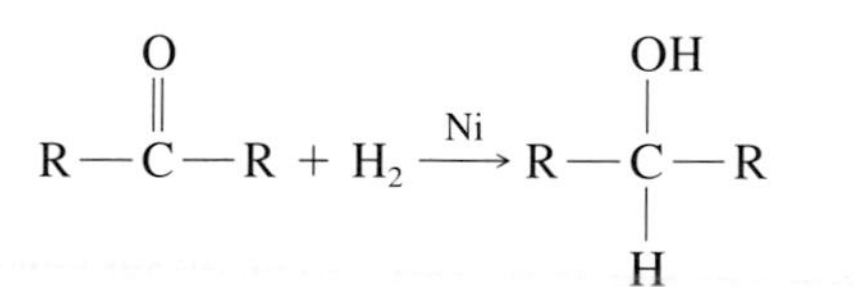

$$R-\overset{\overset{\displaystyle O}{\|}}{C}-R + H_2 \xrightarrow{Ni} R-\underset{\underset{\displaystyle H}{|}}{\overset{\overset{\displaystyle OH}{|}}{C}}-R$$

Ketone — Secondary alcohol

$$CH_3-\overset{\overset{\displaystyle O}{\|}}{C}-CH_3 + H_2 \xrightarrow{Ni} CH_3-\overset{\overset{\displaystyle OH}{|}}{CH}-CH_3$$

Acetone — 2-Propanol

Chapter Summary

12.1 Alcohols and Phenols
The functional group of an alcohol is the hydroxyl group —OH attached to a carbon chain, R—OH. In a phenol, the hydroxyl group is attached to a benzene ring, Ar—OH. Alcohols are classified according to the number of alkyl groups attached to the carbon bonded to the—OH group. In a primary alcohol, there is one alkyl group attached to the carbon atom bonded to the —OH. In a secondary alcohol, there are two alkyl groups, and in a tertiary alcohol there are three alkyl groups attached to the carbon atom with the —OH functional group.

12.2 Names of Alcohols and Phenols
In the IUPAC system, the names of alcohols have *ol* endings, and the location of the —OH group is given by numbering the carbon chain. Simple alcohols are generally named by their common names with the alkyl name preceding the term *alcohol*. A cyclic alcohol is named as a cycloalkanol. An aromatic alcohol is named as a phenol.

12.3 Properties of Alcohols and Phenols
The electronegative oxygen atom makes the —OH functional group very polar. Smaller alcohols (up to four carbons) are soluble in water because the —OH group can hydrogen bond with water molecules. In water, a phenol acts as a weak acid. Derivatives of phenols are found in antiseptics and antioxidants. Thiols are similar to alcohols, except they have an —SH functional group in place of the —OH group, R—SH. Thiols are named by the alkane chain followed by the word *thiol*. Many thiols have strong, disagreeable odors characteristic of garlic, onions, cheeses, and skunk spray.

12.4 Reactions of Alcohols
At high temperatures, alcohols dehydrate in the presence of an acid to yield alkenes. Using an oxidizing agent, primary alcohols oxidize to aldehydes, which usually oxidize further to carboxylic acids. Secondary alcohols are oxidized to ketones, but tertiary alcohols do not oxidize.

12.5 Ethers
In an ether, an oxygen atom is connected by single bonds to two alkyl or aromatic groups, R—O—R. In the common names of ethers, the alkyl groups are listed alphabetically followed by the name *ether*. In the IUPAC name, the smaller alkyl group and the oxygen are named as an *alkoxy group* attached to the longer alkane chain, which is numbered to give the location of the alkoxy group. Less polar than alcohols, only a few shorter-chain ethers are slightly soluble in water. Ethers are widely used as solvents but can be explosive because their vapors are highly flammable.

12.6 Aldehydes and Ketones
Aldehydes and ketones contain a carbonyl group $\left(\rangle C{=}O\right)$, which consists of a double bond between a carbon in the chain and an oxygen atom. In an aldehyde, the carbonyl group appears at the end of a carbon chain attached to at least one hydrogen atom. In a ketone, the carbonyl group occurs between carbon groups and has no hydrogens attached to it.

In the IUPAC system, aldehydes and ketones are named by replacing the *e* in the longest chain containing the carbonyl group with *al* for aldehydes, and *one* for ketones. The location of the carbonyl group in a ketone is given if there are more than four carbon atoms in the chain.

12.7 Properties of Aldehydes and Ketones
The polarity of the carbonyl group makes aldehydes and ketones of one to four carbon atoms soluble in water. Aldehydes are oxidized to carboxylic acids by oxidizing agents in Tollens', Benedict's, and Fehling's tests. Ketones do not oxidize further. Aldehydes and ketones are reduced when hydrogen is added in the presence of a metal catalyst to produce primary or secondary alcohols. In biological systems, many important reactions involve the oxidation and reduction of carbonyl groups.

Glossary of Key Terms

alcohols Organic compounds that contain the hydroxyl (—OH) functional group, attached to a carbon chain, R—OH.

aldehyde An organic compound that has a carbonyl functional group and a hydrogen on the end carbon of the chain,

$$R-\overset{\overset{\displaystyle O}{\|}}{C}-H$$

carbonyl group A functional group that contains a carbon–oxygen double bond $>C=O$.

dehydration A reaction that removes water from an alcohol in the presence of an acid to form alkenes at high temperature, or ethers at lower temperatures.

ether An organic compound, R—O—R, in which an oxygen atom is bonded to two alkyl or two aromatic groups, or a mix of the two.

hydroxyl group The —OH functional group.

ketone An organic compound, $R-\overset{\overset{\displaystyle O}{\|}}{C}-R$ in which the carbonyl functional group is bonded to two alkyl groups.

secondary (2°) alcohol An alcohol that has two alkyl groups attached to the carbon atom bonded to the —OH group.

tertiary (3°) alcohol An alcohol that has three alkyl groups attached to the carbon atom attached to the —OH.

thiols Organic compounds that contain a thiol group (—SH) in place of the —OH group of an alcohol.

oxidation The loss of two hydrogen atoms from a reactant to give a more oxidized compound, e.g., primary alcohols oxidize to aldehydes, secondary alcohols oxidize to ketones. An oxidation can also be the addition of an oxygen atom as in the oxidation of aldehydes to carboxylic acids.

phenol An organic compound that has an —OH group attached to a benzene ring.

primary (1°) alcohol An alcohol that has one alkyl or aryl group attached to the alcohol carbon atom, $R-CH_2OH$.

reduction The addition of hydrogen to a carbonyl bond. Aldehydes are reduced to primary alcohols; ketones to secondary alcohols. Reduction can also be the loss of an oxygen atom.

Chemistry at Home

1. Soak a small piece of paper or cloth in alcohol or rubbing alcohol. Place the sample in a fireproof dish and ignite. What happens to the paper or cloth? Why do you think this was the result?
2. Find a recipe for a dessert that is flamed, e.g., cherries jubilee, bananas Foster, or peach flambé. Get a group of chemistry friends together and prepare your dessert. Explain to your friends how this dramatic presentation works.
3. Read the label of a mouthwash. What are the active ingredients? Write their structures. Use a chemistry reference manual to check.

Answers to Study Checks

12.1 tertiary (3°)

12.2 3-bromo-4-methylcyclopentanol

12.3

$$CH_3CH_2\overset{\overset{\displaystyle OH}{|}}{C}H\underset{\underset{\displaystyle CH_3}{|}}{C}H\overset{\overset{\displaystyle Cl}{|}}{C}HCH_2CH_3$$

12.4 CH_3CH_2SH

12.5 cyclopentene

12.6 $CH_3CH=CH_2 + H_2O \xrightarrow{H^+} CH_3\overset{\overset{\displaystyle OH}{|}}{C}HCH_3 \xrightarrow{[O]} CH_3\overset{\overset{\displaystyle O}{\|}}{C}CH_3$

12.7 ethyl phenyl ether

12.8 2,3-dihydroxypropanal; α, β-dihydroxypropionaldehyde

12.9 ethyl propyl ketone

12.10 1-propanol

Problems

Alcohols and Phenols *(Goal 12.1)*

12.1 Write the structural formula of a primary and secondary alcohol.

12.2 Write the structural formula of a phenol.

12.3 Classify the following alcohols as primary, secondary, or tertiary:

a. 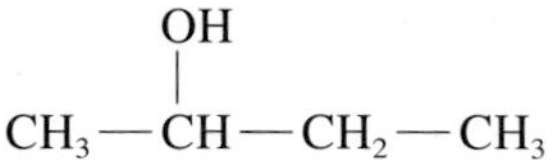 $CH_3-CH(OH)-CH_2-CH_3$

b. $CH_3-CH_2-CH_2-CH_2-CH_2-OH$

c. 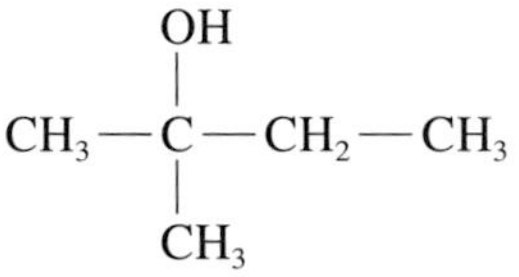 $CH_3-C(OH)(CH_3)-CH_2-CH_3$

d. 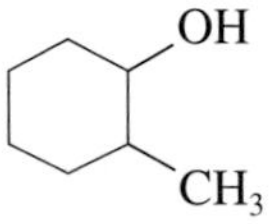 (cyclohexane ring bearing OH and adjacent CH_3)

12.4 Classify the following alcohols as primary, secondary, or tertiary:

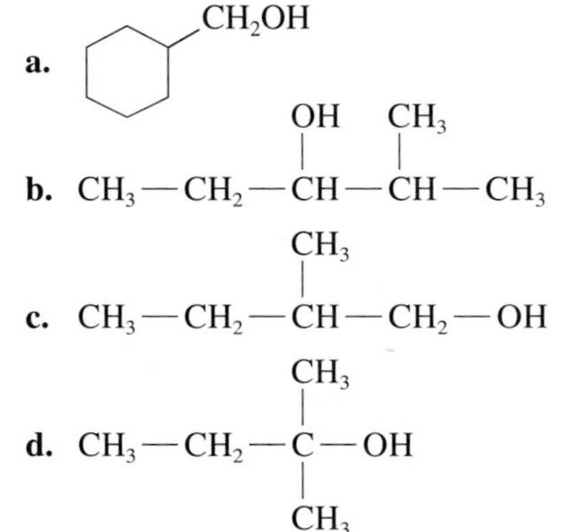

Names of Alcohols and Phenols *(Goal 12.2)*

12.5 Give the IUPAC name for each of the following alcohols:

a. CH_3CH_2OH

b. 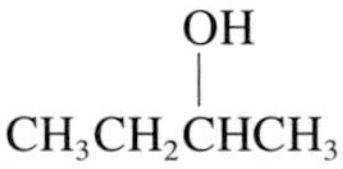 $CH_3CH_2CH(OH)CH_3$

c. 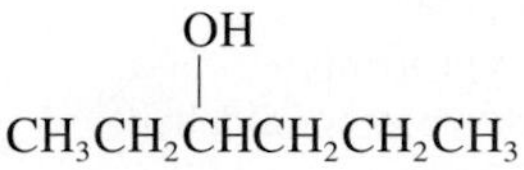 $CH_3CH_2CH(OH)CH_2CH_2CH_3$

d. 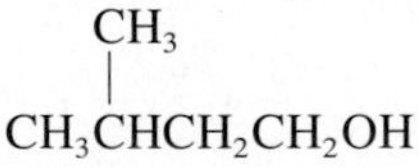 $CH_3CH(CH_3)CH_2CH_2OH$

e. 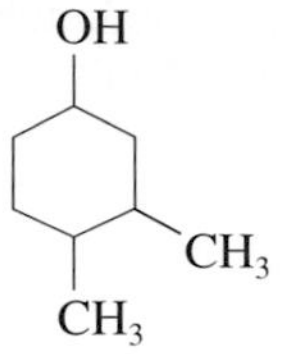 (cyclohexane ring bearing OH, with CH_3 groups on the two adjacent ring carbons)

f. 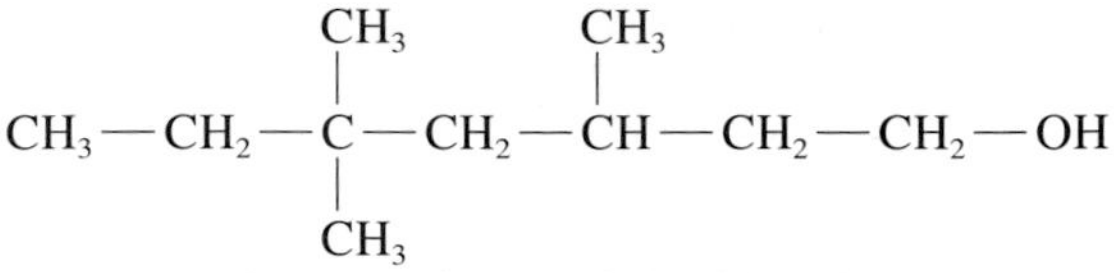 $CH_3-CH_2-C(CH_3)_2-CH_2-CH(CH_3)-CH_2-CH_2-OH$

12.6 Give the IUPAC name of each of the following alcohols:

a. 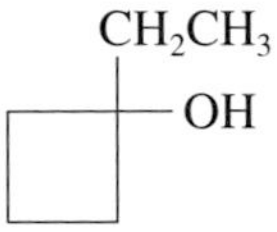 (cyclobutane ring bearing CH_2CH_3 and OH on the same carbon)

b. 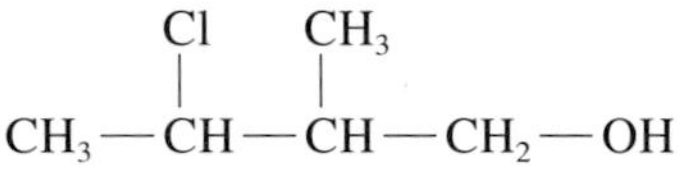 $CH_3-CH(Cl)-CH(CH_3)-CH_2-OH$

c. 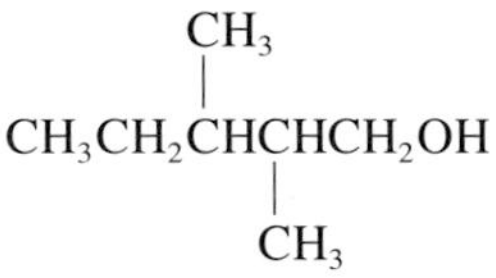 $CH_3CH_2CH(CH_3)CH(CH_3)CH_2OH$

d. 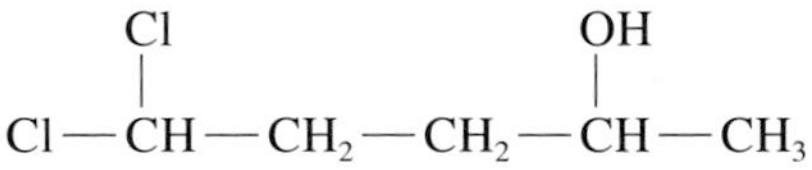 $Cl-CH(Cl)-CH_2-CH_2-CH(OH)-CH_3$

e. 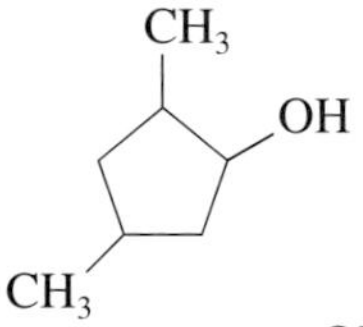 (cyclopentane ring bearing OH, with CH_3 on the adjacent carbon and another CH_3 on the ring)

f. 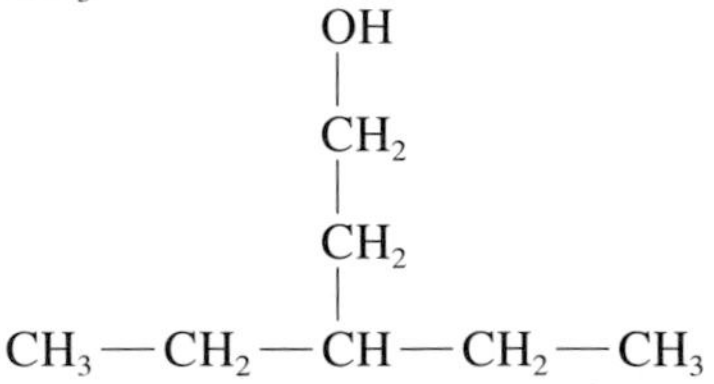 $CH_3-CH_2-CH(CH_2CH_2OH)-CH_2-CH_3$

12.7 Write the condensed structural formula of each of the following alcohols:

a. 1-propanol **b.** methyl alcohol

c. 3-pentanol **d.** 2-methyl-2-butanol

e. cyclohexanol **f.** 1,4-butanediol

12.8 Write the condensed structural formula of each of the following alcohols:

a. ethyl alcohol

b. 3-methyl-1-butanol

c. 2,4-dichlorocyclohexanol

d. propyl alcohol

e. 1,3-cyclopentanediol

f. 2,2,4-trimethyl-3-hexanol

12.9 Name each of the following phenols:

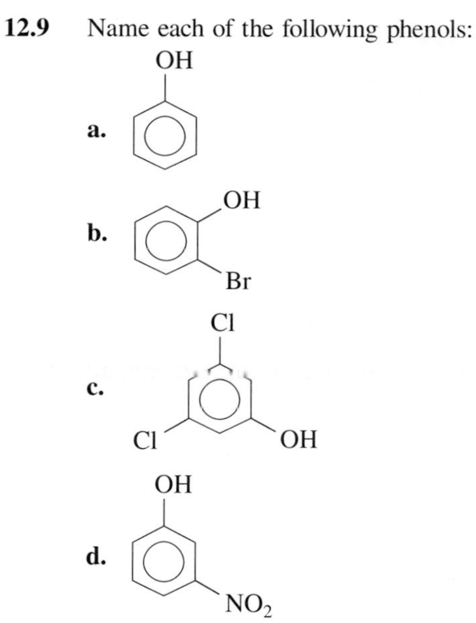

12.10 Name each of the following phenols:

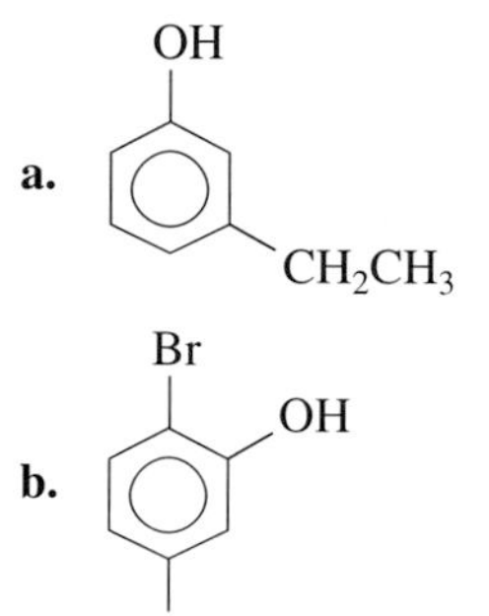

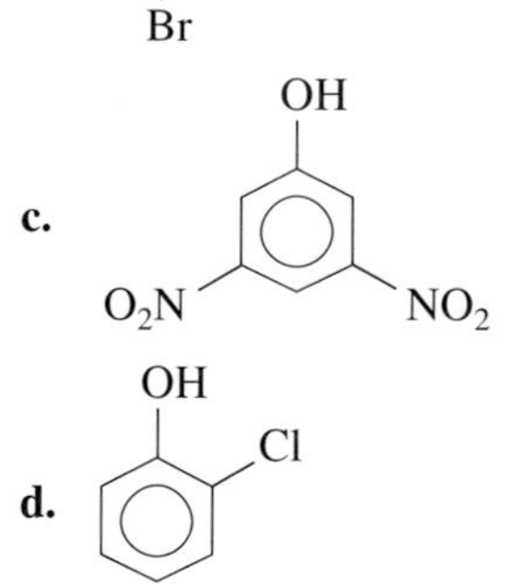

12.11 Write the condensed structural formula of each of the following phenols:

a. *m*-bromophenol
b. *p*-nitrophenol
c. 2,5-dichlorophenol
d. *o*-phenylphenol
e. 4-ethylphenol

12.12 Write the condensed structural formula of each of the following phenols:

a. *o*-ethylphenol
b. 2,4-dichlorophenol
c. 2,4-dimethylphenol
d. 2-ethyl-5-methylphenol
e. *m*-phenylphenol

Properties of Alcohols and Phenols *(Goal 12.3)*

12.13 Why is ethanol soluble in water, but ethane is not?

12.14 Why is propanol more soluble in water than hexanol?

12.15 Write an equation for phenol acting as a weak acid.

12.16 Write an equation for the neutralization reaction of phenol and NaOH.

12.17 Give the IUPAC name for each of the following thiols:

a. $CH_3—SH$

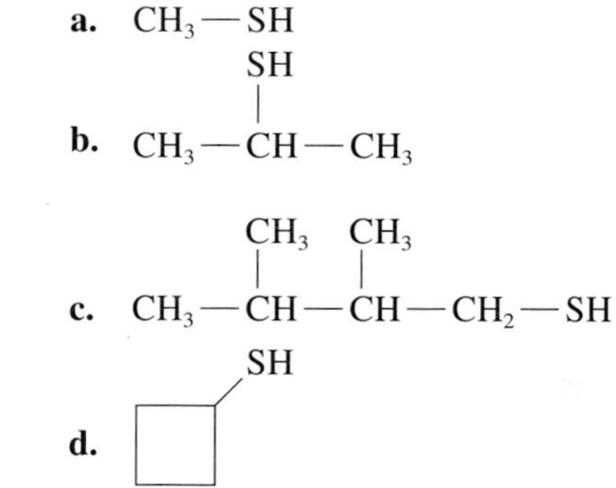

12.18 Give the IUPAC name for each of the following thiols:

a. $CH_3CH_2CH_2—SH$

b. 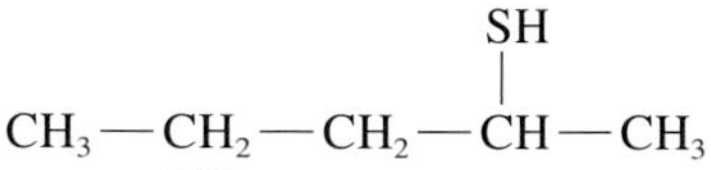 $CH_3—CH_2—CH_2—CH(SH)—CH_3$

c. 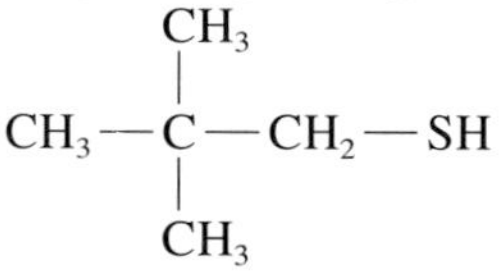 $CH_3—C(CH_3)_2—CH_2—SH$

d. 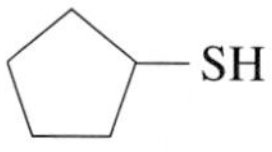 cyclopentyl—SH

Reactions of Alcohols *(Goal 12.4)*

12.19 Draw the condensed structural formula of the organic compounds produced by each of the following dehydration reactions:

a. $CH_3—CH_2—CH_2—CH_2—OH \xrightarrow[\text{Heat}]{H^+}$

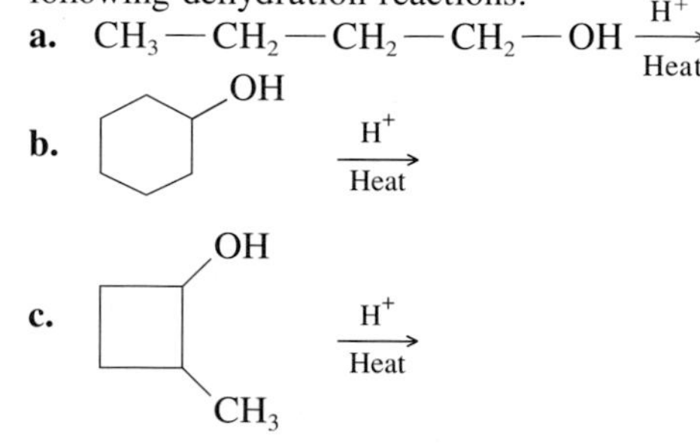

d. $CH_3—CH_2—CH_2—\underset{}{\overset{OH}{\overset{|}{C}}}H—CH_3 \xrightarrow[Heat]{H^+}$

12.20 Draw the condensed structural formula of the organic compounds produced by each of the following dehydration reactions:

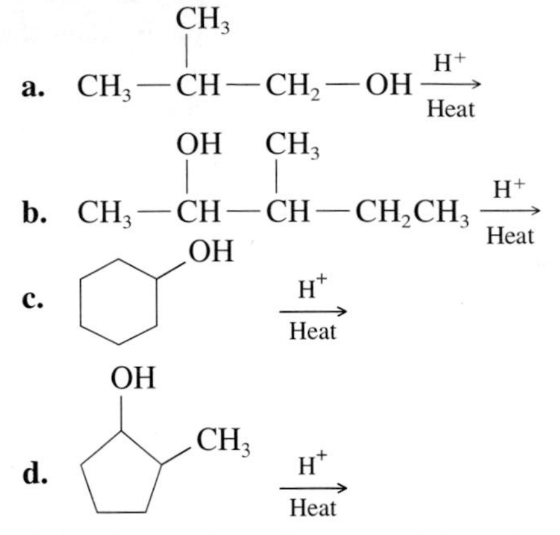

12.21 Draw the condensed structural formula of the organic product when each of the following alcohols is oxidized [O] (if no reaction, write *none*):

a. $CH_3CH_2CH_2CH_2CH_2OH$

b. $CH_3CH_2\overset{OH}{\overset{|}{C}}HCH_3$

c. 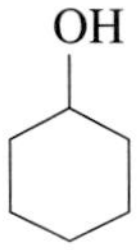 (cyclohexanol: OH on a cyclohexane ring)

d. $CH_3—\overset{OH}{\overset{|}{C}}H—CH_2—\overset{CH_3}{\overset{|}{C}}H—CH_3$

e. $CH_3—\overset{CH_3}{\overset{|}{C}}H—CH_2—CH_2—OH$

12.22 Draw the condensed structural formula of the organic product when each of the following alcohols is oxidized [O] (if no reaction, write *none*):

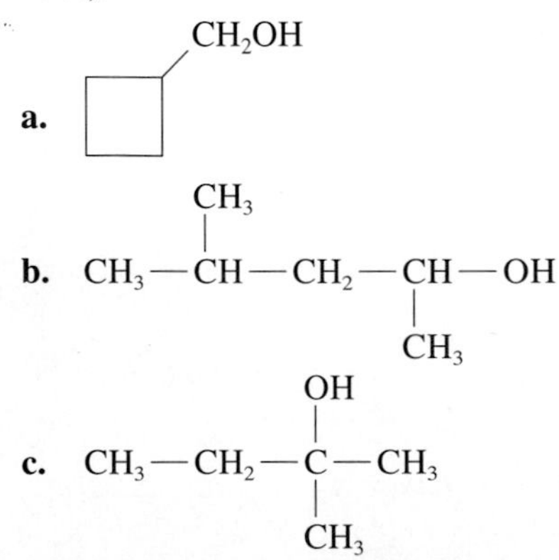

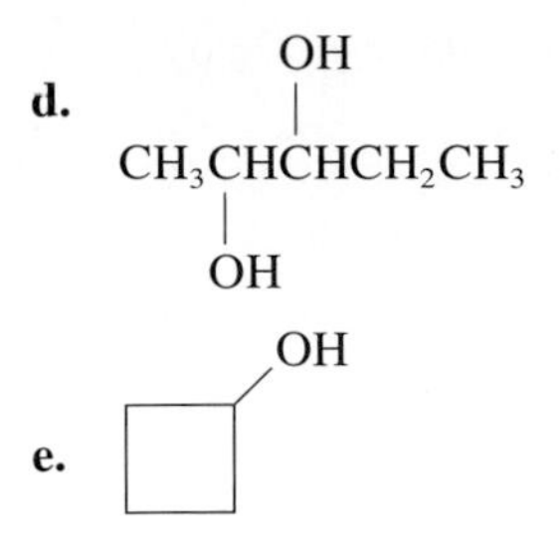

Ethers *(Goal 12.5)*

12.23 Give the IUPAC name and a common name for each of the following ethers:

a. $CH_3—O—CH_2CH_3$

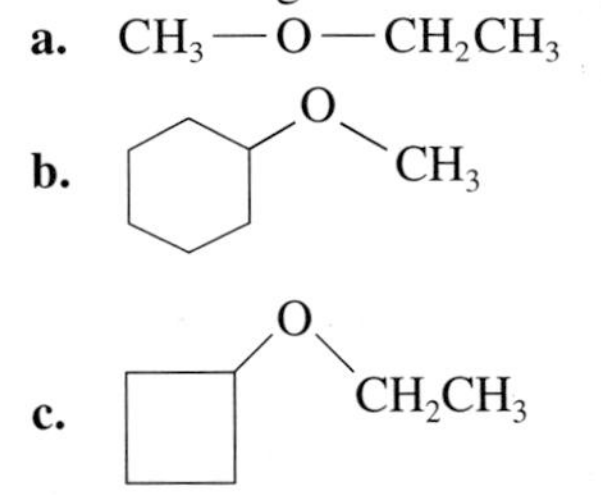

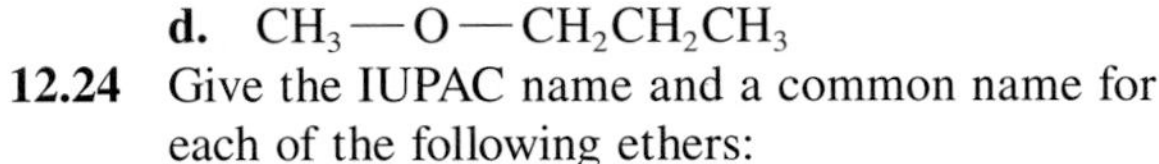

d. $CH_3—O—CH_2CH_2CH_3$

12.24 Give the IUPAC name and a common name for each of the following ethers:

a. $CH_3CH_2—O—CH_2CH_2CH_3$

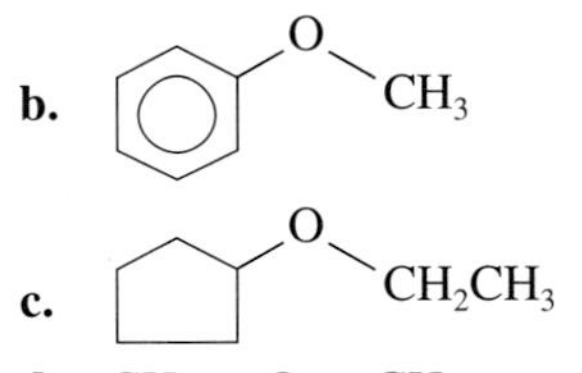

d. $CH_3—O—CH_3$

12.25 Write the condensed structural formula for each of the following ethers:

a. methyl propyl ether
b. ethyl cyclopropyl ether
c. methoxycyclopentane
d. 1-ethoxy-2-methylbutane
e. 2,3-dimethoxypentane

12.26 Write the condensed structural formula for each of the following ethers:

a. diethyl ether
b. diphenyl ether
c. ethoxycyclohexane
d. 2-methoxy-2,3-dimethylbutane
e. 1,2-dimethoxybenzene

Aldehyde and Ketones *(Goal 12.6)*

12.27 Write the IUPAC and common name (if any) for each of the following:

a. $CH_3—\overset{O}{\overset{||}{C}}—H$

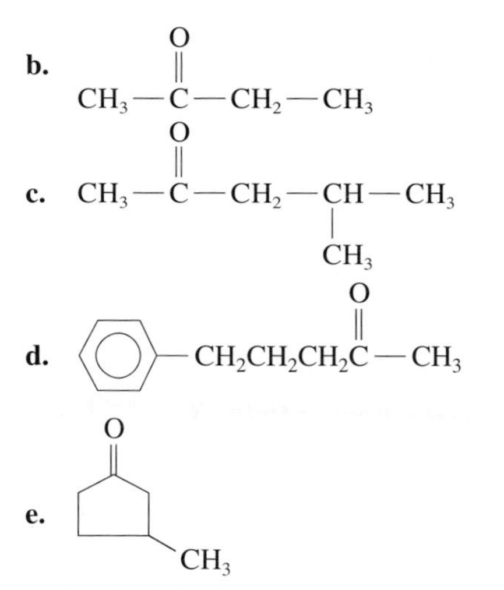

12.28 Write the IUPAC and common name (if any) for each of the following:

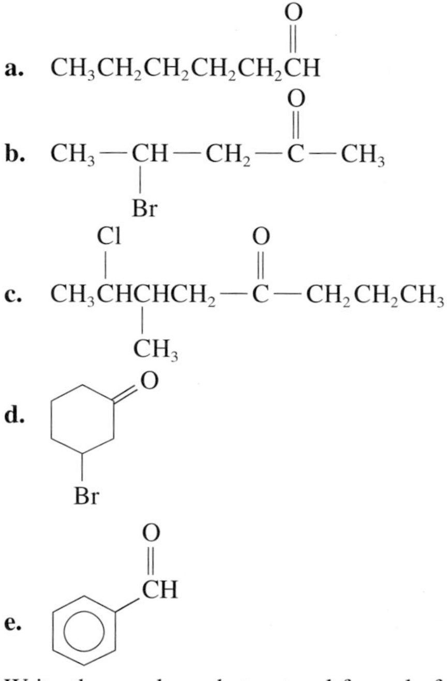

12.29 Write the condensed structural formula for each of the following compounds:

a. propionaldehyde
b. formaldehyde
c. 3-methyl-2-pentanone
d. β-chlorobutyraldehyde
e. benzaldehyde
f. ethyl phenyl ketone

12.30 Write the condensed structural formulas for each of the following compounds:

a. 3-methylpentanal
b. acetaldehyde
c. α-chlorobutyraldehyde
d. 4-bromocyclohexanone
e. 3-hydroxypropanal
f. methyl isopropyl ketone

12.31 Vanillin, a naturally occurring compound in vanilla beans and potato parings, has the following structural formula. If it is named as an aldehyde, give its name.

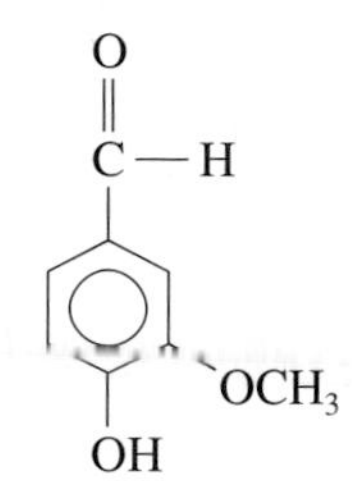

12.32 Veratraldehyde is prepared by adding a methyl group to the hydroxyl group of vanillin. If it is named as an aldehyde, give its name.

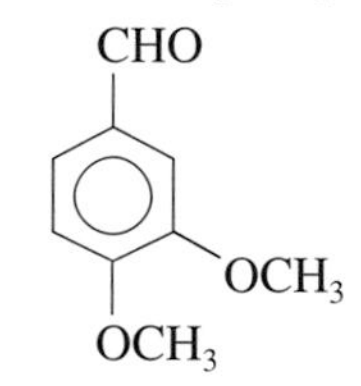

12.33 Draw the condensed structural formula of the alcohol needed to give each of the following oxidation products:

a. formaldehyde
b. cyclopentanone
c. butanone
d. benzaldehyde
e. 3-methylcyclohexanone

12.34 Draw the condensed structural formula of the alcohol needed to give each of the following oxidation products:

a. acetaldehyde
b. 2-methylbutanone
c. cyclohexanone
d. propionaldehyde
e. 3-methylbutanal

Properties of Aldehydes and Ketones *(Goal 12.7)*

12.35 Which compound in each of the following pairs would be more soluble in water? Explain.

a. $CH_3CH_2CH_3$ or $CH_3\overset{\overset{O}{\|}}{C}CH_3$

b. $CH_3\overset{\overset{O}{\|}}{C}H$ or $CH_3CH_2CH_2CH_2\overset{\overset{O}{\|}}{C}H$

12.36 Which compound in each of the following pairs would be more soluble in water? Explain.

a. $CH_3-\overset{\overset{O}{\|}}{C}-CH_2CH_3$ or $CH_3-\overset{\overset{O}{\|}}{C}-\overset{\overset{O}{\|}}{C}-CH_3$

b. $CH_3-\overset{\displaystyle O}{\overset{\|}{C}}-CH_3$ or $CH_3-\overset{\displaystyle OH}{\overset{|}{CH}}-CH_3$

12.37 Indicate the compound in each of the following pairs that will form a brick-red precipitate with Benedict's reagent:
a. pentane or pentanal
b. propanone or propionaldehyde
c. cyclopentanone or cyclopentanal

12.38 Indicate the compound in each of the following pairs that will form a silver mirror in the Tollens' test:
a. hexanone or hexanal
b. butane or butyraldehyde
c. 1-pentanol or pentanal

12.39 Give the condensed structural formula of the organic product formed when each of the following is reduced by hydrogen in the presence of a nickel catalyst:
a. butyraldehyde
b. acetone
c. 3-bromohexanal
d. 2-methyl-3-pentanone

12.40 Give the condensed structural formula of the organic product formed when each of the following is reduced by hydrogen in the presence of a nickel catalyst:
a. ethyl propyl ketone
b. formaldehyde
c. 3-chlorocyclopentanone
d. 2-pentanone

Challenge Problems

The following groups of problems include functional groups, physical and chemical properties, and reactions from all of the organic chapters.

12.41 Give the IUPAC and common names (if any) for each of the following compounds:

a. $CH_3\overset{\displaystyle CH_3}{\overset{|}{\underset{\underset{\displaystyle CH_3}{|}}{C}}}CH_2CH_2\overset{\displaystyle CH_3}{\overset{|}{C}}HCH_3$

b. $CH_3\overset{\displaystyle Br}{\overset{|}{C}}HCH_2\overset{\displaystyle OH}{\overset{|}{C}}HCH_2CH_3$

c. $CH_3CH_2\overset{\displaystyle O}{\overset{\|}{C}}H$

d. $\underset{H}{\overset{CH_3}{}}\!\!>C=C<\!\!\underset{H}{\overset{CH_2CH_3}{}}$

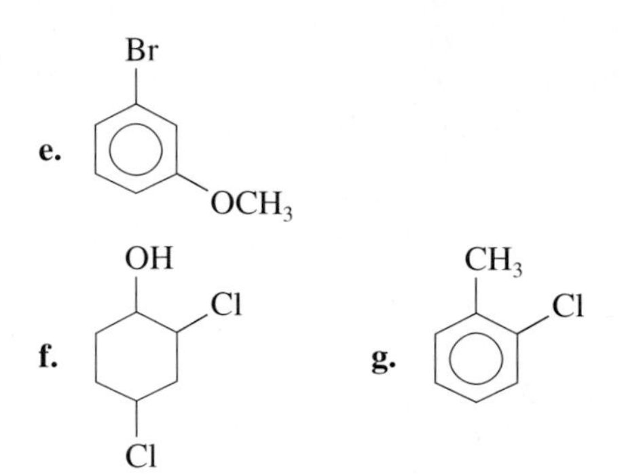

12.42 Give the IUPAC and common names (if any) for each of the following compounds:

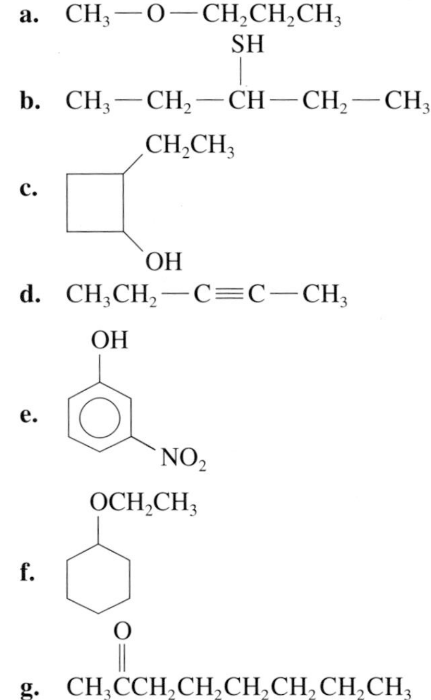

g. $CH_3\overset{\displaystyle O}{\overset{\|}{C}}CH_2CH_2CH_2CH_2CH_2CH_3$

12.43 Draw the condensed structural formula of each of the following:
a. 3-methylcyclopentanone
b. *p*-dichlorobenzene
c. β-chloropropionaldehyde
d. *m*-bromophenol
e. diethyl ether

12.44. Draw the structural formula of each of the following:
a. methanethiol
b. β-chlorobutyraldehyde
c. 3-methyl-1-butanol
d. 3-methoxypentanal
e. *m*-bromotoluene

12.45 Which of the following compounds are soluble in water?
a. $CH_3CH_2CH_2CH_2CH_3$
b. CH_3OH
c. $CH_3CH_2-O-CH_2CH_3$

d. $CH_3\overset{\overset{\displaystyle O}{\|}}{C}CH_3$

e. CH_3CH_2Cl

f. $CH_3CH_2CH_2OH$

g. $CH_3CH_2\overset{\overset{\displaystyle OH}{|}}{C}HCH_2CH_2CH_3$

12.46 Which of the following compounds are soluble in water?

a. $HO—CH_2CH_2CH_2—OH$

b. $CH_3CH_2CH_2\overset{\overset{\displaystyle CH_3}{|}}{C}HCH_3$

c. $CH_2{=}CH—CH_3$

d. $CH_3CH_2\overset{\overset{\displaystyle O}{\|}}{C}CH_3$

e. $CH_3CH_2\overset{\overset{\displaystyle O}{\|}}{C}H$

f. (cyclobutane ring) $—CH_2CH_3$

g. $\underset{\underset{\displaystyle OH}{|}}{CH_2}—\underset{\underset{\displaystyle OH}{|}}{CH}—\underset{\underset{\displaystyle OH}{|}}{CH_2}$

12.47 Draw the structural formulas of all the alcohols with a molecular formula $C_4H_{10}O$.

12.48 Draw the structural formulas of all the ethers with a molecular formula $C_5H_{12}O$.

12.49 Draw the structural formulas of an alcohol, ether, aldehyde, and ketone, all with the formula $C_6H_{12}O$.

12.50 Draw the structural formulas of two ketones and two aldehydes with the formula $C_5H_{10}O$.

12.51 Sometimes, several steps are needed to prepare a compound. Using a combination of the reactions we have studied, indicate how you might prepare the following from the starting substance given. (For example, 2-propanol could be prepared from 1-propanol by first dehydrating the alcohol to give propene and then hydrating it again to give 2-propanol according to Markovnikov's rule.)

$$CH_3CH_2CH_2OH \xrightarrow[\text{Heat}]{H^+} CH_3CH{=}CH_2 + H_2O$$

$$\xrightarrow{H^+} CH_3\overset{\overset{\displaystyle OH}{|}}{C}HCH_3$$

a. acetaldehyde from ethene

b. 2-methylpropane from 2-methyl-2- propanol

c. butanone from 1-butene

12.52 As in problem 12.51, indicate how you might prepare the following from the starting substance given:

a. 1-pentene from pentanal.

b. butanone from butanal.

c. cyclopentene from cyclopentanone.

d. 1,2-dibromobutane from butanal.

12.53 Identify the functional groups in the following molecules:

a.

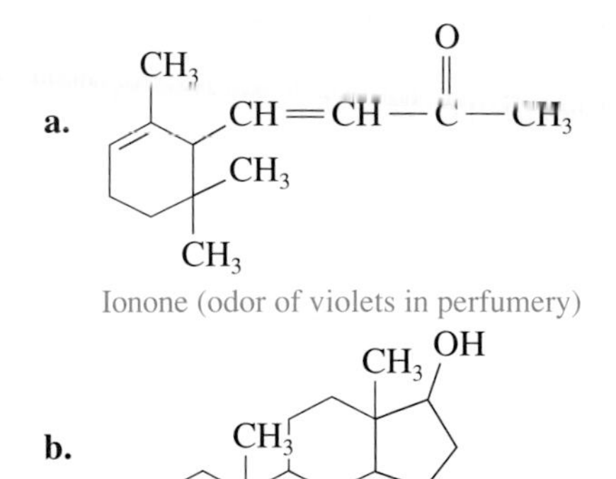

Ionone (odor of violets in perfumery)

b.

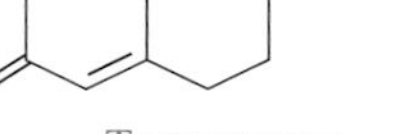

Testosterone

12.54 Identify the functional groups in the following molecules:

a.

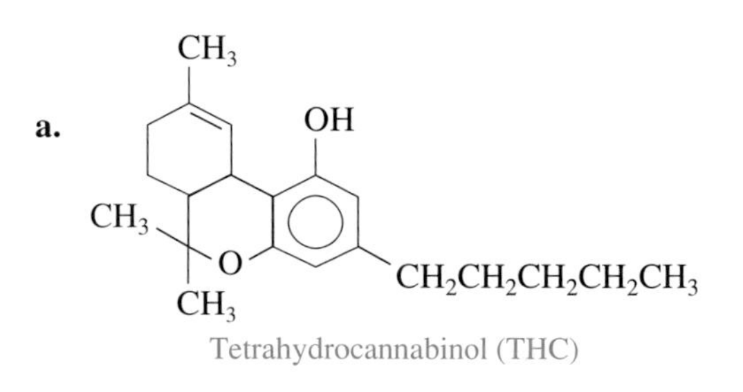

Tetrahydrocannabinol (THC)

b.

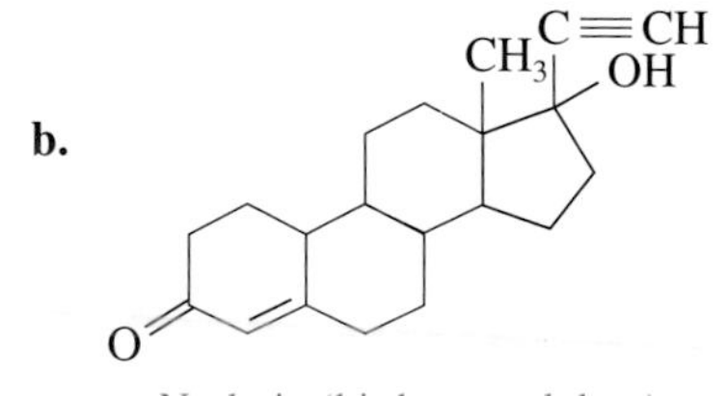

Norlutin (birth control drug)

12.55 Hexylresorcinal, an antiseptic ingredient used in mouthwashes and throat lozenges, has the IUPAC name of 4-hexyl-1,3-benzenediol. Draw its condensed structural formula.

12.56 Menthol, which has a minty flavor, is used in throat sprays and lozenges. Thymol is used as a topical antiseptic to destroy mold. Give each of their IUPAC names. What is similar and different about their structures?

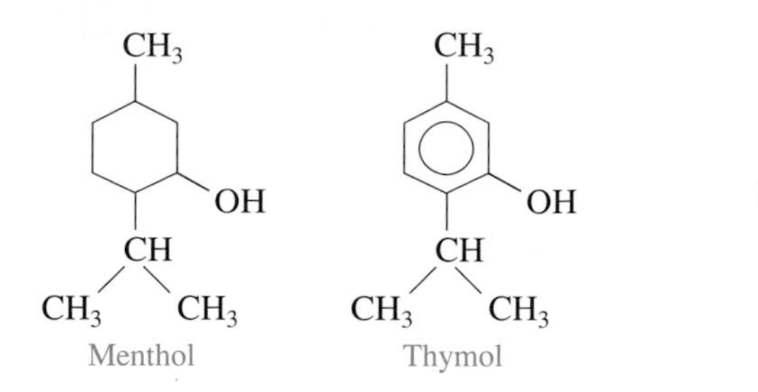

12.57 Write the condensed structural formulas for each of the following naturally occurring compounds:

a. 2-heptanone, an alarm pheromone of bees
b. *trans*-2-hexenal, an alarm pheromone of a myrmicine ant
c. 2,5-dichlorophenol, a defense pheromone of a grasshopper
d. pinacolone, which is 3,3-dimethyl-2-butanone, odor of peppermint or camphor

12.58 Dimethyl ether and ethyl alcohol both have the molecular formula C_2H_6O. One has a boiling point of $-24°C$, and the other, $79°C$. Draw the condensed structural formulas of each compound. Decide which boiling point goes with which compound and explain. Check the boiling points in a chemistry handbook.

Chapter 13

Carboxylic Acids, Esters, Amines, and Amides

The caffeine from these coffee beans is an alkaloid, which is a nitrogen-containing compound that has physiological effects on the body.

Looking Ahead

Learning Goals

13.1 Give the common names, IUPAC names, and condensed structural formulas of carboxylic acids.

13.2 Describe the solubility and ionization of carboxylic acids in water.

13.3 Write equations for the neutralization of carboxylic acids and the preparation of esters.

13.4 Give the common names, IUPAC names, and condensed structural formulas of esters.

13.5 Write equations for the hydrolysis and saponification of esters.

13.6 Classify amines; give their names and condensed structural formulas.

13.7 Write equations for the reactions of amines.

13.8 Write the common names, IUPAC names, and condensed structural formulas of amides.

13.9 Write equations for the hydrolysis of amides.

QUESTIONS TO THINK ABOUT

1. What are the carboxylic acids in vinegar and citrus fruits that make them taste sour?
2. What is the carboxylic acid in the sting of a red ant?
3. What gives the pleasant odors to perfumes, flowers and fruits?
4. What are the chemical groups in fats?
5. How does a soap work? Why are fats used to make soap?
6. What do the codes 1-PETE on a soft drink bottle, 2-PS on an aspirin container, and 3-HDPE on a bottle of antioxidant capsules mean?
7. What makes fish smell "fishy"?
8. What functional group is present in many tranquilizers and hallucinogens?
9. What reaction takes place during the digestion of proteins?

Carboxylic acids are in some ways similar to the inorganic acids we studied in Chapter 9. They have a sour or tart taste, produce hydronium ions in water, and neutralize bases. Many of the carboxylic acids are naturally occurring. A solution of acetic acid and water is the vinegar you use in your salad. The tartness of citrus fruits is due to citric acid. The burning sting of a red ant is due to the carboxylic acid called formic acid. The lactic acid produced in our muscles when we exercise is converted to pyruvic acid as oxygen returns to the muscle.

When a carboxylic acid combines with an alcohol, an ester is produced. Aspirin is an ester. Esters give pleasant aromas and flavors to fruits and flowers. Fats and oils are esters of glycerol and carboxylic acids called fatty acids.

Organic compounds that contain nitrogen include the amines and amides. Many, like adrenaline and amphetamine, are physiologically active. Alkaloids, such as curare, belladonna, and digitalis, are nitrogen-containing compounds obtained from plants. When a carboxylic acid combines with an amine, an amide is produced. Acetaminophen (Tylenol) is an amide. Some amides, such as phenobarbital, are used as sedatives and in anticonvulsant medications. Biologically, proteins are polymers held together by amide linkages between the carboxyl and amine groups of amino acids.

13.1 Carboxylic Acids

Learning Goal

Give the common names, IUPAC names, and condensed structural formulas of carboxylic acids.

In Chapter 12, we described the carbonyl group (C=O) as the functional group in aldehydes and ketones. In a **carboxylic acid,** a hydroxyl group is attached to the carbonyl group forming a **carboxyl group.** The carboxyl functional group may be attached to an alkyl group or an aromatic (Ar) group.

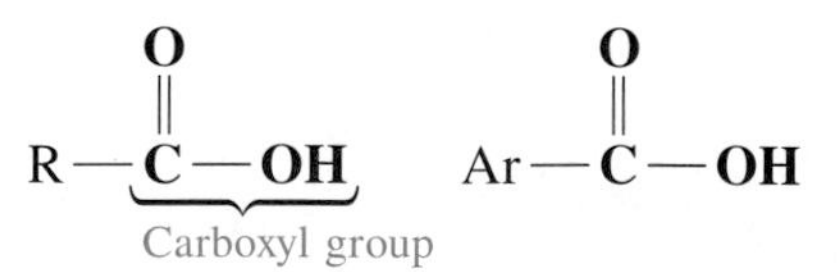

The carboxyl group can be written in several different ways. For example, the condensed structural formula for the carboxylic acid known as acetic acid can be written as follows:

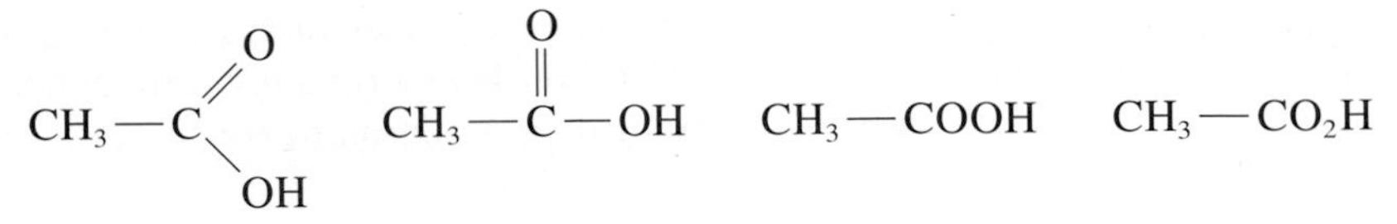

Some condensed structural formulas for acetic acid

Figure 13.1
Formic acid is the substance in the sting of a bee that irritates the skin.

Names of Carboxylic Acids

Using the IUPAC rules, carboxylic acids are named by replacing the *e* in the alkane parent name with *oic acid.* However, the common names of the simple carboxylic acids, which are derived from their natural sources, are used extensively. In Chapter 12, we named aldehydes using the prefixes from the common names of the carboxylic acids. Formic acid is released under the skin from bee or red ant stings and other insect bites. (See Figure 13.1.) Acetic acid is the oxidation product of the ethanol in wines and apple cider. The resulting solution of acetic acid and water is known as vinegar. Butyric acid gives the foul odor to rancid butter. (See Table 13.1.) Some ball-and-stick models of carboxylic acids are shown in Figure 13.2.

Naming Carboxylic Acids $R{-}\overset{\displaystyle O}{\overset{\|}{C}}{-}OH$

Functional group: $-\overset{\displaystyle O}{\overset{\|}{C}}-OH$
IUPAC name: alkanoic acid, benzoic acid
Common names: formic acid, acetic acid, propionic acid, butyric acid

Table 13.1 ***Names and Natural Sources of Carboxylic Acids***

Condensed Structural Formulas	IUPAC Name	Common Name	Occurs In
$H\overset{\displaystyle O}{\overset{\|}{C}}OH$	Methanoic acid	Formic acid	Ant stings (Lat. *formica,* "ant")
$CH_3\overset{\displaystyle O}{\overset{\|}{C}}OH$	Ethanoic acid	Acetic acid	Vinegar (Lat. *acetum,* "sour")
$CH_3CH_2\overset{\displaystyle O}{\overset{\|}{C}}OH$	Propanoic acid	Propionic acid	Dairy products (Greek, *pro,* "first," *pion,* "fat")
$CH_3CH_2CH_2\overset{\displaystyle O}{\overset{\|}{C}}OH$	Butanoic acid	Butyric acid	Rancid butter (Lat. *butyrum,* "butter")

Figure 13.2
Ball-and-stick models of formic acid, acetic acid, and propionic acid.

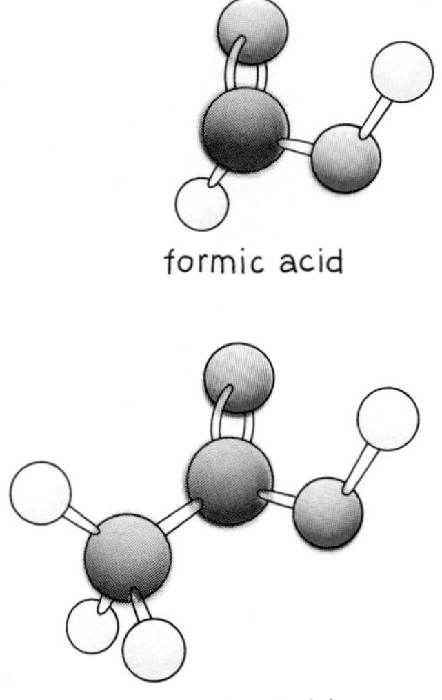

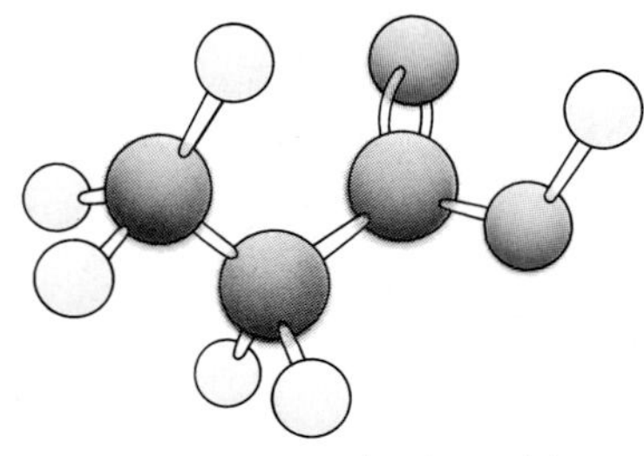

Carboxylic acids are named following IUPAC rules much as we have done for other families of organic compounds. Suppose you want to write the IUPAC and common name for the following carboxylic acid:

$$CH_3{-}\underset{\displaystyle CH_3}{CH}{-}CH_2{-}\overset{\displaystyle O}{\overset{\|}{C}}OH$$

Step 1 Identify the longest carbon chain containing the carboxyl group and replace the *e* of the alkane name by *oic acid;* smaller carboxylic acids are often named by their common names.

$$\mathbf{CH_3{-}CH{-}CH_2{-}\overset{O}{\overset{\|}{C}}OH}$$ (with CH_3 on CH) Butanoic acid (butyric acid)

Step 2 Number the carbon chain beginning with the carboxyl carbon as carbon 1. In the common names, the Greek letters alpha (α), beta (β), and gamma (γ) are assigned to the carbons adjacent to the carboxyl carbon.

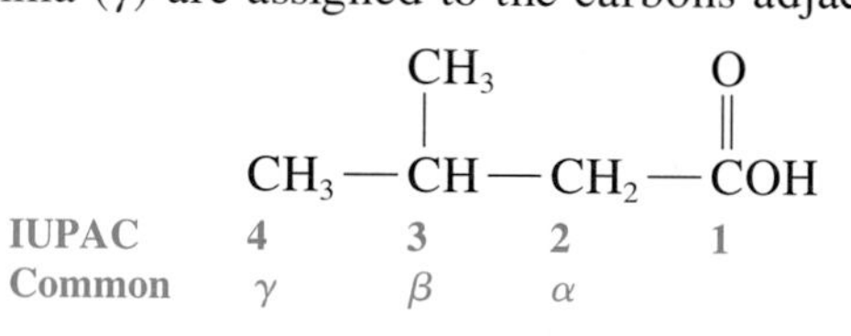

Step 3 Give the location and names of substituents on the parent chain. Use numbers in the IUPAC name and Greek letters in the common name.

$$CH_3{-}CH{-}CH_2{-}\overset{O}{\overset{\|}{C}}OH$$ (with CH_3 on CH) 3-Methylbutanoic acid (β-methylbutyric acid)

We can break the IUPAC name down to check our naming.

3	methyl	butan	oic acid
Carbon 3 has a	$—CH_3$ group	4 C in chain	Carboxyl functional group

Some more examples of naming carboxylic acids are shown below.

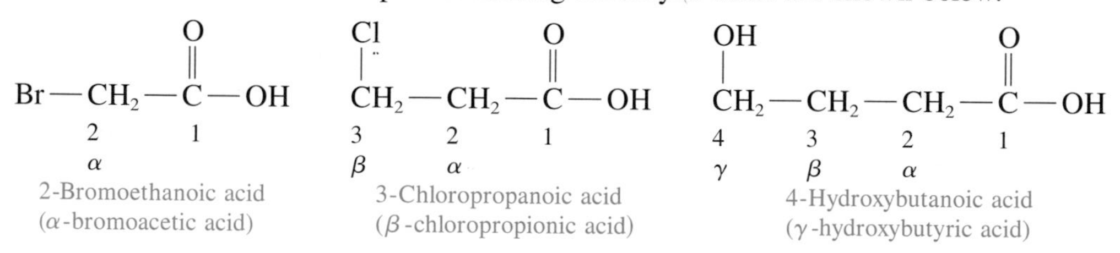

The aromatic carboxylic acid is called benzoic acid.

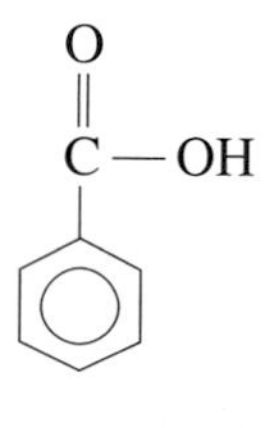

Benzoic acid

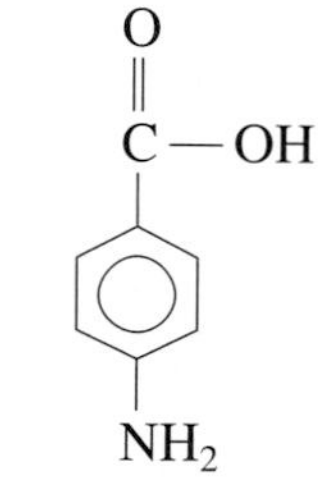

4-Aminobenzoic acid
(*p*-aminobenzoic acid)

O ‖ C—OH (benzene ring with Cl at 3 and 4 positions)

3,4-Dichlorobenzoic acid

Sample Problem 13.1
Naming Carboxylic Acids

Give the IUPAC and common name for each of the following carboxylic acids:

a. $CH_3\overset{O}{\overset{\|}{C}}OH$

b. $CH_3\overset{CH_3}{\overset{|}{C}}H-\overset{O}{\overset{\|}{C}}OH$

Solution

a. This carboxylic acid has two carbon atoms. In the IUPAC system, the *e* in ethane is replaced by *oic acid,* to give the name, *ethanoic acid.* Its common name is *acetic acid.*

b. This carboxylic acid has a methyl group on the second carbon. It has the IUPAC name *2-methylpropanoic acid.* In the common name, the Greek letter α specifies the carbon atom next to the carboxyl carbon, *α-methylpropionic acid.*

Study Check

Write the condensed structural formula of 3-phenylpropanoic acid.

Preparation of Carboxylic Acids

Carboxylic acids can be prepared from primary alcohols or aldehydes. As we saw in Chapter 12, the loss of two hydrogens from a primary alcohol produces an aldehyde. Oxidation continues easily as another oxygen is added to yield a carboxylic acid.

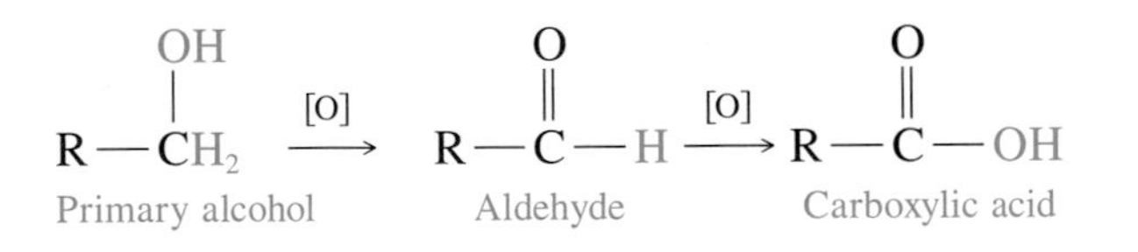

Figure 13.3
Vinegar is a 5% solution of acetic acid and water.

For example, when ethyl alcohol in wine comes in contact with the oxygen in the air, vinegar is produced. The oxidation process converts the ethyl alcohol (primary alcohol) to acetaldehyde, and then to acetic acid, the carboxylic acid in vinegar. (See Figure 13.3.)

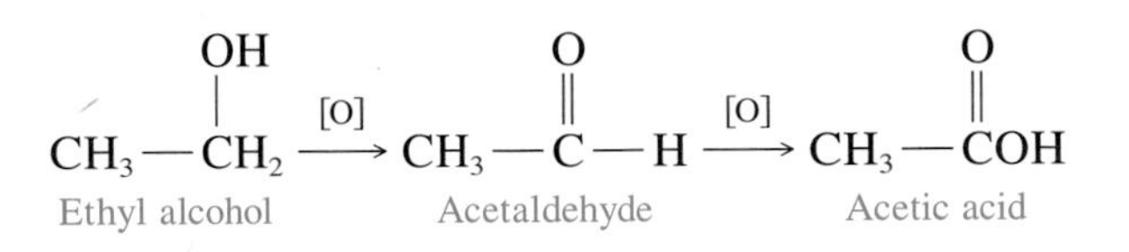

Sample Problem 13.2
Preparation of Carboxylic Acids

Write an equation for the oxidation of 1-propanol and name each product.

Solution

A primary alcohol will oxidize to an aldehyde, which can oxidize further to a carboxylic acid.

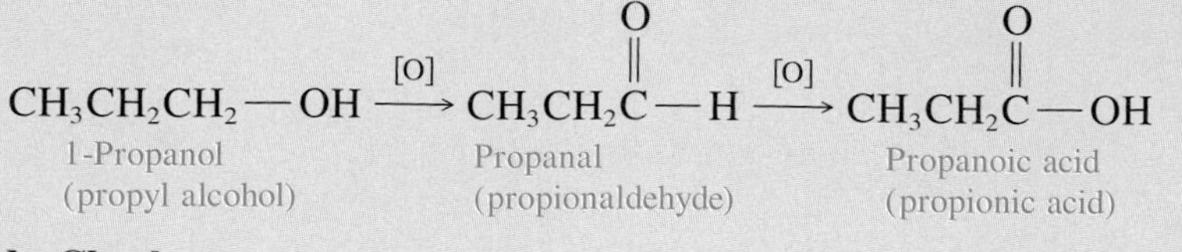

Study Check

Write the condensed structural formula of the carboxylic acid produced by the oxidation of both hydroxyl groups in 1,3-propanediol.

HEALTH NOTE *Alpha Hydroxy Acids*

During strenuous exercise, lactic acid is formed in the muscles when oxygen is used up. Lactic acid is also produced as milk spoils, which turns the milk "sour." (See Figure 13.4.) Cleopatra reportedly bathed in sour milk to reap the benefits of lactic acid.

Other alpha hydroxy acids include tartaric, malic, citric, and glycolic acids. Also known as fruit acid, tartaric acid occurs naturally in grapes and adds a tart taste to wines, malic acid is found in apples and grapes, citric acid in fruits such as lemons, oranges, and grapefruit, and glycolic acid is found in sugarcane and sugar beets. These alpha hydroxy acids are being added to skin formulations to help soften the skin and reduce the appearance of fine lines and wrinkles. Some speculate that the acids enhance the natural exfoliation process.

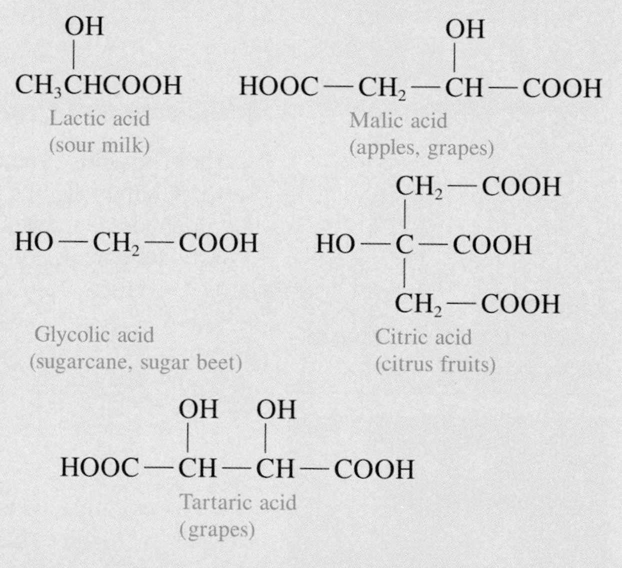

Figure 13.4
Alpha hydroxy acids, including glycolic, lactic, malic, tartaric, and citric acid, are used at concentrations of 4–10% in skin care products to soften skin cells and lessen wrinkles.

13.2 Properties of Carboxylic Acids

Learning Goal
Describe the solubility and ionization of carboxylic acids in water.

Carboxylic acids are among the most polar organic compounds because they contain two polar groups: a hydroxyl (—OH) group and a carbonyl (C=O) group. The smaller carboxylic acids (one to four carbons) are very soluble in water because the carboxyl group can hydrogen bond with several water molecules. However, as the length of the carbon chain increases, the hydrocarbon portion of the molecules affects solubility. Carboxylic acids having five or more carbon atoms are only slightly soluble in water. (See Table 13.2.)

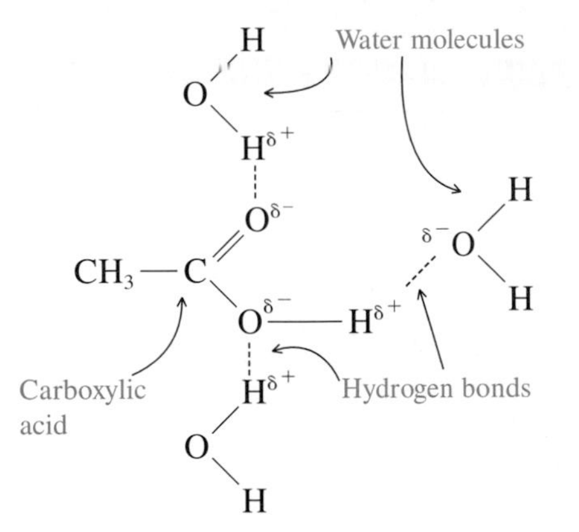

Hydrogen bonds form between the carboxylic acid and water molecules.

Carboxylic Acids Are Weak Acids

In water, carboxylic acids ionize by donating a proton to water to form hydronium ions, H_3O^+. However, carboxylic acids are weak acids, although

Table 13.2 ***Solubility of Some Carboxylic Acids in Water***

Carboxylic Acid	Condensed Structural Formula	Solubility in Water (g/100 g)
Formic acid	$HC(=O)OH$ (HCOH with C=O)	Any amount
Acetic acid	$CH_3C(=O)OH$	Any amount
Propionic acid	$CH_3CH_2C(=O)OH$	Any amount
Butyric acid	$CH_3CH_2CH_2C(=O)OH$	Any amount
Pentanoic acid	$CH_3CH_2CH_2CH_2C(=O)OH$	4
Hexanoic acid	$CH_3CH_2CH_2CH_2CH_2C(=O)OH$	1

they are more acidic than phenols. In solution, only a few molecules of the acid ionize; most of the acid is not ionized. The organic ion that forms is called a *carboxylate ion,* named by replacing the *ic acid* ending with *ate.*

Ionization of a Carboxylic Acid in Water

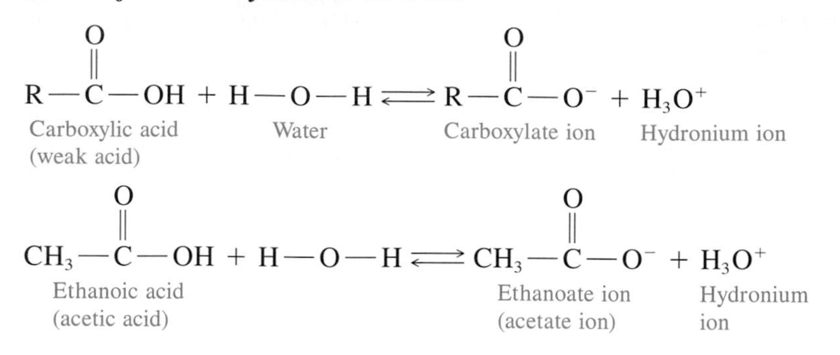

Sample Problem 13.3
Ionization of Carboxylic Acids in Water

Write the equation for the ionization of propionic acid in water.

Solution
The ionization of propionic acid produces a carboxylate ion and an hydronium ion.

$$CH_3CH_2-\overset{\overset{\displaystyle O}{\|}}{C}-OH + H-O-H \rightleftharpoons CH_3CH_2-\overset{\overset{\displaystyle O}{\|}}{C}-O^- + H_3O^+$$

Study Check
Write an equation for the ionization of formic acid in water.

13.3 Reactions of Carboxylic Acids

Learning Goal
Write equations for the neutralization of carboxylic acids and the preparation of esters.

When a carboxylic acid is neutralized by a base, the products are the salt of the carboxylic acid and water. The salt is named by giving the name of the metal ion followed by the name of the carboxylate ion. Because salts of carboxylic acids are ionic, they are more soluble in water than the corresponding acids.

Neutralization of a Carboxylic Acid

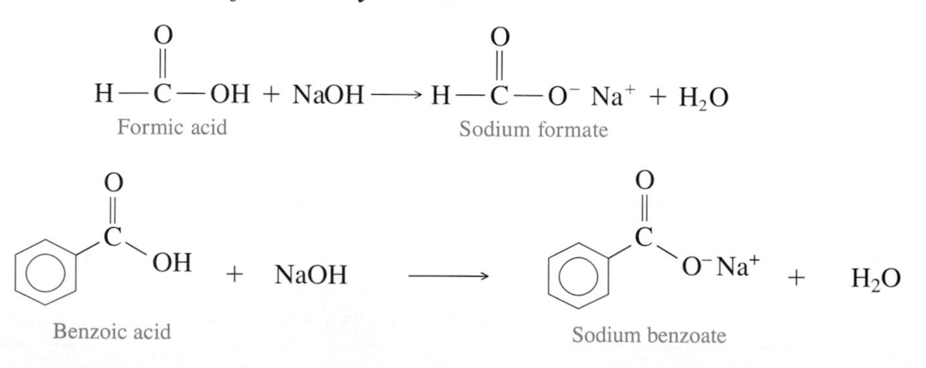

Figure 13.5
Preservatives and flavor enhancers in soups and seasonings are carboxylic acids or their salts.

Sodium propionate, a preservative, is added to bread, cheeses, and bakery items to inhibit the spoilage of the food by microorganisms. Sodium benzoate, an inhibitor of mold and bacteria, is added to juices, margarine, relishes, salads, and jams. Monosodium glutamate (MSG) has been added to meats, fish, vegetables, and bakery items to enhance flavor, although it causes headaches in some people. (See Figure 13.5.)

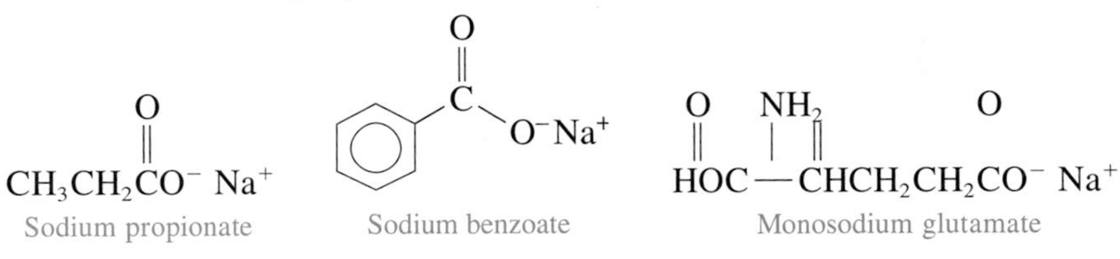

Sample Problem 13.4
Neutralization of a Carboxylic Acid

Write the equation for the neutralization of propionic acid with sodium hydroxide.

Solution
The neutralization of an acid with a base produces the salt of the acid and water.

$$\underset{\text{Propionic acid}}{CH_3CH_2\overset{\overset{O}{\|}}{C}OH} + NaOH \longrightarrow \underset{\text{Sodium propionate}}{CH_3CH_2\overset{\overset{O}{\|}}{C}O^- \ Na^+} + H_2O$$

Study Check
What carboxylic acid will give potassium butyrate when it is neutralized by KOH?

Formation of Esters

In a reaction called **esterification,** a carboxylic acid reacts with an alcohol when heated in the presence of an acid catalyst (usually H_2SO_4) to produce an ester. In the reaction, water formed by an —OH from the carboxylic acid and —H from the alcohol is removed from the reacting molecules. In an ester, we can think of the hydrogen of the —OH group in the carboxyl group as replaced by an alkyl or aromatic group.

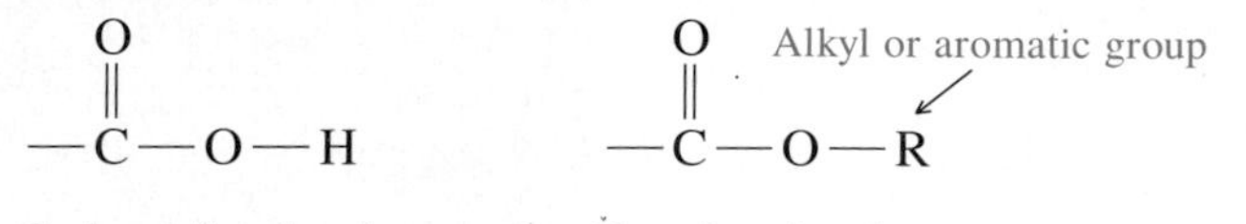

Carboxyl functional group Ester functional group

Esterification: A Reaction of a Carboxylic Acid and an Alcohol

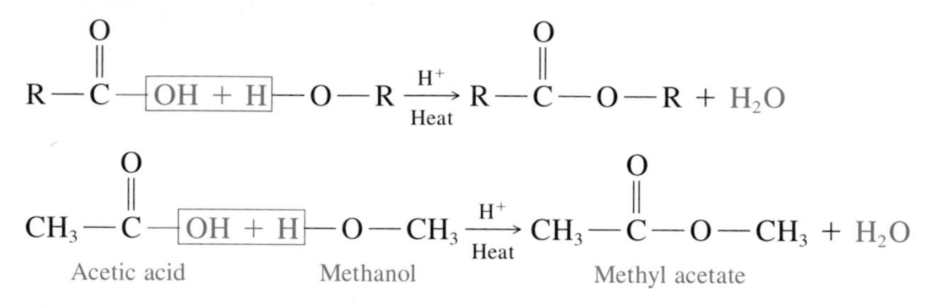

If we change the alcohol to 1-propanol, we can write an equation for the formation of the ester that is responsible for the flavor and odor of pears.

$$CH_3—\overset{\overset{\large O}{\|}}{C}—\boxed{OH + H}—O—CH_2CH_2CH_3 \xrightarrow[\text{Heat}]{H^+} CH_3—\overset{\overset{\large O}{\|}}{C}—O—CH_2CH_2CH_3 + H_2O$$

Acetic acid 1-Propanol Propyl acetate (pears)

Sample Problem 13.5
Writing Esterification Equations

The ester that gives the flavor and odor of apples can be synthesized from butyric acid and methyl alcohol. What is the equation for the formation of the ester in apples?

Solution

$$CH_3CH_2CH_2—\overset{\overset{\large O}{\|}}{C}—\boxed{OH + H}—O—CH_3 \xrightarrow[\text{Heat}]{H^+} CH_3CH_2CH_2—\overset{\overset{\large O}{\|}}{C}—O—CH_3 + H_2O$$

Study Check
What carboxylic acid and alcohol are needed to form the ester shown below that gives the flavor and odor to apricots?

$$CH_3CH_2—\overset{\overset{\large O}{\|}}{C}—O—CH_2CH_2CH_2CH_2CH_3$$

ENVIRONMENTAL NOTE *Plastics and Recycling*

Terephthalic acid (an acid with two carboxyl groups) is produced in large quantities for the manufacture of polyesters such as Dacron, and plastics. When terephthalic acid reacts with ethylene glycol (a diol) ester bonds can form on both ends of the molecules, allowing many molecules to combine until they have formed a long polymer known as a *polyester*.

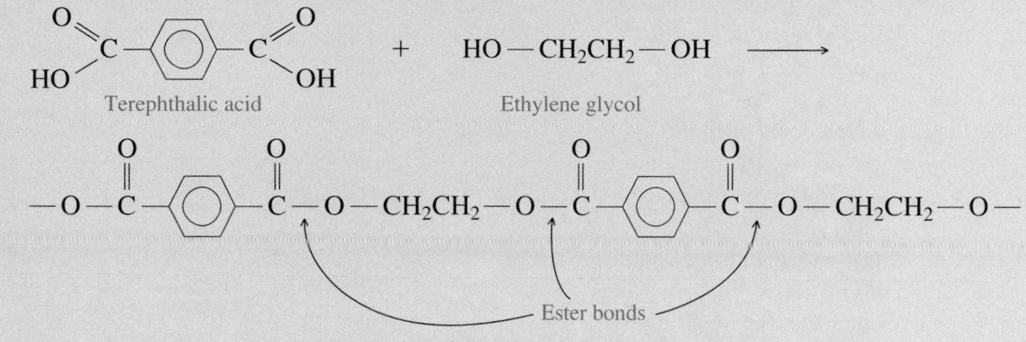

Dacron polyester is used to make permanent press fabrics, carpets, and clothes. In medicine, artificial blood vessels and valves are made of Dacron, which is biologically inert and does not clot the blood. The polyester can also be made as a film called Mylar and as a plastic known as PETE (**p**oly**e**thylene**te**rphthalate). PETE is used for plastic soft drink bottles as well as for containers of salad dressings, shampoos, and dishwashing liquids.

Today PETE is the most widely recycled of all the plastics. (See Figure 13.6.) In 1992, 365 million pounds (166 million kilograms) of PETE were recycled. After it is separated from other plastics, PETE can be changed into other useful items including polyester fabric for T-shirts and coats, fill for sleeping bags, door mats, and tennis ball containers.

You can identify the type of polymer used to make a plastic item by looking at the bottom. A number inside the recycle symbol (arrows in a triangle) corresponds to the plastic. A 1 inside the triangle identifies it as a PETE plastic. (See Figure 13.7.)

1	2	3	4	5	6
PETE	HDPE	PV	LDPE	PP	PS
Polyethylene terephthalate	High-density polyethylene	Polyvinyl chloride	Low-density polyethylene	Polypropylene	Polystyrene

Figure 13.6
Plastic containers are used to hold soft drinks, juices, salad oils and dressings, tennis balls, and mouthwash.

Figure 13.7
Recycling information on a plastic bottle includes a triangle with a code number that identifies the type of polymer used to produce the plastic.

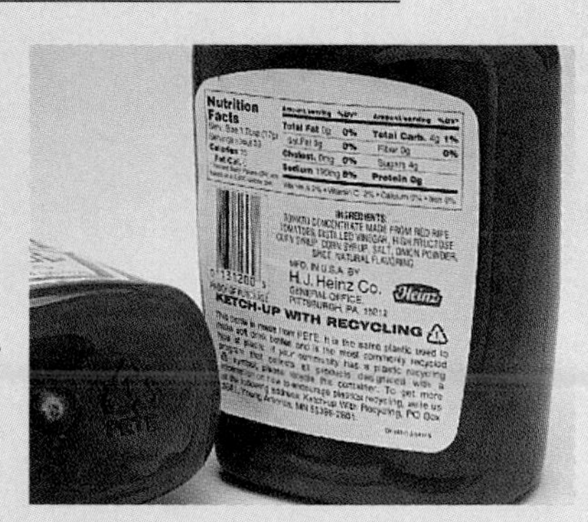

HEALTH NOTE *Salicylic Acid and Aspirin*

Chewing on a piece of willow bark had been a way of relieving pain for many centuries. By the 1800s, chemists discovered that salicylic acid was the agent in the bark responsible for the relief of pain. However, salicylic acid, which has both a carboxylic group and a hydroxyl group, used by itself, irritates the stomach lining. An ester of salicylic acid and acetic acid called acetylsalicylic acid or "aspirin" that is less irritating was prepared by the Bayer chemical company in Germany. In some aspirin preparations, a buffer is added to neutralize the carboxylic acid group and lessen its irritation of the stomach. Today aspirin is used as an analgesic (pain reliever), antipyretic (fever reducer), and antiinflammatory agent.

Figure 13.8
Products to relieve pain and fever made from salicylic acid.

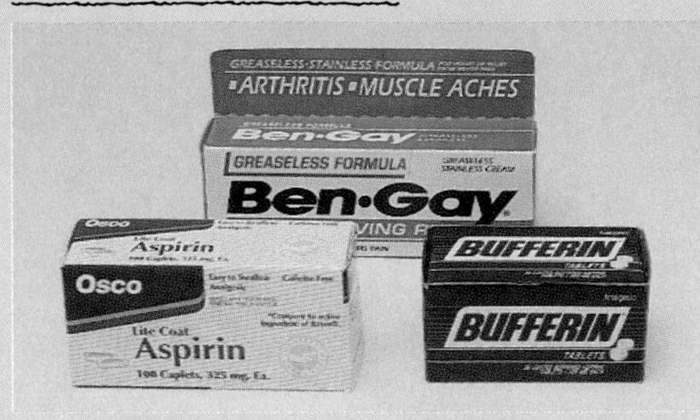

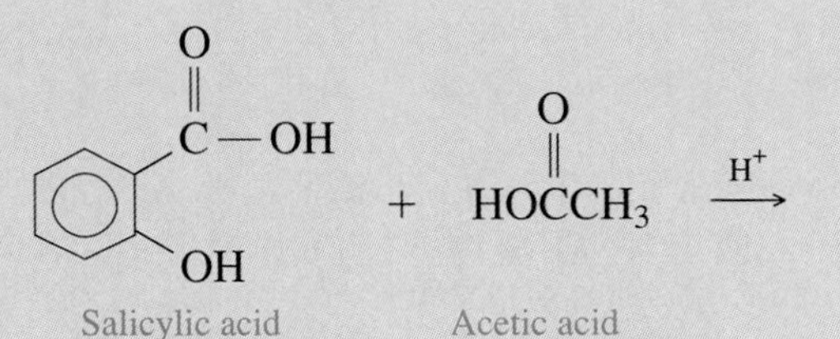

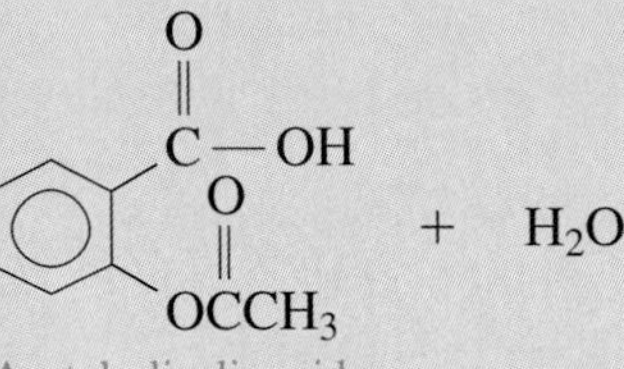

Oil of wintergreen, or methyl salicylate, has a spearmint odor and flavor. It is used in mint flavorings and in skin ointments where it acts as a counterirritant, producing heat to soothe sore muscles. (See Figure 13.8.)

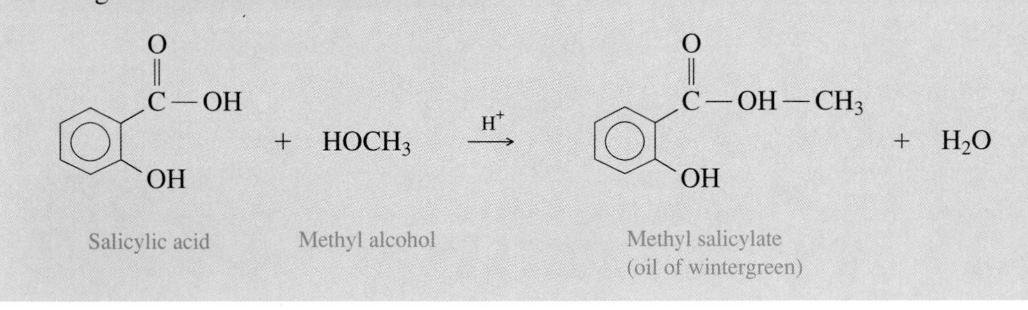

13.4 Esters

Learning Goal
Give the common names, IUPAC names, and condensed structural formulas of esters.

In an **ester** the functional group looks like a carboxylic acid, except the hydrogen in the carboxyl group is replaced by an alkyl or aromatic group. (See Figure 13.9.)

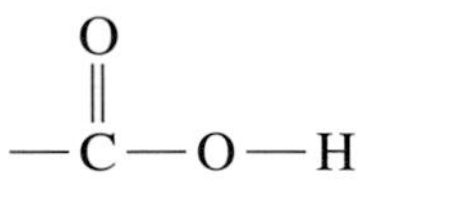

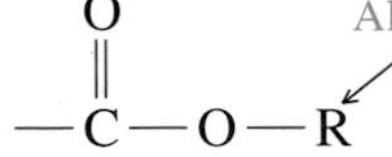

Figure 13.9
A ball-and-stick model of methyl acetate, an ester.

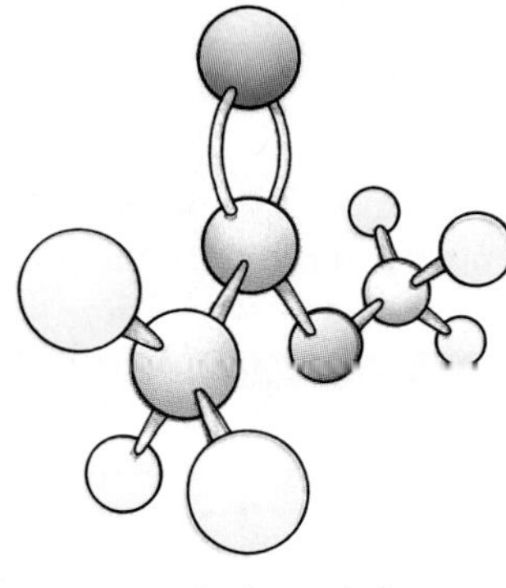

The name of an ester consists of two parts, one from the alcohol and the other from the carboxylic acid.

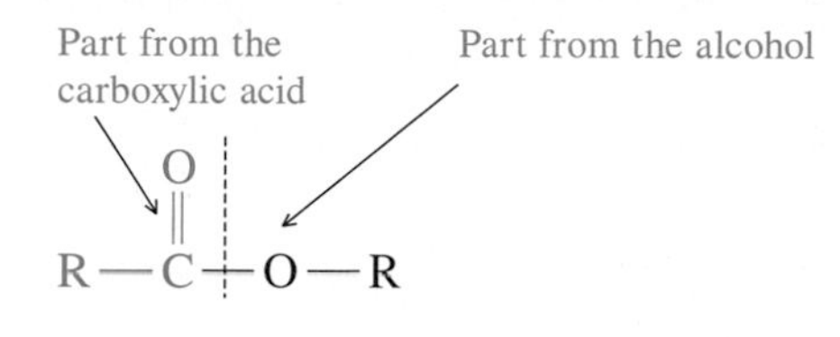

The first part of the ester name is from the alkyl group of the alcohol. The second part is from the name of the carboxylic acid with the ending of *ic acid* replaced by *ate*. In the common names of esters, the common names of the carboxylic acids are used.

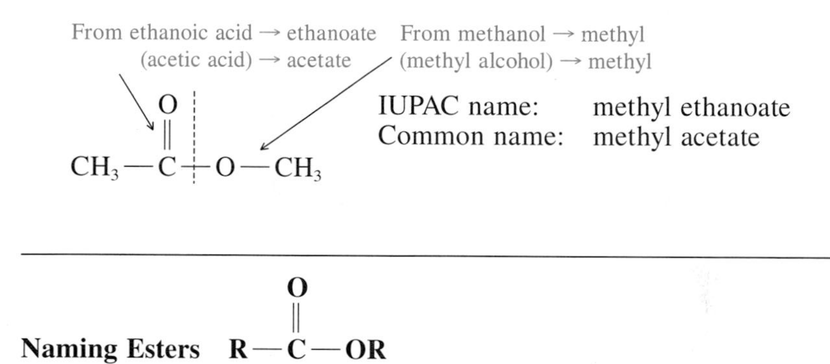

Naming Esters $R-\overset{\overset{O}{\|}}{C}-OR$

Functional group:	$-\overset{\overset{O}{\|}}{C}-O-$
IUPAC name:	alkyl carboxylate (IUPAC name of acid)
Common name:	alkyl carboxylate (common name of acid)

Examples:

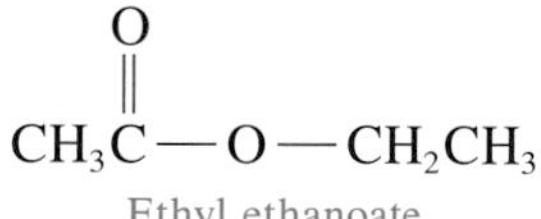

Ethyl ethanoate
(ethyl acetate)

$CH_3CH_2\overset{\overset{O}{\|}}{C}-O-CH_3$

Methyl propanoate
(methyl propionate)

$C_6H_5-\overset{\overset{O}{\|}}{C}-O-CH_2CH_3$

Ethyl benzoate

Sample Problem 13.6
Naming Esters

Write the IUPAC and common names of the following ester:

$$CH_3CH_2-\overset{\overset{O}{\|}}{C}-O-CH_2CH_2CH_3$$

Solution

The alcohol part of the ester is propyl, and the carboxylic acid part is propanoic (propionic) acid.

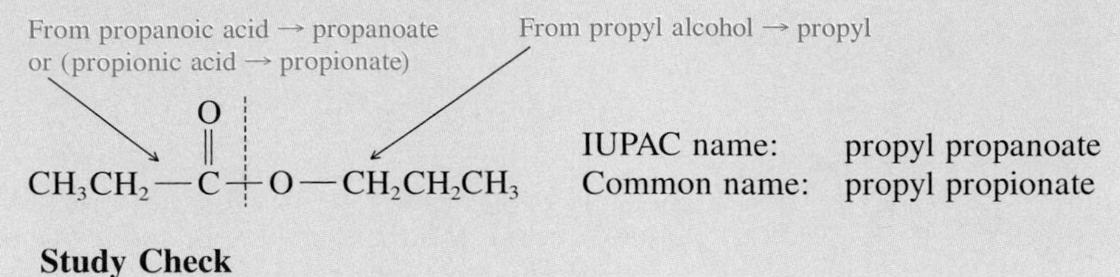

Study Check

Draw the condensed structural formula of pentyl acetate.

Environmental Note *Fragrant Esters*

Many of the fragrances of perfumes and flowers and the flavors of fruits are due to esters. Small esters are volatile so we can smell them and soluble in water so we can taste them. (See Figure 13.10.) Several of these are listed in Table 13.3.

Figure 13.10
Esters are responsible for part of the odor and flavor of oranges, bananas, pears, pineapples, and strawberries.

Table 13.3 *Some Naturally Occurring Esters in Fruits and Flavorings*

Condensed Structural Formula	Name	Flavor/Odor
$HC(=O)—O—CH_2CH_3$	Ethyl methanoate (ethyl formate)	Rum
$HC(=O)—O—CH_2CH(CH_3)CH_3$	Isobutyl methanoate (isobutyl formate)	Raspberries
$CH_3C(=O)—O—CH_2CH_2CH_3$	Propyl ethanoate (propyl acetate)	Pears

Table 13.3 *Some Naturally Occurring Esters in Fruits and Flavorings* (*cont.*)

Condensed Structural Formula	Name	Flavor/Odor
$\mathrm{CH_3\overset{O}{\overset{\|}{C}}-O-CH_2CH_2CH_2CH_2CH_3}$	Pentyl ethanoate (pentyl acetate)	Bananas
$\mathrm{CH_3\overset{O}{\overset{\|}{C}}-O-CH_2CH_2CH_2CH_2CH_2CH_2CH_2CH_3}$	Octyl ethanoate (Octyl acetate)	Oranges
$\mathrm{CH_3CH_2CH_2\overset{O}{\overset{\|}{C}}-O-CH_2CH_3}$	Ethyl butanoate (ethyl butyrate)	Pineapples
$\mathrm{CH_3CH_2CH_2\overset{O}{\overset{\|}{C}}-O-CH_2CH_2CH_2CH_2CH_3}$	Pentyl butanoate (pentyl butyrate)	Apricots
$\mathrm{CH_3CH_2CH_2\overset{O}{\overset{\|}{C}}-SCH_3}$	Methyl thiobutanoate (methyl thiobutyrate)	Strawberries

HEALTH NOTE *Esters in Fats and Oils*

Waxes are high molecular weight esters formed from long-chain carboxylic acids and alcohols. Beeswax is used to make candles and polishes, and carnauba wax is used in car and floor waxes.

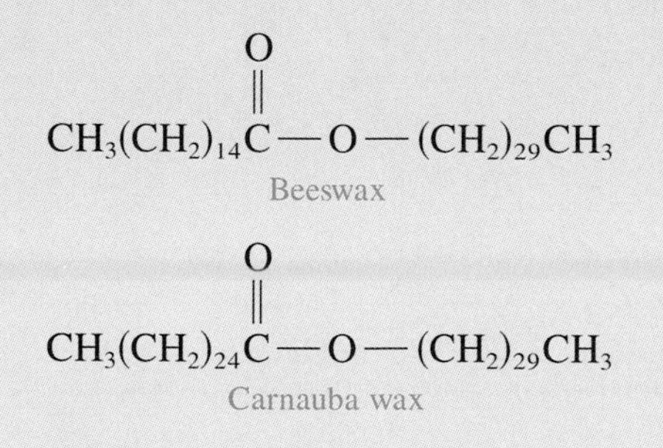

Fats and oils are esters of glycerol, a trihydroxy alcohol, and fatty acids, which are long-chain carboxylic acids. The fatty acids in fats and oils typically contain an even number of carbon atoms from 10 to 18. Most of the fatty acids in animal sources are saturated. In oils from vegetables, more of the fatty acids contain double bonds. A more extensive list of fatty acids is found in Chapter 15.

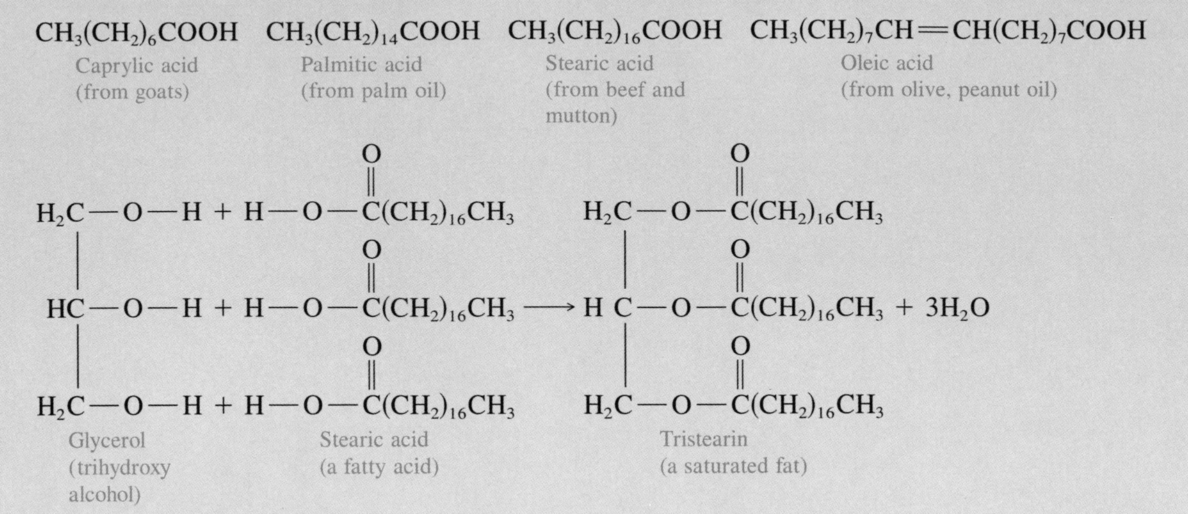

13.5 Reactions of Esters

Learning Goal

Write equations for the hydrolysis and saponification of esters.

In hydrolysis, esters are split apart by a reaction with water in the presence of an acid or base catalyst. The products of ester hydrolysis are the carboxylic acid and alcohol that combined to form the ester. Hydrolysis is an important reaction in the body too. For example, during digestion, the ester bonds in our dietary fats are hydrolyzed (split apart) by enzymes (biological catalysts called hydrolases) to produce glycerol and fatty acids.

Acid Hydrolysis

The hydrolysis of an ester using a strong acid such as H_2SO_4 or HCl and heat, or an esterase enzyme, is the reverse of the reaction that formed the ester. The H of the water molecule becomes part of the alcohol, and the —OH part forms the carboxyl group in the acid.

Acid Hydrolysis of an Ester

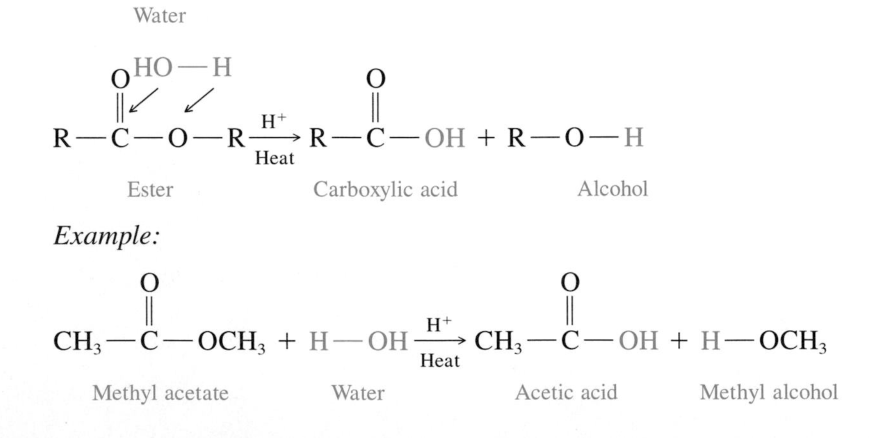

Sample Problem 13.7
Acid Hydrolysis of Esters

Aspirin that has been stored for a long time may undergo hydrolysis in the presence of water and heat. What are the hydrolysis products of aspirin? Why does a bottle of old aspirin smell like vinegar?

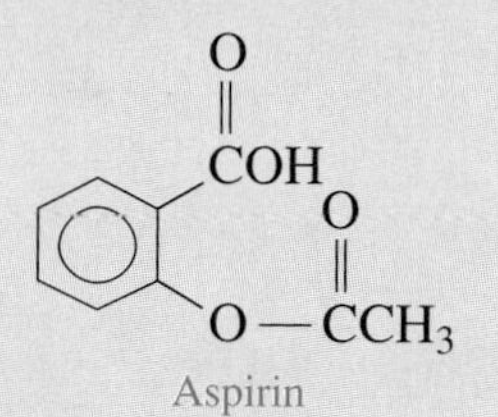

Aspirin

Solution

To write the hydrolysis products, separate the compound at the ester bond. Complete the formula of the carboxylic acid by adding —OH (from water) to the carbonyl group and the —H to complete the alcohol. The acetic acid in the products gives the vinegar odor to a sample of aspirin that has hydrolyzed.

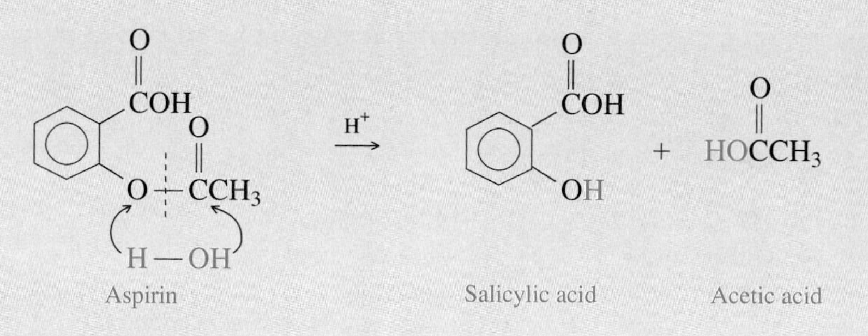

Aspirin Salicylic acid Acetic acid

Study Check

What are the names of the products from the acid hydrolysis of ethyl propionate?

Base Hydrolysis (Saponification)

Saponification is the hydrolysis of an ester using a strong base, NaOH or KOH, and heat, to give the salt of the carboxylic acid and an alcohol.

Base Hydrolysis of an Ester

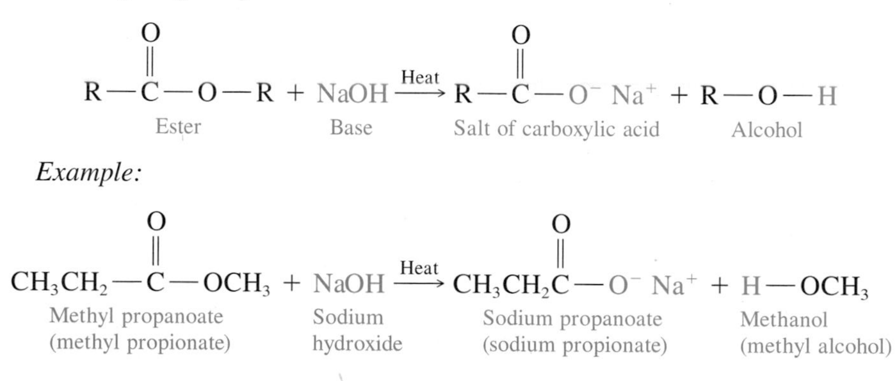

Sample Problem 13.8
Base Hydrolysis of Esters

Ethyl acetate is a solvent widely used for fingernail polish, plastics, and lacquers. Write the equation of the hydrolysis of ethyl acetate by NaOH.

Solution
The hydrolysis of ethyl acetate by NaOH gives the salt of acetic acid and ethyl alcohol.

$$\underset{\text{Ethyl acetate}}{CH_3-\overset{\overset{\displaystyle O}{\|}}{C}-OCH_2CH_3} + NaOH \xrightarrow{\text{Heat}} \underset{\text{Sodium acetate}}{CH_3-\overset{\overset{\displaystyle O}{\|}}{C}-O^-\,Na^+} + \underset{\text{Ethyl alcohol}}{HOCH_2CH_3}$$

Study Check
Write the condensed structural formulas of the products from the hydrolysis of methyl benozate by KOH.

Health Note *Fats and Soaps*

Our diets include fats and oils (also called triglycerides), which are esters of glycerol and long-chain carboxylic acids called fatty acids. The most abundant fatty acids contain 18 carbon atoms and may be saturated or unsaturated. Solid esters of glycerol such as lard and butter are fats that contain mostly saturated fatty acids. Oils are liquid esters such as safflower or corn oil, which contain mostly unsaturated fatty acids. During digestion in the small intestine, enzymes from the pancreas (lipases and esterases) mix with the fats and oils to hydrolyze them to fatty acids and glycerol. (See Figure 13.11.)

Figure 13.11
Liquid vegetable oils such as corn oil contain mostly unsaturated fatty acids.

$$\underset{\text{Glyceryl tripalmitin (in palm oil)}}{\begin{array}{l} H_2C-O-\overset{\overset{\displaystyle O}{\|}}{C}(CH_2)_{14}CH_3 \\ \;|\\ HC-O-\overset{\overset{\displaystyle O}{\|}}{C}(CH_2)_{14}CH_3 \\ \;| \\ H_2C-O-\overset{\overset{\displaystyle O}{\|}}{C}(CH_2)_{14}CH_3 \end{array}} + 3\,H_2O \longrightarrow \underset{\text{Glycerol}}{\begin{array}{l} H_2C-O-H \\ \;| \\ HC-O-H \\ \;| \\ H_2C-O-H \end{array}} \begin{array}{l} + \\ \\ + \\ \\ + \end{array} \underset{\text{3 Palmitic acids (a fatty acid)}}{\begin{array}{l} HO-\overset{\overset{\displaystyle O}{\|}}{C}(CH_2)_{14}CH_3 \\ \\ HO-\overset{\overset{\displaystyle O}{\|}}{C}(CH_2)_{14}CH_3 \\ \\ HO-\overset{\overset{\displaystyle O}{\|}}{C}(CH_2)_{14}CH_3 \end{array}}$$

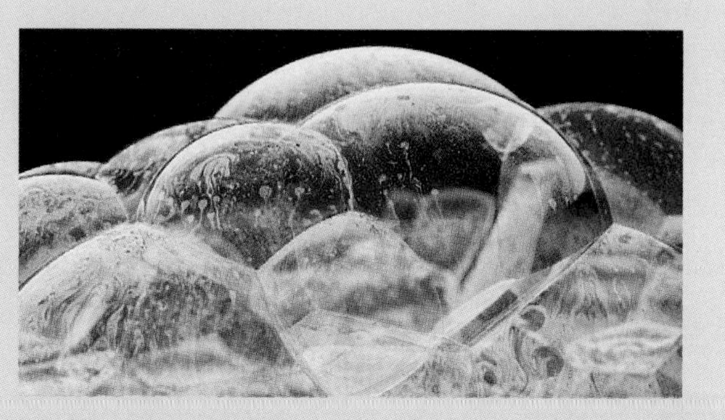

Figure 13.12
Soaps, sodium or potassium salts of long-chain carboxylic acids, have a nonpolar, hydrocarbon end that attracts fats and grease, and removes them from clothes or dishes because the polar end dissolves in water.

When a fat or oil is saponified, the sodium or potassium salt formed from the long-chain fatty acids is called a **soap.** For many centuries, soaps were made by heating a mixture of animal fats (tallow) with lye, a basic solution obtained from wood ashes. Today soaps are also prepared from fats such as coconut oil. Perfumes and dyes are added to give a pleasant smelling soap. (See Figure 13.12.)

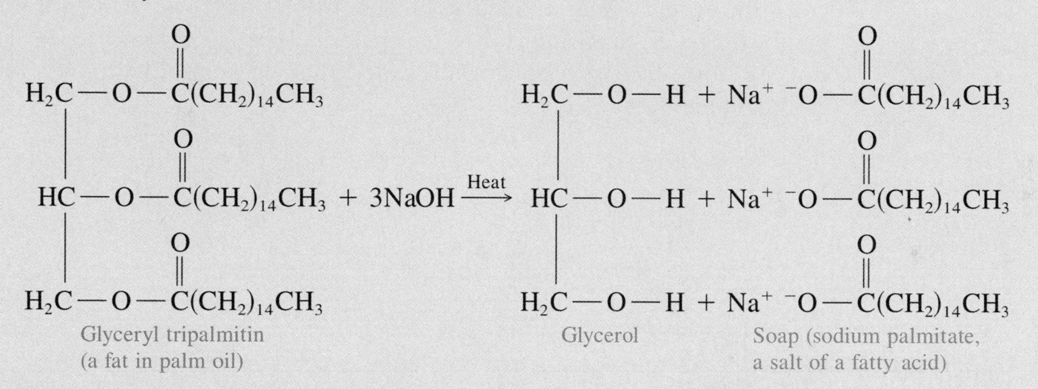

Within a soap molecule, there is a polar part that is hydrophilic or "water-loving," and a long, nonpolar hydrocarbon chain that is hydrophobic, "water-fearing." When soap molecules are mixed with grease or dirt, the nonpolar ends are attracted to the nonpolar fat or oil particles. The grease or dirt is pulled into the water by the attraction of polar (salt) ends of the soap molecules to water.

13.6 Amines

Learning Goal
Classify amines; give their names and condensed structural formulas.

The element nitrogen plays an important role in the physiological activity of compounds such as antibiotics, amphetamines, alkaloids, and tranquilizers. **Amines** are organic compounds in which a nitrogen (N) atom is attached to one, two, or three alkyl or aromatic groups. (See Figure 13.13.)

Amines are classified as *primary (1°)* amines if a nitrogen atom is attached to one alkyl group. An amine is a *secondary (2°)* amine if two carbon groups are attached to nitrogen and a *tertiary (3°)* amine when there are three groups attached.

Figure 13.13
Ball-and-stick models of amines: methylamine and dimethylamine.

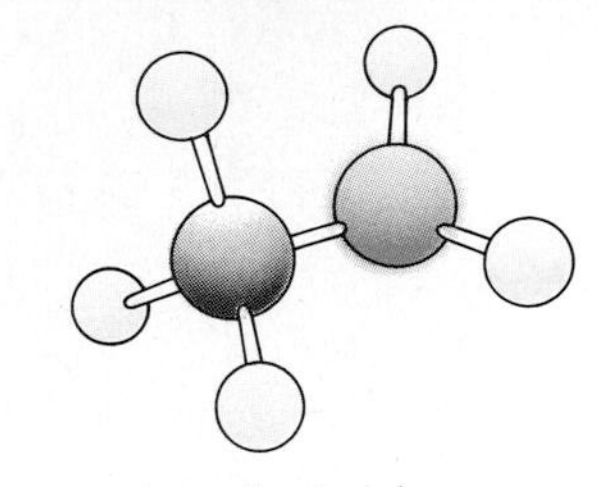

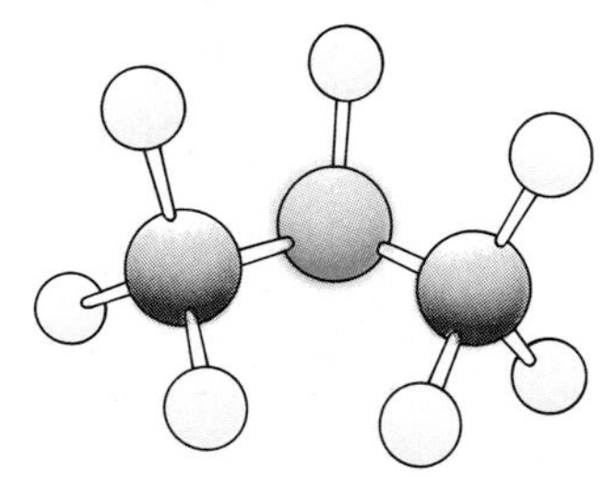

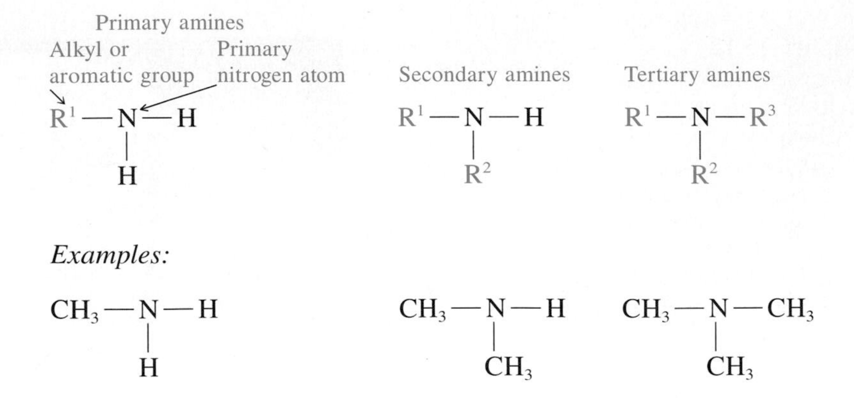

Solubility of Amines

The electronegativity of the nitrogen atom in an amine makes an N—H bond polar. In primary and secondary amines, the N—H bond can hydrogen-bond with water. In all amines, the lone pair of electrons on the nitrogen atom can also form hydrogen bonds with water, making amines with one to five carbon atoms water soluble.

O—H
|
$H^{\delta+}$
⋮ ← The formation of hydrogen bonds in a primary amine
⋮ $\bullet\bullet\delta-$ ↓
CH_3—N—$H^{\delta+}$ ········ $^{\delta-}$O—H
| |
H H

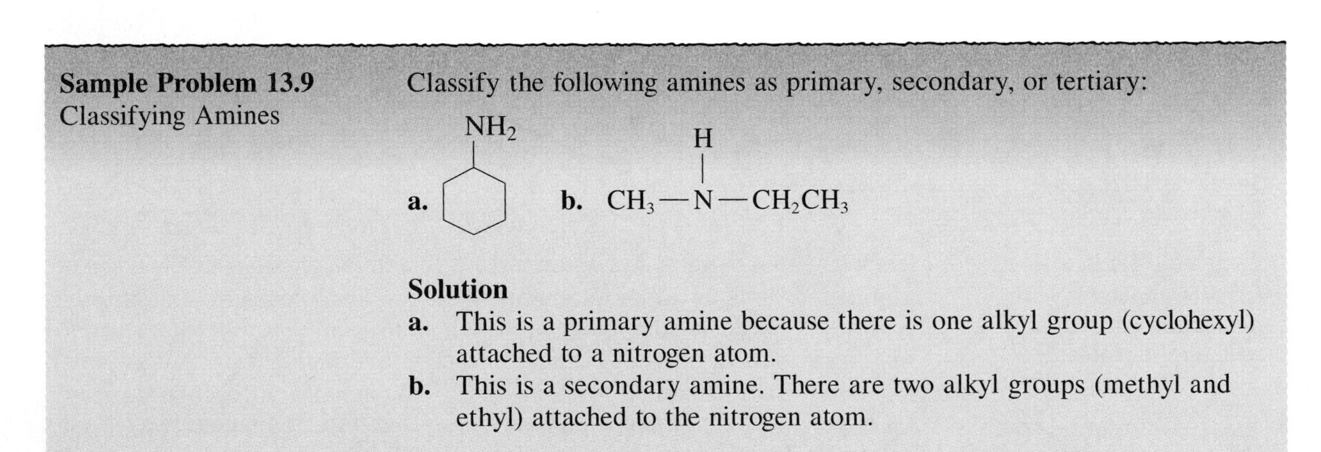

Sample Problem 13.9
Classifying Amines

Classify the following amines as primary, secondary, or tertiary:

a. (cyclohexane ring with NH_2) **b.** CH_3—N(H)—CH_2CH_3

Solution

a. This is a primary amine because there is one alkyl group (cyclohexyl) attached to a nitrogen atom.

b. This is a secondary amine. There are two alkyl groups (methyl and ethyl) attached to the nitrogen atom.

Study Check

Classify the following amine as a primary, secondary, or tertiary amine.

$$CH_3CH_2-\underset{\displaystyle CH_3}{\underset{|}{N}}-CH_2CH_3$$

Naming Amines

Amines are named by both their common name and IUPAC name. Simple amines are most often named as one word using the alkyl group followed by the suffix *amine*. In the IUPAC system, the NH_2 group is named an *amino* group on an alkane chain.

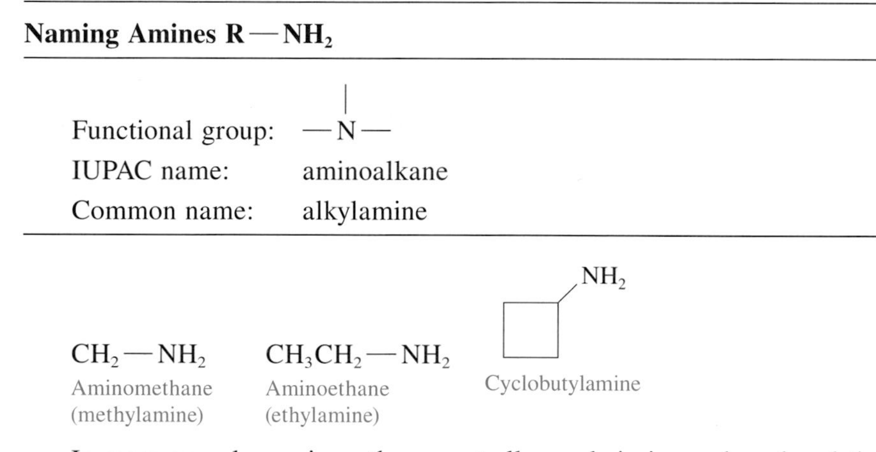

Naming Amines R—NH_2

Functional group:	$-\overset{|}{N}-$
IUPAC name:	aminoalkane
Common name:	alkylamine

Aminomethane (methylamine) · Aminoethane (ethylamine) · Cyclobutylamine

In more complex amines, the parent alkane chain is numbered and the substituents including the amino group are named.

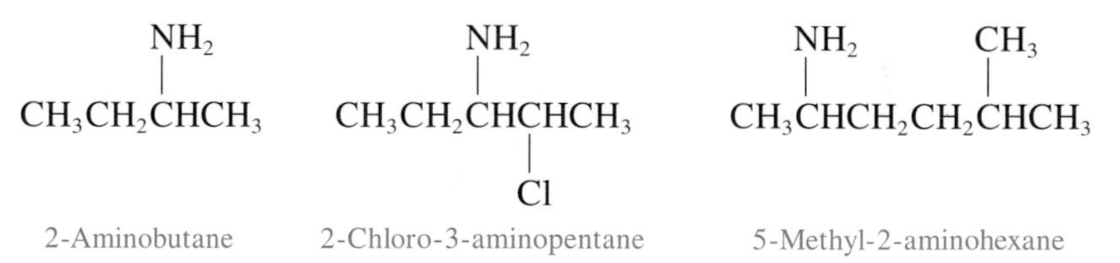

2-Aminobutane · 2-Chloro-3-aminopentane · 5-Methyl-2-aminohexane

Amines are among the substances responsible for the fishy odors of seafood and the putrid odors of the decay products of animal proteins.

$NH_2-CH_2CH_2CH_2CH_2-NH_2$
1,4-Diaminobutane (putresine)

$NH_2-CH_2CH_2CH_2CH_2CH_2-NH_2$
1,5-Diaminopentane (cadaverine)

Naming Substituents on the Amino Group

In the common names for amines, the prefixes *di* and *tri* are used to indicate two or three identical alkyl groups attached to the N atom. If the alkyl groups are not identical, they are listed in alphabetical order. In the IUPAC system, alkyl groups attached to the N atom of the amino group NH_2 are indicated by the prefix *N*-.

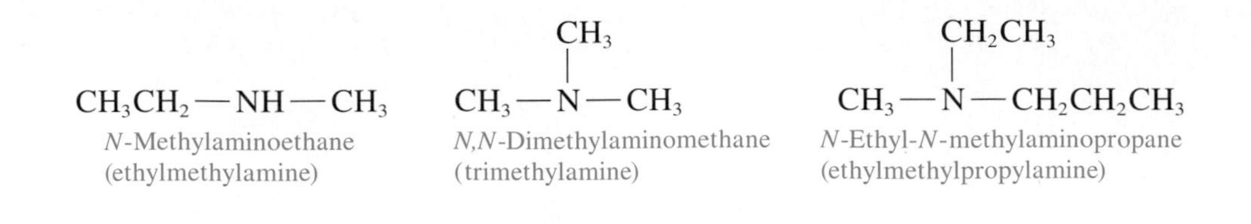

Aromatic Amines. The name *aniline* has been approved as the IUPAC name of the amine of benzene.

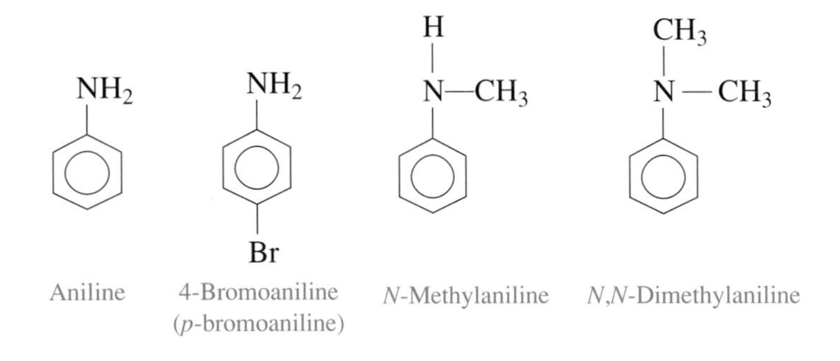

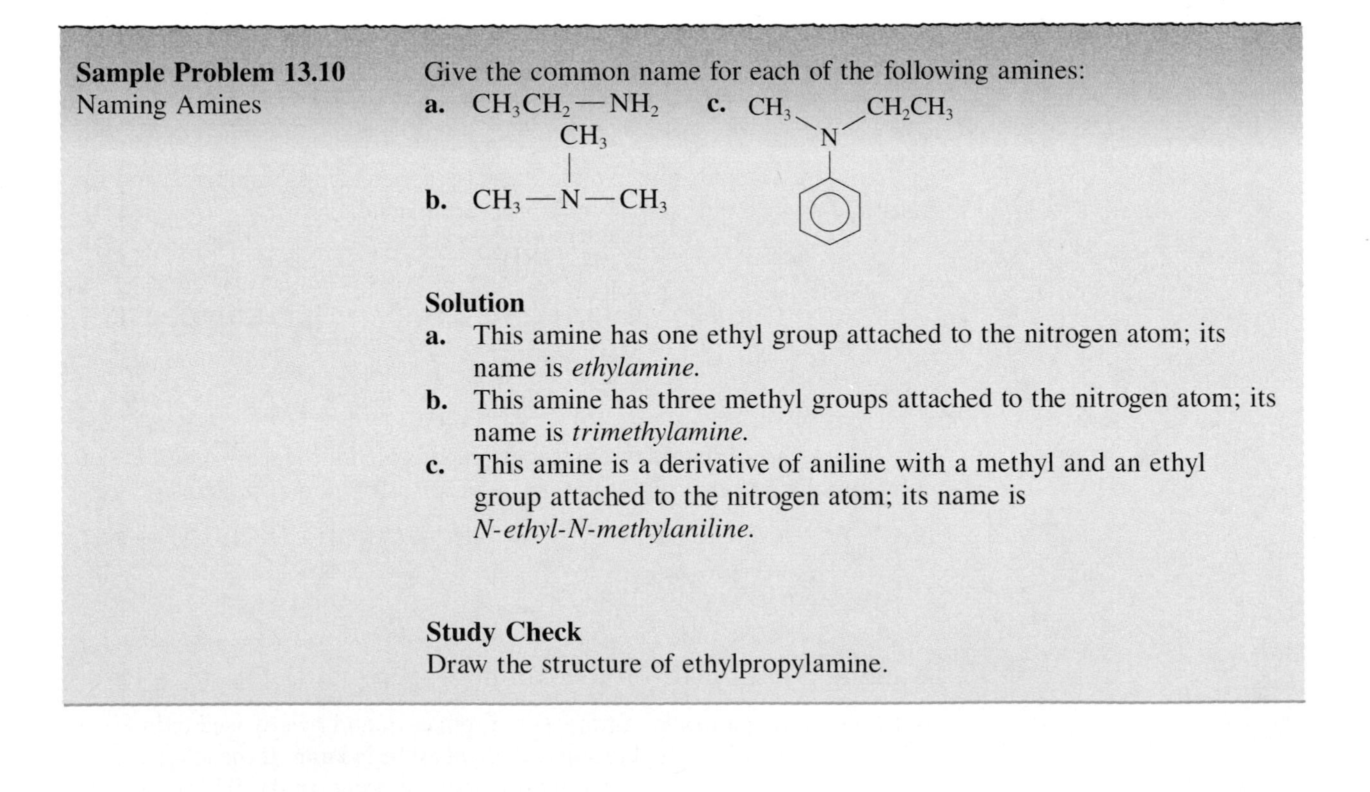

Sample Problem 13.10
Naming Amines

Give the common name for each of the following amines:

Solution

a. This amine has one ethyl group attached to the nitrogen atom; its name is *ethylamine.*

b. This amine has three methyl groups attached to the nitrogen atom; its name is *trimethylamine.*

c. This amine is a derivative of aniline with a methyl and an ethyl group attached to the nitrogen atom; its name is *N-ethyl-N-methylaniline.*

Study Check

Draw the structure of ethylpropylamine.

HEALTH NOTE *Amines in Health and Medicine*

In response to allergic reactions or injury to cells, the body increases the production of histamine, which causes blood vessels to dilate and increases the permeability of the cells. Redness and swelling occur in the area. Administering an antihistamine such as diphenylhydramine helps block the effects of histamine.

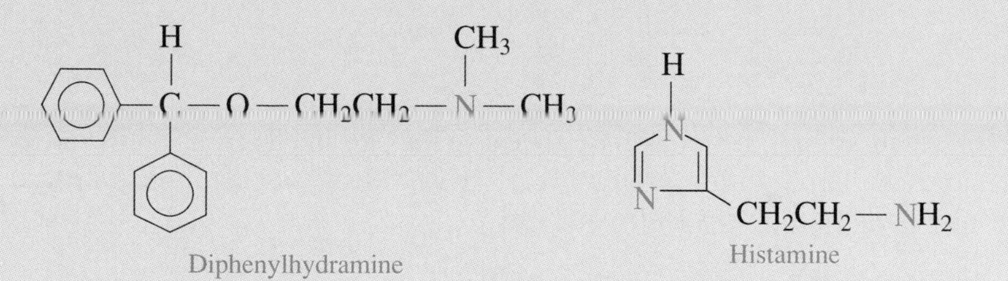

Diphenylhydramine   Histamine

In the body, hormones called biogenic amines carry messages between the central nervous system and nerve cells. Epinephrine (adrenaline) and norepinephrine (noradrenaline) are released by the adrenal medulla in "fight or flight" situations to raise the blood glucose level and move the blood to the muscles. Used in remedies for colds, hay fever, and asthma, the norepinephrine contracts the capillaries in the mucous membranes of the respiratory passages. Parkinson's disease is a result of a deficiency in another biogenic amine called dopamine.

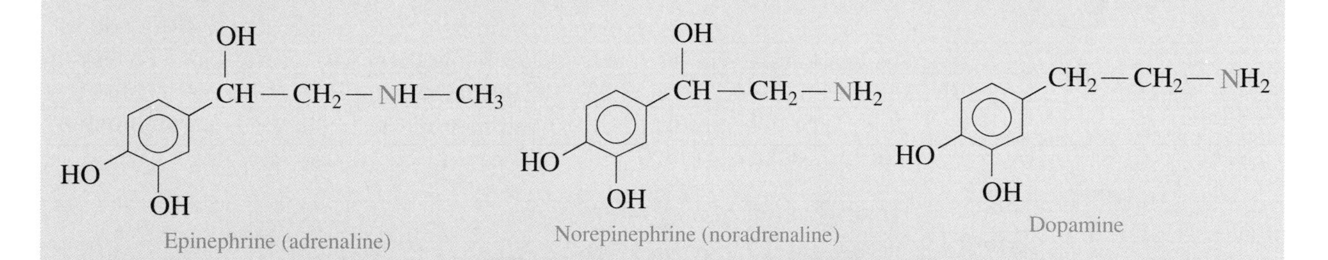

Epinephrine (adrenaline)   Norepinephrine (noradrenaline)   Dopamine

Produced synthetically, amphetamines (known as "uppers") are stimulants of the central nervous system much like adrenaline, but they also increase cardiovascular activity and depress the appetite. They are sometimes used to bring about weight loss, but they can cause chemical dependency. Benzedrine and Neo-Synephrine (phenylephrine) are used in medications to reduce respiratory congestion from colds, hay fever, and asthma. Sometimes, benzedrine is taken internally to combat the desire to sleep, but it has side effects. Methedrine is used to treat depression and in the illegal form is known as "speed" or "crank."

Benzedrine, amphetamine   Neo-Synephrine, phenylephrine   Methamphetamine (methedrine)

13.7 Reactions of Amines

Learning Goal
Write equations for the reactions of amines.

In Chapter 9, we saw that ammonia NH_3 is a weak base that accepts a proton (H^+) from water to produce ammonium ion and hydroxide ion (OH^-).

$$H{-}\underset{\underset{H}{|}}{\overset{\overset{H}{|}}{N}} + HOH \rightleftharpoons H{-}\underset{\underset{H}{|}}{\overset{\overset{H}{|}}{N^+}}{-}H + OH^-$$

Ammonia — Ammonium hydroxide

In water, amines are weak bases because the pair of electrons on the nitrogen atom can bond with a proton (H^+). The products of this ionization are an ammonium ion and a hydroxide ion.

Ionization of an Amine in Water

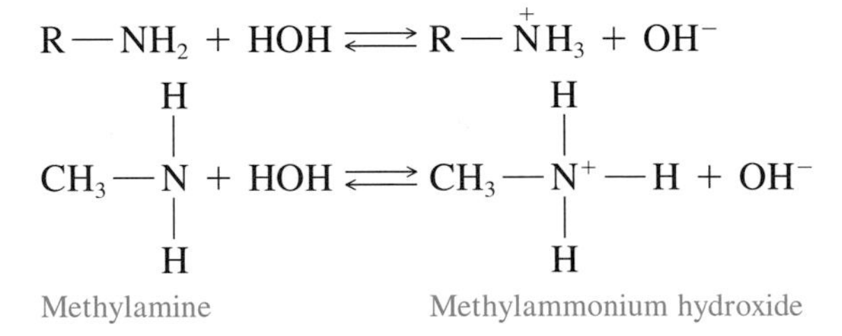

When you squeeze lemon juice on fish, you counteract the fishy odor of the basic amines in the fish by neutralizing them with citric acid. The proton (H^+) from the acid bonds with the unshared pair of electrons on the nitrogen atom. The only product of the neutralization of an amine is an ammonium salt; no water is formed.

Neutralization of an Amine

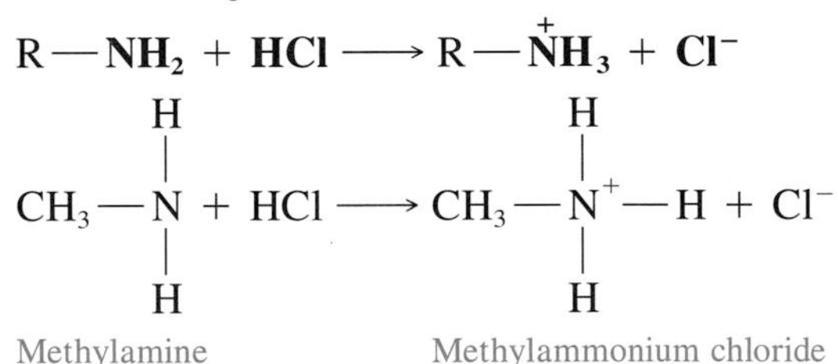

Sample Problem 13.11
Reactions of Amines

Write an equation that shows ethylamine

a. as a weak base in water
b. reacting with HCl

Solution

a. In water, ethylamine acts as a weak base by accepting a proton from water to produce ethylammonium hydroxide.

$$CH_3CH_2{-}NH_2 + H{-}OH \rightleftharpoons CH_3CH_2{-}NH_3^+ + OH^-$$

b. $CH_3CH_2{-}NH_2 + HCl \longrightarrow CH_3CH_2{-}NH_3^+ + Cl^-$

Study Check
What is the condensed structural formula of the salt formed by the reaction of trimethylamine and HCl?

Amidation

When a carboxylic acid reacts with ammonia or an amine, an amide is produced. In **amidation,** a molecule of water is eliminated, and the fragments of the carboxylic acid and amine molecules join to form the amide, much like ester formation. Because a hydrogen must be lost from the amines, only primary and secondary amines will react.

Amidation

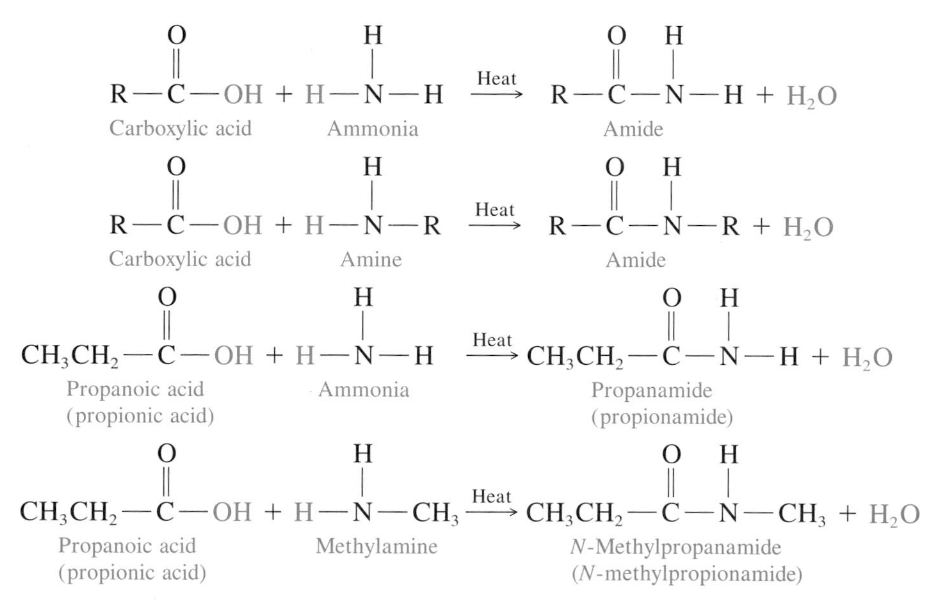

Sample Problem 13.12
Amidation Products

Give the structural formula of the amide product in each of the following reactions:

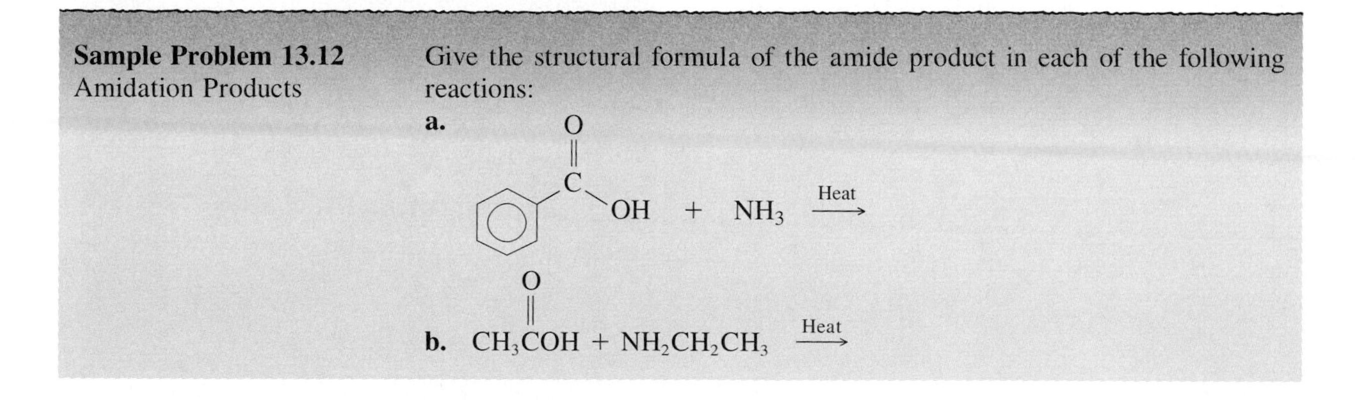

Solution

a. The structural formula of the amide product can be written by attaching the carbonyl group from the acid to the nitrogen atom of the amine. The —OH is removed from the acid and an —H from the amine to form water.

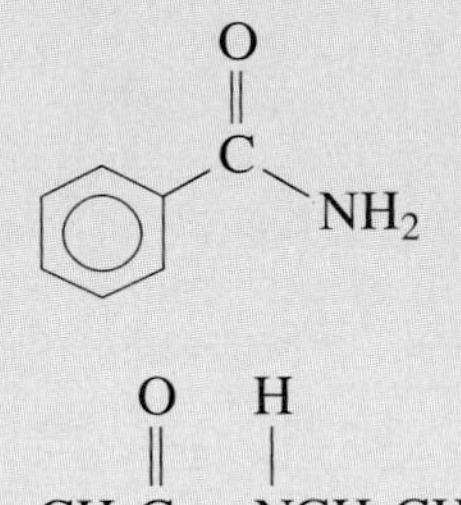

b. $CH_3C(=O)—N(H)CH_2CH_3$

Study Check

What are the condensed structural formulas of the carboxylic acid and amine needed to prepare the following amide?

$H—C(=O)—N(CH_3)—CH_3$

HEALTH NOTE *Alkaloids*

Alkaloids are physiologically active nitrogen-containing compounds produced by plants. The term *alkaloid* refers to the basic (alkali-like) characteristics of amines. Certain alkaloids are used in anesthetics, in antidepressants, and as stimulants, although many are habit forming. (See Figure 13.14.)

Quinine obtained from the bark of the cinchona tree has been used in the treatment of malaria since the 1600s:

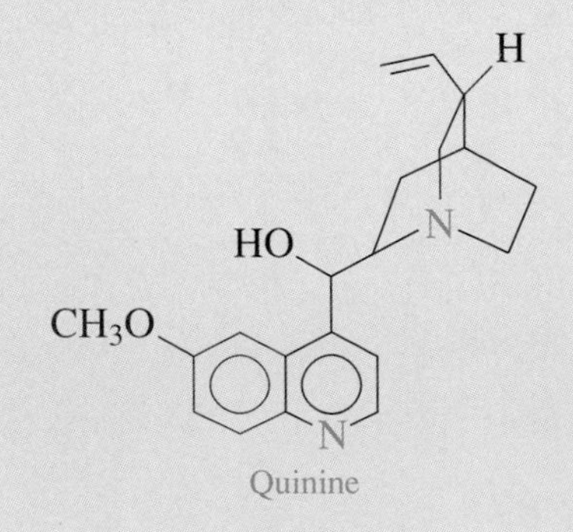

Figure 13.14
Nicotine and caffeine are alkaloids, nitrogen-containing compounds obtained from plants having physiological activity.

Nicotine from the leaves of the tobacco plant, and coniine from hemlock, are extremely toxic alkaloids. Caffeine, a compound found in coffee beans and tea, is a stimulant of the central nervous system.

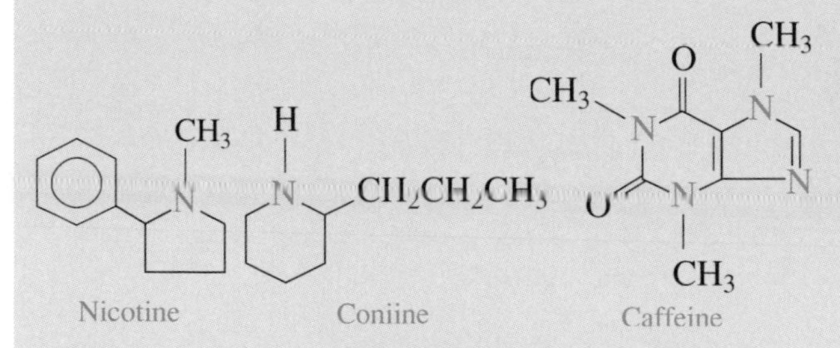

Atropine from belladonna and cocaine from the coca plant are used in low concentrations as anesthetics for eye and sinus procedures. However, higher doses produce euphoria followed by depression and a desire for additional quantities of the drug.

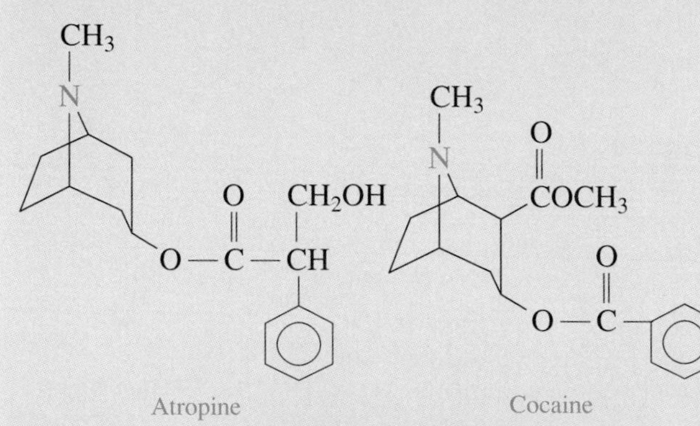

Chemists have altered the structures of substances such as atropine and cocaine to develop synthetic alkaloids such as procaine and xylocaine, which are used for local anesthesia. The synthetic products retain the anesthetic qualities of the natural alkaloid without the addictive side effects.

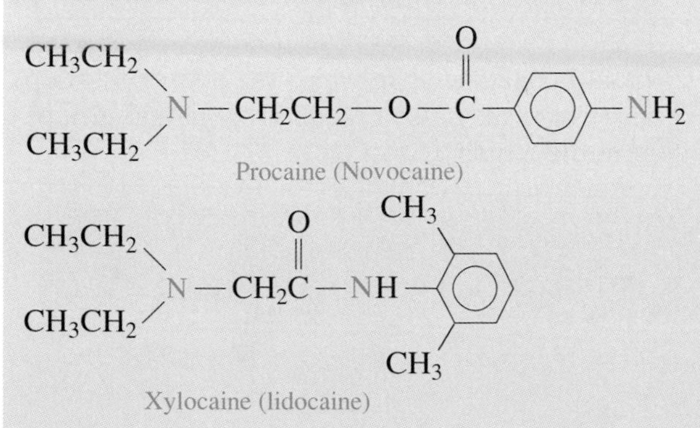

For many centuries, morphine, an alkaloid found in the oriental poppy plant, has been used as a painkiller. However, it also has strong hallucinogenic and addictive side effects. Codeine, which is similar to morphine, is used in prescription painkillers. Heroin, an opiate similar to morphine, is strongly addicting and not used medically. One area of research in pharmacology is the synthesis of morphine-like compounds that can be used safely as effective painkillers with no side effects. A synthetic alkaloid, meperidine, or Demerol, was developed as a painkiller with a chemical structure similar to morphine but with reduced side effects.

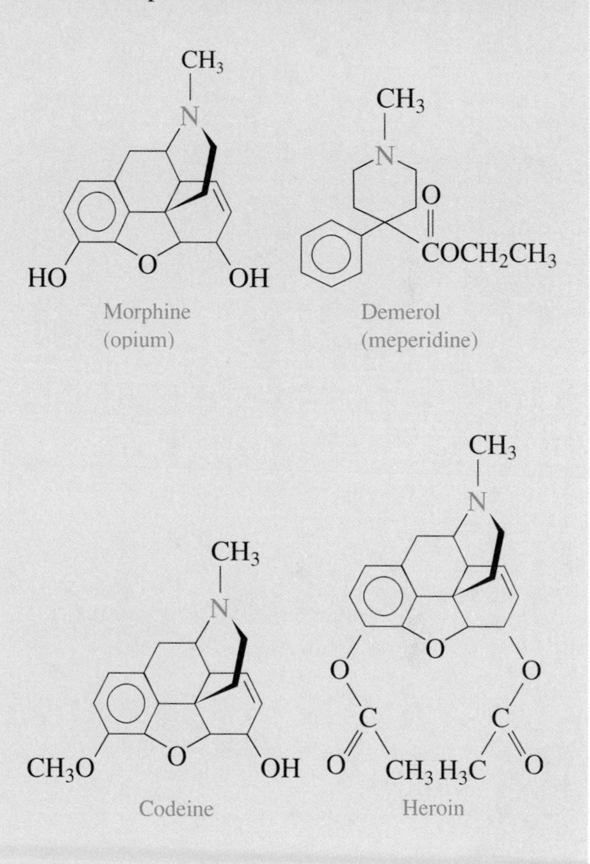

Alkaloids are prevalent among the compounds known as hallucinogens. Examples include mescaline, from the peyote cactus, and LSD (lysergic acid diethylamide), prepared from lysergic acid that is produced by a fungus that grows on rye.

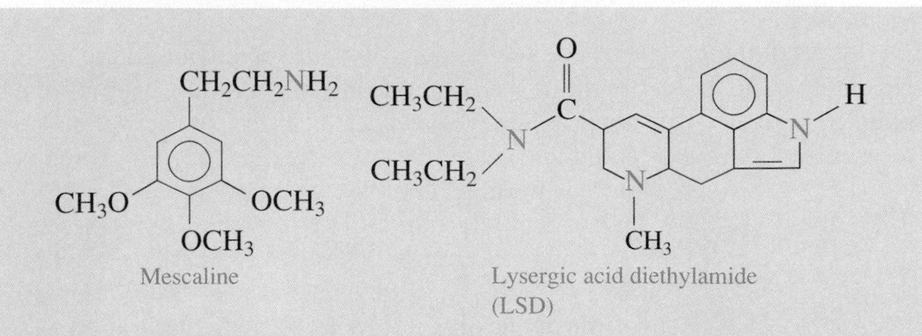

Mescaline

Lysergic acid diethylamide (LSD)

Other heterocyclic amines act as tranquilizers by reducing the transmission of nerve impulses to the brain. Low levels of serotonin in the brain appear to be associated with depressed states. Reserpine has been used as a sedative for psychotic patients, and thorazine has been effective as a medication for schizophrenia.

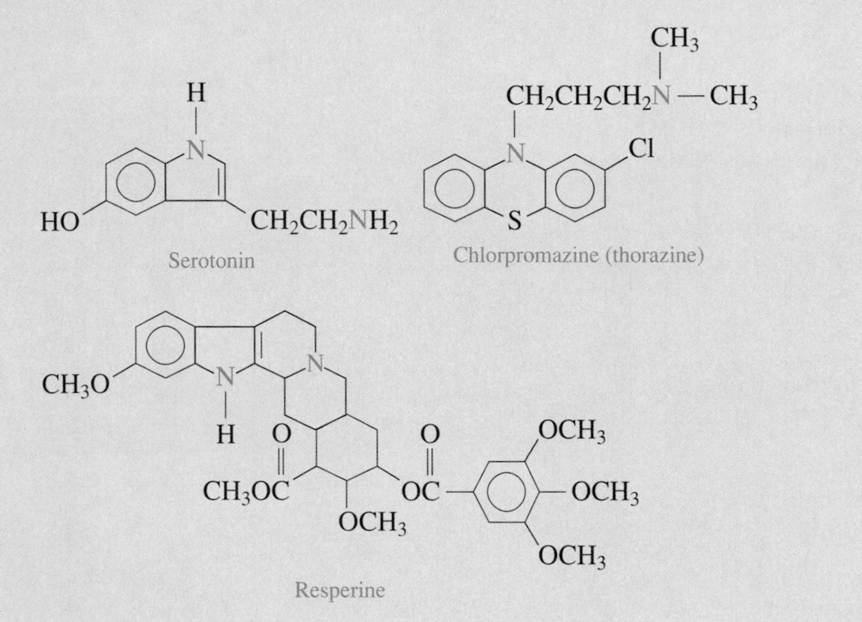

Serotonin

Chlorpromazine (thorazine)

Resperine

13.8 Amides

Learning Goal

Write the common names, IUPAC names, and condensed structural formulas of amides.

The **amides** are derivatives of carboxylic acids in which the hydroxyl group is replaced by an amino group. There may also be one or two alkyl groups attached to the nitrogen atom in the amide. (See Figure 13.15.)

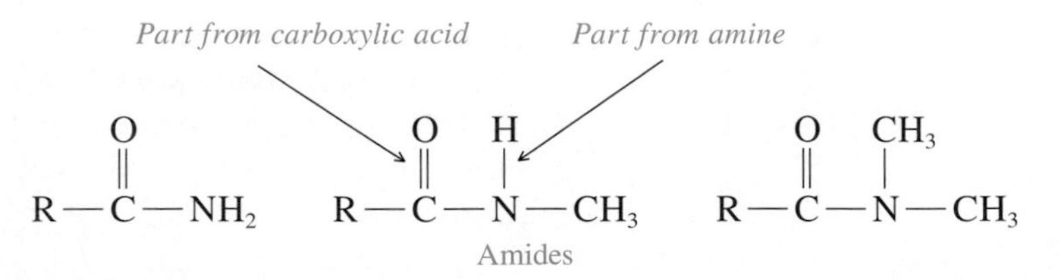

Amides

Figure 13.15
A ball-and-stick model of ethanamide (acetamide), an amide.

ethanamide

Naming Amides

In both the common and IUPAC name, amides are named by dropping the *ic acid* or *oic acid* from the carboxylic acid names (IUPAC or common) and adding the suffix *amide.*

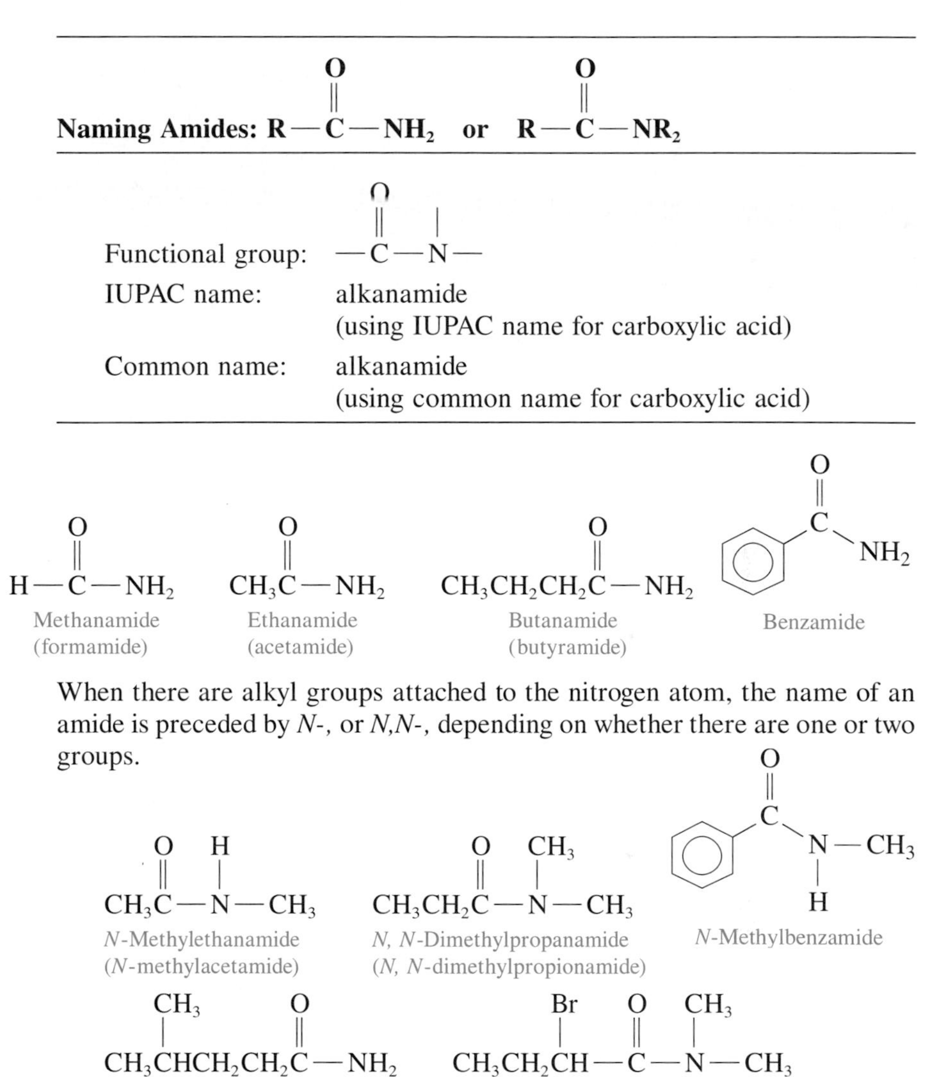

When there are alkyl groups attached to the nitrogen atom, the name of an amide is preceded by *N*-, or *N,N*-, depending on whether there are one or two groups.

Peptide Bonds in Proteins

Amino acids contain both a carboxylic acid group and an alpha (α) amino group in the same molecule. About 20 amino acids are commonly found in proteins. To form a protein, the amino group from one amino acid forms an amide bond with the carboxylic acid group of another amino acid. This is the same reaction we studied in the last section on amidation. In proteins, the amide linkages are called peptide bonds. We will discuss amino acids and proteins further in Chapter 16.

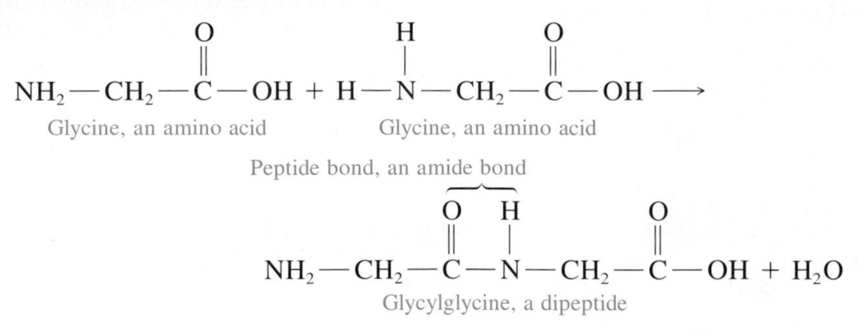

Sample Problem 13.13
Naming Amides

Give the common and IUPAC names for each of the following amides:

a. $CH_3-CH_2-C(=O)-NH_2$

b. $CH_3-CH(Cl)-C(=O)-NH-CH_2CH_3$

Solution

a. The IUPAC name of the carboxylic acid is propanoic acid; the common name is propionic acid. Replacing the *oic acid* or *ic acid* ending with *amide* gives the IUPAC name of *propanamide* and common name of *propionamide.*

b. Using the same carboxylic acid as in part (a), the ethyl group attached to the nitrogen atom is named *N-ethyl.* The amide is named *N-ethyl-2-chloropropanamide* (IUPAC), and *N-ethyl-α-chloropropionamide* (common).

Study Check

Draw the condensed structural formula of *N*,*N*-dimethyl-*p*-chlorobenzamide.

HEALTH NOTE *Amides in Health and Medicine*

The simplest natural amide is urea, an end product of protein metabolism in the body. The kidneys remove urea from the blood and provide for its excretion in urine. If the kidneys malfunction, urea is not removed and builds to a toxic level, a condition called uremia. Urea is also a component of fertilizer, used to increase the nitrogen in the soil.

$NH_2-C(=O)-NH_2$ Urea

Synthetic amides are used as substitutes for sugar and aspirin. Saccharin is a very powerful sweetener and is used as a sugar substitute.

Aspirin substitutes contain phenacetin or acetaminophen, which is used in Tylenol. Like aspirin, acetaminophen is an analgesic and antipyretic, but it has little antiinflammatory effect. (See Figure 13.16.)

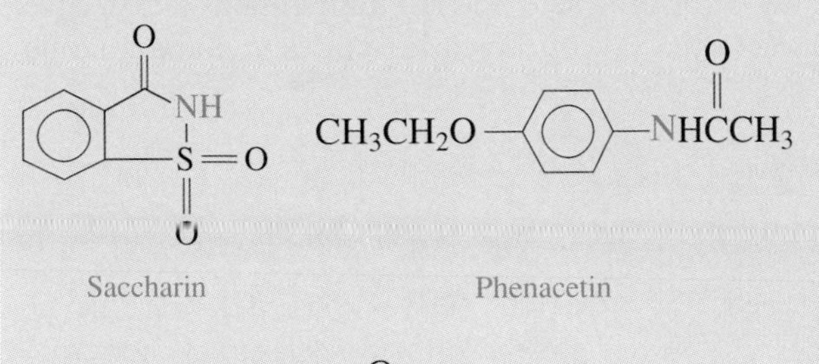

Saccharin Phenacetin

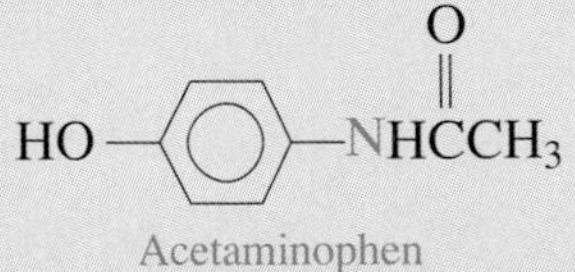

Acetaminophen

Figure 13.16
Sugar and aspirin substitutes contain amide compounds.

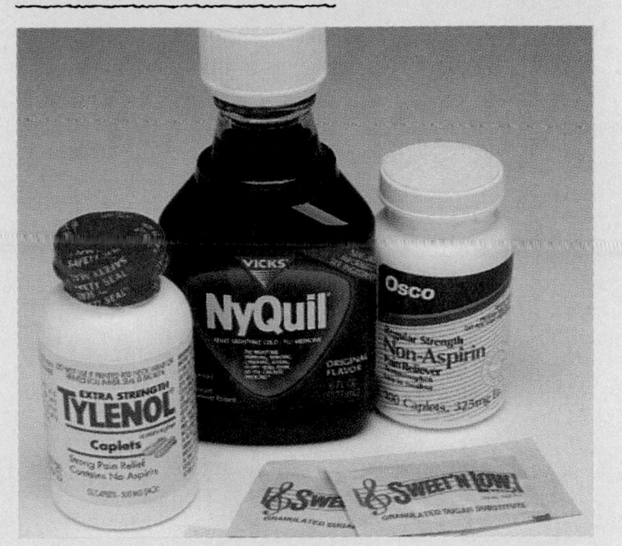

Many barbiturates are cyclic amides of barbituric acid that act as sedatives in small dosages or sleep inducers in larger dosages. They are often habit forming. Barbiturate drugs include phenobarbital (Luminal), pentobarbital (Nembutal), and secobarbital (Seconal).

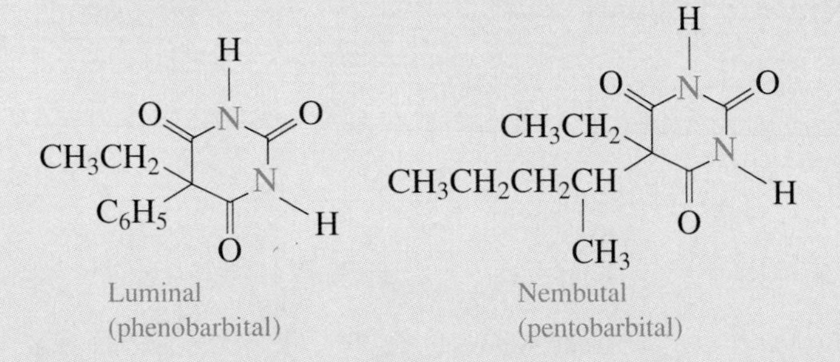

Luminal (phenobarbital) Nembutal (pentobarbital)

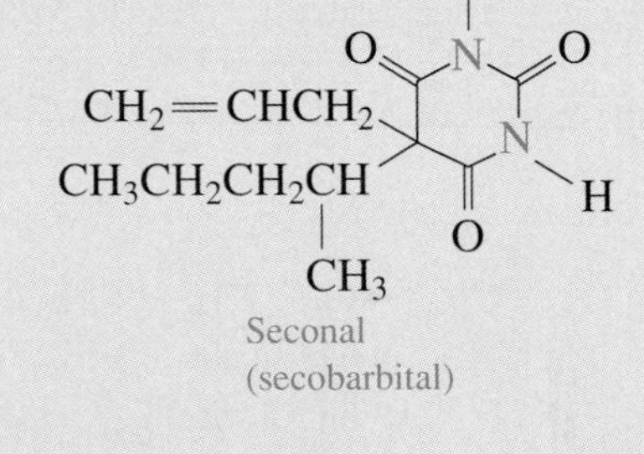

Seconal (secobarbital)

Another group of amides such as diazepam (Valium) and meprobamate (Equanil) are used to reduce anxiety.

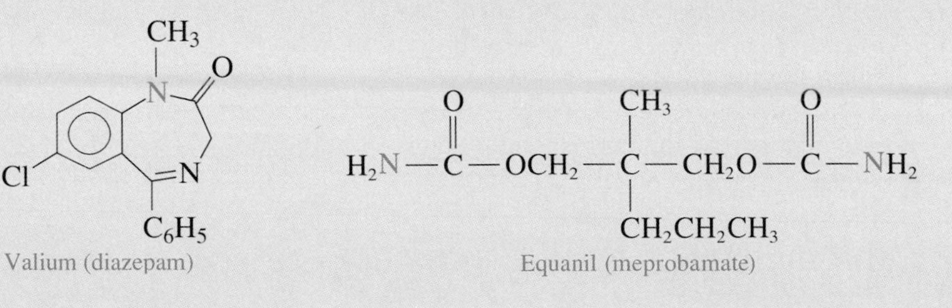

Valium (diazepam) Equanil (meprobamate)

13.9 Reactions of Amides

Learning Goal
Write equations for the hydrolysis of amides.

As we have seen, amide bonds are formed by the elimination of water. The reverse reaction called hydrolysis occurs when water is added back to the amide bond to split the molecule. When an acid is used, the hydrolysis products of an amide are the carboxylic acid and the ammonium salt. In base hydrolysis, the amide produces the salt of the carboxylic acid and ammonia or amine.

Acid Hydrolysis of Amides

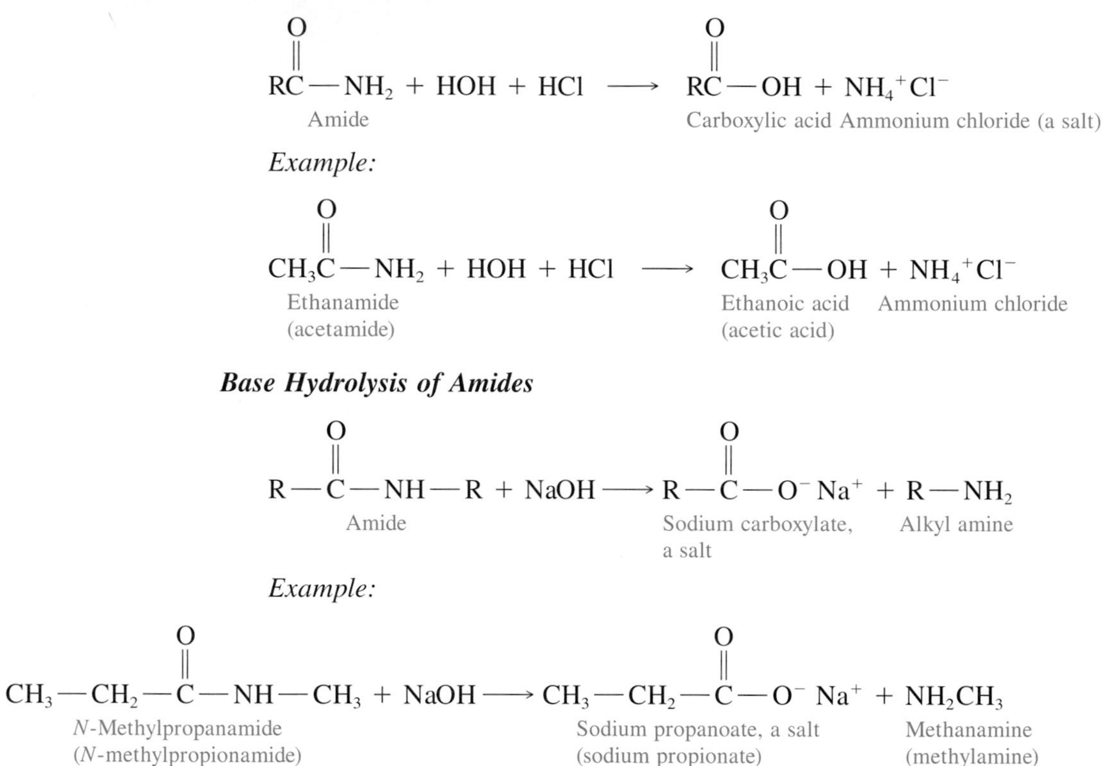

Sample Problem 13.14
Hydrolysis of Amides

Write the structural formulas for the products for the hydrolysis of *N*-methylpentamide with NaOH.

Solution
Hydrolysis of the amide with a base produces a carboxylate salt (sodium pentanoate) and the corresponding amine (methylamine).

$$CH_3CH_2CH_2CH_2\overset{\overset{\displaystyle O}{\|}}{C}O^-Na^+ + NH_2CH_3$$

Study Check
What are the structures of the products from the hydrolysis of *N*-methylbutyramide with HBr?

HEALTH NOTE *Amide Hydrolysis in Nature*

In biological systems, the peptide bonds of proteins are hydrolyzed by enzymes called proteases during the digestion of proteins or when they are broken down in cells. For example, the hydrolysis of a dipeptide to produce amino acids can be shown as follows:

Peptide bond, an amide bond

$$NH_2{-}CH_2{-}\overset{\overset{\displaystyle O}{\|}}{C}{-}\overset{\overset{\displaystyle H}{|}}{N}{-}CH_2{-}\overset{\overset{\displaystyle O}{\|}}{C}{-}OH \xrightarrow{\text{Protease}} NH_2{-}CH_2{-}\overset{\overset{\displaystyle O}{\|}}{C}{-}OH + H{-}\overset{\overset{\displaystyle H}{|}}{N}{-}CH_2{-}\overset{\overset{\displaystyle O}{\|}}{C}{-}OH$$

Glycylglycine, a dipeptide — Glycine, an amino acid — Glycine, an amino acid

In 1989, a controversy arose concerning a hormone, Alar, that was being applied to apples to keep them on the trees longer and allow larger fruit to develop. Upon hydrolysis, the Alar produces an amine, UDMH, suspected of being a carcinogen. UDMH is the abbreviation for unsymmetrical dimethylhydrazine. Of particular concern was apple juice because high temperatures used during the processing of the apples increase the hydrolysis of Alar. Eventually, sale of Alar was discontinued in the United States.

$$HO\overset{\overset{\displaystyle O}{\|}}{C}CH_2CH_2\overset{\overset{\displaystyle O}{\|}}{C}{-}NH{-}\overset{\overset{\displaystyle CH_3}{|}}{N}{-}CH_3 + H{-}OH \xrightarrow{\text{Heat}} HO\overset{\overset{\displaystyle O}{\|}}{C}CH_2CH_2\overset{\overset{\displaystyle O}{\|}}{C}{-}OH + NH_2{-}\overset{\overset{\displaystyle CH_3}{|}}{N}{-}CH_3$$

Alar — Succinic acid — UDMH

Summary of Naming

Family	Condensed Structural Formula	IUPAC Name	Common Name
Carboxylic acid	$CH_3\overset{\overset{O}{\|}}{C}{-}OH$	Ethanoic acid	Acetic acid
Carboxylate salt	$CH_3\overset{\overset{O}{\|}}{C}{-}O^-\ Na^+$	Sodium ethanoate	Sodium acetate
Ester	$CH_3\overset{\overset{O}{\|}}{C}{-}OCH_3$	Methyl ethanoate	Methyl acetate
Amine	$CH_3CH_2{-}NH_2$	Aminoethane	Ethylamine
Amide	$CH_3\overset{\overset{O}{\|}}{C}{-}NH_2$	Ethanamide	Acetamide

Summary of Reactions

Oxidation of Aldehydes to Carboxylic Acids

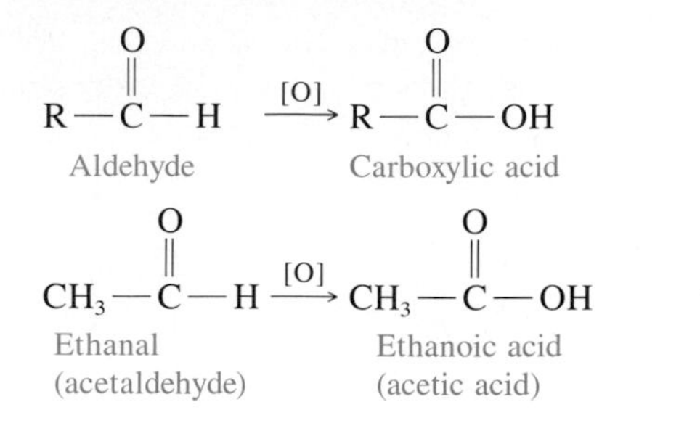

Ionization of a Carboxylic Acid in Water

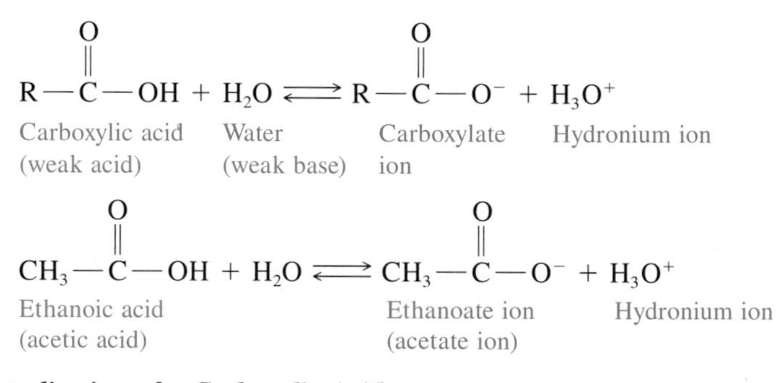

Neutralization of a Carboxylic Acid

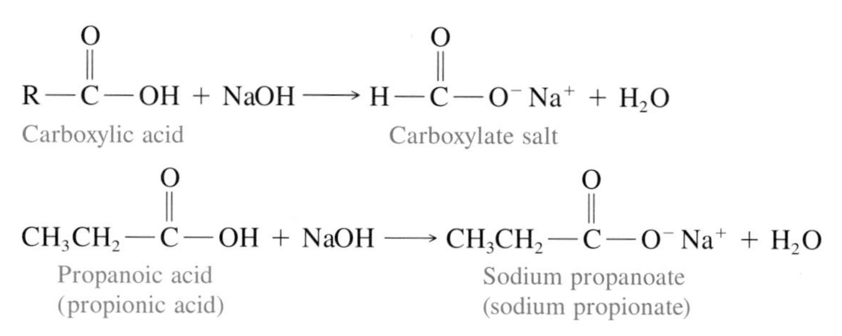

Esterification: Carboxylic Acid and an Alcohol

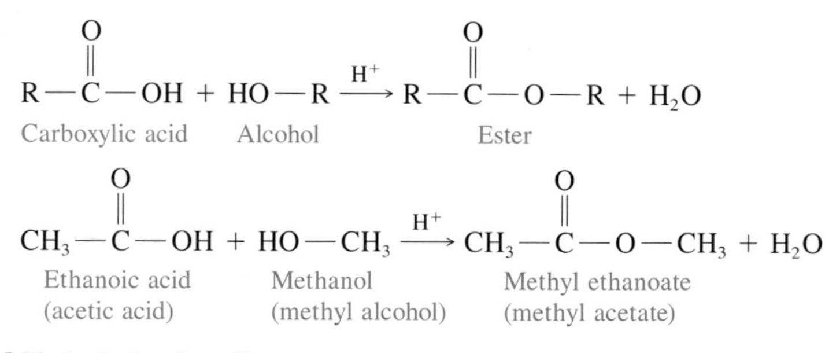

Acid Hydrolysis of an Ester

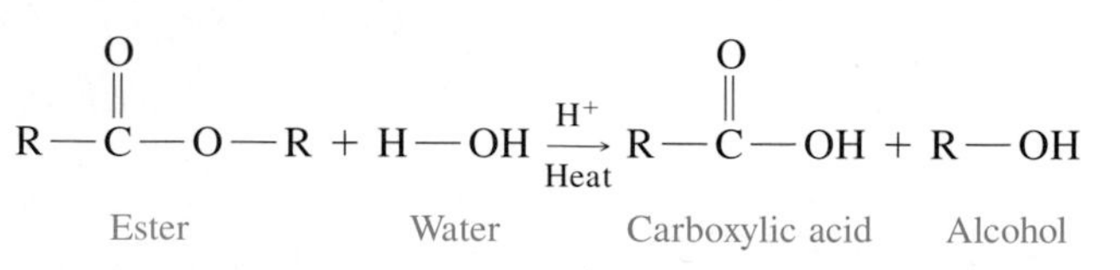

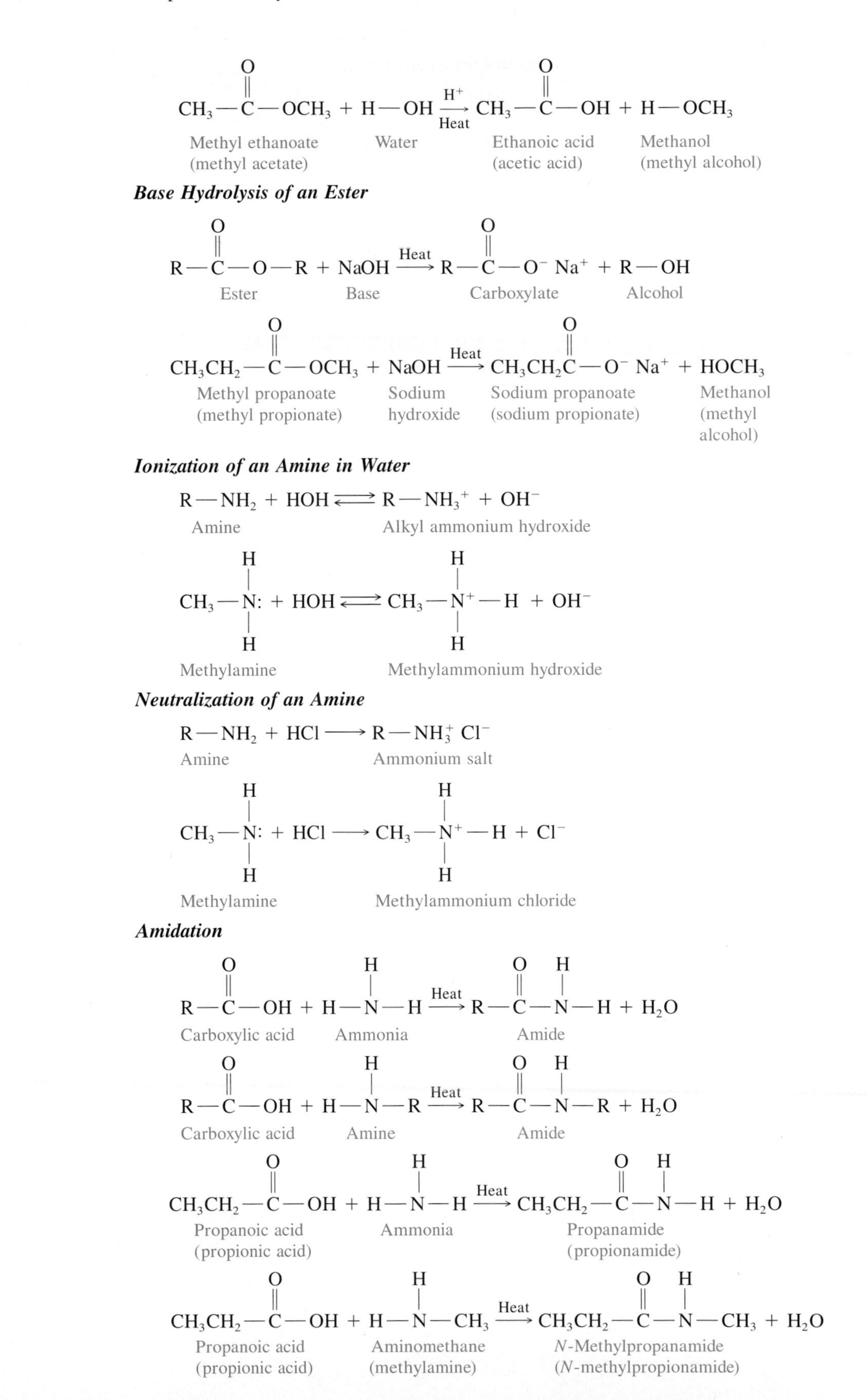
CH3—C(=O)—OCH3 + H—OH —H+, Heat→ CH3—C(=O)—OH + H—OCH3
Methyl ethanoate (methyl acetate)
Water
Ethanoic acid (acetic acid)
Methanol (methyl alcohol)
Base Hydrolysis of an Ester
R—C(=O)—O—R + NaOH —Heat→ R—C(=O)—O⁻ Na⁺ + R—OH
Ester
Base
Carboxylate
Alcohol
CH3CH2—C(=O)—OCH3 + NaOH —Heat→ CH3CH2C(=O)—O⁻ Na⁺ + HOCH3
Methyl propanoate (methyl propionate)
Sodium hydroxide
Sodium propanoate (sodium propionate)
Methanol (methyl alcohol)
Ionization of an Amine in Water
R—NH2 + HOH ⇌ R—NH3⁺ + OH⁻
Amine
Alkyl ammonium hydroxide
CH3—NH2: + HOH ⇌ CH3—N⁺H3 + OH⁻
Methylamine
Methylammonium hydroxide
Neutralization of an Amine
R—NH2 + HCl → R—NH3⁺ Cl⁻
Amine
Ammonium salt
CH3—NH2: + HCl → CH3—N⁺H3 + Cl⁻
Methylamine
Methylammonium chloride
Amidation
R—C(=O)—OH + H—NH—H —Heat→ R—C(=O)—NH—H + H2O
Carboxylic acid
Ammonia
Amide
R—C(=O)—OH + H—NH—R —Heat→ R—C(=O)—NH—R + H2O
Carboxylic acid
Amine
Amide
CH3CH2—C(=O)—OH + H—NH—H —Heat→ CH3CH2—C(=O)—NH—H + H2O
Propanoic acid (propionic acid)
Ammonia
Propanamide (propionamide)
CH3CH2—C(=O)—OH + H—NH—CH3 —Heat→ CH3CH2—C(=O)—NH—CH3 + H2O
Propanoic acid (propionic acid)
Aminomethane (methylamine)
N-Methylpropanamide (N-methylpropionamide)

Hydrolysis of Amides

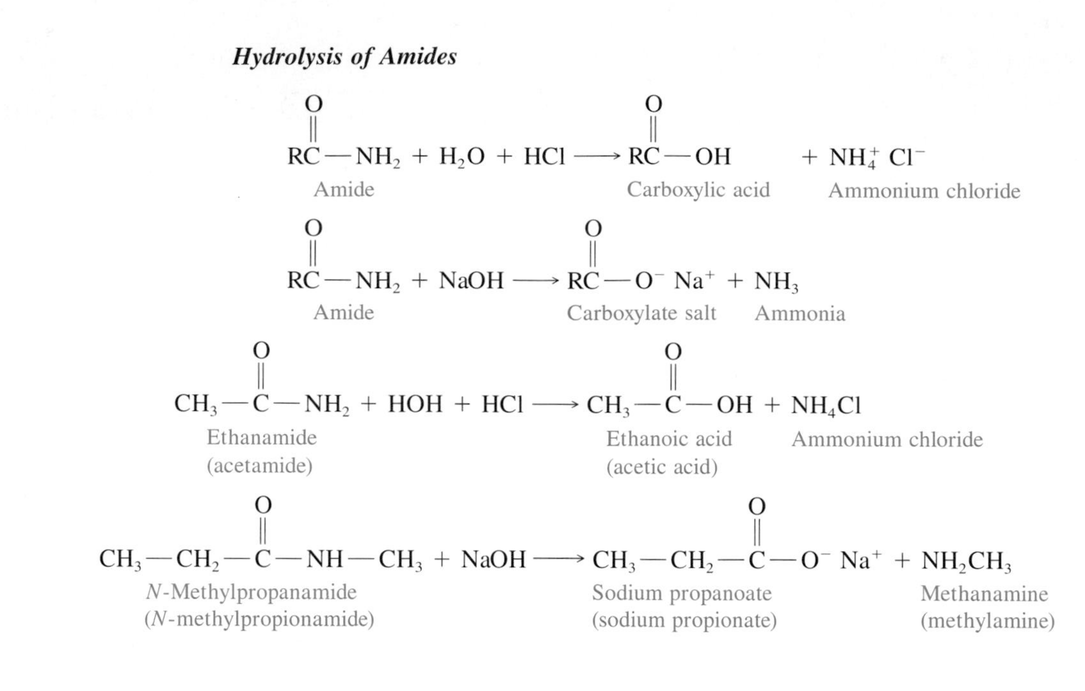

Chapter Summary

13.1 Carboxylic Acids
A carboxylic acid contains the carboxyl functional group which is a hydroxyl group connected to the carbonyl group. A carboxylic acid is formed by the oxidation of a primary alcohol or an aldehyde.

13.2 Properties of Carboxylic Acids
The carboxyl group contains polar bonds of O—H and C=O, which makes a carboxylic acid with one to four carbon atoms very soluble in water. As weak acids, carboxylic acids ionize slightly by donating a proton to water to form carboxylate and hydronium ions.

13.3 Reactions of Carboxylic Acids
Carboxylic acids are neutralized by a base producing the carboxylate salt and water. In the presence of a strong acid, a carboxylic acid reacts with an alcohol to produce an ester. A molecule of water is removed, the —OH from the carboxylic acid and the —H from the alcohol molecule.

13.4 Esters
In an ester, an alkyl or aromatic group has replaced the H of the hydroxyl group of a carboxylic acid. The names of esters consist of two words, one from the alcohol and the other from the carboxylic acid with the *ic* ending replaced by *ate.*

13.5 Reactions of Esters
Esters undergo acid hydrolysis by adding water to yield the carboxylic acid and alcohol (or phenol). Base hydrolysis or saponification of an ester produces the carboxylate salt and an alcohol. Soaps are the salts of long-chain fatty acids produced when fats are saponified.

13.6 Amines
A nitrogen atom attached to one, two, or three alkyl or aromatic groups forms a primary, secondary, or tertiary amine. Amines are prevalent in compounds, synthetic and naturally occurring, that have physiological activity. Amines are usually named by common names in which the names of the alkyl group are listed alphabetically preceding the suffix *amine.* In the IUPAC system, the word *amino* is used to give the position of the NH_2 group on an alkane chain.

13.7 Reactions of Amines
In water, amines act as weak bases by accepting protons from water to produce ammonium and hydroxide ions. Amines react with acids to produce ammonium salts. When a carboxylic acid reacts with ammonia or an amine, an amide is produced.

13.8 Amides
Amides are derivatives of carboxylic acids in

which the hydroxyl group is replaced by an amino or *N*-alkyl (NH_2 or NHR) group. They are named by replacing the *ic acid* or *oic acid* ending with *amide.*

13.9 Reactions of Amides
Amides undergo hydrolysis (in acid or base) to produce the carboxylic acid and the amine (or ammonia) or their salts.

Glossary of Key Terms

alkaloids Amines having physiological activity that are produced in plants.

amidation The formation of an amide from a carboxylic acid and ammonia or an amine.

amides Organic compounds containing the carbonyl group attached to an amino group or a substituted nitrogen atom:

$$R-\overset{\overset{\displaystyle O}{\|}}{C}-NH_2 \qquad R-\overset{\overset{\displaystyle O}{\|}}{C}-\overset{\overset{\displaystyle H}{|}}{N}-R$$

amines Organic compounds containing a nitrogen atom attached to one, two, or three hydrocarbon groups.

carboxyl group A functional group found in carboxylic acids composed of carbonyl and hydroxyl groups.

$$-\overset{\overset{\displaystyle O}{\|}}{C}-OH \quad \text{Carboxyl group}$$

carboxylic acids A family of organic compounds containing the carboxyl group.

$$R-\overset{\overset{\displaystyle O}{\|}}{C}-OH \quad \text{Carboxylic acid}$$

esters A family of organic compounds in which an alkyl group replaces the hydrogen atom in a carboxylic acid.

$$R-\overset{\overset{\displaystyle O}{\|}}{C}-O-R \quad \text{Ester}$$

esterification The formation of an ester from a carboxylic acid and an alcohol with the elimination of a molecule of water in the presence of an acid catalyst.

hydrolysis The splitting of a molecule by the addition of water. Esters hydrolyze to produce a carboxylic acid and an alcohol; amides to yield the corresponding carboxylic acid and amine or their salts.

saponification The hydrolysis of an ester with a strong base to produce a salt of the carboxylic acid and an alcohol.

soap A salt of a long-chain fatty acid produced by the saponification of fats.

Chemistry at Home

1. Find the identification codes on the bottom of plastic bottles. From the code list for plastics, determine the type of polymer that was used in producing that plastic.
2. Read the labels on shampoos, cosmetics, and foods and list the carboxylic acids, carboxylate salts, esters, amines, and amides that you find. Draw their structural formulas. Use a chemistry handbook or other reference, if necessary.

Answers to Study Checks

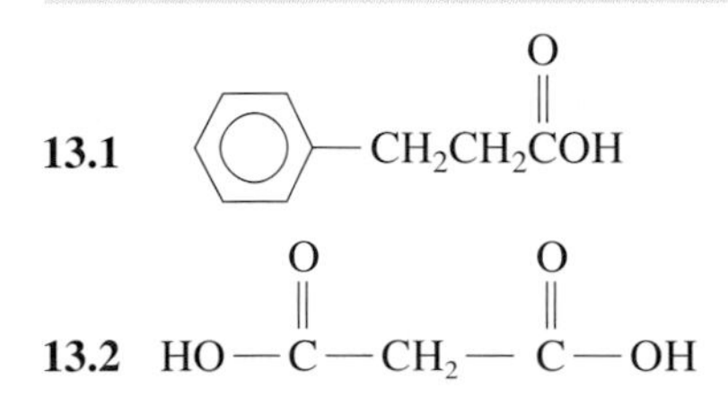

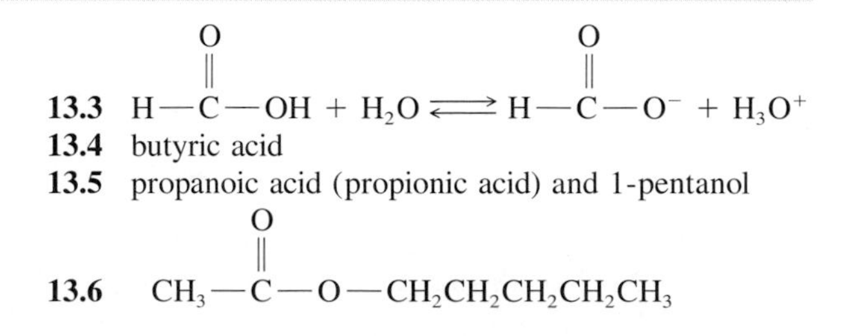

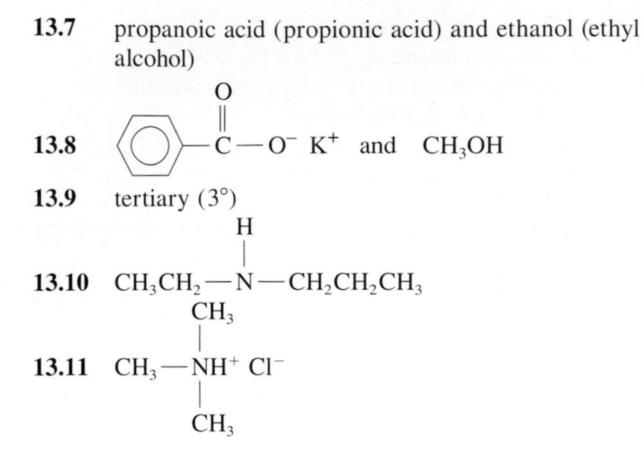

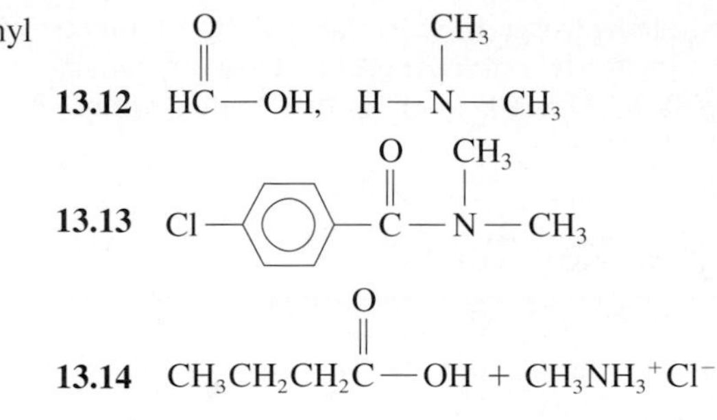

13.14 $CH_3CH_2CH_2C(=O)—OH + CH_3NH_3^+Cl^-$

Problems

Carboxylic Acids *(Goal 13.1)*

13.1 Explain the differences in the condensed structural formulas of propanal and propanoic acid.

13.2 Explain the differences in the condensed structural formulas of benzaldehyde and benzoic acid.

13.3 Give the IUPAC and common names (if any) for the following carboxylic acids:

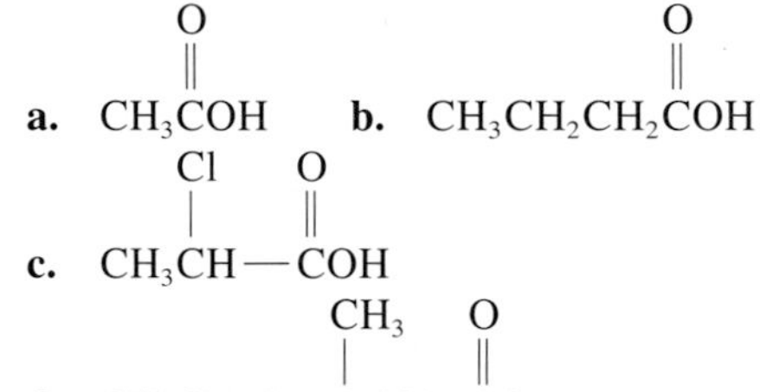

13.4 Give the IUPAC and common names (if any) for the following carboxylic acids:

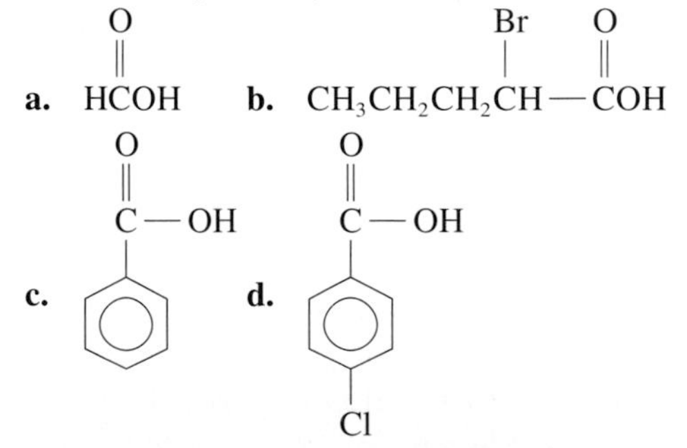

13.5 Draw the condensed structural formulas of each of the following carboxylic acids:
a. propionic acid **b.** benzoic acid
c. 2-chloroethanoic acid
d. 3-hydroxypropanoic acid
e. β-methylbutyric acid
f. 3,5-dibromoheptanoic acid

13.6 Draw the condensed structural formulas of each of the following carboxylic acids:
a. butyric acid **b.** 3-ethylbenzoic acid
c. α-hydroxyacetic acid
d. 2,4-dibromobutanoic acid
e. *m*-methylbenzoic acid
f. 4,4-dibromohexanoic acid

13.7 Draw the condensed structural formulas of the carboxylic acids formed by the oxidation of each of the following:

a. CH_3OH **b.** CH_3CH (with C=O)

c. $CH_3CH(CH_3)CH_2CH_2OH$ **d.** cyclopentyl—CH_2OH

13.8 Draw the condensed structural formulas of the carboxylic acids formed by the oxidation of each of the following:

a. $CH_3CH_2CH_2CH_2CH_2CH_2OH$

b. $CH_3CH_2CH_2CH_2CH$ (with C=O)

c. $CH_3CH(CH_3)CH_2CH$ (with C=O) **d.** phenyl—CH_2CH_2OH

Properties of Carboxylic Acids *(Goal 13.2)*

13.9 In the following pairs, which compound would be more soluble in water? Explain.

a. $CH_3CH_2CH_3$ or $CH_3CH_2\overset{O}{\overset{\|}{C}}OH$

b. $CH_3CH_2CH_2CH_2OH$ or $CH_3CH_2CH_2\overset{O}{\overset{\|}{C}}OH$

c. $CH_3\overset{O}{\overset{\|}{C}}OH$ or $CH_3CH_2CH_2CH_2\overset{O}{\overset{\|}{C}}OH$

13.10 In the following pairs, which compound would be more soluble in water? Explain.

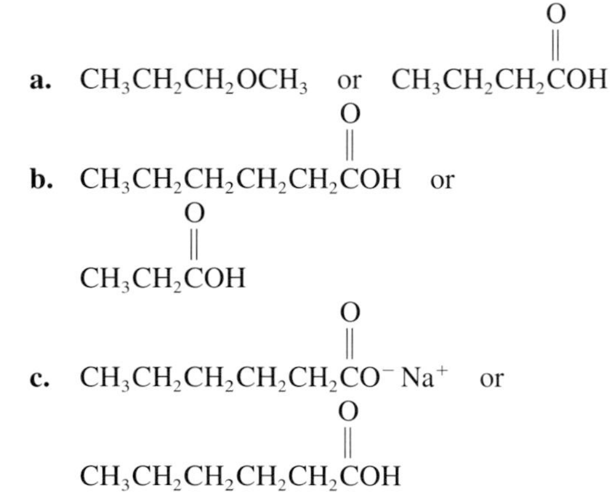

13.11 Write equations for the ionization for each of the following carboxylic acids in water:

a. $H{-}\overset{O}{\overset{\|}{C}}{-}OH$ **b.** $CH_3CH_2\overset{O}{\overset{\|}{C}}OH$

c. acetic acid

13.12 Write equations for the ionization for each of the following carboxylic acids in water:

a. $CH_3\underset{CH_3}{\underset{|}{C}}H\overset{O}{\overset{\|}{C}}OH$ **b.** α-hydroxyacetic acid

c. butanoic acid

Reactions of Carboxylic Acids *(Goal 13.3)*

13.13 Write equations for the reaction of each of the following carboxylic acids with NaOH:
a. formic acid
b. propanoic acid
c. benzoic acid

13.14 Write equations for the reaction of each of the following carboxylic acids with KOH:
a. acetic acid
b. 2-methylbutanoic acid
c. *p*-chlorobenzoic acid

13.15 Write the structural formulas of the esters formed by reacting the following carboxylic acids with methyl alcohol:
a. acetic acid
b. butyric acid
c. benzoic acid

13.16 Write the condensed structural formulas of the esters formed by reacting each of the following carboxylic acids with ethyl alcohol:
a. formic acid
b. propionic acid
c. 2-methylpentanoic acid

13.17 Draw the condensed structural formulas of the esters formed by reacting each of the following carboxylic acids and alcohols:

a. $CH_3\overset{O}{\overset{\|}{C}}{-}OH + HOCH_2CH_2CH_3 \xrightarrow{H^+}$

b. $CH_3CH_2CH_2CH_2\overset{O}{\overset{\|}{C}}OH + HO\overset{CH_3}{\overset{|}{C}}HCH_3 \xrightarrow{H^+}$

13.18 Draw the condensed structural formulas of the esters formed by reacting each of the following carboxylic acids and alcohols:

a. $CH_3{-}CH_2{-}\overset{O}{\overset{\|}{C}}{-}OH + HOCH_3 \xrightarrow{H^+}$

b. $C_6H_5\overset{O}{\overset{\|}{C}}OH$ (benzene ring bearing $\overset{O}{\overset{\|}{C}}OH$) $+ HOCH_2CH_2CH_2CH_3 \xrightarrow{H^+}$

Esters *(Goal 13.4)*

13.19 Identify the following as aldehydes, ketones, carboxylic acids or esters:

a. $CH_3\overset{O}{\overset{\|}{C}}H$ **b.** $CH_3\overset{O}{\overset{\|}{C}}{-}OCH_3$

c. $CH_3CH_2CH_2\overset{O}{\overset{\|}{C}}CH_3$

d. $CH_3\overset{CH_3}{\overset{|}{C}}HCH_2\overset{O}{\overset{\|}{C}}{-}OH$

13.20 Identify the following as aldehydes, ketones, carboxylic acids, or esters:

a. $CH_3\overset{O}{\overset{\|}{C}}OH$ **b.** $CH_3\overset{O}{\overset{\|}{C}}CH_3$

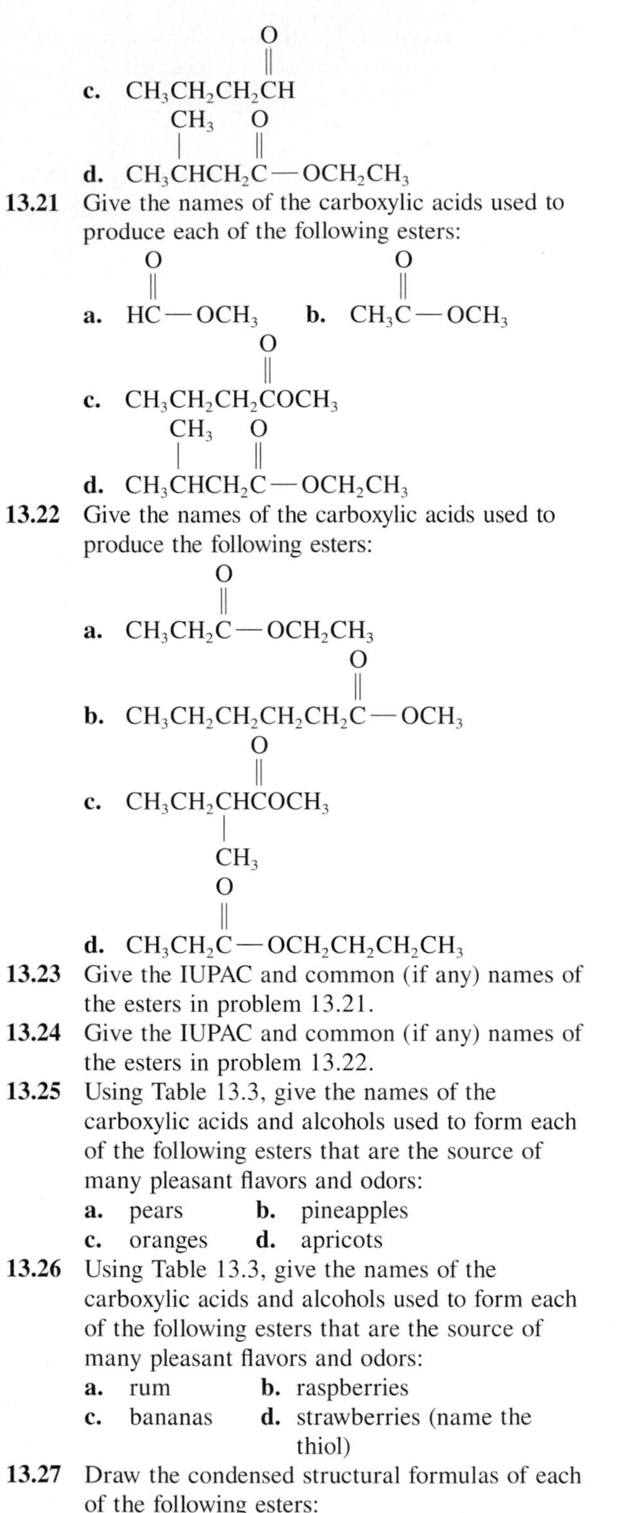

c. $CH_3CH_2CH_2CH$ (with C=O)

d. $CH_3CHCH_2C{-}OCH_2CH_3$ (with CH_3 branch and C=O)

13.21 Give the names of the carboxylic acids used to produce each of the following esters:

a. $HC{-}OCH_3$ (C=O) b. $CH_3C{-}OCH_3$ (C=O)

c. $CH_3CH_2CH_2COCH_3$ (C=O)

d. $CH_3CHCH_2C{-}OCH_2CH_3$ (with CH_3 branch and C=O)

13.22 Give the names of the carboxylic acids used to produce the following esters:

a. $CH_3CH_2C{-}OCH_2CH_3$ (C=O)

b. $CH_3CH_2CH_2CH_2CH_2C{-}OCH_3$ (C=O)

c. $CH_3CH_2CHCOCH_3$ (C=O, with CH_3 branch)

d. $CH_3CH_2C{-}OCH_2CH_2CH_2CH_3$ (C=O)

13.23 Give the IUPAC and common (if any) names of the esters in problem 13.21.

13.24 Give the IUPAC and common (if any) names of the esters in problem 13.22.

13.25 Using Table 13.3, give the names of the carboxylic acids and alcohols used to form each of the following esters that are the source of many pleasant flavors and odors:

a. pears **b.** pineapples
c. oranges **d.** apricots

13.26 Using Table 13.3, give the names of the carboxylic acids and alcohols used to form each of the following esters that are the source of many pleasant flavors and odors:

a. rum **b.** raspberries
c. bananas **d.** strawberries (name the thiol)

13.27 Draw the condensed structural formulas of each of the following esters:

a. methyl acetate **b.** butyl formate
c. ethyl pentanoate
d. 2-bromopropyl propanoate

13.28 Draw the condensed structural formulas of each of the following esters:

a. hexyl acetate
b. propyl propionate
c. ethyl 2-hydroxybutanoate
d. methyl benzoate

Reactions of Esters *(Goal 13.5)*

13.29 What are the products of the acid hydrolysis of an ester?

13.30 What are the products of the base hydrolysis of an ester?

13.31 Draw the condensed structural formulas of the products from the acid or base catalyzed hydrolysis of each of the following compounds:

a. $CH_3CH_2{-}C({=}O){-}O{-}CH_3 + NaOH \longrightarrow$

b. $CH_3{-}C({=}O){-}O{-}CH_2CH_2CH_3 + H_2O \xrightarrow{H^+}$

c. $CH_3CH_2CH_2{-}C({=}O){-}O{-}CH_2CH_3 + H_2O \xrightarrow{H^+}$

d. $C_6H_5{-}C({=}O){-}O{-}CH_2CH_3 + H_2O \xrightarrow{H^+}$

e. $C_6H_5{-}C({=}O){-}O{-}CH_2CH_3 + NaOH \longrightarrow$

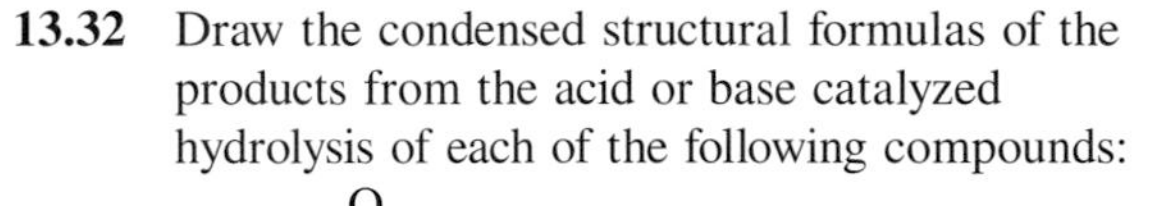

13.32 Draw the condensed structural formulas of the products from the acid or base catalyzed hydrolysis of each of the following compounds:

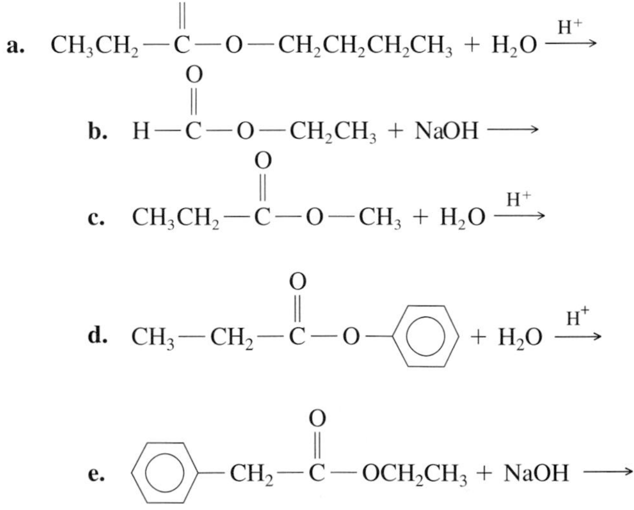

a. $CH_3CH_2{-}C({=}O){-}O{-}CH_2CH_2CH_2CH_3 + H_2O \xrightarrow{H^+}$

b. $H{-}C({=}O){-}O{-}CH_2CH_3 + NaOH \longrightarrow$

c. $CH_3CH_2{-}C({=}O){-}O{-}CH_3 + H_2O \xrightarrow{H^+}$

d. $CH_3{-}CH_2{-}C({=}O){-}O{-}C_6H_5 + H_2O \xrightarrow{H^+}$

e. $C_6H_5{-}CH_2{-}C({=}O){-}OCH_2CH_3 + NaOH \longrightarrow$

Amines *(Goal 13.6)*

13.33 What is a primary amine?

13.34 What is a tertiary amine?

13.35 Classify the following amines as primary, secondary, or tertiary:

a. $CH_3CH_2CH_2CH_2—NH_2$

b. $CH_3—N(CH_3)—CH_2CH_3$

c. (piperidine ring with N–H)

d. (benzene ring with NH_2)

13.36 Classify the following amines as primary, secondary, or tertiary:

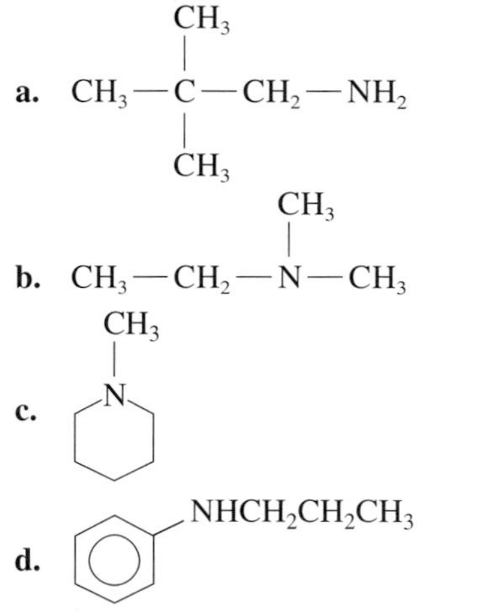

13.37 Give the common and IUPAC names for each of the following amines:

a. $CH_3CH_2—NH_2$

b. $CH_3—NH—CH_2CH_2CH_3$

c. $CH_3CH_2—N(CH_2CH_3)—CH_2CH_2$

d. (benzene ring with $NHCH_3$)

13.38 Give the common and IUPAC names for each of the following amines:

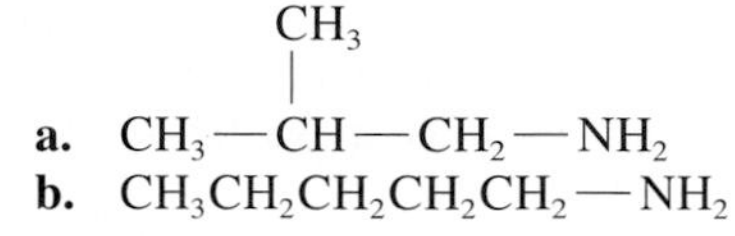

b. $CH_3CH_2CH_2CH_2CH_2—NH_2$

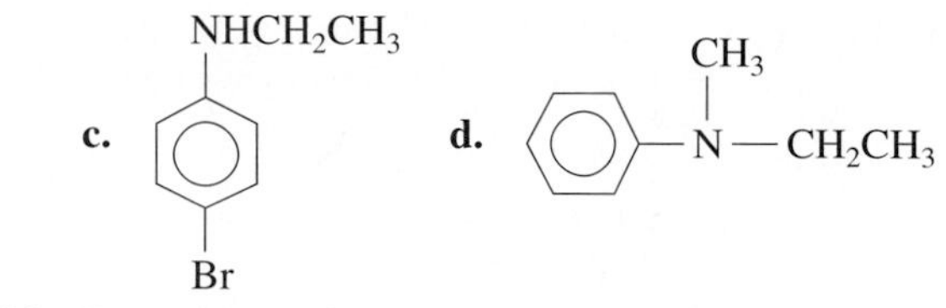

13.39 Draw the condensed structural formulas for each of the following amines:

a. ethylamine **b.** *N*-methylaniline

c. butylpropylamine **d.** 2-aminopentane

13.40 Draw the condensed structural formulas for each of the following amines:

a. dimethylamine **b.** *p*-chloroaniline

c. *N,N*-diethylaniline

d. ethyldimethylamine

Reactions of Amines *(Goal 13.7)*

13.41 Write an equation for the ionization of each of the following amines in water:

a. methylamine

b. dimethylamine

c. aniline

13.42 Write an equation for the ionization of each of the following amines in water:

a. ethylamine

b. propylamine

c. *N*-methylaniline

13.43 Write the condensed structural formula of the salts obtained from the reactions of each of the amines in problem 13.41 with HCl.

13.44 Write the condensed structural formula of the salts obtained from the reactions of each of the amines in problem 13.42 with HBr.

13.45 Draw the condensed structural formulas of the amides formed in each of the following reactions:

a. $CH_3\overset{O}{\overset{\|}{C}}OH + NH_3 \xrightarrow{\text{Heat}}$

b. $CH_3\overset{O}{\overset{\|}{C}}OH + NH_2CH_2CH_3 \xrightarrow{\text{Heat}}$

c. (benzene ring with $\overset{O}{\overset{\|}{C}}OH$) $+ NH_2CH_2CH_2CH_3 \xrightarrow{\text{Heat}}$

13.46 Draw the condensed structural formulas of the amides formed in each of the following reactions:

a. $CH_3CH_2CH_2CH_2\overset{O}{\overset{\|}{C}}OH + NH_3 \xrightarrow{\text{Heat}}$

b. $CH_3\underset{}{\overset{CH_3}{\overset{|}{C}}}HCH_2\overset{O}{\overset{\|}{C}}OH + NH_2CH_2CH_2CH_3 \xrightarrow{\text{Heat}}$

c. $CH_3CH_2\overset{O}{\overset{\|}{C}}OH + C_6H_5NH_2 \xrightarrow{\text{Heat}}$

Amides *(Goal 13.8)*

13.47 Give the IUPAC and common names (if any) for each of the following amides:

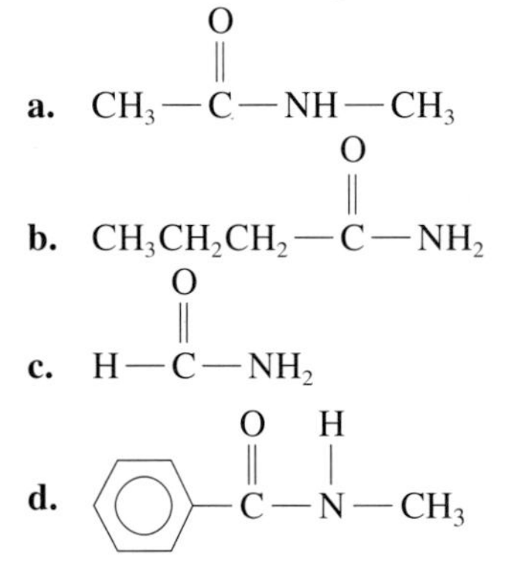

13.48 Give the IUPAC and common names (if any) for each of the following amides:

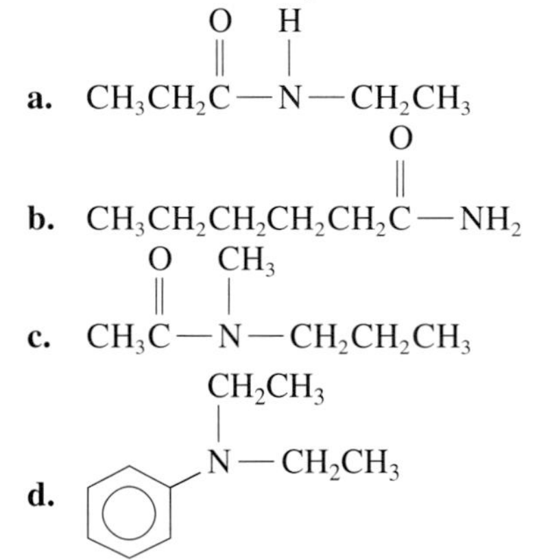

13.49 Draw the condensed structural formulas for each of the following amides:

a. propionamide **b.** 2-methylpentamide
c. methanamide **d.** *N*-ethylbenzamide
e. *N*-ethylbutyramide

13.50 Draw the condensed structural formulas for each of the following amides:

a. formamide
b. *N,N*-dimethylbenzamide
c. 3-methylbutyramide
d. 2,2-dichlorohexamide
e. *N*-propyl-3-chloropentamide

Reactions of Amides *(Goal 13.9)*

13.51 Write the condensed structural formulas for the products of the acid hydrolysis of each of the following amides with HCl:

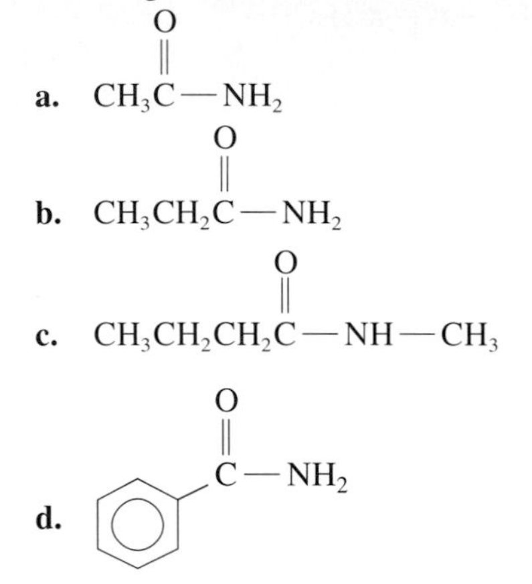

e. *N*-ethylpentanamide

13.52 Write the condensed structural formulas for the products of the base hydrolysis of each of the following amides with NaOH:

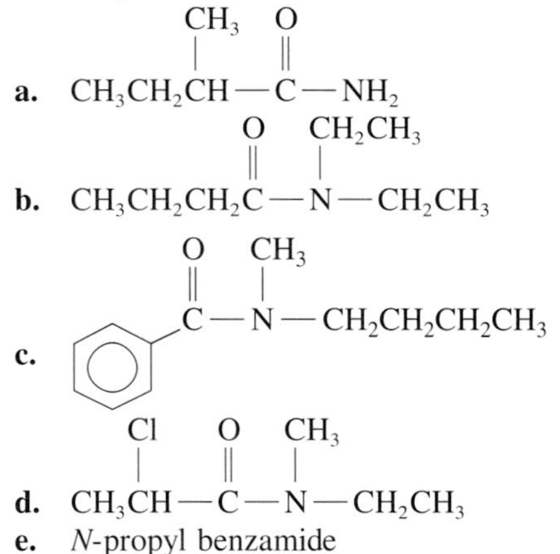

e. *N*-propyl benzamide

Challenge Problems

13.53 Methyl benzoate is not soluble in water; but when it is heated with KOH, it dissolves. Write an equation for the reaction and explain what happens. When HCl is added to the solution, a white solid forms. What is the solid?

13.54 Hexanoic acid is soluble in NaOH solution, but hexanal is not. Explain.

13.55 Salicylic acid could be named *o*-hydroxybenzoic acid.

a. What two reactive functional groups are present?

b. Draw the structure of the ester product that forms when the hydroxyl group of salicylic acid reacts with acetic acid.

c. Draw the structure of methyl salicylate, oil of wintergreen, formed when salicylic acid forms an ester with methyl alcohol.

13.56 What volume of 0.100 *M* NaOH is needed to neutralize 3.00 g of solid benzoic acid?

13.57 Draw the condensed structural formulas of the products produced by the hydrolysis of each of the following compounds:

a. $CH_3O{-}\overset{O}{\overset{\|}{C}}{-}CH_2CH_2{-}\overset{O}{\overset{\|}{C}}{-}OCH_3 + KOH \longrightarrow$

b. $CH_3{-}\overset{CH_3}{\overset{|}{C}H}{-}O{-}\overset{O}{\overset{\|}{C}}{-}\overset{CH_3}{\overset{|}{C}H}{-}CH_3 + H_2O \xrightarrow{H^+}$

13.58 There are four amine isomers with the molecular formula C_3H_9N. Draw their condensed structural formulas. Name and classify each as a primary, secondary, or tertiary amine.

13.59 Starting with 1-propanol as your only organic compound, write reactions to convert the alcohol into each of the following:

a. propionic acid b. 1-propyl propanoate
c. propane d. 2-propanol
e. acetone (propanone)
f. propanamide

13.60 Toradol is used in dentistry to relieve pain. Name the functional groups in this molecule.

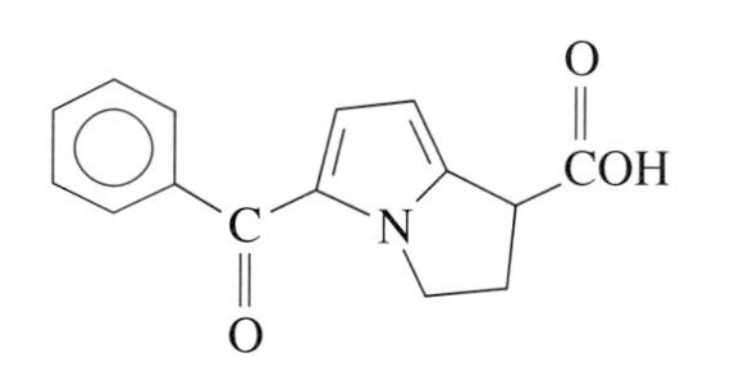

13.61 Voltaren is indicated for acute and chronic treatment of the symptoms of rheumatoid arthritis. Name the functional groups in this molecule.

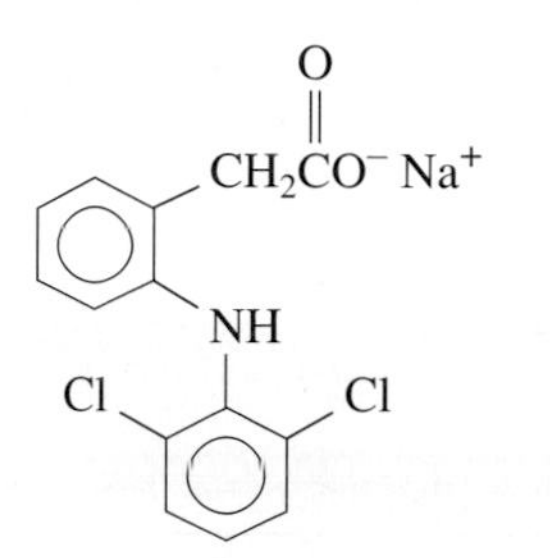

13.62 Many amine-containing drugs are given to patients in the form of their salt, e.g., hydrochloride, sulfate, and so on. What might be the reason?

13.63 Using a reference book such as the *Merck Index* or *Physicians' Desk Reference,* look up the structural formula of the following medicinal drugs and list the functional groups in the compounds. You may need to refer to the cross-index of names in the back of the book.

a. Keflex, an antibiotic
b. Inderal, a β-channel blocker used to treat heart irregularities
c. Ibuprofen, an antiinflammatory agent
d. Aldomet (methyldopa)
e. Percodan, a narcotic pain reliever
f. Triamterene, a diuretic

Chapter 14

Carbohydrates

Complex carbohydrates found in grains such as wheat, serve as a major source of energy that maintains a constant level of glucose in the blood.

Looking Ahead

Learning Goals

14.1 Classify carbohydrates as monosaccharides, disaccharides, and polysaccharides.

14.2 Classify a monosaccharide as an aldose or ketose and indicate the number of carbon atoms.

14.3 Draw and identify Fischer projections for carbohydrate molecules.

14.4 Draw the open-chain structures for D-glucose, D-galactose, and D-fructose.

14.5 Draw and identify the cyclic structures of monosaccharides.

14.6 Describe the monosaccharide units and linkages in disaccharides.

14.7 Describe the structural features of amylose, amylopectin, glycogen, and cellulose.

QUESTIONS TO THINK ABOUT

1. What elements are found in carbohydrates?
2. What is the relationship of photosynthesis and respiration?
3. What are the carbohydrates in table sugar, milk, and wood?
4. What is the storage form of carbohydrates in plants? In humans?
5. What is meant by a "high-fiber" diet?

Bread, pasta, potatoes, rice, and many other foods in our diets contain carbohydrates. A carbohydrate is also called a **saccharide,** a word that comes from the Latin term *saccharum,* "sugar." Digestion breaks the bonds of the polysaccharides in these foods to give monosaccharides, primarily glucose, a carbohydrate that provides most of our cellular energy. Cellulose is a polysaccharide in plants that forms rigid cell walls. Although cellulose is not digestible by humans, it does play an important role in our diets by providing fiber. Our diets may also include sucrose (table sugar) and lactose (milk sugar).

The National Academy of Sciences has recommended that one-half of our calories be obtained from complex carbohydrates found in foods such as potatoes, corn, rice, and wheat. Because complex carbohydrates are digested more slowly, a constant level of glucose is maintained in the blood over a longer period of time.

14.1 Classification of Carbohydrates

Learning Goal

Classify carbohydrates as monosaccharides, disaccharides, and polysaccharides.

Carbohydrates such as table sugar, lactose, and cellulose are made of carbon, hydrogen, and oxygen. Simple sugars such as glucose or fructose, which have formulas of $(CH_2O)_6$, were once thought to be hydrates of carbon, thus the name *carbohydrate.* In the process of photosynthesis, carbohydrates are produced in the green leaves of plants including algae in lakes and oceans. Energy from sunlight is used to combine carbon dioxide (CO_2) and water into simple carbohydrates such as glucose.

Carbon cycle

$$6CO_2 + 6H_2O + \text{energy} \underset{\text{Respiration}}{\overset{\text{Photosynthesis}}{\rightleftharpoons}} \underset{\text{Glucose}}{C_6H_{12}O_6} + 6O_2$$

In our body tissues, glucose is oxidized in metabolic reactions known as respiration and chemical energy is released to do work in the cells. Carbon dioxide and water are formed. The combination of photosynthesis and respiration is called the carbon cycle in which energy from the sun is stored by photosynthesis and made available to our cells when carbohydrates are metabolized. (See Figure 14.1.)

Types of Carbohydrates

The simplest carbohydrates are the **monosaccharides. Disaccharides** consist of two monosaccharide units joined together, whereas **polysaccharides** are complex carbohydrates that may contain thousands of monosaccharide units.

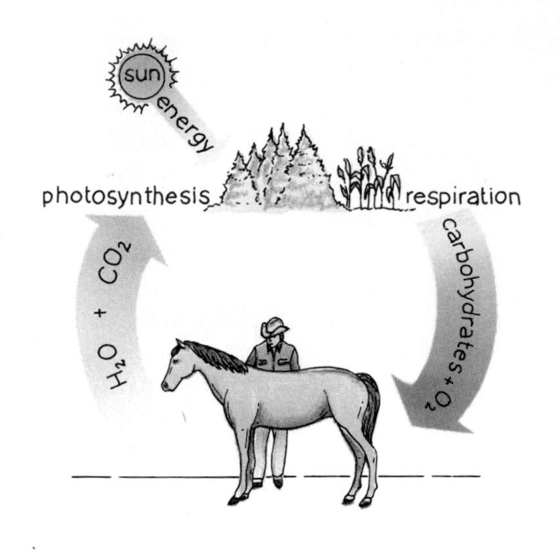

Figure 14.1
The carbon cycle in nature depicts the interdependence of photosynthesis and respiration.

When disaccharides or polysaccharides react with water, they can be completely hydrolyzed to yield monosaccharides.

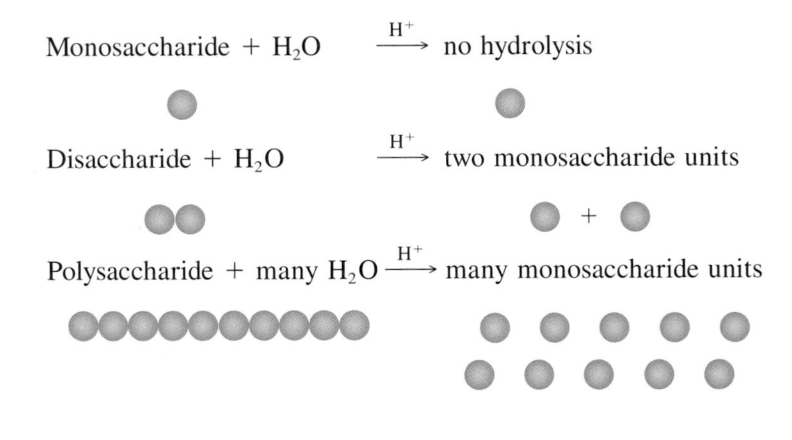

Sample Problem 14.1
Types of Carbohydrates

Classify the following carbohydrates as mono-, di-, or polysaccharides:

a. When lactose, milk sugar, is hydrolyzed, two monosaccharide units are produced.
b. Cellulose, a carbohydrate in cotton, yields thousands of monosaccharide units when completely hydrolyzed.

Solution

a. A disaccharide contains two monosaccharide units.
b. A polysaccharide contains many monosaccharide units.

Study Check

Fructose found in fruits does not undergo hydrolysis. What type of carbohydrate is it?

14.2 Monosaccharides

Learning Goal

Classify a monosaccharide as an aldose or ketose and indicate the number of carbon atoms.

Monosaccharides are polyhydroxy (many —OH groups) compounds with a carbonyl group (>C=O). A monosaccharide that is an *ald*ehyde is classified as an **aldose;** a **ketose** contains the *ket*one carbonyl group.

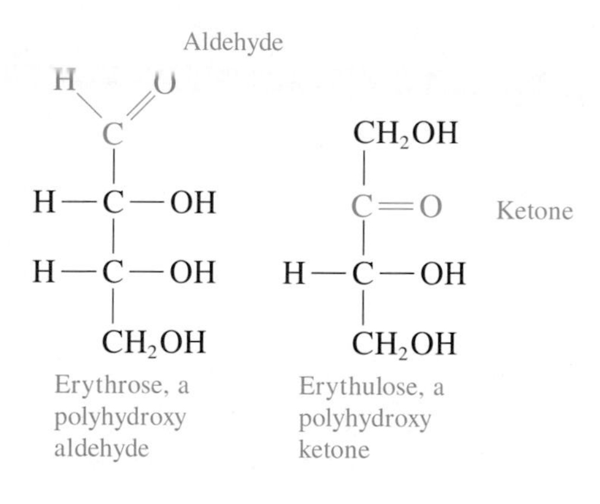

Erythrose, a polyhydroxy aldehyde

Erythulose, a polyhydroxy ketone

Most of the naturally occurring **monosaccharides** contain a chain of three to six carbon atoms. A monosaccharide with three carbon atoms is a *triose,* one with four carbon atoms is a *tetrose,* a *pentose* has five carbons, and a *hexose* contains six carbons. We can use both classification systems to indicate the type of carbonyl group and the number of carbon atoms. An aldopentose is a five-carbon monosaccharide that is an aldehyde; a ketohexose would be a six-carbon monosaccharide that is a ketone. Some examples are given below:

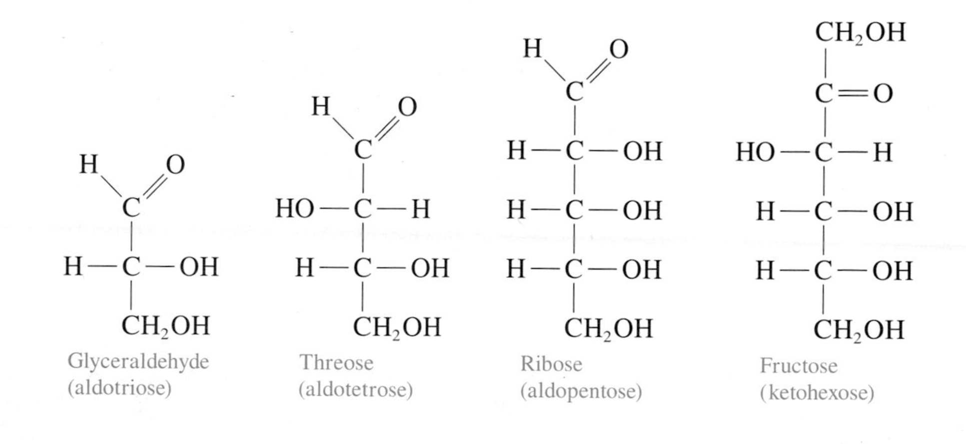
Glyceraldehyde (aldotriose)

Threose (aldotetrose)

Ribose (aldopentose)

Fructose (ketohexose)

Sample Problem 14.2
Monosaccharides

Classify each of the following monosaccharides to indicate their carbonyl group and number of carbon atoms:

a. **b.**

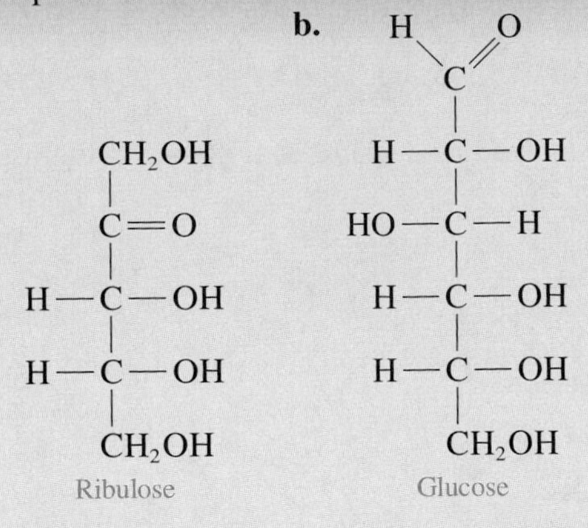

Solution

a. The structural formula has a ketone group; ribulose is a ketose. Because there are five carbon atoms, it is a pentose. Combining these classifications makes it a ketopentose.

b. The structural formula has an aldehyde group; glucose is an aldose. Because there are six carbon atoms, it is an aldohexose.

Study Check

The simplest ketose is a triose named dihydroxyacetone. Draw its structural formula.

14.3 Chiral Carbon Atoms

Learning Goal
Draw and identify Fischer projections for carbohydrate molecules.

If you look at your hands, each has a palm, a back, and the same number of fingers and thumbs. When the palms are up, your thumbs are on opposite sides. This is the reason you cannot put a left-handed glove on your right hand.

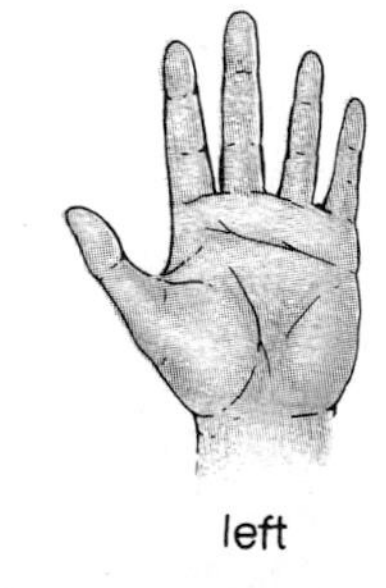

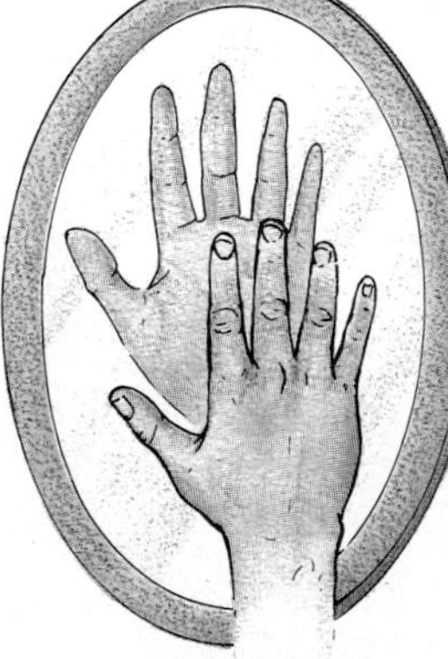

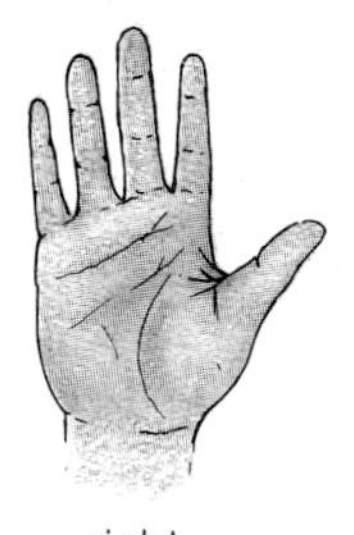

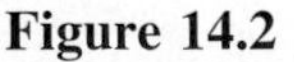

Figure 14.2
A pair of hands are chiral because they are mirror images of each other that are not superimposable.

Figure 14.3
(a) Objects that are chiral.
(b) Objects that are achiral.

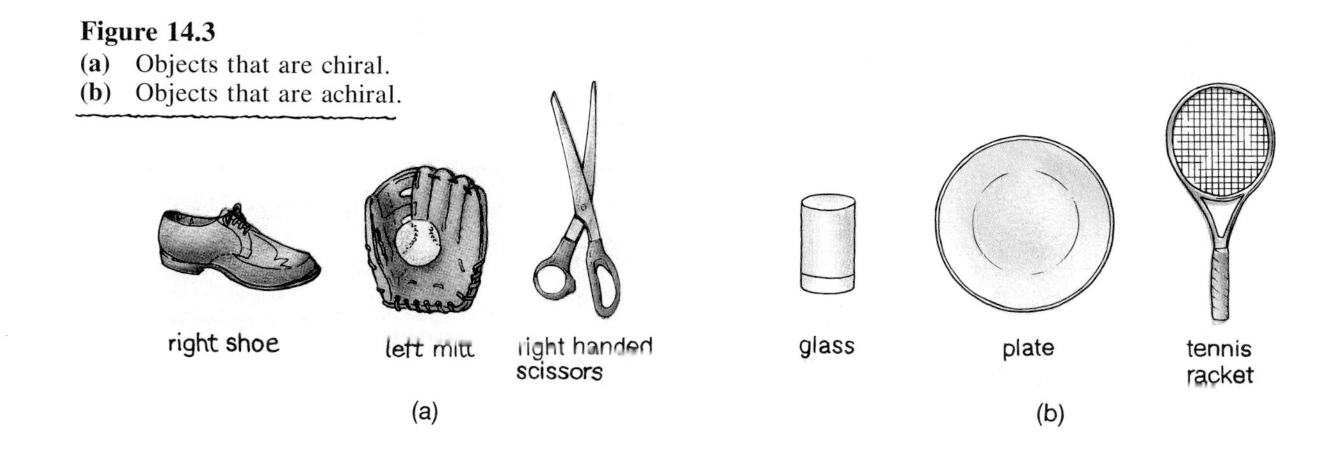

If you turn your palms toward each other, they are mirror images. Our left hand is the mirror image of the right hand. (See Figure 14.2.) We say that our hands are nonsuperimposable. Objects like hands that are nonsuperimposable mirror images are **chiral** objects. Left and right shoes are chiral; left- and right-handed golf clubs are chiral. They are mirror images of each other that cannot be superimposed. Objects such as a glass or a ball whose mirror images can be superimposed on each other are achiral. (See Figure 14.3.)

Sample Problem 14.3
Chiral and Achiral

Objects

Classify each of the following objects as chiral or achiral:
a. left ear **b.** a Styrofoam coffee cup **c.** a left-handed mitt

Solution

a. Chiral; the left ear is the mirror image of the right ear.
b. Achiral; the mirror images of a cup can be superimposed on each other.
c. Chiral; a left-handed mitt cannot be superimposed on a right-handed mitt.

Study Check

Would a bowling pin be chiral or achiral?

Chiral Carbon Atoms

A carbon compound is chiral if it has a carbon atom (a chiral carbon) attached to four different atoms or groups. It is chiral because four different atoms or groups can be attached to a carbon atom in two different ways. The resulting structures are mirror images of that molecule and nonsuperimposable. Compounds that have the same structural formula, but different spatial arrangements of the attached atoms are called **stereoisomers.** When stereoisomers are mirror images of each other, they are **enantiomers.** In Figure 14.4, we have drawn a chiral molecule and its mirror image. If we line up the hydrogen and iodine atoms, the bromine and chlorine atoms appear on opposite sides.

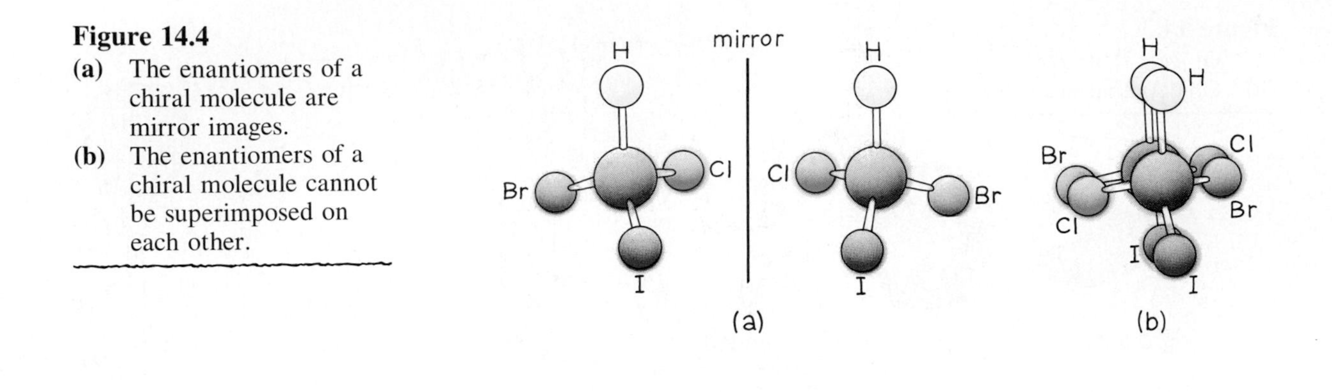

Figure 14.4
(a) The enantiomers of a chiral molecule are mirror images.
(b) The enantiomers of a chiral molecule cannot be superimposed on each other.

Sample Problem 14.4
Chiral Carbon Atoms

Indicate whether the central carbon atom in each of the following compounds is chiral or achiral; if chiral draw its mirror image.

a. Cl (top), H—C—H, Br (bottom)

b. CH_3 (top), H—C—OH, Cl (bottom)

c. CH_3 (top), HO—C—H, CH_2OH (bottom)

Solution

a. This molecule is achiral because two of the atoms attached to the carbon atom are the same (two hydrogens).

b. This molecule is chiral because there are four different atoms or groups attached to the central carbon. The mirror image would be written as:

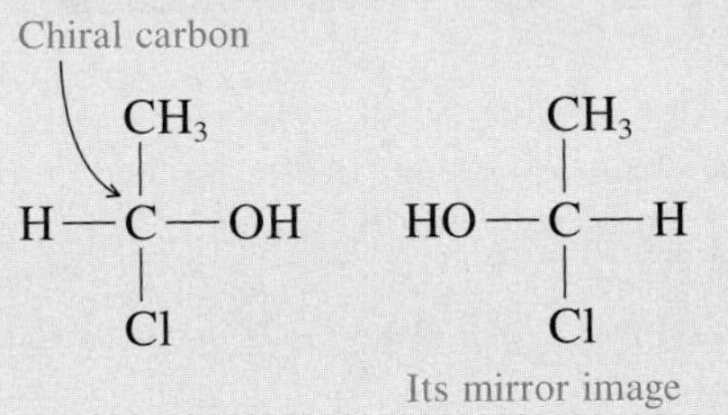

c. The central carbon atom is chiral because there are four different groups attached. Its mirror image is written as:

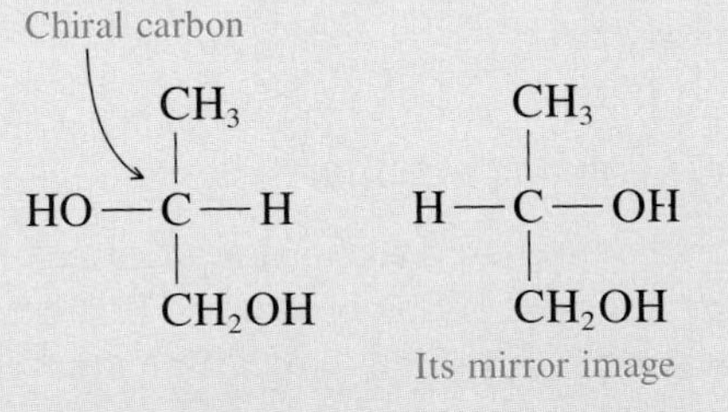

Study Check
Identify the chiral carbon atom in 2-butanol.

Fischer Projections

In a three-dimensional view of glyceraldehyde, the simplest sugar, the bonds that extend back from the chiral carbon are shown as dotted wedges; bonds that extend to the front are solid wedges. In a two-dimensional view called a **Fischer projection,** the bonds that extend to the front are drawn as horizontal lines to the left and right sides of the chiral atoms. The bonds that extend back are drawn as vertical lines. The chiral carbon is the intersection of the horizontal and vertical lines.

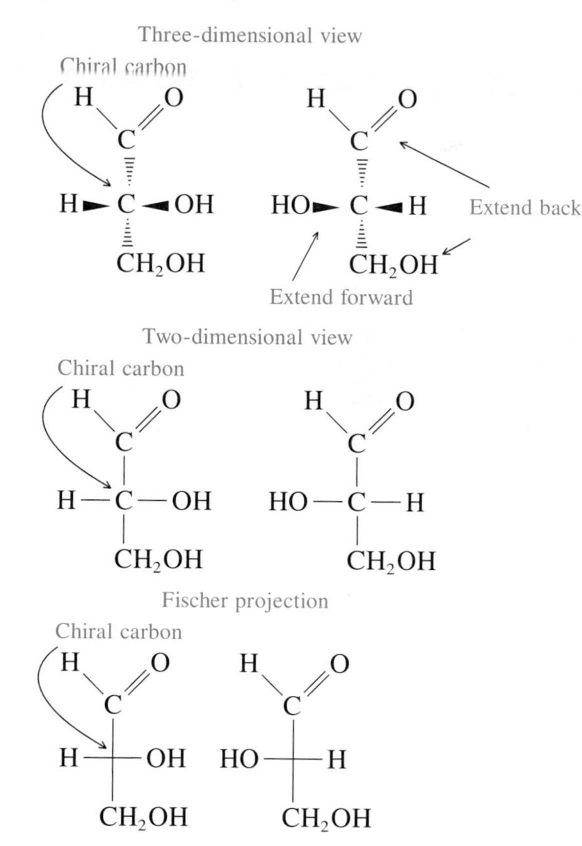

D and L Notation for Enantiomers

One way to distinguish between enantiomers is to label the molecules as the left- or right-handed form. By convention, the Fischer projection of glyceraldehyde is written with the carbonyl group at the top (carbon 1) and the —CH_2OH group at the bottom. The letter L is assigned to the stereoisomer if the —OH group is on the left of the chiral carbon. In D-glyceraldehyde, the —OH is on the right.

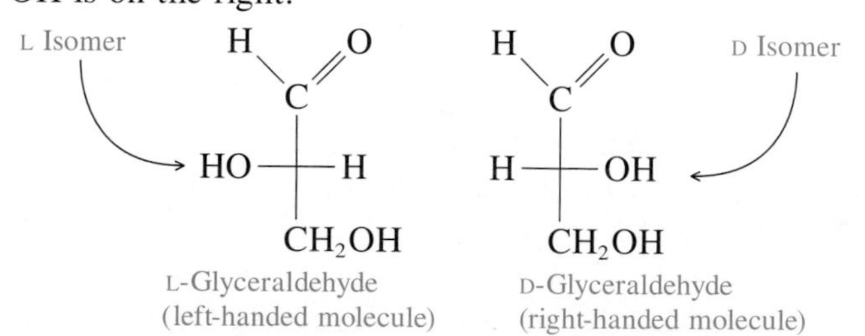

L-Glyceraldehyde (left-handed molecule)

D-Glyceraldehyde (right-handed molecule)

Most carbohydrates have longer carbon chains and several chiral carbon atoms. Then the chiral carbon furthest from the carbonyl group determines the D or L isomer. For example, in glucose, carbon 5 is the reference chiral carbon in the chain (carbon 6 in $-CH_2OH$ is not chiral).

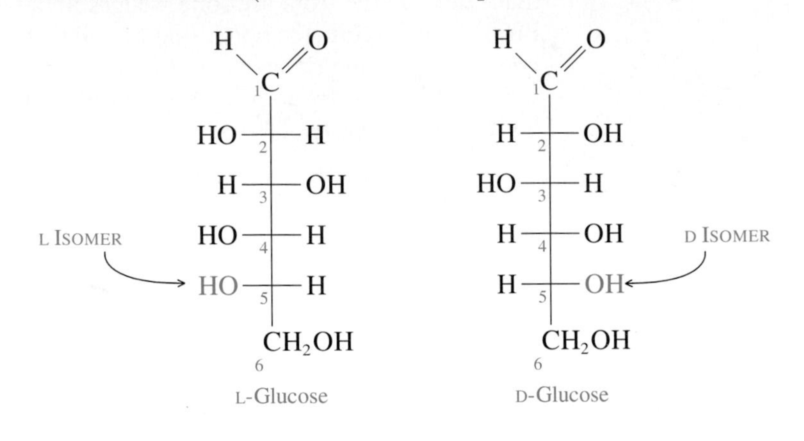

Sample Problem 14.5
Identifying D and L Isomers of Sugars

Is the following structure of ribose the D or L enantiomer?

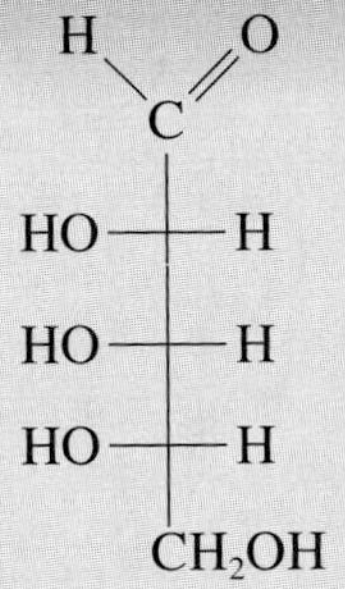

H, O, C
HO—H
HO—H
HO—H
CH_2OH

Solution

In ribose, carbon 4 is the chiral atom furthest from the carbonyl group. Because the hydroxyl group on carbon 4 is on the left, this enantionmer is L-ribose.

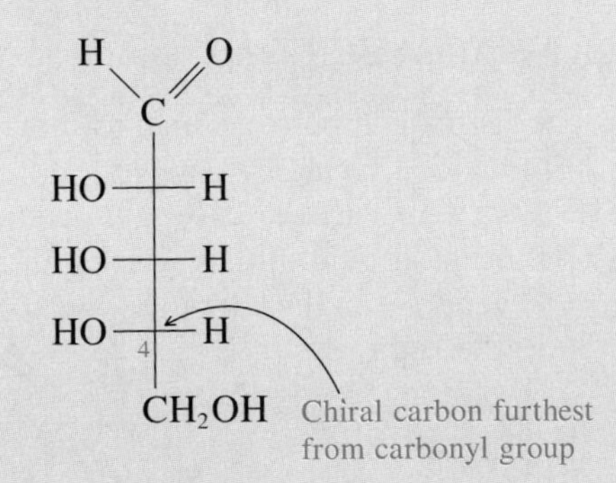

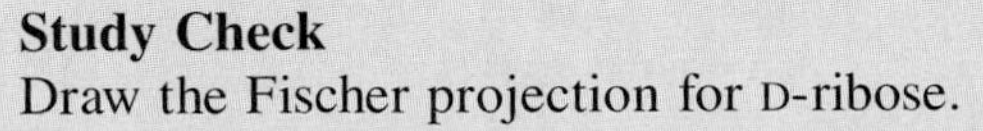

Study Check

Draw the Fischer projection for D-ribose.

HEALTH NOTE *Chiral Compounds in Nature*

Most stereoisomers that are active in biological systems consist of only one of the enantiomers. Rarely are both the D and L forms of a biomolecule biologically active. This happens because our enzymes and cell surface receptors on which metabolic reactions take place also have "handedness." Thus, only one of the enantiomers of a reactant or a drug interacts with those enzymes or receptors; the other is inactive. For example, only the D isomers of the monosaccharides of glucose, galactose, and fructose are active. Among the amino acids, only the L isomers such as L-alanine, or L-lysine are used to make proteins.

A substance called carvone exists as D and L isomers. The L-carvone isomer gives the flavor to spearmint oil, and its enantiomer D-carvone is the flavor of caraway seed oil. Thus, our senses of taste and smell are sensitive to the chirality of molecules. In the brain, D-LSD causes hallucinations because it affects the production of serotonin, a chemical that is important in sensory perception. However, its enantiomer L-LSD produces little effect in the brain.

In the past, a drug contained both enantiomers, even though one was not useful, or might even have been toxic. For example, in the late 1950s, a drug called thalidomide was prescribed as a tranquilizer for pregnant women in some countries (not the United States). Its tragic effects on developing embryos was soon apparent as babies were born with deformed limbs. Scientists have determined that only one enantiomer of thalidomide produced the birth defects, but its mirror image was useful as a tranquilizer.

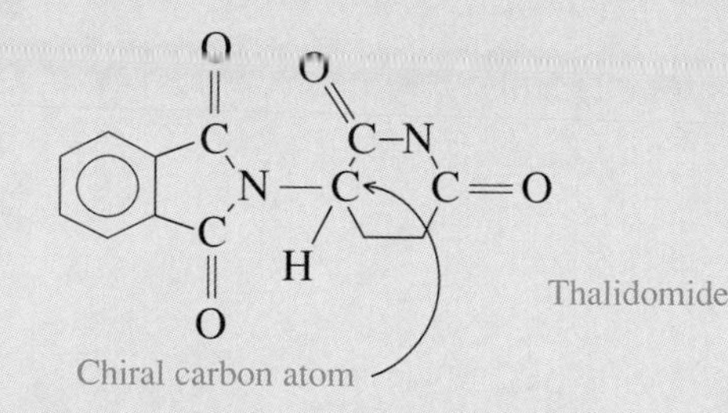

Today, drug researchers are beginning to apply *chiral technology* to produce the active enantiomers or chiral drugs. The benefits of producing the active enantiomer include reducing the dose, enhancing activity, and reducing interactions with other drugs and harmful side effects. A substance used to treat Parkinson's disease is L-dopamine; its enantiomer D-dopamine has no biological effect. Even the active enantiomer of the popular analgesic ibuprofen, the active ingredient in Advil, Motrin, and Nuprin, is expected to appear on the market soon.

Optical Activity

The different spatial arrangements of stereoisomers make them react in different ways with plane-polarized light. When a beam of light passes through a polarizer, the emerging polarized light vibrates in only one plane. When the polarized light passes through a solution of a chiral compound, the plane of polarized light rotates and emerges at an angle to the original plane. (See Figure 14.5.) Called an **optical isomer,** one chiral compound can rotate polarized light to the right (dextrorotatory), whereas its enantiomer rotates light to the left (levorotatory).

Figure 14.5
When plane-polarized light passes through a sample of an optical isomer (a chiral molecule), it emerges at a different angle.

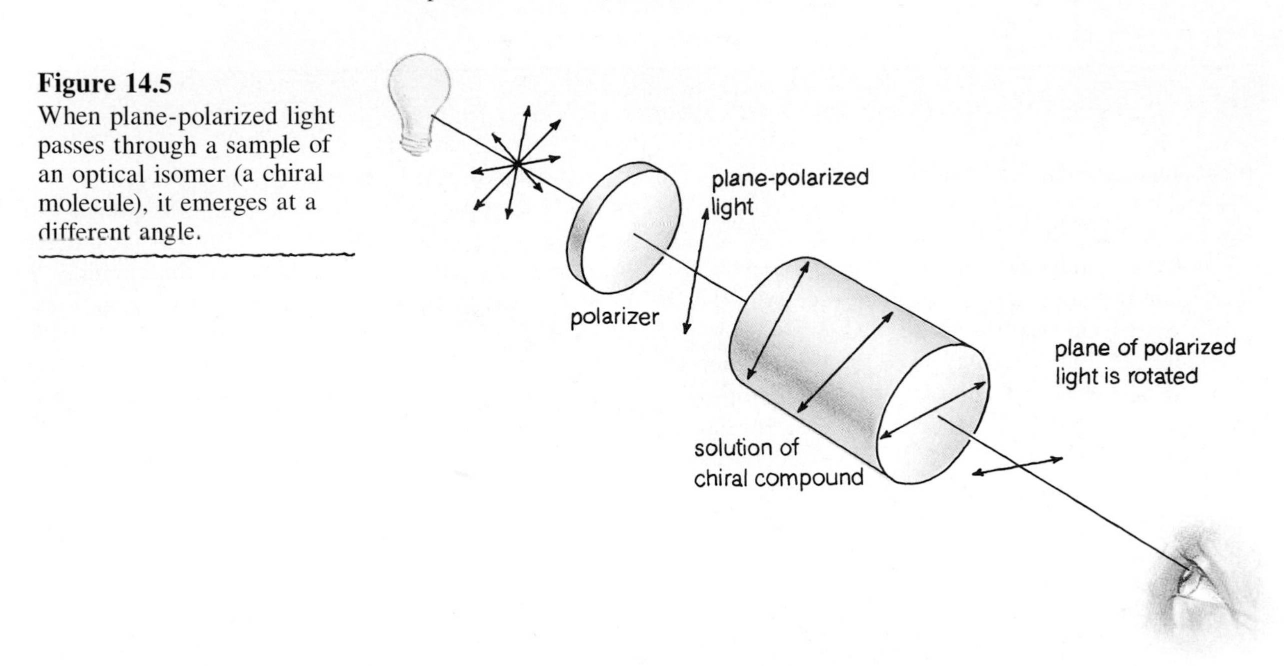

14.4 Structures of Some Important Monosaccharides

Learning Goal
Draw the open-chain structures for D-glucose, D-galactose, and D-fructose.

The hexoses glucose, galactose, and fructose are important monosaccharides. Although structural formulas for both the D and L isomers can be written for these hexoses, only the D isomers shown below occur naturally.

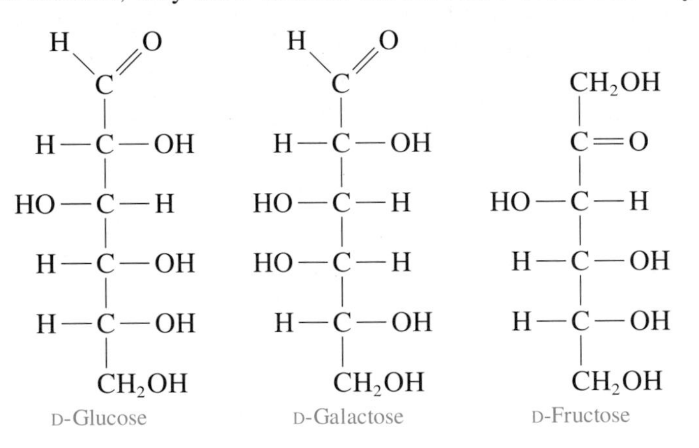

D-Glucose

The most important hexose is D-glucose, which has the molecular formula $C_6H_{12}O_6$. Also known as dextrose, **glucose** is found in fruits, vegetables, corn syrup, and honey. It is a building block of the disaccharides sucrose, lactose, and maltose and is the monosaccharide unit in polysaccharides such as starch, cellulose, and glycogen.

In the body, glucose normally has a concentration of 70–90 mg/dL (100 mL) of blood. However, the amount of glucose depends on the time that has passed since eating. In the first hour after a meal, the level of glucose rises to about 130 mg/dL (100 mL) of blood, and then it decreases over the next

2–3 hours as it is used in the tissues. Some glucose is converted to glycogen and stored in the liver and muscle.

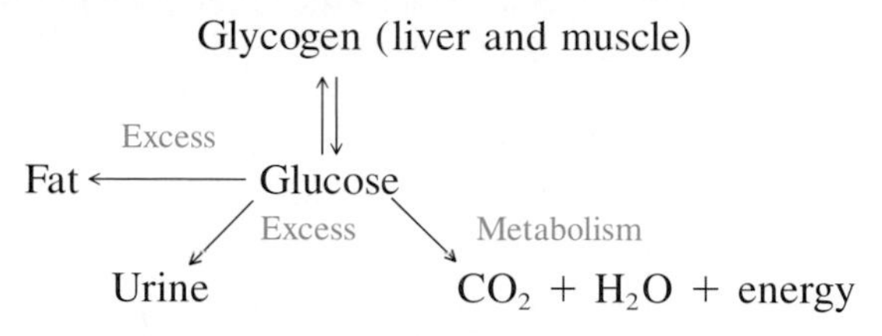

HEALTH NOTE *Hyperglycemia and Hypoglycemia*

A doctor may order a glucose tolerance test to evaluate the body's ability to return to normal glucose concentration in response to the ingestion of a specified amount of glucose. The patient fasts for 12 hours and then drinks a solution containing 100 g of glucose. A blood sample is taken immediately, followed by more blood samples each half-hour for 2 hours, and then every hour for a total of 5 hours. If the blood glucose exceeds 140 mg/dL (1 dL = 100 ml) plasma and remains high, hyperglycemia may be indicated. An example of a disease that can cause hyperglycemia is diabetes mellitus, which occurs when the pancreas is unable to produce sufficient quantities of insulin. As a result, glucose levels in the body fluids can rise as high as 350 mg/dL plasma. Symptoms of diabetes in people under the age of 40 include thirst, excessive urination, increased appetite, and weight loss. In older persons, diabetes is sometimes a consequence of excessive weight gain.

When a person is hypoglycemic, the blood glucose level rises and then decreases rapidly to levels as low as 40 mg/dL plasma. In some cases, hypoglycemia is caused by overproduction of insulin by the pancreas. Low blood glucose can cause dizziness, general weakness, and muscle tremors. A diet may be prescribed that consists of several small meals high in protein and low in carbohydrate. Some hypoglycemic patients are finding success with diets that include more complex carbohydrates rather than simple sugars. (See Figure 14.6.)

Figure 14.6
(a) In a glucose tolerance test, 100 g of glucose is given to a patient.
(b) Typical blood glucose levels for normal, hyperglycemic, and hypoglycemic conditions.

(a)

(b)

D-Galactose

Galactose is an aldohexose that does not occur free in nature. It is obtained as a hydrolysis product of the disaccharide lactose, a sugar found in milk and milk products. Galactose is the prevalent monosaccharide in the cellular membranes of the brain and nervous system. The only difference in the structures of D-glucose and D-galactose is the arrangement of the —OH group on carbon 4.

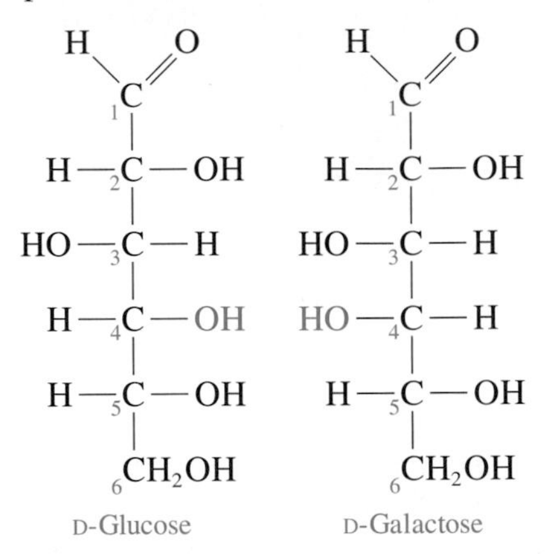

In a condition called *galactosemia,* an enzyme needed to convert galactose to glucose is missing. The accumulation of galactose in the blood and tissues leads to cataracts, mental retardation, and cirrhosis. The treatment for galactosemia is the removal of all galactose-containing foods, mainly milk and milk products, from the diet. If this is done for an infant immediately after birth, no ill effects occur.

D-Fructose

In contrast to glucose and galactose, **fructose** is a ketohexose. The structure of fructose differs from glucose at carbons 1 and 2 by the location of the carbonyl group.

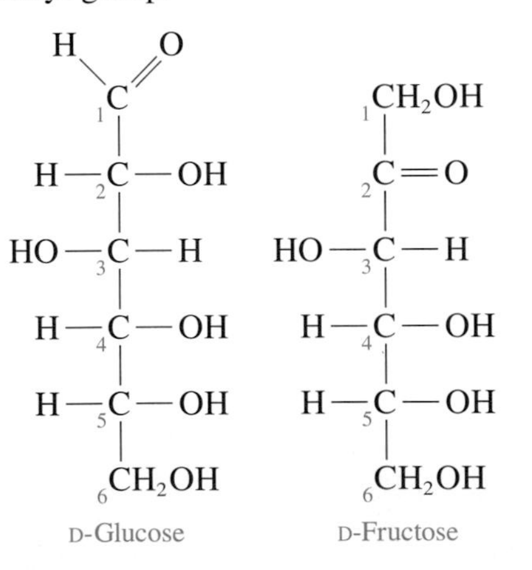

Fructose is the sweetest of the carbohydrates, twice as sweet as sucrose (table sugar). This makes fructose popular with dieters since less fructose, and therefore fewer calories, are needed to provide a pleasant taste. After

Figure 14.7
Some common carbohydrates in our diets are fructose in fruit and honey, glucose in corn syrup, lactose in milk and milk products, and sucrose in table sugar.

fructose enters the bloodstream, it is converted to its isomer, glucose. Fructose is found in fruit juices and honey; it is also called levulose and fruit sugar. Fructose is also obtained as a hydrolysis product of sucrose, the disaccharide known as table sugar. (See Figure 14.7.)

14.5 Cyclic Structures of Monosaccharides

Learning Goal
Draw and identify the cyclic structures of monosaccharides.

When an aldehyde or ketone reacts with an alcohol, a **hemiacetal** or **hemiketal** is formed.

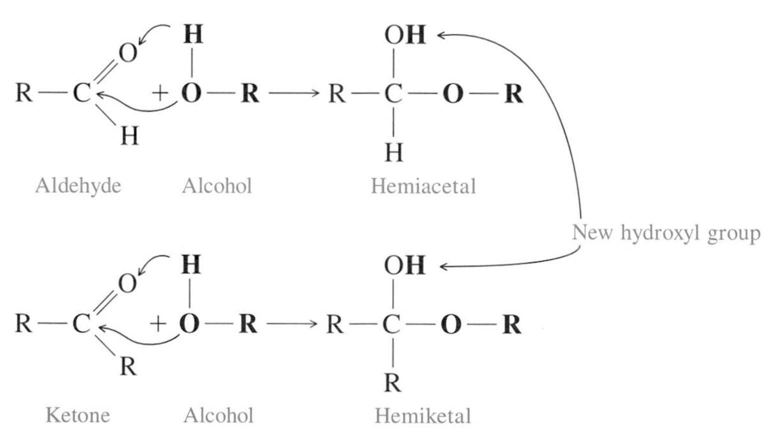

Up to now in this chapter, we have represented monosaccharides by their open-chain linear structural formulas. However, most of the time, chains of five- or six-carbon atoms exist as stable ring structures. In an aldohexose, the hydroxyl group on carbon 5 reacts easily with the aldehyde group at the other end of the molecule. The product is a cyclic hemiacetal structure with six atoms, the most prevalent form of the monosaccharides. Let us look at the formation of the cyclic structures of glucose.

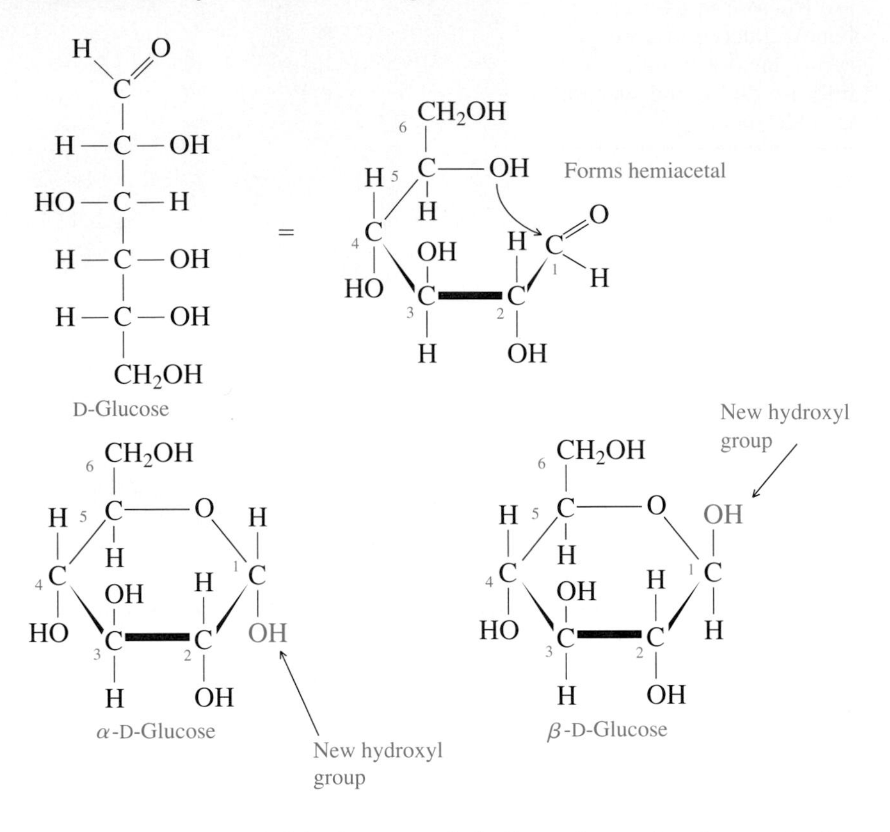

The structure at the left is the open-chain linear form of D-glucose. In the second structure, the open-chain folds into a hexagon to bring the hydroxyl group on carbon 5 close to carbon 1 in the aldehyde group. In this form, also called a **Haworth structure,** all the —OH groups drawn on the right of the linear form appear below the ring. The intramolecular reaction between the alcohol group on carbon 5 and the aldehyde group forms the cyclic hemiacetal. Sometimes, the cyclic structure is simplified to show only the position of the hydroxyl groups on the six-atom ring structure.

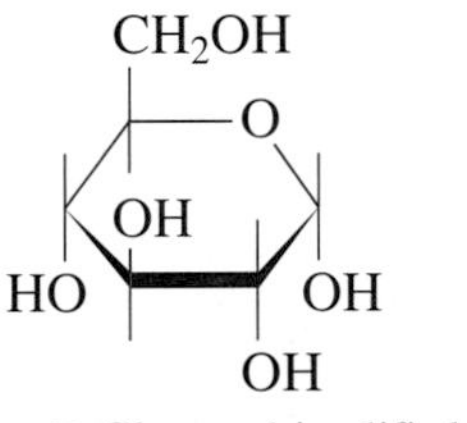

α-D-Glucose (simplified structure)

α and β Anomers

In the cyclic hemiacetal structures, a new hydroxyl group appears on carbon 1. Because the new —OH at carbon 1 can be up or down, two isomers called

anomers are formed. When the —OH on carbon 1 is down, the isomer is the α form of the sugar. In the β anomer, the —OH on carbon 1 is up. Such differences in structural forms may seem trivial. However, we can digest starch products such as pasta to obtain glucose because the polysaccharide contains the α isomers of glucose. We cannot digest paper or wood because cellulose consists of only β-D-glucose units. Humans have an α-amylase for starches but not a β-amylase for cellulose.

Mutarotation

In solution, the α-glucose can convert to β-glucose and back again. In a process called **mutarotation,** the hemiacetal of each isomer converts from the closed ring to the open chain and back again. As the ring opens and closes, the hydroxyl (—OH) group on carbon 1 can shift between the α and the β position. Although the open chain is an essential part of mutarotation, only a small amount is present at any given time.

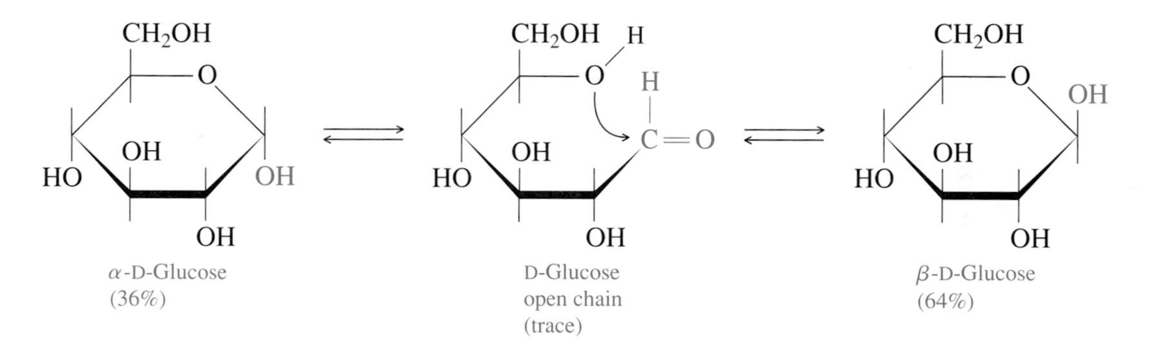

α-D-Glucose (36%)

D-Glucose open chain (trace)

β-D-Glucose (64%)

Cyclic Structures of Galactose

Galactose is an aldohexose like glucose, differing only in the arrangement of the —OH group on carbon 4. Thus, its cyclic structure is also similar to glucose, except that in galactose the —OH on carbon 4 is up. With the formation of a new hydroxyl group on carbon 1, galactose also exists as α and β anomers and undergoes mutarotation with the open chain form in solution.

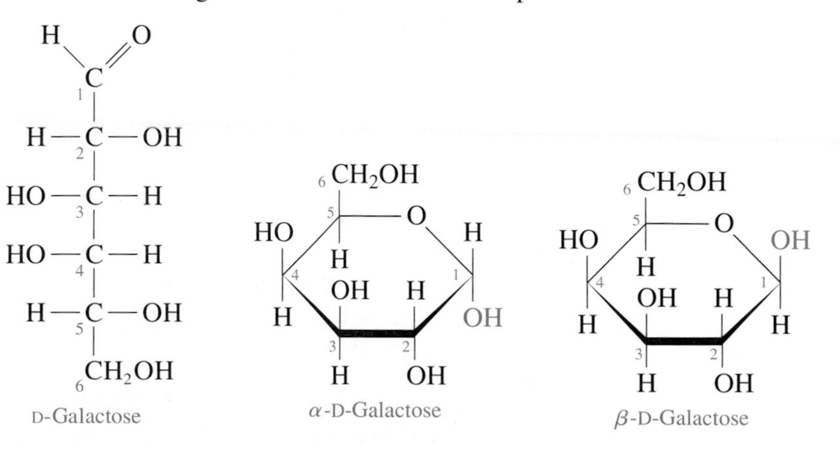

D-Galactose

α-D-Galactose

β-D-Galactose

Cyclic Structures of Fructose

In contrast to glucose and galactose, fructose is a ketohexose. It forms a hemiketal when a hydroxyl group on carbon 5 reacts with the ketone group. The cyclic structure for fructose is a five-atom ring with carbon 2 at the right corner. There is a new hydroxyl group on carbon 2 along with the carbon 1 in the CH_2OH group. There are also α and β anomers of fructose that undergo mutarotation in solution.

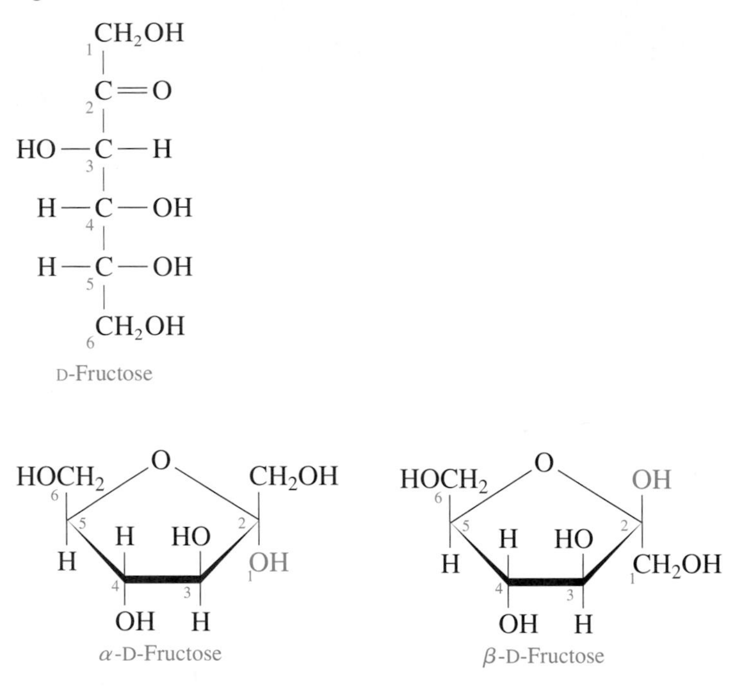

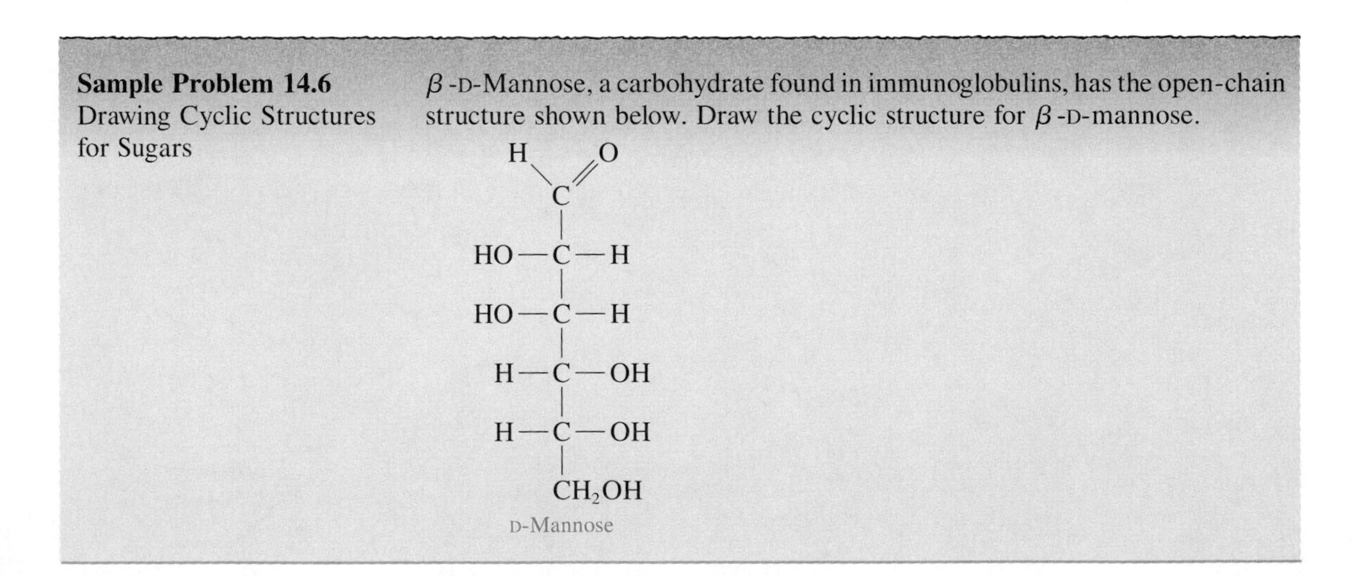

Sample Problem 14.6
Drawing Cyclic Structures for Sugars

β-D-Mannose, a carbohydrate found in immunoglobulins, has the open-chain structure shown below. Draw the cyclic structure for *β*-D-mannose.

Solution

First, number the carbon atoms in the open chain starting at the aldehyde group.

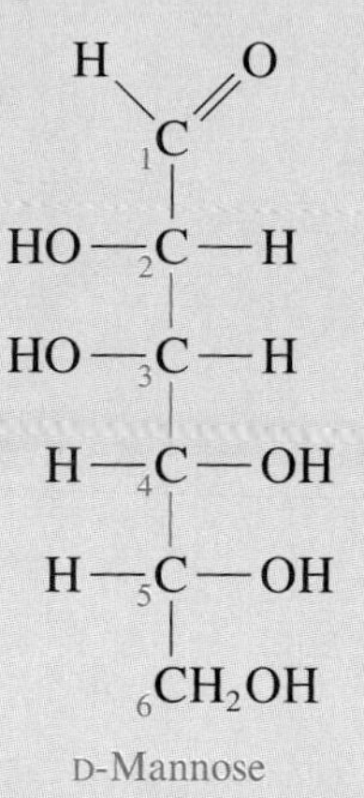

D-Mannose

Draw a six-atom hexagon containing carbon atoms 1–5 and an oxygen atom. Number the carbon atoms in the cyclic structure from the right corner in clockwise order. Place the CH_2OH group (carbon 6) above carbon 5.

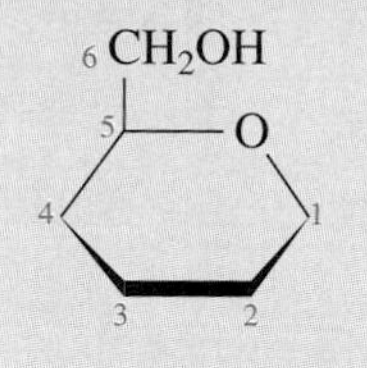

Draw the —OH groups on the left of the open chain above the ring, and the —OH groups on the right below. At carbon 1, write the new hydroxyl group upward to make the β-D-mannose anomer.

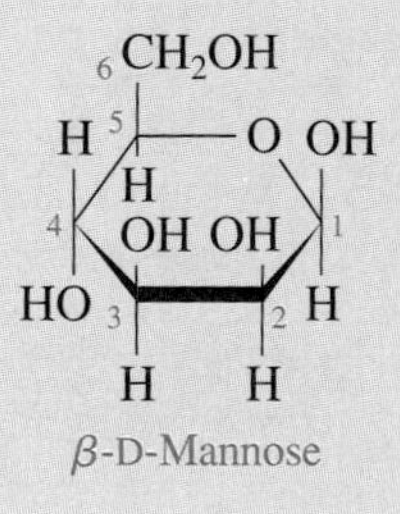

β-D-Mannose

Study Check

Draw the cyclic structure for α-D-glucose.

Oxidation of Monosaccharides

Monosaccharides are also called **reducing sugars** because they easily oxidize to yield carboxylic acids. Although they exist primarily in the cyclic forms, the aldehyde group of the open chain reacts with oxidizing agents such as Benedict's reagent. (See Chapter 13.) All monosaccharides are reducing sugars.

When Benedict's solution is used, the Cu^{2+} ions are reduced to Cu^{+} in Cu_2O.

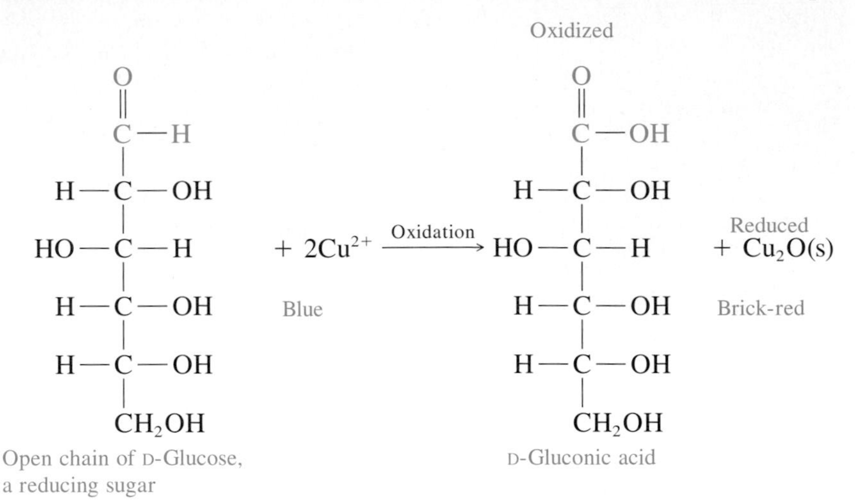

Open chain of D-Glucose, a reducing sugar
D-Gluconic acid

Fructose, a hemiketal, is also a reducing sugar. In the open-chain form, a rearrangement between the hydroxyl group on carbon 1 and the ketone group provides an aldehyde group that can be oxidized.

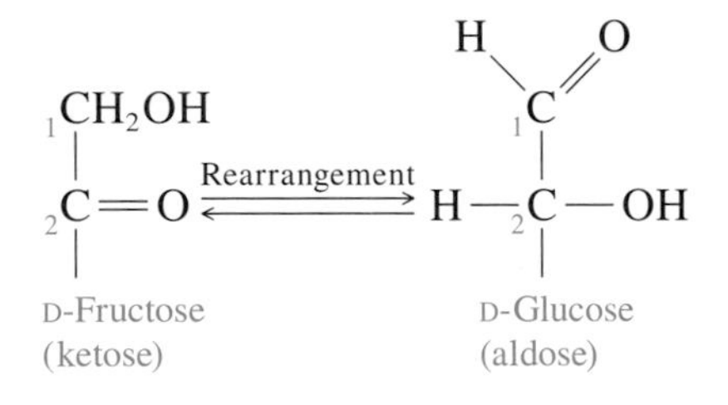

D-Fructose (ketose)
D-Glucose (aldose)

HEALTH NOTE *Testing for Glucose in Urine*

Normally, blood glucose flows through the kidneys and is reabsorbed into the bloodstream. However, if the blood level exceeds about 160 mg of glucose/dL (100 mL) of blood, the kidneys cannot reabsorb it all, and glucose spills over into the urine, a condition known as glucosuria. A symptom of diabetes mellitus is a high level of glucose in the urine.

Benedict's test can be used to determine the presence of glucose in urine. The amount of cuprous oxide (Cu_2O) formed is proportional to the amount of reducing sugar present in the urine. Low to moderate levels of reducing sugar turn the solution green; solutions with high glucose levels turn Benedict's yellow or bright orange. Table 14.1 lists some colors associated with the concentration of glucose in the urine.

Table 14.1 *Glucose Test Results*

	Glucose Present in Urine	
Color	*%*	*mg/dL*
Blue	0	0
Blue-green	0.25	250
Green	0.50	500
Yellow	1.00	1000
Orange	2.00	2000

In another clinical test that is more specific for glucose, the enzyme glucose oxidase is used. The oxidase enzyme converts glucose to gluconic acid and hydrogen peroxide, H_2O_2. The peroxide produced reacts with a dye in the urine test strip to give different colors. The level of glucose present in the urine is found by matching the color produced to a color chart on the container. (See Figure 14.8.)

Figure 14.8
Blood glucose test: blood glucose level is determined by comparing the color of a urine test strip with the color chart on the container.

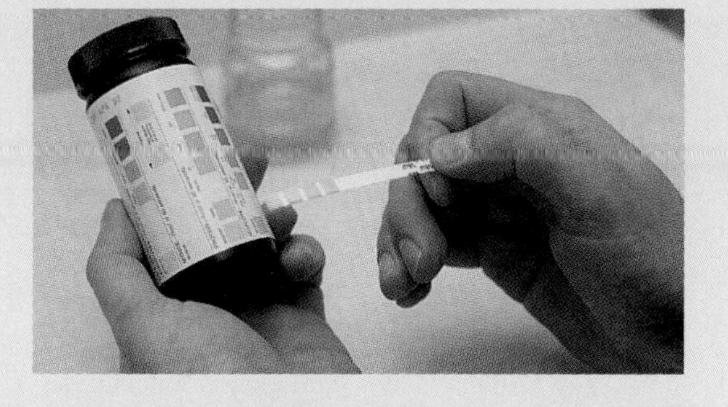

Sample Problem 14.7
Reducing Sugars

Why is D-glucose called a *reducing sugar?*

Solution
D-glucose is easily oxidized by Benedict's or Tollens' solution. A carbohydrate that reduces a metal ion (Cu^{2+} to Cu^+ or Ag^+ to Ag) is called a reducing sugar.

Study Check
A test using Benedict's reagent turns orange with a urine sample. According to Table 14.1, what might this result indicate?

14.6 Disaccharides

Learning Goal
Describe the monosaccharide units and linkages in disaccharides.

A disaccharide is composed of two monosaccharides linked together. The most common disaccharides are maltose, lactose, and sucrose. Their hydrolysis gives the following monosaccharides.

$$\text{Maltose} + H_2O \longrightarrow \text{glucose} + \text{glucose}$$

$$\text{Lactose} + H_2O \longrightarrow \text{glucose} + \text{galactose}$$

$$\text{Sucrose} + H_2O \longrightarrow \text{glucose} + \text{fructose}$$

Glycosides (Acetals)

In their cyclic forms, the hydroxyl group on the anomeric carbon of a monosaccharide reacts with an alcohol to give an **acetal.** These acetal products of monosaccharides are called **glycosides,** and the linkage between the hydroxyl groups is called a **glycosidic bond.** The glycosidic bond can be in the α or β position, depending on the anomer that reacted.

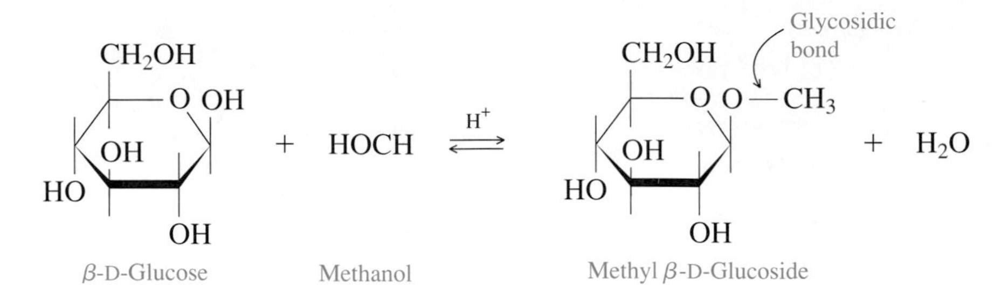

In the acetal, the glycosidic bond does not open. Thus, neither mutarotation nor oxidation can occur.

Maltose

Maltose, or malt sugar, is a disaccharide obtained from the hydrolysis of starch. When maltose in barley and other grains is hydrolyzed by yeast enzymes, glucose is obtained that can undergo fermentation to give ethanol. Maltose is used in cereals, candies, and the brewing of beverages.

In maltose, the hydroxyl group on carbon 1 of one glucose molecule reacts with the hydroxyl group on carbon 4 of the second glucose molecule. A glycosidic bond joins the two glucose molecules and a molecule of water is eliminated. The glycosidic bond is designated as an α-1,4- to show that —OH on carbon 1 of the α anomer is joined to the —OH on carbon 4 of the second glucose. Because the second glucose molecule has a free —OH on the anomeric carbon, there are α and β anomers of maltose. This anomeric carbon also opens up to give a free aldehyde group to oxidize, which means that maltose is also a reducing sugar.

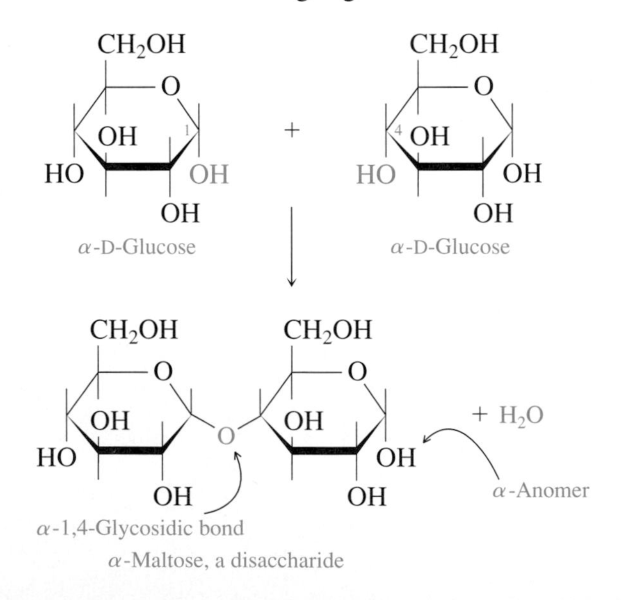

Lactose

Lactose, milk sugar, is a disaccharide found in milk and milk products. It makes up 6–8% of human milk and about 4–5% of cow's milk and is used in products that attempt to duplicate mother's milk. Some people do not produce sufficient quantities of the enzyme needed to hydrolyze lactose, and the sugar remains undigested, causing abdominal cramps and diarrhea. In some commercial milk products, an enzyme called lactase is added to assist the hydrolysis of lactose.

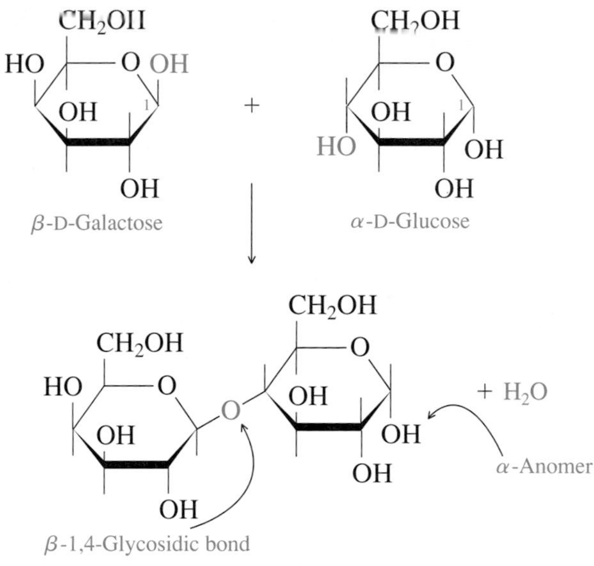

α-Lactose, a disaccharide

Lactose consists of β-D-galactose and D-glucose joined by a β-1,4-glycosidic bond. The free hydroxyl group on the anomeric carbon of the glucose gives α- and β-lactose. Although the glycosidic bond ties up the anomeric carbon in galactose, the anomeric carbon of the glucose can open to give an aldehyde group. Thus, lactose is a reducing sugar.

Figure 14.9
Sucrose, one of the most abundant carbohydrates, is obtained from sugar cane and sugar beets.

Sucrose

The disaccharide **sucrose,** table sugar, is a product of sugar cane and sugar beets. See Figure 14.9. Some estimates indicate that each person in the United States consumes an average of 100 lb of sucrose every year either by itself or in a variety of food products.

Sucrose consists of α-D-glucose and β-D-fructose molecule joined by an α,β-1,2-glycosidic bond. The structure of sucrose differs from the other disaccharides because the glycosidic bond ties up the anomeric carbons of both monosaccharide units. There are no isomers for sucrose. Without a free aldehyde group, it cannot react with Benedict's or Tollens' reagent; sucrose is not a reducing sugar.

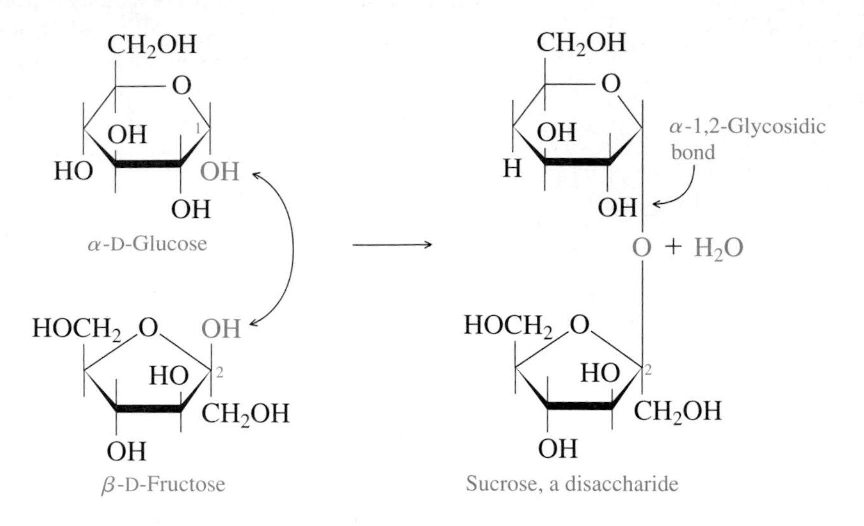

HEALTH NOTE *How Sweet Is My Sweetener?*

Although many of the monosaccharides and disaccharides taste sweet, there is considerable difference in the degree of sweetness. Dietetic foods contain sweeteners that are noncarbohydrate or carbohydrates that are sweeter. Some examples of sweeteners compared with sucrose are shown in Table 14.2.

Table 14.2 ***Relative Sweetness of Sugars and Artificial Sweeteners***

	Sweetness Relative to Sucrose (= 100)
Monosaccharides	
Galactose	30
Glucose	75
Fructose	175
Disaccharides	
Lactose	16
Maltose	33
Sucrose	100 ⟵ reference standard
Artificial Sweeteners (noncarbohydrate)	
Aspartame	18,000
Saccharin	45,000

Fermentation

When treated with yeast, the hexoses glucose and fructose (but not galactose) will undergo **fermentation** to produce ethanol and carbon dioxide gas.

$$\underset{\text{Glucose or fructose}}{C_6H_{12}O_6} \xrightarrow{\text{Yeast enzymes}} \underset{\text{Ethanol}}{2C_2H_5OH} + 2\,CO_2(g)$$

The disaccharides maltose and sucrose also undergo fermentation because yeast contains enzymes for their hydrolysis to give glucose and fructose. However, lactose will not ferment because the enzyme lactase required for its hydrolysis is not present in yeast. Fermentation does not occur with polysaccharides.

Sample Problem 14.8
Glycosidic Bonds in Disaccharides

Melebiose is a disaccharide that has a sweetness of about 30 compared with sucrose (= 100).

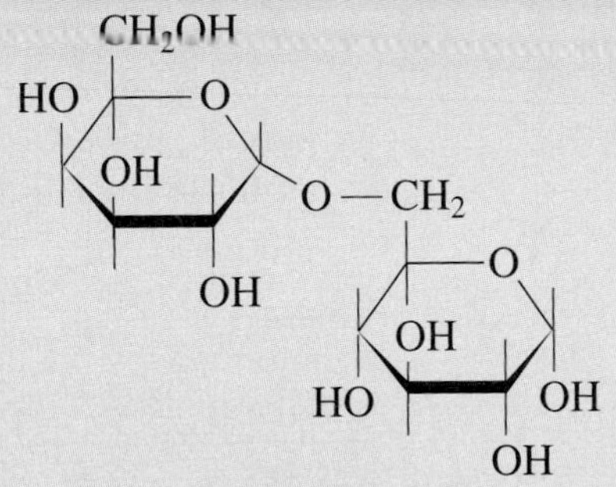

Melebiose

a. What are the monosaccharide units in melebiose?
b. How is the glycosidic bond designated?
c. Is the compound α- or β-melebiose?

Solution

a. The monosaccharide on the left side is α-D-galactose; on the right is α-D-glucose.
b. The monosaccharide units are linked by a α-1,6-glycosidic bond.
c. The downward position of the anomeric OH makes it α melebiose.

Study Check

Cellobiose is a disaccharide composed of two glucose molecule linked by a β-1,4-glycosidic linkage. Draw a structural formula for β-cellobiose.

HEALTH NOTE *Blood Types and Carbohydrates*

Every individual's blood can be typed as one of four blood groups, A, B, AB, and O. A group of three or four monosaccharides on the surface of a red blood cell acts as a marker, which allows the body to know its own cells and to produce antibodies that attack foreign cells.

Three monosaccharides on the surface of red blood cells characterize the type O blood group. In type A blood, a fourth monosaccharide, *N*-acetylgalactosamine, is bonded to galactose. In type B blood, there is a second unit of galactose. Type AB blood contains both. The genetic markers that determine blood type also appear in saliva and semen and are used to establish paternity or the presence of suspects from blood at the scene of a crime. The structures of the monosaccharides that

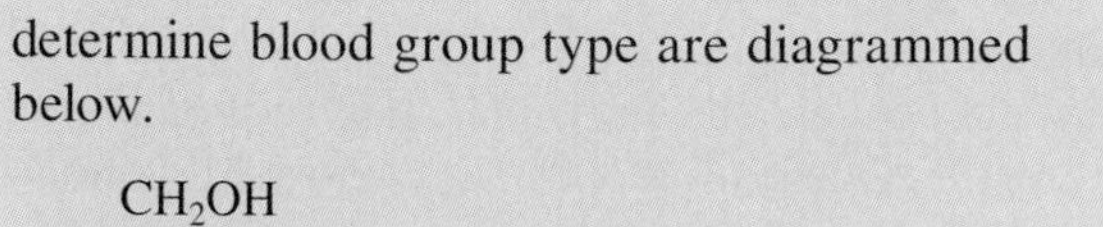
determine blood group type are diagrammed below.

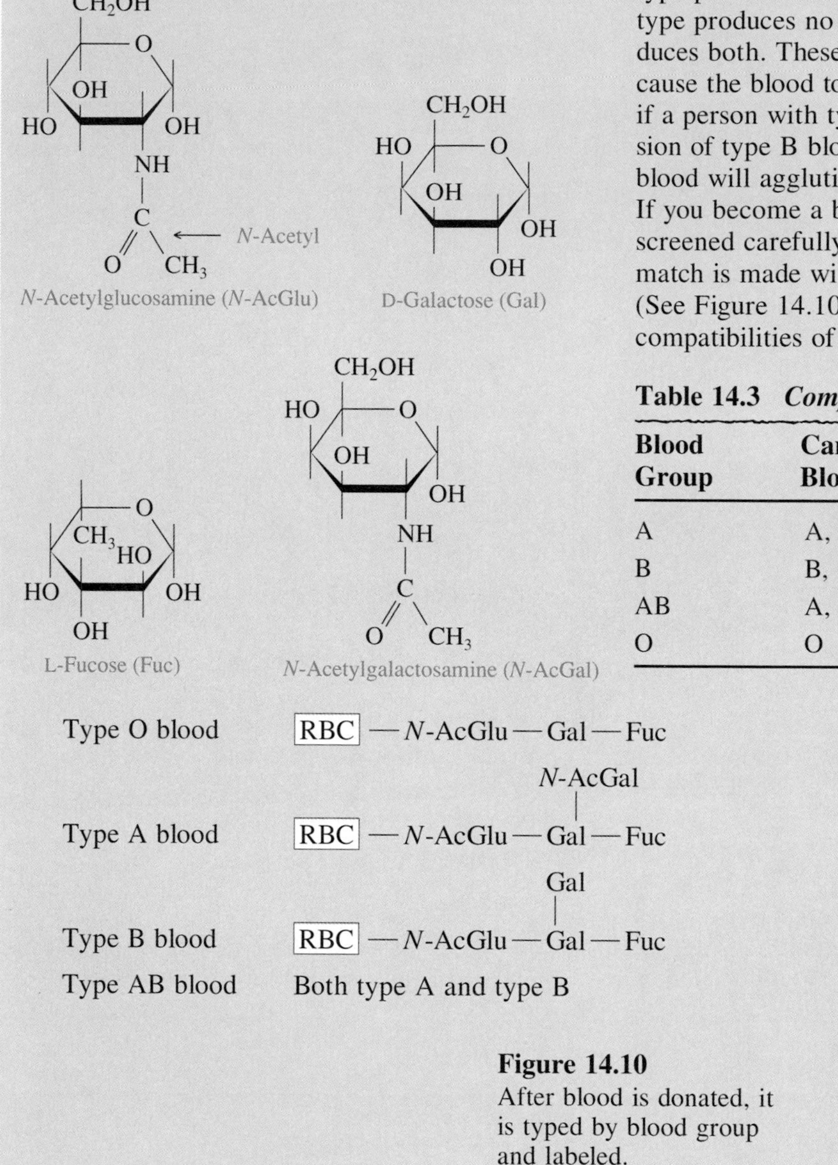

A person with A blood type produces antibodies against B type, whereas a person with B blood type produces antibodies against A. An AB blood type produces no antibodies, whereas O type produces both. These antibodies called agglutinins cause the blood to coagulate or agglutinate. Thus, if a person with type A blood receives a transfusion of type B blood, factors in the recipient's blood will agglutinate the donor's red blood cells. If you become a blood donor, your blood is screened carefully to make sure that an exact match is made with the blood type of the recipient. (See Figure 14.10.) Table 14.3 summarizes the compatibilities of blood groups for transfusion.

Table 14.3 ***Compatibility of Blood Groups***

Blood Group	Can Receive Blood Types	Cannot Receive Blood Types
A	A, O	B, AB
B	B, O	A, AB
AB	A, B, AB, O	None
O	O	A, B, AB

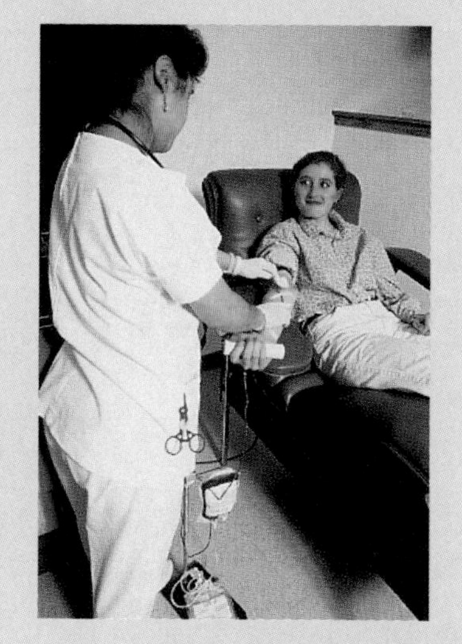

Figure 14.10
After blood is donated, it is typed by blood group and labeled.

14.7 Polysaccharides

Learning Goal
Describe the structural features of amylose, amylopectin, glycogen, and cellulose.

A polysaccharide is a polymer of many monosaccharides joined together. Three important polysaccharides—starch, cellulose, and glycogen—are all polymers of D-glucose, which differ only in the type of glycosidic bonds and the amount of branching in the molecule.

Plant Starch: Amylose and Amylopectin

Starch, a storage form of glucose in plants, is found as insoluble starch granules in rice, wheat, potatoes, beans, and cereals. Starch is composed of two kinds of polysaccharides, amylose and amylopectin. **Amylose,** which makes up about 20% of starch, consists of α-D-glucose molecules connected by α-1,4-glycosidic bonds in a continuous chain. A typical polymer of amylose may contain from 250 to 4000 glucose units. Sometimes called a straight-chain polymer, polymers of amylose are actually coiled in helical fashion.

Amylopectin, which makes up as much as 80% of starch, is a branched-chain polysaccharide. Like amylose, the glucose molecules are connected by α-1,4-glycosidic bonds. However, at about every 25 glucose units, there is a branch of glucose molecules attached by an α-1,6-glycosidic bond between carbon 1 of the branch and carbon 6 in the main chain. See Figure 14.11.

Figure 14.11
The structures of
(a) amylose, a straight-chain polysaccharide of glucose units, and
(b) amylopectin, a branched chain.

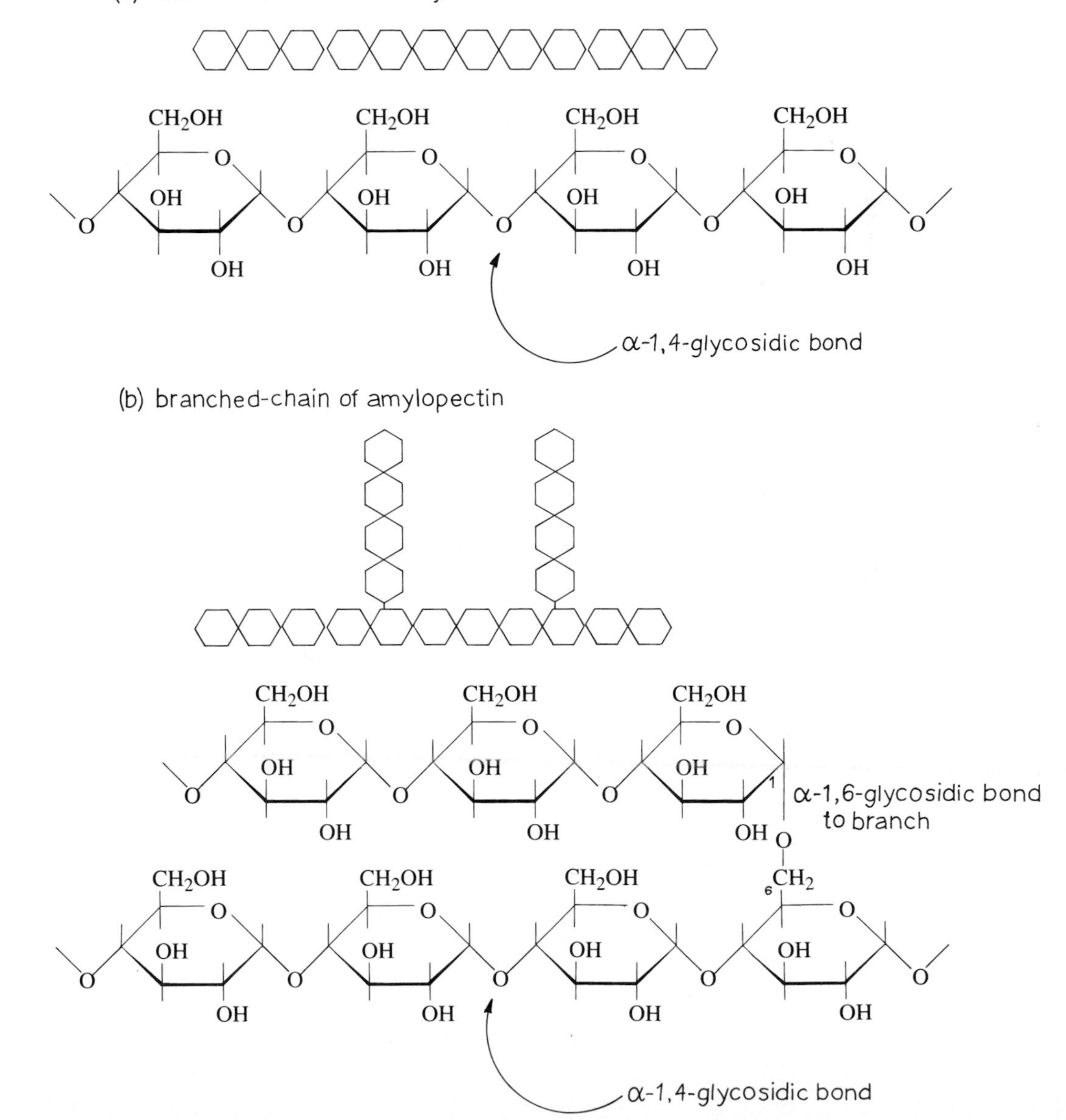

The starches hydrolyze easily in water and acid to give smaller saccharides called dextrins, which then hydrolyze to maltose and finally glucose. In our bodies, these complex carbohydrates are digested by the enzymes amylase and maltase. The glucose obtained provides about 50% of our nutritional calories.

$$\text{Amylose, amylopectin} \xrightarrow{\text{H}^+ \text{ or amylase}} \text{dextrins} \xrightarrow{\text{H}^+ \text{ or amylase}} \text{maltose} \xrightarrow{\text{H}^+ \text{ or maltase}} \text{many D-glucose units}$$

Iodine Test

In the **iodine test,** iodine (I_2) is used to test for the presence of starch. The unbranched helical shape of the polysaccharide amylose in starch reacts strongly with iodine to form a deep blue-black complex. Amylopectin, cellulose, and glycogen produce reddish-purple and brown colors. Such colors do not develop when iodine is added to samples of mono- or disaccharides.

Animal Starch: Glycogen

Glycogen, or animal starch, is a polymer of glucose that is stored in the liver and muscle of animals. It is hydrolyzed in our cells at a rate that maintains the blood level of glucose and provides energy between meals. The structure of glycogen is very similar to that of amylopectin found in plants except that glycogen is more highly branched. In glycogen, the glucose units are joined by α-1,4-glycosidic bonds, and branches occurring about every 10–15 glucose units are attached by α-1,6-glycosidic bonds.

Structural Polysaccharide: Cellulose

Cellulose is the major structural material of wood and plants. Cotton is almost pure cellulose. In **cellulose,** glucose molecules form a long unbranched chain similar to that of amylose. However, the glucose units in cellulose are linked by β-1,4-glycosidic bonds. The β isomers do not form coils like the α isomers but are aligned in parallel rows that are held in place by hydrogen bonds between hydroxyl groups in adjacent chains. This gives a rigid structure to the cell walls in wood and fiber that is more resistant to hydrolysis than the starches. See Figure 14.12.

Enzymes in our saliva and pancreatic juices hydrolyze the α-1,4-glycosidic bonds of the starches. However, there are no enzymes in humans that hydrolyze the β-1,4-glycosidic bonds of cellulose; we cannot digest cellulose. Some animals such as goats and cows and insects like termites are able to obtain glucose from cellulose. Their digestive systems contain bacteria with enzymes that can hydrolyze β-1,4-glycosidic bonds.

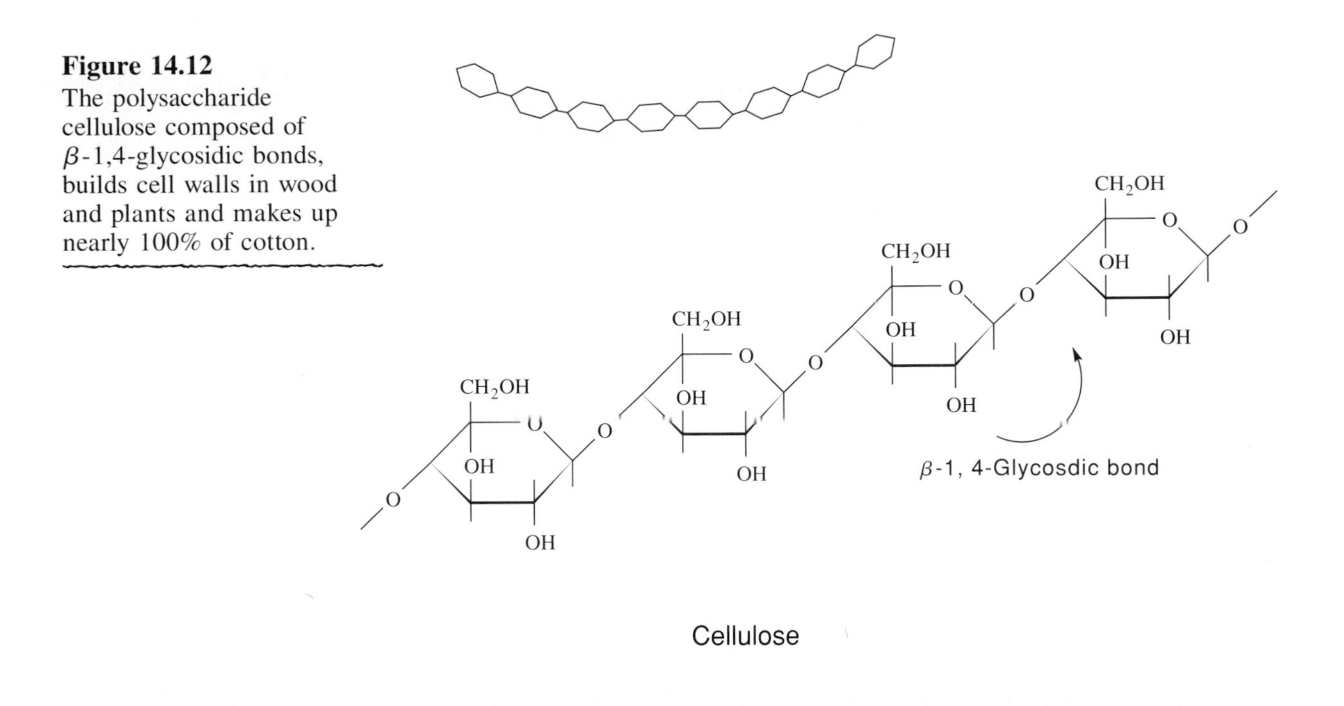

Figure 14.12
The polysaccharide cellulose composed of β-1,4-glycosidic bonds, builds cell walls in wood and plants and makes up nearly 100% of cotton.

HEALTH NOTE *Fiber in the Diet*

The term *dietary fiber* includes all plant material that is not digestible by humans, including cellulose. Food sources of dietary fiber include whole grains, bran, fruits, and vegetables. (See Figure 14.13.)

In the large intestine, the retention of water by dietary fiber has a laxative effect. As a result, wastes pass through the intestinal tract faster, which lessens the time that bacteria and any carcinogens come in contact with the intestine. Some studies indicate that a high-fiber diet may lower the incidence of colon cancer, diverticulitis (inflammation of the colon), heart disease, and hemorrhoids.

The absorptive effects of fiber may also be beneficial in weight maintenance. Fiber increases the bulk in the stomach and intestines without contributing to caloric intake. Fiber may also absorb some of the carbohydrate and cholesterol from the diet, thus decreasing the quantity that diffuses through the intestinal walls.

Figure 14.13
High-fiber foods include beans, peas, cauliflower, carrots, fruits, wheat, and brown rice.

Sample Problem 14.9
Structures of Polysaccharides

Identify the polysaccharide described by each of the following statements:

a. A polysaccharide that is stored in the liver and muscle tissues.
b. An unbranched polysaccharide containing β-1,4-glycosidic bonds.
c. A starch containing α-1,4- and α-1,6-glycosidic bonds.

Solution

a. glycogen **b.** cellulose **c.** amylopectin, glycogen

Study Check

Cellulose and amylose are both unbranched glucose polymers. How do they differ?

ENVIRONMENTAL NOTE *Acidic and Amino Polysaccharides Found in Nature*

When carbon 6 of glucose is oxidized, it forms a carboxylic acid called glucuronic acid. In glucosamine and *N*-acetylglucosamine, a hydroxyl group on carbon 2 of glucose has been replaced by an amino or *N*-acetylamino group. The oxidation of glucose or the addition of an amino or amide group gives monosaccharide units that form polysaccharides that are important in cellular structure, connective tissues, anticoagulants, and antibiotics.

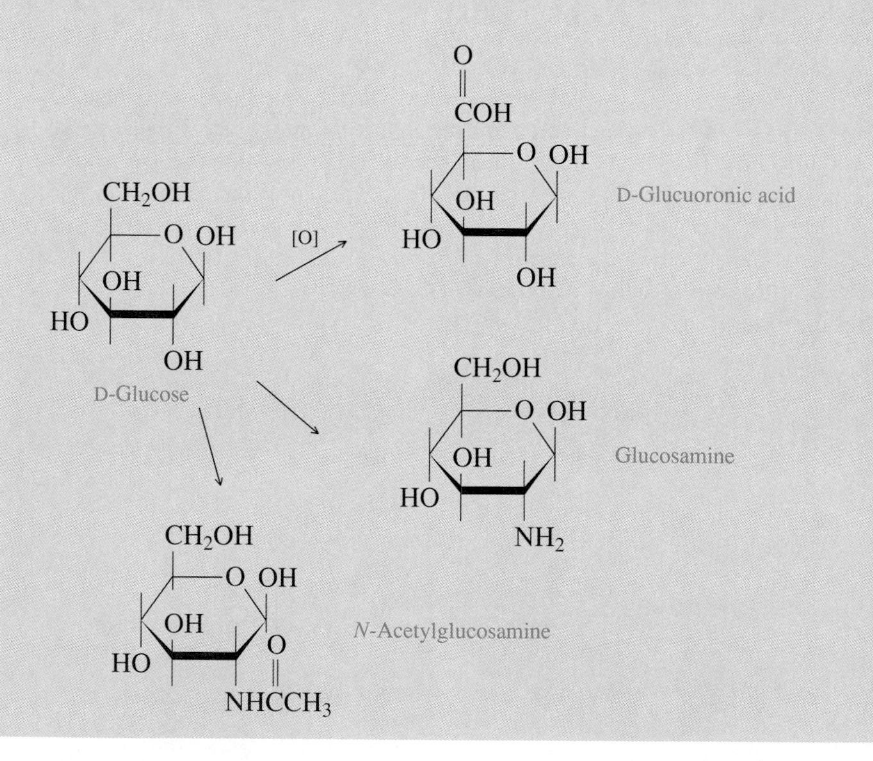

Chitin (pronounced *kité in*) is the polysaccharide that gives the rigid structure to the skeletons of insects, and forms the shells of crabs, lobsters, and shrimp. Similar to cellulose, chitin is an unbranched polymer of *N*-acetylglucosamine units joined by β-1,4-glycosidic bonds. (See Figure 14.14.)

The gellike viscous material that bathes connective tissues, lubricates bone joints, and maintains the shape of the eyeball contains mucopolysaccharides such as hyaluronic acid. Hyaluronic acid is a polysaccharide consisting of alternating units of glucuronic acid and *N*-acetylglucosamine.

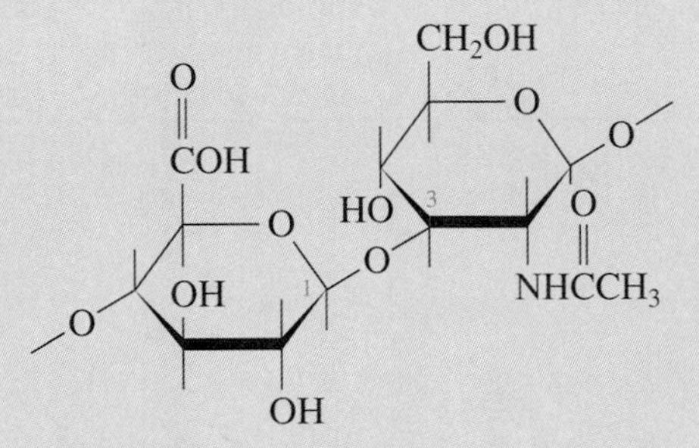

Hyaluronic acid, a repeating disaccharide unit in bone joints and connective tissues

Heparin, an acidic polysaccharide found in the cytoplasm, is an anticoagulant. It is often used in open-heart surgery to prevent the clotting of the blood. Heparin consists of repeating units of glucuronic acid and *N*-acetylglucosamine joined by α-1,4-glycosidic bonds.

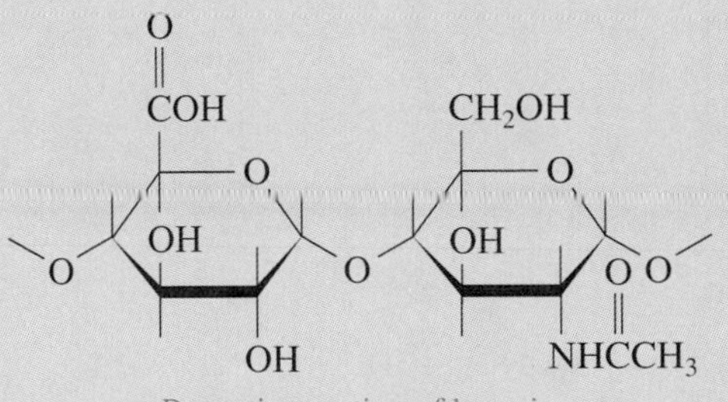

Repeating section of heparin

Figure 14.14
Chitin, a polymer of *N*-acetylglucosamine, is the structural polysaccharide in the shells of shrimp, crab, and lobsters.

Summary of Carbohydrates

Carbohydrate	Food Sources	Monosaccharides
Monosaccharides		
Glucose	Fruit juices, honey, corn syrup	Glucose
Galactose	Lactose hydrolysis	Galactose
Fructose	Fruit juices, honey, sucrose hydrolysis	Fructose
Disaccharides		
Maltose	Germinating grains, starch hydrolysis	Glucose + glucose
Lactose	Milk, yogurt, ice cream	Glucose + galactose

Summary of Carbohydrates (continued)

Carbohydrate	Food Sources	Monosaccharides
Sucrose	Sugar cane, sugar beets	Glucose + fructose
Polysaccharides		
Amylose	Rice, wheat, grains, cereals	Unbranched polymer of glucose joined by α- 1,4-glycosidic bonds
Amylopectin	Rice, wheat, grains, cereals	Branched polymer of glucose joined by α- 1,4- and α- 1,6-glycosidic bonds
Glycogen	Liver, muscles	Highly branched polymer of glucose joined by α- 1,4- and α- 1,6- glycosidic bonds
Cellulose	Plant fiber, bran, beans, celery	Unbranched polymer of glucose joined by β- 1,4-glycosidic bonds

Summary of Reactions

Hemiacetal Formation

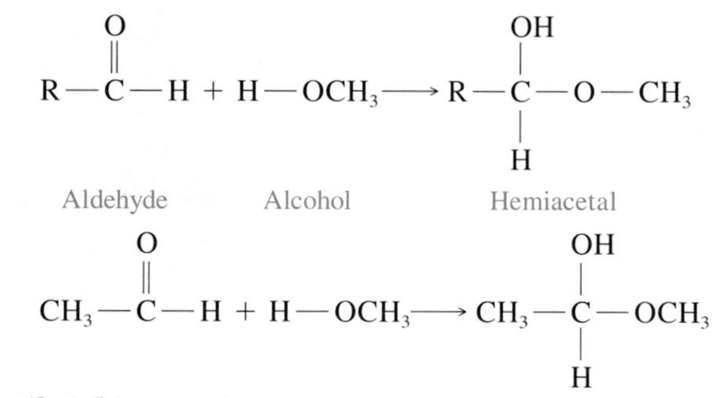

Hemiketal Formation

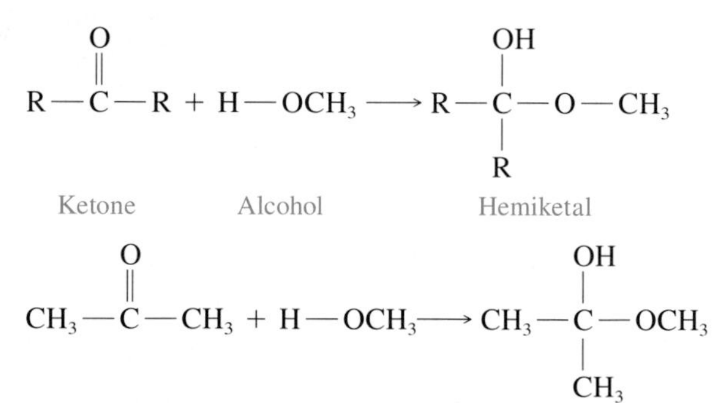

Glycoside (acetal) Formation

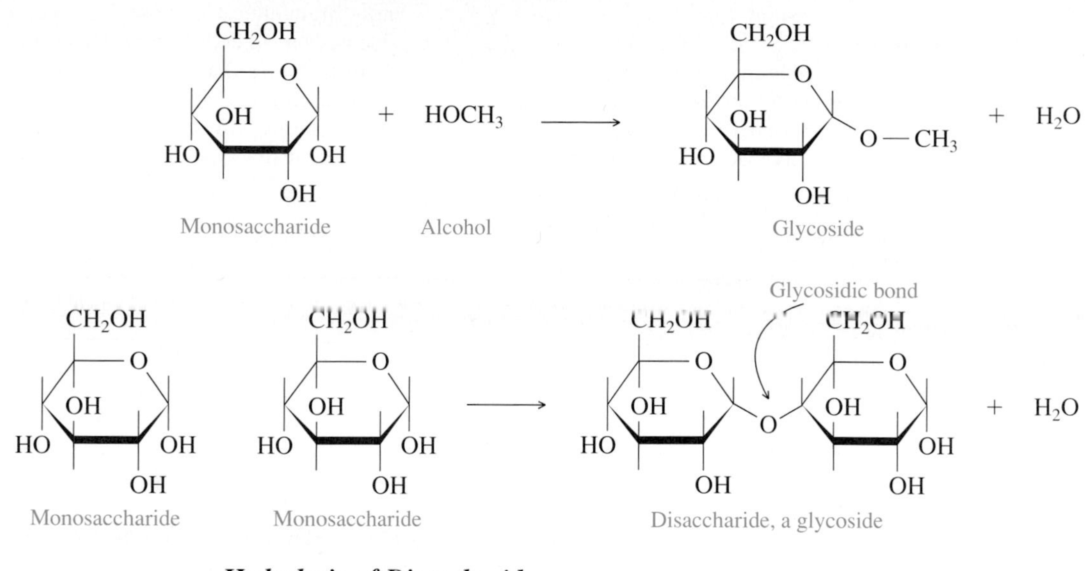

Hydrolysis of Disaccharides

Sucrose + H_2O ⟶ glucose + fructose

Lactose + H_2O ⟶ glucose + galactose

Maltose + H_2O ⟶ glucose + glucose

Chapter Summary

14.1 Classification of Carbohydrates
Carbohydrates are classified as monosaccharides (simple sugars), disaccharides (two monosaccharide units), and polysaccharides (many monosaccharide units).

14.2 Monosaccharides
Monosaccharides are polyhydroxy aldehydes (aldoses) or ketones (ketoses). Monosaccharides are also classified by their number of carbon atoms: *triose, tetrose, pentose,* or *hexose.*

14.3 Chiral Carbon Atoms
Chiral molecules have mirror images that cannot be superimposed on each other. In a chiral molecule, there is one or more carbon atoms attached to four different atoms or groups. The mirror images of a chiral molecule represent two different molecules called enantiomers. In a Fischer projection (straight chain), the prefixes D- and L- are used to distinguish between the mirror images. In D-glyceraldehyde, the —OH is on the right of the chiral carbon; it is on the left in L-glyceraldehyde. Enantiomers are also called optical isomers because they rotate polarized light in opposite directions.

14.4 Structures of Some Important Monosaccharides
Important monosaccharides are the aldohexoses glucose and galactose and the ketohexose fructose.

14.5 Cyclic Structures of Monosaccharides
The predominant form of monosaccharides is the cyclic form of five or six atoms. The cyclic structure forms by a reaction between an OH (usually the one on carbon 5 in hexoses) with the carbonyl group of the same molecule. The formation of a new hydroxyl group on carbon 1 (or 2 in fructose) gives α and β anomers of the cyclic monosaccharide. Because the molecule opens and closes continuously (mutarotation) while in solution, both anomers are present. Monosaccharides are reducing sugars because the aldehyde group (also available in ketoses) is oxidized by a metal ion such as Cu^{2+} which is reduced.

14.6 Disaccharides
Disaccharides are glycosides of two monosaccharide units joined together by a glycosidic bond. In the most common

disaccharides, maltose, lactose, and sucrose, there is at least one glucose unit.

14.7 Polysaccharides
Polysaccharides are polymers of monosaccharide units. Starches consist of amylose, an unbranched chain of glucose; amylopectin is a branched polymer of glucose. Glycogen, the storage form of glucose in animals, is similar to amylopectin with more branching. Cellulose is also a polymer of glucose, but in cellulose the glycosidic bonds are β bonds rather than α bonds as in the starches. Humans can digest starches, but not cellulose, to obtain energy. However, cellulose is important as a source of fiber in our diets.

Glossary of Key Terms

acetal The product of the reaction whereby a monosaccharide forms a glycosidic bond with an —OH of a second monosaccharide.

aldose Monosaccharides that contain an aldehyde group.

amylopectin A branched-chain polymer of starch composed of glucose units joined by α-1,4- and α-1,6-glycosidic bonds.

amylose An unbranched polymer of starch composed of glucose units joined by α-1,4-glycosidic bonds.

anomers The isomers that occur when the formation of a hemiacetal or hemiketal of monosaccharides produces a new hydroxyl group on carbon 1 (or carbon 2). In the α anomer, the OH is drawn downward; in the β isomer the OH is up.

Benedict's test A chemical test for reducing sugars in which blue Cu^{2+} is reduced to give brick-red $Cu_2O(s)$.

carbohydrate A simple or complex sugar composed of carbon, hydrogen, and oxygen.

cellulose An unbranched polysaccharide composed of glucose units linked by β-1,4-glycosidic bonds that cannot be hydrolyzed by humans.

chiral A carbon atom that is attached to four different atoms or groups of atoms; the mirror images of a chiral molecule are not superimposable.

disaccharides Carbohydrates composed of two monosaccharides joined by a glycosidic bond.

enantiomers Stereoisomers that are mirror images of each other.

fermentation A reaction of glucose, fructose, maltose, or sucrose, in which the sugar reacts with enzymes in yeast to give ethanol and carbon dioxide gas.

Fischer projection A two-dimensional straight-chain representation of a molecule.

fructose A monosaccharide found in honey and fruit juices; it is combined with glucose in sucrose. Also called levulose and grape sugar.

galactose A monosaccharide that occurs combined with glucose in lactose.

glucose The most prevalent monosaccharide in the diet. An aldohexose that is found in fruits, vegetables, corn syrup, and honey. Also known as blood sugar and dextrose. Combines in glycosidic bonds to form most of the polysaccharides.

glycogen A polysaccharide formed in the liver and muscles for the storage of glucose as an energy reserve. It is composed of glucose in a highly branched polymer joined by α-1,4- and α-1,6-glycosidic bonds.

glycosides Acetal products of a monosaccharide reacting with an alcohol or another sugar.

glycosidic bond The acetal bond that forms when an alcohol or a hydroxyl group of a monosaccharide adds to a hemiacetal. It is the type of bond that links monosaccharide units in di- or polysaccharides.

Haworth structure The cyclic structure that represents the closed chain of a monosaccharide.

hemiacetal The product of the addition of an alcohol to the carbonyl group of an aldehyde.

hemiketal The product of the addition of an alcohol to the carbonyl group of a ketone.

iodine test A test for amylose that forms a blue-black color after iodine is added to the sample.

ketose A monosaccharide that contains a ketone group.

lactose A disaccharide consisting of glucose and galactose found in milk and milk products.

maltose A disaccharide consisting of two glucose units; it is obtained from the hydrolysis of starch and in germinating grains.

monosaccharide A polyhydroxy compound that contains an aldehyde or ketone group.

mutarotation The conversion between α and β anomers.

optical isomers Mirror images of a chiral compound, which rotate plane-polarized light in opposite directions.

polysaccharides Polymers of many monosaccharide units, usually glucose. Polysaccharides differ in the types of glycosidic bonds and the amount of branching in the polymer.

reducing sugar A carbohydrate with a free aldehyde group capable of reducing the Cu^{2+} in Benedict's reagent.

saccharide A term from the Latin word *saccharum,* meaning "sugar"; it is used to describe the carbohydrate family.

stereoisomers Molecules with the same structural formulas and attached atoms but having different three-dimensional arrangements of the atoms.

sucrose A disaccharide composed of glucose and fructose; a nonreducing sugar, commonly called table sugar or "sugar."

Chemistry at Home

1. Carefully wrap four or five leaves of a house plant with aluminum foil and masking tape. Place the plant in the sun for 1 week. What do you think will happen to the leaves that are covered? Unwrap the leaves. What differences are there between leaves that were wrapped and not wrapped? What explanation would you give?
2. Read the nutrition facts label on a cereal box. How are carbohydrates listed? What is dietary fiber? What carbohydrates are listed in the ingredients?
3. The major ingredient in crackers is flour, a starch. What type of carbohydrate is starch? Chew on a cracker for 4–5 minutes. There is an enzyme (amylase) in your saliva that breaks apart the bonds in the starch. How does its taste change? Why? How are you getting energy from the cracker?
4. Add some sugar to a glass of water and stir. What happens? If you have other carbohydrates such as fructose, cornstarch, arrowroot, or flour, add some of each to separate glasses of water and stir. Do they dissolve like the sugar? Why?
5. If you have some iodine solution for cuts, you can use it to test for starch. Collect samples of food—crackers, bread, pasta, cereals, candy, and so on—an aspirin, and an envelope. Put a few drops of iodine on each. If a blue-black color develops, there is starch present. Is there starch in an aspirin tablet? On the glue of the envelope flap? Throw all your samples away! Iodine is toxic.

Answers to Study Checks

14.1 a monosaccharide

14.2

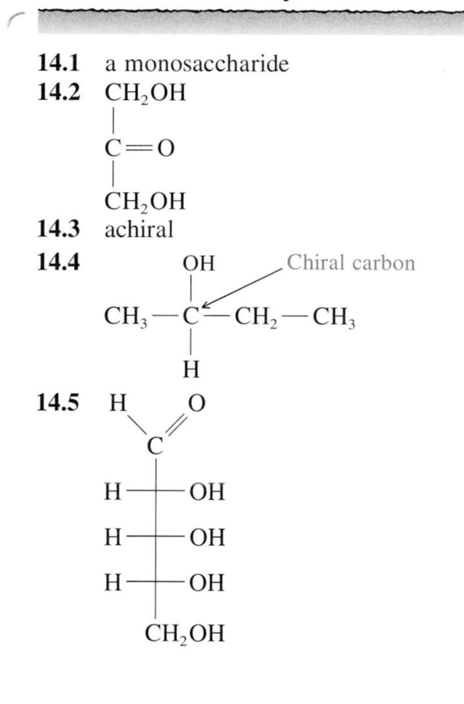

14.3 achiral

14.4

14.5

14.6

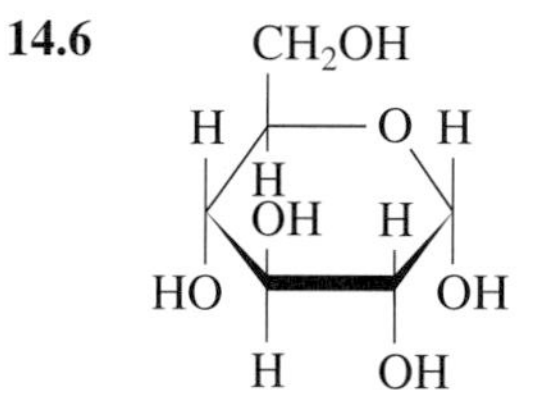

14.7 This indicates a high level of reducing sugar (probably glucose) in the urine. One common cause of this condition is diabetes mellitus.

14.8

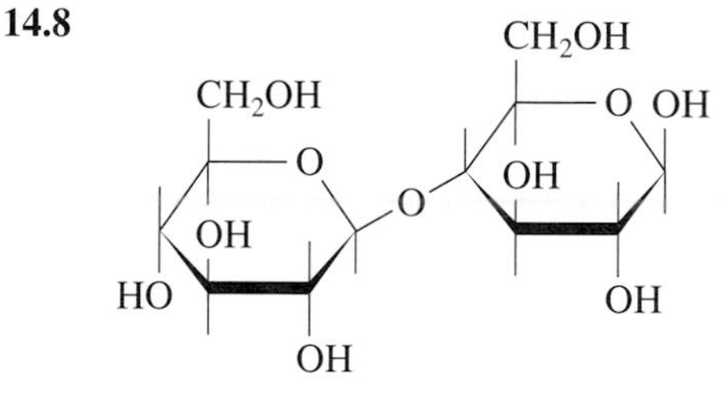

14.9 Cellulose contains glucose units connected by β- 1,4-glycosidic bonds, whereas the glucose units in amylose are connected by α- 1,4-glycosidic bonds.

Problems

Classification of Carbohydrates *(Goal 14.1)*

14.1 What is a monosaccharide?
14.2 What is a polysaccharide?

Monosaccharides *(Goal 14.2)*

14.3 What functional groups are found in all monosaccharides?
14.4 What is the difference between an aldose and a ketose?
14.5 What are the functional groups and number of carbons in a ketopentose?
14.6 What are the functional groups and number of carbons in an aldohexose?
14.7 Classify each of the following monosaccharides as an aldose or ketose.

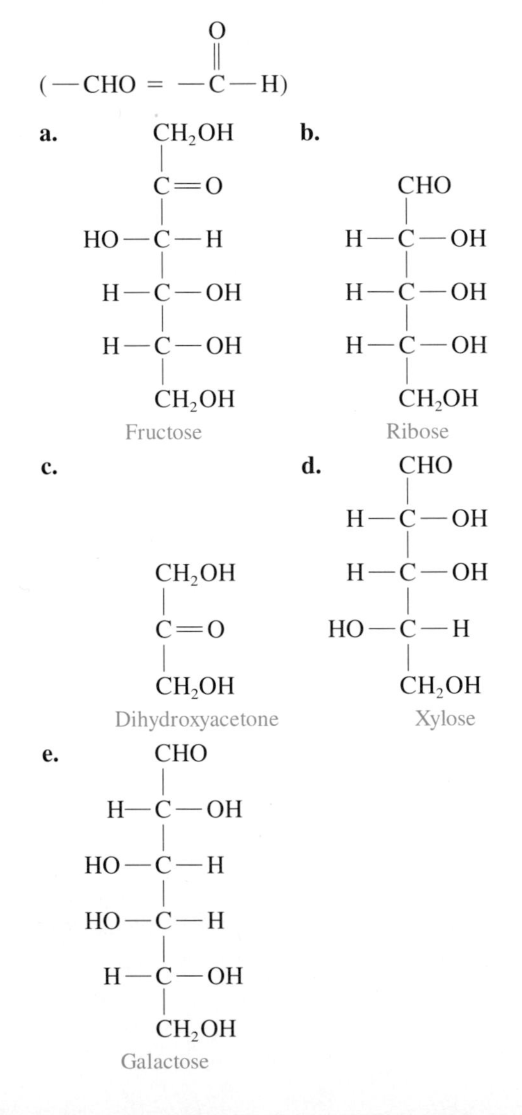

a. Fructose b. Ribose c. Dihydroxyacetone d. Xylose e. Galactose

14.8 Classify each of the monosaccharides in problem 14.7 according to the number of carbon atoms in the chain.

Chiral Carbon Atoms *(Goal 14.3)*

14.9 Indicate whether the following objects are chiral or not chiral. Consider each object to be without markings or printing.
a. a Ping-Pong ball for table tennis
b. a tennis shoe
c. an unsharpened pencil
d. a new piece of chalk

14.10 Indicate whether the following objects are chiral or not chiral. Consider each object to be without markings or printing.
a. a boxing glove
b. a blank sheet of typing paper
c. a golf club
d. a golf tee

14.11 Indicate whether the starred carbon atom in each of the following compounds is a chiral carbon or not:

a. H—C*(H)(H)—Br

b. CH_3—C*(H)(Cl)—Cl

c. H—C*(CHO)(CH_2OH)—OH

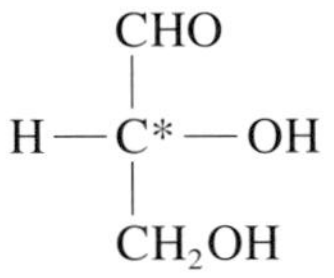

d. CH_3CH_2—C*(COOH)(OH)—CH_3

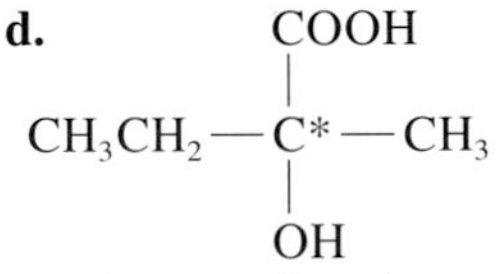

14.12 Indicate whether the central (starred) carbon atom in each of the following compounds is a chiral carbon or not:

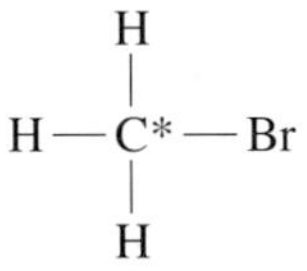

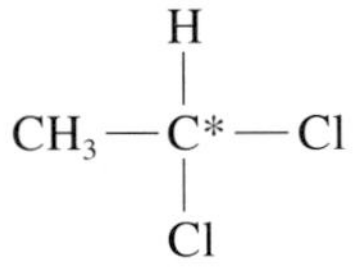

a. H—C*(CHO)(H)—CH_2CH_3

b. H—C*(COOH)(CH_3)—OH

Lactic acid

c.

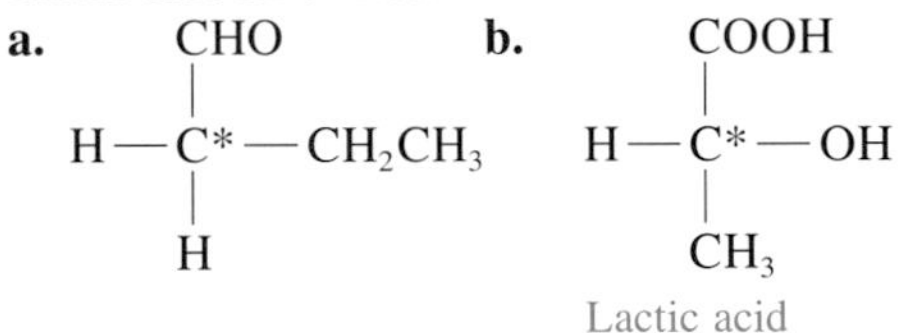

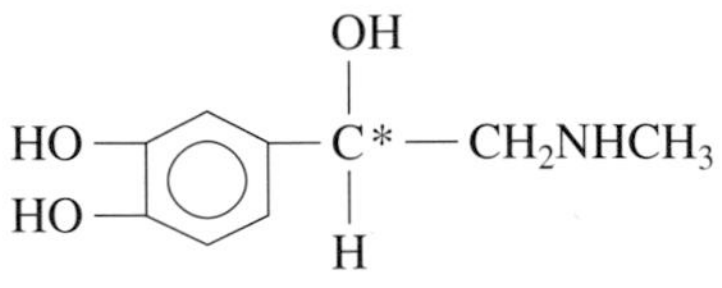

Epinephrine

d. CH_3—C*(CH_3)(Cl)—OH

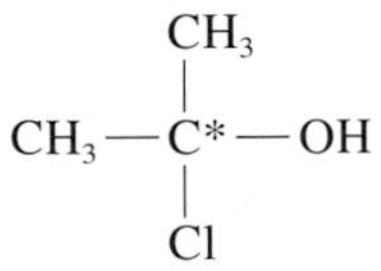

14.13 What is a Fischer projection?

14.14 Write the projection formula for D-glyceraldehyde and L-glyceraldehyde.

14.15 For each of the following sugars state whether each is the D or L isomer:

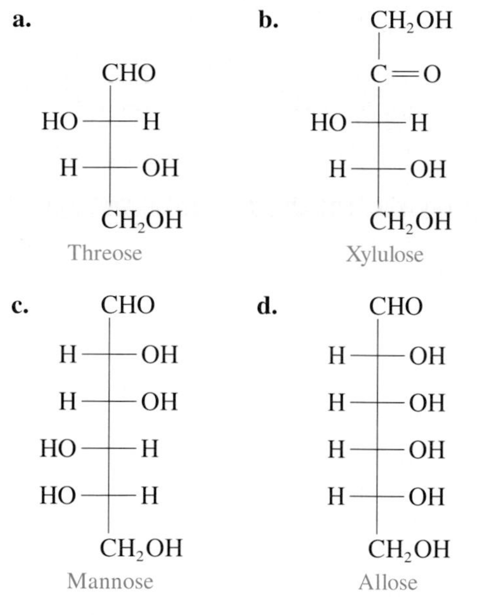

Threose Xylulose Mannose Allose

14.16 For each of the following sugars state whether each is the D or L isomer:

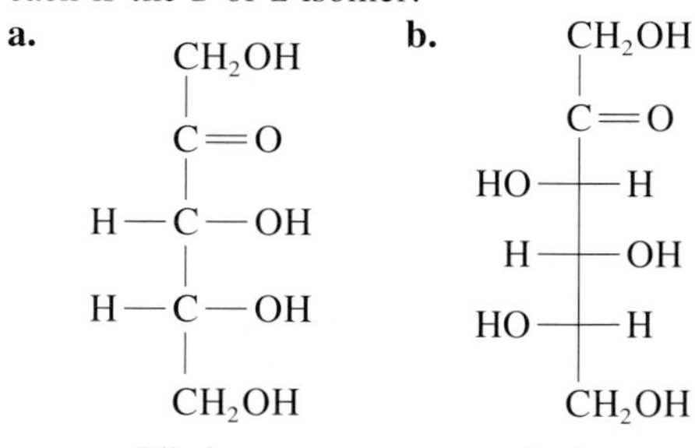

Ribulose Sorbose

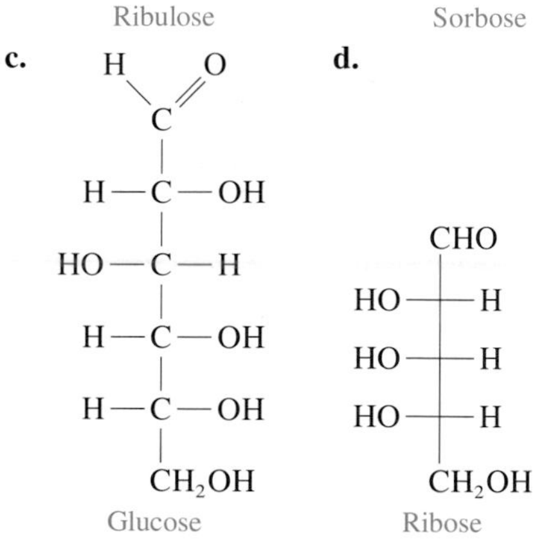

Glucose Ribose

14.17 D-lactic acid is optically active. What does that mean?

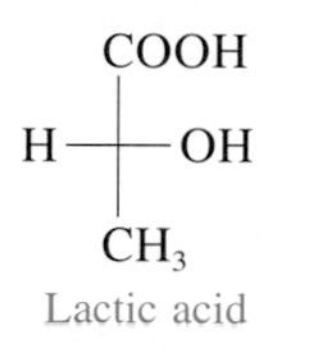

Lactic acid

14.18 What is the projection formula of L-lactic acid (see problem 14.17)? How does it rotate polarized light compared with D-lactic acid?

Structures of Some Important Monosaccharides *(Goal 14.4)*

14.19 Draw the open-chain structure of D-glucose and L-glucose.

14.20 Draw the open-chain structure of D-fructose and L-fructose.

14.21 How does the open-chain structure of D-galactose differ from D-glucose?

14.22 How does the open-chain structure of D-fructose differ from D-glucose?

14.23 Identify a monosaccharide that fits the following description:

a. is also called blood sugar
b. is not metabolized in galactosemia
c. is also called fruit sugar

14.24 Identify a monosaccharide that fits the following description:

a. high blood levels in diabetes
b. obtained as a hydrolysis product of lactose
c. is the sweetest of the monosaccharides

Cyclic Structures of Monosaccharides *(Goal 14.5)*

14.25 What are the kind and number of atoms in the cyclic structure of glucose? Why?

14.26 What are the kind and number of atoms in the cyclic structure of fructose? Why?

14.27 Draw the cyclic structures for the α and β anomers of D-glucose.

14.28 Draw the cyclic structures for the α and β anomers of D-fructose.

14.29 Identify each of the following cyclic structures as a hemiacetal or a hemiketal:

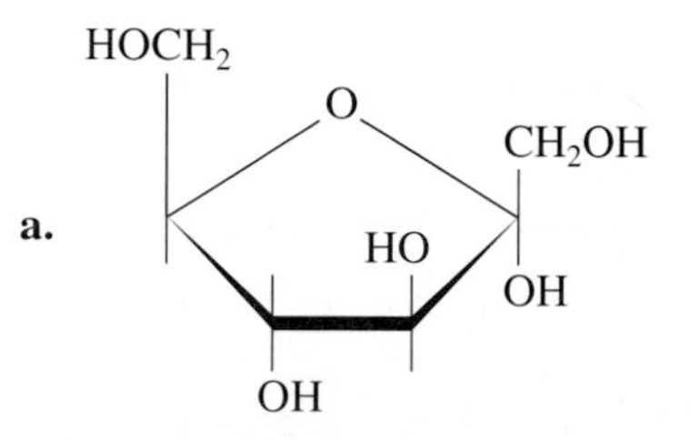

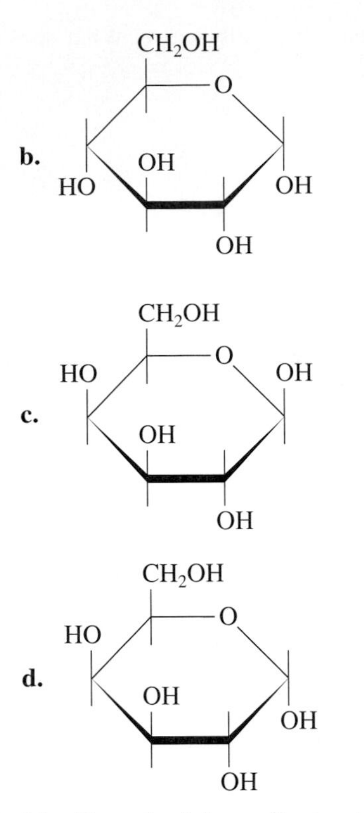

14.30 Identify each of the cyclic structures in problem 14.29 as the α or β anomer.

14.31 What is a reducing sugar?

14.32 What are the products when D-galactose reacts with Benedict's reagent?

Disaccharides *(Goal 14.6)*

14.33 For each of the following disaccharides, give the monosaccharide units produced by hydrolysis, the type of glycosidic bond, and the identity of the disaccharide including the α or β anomer:

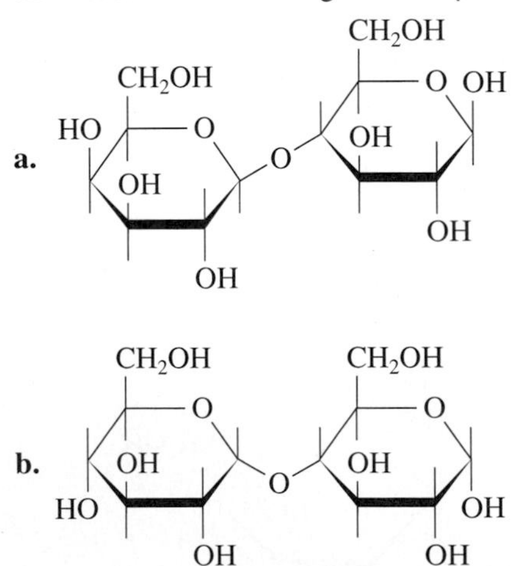

14.34 For each of the following disaccharides, give the monosaccharide units produced by hydrolysis, the type of glycosidic bond, and the identity of the disaccharide including the α or β anomer:

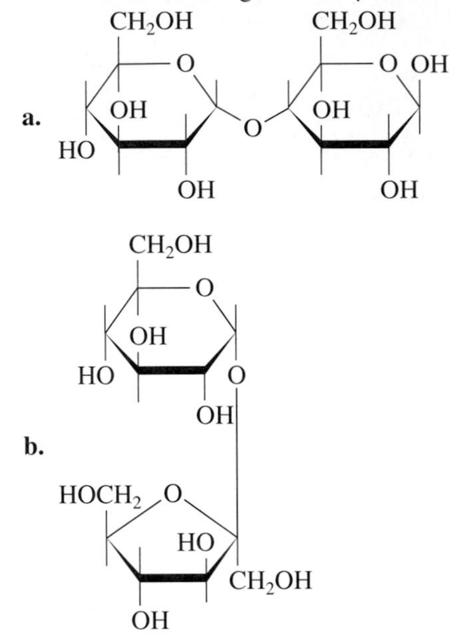

14.35 Identify a disaccharide that fits the following description:

a. is ordinary table sugar
b. found in milk and milk products
c. is also called *malt sugar*
d. Hydrolysis gives galactose and glucose.

14.36 Identify a disaccharide that fits the following description:

a. is not a reducing sugar
b. is composed of two glucose units
c. is also called *milk sugar*
d. Hydrolysis gives glucose and fructose.

Polysaccharides *(Goal 14.7)*

14.37 Describe the similarities and differences in the following polysaccharides:

a. amylose and amylopectin
b. amylopectin and glycogen

14.38 Describe the similarities and differences in the following polysaccharides:

a. amylose and cellulose
b. cellulose and chitin

14.39 Give the name of one or more polysaccharides that matches each of the following descriptions:

a. is not digestible by humans
b. is the storage form of carbohydrates in plants
c. contains only α-1,4-glycosidic bonds
d. is the most highly branched polysaccharide

14.40 Give the name of one or more polysaccharides that matches each of the following descriptions:

a. is the storage form of carbohydrates in animals

b. contains only β- 1,4-glycosidic bonds

c. contains both α- 1,4-and α- 1,6-glycosidic bonds

d. produces maltose during digestion

Challenge Problems

14.41 D-Sorbitol, a sweetener found in seaweed and berries, contains only hydroxyl functional groups. When D-sorbitol is oxidized, it forms D-glucose. What is the structural formula of D-sorbitol?

14.42 Raffinose is a trisaccharide found in Australian manna and in cottonseed meal. It is composed of three different monosaccharides and is not a reducing sugar. Identify the monosaccharides in raffinose.

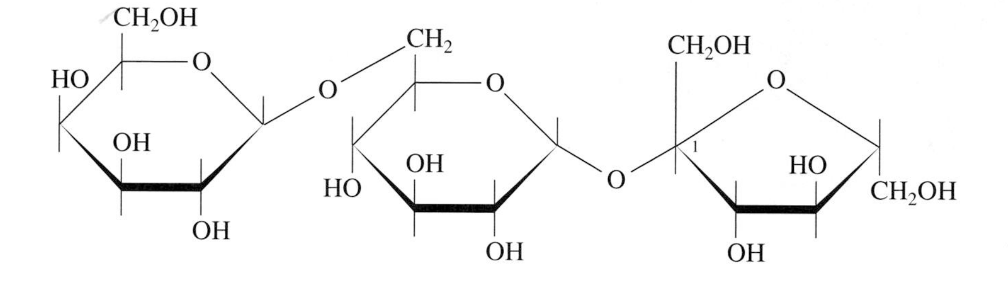

14.43 If α-galactose is dissolved in water, β-galactose is eventually present. Explain how this occurs.

14.44 Why are lactose and maltose reducing sugars, but sucrose is not?

14.45 β-Cellobiose is a disaccharide obtained from the hydrolysis of cellulose. It is quite similar to maltose except it has a β- 1,4-glycosidic bond. What is the structure of β-cellobiose?

14.46 The disaccharide trehalose found in mushrooms is composed of two α-D-glucose molecules joined by α- 1,1-glycosidic bond. Draw the structure of trehalose.

Chapter 15

Lipids

High levels of serum cholesterol have been associated in many studies with the accumulation of lipid deposits within the coronary arteries, restricting the flow of blood to the tissue.

Looking Ahead

Learning Goals

15.1 Describe the classes of lipids.

15.2 Identify a fatty acid as saturated or unsaturated.

15.3 Write the structural formula of a wax, fat, or oil produced by the reaction of a fatty acid and an alcohol or glycerol.

15.4 Draw the structure of the product from the reaction of a triglyceride with hydrogen, an acid or base, or an oxidizing agent.

15.5 Describe the components of phosphoglycerides, sphingolipids, and glycolipids.

15.6 Describe the structure of a steroid and cholesterol.

15.7 Describe the steroid hormones.

Questions to Think About

1. What are some functions of fats in the body?
2. What foods in your diet are high in fats?
3. What is the difference between a fat and an oil?
4. How are solid margarines produced?
5. What kind of lipid is cholesterol?
6. What is the difference between HDL and LDL lipids?

When we talk of fats, waxes, vegetable oils, steroids, or fat-soluble vitamins, we are discussing lipids. In the body, that ever-present adipose tissue known as body fat is a storage form of energy that provides more than twice as much energy (9 kcal/g) as carbohydrates (4 kcal/g). Fats also provide protection and insulation for internal organs.

Many people are concerned about the amounts of saturated fats and cholesterol in our diets. Researchers suggest that saturated fats and cholesterol are associated with diseases such as diabetes, cancers of the breast, pancreas, and colon, and arteriosclerosis, a condition in which deposits of lipid materials accumulate in the coronary blood vessels. These plaques restrict the flow of blood to the tissue, causing necrosis (death) of the tissue. In the heart, this could result in a *myocardial infarction* (heart attack).

The American Institute of Cancer Research has recommended that our diets contain more fiber and starch by adding more vegetables, fruits, and whole grains with moderate amounts of foods with low levels of fat and cholesterol such as fish, poultry, lean meats, and low-fat dairy products. They also suggest that we limit our intake of foods high in fat and cholesterol such as eggs, nuts, fatty or organ meats, cheeses, butter, and coconut and palm oil.

15.1 Types of Lipids

Learning Goal
Describe the classes of lipids.

Lipids are a family of biomolecules that have the common property of being soluble in organic solvents but not in water. The word "lipid" comes from the Greek word *lipos,* meaning fat or lard. Typically, the lipid content of a plant cell can be extracted using an organic solvent such as ether, chloroform, or acetone. Lipids are an important feature in cell membranes, fat-soluble vitamins, and steroid hormones.

Within the lipid family, there are distinct structures that distinguish the different types of lipids as shown in Table 15.1. Lipids such as waxes, fats,

Table 15.1 ***Classes of Lipid Molecules***

Lipids	Composition
Waxes	Fatty acid and long-chain alcohol
Fats and oils (triglycerides)	Fatty acids and glycerol
Phosphoglycerides	Fatty acids, glycerol, phosphate, amino alcohol
Sphingolipids	Fatty acids, sphingosine, phosphate, amino alcohol
Glycolipids	Fatty acids, glycerol or sphingosine, one or more monosaccharides
Steroids	A fused structure of three cyclohexanes and a cyclopentane

oils, and phospholipids are esters that can be hydrolyzed to give fatty acids along with other products including an alcohol. Sphingolipids contain an alcohol called sphingosine and glycolipids contain a carbohydrate. Steroids are characterized by the steroid nucleus of four fused carbon rings. They do not contain fatty acids and cannot be hydrolyzed. Figure 15.1 illustrates the general structure of some classes of lipids.

The structural formulas of some typical lipids are shown below:

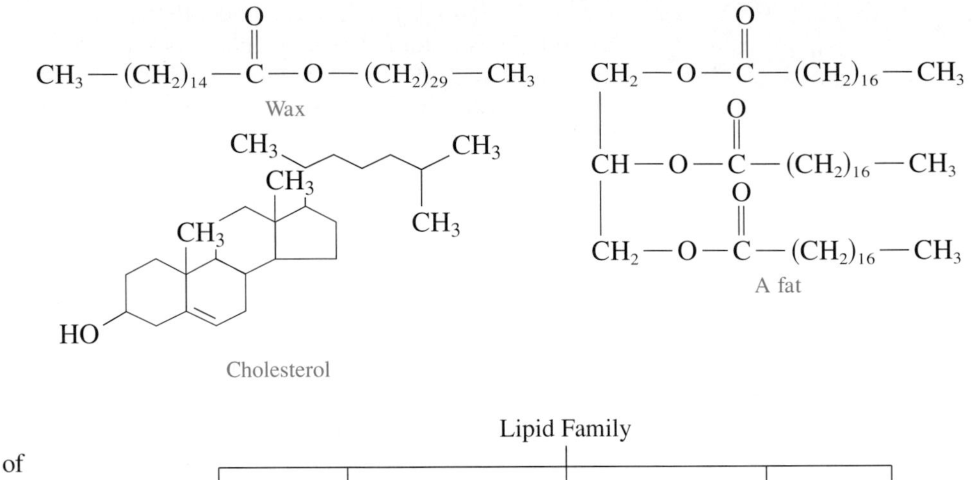

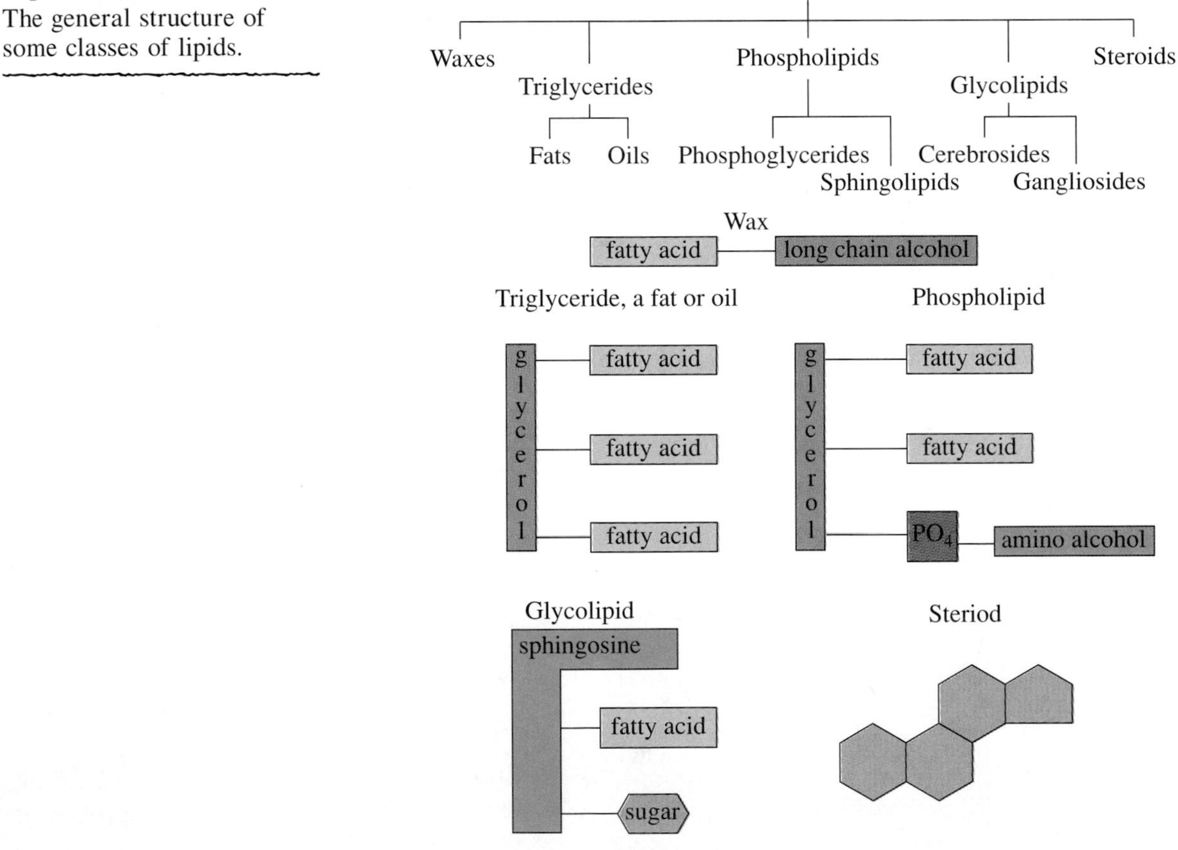

Figure 15.1
The general structure of some classes of lipids.

Sample Problem 15.1
Classes of Lipids

What type of lipid does not contain fatty acids?

Solution

The steroids are a group of lipids with no fatty acids.

Study Check

What type of lipid contains a carbohydrate?

15.2 Fatty Acids

Learning Goal

Identify a fatty acid as saturated or unsaturated.

Fatty acids are long-chain carboxylic acids that are insoluble in water. The fatty acids in fats and oils typically have an even number of carbon atoms; 12–18 carbon atoms being the most common in biological systems. An example is lauric acid, a 12-carbon acid, found in coconut oil. The structural formula of a fatty acid may also be written in a zigzag form or in the condensed form as shown below.

Some Structures of Lauric Acid

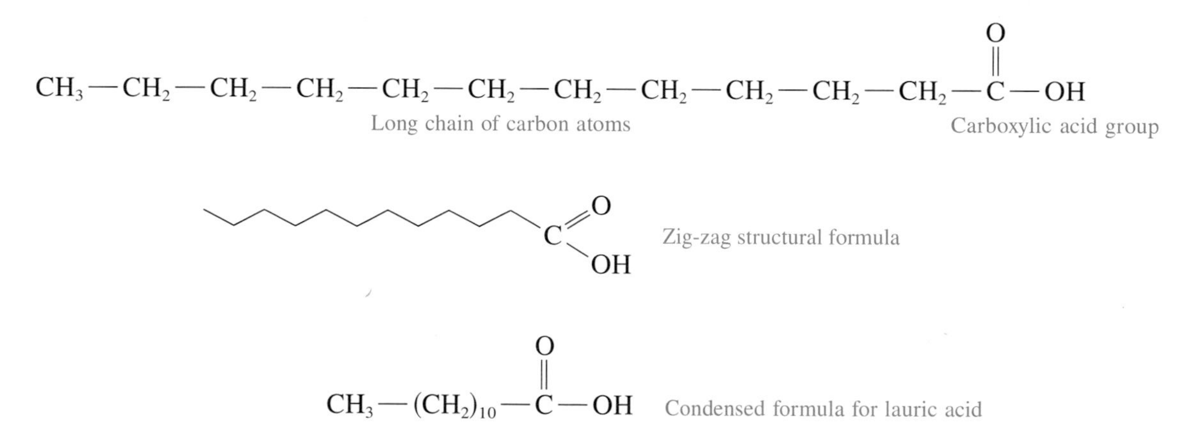

Saturated fatty acids such as lauric acid contain only single bonds between carbons. *Monounsaturated* fatty acids have one double bond in the carbon chain, and *polyunsaturated* fatty acids have two or more double bonds. Table 15.2 lists some of typical fatty acids in lipids. An example of the structural, zigzag, and condensed formulas of oleic acid, a monounsaturated fatty acid found in olives and corn, is shown below. In the full structure, the bend in the carbon chain is caused by the double bond that has a cis configuration. The double bonds in most **unsaturated** fatty acids are cis causing one of or more bends in the carbon chain.

Table 15.2 ***Formulas, Melting Points, and Sources of Some Fatty Acids***

Number of Carbon Atoms	Structural Formula	Melting Point (°C)	Common Name	Source
Saturated Fatty Acids (Single Carbon–Carbon Bonds)				
12	$CH_3(CH_2)_{10}COOH$	44	Lauric	Cocount
14	$CH_3(CH_2)_{12}COOH$	54	Myristic	Nutmeg
16	$CH_3(CH_2)_{14}COOH$	63	Palmitic	Palm
18	$CH_3(CH_2)_{16}COOH$	70	Stearic	Animal fat
Monounsaturated Fatty Acids (One cis Double Bond)				
16	$CH_3(CH_2)_5CH{=}CH(CH_2)_7COOH$	−1	Palmitoleic	Butter fat
18	$CH_3(CH_2)_7CH{=}CH(CH_2)_7COOH$	14	Oleic	Olives, corn
Polyunsaturated Fatty Acids (Two or More cis Double Bonds)				
18	$CH_3(CH_2)_4CH{=}CHCH_2CH{=}CH(CH_2)_7COOH$	−5	Linoleic	Soybean, safflower, sunflower
18	$CH_3CH_2CH{=}CHCH_2CH{=}CHCH_2CH{=}CH(CH_2)_7COOH$	−11	Linolenic	Corn

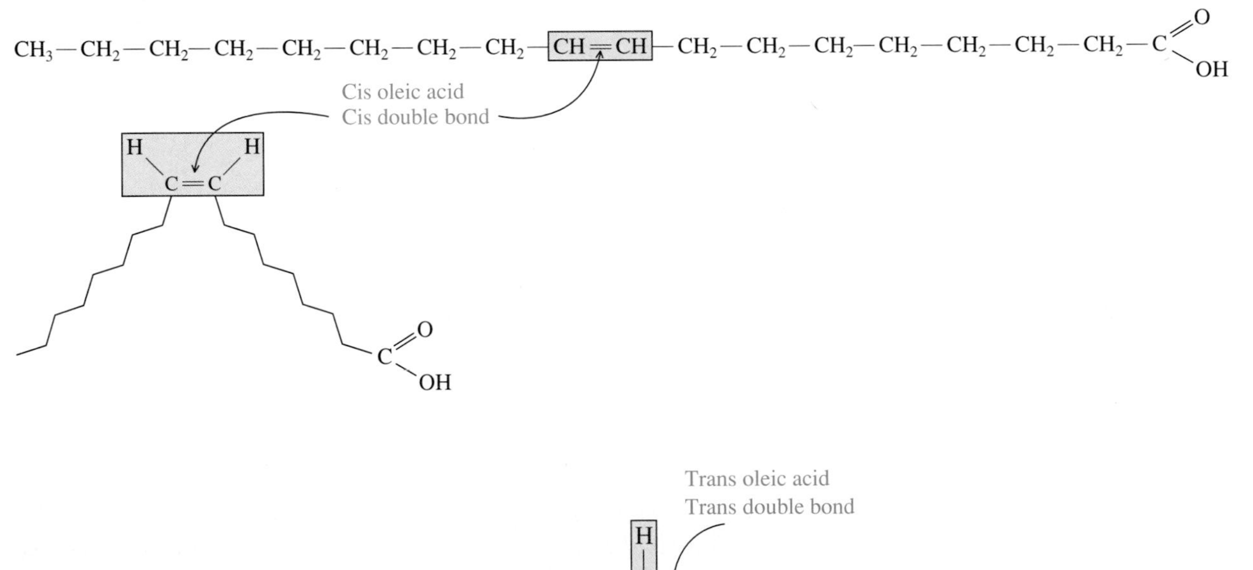

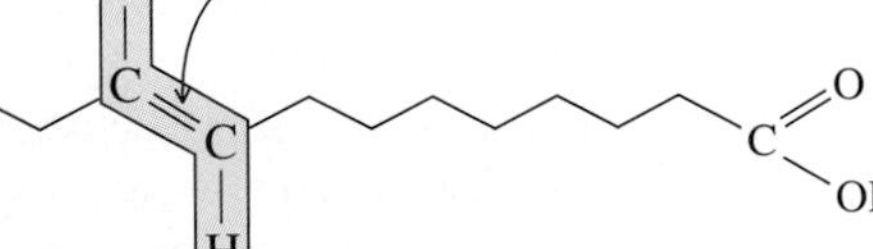

The human body is capable of synthesizing most fatty acids from carbohydrates or other fatty acids. However, humans and other mammals cannot synthesize fatty acids that have more than one double bond, such as linoleic and linolenic acid. These fatty acids are called *essential* fatty acids because they must be provided by the diet.

A deficiency of essential fatty acids can cause skin dermatitis in infants. However, the role of fatty acids in adult nutrition is not well understood. Adults do not usually have a deficiency of essential fatty acids.

Physical Properties of Fatty Acids

The saturated fatty acids have straight-chain structures that allow their molecules to fit closely together and form strong attractions between the carbon chains. (See Figure 15.2.) As a result, they have high melting points because energy is needed to break the bonds between the carbon chains to melt the fatty acid. Saturated fats are usually solids at room temperatures.

In contrast, most unsaturated fatty acids contain one or more cis double bonds that cause one or more bends in the carbon chain. As a result, unsaturated fatty acids have an irregular shape, do not fit closely together, and form fewer bonds between carbon chains. This makes their melting points so low that most are liquids at room temperature. In general, the melting point of an unsaturated fatty acid is lower as the number of double bonds increases.

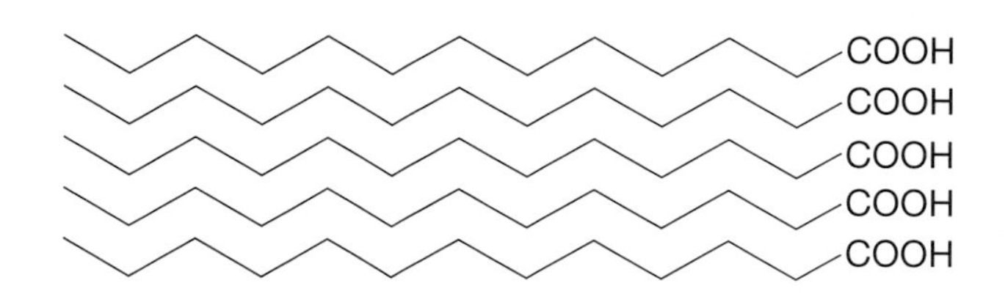

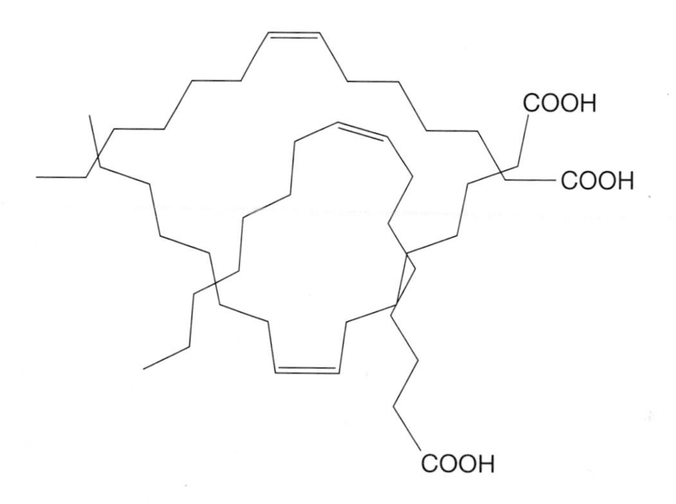

Figure 15.2
(a) A saturated fatty acid in which molecules fit closely together has a high melting point.
(b) An unsaturated fatty acid in which molecules are further apart has a low melting point.

HEALTH NOTE *Prostaglandins*

Prostaglandins are hormonelike substances found in very low amounts in most tissues and fluids of the body. The parent structure, prostanoic acid, which has 20 carbon atoms, is formed from an unsaturated fatty acid called arachidonic acid. Small changes in the structure of prostanoic acid produce the different prostaglandins.

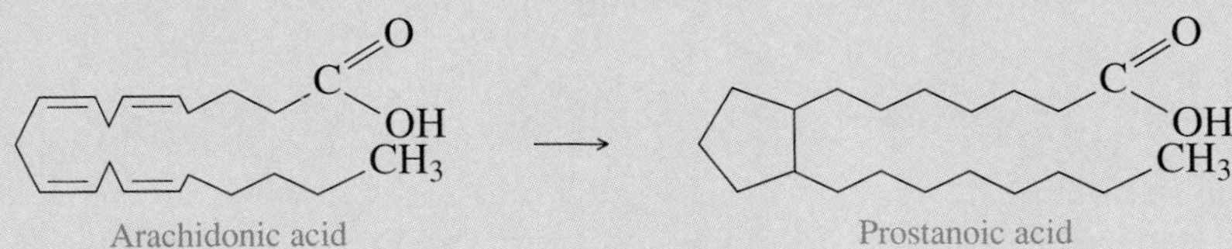

Prostaglandins produced in the tissues enter the blood, where they regulate several physiological activities. Some increase blood pressure as vasoconstrictors, and others lower blood pressure as vasodilators. There are also prostaglandins that stimulate contraction and relaxation of smooth muscle. Still others play a role in reproduction, respiration, fat metabolism, and nerve impulse transmission. Some also cause inflammation, pain, and fever. These effects are inhibited by anti-inflammatory agents such as aspirin and acetaminophen, which act by preventing the formation of the prostaglandins in the tissues.

The prostaglandins PGE_2 and $PGF_{2\alpha}$ are used medically to treat hypertension, to increase uterine contractions, and to relieve bronchial asthma.

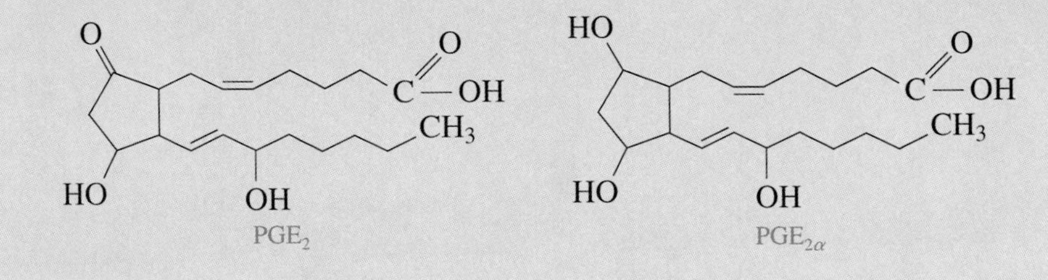

Sample Problem 15.2
Structures and Properties of Fatty Acids

Consider the structural formula of oleic acid.

$$CH_3(CH_2)_7CH{=}CH(CH_2)_7\overset{\overset{\large O}{\|}}{C}OH$$

a. Why is the substance called an acid?
b. How many carbon atoms are in oleic acid?
c. Is it a saturated or unsaturated fatty acid?
d. Is it most likely to be solid or liquid at room temperature?
e. Would it be soluble in water?

Solution

a Oleic acid contains a carboxylic acid group.
b. It contains 18 carbon atoms.
c. It is an unsaturated fatty acid.
d. It is liquid at room temperature.
e. No, its long hydrocarbon chain makes it insoluble in water.

Study Check
Palmitoleic acid is a fatty acid with the following formula:

$$CH_3(CH_2)_5CH{=}CH(CH_2)_7\overset{\overset{\displaystyle O}{\|}}{C}OH$$

a. How many carbon atoms are in palmitoleic acid?
b. It is a saturated or unsaturated fatty acid?
c. Is it most likely to be solid or liquid at room temperature?

15.3 Waxes, Fats, and Oils

Learning Goal
Write the structural formula of a wax, fat, or oil produced by the reaction of a fatty acid and an alcohol or glycerol.

Waxes are found in many plants and animals. Wax coatings on fruits and the leaves and stems of plants help to prevent loss of water and damage from pests. Waxes on the skin, fur, and feathers of animals and birds provide a waterproof coating. A **wax** is an ester of a saturated fatty acid and a long-chain alcohol, each containing from 14 to 30 carbon atoms.

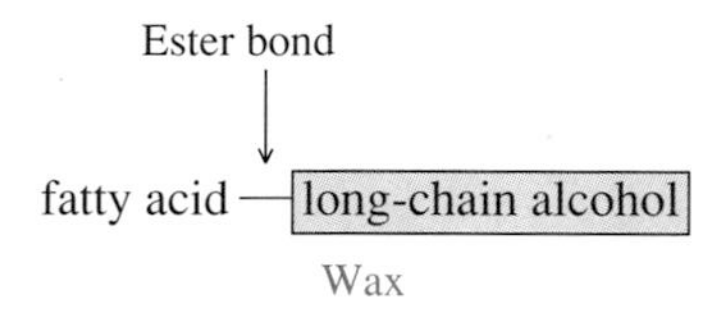

The formulas of some common waxes are given in Table 15.3. Beeswax obtained from honeycombs, and carnauba wax obtained from palm trees, are used to give a protective coating to furniture, cars, and floors. Jojoba wax is used in making candles and cosmetics such as lipstick. Lanolin, a mixture of waxes obtained from wool, is used in hand and facial lotions to aid retention of water, which softens the skin. (See Figure 15.3.)

Fats and Oils

Fats and *oils* are the most prevalent forms of lipids. Also known as **triglycerides** or *triacylglycerols,* fats and oils are esters of glycerol and three fatty

Table 15.3 *Some Typical Waxes*

Type	Structural Formula	Source	Uses
Beeswax	$CH_3(CH_2)_{14}{-}\overset{\overset{\displaystyle O}{\|}}{C}{-}O{-}(CH_2)_{29}CH_3$	Honeycomb	Candles, shoe polish, wax paper
Carnauba wax	$CH_3(CH_2)_{24}{-}\overset{\overset{\displaystyle O}{\|}}{C}{-}O{-}(CH_2)_{29}CH_3$	Brazilian palm tree	Waxes for furniture, cars, floors, shoes
Jojoba wax	$CH_3(CH_2)_{18}{-}\overset{\overset{\displaystyle O}{\|}}{C}{-}O{-}(CH_2)_{19}CH_3$	Jojoba	Candles, soaps, cosmetics

Figure 15.3
Waxes are used in candles, furniture, car and shoe polishes, lotions, and cosmetics.

acids. Glycerol is a trihydroxy alcohol that can form ester links with carboxyl groups of up to three fatty acids.

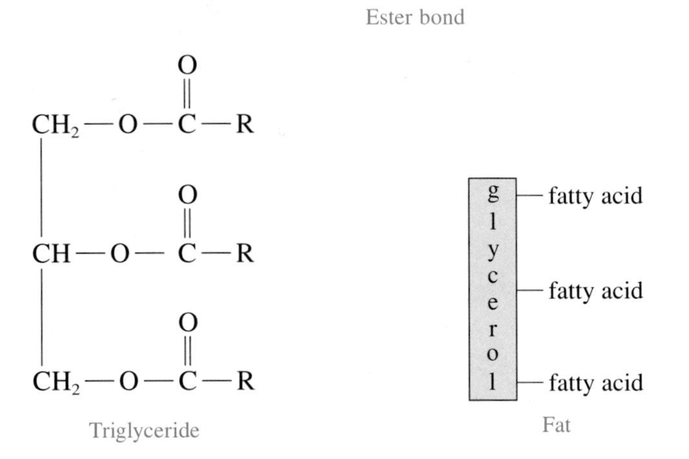

For example, in tristearin, a simple triglyceride, three stearic acid molecules form ester bonds with glycerol.

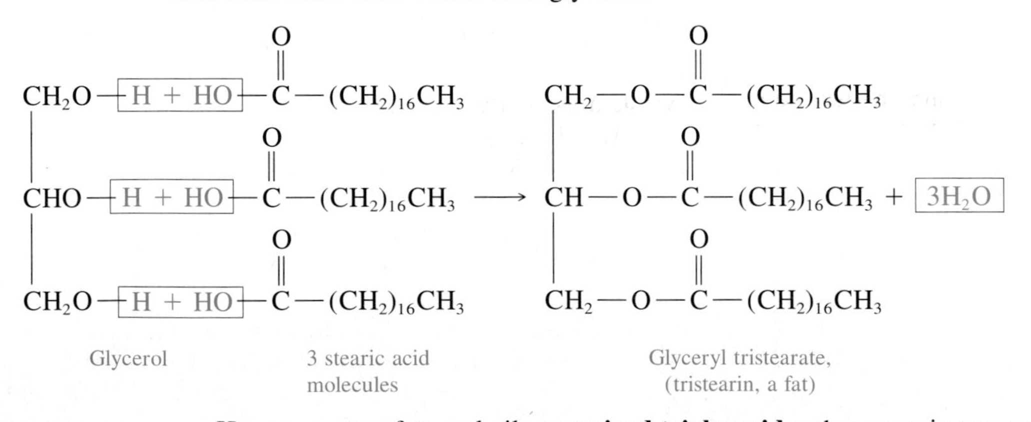

However, most fats and oils are **mixed triglycerides** that contain two or three different fatty acids. For example, a mixed triglyceride might be made from lauric acid, myristic acid, and palmitic acid. One possible structure for the mixed triglyceride follows:

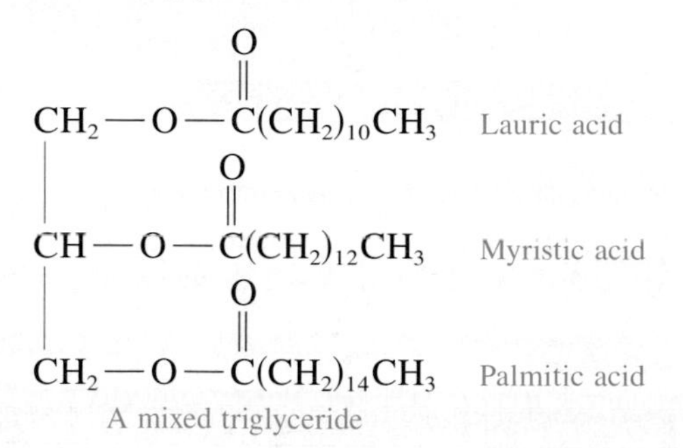

A mixed triglyceride

Sample Problem 15.3
Writing Structures for Triglycerides

Draw the structural formula of triolein, a simple triglyceride.

Solution

Triolein is the triglyceride of glycerol and three oleic acid molecules. Each fatty acid is attached by an ester bond to one of the hydroxyl groups in glycerol.

$$\begin{array}{l} CH_2-O-\overset{\overset{\displaystyle O}{\|}}{C}(CH_2)_7CH=CH(CH_2)_7CH_3 \\ | \\ CH-O-\overset{\overset{\displaystyle O}{\|}}{C}(CH_2)_7CH=CH(CH_2)_7CH_3 \\ | \\ CH_2-O-\overset{\overset{\displaystyle O}{\|}}{C}(CH_2)_7CH=(CH_2)_7CH_3 \end{array}$$

Glyceryl trioleate (triolein)

Study Check

Write the structure of trimyristin.

Figure 15.4
Vegetable oils such as olive oil, corn oil, and safflower oil contain unsaturated fats.

Melting Points of Fats and Oils

A **fat** is a triglyceride that is solid at room temperature, such as fats in meat, whole milk, butter, and cheese. Most fats come from animal sources.

An **oil** refers to a triglyceride that is a liquid at room temperature. (See Figure 15.4.) The most commonly used oils come from plant sources. However, some oils such as fish oils come from animals. Olive oil and peanut oil are called monounsaturated because they contain large amounts of oleic acid. Oils from corn, cottonseed, safflower, and sunflower are polyunsaturated because they contain large amounts of fatty acids with two or more double bonds. However, a few oils such as palm oil and coconut oil are solid at room temperature because they consist mostly of saturated fatty acids.

A comparison of the saturated, monounsaturated, and polyunsaturated fatty acids in some typical fats and oils is shown in Figure 15.5. Animal fats usually contain more saturated fatty acids than do vegetable oils. Because

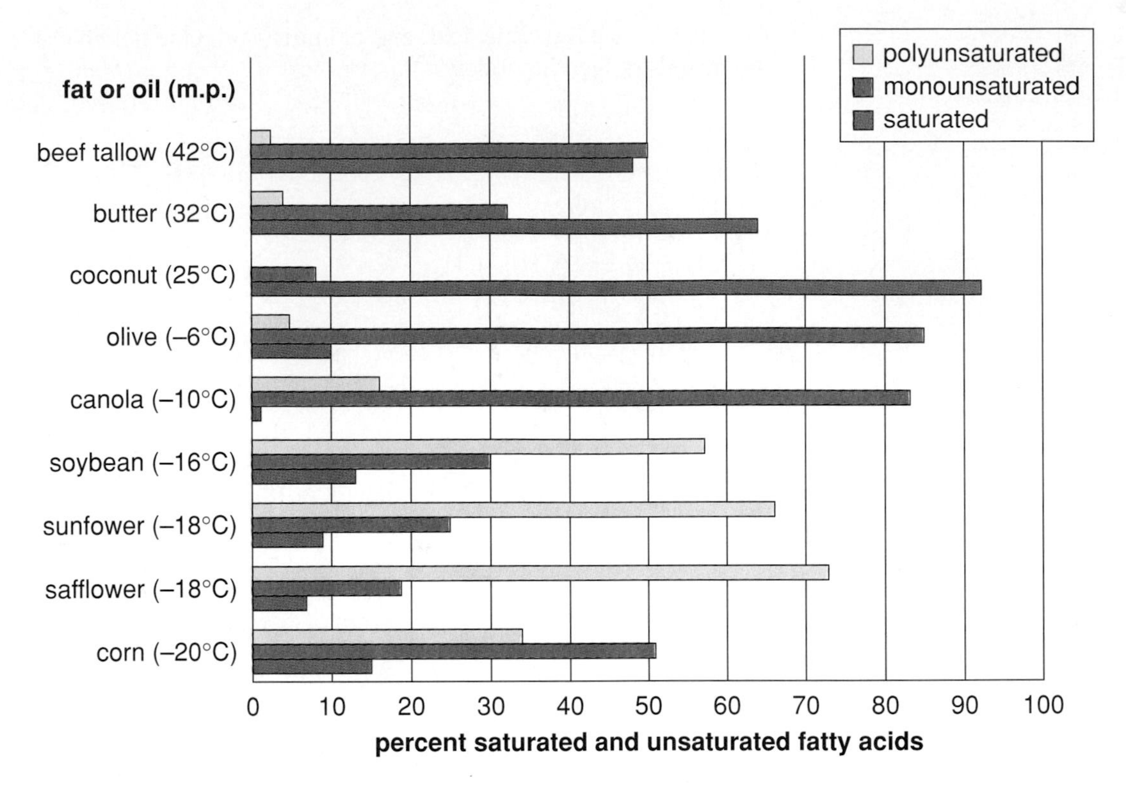

Figure 15.5
The melting points and polyunsaturation in some fats and oils.

saturated fatty acids have higher melting points than unsaturated fatty acids, the melting points of saturated fats are also higher. They are usually solid at room temperature. Because an oil contains more unsaturated fatty acids, it is usually a liquid at room temperature.

15.4 Chemical Properties of Triglycerides

Learning Goal
Draw the structure of the product from the reaction of a triglyceride with hydrogen, an acid or base, or an oxidizing agent.

The chemical reactions of the triglycerides (fats and oils) are the same as we discussed in the organic chapters for the reactions of alkenes, carboxylic acids, and esters. We will look at the addition of hydrogen to the double bonds, the oxidation of double bonds, and the acid and base hydrolysis of ester bonds.

Hydrogenation

converts double to single bond by adding hydrogen

The **hydrogenation** of unsaturated fats converts carbon–carbon double bonds to single bonds by the addition of hydrogen. The hydrogen gas is bubbled through the heated oil in the presence of a nickel catalyst.

$$-\mathrm{CH}{=}\mathrm{CH}- \; + \; \mathrm{H_2} \xrightarrow{\mathrm{Ni}} -\overset{\displaystyle \mathrm{H}}{\underset{\displaystyle \mathrm{H}}{\overset{|}{\underset{|}{\mathrm{C}}}}}-\overset{\displaystyle \mathrm{H}}{\underset{\displaystyle \mathrm{H}}{\overset{|}{\underset{|}{\mathrm{C}}}}}-$$

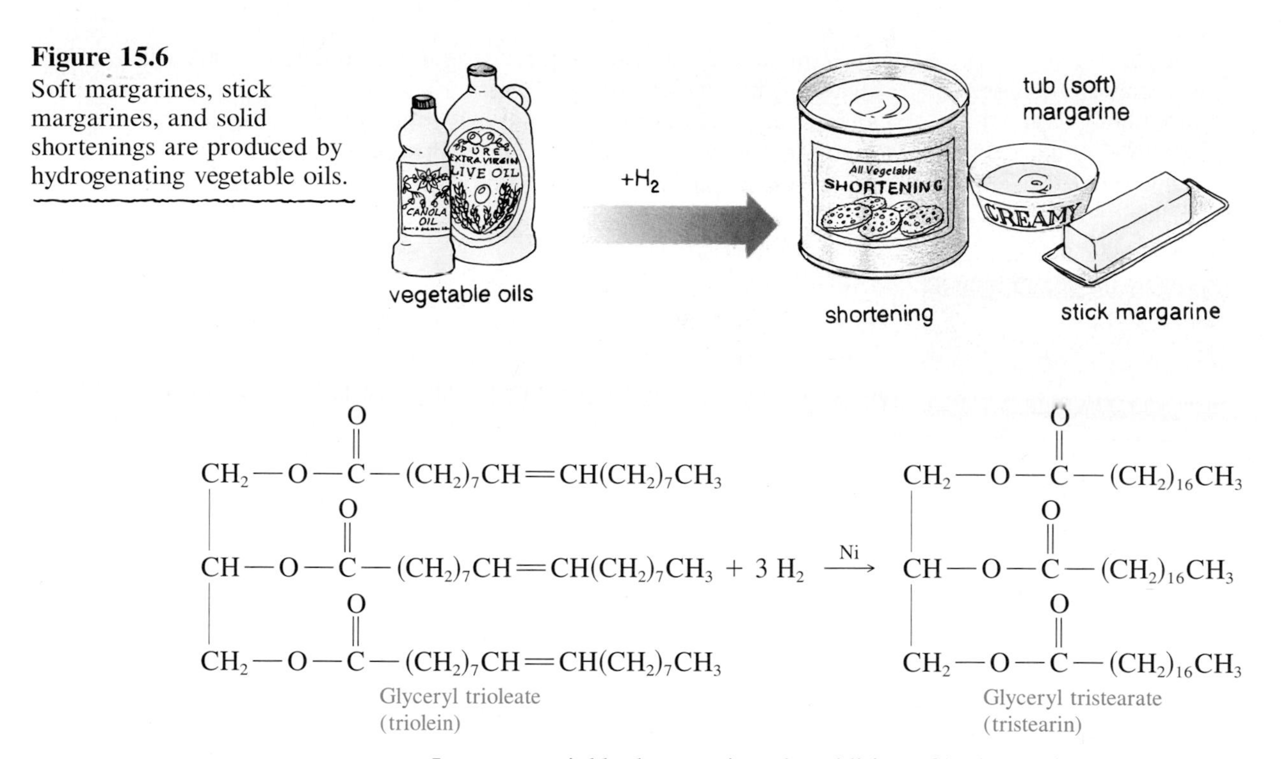

Figure 15.6
Soft margarines, stick margarines, and solid shortenings are produced by hydrogenating vegetable oils.

In commercial hydrogenation, the addition of hydrogen is stopped before all the double bonds in an oil become completely saturated. Complete hydrogenation gives a very brittle product, whereas the partial hydrogenation of a liquid vegetable oil changes it to a soft, semisolid fat. As the oil becomes more saturated, the melting point increases. Control of the degree of hydrogenation gives the various types of partially hydrogenated vegetable oil products on the market today—soft margarines, solid stick margarines, and solid shortenings. (See Figure 15.6.) Although these products now contain more saturated fatty acids than the original oils, they contain no cholesterol, unlike similar products from animal sources, such as butter and lard.

Oxidation

A fat or oil becomes rancid when its double bonds are oxidized in the presence of oxygen and microorganisms. The products are short-chain fatty acids and aldehydes that have very disagreeable odors.

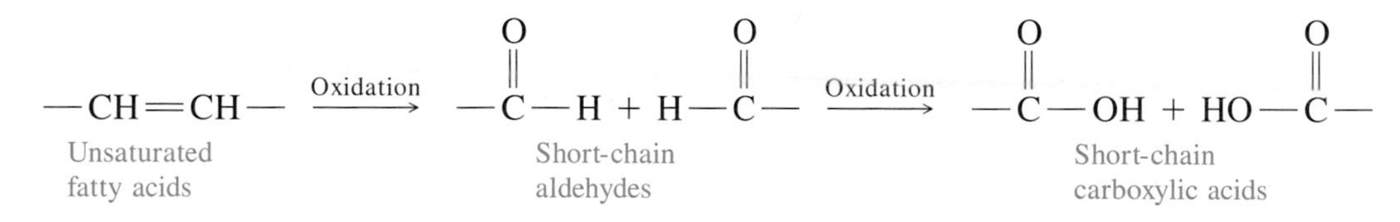

If a vegetable oil does not contain an antioxidant, it will oxidize easily. You can detect an oil that has become rancid by its unpleasant odor. If an oil is covered tightly and stored in a refrigerator, the process of oxidation can be slowed down and the oil will last longer.

Figure 15.7
After strenuous exercise, increased body temperature and perspiration promote the oxidation of oils that accumulate on the surface of the skin.

Oxidation also occurs in the oils that accumulate on the surface of the skin during heavy exercise. (See Figure 15.7.) At body temperature, microorganisms on the skin promote rapid oxidation of the oils as they are exposed to oxygen and water. The resulting short-chain aldehydes and fatty acids account for the body odor associated with workout and heavy perspiration.

Hydrolysis

Triglycerides are hydrolyzed (split by water) in the presence of strong acids or digestive enzymes called *lipases*. The products of **hydrolysis** of the ester bonds are glycerol and three fatty acids. The polar glycerol is soluble in water, but the fatty acids with their long hydrocarbon chains are not.

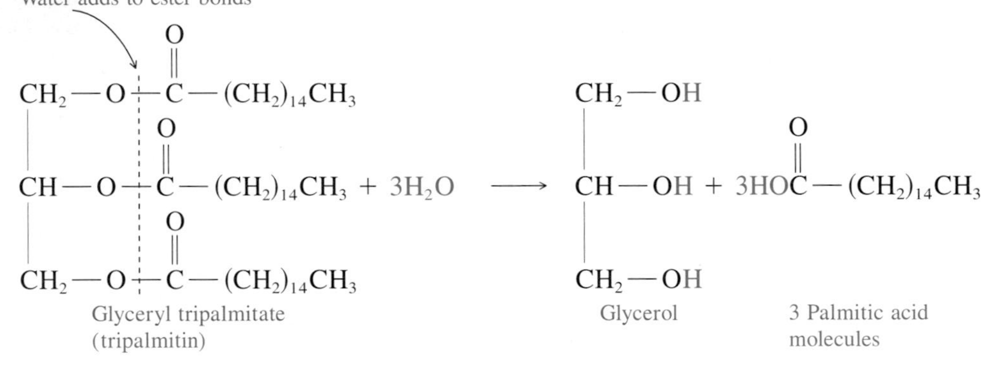

Saponification

When a fat is heated with a strong base such as sodium hydroxide, **saponification** of the fat gives glycerol and the sodium salts of the fatty acids, which are soaps. When NaOH is used, a solid soap is produced that can be molded into a desired shape; KOH produces a softer, liquid soap. Oils that are polyunsaturated produce softer soaps. Names like "coconut" or "avocado shampoo" tell you the sources of the oil used in the hydrolysis reaction.

Fat or oil + strong base → glycerol + salts of fatty acids (soaps)

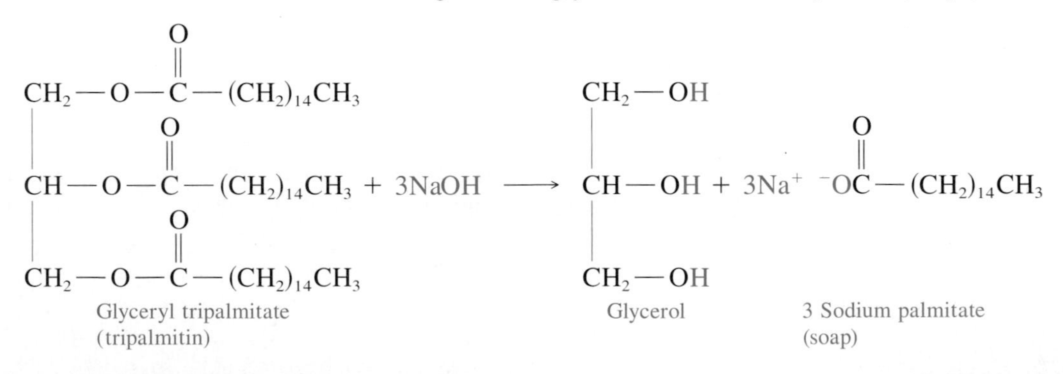

Sample Problem 15.4
Reactions of Lipids

Write the equation for the reaction of the enzyme lipase that hydrolyzes trilaurin during the digestion process.

Solution

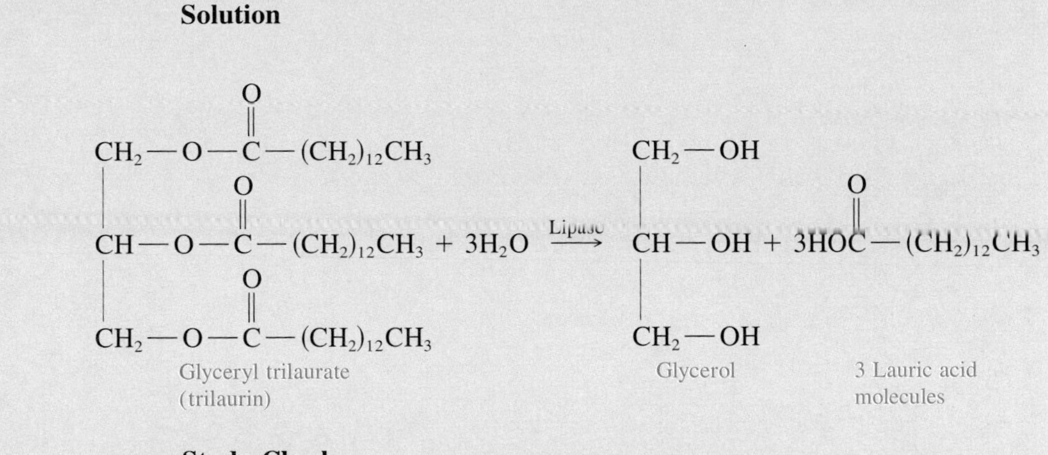

Study Check
What is the name of the product formed when a triglyceride containing oleic acid and linoleic acid is completely hydrogenated?

ENVIRONMENTAL NOTE *Cleaning Action of Soaps*

For many centuries, soaps were made by heating a mixture of animal fats (tallow) with lye, a basic solution obtained from wood ashes. Today soaps are also prepared from fats such as coconut oil. Perfumes are added to give a pleasant smelling soap. Because a soap is the salt of a long-chain fatty acid, the two ends of a soap molecule have different polarities. The long carbon chain end is nonpolar and *hydrophobic* (water-fearing). It is soluble in nonpolar substances such as oil or grease; but it is not soluble in water. The carboxylate salt end is ionic and *hydrophilic* (water-loving). It is very soluble in water but not in oils or grease.

The Dual Polarity of a Soap (Salt of a Fatty Acid)

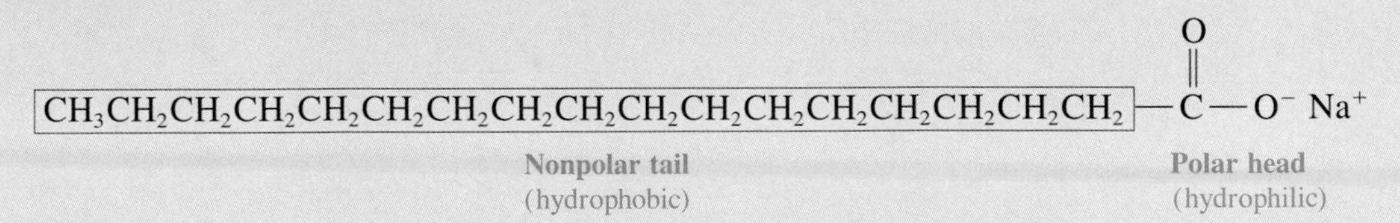

When a soap is used to clean grease or oil, the nonpolar ends of the soap molecules dissolve in nonpolar fats and oils that accompany dirt. The water-loving salt ends of the soap molecules extend outside where they can dissolve in water. The soap molecules coat the oil- or grease-forming cluster called *micelles.* The ionic ends of the soap molecules provide polarity to the micelles, which makes them soluble in water. As a result, small globules of oil and fat coated with soap molecules are pulled into the water layer and rinsed away. (See Figure 15.8.)

Figure 15.8
The cleaning action of soap. The nonpolar portion of the soap molecule dissolves in the grease and oil accompanying dirt on clothing and dishes. The polar ends of the soap molecules are attracted to water and pull the grease or oil into the aqueous solution.

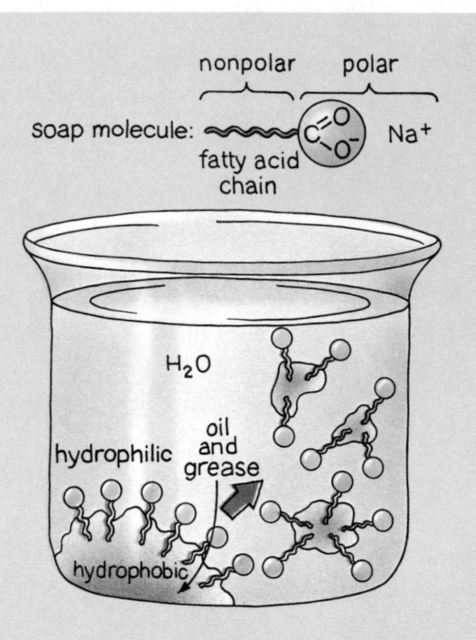

One of the problems of using soaps is that the carboxylate end reacts with ions in water such as Ca^{2+} and Mg^{2+} and forms insoluble substances.

$$2CH_3(CH_2)_{16}COO^- + Mg^{2+} \longrightarrow$$
Stearate ion — Magnesium ion

$$[CH_3(CH_2)_{16}COO^-]_2\ Mg^{2+}$$
Magnesium stearate (insoluble)

These insoluble products produce a dull coating or scum on fabrics and dishes. To solve this problem, detergents that do not form insoluble products with ions in water were developed. One type of detergent found in shampoos consists of long nonpolar hydrocarbon chains from fatty acids such as lauric acid attached to sulfate. The calcium and magnesium salts of these detergents are soluble in water. They are also biodegradable, which means they break down into nonpolluting products in the environment as do soaps.

Sodium lauryl sulfate, a detergent

$$CH_3CH_2CH_2CH_2CH_2CH_2CH_2CH_2CH_2CH_2CH_2CH_2-O-\overset{\overset{\displaystyle O}{\|}}{\underset{\underset{\displaystyle O}{\|}}{S}}-O^-\ Na^+$$

Hydrophobic part — Hydrophilic part

15.5 Phospholipids and Glycolipids

Learning Goal
Describe the components of the phosphoglycerides, sphingolipids, and glycolipids.

The phospholipids or phosphatides are a family of lipids containing glycerol, fatty acids, and a phosphate group. Unlike most other lipids, phospholipids have polar and nonpolar sections, which enables them to play a role in the transport of less soluble lipids through the bloodstream. They are also major components in the structure of cell membranes.

Phosphoglycerides

Phosphoglycerides are the most abundant lipids in cell membranes, where they play an important role in cellular permeability. They make up much of the myelin sheath that protects the nerve cells. In the body fluids, they combine with the less polar triglycerides and cholesterol to make them more soluble as they are transported in the body.

Like the triglycerides, **phosphoglycerides** contain glycerol and fatty acids, both saturated and unsaturated. They also contain a phosphate group and an amino alcohol that ionize. The resulting polar section makes phospholipids more soluble in water than most lipids.

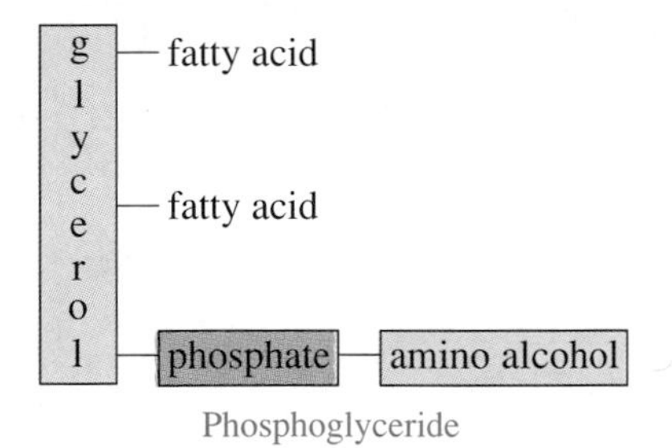

Phosphoglyceride

Some of the amino alcohols found in phosphoglycerides are listed below:

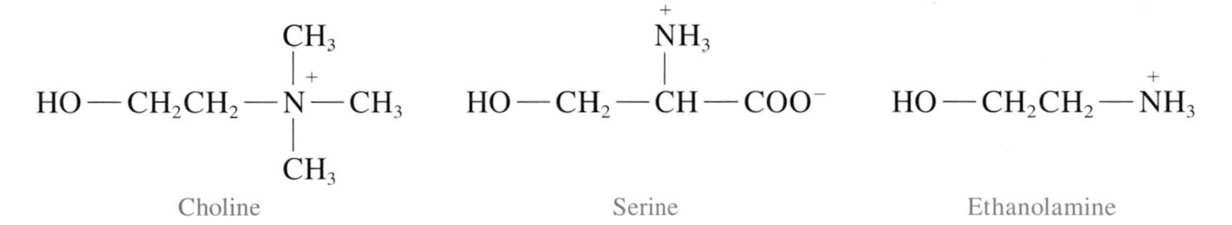

Lecithins and **cephalins** are two types of phosphoglycerides that are particularly abundant in brain and nerve tissues as well as egg yolks, wheat germ, and yeast.

Lecithins contain choline, and cephalins contain ethanolamine, and sometimes serine. In the following structural formulas, palmitic and stearic acids are used. The ionization of the phosphate and amino group provides a polar section that makes phosphoglycerides such as lecithin and cephalin soluble in water.

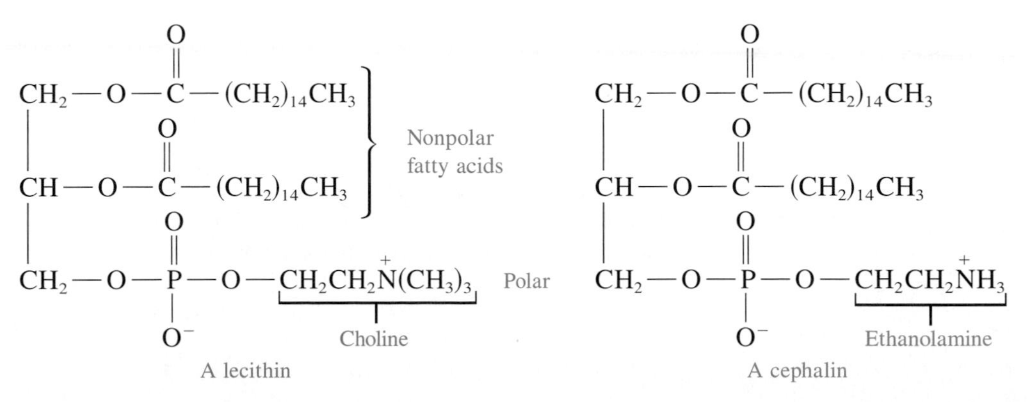

Phospholipids in Cell Membranes

One of the functions of phospholipids in cellular membranes is to separate one fluid compartment from another. Every phospholipid has both a polar and nonpolar portion, which means that one end is attracted to water and the other end is repelled. The widely accepted **fluid mosaic model** suggests that a cellular membrane consists of two layers of phospholipid molecules. In this *bilipid layer,* the hydrophobic, nonpolar portions of the phospholipids are at the center. The hydrophilic, polar groups are arranged along the outer surfaces of the membrane where they can hydrogen bond with water.

The cell membrane acts as a wall that separates the contents inside a cell from the surrounding fluids. The lipid portion allows nonpolar molecules such as oxygen and carbon dioxide to pass through but is impermeable to ions and polar molecules.

Polar molecules such as water and ions enter a cell through protein tunnels that are embedded throughout the bilipid layer. Proteins on the surface also act as receptors for hormones, neurotransmitters, antibiotics, and other chemicals that modify cellular activities. (See Figure 15.9.)

Sample Problem 15.5
Drawing Phospholipid Structures

Draw the structure of cephalin, using stearic acid for the fatty acids and serine for the amino alcohol. Describe each of the components in the phosphoglyceride.

Solution

In general, phosphoglycerides are composed of a glycerol molecule in which two carbon atoms are attached to fatty acids such as stearic acid. The third carbon atom is attached in an ester bond to phosphate linked to an amino alcohol. In this example, the amino alcohol is serine.

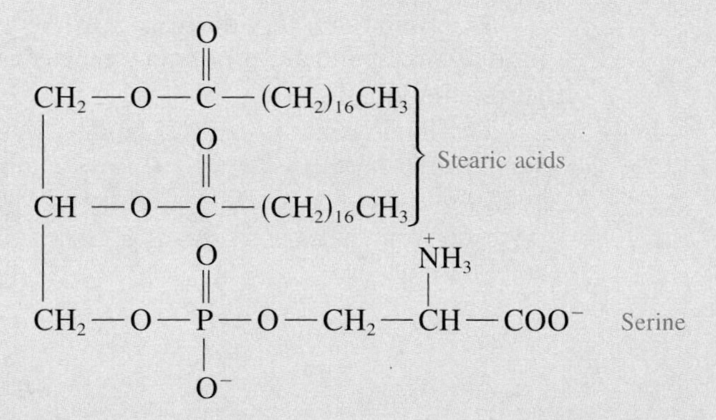

Study Check

Phospholipids containing saturated fatty acids can pack together tightly to form rigid membranes, whereas phospholipids with unsaturated fatty acids form more flexible membranes. Explain.

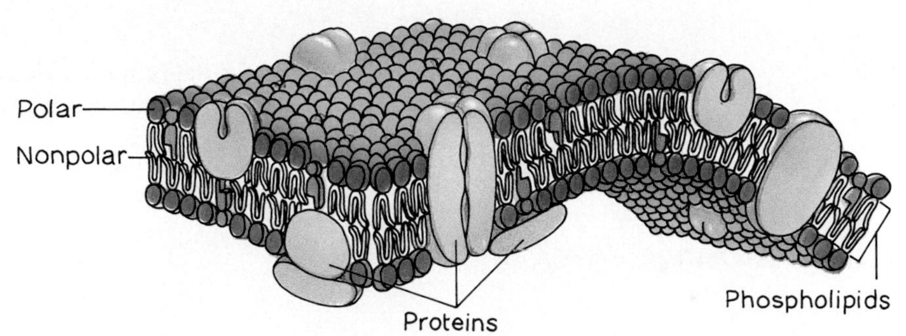

Figure 15.9
Fluid mosaic model of a cell membrane. A double row of phospholipids forms a barrier with polar ends in contact with fluids and nonpolar ends at the center away from the water. Proteins embedded in the bilipid layer permit passage of polar substances and act as receptors.

Sphingolipids

The *sphingolipids* are a group of phospholipids that are abundant in brain and nerve tissues. The **sphingolipids** are esters of an 18-carbon alcohol called sphingosine instead of glycerol.

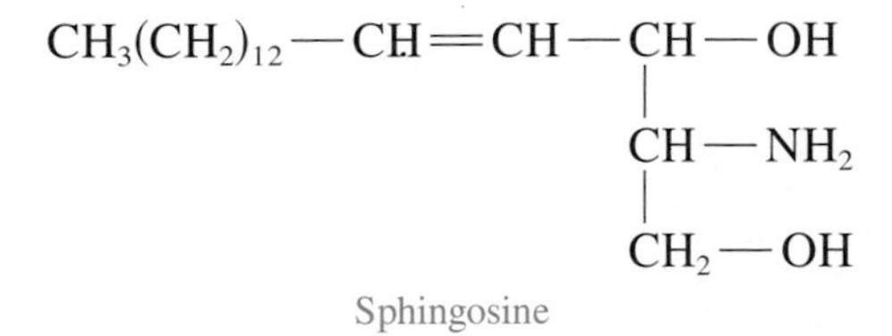

Sphingosine

One of the most abundant sphingolipids is sphingomyelin. It is the white matter of the myelin sheath, a coating surrounding the nerve cells that increases the speed of nerve impulses and insulates and protects the nerve cells. In sphingomyelin, the sphingosine is linked by an amide bond to a fatty acid, and to a phosphate ester of choline, an amino alcohol.

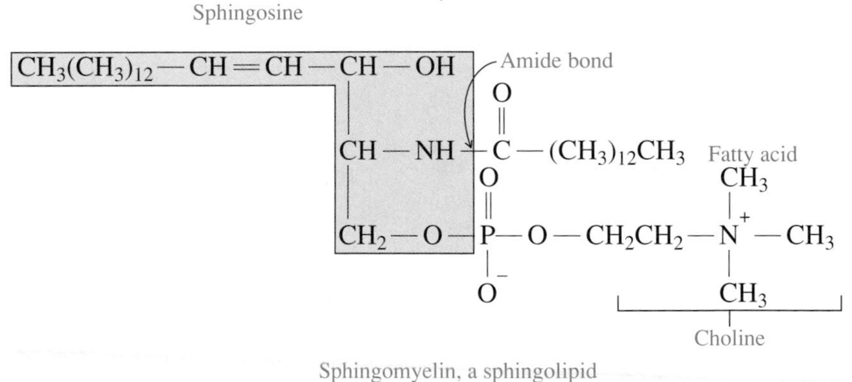

Sphingomyelin, a sphingolipid

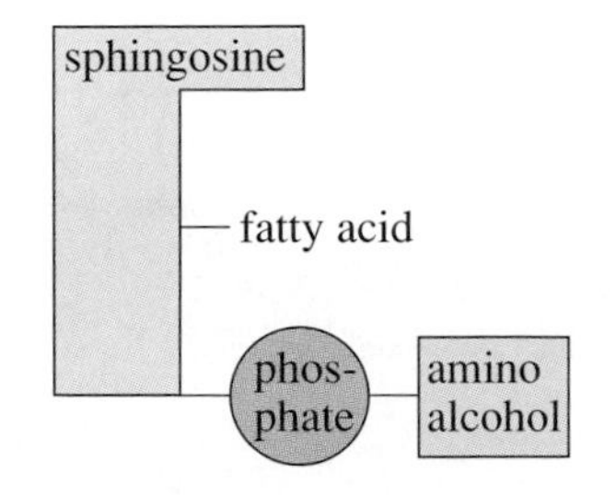

Sphingolipid

Glycolipids

Glycolipids are abundant in the brain and in the myelin sheaths of nerves. Some **glycolipids** contain glycerol attached to two fatty acids and a monosaccharide, usually galactose.

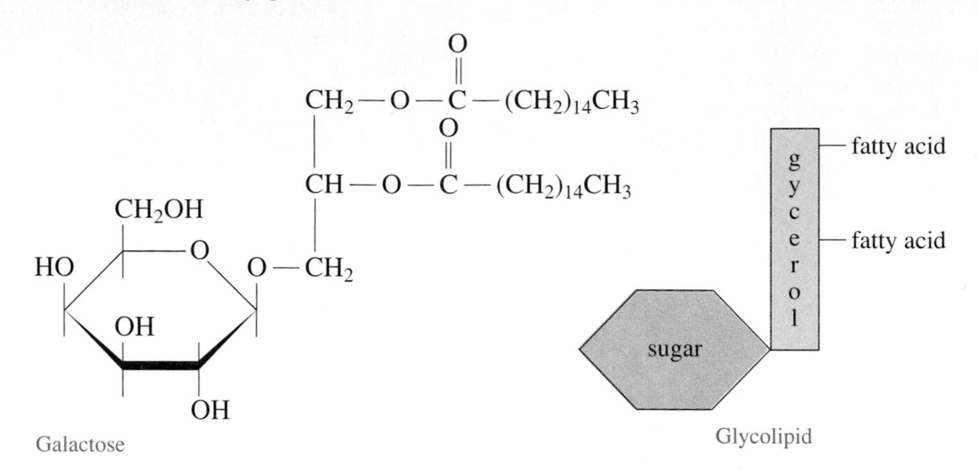

Cerebrosides, another type of glycolipid, are similar to the sphingolipids. **Cerebrosides** contain sphingosine, a fatty acid with 24 carbon atoms, and a monosaccharide, which is usually galactose and sometimes glucose. Gangliosides are similar to cerebrosides but contain two or more monosaccharides, such as glucose and galactose. Gangliosides are important in the membranes of neurons and act as receptors for hormones, viruses, and several drugs.

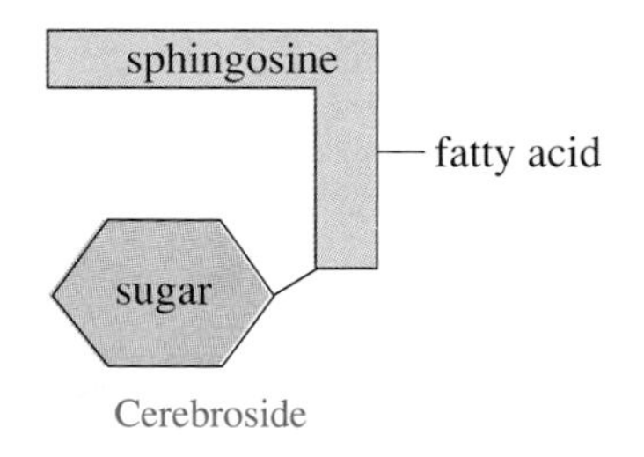

Cerebroside

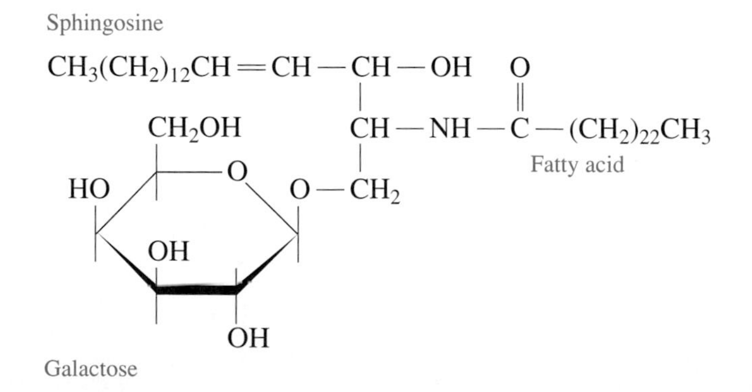

Galactocerebroside

HEALTH NOTE *Lipid Diseases*

Many lipid diseases **(lipidoses)** involve the excessive accumulation of a sphingolipid or glycolipid because an enzyme needed for its breakdown is deficient or absent. The accumulation of these glycolipids may enlarge the spleen, liver, and bone marrow cells (Gaucher's disease) and cause mental retardation, seizures, blindness, and death in early infancy. Some lipid storage diseases are listed in Table 15.4.

Table 15.4 ***Lipid Diseases***

Names of Disease	Lipid Stored	Type	Enzyme Absent
Fabry's	Gal-gal-glucosylceramide	Ganglioside	α-Galactosidase
Gaucher's	Glucosylceramide	Cerebroside	β-Glucosidase
Niemann-Pick	Sphingomyelin	Sphingolipid	Sphingomyelinase
Tay-Sachs	G_{m2} ganglioside	Ganglioside	Hexosaminidase A

In multiple sclerosis, sphingomyelins are lost from the myelin sheath, which is the protective membrane surrounding the neurons in the brain and spinal cord. (See Figure 15.10.) As the disease progresses, the myelin sheath deteriorates. Scars form on the neurons, and impair the transmission of nerve signals. The symptoms of multiple sclerosis include various levels of muscle weakness and loss of coordination and vision depending on the amount of damage. The cause of multiple sclerosis is not yet known although some researchers suggest that a virus is involved.

Figure 15.10
The myelin sheath with its high lipid content forms a protective membrane around the nerve axons in the brain and spinal cord.

Sample Problem 15.6
Glycolipids

In Fabry's disease, the trihexosylceramide shown below accumulates due to a deficiency of α-galactosidase. Identify the components A–E in this glycolipid.

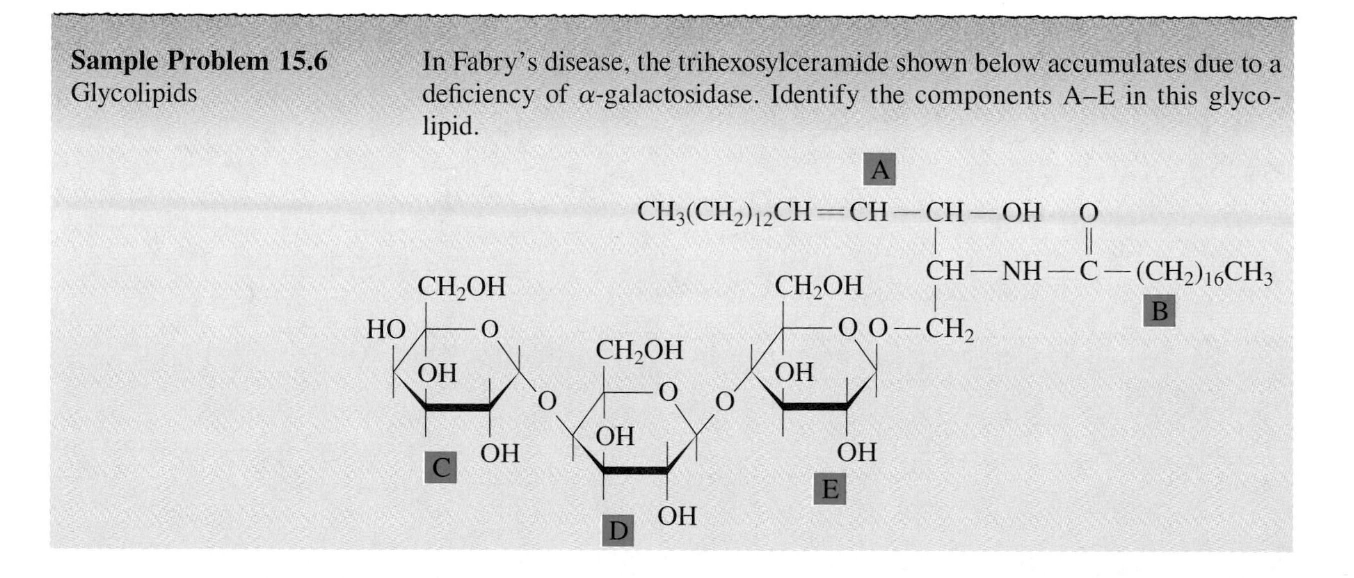

Solution

In this glycolipid, the components are sphingosine (A), stearic acid (B), an 18-carbon fatty acid, two galactose units (C, D), and one glucose (E).

Study Check

How do we know that this glycolipid is a breakdown product of a ganglioside rather than a cerebroside?

15.6 Steroids and Cholesterol

Learning Goal

Describe the structure of a steroid and cholesterol.

The word steroid comes from the Latin word *stereos,* meaning "solid." **Steroids** are compounds containing the steroid nucleus, which consists of three cyclohexane rings and a cyclopentane fused together.

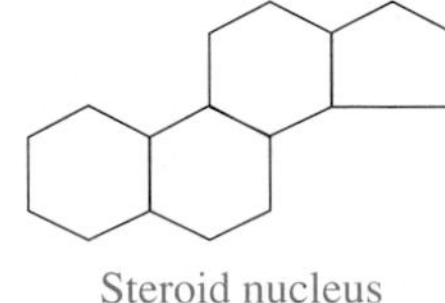

Steroid nucleus

By attaching various groups to the steroid nucleus, many steroid compounds are formed including several that are of medical and biochemical importance. These include cholesterol, bile salts, steroid hormones, male and female sex hormones, and anabolic steroids.

Cholesterol

Cholesterol is the most important and abundant steroid compound in the body. It is called a *sterol* because it contains an alcohol group (—OH) on the first ring. In cholesterol, methyl groups and a branched chain of carbon atoms have been added to the steroid nucleus.

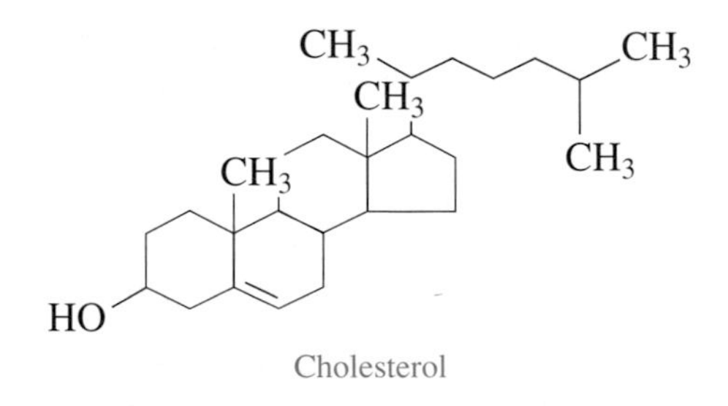

Cholesterol

Sample Problem 15.7
Cholesterol

Observe the structure of cholesterol for the following questions:

a. What part of cholesterol is the steroid nucleus?

b. What features have been added to the steroid nucleus in cholesterol?

c. What classifies cholesterol as a sterol?

Solution

a. The four fused rings form the steroid nucleus.

b. The cholesterol molecule contains an alcohol group ($-OH$) on the first ring, one double bond in the second ring, two methyl groups ($-CH_3$), and a branched carbon chain.

c. The alcohol group determines the sterol classification.

Study Check

Why is cholesterol in the lipid family?

HEALTH NOTE *Cholesterol in the Body*

Cholesterol is a component of cellular membranes, myelin sheath, and brain and nerve tissue. It is also found in the liver, bile salts, and skin, where it forms vitamin D. In the adrenal gland, it is used to synthesize steroid hormones. Whereas cholesterol in the body is obtained from eating meats, milk, and eggs, it is also synthesized by the liver from fats, carbohydrates, and proteins. There is no cholesterol in vegetable and plant products.

If a diet is high in cholesterol, the liver produces less. A typical daily American diet includes 400–500 mg of cholesterol, one of the highest in the world. The American Heart Association has recommended that we consume no more than 300 mg of cholesterol a day. The cholesterol contents of some typical foods are listed in Table 15.5.

Table 15.5 *Cholesterol Content of Some Foods*

Food	Serving Size	Cholesterol (mg)
Liver (beef)	3 oz	370
Egg	1	250
Lobster	3 oz	175
Fried chicken	$3\frac{1}{2}$ oz	130
Hamburger	3 oz	85
Chicken (no skin)	3 oz	75
fish (salmon)	3 oz	40
Butter	1 tablespoon	30
Whole milk	1 cup	35
Skim milk	1 cup	5
Margarine	1 tablespoon	0

When cholesterol exceeds its saturation level in the bile, gallstones may form. Gallstones are composed of almost 100% cholesterol with some calcium salts, fatty acids, and phospholipids. High levels of cholesterol are associated with the accumulation of lipid deposits (plaques) that line and narrow the coronary arteries. (See Figure 15.11.) Clinically, cholesterol levels are considered elevated if the total plasma cholesterol level exceeds 200–220 mg/dL.

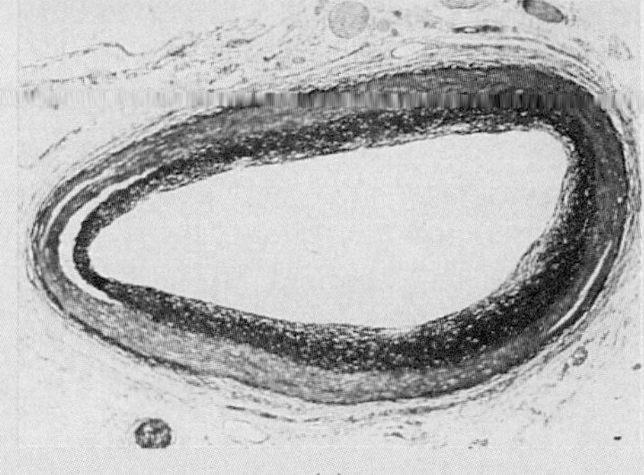

(a)

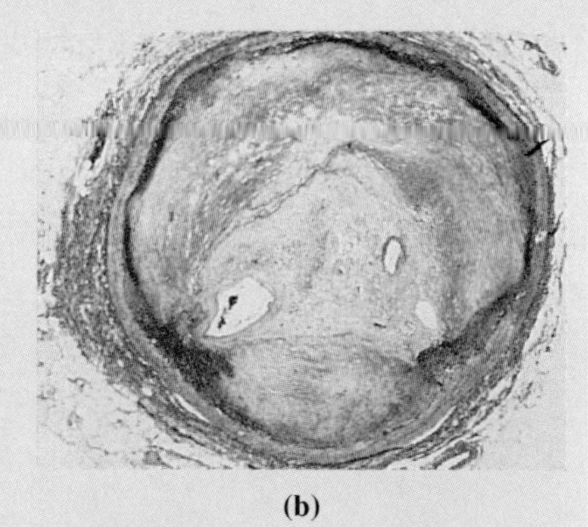

(b)

Figure 15.11
Excess cholesterol forms plaque that eventually can block an artery, resulting in a heart attack.
(a) A normal, open artery shows no buildup of plaque.
(b) An artery that is almost completely clogged by atherosclerotic plaque.

Some research indicates that saturated fats in the diet may stimulate the production of cholesterol by the liver. A diet that is low in foods containing cholesterol and saturated fats appears to be helpful in reducing the serum cholesterol level. Other factors that may also increase the risk of heart disease are family history, lack of exercise, smoking, obesity, diabetes, gender, and age.

Bile Salts

The *bile* salts are synthesized in the liver from cholesterol and stored in the gallbladder. When bile is secreted into the small intestine, the bile salts mix with the water-insoluble fats and oils in our diets. The bile salts with their nonpolar and polar regions act much like soaps, breaking apart and emulsifying large globules of fat. The emulsions that form have a larger surface area for the lipases, enzymes that digest fat. The bile salts also help in the absorption of cholesterol into the intestinal mucosa.

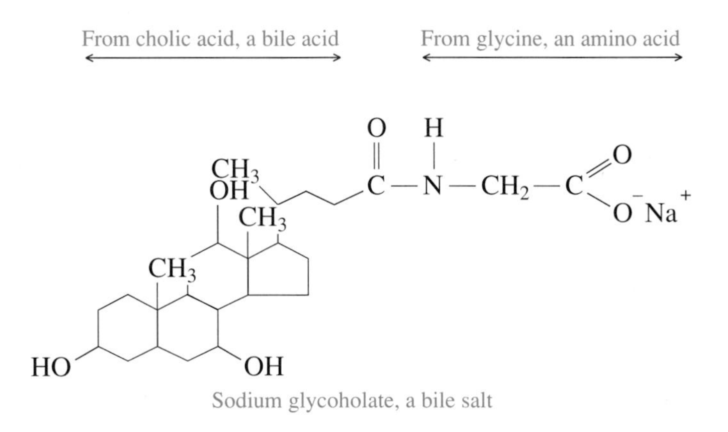

Sodium glycoholate, a bile salt

Fat-Soluble Vitamins

The fat-soluble vitamins A, D, E, and K are all nonpolar, which makes them soluble in body fat, where they are stored. An excess of a fat-soluble vitamin can result in hypervitaminosis and damage to the body. Vitamin D, a fat-soluble vitamin related to cholesterol, is produced in the skin from reactions of sunlight and is found in sardines, salmon, and egg yolks. In the body, vitamin D increases the absorption of calcium from the intestinal tract and regulates the amount of calcium deposited in the bones and teeth. The other fat-soluble vitamins are unsaturated hydrocarbons. Vitamin A is important in the formation of visual pigments for night vision. Vitamin K is important in blood clotting. Table 15.6 lists the fat-soluble vitamins.

Table 15.6 *Fat-Soluble Vitamins*

Structure	Biological Function	Deficiency	Symptoms of Excess
Retinol (vitamin A)	Formation of visual pigment; development of epithelial cells	Night blindness; dry, scaly skin; bacterial infections	Hypervitaminosis, irritability, sloughing of skin, joint pain, weight loss, liver enlargement
Calciferol (vitamin D)	Absorption of calcium and phosphate; deposition of calcium and phosphate in bone	Rickets; bone decalcification; bone deformities	Hypervitaminosis D; muscle weakness, nausea, diarrhea, anorexia; calcification in tissue, kidney damage
α-Tocopherol (vitamin E)	May prevent oxidation of unsaturated fatty acids in cell membranes	Hemolysis of red blood cells; infertility in rats	
Phylloquinone (vitamin K)	Synthesis of prothrombin for blood clotting	Bruising, hemorrhaging from minor cuts, longer clotting time	

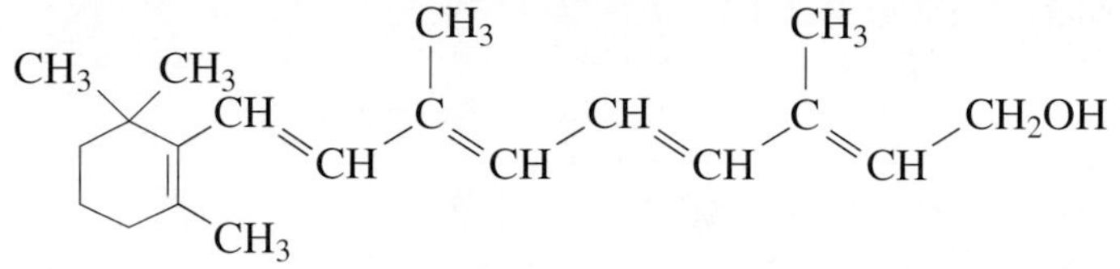

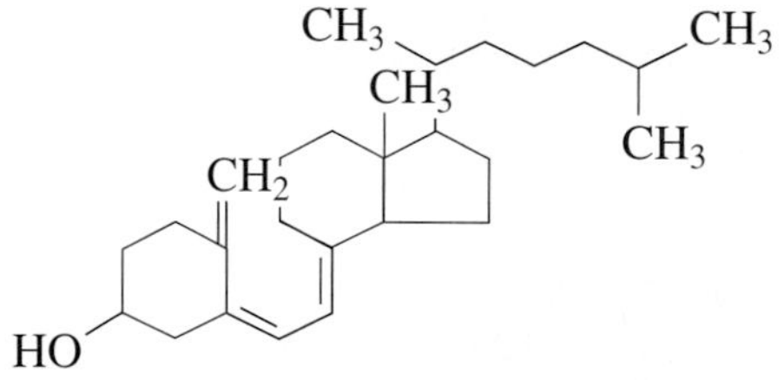

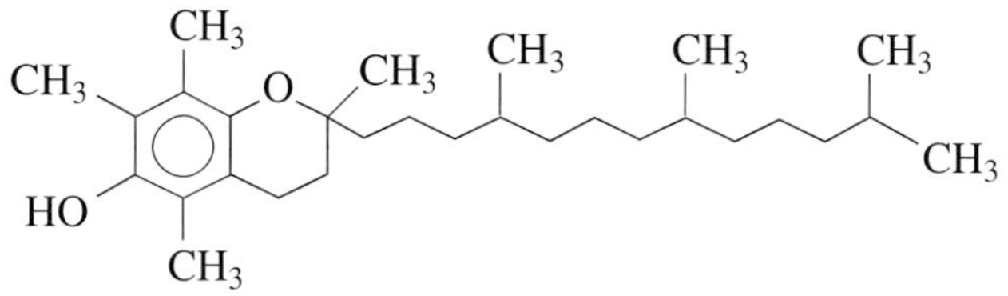

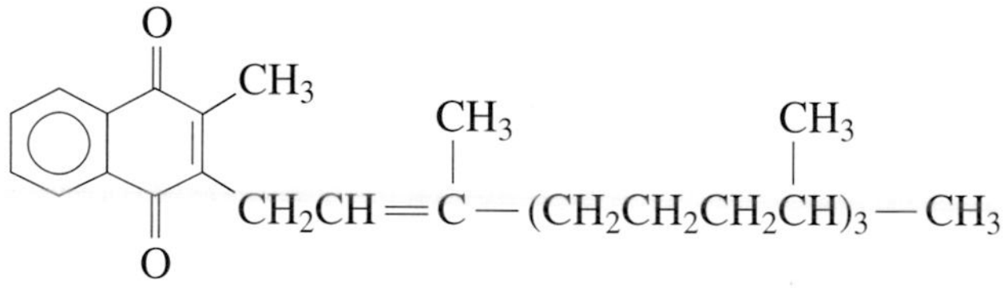

Retinol structure: CH_3, CH_3 on ring; ring–$CH{=}CH{-}C(CH_3){=}CH{-}CH{=}CH{-}C(CH_3){=}CH{-}CH_2OH$; ring CH_3

Calciferol structure: HO, CH_2, CH_3, CH_3, CH_3, CH_3

α-Tocopherol structure: CH_3, CH_3, O, HO, CH_3, CH_3, CH_3, CH_3, CH_3, CH_3

Phylloquinone structure: O, O, CH_3, $CH_2CH{=}C(CH_3){-}(CH_2CH_2CH_2CH(CH_3))_3{-}CH_3$

HEALTH NOTE *Lipoproteins: Transporting Lipids in Solution*

Because trigylcerides and cholesterol are nonpolar, they are insoluble in blood and are not transported easily. To improve their solubility, they attach to proteins and phospholipids, forming a polar complex called a **lipoprotein.** (See Figure 15.12.)

There are four types of lipoproteins of varying densities and lipid composition. The chylomicrons and the VLDLs (very-low density lipoproteins) are the major carriers of triglycerides between the intestines or the liver and the cells in the body where they are used for energy or stored as adipose tissue.

The LDLs (low-density lipoproteins) carry cholesterol to the tissues to be used for the synthesis of cell membranes and steroid hormones. If the need for cholesterol is exceeded by the LDLs, excess cholesterol can accumulate in the cells and the arteries. The HDLs (high-density lipoproteins) carry cholesterol from the tissues to the liver where it produces bile salts, which are excreted. (See Table 15.7.)

Figure 15.12
A lipoprotein particle; a form in which nonpolar lipids are carried through the bloodstream.

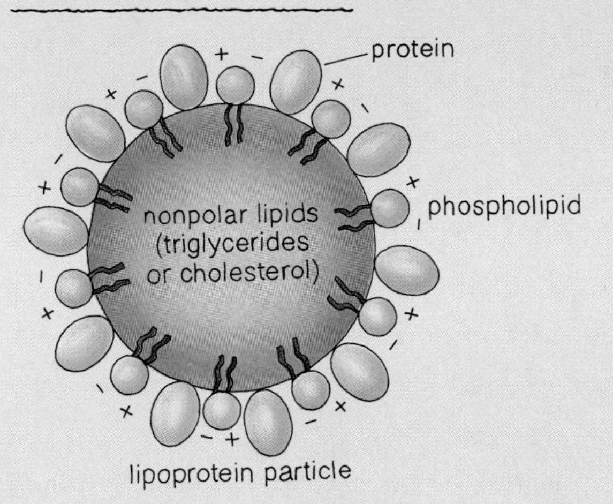

Table 15.7 *Some Properties of Plasma Lipoproteins*

Type	Density (g/mL)	Protein (%)	Phospholipid (%)	Other Lipid[a] (%)
Chylomicrons	0.94	2	9	84
VLDL	0.940–1.006	10	18	72
LDL	1.006–1.063	25	20	55
HDL	1.063–1.210	55	25	20

[a] Other lipid includes cholesterol and triglycerides.

Because high cholesterol levels have been associated with the onset of atherosclerosis and heart disease, the measurement of the serum levels of LDL and HDL is of much interest to medical researchers. When the LDL level is high, there is an increase in the levels of triglycerides and cholesterol in the blood serum. The excess cholesterol can deposit in the blood vessels, restricting blood flow and increasing the risk of developing heart disease and myocardial infarctions (heart attacks).

Low levels of LDL and high levels of HDL are associated with decreased risk of myocardial infarctions because HDLs remove cholesterol from the cells. A lower level of serum cholesterol decreases the risk of plaque formation within the blood vessels and, therefore, heart disease. Higher HDL levels are found in persons who exercise regularly and eat less saturated fats. They are higher in women whose estrogen levels are adequate.

15.7 Steroid Hormones

Learning Goal
Describe the steroid hormones.

The word *hormone* comes from the Greek "to arouse" or "to excite." Hormones are chemical messengers that serve as a kind of communication system from one part of the body to another. The *steroid* hormones, which include the sex hormones and the adrenocortical hormones, are closely related in structure to cholesterol and depend on cholesterol for their synthesis.

Sex Hormones

Two important male sex hormones, *testosterone* and *androsterone*, promote the growth of muscle, facial hair, and the maturation of the male sex organs and of sperm. Table 15.8 gives the structures of some of the steroid hormones and their physiological effects.

Table 15.8 ***Examples of Some Steroid Hormones Derived from Cholesterol***

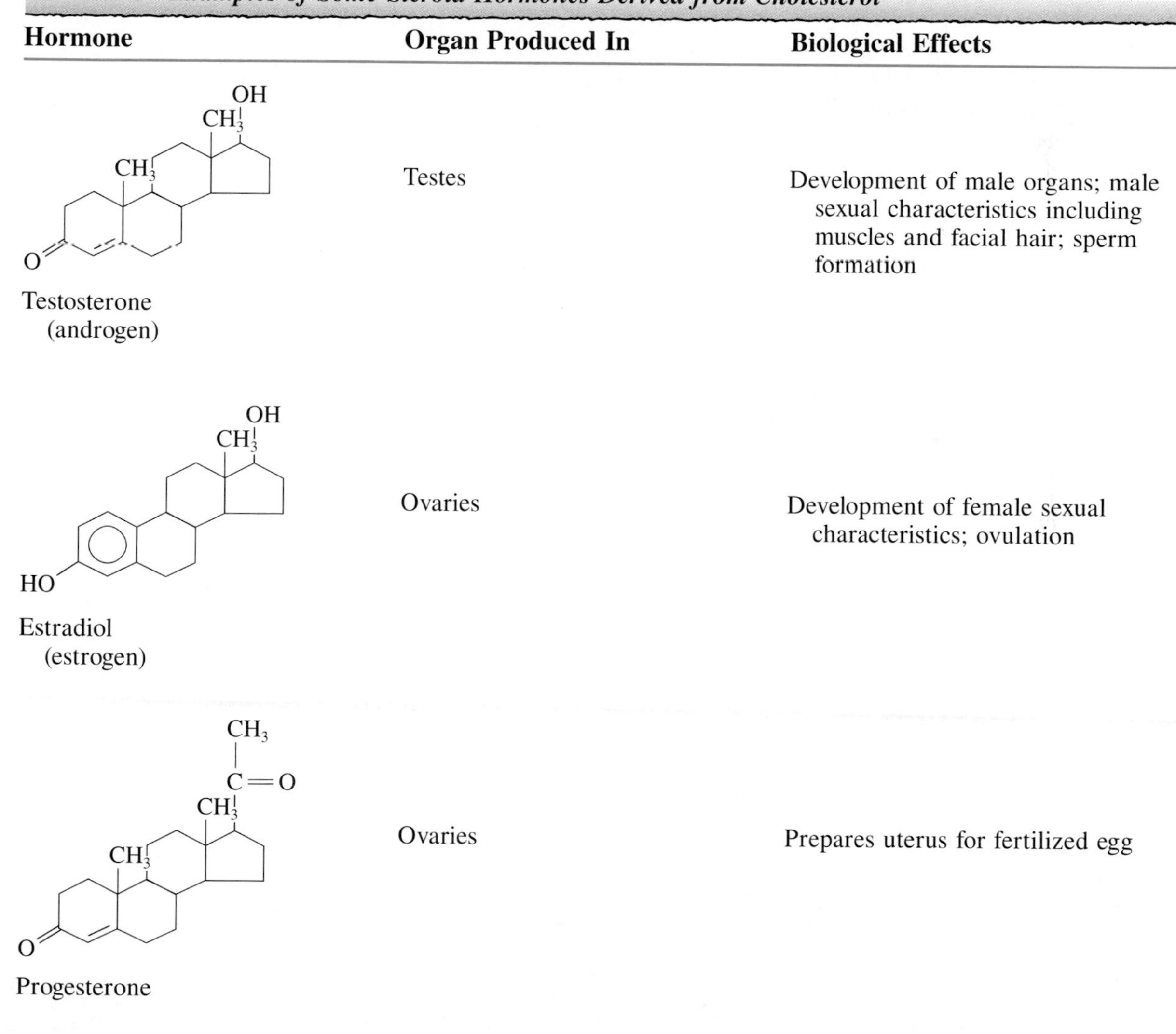

Hormone	Organ Produced In	Biological Effects
Testosterone (androgen)	Testes	Development of male organs; male sexual characteristics including muscles and facial hair; sperm formation
Estradiol (estrogen)	Ovaries	Development of female sexual characteristics; ovulation
Progesterone	Ovaries	Prepares uterus for fertilized egg

Table 15.8 Examples of Some Steroid Hormones Derived from Cholesterol *(cont.)*

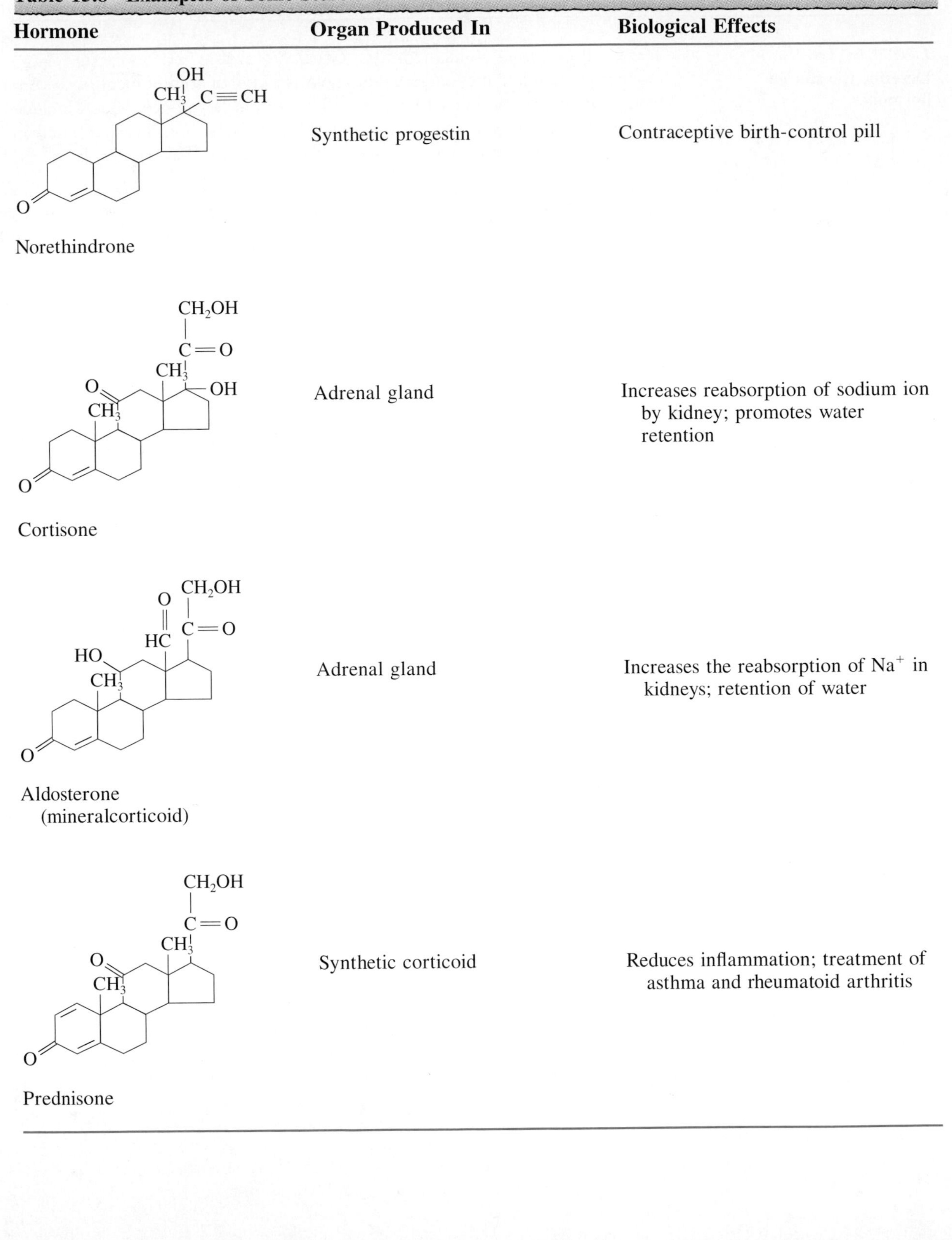

Hormone	Organ Produced In	Biological Effects
Norethindrone	Synthetic progestin	Contraceptive birth-control pill
Cortisone	Adrenal gland	Increases reabsorption of sodium ion by kidney; promotes water retention
Aldosterone (mineralcorticoid)	Adrenal gland	Increases the reabsorption of Na^+ in kidneys; retention of water
Prednisone	Synthetic corticoid	Reduces inflammation; treatment of asthma and rheumatoid arthritis

The *estrogens,* a group of female sex hormones, direct the development of female sexual characteristics: the uterus increases in size, fat is deposited in the breasts, and the pelvis broadens. *Progesterone* prepares the uterus for the implantation of a fertilized egg. If an egg is not fertilized, the levels of progesterone and estrogen drop sharply, and menstruation follows. Synthetic forms of the female sex hormones are used in birth-control pills. As with other kinds of steroids, side effects include weight gain and a greater risk of forming blood clots.

Adrenal Corticosteroids

The adrenal glands, located on the top of each kidney, produce the corticosteroids. *Aldosterone,* a mineralcorticoid, is responsible for electrolyte and water balance by the kidneys. *Cortisone,* a glucocorticoid, increases the blood glucose level and stimulates the synthesis of glycogen in the liver from amino acids. Synthetic corticoids such as *prednisone* are derived from cortisone and used medically for reducing inflammation and treating asthma and rheumatoid arthritis, although precautions are given for long-term use.

HEALTH NOTE *Anabolic Steroids*

Some of the physiological effects of testosterone are to increase muscle mass and decrease body fat. Derivatives of testosterone called *anabolic steroids* that enhance these effects have been synthesized. Although they have some medical uses, anabolic steroids have been used in rather high dosages by some athletes in an effort to increase muscle mass. Such use is illegal.

Use of anabolic steroids in attempting to improve athletic strength can cause several side effects including hypertension, fluid retention, increased hair growth, sleep disturbances, and acne. Over a long period of time, their use can be devastating and may cause irreversible liver damage and decreased sperm production.

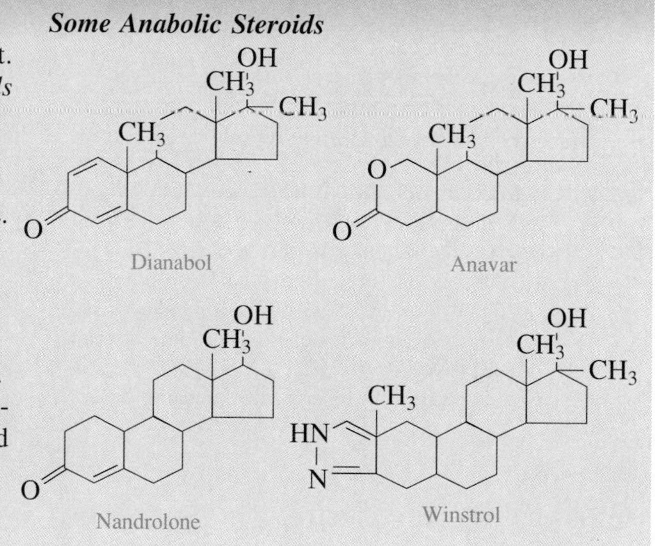

Sample Problem 15.8
Steroid Hormones

What are the functional groups on the steroid nucleus in the sex hormones estradiol and testosterone?

Solution

Estradiol contains a benzene ring, a methyl group, and an alcohol group. Testosterone contains a ketone group, a double bond, two methyl groups, and an alcohol group.

Study Check

What are the similarities and differences in the structures of testosterone and the anabolic steroid Nandrolone?

Chapter Summary

15.1 Types of Lipids
Lipids are nonpolar compounds that are not soluble in water. Classes of lipids include waxes, fats and oils, phospholipids, and steroids.

15.2 Fatty Acids
Fatty acids are unbranched carboxylic acids that typically contain an even number (12–18) carbon atoms. Fatty acids may be saturated, monounsaturated with one double bond, or polyunsaturated with two or more double bonds. The double bonds in unsaturated fatty acids are almost always cis.

15.3 Waxes, Fats, and Oils
A wax is an ester of a long-chain fatty acid and a long-chain alcohol. The triglycerides of fats and oils are esters of glycerol with three long-chain fatty acids. Fats contain more saturated fatty acids and have higher melting points than most vegetable oils.

15.4 Chemical Properties of Triglycerides
The hydrogenation of unsaturated fatty acids converts double bonds to single bonds. The oxidation of unsaturated fatty acids produces short-chain fatty acids with disagreeable odors. The hydrolysis of the ester bonds in fats or oils produces glycerol and fatty acids. In saponification, a fat heated with a strong base produces glycerol and the salts of the fatty acids or soaps. The dual polarity of a soap permits its solubility in both water and fats.

15.5 Phospholipids and Glycolipids
Phosphoglycerides are esters of glycerol with two fatty acids and a phosphate group attached to an amino alcohol. In sphingolipids, the alcohol sphingosine forms an ester bond with one fatty acid and the phosphate–amino alcohol group. In glycolipids, glycerol or sphingosine is bonded to fatty acids and one or more monosaccharides. Phospholipids are the major components of the lipid bilayer of cellular membranes.

15.6 Steroids and Cholesterol
Steroids are lipids containing the steroid nucleus, which is a fused structure of four rings. Steroids include cholesterol, bile salts, and vitamin D.

15.7 Steroid Hormones
The steroid hormones are closely related in structure to cholesterol and depend on cholesterol for their synthesis. The sex hormones such as estrogen and testosterone are responsible for sexual characteristics and reproduction. The adrenal corticosteroids such as aldosterone and cortisone regulate water balance and glucose levels in the cells.

Glossary of Key Terms

cephalins Phospholipids found in brain and nerve tissues that incorporate the amino alcohol serine or ethanolamine.

cerebroside A glycolipid consisting of sphingosine, a fatty acid with 24 carbon atoms, and a monosaccharide (usually galactose).

cholesterol The most prevalent of the steroid compounds found in cellular membranes; needed for the synthesis of vitamin D, hormones, and bile acids.

fat Another term for solid triglycerides.

fatty acids Long-chain carboxylic acids found in fats.

fluid mosaic model A model of a cell membrane in which phospholipids are arranged as a bilipid layer interspersed with proteins arranged at different depths.

glycolipid A phospholipid that contains one or more monosaccharides.

hydrogenation The addition of hydrogen to unsaturated fats.

hydrolysis The splitting of bonds by water, usually in the presence of acid or enzymes; in fats, hydrolysis breaks apart the ester bonds between glycerol and fatty acids.

lecithins Phospholipids containing choline as the amino alcohol.

lipidoses Genetic diseases in which a deficiency of an enzyme for the hydrolysis of a lipid causes the accumulation of that lipid to toxic levels.

lipids A family of compounds that is nonpolar in nature and not soluble in water; includes fats, waxes, phospholipids, and steroids.

lipoprotein A combination of nonpolar lipids with proteins to form a polar complex that can be transported through body fluids.

mixed triglycerides Fats or oils that contain two or more different fatty acids.

oil Another term for liquid triglycerides.

phosphoglycerides Polar lipids of glycerol attached to two fatty acids and a phosphate group connected to an amino group such as choline, serine, or ethanolamine.

prostaglandins A number of compounds derived from arachidonic acid that regulate several physiological processes.

saponification The reaction of fats with strong bases producing salts of the fatty acids, known as soaps, and glycerol.

saturated Saturated lipids composed of saturated fatty acids that have no double bonds; they have higher melting points than unsaturated lipids and are usually solid at room temperatures.

sphingolipids Phospholipids in which sphingosine has replaced glycerol.

steroids Types of lipid composed of a multicyclic ring system.

triglycerides A family of lipids composed of three fatty acids bonded through ester bond to glycerol, a trihydroxy alcohol.

unsaturated A lipid or fatty acid that contains double bonds.

wax The ester of a long-chain alcohol and a long-chain saturated fatty acid.

Chemistry at Home

1. Read the labels on butter, margarines, and vegetable oils. Describe the terms used such as saturated, unsaturated, cholesterol or no cholesterol, and partially hydrogenated. Which type of product, plant or animal, contains no cholesterol?
2. Read the nutrition facts label on some packaged foods such as cereals, potato chips, and milk. How many grams of fat are in one serving? How many calories (kilocalories) come from the fat (9 kcal/g)? What is the percentage of fat in the total calories for one serving?
3. Place some nuts (walnuts, pecans, and so on) in a plastic bag or on a piece of paper. Carefully smash the nuts with a hammer or rolling pin. The oil from the nuts should appear as it is pressed out of the nuts. This is an example of how peanut oil or walnut oil is obtained. How would you obtain oil from olives?
4. Place some water in a small bowl or dish. Add a drop of vegetable oil. Then add a second, and a third. Do the drops of oil mix with the water? Do they stay separated? Explain. Add a few drops of soap and mix. What happens to the oil layer? Explain.
5. In mayonnaise, oil and water stay mixed in emulsions because egg yolks are added. The lipids in egg yolk are emulsifying agents that coat the oil particles so the oil particles don't come together to form a separate layer. To make mayonnaise, place 1 egg yolk, $\frac{1}{4}$ teaspoon salt (if desired), and a $\frac{1}{2}$ teaspoon of vinegar in a bowl and use an electric beater to mix. Add $\frac{1}{2}$ cup of oil, slowly at first, drop by drop, to the egg yolk mixture, and faster as the mixture thickens. This allows small amounts of oil to be coated by the emulsifying agent in the egg yolk. Add a teaspoon of vinegar. The product should be yellow and shiny with no separate layers. You can use it, but with no preservatives, it may spoil quickly.

Answers to Study Checks

15.1 A glycolipid

15.2 **a.** 16 **b.** unsaturated **c.** liquid

15.3

$$\begin{array}{l} \quad\quad\quad\quad\quad O \\ \quad\quad\quad\quad\quad \| \\ CH_2-O-C-(CH_2)_{12}-CH_3 \\ | \quad\quad\quad\quad\; O \\ | \quad\quad\quad\quad\; \| \\ CH-O-C-(CH_2)_{12}-CH_3 \\ | \quad\quad\quad\quad\; O \\ | \quad\quad\quad\quad\; \| \\ CH_2-O-C-(CH_2)_{12}-CH_3 \end{array}$$

15.4 tristearin

15.5 Phospholipids containing saturated fatty acid molecules with their regular structure would line up close together. Unsaturated fatty acids with their nonregular, bent shape would be found in phospholipids that form looser, more flexible membranes.

15.6 Cerebrosides contain only one monosaccharide, and gangliosides contain two or more monosaccharide units.

15.7 Cholesterol is not soluble in water; it is classified with the lipid family.

15.8 Testosterone and Nandrolone both contain a steroid nucleus with one double bond and ketone group in the first ring, a methyl and alcohol group on the five-carbon ring. Nandrolone does not have the second methyl group at the first and second ring fusion that is seen in the structure of testosterone.

Problems

Types of Lipids *(Goal 15.1)*

15.1 What is a lipid?

15.2 Which of the following solvents might be used to dissolve an oil stain?
a. water **b.** CCl_4
c. diethyl ether **d.** benzene
e. NaCl solution

Fatty Acids *(Goal 15.2)*

15.3 Describe some similarities and differences in the structures of a saturated fatty acid and an unsaturated fatty acid.

15.4 Stearic acid and linoleic acid both have 18 carbon atoms. Why does stearic acid melt at 70°C, but linoleic acid melts at −5°C?

15.5 Write the structure of the following fatty acids:
a. palmitic acid **b.** oleic acid

15.6 Write the structure of the following fatty acids:
a. stearic acid **b.** linoleic acid

15.7 Which of the following fatty acids are saturated, and which are unsaturated?
a. lauric acid **b.** linolenic
c. palmitoleic acid **d.** stearic acid

15.8 Which of the following fatty acids are saturated, and which are unsaturated?
a. linolenic acid **b.** palmitic acid
c. myristic acid **d.** oleic acid

Waxes, Fats, and Oils *(Goal 15.3)*

15.9 Draw the structure of a component of beeswax that hydrolyzes to yield myricyl alcohol, $CH_3(CH_2)_{29}OH$, and palmitic acid.

15.10 Draw the structure of a component of jojoba wax that hydrolyzes to yield arachidic acid, a 20-carbon saturated fatty acid, and 1-docosanol, $CH_3(CH_2)_{21}OH$.

15.11 Draw the structure of a triglyceride that forms stearic acid and glycerol upon hydrolysis.

15.12 A mixed triglyceride contains two palmitic acid molecules to every one oleic acid molecule. Write two possible structures (isomers) for the compound.

15.13 Draw the structure of tripalmitin.

15.14 Draw the structure of triolein.

15.15 Safflower oil is called a polyunsaturated oil, whereas olive oil is a monounsaturated oil. Explain.

15.16 Why does olive oil have a lower melting point than butter fat?

15.17 Why does coconut oil, a vegetable oil, have a melting point similar to fats from animal sources?

15.18 A label on a bottle of 100% sunflower seed oil states that it is lower in saturated fats than all the leading oils.
a. How does the percentage of saturated fats in sunflower seed oil compare to that of safflower, corn, and canola oil? (See Figure 15.5.)
b. Is the claim valid?

Chemical Properties of Triglycerides *(Goal 15.4)*

15.19 Write an equation for the hydrogenation of glyceryl trioleate, a fat containing glycerol and three oleic acid units.

15.20 Write an equation for the hydrogenation of glyceryl trilinolenate, a fat containing glycerol and three linolenic acid units.

15.21 A label on a container of margarine states that it contains partially hydrogenated corn oil.
a. How has the liquid corn oil been changed?
b. Why is the margarine product solid?

15.22 Why should a bottle of vegetable oil that has no preservatives be tightly covered and refrigerated?

15.23 **a.** Write an equation for the acid hydrolysis of glyceryl trimyristate (trimyristin).
b. Write an equation for the NaOH saponification of glyceryl trimyristate (trimyristin).

15.24 **a.** Write an equation for the acid hydrolysis of glyceryl trioleate (triolein).
b. Write an equation for the NaOH saponification of glyceryl trioleate (triolein).

Phospholipids and Glycolipids *(Goal 15.5)*

15.25 Describe the differences between the following pairs:
a. a fat and a phosphoglyceride
b. a phosphoglyceride and a sphingolipid

15.26 Describe the differences between the following pairs:
a. a lecithin and a cephalin
b. a cerebroside and a ganglioside

15.27 Draw the structure of a phosphoglyceride containing two molecules of palmitic acid, and ethanolamine. What is another name for this type of phospholipid?

15.28 What amino alcohol is found in sphingomyelin? Draw the structure of a sphingomyelin containing palmitic acid.

15.29 Draw the structure of a cerebroside containing palmitic acid and galactose.

15.30 Draw the structure of a phospholipid that contains choline and palmitic acids.

15.31 Identify the following phospholipid and list its components:

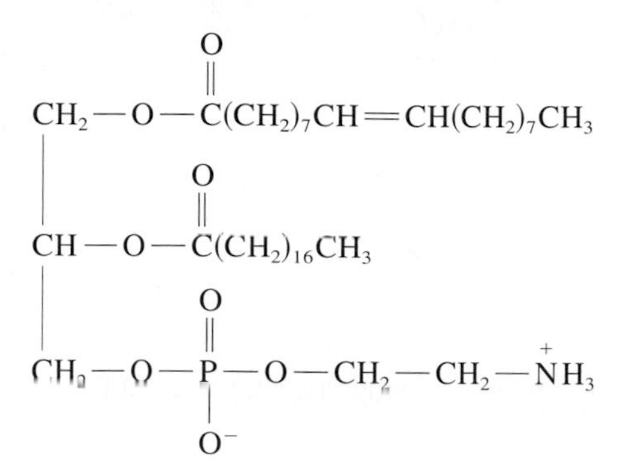

15.32 Identify the following phospholipid and list its components:

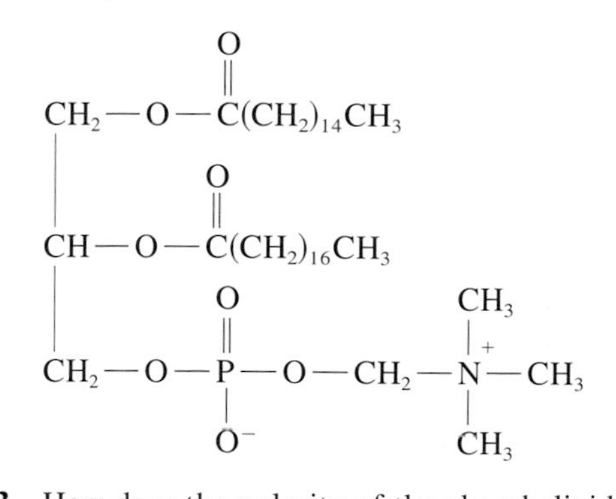

15.33 How does the polarity of the phospholipids contribute to their function in cell membranes?

15.34 How are phospholipids similar to soaps?

Steroids and Cholesterol *(Goal 15.6)*

15.35 Draw the structure for the steroid nucleus.

15.36 Which of the following compounds are derived from cholesterol?
a. glyceryl tristearate **b.** cortisone
c. bile salts **d.** cholesterol
e. estradiol

15.37 What is the function of bile salts digestion?

15.38 Why are gallstones composed of cholesterol?

Steroid Hormones *(Goal 15.7)*

15.39 What are the similarities and differences between the sex hormones estradiol and testosterone?

15.40 What are the similarities and differences between the adrenal hormone cortisone and the synthetic corticoid prednisone?

15.41 Which of the following are male sex hormones?
a. cholesterol **b.** aldosterone
c. estrogen **d.** testosterone
e. choline

15.42 Which of the following are adrenal steroids?
a. cholesterol **b.** aldosterone
c. estrogen **d.** testosterone
e. choline

Challenge Problems

15.43 Among the ingredients of a lipstick are beeswax, carnauba wax, hydrogenated vegetable oils, capric triglyceride, and tocopherol. What types of lipids have been used? Draw the structure of capric triglyceride (capric acid is the saturated 10-carbon fatty acid).

15.44 Because peanut oil floats on the top of peanut butter, many brands of peanut butter are hydrogenated. A solid product then forms that is mixed into the peanut butter and does not separate. If a triglyceride in peanut oil that contains one palmitic acid, one oleic acid, and one linoleic acid is completely hydrogenated, what is the product?

15.45 A recent report states that trans fats are produced during the hydrogenation of polyunsaturated oils to make them more solid and easier to use. Studies report that trans fats tend to increase the levels of low-density lipoprotein, LDL, and to decrease the high-density lipoprotein, HDL. Although "cholesterol free," trans fats may contribute to heart disease.
a. What is the typical configuration of a monounsaturated fatty acid?
b. How does a trans fat compare with a cis fat?
c. Draw the structure of trans oleic acid.

15.46 One mole of triolein is completely hydrogenated. What is the product? How many moles of hydrogen are required? How many grams of hydrogen? How many liters of hydrogen are needed if the reaction is run at STP?

15.47 On the list of ingredients of a makeup are glyceryl stearate and lecithin. What are the structures of these compounds?

15.48 Some typical meals at fast-food restaurants are listed here. Calculate the number of kilocalories from fat and the percentage of total kilocalories due to fat (1 grams of fat = 9 kcal). Would you expect the fats to be mostly saturated or unsaturated? Why?
a. a chicken dinner, 830 kcal, 46 g of fat
b. a quarter-pound cheeseburger 518 kcal, 29 g of fat
c. pepperoni pizza (three slices), 560 kcal, 18 g of fat
d. beef burrito, 466 kcal, 21 g of fat
e. deep-fried fish (three pieces), 477 kcal, 28 g of fat

15.49 Identify the following as fatty acids, soaps, triglycerides, wax, phosphoglyceride, sphingolipid, or steroid:
a. beeswax **b.** cholesterol
c. lecithin
d. glyceryl palmitate (tripalmitin)
e. sodium stearate **f.** safflower oil
g. sphingomyelin **h.** whale blubber
i. adipose tissue **j.** progesterone
k. cortisone **l.** stearic acid

Chapter 16

Amino Acids, Proteins, and Enzymes

Athletes need protein to build their muscles, to produce the hemoglobin that carries oxygen in the bloodstream, and to build enzymes and hormones that direct metabolic activities.

Looking Ahead

Learning Goals

16.1 Classify proteins by their functions in the cells.

16.2 Draw the structure for an amino acid in neutral or zwitterion form.

16.3 Describe a peptide bond; draw the structure for a peptide.

16.4 Distinguish between the primary and secondary structures of a protein.

16.5 Distinguish between the tertiary and quaternary structures of a protein.

16.6 Describe enzymes and their functions.

16.7 Describe the role of an enzyme in an enzyme-catalyzed reaction.

16.8 Discuss the effect of cofactors, temperature, and pH on enzyme-catalyzed reactions.

16.9 Describe the effect of an inhibitor upon enzyme activity.

Questions to Think About

1. What are some uses of protein in the body?
2. What are some sources of protein in your diet?
3. What elements are in proteins?
4. What are the units that form a protein called?
5. Why is heat or alcohol used to sterilize hospital equipment?
6. What is the purpose of adding lactase (enzyme) to milk and milk products?
7. Why are minerals such as copper, manganese, and zinc included in a daily vitamin supplement?

The word "protein" is derived from the Greek word *proteios,* meaning "first." Made of amino acids, proteins provide structure in membranes, build cartilage and connective tissue, transport oxygen in blood and muscle, direct biological reactions as enzymes, defend the body against infection, and control metabolic processes as hormones. They can even be a source of energy.

Compared with many of the compounds we have studied, protein molecules can be gigantic. Insulin has a molecular weight of 5,700, and hemoglobin has a molecular weight of about 64,000. Some virus proteins are still larger, having molecular weights of more than 40 million. Yet all proteins are polymers made up of about 20 different kinds of amino acids. Each kind of protein is composed of amino acids arranged in a specific order that determines the characteristics of the protein and its biological action.

Enzymes are proteins that act as biological catalysts in the cells of our body. Enzymes help us digest our food, contract our muscles, and produce the biomolecules and energy we need for survival.

16.1 Types of Proteins

Learning Goal
Classify proteins by their functions in the cells.

There are many kinds of proteins that perform many different functions in the body. Within each cell, there are proteins that form structural components such as cartilage, muscles, hair, and nails. Wool, silk, feathers, and horns are some other proteins made by animals. In the body, proteins called enzymes regulate biological reactions such as digestion and cellular metabolism. Still other proteins, hemoglobin and myoglobin, carry oxygen in the blood and muscle. Table 16.1 gives examples of proteins that are classified by their functions in biolgical systems.

Table 16.1 ***Classification of Some Proteins and their Function***

Type of Protein	Function	Example
Structural	Builds tendons and cartilage	Collagen
	Forms hair, skin, wool, and nails	Keratin
Contractile	Contracts muscle	Myosin, actin
Transport	Transports oxygen	Hemoglobin
	Transports lipids	Lipoprotein
Storage	Stores protein as milk	Casein
	Stores iron in liver for the production of red blood cells	Ferritin

Table 16.1 ***Classification of Some Proteins and their Functions (cont.)***

Type of Protein	Function	Example
Hormone	Increases glucose metabolism	Insulin
	Regulates body growth	Growth hormone
Enzyme	Catalyzes hydrolysis of sucrose	Sucrase
	Catalyzes hydrolysis of protein	Trypsin
	Converts ethanol to acetaldehyde	Dehydrogenase
Antibody	Recognizes and destroys bacterial or viral antigens.	Immunoglobulins

Sample Problem 16.1
Classifying Proteins by Function

Give the type of protein that would perform each of the following functions.

a. catalyzes metabolic reactions of lipids
b. carries oxygen in the bloodstream
c. stores amino acids in milk

Solution

a. Enzymes catalyze metabolic reactions.
b. Transport proteins carry substances such as oxygen through the bloodstream.
c. A storage protein stores nutrients such as amino acids in milk.

Study Check

What proteins are known as chemical messengers?

16.2 Amino Acids

Learning Goal

Draw the structure for an amino acid in neutral or zwitterion form.

Proteins are composed of molecular building blocks called amino acids. An **amino acid** contains two functional groups, an amino group (—NH_2) and a carboxylic acid group (—COOH). In most of the amino acids found in proteins, the amino group, the carboxylic group and a hydrogen atom are bonded to a central carbon atom. Amino acids with this structure are called *α- (alpha)-amino acids.* Different side groups (R) are attached to the center (α) carbon, giving each amino acid unique characteristics.

General Structure of an α-Amino Acid

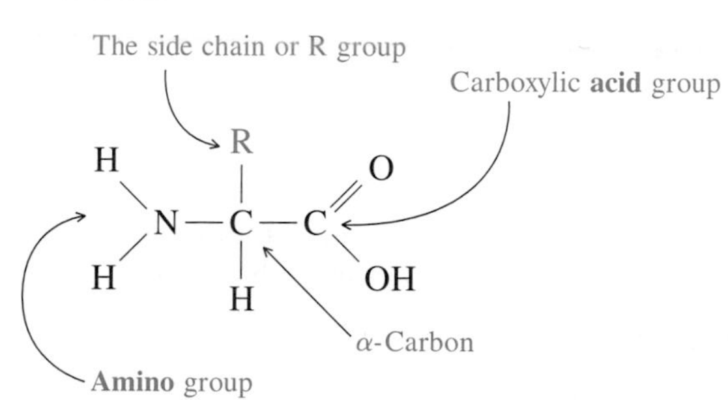

Table 16.2
Essential Amino Acids

Arginine	Methionine
Histidine	Phenylalanine
Isoleucine	Threonine
Leucine	Tryptophan
Lysine	Valine

Of the 20 amino acids used to build all the proteins in our body, 10 can be synthesized in the body. However, 10 amino acids, listed in Table 16.2,

Figure 16.1
In our diet, amino acids are obtained from the proteins of eggs, milk, fish and poultry, corn, rice, beans, and nuts.

cannot be synthesized. They are **essential amino acids** and must be obtained from the proteins in the diet. All of the essential amino acids are found in the proteins of animal products such as eggs, milk, meat, fish, and poultry. Because the proteins in grains, beans, and nuts are deficient in one or more essential amino acids, different proteins must be eaten together. (See Figure 16.1.)

In a condition known as kwashiorkor, the number of calories in the diet of corn meal is sufficient, but corn is deficient in lysine and tryptophan. (See Table 16.3.) As a result, a child fails to grow, becomes lethargic, and has an enlarged abdomen from hypoprotein edema due to the inability to properly build proteins.

Table 16.3 *Amino Acids Missing in Selected Vegetables and Grains*

Food Source	Amino Acid Deficiency
Eggs, milk, meat, fish, poultry	None
Wheat, rice, oats	Lysine
Corn	Lysine, tryptophan
Beans	Methionine, tryptophan
Peas	Methionine
Almonds, walnuts	Lysine, tryptophan

Classification of Amino Acids

Amino acids are classified according to their side (R) groups. **Nonpolar amino acids** contain a hydrocarbon or aromatic side chain, which makes them **hydrophobic. Polar amino acids** are **hydrophilic** because their side (R) groups contain atoms such as —OH that are attracted to water. Some of the polar amino acids are *acidic* if the R contains a carboxylic acid group, and *basic* if the R group contains an amino group. The structures of the amino acids, their R side groups, common names, and their abbreviations (one or three letters) are listed in Table 16.4.

Table 16.4 *The 20 Amino Acids Found in Proteins*

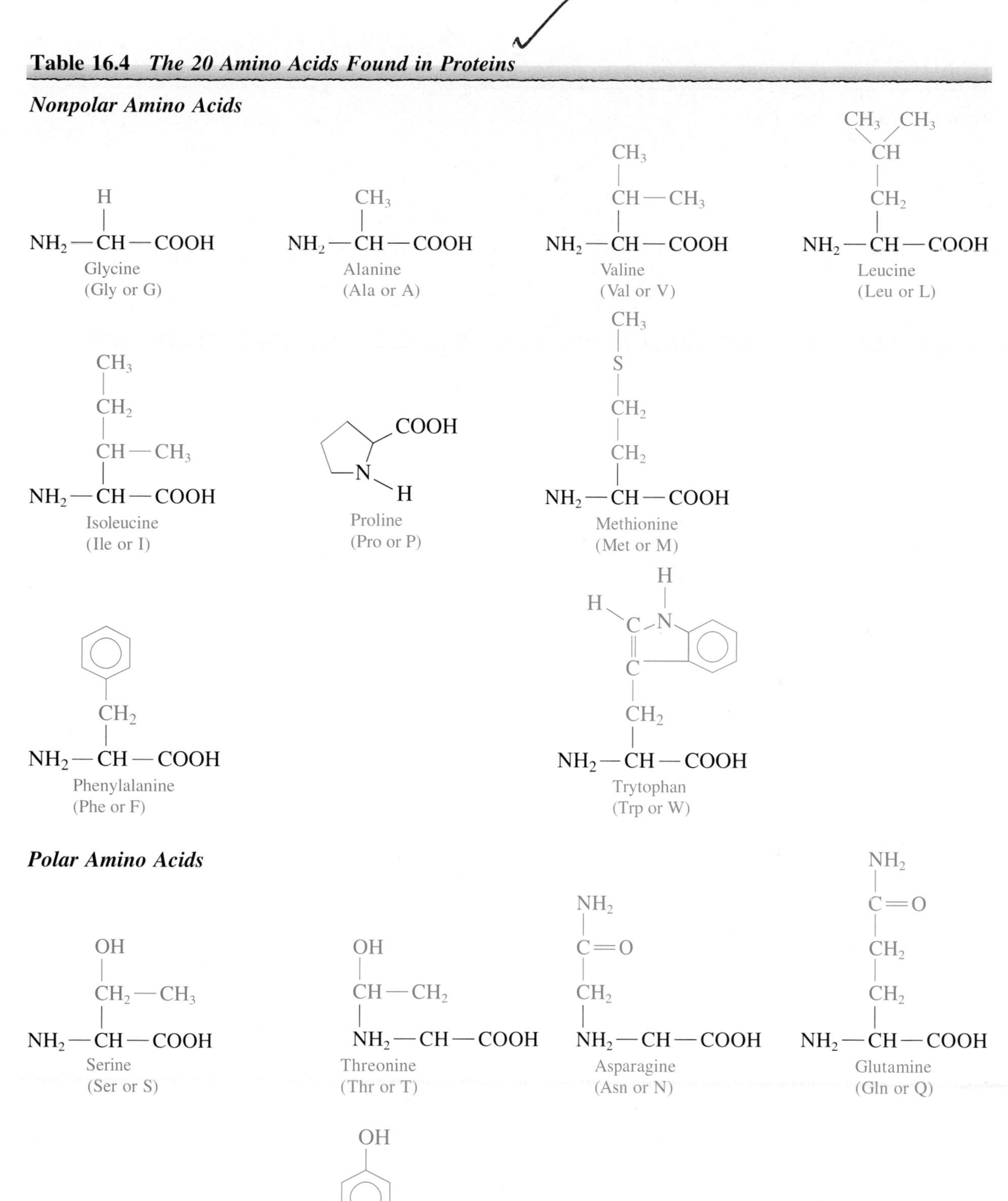

Table 16.4 *The 20 Amino Acids Found in Proteins (cont.)*

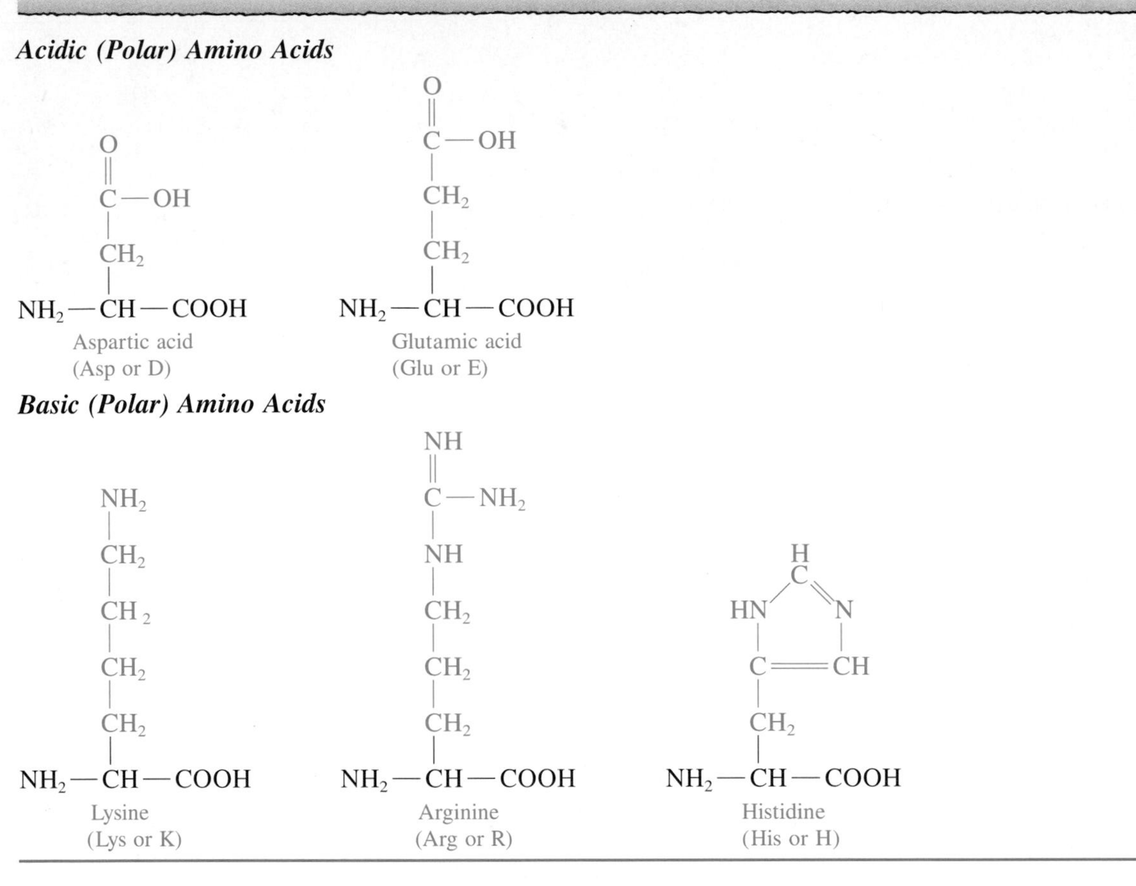

Sample Problem 16.2
Structural Formulas of Amino Acids

Write the structural formulas and abbreviations for the following amino acids:
a. alanine (R = $-CH_3$) **b.** serine (R = $-CH_2OH$)

Solution

a. The structure of the amino acids is written by attaching the side group (R) to the central carbon atom of the general structure of an amino acid.

$$\boxed{CH_3} \leftarrow \text{R group}$$

alanine (ala or A) $NH_2-CH-COOH$

b. serine (ser or S)

$$\boxed{OH-CH_2}$$

$NH_2-CH-COOH$

Study Check
Classify the amino acids in the above question as polar or nonpolar.

Ionization of Amino Acids

An amino acid can ionize when the carboxyl group donates a proton and when the lone pair of electrons on the amino group attracts a proton. Then the carboxyl group has a negative charge, and the amino group has a positive charge. This ionized form of the amino acid, called a **zwitterion** or *dipolar ion,* has a net charge of zero.

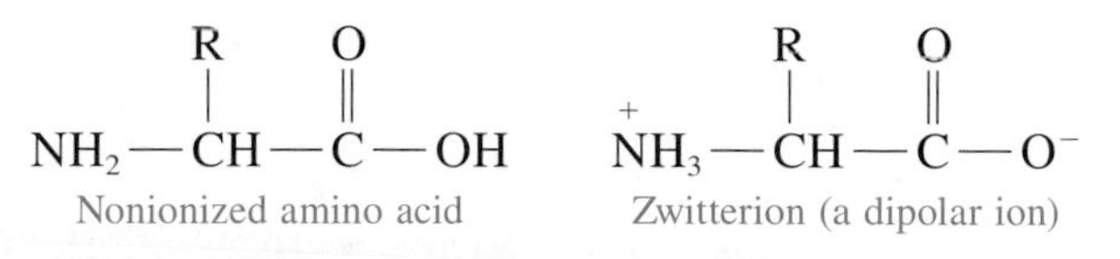

Most of the amino acids exist in their zwitterion form in solutions that have pH values around 6 or 7, close to neutral. For example, alanine has the following zwitterion form at a pH of 6.1:

$$\overset{+}{N}H_3—CH(CH_3)—C(=O)—O^-$$

Zwitterion of alanine

When placed in an acidic solution (low pH), the ($—COO^-$) of the zwitterion accepts a proton (H^+), forming an ion with a positive charge. When placed in a basic solution (high pH), the ($—NH_3^+$) of the zwitterion donates a proton (H^+), forming an ion with a negative charge. This is illustrated below using alanine.

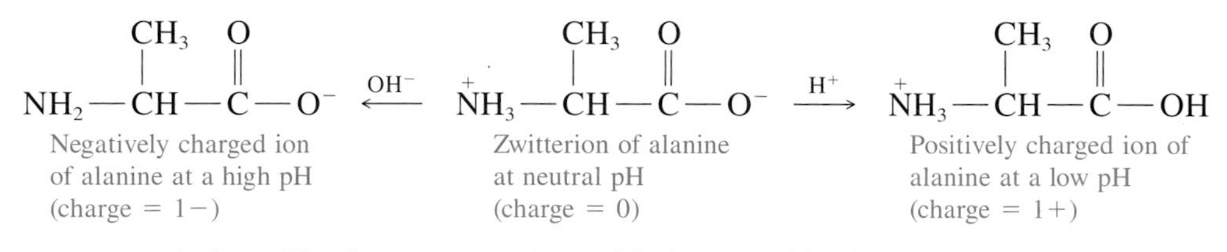

At low pH values, most amino acids form positive ions; at neutral pH values, most exist as neutral zwitterions; and at high pH values, most form negative ions.

Sample Problem 16.3
Amino Acids in Acid or Base

Serine exists in its zwitterion form at a pH of 5.7. Draw the structural formula for the zwitterion of serine.

Solution

As a zwitterion, both the carboxylic acid group and the amino group are ionized.

$$\overset{+}{N}H_3—CH(CH_2OH)—C(=O)—O^-$$

Zwitterion of serine at neutral pH

Study Check

At a pH of 2.0 serine is a positively charged ion. Draw its structure.

16.3 The Peptide Bond

Learning Goal

Describe a peptide bond; draw the structure for a peptide.

A **peptide bond,** the link between amino acids, is formed between the carboxyl group of one amino acid and the amino group of another. In the reaction, a molecule of water is released.

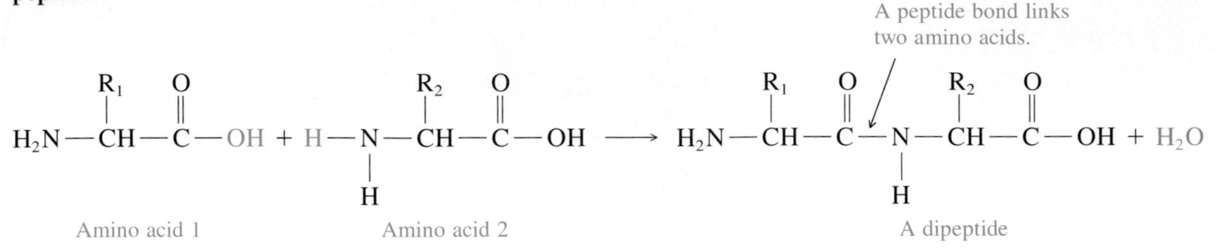

The linking of two or more amino acids forms a **peptide.** Two amino acids are linked by a peptide bond in a *dipeptide.* In a *tripeptide,* there are three amino acids linked by peptide bonds. In any peptide, the amino acid on one end that has a free amino group ($—NH_2$) is called the N-terminus. On the other end of the peptide, there is an amino acid with a free carboxyl group (—COOH) called the C-terminus.

The amino acids in a peptide are indicated by drawing their structural formulas or by writing the abbreviations of their names. To name a peptide, each of the amino acids beginning at the N-terminus is named with a *yl* ending followed by the full name of the amino acid at the C-terminus.

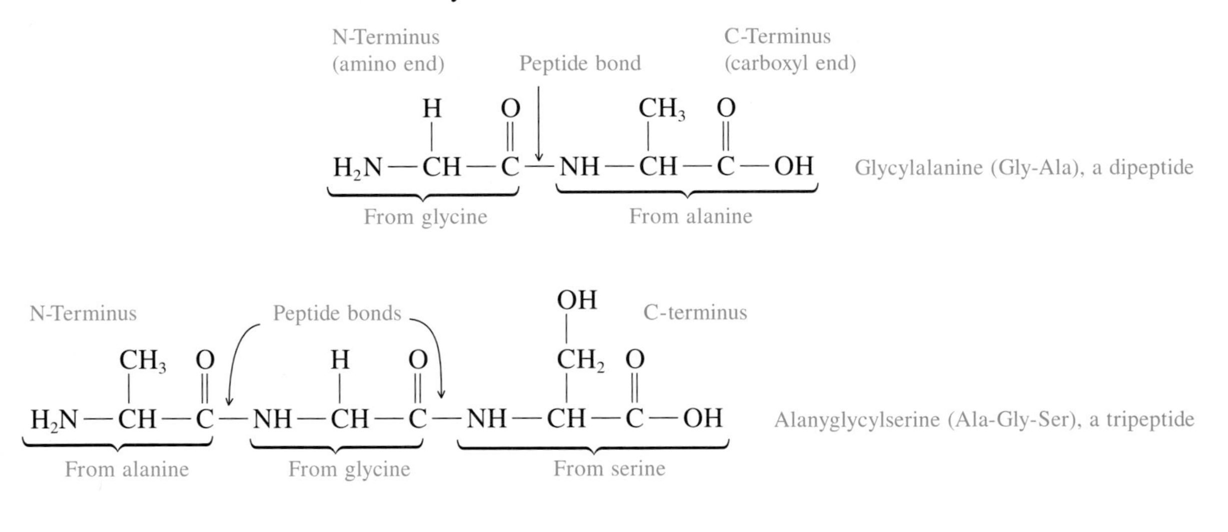

Hydrolysis of Peptides

Peptides are broken apart by hydrolysis, a reaction that adds water to the peptide bonds in the presence of an acid or enzyme such as pepsin or trypsin.

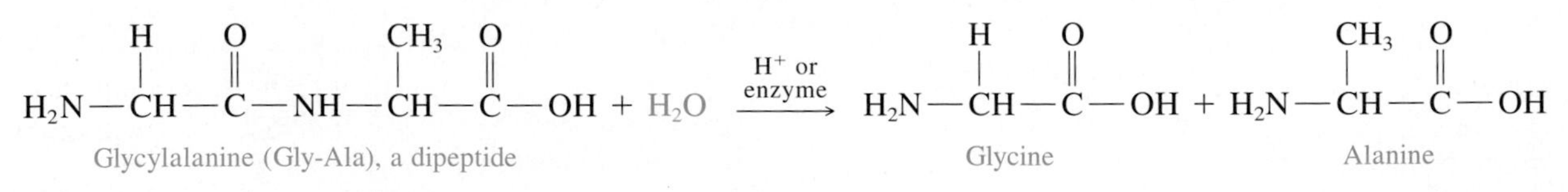

Sample Problem 16.4
Writing Structures of Dipeptides

Write a structural formula for the dipeptide valylserine.

Solution

The N-terminus, valine, is joined to serine by a peptide bond.

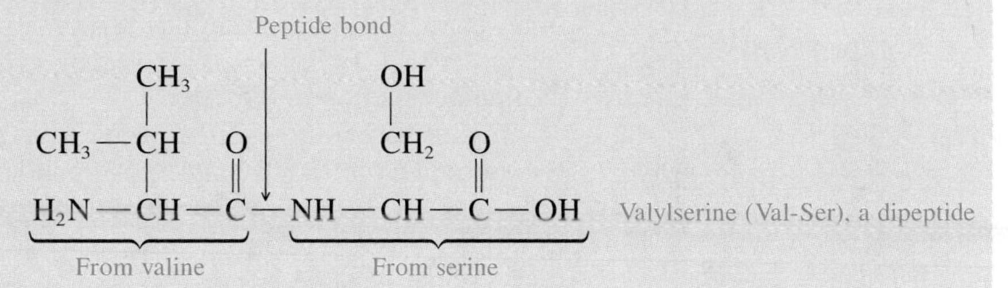

Valylserine (Val-Ser), a dipeptide

Study Check

Aspartame, an artificial sweetener 200 times sweeter than sucrose, is a methyl ester of aspartylphenylalanine, a dipeptide. What is the structure of the dipeptide in aspartame?

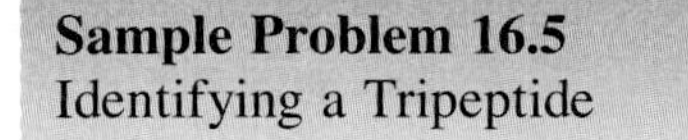

Sample Problem 16.5
Identifying a Tripeptide

Consider the following tripeptide.

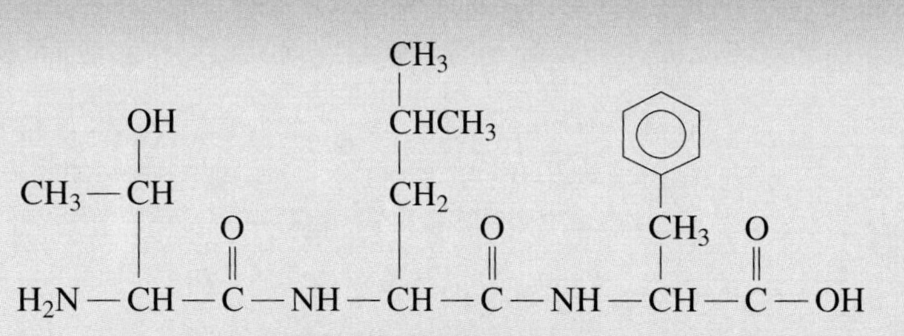

a. What is the N-terminus amino acid? C-Terminus amino acid?

b. Using the three-letter abbreviations, what is the order of amino acids in the peptide?

Solution

a. The N-terminus is threonine, the amino acid with the free amino group; phenylalanine, the C-terminus has the free carboxyl group.

b. Thr-Leu-Phe

Study Check

What is the name of the tripeptide?

16.4 Primary and Secondary Protein Structure

Learning Goal

Distinguish between the primary and secondary structures of a protein.

Long chains of amino acids linked by peptide bonds are called **polypeptides.** When there are more than 50 amino acids in a chain, the polypeptide is usually called a **protein.**

Figure 16.2
Primary structure of human insulin.

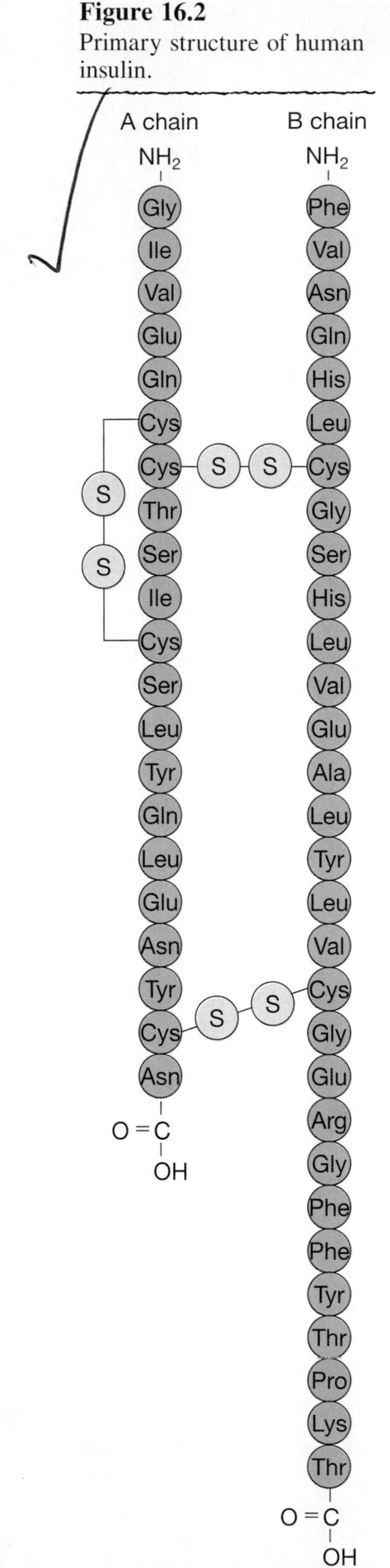

Primary Structure

The **primary structure** of a protein is the sequence of the amino acids that are linked together. For example, the primary structure of a hormone that stimulates the thyroid to release thyroxine is the tripeptide Glu-His-Pro. Although other sequences are possible for these three amino acids, only this order is active. Each protein has certain amino acids in a unique order. It is this order that determines all the other structural levels of a protein and its particular function in the cells.

Figure 16.2 illustrates the primary structure of human insulin, which is the order of amino acids in two polypeptide chains held together by disulfide cross-links. There are 21 amino acids in chain A and 30 amino acids in chain B. For many years, the bovine insulin produced by cows, horses, and pigs was used to treat diabetes because its primary structure is such a close match to human insulin. Today, most of the insulin used to treat diabetes is human insulin produced through genetic engineering methods.

Secondary Structure

In the **secondary structure** of a protein, hydrogen bonds between portions of the peptide chain fold the protein into a characteristic shape. The three most common patterns of secondary structure are the α-helix, the β-pleated sheet, and the triple helix.

The α-Helix

The corkscrew shape of an α **(alpha)-helix** is held in place by hydrogen bonds between every N—H group and the oxygen of a C=O group in the next turn of the helix, four amino acids down the chain. (See Figure 16.3.) Because

Figure 16.3
The α-helix acquires a coiled shape through the hydrogen bonding between the N—H and C=O of the next loop.

Figure 16.4
In hair and wool, which are fibrous proteins of α-keratins, many α-helixes are wrapped together to form fibrils.

there are many hydrogen bonds between the loops of the peptide chain, the protein takes the shape of a strong, tight coil that looks like a telephone cord or a Slinky toy, with the R groups lying on the outside of the helix.

Fibrous proteins are insoluble in water. One type of fibrous protein is the α-keratins, the proteins that make up hair, wool, skin, and nails. For example, in hair or wool, three α-helical chains coil together like a rope to form a fibril. Many of these fibrils are used to form one strand of hair or wool. (See Figure 16.4.) In hair, the α-helixes are held together by disulfide linkages between the side groups of the many cysteines in hair protein. When wet, the fibrils in wool stretch as hydrogen bonds are broken and the return to the original shape as the wool dries.

HEALTH NOTE *Natural Opiates in the Body*

Painkillers known as enkephalins and endorphins are produced naturally in the body. They are polypeptides that bind to receptors in the brain to give relief from pain. This effect appears to be responsible for the runner's high, the temporary loss of pain when severe injury occurs, and the analgesic effects of acupuncture.

The *enkephalins,* found in the thalamus and the spinal cord, are pentapeptides, the smallest molecules with opiate activity. The amino acid sequence of an enkephalin is found in the longer amino acid sequence of the endorphins.

Four groups of *endorphins* have been identified: α-endorphin contains 16 amino acids, β-endorphin contains 31 amino acids, γ-endorphin has 17 amino acids, and δ-endorphin has 27 amino acids. Endorphins may produce their sedating effects by preventing the release of substance P, a polypeptide with 11 amino acids, which has been found to transmit pain impulses to the brain.

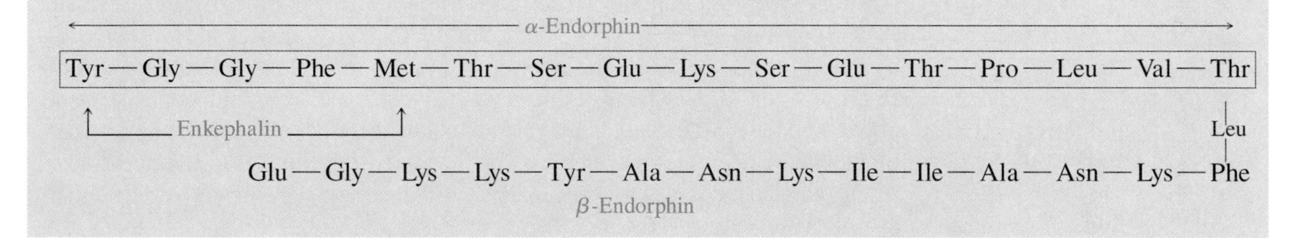

HEALTH NOTE *Hair Permanents*

Much of the protein in the hair is cross-linked by disulfide bonds between R groups of cysteine, an amino acid that is prevalent in hair.

$$\begin{array}{l} \quad\ \ CH_2—\mathbf{SH} \longleftarrow \text{Thiol} \\ \quad\ \ | \\ H_2N—CH—COOH \quad \text{Cysteine} \end{array}$$

In a disulfide, two sulfur atoms are linked together to give an —S—S— group.

$$CH_3—S—S—CH_3$$

Dimethyl disulfide

Disulfides are formed by the reaction of two thiols.

$$CH_3—S—H + H—S—CH_3 \xrightarrow{[O]} CH_3—S—S—CH_3 + H_2O$$

Methanethiol Dimethyl disulfide

The shape of the hair, curly or straight, is determined by the way the proteins are cross-linked by disulfide bonds formed between molecules of cysteine.

When a permanent is given, a reducing substance is put on the hair to break apart the disulfide bonds. Then, the hair is wrapped around curlers and an oxidizing substance is used that causes the disulfide bonds to form between protein strands that take the shape of the curlers. (See Figure 16.5.)

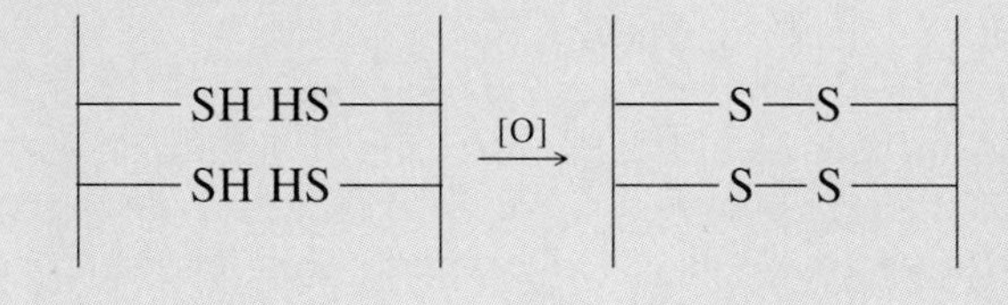

Figure 16.5
In a permanent, the disulfide bonds in hair protein are reduced so they break apart and then oxidized so they reform disulfide bonds to give a new shape.

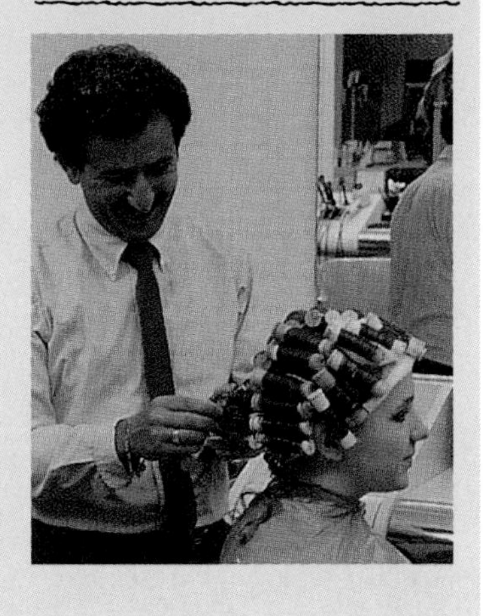

β-Pleated Sheet

Another type of secondary structure is known as the β **(beta)-pleated sheet.** In a β-pleated sheet, polypeptide chains are held together side by side by hydrogen bonds between the peptide bonds. In a β-pleated sheet, the small R groups of the prevalent amino acids, glycine, alanine, and serine, extend above and below the sheet. This results in a series of β-pleated sheets that can be stacked close together. The hydrogen bonds holding the β-pleated sheets tightly in place account for the strength and durability of fibrous proteins such as silk. (See Figure 16.6.)

Figure 16.6
In a β-pleated sheet, hydrogen bonds form between the peptide bonds in adjacent rows of polypeptide chains.

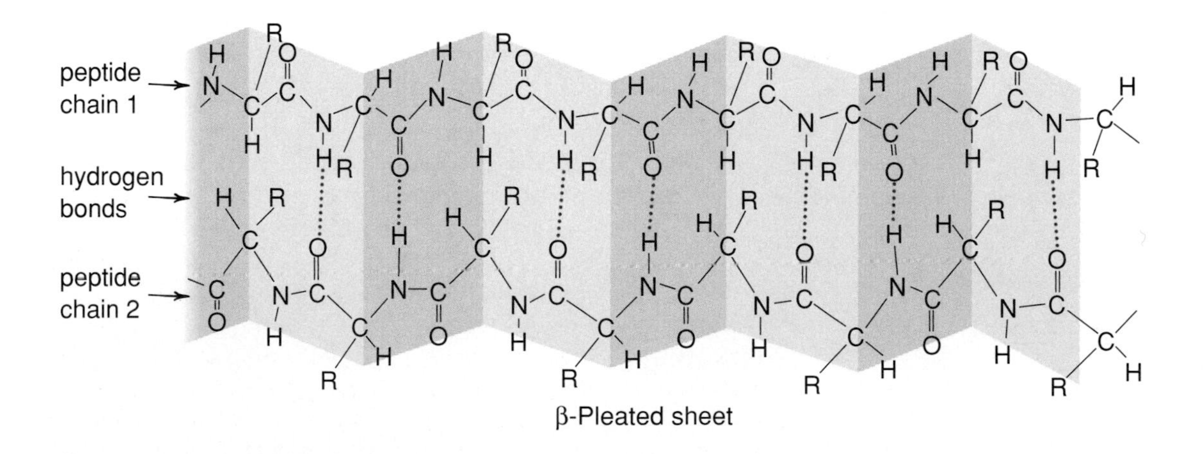

Figure 16.7
The triple helix in collagen.

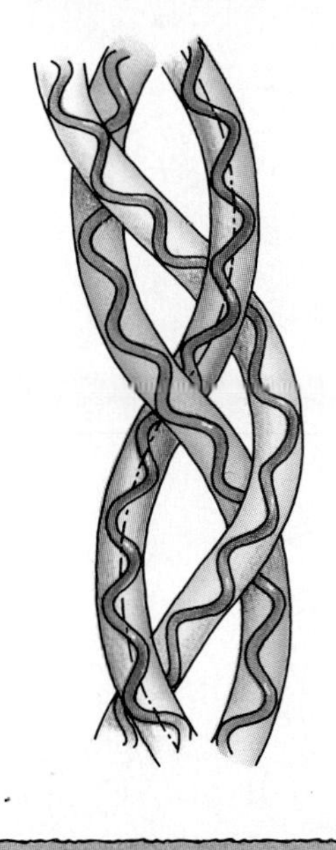

Most proteins in the cells are globular proteins, which are soluble in water. In globular proteins, only certain parts of the molecule have the α-helix secondary structure. Other parts of a globular protein take a random shape, and sometimes contain a β-pleated sheet secondary structure.

Collagen: A Triple Helix

Collagen makes up as much as one-third of all the protein in animals. It is prevalent in connective tissue, skin, tendons, ligaments, the cornea of the eye, and cartilage. The strong structure of collagen is a result of three peptide chains woven together like a braid to form a **triple helix,** seen in Figure 16.7. Collagen has a high percentage of glycine (35%) and proline (40%) and smaller amount of hydroxyproline and hydroxylysine, forms of proline and lysine with —OH groups. Hydrogen bonding between these hydroxyl groups gives strength to the collagen triple helix. When several triple helixes wrap together, they form the fibrils of collagen that make up connective tissues.

Sample Problem 16.6
Identifying Secondary Structures

Indicate the type of secondary structure (α-helix, β-pleated sheet, or triple helix) described in each of the following statements:

a. a coiled peptide chain held in place by hydrogen bonding between peptide bonds in the same chain

b. a structure that has hydrogen bonds between polypeptide chains arranged side by side.

Solution

a. α-helix **b.** β-pleated sheet

Study Check

What is the type of secondary structure in collagen?

16.5 Tertiary and Quaternary Protein Structure

Learning Goal
Distinguish between the tertiary and quarternary structures of a protein.

Globular proteins such as insulin, hemoglobin, enzymes, and antibodies have a compact, three-dimensional shape called the *tertiary structure*. With the nonpolar amino acid groups pushed to the inside and the more polar groups on the surface, globular proteins are soluble in aqueous solutions. The polar side groups, which are hydrophilic (water-loving), move to the outside of the protein structure where they can hydrogen bond with water.

In a **tertiary structure,** a polypeptide chain with all of its regions of secondary structures (α-helix and β-pleated) has folded over upon itself. The polypeptide folds spontaneously as the side groups of the amino acids along the chain interact with water or other R groups in different regions of the chain. Hydrophobic attractions between nonpolar R groups form a nonpolar

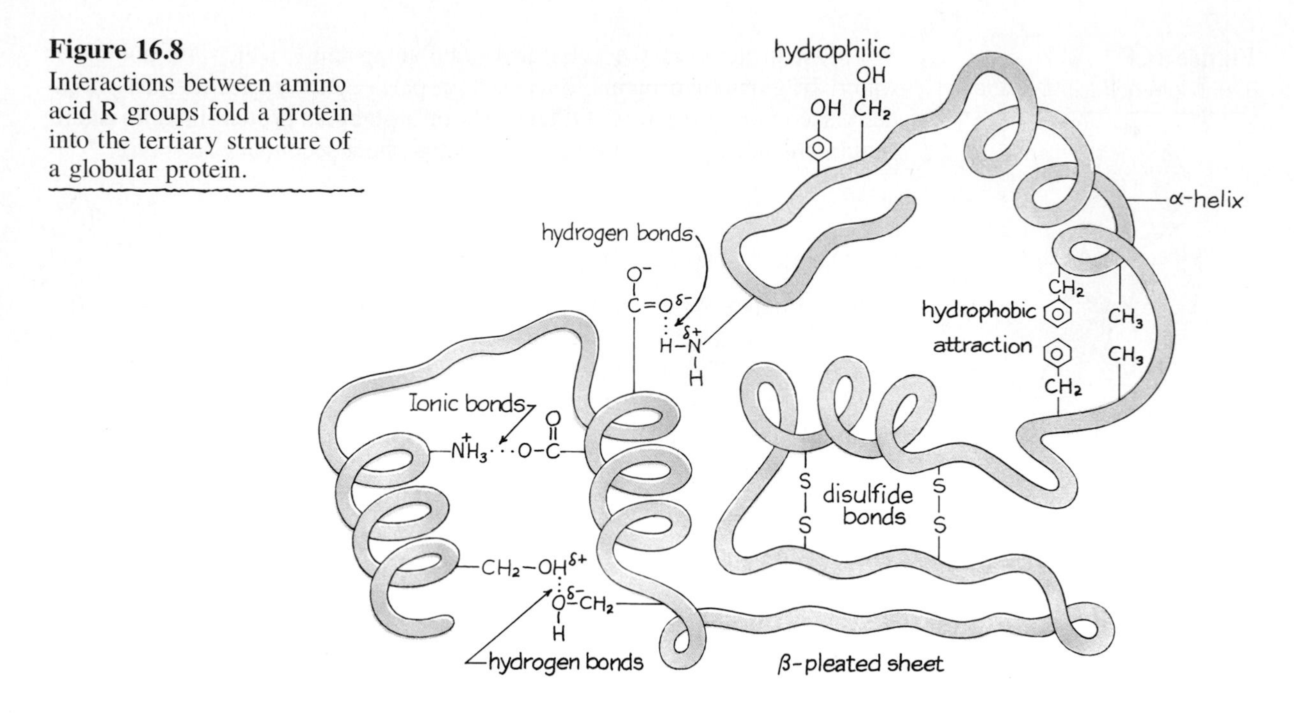

Figure 16.8
Interactions between amino acid R groups fold a protein into the tertiary structure of a globular protein.

center away from the water. Disulfide (—S—S—) covalent bonds form between the sulfur atoms of two close cysteine R groups. Ionic bonds, also known as salt bridges, form between the COO^- of the R group of acidic amino acids, and the $—NH_3^+$ of the R group of basic amino acids. Hydrogen bonds can form between a carbonyl group (C═O) and a hydroxyl (—OH) or amino ($—NH_2$) in another R group. (See Figure 16.8.) All of these interactions between the R groups stabilize the tertiary structure of both the α-helix and β-pleated sheet regions of the protein.

Sample Problem 16.7
Cross-Links in Tertiary Structures

What type of interaction would you expect between the R groups of the following amino acids?

a. cysteine and cysteine **b.** glutamic acid and lysine

Solution

a. Because cysteine has an R group with an —SH, a disulfide bond will form.

b. An ionic bond (salt bridge) can form by the interaction of $—COO^-$ of glutamic acid and the $—NH_3^+$ of lysine.

Study Check

Would you expect to find valine and leucine in a globular protein on the outside or the inside of the tertiary structure? Why?

Myoglobin, a Tertiary Protein Structure

Myoglobin is a globular protein that stores oxygen in skeletal muscle. High concentrations of myoglobin have been found in the muscles of sea mammals,

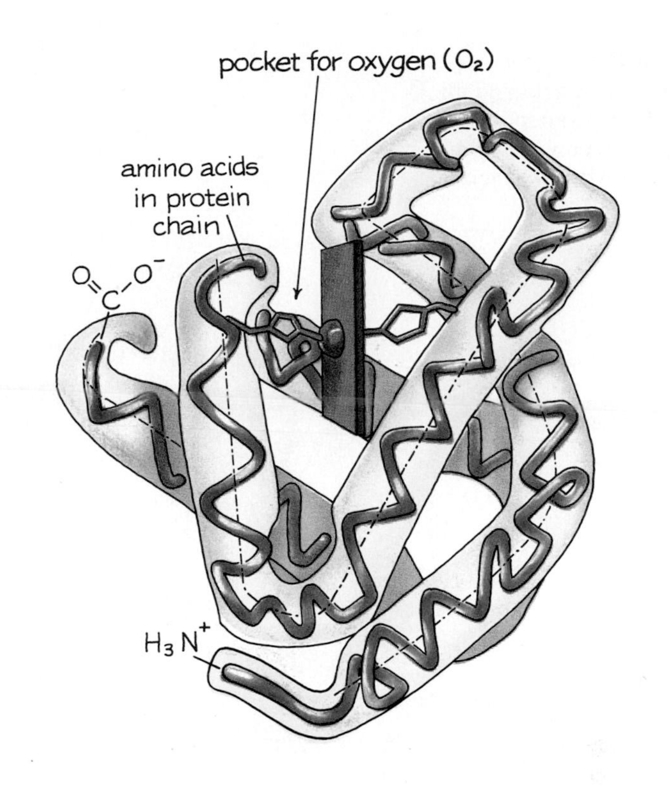

Figure 16.9
In the tertiary structure of myoglobin, a heme group and surrounding nonpolar amino acids form a nonpolar pocket for storing oxygen.

such as seals and whales, that stay under the water for long periods of time. Myoglobin contains 153 amino acids in a single polypeptide chain with about three-fourths of the chain in the α-helix secondary structure. The polypeptide chain, including its helical regions, forms a compact tertiary structure by folding upon itself. The polar R groups are found on the outside of the molecule, and the nonpolar R groups are found on the inside. (See Figure 16.9.) Within the tertiary structure, a nonpolar pocket of amino acids and a heme group binds and stores oxygen (O_2).

Quaternary Structure: Hemoglobin

When a biologically active protein consists of two or more polypeptide subunits, the structural level is referred to as a **quaternary structure.** Hemoglobin, a globular protein that transports oxygen in blood, consists of four polypeptide chains or subunits, two α-chains and two β-chains. The subunits are held together in the quaternary structure by the same interactions that stabilize the tertiary structure, such as hydrogen and ionic bonds between side groups, disulfide links, and hydrophobic attractions. (See Figure 16.10.) Each subunit of the hemoglobin contains a heme group that binds oxygen. In the hemoglobin molecule, all four subunits ($\alpha_2\beta_2$) must be combined for the hemoglobin to properly function as an oxygen carrier. Therefore, the complete quaternary structure of hemoglobin can bind and transport four molecules of oxygen.

Hemoglobin and myoglobin have similar biological functions. Hemoglobin carries oxygen in the blood, whereas myoglobin carries oxygen in muscle. Myoglobin, a single polypeptide chain with a molecular weight of 17,000, has about one-fourth the molecular weight of hemoglobin (64,000).

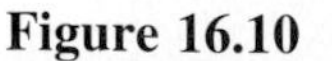

Figure 16.10
The structure of hemoglobin contains four polypeptide subunits, each containing a heme group where oxygen is stored.

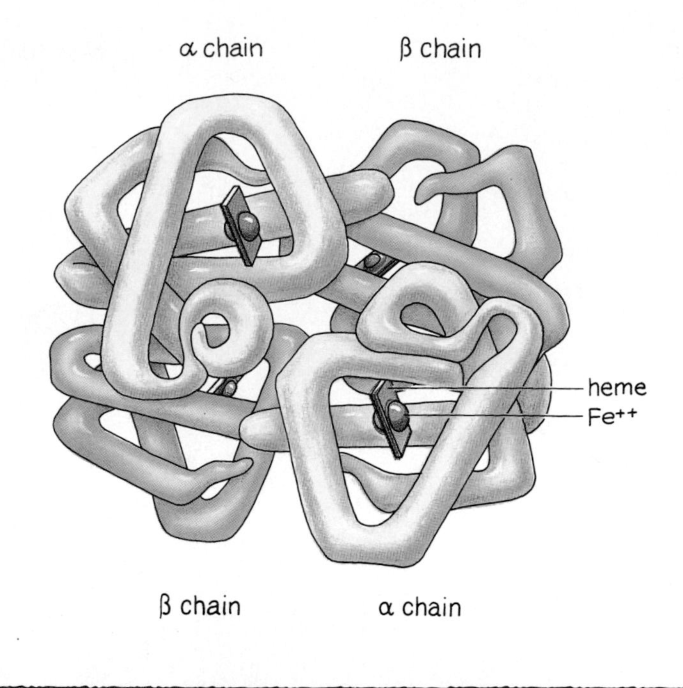

HEALTH NOTE *Sickle-Cell Anemia*

Sickle-cell anemia is a disease caused by an abnormality in the shape of one of the subunits of the hemoglobin protein. In the β-chain, the sixth amino acid, glutamic acid, which is polar, is replaced by valine, a nonpolar amino acid. An essential ionic cross-link no longer forms causing a severe change in the shape and function of the tertiary structure of the hemoglobin. The affected red blood cells change from a rounded shape to a crescent shape, like a sickle, which interferes with their ability to transport adequate quantities of oxygen. (See Figure 16.11.)

Sickled cells are removed from circulation more rapidly than normal red blood cells, causing anemia and low oxygen concentration in the tissues. Clumps of sickled cells clog the capillaries, where they cause great pain and organ damage. Critically low oxygen levels may occur in the affected tissues.

Figure 16.11
Comparison of the shape of sickled cells with normal red blood cells.

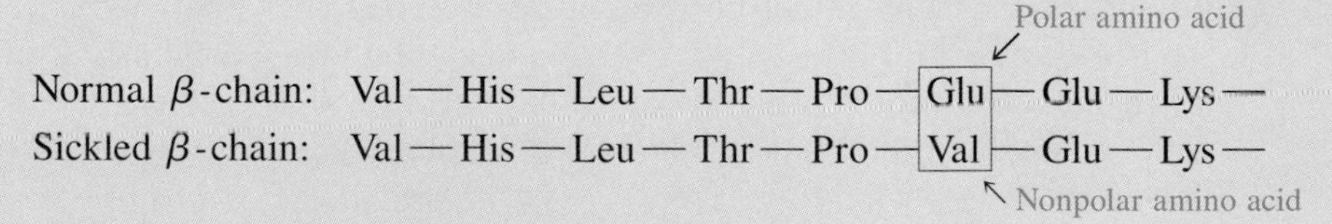

In sickle cell anemia, both genes for the altered hemoglobin must be inherited. However, there are a few sickle-cells found in persons who carry one gene for sickle-cell hemoglobin, a condition that is also known to provide protection from malaria.

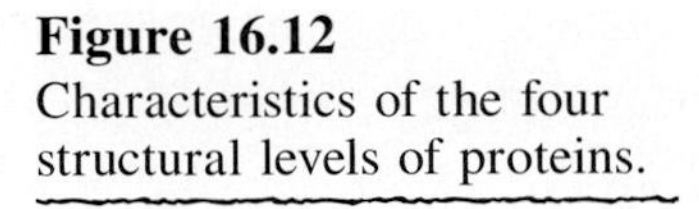

Figure 16.12
Characteristics of the four structural levels of proteins.

primary

-Gly-Ala-Tyr-His-Phe-Iso-Arg-Leu-

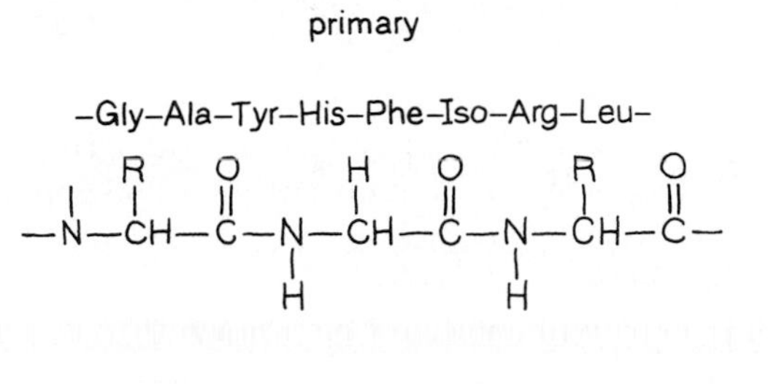

secondary

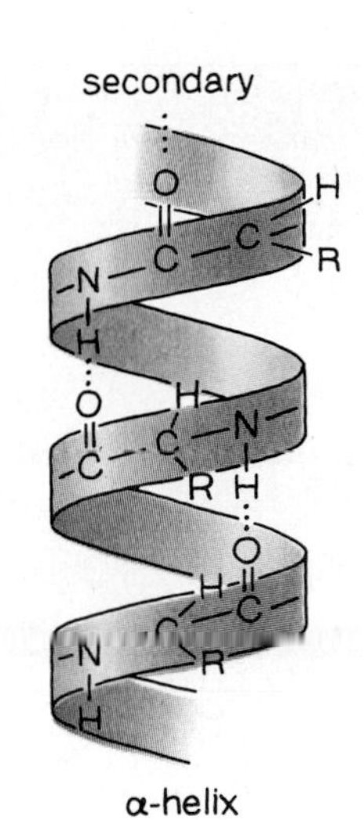

α-helix

tertiary

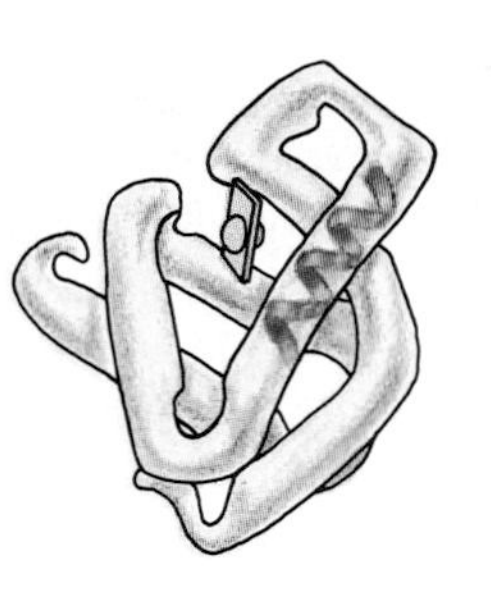

myoglobin

quaternary

hemoglobin

Table 16.5 ***Summary of Structural Levels in Proteins***

Structural Level	Characteristics
Primary	The sequence of amino acids.
Secondary	The coiled α-helix, β-pleated sheet, or a triple helix formed by hydrogen bonding between peptide bonds along the chain.
Tertiary	A folding of the protein into a compact, three-dimensional shape stabilized by interactions between side R groups of amino acids.
Quaternary	A combination of two or more protein subunits to form a larger, biologically active protein.

In addition, the tertiary structure of the single polypeptide myoglobin is almost identical to the tertiary structure of each of the subunits of hemoglobin. Myoglobin stores just one molecule of oxygen, just as each subunit of hemoglobin carries one oxygen molecule. The similarity in tertiary structures allows each protein to bind and release oxygen in a similar manner. Table 16.5 and Figure 16.12 summarize the structural levels of proteins.

Sample Problem 16.8
Identifying Protein Structure

Indicate whether the following conditions are responsible for primary, secondary, tertiary, or quaternary protein structures:

a. Disulfide bonds form between portions of a protein chain.
b. Peptide bonds form a chain of amino acids.

Solution

a. Disulfide bonds help to stabilize the tertiary structure of a protein.
b. The sequence of amino acids in a polypeptide is a primary structure.

Study Check

What structural level is represented by the grouping of two subunits in insulin?

Electrophoresis

At a pH called the **isoelectric point,** there are equal numbers of positive and negative charges in the R groups of a protein. The protein is neutral and not soluble in water. The isoelectric point of egg albumin and milk casein is pH 4.6, insulin is pH 5.3, myoglobin is pH 7.0, and chymotrypsin is pH 9.5.

In the hospital laboratory, **electrophoresis** is used to separate proteins. An electric current applied to a protein mixture shown in Figure 16.13 causes the positively charged proteins to move toward the negative electrode and negatively charged proteins to move toward the positive electrode. The proteins are separated because they move at different rates toward the electrode depending on their ionic charge. A protein at its isoelectric point is neutral and does not move during electrophoresis. The proteins are identified by their direction and rate of migration in the electric field. Electrophoresis is also used to separate other charged molecules, most notably DNA in DNA fingerprinting techniques.

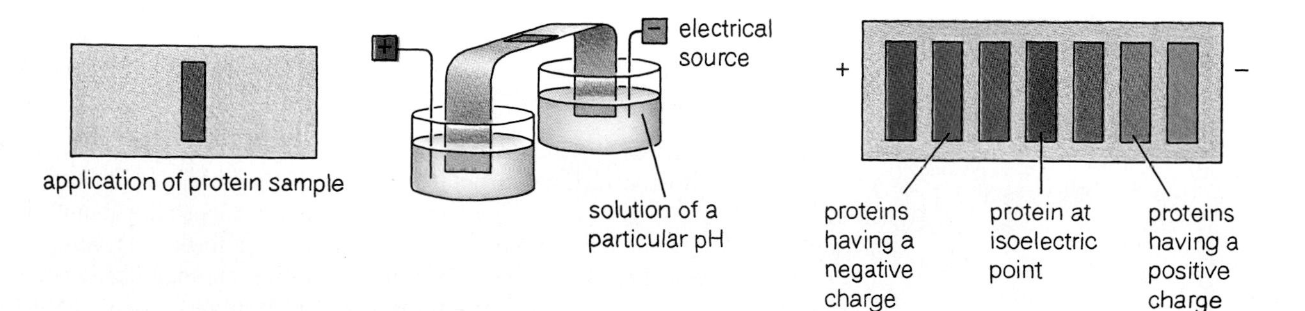

Figure 16.13
A positively charged protein moves toward the negative electrode; a negatively charged protein moves toward the positive electrode; a protein at its isoelectric point does not migrate.

Figure 16.14
Denaturation of a protein occurs when heat, acid, base, or heavy metal salts disrupt the bonds of the tertiary structure, destroy the shape, and render the protein biologically inactive.

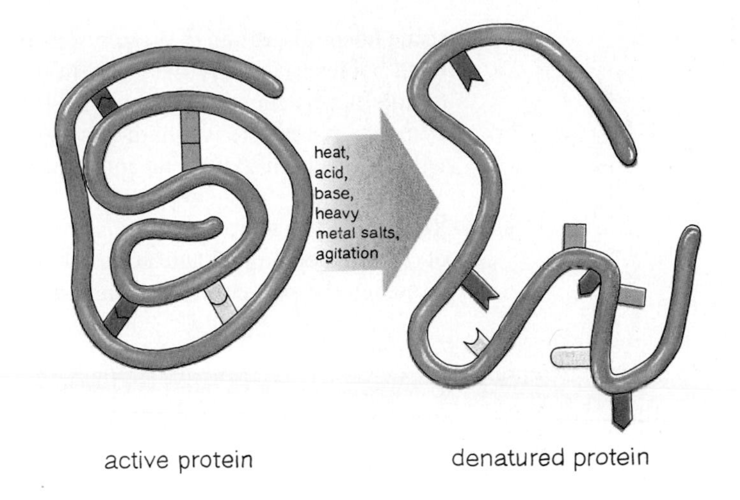

Denaturation of Protein

Denaturation occurs when there is a disruption of the bonds that stabilize the secondary, tertiary, or quaternary structures of proteins. When the interactions between the R groups are undone or altered, a globular protein unfolds like a loose piece of spaghetti. With the loss of its overall shape, the protein is no longer biologically active. (See Figure 16.14.)

Denaturing agents include heat, acids and bases, organic compounds, heavy metal ions, and mechanical agitation.

Heat. Heat denatures proteins by breaking apart hydrogen bonds and the hydrophobic attraction between nonpolar side groups. Few proteins can remain biologically active above 50°C. Whenever you cook food, you are using heat to denature protein. The nutritional value of the proteins in food is not changed, but they are made more digestible. High temperatures are also used to disinfect surgical instruments and gowns by denaturing the proteins of any bacteria present.

Acids and Bases. Placing a protein in an acid or base affects the acidic and basic R groups and disrupts the ionic bonds (salt bridges). In the preparation of yogurt and cheese, a bacteria that produces lactic acid is added to denature the milk protein and produce solid casein. Tannic acid, a weak acid used in burn ointments, coagulates proteins at the site of the burn, forming a protective cover and preventing further loss of fluid from the burn area.

Organic Compounds. Ethanol and isopropyl alcohol act as disinfectants by forming their own hydrogen bonds with a protein and disrupting the original hydrogen bonds. An alcohol swab is used to clean wounds or to prepare the skin for an injection because the alcohol passes through the cell walls of bacteria and coagulates the proteins inside the cell.

Heavy Metal Ions. When heavy metal ions like Ag^+, Pb^{2+}, and Hg^{2+}, react with the disulfide ($-S-S-$) bonds, the denatured protein solidifies. In some hospitals, dilute (1%) solutions of $AgNO_3$ are placed in the eyes of newborn babies to destroy the bacteria that causes gonorrhea. If ingested, heavy metals act as poisons by severely disrupting body proteins, especially enzymes. An antidote is a high-protein food such as milk, eggs, or cheese that will tie up the heavy metal ions until the stomach can be pumped.

Agitation. The whipping of cream and the beating of egg whites are examples of using mechanical agitation to denature protein. The whipping action stretches the polypeptide chains until the bonds break apart.

Sample Problem 16.9
Effects of Denaturation

What happens to the tertiary structure of a globular protein when it is placed in an acidic solution?

Solution
An acid causes denaturation by disrupting the ionic bonds between the $-COO^-$ and the NH_3^+ of the R groups that formed salt bridges. A loss in ionic interactions causes the tertiary structure to lose stability. As the protein unfolds, both the shape and biological function are lost.

Study Check
Why is a dilute solution of $AgNO_3$ used to disinfect the eyes of newborn infants?

16.6 Enzymes

Learning Goal
Describe enzymes and their functions.

In the laboratory, we can break down polysaccharides, fats, or proteins by using a strong acid or base, high temperatures, and long reaction times. In the cells of our body, the same reactions occur at incredibly fast rates under mild conditions close to pH 7 and a body temperature of 37°C. In our cells almost all the chemical reactions are catalyzed by proteins known as **enzymes.** At this moment, in each of your cells, there are as many as 1500 different types of enzymes that catalyze the chemical reactions needed by your body.

In our discussion of chemical reactions in Chapter 5, we described a catalyst as a substance that lowers the activation energy of a reaction without being changed itself. Enzymes are catalysts that speed up chemical reactions in our cells by lowering the energy required for the reaction. The same reactions would eventually occur, but not at a rate fast enough for survival. For example, the hydrolysis of protein in our diet would occur without a catalyst, but not fast enough to meet the body's requirements for amino acids. However, when a digestive enzyme such as pepsin or trypsin is present, the hydrolysis occurs at a much faster rate.

In the blood, an enzyme called carbonic anhydrase converts large amounts of carbon dioxide and water to carbonic acid. Each molecule of anhydrase catalyzes about 35 million reactant molecules every minute. Typ-

Figure 16.15
The energy needed for the reaction of CO_2 and H_2O is lowered by the enzyme carbonic anhydrase.

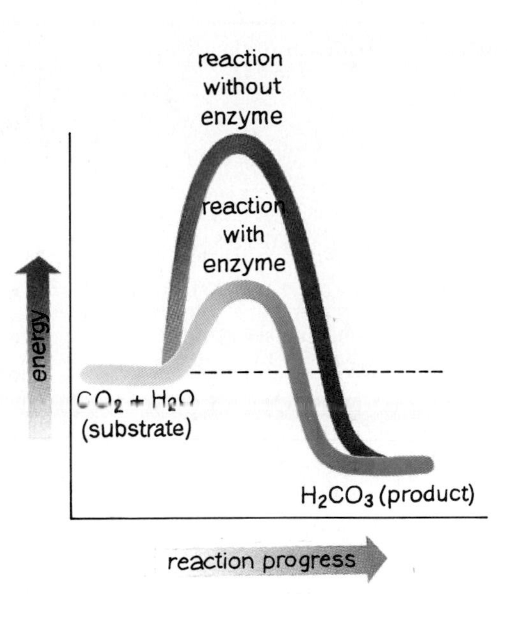

ically, an enzyme speeds up a reaction by as much as 10 million. (See Figure 16.15.)

Classification and Names of Enzymes

There are many different enzymes because each one catalyzes one particular reaction. As enzymes were discovered, many were named by adding the suffix *ase* to the names of the reactants. The enzyme that hydrolyzes sucrose is *sucrase,* and the enzymes that hydrolyze lipids are *lipases.* However, the names of some digestion enzymes, pepsin, renin, and trypsin, have been retained.

More recently, a systematic method of classifying and naming enzymes has been established. The name and class of each indicates the type of reaction it catalyzes. There are six classes of enzymes as described in Table 16.6.

Sample Problem 16.10
Enzyme Names

What chemical reaction would be catalyzed by the following enzymes?
a. amino transferase **b.** lactate dehydrogenase

Solution
a. catalyzes the transfer of an amino group
b. catalyzes the removal of hydrogen from lactate

Study Check
What is the class of lipase, an enzyme that catalyzes the hydrolysis of ester bonds in triglycerides?

Table 16.6 *Classification of Enzymes*

Class	General Reaction	Examples	Reactions Catalyzed
Oxidoreductases	Oxidation/reduction	Oxidases	Oxidation
		Reductases	Reduction
		Dehydrogenases	Remove hydrogen to form double bonds
$CH_3CH_2OH + NAD^+ \xrightarrow{\text{Alcohol dehydrogenase}} CH_3C(=O)—H + NADH + H^+$ (Reduced) → (Oxidized)			
Transferases	Transfer of a group	Transaminases	Transfer amino groups
		Kinases	Transfer phosphate
$CH_3—CH(NH_2)—COOH + HOOC—C(=O)—CH_2CH_2—COOH \xrightarrow{\text{Transaminase}} CH_3—C(=O)—COOH + HOOC—CH(NH_2)—CH_2CH_2—COOH$ (Alanine + α-Ketoglutaric acid → Pyruvic acid + Glutamic acid)			
Hydrolases	Hydrolysis of bonds	Peptidases	Hydrolyze peptide bonds
		Lipases	Hydrolyze ester bonds in triglycerides
		Amylase	Splits 1,4-glycosidic bonds in amylose
		Phosphatases	Hydrolyze phosphate
$\text{Triglyceride} + 3\ H_2O \xrightarrow{\text{Lipase}} \text{glycerol} + 3\ \text{fatty acids}$ (Lipid)			
Lyases	Remove/add small groups (usually with double bonds)	Decarboxylases	Remove CO_2
		Dehydrases	Remove water (H_2O)
		Deaminases	Remove NH_3
$CH_3—C(=O)—COOH \xrightarrow{\text{Pyruvic decarboxylase}} CH_3—C(=O)—H + CO_2$ (Three-carbon molecule → Two-carbon molecule + Carbon dioxide)			
Isomerases	Rearrange the atoms in a molecule	Isomerases	Convert between cis and trans; ketose and aldose
		Epimerases	Convert between D- and L-isomers
$\text{D-Fructose} \xrightarrow{\text{Fructose isomerase}} \text{D-glucose}$			
Ligases	Synthesize larger molecules using ATP	Synthetases	Combine two molecules
		Carboxylases	Add CO_2
$CH_3—C(=O)—COOH + CO_2 + ATP \xrightarrow{\text{Pyruvic carboxylase}} CH_3—C(=O)—CH_2—COOH + ADP + P_i$ (Three-carbon molecule → Four-carbon molecule)			

16.7 Enzyme Action

Learning Goal
Describe the role of an enzyme in an enzyme-catalyzed reaction.

An enzyme is a globular protein that has a unique three-dimensional shape. In many enzyme-catalyzed reactions, an enzyme catalyzes just one type of reaction for one specific substance known as a **substrate.** The enzyme is said to be substrate specific. An enzyme recognizes and binds to a substrate because the tertiary (or quaternary) structure of the enzyme complements the structure of a substrate. The combination of an *enzyme (E)* with a *substrate (S)* is called the **enzyme–substrate complex (ES).**

$$E + S \rightleftharpoons ES$$

The part of the enzyme that binds the substrate is called the **active site.** It is a pocket that fits the shape of a specific substrate formed by the side groups of a few of the amino acids in the protein. Because the substrate fits into the active site like a key fits a particular lock, this combination of enzyme and substrate is known as the **lock-and-key model.** (See Figure 16.16.)

However, the lock and key model assumes that the active site is a rigid, geometric structure. We now know that some enzymes react with more than one substrate; they have a broader range of specificity. In the **induced-fit model,** the active site is flexible, which allows it to accommodate the shapes of several different substrates.

Within the enzyme–substrate complex, the reaction of the substrate is catalyzed. In the active site, bonds of the substrate are broken and new bonds

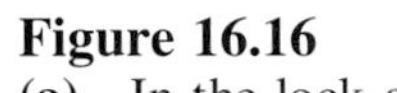

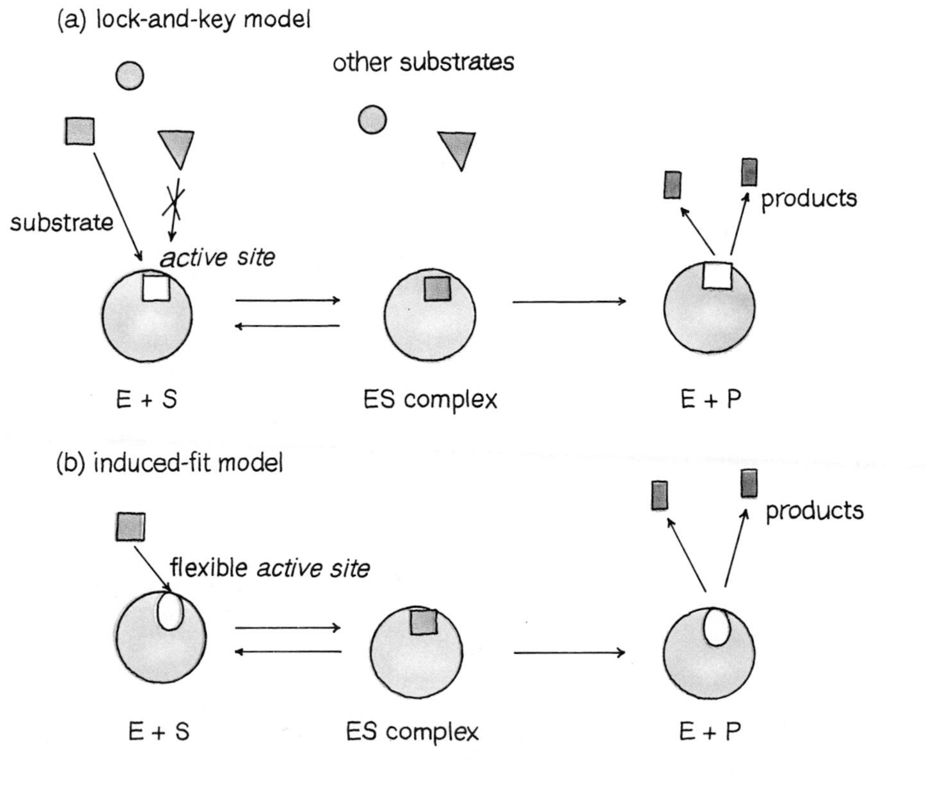

Figure 16.16
(a) In the lock-and-key model of enzyme action, a substrate that fits the shape of the active site binds to the enzyme.
(b) In the induced-fit model, the flexible active site adapts to the shape of the substrate. After catalysis occurs in the ES complex, the products disperse and the enzyme is free to bind another molecule of substrate.

formed to yield a product. Because the shape of the product is different from that of the substrate, the molecules of product detach from the enzyme. The active site on the enzyme is available again to bind and catalyze the reaction of another substrate molecule. (See Figure 16.16.) The overall enzyme-catalyzed reaction can be written as:

$$E + S \rightleftarrows E{-}S \longrightarrow E + P$$

Enzyme + substrate | Enzyme–substrate complex | Enzyme + product

Sample Problem 16.11
Enzyme Action

The enzyme sucrase catalyzes the hydrolysis of sucrose to give fructose and glucose. Using the equation for an enzyme-catalyzed reaction, describe the hydrolysis of sucrose.

Solution

Sucrose and the enzyme form an enzyme–substrate complex. At the active site, the glycosidic bond of sucrose is broken to give glucose and fructose.

$$E + S \rightleftarrows E{-}S \longrightarrow E + P$$

Sucrase + sucrose | Sucrase–sucrose complex | Sucrase + fructose + glucose

Study Check

How is the active site described by the lock-and-key model and the induced-fit model?

HEALTH NOTE *Enzymes as Diagnostic Tools*

Enzymes that catalyze the same reactions but vary slightly in structure are called **isoenzymes.** Different cells in the body produce isoenzymes for the same type of reactions. For example, there are five isoenzymes for lactate dehydrogenase (LDH), an enzyme that converts pyruvic acid to lactic acid. Each isoenzyme consists of a quaternary structure containing four subunits. In heart muscle, the most prevalent subunit is designated as H. In skeletal muscle, the major subunit is designated as M. With different combinations of subunits, isoenzymes can be separated by electrophoresis.

In healthy tissues, enzymes are contained within cellular membranes. However, if a particular organ is damaged, the contents of the cells, including the enzymes, enter the blood. By identifying the isoenzyme that is elevated in the blood serum, the damaged organ can be identified. For example, liver diseases are indicated by a rise in the serum LDH_5 level. When myocardial infarction (MI), or heart attack, damages the cells in heart muscle, there is an increase in the serum LDH_2 level.

Isoenzyme	LDH_1	LDH_2	LDH_3	LDH_4	LDH_5
Subunits	H_4	H_3M	H_2M_2	HM_3	M_4
Abundant in	Heart, kidneys	Heart, kidneys, brain, red blood cells	Kidneys, lungs	Spleen	Liver, skeletal muscle

Another isoenzyme used diagnostically is creatine kinase (CK), which consists of two types of subunits. One subunit is prevalent in the brain (B) and the other predominates in skeletal muscle (M). Normally only the CK_3 is present in low amounts in the blood serum. However, in a patient who has suffered an MI, the levels of CK_2 will be elevated soon after the heart attack. (See Figure 16.17.)

Isoenzyme	CK_1	CK_2	CK_3
Subunits	BB	MB	MM
Abundant in	Brain, lung	Heart	Skeletal muscle, red blood cells

Table 16.7 lists some enzymes used to diagnose diseases of certain organs.

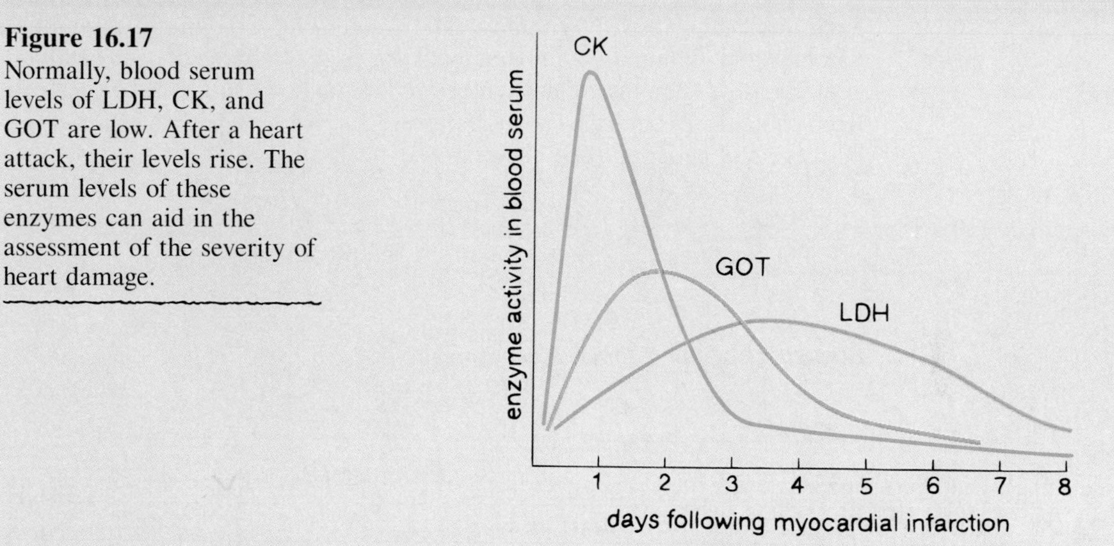

Figure 16.17
Normally, blood serum levels of LDH, CK, and GOT are low. After a heart attack, their levels rise. The serum levels of these enzymes can aid in the assessment of the severity of heart damage.

Table 16.7 ***Serum Enzymes Used in Diagnosis of Tissue Damage***

Organ	Condition	Diagnostic Enzymes Elevated
Heart	Myocardial infarction (MI)	Lactate dehydrogenase (LDH_1) Creatine phosphokinase (CK_2) Glutamic oxaloacetic transaminase (GOT)
Liver	Cirrhosis, carcinoma, hepatitis	Glutamic pyruvic transaminase (GPT) Lactate dehydrogenase (LDH_5) Alkaline phosphatase (ALP) Glutamic oxaloacetic transaminase (GOT)
Bone	Rickets, carcinoma	Alkaline phosphatase (ALP)
Pancreas	Pancreatic diseases	Amylase Cholinesterase Lipase (LPS)
Prostate	Carcinoma	Acid phosphatase (ACP)

16.8 Factors Affecting Enzyme Action

Learning Goal

Discuss the effect of cofactors, temperature, and pH on enzyme-catalyzed reactions.

A functional enzyme that consists of only polypeptides is called a simple enzyme. However, many enzymes require the addition of a nonprotein portion to catalyze a reaction. Then the protein part of an enzyme is called the **apoenzyme** and the nonprotein, a **cofactor.** A cofactor may be a metal ion such as copper (Cu^{2+}), iron (Fe^{2+}), magnesium (Mg^{2+}), manganese (Mn^{2+}), and zinc (Zn^{2+}). For example, carbonic anhydrase requires a zinc ion cofactor for catalytic activity, and pyruvate decarboxylase needs manganese ion. Many of these metal ions are trace elements that must be obtained in our diet.

Other enzymes require cofactors called *coenzymes,* which are small organic molecules, usually vitamins. The water-soluble vitamins, vitamin B complex and vitamin C, supply many of the coenzymes needed for catalytic activity. Most vitamins are not synthesized in the body and must be obtained from the diet. Deficiencies of vitamins are known to cause scurvy, rickets, beriberi, and pellagra. (See Table 16.8.)

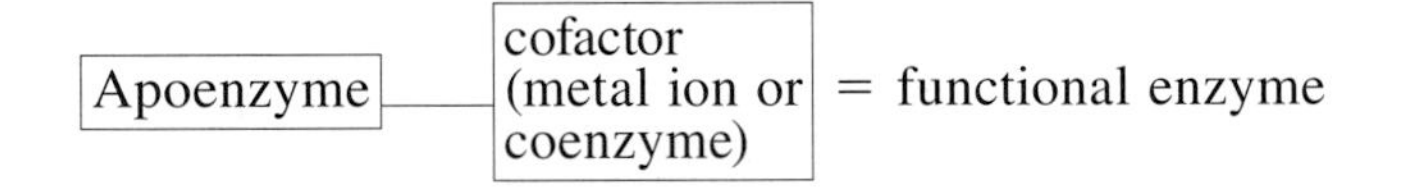

Structures of the Vitamin Coenzymes

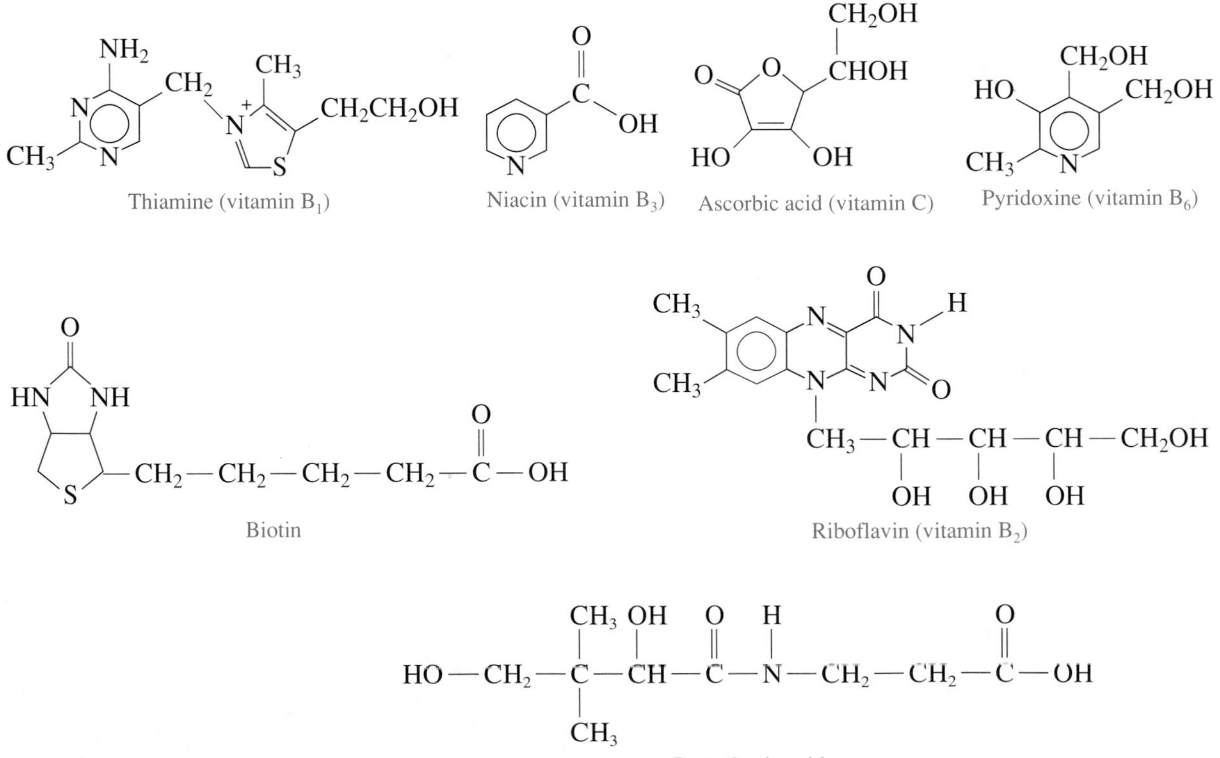

Table 16.8 ***The Water-Soluble Vitamins as Coenzymes***

Vitamin	Coenzyme Function	Deficiency Symptoms
Thiamine (vitamin B_1)	Decarboxylation of α-keto acids; thiamine pyrophosphate	Beriberi (fatigue, anorexia, nerve degeneration, heart failure)
Riboflavin (vitamin B_2)	Biological oxidation; FAD (flavin adenine dinucleotide) and FMN (flavin mononucleotide)	Dermatitis, glossitis (tongue inflammation), cataracts
Niacin (vitamin B_3)	Biological oxidation NADH (nicotinamide adenine dinucleotide)	Pellagra (scaly skin, muscle fatigue, diarrhea, mouth sores, mental disorder)
Pyridoxine (vitamin B_6)	Transamination reactions; pyridoxal phosphate	Dermatitis, fatigue, anemia, irritability
Cobalamin (vitamin B_{12})	Transfer of methyl groups in biosynthesis of red blood cells, methionine, choline, purines	Pernicious anemia, malformed red blood cells, mental disorders
Ascorbic acid (vitamin C)	Synthesis of collagen, protein metabolism, iron absorption, healing of wounds	Scurvy (bleeding gums, slow-healing wounds, muscle pain, anemia)
Biotin	Carboxylation of pyruvic acid to oxaloacetic acid, fatty acid and amino acid metabolism	Dermatitis, fatigue, anemia, nausea, depression
Folic acid	Transfer of methyl groups in biosynthesis reactions; tetrahydrofolate	Abnormal red and white blood cells, intestinal tract disturbances
Pantothenic acid	Transfer of acetyl groups; coenzyme A	Fatigue, anemia

Sample Problem 16.12
Enzymes and Cofactors

Indicate whether each of the following is a simple enzyme or one that requires a cofactor.

a. a polypeptide that needs Mg^{2+} for catalytic activity
b. a functional enzyme composed only of a polypeptide chain
c. an enzyme that consists of a protein portion attached to vitamin B_6

Solution

a. The metal ion Mg^{2+} is a cofactor.
b. A functional enzyme consisting only of a polypeptide chain is a simple enzyme.
c. The vitamin B_6 is a cofactor.

Study Check

Which of the nonprotein portions of the enzymes in the above question is a coenzyme?

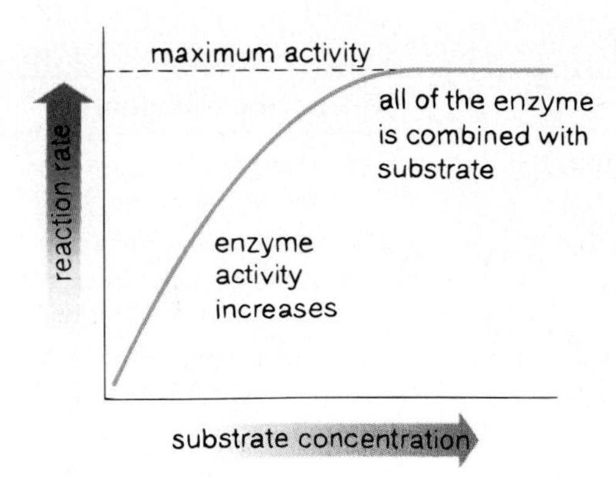

Figure 16.18
An increase in substrate concentration increases the rate of an enzyme-catalyzed reaction until all of the available enzyme has combined with substrate.

Effect of Substrate Concentration

When the amount of enzyme is fixed and we add more substrate, the catalyzed reaction goes faster. However, at some point, all of the active sites on the enzymes are filled with substrate. When enzymes are saturated with substrate, there can be no further increase in the rate at which substrate is catalyzed. The reaction proceeds at its maximum rate, even if more substrate is added. (See Figure 16.18.)

Effect of Temperature

At low temperatures, most enzymes show little activity because there is not a sufficient amount of energy for the catalyzed reaction to take place. As the temperature increases, the rate of a reaction also increases. Enzymes are most active at an *optimum temperature,* usually around 37°C. At temperatures above the optimum, the protein of an enzyme begins to denature. Higher temperatures alter the structure of the enzyme further, leading to a rapid loss of catalytic activity. (See Figure 16.19.)

Effect of pH

Enzymes are most active at their **optimum pH,** the pH that maintains the proper tertiary and quaternary structure of the protein. (See Figure 16.20.) At pH values above or below the optimum, changes occurring in the ionic side

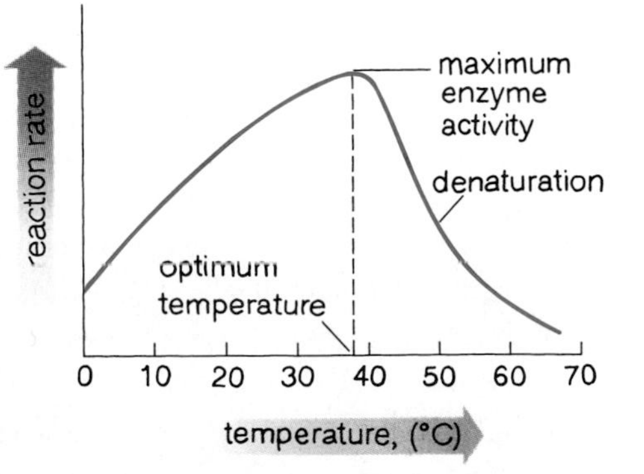

Figure 16.19
An increase in temperature increases enzyme activity to its optimum temperature. Higher temperatures cause denaturation and loss of catalytic activity.

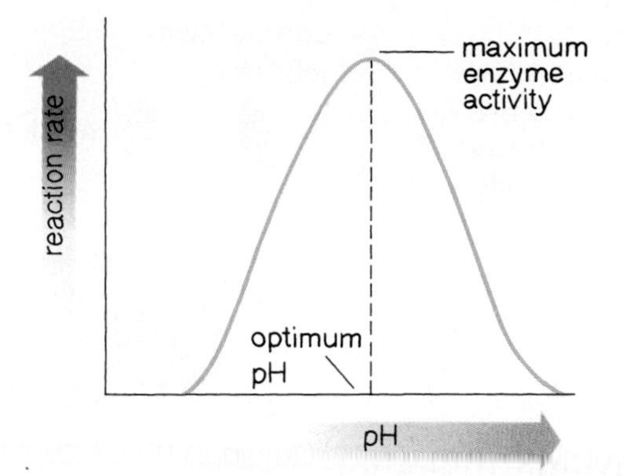

Figure 16.20
Enzymes are most active at their optimum pH. At higher or lower pH, denaturation of the enzyme causes a loss of catalytic activity.

groups of the enzyme disrupt the active site until it no longer binds with the substrate. For example, the digestive enzymes in the stomach, such as pepsin, have an optimum pH of 2. If pH rises to pH 4 or 5, pepsin shows little or no digestive activity. At extreme pH values, the enzyme may be completely denatured. Table 16.9 lists the optimum pH values for some enzymes.

Table 16.9 ***Optimum pH for Selected Enzymes***

Enzyme	Location	Substrate	Optimum pH
Pepsin	Stomach	Protein	2.0
Urease	Liver	Urea	5.0
Sucrase	Small intestine	Sucrose	6.2
Pancreatic amylase	Pancreas	Amylose	7.0
Trypsin	Small intestine	Polypeptides	8.0

Sample Problem 16.13
Factors Affecting Enzymatic Activity

Describe the effect of each of the following on the rate of the reaction catalyzed by urease.

$$H_2N-\overset{\overset{\displaystyle O}{\|}}{C}-NH_2 + H_2O \xrightarrow{\text{Urease}} 2NH_3 + CO_2$$

Urea

a. increasing the urea concentration
b. lowering the temperature to 10°C

Solution

a. An increase in urea concentration will increase the rate of reaction until all of the enzyme molecules are bound to the urea substrate. No further increase in rate occurs.
b. Because 10°C is lower than the optimum temperature of 37°C, the lower temperature will decrease the rate of the reaction.

Study Check
If urease has an optimum pH of 5, what is the effect of lowering the pH to 3?

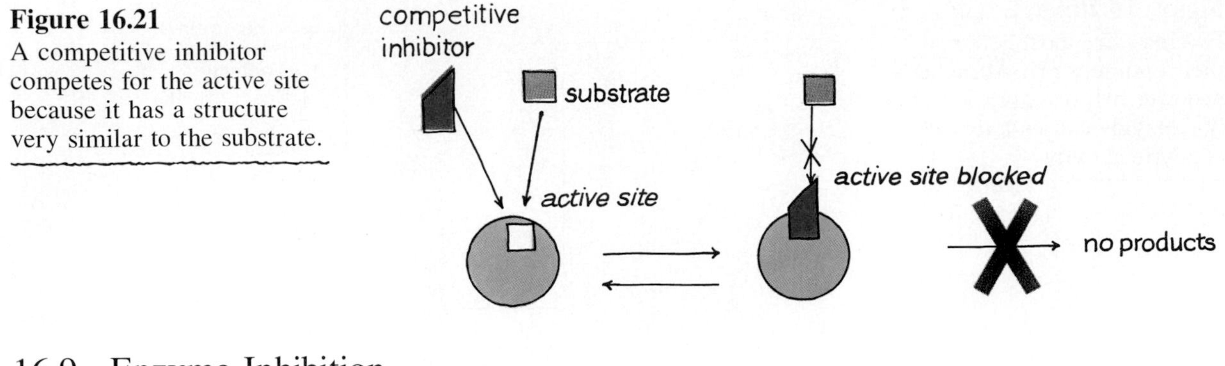

Figure 16.21
A competitive inhibitor competes for the active site because it has a structure very similar to the substrate.

16.9 Enzyme Inhibition

Learning Goal
Describe the effect of an inhibitor upon enzyme activity.

Inhibition is a loss of catalytic activity by enzymes. Chemical compounds called **inhibitors** combine with enzymes causing a change in the protein structure. The active site is prevented from reacting with its proper substrate and the enzyme is inactive. For example, many antibiotics stop bacterial infections because they inhibit enzymes essential to the growth process of bacteria.

Competitive Inhibition

A **competitive inhibitor** has a structure similar to the substrate. Therefore, the inhibitor fits into the active site on the enzyme just like the substrate. As long as the inhibitor occupies the active site, no reaction of the substrate can take place. However, the attractions for competitive inhibitors are weak, and the inhibitor is easily released from the active site. When the active site is open, a substrate molecule can bind to the enzyme. As a result, the substrate and its inhibitor are competing for the active site on the enzyme. As the number of substrate molecules increases, the active sites are more likely to be filled with substrate rather than inhibitor molecules. Increasing the substrate concentration can reverse competitive inhibition. (See Figure 16.21.)

Malonic acid is a competitive inhibitor of the enzyme succinic dehydrogenase. Because malonic acid has a structure similar to the substrate, succinic acid, the two substances compete for the active site on the dehydrogenase. As long as the inhibitor malonic acid occupies the active site on the enzyme, no reaction occurs. When more succinic acid is added, more active sites will fill with substrate, and there will be less inhibition.

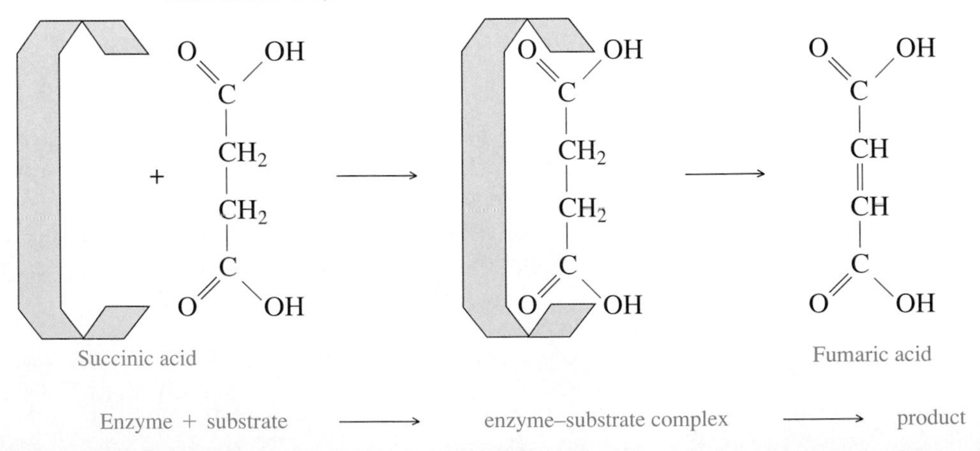

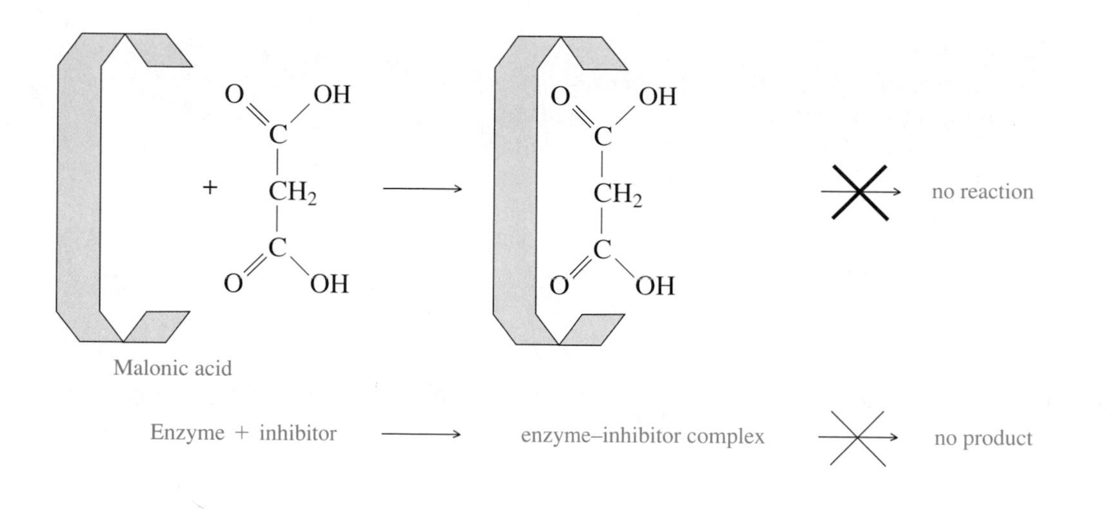

HEALTH NOTE *Competitive Inhibitors in Medicine*

Some bacterial infections are treated with competitive inhibitors called **antimetabolites.** Sulfanilamide, one of the first sulfa drugs, competes with PABA, a substrate (metabolite), in the growth cycle of bacteria.

Substrate of Bacterial Growth *Inhibitor*

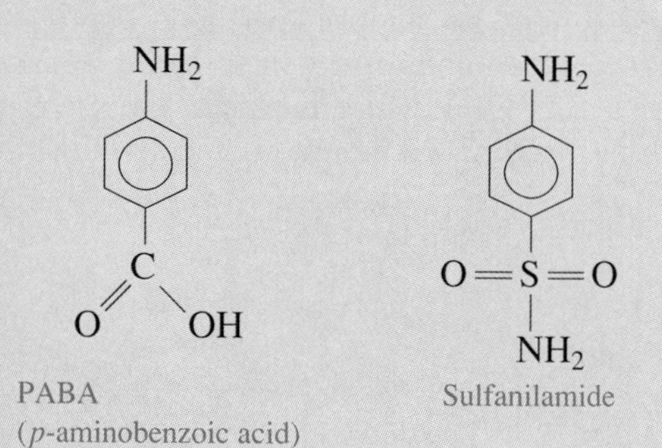

PABA
(p-aminobenzoic acid)

Sulfanilamide

Antibiotics produced by bacteria, mold, or yeast are used to inhibit bacterial growth. For example, penicillin inhibits an enzyme needed for the formation of cell walls in bacteria, but not human cell membranes. With an incomplete cell wall, bacteria cannot survive, and the infection is stopped.

However, some bacteria are resistant to penicillin because they contain penicillinase, an enzyme that breaks down penicillin. New kinds of penicillin to which bacteria have not yet become resistant have been produced.

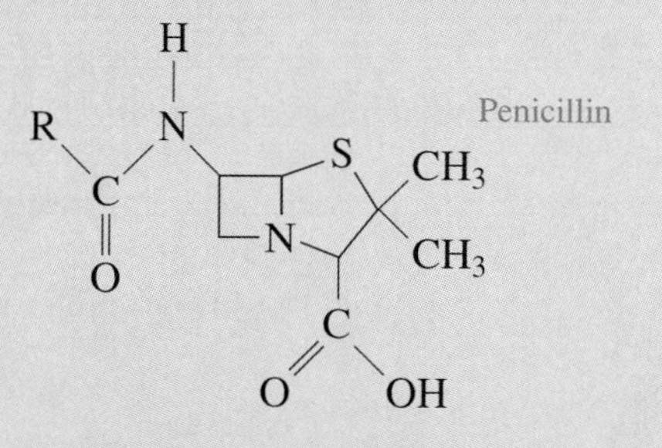

Penicillin

R Groups for Penicillin Derivatives

Penicillin G

Penicillin V

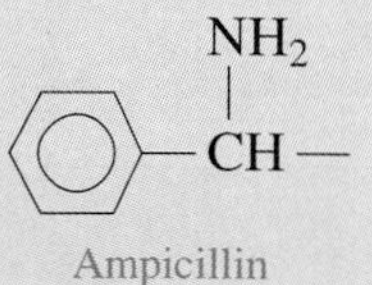

Ampicillin

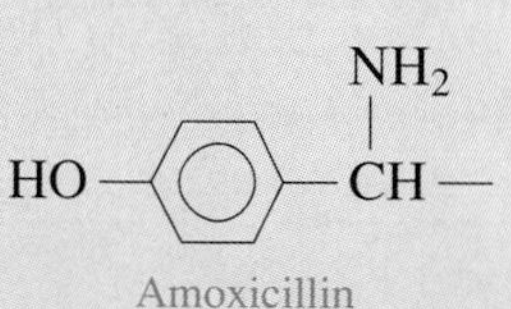

Amoxicillin

Another group of antibiotics, tetracycline, streptomycin, chloramphenicol, and aureomycin, stops bacterial infections by inhibiting the synthesis of protein in bacteria. They are also used to fight bacterial infections in people allergic to the penicillins.

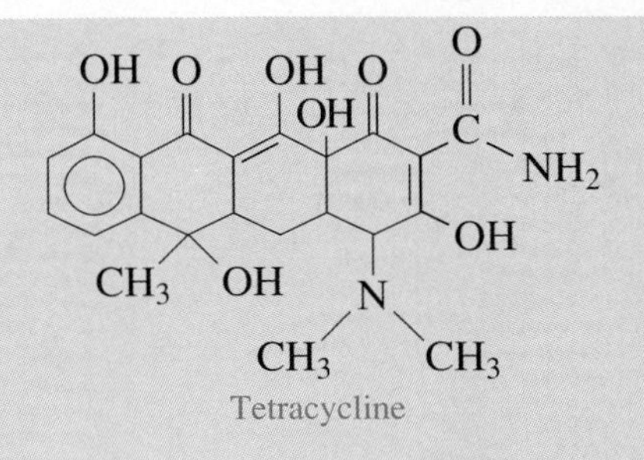

Tetracycline

Noncompetitive Inhibition

The structure of a **noncompetitive inhibitor** does not resemble the substrate. Therefore, it does not compete with the substrate for the active site and usually binds to a part of an enzyme that is not the active site. However, when a noncompetitive inhibitor attaches to the enzyme, it changes the shape of the enzyme as well as the shape of the active site. Inhibition occurs because the substrate cannot fit into the active site and no reaction takes place. (See Figure 16.22.) Examples of noncompetitive inhibitors are the heavy metal ions Pb^{2+}, Ag^{+}, and Hg^{2+} that bond with side groups such as —SH, $—COO^{-}$, or —OH. Catalytic activity is restored when the inhibitors leave the enzyme or are removed by chemical reagents. Because there is no competition for the active site, the addition of more substrate does not reverse this type of inhibition.

A noncompetitive inhibition can be irreversible if the inhibitor is tightly bonded at the active site and cannot be removed without destroying the protein. Nerve gases act as irreversible inhibitors by forming a covalent bond with the side group of serine ($—CH_2OH$), which is part of the active site in acetylcholinesterase. When this enzyme is inhibited, the transmission of nerve impulses is blocked.

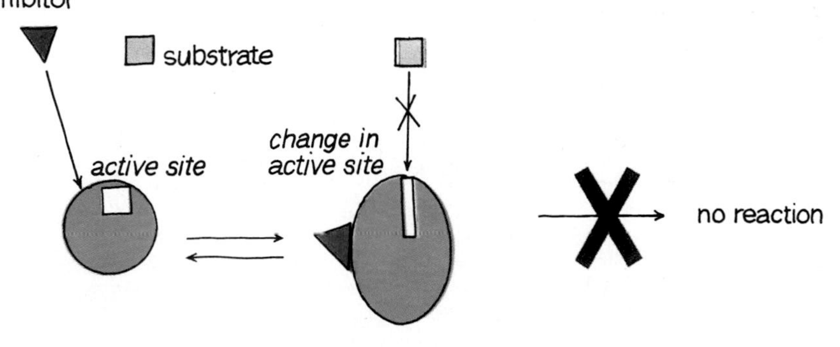

Figure 16.22
When a noncompetitive inhibitor attaches to the enzyme away from the active site, it changes the shape of the enzyme, which alters the active site and prevents the binding of the substrate.

Sample Problem 16.14
Enzyme Inhibition

State the type of inhibition in the following:

a. The inhibitor has a structure that is similar to the substrate.
b. This inhibitor binds to the surface of the enzyme, changing its shape in such a way that it cannot bind to substrate.

Solution

a. competitive inhibition **b.** noncompetitive inhibition

Study Check

Hydrogen cyanide (HCN) forms strong bonds with catalase, an enzyme that contains iron (Fe^{3+}). What type of inhibitor is HCN?

Chapter Summary

16.1 Types of Proteins
Some proteins are enzymes or hormones, whereas others are important in structure, transport, protection, storage, and contraction of muscles.

16.2 Amino Acids
A group of 20 amino acids provides the molecular building blocks of proteins. Attached to the central (alpha) carbon of each amino acid is an amino group, a carboxyl group, and a unique side group (R). The R group gives an amino acid the property of being nonpolar, polar, acidic, or basic. Amino acids exist as dipolar ions called zwitterions, positive ions at low pH, and negative ions at high pH levels.

16.3 The Peptide Bond
Peptide bonds join amino acids to form peptides by forming an amide bond between the carboxyl group of one amino acid and the amino group of the second. Long chains of amino acids are called proteins.

16.4 Primary and Secondary Protein Structure
The primary structure of a protein is its sequence of amino acids. In the secondary structure, hydrogen bonds between peptide bonds produce a characteristic shape such as an α-helix, β-pleated sheet, or a triple helix.

16.5 Tertiary and Quaternary Protein Structure
In globular proteins, the polypeptide chain including its α-helical and β-pleated sheet regions folds upon itself to form a tertiary structure. A tertiary structure is stabilized by interactions that move hydrophobic R groups to the inside, hydrophilic R groups to the outside surface, and attractions between R groups that form hydrogen, disulfide, and ionic bonds. In a quaternary structure, two or more tertiary subunits must combine for biological activity. They are held together by the same interactions found in tertiary structures. Denaturation of a protein occurs when heat or other denaturing agents destroy the structure of the protein (but not the primary structure) until biological activity is lost.

16.6 Enzymes
Enzymes are globular proteins that act as biological catalysts. The names of most enzymes are indicated by their *ase* endings. Enzymes are classified by the type of reaction they catalyze such as oxidoreductase, hydrolase, or isomerase.

16.7 Enzyme Action
Within the structure of the enzyme, there is a small pocket called the active site, which has a specific shape that fits a specific substrate. In the lock-and-key model or the induced-fit model, an enzyme and substrate form an enzyme–substrate complex so the reaction of the substrate can be catalyzed at the active site.

16.8 Factors Affecting Enzyme Action
Simple enzymes are biologically active as a protein only, whereas other enzymes require both a protein portion (apoenzyme) and a cofactor. A cofactor may be a metal ion such as Cu^{2+} or Fe^{2+}, or an organic compound called a coenzyme, usually a vitamin.

An increase in substrate concentration increases the reaction rate of an enzyme-catalyzed reaction until all the enzyme molecules combine with substrate. Then additional substrate will not increase the rate further. Enzymes are most effective at optimum temperature and pH. The rate of an enzyme reaction decreases considerably at temperatures and pH above or below the optimum.

16.9 Enzyme Inhibition
An enzyme can be made inactive by changes in pH, temperature, or by chemical compounds called inhibitors. A competitive inhibitor has a stucture similar to the substrate and competes for the active side. When the active site is occupied, the enzyme cannot catalyze the reaction of the substrate. A noncompetitive inhibitor attaches elsewhere on the enzyme changing the shape of both the enzyme and the active site. As long as the noncompetitive inhibitor is attached to the enzyme, the altered active site cannot bind with substrate.

Glossary of Key Terms

active site The region of an enzyme that binds and catalyzes a reaction of substrate.

α (alpha)-helix A secondary level of protein structure, in which hydrogen bonds connect the NH of one peptide bond with the C=O of a peptide bond later in the chain to form a coiled or corkscrew structure.

amino acid The building block of proteins, consisting of an amino group, a carboxylic acid group, and a unique side group attached to the alpha carbon.

antibiotic An inhibitor produced by a bacteria, mold, or yeast that is toxic to other bacteria.

apoenzyme The protein portion that along with a cofactor provides a functional enzyme.

β (beta)-pleated sheet A secondary level of protein structure that consists of hydrogen bonds between peptide links in parallel polypeptide chains.

cofactor The nonprotein portion, a metal ion or an organic compound, that is necessary for a biologically functional enzyme.

competitive inhibitor A molecule that has a structure similar to the substrate and inhibits enzyme action by competing for the active site.

denaturation The loss of secondary and tertiary protein structure caused by heat, acids, bases, organic compounds, heavy metals, and/or agitation.

electrophoresis The use of electrical current to separate proteins or other charged molecules with different isoelectric points.

enzymes Proteins that catalyze biological reactions.

enzyme–substrate complex An intermediate in an enzyme-catalyzed reaction in which the substrate combines with the enzyme.

essential amino acids Amino acids that must be supplied by the diet because they are not synthesized by the body.

fibrous protein A protein that is insoluble in water; consists of polypeptide chains with α-helixes, or β-pleated sheet, that make up the fibers of hair, wool, skin, nails, and silk.

globular proteins Proteins that acquire a compact shape from attractions between the R group of the amino acid residues in the protein.

hydrophilic amino acid An amino acid having polar, acidic, or basic R groups that are attracted to water; "water-loving."

hydrophobic amino acid A nonpolar amino acid with hydrocarbon R groups; "water-fearing."

induced-fit model A model of enzyme action that describes a flexible active site that adapts to the structure of a substrate to form the enzyme–substrate complex.

inhibitors Substances that make an enzyme inactive by interfering with its ability to react with a substrate.

isoelectric point The pH at which a protein is electrically neutral.

lock-and-key model A model of enzyme action that represents the substrate as a key fitting the correct lock which is the specific shape of the active site in an enzyme.

noncompetitive inhibitor A substance that changes the shape of the enzyme and the active site, preventing the active site from binding with the substrate.

nonpolar amino acids Amino acids that are not soluble in water because they contain a nonpolar R group.

optimum pH The pH at which an enzyme is most active.

peptide The combination of two or more amino acids joined by peptide bonds; dipeptide, tripeptide, and so on.

peptide bond The amide bond that joins amino acids in polypeptides and proteins.

polar amino acids Amino acids that are soluble in water because their R group is polar: hydroxyl (OH), thiol (SH), carbonyl (C=O), amino (NH_2), or carboxyl (COO^-).

polypeptides Polymers of amino acids joined by peptide bonds.

primary structure The sequence of the amino acids in a protein.

protein A term used for biologically active polypeptides that have many amino acids linked together by peptide bonds.

quaternary structure A protein structure in which two or more protein subunits form an active protein.

secondary structure The formation of an α-helix, β-pleated sheet or triple helix.

substrate The substance that reacts at the active site in an enzyme-catalyzed reaction.

tertiary structure The folding of the secondary structure of a protein into a compact structure that is stabilized by the interactions of R groups such as ionic and disulfide bonds.

triple helix The protein structure found in collagen consisting of three polypeptide chains woven together like a braid.

zwitterion The dipolar form of an amino acid consisting of two oppositely charged ionic regions, $-NH_3^+$ and $-COO^-$.

Chemistry at Home

1. Read the nutrition facts on some food products. What is the quantity of a serving size? How many servings are in the container? How many grams of fat, carbohydrate, and protein are in one serving? Using the guide near the bottom of the Facts list, Calories per gram, calculate the Calories from each food type, and the total Calories in one serving. How do your calculations compare with the stated number of Calories at the top of the Facts list? What is the percentage of the Calories obtained from protein? Fat? Carbohydrate?
2. Place some milk in four glasses. To the first milk sample, begin adding vinegar, drop by drop. Stir. What changes do you observe? Explain. To the next milk sample, add $\frac{1}{2}$ teaspoon of meat tenderizer. Stir. What enzyme is listed on the package label? Provide an explanation for your observation. To the third milk sample, add 1 teaspoon of fresh pineapple juice. (Canned juice has been heated.) If pineapple juice contains the same enzyme as tenderizer, how would you explain your observations? To the fourth milk sample, add 1 teaspoon of pineapple juice that you have heated to boiling. What is its effect compared with the fresh juice? Explain.
3. The enzymes on the surface of a freshly cut apple or avocado react with oxygen in the air to turn the surface brown. An antioxidant such as vitamin C in lemon juice prevents the oxidation reaction. Cut an apple into several large slices. Place one slice in a plastic zip locked bag, squeeze out the air, and close the zipper lock. Place three apple slices on a plate. Leave one alone. Dip one in lemon juice. Sprinkle the third slice with a crushed vitamin C tablet. Record your observations immediately, and then every hour for 4 hours. Which slice(s) had the most oxidation (turned brown)? Which had the least? Explain.
4. Some detergents contain enzymes to help clean fabrics. Obtain a detergent with enzymes on the contents, and one without enzymes. Obtain two large glasses or jars and prepare two hard boiled eggs. Place equal amounts of water in each glass or jar. Add 2–3 tablespoons of enzyme-containing detergent in one and the same amount of nonenzyme detergent in the other. Remove the egg shells and place a hard-boiled egg in each solution. Put the jars and contents aside and observe each day for one week. One detergent claims it has a "grease releaser." What might that be?
5. Read the label on a permanent wave. What are some of the chemicals used to reduce and oxidize the disulfide bonds in hair proteins?

Answers to Study Checks

16.1 hormones

16.2 **a.** nonpolar **b.** polar

16.3

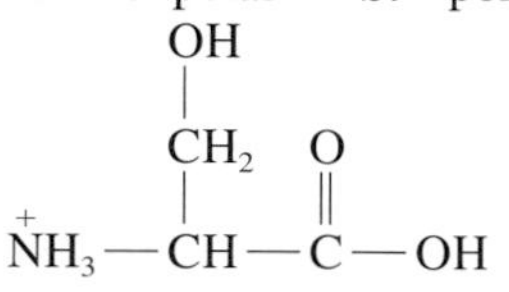

$$\overset{+}{N}H_3-\underset{\underset{\displaystyle OH}{|}}{\overset{}{CH}}\!\!\!\!\!$$

$$\begin{array}{l} \quad\ \ OH \\ \quad\ \ | \\ \quad\ \ CH_2\ \ O \\ \quad\ \ |\qquad \| \\ \overset{+}{N}H_3-CH-C-OH \end{array}$$

16.4

$$\begin{array}{l} \qquad O\ \ \ OH \\ \qquad\ \ \backslash\!\!\backslash\ / \\ \qquad\quad C \qquad\qquad\qquad\quad C_6H_5 \\ \qquad\quad | \qquad\qquad\qquad\quad\ | \\ \qquad\ CH_2\ \ O \qquad\qquad\ CH_2\ \ O \\ \qquad\quad |\qquad \| \qquad\qquad\quad |\qquad \| \\ NH_2-CH-C-NH-CH-C-O-H \end{array}$$

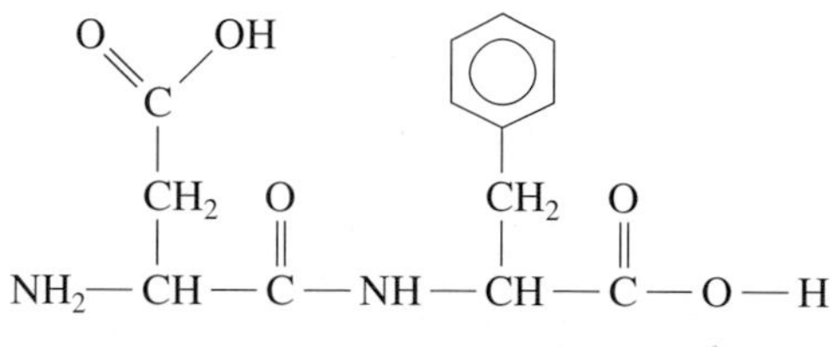

16.5 threonylleucylphenyalanine
16.6 a triple helix
16.7 Both are nonpolar and would be found on the inside of the tertiary structure.
16.8 quaternary
16.9 The heavy metal Ag^+ denatures the protein in bacteria that cause gonorrhea.
16.10 hydrolase
16.11 The active site has a rigid shape in the lock-and-key model but is flexible in the induced-fit and adapts to the shape of the substrate.
16.12 The vitamin B_6 in part (c).
16.13 Urease will lose enzymatic activity as denaturation of its protein occurs.
16.14 noncompetitive inhibitor; irreversible

Problems

Types of Proteins *(Goal 16.1)*

16.1 Classify each of the following proteins according to its function:
a. hemoglobin, oxygen carrier in the blood
b. collagen, a major component of tendons and cartilage
c. keratin, a protein found in hair
d. amylase, an enzyme that hydrolyzes starch

16.2 Classify each of the following proteins according to its function:
a. insulin, a hormone needed for glucose utilization
b. antibodies, proteins that disable foreign proteins
c. casein, milk protein
d. lipases, enzymes that hydrolyze lipids

Amino Acids *(Goal 16.2)*

16.3 Describe the functional groups found in all amino acids.

16.4 What is the difference between a polar and a nonpolar side (R) group?

16.5 Draw the structural formula of each of the following amino acids:
a. alanine **b.** threonine
c. glutamic acid **d.** phenylalanine

16.6 Draw the structural formula of each of the following amino acids:
a. lysine **b.** aspartic acid
c. leucine **d.** tyrosine

16.7 Describe of the amino acids in problem 16.5 as nonpolar or polar.

16.8 Describe each of the amino acids in problem 16.6 as nonpolar, or polar.

16.9 Give the name of the amino acid represented by each of the following three-letter abbreviations:
a. Ala **b.** Val **c.** Lys **d.** Cys

16.10 Give the name of the amino acid represented by each of the following three-letter abbreviations:
a. Trp **b.** Met **c.** Pro **d.** Gly

16.11 What is an essential amino acid? Which of the amino acids listed in problem 16.9 are essential amino acids?

16.12 How are essential amino acids obtained? Which of the amino acids listed in problem 16.10 are essential amino acids?

16.13 Write the zwitterion of each of the following amino acids:
a. glycine **b.** cysteine
c. serine **d.** alanine

16.14 Write the zwitterion of each of the following amino acids:
a. phenylalanine **b.** methionine
c. leucine **d.** valine

16.15 Write the ionized form of the amino acids in problem 16.13 at low pH.

16.16 Write the ionized form of the amino acids in problem 16.14 at high pH.

The Peptide Bond *(Goal 16.3)*

16.17 Draw the structural formulas of the following peptides:
a. Ala-Cys **b.** serylphenylalanine
c. Gly-Ala-Val **d.** Val-Ile-Trp

16.18 Draw the structural formulas of the following peptides:
a. Met-Asp **b.** alanyltyrosine
c. Met-Gln-Lys **d.** His-Gly-Glu-Ala

Primary and Secondary Protein Structure *(Goal 16.4)*

16.19 What is the primary structure of a protein?

16.20 Distinguish between the α-helix structure of a protein and the β-pleated sheet.

Tertiary and Quaternary Protein Structure *(Goal 16.5)*

16.21 What type of interaction would you expect from the following R groups in a tertiary structure?
- **a.** two cysteine residues
- **b.** glutamic acid and lysine
- **c.** serine and aspartic acid
- **d.** two leucine residues

16.22 In myoglobin, about one-half of the 153 amino acids have nonpolar side chains.
- **a.** Where would you expect those amino acids to be located in the tertiary structure?
- **b.** Where would you expect the polar side chains to be?
- **c.** Why is myoglobin more soluble in water than silk or wool?

16.23 A portion of a polypeptide chain contains the following sequence of amino acid residues:

-Leu-Val-Cys-Asp-

- **a.** Which R groups can form a disulfide cross-link?
- **b.** Which amino acid residues are likely to be found on the inside of the protein structure? Why?
- **c.** Which amino acid residues would be found on the outside of the protein? Why?
- **d.** How does the primary structure of a protein affect its tertiary structure?

16.24 State whether the following statements apply to primary, secondary, tertiary or quaternary protein structure:
- **a.** Side groups interact to form disulfide bonds or ionic bonds.
- **b.** Peptide bonds join amino acids in a polypeptide chain.
- **c.** Several polypeptides are held together by hydrogen bonds between adjacent chains.
- **d.** Hydrogen bonding between carbonyl oxygen atoms and nitrogen atoms of amide groups causes a polypeptide to coil.
- **e.** Hydrophobic side chains seeking a nonpolar environment move toward the inside of the folded protein.
- **f.** Protein chains of collagen form a triple-helix.
- **g.** An active protein contains four tertiary subunits.

16.25 Indicate the changes that occur under the following conditions:
- **a.** An egg placed in hot water at 100°C is soft-boiled in about 3 min.
- **b.** Prior to an injection, the skin is cleansed with an alcohol swab.
- **c.** Surgical instruments are placed in a 120°C autoclave.

16.26 Indicate the changes that occur under the following conditions:
- **a.** Tannic acid is placed on a burn.
- **b.** Milk is heated to 60°C to make yogurt.
- **c.** Seeds are treated with an $HgCl_2$ solution.

Enzymes *(Goal 16.6)*

16.27 What class of enzyme catalyzes each of the following reactions?
- **a.** hydrolysis of sucrose
- **b.** addition of oxygen
- **c.** isomerization of glucose to fructose
- **d.** transfer of an amino group

16.28 What class of enzyme catalyzes each of the following reactions?
- **a.** addition of water to a double bond
- **b.** removal of hydrogen
- **c.** hydrolysis of ester bonds in triglycerides
- **d.** decarboxylation of pyruvate

16.29 What is the substrate of each of the following enzymes?
a. galactase **b.** lipase **c.** aspartase

16.30 What is the substrate of each of the following enzymes?
a. peptidase **b.** cellulase **c.** lactase

Enzyme Action *(Goal 16.7)*

16.31 Match the terms, (1) enzyme–substrate complex, (2) enzyme, and (3) substrate, with the following phrases:
- **a.** has a tertiary structure that recognizes the substrate
- **b.** the combination of an enzyme with the substrate
- **c.** has a structure that complements the structure of the enzyme

16.32 Match the terms, (1) active site, (2) lock-and-key model, and (3) induced-fit model with the following phrases:
- **a.** the portion of an enzyme where catalytic activity occurs
- **b.** an active site that adapts to the shape of a substrate
- **c.** an active site having a rigid shape

16.33 Write an equation that represents an enzyme-catalyzed reaction.
- **a.** Identify each of the symbols used in the equation.
- **b.** How is the active site different from the whole enzyme structure?
- **c.** After the products have formed, what happens to the enzyme?

16.34 Lactase is an enzyme that hydrolyzes lactose to glucose and galactose.
- **a.** What are the reactants and products of the reaction?
- **b.** Draw an energy diagram for the reaction with and without the enzyme.
- **c.** Why does the enzyme make the reaction go faster?

Factors Affecting Enzyme Activity *(Goal 16.8)*

16.35 Is the enzyme described in each of the following statements simple or one that contains a cofactor?
- **a.** contains a sugar portion
- **b.** requires Zn^{2+} for activity
- **c.** contains only protein

16.36 Is the enzyme described in each of the following statements simple or one that contains a cofactor?
- **a.** requires thiamine
- **b.** is composed of 155 amino acids
- **c.** produces Cu^{2+} upon hydrolysis

16.37 Trypsin, a peptidase that hydrolyzes polypeptides, functions in the small intestine at an optimum pH of 8. How is the rate of a trypsin-catalyzed reaction affected by each of the following conditions?
- **a.** lowering the concentration of polypeptides
- **b.** changing the pH to 3
- **c.** running the reaction at 75°C

16.38 Pepsin, a peptidase that hydrolyzes proteins, functions in the stomach at an optimum pH of 2. How is the rate of a pepsin-catalyzed reaction affected by each of the following conditions?
- **a.** increasing the concentration of proteins
- **b.** changing the pH to 5
- **c.** running the reaction at 0°C

16.39 The following graph shows the curves for pepsin, urease, and oxidase. Estimate the optimum pH for each.

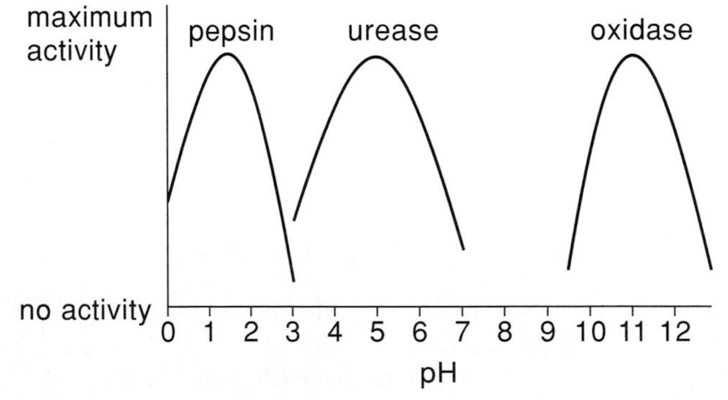

16.40 Refer to the graph in problem 16.39 to determine if the reaction rate in each condition will be at its optimum rate or not.
- **a.** oxidase, pH 6 **b.** urease, pH 5
- **c.** pepsin, pH 4 **d.** urease, pH 8

Enzyme Inhibition *(Goal 16.9)*

16.41 Indicate whether the following describe a competitive or a noncompetitive enzyme inhibitor:
- **a.** The inhibitor has a structure similar to the substrate.
- **b.** The effect of the inhibitor cannot be reversed by adding more substrate.
- **c.** The inhibitor competes with the substrate for the active site.
- **d.** The structure of the inhibitor is not similar to the substrate.
- **e.** The addition of more substrate reverses the inhibition.

16.42 Oxaloacetic acid is an inhibitor of succinic dehydrogenase:

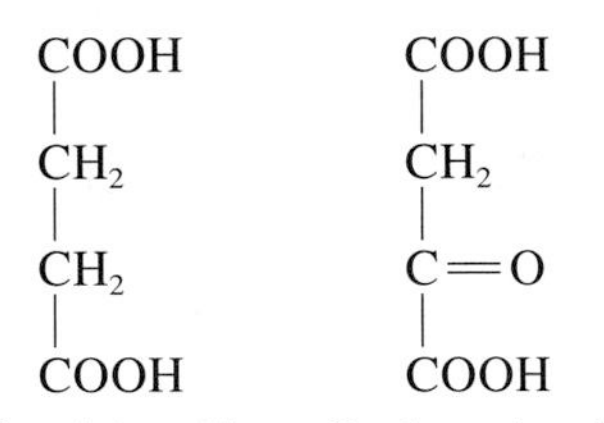

Succinic acid Oxaloacetic acid

- **a.** Would you expect oxaloacetic acid to be a competitive or a noncompetitive inhibitor? Why?
- **b.** Would oxaloacetic acid bind to the active site or elsewhere on the enzyme?
- **c.** How would you reverse the effect of the inhibitor?

Challenge Problems

16.43 Seeds and vegetables are often deficient in one or more essential amino acids. Using the following table, state whether the following combinations would provide all the essential amino acids:

Source	Lysine	Tryptophan	Methionine
Oatmeal	No	Yes	Yes
Rice	No	Yes	Yes
Garbanzo beans	Yes	No	Yes
Lima beans	Yes	No	No
Cornmeal	No	Yes	Yes

a. rice and garbanzo beans
b. lima beans and cornmeal
c. a salad of garbanzo beans and lima beans
d. rice and lima beans
e. rice and oatmeal
f. cornmeal and lima beans

16.44 How does denaturation of a protein differ from its hydrolysis?

16.45 What are some differences between the following pairs:
a. secondary and tertiary protein structures
b. essential and nonessential amino acids
c. polar and nonpolar amino acids
d. di- and tripeptides
e. an ionic bond (salt bridge) and a disulfide bond
f. fibrous and globular proteins
g. α-helix and β-pleated sheet
h. tertiary and quaternary structures of proteins

16.46 The proteins placed on a gel for electrophoresis have the following isoelectric points: albumin, 4.9, hemoglobin, pH 6.8, and lysozyme, 11.0. A buffer of pH 6.8 is placed on the gel.
a. Which protein will migrate toward the positive electrode?
b. Which protein will migrate toward the negative electrode?
c. Which protein will remain at the same place it was originally placed?

16.47 What is the class of the enzyme that would catalyze each of the following reactions?

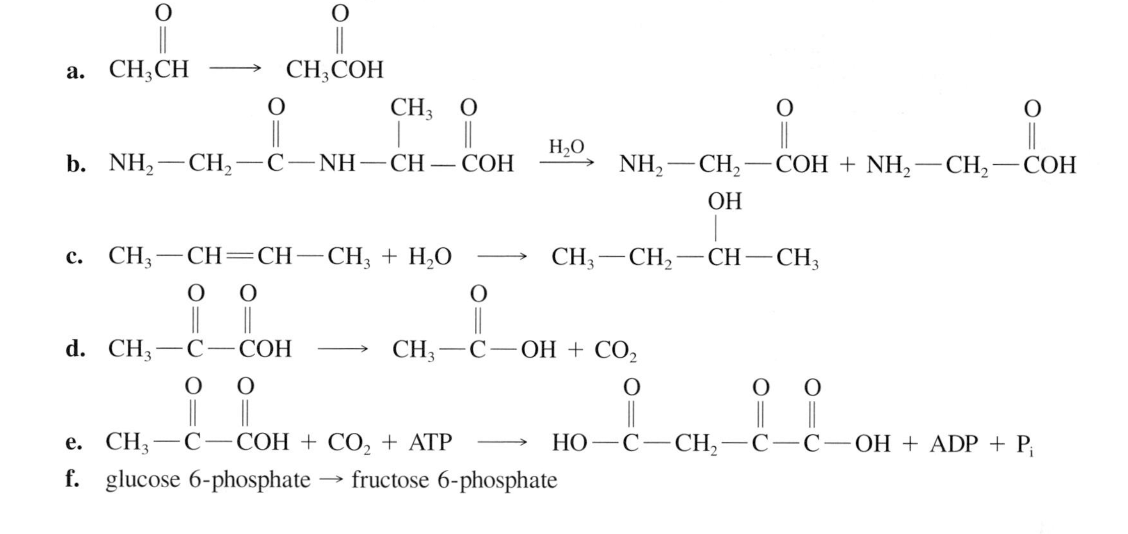

a. $CH_3CHO \longrightarrow CH_3COOH$

b. $NH_2-CH_2-C(=O)-NH-CH(CH_3)-COOH \xrightarrow{H_2O} NH_2-CH_2-COOH + NH_2-CH_2-COOH$

c. $CH_3-CH{=}CH-CH_3 + H_2O \longrightarrow CH_3-CH_2-CH(OH)-CH_3$

d. $CH_3-C(=O)-COOH \longrightarrow CH_3-C(=O)-OH + CO_2$

e. $CH_3-C(=O)-COOH + CO_2 + ATP \longrightarrow HO-C(=O)-CH_2-C(=O)-C(=O)-OH + ADP + P_i$

f. glucose 6-phosphate → fructose 6-phosphate

Chapter 17

Nucleic Acids

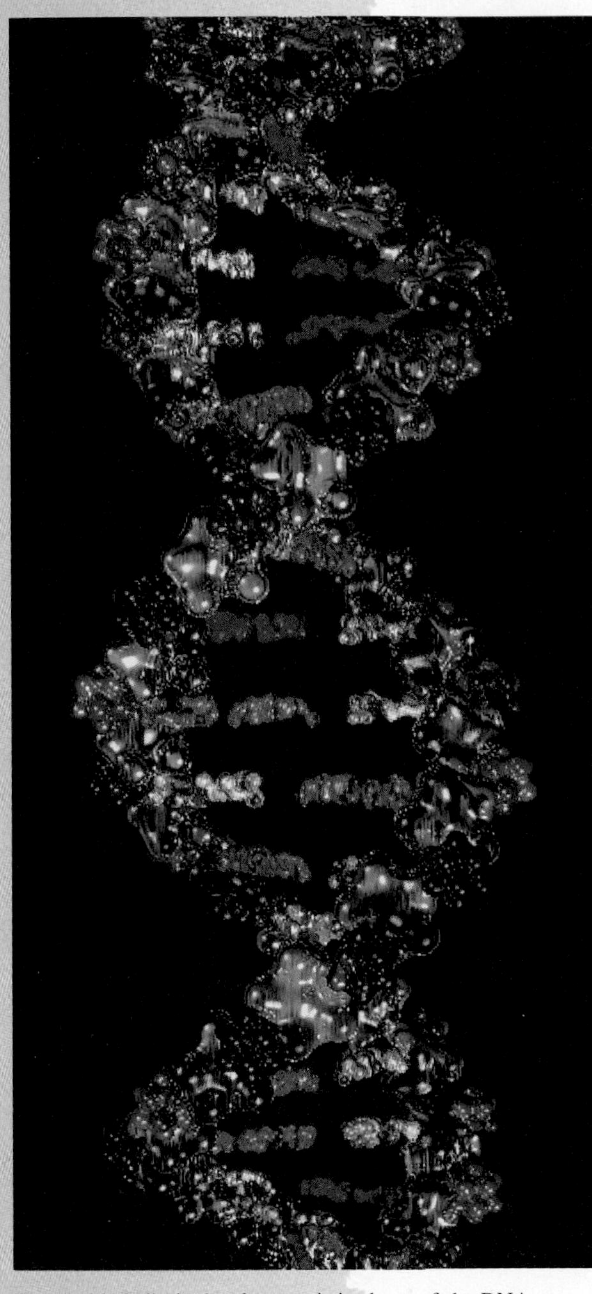

The double helix is the characteristic shape of the DNA molecules that transmit genetic information from one generation to the next.

Looking Ahead

Learning Goals

17.1 Describe the nucleotides contained in DNA and RNA.

17.2 Describe the structures of RNA and DNA; show the relationship between the bases in the double helix.

17.3 Explain the process of DNA replication.

17.4 Describe the structures and characteristics of the three types of RNA.

17.5 Describe the synthesis of mRNA (transcription).

17.6 Describe the function of the codons in the genetic code.

17.7 Describe the process of protein synthesis (translation).

17.8 Describe some ways in which DNA is altered to cause mutations.

17.9 Explain cellular regulation of protein

QUESTIONS TO THINK ABOUT

1. Where is DNA found in your cells?
2. How does DNA determine the color of your hair or eyes?
3. When a cell divides, how is DNA passed to the new cells?
4. What is the genetic code?
5. How do mutations occur?

In the preceding chapters we discussed the digestion and metablism of carbohydrates, lipids, and proteins essential to the survival of living organisms. There is one more group of biomolecules that must be included in our discussion of life—the nucleic acids.

Nucleic acids are the molecules in our cells that store and direct information for cellular growth and reproduction. The nucleic acid deoxyribonucleic acid (DNA), the genetic material in the nucleus of the cell, contains all the information needed for the development of a complete living system. The way you grow, your hair, your eyes, your physical appearance, and the activities of the cells in your body are all determined by a set of directions contained in the DNA of each of your cells. DNA controls the functioning of the cells by coding for the amino acid sequences of enzymes. Another type of nucleic acid, RNA, carries the genetic information from the DNA to the ribosomes, the cellular factories for the synthesis of protein. When the genetic blueprint is altered, incorrect information is transmitted to the ribosomes, leading to the formation of defective proteins and malfunctioning enzymes.

17.1 Nucleic Acids

Learning Goal
Describe the nucleotides contained in DNA and RNA.

The **nucleic acids**, large molecules first discovered in the nuclei of cells, contain all the information to direct the activities of a cell and its reproduction. There are two closely related types of nucleic acids: *deoxyribonucleic acid* (**DNA**), and *ribonucleic acid* (**RNA**). Both are linear polymers of *nucleotides.* A DNA molecule may contain several million nucleotides; smaller RNA molecules may contain up to several thousand.

Composition of Nucleotides

In a DNA or RNA molecule, there are four **nucleotides.** As shown in Figure 17.1, a nucleotide has the following three parts: a nitrogen base, a sugar, and a phosphate group.

There are five kinds of **nitrogen bases** in nucleic acids, all derived from *pyrimidine* or *purine.* The **pyrimidines** are single ring structures of carbon and nitrogen atoms; **purines** have fused ring structures. The bases, adenine (abbreviation: A), cytosine (C), and guanine (G) are found in the nucleotides of both DNA and RNA; thymine (T) is found only in DNA, and uracil (U) only in RNA. (See Figure 17.2.)

In RNA, the sugar is *ribose,* a five-carbon sugar, *deoxyribose* is found in DNA. Deoxyribose is similar to ribose except that there is no hydroxyl (—OH) on carbon 2, thus the prefix *deoxy*. The acid part of the name nucleic acid comes from phosphoric acid that ionizes to yield a phosphate group.

Figure 17.1
A diagram of the general structure of a nucleotide.

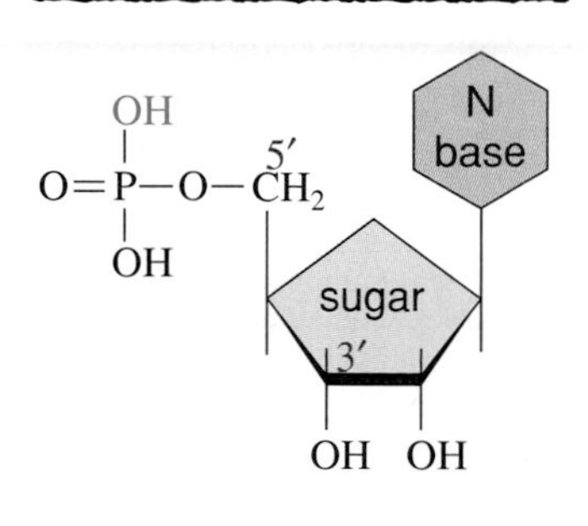

Figure 17.2
A nucleotide consists of a nitrogen base, a ribose or deoxyribose sugar, and a phosphate group.

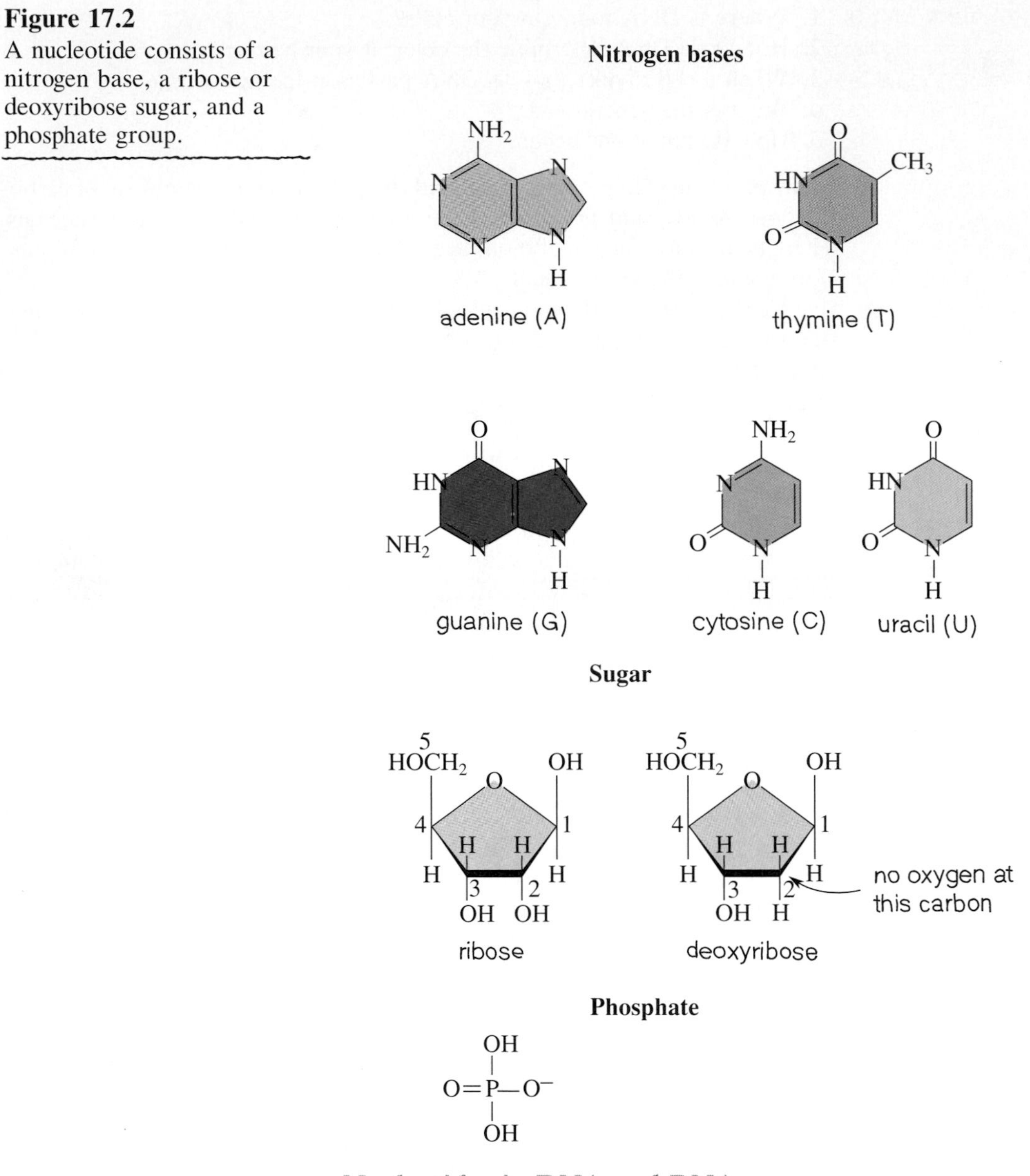

Nucleotides in DNA and RNA

The bonding of a nitrogen base with a sugar, ribose (RNA) or deoxyribose (DNA), yields a **nucleoside**. When there is a phosphate group attached to the sugar portion of the nucleoside, a *nucleotide* forms. (See Figure 17.3.) A nucleotide in DNA is named a *deoxynucleoside phosphate*; in RNA, a *nucleoside phosphate*. The names of the nucleosides and nucleotides in DNA and RNA are listed in Table 17.1. Although the letters A, G, C, U and T represent the nitrogen bases, they are often used to represent their respective nucleotides as well.

$$\text{Sugar} + \text{nitrogen base} \longrightarrow \text{nucleoside}$$

$$\text{Sugar} + \text{nitrogen base} + \text{phosphate} \longrightarrow \text{nucleotide}$$

Figure 17.3
The structures of the nucleotides of DNA.

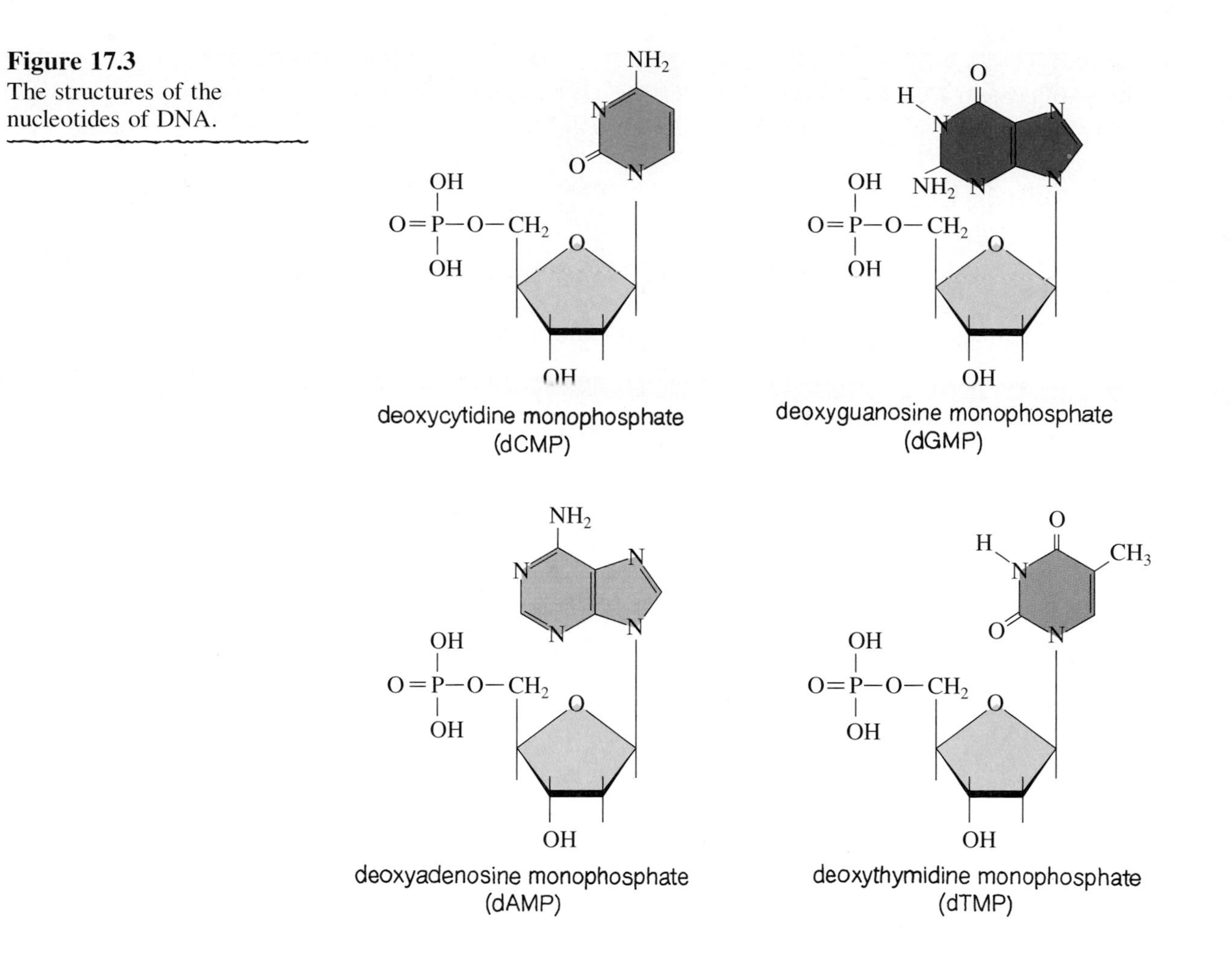

Table 17.1 ***Names of Nucleosides and Nucleotides in DNA and RNA***

		RNA
Base	**Nucleosides**	**Nucleotides**
Adenine (A)	Adenosine (A)	Adenosine monophosphate (AMP)
Guanine (G)	Guanosine (G)	Guanosine monophosphate (GMP)
Cytosine (C)	Cytidine (C)	Cytidine monophosphate (CMP)
Uracil (U)	Uridine (U)	Uridine monophosphate (UMP)
		DNA
Base	**Nucleosides**	**Nucleotides**
Adenine (A)	Deoxyadenosine (A)	Deoxyadenosine monophosphate (dAMP)
Guanine (G)	Deoxyguanosine (G)	Deoxyguanosine monophosphate (dGMP)
Cytosine (C)	Deoxycytidine (C)	Deoxycytidine monophosphate (dCMP)
Thymine (T)	Deoxythymidine (T)	Deoxythymidine monophosphate (dTMP)

Sample Problem 17.1
Nucleotides

Identify the nucleic acid (DNA or RNA) that contains each of the following nucleotides; state the components of each nucleotide:

a. deoxyguanosine monophosphate (dGMP)
b. adenosine monophosphate (AMP)

Solution

a. This DNA nucleotide consists of the sugar deoxyribose, a nitrogen base guanosine, and phosphate.
b. This RNA nucleotide contains the sugar ribose, nitrogen base adenine, and phosphate.

Study Check

What is the name of the DNA nucleotide of cytosine?

17.2 Structures of Nucleic Acids

Learning Goal

Describe the structures of RNA and DNA; show the relationship between the bases in the double helix.

A nucleic acid is a polymer formed by joining the nucleotides. A hydroxyl group of the sugar group of one nucleotide attaches to the phosphate group of the next nucleotide to yield an ester bond and a molecule of water as shown in Figure 17.4.

In the formation of a strand of DNA or RNA, the hydroxyl group on carbon 3 (3′ carbon) of the sugar in one nucleotide bonds to the phosphate group on the 5′ carbon of the sugar of the next nucleotide. As more nucleotides are added, a backbone forms with alternating sugar and phosphate groups joined by 3′, 5′-phosphodiester linkages. The bases attached to each

Figure 17.4
Nucleotides are joined by forming an ester bond between the 3′ hydroxyl group of one nucleotide and the phosphate group of the next nucleotide.

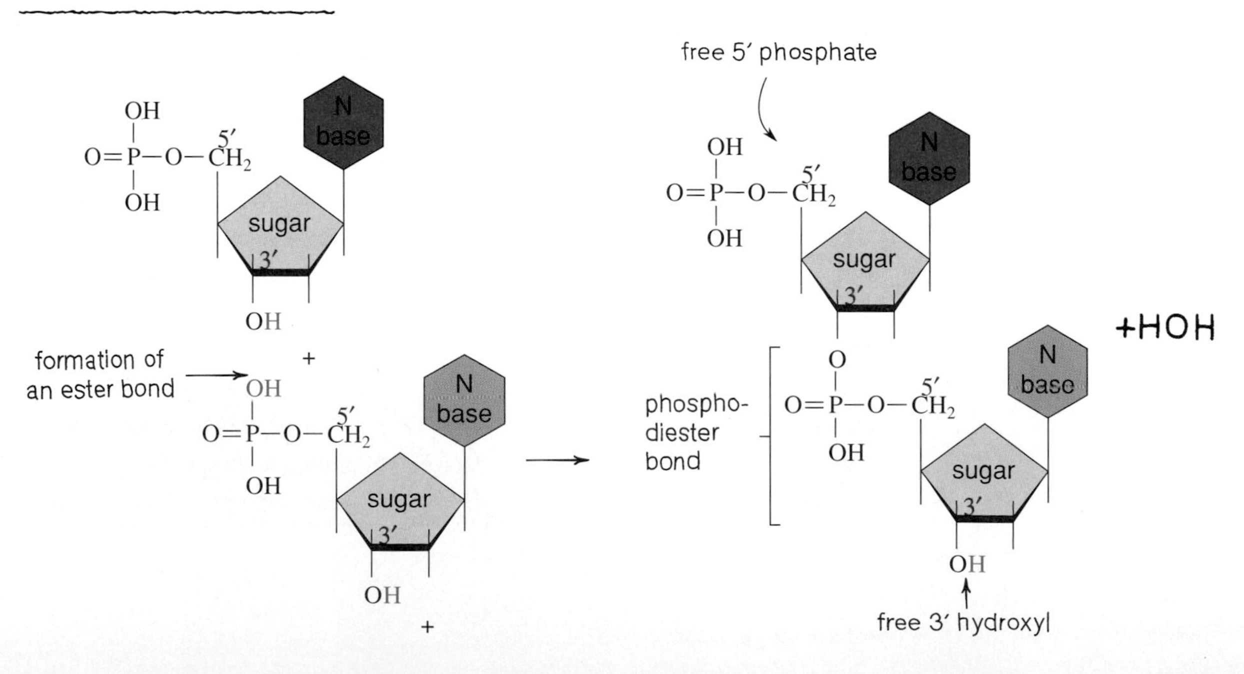

of the sugars extend out from the backbone. In every nucleic acid chain, the 5′ carbon in the sugar at one end has an unbonded phosphate group; the sugar at the other end has a free hydroxyl group on the 3′ carbon.

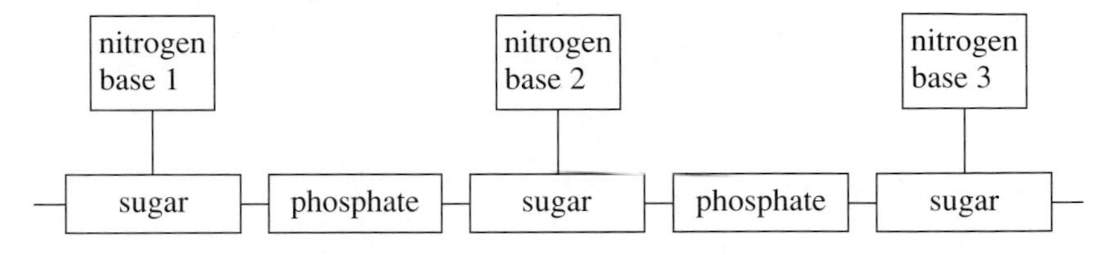

A sequence of four nucleotides in RNA is shown in Figure 17.5. In a strand of DNA, the sugar would be deoxyribose; the nitrogen base thymine replaces uracil.

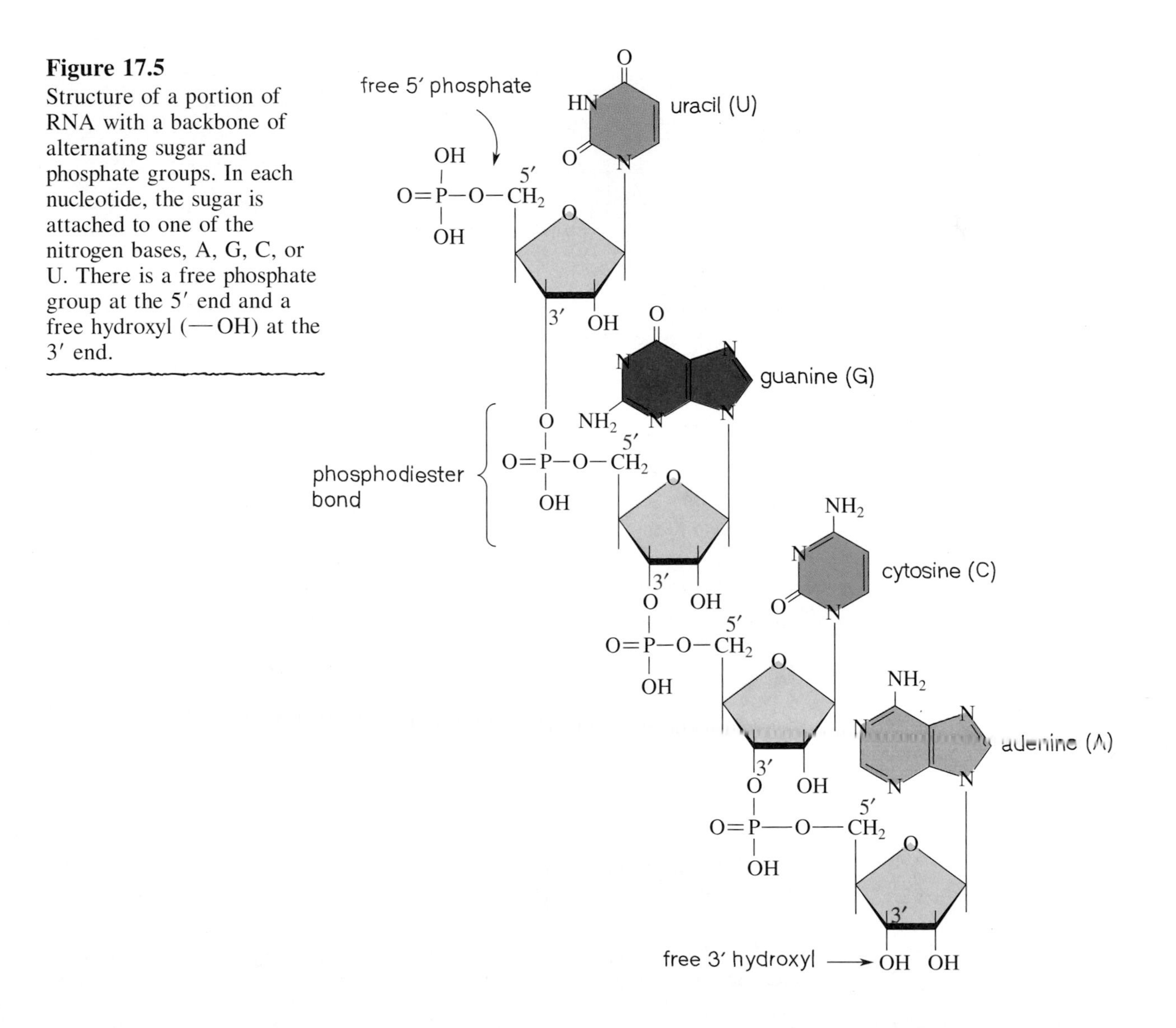

Figure 17.5
Structure of a portion of RNA with a backbone of alternating sugar and phosphate groups. In each nucleotide, the sugar is attached to one of the nitrogen bases, A, G, C, or U. There is a free phosphate group at the 5′ end and a free hydroxyl (—OH) at the 3′ end.

Sample Problem 17.2
Bonding of Nucleotides

Draw the structure of an RNA dinucleotide formed by two cytidine monophosphates.

Solution

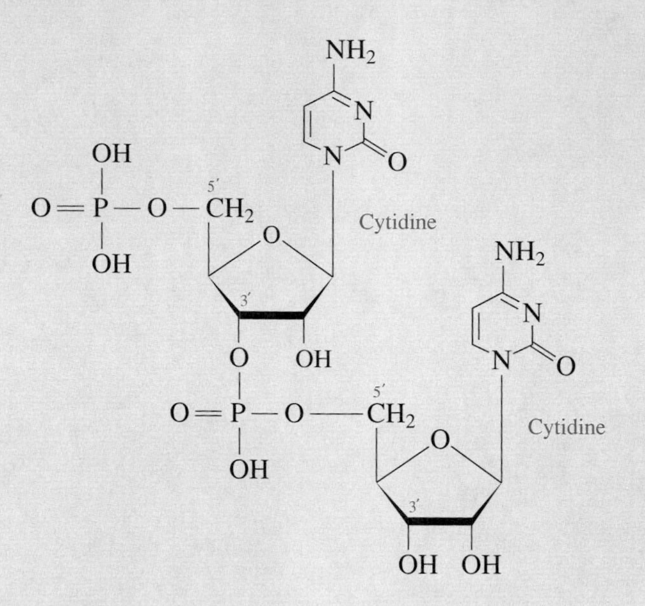

Study Check
In the above dinucleotide of cytidine, identify the free 5′ phosphate group and the free 3′ hydroxyl (—OH) group.

Double Helix of DNA

In the early 1950s, James Watson and Francis Crick determined that DNA had a double-helical structure with two nucleotide strands winding about each other like a spiral staircase. In the DNA **double helix,** the sugar–phosphate backbones are like the railings. Holding the two strands together are hydrogen bonds that form between the bases on one strand with the bases from the other, much like stair steps in a spiral staircase. An important feature of these base pairs is they always pair the same way, adenine is always opposite thymine (A—T or T—A), and guanine is always opposite cytosine (G—C or C—G). The two strands of DNA are not identical to each other, rather they are complementary. Thus, the pairs of bases (A—T and G—C) are called **complementary base pairs**. (See Figure 17.6.) There are three hydrogen bonds between guanine and cytosine, and two hydrogen bonds are between adenine and thymine. Because other pair combinations do not fit together well, there are no C—T, C—A, G—T, or G—A pairs. As we shall see, complementary base pairing plays a crucial role in cell replication and the transfer of hereditary information.

X-ray diffraction patterns of DNA indicate that the sugar–phosphate backbone in most DNA is a right-handed helix. In one complete turn, there are about 10 pairs of nucleotides. (See Figure 17.7.) In mitochondria,

Figure 17.6
Complementary base pairs hold the DNA strands of the double helix together.

Figure 17.7
Two representations of the double helix of DNA:
(a) a diagram of the double helix of DNA,
(b) a computer-designed model of the DNA molecule.

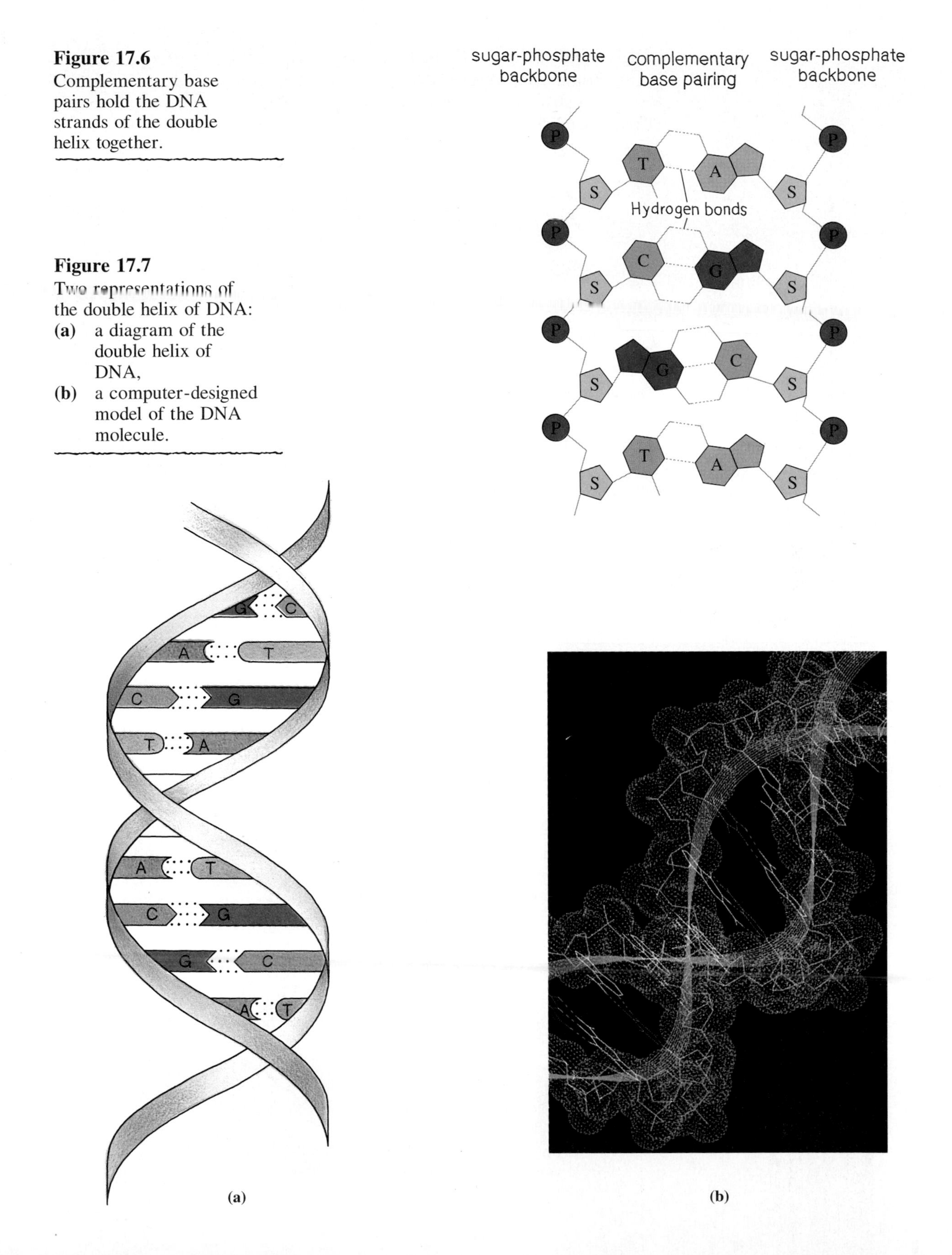

bacteria, and viruses, DNA molecules are compact, highly coiled molecules. In the chromosomes, these DNA strands are wrapped around proteins called *histones,* a structure that provides the most stable arrangement for the long DNA molecules.

Sample Problem 17.3
Complementary Base Pairs

Write the base sequence for the complementary portion for the following portion of a strand of DNA.

—A—C—G—A—T—C—T—

Solution

In the complementary strand of DNA, the base A pairs with T, and G pairs with C.

Given segment of DNA	—A—C—G—A—T—C—T—
	: : : : : : :
Complementary segment	—T—G—C—T—A—G—A—

Study Check

What is the sequence of bases that is complementary to a portion of DNA with a base sequence of —G—G—T—T—A—A—C—C—?

17.3 DNA Replication

Learning Goal
Explain the process of DNA replication.

All of our body cells have identical strands of DNA (egg and sperm cells contain half as much). DNA, along with some proteins (histones), makes up the material of the chromosomes, the threadlike structures observed in the nucleus during cellular division. In humans, every body cell contains 23 pairs of chromosome, or 46 chromosomes in all (except egg and sperm cells, in which each contains 23). (See Figure 17.8.)

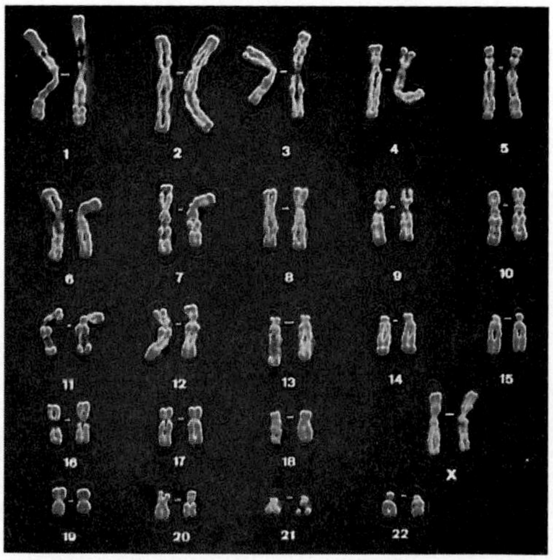

Figure 17.8
A set of 23 chromosomes stained in metaphase.

When a cell divides, DNA *replication* must provide the same genetic information for each new daughter cell. Therefore, the order of nucleotides in the parent DNA must be duplicated exactly. Central to the process of DNA **replication** is the concept of complementary base pairing. The replication process begins when hydrogen bonds between the base pairs are broken and a portion of the double helix unwinds, much like opening a zipper (See Figure 17.9.) Then DNA polymerases begin to copy each of the open strands by pairing each base on the DNA with its complementary base: A with T, and G with C. For example, a thymine in the parent DNA bonds to a nucleotide with adenine, not to any of the other nucleotides. As each base bonds to its complementary base, a phosphodiester bond joins the new nucleotide to the backbone of the growing DNA daughter strand.

During replication, DNA polymerase catalyzes only ester bonds between a 5′ phosphate of one nucleotide and a 3′ —OH of another. That means that DNA polymerases are moving in opposite directions along the separated strands of the DNA. On one of the strands called the template, a new DNA is synthesized continuously. On the other strand, new DNA is synthesized in the opposite direction in short segments of DNA. Several segments are synthesized at the same time by many DNA polymerases and connected later by a DNA ligase enzyme.

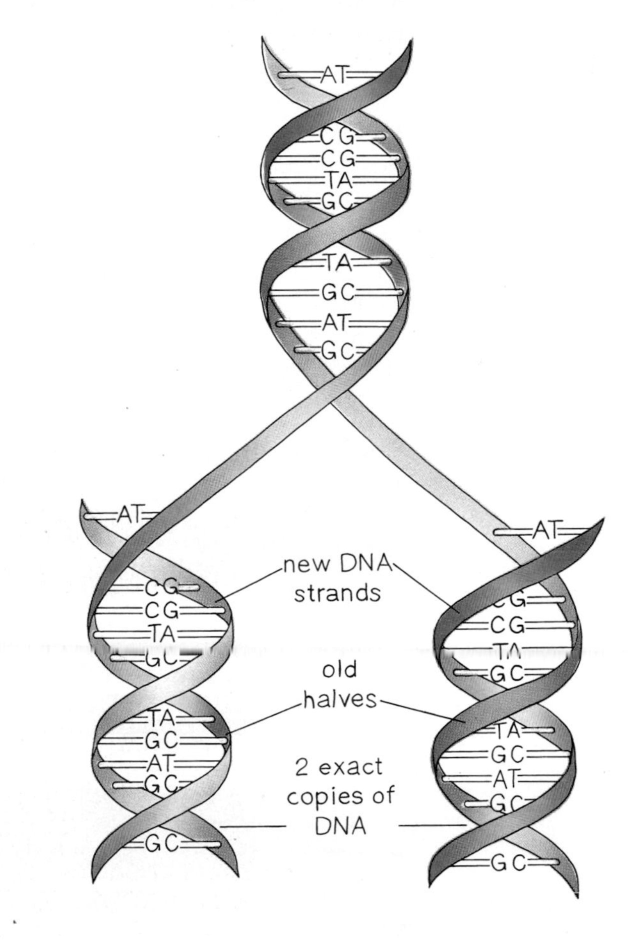

Figure 17.9
In DNA replication, two exact copies of DNA are produced. As the strands in the double helix of the original (parent) unwind and separate, new strands of daughter DNA are synthesized that contain complementary base pairs to the original DNA.

Eventually the entire double helix of the original DNA is unwound and copied. In each new DNA, one strand of the helix is from the original DNA and one is a newly synthesized strand. This process, called semiconservative replication, produces two new daughter DNAs that are identical to each other and exact copies of the original parent DNA. In the process of DNA replication, complementary base pairing ensures the correct placement of bases for the new identical DNA strands.

Sample Problem 17.4
DNA Replication

In an original DNA strand, there is a portion with the bases A—G—T. What nucleotides are placed in a growing DNA daughter strand for this sequence?

Solution
Complementary base pairing allows only one possible nucleotide to pair with each base on the original strand. Thymine only will pair with adenine, cytosine with guanine, and adenine with thymine to give the sequence T—C—A.

Study Check
After a new base is paired with a base in an original DNA strand, how is the backbone formed?

17.4 Structure and Types of RNA

Learning Goal
Describe the structures and characteristics of the three types of RNA.

Ribonucleic acid, RNA, makes up most of the nucleic acid found in the cell. (See Figure 17.10.) Similar to DNA, RNA molecules are linear polymers of nucleotides. However, there are several important differences. The sugar in RNA is ribose rather than the deoxyribose found in DNA. The nitrogen base uracil is used in place of thymine. RNA molecules are single strands of nucleotides, not double, and are much smaller than DNA molecules.

Types of RNA

There are three types of RNA that differ in their functions in the cells: one type of RNA carries the genetic information to synthesize protein, whereas others are involved in the construction of proteins. The different types of RNA molecules found in a cell are classified according to their location and function. (See Table 17.2.)

Ribosomal RNA (**rRNA**), the most abundant RNA, makes up 60% of the structural material of the ribosomes. *Ribosomes*, the sites for protein synthesis, are small, complex structures in the cytoplasm composed of rRNA and protein.

Messenger RNA (**mRNA**) carries genetic information from the DNA in the nucleus to the ribosomes for protein synthesis. Each gene in a DNA produces a separate mRNA when a certain protein is required by the cell. The size of an mRNA depends on the number of nucleotides in that particular gene.

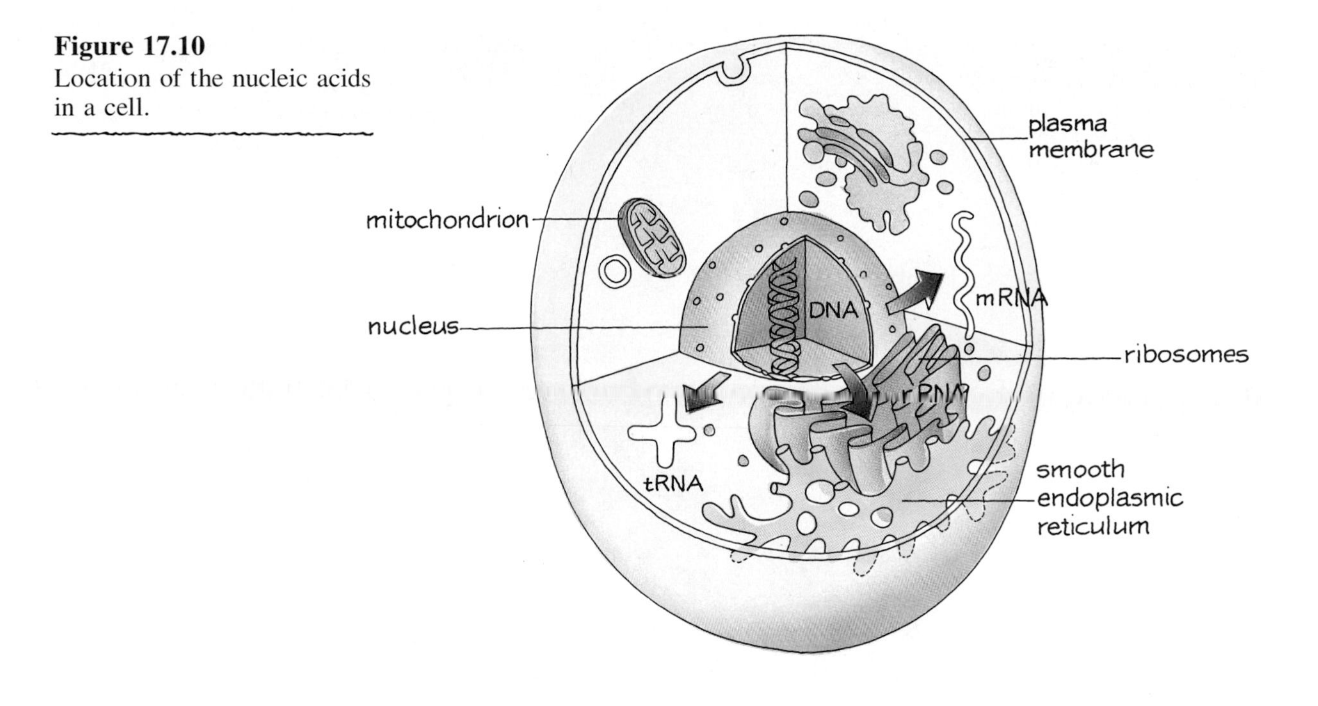

Figure 17.10
Location of the nucleic acids in a cell.

Table 17.2 *Types of RNA Molecules*

Type	Abbreviation	Percentage of Total RNA	Function in the Cell
Ribosomal RNA	rRNA	75	Major component of the ribosomes
Messenger RNA	mRNA	5–10	Carries information for protein synthesis from the DNA in the nucleus to the ribosomes
Transfer RNA	tRNA	10–15	Brings amino acids to the ribosomes for protein synthesis

Transfer RNA (**tRNA**), the smallest RNA molecule, interprets the genetic code and brings specific amino acids to the ribosome for protein synthesis. They are the only RNAs that link the genetic code to the amino acids in the proteins. There are different tRNAs for all of the amino acids.

Transfer RNAs Are Adapters

The structures of the transfer RNAs are similar. Each tRNA consists of a single chain of about 70–90 nucleotides. Hydrogen bonds between some of the complementary nitrogen bases in the chain produce loops that give an overall cloverleaf structure. (See Figure 17.11.)

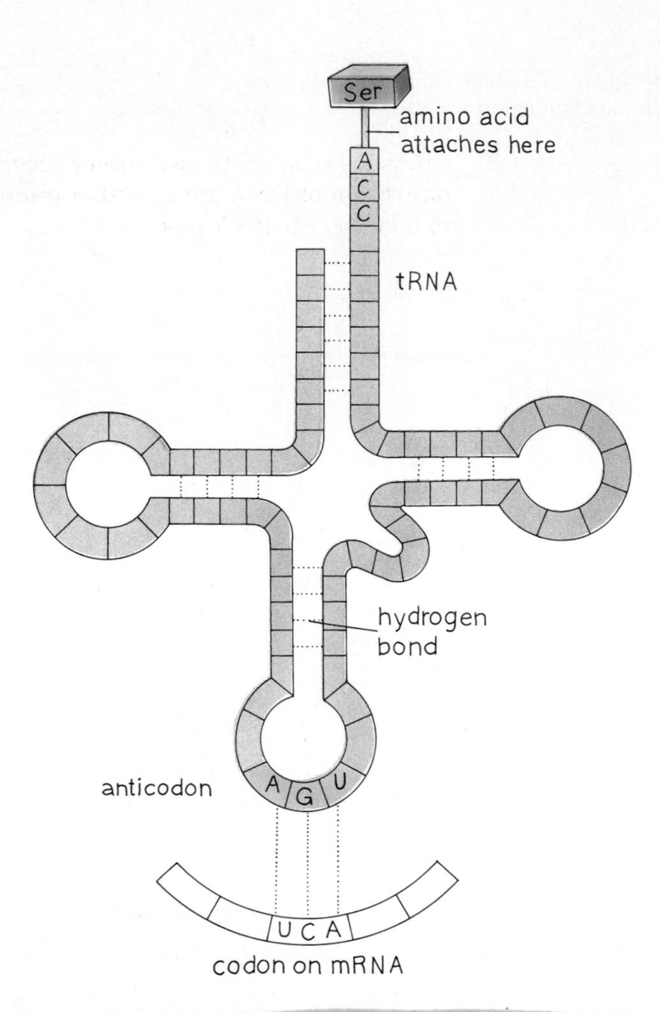

Figure 17.11
The cloverleaf structure of tRNA picks up an amino acid and carries it to an mRNA codon that is complementary to the tRNA anticodon loop.

There are two important sections on a tRNA that make it an adapter molecule, one that recognizes a particular amino acid and one that places the amino acid in the correct place in a protein. Each tRNA binds to its specific amino acid by an ester bond with the adenine at the end of the tRNA. At the center loop, there are three bases known as the **anticodon** that are complementary to three bases on an mRNA codon. As an adapter molecule, a tRNA brings an amino acid to the ribosome and attaches to the codon that is complementary to its anticodon. As a result, the amino acid is in position to be incorporated into a growing peptide chain.

Sample Problem 17.5
Types of RNA

What are the functions of mRNA and tRNA in a cell?

Solution
mRNA carries the instructions for the synthesis of a protein from the DNA to the ribosomes. tRNA brings specific amino acids to the ribosomes during the synthesis of a protein.

Study Check
Why is tRNA called an adapter molecule?

17.5 Synthesis of RNA: Transcription

Learning Goal

Describe the synthesis of mRNA (transcription).

In a process called **transcription** occuring in the nucleus, the nucleotides on one strand of DNA are copied or transcribed as their complementary bases in a messenger RNA molecule.

$$\text{DNA} \xrightarrow{\text{Transcription (nucleus)}} \text{mRNA}$$

As in DNA replication, a portion of the DNA is unwound, this time by the enzyme RNA polymerase. In RNA synthesis, only one of the separated DNA strands is used as a pattern or template for the synthesis of an mRNA. As the RNA polymerase moves along the separated DNA template strand, each ribonucleotide is paired with its complementary base. In RNA, uracil is paired with each adenine in the DNA template strand. (See Figure 17.12.) Each new ribonucleotide is attached to the growing RNA strand by a phosphoester bond. When the polymerase reaches a termination sequence, transcription ends and the newly formed RNA strand dissociates from the DNA template.

Bases on a DNA template strand:	—G—T—A—C—
	↓ ↓ ↓ ↓
Complementary base sequence in mRNA:	—C—A—U—G—

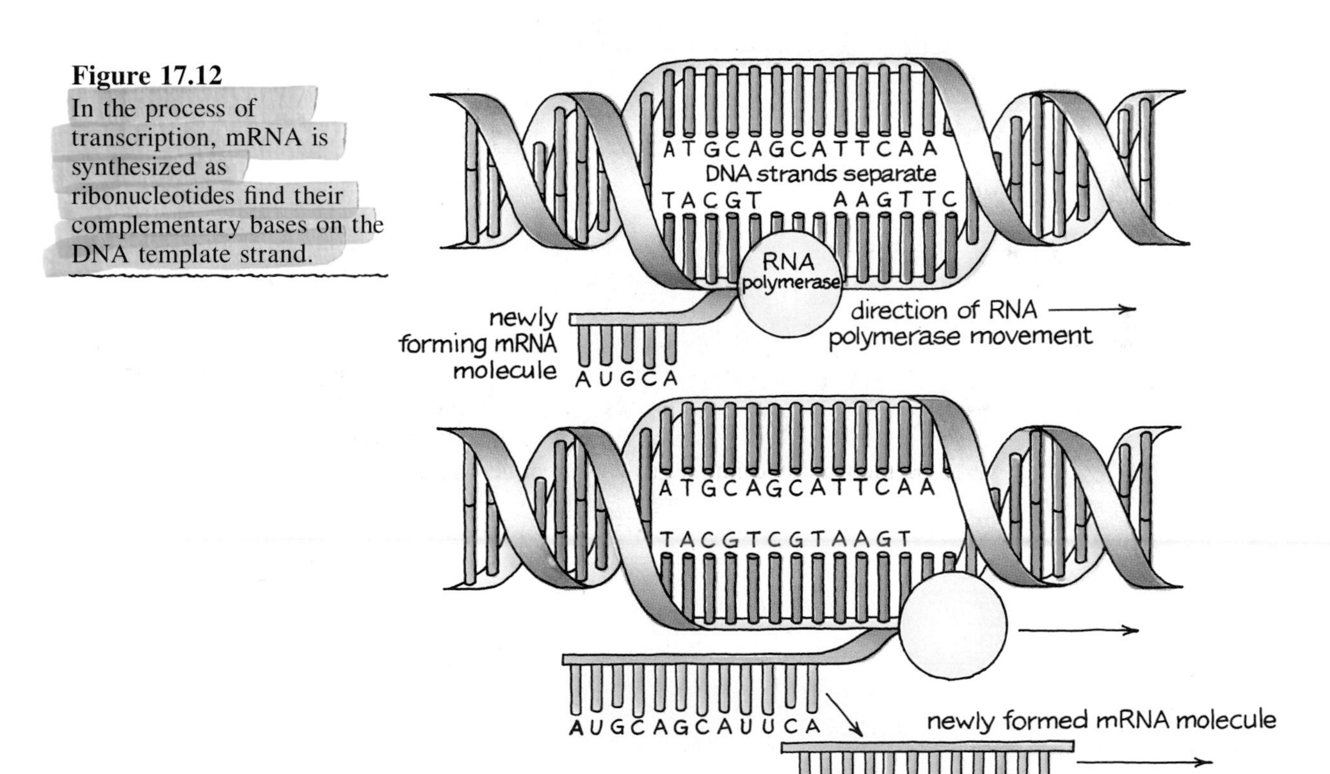

Figure 17.12
In the process of transcription, mRNA is synthesized as ribonucleotides find their complementary bases on the DNA template strand.

Sample Problem 17.6
RNA Synthesis

The sequence of bases in a part of the DNA template for mRNA is CGATCA. What is the corresponding section of the mRNA produced?

Solution
The nucleotides in DNA pair up with the ribonucleotides, G → C, C → G, T → A, and A → U.

Portion of DNA template strand:

Complementary bases in mRNA section:

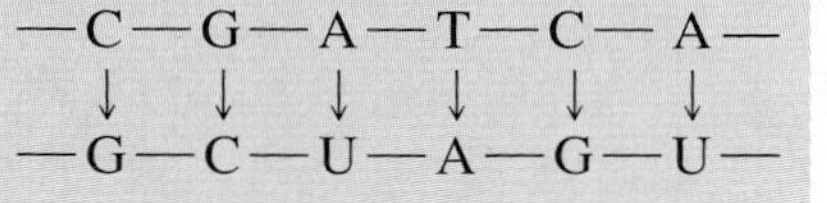

Study Check
What is the DNA template section that codes for an RNA section having the ribonucleotide sequence of GGGUUUAAA?

17.6 The Genetic Code

Learning Goal
Describe the function of the codons in the genetic code.

The overall function of the RNAs in the cell is to facilitate the task of synthesizing protein. After the genetic information encoded in DNA is transcribed into mRNA molecules, the mRNAs move out of the nucleus to the ribosomes in the cytoplasm. At the ribosomes, **translation** by the transfer RNAs converts the genetic information in the mRNAs into a sequence of amino acids in protein. (See Figure 17.13.)

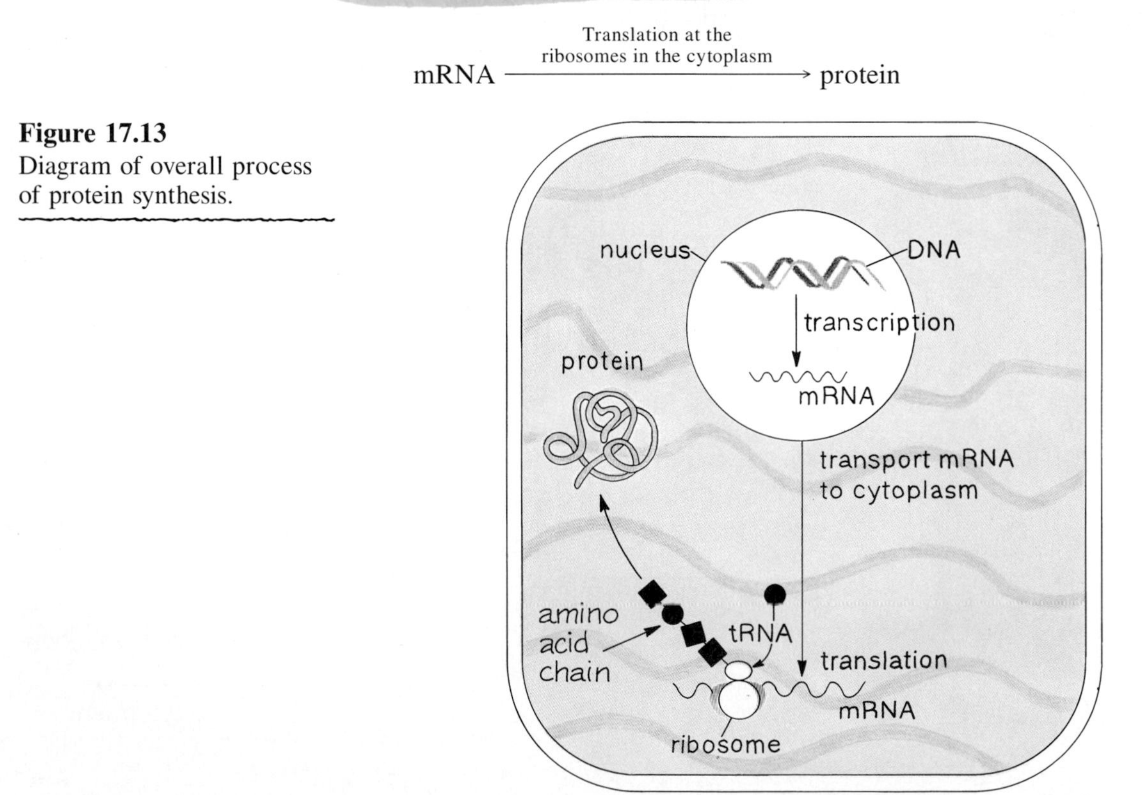

Figure 17.13
Diagram of overall process of protein synthesis.

Genetic Code

Genetic information from DNA is encoded in the mRNA as a sequence of nucleotides. In the **genetic code,** a sequence of three bases (triplet), called a **codon,** specifies each amino acid in the protein. Early work on protein synthesis showed repeating triplets of uracil (UUU) produced a polypeptide that contained only phenylalanine. Therefore, a sequence of UUU—UUU—UUU codes are for three phenylalanines.

Codons (base sequence) in mRNA	UUU—UUU—UUU
Translation	↓ ↓ ↓
Amino acid sequence	Phe — Phe — Phe

The codons have now been determined for all 20 amino acids. A total of 64 codons are possible from the triplet combinations of A, G, C and U. Three of these, UGA, UAA, and UAG, are stop signals that code for the termination of protein synthesis. All the other three-base codons shown in Table 17.3 specify amino acids, which means that one amino acid can have several codons. For example, glycine is the amino acid coded by the codons, GGU, GGC, GGA, and GGG. The triplet AUG has two roles in protein synthesis. At the beginning of an mRNA, the codon AUG signals the start of protein synthesis. In the middle of a series of codons, the AUG codon specifies the amino acid methionine.

Table 17.3 ***mRNA Codons: The Genetic Code for Amino Acids***

First Letter ↓	Second Letter: U		C		A		G		Third Letter ↓
U	UUU	Phe	UCU	Ser	UAU	Tyr	UGU	Cys	U
	UUC		UCC		UAC		UGC		C
	UUA	Leu	UCA		UAA	STOP	UGA	STOP	A
	UUG		UCG		UAG	STOP	UGG	Trp	G
C	CUU	Leu	CCU	Pro	CAU	His	CGU	Arg	U
	CUC		CCC		CAC		CGC		C
	CUA		CCA		CAA	Gln	CGA		A
	CUG		CCG		CAG		CGG		G
A	AUU	Ile	ACU	Thr	AAU	Asn	AGU	Ser	U
	AUC		ACC		AAC		AGC		C
	AUA		ACA		AAA	Lys	AGA	Arg	A
	[a]AUG	Met	ACG		AAG		AGG		G
G	GUU	Val	GCU	Ala	GAU	Asp	GGU	Gly	U
	GUC		GCC		GAC		GGC		C
	GUA		GCA		GAA	Glu	GGA		A
	GUG		GCG		GAG		GGG		G

[a] Codon that signals the start of a peptide chain.
STOP codon signals the end of a peptide chain.

Sample Problem 17.7
Codons

What is the sequence of amino acids coded by the following codons in mRNA?

GUC—AGC—CCA

Solution
According to the codon given in Table 17.3, GUC codes for valine, AGC for serine, and CCA for proline. The sequence of amino acids is Val-Ser-Pro.

Study Check
The codon UGA does not code for an amino acid. What is its function in mRNA?

17.7 Protein Synthesis

Learning Goal
Describe the process of protein synthesis (translation).

After an mRNA is synthesized, it migrates out of the nucleus into the cytoplasm. During *translation*, ribosomes, tRNAs, and their amino acids, along with some enzymes, convert the three-base sequence of the mRNA into a protein. Protein synthesis takes place on a ribosome at two adjacent sites. Amino acids are attached to the A site by the tRNAs and peptide bonds are formed at the P site. We will discuss protein synthesis (translation) as a series of the following three steps: initiation, elongation, and termination.

Step 1: Initiation

Protein synthesis begins when an mRNA attaches to a binding site on a ribosome. The first codon in mRNA, AUG, is placed in the P site. At the first codon, AUG is read as a start codon. A tRNA carrying methionine forms hydrogen bonds with the AUG codon. Although other tRNAs might try to bind to the AUG codon, the tRNA with methionine is the only tRNA with the anticodon (UAC) that is complementary to the mRNA codon. (See Figure 17.14.)

Step 2: Elongation

The second codon on the mRNA at the A site is now ready for translation. Again, there is only one tRNA that has the three-base anticodon to complement the second codon. Suppose that the second codon is UCC, the codon for the amino acid serine. Only the serine-carrying tRNAs with the anticodon AGG can bond with the codon at the A site. While the two amino acids methionine and serine are held in the P and A sites, a peptide bond is formed. Then the ribosome shifts to the next codon (three bases) on the mRNA in a process called **translocation.** The first tRNA detaches from the ribosome, and the second tRNA along with the peptide chain moves into the P site. This shift places a new codon in the A site and the cycle repeats.

Figure 17.14
Process of translation. An mRNA attaches to the binding sites on a ribosome. A tRNA carrying methionine binds at the P site. A second tRNA with the anticodon for the second codon on mRNA binds at the A site and a peptide bond forms between the two amino acids. The ribosome shifts down one codon and a new tRNA with an amino acid binds at the A site. As the ribosome translocates, the cycle repeats and a peptide chain is formed.

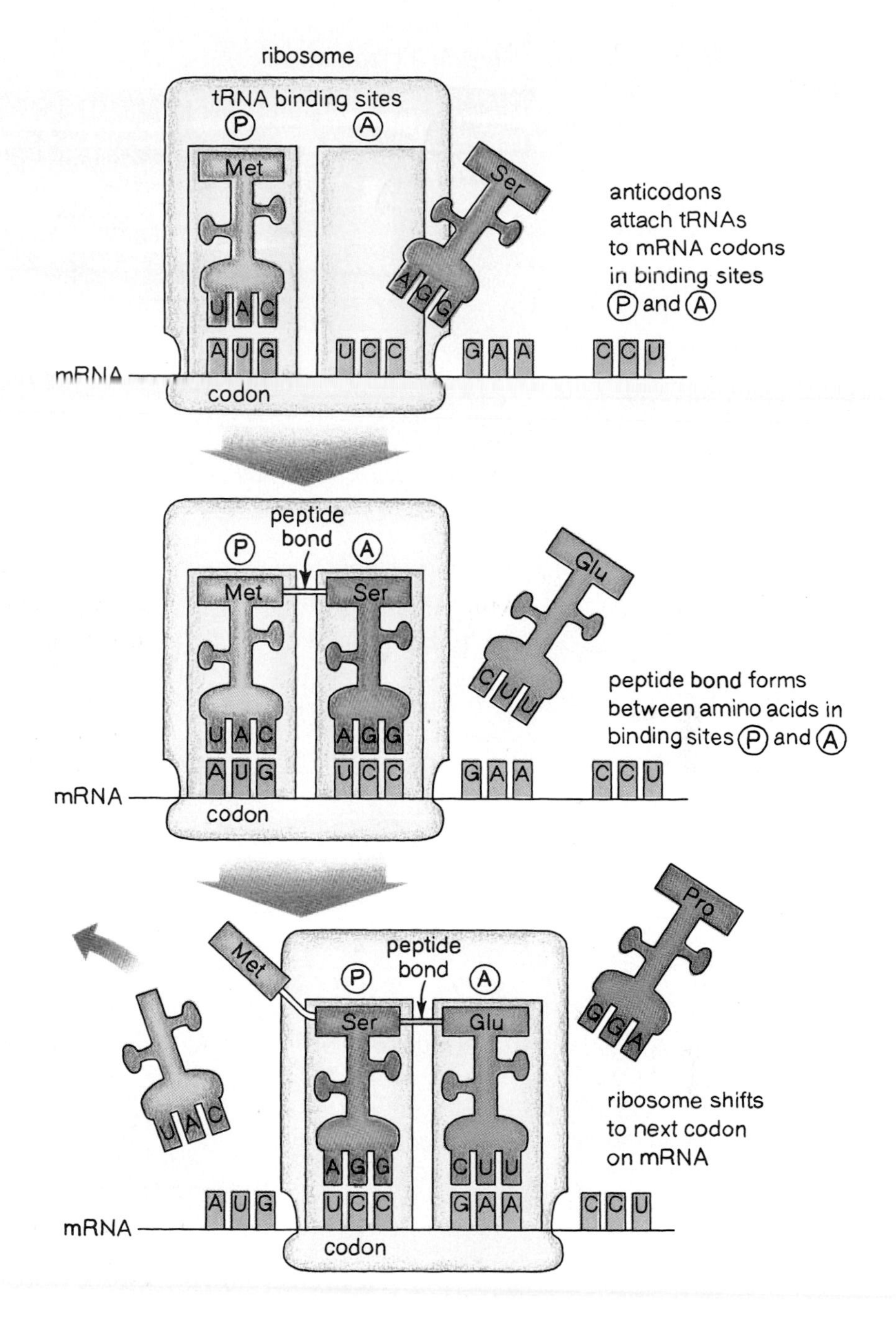

The process of elongation continues as the ribosome moves along (translocates) the mRNA, one codon at a time. Sometimes, several ribosomes (a polysome) translate the same strand of mRNA at the time. For each new codon at the A site, an amino acid is brought by the appropriate tRNA to the A site. At the P site, a peptide bond attaches the new amino acid to the growing peptide chain.

Step 3: Termination

Protein synthesis stops when a ribosome encounters the codons UGA, UAA or UAG. Because there are no tRNAs to complement the termination codons, translation ends. An enzyme releases the completed polypeptide chain from the ribosome. The free peptide chain forms secondary and tertiary cross-links and binds to coenzymes, if needed, to make the protein biologically active. If the functional protein requires a quaternary structure, different mRNAs simultaneously produce the subunits that combine as soon as they are released from the ribosomes.

Sample Problem 17.8
Protein Synthesis

What order of amino acids would you expect in a peptide for the mRNA sequence of UCA—AAA—GCC—CUU?

Solution
Each of the three base codons specifies a particular amino acid. Using Table 17.3, we write a peptide with the following amino acid sequence:

RNA codons	UCA — AAA — GCC — CUU
	↓ ↓ ↓ ↓
Amino acid sequence	Ser — Lys — Ala — Leu

Study Check
Where would protein synthesis stop in the following series of bases in an mRNA: GGG—AGC—AGU—UAG—GUU?

HEALTH NOTE *Many Antibiotics Inhibit Protein Synthesis*

Several antibiotics stop bacterial infections by interfering with the synthesis of proteins needed by the bacteria. Some antibiotics act only on bacterial cells by binding to the ribosomes in bacteria but do not act on human cells. A description of some of these antibiotics is given in Table 17.4.

Table 17.4 ***Antibiotics that Inhibit Protein Synthesis in Bacterial Cells***

Antibiotic	Effect on Ribosomes to Inhibit Protein Synthesis
Chloramphenicol	Inhibits peptide bond formation and prevents the binding of tRNA
Erythromycin	Inhibits peptide chain growth by preventing the translocation of the ribosome along the mRNA
Puromycin	Causes release of an incomplete protein by ending the growth of the polypeptide early
Streptomycin	Prevents the proper attachment of tRNAs
Tetracycline	Prevents the binding of tRNAs

17.8 Genetic Mutations

Learning Goal

Describe some ways in which DNA is altered to cause mutations.

A **mutation** is a change in the DNA base sequence that alters the structure and function of a protein in the cell. Some mutations are known to result from X rays, overexposure to sun (ultraviolet or UV light), chemicals called mutagens, and possibly some viruses. If a change in DNA occurs in a somatic cell, a cell other than a reproductive cell, the altered DNA will be limited to that cell and its daughter cells. If there is uncontrolled growth, the mutation could lead to cancer. If the mutation occurs in germ cell DNA (egg or sperm), then all the DNA produced in a new individual will contain the same genetic error. If the genetic error greatly affects the catalysis of metabolic reactions or the formation of important structural proteins, the new cells may not survive or the person may exhibit a genetic disease. (See Figure 17.15.)

Consider a triplet of bases such as CCC, which produces the codon GGG in the mRNA. At the ribosome, a tRNA would place the amino acid glycine in the peptide chain. Now, suppose the DNA sequence is changed to CAC. The codon in the resulting mRNA is altered to give GUG, which codes for valine not glycine. The replacement of one base in DNA with another is called **substitution.** The change in the codon may cause a different amino acid to be inserted at that point in the polypeptide. Substitution is the most common way in which mutations occur. (See Table 17.5.)

Table 17.5 ***Effect of Point Mutations on Amino Acid Sequence***

Substance	Normal Sequence	Effect of Mutation
DNA	ACA—**CCC**—AGG—TTT	Possible change in base ↓ ACA—**CAC**—AGG—TTT
mRNA	UGU—**GGG**—UCC—AAA	Change in codon UGU—**GUG**—UCC—AAA
Amino acid sequence	Cys—**Gly**—Ser—Lys	Change in amino acid order Cys—**Val**—Ser—Lys

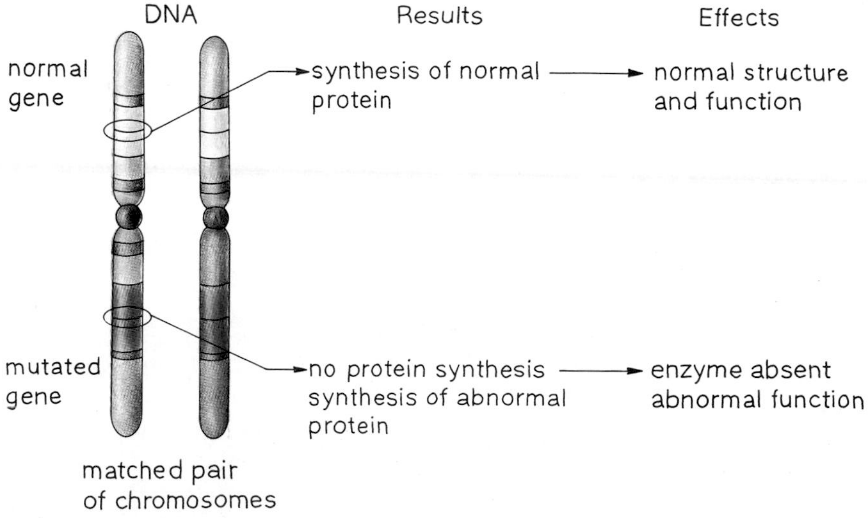

Figure 17.15
Most genetic diseases are a result of gene mutations that produce no protein or abnormal protein.

In a **frame shift mutation,** a base is added to or deleted from the normal order of bases in the DNA. Then there is a change in the codon that lost or gained a base, and in all the codons that follow. Table 17.6 illustrates addition and deletion mutations.

Table 17.6 *Addition and Deletion Mutations*

	Normal Sequence	Frame Shift Effect of Mutation	
		A added ↓	A deleted
DNA	—CCC—A**GG**—TTT—	—CCC—**AAG**—**GTT**—T	—CCC—**GGT**—TT
		Shift in codons	
mRNA	—GGG—**UCC**—AAA—	—GGG—**UUC**—CAA—A	—GGG—**CCA**—**AA**
		Incorrect order	
Amino acid sequence	—Gly—**Ser**—Lys—	—Gly—**Phe**—**Gln**—	—Gly—**Pro**—

Effect of Mutations

When a mutation causes a change in the amino acid sequence, the structure of the resulting protein can be altered severely. If a polypeptide contains a large section of incorrect amino acids, it may lose biological activity. If the protein is an enzyme, it no longer binds to its substrate or reacts with substrate at the active site. When an enzyme cannot catalyze a reaction, certain substances may accumulate until they act as poisons in the cell. If a defective enzyme occurs in a major metabolic pathway or in the building of a cell membrane, the mutation can be lethal. If the defect is in a structural protein, it can lead to detrimental anatomical changes. When a deficiency of a protein is genetic, the condition is called a **genetic disease.**

DNA —(X rays, UV sunlight, Mutagens, Viruses)→ alteration of DNA → defective protein → genetic disease (germ cells) or cancer (somatic cells)

Sample Problem 17.9
Mutations

An RNA has a sequence of codons of CCC—AGA—GGG. If a base substitution in the DNA changes the m-RNA codon of AGA to ACA, how is the amino acid sequence affected in the resulting protein?

Solution

The initial mRNA sequence of CCC—AGA—GGG codes are for the amino acids proline, arginine, and glycine. When the mutation occurs, the new sequence of mRNA codons CCC—ACA—GGG are for proline, threonine, and glycine. The amino acid serine is replaced by threonine.

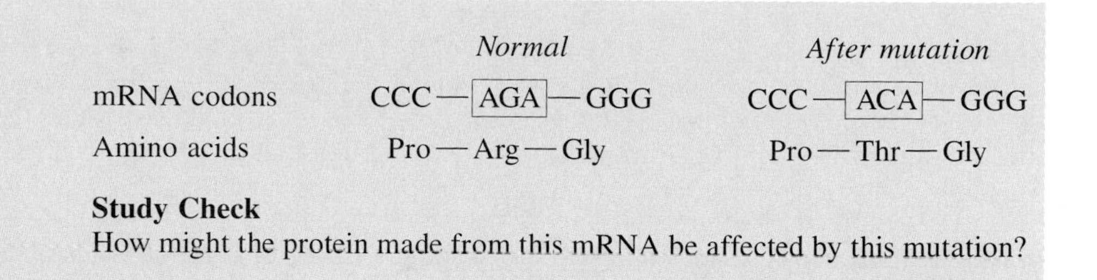

	Normal	*After mutation*
mRNA codons	CCC—AGA—GGG	CCC—ACA—GGG
Amino acids	Pro—Arg—Gly	Pro—Thr—Gly

Study Check

How might the protein made from this mRNA be affected by this mutation?

HEALTH NOTE *Recombinant DNA*

Over the past 25 years, geneticists have been cutting, splicing, and rejoining DNA from different genes to form a new, synthetic DNA, called *recombinant* DNA. Most of this experimentation has been done with a bacterium called *Escherichia coli* (*E. coli*), which contains a single chromosome and several small cyclic DNA particles called *plasmids.*

When the cells are soaked in a detergent solution, they break open, releasing the plasmids, which are collected. Restriction enzymes are used to remove a specific section of DNA from the plasmid. Then, a piece of DNA from another organism, which might be the gene that produces insulin or growth hormone, is placed in the cut region of the plasmid. The ends of the inserted DNA piece and the plasmid are joined by a DNA ligase. This process of gene splicing forms a hybrid DNA or recombinant DNA that can now be reabsorbed by the bacterium. The altered *E. coli* cells are then cloned. As the *E. coli* reproduce, the recombinant DNA in the plasmids is replicated. The new cells now produce the new protein for use in medicine, agriculture, or industry. (See Figure 17.16.)

Recombinant DNA has caused much excitement because of its potential medical uses. Already, recombinant DNA methods have produced enzymes, hormones, antibodies, clotting factors, and other compounds that some people cannot

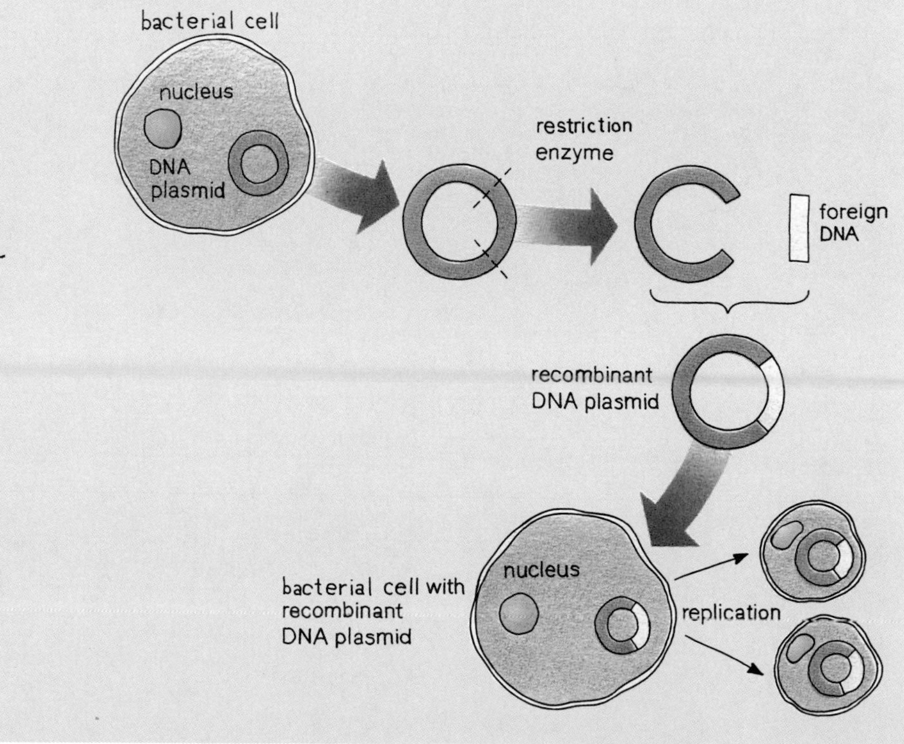

Figure 17.16
Formation of recombinant DNA by placing new DNA in plasmid DNA of bacterium. The bacterium can now produce a nonbacterial protein such as insulin or growth hormone.

produce in their cells. Recently, the DNA that synthesizes human insulin has been successfully incorporated into *E. coli* plasmids. The protein interferon, which helps fight viral infections and possibly cancer, has been produced by recombinant DNA technology in greater quantity than was previously available. Perhaps in the future, defective genes that cause genetic disease can be corrected by recombinant DNA methods.

There has been concern over recombinant DNA because mistakes could produce toxic strains of bacteria against which there is no biological defense. Guidelines to control such possibilities have been established by the National Institutes of Health. As a control measure, the *E. coli* strain used for research is made deficient in one amino acid, so that it is dependent on artificial conditions for survival.

HEALTH NOTE *Genetic Diseases*

Today, more than 300 metabolic diseases are known to be inherited. A genetic disease is the result of a defective enzyme caused by a mutation in its genetic code. For example, phenylketonuria, PKU, results when DNA cannot produce the enzyme phenylalanine hydroxylase, required for the conversion of phenylalanine to tyrosine. In an attempt to break down the phenylalanine, other enzymes in the cells convert it to phenylpyruvic acid. The accumulation of phenylalanine and phenylpyruvic acid in the blood can lead to severe brain damage and mental retardation. If PKU is detected in a newborn baby, a diet is prescribed that eliminates all the foods that contain phenylalanine. Preventing the buildup of the phenylpyruvic acid ensures normal growth and development.

The amino acid tyrosine is needed in the formation of melanin, the pigment that gives the color to our skin and hair. If the enzyme that converts tyrosine to melanin is defective, no melanin is produced, a genetic disease known as albinism. Persons and animals with no melanin have no skin or hair pigment. (See Figure 17.17.) Table 17.7 lists some other common genetic deseases and the type of metabolism or area affected.

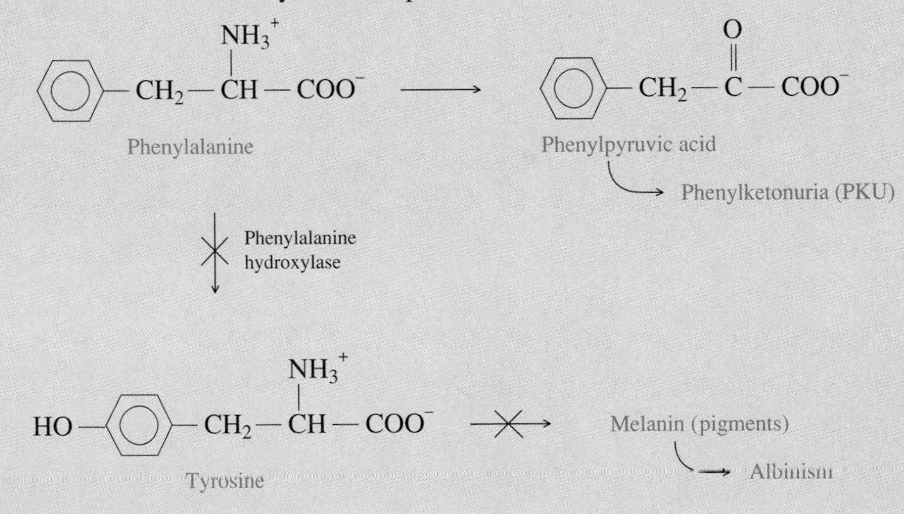

Figure 17.17
This peacock with albinism does not produce the melanin for the bright colors of its feathers.

Table 17.7 *Some Genetic Diseases*

Genetic Disease	Result
Galactosemia	Absence of the transferase enzyme required for the metabolism of galactose-1-phosphate. Accumulation of Gal-1-P leads to cataracts and mental retardation.
Cystic fibrosis	The most common inherited disease, thick mucous secretions make breathing difficult, and block pancreatic function.
Down's syndrome	The leading cause of mental retardation, occurring in about 1 of every 800 live births. Mental and physical problems including heart and eye defects are the result of the formation of three chromosomes, usually chromosome 21, instead of a pair.
Familial hypercholesterolemia	A mutation of a gene on chromosome 19 characterized by high cholesterol levels that lead to early coronary heart disease in 30–40 year-old persons.
Muscular dystrophy (Duchenne)	One of 10 forms of MD caused by a mutation in the X chromosome occurring in about 1 of 10,000 males due to the low or abnormal production of *dystrophin* by the X gene; a muscle-destroying disease beginning at about age 5 with death by age 20.
Huntington's disease (HD)	A genetic disorder appearing in middle age affecting the nervous system that leads to total physical impairment; result of a mutation in a gene on chromosome 4, which can now be mapped to test people in families with HD.

Table 17.7 ***Some Genetic Diseases (cont.)***

Genetic Disease	Result
Sickle-cell anemia	In the disorder, a defective hemoglobin from a mutation in a gene on chromosome 11 decreases the oxygen-carrying ability of red blood cells that take on a sickled shape, causing anemia and plugged capillaries from red blood cell aggregation.
Hemophilia	One or more defective blood clotting factors leading to poor coagulation, excessive bleeding, and internal hemorrhages.
Tay-Sachs disease	A defective hexosaminidase A, causing an accumulation of gangliosides resulting in mental retardation, loss of motor control, and early death.

17.9 Cellular Regulation

Learning Goal

Explain cellular regulation of protein synthesis.

If a cell is to maximize available energy, it must operate efficiently. Therefore, materials are produced in a cell only as they are needed and degraded when they are not needed. This means that when a substance enters a cell, the enzyme required to catalyze its reactions is also produced via protein synthesis. We also need different proteins at different times in our lives. Different organs and tissues in the body have different functions, and therefore require the synthesis of different proteins. mRNA is not randomly synthesized by DNA; rather, it is synthesized in response to cellular needs for certain materials. That means that a gene can be turned on to synthesize a particular protein, and turned off again when the cell no longer requires that protein.

Operon Model

In 1961, the French biologists François Jacob and Jacques Monod proposed the *operon model* of cellular control to explain how genes in bacteria regulate protein synthesis. In this model, groups of nucleotides in DNA act to control the synthesis of mRNA and, therefore, the synthesis of protein. Each group of related proteins is regulated by an **operon,** a section of DNA containing a control site and structural genes. The **structural genes** are the DNA sections that code for proteins. The **promotor** and the **operator** make up the **control site,** which is the section of DNA ahead of the structural genes. Preceding the control site is a **regulatory gene.** (See Figure 17.18.)

In many cases, one operator controls several structural genes. When the operator is turned on, its structural genes produce mRNA for a specific group of proteins. Because mRNA is broken down rapidly, there must be a continuous formation (transcription) of mRNA as long as the protein is synthesized.

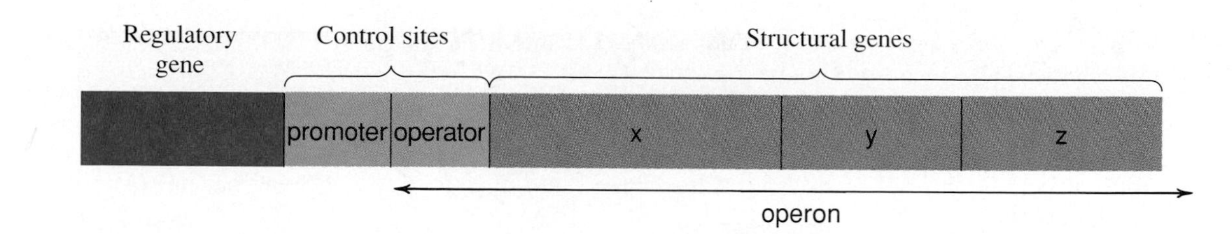

Figure 17.18
Diagram of an operon and regulatory gene. The structural genes x, y, and z represent structural genes that code for proteins x, y, and z.

When those proteins are no longer required by the cell, their structural genes are turned off by the operator gene and no mRNA is produced. Thus, resources are conserved by not producing enzymes when the substrates for those enzymes are not present.

Enzyme Induction

In **enzyme induction,** a substrate induces the synthesis of the enzymes for its metabolism. Suppose some bacteria are grown on glucose, and some on lactose. The bacteria grown on glucose will produce enzymes that metabolize glucose but not lactose. The bacteria grown on lactose will produce enzymes that metabolize lactose but not glucose. The enzymes synthesized were *induced* by the presence of a particular substrate.

In enzyme induction, the **regulatory gene** produces a protein called a **repressor** that binds to the operator. (See Figure 17.19.) As long as the repressor remains bound, RNA polymerase cannot read the structural genes and no mRNA is produced. For example, if there is no lactose, the structural genes that produce enzymes for the metabolism of lactose are blocked by a repressor. As long as the operator is blocked, the process of transcription is turned off. In fact, at any given time, most of the structural genes in a DNA are not operating.

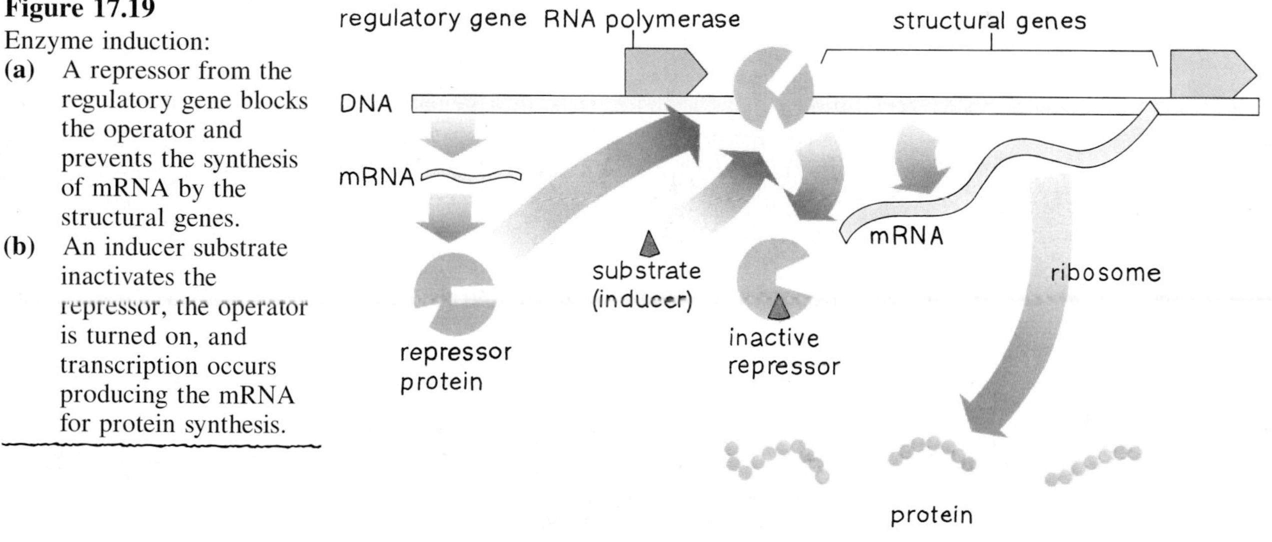

Figure 17.19
Enzyme induction:
(a) A repressor from the regulatory gene blocks the operator and prevents the synthesis of mRNA by the structural genes.
(b) An inducer substrate inactivates the repressor, the operator is turned on, and transcription occurs producing the mRNA for protein synthesis.

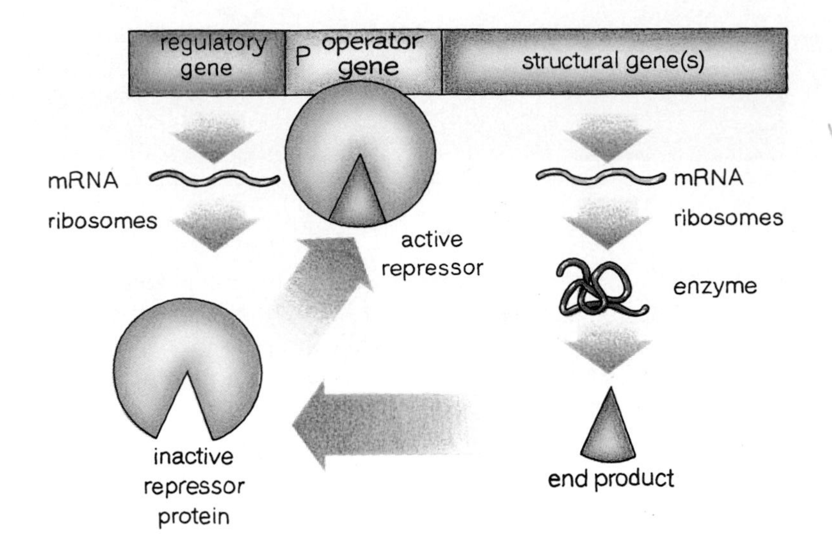

Figure 17.20
Enzyme repression. When the concentration of end products of an enzyme is elevated, the end-product molecules form an active repressor that blocks the operator and stops protein synthesis.

Suppose now that some lactose enters the cell. The lactose binds to the repressor, which alters its shape, and causes the repressor to detach from the operator gene. With an open operator, the RNA polymerase transcribes the structural genes and protein is produced. In this case, lactose initiates (induces) the production of its own enzymes. As the enzymes metabolize the lactose, the concentration of lactose decreases. Eventually there is no longer any lactose to bind to the repressor. Then, the repressor binds again to the operator gene and stops the synthesis of protein.

Enzyme Repression

In **enzyme repression,** it is the end products of an enzyme-catalyzed reaction, not the substrate, that turn off (repress) the synthesis of their enzymes. In the repression model, the operator gene is on and the structural genes are producing mRNA. (See Figure 17.20.) As long as the concentration of end product stays low, the structural genes continue to direct the synthesis of protein. However, when the amount of end product exceeds the needs of the cell, the end-product molecule combines with a protein produced by the regulator gene. The combination becomes an active *repressor* that binds to the operator gene and blocks the RNA polymerase. The operator is turned off and there is no further synthesis of mRNA. The end product has repressed the synthesis of the enzymes needed to produce that end product.

Sample Problem 17.10
Cellular Regulation

Indicate whether the following statements are *true* or *false:*

a. In enzyme induction, the operon is normally repressed and must be turned on by the substrate.

b. In enzyme repression, an active repressor forms when the concentration of the end product of an enzyme series is elevated.

Solution

a. True. A substrate binds with a repressor, causing the repressor to detach from the operator gene. With the operator gene turned on, the structural genes transcribe mRNA for the synthesis of protein.

b. True. When the concentration of the end product of an enzyme rises, there is no further need for that enzyme. An active repressor formed by the binding of end product with a protein turns off the operator and protein synthesis stops.

Study Check

When there is no sucrose present in the cell, why is there no sucrase enzyme?

HEALTH NOTE *Viruses and AIDS*

Viruses are small particles of 3 to 200 genes that cannot replicate without a host cell. A typical virus contains a nucleic acid, DNA or RNA, but not both, inside a protein coat. Virions are infectious viruses that invade living cells and cause viral infections. (See Figure 17.21.) Some infections caused by viruses are listed in Table 17.8.

From about 1981, a new disease called AIDS (acquired immune deficiency syndrome) began to claim an alarming number of lives. A virus called HIV-1 (human immunodeficiency virus type 1) is now known as the AIDS-causing agent. (See Figure 17.22.) HIV is a retrovirus that infects and destroys the T4 lymphocyte cells. T4 cells are helper cells on the surface of a membrane that are involved in the immune response. The gradual depletion of T4 cells reduces the ability of the immune system to destroy harmful organisms. The AIDS syndrome is characterized by opportunistic

Figure 17.21
Action of a virus.

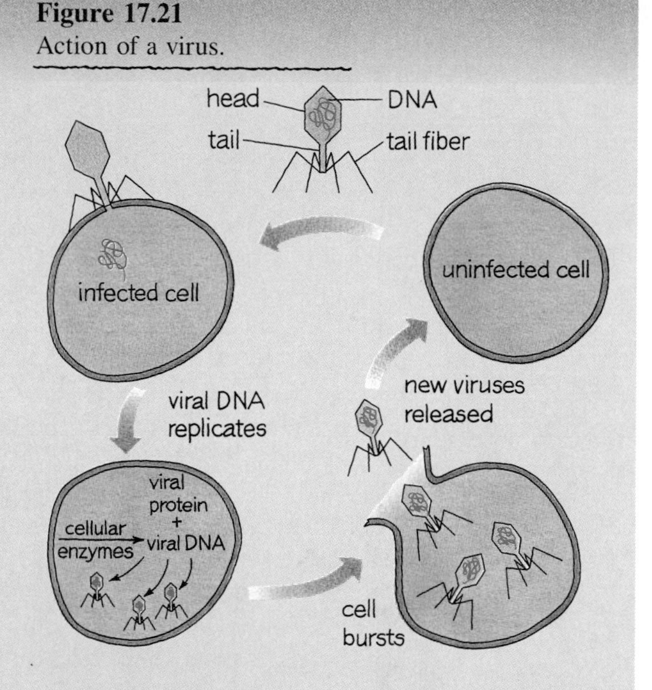

Table 17.8 ***Some Diseases Caused by Viral Infection***

Disease	Virus
Common cold	Coronavirus (over 100 types)
Influenza	Orthomyxovirus
Warts	Papovavirus
Herpes	Herpesvirus
Leukemia, cancers, AIDS	Retrovirus
Hepatitis	Hepatitis A virus (HAV), hepatitis B virus (HBV)
Mumps	Paramyxovirus
Epstein-Barr	Epstein-Barr virus (EBV)

Figure 17.22
A diagram of HIV-1.

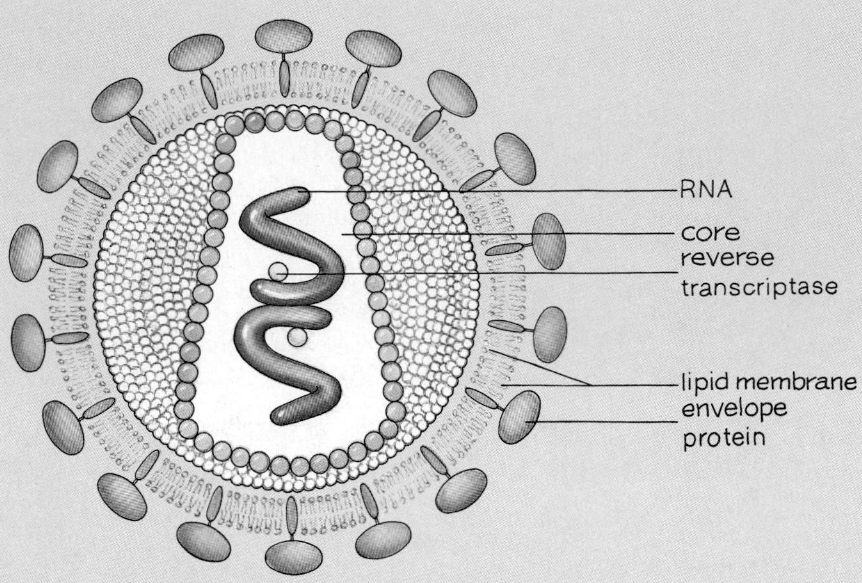

RNA
core
reverse
transcriptase
lipid membrane
envelope
protein

Figure 17.23
After the HIV virus binds to T4 cells, viral RNA is injected. The viral RNA is copied into a DNA provirus by reverse transcription. The provirus joins the DNA in the nucleus. When the cell reproduces, the provirus DNA duplicates the viral RNA needed to produce more HIV virus particles. More T4 cells are infected until all T4 cells are depleted.

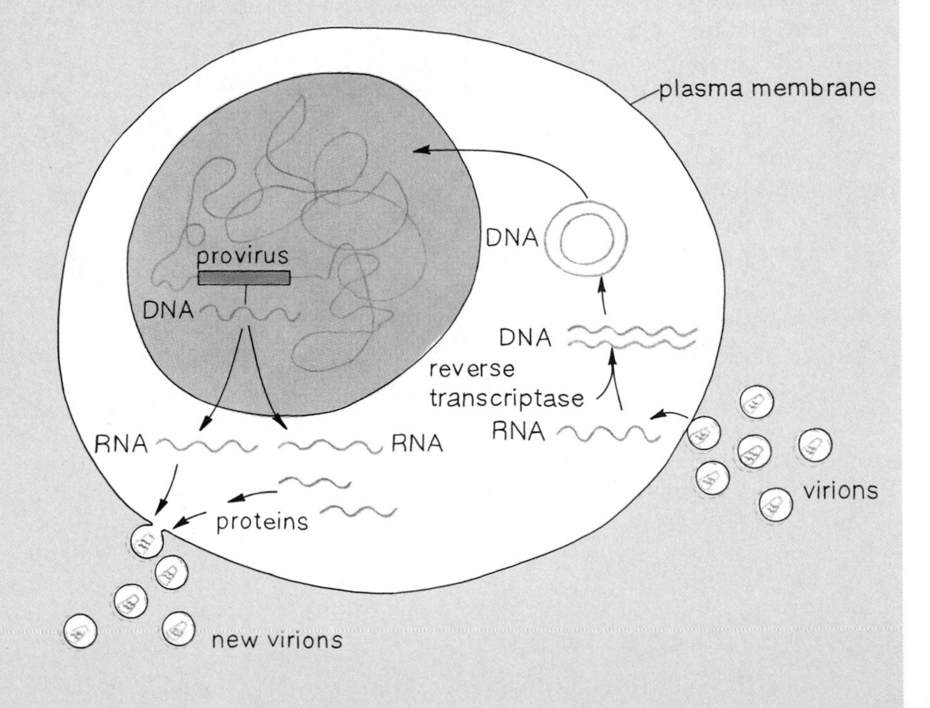

infections such as *pneumocystis carinii,* which causes a pneumonia, and *Karposi's sarcoma,* a skin cancer.

The HIV virus is a retrovirus that contains a single strand of RNA as its genetic material. The T4 cell is infected when the HIV virus binds to receptors on the surface in lock-and-key fashion and injects viral RNA into the host cell. Then the viral RNA is copied by reverse transcriptase into double-helical DNA using the nucleotides present in the host cell. (See Figure 17.23.)

The viral DNA, called a provirus, joins the DNA in the nucleus where it remains inactive within the cell until the cell begins to reproduce. Then the genes of the provirus DNA direct the formation of viral RNA for more HIV viruses. The new HIV viruses leave the host cell and infect more T4 cells. Health may remain adequate for up to 5 years after infection. Eventually immune response is so reduced that opportunistic infections and other diseases take hold.

As of yet, there is no cure for HIV infection. There is an effort to try and block the reverse transcriptase process with antiviral vaccines, but none has yet been successful. Although not a cure, AZT has been made available for late-stage HIV to extend life a few years. However, AZT does have toxic side effects. A polarized optical micrograph of the AIDS antiviral drug zidovudine or azidothymidine (AZT) is shown in Figure 17.24. There are also efforts to block the binding of the HIV virus to the cell, and to prevent the newly formed viral RNA from being incorporated into mature retroviral material.

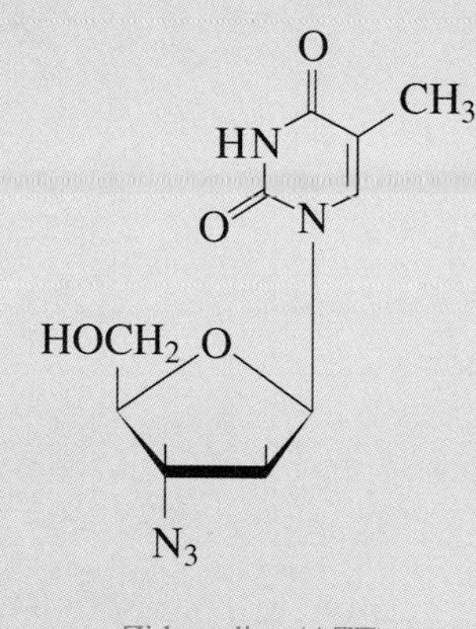

Zidovudine (AZT)

Figure 17.24
A micrograph of crystallites of the anti-AIDS drug azidothymidine (AZT). In reverse transcription, AZT causes a premature termination of the DNA chain, which prevents the DNA from combining with the host DNA.

HEALTH NOTE *Cancer*

In an adult body, most cells do not continue to reproduce. When cells in the body begin to grow and multiply without control, they invade neighboring cells and appear as a tumor or growth (neoplasm). When tumors interfere with normal functions of the body, they are cancerous. If they are limited, they are benign. Cancer can be caused by chemical and environmental substances, by radiation, or by oncogenic viruses.

Some reports estimate that 70–80% of all human cancers are initiated by chemical and environmental substances. A carcinogen is any substance that increases the chance of inducing a tumor. Known carcinogens include dyes, cigarette smoke, and asbestos. More than 90% of all persons with lung cancer are smokers. A carcinogen causes cancer by reacting with molecules in a cell, probably DNA, and altering the growth of that cell. Some known carcinogens are listed in Table 17.9.

Table 17.9 *Some Chemical and Environmental Carcinogens*

Carcinogen	Tumor Site
Asbestos	Lung, respiratory tract
Arsenic	Skin, lung
Cadmium	Prostate, kidneys
Chromium	Lung
Nickel	Lung, sinuses
Aflatoxin	Liver
Nitrites	Stomach
Aniline dyes	Bladder
Vinyl chloride	Liver

Radiant energy from sunlight or medical radiation is another type of environmental factor. Skin cancer has become one of the most prevalent forms of cancer. It appears that DNA damage in the exposed areas of the skin causes mutations. The cells lose their ability to control protein synthesis, and uncontrolled cell division leads to cancer. The incidence of malignant melanoma, one of the most serious skin cancers, has been rapidly increasing. Some possible factors for this increase may be the popularity of suntanning as well as the reduction of the ozone layer, which absorbs much of the harmful radiation from sunlight.

Oncogenic viruses cause cancer when cells are infected. Several viruses associated with human cancers are listed in Table 17.10. Some cancers such as retinoblastoma and breast cancer appear to occur more frequently in families. There is some indication that a missing or defective gene may be responsible.

Table 17.10 *Human Cancers Caused by Oncogenic Viruses*

Virus	Disease
RNA viruses	
Human T-cell lymphotropic virus-type I (HTLV-I)	Leukemia
DNA viruses	
Epstein-Barr virus (EBV)	Burkitt's lymphoma (cancer of white blood B cells), Nasopharyngeal carcinoma, Hodgkin's disease
Hepatitis B virus (HBV)	Liver cancer
Herpes simplex virus (type 2)	Cervical and uterine cancer
Papilloma virus	Cervical and colon cancer, genital warts

Chapter Summary

17.1 Nucleic Acids
Nucleic acids, deoxyribonucleic acid (DNA) and ribonucleic acid (RNA), are linear polymers of nucleotides. A nucleotide is composed of three parts: a nitrogen base, a sugar, and a phosphate group. In DNA, the sugar is deoxyribose and the nitrogen base can be adenine, thymine, guanosine, or cytosine. In RNA, the sugar is ribose and uracil replaces thymine.

17.2 Structures of Nucleic Acids
A DNA molecule consists of two strands of nucleotides that are wound around each other like a spiral staircase. The two strands are held together by hydrogen bonds between complementary base pairs, A with T, and G with C.

17.3 DNA Replication
During DNA replication, DNA polymerase makes new DNA strands along each of the original DNA strands that serve as templates. Complementary base pairing ensures the correct pairing of bases to give identical copies of the original DNA.

17.4 Structure and Types of RNA
The three types of RNA differ by function in the cell: ribosomal RNA makes up most of the structure of the ribosomes, messenger RNA carries genetic information from the DNA to the ribosomes, and transfer RNA places the correct amino acids in the protein.

17.5 Synthesis of RNA: Transcription
Transcription is the process by which RNA polymerase produces mRNA from one strand of DNA. The bases in the mRNA are complementary to the DNA, except U is paired with A in DNA.

17.6 The Genetic Code
The genetic code consists of a sequence of three bases (triplet) that specifies the order for the amino acids in a protein. There are 64 codons for the 20 animo acids, which means there are several codons for some amino acids. The codon AUG signals the start of transcription and codons UAG, UGA, and UAA signal stop.

17.7 Protein Synthesis
Proteins are synthesized at the ribosomes in a translation process that includes three steps: initiation, elongation, and termination. During translation, tRNAs bring the appropriate amino acids to the ribosome and peptide bonds form. When the polypeptide is released, it takes on its secondary and tertiary structures and becomes a functional protein in the cell.

17.8 Genetic Mutations
A genetic mutation is a change of one or more bases in the DNA sequence that alters the structure and ability of the resulting protein to function properly. In a substitution, one base may be altered, and a frame shift mutation inserts or deletes a base and changes all the codons after the base change.

17.9 Cellular Regulation
The production of mRNA occurs when certain proteins are needed in the cell. In enzyme induction, the appearance of a substrate in a cell removes a repressor, which allows RNA polymerase to produce mRNA at the structural genes. In enzyme repression, the end product of an enzyme-catalyzed sequence activates a repressor to inhibit the further production of mRNA.

Glossary of Key Terms

anticodon The triplet of bases in the center loop of tRNA that is complementary to a codon on mRNA.

codon A sequence of three bases in mRNA that specifies a certain amino acid to be placed in a protein. A few codons signal the start or stop of transcription.

complementary base pairs In DNA, adenine is always paired with thymine (A–T or T–A), and guanine is always paired with cytosine (G–C or C–G). In forming RNA, adenine is paired with uracil (A–U).

control site A section of DNA composed of a promotor and operator that regulates protein synthesis.

DNA Deoxyribonucleic acid; the genetic material of all cells containing nucleotides with deoxyribose sugar, phosphate, and the four nitrogenous bases adenine, thymine, guanine, and cytosine.

double helix The helical shape of the double chain of DNA that is like a spiral staircase with a sugar–phosphate backbone on the outside and base pairs like stair steps on the inside.

enzyme induction A model of cellular regulation in which protein synthesis is induced by a substrate.

enzyme repression A model of cellular regulation in which protein synthesis is repressed by elevated levels of end product.

frame shift A mutation that inserts or deletes a base in a DNA sequence.

gene A section of DNA that is associated with the synthesis of a particular protein.

genetic code The sequence of codons in mRNA that specifies the amino acid order for the synthesis of protein.

genetic disease A physical malformation or metabolic dysfunction caused by a mutation in the genetic code.

mRNA Messenger RNA; produced in the nucleus by DNA, that carries the genetic information to the ribosomes for the construction of a protein.

mutation A change in the DNA base sequence that alters the formation of a protein in the cell.

nitrogen base Nitrogen-containing compounds found in DNA and RNA: adenine (A), thymine (T), cytosine (C), guanine (G), and uracil (U).

nucleic acids Large molecules, composed of nucleotides, found as a double helix in DNA, or the single strands of RNA.

nucleoside The combination of a pentose sugar and a nitrogen base.

nucleotides Building blocks of a nucleic acid consisting of a nitrogen base, a pentose sugar (ribose or deoxyribose), and a phosphate group.

operator The section of DNA that precedes the structural genes and controls the synthesis of an mRNA

operon A group of genes, control site and structural genes, whose transcription is controlled by the same regulatory gene.

promotor A component of the control site on an operon that regulates DNA synthesis.

purines The double-ringed nitrogen-containing compounds of adenine and guanine found in nucleic acids.

pyrimidines The single-ring nitrogen-containing compounds of cytosine, uracil, and thymine found in nucleic acids.

RNA Ribonucleic acid, a type of nucleic acid that is a single strand of nucleotides containing adenine, cytosine, guanine, and uracil.

rRNA Ribosomal RNA; the most prevalent type of RNA; a major component of the ribosomes.

regulatory gene A gene in front of the control site that produces a repressor.

replication The process of duplicating DNA by pairing the bases on each parent strand with their complementary base.

repressor A protein that interacts with the operator gene in an operon to prevent the transcription of mRNA.

structural genes The sections of DNA that code for the synthesis of proteins.

substitution A mutation that replaces one base in a DNA with a different base.

transcription The transfer of genetic information from DNA by the formation of mRNA.

translation The interpretation of the codons in mRNA as amino acids in a peptide.

translocation The shift of a ribosome along mRNA from one codon (three bases) to the next codon during translation.

tRNA Transfer RNA; an RNA that places a specific amino acid into a peptide chain at the ribosome. There is one or more tRNA for each of the 20 different amino acids.

Chemistry at Home

1. Cut out 16 rectangular pieces of paper. For strand 1, write on two pieces each the symbols for the four nucleotides with hydrogen bonds: A≡, T≡, G═, and C═. For strand 2, write on two pieces each: ≡A, ≡T, ═G, and ═C. Mix up the pieces for strand 1 and randomly place in vertical order. From the second group, strand 2 nucleotides, select the complementary base to complete this section of DNA.

2. Using the nucleotides from experiment 1, make a strand 1 of A—T—T—G—C—C. Arrange the corresponding bases to complete the DNA section. In the original column, change the G to an A. What does this do to the complementary strand? How could this change result in a mutation?

Answers to Study Checks

17.1 deoxycytidine monophosphate

17.2

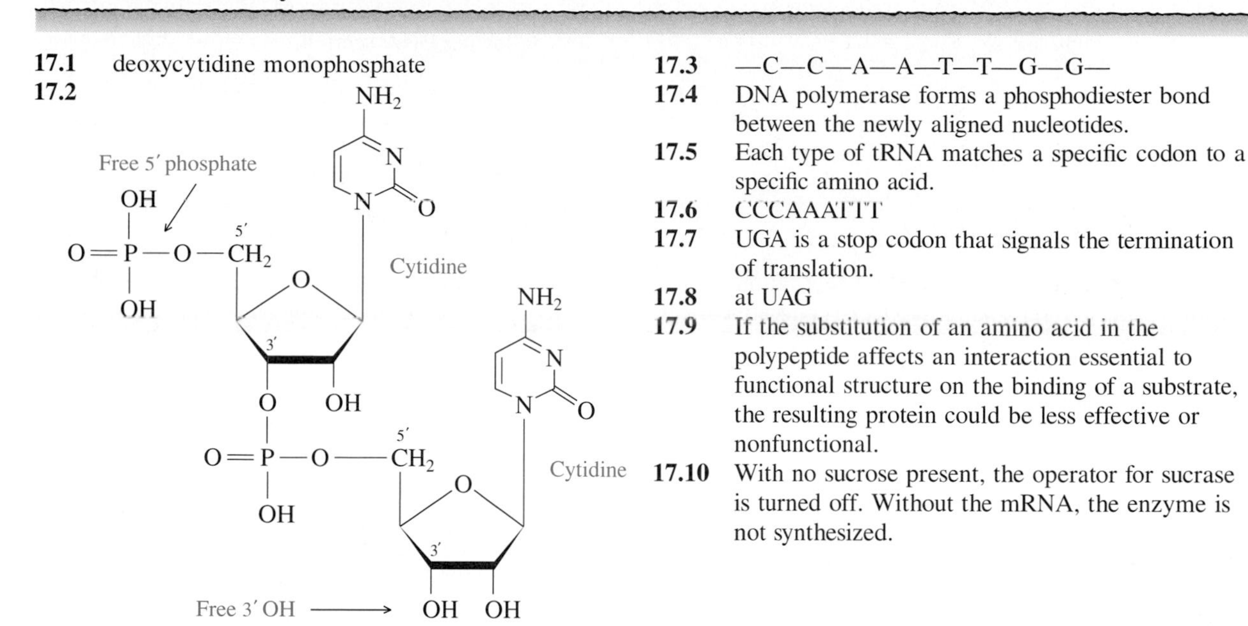

17.3 —C—C—A—A—T—T—G—G—

17.4 DNA polymerase forms a phosphodiester bond between the newly aligned nucleotides.

17.5 Each type of tRNA matches a specific codon to a specific amino acid.

17.6 CCCAAATTT

17.7 UGA is a stop codon that signals the termination of translation.

17.8 at UAG

17.9 If the substitution of an amino acid in the polypeptide affects an interaction essential to functional structure on the binding of a substrate, the resulting protein could be less effective or nonfunctional.

17.10 With no sucrose present, the operator for sucrase is turned off. Without the mRNA, the enzyme is not synthesized.

Problems

Nucleic Acids *(Goal 17.1)*

17.1 What are the nitrogen bases and sugar in the nucleotides of DNA?

17.2 What are the nitrogen bases and sugar in the nucleotides of RNA?

17.3 What are the four nucleotides in DNA?

17.4 What are the four nucleotides in RNA?

17.5 Identify each of the following as a nucleoside or nucleotide:
- **a.** adenosine
- **b.** deoxycytidine
- **c.** uridine
- **d.** cytidine monophosphate

17.6 Identify each of the following as a nucleoside or nucleotide in DNA or RNA:
- **a.** deoxythymidine
- **b.** guanosine
- **c.** adenosine
- **d.** uridine monophosphate

17.7 Draw the structure of deoxyadenosine monophosphate (dAMP).

17.8 Draw the structure of uridine monophosphate (UMP).

Structures of Nucleic Acids *(Goal 17.2)*

17.9 How are nucleotides held together in a nucleic acid chain?

17.10 How are the two strands of nucleic acid in DNA held together?

17.11 To what part of the backbone are the nitrogen bases attached?

17.12 What is meant by complementary base pairing?

17.13 Write the structure of the dinucleotide GC in RNA.

17.14 Write the structure of the dinucleotide AT in DNA.

17.15 Complete the base sequences in a second DNA strand if a portion of one strand has the following base sequence:
- **a.** AAAAAA
- **b.** GGGGGG
- **c.** AGTCCAGGT
- **d.** CTGTATACGTTA

17.16 Complete the base sequences in a second DNA strand if a portion of one strand has the following base sequence:
a. TTTTTT
b. CCCCCCCCC
c. ATGGCA
d. ATATGCGCTAAA

DNA Replication *(Goal 17.3)*

17.17 Describe the process of DNA replication.
17.18 How does replication of DNA produce identical copies?
17.19 Describe the replication of the following portion of a DNA molecule:

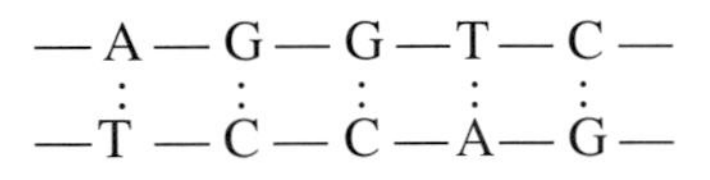

17.20 Where are the DNA strands of the original DNA found in the α-helixes of the new daughter DNA?

Structure and Types of RNA *(Goal 17.4)*

17.21 List the three types of RNA.
17.22 Where are the nucleic acids DNA and RNA found in the cell?
17.23 What is the general structure of a tRNA molecule?
17.24 Why are tRNA molecules called adapter molecules?

Synthesis of RNA: Transcription *(Goal 17.5)*

17.25 What is meant by the term "transcription"?
17.26 What bases in mRNA are used to complement the bases in DNA?
17.27 Write the corresponding section of an mRNA produced from the template strand of the following DNA section:

—C—C—G—A—A—G—G—T—T—C—A—C—
section of DNA template

17.28 Write the corresponding section of an mRNA produced from the template strand of the following DNA section:

—T—A—C—G—G—C—A—A—G—C—T—A—
section of DNA template

The Genetic Code *(Goal 17.6)*

17.29 What is a codon?
17.30 What is the genetic code?
17.31 What amino acid is coded for by each codon?
a. CUU
b. UCA
c. GGU
d. AGG
17.32 What amino acid is coded for by each codon?
a. AAA
b. GUC
c. CGG
d. GCA
17.33 When does the codon AUG signal the start of a protein, and when does it code for the amino acid methionine?
17.34 The codons UGA, UAA, and UAG do not code for amino acids. What is their role as codons in mRNA?

Protein Synthesis *(Goal 17.7)*

17.35 What is the difference between a *codon* and an *anticodon?*
17.36 Why are there at least 20 different tRNAs?
17.37 What are the three steps of translation?
17.38 Where does protein synthesis take place?
17.39 What amino acid sequence would you expect from each of the following mRNA segments?
a. —A—A—A—A—A—A—A—A—A—
b. —U—U—U—C—C—C—U—U—U—C—C—C
c. —U—A—C—G—G—G—A—G—A—U—G—U
17.40 What amino acid sequence would you expect from each of the following mRNA segments?
a. —A—A—A—C—C—C—U—U—G—G—C—C
b. —C—C—U—C—G—A—A—G—C—C—C—A—U—G—A
c. —A—U—G—C—A—C—A—A—A—G—A—A—G—U—A—C—U—U—

17.41 How is a peptide chain extended at sites A and P?

17.42 What is meant by "translocation"?

17.43 The following portion of DNA is in the template DNA strand:

—G—C—T—T—T—T—C—A—A—A—A—A—

- **a.** What is the corresponding mRNA section?
- **b.** What are the anticodons of the tRNAs?
- **c.** What amino acids will be placed in the peptide chain?

17.44 The following portion of DNA is in the template DNA strand:

—T—G—T—G—G—G—G—T—T—A—T—T—

- **a.** What is the corresponding mRNA section?
- **b.** What are the anticodons of the tRNAs?
- **c.** What amino acids will be placed in the peptide chain?

Genetic Mutations *(Goal 17.8)*

17.45 What is a substitution mutation?

17.46 How does a substitution mutation in the genetic code for an enzyme affect the order of amino acids in that protein?

17.47 What is the effect of a mutation on the amino acids in the polypeptide?

17.48 How can a mutation decrease the activity of a protein?

17.49 How is protein synthesis affected if the normal base sequence —T—T—T— in the DNA template is changed to —T—T—C—?

17.50 How is protein synthesis affected if the normal base sequence —C—C—C— is changed to —A—C—C?

17.51 Consider the following portion of mRNA produced by the normal order of DNA nucleotides:

—ACA—UCA—CGG—GUA—

- **a.** What is the amino acid order produced for normal DNA?
- **b.** What is the amino acid order if a mutation changes UCA to ACA?
- **c.** What is the amino acid order if a mutation changes CGG to GGG?
- **d.** What happens to protein synthesis if a mutation changes UCA to UAA?

17.52 Consider the following portion of mRNA produced by the normal order of DNA nucleotides:

—CUU—AAA—CGA—GUU—

- **a.** What is the amino acid order produced for normal DNA?
- **b.** What is the amino acid order if a mutation changes CUU to CCU ?
- **c.** What is the amino acid order if a mutation changes CGA to AGA?
- **d.** What happens to protein synthesis if a mutation changes AAA to UAA?

17.53 **a.** A base substitution changes a codon for an enzyme from GCC to GCA. Why is there no change in the amino acid order in the protein?

b. In sickle-cell anemia, a base substitution in hemoglobin replaces glutamine (a polar amino acid) by valine. Why does the replacement by one amino acid cause such a drastic change in biological function?

17.54 **a.** A base substitution for an enzyme replaces leucine (a nonpolar amino acid) by alanine. Why does this change in amino acids have little effect on the biological activity of the enzyme?

b. A base substitution replaces cytosine in the codon UCA by adenine. How would this substitution affect the amino acids in the protein?

Cellular Regulation *(Goal 17.9)*

17.55 What is an operon?

17.56 Does the operon model control protein synthesis at transcription or translation? Why?

17.57 How is protein synthesis repressed in enzyme induction?

17.58 How is protein synthesis repressed in enzyme repression?

17.59 A bacterium grown in glucose is switched to lactose. Soon enzymes are produced that metabolize lactose. Why?

17.60 In enzyme repression, how is the active repressor formed?

Challenge Problems

17.61 Which amino acids have the fewest number of codons? Which have the highest number of codons?

17.62 A protein contains 35 amino acids. How many nucleotides would be found in the mRNA for this protein?

17.63 Leu-enkephalin is a brain pentapeptide. What is a possible mRNA if it has an amino acid sequence of Tyr-Gly-Gly-Phe-Leu?

17.64 Why does the cell regulate protein synthesis from DNA?

Chapter 18

Metabolic Pathways and Energy Production

During physical activity, ATP energy from the metabolism of glucose stored in muscle cells is used to contract the muscles.

Looking Ahead

Learning Goals

18.1 Describe the role of ATP in catabolic and anabolic reactions.

18.2 Give the sites and products of digestion for carbohydrates, lipids, and proteins.

18.3 Describe the conversion of glucose to pyruvate in glycolysis.

18.4 Give the conditions for the conversion of pyruvate to lactate, ethanol, and acetyl coenzyme A.

18.5 Describe the oxidation of acetyl CoA in the citric acid cycle.

18.6 Describe the synthesis of ATP from NADH and $FADH_2$ by the electron transport chain.

18.7 Account for the ATP produced by the complete oxidation of glucose.

18.8 Describe the role of oxidation of fatty acids in ATP production.

18.9 Describe some metabolic pathways for amino acids.

Questions to Think About

1. What is the purpose of digestion?
2. What products are obtained when carbohydrates, fats, and proteins are digested?
3. What is the importance of ATP in the cells?
4. How do you obtain energy to walk or run?
5. When are fats oxidized for energy?
6. Why is it dangerous to use body proteins for energy?

All the chemical reactions in living cells that break down or build molecules are known as *metabolism.* Many reactions, each catalyzed by a specific enzyme, link together to form a metabolic pathway. Some of the pathways produce energy for the cell; others use energy.

When we digest food, products are obtained that are small enough to be absorbed into the cells of the body. Within our cells, the oxidation of these food products releases energy that is stored in a high-energy compound called adenosine triphosphate (ATP). Our cells need ATP energy to do work such as contracting muscles, synthesizing larger molecules, sending nerve impulses, and moving substances across cell membranes.

18.1 ATP: The Energy Storehouse

Learning Goal

Describe the role of ATP in catabolic and anabolic reactions.

A 25-year-old college student requires a daily diet of 2000 Calories (kcal), an aerobics instructor needs a daily diet of 3000 Calories, and a bicyclist preparing for an international race needs 6000 Calories a day. If there is no change in weight, we know that the amount of energy expended is balanced by the amount of energy obtained from the diet.

In **metabolism,** thousands of chemical reactions are taking place in our cells. **Catabolic** (energy producing) **reactions** break apart complex compounds releasing energy to be stored as ATP. **Anabolic** (energy requiring) **reactions** use the energy from ATP hydrolysis to do work in the cells.

ATP

In our cells, the energy released from the oxidation of the food we eat is used to form a compound called *adenosine triphosphate* (**ATP**). The ATP molecule is composed of adenine, ribose, and three phosphate groups. (See Figure 18.1.) ATP is like a battery in the cell: it provides energy to run reactions, but it runs down if not recharged.

Hydrolysis of ATP Yields Energy

We can think of ATP as the storage form of energy within the cell that connects the energy-producing reactions with the energy-requiring reactions.

Figure 18.1
The structure of adenosine triphosphate, ATP, a compound that stores energy in the cell.

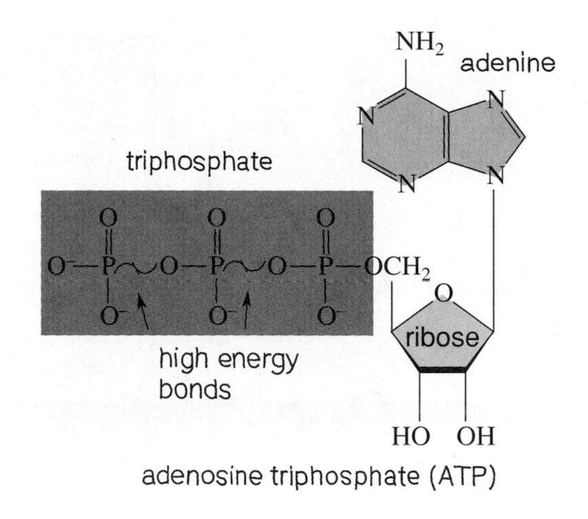

When the bond between two of the phosphate groups in ATP is hydrolyzed, 7.3 kcal/mol of energy is released. This energy, which varies with conditions, is used to drive energy-requiring processes. (See Figure 18.2.) The other products of ATP hydrolysis are **ADP** and inorganic phosphate ion (P_i).

$$ATP + H_2O \longrightarrow ADP + P_i + 7.3 \text{ kcal/mol}$$

Every time we contract muscles, move substances across cellular membranes, send nerve signals, or synthesize an enzyme, we use some energy from ATP hydrolysis. In a cell that is doing a lot of work, there may be 1–2 million ATPs hydrolyzed in a second. (See Figure 18.3.) The amount of ATP hydrolyzed in one day can be as much as our body mass, even though there is only about 1 gram of ATP in all our cells at any given time.

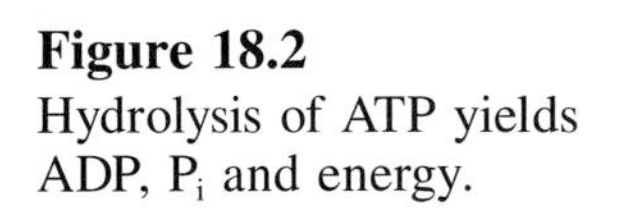

Figure 18.2
Hydrolysis of ATP yields ADP, P_i and energy.

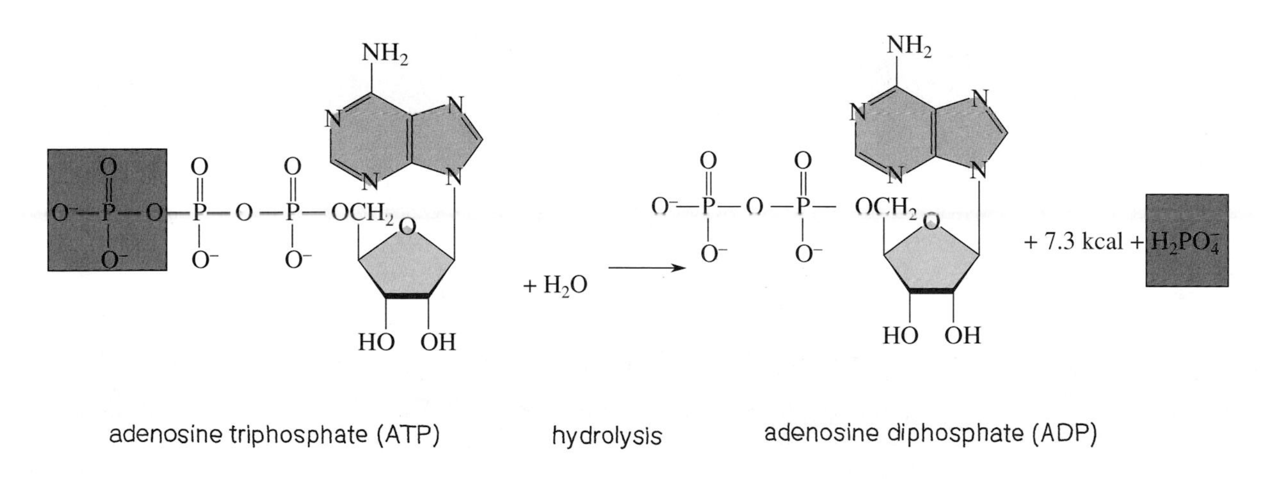

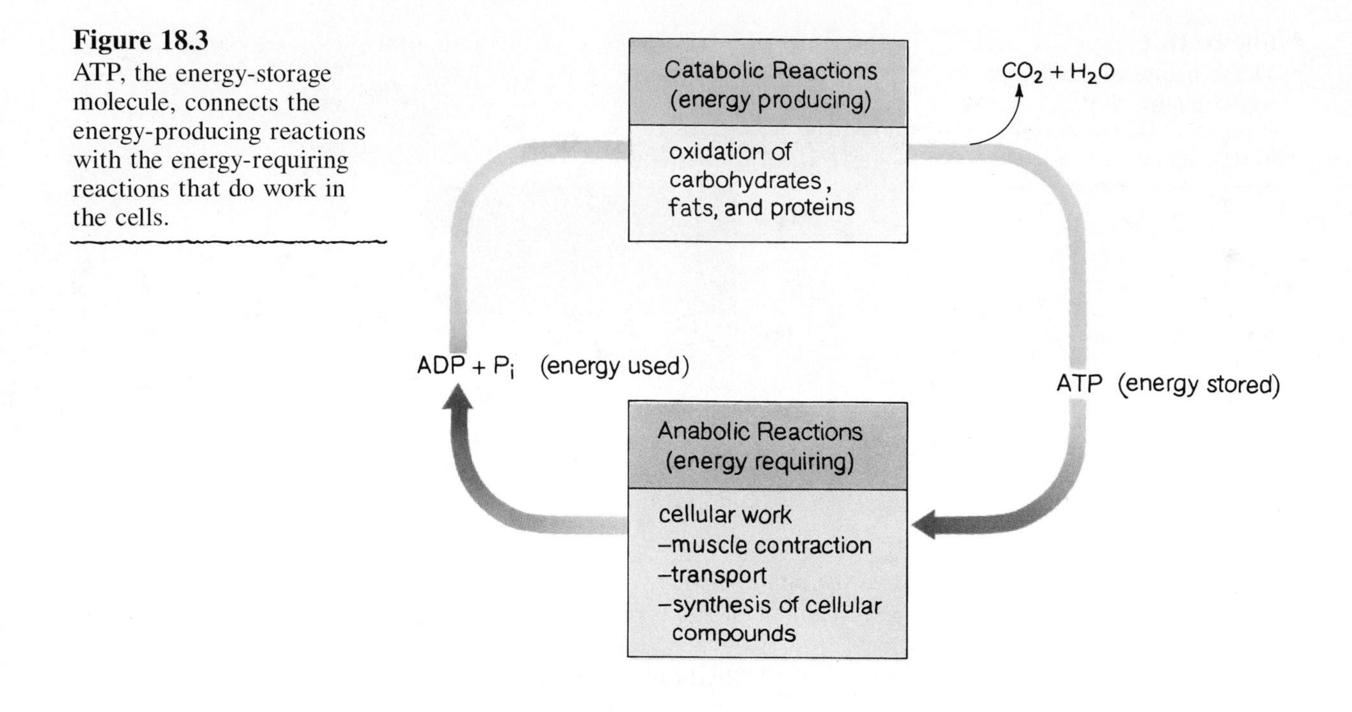

Figure 18.3
ATP, the energy-storage molecule, connects the energy-producing reactions with the energy-requiring reactions that do work in the cells.

Overview of Catabolic Reactions

Catabolic reactions begin with digestion, the process of breaking down carbohydrates, fats, and proteins in the intestinal tract. The products, amino acids, fatty acids, and monosaccharides, are absorbed into our cells where they are oxidized when we need energy. The energy released by catabolism is linked with the synthesis of ATP, which stores energy. Although different metabolic pathways are used to extract energy, their carbon, hydrogen, and oxygen atoms are eventually found in the oxidation products CO_2 and H_2O. (See Figure 18.4.)

Sample Problem 18.1
Hydrolysis of ATP

Write an equation for the hydrolysis of ATP.

Solution
The hydrolysis of ATP produces ADP, P_i, and energy.

$$\text{ATP} + H_2O \longrightarrow \text{ADP} + P_i + 7.3 \text{ kcal/mol energy}$$

Study Check
How much energy is needed to form 1 mol of ATP from 1 mol of ADP and P_i?

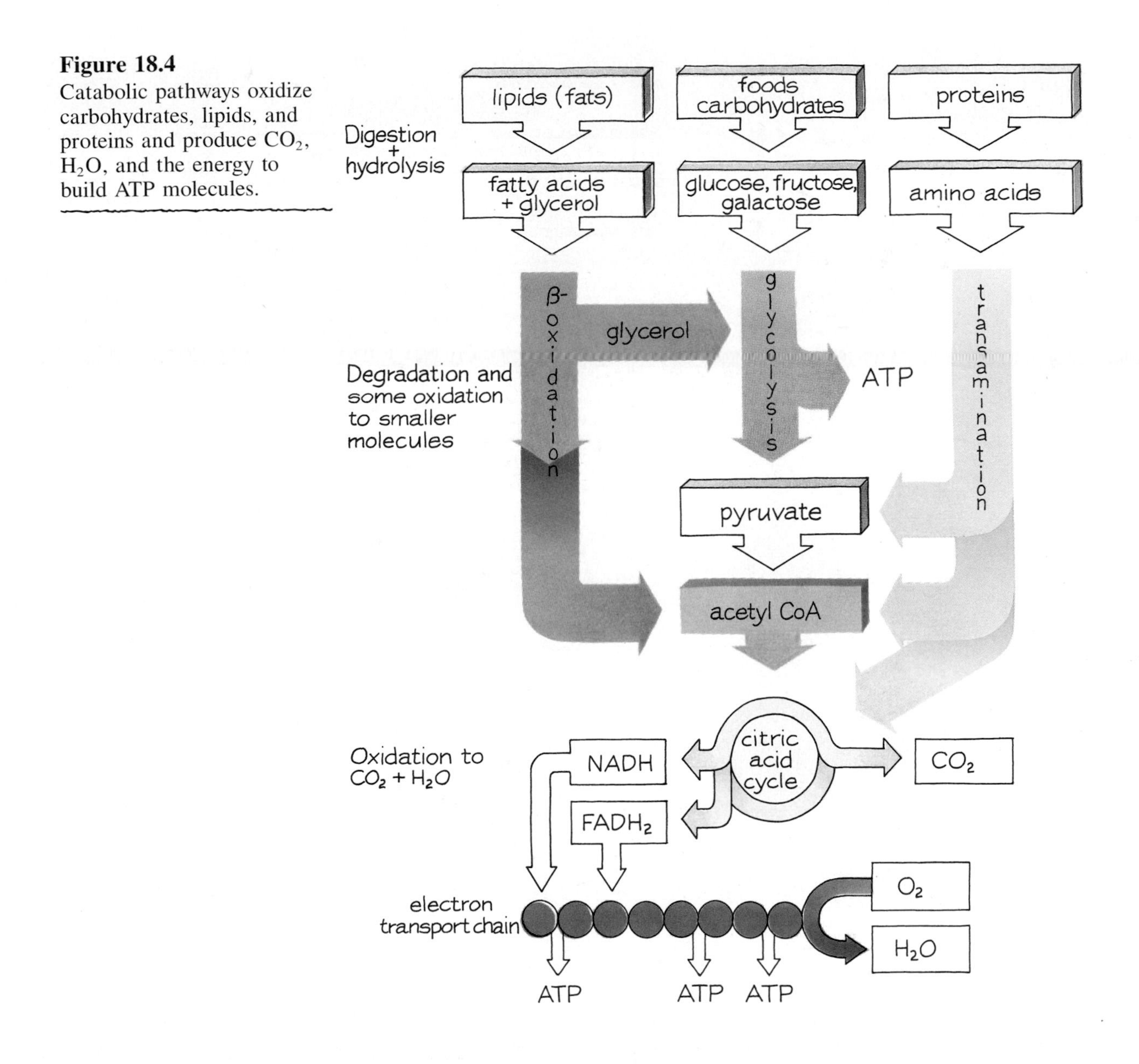

Figure 18.4
Catabolic pathways oxidize carbohydrates, lipids, and proteins and produce CO_2, H_2O, and the energy to build ATP molecules.

HEALTH NOTE *ATP Energy Needed to Contract Muscles*

A muscle consists of thousands of parallel fibers. Within these muscle fibers are fibrils composed of two kinds of proteins called filaments. Arranged in alternating rows, the thick filaments of myosin overlap the thin filaments containing actin. (See Figure 18.5.)

In relaxed muscle, the Ca^{2+} concentration is low and the ATP concentration is high. When a

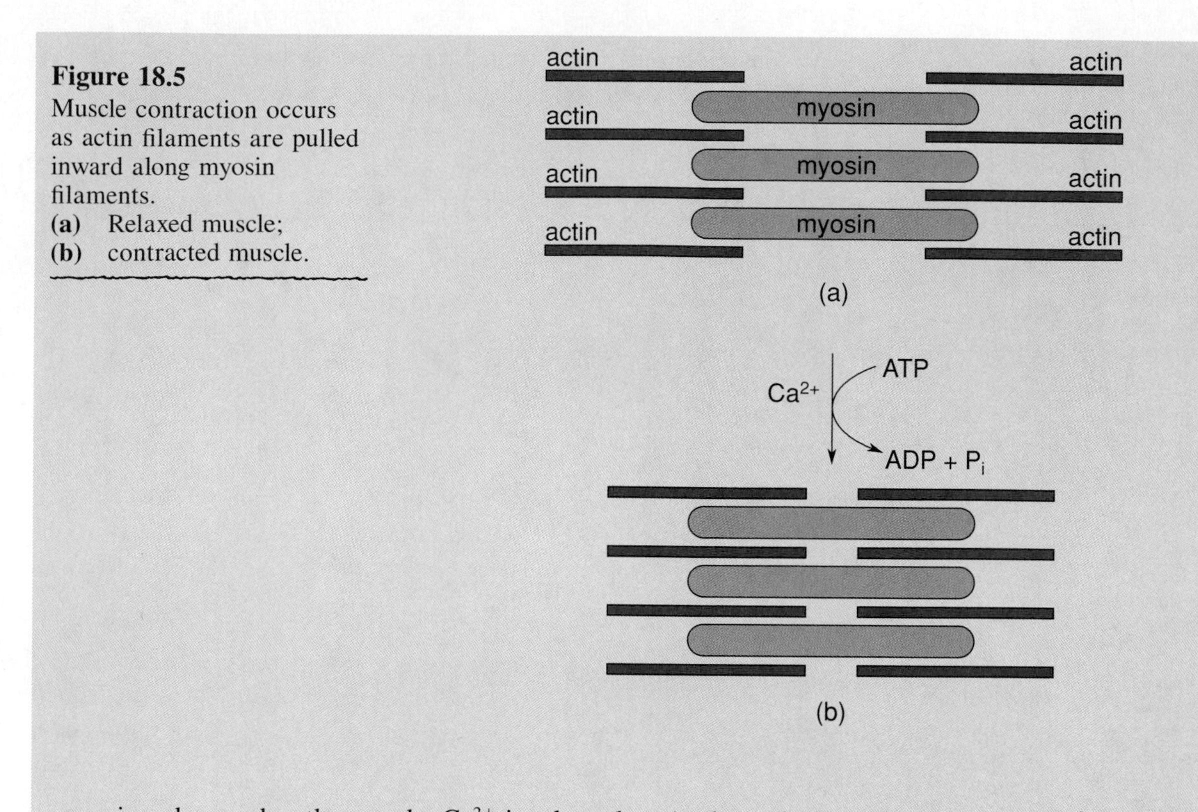

Figure 18.5
Muscle contraction occurs as actin filaments are pulled inward along myosin filaments.
(a) Relaxed muscle;
(b) contracted muscle.

nerve impulse reaches the muscle, Ca^{2+} is released, causing the hydrolysis of ATP. Muscle contraction occurs as the ATP energy pulls the actin filaments inward along the myosin fibers.

$$\text{Relaxed muscle} + \text{ATP} \xrightarrow{Ca^{2+}} \text{contracted muscle} + \text{ADP} + P_i$$

At the end of the nerve impulse, Ca^{2+} level drops and ATP concentration rises. The myosin and actin filaments return to their stretched, relaxed state. With the next nerve impulse, Ca^{2+} is released to hydrolyze more ATP, and the muscle contracts again.

18.2 Digestion of Foods

Learning Goal
Give the sites and products of digestion for carbohydrates, lipids, and proteins.

In the first step of catabolism, our foods must undergo **digestion,** a process that converts large molecules to smaller ones that can be absorbed by the body. (See Figure 18.6.)

Digestion of Carbohydrates

We begin to digest carbohydrates as soon as we chew food. An enzyme in the saliva, α-**amylase,** breaks apart some of the α-glycosidic bonds in the starches producing smaller polysaccharides called dextrins (three to eight glucose units), maltose, and some glucose. After swallowing, the partially digested starches encounter the highly acidic environment of the stomach, where little carbohydrate digestion occurs.

Figure 18.6
Digestion of foods along the gastrointestinal tract.

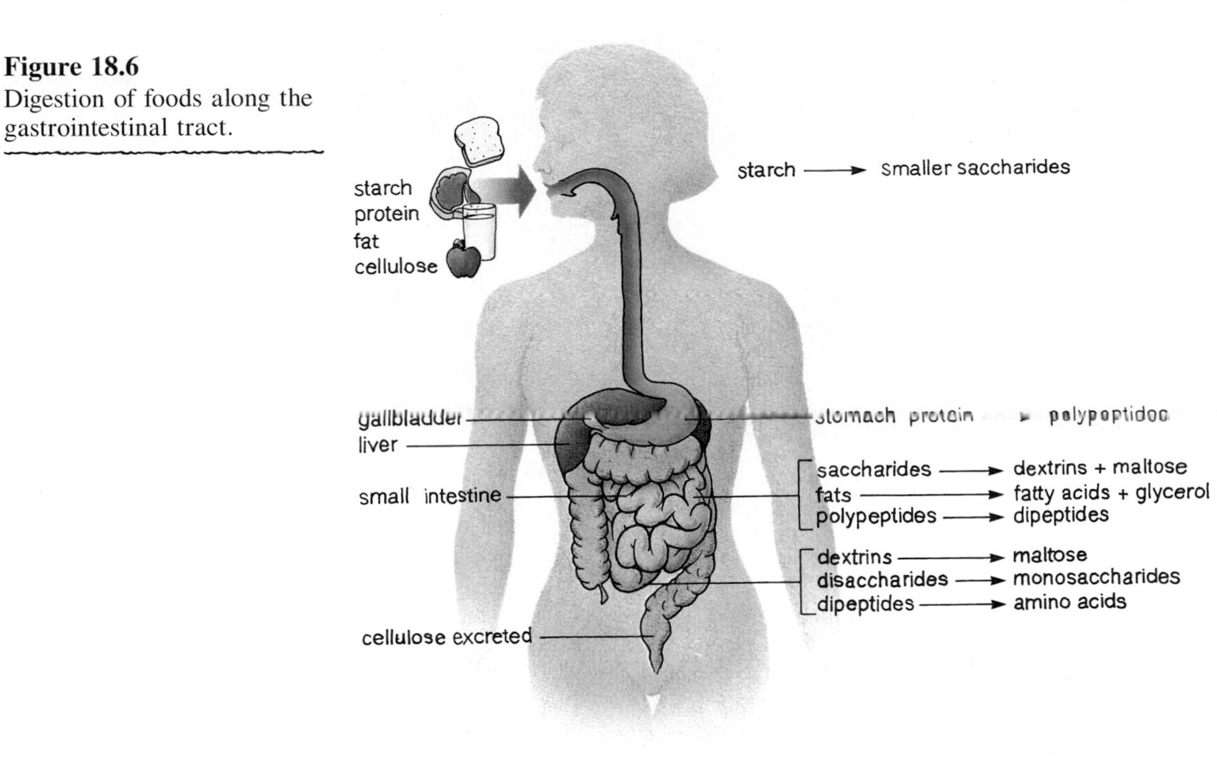

In the small intestine, which has a pH of about 8, amylase secreted by the pancreas converts the remaining polysaccharides to maltose. Finally, the maltose and any sucrose, and lactose from the diet are hydrolyzed by their enzymes maltase, sucrase, and lactase. (See Figure 18.7.) The resulting monosaccharides, glucose, fructose, and galactose, move through the intestinal wall and enter the bloodstream. In the body, they may be used to synthesize ATP or stored as glycogen in the muscle and liver.

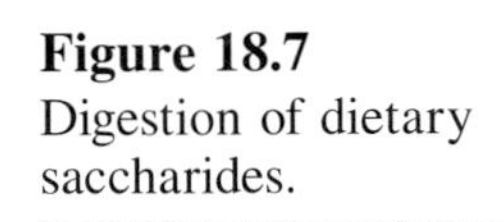
Figure 18.7
Digestion of dietary saccharides.

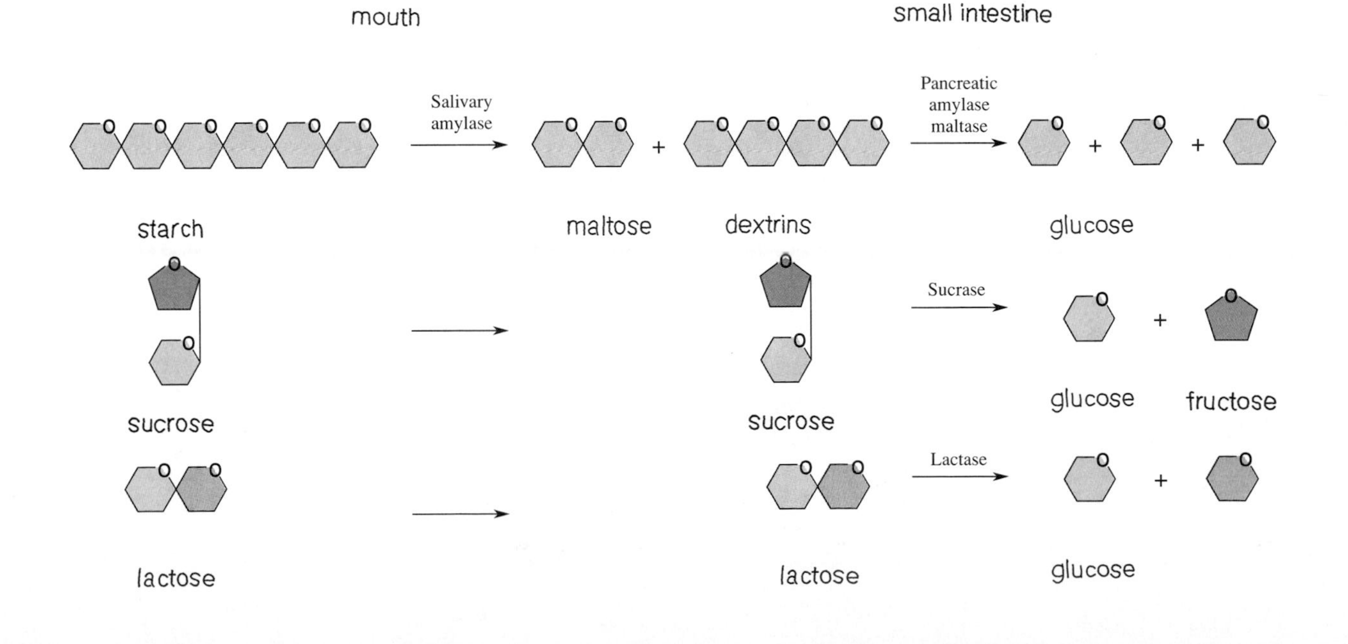

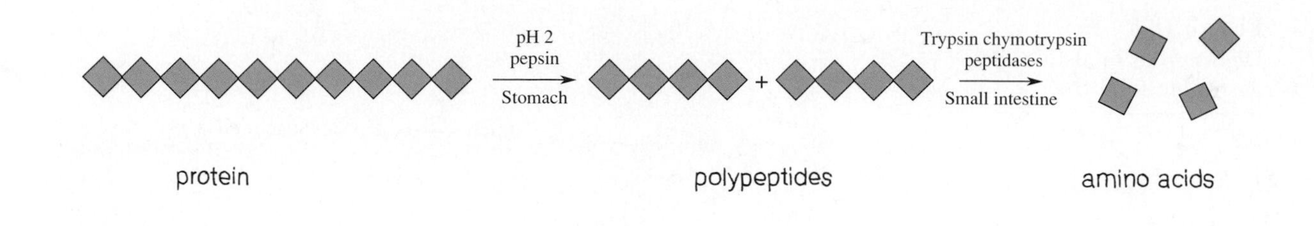

Figure 18.8
Digestion of dietary proteins.

Digestion of Proteins

The digestion of proteins begins in the stomach, where hydrochloric acid (HCl) at pH 2 denatures the protein and pepsin hydrolyzes many of the peptide bonds. In the basic environment (pH 8) of the small intestine, proteases such as trypsin and chymotrypsin complete the hydrolysis of the polypeptides to amino acids. The amino acids are absorbed through the intestinal walls into the bloodstream. (See Figure 18.8.)

Digestion of Lipids

The lipids in our diets undergo little digestion until they enter the small intestine where they are mixed with bile salts from the gallbladder and lipases from the pancreas. The bile salts break the fats globule into smaller droplets called micelles. In the small intestine, lipases hydrolyze the triglycerides on the surface of the micelles to yield fatty acids, monoglycerides, and glycerol. Water-soluble glycerol and short-chain fatty acids move through the intestinal lining into the bloodstream.

The insoluble fatty acids and triglycerides as well as cholesterol are absorbed into the intestinal lining where they are coated with proteins. The resulting lipoproteins are the soluble forms that carry lipids through the

Figure 18.9
Digestion of dietary fats.

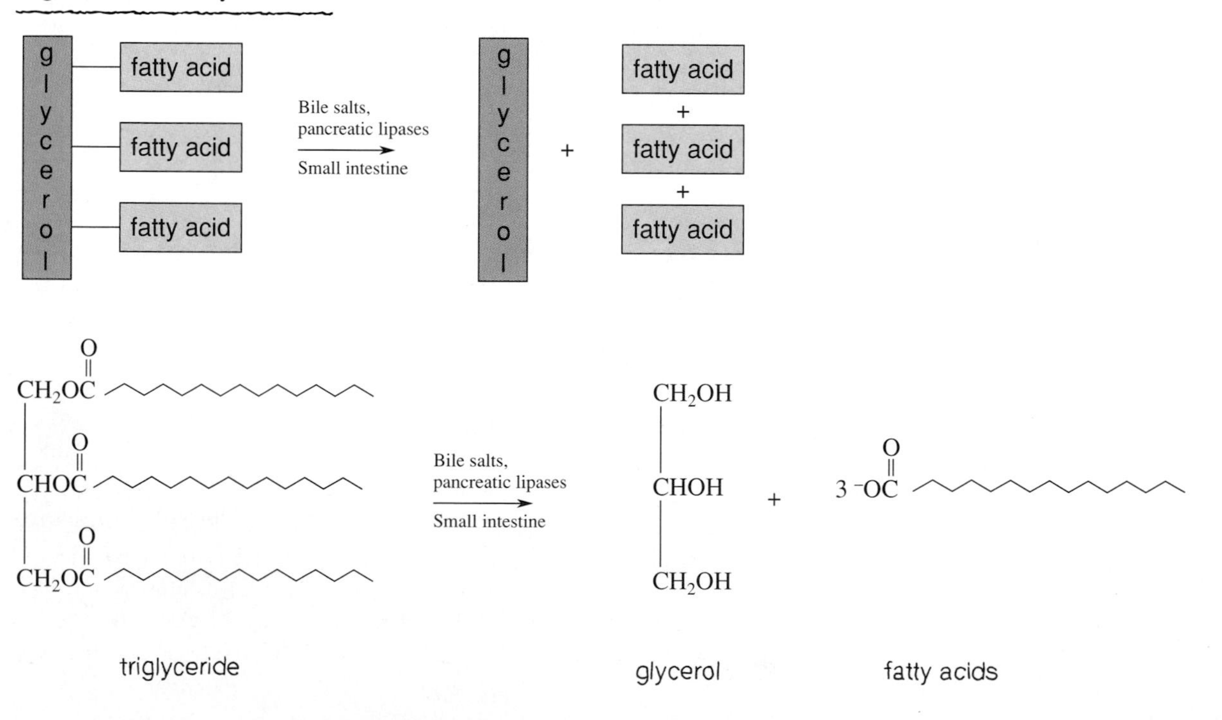

bloodstream to the liver where they can be used for energy or stored in fat cells. (See Figure 18.9.)

Sample Problem 18.2
Digestion of Foods

Describe the role of bile salts and pancreatic lipase in fat digestion.

Solution
In the small intestine, bile salts released from the gallbladder break apart fat globules to yield micelles. On the surface of the micelles, pancreatic lipase hydrolyzes triglycerides.

Study Check
Describe the digestion of amylose, a polymer of glucose molecules joined by α-glycosidic bonds.

18.3 Glycolysis: Oxidation of Glucose

Learning Goal
Describe the conversion of glucose to pyruvate in glycolysis.

The glucose we obtain from the digestion of carbohydrates is our major source of energy. In a metabolic pathway called **glycolysis** (glyco = "sugar," lysis = "a breaking apart"), a six-carbon glucose molecule breaks down step-by-step to yield two three-carbon molecules of pyruvate. (See Figure 18.10.) The product pyruvate is written as an anion, which is the prevalent form of pyruvic acid in physiological solutions (pH 7.4). A total of 10 steps are required because energy must be released in small amounts. If a lot of energy were released at one time, it would raise the temperature in the cells so much that the proteins would be destroyed. Scientists believe that glycolysis was one of the first biological pathways that produced energy, even before there was oxygen in the atmosphere.

$$\underset{\text{Glucose}}{C_6H_{12}O_6} \xrightarrow{\text{10 steps of glycolysis}} \underset{\text{Pyruvate (anion of pyruvic acid in physiological solutions)}}{2CH_3-\overset{\overset{O}{\|}}{C}-\overset{\overset{O}{\|}}{C}-O^-}$$

As we describe the steps in glycolysis, you may want to refer to Figure 18.10. Although many details are given for the pathway, it should not be necessary for you to memorize all the steps. However, it is important that you gain an understanding of the overall process.

A Summary of Reactions in Glycolysis

***Phosphorylation* (*Steps 1, 2, and 3*).** When glucose from the bloodstream enters a cell, it goes into the cytoplasm, a part of the cell where there is no oxygen. Glycolysis is an **anaerobic** process. In the first three steps, a molecule of glucose is prepared to be split apart. In Step 1, a *hexokinase* uses one ATP to add a phosphate to glucose, producing glucose 6-phosphate. This step is accelerated by insulin produced when the glucose level is high within the cell.

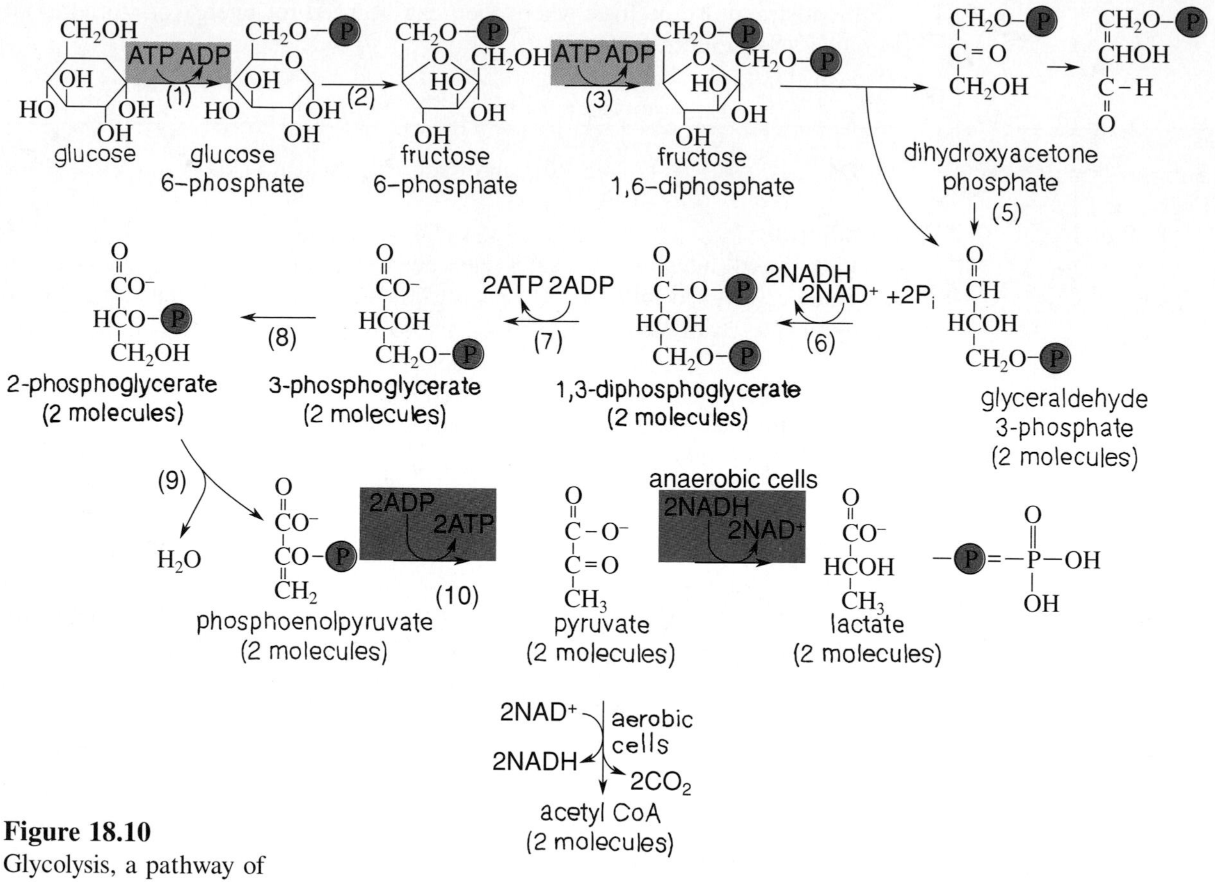

Figure 18.10
Glycolysis, a pathway of reactions that oxidize glucose to pyruvate.

In Step 2, an *isomerase* converts glucose 6-phosphate to fructose 6-phosphate. In Step 3, another phosphate group from ATP is transferred to yield fructose 1,6-diphosphate.

***The Six-carbon Sugar Splits Apart* (*Steps 4 and 5*).** In Step 4, the enzyme *aldolase* splits the six-carbon molecule fructose 1,6-diphosphate into two triose molecules, an aldose phosphate and a ketose phosphate. Because the ketose molecule cannot be oxidized further, an *isomerase* converts it, in Step 5, to a second molecule of glyceraldehyde 3-phosphate. Up to this point in glycolysis, one glucose molecule has been phosphorylated, isomerized, and split to yield two aldotriose phosphate molecules. Two molecules of ATP have been hydrolyzed; but so far, no ATP energy has been generated.

***Oxidation–Reduction* (*Step 6*).** In Step 6, the glyceraldehyde 3-phosphate undergoes both an oxidation and a phosphorylation catalyzed by a *dehydrogenase.* The **oxidation** (loss of 2H) of the aldehyde to a carboxylate ion is coupled with the **reduction** (gain of 2H) of NAD^+ (nicotinamide adenine dinucleotide) to NADH + H^+. The addition of an inorganic phosphate group (P_i) yields 1,3-diphosphoglycerate, which has a high-energy phosphate bond.

***Formation of ATP* (*Step 7*).** In Step 7, the first molecule of ATP forms when the phosphate group from the energy-rich 1,3-diphosphoglycerate is transferred directly to an ADP. This process, known as a **direct substrate–phosphorylation,** is catalyzed by a *phosphokinase.* Because two triose molecules are obtained from one glucose, this step yields two ATP molecules. At this point, glycolysis pays back the debt of two ATP used in the earlier steps.

***Isomerization and Dehydration* (*Steps 8 and 9*).** The 3-phosphoglycerate is converted into a high-energy compound when a *mutase* catalyzes the isomerization of its phosphate and hydroxyl groups to yield 2-phosphoglycerate. Then, an *enolase* catalyzes the removal of water from 2-phosphoglycerate to yield phosphoenolpyruvate, a high-energy phosphate compound.

***Formation of more ATP* (*Step 10*).** In Step 10, a *kinase* catalyzes the direct transfer of the remaining phosphate group to ADP to yield ATP and pyruvate. In this step, there is a yield of two more ATP from the original glucose.

ATP and NADH Production in Glycolysis

Overall glycolysis produces two pyruvates, two NADH, and two ATP molecules. In Steps 7 and 10, four ATP are produced by **direct substrate phosphorylation.** However, the initial hydrolysis of two ATP gives a net yield of two ATP from the degradation of one glucose.

$$\underset{\text{Glucose}}{C_6H_{12}O_6} + 2\ ADP + 2\ P_i + 2\ NAD^+ \xrightarrow{\text{10 steps of glycolysis}} \underset{\text{Pyruvate}}{2CH_3\overset{O}{\overset{\|}{C}}\overset{O}{\overset{\|}{C}}O^-} + 2\ ATP + 2\ NADH + 2H^+ + 2H_2O$$

Sample Problem 18.3
Glycolysis

What are the steps in glycolysis that generate ATP?

Solution
ATP is produced when phosphate groups are transferred directly to ADP from 1,3-diphosphoglycerate (Step 7) and from phosphoenolpyruvate (Step 10).

Study Check
If four ATP molecules are produced in glycolysis, why is there a net yield of two ATP?

18.4 Pathways for Pyruvate

Learning Goal
Give the conditions for the conversion of pyruvate to lactate, ethanol, and acetyl coenzyme A.

The pyruvate produced from glucose can now enter one of three pathways that continue to extract energy. The available pathway depends on whether there is sufficient oxygen in the cell. In aerobic conditions, oxygen is available to convert pyruvate to acetyl coenzyme A(CoA). In anaerobic conditions, oxygen levels are low and pyruvate is reduced to lactate. In yeast cells, which are anaerobic, pyruvate is converted to ethanol.

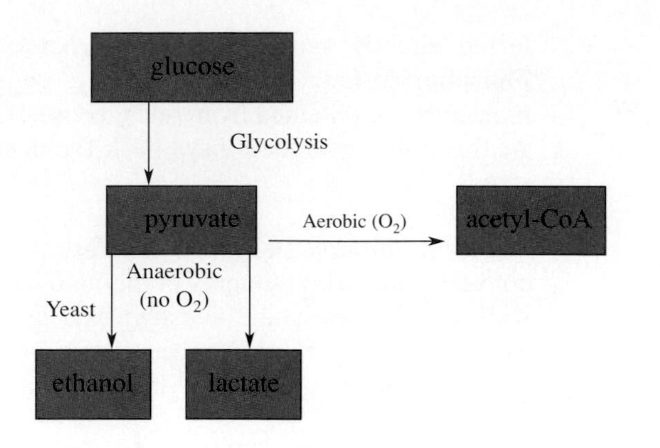

Formation of Acetyl CoA (Aerobic)

The greatest amount of the energy is obtained from glucose when the oxygen levels are high in the cells. Under **aerobic** conditions, pyruvate moves from the cytoplasm (where glycolysis took place) into the mitochondria for oxidation. In a series of reactions catalyzed by pyruvate dehydrogenase, a carbon is removed as CO_2 and the resulting two-carbon compound is oxidized using NAD^+ to accept the hydrogen. The resulting acetyl group is attached to CoA producing **acetyl CoA,** an important intermediate in many metabolic pathways. Acetyl CoA links glycosis and the citric acid cycle and supplies carbons for fatty acid and steroid synthesis.

$$\underset{\text{Pyruvate}}{CH_3-\overset{\overset{O}{\|}}{C}-\overset{\overset{O}{\|}}{C}-O^-} + \boxed{NAD^+} + CoA \xrightarrow{\text{Pyruvate dehydrogenase}} \underset{\text{Acetyl CoA}}{CH_3-\overset{\overset{O}{\|}}{C}-CoA} + CO_2 + \boxed{NADH}$$

CoA, the carrier for acetyl groups, is composed of the B vitamin pantothenic acid, a molecule of ATP, and the amino acid cysteine, which has an —SH that joins the CoA to the acetyl group. (See Figure 18.11.)

Reduction of Pyruvate to Lactate (Anaerobic)

When we engage in strenuous exercise, the oxygen stored in our muscle cells is quickly depeleted. With low oxygen levels, pyruvate produces ATP only by glycolysis. However, glycolysis requires NAD^+, which is produced in the cytoplasm by the reduction of pyruvate to lactate. (See Figure 18.12.) Then NAD^+ becomes available for the oxidation of glyceraldehyde 3-phosphate in glycolysis, which produces a small, but needed amount of ATP.

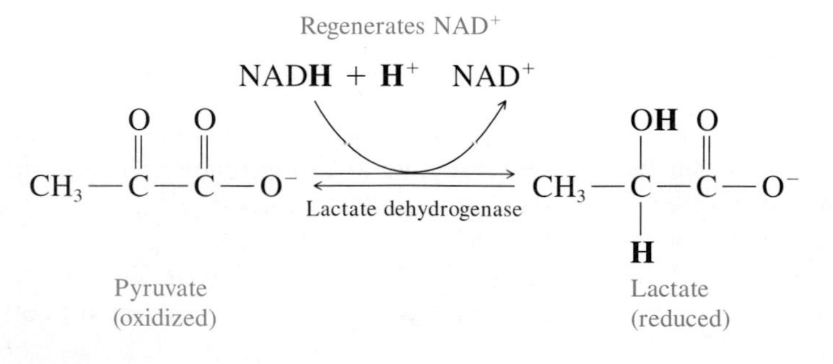

Figure 18.11
Structure of acetyl coenzyme A (CoA).

Figure 18.12
Reactions of pyruvate under aerobic and anaerobic conditions.

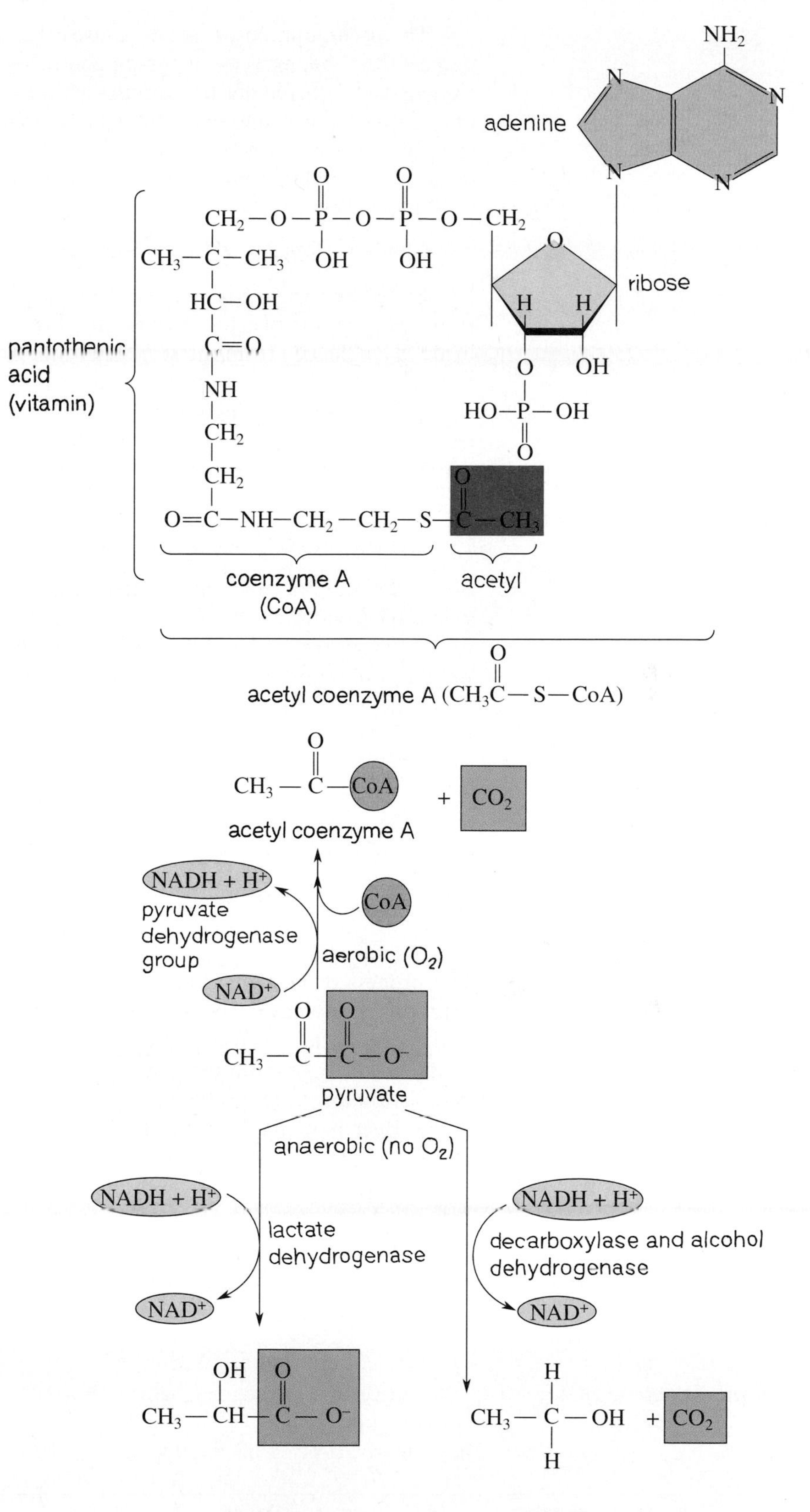

The accumulation of lactate causes the muscles to tire rapidly and become sore. After exercise, a person continues to breathe rapidly to repay the oxygen debt incurred during exercise. Most of the lactate is carried to the liver and converted back into pyruvate, which can undergo further reaction. Under strictly anaerobic conditions, the only ATP production in glycolysis occurs during the steps that phosphorylate ADP directly, giving a net gain of two ATP molecules:

$$\text{Glucose} + 2\text{ADP} + 2\text{P}_i \longrightarrow 2\text{ lactate} + 2\text{ATP} + 2\text{H}_2\text{O}$$

Bacteria can also convert pyruvate to lactate under anaerobic conditions. In the preparation of kimchee and sauerkraut, cabbage is covered with a salt brine. The glucose from the starches is converted to lactate. This acid environment acts as a preservative because it prevents the growth of other bacteria. The pickling of olives and cucumbers gives similar products. When cultures of bacteria that produce lactate are added to milk, the change in pH denatures the milk proteins to give sour cream and yogurt.

Fermentation of Pyruvate to Ethanol (Anaerobic)

Some microorganisms, particularly yeast, convert sugars to ethanol under anaerobic conditions by a process called **fermentation.** After pyruvate is formed in glycolysis, a carbon atom is removed in the form of CO_2 **(decarboxylation).** The NAD^+ for continued glycolysis is regenerated from NADH when the acetaldehyde is reduced to ethanol.

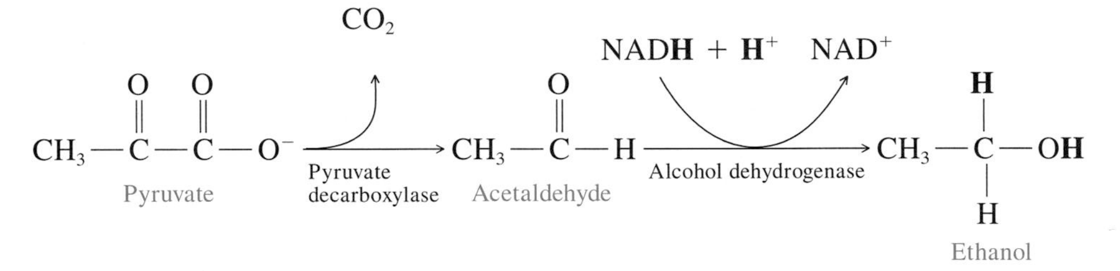

The process of fermentation by yeast is one of the oldest known chemical reactions. Enzymes in the yeast convert the sugars in a variety of carbohydrate sources to glucose and then to ethanol. The evolution of CO_2 gas produces the bubbles in beer, sparkling wines, and champagne. The type of carbohydrate determines the taste associated with a particular alcoholic beverage. Beer is made from the fermentation of barley malt, wine from the sugars in grapes, vodka from potatoes, sake from rice, and whiskeys from corn or rye. Fermentation produces solutions up to about 15% alcohol by volume. At this concentration, the alcohol denatures the yeast, and fermentation stops. Higher concentrations are obtained by distilling and concentrating the alcohol.

Sample Problem 18.4
Fates of Pyruvate

Is each of the following products from pyruvate produced under anaerobic or aerobic conditions?

a. acetyl CoA **b.** lactate

Solution
a. aerobic conditions **b.** anaerobic conditions

Study Check
After strenuous exercise, some lactate is oxidized back to pyruvate by lactate dehydrogenase using NAD^+. Write an equation to show this reaction.

18.5 Citric Acid Cycle

Learning Goal:
Describe the oxidation of acetyl CoA in the citric acid cycle.

The **citric acid cycle** (or Krebs cycle) is a series of reactions that convert acetyl CoA to CO_2 in the presence of oxygen. All of the enzymes for the citric acid cycle are located in the mitochondria considered as the powerhouse of the cell because most of the energy from glucose is extracted there.

In a mitochondrion, there are two different types of membranes, an outer membrane and an inner membrane. Small molecules in the cytoplasm such as pyruvate can move through the outer membrane into the mitochondria. (See Figure 18.13.) The compartment between the outer and inner membrane is called the intermembrane space and the central compartment inside the inner membrane is the matrix space. Most of the enzymes and coenzymes that catalyze the oxidation of pyruvate and fatty acids are found along the inner membrane and in the matrix space of the mitochondria.

In the citric acid cycle, the oxidation of metabolites requires the coenzymes NAD^+ or FAD, which accept hydrogen to yield their reduced forms NADH or $FADH_2$. Both $\mathbf{NAD^+}$ (nicotinamide adenine dinucleotide) and **FAD** (flavin adenine dinucleotide) are derived from ADP and a B vitamin: niacin in NAD^+ and riboflavin in FAD. (See Figure 18.14.) NAD^+ is the

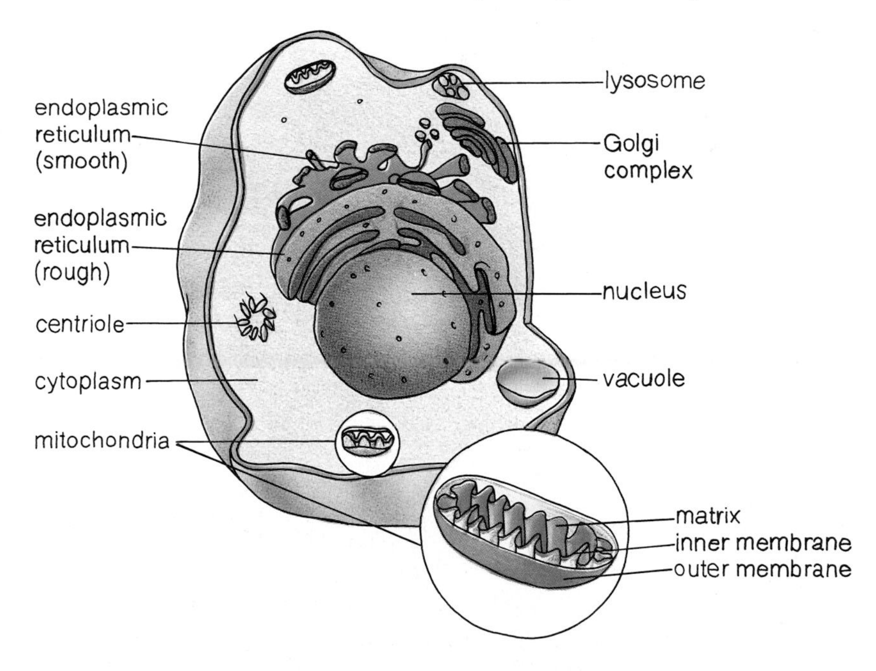

Figure 18.13
A simplified drawing of a cell. Some degradative pathways such as glycolysis occur in the cytoplasm, but most of the oxidation and energy production occurs within the mitochondria, the powerhouse of the cell.

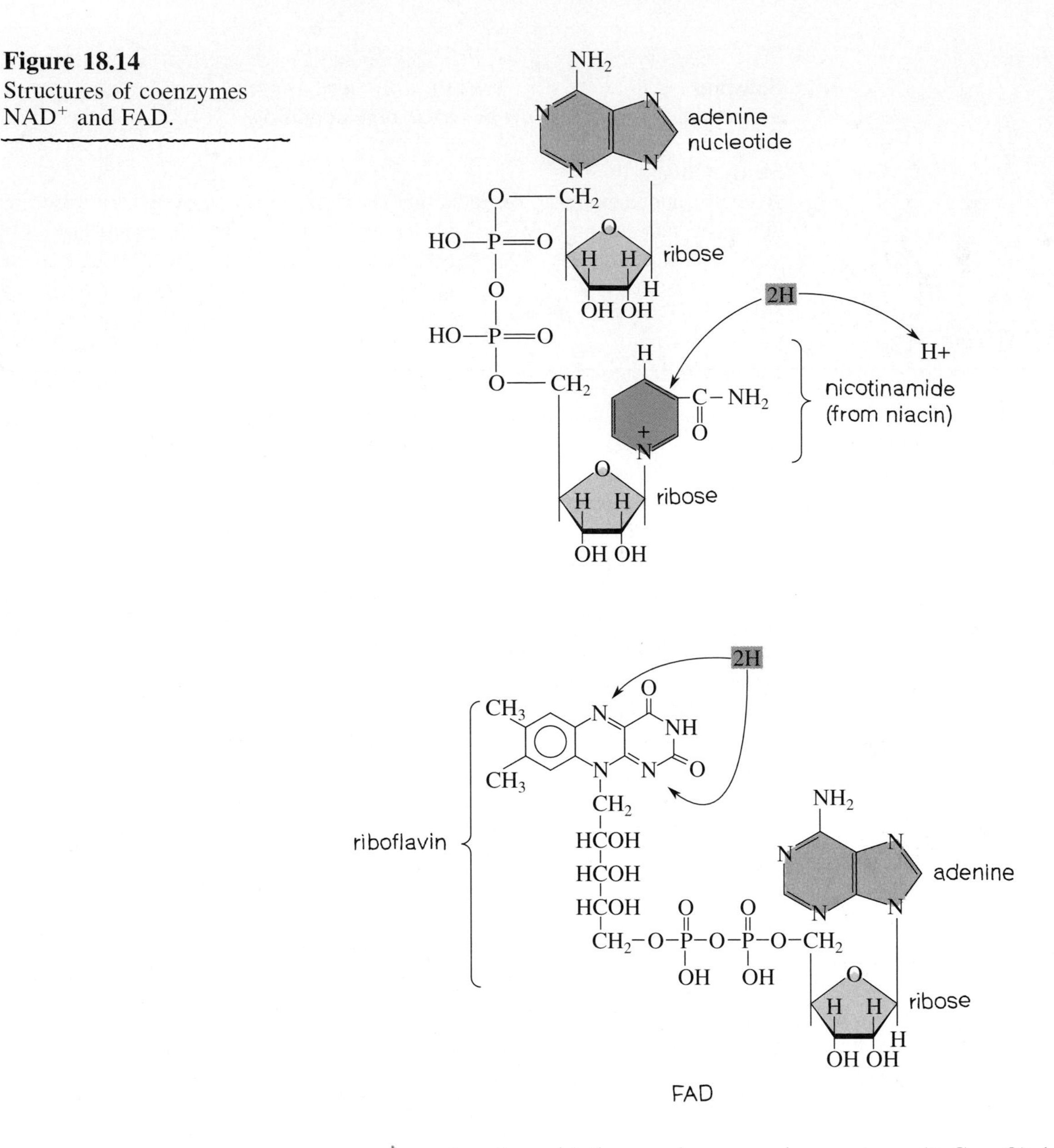

Figure 18.14
Structures of coenzymes NAD^+ and FAD.

coenzyme when the oxidation produces a carbon–oxygen (>C=O) double bond. FAD is the hydrogen acceptor when an oxidation produces a carbon–carbon (>C=C<) double bond, as in the case of succinate in the citric acid cycle.

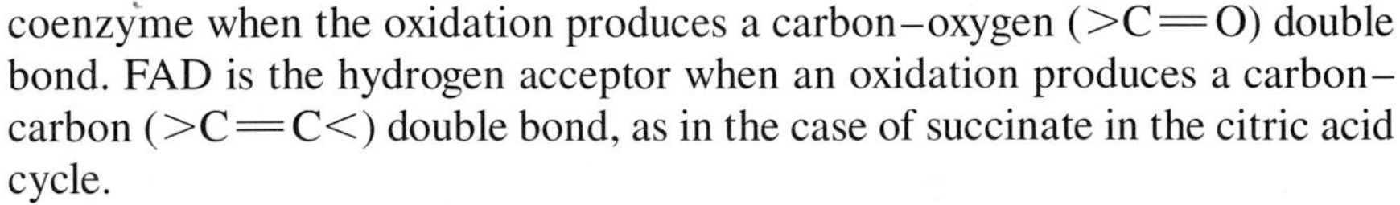

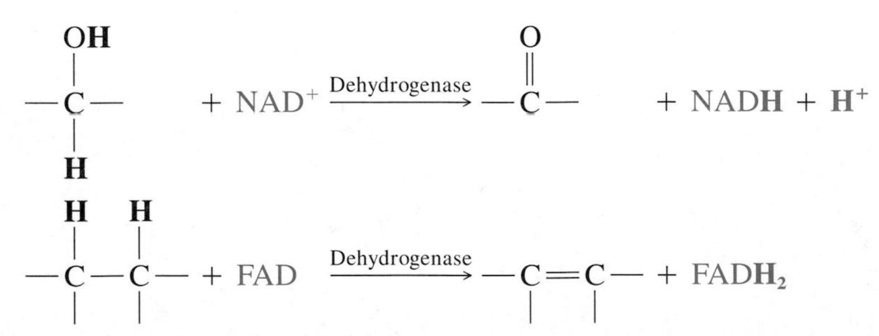

The chemical energy of the NADH and $FADH_2$ produced in the citric acid cycle is linked with a pathway that produces energy-rich ATP. We will see how this is done when we discuss the electron transport chain later in this chapter.

Summary of the Reactions of the Citric Acid Cycle

As we discuss the steps of the citric acid cycle, refer to the diagram in Figure 18.15 for the structures of the compounds that undergo reaction. In the citric acid cycle, the compounds are shown as anions, the prevalent form of these di- and tricarboxylic acids at physiological pH (7.4). For example, citrate ion is the anion for citiric acid at this pH.

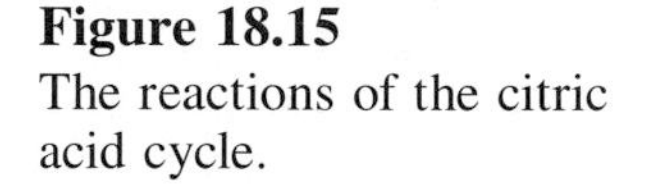

Figure 18.15
The reactions of the citric acid cycle.

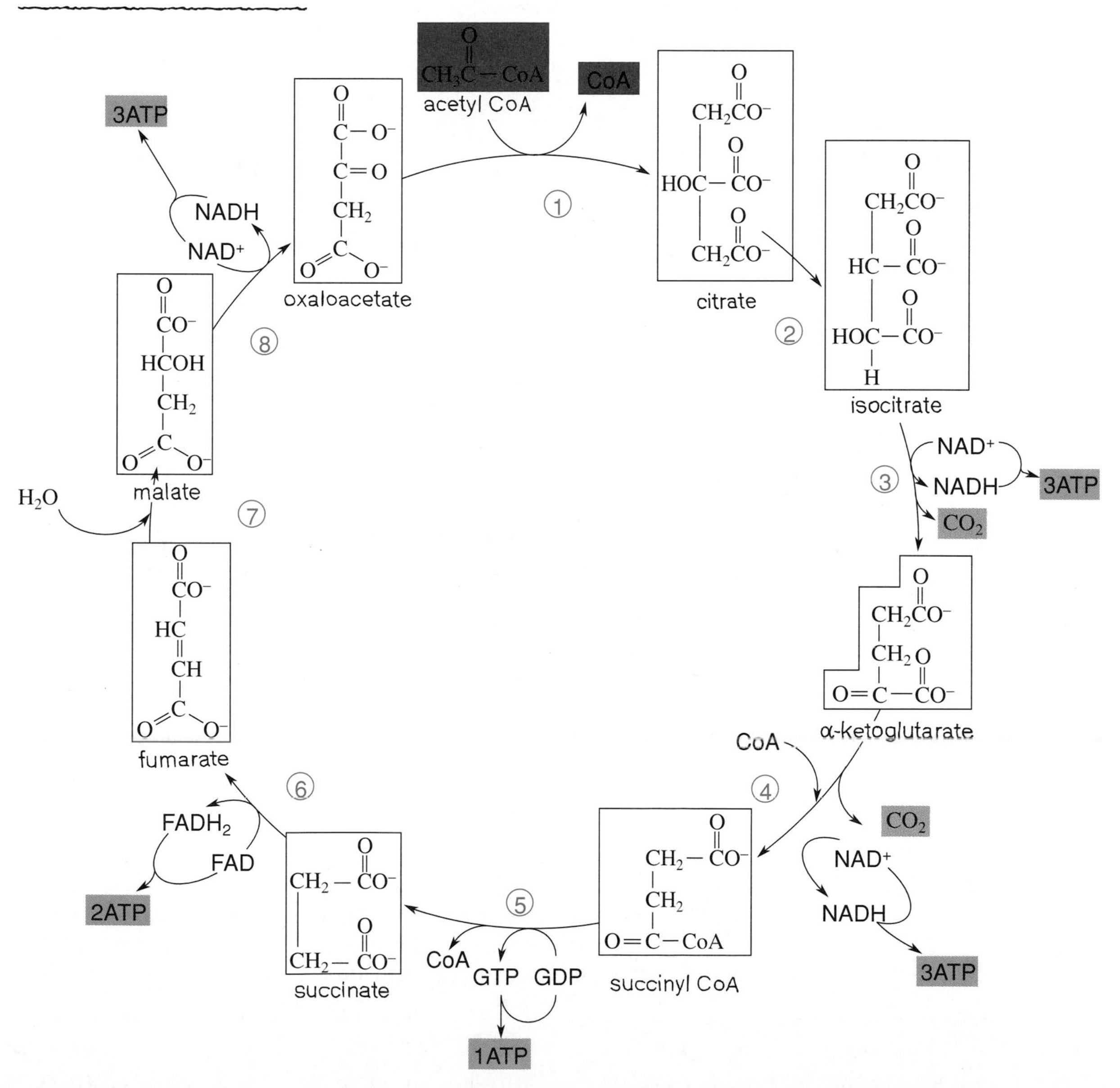

***Condensation Forms Citrate* (*Step 1*).** In this step, the two-carbon acetyl CoA enters the citric acid cycle by combining with a four-carbon molecule, oxaloacetate. Catalyzed by *citrate synthetase,* the condensation yields a six-carbon compound, citrate.

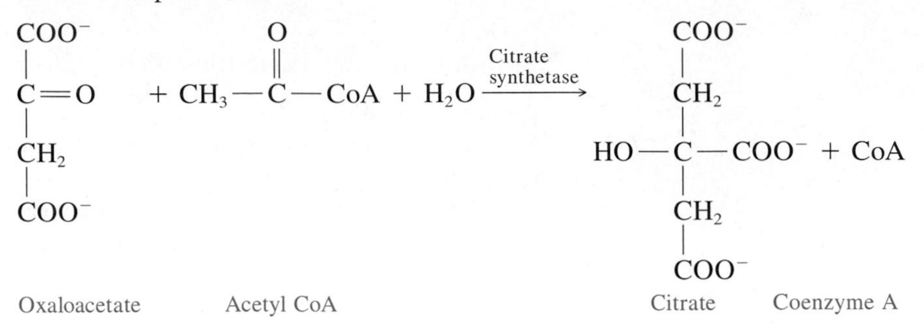

***Isomerization to Isocitrate* (*Step 2*).** The enzyme aconitase catalyzes the isomerization of citrate to isocitrate. The tertiary alcohol group (—OH) in citrate is converted to a secondary alcohol group in isocitrate that can now be oxidized.

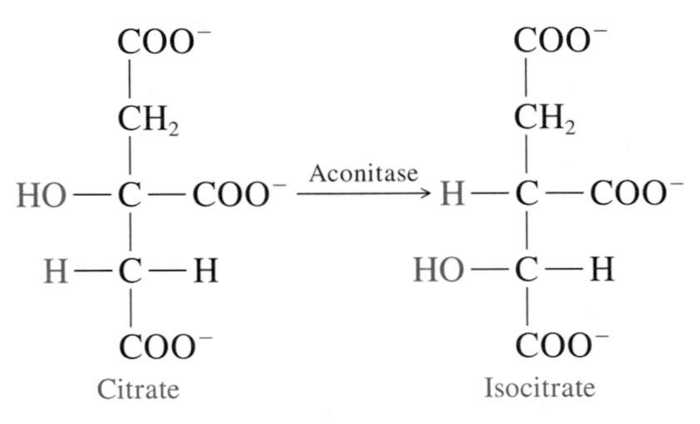

***Oxidative Decarboxylation* (*Step 3*).** At this point in the citric acid cycle, *isocitrate dehydrogenase complex* catalyzes several reactions collectively called oxidative decarboxylation. The hydroxyl (—OH) group of isocitrate is oxidized to a ketone and the hydrogen used to reduce NAD^+ to NADH. The carbon chain is also shortened by the loss of CO_2 to yield the five-carbon compound, α-ketoglutarate.

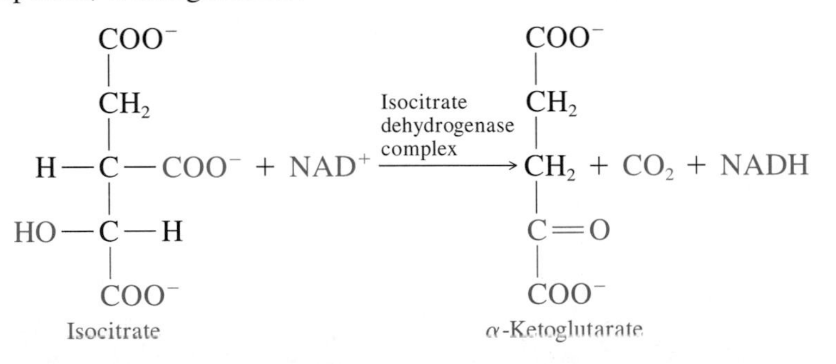

***Second Oxidative Decarboxylation* (*Step 4*).** Another oxidative decarboxylation is catalyzed by the enzymes and cofactors of the *α-ketoglutarate dehydrogenase complex.* The loss of another carbon as CO_2 yields a four-carbon chain. A reaction with CoA produces a high-energy bond in succinyl CoA and reduces NAD^+ to NADH.

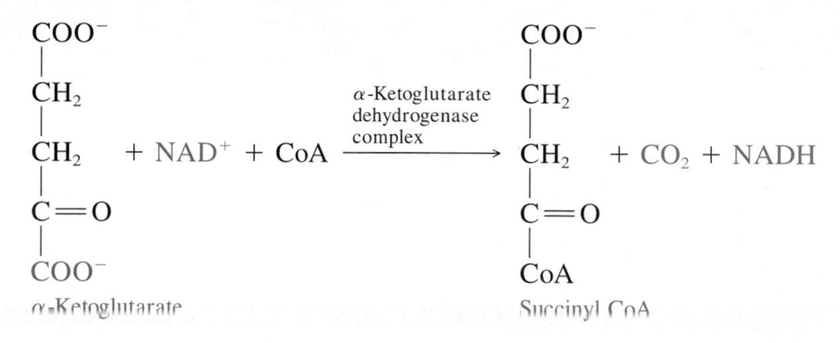

***Hydrolysis* (*Step 5*).** When the high-energy bond in succinyl CoA is hydrolyzed by *succinyl CoA synthetase,* sufficient energy is released to add inorganic phosphate to guanosine diphosphate (GDP). The product, guanosine triphosphate (GTP), is a high-energy compound similar to ATP.

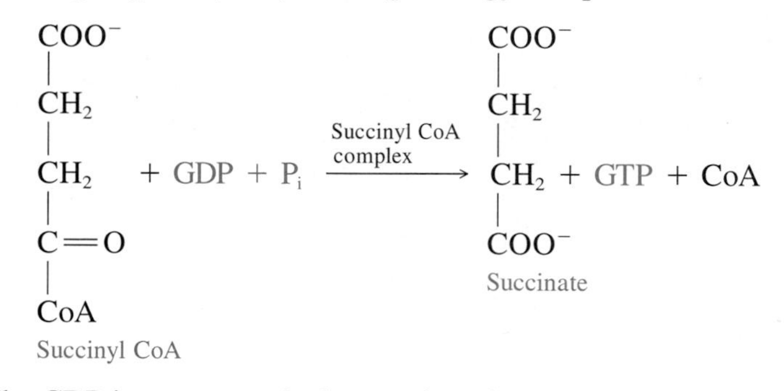

The GDP is regenerated when *nucleotide phosphokinase* transfers the phosphate group from GTP to ADP to yield ATP. This is the only place in the citric acid cycle where ATP is formed by a direct substrate phosphorylation.

$$\text{GTP} + \text{ADP} \underset{}{\overset{\text{Nucleotide phosphokinase}}{\rightleftharpoons}} \text{GDP} + \text{ATP}$$

***Third Oxidation* (*Step 6*).** In this step, succinate is oxidized by *succinate dehydrogenase* to yield fumarate. The hydrogen removed to form a carbon–carbon double bond of fumarate reduces the coenzyme FAD to $FADH_2$. This is the only reaction in the citric acid cycle that produces $FADH_2$ instead of NADH.

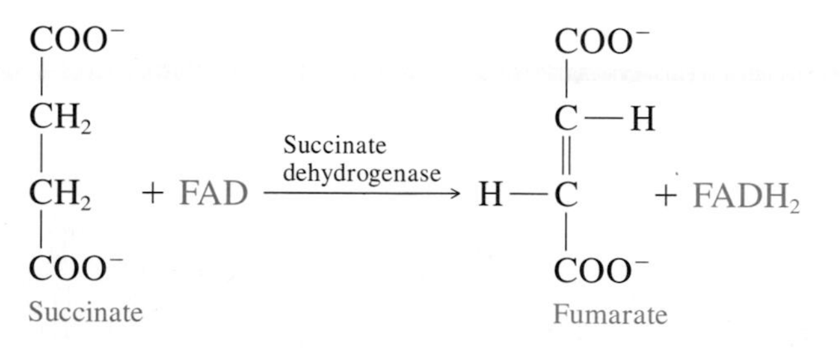

Hydration (Step 7). In a hydration step catalyzed by *fumarase,* water is added to the double bond of fumarate to yield malate.

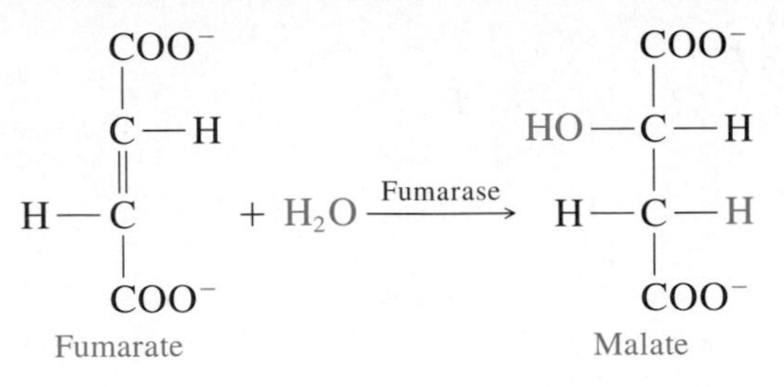

Fourth Oxidation (Step 8). In the last step of the citric acid cycle, *malate dehydrogenase* catalyzes the oxidation of the hydroxyl (—OH) group in malate to a ketone group. Oxaloacetate is formed as NAD^+ is reduced to NADH.

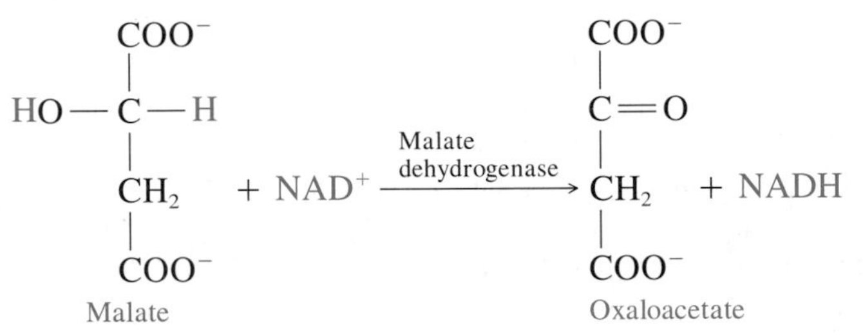

These eight reactions make up one full turn of the citric acid cycle. The cycle starts all over again when oxaloacetate picks up another acetyl CoA. In one turn of the citric acid cycle, one acetyl CoA produced two molecules of CO_2, three molecules of NADH, one molecule of $FADH_2$, and one molecule of GTP, which was converted to ATP. We can summarize the citric acid cycle by writing the following equation:

$$\underset{\text{Acetyl CoA}}{CH_3-\overset{\overset{\Large O}{\|}}{C}-CoA} + 3\,NAD^+ + FAD + ADP \xrightarrow{\text{Citric acid cycle}} 2CO_2 + 3\,NADH + FADH_2 + ATP$$

Sample Problem 18.5
Citric Acid Cycle

When one acetyl CoA completes the citric acid cycle, how many of each of the following are produced?

a. NADH **b.** ketone group **c.** CO_2

Solution

a. One turn of the citric acid cycle produces three molecules of NADH.
b. Two ketone groups form when the secondary alcohol groups in isocitrate and malate are oxidized by NAD^+.
c. Two molecules of CO_2 are produced by the decarboxylation of isocitric acid and α-ketoglutarate.

Study Check
What substance is a substrate in the first step of the citric acid cycle and a product in the last step?

18.6 The Electron Transport Chain

Learning Goal

Describe the synthesis of ATP from NADH and $FADH_2$ by the electron transport chain.

As we study the catabolism of glucose and fatty acids, we find that several oxidation reactions produce NADH or $FADH_2$. The electrons from these reduced coenzymes are transferred to the **electron transport chain,** which is a series of electron carriers composed of flavins, iron-sulfur clusters, and heme groups. (See Figure 18.16.) Within the electron transport chain, each type of carrier is reduced when it accepts the electrons, and oxidized when it passes the electrons to the next carrier. Eventually, the electrons combine with molecular oxygen (O_2), the final acceptor in the chain. As the electrons pass from one electron carrier to next, the energy level drops. When the energy drop is great enough, the energy released is used to pump protons (H^+) across the inner membrane and to synthesize ATP. During the process of electron transfer, one mole of NADH is oxidized, H_2O is formed and energy is released, 53 kcal for one mole NADH and 43 kcal for one mole $FADH_2$. The energy made available from the oxidation of NADH is used to synthesize 3 ATP, and the energy from $FADH_2$ is used to synthesize 2 ATP. The overall oxidation of NADH + H^+ and $FADH_2$ is

$$NADH + H^+ + \tfrac{1}{2}O_2 \longrightarrow NAD^+ + H_2O + \text{energy for 3 ATP}$$

$$FADH_2 + \tfrac{1}{2}O_2 \longrightarrow FAD + H_2O + \text{energy for 2 ATP}$$

Components of the Electron Transport Chain

Most of the electron carriers are found in protein clusters that are located along the inner mitochondrial membrane. Two of the electron carriers, **coenzyme Q** and cytochrome *c* move freely between the clusters to carry electrons from one cluster to the next.

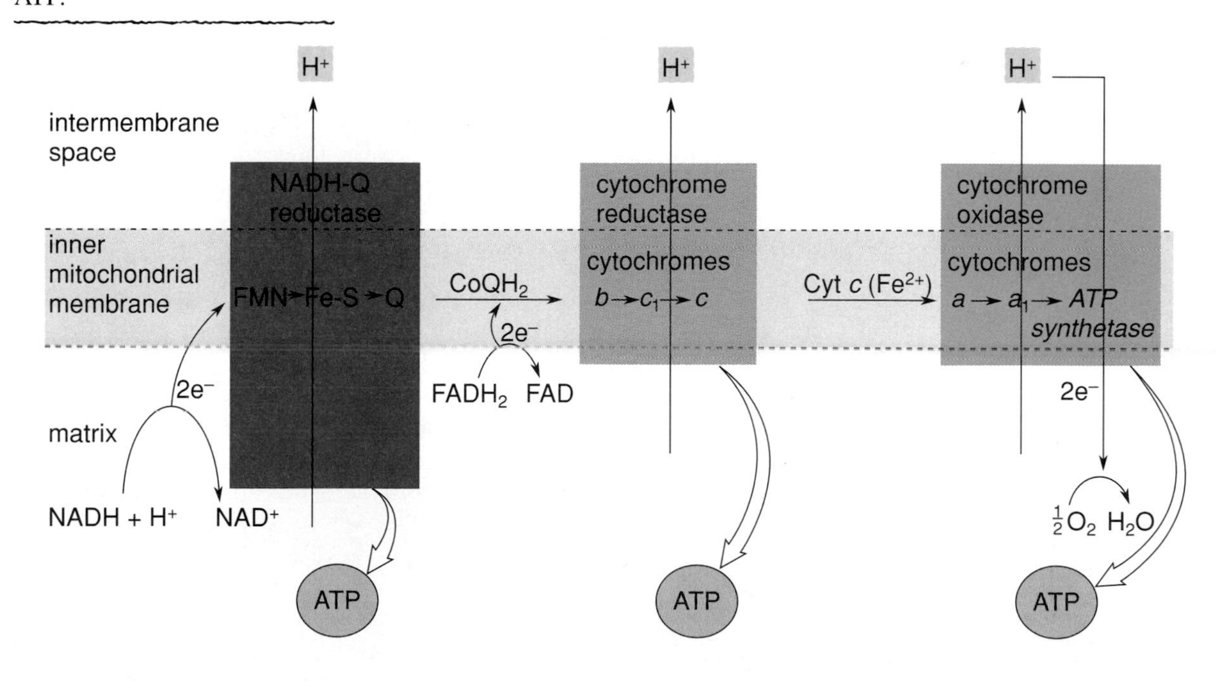

Figure 18.16
The electron transport chain. As electrons flow from NADH, and $FADH_2$ to O_2, protons pumped across the inner membrane at three different sites establish a proton gradient that supplies energy for the synthesis of ATP.

NADH-Q Reductase

At the first stop along the electron chain known as the NADH-Q reductase, electrons from NADH are accepted by a *flavin mononucleotide* (FMN).

$$NADH + H^+ + FMN \longrightarrow FMNH_2 + NAD^+$$

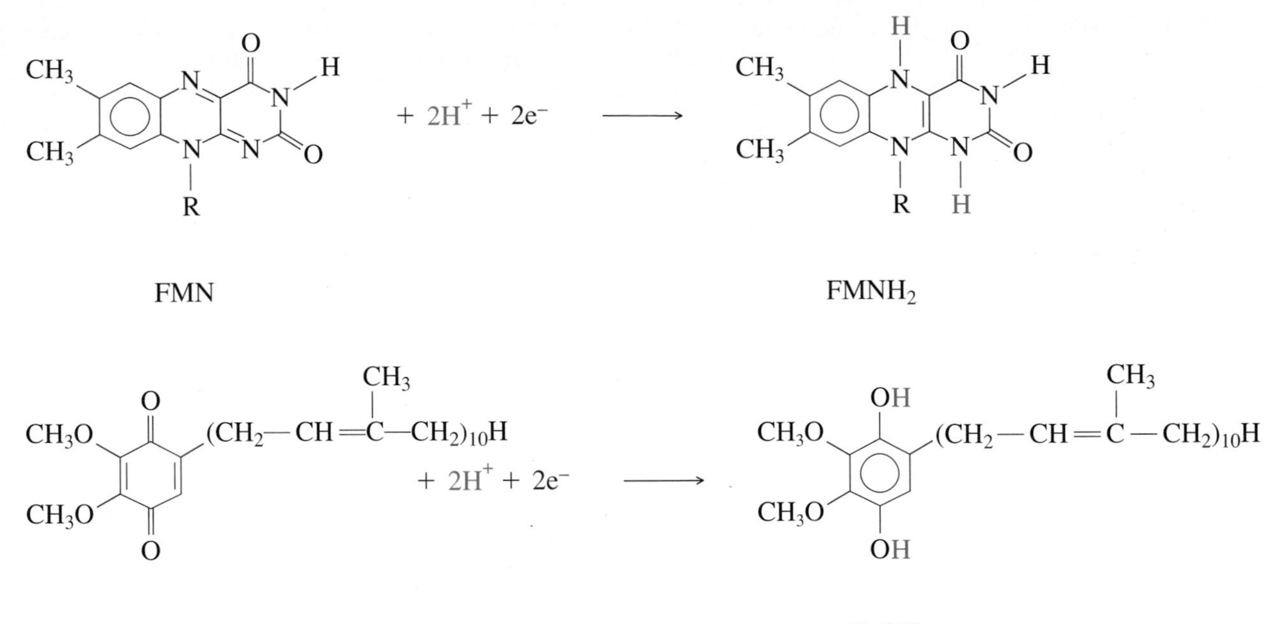

In a second step within this cluster the electrons are transferred to a series of Fe • S proteins that reduce and oxidize between Fe^{2+} and Fe^{3+}. In the final step at this cluster, the electrons are transferred to coenzyme Q (**CoQ** or ubiquinone). Coenzyme Q is a mobile carrier that shuttles the electrons to the next cluster, cytochrome reductase.

Some of the energy released by the flow of electrons through the NADH-Q reductase is used to drive protons across the inner membrane.

Oxidation of $FADH_2$ by Coenzyme Q

Coenzyme Q is also the electron acceptor for the electrons from $FADH_2$. $FADH_2$ is the reduced coenzyme produced when succinate is oxidized to fumarate in the citric acid cycle. The $FADH_2$ enters the electron transport chain at a later point by transferring its electrons directly to CoQ, and not to FMN as in the case of $NADH_2$. The energy change for the transfer of electrons is not sufficient to drive protons across the inner mitochondria membrane and $FADH_2$ does not contribute to ATP synthesis at this point in the electron transport chain.

$$FADH_2 + CoQ \longrightarrow FAD + CoQH_2$$

Cytochrome Reductase

The electrons collected from NADH and $FADH_2$ are now carried by reduced coenzyme Q ($CoQH_2$) to the next cluster known as cytochrome reductase. In this cluster, electrons are passed to cytochromes *b* and *c*—much like buckets of water in a fire brigade. As each cytochrome accepts an electron, the iron in the center of the cytochrome is reduced to Fe^{2+}. When that electron is transferred to the next cytochrome, the iron oxidizes again to Fe^{3+}. The electrons in the cytochrome reductase are finally transferred to cytochrome *c*, another mobile carrier that shuttles the electrons to the next cluster.

$$\text{cytochrome-Fe}^{2+}\text{ (reduced)} \longrightarrow \text{cytochrome-Fe}^{3+}\text{ (oxidized)} + e^-$$

$$CoQH_2 + 2\text{cyt } b(Fe^{3+}) \longrightarrow CoQ + 2\text{ cyt } b\ (Fe^{2+}) + 2H^+$$

Cytochrome Oxidase

The last cluster is known as cytochrome oxidase because its carriers accept electrons from the mobile carrier cytochrome *c* and transfer them to cytochromes a and a_3, and finally to oxygen (O_2) to form H_2O. This is the final product formed by the electrons obtained from the oxidation pathways of the molecules we digested from our food.

$$4\text{ cyt c }(Fe^{2+}) + 4\ H^+ + O_2 \longrightarrow 4\text{ cyt c }(Fe^{3+}) + 2\ H_2O$$

As the electrons are transferred to O_2, the energy released drives more protons across the inner mitochondrial membrane.

Oxidative Phosphorylation

In this model of electron transport, electrons are collected by coenzyme Q from NADH and $FADH_2$ and carried from NADH-Q reductase or from $FADH_2$ to cytochrome reductase. Then, from cytochrome reductase, electrons are shuttled by cytochrome *c* to cytochrome oxidase and O_2. During the process, energy that drives protons across the inner membrane is released.

In 1961, Peter Mitchell (Nobel Prize 1978) proposed that the accumulation of protons on the inner membrane stores energy used in the synthesis of ATP. In **Mitchell's chemiosmotic theory,** the reactions in each of the clusters of the electron transport chain drive protons from the matrix into the intermembrane space. As the protons accumulate, a high-energy proton gradient results. The change in pH as the H^+ leaves creates an electrical gradient across the inner membrane. The protons then flow down this electrical gradient through a protein tunnel called ATP synthetase. During the process, energy is released and coupled with the synthesis of ATP from ADP and P_i, a process called **oxidative phosphorylation.** (See Figure 18.17.)

$$ADP + P_i \xrightarrow{\text{ATP synthetase}} ATP$$

The overall process of oxidative phosphorylation links the oxidation energy of glucose and other substrates such as fatty acids to the formation of energy-rich ATP. Because there is sufficient energy released at each cluster, the oxidation of NADH by the elecron transport chain produces three ATP molecules.

$$NADH + H^+ + 3\ ADP + 3\ P_i + \tfrac{1}{2}O_2 \longrightarrow NAD^+ + 3\ ATP + H_2O$$

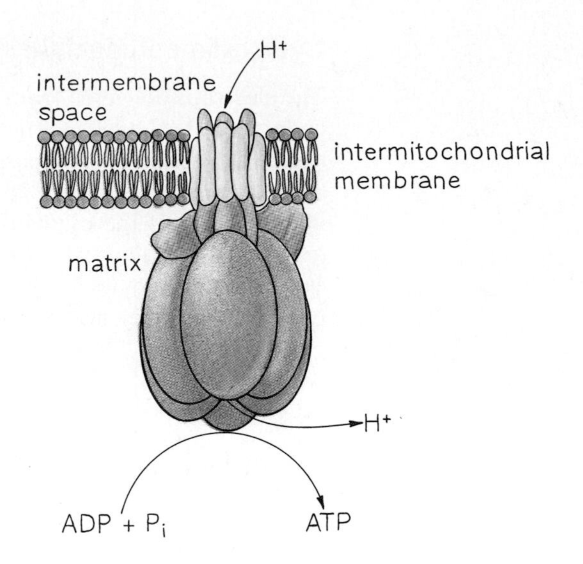

Figure 18.17
The energy of the H^+ gradient is used by ATP synthetase to produce ATP from ADP and P_i.

Because $FADH_2$ enters the electron transport chain after the NADH-Q reductase cluster, the electrons from $FADH_2$ produce energy at only the second and third clusters. Thus $FADH_2$ provides sufficient energy to synthesize two ATP molecules.

$$FADH_2 + 2\ ADP + 2\ P_i + \tfrac{1}{2}O_2 \longrightarrow FAD + 2\ ATP + H_2O$$

Sample Problem 18.6
Electron Transport System

Why does the oxidation of NADH by the electron transport chain provide energy for the formation of three ATP whereas $FADH_2$ produces two ATP?

Solution
Electrons from the oxidation of NADH enter the electron chain at an earlier site than the $FADH_2$, providing energy to pump three pairs of protons to the intermembrane. However, $FADH_2$ transfers electrons to Q that go through two sites that pump protons and therefore provide energy for two ATP.

Study Check
Where is water (H_2O) formed in the electron chain?

18.7 ATP Energy from Glucose

Learning Goal
Account for the ATP produced by the complete oxidation of glucose.

We can now determine the total ATP production for one glucose molecule when it undergoes complete oxidation to carbon dioxide and water using the pathways of glycolysis, formation of acetyl CoA, and the citric acid cycle. (See Figure 18.18.)

ATP from Glycolysis

In glycolysis, a glucose molecule produces two molecules of pyruvate and two ATPs from direct substrate phosphorylation. There are also two molecules of

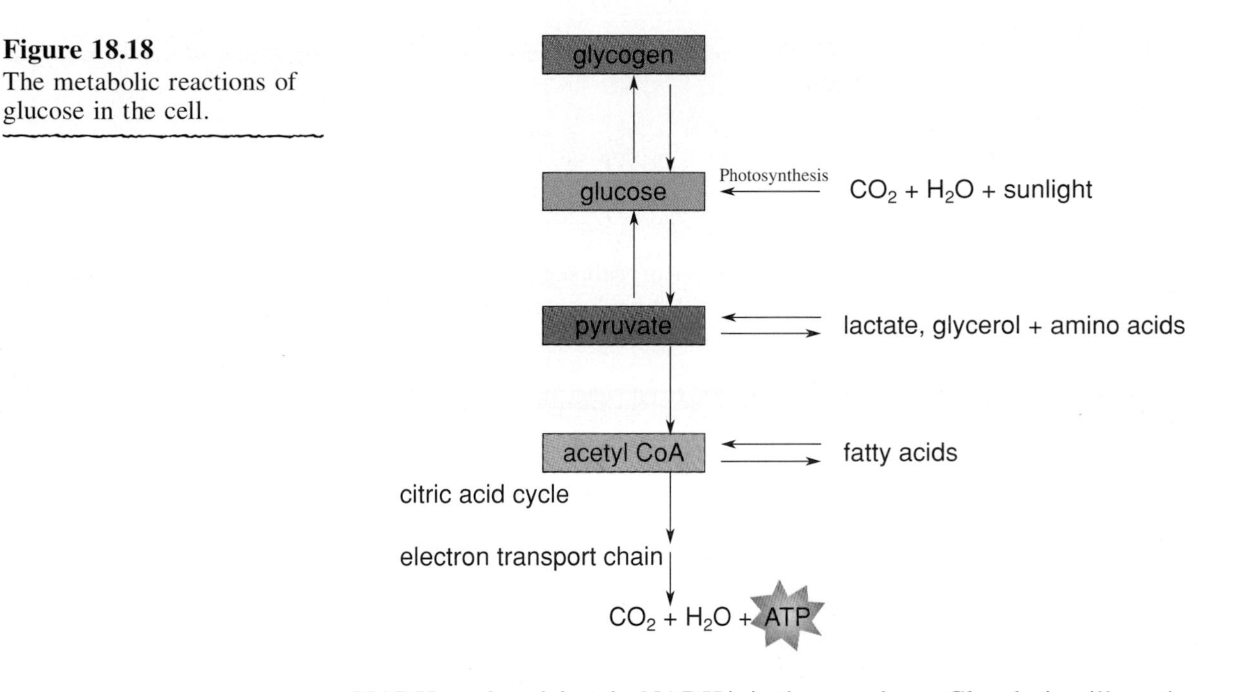

Figure 18.18
The metabolic reactions of glucose in the cell.

NADH produced, but the NADH is in the cytoplasm. Glycolysis will continue only if NAD^+ is regenerated. However, neither NADH or NAD^+ can pass through the inner membrane of the mitochondria. Instead, a shuttle system transports the electrons from the NADH in the cytoplasm to the electron transport chain but at a cost of one ATP. Thus, an NADH from glycolysis provides a net yield of two ATP (not three). In glycolysis, the oxidation of one glucose forms two ATPs from direct substrate phosphorylation and two molecules of NADH that yield four ATP for a total of six ATP.

Glycolysis (aerobic)

$$\text{Glucose} + 6\text{ ADP} + 6\text{ P}_i \longrightarrow 2\text{ pyruvate} + 6\text{ ATP} + 2H_2O$$

ATP from Oxidation of Pyruvate

Almost all the ATP in the mitochondria is formed under aerobic conditions by the further oxidation of the two pyruvates. After pyruvate crosses the inner membrane into the mitochondrion, it is converted to acetyl CoA and CO_2, an oxidation that produces one molecule of NADH. For every NADH oxidized by the electron transport chain, three ATP are produced.

Formation of Acetyl CoA from Pyruvate

$$\text{Pyruvate} + 3\text{ ADP} + 3\text{ P}_i \longrightarrow \text{acetyl CoA} + CO_2 + 3\text{ ATP}$$

When acetyl CoA enters the citric acid cycle, it goes through a series of oxidation reactions to yield two CO_2, three NADH, one $FADH_2$, and one GTP, which is converted to ATP. In the electron transport chain, the electrons from the oxidation of NADH or $FADH_2$ produce a proton gradient that supplies energy to synthesize three ATPs for each NADH oxidized; two ATPs for each $FADH_2$. In one turn, the citric acid cycle provides a total of 12 ATP.

ATP Produced from the Electron Carriers in One Turn of the Citric Acid Cycle

$$
\begin{array}{r}
3\ \text{NADH} \times 3\ \text{ATP} = \ \ 9\ \text{ATP} \\
1\ \text{FADH}_2 \times 2\ \text{ATP} = \ \ 2\ \text{ATP} \\
1\ \text{GTP} \times 1\ \text{ATP} = \ \ 1\ \text{ATP} \\
\hline
\text{Total ATP} = 12\ \text{ATP}
\end{array}
$$

We can write an overall equation for the amount of ATP produced by the oxidation of acetyl CoA in the citric acid cycle.

Citric Acid Cycle

$$\text{Acetyl CoA} + 12\ \text{ADP} + 12\ \text{P}_i \longrightarrow 2\text{CO}_2 + 12\ \text{ATP} + 2\text{H}_2\text{O}$$

Figure 18.19
The complete oxidation of glucose yields 36 ATP.

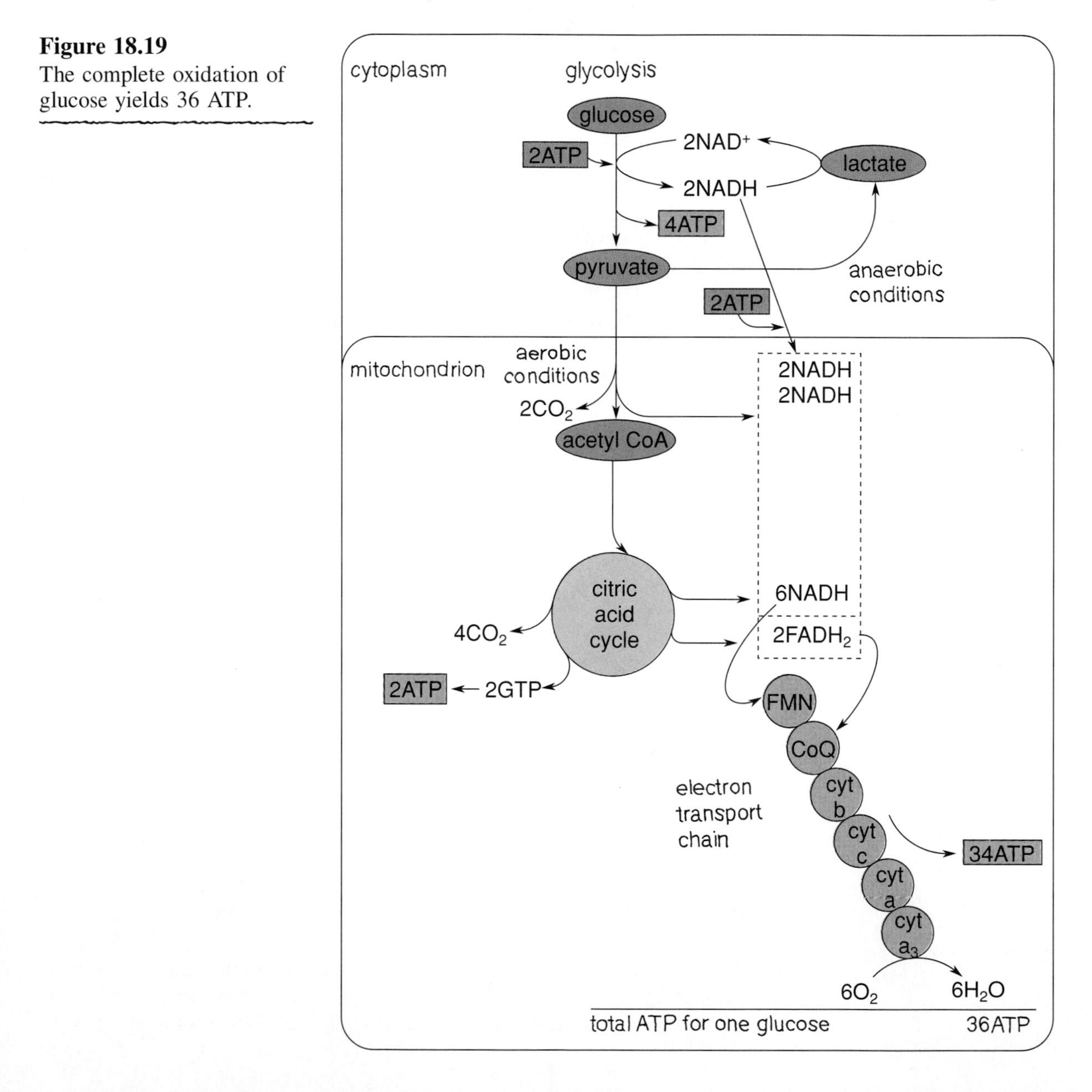

ATP from the Complete Oxidation of Glucose

When glucose is completely oxidized to CO_2 and H_2O, a total of 36 ATP molecules are synthesized. Recall that in glycolysis, glucose yields two pyruvate molecules that are oxidized to two acetyl CoA molecules and two CO_2. When oxidized further, the two acetyl CoA molecules will complete two turns of the citric acid cycle. The total amount of ATP from direct substrate phosphorylation and the oxidation of NADH or $FADH_2$ in the electron chain can be calculated as shown in Table 18.1. The oxidation of glucose is summarized in Figure 18.19.

Table 18.1 *ATP Produced by the Complete Combustion of Glucose*

Reaction Pathway	ATP for One Glucose
ATP from Glycolysis	
Activation of glucose	−2 ATP
Oxidation of glyceraldehyde 3-phosphate (2 NADH)	6 ATP
Transport of 2 NADH across membrane	−2 ATP
Direct ADP phosphorylation (two triose phosphate)	4 ATP
Summary: $\underset{\text{Glucose}}{C_6H_{12}O_6} \longrightarrow$ 2 pyruvate + $2H_2O$	6 ATP
ATP from Acetyl CoA	
2 pyruvate $\longrightarrow$ 2 acetyl CoA (2 NADH)	6 ATP
ATP from Citric Acid Cycle	
Oxidation of 2 isocitrate (2 NADH)	6 ATP
Oxidation of 2 α-ketoglutarate (2 NADH)	6 ATP
2 direct substrate phosphorylations (2 GTP)	2 ATP
Oxidation of 2 succinate (2 $FADH_2$)	4 ATP
Oxidation of 2 malate (2 NADH)	6 ATP
Summary: 2 Acetyl CoA $\longrightarrow 4CO_2 + 2H_2O$	24 ATP
Overall ATP Production for One Glucose	
$\underset{\text{Glucose}}{C_6H_{12}O_6} + 6O_2 + 36\text{ ADP} + 36\text{ P}_i \longrightarrow 6CO_2 + 6H_2O + 36\text{ ATP}$	

Sample Problem 18.7
ATP Production

Indicate the amount of ATP produced by each of the following oxidation reactions:

a. pyruvate to acetyl CoA **b.** glucose to acetyl CoA

Solution

a. The oxidation of pyruvate to acetyl CoA produces one NADH, which yields three ATP.

b. Six ATP are produced from the oxidation of glucose to two pyruvate molecules. Another six ATP result from the oxidation of two pyruvate molecules to two acetyl CoA molecules.

Study Check

Why does the NADH produced in the cytoplasm by glycolysis only yield two ATP?

Glycogen Stores Energy

When glucose is not immediately used by the cells for energy, it is stored as glycogen in the liver and muscles. When the levels of glucose in the brain or blood become low, the glycogen reserves are hydrolyzed and glucose is released into the blood. If glycogen stores are depleted, some glucose can be synthesized from noncarbohydrate sources. It is the balance of all these reactions that maintains the necessary blood glucose level available to our cells and provides the necessary amount of ATP for our energy needs.

18.8 Oxidation of Fatty Acids

Learning Goal

Describe the role of oxidation of fatty acids in ATP production.

We store most of our energy in the form of triglycerides (lipids) in adipose (fat) cells. (See Figure 18.20.) If our reserves of glucose and glycogen are depleted, we obtain energy by hydrolyzing the triglycerides to glycerol and free fatty acids. Glycerol, a triose, is converted to dihydroxyacetone phosphate, a compound that is oxidized by the glycolysis pathway. The fatty acids are oxidized by our tissues for energy. The oxidation of triglycerides, which produces as much as six times the ATP from glucose, is an important reserve source of ATP in the body.

Figure 18.20
Photomicrograph of an adipose cell filled with fat.

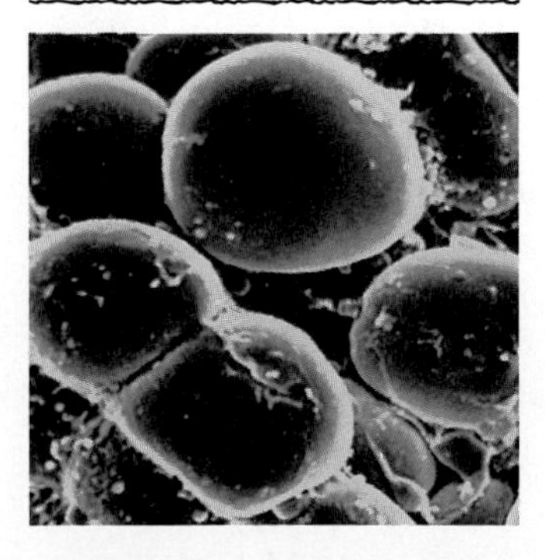

Fatty Acid Activation

In the cytoplasm, a fatty acid is joined to CoA using energy from the hydrolysis of ATP to adenosine monophosphate (AMP) and inorganic phosphate (2 P_i), an energy cost that is equivalent to the hydrolysis of two ATPs. The activated fatty acyl CoA is transported across the inner membrane into the mitochondria where the machinery for its oxidation is located.

$$\underset{\text{Fatty acid}}{R-CH_2-CH_2-\overset{\overset{\displaystyle O}{\|}}{C}-OH} + CoA + ATP \xrightarrow[\text{activation}]{\text{Fatty acid}} \underset{\text{Fatty acyl CoA}}{R-CH_2-CH_2-\overset{\overset{\displaystyle O}{\|}}{C}-CoA} + AMP + 2\,P_i$$

Beta (β)-Oxidation of Fatty Acids

Beta (β)-Oxidation involves a cycle of four reactions that removes two-carbon segments from a fatty acid in sequential order. In each cycle, the β carbon in the fatty acid chain is oxidized, yielding an acetyl CoA unit and a shortened fatty acid chain.

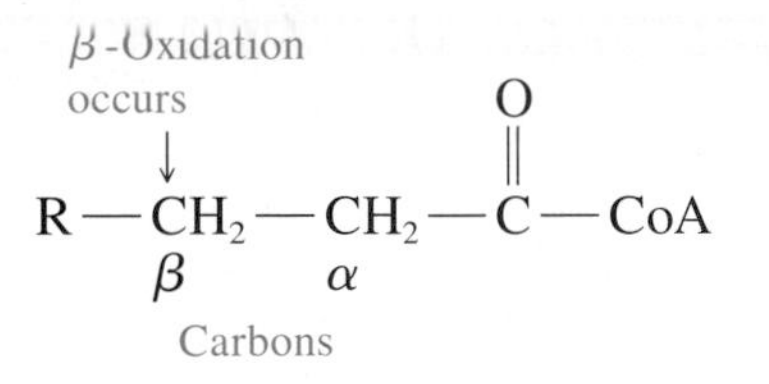

The cycle repeats until the original fatty acid is entirely degraded into two-carbon acetyl CoA units. Each acetyl CoA unit is then processed through the citric acid cycle in the same way as the acetyl CoA units derived from glucose. The number of carbon atoms in a fatty acid determines the number of times the cycle is repeated and the number of acetyl units produced. We will now take a look at the individual steps of β-oxidation as shown in Figure 18.21.

Summary of Reactions in the Oxidation of Fatty Acids

***First Oxidation* (*Step 1*).** In an oxidation reaction, a double bond forms when a *dehydrogenase* removes two hydrogen atoms from the α and β carbons of the fatty acid. FAD picks up the hydrogen to yield $FADH_2$.

***Hydration* (*Step 2*).** Water adds across the double bond placing a hydroxyl group (—OH) on the β carbon.

***Second Oxidation* (*Step 3*).** A *dehydrogenase* oxidizes the hydroxyl group on the β carbon to a ketone and NAD^+ reduces to NADH.

***Release of Acetyl CoA* (*Step 4*).** In the final step, a unit of acetyl CoA separates from the fatty acid chain, and another molecule of CoA bonds to the β carbon. The remaining fatty acetyl CoA is two carbons shorter than the fatty acid chain that started the cycle. As long as the fatty acetyl CoA contains four or more carbons, it recycles through β-oxidation again.

β-Oxidation Produces Energy for ATP

To determine the number of ATP produced from β-oxidation, we count the number of two-carbon units in the carbon chain. For example, a fatty acid with eight carbon (caprylic acid) atoms can be divided into four acetyl units. A fatty acid repeats the cycle one time less than its number of acetyl CoA units. Thus, the carbon chain from caprylic acid cycles through β-oxidation

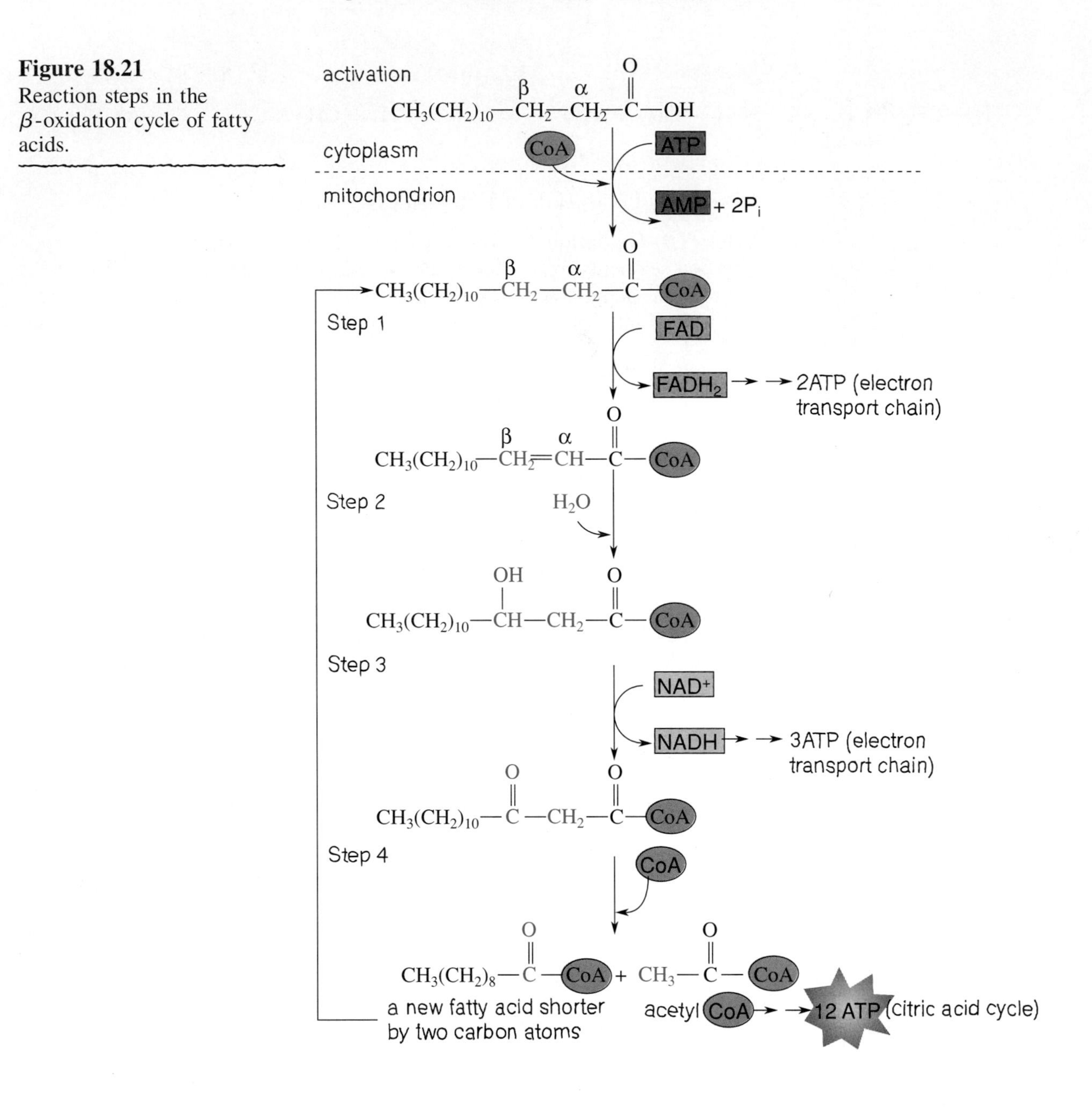

Figure 18.21
Reaction steps in the β-oxidation cycle of fatty acids.

three times. Each time, one NADH, one $FADH_2$, and one acetyl CoA are produced. The NADH and $FADH_2$ from β-oxidation are oxidized further in the electron transport chain to produce energy for ATP synthesis: NADH yields three ATP, and $FADH_2$ yields two ATP.

The acetyl CoA is oxidized by the citric acid cycle producing 12 ATP. A large amount of ATP is produced although two ATP must be subtracted from the total ATP for the initial activation of the fatty acid. Table 18.2 summarizes the production of ATP from caprylic acid, an eight-carbon atom fatty acid.

Table 18.2 ***Total ATP Produced by Caprylic Acid (Eight Carbon Atoms)***

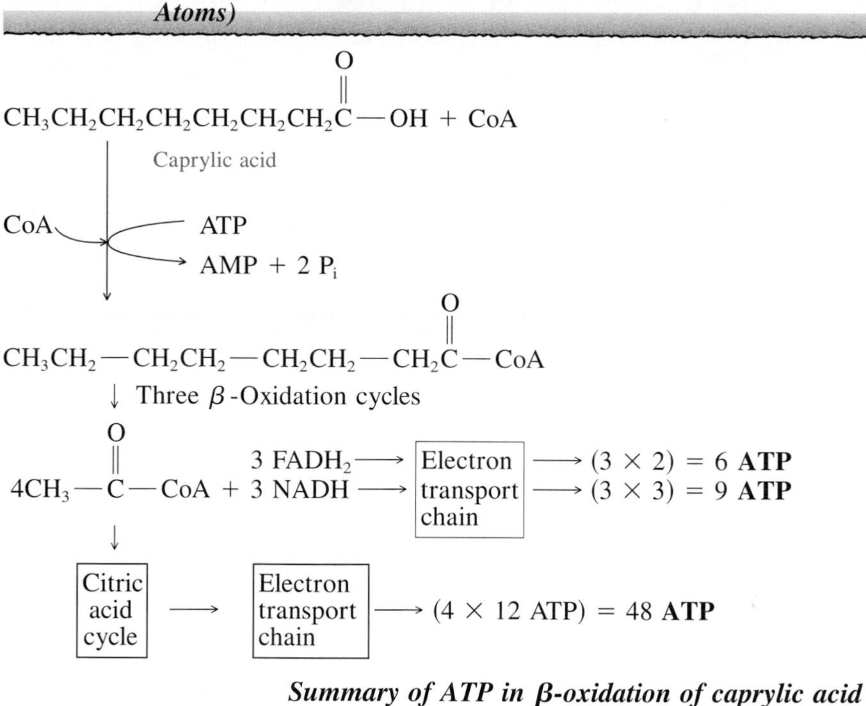

Summary of ATP in β-oxidation of caprylic acid

Total ATP:	63 **ATP**
Activation:	−2 **ATP**
	61 **ATP**

Sample Problem 18.8
ATP Production from β-Oxidation

How much ATP will be produced from the β-oxidation of myristic acid, a 14-carbon fatty acid?

Solution

A 14-carbon fatty acid will produce seven acetyl CoA units and go through six β-oxidation cycles. Each acetyl CoA will produce 12 ATP by the citric acid cycle for a total of 84 ATP. Each β-oxidation cycle will produce one NADH and one $FADH_2$ for a total of six NADH and six $FADH_2$. In the electron transport chain, each NADH will produce energy for three ATP; each $FADH_2$ for two ATP.

ATP Production for Myristic Acid		
Activation		−2ATP
7 acetyl CoA × 12 ATP	(Citric acid cycle)	84 ATP
6 $FADH_2$ × 2 ATP	(Electron transport chain)	12 ATP
6 NADH × 3 ATP	(Electron transport chain)	18 ATP
	Total ATP by β-oxidation of myristic acid	112 ATP

Study Check

How much ATP is produced if stearic acid (an 18-carbon acid) goes through only two turns of the β-oxidation cycle?

HEALTH NOTE *Ketone Bodies*

When carbohydrates are not available to meet energy needs, the body breaks down large quantities of body fat. The oxidation of fatty acids produces so many units of acetyl CoA that the citric acid cycle is overwhelmed. At the same time, less oxaloacetate is available for the citric acid cycle because it is used to synthesize glucose. Without oxaloacetate, acetyl CoA from β-oxidation cannot be used completely by the citric acid cycle. Instead, acetyl CoA enters a pathway called *ketogenesis* in which four-carbon *ketone bodies,* acetoacetate and β-hydroxybutyrate, are produced from two acetyl CoA units. Acetoacetate is then converted to acetone, another ketone body.

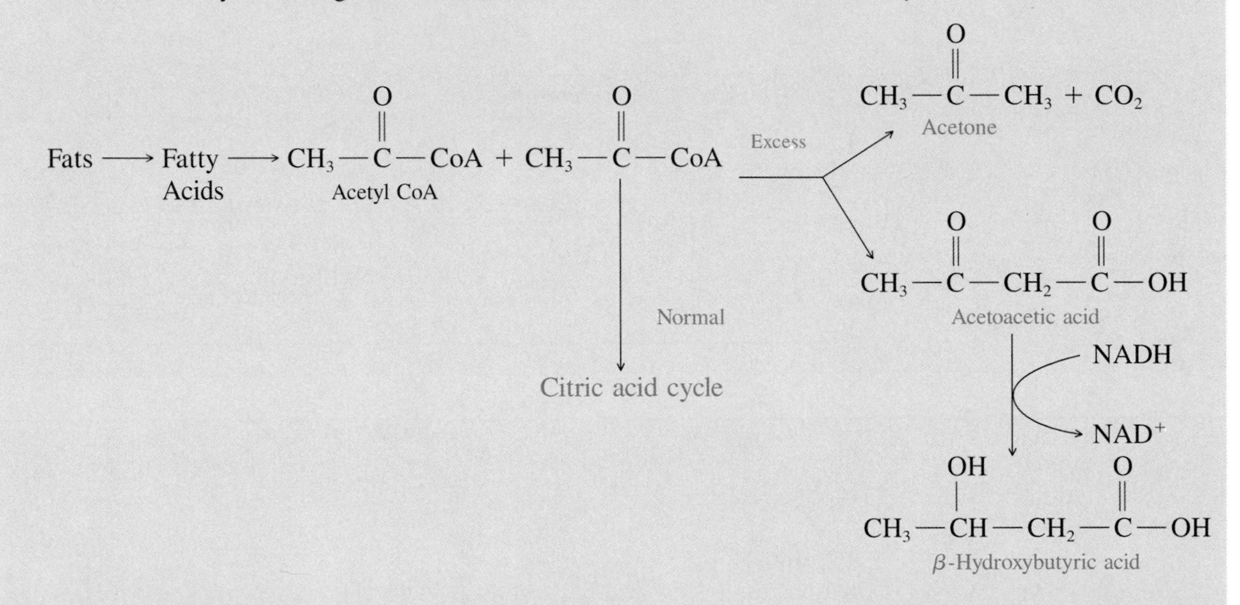

In the body, a small amount of energy is derived from ketone bodies by converting them back to acetyl CoA. When depleted of glucose, the brain uses ketone bodies for energy. However, in **ketosis,** the level of ketone bodies becomes so high that they are not completely metabolized by the body. This condition occurs in severe diabetes, diets that are high in fat and low in carbohydrate, and starvation. The fruity odor of acetone can often be detected on the breath of a person with ketosis.

Acidosis is also associated with ketosis because the acidity of the ketone bodies lowers the pH of the blood. A severe drop in blood pH can interfere with the ability of the blood to carry oxygen. As a result, breathing becomes strained. As the body tries to excrete ketone bodies in the urine, large amounts of sodium are removed from body fluids. As sodium supply is depleted, the output of urine becomes high accompanied with a strong sensation of thirst.

Fatty Acid Synthesis

When the body has met all its energy needs and glycogen stores are full, excess acetyl CoA is converted to fatty acids. In a series of reactions known as *lipogenesis,* two-carbon units obtained from acetyl CoA are linked together to form a fatty acid. Several of the reactions are essentially the reverse of the reactions described in β-oxidation. However, the enzymes for fatty acid synthesis are located in the cytoplasm, and not in the mitochondria where β-oxidation occurs. Also, the reduced coenzyme in fatty acid synthesis is NADPH, not NADH or $FADH_2$, the coenzymes used for degradative reactions.

Reactions in Fatty Acid Synthesis

Carboxylation of Acetyl CoA. Energy from the hydrolysis of an ATP is used to add CO_2 to acetyl CoA yielding a three-carbon compound called malonyl CoA.

$$\underset{\text{Acetyl CoA}}{CH_3-\overset{\overset{\displaystyle O}{\|}}{C}-CoA} + CO_2 + ATP \xrightarrow{\text{Acetyl CoA carboxylase}} \underset{\text{Malonyl CoA}}{HO-\overset{\overset{\displaystyle O}{\|}}{C}-CH_2-\overset{\overset{\displaystyle O}{\|}}{C}-CoA} + ADP + P_i$$

Activation. The malonyl CoA and another acetyl CoA are each attached to a protein called *acetyl carrier protein (ACP).*

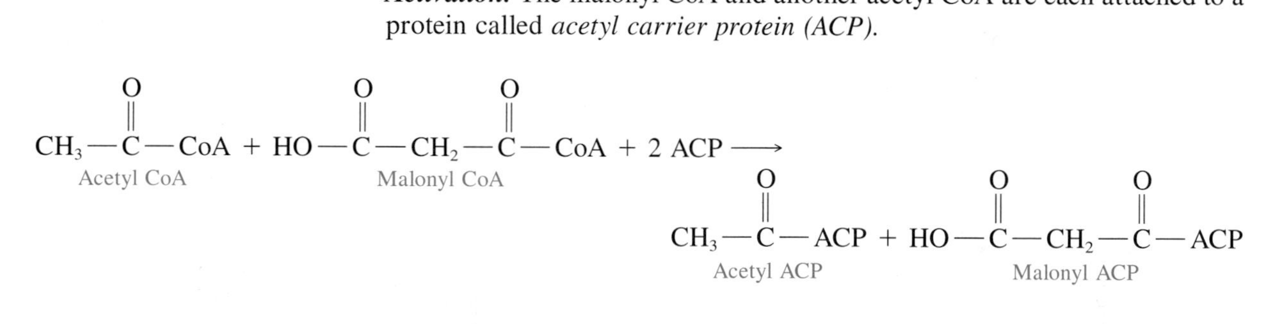

Condensation. The two-carbon acetyl and three-carbon malonyl ACP groups combine to form a four-carbon acetoacetyl ACP, accompanied by a loss of CO_2:

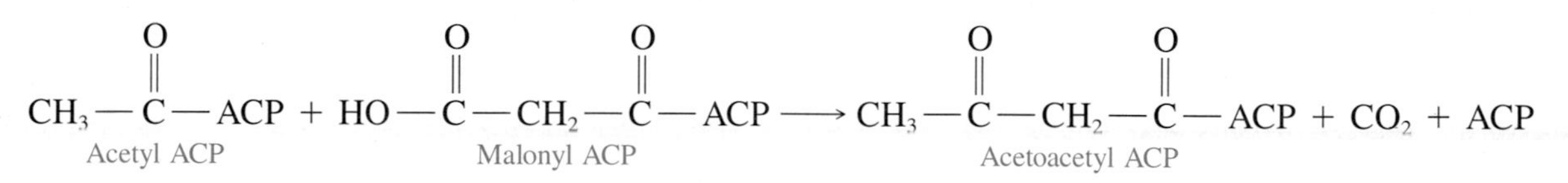

Reduction, Dehydration, and Reduction. In these three steps, hydrogen from the reduced coenzymes NADPH reduces the β-ketone group to a hydroxyl group. The hydroxy compound is dehydrated to yield a carbon–carbon double bond. The reduction of the double bond yields a saturated carbon chain that is longer by two carbon atoms. The chain is lengthened when the four-carbon chain combines with another malonyl ACP and the last three steps are repeated. The typical fatty acid product of lipogenesis is palmitic acid, a

16-carbon chain, from which many other fatty acids are derived. After synthesis, fatty acids are converted to triglycerides (fats) and stored in the adipose cells.

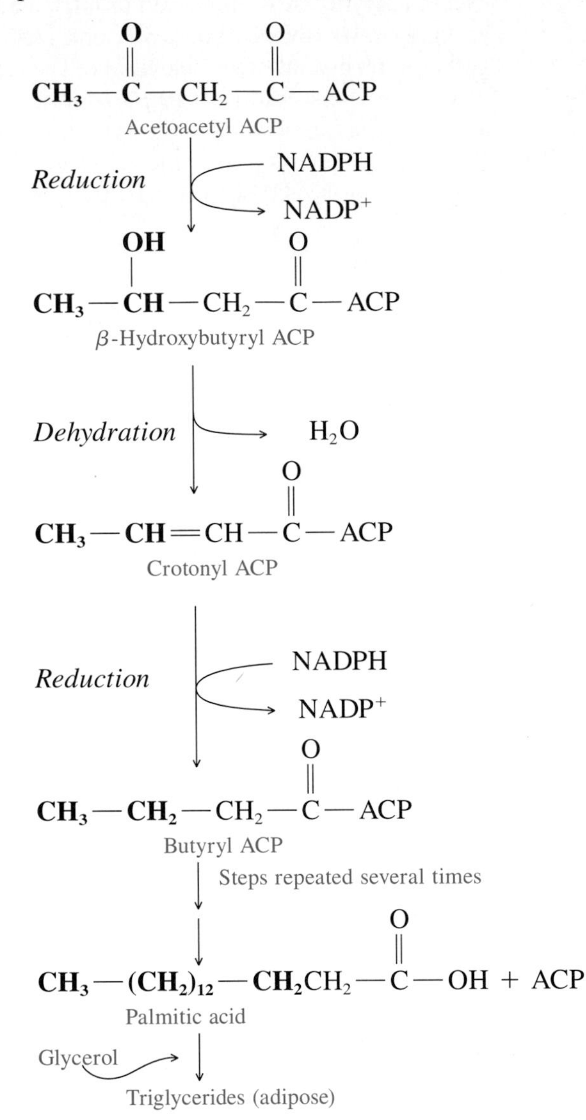

18.9 Metabolic Pathways for Amino Acids

Learning Goal
Describe some metabolic pathways for amino acids.

Amino acids obtained from the digestion of proteins are absorbed by the bloodstream and carried to the cells in the body, where most are used to build proteins. Our bodies are constantly replacing old proteins with new ones. Although amino acids are not a major energy source, energy can be extracted from amino acids if glycogen and fat reserves have been depleted. If amino acids become the only source of energy as in starvation, the breakdown of the body's own proteins eventually leads to a destruction of essential body tissues.

Transamination

Enzymes known as amino transferases, or **transaminases,** move amino groups from amino acids to α-keto acids, usually α-ketoglutaric acid, an intermediate in the citric acid cycle. The following equation illustrates the transfer of the amino group of alanine to yield glutamic acid and pyruvic acid:

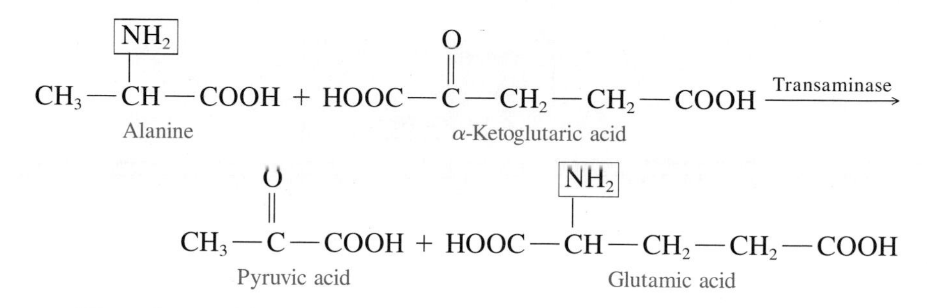

The α-keto acids that result from removing the amino group can become energy sources as intermediates of the citric acid cycle. Many of the three-carbon amino acids are converted to pyruvic acid; four-carbon amino acids yield oxaloacetic acid; five-carbon chains are converted to α-ketoglutaric acid. (See Figure 18.22.) By entering the citric acid cycle, these derivatives of the amino acids can undergo oxidation to form ATP or be used in the synthesis of glucose.

Transamination is also used to prepare the nonessential amino acids in the body by transferring an amino group from glutamic acid to several of the α-keto acids obtained from the citric acid cycle or glycolysis.

$$\text{Pyruvate} \xrightarrow{\text{Transamination}} \text{alanine, valine, leucine, serine, cysteine, glycine}$$

$$\text{Oxaloacetate} \xrightarrow{\text{Transamination}} \text{aspartic acid, threonine, lysine, isoleucine}$$

$$\alpha\text{-Ketoglutarate} \xrightarrow{\text{Transamination}} \text{glutamine, proline}$$

Oxidative Deamination

In many transamination reactions, the amino group is transferred to α-ketoglutaric acid to yield glutamic acid. To regenerate α-ketoglutaric acid, glutamic acid is oxidized and the amino group is removed as NH_3, a process called **oxidative deamination.** Because the NH_3 product is highly toxic to the body, another series of reactions, the urea cycle, combines NH_3 with CO_2 to yield urea, a nontoxic substance that is excreted in the urine.

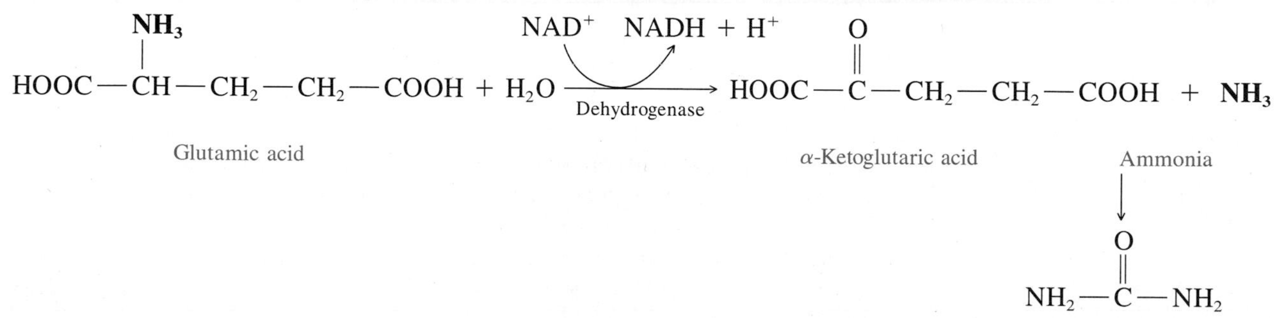

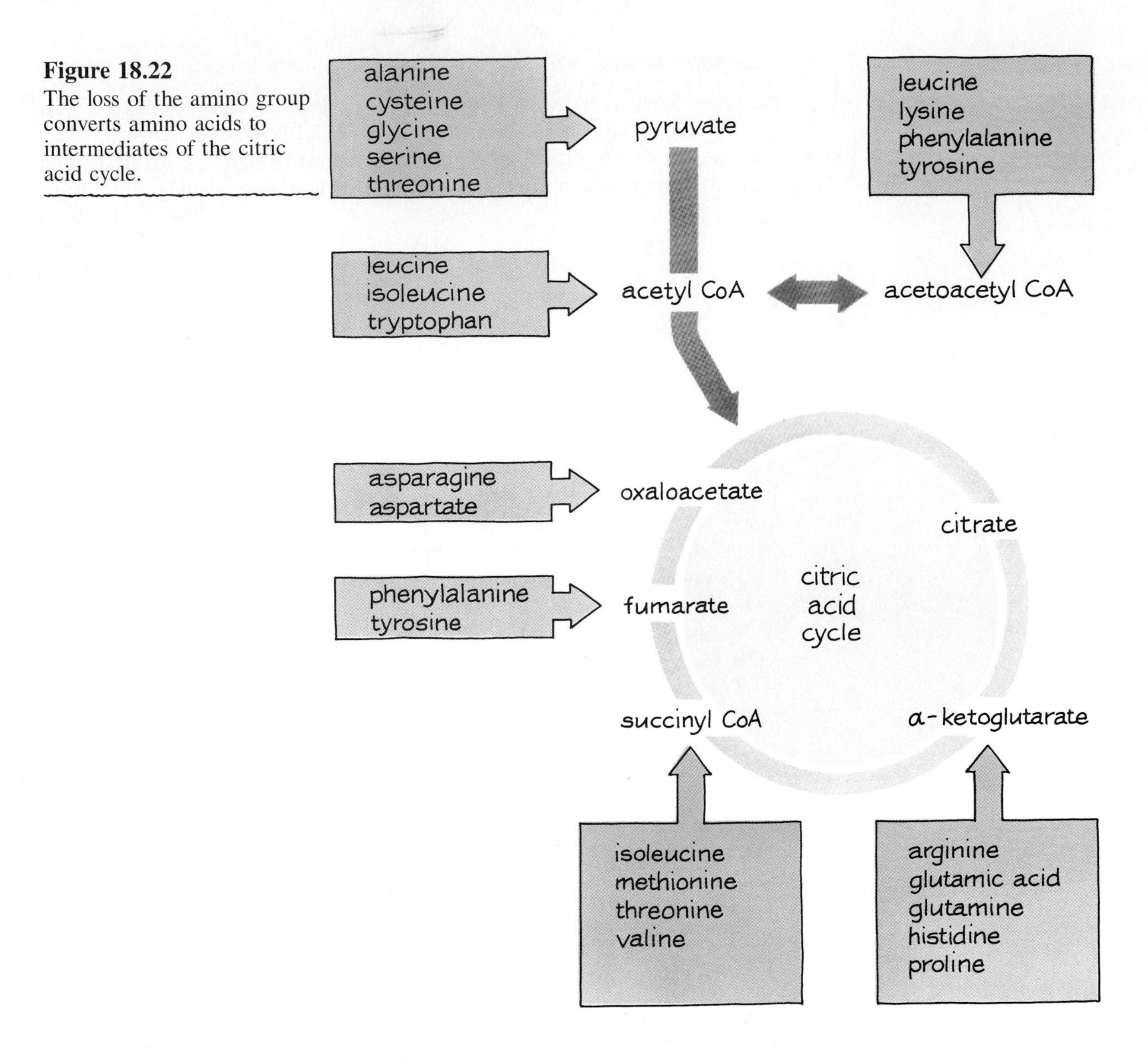

Figure 18.22
The loss of the amino group converts amino acids to intermediates of the citric acid cycle.

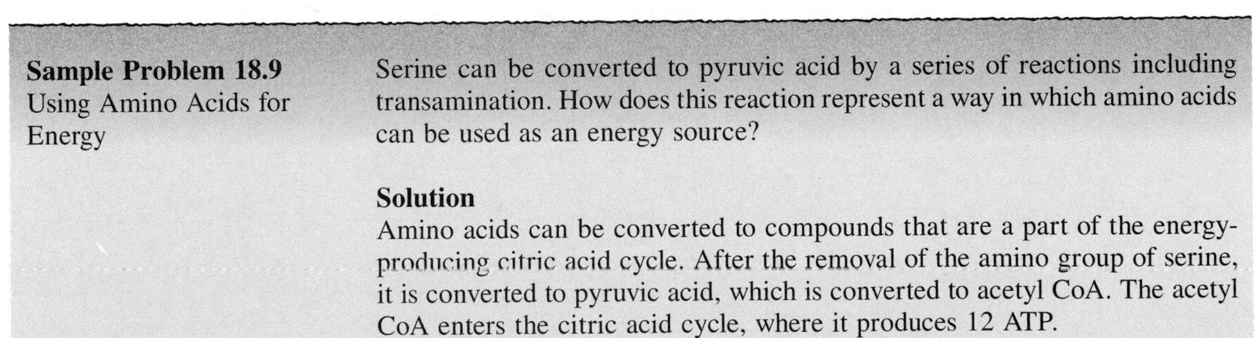

Sample Problem 18.9
Using Amino Acids for Energy

Serine can be converted to pyruvic acid by a series of reactions including transamination. How does this reaction represent a way in which amino acids can be used as an energy source?

Solution
Amino acids can be converted to compounds that are a part of the energy-producing citric acid cycle. After the removal of the amino group of serine, it is converted to pyruvic acid, which is converted to acetyl CoA. The acetyl CoA enters the citric acid cycle, where it produces 12 ATP.

Study Check
What α-keto acid from the citric acid cycle would undergo reactions including transamination to yield asparate acid?

HEALTH NOTE *Why Do I Get Hungry?*

When you say "I'm hungry," you probably decide to find something to eat. You are responding to a stimulation of the hunger center of the brain. When you eat, the satiety center near by is stimulated, which makes you feel full, and you stop eating. The hunger center is operational all the time, but it is inhibited by the satiety center.

In the glucostatic theory, a low blood glucose level decreases the activity of the satiety center so it no longer inhibits the hunger center. We respond by eating a meal, which causes our blood glucose levels to rise until we are no longer hungry. An increase in the level of fatty acids and amino acids in the blood also depresses the hungry feeling. They act in a fashion similar to glucose and cause an inhibition of the hunger center. In general, we feel hungry and eat when the nutrient levels in the body are low.

Our body temperature also affects our food intake. In a cold environment, food intake increases, which accelerates metabolic reactions, increases body temperature, and provides more body fat for insulation. When we are hot, less body fat is needed and we tend to eat less.

Overview of Metabolism

The oxidative pathways degrade large molecules to small molecules that can be used for energy production via the citric acid cycle and the electron transport chain. These same molecules can also enter pathways that lead to the synthesis of larger molecules in the cell. Such synthesis is energy consuming and depends on a supply of ATP. (See Figure 18.23.)

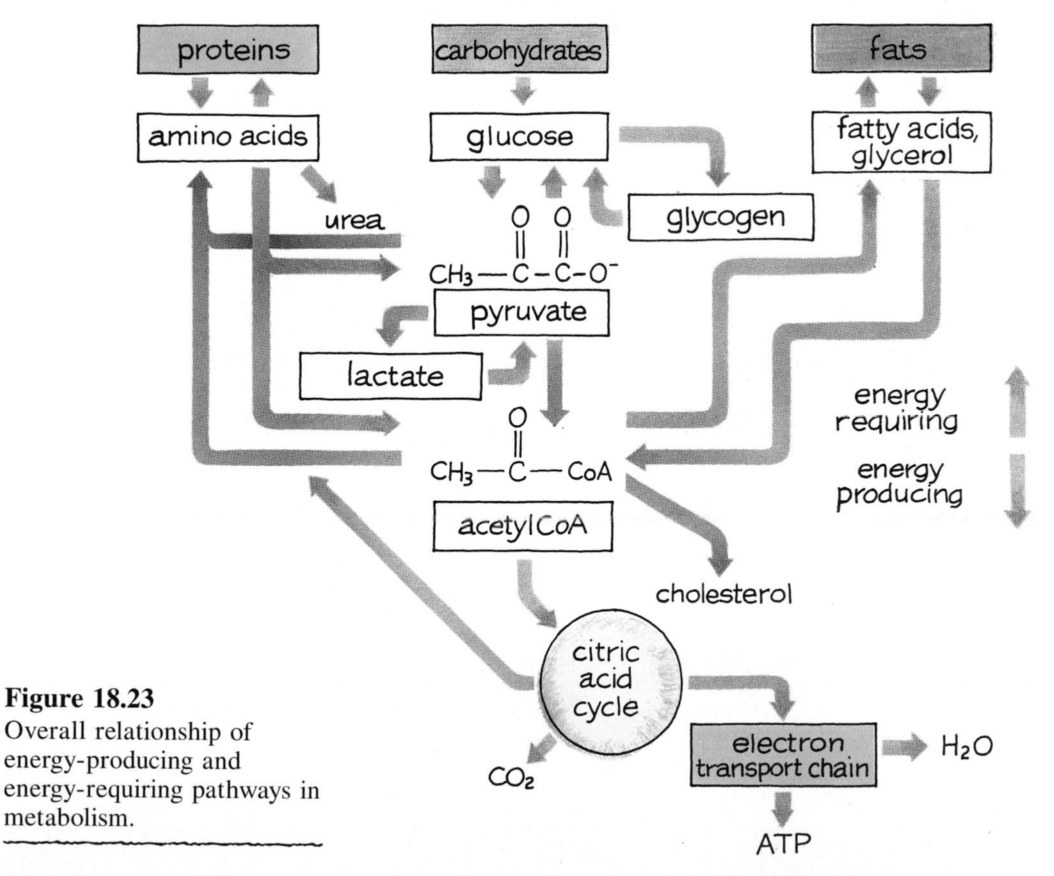

Figure 18.23
Overall relationship of energy-producing and energy-requiring pathways in metabolism.

The pathways in metabolism have several branch points from which compounds may be degraded for energy or used to synthesize larger molecules. For example, glucose can be converted to glycogen for storage, or degraded to acetyl CoA for the citric acid cycle to produce energy. When glycogen stores are depleted, fatty acids can be oxidized for energy. Amino acids are normally used to synthesize proteins for the cells, but they can be converted to intermediates of the citric acid cycle in an energy crisis. The α-keto acids of the citric acid may be used in the synthesis of several nonessential amino acids by transamination reactions.

Summary of Key Equations

Hydrolysis of ATP

$$\text{ATP} + \text{H}_2\text{O} \longrightarrow \text{ADP} + \text{P}_i + 7.3\ \text{kcal/mol}$$

Glycolysis

$$\text{glucose} + 2\ \text{ADP} + 2\ \text{P}_i + \text{NAD}^+ \longrightarrow \text{pyruvate} + 2\ \text{ATP} + 2\ \text{NADH} + 2\ \text{H}^+$$

Formation of Lactate

$$\text{pyruvate} + \text{NADH} + \text{H}^+ \xrightarrow{\text{Anaerobic}} \text{lactate} + \text{NAD}^+$$

Formation of Acetyl CoA

$$\text{pyruvate} + \text{NAD}^+ + \text{CoA} \xrightarrow{\text{Aerobic}} \text{acetyl CoA} + \text{NADH} + \text{CO}_2$$

Citric Acid Cycle

$$\text{acetyl CoA} + 3\ \text{NAD}^+ + \text{FAD} + \text{ADP} \longrightarrow 2\ \text{CO}_2 + 3\ \text{NADH} + \text{FADH}_2 + \text{ATP}$$

Electron Transport Chain

$$\text{NADH} + 3\ \text{ADP} + 3\ \text{P}_i + \tfrac{1}{2}\text{O}_2 \longrightarrow \text{NAD}^+ + 3\ \text{ATP} + \text{H}_2\text{O}$$
$$\text{FADH}_2 + 2\ \text{ADP} + 2\ \text{P}_i + \tfrac{1}{2}\text{O}_2 \longrightarrow \text{FAD} + 2\ \text{ATP} + \text{H}_2\text{O}$$

Phosphorylation of ADP

$$\text{ADP} + \text{P}_i \longrightarrow \text{ATP} + \text{H}_2\text{O}$$

Complete Combustion of Glucose

$$\text{glucose} + 36\ \text{ADP} + 36\ \text{P}_i \longrightarrow 6\ \text{CO}_2 + 36\ \text{ATP} + 6\ \text{H}_2\text{O}$$

Chapter Summary

18.1 ATP: The Energy Storehouse

In metabolism, catabolic reactions break down large molecules into small molecules in processes that provide energy. The energy is stored in ATP, a high-energy compound, that is hydrolyzed when energy is required by the anabolic reactions that do work in the cells.

18.2 Digestion of Foods

Digestion is a series of reactions that break down large food molecules of carbohydrates, lipids, and proteins into smaller molecules that can be absorbed and used by the cells. The end products of digestion include monosaccharides from carbohydrates, fatty acids and glycerol from triglycerides, and amino acids from proteins.

18.3 Glycolysis: Oxidation of Glucose
Glycolysis is the primary anaerobic pathway for the degradation of glucose to yield pyruvic acid. Glucose is phosphorylated and isomerized to fructose 1,6-diphosphate that is split into two triose phosphate molecules. The oxidation of the three-carbon sugars yields the reduced coenzyme 2 NADH and 2 ATP.

18.4 Pathways for Pyruvate
In the absence of oxygen, pyruvate is reduced to lactate and NAD^+ is regenerated for the continuation of glycolysis. Under aerobic conditions, pyruvate is oxidized in the mitochondria to acetyl CoA, which enters the citric acid cycle.

18.5 Citric Acid Cycle
In a sequence of reactions called the citric acid cycle, acetyl CoA combines with oxaloacetate to yield citrate. Citrate undergoes oxidation and decarboxylation to yield two CO_2, GTP, three NADH, and $FADH_2$. The phosphorylation of ADP by GTP yields ATP.

18.6 The Electron Transport Chain
The reduced coenzymes from glycolysis and the citric acid cycle are oxidized to NAD^+ and FAD by transferring protons and electrons to the electron transport chain. In the electron chain, electrons are transferred to electron carriers including flavins, iron-sulfur proteins, and cytochromes with Fe^{3+}/Fe^{2+}. The flow of electrons along the electron chain pumps protons across the inner membrane, which produces a high-energy proton gradient that provides energy for the synthesis of ATP. The final acceptor, O_2, combines with protons and electrons to yield H_2O. The process of using the energy of the electron transport chain to synthesize ATP is called oxidative phosphorylation. The oxidation of NADH yields three ATP molecules, and $FADH_2$ yields two ATP.

18.7 ATP Energy from Glucose
The complete oxidation of one glucose yields a total of 36 ATP from direct phosphorylation and the oxidation of the reduced coenzymes NADH and $FADH_2$ by the electron transport chain and oxidative phosphorylation.

18.8 Oxidation of Fatty Acids
As an energy source, fatty acids are linked to coenzyme A and transported into the mitochondria where they undergo β-oxidation. The fatty acetyl chain is oxidized to yield a shortened fatty acid, acetyl CoA, and the reduced coenzymes NADH and $FADH_2$. Although the energy from a particular fatty acid depends on its length, each cycle of β-oxidation yields 17 ATP. When there is an excess of acetyl CoA in the cell, the two-carbon units link together to synthesize fatty acids that are stored in the adipose tissue.

18.9 Metabolic Pathways for Amino Acids
Amino acids are normally used for protein synthesis. The nonessential amino acids can be prepared by transferring an amino group to an α-keto acid to yield a different amino acid and α-keto acid. Some α-keto acids that are intermediates of the citric acid cycle can be used in the synthesis of lipids or glucose or oxidized further for energy. Oxidative transamination removes an amino group as NH_3 and converts it to urea.

Glossary of Key Terms

acetyl CoA A two-carbon acetyl unit attached to coenzyme A. A product of glucose, fatty acid, and protein metabolism, acetyl CoA is oxidized to yield energy via the citric acid cycle or used as a substrate to synthesize fatty acids or glucose.

ADP Adenosine diphosphate; a compound of adenine, a ribose sugar, and two phosphate groups, it is formed by the hydrolysis of ATP.

aerobic An oxygen-containing environment in the cells.

anabolic reactions Metabolic reactions that are energy requiring.

anaerobic The absence of oxygen in the cells.

ATP Adenosine triphosphate; a high-energy compound that stores energy in the cells and consists of adenine, a ribose sugar, and three phosphate groups.

beta (β)-oxidation A reaction cycle that removes acetyl CoA to shorten a fatty acid and yields the reduced coenzymes NADH and $FADH_2$.

catabolic reactions Metabolic reactions that produce energy for the cell by the degradation and oxidation of glucose, fatty acids, and amino acids.

chemiosmotic theory The conservation of energy from transfer of electrons in the electron chain by pumping protons into the intermembrane space to produce a high-energy gradient that provides the energy for the phosphorylation of ADP to yield ATP.

citric acid cycle A series of oxidation reactions in the mitochondria that convert acetyl CoA to CO_2 and yield NADH and $FADH_2$. It is also called the Krebs cycle.

coenzyme Q A mobile carrier that accepts electrons from NADH-Q reductase or $FADH_2$ in the electron transport chain.

cytochromes Iron-containing proteins (Fe^{3+}, Fe^{2+}) in the electron transport chain that transfer electrons from $CoQH_2$ to oxygen.

decarboxylation The loss of a carbon atom in the form of CO_2.

digestion The processes that break down large food molecules to small ones that can be absorbed by cells of the body.

direct substrate phosphorylation The transfer of inorganic phosphate (P_i) directly from a high-energy phosphate bond to a substrate.

electron transport chain A mitochondrial system that transfers electrons from the reduced coenzymes NADH and $FADH_2$ to clusters of electron carriers such as flavins, iron-sulfur proteins, and cytochromes. The final electron acceptor, O_2, yields H_2O. Energy changes during some of the transfers provide energy to phosphorylate ADP to yield ATP.

FAD A coenzyme (flavin adenine dinucleotide) for dehydrogenase enzymes that form carbon–carbon double bonds. The reduced form, $FADH_2$, transfers electrons to CoQ in the electron transport chain to regenerate FAD.

fermentation The anaerobic conversion of sugars by enzymes in yeast to yield alcohol and CO_2.

FMN The initial electron acceptor (flavin mononucleotide) of the electron transport chain for NADH.

glycolysis The oxidation reactions of glucose that yield two pyruvate molecules.

ketosis A condition in which the level of ketone bodies is elevated.

metabolism All the chemical reactions in living cells that carry out molecular and energy transformations.

NAD^+ The hydrogen acceptor in oxidation reactions that forms carbon–oxygen double bonds. The reduced coenzyme NADH transfers electrons to FMN in the electron transport chain to yield three ATP molecules and regenerate NAD^+.

oxidation The loss of hydrogen atoms or electrons by a reactant. In metabolism, it is generally associated with the degradation of large molecules and the production of energy.

oxidative deamination The removal of an amino group as NH_3.

oxidative phosphorylation The synthesis of ATP in the mitochondria by combining ADP and P_i using energy from the oxidation reactions in the electron transport chain.

reduction The gain of hydrogen or electrons by a reactant.

transamination The transfer of an amino group from an α-amino acid to produce α-keto acids that are intermediates in the citric acid cycle.

Chemistry at Home

1. Some milk products contains Lactaid, an enzyme that digests lactose in milk. What is the chemical name of the enzyme? Write the equation for the digestion of lactose. Where in the intestinal tract does this reaction occur?
2. Chew a cracker or piece of bread for 5 minutes. How does its taste change? What part of carbohydrate digestion occurs in the mouth?
3. Place some oil in a glass of water. What happens? Add a few drops of soap and shake well. What happens to the size of the oil droplets? How is soap like the bile salts in digestion? Where does this part of fat digestion occur in the body? People without gallbladders take lipase medication. Why is this necessary?

Answers to Study Checks

18.1 7.3 kcal/mol ATP

18.2 The digestion of amylose begins in the mouth when salivary amylase hydrolyzes some of the α-1,4-glycosidic bonds. In the small intestine, pancreatic amylase hydrolyzes more glycosidic bonds, and finally maltose is hydrolyzed by maltase to yield glucose.

18.3 In the initial steps of glycolysis, two ATP are used to convert glucose to fructose 1,6-diphosphate.

18.4

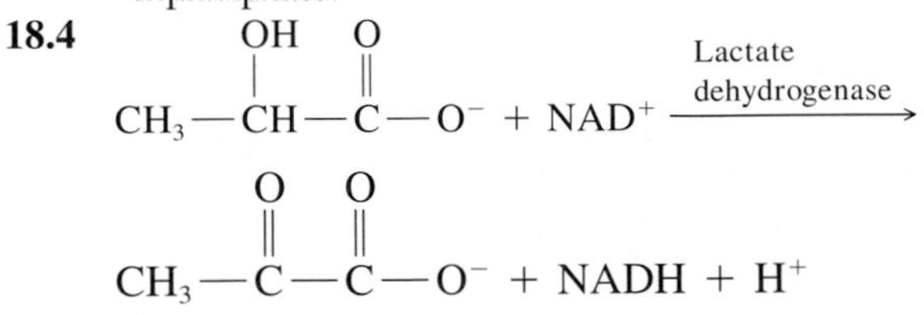

$$CH_3{-}CH(OH){-}C(=O){-}O^- + NAD^+ \xrightarrow{\text{Lactate dehydrogenase}} CH_3{-}C(=O){-}C(=O){-}O^- + NADH + H^+$$

18.5 oxaloacetate

18.6 When the final acceptor, O_2, combines with protons and electrons from cytochrome a_3.

18.7 The process that transports NADH from the cytoplasm (site of glycolysis) to the mitochondria requires the energy of 1 ATP.

18.8 activation (−2 ATP); 2 acetyl CoA (24 ATP); 2 NADH (6 ATP); and 2 $FADH_2$ (4 ATP) = 32 ATP

18.9 oxaloacetate

Problems

ATP: The Energy Storehouse *(Goal 18.1)*

18.1 What is a catabolic reaction in metabolism?

18.2 What is an anabolic reaction in metabolism?

18.3 How does ADP differ from ATP? Write an equation for the formation of ATP from ADP and P_i.

18.4 Why is ATP considered a high-energy compound? Write an equation to represent the hydrolysis of ATP.

18.5 How much energy is needed to convert 2 mol ADP to 2 mol ATP?

18.6 How much energy is released if 5 mol ATP were hydrolyzed to 5 mol ADP?

Digestion of Foods *(Goal 18.2)*

18.7 What is the purpose of digestion?

18.8 What are the major sites of digestion?

18.9 Give the site, enzyme(s), and end products for the digestion of the following:
a. sucrose **b.** proteins **c.** amylose

18.10 Give the site, enzyme(s), and end products for the digestion of the following:
a. triglycerides **b.** peptides **c.** lactose

Glycolysis: Oxidation of Glucose *(Goal 18.3)*

18.11 What is the starting product of glycolysis?

18.12 What is the end product of glycolysis?

18.13 How is ATP used in the initial steps of glycolysis?

18.14 How many ATP molecules are used in the initial steps of glycolysis?

18.15 What trioses are obtained when fructose 1,6-diphosphate splits?

18.16 Why does one of the triose products undergo isomerization?

18.17 How does direct phosphorylation account for the production of ATP in glycolysis?

18.18 Why are there two ATP molecules formed for one molecule of glucose?

18.19 Indicate the enzyme that catalyzes the following reactions in glycolysis:
a. phosphorylation
b. direct transfer of a phosphate group

18.20 Indicate the enzyme that catalyzes the following reactions in glycolysis:
a. isomerization
b. formation of a ketotriose and an aldotriose

18.21 How many ATP or NADH are produced (or required) in each of the following steps in glycolysis?
a. glucose to glucose 6-phosphate
b. glyceraldehyde 3-phosphate to 1,3-diphosphoglycerate
c. glucose to pyruvate

18.22 How many ATP or NADH are produced (or required) in each of the following steps in glycolysis?
a. 1,3-diphosphoglycerate to 3-phosphoglycerate
b. fructose 6-phosphate to fructose 1,6-diphosphate
c. phosphoenolpyruvate to pyruvate

Pathways for Pyruvate *(Goal 18.4)*

18.23 Write an equation for the conversion of pyruvate to acetyl CoA.

18.24 Write an equation for the conversion of pyruvate to lactate.

18.25 Why is glucose converted to lactate under anaerobic conditions?

18.26 What is fermentation?

Citric Acid Cycle *(Goal 18.5)*

18.27 Where in the cell are the enzymes and coenzymes for the citric acid cycle?

18.28 What conditions are necessary for the operation of the citric acid cycle?

18.29 NAD^+ is the coenzyme for some dehydrogenases in the citric acid cycle.
 a. What are the components of NAD^+?
 b. What type of group does NAD^+ oxidize? What type of product forms?
 c. When a reactant is oxidized, what happens to NAD^+?

18.30 FAD is the coenzyme for some dehydrogenases in the citric acid cycle.
 a. What are the components of FAD?
 b. What type of group does FAD oxidize? What type of product forms?
 c. When a reactant is oxidized, what happens to FAD?

18.31 Refer to the diagram of the citric acid cycle to answer each of the following:
 a. What are the six-carbon compounds?
 b. How is the number of carbon atoms decreased?
 c. How many NADH molecules form?
 d. What are the five-carbon compounds?
 e. Which steps are oxidation reactions?
 f. In which steps are secondary alcohols oxidized?

18.32 Refer to the diagram of the citric acid cycle to answer each of the following:
 a. What is the yield of CO_2 molecules?
 b. How many $FADH_2$ molecules form?
 c. What are the four-carbon compounds?
 d. What is the yield of GTP molecules?
 e. What are the decarboxylation steps?
 f. Where does a hydration occur?

18.33 Indicate the name of the enzyme for each of the following reactions in the citric acid cycle:
 a. joins acetyl CoA to oxaloacetate
 b. forms a carbon–carbon double bond
 c. adds water to fumarate

18.34 Indicate the name of the enzyme for each of the following reactions in the citric acid cycle:
 a. isomerizes citrate
 b. oxidizes and decarboxylates α-ketoglutarate
 c. adds P_i to GDP

18.35 State the acceptor for hydrogen or phosphate in each of the following reactions:
 a. isocitrate $\longrightarrow$ α-ketoglutarate
 b. succinyl CoA $\longrightarrow$ succinate
 c. succinate $\longrightarrow$ fumarate

18.36 State the acceptor for hydrogen or phosphate in each of the following reactions:
 a. malate $\longrightarrow$ oxaloacetate
 b. α-ketoglutarate $\longrightarrow$ succinyl CoA
 c. pyruvate $\longrightarrow$ acetyl CoA

The Electron Transport Chain *(Goal 18.6)*

18.37 What reduced coenzymes provide the electrons for the electron transport chain?

18.38 What happens to the energy level as electrons are passed down the electron transport chain?

18.39 What are the oxidized and reduced forms of each of the following?
 a. NADH **b.** FMN

18.40 What are the oxidized and reduced forms of each of the following?
 a. cytochrome *c* **b.** O_2

18.41 In what order do the following appear in the electron chain: cytochrome *c*, FAD, and coenzyme Q?

18.42 In what order do the following appear in the electron chain: O_2, NAD^+, ATP synthetase?

18.43 How is NADH oxidized by the electron chain? What is the ATP yield?

18.44 How is $FADH_2$ oxidized by the electron chain? What is the ATP yield?

18.45 What is meant by oxidative phosphorylation?

18.46 How is the proton gradient established?

18.47 According to the chemiosmotic theory, how does the proton gradient provide energy to synthesize ATP?

18.48 How does the phosphorylation of ADP occur?

18.49 How are glycolysis and the citric acid cycle linked to the production of ATP by the electron transport chain?

18.50 Why does $FADH_2$ have a yield of 2 ATP via the electron chain, but NADH yields 3 ATP?

ATP Energy from Glucose *(Goal 18.7)*

18.51 Indicate the amount of ATP associated with each of the following:
 a. NADH $\longrightarrow$ NAD^+
 b. glucose $\longrightarrow$ 2 pyruvate
 c. 2 pyruvate $\longrightarrow$ 2 acetyl CoA + $2CO_2$
 d. acetyl CoA $\longrightarrow$ $2CO_2$

18.52 Indicate the amount of ATP associated with each of the following:
 a. acetyl CoA $\longrightarrow$ CO_2 (citric acid cycle)
 b. $FADH_2$ $\longrightarrow$ FAD
 c. glucose + $6O_2$ $\longrightarrow$ $6CO_2$ + $6H_2O$
 d. glucose $\longrightarrow$ 2 lactate
 e. pyruvate $\longrightarrow$ lactate

Oxidation of Fatty Acids *(Goal 18.8)*

18.53 Consider the activation of a fatty acid.
- **a.** Triglycerides provide fatty acids for β-oxidation. What happens to the glycerol?
- **b.** Where in the cells are fatty acids activated for β-oxidation?
- **c.** What is the energy cost in ATP for activation of fatty acids?
- **d.** What is the purpose of activating fatty acids?

18.54 Consider the β-oxidation pathway for fatty acids.
- **a.** Why is the oxidation of fatty acids called β-oxidation?
- **b.** What coenzymes are required?
- **c.** When is each coenzyme used?
- **d.** What is the yield in ATP molecules for one cycle of β-oxidation?

18.55 Lauric acid, $CH_3—(CH_2)_{10}—COOH$, is a fatty acid.
- **a.** Write the formula of the activated lauric acid.
- **b.** Indicate the α- and β-carbon atoms in lauric acid.
- **c.** Write the equations for one cycle of β-oxidation for lauric acid.
- **d.** How many acetyl CoA units will be produced when β-oxidation of lauric acid is complete?
- **e.** How many cycles of β-oxidation are needed to complete the oxidation?
- **f.** Account for the total ATP yield from β-oxidation of lauric acid by completing the following table:

Units Produced from Lauric Acid

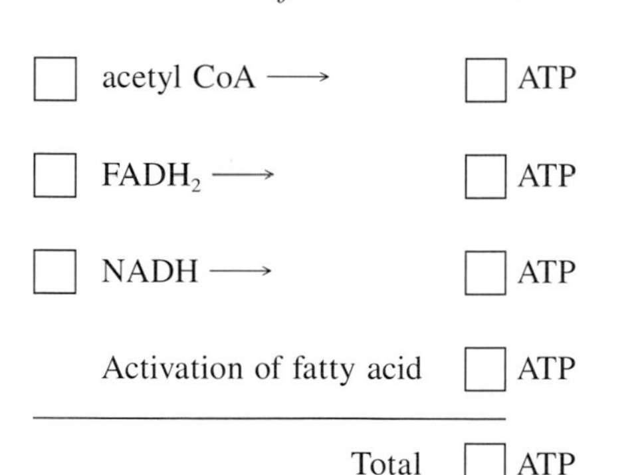

18.56 Stearic acid, $CH_3—(CH_2)_{16}—COOH$, is a fatty acid in butter fat and beef fat.
- **a.** Write the formula of the activated stearic acid.
- **b.** Indicate the α- and β-carbon atoms in stearic acid.
- **c.** Write the equations for one cycle of β-oxidation for stearic acid.
- **d.** How many acetyl CoA units will be produced when β-oxidation of stearic acid is complete?
- **e.** How many cycles of β-oxidation are needed to complete the oxidation?
- **f.** Account for the total ATP yield from β-oxidation of stearic acid by completing the following table:

Units Produced from Stearic Acid

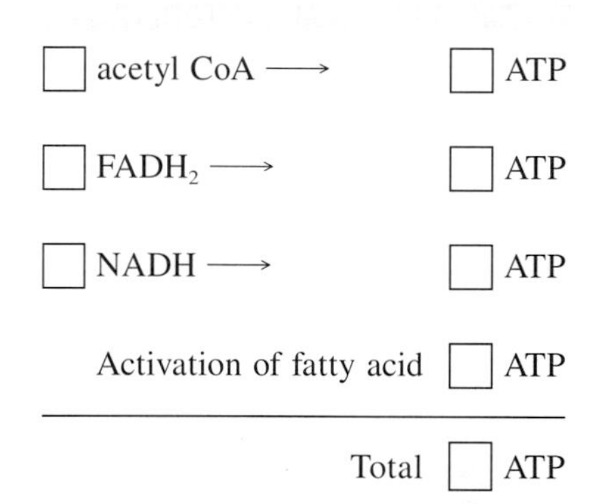

18.57 How does the breakdown of large amounts of fats lead to the formation of excess ketone bodies? Under what conditions does ketosis occur? What are some symptoms of high levels of ketone bodies?

18.58 Consider the lipogenesis of myristyl CoA.
- **a.** How many malonyl CoA units are needed?
- **b.** How many NADPH are used?

Metabolic Pathways for Amino Acids *(Goal 18.9)*

18.59 What are the products of transamination of amino acids?

18.60 What are the products of oxidative deamination of amino acids?

18.61 Write equations for the transamination of each of the following amino acids and α-ketoglutarate:
a. alanine **b.** aspartate

18.62 Write the equation for the oxidative deamination of glutamic acid.

18.63 If necessary, how are amino acids used as a fuel in the cells?

18.64 If a nonessential amino acid is low in a cell, how might it be prepared from intermediates of the citric acid cycle?

Challenge Problems

18.65 Why is NAD^+ used for the oxidation of isocitrate and malate, whereas FAD is the coenzyme used for the oxidation of succinate?

18.66 What would be the ATP yield for the complete oxidation of maltose?

18.67 At the gym, you expend 300 kcal riding the stationary bicycle for 1 hr. Using the value of 7.3 kcal/mol for the hydrolysis of ATP, how many grams of glucose (molar mass, 180 g/mol) would you need to oxidize?

18.68 Why are the degradative pathways carried out in a series of steps, and not in one step?

18.69 If the citric acid cycle contains no reactions with oxygen, O_2, why does it only operate under aerobic conditions?

18.70 What is the ratio of ATP produced by the oxidation of glucose under aerobic versus anaerobic conditions?

18.71 When alcohol is consumed, it is oxidized to acetaldehyde and then to acetate. How does alcohol provide calories for the body?

18.72 Your friend has gone on a low calorie diet and has developed "fruity" breath. What would you tell your friend about his diet?

Appendix A Scientific (Exponential) Notation

Powers of 10

When numbers are very large or very small, it is more convenient to use scientific notation to represent those numbers in **powers of 10.** For example, 1000 is the same as 10 multiplied three times. The three is written as a power of 10 (exponent). Some examples of powers of 10 are listed in Table A.1.

$$1000 = 10 \times 10 \times 10 = 10^3 \quad \text{(3 = Exponent)}$$

Table A.1 *Some Examples of Positive Powers of 10*

$10^5 = 10 \times 10 \times 10 \times 10 \times 10 = 100{,}000$
$10^4 = 10 \times 10 \times 10 \times 10 = 10{,}000$
$10^3 = 10 \times 10 \times 10 = 1000$
$10^2 = 10 \times 10 = 100$
$10^1 = 10$
$10^0 = 1$

We can represent 2400 in scientific notation by converting the number to 2.4 (a value between 1 and 10) and multiplying by 10 three times.

$$2400 = 2.4 \times 1000 = \underbrace{2.4}_{\text{Number between 1 and 10}} \times \underbrace{10^3}_{\text{Power of 10}}$$

To convert a large number into scientific notation, the decimal point is moved to the left until it is located after the first digit. The number of places the decimal point was moved becomes the power of 10 (exponent). Suppose we want to write the number 18,000 in scientific notation. We move the decimal point to give 1.8. (Only the significant numbers are retained.)

$$1_{\wedge}8_{\smile}0_{\smile}0_{\smile}0.$$
4 3 2 1

Because we had to move the decimal point four places to the left, the power of 10 will be 4.

$$18\ 000 = 1.8 \times 10^4$$

Sample Problem A.1
Writing Large Numbers in Scientific Notation

Express the following numbers in scientific notation:
a. 35,200 **b.** 5,000,000 **c.** 8100

Solution
For numbers larger than 10, we need to move the decimal point from the end of the number to the left until it is located behind the first number.

Decimal point moves four places to the left

a. 3 . 5 2 0 0. 3.52×10^4

b. 5, 0 0 0, 0 0 0. 5×10^6

c. 8 . 1 0 0. 8.1×10^3

Study Check
Convert the following numbers to their decimal form:
a. 5.8×10^3 **b.** 2×10^5

Small numbers can also be written using scientific notation. For example, 0.01 or 1/100 is the same as 10^{-2} and 0.001 or 1/1000 is the same as 10^{-3}.

$$0.01 = \frac{1}{100} = \frac{1}{(10)(10)} = \frac{1}{10^2} = 10^{-2}$$

$$0.001 = \frac{1}{1000} = \frac{1}{(10)(10)(10)} = \frac{1}{10^3} = 10^{-3}$$

Some examples of negative powers of 10 are given in Table A.2.

Table A.2 *Some Examples of Negative Powers of 10*

$$10^{-1} = \frac{1}{10} = \frac{1}{10^1} = 0.1$$

$$10^{-2} = \frac{1}{100} = \frac{1}{(10)(10)} = \frac{1}{10^2} = 0.01$$

$$10^{-3} = \frac{1}{1000} = \frac{1}{(10)(10)(10)} = \frac{1}{10^3} = 0.001$$

$$10^{-4} = \frac{1}{10{,}000} = \frac{1}{(10)(10)(10)(10)} = \frac{1}{10^4} = 0.0001$$

$$10^{-5} = \frac{1}{100{,}000} = \frac{1}{(10)(10)(10)(10)(10)} = \frac{1}{10^5} = 0.00001$$

To convert a small number into scientific notation, the decimal point must be moved to the right until it is located after the first nonzero digit. For example, to write 0.00086 in scientific notation, the decimal point must be moved to the right until it is between the 8 and the 6, 8.6.
example, to write 0.00086 in scientific notation, the decimal point must be moved to the right until it is between the 8 and the 6, 8.6.

0.0 0 0 8 6 — Decimal point moves four places to the right
1 2 3 4

When the decimal point moves four places to the right, it is the same as dividing by 10 four times. This is expressed as a negative exponent in the power of 10.

$$0.00086 = 8.6 \times \frac{1}{10^4} = 8.6 \times 10^{-4}$$

Sample Problem A.2
Writing Small Numbers in Scientific Notation

Write the following numbers in scientific notation:
a. 0.0042 **b.** 0.00000255 **c.** 0.000108

Solution
Moving the decimal to the right converts each number to scientific notation form with negative powers of 10:

a. 0.0 0 4 2 — 4.2×10^{-3}

b. 0.0 0 0 0 0 2 5 5 — 2.55×10^{-6}

c. 0.0 0 0 1 0 8 — 1.08×10^{-4}

Study Check
Express the following numbers in decimal form:
a. 8.5×10^{-2} **b.** 2×10^{-5}

Appendix B
Some Useful Equalities for Conversion Factors

Length: *metric (SI):* meter (m)

1 kilometer (km) = 1000 meters

1 meter = 1000 millimeters (mm)

= 100 centimeters (cm)

metric–American:

1 inch (in.) = 2.54 centimeters

1 mile = 1.61 kilometers

Volume: *metric:* liter (L); *SI:* cubic meter (m^3)

1 m^3 = 1000 liters

1 liter = 1000 milliliters (mL)

= 1000 cubic centimeters (cm^3)

= 10 deciliters (dL)

1 milliliter = 1 cubic centimeter

metric–American:

1 liter = 1.06 quarts

1 quart = 0.946 liter = 946 milliliters

Mass: *metric:* gram (g); *SI:* kilogram (kg)

1 kilogram = 1000 grams

1 gram = 1000 milligrams (mg)

1 kilogram = 1,000,000 milligrams (mg)

metric–American:

1 pound = 454 grams

1 kilogram = 2.20 pounds

Density: mass/volume (g/mL or g/L)

density of water (4°C) = 1.00 g/mL

Temperature: *metric:* Celsius (°C); *SI:* Kelvin (K)

$°F = 1.8°C + 32°$

$$°C = \frac{°F - 32}{1.8}$$

$K = °C + 273°C$

Heat and energy: *metric:* calorie (cal); *SI:* joule (J)

1 calorie = 4.18 joules

1 kilocalorie (kcal) = 1000 calories

$$\text{specific heat of water} = \frac{1\text{ calorie}}{(1\text{ g})(1°C)}$$

heat of fusion for water = 80 cal/g

heat of vaporization for water = 540 cal/g

Pressure: *metric:* atmosphere (atm); *SI:* pascal (Pa)

1 atmosphere = 760 mm Hg (torr)

1 mole gas (STP) = 22.4 liters

metric–American:

1 atmosphere = 14.7 lb/in^2

Acids and bases:

$K_W = [H_3O^+][OH^-] = 1 \times 10^{-14}$

$pH = -\log[H_3O^+]$

Appendix C
Using Your Calculator

A calculator makes it possible to carry out mathematical operations quickly. Although there are different kinds of calculators, most have similar procedures. Sometimes, the steps are a little different from what you remember from basic math. Learning to use your calculator correctly can help you be efficient while avoiding mistakes.

Adding and Subtracting

When you need to add or subtract, you may follow these steps:

1. Use the number keys to enter the first number.
2. Press the function key for the correct operation, [+] to add, [−] to subtract.
3. Enter the second number.
4. Press the equals key [=].
5. Read your answer in the display area.

If there are additional numbers to add or subtract, repeat steps 2, 3, and 4. To give the correct answer, be sure to round off the display answer to give the correct number of significant figures. For addition and subtraction, round off to the last digit where all the numbers have a significant figure.

Sample Problem C.1
Adding and Subtracting on a Calculator

Give the correct answer for the following calculations:

a. 4.82 + 25.3
b. 5.18 − 0.926

Solution

a.

Press Key(s)	*Display Reads*	
[4] [.] [8] [2]	4.82	
[+]	4.82	
[2] [5] [.] [3]	25.3	
[=]	30.12	Final answer = 30.1

Because one of the numbers has significant figures only to the tenths (0.1) place, the correct answer, 30.1, is obtained by rounding off the display answer.

b.

Press Key(s)	*Display Reads*	
[5] [.] [1] [8]	5.18	
[−]	5.18	
[.] [9] [2] [6]	0.926	
[=]	4.254	Final answer = 4.25

The final answer, 4.25, is obtained by rounding off the display answer.

Study Check
Carry out the following calculations:
a. 5.104 + 25.2 − 14.38 **b.** 37 − 85.42

Multiplying and Dividing

When you need to multiply or divide, you may follow these steps:

1. Use the number keys to enter the first number.
2. Press the function key for the correct operation, [×] to multiply, [÷] to divide.
3. Enter the second number.
4. Press the equals key [=].
5. Read your answer in the display area.

To give the correct answer, be sure to round off the display answer to give the correct number of significant figures. For multiplication and division, round off to give as many digits as the number with the fewest significant figures.

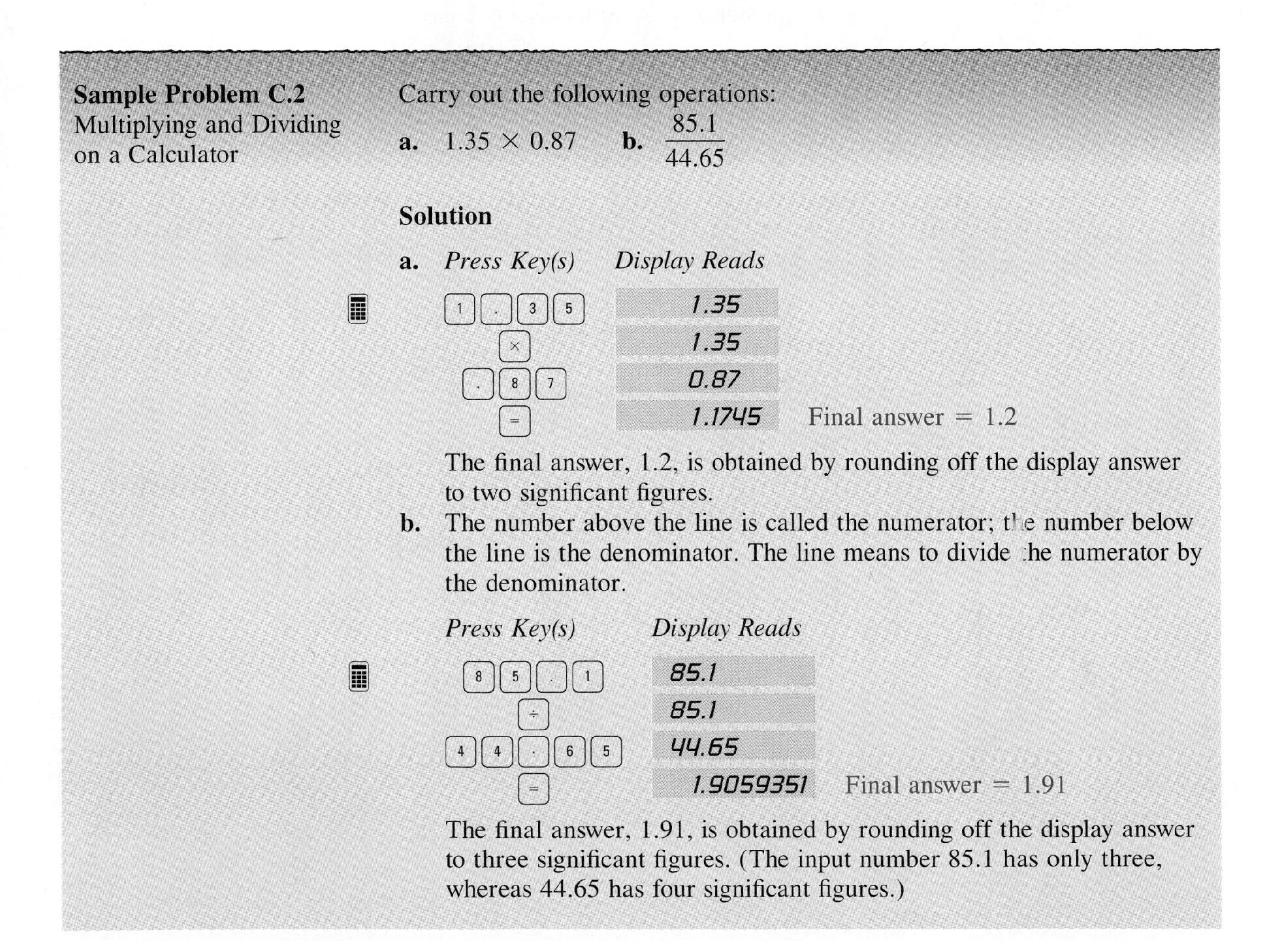

Sample Problem C.2
Multiplying and Dividing on a Calculator

Carry out the following operations:

a. 1.35×0.87 **b.** $\frac{85.1}{44.65}$

Solution

a.

Press Key(s)	*Display Reads*	
[1] [.] [3] [5]	1.35	
[×]	1.35	
[.] [8] [7]	0.87	
[=]	1.1745	Final answer = 1.2

The final answer, 1.2, is obtained by rounding off the display answer to two significant figures.

b. The number above the line is called the numerator; the number below the line is the denominator. The line means to divide the numerator by the denominator.

Press Key(s)	*Display Reads*	
[8] [5] [.] [1]	85.1	
[÷]	85.1	
[4] [4] [.] [6] [5]	44.65	
[=]	1.9059351	Final answer = 1.91

The final answer, 1.91, is obtained by rounding off the display answer to three significant figures. (The input number 85.1 has only three, whereas 44.65 has four significant figures.)

Study Check
Carry out the following calculations:

a. $\dfrac{185 - 22}{34}$

b. $(2 \times 18.0) + (4 \times 30.1)$; the 2 and 4 are exact.

Positive and Negative Numbers

A negative number has the same magnitude as a positive number but has a negative sign in front. Most positive numbers are written without the positive sign.

Positive twenty-five	$+25$ or 25
Negative four point eight	-4.8

To multiply numbers, the rule of signs is used.

$$(+1) \times (+1) = +1$$
$$(+1) \times (-1) = -1$$
$$(-1) \times (+1) = -1$$
$$(-1) \times (-1) = +1$$

Suppose we want to solve 1.8×-14. To enter a number with a negative sign, use the change sign key, [+/−]. DO NOT USE THE SUBTRACT OPERATION KEY. This is a multiplication of a negative number, not a subtraction.

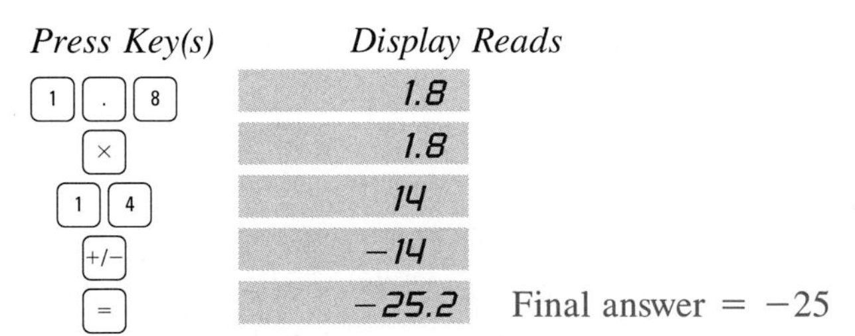

Press Key(s)	*Display Reads*	
[1] [.] [8]	1.8	
[×]	1.8	
[1] [4]	14	
[+/−]	−14	
[=]	−25.2	Final answer = −25

The display answer with its negative sign is rounded to two significant figures to give −25, the final answer.

Chain Calculations

The advantage of the calculator is that you can carry out several mathematical operations in a sequence. For example, suppose you want to solve the following calculation:

$$\frac{(18.6)(2.1)}{4.6}$$

The multiplication and division can all be done without writing down any intermediate answer, only the final answer. After each operation press [=] to review the current progress.

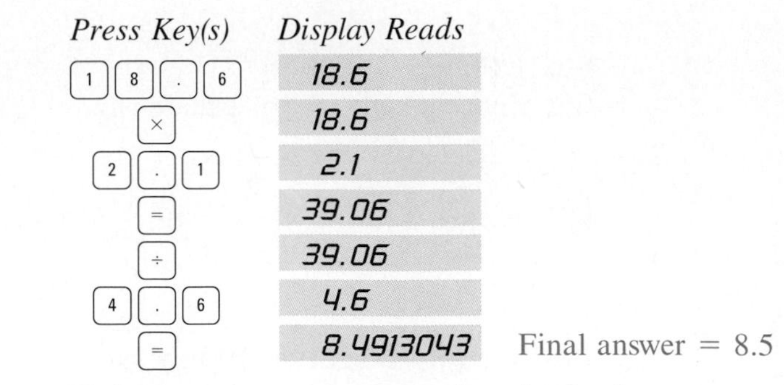

Press Key(s)	*Display Reads*	
1 8 . 6	18.6	
×	18.6	
2 . 1	2.1	
=	39.06	
÷	39.06	
4 . 6	4.6	
=	8.4913043	Final answer = 8.5

If there are two or more numbers in the denominator, be sure to divide by each, one at a time. Consider the following:

$$\frac{(1.45)(26.8)}{(16.8)(4.56)}$$

The problem should be read as 1.45 multiplied by 26.8 divided by 16.8 divided by 4.56. The steps are as follows:

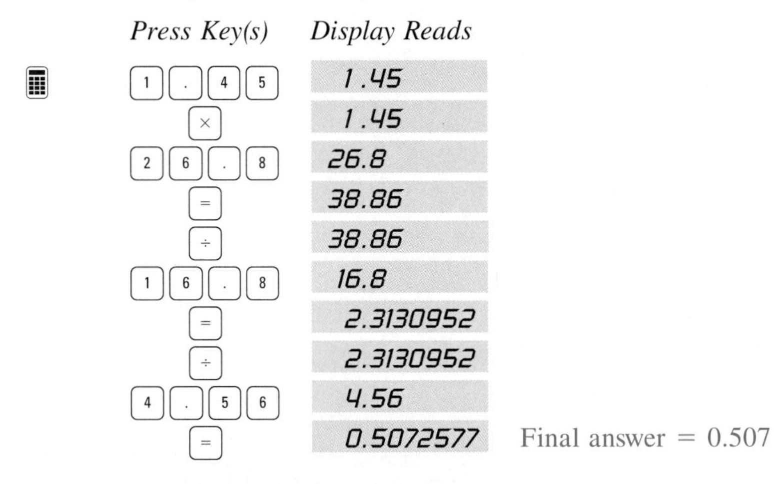

Press Key(s)	*Display Reads*	
1 . 4 5	1.45	
×	1.45	
2 6 . 8	26.8	
=	38.86	
÷	38.86	
1 6 . 8	16.8	
=	2.3130952	
÷	2.3130952	
4 . 5 6	4.56	
=	0.5072577	Final answer = 0.507

Sample Problem C.3
Chain Calculations with a Calculator

Perform the following calculations using the calculator:

a. $\frac{(1.6)(85.2)}{125}$

b. $\frac{(68)(1.52)}{(24)(5.16)}$

Solution

a.

Press Key(s)	*Display Reads*	
1 . 6	1.6	
×	1.6	
8 5 . 2	85.2	
=	136.32	
÷	136.32	
1 2 5	125	
=	1.09056	Final answer = 1.1

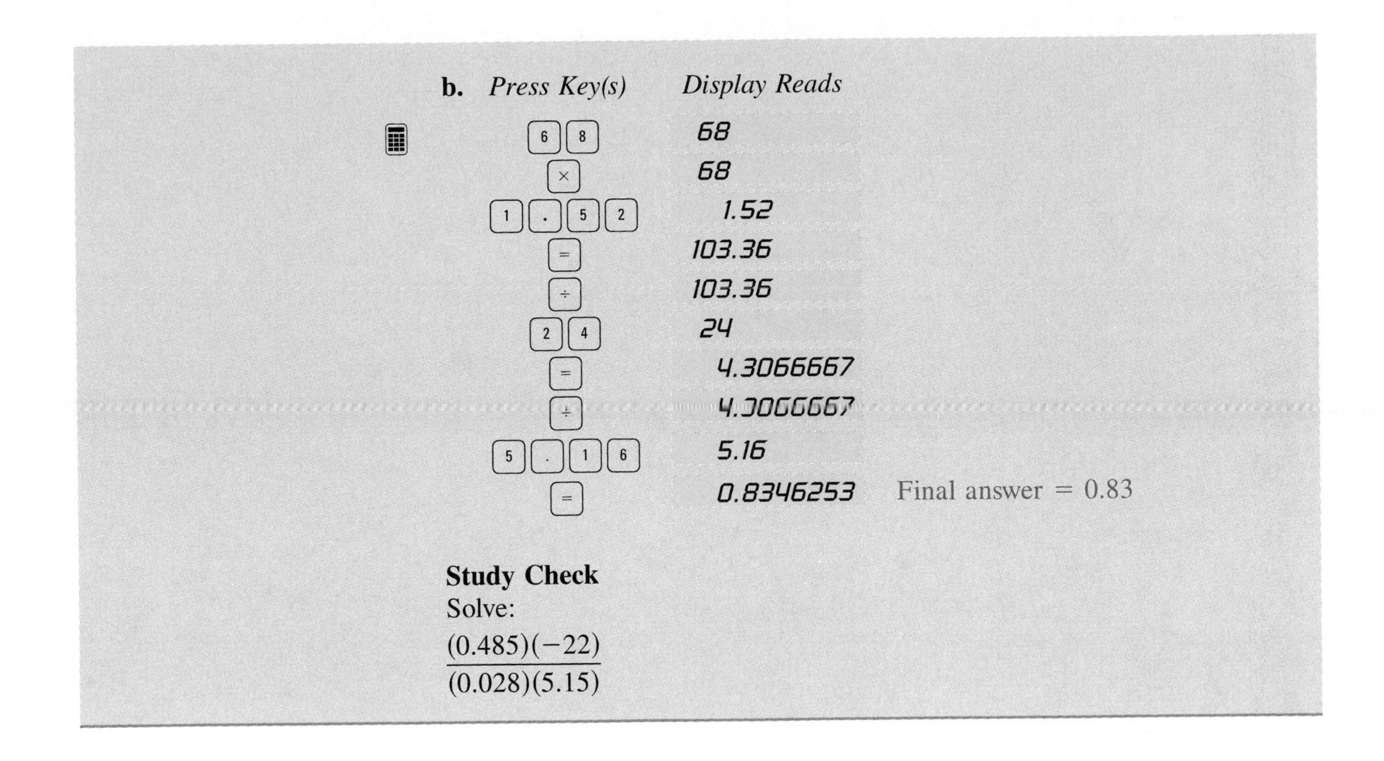

b.

Press Key(s)	*Display Reads*
6 8	68
×	68
1 . 5 2	1.52
=	103.36
÷	103.36
2 4	24
=	4.3066667
÷	4.3066667
5 . 1 6	5.16
=	0.8346253 Final answer = 0.83

Study Check

Solve:

$$\frac{(0.485)(-22)}{(0.028)(5.15)}$$

Numbers in Scientific Notation

Some calculators will convert an answer that is a large number or small number into scientific notation. For example, in the display area you may see the following numbers separated by a space or raised:

7.5 04 and 8^{-03}

These numbers have been given in scientific notation. The second part of the number is the power of 10. To write them in proper exponential form, the term × 10 must be inserted.

Display Reads		*Scientific Notation*
7.5 04	=	7.5×10^4
8^{-03}	=	8×10^{-3}

To enter numbers in scientific notation, use the EE key or the EXP key. For example, the steps to enter the number 2.5×10^4 are:

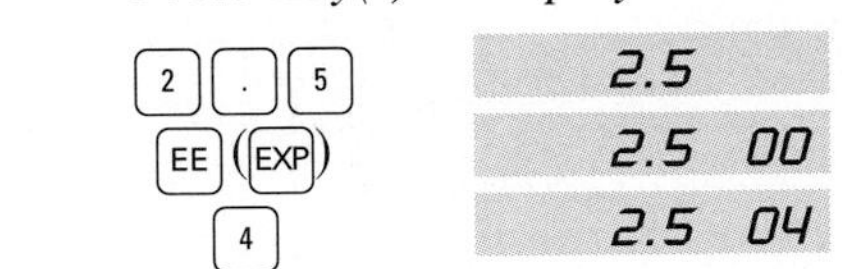

Press Key(s)	*Display Reads*
2 . 5	2.5
EE (EXP)	2.5 00
4	2.5 04

If the number has a negative exponent, use the +/− key to change its sign. For example, 1.8×10^{-6} is entered by:

Press Key(s)	*Display Reads*
1 . 8	1.8
EE (EXP)	1.8 00
6	1.8 06
+/−	1.8 −06

Sample Problem C.4
Scientific Notation in Calculations

Solve the following using the calculator:
a. $(2.7 \times 10^5) \times (3.6 \times 10^2)$
b. $(3.1 \times 10^{-2}) \times (1.8 \times 10^4)$

Solution

a.

Press Key(s)	*Display Reads*	
2 . 7	2.7	
EE (EXP)	2.7 00	
5	2.7 05	
×	2.7 05	
3 . 6	3.6	
EE (EXP)	3.6 00	
2	3.6 02	
=	9.72 07	Final answer = 9.7×10^7

b.

Press Key(s)	*Display Reads*	
3 . 1	3.1	
EE (EXP)	3.1 00	
2	3.1 02	
+/−	3.1 −02	
×	3.1 −02	
1 . 8	1.8	
EE (EXP)	1.8 00	
4	1.8 04	
=	5.58 02	Final answer = 5.6×10^2

Study Check
Solve:

$$\frac{5.2 \times 10^{-6}}{3.8 \times 10^3}$$

Appendix D Using Percent in Calculations

The term percent means parts per hundred. It represents a comparison of each item to the total items in the group. For example, suppose you have a basket of fruit containing 10 apples and 30 oranges. There is a total number of 40 items (apples and oranges) in this group of items called the whole. To find the percentage of apples in the basket, we compare the number of apples to the whole by dividing the number of apples by the whole and multiplying by 100. The percentage of oranges is found in the same way.

Percent Apples

$$\frac{\text{part}}{\text{whole}} = \frac{\text{number of apples}}{\text{total number of fruit}} = \frac{10 \text{ apples}}{40 \text{ fruit}} \times 100 = 25\% \text{ apples}$$

Percent Oranges

$$\frac{\text{part}}{\text{whole}} = \frac{\text{number of oranges}}{\text{total number of fruit}} = \frac{30 \text{ oranges}}{40 \text{ fruit}} \times 100 = 75\% \text{ oranges}$$

Note that the total percentages of all the items is equal to 100% (25% + 75%).

Sample Problem D.1
Calculating a Percentage

In one evening at a restaurant, 42 people ordered pasta, 54 people ordered salmon, and 24 people ordered steak. What was the percentage of salmon dinners ordered?

Solution
The total number of orders was 120 (42 + 54 + 24). The part of the whole that ordered salmon was 54. The percentage is calculated as:

$$\frac{\text{part}}{\text{whole}} = \frac{\text{number of salmon dinners}}{\text{total dinners ordered}} = \frac{54}{120} \times 100 = 45\%$$

That night, 45% of all the dinners ordered were salmon dinners.

Study Check
In the above problem, what were the percentage of pasta dinners and the percentage of steak dinners?

Using Percents as Conversion Factors

Sometimes, percentages are given in a problem. To work with a percent, it is convenient to write it as a conversion factor. To do this, we choose units from the problem to express the numerical relationship of the part to

100 parts of the whole. For example, an athlete might have 18% body fat by weight. We can use a choice of units for weight or mass to write the following conversion factors:

$$\frac{18\text{ kg body fat}}{100\text{ kg body mass}} \qquad \frac{18\text{ lb body fat}}{100\text{ lb body weight}}$$

If a question is asked about the number of kilograms of body fat in a marathon runner with a total body mass of 48 kg, we would state the units in the factor in kg body fat. Note that the units in the factor for the percentage cancel the given.

$$\underset{\text{Given}}{48\ \cancel{\text{kg body mass}}} \times \underset{\text{Percent factor}}{\frac{18\text{ kg body fat}}{100\ \cancel{\text{kg body mass}}}} = 8.6\text{ kg body fat}$$

If the problem asked about the pounds of body fat in a dancer weighing 126 lb, we would state the factor in lb body fat.

$$126\ \cancel{\text{lb body weight}} \times \frac{18\text{ lb body fat}}{100\ \cancel{\text{lb body weight}}} = 23\text{ lb body fat}$$

Sample Problem D.2
Using Percent Factors

A wine contains 13% alcohol by volume. How many milliliters of alcohol are contained in a glass containing 125 mL of the wine?

Solution

The factor for 13% alcohol can be written using mL units as:

$$\frac{13\text{ mL alcohol}}{100\text{ mL wine}}$$

The factor is then used with the given to set up the problem.

$$125\ \cancel{\text{mL}} \times \frac{13\text{ mL alcohol}}{100\ \cancel{\text{mL}}\text{ wine}} = 16\text{ mL alcohol}$$

Study Check

There are 85 students in a chemistry class. On the last test, 20% received grades of A. How many As were given on the test?

Answers to Study Checks

A.1 **a.** $5.8 \times 1000 = 5800$
b. $2 \times 100{,}000 = 200{,}000$

A.2 **a.** $8.5 \times \frac{1}{100} = 0.085$
b. $3 \times \frac{1}{100{,}000} = 0.00003$

C.1 **a.** 15.9 **b.** −48

C.2 **a.** 4.8 **b.** 156

C.3 −74

C.4 Display reads: 1.3684211 −09
Final answer = 1.4×10^{-9}

D.1 pasta dinners: 35%; steak dinners: 20%

D.2 17 A grades

Appendix E
Answers to Odd-Numbered Problems

Chapter 1

1.1 The base unit used to measure length in the metric system is the meter (m). In the American system, several units are used, including the inch (in.), foot (ft), and mile (mi).

1.3 **a.** meter; length **b.** gram; mass **c.** milliliter; volume **d.** meter; length

1.5 The prefix kilo means to multiply by 1000.

1.7 **a.** mg **b.** dL **c.** km **d.** kg **e.** μL

1.9 **a.** 0.01 **b.** 1000 **c.** 0.001 **d.** 0.1 **e.** 1,000,000

1.11 **a.** 100 cm **b.** 1000 m **c.** 0.001 m **d.** 1000 mL

1.13 **a.** kilogram **b.** centiliter **c.** km

1.15 A conversion factor can be inverted to give the second conversion factor.

1.17 **a.** 3 ft = 1 yd; 3 ft/1 yd, 1 yd/3 ft
b. 1 mile = 5280 feet; 5280 ft/1 mi, 1 mi/5280 ft
c. 1 min = 60 sec; 60 sec/1 min, 1 min/60 sec
d. 1 gal = 27 mi; 1 gal/27 mi, 27 mi/1 gal

1.19 **a.** 100 cm = 1 m; 100 cm/1 m, 1 m/100 cm
b. 1000 mg = 1 g; 1000 mg/1 g, 1 g/1000 mg
c. 1 L = 1000 mL; 1000 mL/1 L, 1 L/1000 mL
d. 1 dL = 100 mL; 100 mL/1 dL, 1 dL/100 mL

1.21 Significant figures must be counted in measured numbers used in calculations.

1.23 **a.** measured **b.** exact **c.** exact **d.** measured

1.25 Zeros are significant when they are part of a measured number. When they come before a recorded number or at the end of a large number (placeholders), they are not significant.

1.27 **a.** 5 **b.** 2 **c.** 2 **d.** 5

1.29 **a.** 1.85 **b.** 184 **c.** 0.00474 **d.** 8810 **e.** 1.83

1.31 **a.** 1.6 **b.** 0.01 **c.** 27.6 **d.** 3.5

1.33 **a.** 53.54 cm **b.** 127.6 g **c.** 121.5 mL **d.** 0.50 L

1.35 Conversion factors are used to solve problems. They are derived when an equality is expressed as a fraction.

1.37 **a.** 8.0 yd **b.** 900 min **c.** 0.88 gal

1.39 **a.** 1.75 m **b.** 5.5 L **c.** 55 g

1.41 **a.** 710. mL **b.** 75.0 kg **c.** 495 mm **d.** 9.5 kg

1.43 **a.** 66 gal **b.** 3 tablets **c.** 1800 mg

1.45 **a.** 1.20 g/mL **b.** 4.4 g/ml **c.** 3.10 g/mL

1.47 **a.** 210 g **b.** 575 g **c.** 62 oz

1.49 **a.** 1.030 **b.** 1.13 **c.** 0.85 g/mL

1.51 **a.** 31 kg **b.** $300.77

1.53 9 onions

1.55 **a.** 96 crackers **b.** 0.23 oz of fat **c.** 112 g of sodium

1.57 Dear Cousin, If you want to find the density of a diamond you will need a balance and a graduated cylinder. You can measure the mass of the diamond on the balance. Because its shape is probably irregular, you would first place some water in the cylinder and record its level. Then place the diamond under water and record the new water level. Subtract the first volume measurement from the second to find the volume of the diamond. Don't forget the units! Density is calculated by dividing the mass of the diamond by its volume. Your answer will be in units of g/cm^3. If you get 3.51 g/cm^3, you probably have a real diamond and you should take it to a jewelry store. If the density doesn't match, it may just be a piece of glass. Good luck, Cousin Bill.

1.59 720 cm^3

Chapter 2

2.1 When the car is at the top of the hill, it has all potential energy. As it descends, potential energy changes to kinetic energy. At the bottom of the hill, all the energy is kinetic.

2.3 **a.** potential **b.** kinetic **c.** potential **d.** potential

2.5 **a.** electrical energy changes to heat energy
b. electrical energy changes to kinetic energy
c. chemical energy changes to kinetic energy
d. radiant energy changes to heat energy

2.7 In the United States, we still use the Fahrenheit temperature scale. In °F, normal body temperature is 98.6. On the Celsius scale, her temperature would be 37.7°C, a mild fever.

2.9 **a.** 98.6°F **b.** 18.5°C **c.** 246 K **d.** 335 K **e.** 43°C **f.** 295 K

2.11 **a.** 41°C
b. No. The temperature is equivalent to 39°C.

2.13 Metals (aluminum, iron, or copper) conduct heat easily and get hot quickly.

2.15 **a.** 250 cal **b.** 230 cal **c.** 47,000 J **d.** 110 kJ

2.17 130 cal

2.19 **a.** 5 kcal **b.** 210 kcal **c.** 25 kcal

2.21 **a.** 68 kcal **b.** 18 g

2.23 210 kcal

2.25 **a.** gas **b.** gas **c.** solid

2.27 **a.** evaporation **b.** freezing **c.** boiling **d.** sublimation

2.29 **a.** physical property **b.** physical change **c.** physical change **d.** physical change

2.31 **a.** The water in perspiration changes to vapor. The heat needed for the change is lost by the skin.
b. On a hot day, there are more molecules with sufficient energy to become water vapor.
c. In a closed bag, some molecules evaporate; but they cannot escape and will condense back to liquid, and the clothes will not dry.

2.33

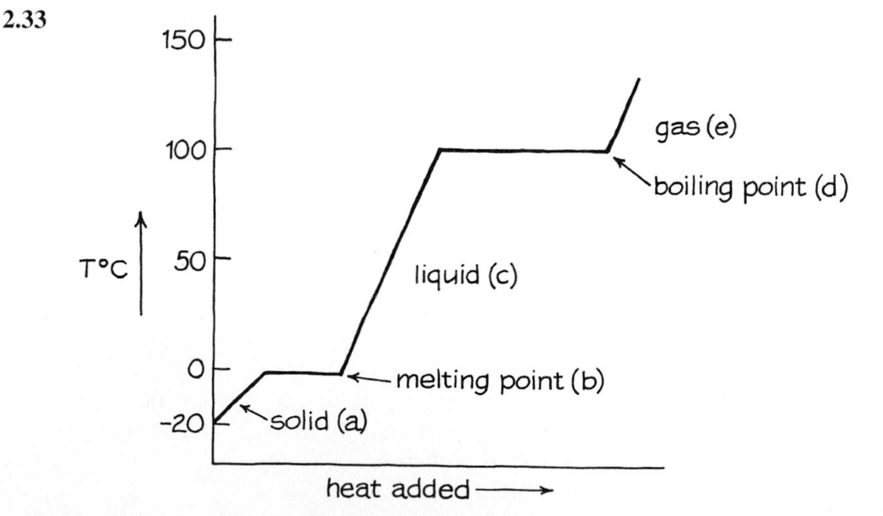

2.35 **a.** 9200 cal **b.** 640,000 cal **c.** 5400 cal **d.** 27,000 cal

2.37 **a.** 1100 cal **b.** 7300 cal **c.** 64,000 cal **d.** 58,000 cal

2.39 22 kcal, 93 kJ

2.41 60°C

2.43 **a.** 38 g of protein, 110 g of carbohydrate, 44 g of fat
b. 68 g of protein, 200 g of carbohydrate, 80 g of fat
c. 98 g of protein, 290 g of carbohydrate, 120 g of fat

2.45 3500 kcal, 15,000 kJ

2.47 **a.** As water freezes, heat is released into the air. The temperature will not go below 0°C until all the water has frozen.
b. 480 kcal

2.49 120,000 J; 1.2×10^5 J

2.51 320 g of ice melt

Chapter 3

3.1 **a.** Cu **b.** Si **c.** K **d.** N **e.** Fe **f.** Ba **g.** Pb **h.** Sr

3.3 **a.** carbon **b.** chlorine **c.** iodine **d.** mercury **e.** fluorine **f.** argon **g.** zinc **h.** nickel

3.5 **a.** electron **b.** proton **c.** electron **d.** neutron

3.7 **a.** atomic number **b.** both **c.** mass number **d.** atomic number

3.9 **a.** lithium, Li **b.** fluorine, F **c.** calcium, Ca **d.** zinc, Zn **e.** neon, Ne **f.** silicon, Si **g.** iodine, I **h.** oxygen, O

3.11 **a.** 12 **b.** 30 **c.** 53 **d.** 19

3.13

Name of Element	Symbol	Atomic Number	Mass Number	Number of Protons	Number of Neutrons	Number of Electrons
Aluminum	Al	13	27	13	14	13
Magnesium	Mg	12	24	12	12	12
Potassium	K	19	39	19	20	19
Sulfur	S	16	31	16	15	16
Iron	Fe	26	56	26	30	26

3.15 **a.** 13 protons, 14 neutrons, 13 electrons
b. 24 protons, 28 neutrons, 24 electrons
c. 16 protons, 18 neutrons, 16 electrons
d. 26 protons, 30 neutrons, 26 electrons

3.17 **a.** $^{31}_{15}P$ **b.** $^{80}_{35}Br$ **c.** $^{27}_{13}Al$ **d.** $^{35}_{17}Cl$

3.19 **a.** $^{32}_{16}S$ $^{33}_{16}S$ $^{34}_{16}S$ $^{36}_{16}S$
b. They all have the same number of protons and electrons.
c. They have different numbers of neutrons, which gives them different mass numbers.
d. The atomic weight of S is the weighted average mass of all the isotopes.

3.21 **a.** Period 2 **b.** Group 8A **c.** Group 1A
d. Period 2

3.23 **a.** alkaline earth **b.** transition element
c. noble gas **d.** alkali metal **e.** halogen

3.25 **a.** C **b.** He **c.** Na **d.** Ca **e.** Al

3.27 **a.** metal **b.** nonmetal **c.** metal
d. nonmetal **e.** nonmetal

3.29 **a.** nonmetal **b.** metal **c.** metal
d. nonmetal **e.** nonmetal **f.** nonmetal
g. nonmetal **h.** metal

3.31 **a.** 2,4 **b.** 2, 8, 8 **c.** 2, 8, 6 **d.** 2, 8, 8, 1
e. 2, 8, 3 **f.** 2, 5

3.33 **a.** Li **b.** Mg **c.** H **d.** Cl **e.** O

3.35 **a.** absorb **b.** give off (emit)

3.37 **a.** B 2, 3 Al 2, 8, 3 **b.** 3 **c.** 3A

3.39 **a.** 2 e^-, 2A **b.** 7 e^-, 7A **c.** 6 e^-, 6A
d. 5 e^-, 5A

3.41 Yes. They each have two electrons in their outer shells.

3.43 **a.** 2 **b.** 6 **c.** 8 **d.** 2

3.45 **a.** $1s^2 2s^2 2p^6 3s^2$ **b.** $1s^2 2s^2 2p^6 3s^2 3p^3$
c. $1s^2 2s^2 2p^6 3s^2 3p^6$ **d.** $1s^2 2s^2 2p^6 3s^2 3p^4$
e. $1s^2 2s^2 2p^6 3s^2 3p^5$

3.47 **a.** H **b.** N **c.** Na **d.** Ne

3.49 **a.** An isotope is a specific form of an atom.
b. Atomic number gives the number of protons in an atom; mass number gives the total number of protons and neutrons.

3.51 **a.** The proton is a positive particle.
b. Electrons are found in the outer part of the atom.
c. The nucleus is a very small part of the atom.
d. The electron has a negative charge.
e. Most of the mass of the atom is due to its protons and neutrons.

3.53 **a.** 26 protons, 30 neutrons, 26 electrons
b. $^{51}_{26}Fe$ **c.** $^{51}_{24}Cr$

3.55 63.5

3.57 8.09×10^7 sodium atoms

Chapter 4

4.1
a. 2, 5 five valence electrons
b. 2, 6 six valence electrons
c. 2, 8, 8 eight valence electrons
d. 2, 8, 8, 1 one valence electron
e. 2, 8, 6 six valence electrons

4.3 a. 6A :S̤̈·
b. 5A ·N̤̈·
c. 2A Ca·
d. 1A Na·
e. 1A K·

4.5 **a.** :N̤̈e: stable
b. :Ö· will form a compound
c. Li· will form a compound
d. :Ä̤r: stable

4.7 **a.** Li^+ **b.** F^- **c.** Mg^{2+} **d.** Fe^{3+} **e.** Zn^{2+}

4.9 **a.** lose $2e^-$ **b.** gain $3e^-$ **c.** gain $1e^-$ **d.** lose $1e^-$ **e.** lose $3e^-$

4.11 **a.** Cl^- **b.** K^+ **c.** O^{2-} **d.** Al^{3+}

4.13 **a.** potassium ion **b.** sulfide ion **c.** calcium ion **d.** nitride ion

4.15 **a.** iron(II) or ferrous ion **b.** copper(II) or cupric ion **c.** zinc ion **d.** lead(IV) or plumbic ion

4.17 **a.** K· ·$\ddot{\underset{..}{Cl}}$: ⟶ K^+ :$\ddot{\underset{..}{Cl}}$:$^-$ ⟶ KCl

b. $\dot{Mg}$· ·$\ddot{\underset{..}{Cl}}$: ·$\ddot{\underset{..}{Cl}}$: ⟶ Mg^{2+} 2[:$\ddot{\underset{..}{Cl}}$: $^-$] ⟶ $MgCl_2$

c. Na· Na· Na· → ·$\ddot{N}$· ⟶ 3 Na^+ :$\ddot{\underset{..}{N}}$:$^{3-}$ ⟶ Na_3N

4.19 **a.** Na_2O **b.** $AlBr_3$ **c.** $BaCl_2$ **d.** $ZnCl_2$ **e.** Al_2S_3

4.21 **a.** Na_2S **b.** K_3N **c.** AlI_3 **d.** Li_2O

4.23 **a.** HCO_3^- **b.** NH_4^+ **c.** PO_4^{3-} **d.** HSO_4^-

4.25 **a.** sulfate ion **b.** carbonate ion **c.** phosphate ion **d.** nitrate ion

4.27

	OH^-	NO_2^-	CO_3^{2-}	HSO_4^-	PO_4^{3-}
Li^+	$LiOH$	$LiNO_2$	Li_2CO_3	$LiHSO_4$	Li_3PO_4
Cu^{2+}	$Cu(OH)_2$	$Cu(NO_2)_2$	$CuCO_3$	$Cu(HSO_4)_2$	$Cu_3(PO_4)_2$
Ba^{2+}	$Ba(OH)_2$	$Ba(NO_2)_2$	$BaCO_3$	$Ba(HSO_4)_2$	$Ba_3(PO_4)_2$

4.29 **a.** aluminum oxide **b.** calcium chloride **c.** sodium oxide **d.** magnesium nitride **e.** potassium iodide

4.31 **a.** 2+ **b.** 2+ **c.** 1+ **d.** 2+

4.33 **a.** iron(II) chloride; ferrous chloride **b.** copper(II) oxide; cupric oxide **c.** iron(III) sulfide; ferric sulfide **d.** aluminum phosphide

4.35 **a.** $MgCl_2$ **b.** Na_2S **c.** Cu_2O **d.** Zn_3P_2 **e.** AuN

4.37 **a.** $Na_2(CO_3)$ sodium carbonate

b. $(NH_4)Cl$ ammonium chloride

c. $Li_3(PO_4)$ lithium phosphate

d. $Cu(NO_2)_2$ copper(II) nitrite; cupric nitrite

e. $Fe(SO_3)$ iron(II) sulfite; ferrous sulfite

4.39 **a.** $Ba(OH)_2$ **b.** Na_2SO_4 **c.** $Fe(NO_3)_2$ **d.** $Zn_3(PO_4)_2$ **e.** $Fe_2(CO_3)_3$

4.41 **a.** :$\ddot{\underset{..}{Br}}$:$\ddot{\underset{..}{Br}}$: **b.** H:$\ddot{\underset{..}{S}}$: with H below S **c.** H:$\ddot{\underset{..}{F}}$: **d.** :$\ddot{\underset{..}{F}}$:$\ddot{\underset{..}{O}}$: with :$\ddot{\underset{..}{F}}$: below O

4.43 **a.** :$\ddot{O}$: :$\ddot{S}$:$\ddot{\underset{..}{O}}$: with :$\underset{..}{O}$: below S

b. H :$\ddot{C}$: H with :$\ddot{O}$: below C

c. :$\ddot{\underset{..}{O}}$:$\ddot{S}$: :$\ddot{O}$:

4.45 **a.** dihydrogen sulfide **b.** carbon tetrabromide **c.** sulfur dichloride **d.** hydrogen fluoride **e.** nitrogen triiodide

4.47 **a.** CCl_4 **b.** CO **c.** PCl_3 **d.** N_2O_4

4.49 **a.** aluminum sulfate **b.** calcium carbonate **c.** dinitrogen monoxide **d.** sodium phosphate **e.** ammonium sulfate

4.51 **a.** $H^{\delta+}—F^{\delta-}$ **b.** $C^{\delta+}—Cl^{\delta-}$ **c.** $N^{\delta+}—O^{\delta-}$ **d.** $N^{\delta+}—F^{\delta-}$

4.53 **a.** covalent **b.** none **c.** covalent **d.** ionic **e.** polar covalent

4.55 **a.** P **b.** Na **c.** Al **d.** Si

4.57 **a.** 2, 8 **b.** 2, 8 **c.** 2, 8 **d.** 2, 8 **e.** 2

4.59 **a.** 2A **b.** $\dot{Z}$· **c.** Be

4.61 **a.** Sn^{4+} **b.** 50 protons, 46 electrons **c.** SnO_2 **d.** $Sn_3(PO_4)_4$

Chapter 5

5.1 **a.** one atom of iron, one atom of sulfur, and four atoms of oxygen
b. two atoms of aluminum and three atoms of oxygen
c. one atom aluminum, three atoms oxygen, and three atoms of hydrogen
d. two atoms of nitrogen, eights atoms of hydrogen, one atom of carbon, and three atoms of oxygen
e. seven atoms of carbon, five atoms of hydrogen, one atom of nitrogen, three atoms of oxygen, and one sulfur atom
f. 22 atoms of carbon, 25 atoms of hydrogen, 1 atom of chlorine, 2 atoms of nitrogen, and 8 atoms of oxygen

5.3 **a.** 151.9 amu **b.** 102.0 amu **c.** 78.0 amu **d.** 96.0 amu **e.** 183.1 amu **f.** 481 amu

5.5 A mole is the quantity of a substance that contains 6.02×10^{23} particles of the substance.

5.7 **a.** 6.02×10^{23} atoms C
b. 6.02×10^{23} atoms Fe
c. 6.02×10^{23} molecules SO_2
d. 6.02×10^{23} molecules HCl

5.9 A mole of iron atoms is heavier than a mole of sodium atoms because an atom of iron is heavier than an atom of sodium.

5.11 **a.** 23.0 amu **b.** 35.5 amu **c.** 12.0 amu **d.** 207.2 amu **e.** 63.6 amu

5.13 **a.** 42.0 g **b.** 78.0 g **c.** 80.0 g **d.** 355 g **e.** 322 g

5.15 **a.** 34.5 g **b.** 32 g **c.** 14.8 g **d.** 398 g **e.** 0.48 g

5.17 **a.** 8.50 g **b.** 109 g **c.** 8.52 g **d.** 176 g **e.** 1.5 g

5.19 **a.** 19.6 g **b.** 35.5 g **c.** 53.0 g **d.** 2.00 g **e.** 49.0 g

5.21 **a.** 600 g **b.** 11 g

5.23 **a.** 0.463 mol **b.** 0.629 mol **c.** 0.0167 mol **d.** 0.882 mol **e.** 1.17 mol

5.25 **a.** 0.568 mol **b.** 0.625 mol **c.** 0.262 mol **d.** 0.543 mol **e.** 0.467 mol

5.27 12 mol

5.29 **a.** physical **b.** chemical **c.** physical **d.** chemical **e.** physical **f.** chemical

5.31 **a.** Two molecules of nitrogen monoxide react with one molecule of oxygen to produce two molecules of nitrogen dioxide.
b. Two molecules of dihydrogen sulfide react with three molecules of oxygen to produce two molecules of sulfur dioxide and two molecules of water.
c. One molecule of ethanol reacts with three molecules of oxygen to produce two molecules of carbon dioxide and three molecules of water.

5.33 **a.** $4NH_3 + 3O_2 \longrightarrow 2N_2 + 6H_2O$
b. $4Fe + 3O_2 \longrightarrow 2Fe_2O_3$

5.35 **a.** 2Na, 2Cl **b.** 2N, 8H, 4O
c. 4P, 16O, 12H **d.** 3C, 8H, 10O

5.37 **a.** $N_2 + O_2 \longrightarrow 2NO$
b. $2HgO \longrightarrow 2Hg + O_2$
c. $4Fe + 3O_2 \longrightarrow 2Fe_2O_3$
d. $2Na + Cl_2 \longrightarrow 2NaCl$
e. $2Cu_2O + O_2 \longrightarrow 4CuO$

5.39 **a.** $Mg + 2AgNO_3 \longrightarrow Mg(NO_3)_2 + 2Ag$
b. $CuCO_3 \longrightarrow CuO + CO_2$
c. $2Al + 3CuSO_4 \longrightarrow 3Cu + Al_2(SO_4)_3$
d. $Pb(NO_3)_2 + 2NaCl \longrightarrow PbCl_2 + 2NaNO_3$
e. $2Al + 6HCl \longrightarrow 2AlCl_3 + 3H_2$

5.41 **a.** The energy of activation is the energy required to break the bonds of the reacting molecules.
b. A catalyst lowers the activation energy and speeds up a reaction.
c. In exothermic reactions, the energy of the products is lower than the reactants.
d.

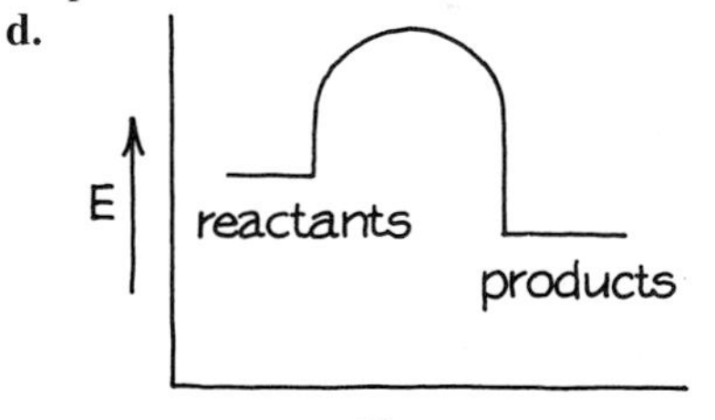

5.43 **a.** exothermic **b.** endothermic **c.** exothermic

5.45 **a.** exothermic **b.** endothermic **c.** exothermic

5.47 **a.** The rate of a reaction tells how fast the products are formed.
b. Reactions go faster at higher temperatures.

5.49 **a.** increase **b.** increase **c.** increase **d.** decrease

5.51 **a.** Two moles of SO_2 react with one mole of O_2 to produce two moles of SO_3.
b. Three moles of carbon react with two moles of SO_2 to produce one mole of CS_2 and two moles of CO_2.

5.53 **a.** $2 \times SO_2(128.2\ g) + O_2\ (32.0\ g) = 2 \times SO_3\ (160.2\ g)$
$160.2\ g = 160.2\ g$
b. $3 \times C(36.0\ g) + 2 \times SO_2\ (128.2\ g) = CS_2\ (76.2\ g) + 2 \times CO_2\ (88.0\ g)$
$164.2\ g = 164.2\ g$

5.55 **a.** $\frac{2\ mol\ SO_2}{1\ mol\ O_2}$ and $\frac{1\ mol\ O_2}{2\ mol\ SO_2}$;
$\frac{2\ mol\ SO_2}{2\ mol\ SO_3}$ and $\frac{2\ mol\ SO_3}{2\ mol\ SO_2}$
$\frac{1\ mol\ O_2}{2\ mol\ SO_3}$ and $\frac{2\ mol\ SO_3}{1\ mol\ O_2}$
b. $\frac{3\ mol\ C}{2\ mol\ SO_2}$ and $\frac{2\ mol\ SO_2}{3\ mol\ C}$;
$\frac{2\ mol\ SO_2}{1\ mol\ CS_2}$ and $\frac{1\ mol\ CS_2}{2\ mol\ SO_2}$
$\frac{3\ mol\ C}{2\ mol\ CO_2}$ and $\frac{2\ mol\ CO_2}{3\ mol\ C}$;
$\frac{2\ mol\ SO_2}{2\ mol\ CO_2}$ and $\frac{2\ mol\ CO_2}{2\ mol\ SO_2}$
$\frac{3\ mol\ C}{1\ mol\ CS_2}$ and $\frac{1\ mol\ CS_2}{3\ mol\ C}$;
$\frac{1\ mol\ CS_2}{2\ mol\ CO_2}$ and $\frac{2\ mol\ CO_2}{1\ mol\ CS_2}$

5.57 **a.** 1.0 mol S **b.** 5.0 mol Cu_2S
c. 200 g Cu_2S

5.59 **a.** 6.0 mol O_2 **b.** 84.0 g N_2 **c.** 80.0 g O_2
d. 54 g H_2O

5.61 336 mol Al

5.63 0.009 mol ethanol

5.65 **a.** $C_2H_6O + 3O_2 \longrightarrow 2CO_2 + 3H_2O$
b. 12 mol O_2
c. 96 g O_2
d. The rate of reaction would increase.

Chapter 6

6.1 **a.** Both an α-particle and a helium nucleus have 2 protons and 2 neutrons. However, an α-particle is emitted from a nucleus during radioactive decay.
b. α, ${}^{4}_{2}He$

6.3 **a.** ${}^{39}_{19}K$, ${}^{40}_{19}K$, ${}^{41}_{19}K$
b. They all have 19 protons and 19 electrons, but they differ in the number of neutrons.

6.5

Medical Use	Isotope Symbol	Mass Number	Number of Protons	Number of Neutrons
spleen imaging	${}^{51}_{24}Cr$	51	24	27
malignancies	${}^{60}_{27}Co$	60	27	33
blood volume	${}^{59}_{26}Fe$	59	26	33
hyperthyroidism	${}^{131}_{53}I$	131	53	78
leukemia treatment	${}^{32}_{15}P$	32	15	17

6.7 **a.** α, ${}^{4}_{2}He$ **b.** ${}^{1}_{0}n$ **c.** β, ${}^{0}_{-1}e$ **d.** ${}^{15}_{7}N$ **e.** ${}^{125}_{53}I$

6.9 **a.** β **b.** α **c.** neutron **d.** sodium-24
e. carbon-14

6.11 **a.** Because β-particles move faster than α-particles, they can penetrate further into tissue.
b. Ionizing radiation breaks bonds and forms reactive species that cause undesirable reactions in the cells.
c. X-ray technicians leave the room to increase the distance between them and the radiation. Also, they are shielded by a wall that contains a lead shield.
d. Wearing gloves shields the skin from α and β radiation.

6.13 **a.** ${}^{208}_{84}Po \longrightarrow {}^{204}_{82}Pb + {}^{4}_{2}\alpha$
b. ${}^{232}_{90}Th \longrightarrow {}^{228}_{88}Ra + {}^{4}_{2}\alpha$
c. ${}^{251}_{102}No \longrightarrow {}^{247}_{100}Fm + {}^{4}_{2}\alpha$
d. ${}^{220}_{86}Rn \longrightarrow {}^{216}_{84}Po + {}^{4}_{2}\alpha$

6.15 **a.** ${}^{25}_{11}Na \longrightarrow {}^{25}_{12}Mg + {}^{0}_{-1}e$
b. ${}^{20}_{8}O \longrightarrow {}^{20}_{9}F + {}^{0}_{-1}e$
c. ${}^{92}_{38}Sr \longrightarrow {}^{92}_{39}Y + {}^{0}_{-1}e$
d. ${}^{42}_{19}K \longrightarrow {}^{42}_{20}Ca + {}^{0}_{-1}e$

6.17 **a** ${}^{28}_{14}Si$ **b.** ${}^{87}_{36}Kr$ **c.** ${}^{0}_{-1}e$ **d.** ${}^{238}_{92}U$

6.19 **a.** ${}^{10}_{4}Be$ **b.** ${}^{0}_{-1}e$ **c.** ${}^{27}_{13}Al$

6.21 **a.** When radiation enters the Geiger counter, it ionizes a gas in the detection tube. The electrical charges in the tube produce a burst of current that is detected by the instrument.
b. Ci
c. RAD

6.23 **a.** 2.2×10^{12} disintegrations **b.** 294 μCi

6.25 When pilots are flying at high altitudes, there is less atmosphere to protect them from radiation.

6.27 Because the elements Ca and P are part of bone, their radioactive isotopes will also become part of the bony structures of the body. The radioactive Sr also seeks out bone because Sr acts much like Ca. Once the radioactive isotopes are in the bone, their radiation can be used to diagnose and reduce a bone lesion or bone tumor.

6.29 ^{131}I locates in the thyroid, where its radiation will destroy some of the thyroid cells that produce thyroid hormone.

6.31 Half-life is the time it takes for one-half of a radioactive sample to decay.

6.33 **a.** 40.0 mg **b.** 20.0 mg **c.** 10.0 mg **d.** 5.0 mg

6.35 128 days, 192 days

6.37 Nuclear fission is the splitting of a large atom into smaller fragments and a large amount of energy.

6.39 $^{103}_{42}Mo$

6.41 **a.** fission **b.** fusion **c.** fission **d.** fusion

6.43 **a.** $^{17}_{8}O$ **b.** $^{28}_{14}Si$

6.45 16 μCi/g

Chapter 7

7.1 **a.** Gases are composed of very small particles (atoms or molecules).
b. Particles of a gas move rapidly in all directions.
c. The particles of gas are far apart.
d. There are no attractive forces between gas particles.
e. Gases move faster at high temperatures.

7.3 volume, mL or L; temperature, °C, K; quantity, moles

7.5 atmospheres (atm), mm Hg, torr, lb/in.^{2}

7.7 **a.** 1520 torr **b.** 29.4 lb/in^2 **c.** 1520 mm Hg

7.9 **a.** boiling point **b.** vapor pressure **c.** atmospheric pressure **d.** boiling point

7.11 **a.** On top of a mountain, water boils below 100°C because the atmospheric pressure is less than 1 atm.
b. The pressure inside a pressure cooker is greater than 1 atm. Therefore water boils above 100°C. At a higher temperature, food cooks faster.

7.13 **a.** increases **b.** decreases

7.15 **a.** 328 torr **b.** 1310 torr **c.** 4400 torr **d.** 65,500 torr

7.17 2300 mm Hg

7.19 **a.** 25.3 L **b.** 25 L **c.** 100. L **d.** 44.7 L

7.21 25 L

7.23 **a.** inspiration **b.** expiration **c.** inspiration

7.25 **a.** decreases **b.** increases

7.27 **a.** 2400 mL **b.** 4900 mL **c.** 1800 mL **d.** 1700 mL

7.29 **a.** increases **b.** decreases

7.31 **a.** 770 torr **b.** 1.51 atm

7.33 $\frac{P_iV_i}{T_i} = \frac{P_fV_f}{T_f}$
Boyles, Charles, and Gay–Lussac's laws are combined to make this law.

7.35 **a.** 5.53 atm **b.** 0.948 atm **c.** 47.8 atm **d.** 0.625 atm

7.37 110 mL

7.39 **a.** 5.00 L **b.** 1300 mL **c.** 3.0 mol SO_2

7.41 **a.** 2.00 mol O_2 **b.** 0.179 mol CO_2 **c.** 4.48 L **d.** 55,400 mL

7.43 765 torr

7.45 425 torr

7.47 4.9 atm

7.49 29 mol O_2

7.51 44 L

7.53 12 atm

7.55 −223°C

7.57 He 600 torr O_2 1800 torr

7.59 370 torr

7.61 44.8 L CO_2

Chapter 8

8.1 **a.** $O^{\delta-}$, $^{\delta+}H$, $H^{\delta+}$
The oxygen in the water molecule has a partial negative charge and the hydrogen atoms have partial positive charges.
b.

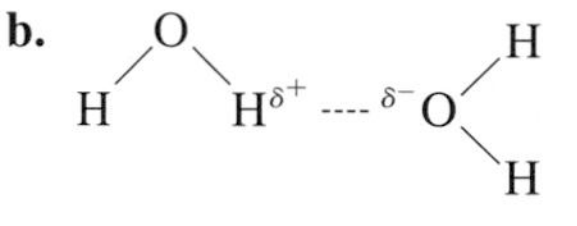

c. HF is polarized, with a partial positive charge on H and partial negative on F.

$H^{\delta+} — F^{\delta-}$

d. CH_4 would not form hydrogen bonds because the C—H bond is not sufficiently polar.

8.3 **a.** NaCl, solute; water, solvent **b.** water, solute; ethanol, solvent **c.** oxygen, solute; nitrogen, solvent

8.5 **a.** oxygen atom **b.** hydration **c.** hydrogen atom

8.7 The K^+ and I^- ions are pulled away from the solid by the polar water molecules into the water where they are hydrated.

8.9 **a.** increase **b.** decrease **c.** decrease **d.** increase

8.11 **a.** water **b.** CCl_4 **c.** water **d.** CCl_4

8.13 **a.** saturated **b.** unsaturated

8.15 **a.** unsaturated **b.** saturated **c.** saturated **d.** unsaturated

8.17 **a.** unsaturated **b.** saturated **c.** 20 g KBr

8.19 **a.** The solubility increases as temperature increases.
b. The solubility of a gas (CO_2) is less at higher temperature.
c. The solubility is less at higher temperature and the CO_2 pressure in the can is increased.

8.21 **a.** soluble **b.** insoluble **c.** insoluble **d.** soluble **e.** soluble

8.23 5% (m/m) is 5 g of glucose in 100 g of solution, whereas 5% (m/v) is 5 g of glucose in 100 mL of solution.

8.25 **a.** 13.3% **b.** 16.0% **c.** 11.1%

8.27 **a.** 30.% **b.** 0.30% **c.** 10.%

8.29 **a.** $\frac{10\text{g NaOH}}{100\text{ g solution}}$ and $\frac{100\text{ g solution}}{10\text{ g NaOH}}$

b. $\frac{0.5\text{ g NaCl}}{100\text{ mL solution}}$ and $\frac{100\text{ mL solution}}{0.5\text{ g NaCl}}$

c. $\frac{15\text{ mL }CH_3OH}{100\text{ mL solution}}$ and $\frac{100\text{ mL solution}}{15\text{ mL }CH_3OH}$

8.31 **a.** 2.5 g KCl **b.** 22.5 g K_2CO_3 **c.** 50. g NH_4Cl **d.** 25.0 mL acetic acid

8.33 **a.** 20 g mannitol **b.** 480 g mannitol

8.35 **a.** 1000 g **b.** 20 mL **c.** 400 mL **d.** 20 mL

8.37 2.0 L

8.39 **a.** 0.036 M KOH **b.** 0.50 M glucose **c.** 1.00 M NaOH **d.** 2.5 M NaCl

8.41 **a.** $\frac{6.00\text{ mol HCl}}{1\text{ L solution}}$ and $\frac{1\text{ L solution}}{6.00\text{ mol HCl}}$

b. $\frac{0.250\text{ mol }NaHCO_3}{1\text{ L solution}}$ and $\frac{1\text{ L solution}}{0.250\text{ mol }NaHCO_3}$

c. $\frac{1.0\text{ mol }H_2SO_4}{1\text{ L solution}}$ and $\frac{1\text{ L solution}}{1.0\text{ mol }H_2SO_4}$

8.43 **a.** 3.0 mol NaCl **b.** 0.40 mol KBr **c.** 0.80 mol NaCl

8.45 **a.** 40. g NaOH **b.** 600 g KCl **c.** 29 g NaCl

8.47 **a.** 1.0 L **b.** 10 L **c.** 2 L

8.49 **a.** 4.0% KCl **b.** 10% mannitol **c.** 2.0 M HCl **d.** 3.0% NaCl

8.51 **a.** solution **b.** colloid **c.** suspension

8.53 An emulsion is a colloid in which a liquid is dispersed in another liquid or a solid.

8.55 **a.** 10% starch **b.** from pure water into the starch **c.** starch

8.57 **a.** B 10% glucose **b.** B 8% albumin **c.** B 10% NaCl

8.59 **a.** hypotonic **b.** hypotonic **c.** isotonic **d.** isotonic **e.** hypertonic

8.61 **a.** NaCl **b.** alanine **c.** NaCl **d.** urea

8.63 **a.** 60 g of amino acids, 380 g of glucose and 100 g of lipids **b.** 2700 kcal

8.65 **a.** 35.0% HNO_3 **b.** 165 mL **c.** 42.4% **d.** 6.73 M

8.67 The solution will dehydrate the flowers because water will flow out of the cells of the flowers into the salt solution.

8.69 Drinking sea water will cause water to flow out of the body cells and further dehydrate a person.

Chapter 9

9.1 In a solution of KF, only the ions of K^+ and F^- are present. In an HF solution, there are a few ions of H^+ and F^- present but mostly dissolved HF molecules.

9.3 **a.** $KCl \xrightarrow{H_2O} K^+ + Cl^-$

b. $CaCl_2 \xrightarrow{H_2O} Ca^{2+} + 2Cl^-$

c. $K_3PO_4 \xrightarrow{H_2O} 3K^+ + PO_4^{3-}$

d. $Fe(NO_3)_3 \xrightarrow{H_2O} Fe^{3+} + 3NO_3^-$

9.5 **a.** mostly molecules and a few ions
b. ions only
c. molecules only

9.7 **a.** $HC_2H_3O_2 \overset{H_2O}{\rightleftharpoons} H^+ + C_2H_3O_2^-$

b. $NaBr \xrightarrow{H_2O} Na^+ + Br^-$

c. $C_6H_{12}O_6(s) \xrightarrow{H_2O} C_6H_{12}O_6\ (aq)$

9.9 **a.** strong electrolyte
b. weak electrolyte
c. nonelectrolyte

9.11 **a.** 1 equiv **b.** 2 equiv **c.** 2 equiv **d.** 6 equiv

9.13 **a.** 0.704 equiv **b.** 0.806 equiv **c.** 0.200 equiv **d.** 1.00 equiv

9.15 0.18 g Mg^{2+}

9.17 3.54 g Na^+, 5.47 g Cl^-

9.19 55 meq/L

9.21 Arrhenius acid: $HBr \xrightarrow{H_2O} H^+ + Br^-$
Brønsted–Lowry acid: $HBr + H_2O \longrightarrow H_3O^+ + Br^-$

9.23 **a.** acid **b.** acid **c.** acid **d.** base **e.** base

9.25 **a.** hydrochloric acid
b. calcium hydroxide
c. lithium hydroxide
d. sulfuric acid

9.27 A strong acid is completely ionized in solution, but a weak acid is only partially ionized.

9.29 HCl, HNO_3, H_2SO_4

9.31 **a.** $HI + H_2O \longrightarrow H_3O^+ + I^-$

b. $Ca(OH)_2 \longrightarrow Ca^{2+} + 2OH^-$

c. $LiOH \longrightarrow Li^+ + OH^-$

d. $HNO_2 + H_2O \rightleftharpoons H_3O^+ + NO_2^-$

9.33 (b) and (d) are neutralization reactions; (a) and (c) are not.

9.35 **a.** $2HCl + Mg(OH)_2 \longrightarrow MgCl_2 + 2H_2O$

b. $H_3PO_4 + 3LiOH \longrightarrow Li_3PO_4 + 3H_2O$

9.37 **a.** $H_2SO_4 + 2NaOH \longrightarrow Na_2SO_4 + 2H_2O$

b. $3HCl + Fe(OH)_3 \longrightarrow FeCl_3 + 3H_2O$

c. $H_2CO_3 + Mg(OH)_2 \longrightarrow MgCO_3 + 2H_2O$

9.39 $[H_3O^+] = [OH^-] = 1 \times 10^{-7}$ M

9.41 **a.** $[H_3O^+] = 1 \times 10^{-6}$ M
b. $[H_3O^+] = 1 \times 10^{-12}$ M
c. $[H_3O^+] = 1 \times 10^{-3}$ M
d. $[H_3O^+] = 5 \times 10^{-10}$ M

9.43 **a.** basic **b.** basic **c.** neutral **d.** acidic **e.** basic **f.** acidic

9.45 **a.** basic **b.** acidic **c.** basic **d.** acidic **e.** neutral

9.47 **a.** 5, 7, 10, 14 **b.** 1, 8, 10, 13 **c.** 1.6, 2.4, 5.5, 8.5 **d.** 3.5, 8.8, 9.7, 11.4

9.49 **a.** 4.0 **b.** 9.0 **c.** 12.0 **d.** 9.0 **e.** 3.0

9.51 **a.** 1.4 **b.** 7.7 **c.** 3.7

9.53

	$[H_3O^+]$	$[OH^-]$
a.	1×10^{-2} M	1×10^{-12} M
b.	1×10^{-7} M	1×10^{-7} M
c.	1×10^{-13} M	1×10^{-1} M
d.	1×10^{-5} M	1×10^{-9} M

9.55

$[H_3O^+]$	$[OH^-]$	pH	Acidic, Basic, Neutral?
1×10^{-8} M	1×10^{-6} M	8	basic
1×10^{-2} M	1×10^{-12} M	2	acidic
1×10^{-5} M	1×10^{-9} M	5	acidic
1×10^{-7} M	1×10^{-7} M	7	neutral

9.57 (b) and (c) are buffer systems. (b) contains the weak acid H_2CO_3 and its salt $NaHCO_3$. (c) contains HF, a weak acid, and its salt KF.

9.59 **a.** A buffer system keeps the pH constant.
b. To neutralize any acid added.
c. The added H^+ reacts with F^- from NaF.
d. The added OH^- is neutralized by the HF.

9.61 To a known volume of ascorbic acid solution, add a few drops of indicator. Place an NaOH solution of known molarity in a buret. Add base to acid until one drop changes the color of the solution. Use the volume and molarity of NaOH to calculate the concentration of the ascorbic acid in the sample.

9.63 **a.** 8.8 M HCl **b.** 0.75 M H_2SO_4 **c.** 0.53 M H_3PO_4

9.65 $[OH^-] = 0.020$ M

9.67 The $Mg(OH)_2$ that dissolves is completely ionized, making it a strong base.

9.69 100 mL of NaOH

9.71 **a.** 7.4 **b.** 1.2 **c.** 2.5 **d.** 10.0

Chapter 10

10.1 Historically, organic chemistry involved the idea of a vital force, a force where a living organism was involved. Now, organic chemistry is the study of compounds containing carbon and hydrogen.

10.3 **a.** inorganic **b.** organic **c.** organic **d.** inorganic **e.** organic

10.5 **a.** propane **b.** propane **c.** KCl **d.** KCl **e.** propane

10.7 **a.** $CH_3CH_2CH_2CH_2CH_3$

b. $CH_3CH(CH_3)CH(CH_3)CH_3$

10.9

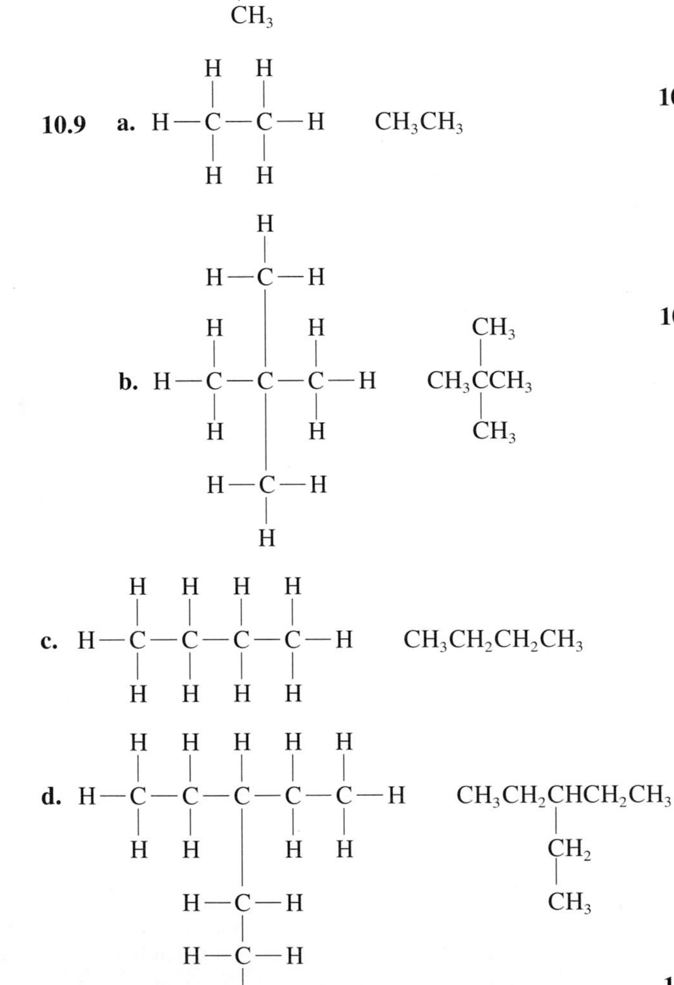

10.11 In an unbranched-chain alkane, the carbon atoms are connected in a row, one after the other.

10.13 **a.** unbranched **b.** unbranched **c.** branched **d.** unbranched

10.15 Structural isomers have the same molecular formula but a different sequence of atoms.

10.17 **a.** not structural isomers
b. structural isomers
c. not structural isomers
d. structural isomers

10.19 **a.** ethane **b.** 2-methylbutane **c.** hexane **d.** octane

10.21 **a.** $-CH_3$ **b.** $-CH_2CH_2CH_3$

10.23 **a.** Methane is a compound with the formula CH_4; a methyl group is an alkyl substituent with a formula $-CH_3$.
b. Ethane is an alkane with the formula CH_3CH_3. The ethyl group is an alkyl substituent that has a formula of $-CH_2CH_3$.

10.25 **a.** 2-methylpropane
b. 2-methylpentane
c. 2,3-dimethylpentane
d. 2,4-dimethylhexane
e. 2,4-dimethylheptane
f. 3-ethyl-2-methylpentane
g. 4-propylheptane

10.27 **a.** $CH_3CH_2CH_2CH_3$

b. $CH_3CH_2C(CH_3)_2CH_2CH_3$

c. $CH_3C(CH_3)_2CH_2CH(CH_3)CH_2CH_2CH_2CH_3$

d. $CH_3CH_2C(CH_3)_2CH_2CH_2CH_3$

e. $CH_3CH(CH_3)CH_2CH(CH(CH_3)CH_3)CH_2CH_2CH_3$

10.29 **a.** These are not isomers.
b. These are isomers.
c. These are not isomers.
d. These are not isomers.

10.31 Butane has four carbon atoms in an unbranched chain. Cyclobutane has four carbon atoms in a cyclic structure.

10.33 **a.** cyclobutane **b.** methylcyclobutane
c. cyclopentane
d. 1-ethyl-3-methylcyclohexane
e. 1,2-dimethylcyclopentane

10.35

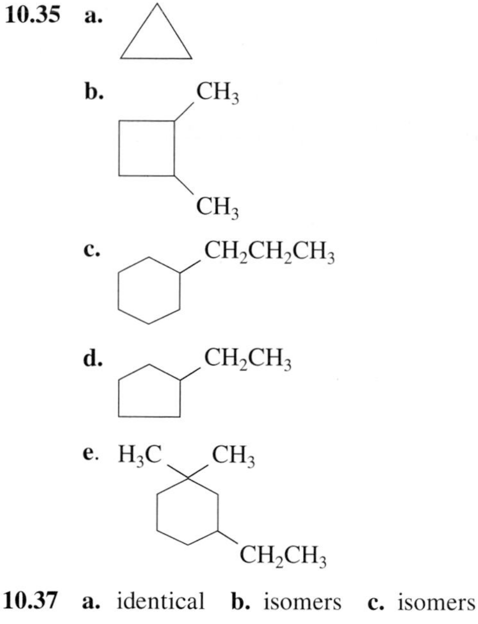

10.37 **a.** identical **b.** isomers **c.** isomers

10.39 **a.** 1,2-dibromopropane
b. 2-bromo-4-chloropentane
c. 1,2-dichlorocyclobutane
d. 2-fluoro-2-methylbutane
e. 2-bromo-4,4-dichloro-2-fluorohexane

10.41 **a.** CH_3CHCH_3 (Cl on the central C)
b. $BrCH_2CHCH_3$ (Cl on the CH)
c. CH_3Br

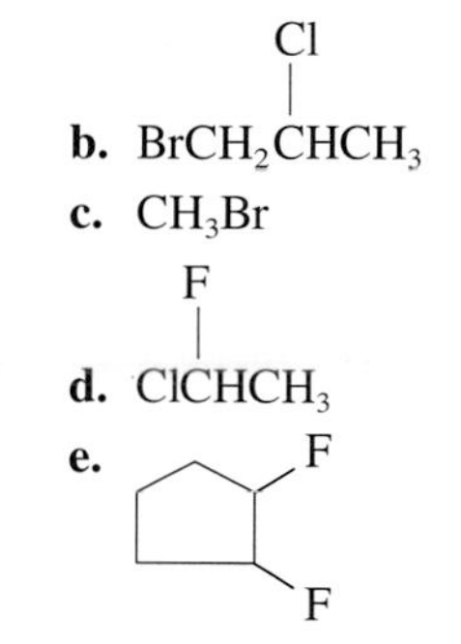

10.43 chloromethane, CH_3Cl
chloroethane, CH_3CH_2Cl

10.45 $ClCH_2CH_2CH_2CH_3$ 1-chlorobutane
$CH_3CHCH_2CH_3$ (Cl on C-2) 2-chlorobutane
CH_3CCH_3 (Cl and CH_3 on central C) 2-chloro-2-methylpropane
$ClCH_2CHCH_3$ (CH_3 on CH) 1-chloro-2-methylpropane

10.47 **a.** $CH_4 + 2O_2 \longrightarrow CO_2 + 2H_2O$
b. $C_3H_8 + 5O_2 \rightarrow 3CO_2 + 4H_2O$
c. $C_4H_8 + 6O_2 \rightarrow 4CO_2 + 4H_2O$

10.49 **a.** $CH_3CH_2CH_2CH_2CH_2CH_2CH_3$
b. liquid
c. No.
d. It will float.
e. $C_7H_{16} + 11O_2 \longrightarrow 7CO_2 + 8H_2O$

10.51 **a.** CH_3CH_2Br
b. $CH_3CH_2CH_2Cl + CH_3CHCH_3$ (Cl on the CH)
c. no reaction
d. CH_3Cl

10.53 $ClCH_2CH_2CH_2CH_2CH_3$
$CH_3CHCH_2CH_2CH_3$ (Cl on the CH)
$CH_3CH_2CHCH_2CH_3$ (Cl on the CH)
$ClCH_2CHCH_2CH_3$ (CH_3 on the CH)
$CH_3CCH_2CH_3$ (CH_3 and Cl on the C)
$CH_3CHCHCH_3$ (CH_3 on C-2, Cl on C-3)
$CH_3CHCH_2CH_2Cl$ (CH_3 on the CH)
CH_3CCH_2Cl (two CH_3 on the C)

10.55 1,2,3,4,5,6-hexachlorocyclohexane

10.57 **a.** $C_5H_{12} + 8O_2 \rightarrow 5CO_2 + 6H_2O$
b. 72.0 g **c.** 28 000 kcal, 120 000 kJ
d. 3700 L of CO_2

10.59 1,1-dichlorocyclohexane
1,2-dichlorocyclohexane
1,3-dichlorocyclohexane
1,4-dichlorocyclohexane

Chapter 11

11.1 Cyclohexane has six carbon atoms in a cyclic structure, all with single bonds. Cyclohexene has six carbon atoms in a cyclic structure, but cyclohexene contains a carbon–carbon double bond.

11.3 **a.** ethene **b.** 3,4-dibromo-3-hexene
c. 3-chloro-2-methyl-1-butene
d. 2-pentene
e. 3-methylcyclohexene

11.5 **a.** $CH_2{=}CHCH_3$

b. $CH_3C(CH_3){=}CHCH_2CH_3$

c. (cyclobutene)

d. Br (3-bromocyclopentene)

e. $ClCH_2C(Cl){=}C(CH_3)CH_2CH_2CH_2CH_3$

11.7 **a.** *trans*-1,2-difluoroethene
b. *trans*-3,4-dibromo-3-heptene
c. *cis*-2-pentene
d. *trans*-3-hexene

11.9 **a.** Br(H)C=C(Br)H

b. H_3C(H)C=C(H)CH_3

c. Br(H)C=C(Br)CH_2CH_3

d. CH_3CH_2(H)C=C(H)$CH_2CH_2CH_3$

11.11 **a.** $CH_3CH_2CH_3$ Propane

b. $ClCH_2CH(Cl)CH(CH_3)CH_3$ 1,2-dichloro-3-methylbutane

c. Br, Br (cyclobutane) 1,2-dibromocyclobutane

d. (cyclopentane) cyclopentane

e. $CH_3C(CH_3)(Cl){-}CH(Cl)CH_3$ 2, 3-dichloro-2-methylbutane

11.13 **a.** $CH_3CH_2CH(Br)CH_3$

b. OH (cyclopentanol)

c. $CH_3CH(Cl)CH_2CH_3$

d. $CH_3CH(CH_3){-}C(I)(CH_3)CH_3$

e. $CH_3CH_2CH_2CH(Br)CH_3$ and $CH_3CH_2CH(Br)CH_2CH_3$

11.15 **a.** $CH_3C(CH_3){=}CH_2 + H_2 \xrightarrow{Pt} CH_3CH(CH_3)CH_3$

b. (cyclopentene) + HCl $\longrightarrow$ (chlorocyclopentane) Cl

c. $CH_3CH{=}CHCH_2CH_3 + Br_2 \longrightarrow CH_3CH(Br)CH(Br)CH_2CH_3$

d. $CH_3CH{=}CH_2 + H_2O \xrightarrow{H^+} CH_3CH(OH)CH_3$

11.17 **a.** ethyne **b.** propyne **c.** 2-pentyne
d. 4,4-dichloro-1-pentyne
e. 4,5-dimethyl-2-hexyne

11.19 **a.** $CH_3C{\equiv}CCH_3$
b. $HC{\equiv}CH$

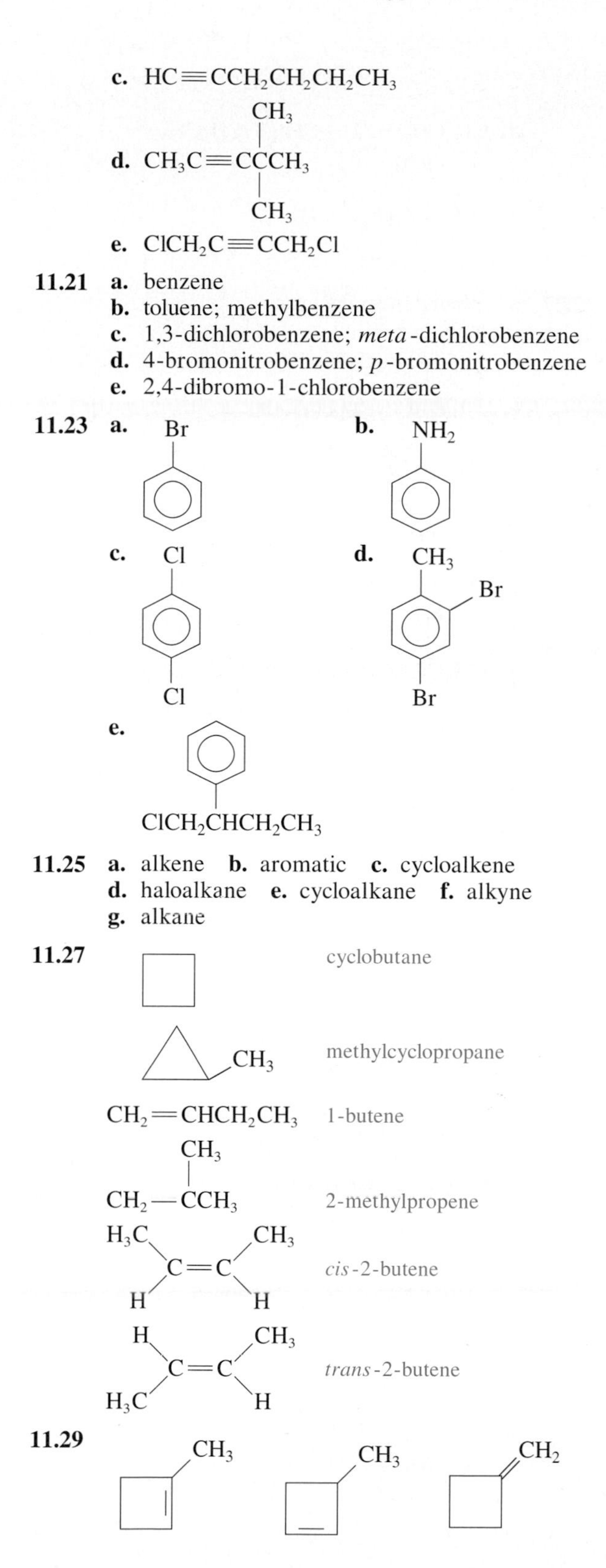

c. $HC{\equiv}CCH_2CH_2CH_2CH_3$

d. $CH_3C{\equiv}CCCH_3$ (with CH_3 above and CH_3 below the quaternary C)

e. $ClCH_2C{\equiv}CCH_2Cl$

11.21 **a.** benzene
b. toluene; methylbenzene
c. 1,3-dichlorobenzene; *meta*-dichlorobenzene
d. 4-bromonitrobenzene; *p*-bromonitrobenzene
e. 2,4-dibromo-1-chlorobenzene

11.23 **a.** Br **b.** NH_2 **c.** Cl, Cl **d.** CH_3, Br, Br **e.** $ClCH_2CHCH_2CH_3$

11.25 **a.** alkene **b.** aromatic **c.** cycloalkene
d. haloalkane **e.** cycloalkane **f.** alkyne
g. alkane

11.27 cyclobutane
CH_3 methylcyclopropane
$CH_2{=}CHCH_2CH_3$ 1-butene
$CH_2{-}CCH_3$ (with CH_3 above) 2-methylpropene
H_3C, CH_3, C=C, H, H *cis*-2-butene
H, CH_3, C=C, H_3C, H *trans*-2-butene

11.29 CH_3 CH_3 CH_2

11.31 **a.** $2C_6H_6 + 15O_2 \longrightarrow 12CO_2 + 6H_2O$
b. 12.0 mol CO_2
c. 336 L O_2

Chapter 12

12.1 CH_3CH_2OH Primary 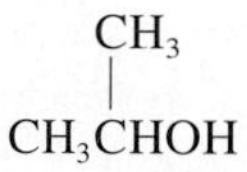
CH_3CHOH (with CH_3 above) Secondary

12.3 **a.** secondary **b.** primary **c.** tertiary
d. secondary

12.5 **a.** ethanol **b.** 2-butanol **c.** 3-hexanol
d. 3-methyl-1-butanol
e. 3,4-dimethylcyclohexanol
f. 3,5,5-trimethyl-1-heptanol

12.7 **a.** $CH_3CH_2CH_2OH$
b. CH_3OH
c. 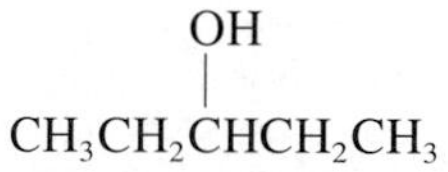 $CH_3CH_2CHCH_2CH_3$ (with OH above)
d. 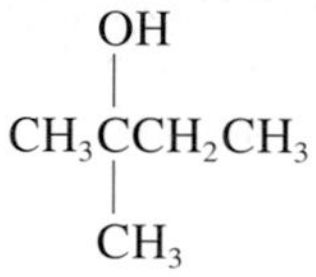 $CH_3CCH_2CH_3$ (with OH above and CH_3 below)
e. 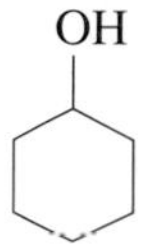 OH (cyclohexanol)
f. $HOCH_2CH_2CH_2CH_2OH$

12.9 **a.** phenol
b. 2-bromophenol, *ortho*-bromophenol
c. 3,5-dichlorophenol
d. 3-nitrophenol, *meta*-nitrophenol

12.11

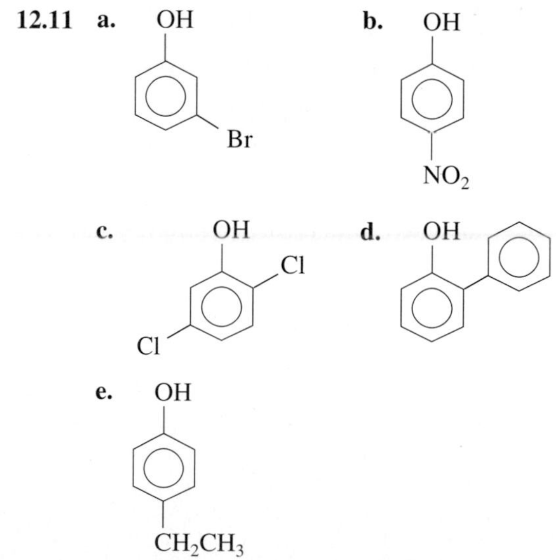

12.13 Ethanol has a hydroxyl (OH) group that can hydrogen bond to water. Ethane does not have any polar groups that could hydrogen bond to water.

12.15

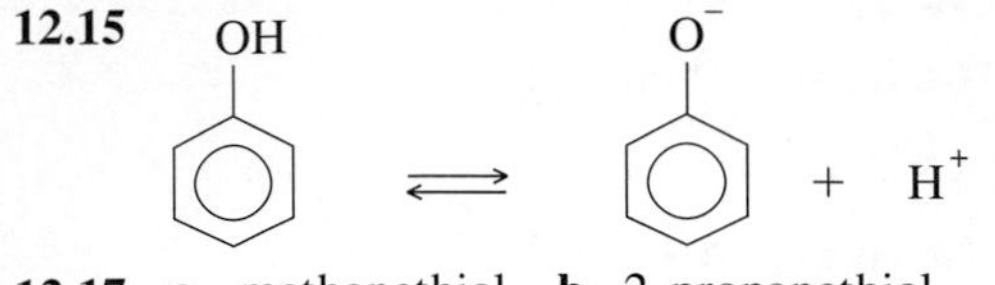

12.17 **a.** methanethiol **b.** 2-propanethiol
c. 2,3-dimethyl-1-butanethiol
d. cyclobutanethiol

12.19 **a.** $CH_3CH_2CH{=}CH_2$
b. 
c.

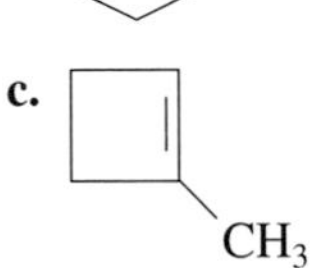

d. $CH_3CH_2CH{=}CHCH_3$

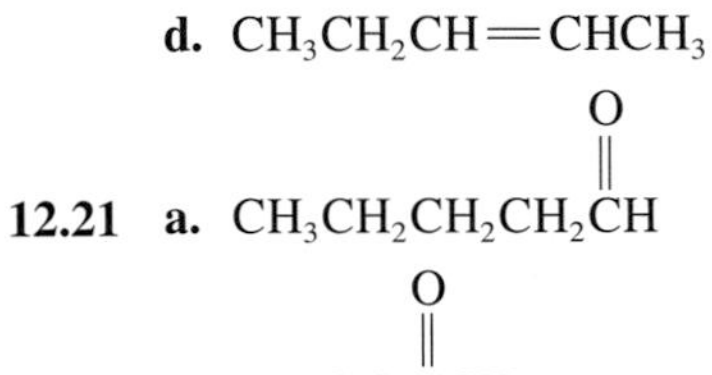

12.21 **a.** $CH_3CH_2CH_2CH_2\overset{O}{\overset{\|}{C}}H$
b. $CH_3CH_2\overset{O}{\overset{\|}{C}}CH_3$
c. O
d. $CH_3\overset{O}{\overset{\|}{C}}CH_2\overset{CH_3}{\overset{|}{C}}HCH_3$
e. $CH_3\overset{CH_3}{\overset{|}{C}}HCH_2\overset{O}{\overset{\|}{C}}H$

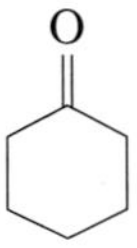

12.23 **a.** methoxyethane, ethyl methyl ether
b. methoxycyclohexane, cyclohexyl methyl ether
c. ethoxycyclobutane, cyclobutyl ethyl ether
d. 1-methoxypropane, methyl propyl ether

12.25 **a.** $CH_3{-}O{-}CH_2CH_2CH_3$
b. $CH_3CH_2{-}O{-}$
c.

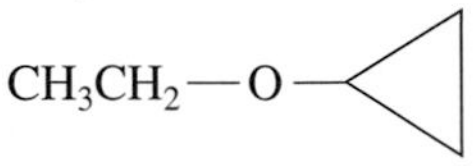

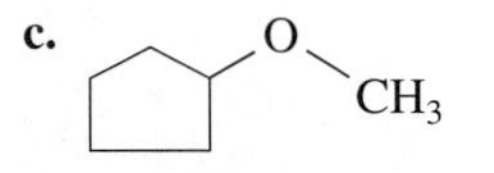

d. $CH_3CH_2{-}O{-}CH_2\overset{CH_3}{\overset{|}{C}}HCH_2CH_3$
e. $CH_3\overset{OCH_3}{\overset{|}{C}}H\underset{OCH_3}{\underset{|}{C}}HCH_2CH_3$

12.27 **a.** ethanal, acetaldehyde
b. butanone, ethyl methyl ketone
c. 4-methyl-2-pentanone
d. 5-phenyl-2-pentanone
e. 3-methylcyclopentanone

12.29 **a.** $CH_3CH_2\overset{O}{\overset{\|}{C}}H$
b. $H{-}\overset{O}{\overset{\|}{C}}{-}H$
c. $CH_3\overset{O}{\overset{\|}{C}}\underset{CH_3}{\underset{|}{C}}HCH_2CH_3$
d. $CH_3\underset{Cl}{\underset{|}{C}}HCH_2\overset{O}{\overset{\|}{C}}H$
e. $\overset{O}{\overset{\|}{C}}H$
f. $CH_3CH_2{-}\overset{O}{\overset{\|}{C}}{-}$

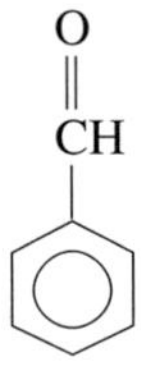

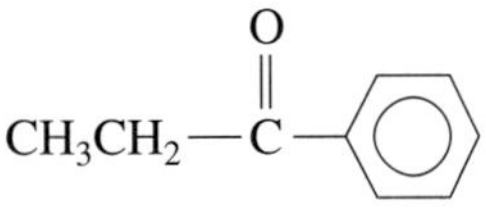

12.31 4-hydroxy-3-methoxybenzaldehyde

12.33 **a.** CH_3OH
b. OH
c. $CH_3CH_2\overset{OH}{\overset{|}{C}}HCH_3$
d. CH_2OH

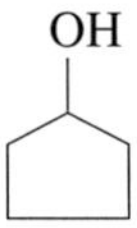

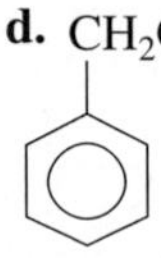

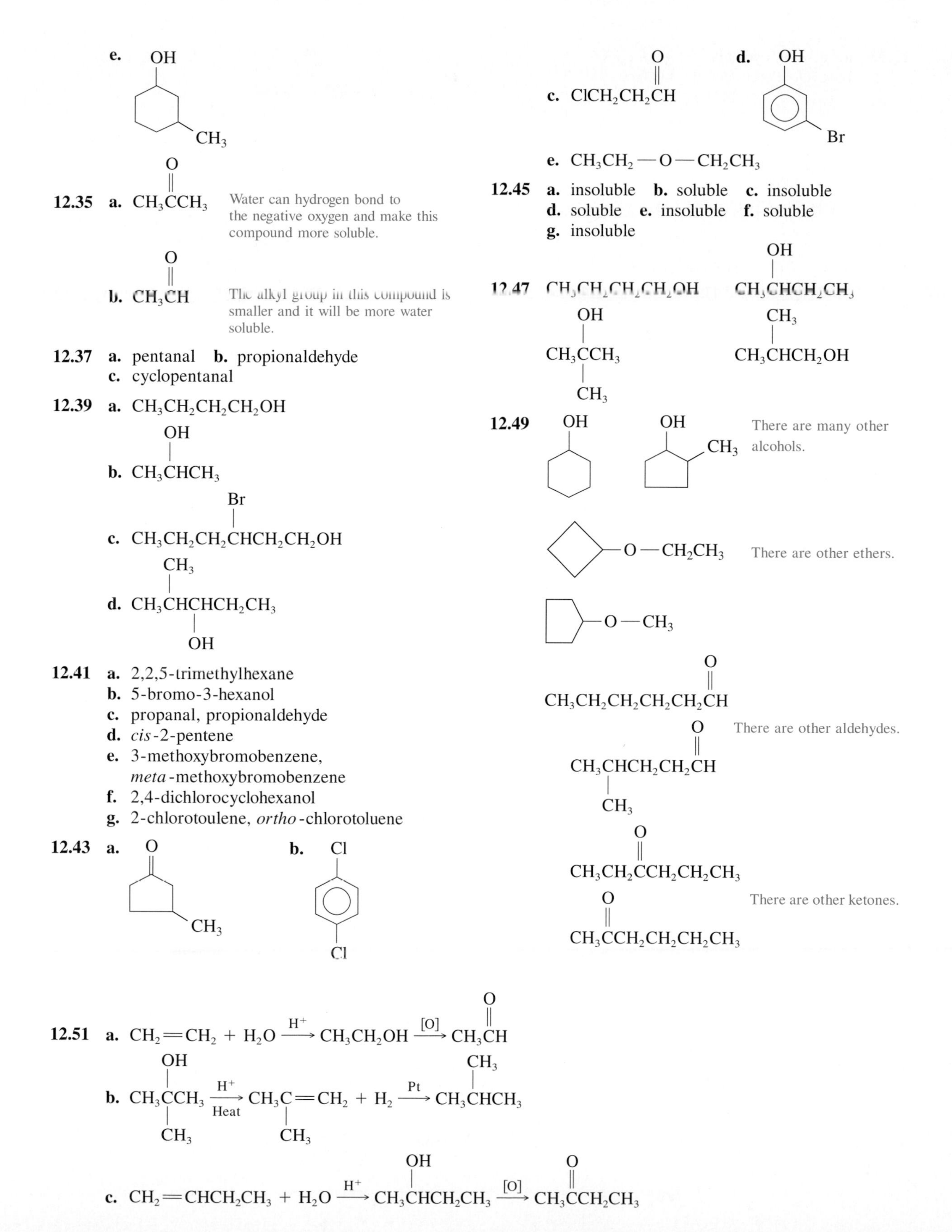

e. (cyclohexane ring with OH and CH_3)

12.35 **a.** CH_3CCH_3 (with C=O) Water can hydrogen bond to the negative oxygen and make this compound more soluble.

b. CH_3CH (with C=O) The alkyl group in this compound is smaller and it will be more water soluble.

12.37 **a.** pentanal **b.** propionaldehyde **c.** cyclopentanal

12.39 **a.** $CH_3CH_2CH_2CH_2OH$

b. CH_3CHCH_3 (OH on C-2)

c. $CH_3CH_2CH_2CHCH_2CH_2OH$ (Br on C-4)

d. $CH_3CHCHCH_2CH_3$ (CH_3 on C-2, OH on C-3)

12.41 **a.** 2,2,5-trimethylhexane
b. 5-bromo-3-hexanol
c. propanal, propionaldehyde
d. *cis*-2-pentene
e. 3-methoxybromobenzene, *meta*-methoxybromobenzene
f. 2,4-dichlorocyclohexanol
g. 2-chlorotoulene, *ortho*-chlorotoluene

12.43 **a.** (cyclopentanone ring with CH_3) **b.** (benzene ring with Cl and Cl)

c. $ClCH_2CH_2CH$ (with C=O) **d.** (benzene ring with OH and Br)

e. $CH_3CH_2—O—CH_2CH_3$

12.45 **a.** insoluble **b.** soluble **c.** insoluble **d.** soluble **e.** insoluble **f.** soluble **g.** insoluble

12.47 $CH_3CH_2CH_2CH_2OH$ $CH_3CHCH_2CH_3$ (OH on C-2)

CH_3CCH_3 (OH and CH_3 on C-2) CH_3CHCH_2OH (CH_3 on C-2)

12.49 (cyclohexanol) (2-methylcyclopentanol: OH, CH_3) There are many other alcohols.

(cyclobutyl)$—O—CH_2CH_3$ There are other ethers.

(cyclopentyl)$—O—CH_3$

$CH_3CH_2CH_2CH_2CH_2CH$ (with C=O)

$CH_3CHCH_2CH_2CH$ (with C=O; CH_3 on C-2) There are other aldehydes.

$CH_3CH_2CCH_2CH_2CH_3$ (with C=O)

$CH_3CCH_2CH_2CH_2CH_3$ (with C=O) There are other ketones.

12.51 **a.** $CH_2{=}CH_2 + H_2O \xrightarrow{H^+} CH_3CH_2OH \xrightarrow{[O]} CH_3CH$ (with C=O)

b. CH_3CCH_3 (OH, CH_3) $\xrightarrow[\text{Heat}]{H^+} CH_3C{=}CH_2$ (CH_3) $+ H_2 \xrightarrow{Pt} CH_3CHCH_3$ (CH_3)

c. $CH_2{=}CHCH_2CH_3 + H_2O \xrightarrow{H^+} CH_3CHCH_2CH_3$ (OH) $\xrightarrow{[O]} CH_3CCH_2CH_3$ (with C=O)

12.53 **a.** alkene, cycloalkene, ketone
b. cycloalkene, ketone, alcohol

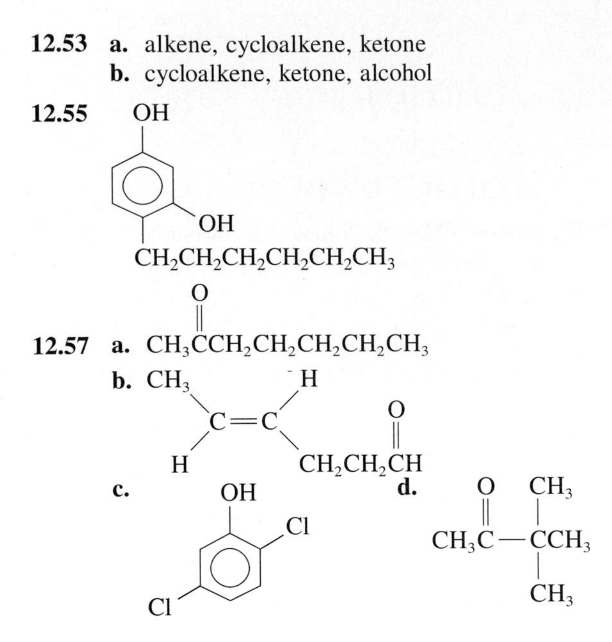

Chapter 13

13.1 Each compound contains three carbon atoms. They differ because propanal, an aldehyde, contains a carbonyl group bonded to a hydrogen. In propanoic acid, the carbonyl group connects to a hydroxyl group.

13.3 **a.** ethanoic acid (acetic acid)
b. butanoic acid (butyric acid)
c. 2-chloropropanoic acid (α-chloropropionic acid)
d. 3-methylhexanoic acid

13.5 **a.** $CH_3CH_2\overset{O}{\overset{\|}{C}}-OH$
b. $C_6H_5-\overset{O}{\overset{\|}{C}}-OH$ (benzene ring bonded to $\overset{O}{\overset{\|}{C}}-OH$)
c. $Cl-CH_2\overset{O}{\overset{\|}{C}}-OH$
d. $HO-CH_2CH_2\overset{O}{\overset{\|}{C}}-OH$
e. $CH_3\underset{}{\overset{CH_3}{\overset{|}{C}}}HCH_2\overset{O}{\overset{\|}{C}}-OH$
f. $CH_3CH_2\overset{Br}{\overset{|}{C}}HCH_2\overset{Br}{\overset{|}{C}}HCH_2\overset{O}{\overset{\|}{C}}OH$

13.7 **a.** $H\overset{O}{\overset{\|}{C}}-OH$
b. $CH_3\overset{O}{\overset{\|}{C}}-OH$
c. $CH_3\overset{CH_3}{\overset{|}{C}}HCH_2\overset{O}{\overset{\|}{C}}-OH$
d. cyclopentane ring bonded to $\overset{O}{\overset{\|}{C}}OH$

13.9 **a.** CH_3CH_2COOH has a polar cabon–oxygen double bond and a polar —OH; both polar groups hydrogen bond to water.
b. Both the carbonyl and the hydroxyl groups in $CH_3CH_2CH_2COOH$ are strongly polar and hydrogen bond to water.
c. CH_3COOH has a smaller carbon chain and will be more soluble.

13.11 **a.** $HCOOH + H_2O \rightleftarrows HCOO^- + H_3O^+$
b. $CH_3CH_2COOH + H_2O \rightleftarrows CH_3CH_2COO^- + H_3O^+$
c. $CH_3COOH + H_2O \rightleftarrows CH_3COO^- + H_3O^+$

13.13 **a.** $HCOOH + NaOH \longrightarrow HCOO^- Na^+ + H_2O$
b. $CH_3CH_2COOH + NaOH \longrightarrow CH_3CH_2COO^- Na^+ + H_2O$
c.

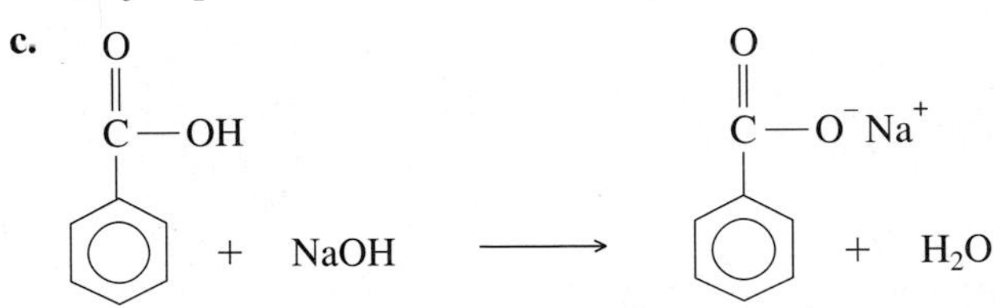

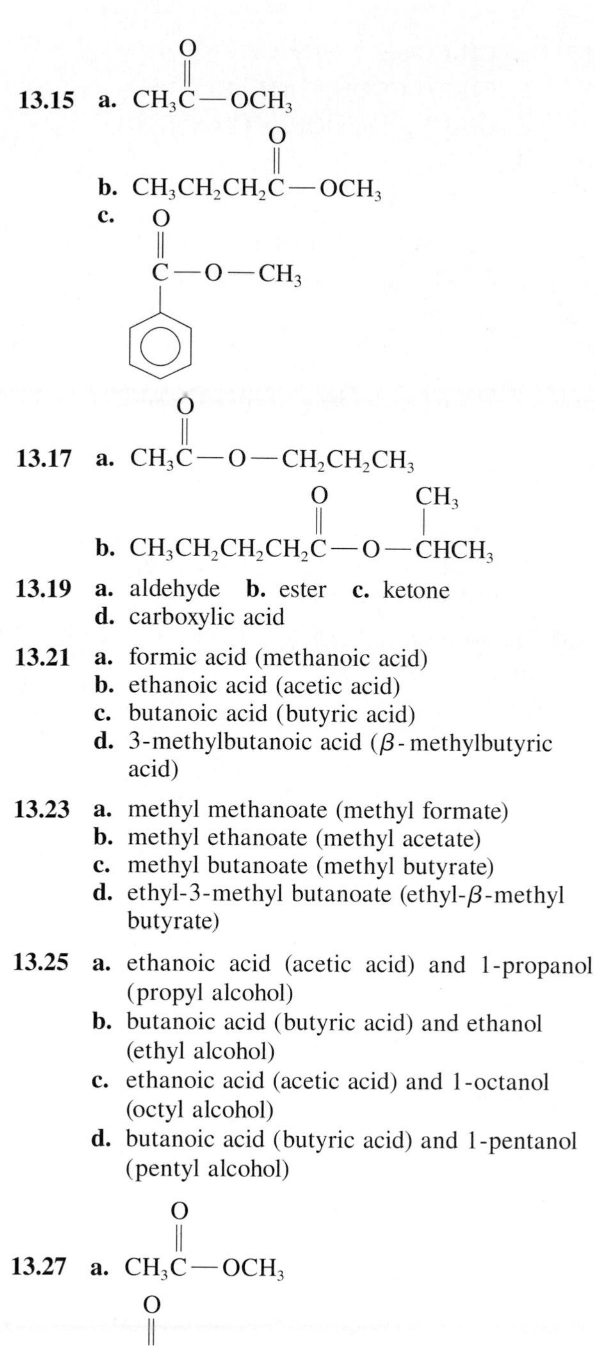

13.15 **a.** $CH_3C(=O)—OCH_3$

b. $CH_3CH_2CH_2C(=O)—OCH_3$

c. $C_6H_5—C(=O)—O—CH_3$

c. $CH_3CH_2CH_2CH_2C(=O)—OCH_2CH_3$

d. $CH_3CH_2C(=O)—OCH_2CH(Br)CH_3$

13.17 **a.** $CH_3C(=O)—O—CH_2CH_2CH_3$

b. $CH_3CH_2CH_2CH_2C(=O)—O—CH(CH_3)CH_3$

13.19 **a.** aldehyde **b.** ester **c.** ketone
d. carboxylic acid

13.21 **a.** formic acid (methanoic acid)
b. ethanoic acid (acetic acid)
c. butanoic acid (butyric acid)
d. 3-methylbutanoic acid (β-methylbutyric acid)

13.23 **a.** methyl methanoate (methyl formate)
b. methyl ethanoate (methyl acetate)
c. methyl butanoate (methyl butyrate)
d. ethyl-3-methyl butanoate (ethyl-β-methyl butyrate)

13.25 **a.** ethanoic acid (acetic acid) and 1-propanol (propyl alcohol)
b. butanoic acid (butyric acid) and ethanol (ethyl alcohol)
c. ethanoic acid (acetic acid) and 1-octanol (octyl alcohol)
d. butanoic acid (butyric acid) and 1-pentanol (pentyl alcohol)

13.27 **a.** $CH_3C(=O)—OCH_3$

b. $HC(=O)O—CH_2CH_2CH_2CH_3$

13.29 The products of the acid hydrolysis of an ester are an alcohol and a carboxylic acid.

13.31 **a.** $CH_3CH_2COO^- Na^+$ and CH_3OH
b. CH_3COOH and $CH_3CH_2CH_2OH$
c. $CH_3CH_2CH_2COOH$ and CH_3CH_2OH
d. $C_6H_5—COOH$ and CH_3CH_2OH
e. $C_6H_5—COO^- Na^+$ and CH_3CH_2OH

13.33 In a primary amine, there is one alkyl group (and two hydrogens) attached to a nitrogen atom.

13.35 **a.** primary **b.** tertiary **c.** secondary
d. primary

13.37 **a.** aminoethane (ethylamine)
b. *N*-methyl-1-aminopropane (methylpropylamine)
c. *N,N*-diethylaminoethane (triethylamine)
d. *N*-methylaniline

13.39 **a.** $CH_3CH_2NH_2$
b. $C_6H_5—NHCH_3$
c. $CH_3CH_2CH_2CH_2—NH—CH_2CH_2CH_3$
d. $CH_3CH(NH_2)CH_2CH_2CH_3$

13.41 **a.** $CH_3NH_2 + H_2O \rightleftarrows CH_3NH_3^+ + OH^-$

b. $CH_3—NH(CH_3) + H_2O \rightleftarrows CH_3—NH_2^+(CH_3) + OH^-$

c. $C_6H_5NH_2 + H_2O \rightleftarrows C_6H_5NH_3^+ + OH^-$

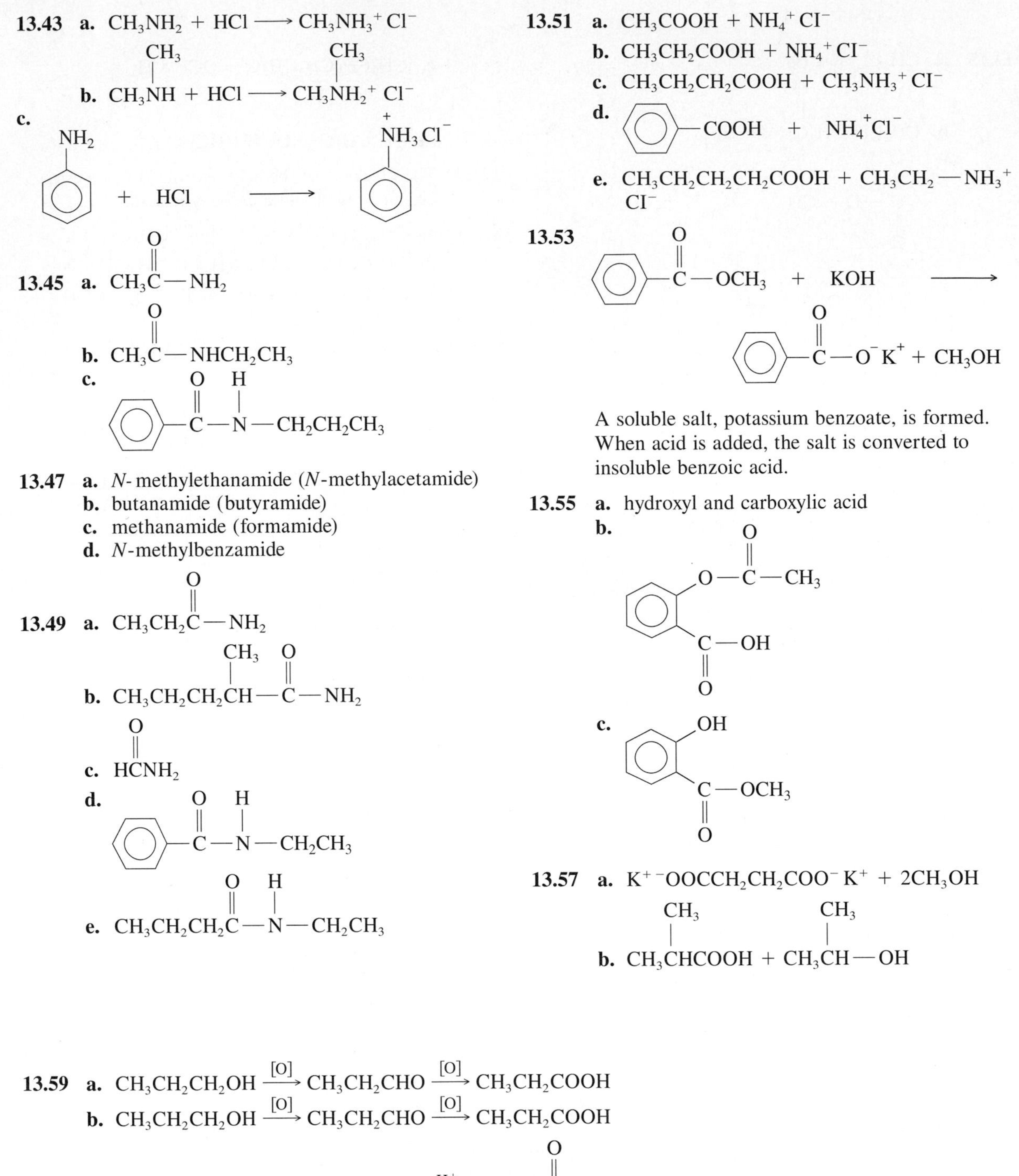

13.43 a. $CH_3NH_2 + HCl \longrightarrow CH_3NH_3^+ Cl^-$

b. $CH_3NH(CH_3) + HCl \longrightarrow CH_3NH_2^+(CH_3)\ Cl^-$

c. $C_6H_5NH_2 + HCl \longrightarrow C_6H_5NH_3^+ Cl^-$

13.45 a. $CH_3C(=O)-NH_2$

b. $CH_3C(=O)-NHCH_2CH_3$

c. $C_6H_5-C(=O)-N(H)-CH_2CH_2CH_3$

13.47 a. *N*-methylethanamide (*N*-methylacetamide)
b. butanamide (butyramide)
c. methanamide (formamide)
d. *N*-methylbenzamide

13.49 a. $CH_3CH_2C(=O)-NH_2$

b. $CH_3CH_2CH_2CH(CH_3)-C(=O)-NH_2$

c. $HC(=O)NH_2$

d. $C_6H_5-C(=O)-N(H)-CH_2CH_3$

e. $CH_3CH_2CH_2C(=O)-N(H)-CH_2CH_3$

13.51 a. $CH_3COOH + NH_4^+ Cl^-$
b. $CH_3CH_2COOH + NH_4^+ Cl^-$
c. $CH_3CH_2CH_2COOH + CH_3NH_3^+ Cl^-$
d. $C_6H_5-COOH + NH_4^+Cl^-$
e. $CH_3CH_2CH_2CH_2COOH + CH_3CH_2-NH_3^+$ Cl^-

13.53 $C_6H_5-C(=O)-OCH_3 + KOH \longrightarrow C_6H_5-C(=O)-O^-K^+ + CH_3OH$

A soluble salt, potassium benzoate, is formed. When acid is added, the salt is converted to insoluble benzoic acid.

13.55 a. hydroxyl and carboxylic acid
b. $C_6H_4(-O-C(=O)-CH_3)(-C(=O)-OH)$
c. $C_6H_4(-OH)(-C(=O)-OCH_3)$

13.57 a. $K^+\ ^-OOCCH_2CH_2COO^-\ K^+ + 2CH_3OH$
b. $CH_3CH(CH_3)COOH + CH_3CH(CH_3)-OH$

13.59 a. $CH_3CH_2CH_2OH \xrightarrow{[O]} CH_3CH_2CHO \xrightarrow{[O]} CH_3CH_2COOH$

b. $CH_3CH_2CH_2OH \xrightarrow{[O]} CH_3CH_2CHO \xrightarrow{[O]} CH_3CH_2COOH$

$CH_3CH_2COOH + CH_3CH_2CH_2OH \xrightarrow{H^+} CH_3CH_2C(=O)-OCH_2CH_2CH_3$

c. $CH_3CH_2CH_2OH \xrightarrow{H^+,\ Heat} CH_3CH{=}CH_2 + H_2 \xrightarrow{Pt} CH_3CH_2CH_3$

d. $CH_3CH_2CH_2OH \xrightarrow{H^+,\ Heat} CH_3CH{=}CH_2 + H_2O \xrightarrow{H^+} CH_3CH(OH)CH_3$

e. $$CH_3CH_2CH_2OH \xrightarrow{H^+,\ Heat} CH_3CH{=}CH_2 + H_2O \xrightarrow{H^+} CH_3\overset{OH}{\overset{|}{C}}HCH_3 \xrightarrow{[O]} CH_3{-}\overset{O}{\overset{\|}{C}}{-}CH_3$$

f. $$CH_3CH_2CH_2OH \xrightarrow{[O]} CH_3CH_2\overset{O}{\overset{\|}{C}}{-}H \xrightarrow{[O]} CH_3CH_2\overset{O}{\overset{\|}{C}}{-}OH + NH_3 \xrightarrow{Heat} CH_3CH_2\overset{O}{\overset{\|}{C}}{-}NH_2$$

13.61 carboxylate salt, aromatic, amine, halide

13.63 **a.** aromatic, amine, amide, carboxylic acid, cycloalkene
b. aromatic, ether, alcohol, amine
c. aromatic, carboxylic acid
d. phenol, amine, carboxylic acid
e. aromatic, ether, alcohol, amine, ketone
f. aromatic, amine

Chapter 14

14.1 A monosaccharide is a simple sugar composed of from three to six carbon atoms.

14.3 Hydroxyl groups are found in all monosaccharides along with a carbonyl on the first or second carbon.

14.5 A ketopentose contains hydroxyl and ketone functional groups and has five carbon atoms.

14.7 **a.** ketose **b.** aldose **c.** ketose **d.** aldose **e.** aldose

14.9 **a.** achiral **b.** chiral **c.** achiral **d.** achiral

14.11 **a.** achiral **b.** achiral **c.** chiral **d.** chiral

14.13 A Fischer projection is a two-dimensional representation of the three-dimensional structure of a molecule.

14.15 **a.** D **b.** D **c.** L **d.** D

14.17 This means that when plane polarized light is passed through a sample of this compound, the plane of the light will be rotated.

14.19

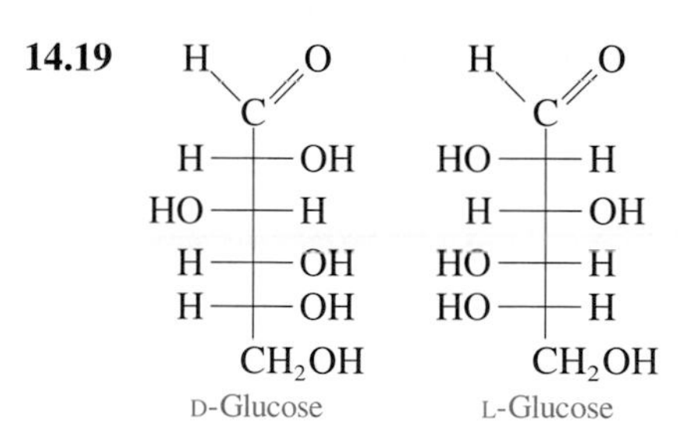

D-Glucose L-Glucose

14.21 In D-galactose the hydroxyl on carbon four extends to the left, in glucose this hydroxyl goes to the right.

14.23 **a.** glucose **b.** galactose **c.** fructose

14.25 In the cyclic structure of glucose, there are five carbon atoms and an oxygen. When one part of the molecule forms a hemiacetal with another part, a C—O—C bond is formed.

14.27

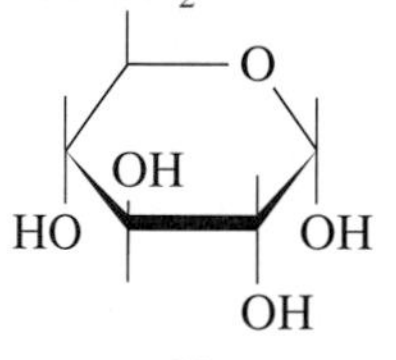

α-D-Glucose

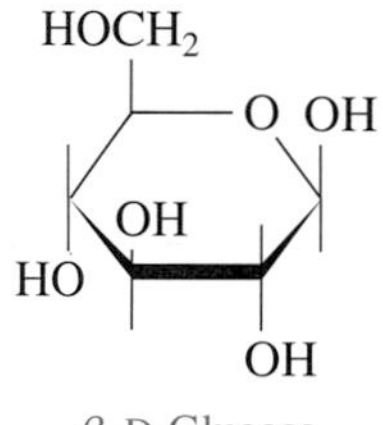

β-D-Glucose

14.29 **a.** hemiketal
b. hemiacetal
c. hemiacetal
d. hemiacetal

14.31 A reducing sugar is one that reacts with Tollens' or Benedict's reagent.

14.33 **a.** galactose and glucose; β-1,4 bond; β-lactose
b. glucose; α-1,4 bond; α-maltose

14.35 **a.** sucrose
b. lactose
c. maltose
d. lactose

14.37 **a.** Amylose is an unbranched polymer of glucose units joined by α-1,4 bonds; amylopectin is a branched polymer of glucose joined by α-1,4 and α-1,6 bonds.
b. Amylopectin is a branched polymer of glucose, joined by α-1,4 and α-1,6 bonds. Glycogen is a highly branched polymer of glucose, joined by α-1,4 and α-1,6 bonds.

14.39 **a.** cellulose **b.** amylose, amylopectin **c.** amylose **d.** glycogen

14.41

H—C(H)—OH
H—C—OH
HO—C—H
H—C—OH
H—C—OH
CH_2OH

14.43 The α-galactose forms an open-chain structure, and when the chain closes, it can form both α- and β-galactose.

14.45

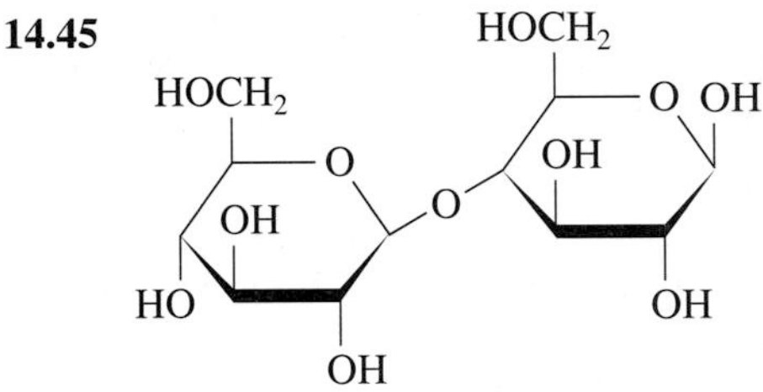

Chapter 15

15.1 A lipid is a biomolecule that is not soluble in water.

15.3 All fatty acids contain a long chain of carbon atoms with a carboxylic acid group. Saturated fats contain only carbon-to-carbon single bonds; unsaturated fats contain one or more double bonds. More saturated fats are found in animal fats, whereas vegetable oils contain more unsaturated fats.

15.5

$CH_3(CH_2)_{14}\overset{O}{\overset{\|}{C}}OH$ — Palmitic acid

$CH_3(CH_2)_7CH{=}CH(CH_2)_7\overset{O}{\overset{\|}{C}}OH$ — Oleic acid

15.7 **a.** saturated
b. unsaturated
c. unsaturated
d. saturated

15.9

$CH_3(CH_2)_{14}\overset{O}{\overset{\|}{C}}O(CH_2)_{29}CH_3$

15.11

$CH_2O\overset{O}{\overset{\|}{C}}(CH_2)_{16}CH_3$
|
$CHO\overset{O}{\overset{\|}{C}}(CH_2)_{16}CH_3$
|
$CH_2O\overset{O}{\overset{\|}{C}}(CH_2)_{16}CH_3$

15.13

$CH_2O\overset{O}{\overset{\|}{C}}(CH_2)_{14}CH_3$
|
$CHO\overset{O}{\overset{\|}{C}}(CH_2)_{14}CH_3$
|
$CH_2O\overset{O}{\overset{\|}{C}}(CH_2)_{14}CH_3$

15.15 Safflower oil contains fatty acids with two or three double bonds; olive oil contains a large amount of oleic acid, which has a single (monounsaturated) double bond.

15.17 Although coconut oil comes from a vegetable, it has large amounts of saturated fatty acids and small amounts of unsaturated fatty acids.

15.19

$CH_2O\overset{O}{\overset{\|}{C}}(CH_2)_7CH{=}CH(CH_2)_7CH_3$
|
$CHO\overset{O}{\overset{\|}{C}}(CH_2)_7CH{=}CH(CH_2)_7CH_3 + H_2 \xrightarrow{Pt}$
|
$CH_2O\overset{O}{\overset{\|}{C}}(CH_2)_7CH{=}CH(CH_2)_7CH_3$

$CH_2O\overset{O}{\overset{\|}{C}}(CH_2)_{16}CH_3$
|
$CHO\overset{O}{\overset{\|}{C}}(CH_2)_{16}CH_3$
|
$CH_2O\overset{O}{\overset{\|}{C}}(CH_2)_{16}CH_3$

15.21 **a.** Some of the double bonds in the unsaturated fatty acids have been converted to single bonds.
b. It is mostly saturated fatty acids.

15.23

a.

$$\begin{array}{l} CH_2OC(=O)(CH_2)_{12}CH_3 \\ | \\ CHOC(=O)(CH_2)_{12}CH_3 \\ | \\ CH_2OC(=O)(CH_2)_{12}CH_3 \end{array} + 3H_2O \xrightarrow{H^+} \begin{array}{l} CH_2OH \\ | \\ CHOH \\ | \\ CH_2OH \end{array} + 3CH_3(CH_2)_{12}COH \ (C{=}O)$$

b.

$$\begin{array}{l} CH_2OC(=O)(CH_2)_{12}CH_3 \\ | \\ CHOC(=O)(CH_2)_{12}CH_3 \\ | \\ CH_2OC(=O)(CH_2)_{12}CH_3 \end{array} + 3\ NaOH \longrightarrow \begin{array}{l} CH_2OH \\ | \\ CHOH \\ | \\ CH_2OH \end{array} + 3\ CH_3(CH_2)_{12}C(=O)O^-\ Na^+$$

15.25 **a.** A fat consists of glycerol and three fatty acids. A phosphoglyceride consists of glycerol, two fatty acids, a phosphate group, and an amino alcohol.
b. A phosphoglyceride consists of glycerol, two fatty acids, a phosphate group, and an amino alcohol. A sphingolipid contains the amino alcohol sphingosine instead of glycerol.

15.27

$$\begin{array}{l} CH_2OC(=O)(CH_2)_{14}CH_3 \\ | \\ CHOC(=O)(CH_2)_{14}CH_3 \\ | \\ CH_2OP(=O)(O^-)OCH_2CH_2NH_3^+ \end{array}$$

This is a cephalin

15.29

$CH_3(CH_2)_{12}CH{=}CHOH$

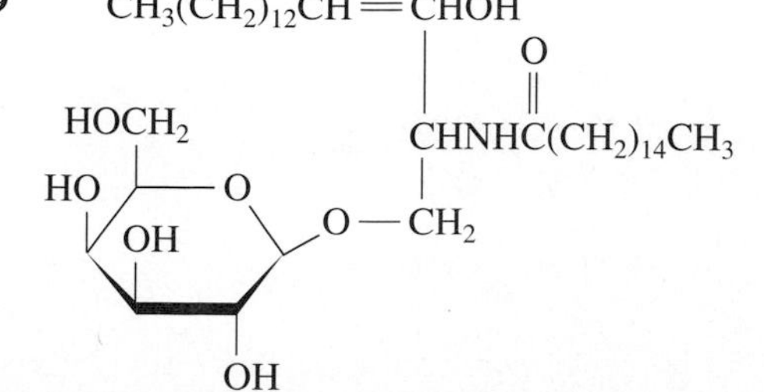

15.31 This phospholipid is a cephalin. It contains glycerol, oleic acid, stearic acid, a phosphate, and ethanolamine.

15.33 In a cell membrane, the phospholipids are arranged in a double row with the polar parts that are hydrophilic on the outside and the nonpolar hydrophobic fatty acids on the inside.

15.35

15.37 Bile salts act to emulsify fat globules, allowing the fat to be more easily digested.

15.39 Both estradiol and testosterone contain the steroid nucleus and a hydroxyl group. Testosterone has a ketone group, a double bond, and an extra methyl group. Estradiol has a benzene ring, an extra hydroxyl group, and a methyl group.

15.41 Testosterone is a male sex hormone.

15.43 Beeswax and carnauba are waxes. Vegetable oil and capric triglyceride are triglycerides. Tocopherol is a vitamin.

$$\begin{array}{l} CH_2OC(=O)(CH_2)_8CH_3 \\ | \\ CHOC(=O)(CH_2)_8CH_3 \\ | \\ CH_2OC(=O)(CH_2)_8CH_3 \end{array}$$

Capric triglyceride

15.45 **a.** A typical fatty acid has a cis double bond.
b. A trans fatty acid has a trans double bond.

c.

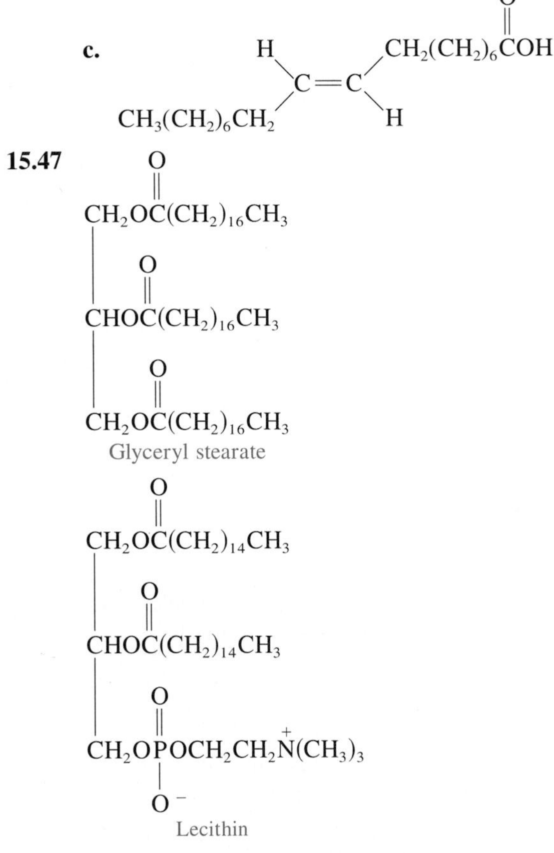

15.47

Glyceryl stearate

Lecithin

15.49 Stearic acid is a fatty acid. Sodium stearate is a soap. Glyceryl tripalmitate, safflower oil, whale blubber, and adipose tissue are triglycerides. Beeswax is a wax. Lecithin is a phosphoglyceride. Spingomyelin is a spingolipid. Cholesterol, progesterone, and cortisone are steroids.

Chapter 16

16.1 **a.** transport **c.** structural **b.** structural **d.** catalytic

16.3 All amino acids contain a carboxylic acid group and an amino group on the α carbon.

16.5

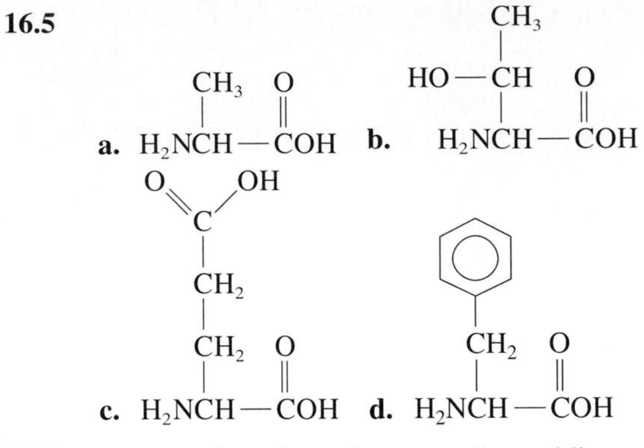

16.7 **a.** nonpolar **b.** polar **c.** polar, acidic **d.** nonpolar

16.9 **a.** alanine **b.** valine **c.** lysine **d.** cysteine

16.11 An essential amino acid is one that cannot be formed in the body. Essential amino acids must be supplied in the diet. The essential amino acids in 16.9 are valine and lysine.

16.13
a. $H_3\overset{+}{N}CH(H)—C(=O)O^-$
b. $H_3\overset{+}{N}CH(CH_2SH)—C(=O)O^-$
c. $H_3\overset{+}{N}CH(CH_2OH)—C(=O)O^-$
d. $H_3\overset{+}{N}CH(CH_3)—C(=O)O^-$

16.15
a. $H_3\overset{+}{N}CH(H)—C(=O)OH$
b. $H_3\overset{+}{N}CH(CH_2SH)—C(=O)OH$
c. $H_3\overset{+}{N}CH(CH_2OH)—C(=O)OH$
d. $H_3\overset{+}{N}CH(CH_3)—C(=O)OH$

16.17

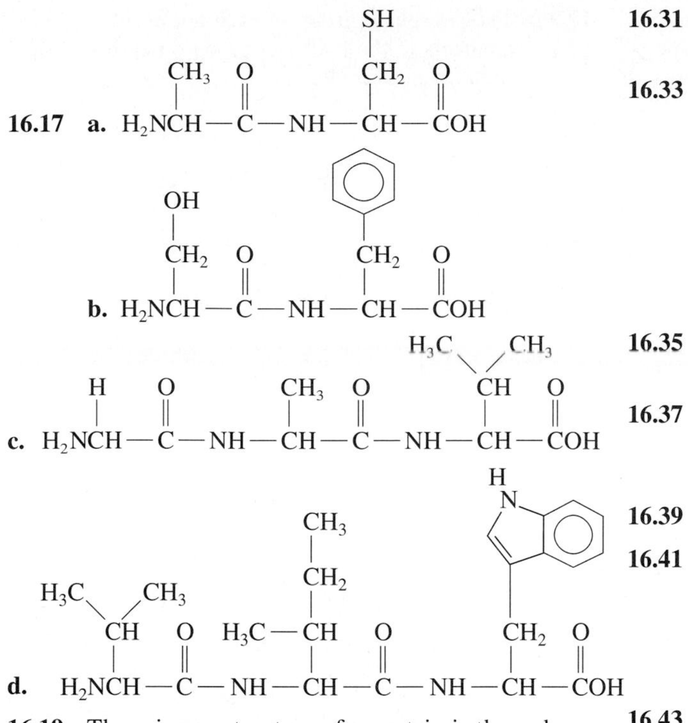

16.19 The primary structure of a protein is the order of amino acids in that protein.

16.21 **a.** a disulfide bond
b. an ionic bond (salt bridge)
c. hydrogen bond
d. hydrophobic interaction

16.23 **a.** cysteine
b. Leucine and valine will be found on the inside of the protein because they are hydrophobic.
c. The cysteine and aspartic acid would be on the outside of the protein because they are polar.
d. The order of the amino acid (the primary structure) provides the R groups, whose interactions determine the tertiary structure of the protein.

16.25 **a.** Placing an egg in boiling water coagulates the protein of the egg.
b. Using an alcohol swab coagulates the protein of any bacteria present.
c. The heat from an autoclave will coagulate the protein of any bacteria on the surgical instruments.

16.27 **a.** hydrolase **b.** oxidoreductase **c.** isomerase **d.** transferase

16.29 **a.** galactose **b.** lipids **c.** aspartic acid

16.31 **a.** enzyme **b.** enzyme–substrate complex **c.** substrate

16.33 $E + S \rightleftharpoons ES \longrightarrow E + P$
a. E = enzyme, S = substrate, ES = enzyme-substrate complex, P = products
b. The active site is the portion of the enzyme which binds to the substrate.
c. After products are formed, the enzyme can catalyze further reactions with other substrate molecules.

16.35 **a.** requires a cofactor **b.** requires a cofactor **c.** simple

16.37 **a.** The rate would decrease.
b. The rate would decrease.
c. The rate would decrease.

16.39 pepsin, pH 1.5; urease, pH 5; oxidase, pH 11

16.41 **a.** competitive inhibitor
b. noncompetitive inhibitor
c. competitive inhibitor
d. noncompetitive inhibitor
e. competitive inhibitor

16.43 **a.** yes **b.** yes **c.** no **d.** yes **e.** no **f.** yes

16.45 **a.** The secondary structure of a protein depends on hydrogen bonds to form a helix or a pleated sheet; the tertiary structure is determined by the interaction of R groups and determines the three-dimensional structure of the protein.
b. Nonessential amino acids can be synthesized by the body; essential amino acids must be supplied by the diet.
c. Polar amino acids have hydrophilic side groups, whereas nonpolar amino acids have hydrophobic side groups.
d. Dipeptides contain two amino acids, whereas tripeptides contain three.
e. An ionic bond is an interaction between a basic and acidic side group; a disulfide bond links the sulfides of two cysteines.
f. Fibrous proteins consist of three to seven α-helixes coiled like a rope. Globular proteins form a compact spherical shape.
g. The α-helix is the secondary shape like a staircase or corkscrew. The β-pleated sheet is a secondary structure that is formed by many protein chains side by side like a pleated sheet.
h. The tertiary structure of a protein is its three-dimensional structure. The quaternary structure involves the grouping of two or more peptide units for the protein to be active.

16.47 **a.** The albumin, which has a negative charge.
b. The lysozyme, which has a positive charge.
c. The hemoglobin, which is neutral because the pH of the buffer is the same as the isoelectric point of hemoglobin.

Chapter 17

17.1 adenine (A), thymine (T), cytosine (C), guanine (G), and deoxyribose

17.3 deoxyadenosine monophosphate, deoxythymidine monophosphate, deoxycytidine monophosphate, and deoxyguanosine monophosphate

17.5 **a.** nucleoside **b.** nucleoside
c. nucleoside **d.** nucleotide

17.7

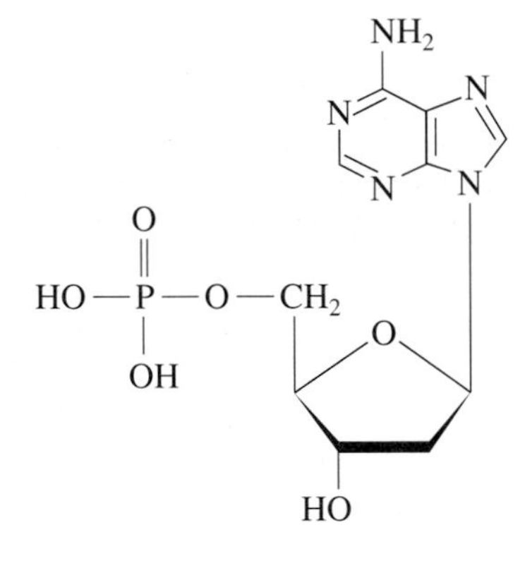

17.9 The phosphate group of each nucleotide connects to the hydroxyl group on carbon 3 of the next nucleotide.

17.11 The sugar.

17.13

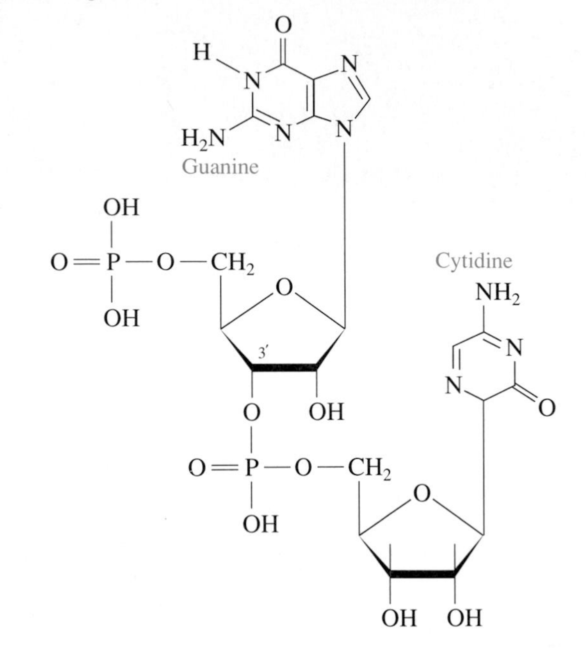

17.15 **a.** TTTTTT **b.** CCCCCC **c.** TCAGGTCCA
d. GACATATGCAAT

17.17 The DNA strands separate and the DNA polymerase pairs each of the bases with its complementary base and produces two exact copies of the original DNA.

17.19 The pairs of bases along the DNA strands are separated and paired with new bases.

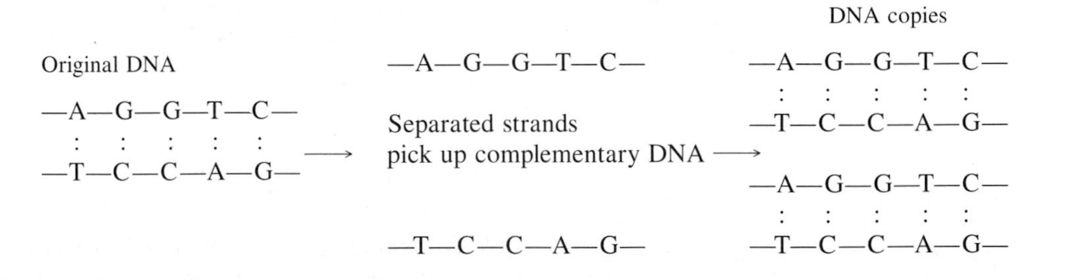

17.21 ribosomal RNA, messenger RNA, and transfer RNA

17.23 In the polypeptide chain of a tRNA, hydrogen bonds between some of the bases produce loops that give a cloverleaf structure.

17.25 In transcription, the sequence of nucleotides on a DNA template (one strand) is used to produce the base sequences of a messenger RNA.

17.27
—G—G—C—U—U—C—C—A—A—G—U—G—

17.29 A three-base sequence in mRNA that codes for a specific amino acid in a protein.

17.31 **a.** leucine
b. serine
c. glycine
d. arginine

17.33 When AUG is the first codon, it signals the start of protein synthesis. Thereafter, AUG codes for methionine.

17.35 A codon is a base triplet in the mRNA template. An anticodon is the complementary triplet on a tRNA for a specific amino acid.

17.37 Initiation, elongation, and termination.

17.39 **a.** —Lys—Lys—Lys—

b. —Phe—Pro—Phe—Pro—

c. —Tyr—Gly—Arg—Cys—

17.41 The amino acid in the A site is joined by a peptide bond to the peptide chain in the P site. The peptide moves to the P site and a new amino acid attaches to the A site.

17.43

a. —C—G—A—A—A—A—G—U—U—U—U—U—

b. GCU, UUU, CAA, AAA

c. Using codons in mRNA: Arg—Lys—Val—Phe

17.45 A base in DNA is replaced by a different base.

17.47 If the resulting codon still codes for the same amino acid, there is no effect. If the new codon codes for a different amino acid, there is a change in the order of amino acids in the polypeptide.

17.49 The normal triplet TTT forms a codon AAA, which codes for lysine. The mutation TTC forms a codon AAG, which also codes for lysine. There is no effect on the amino acid sequence.

17.51 **a.** —Thr—Ser—Arg—Val—

b. —Thr—Thr—Arg—Val—

c. —Thr—Ser—Gly—Val—

d. —Thr—STOP Protein synthesis would terminate early. If this occurs early in the formation of the polypeptide, the resulting protein will probably be nonfunctional.

17.53 **a.** GCC and GCA both code for alanine.

b. A vital ionic cross–link in the tertiary structure of hemoglobin cannot be formed when the polar glutamine is replaced by valine, which is nonpolar. The resulting hemoglobin is malformed and less capable of carrying oxygen.

17.55 An operon is a section of DNA that regulates the synthesis of one or more proteins.

17.57 In enzyme induction, protein synthesis is repressed when the regulatory gene produces a repressor protein that binds to the operator and inhibits the ability of the RNA polymerase to produce mRNA.

17.59 Lactose induced the synthesis of the enzymes for its metabolism.

17.61 Tryptophan (Trp) and methionine (Met) each has only a single codon. Leucine (Leu), serine (Ser), and arginine (Arg) each have six codons.

17.63 An mRNA could be:
UAU—GGU—GGU—UUU—CUU

Chapter 18

18.1 In metabolism, a catabolic reaction breaks apart complex molecules, releasing energy for the synthesis of ATP.

18.3 ATP, adenosine triphosphate, is composed of adenine, ribose, and three phosphate groups. In ADP, there are two phosphate groups.

$\text{ADP} + \text{P}_i + 7.3 \text{ kcal/mol} \rightarrow \text{ATP} + \text{H}_2\text{O}$

18.5 $2 \text{ ADP} + 2 \text{ P}_i + 14.6 \text{ kcal} \longrightarrow 2 \text{ ATP} + 2\text{H}_2\text{O}$

18.7 Digestion breaks down the large molecules in food to smaller compounds that can be absorbed by the body.

18.9 **a.** small intestine, sucrase, glucose, and fructose
b. stomach, pepsin, small intestine, proteases, amino acids
c. mouth, α-amylase, dextrins, maltose; small intestine, pancreatic α-amylase, maltose

18.11 glucose

18.13 ATP is required in phosphorylation steps.

18.15 glyceraldehyde 3-phosphate and dihydroxyacetone phosphate

18.17 ATP is produced in glycolysis by transferring a phosphate from 1,3-phosphoglycerate and from phosphoenol phosphate directly to ADP.

18.19 **a.** hexokinase **b.** phosphokinase

18.21 **a.** 1 ATP required
b. One (1) NADH is produced for each triose.
c. two ATP and two NADH

18.23 $CH_3C(=O)COO^- + NAD^+ + CoA \longrightarrow$
Pyruvate

$CH_3C(=O)—CoA + CO_2 + NADH$
Acetyl CoA

18.25 The formation of lactate makes NAD^+ available for the continuation of glycolysis.

18.27 The enzymes and coenzymes for the citric acid cycle are located in the mitochondria.

18.29 **a.** nicotinamide from niacin, adenine, ribose, and two phosphate groups
b. NAD^+ oxidizes hydroxyl groups to carbon–oxygen double bonds.
c. NAD^+ is converted to $NADH + H^+$

18.31 **a.** citrate and isocitrate
b. In a decarboxylation, a carbon atom is lost as CO_2.
c. 3 NADH
d. α-ketoglutarate
e. isocitrate → α-ketoglutrate; α-ketoglutrate → succinyl CoA; succinate → fumarate; malate → oxaloacetate
f. steps 3, 8

18.33 **a.** citrate synthetase
b. succinate dehydrogenase
c. fumarase

18.35 **a.** NAD^+
b. GDP
c. FAD

18.37 NADH and $FADH_2$

18.39 **a.** reduced NADH, oxidized NAD^+
b. reduced $FMNH_2$, oxidized FMN

18.41 FAD, CoQ, cyt *c*

18.43 Electrons are transferred from NADH to FMN; three ATP are produced.

18.45 In oxidative phosphorylation, the energy from the oxidation reactions in the electron transport system is used to add a phosphate to ADP for ATP production.

18.47 The release of energy from oxidation causes an accumulation of protons that creates an energy-rich proton gradient. As protons interact with ATPase, energy is released to phosphorylate ADP.

18.49 Reactions in glycolysis and the citric acid cycle produce reduced coenzymes NADH and $FADH_2$ that are oxidized by the acceptors in the electron transport chain.

18.51 **a.** 3 ATP **b.** 2 ATP **c.** 6 ATP **d.** 12 ATP

18.53 **a.** Glycerol is converted to dihydroxyacetone phosphate, a compound that can enter the glycolysis pathway for oxidation.
b. Fatty acids are activated for β-oxidation in the cytoplasm.
c. two ATP
d. For transport to the mitochondria for β-oxidation.

18.55 **a.** $CH_3—(CH_2)_{10}—C(=O)—CoA$

b. $CH_3—(CH_2)_8—\underset{\beta}{CH_2}—\underset{\alpha}{CH_2}—C(=O)—CoA$

c. $CH_3—(CH_2)_8—\underset{\beta}{CH_2}—\underset{\alpha}{CH_2}—C(=O)—CoA + FAD \longrightarrow$

$CH_3—(CH_2)_8—\underset{\beta}{CH}=\underset{\alpha}{CH}—C(=O)—CoA + H_2O \longrightarrow$

$CH_3—(CH_2)_8—\underset{\beta}{CH(OH)}—\underset{\alpha}{CH_2}—C(=O)—CoA + NAD^+ \longrightarrow$

$CH_3—(CH_2)_8—C(=O)—CH_2—C(=O)—CoA + CoA \longrightarrow$

$CH_3—(CH_2)_8—C(=O)—CoA + CH_3—C(=O)—CoA$

d. six acetyl CoA

e. five cycles

f.

6 acetyl CoA × 12 ATP/acetyl CoA ⟶	72 ATP
5 $FADH_2$ × 2 ATP/$FADH_2$ ⟶	10 ATP
5 NADH × 3 ATP/NADH ⟶	15 ATP
Activation of fatty acid ⟶	−2 ATP
	95 ATP

18.57 When large quantities of fats are oxidized, the excess acetyl CoA forms four-carbon ketones through ketogenesis. Conditions leading to ketosis include diabetes, high fat and low carbohydrate diets, and starvation. Because ketone bodies are acidic, blood pH is lowered and may hinder the ability of the blood to carry oxygen. A person in ketosis may experience increased urine output, sodium depletion, and a sensation of thirst.

18.59 α-keto acids and other amino acids

18.61

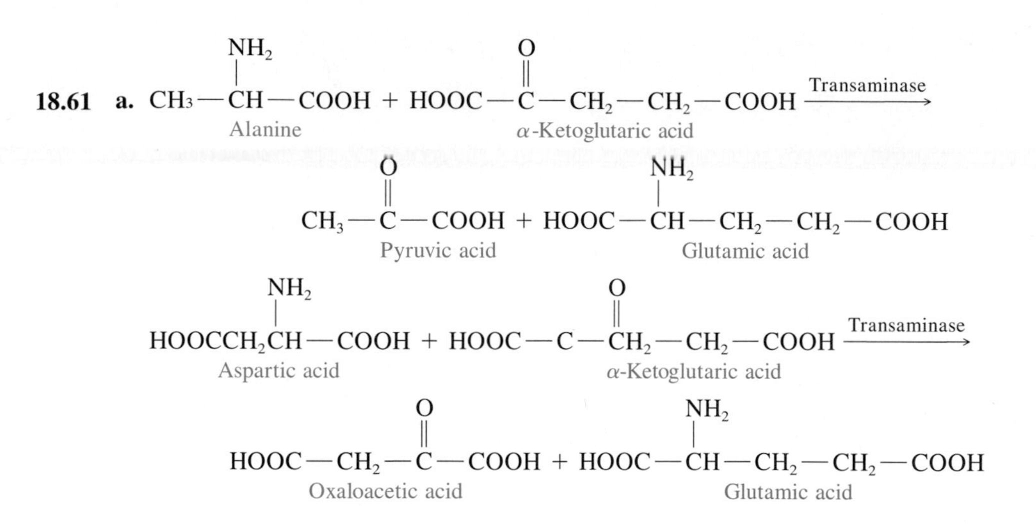

18.63 Amino acids can undergo transamination to α-keto acids such as pyruvate, oxaloacetate, and α-ketoglutarate, intermediates of the citric acid cycle that oxidize to produce ATP.

18.65 NAD^+ accepts H in oxidizations that form C=O bonds, but FAD is used when C=C bonds form.

18.67 300 kcal × 1 mol ATP/7.3 kcal = 41 mol ATP

41 mol ATP × 1 glucose/36 mol ATP = 1.1 mol glucose

1.1 mol glucose × 180 g/mol = 200 g glucose (rounded)

18.69 The reduced coenzymes NADH and $FADH_2$ must be regenerated by the electron transport chain, which operates by transferring electrons to O_2.

18.71 Acetate can be converted to acetyl CoA and enter the citric acid cycle to be oxidized.

Acknowledgments

Illustration Credits

Unless otherwise acknowledged, all illustrations are the property of HarperCollins Publishers.

Figures 2.1, 3.13. From SCOTTFORESMAN SCIENCE: DISCOVER THE WONDER, Grade 5. Copyright © 1993, 1994 Scott, Foresman and Company, Glenview, Illinois. Reprinted by permission.

Figures 15.9, 17.6, 17.9, 17.11, 17.12, 17.13, 17.15, 17.22, 17.23, 18.13, From BIOLOGY: THE NETWORK OF LIFE by Michael C. Mix, Paul Farber, and Keith I. King. Copyright © 1992 by Michael C. Mix, Paul Farbar, and Keith I. King. Reprinted by Permission.

Photo Credits

Unless otherwise acknowledged, all photographs are the property of Scott, Foresman and Company. Pages abbreviations are as follows: (T) top, (C) center, (B) bottom, (L) left, (R) right.

xi: Tom Stewart/The Stock Market
xii: Hans Pfletschinger/Peter Arnold, Inc.
xiv: Jan Kapec/Tony Stone Images
xv: Richard R. Hansen/Photo Researchers
xvi: Al Assid/The Stock Market
xix: Douglas Struthers/Tony Stone Images
xxxi: Courtesy, Karen Timberlake
1: Jay Freis/The Image Bank
4: SIU/Visuals Unlimited
8L: Seth Resnick/Stock Boston
8R: C. C. Duncan/Medical Images, Inc.
30: Tom Pantages
31: David R. Frazier Photolibrary
38: Tom Stewart/The Stock Market
49R: Custom Medical Stock Photo
51: Courtesy of U.S. Dept. of Agriculture
53: Daniel Aubry/The Stock Market
54T: William McCoy/Rainbow
54C and R: Runk/Schoenberger/Grant Heilman Photography
54B: Judi Buie/Bruce Coleman Inc.
57T: F. Stevenson/Photo Researchers
57B: Bruce Barthel/The Stock Market
72: Bob Burch/Bruce Coleman Inc.
77: Lawrence Berkeley Laboratory, Courtesy University of Califormia
87ALL: Andy Washnik for HarperCollins Publishers
94: National Museum of American History/Smithsonian Institution
96: Anderson Cancer Center, University of Texas
107: Dan McCoy/Rainbow
112: Robert E. Lyons/Visuals Unlimited
120L: Barry L. Runk/Grant Heilman Photography
120C: James Pfletschinger/Peter Arnold, Inc.
124ALL: Professor P. Motta/Department of Anatomy/University "La Sapienza," Rome/Science Photo Library/Photo Researchers
148: Bonnie McGrath/Rainbow
153: Andy Washnik for HarperCollins Publishers
161TL: Tom Pantages
161TR: Tom Pantages
161 (all photos except TL and TR): Andy Washnik for HarperCollins Publishers
165: Leonard Lee Rue IV/Bruce Coleman Inc.
183: Howard Sochurek/Medical Images Inc.
188: John Coletti/Stock Boston
195: David Parker/Science Photo Library/Photo Researchers
201ALL: S.I.U./Visuals Unlimited
202: Dan McCoy/Rainbow
203L Hank Morgan/Rainbow
203R: Courtesy of Drs. Michael E. Phelps and John C. Mazziotta U.C.L.A.
204: Yoay Levy/Phototake
210: Jessica Ehler/Bruce Coleman Inc.
217: Uniphoto
222: Matt Meadows/Peter Arnold, Inc.
230: Jan Kapec/Tony Stone Images
242: S.I.U./Visuals Unlimited
250: Roger Tully/Tony Stone Images
251: Tom Pantages
253: George I. Bernard/ANIMALS ANIMALS
257: Andy Washnik for HarperCollins Publishers
261T and C: Andy Washnik for HarperCollins Publishers
261B: CNRI/Science Photo Library/Photo Researchers
262: Andy Washnik for

HarperCollins Publishers
271: Matt Meadows/Peter Arnold, Inc.
279: S.I.U./Bruce Coleman Inc.
287: Raymond G. Barnes/Tony Stone Images
309L: NYC Parks Photo Archive/ Fundamental Photographs
309R: Kristen Brochnann/ Fundamental Photographs
314: Andy Washnik for HarperCollins Publishers
322: Uniphoto
326: A. Limont/Bruce Coleman Inc.
343: Stephen Kline/Bruce Coleman Inc.
345: Richard R. Hansen/Photo Researchers
346: NASA
347: Michael Newman/Photo Edit
359: Charles Thatcher/Tony Stone Images
368: Runk/Schoenberger/Grant Heilman Photography
372ALL: Tom Pantages
376L: S.I.U./Visuals Unlimited
376R: Ellen Harmon/Medical Images, Inc.
379: L. L. T. Rhodes/Tony Stone Images
384: Dianora Niccolini/Medical Images, Inc.
391: Uniphoto
403: Al Assid/The Stock Market
406: Jim Pickerell/Tony Stone Images
412: Tom Pantages
414: Frank Siteman/Uniphoto
419: Tom Pantages
431: Ken Reid/FPG
433: DonaldSpecker/ANIMALS ANIMALS
448: Grant Heilman Photography
449: Matt Meadows/Peter Arnold, Inc.
456: Ed Drews/Photo Researchers
474: Frank Siteman/Uniphoto
485: Tom Pantages
493: Custom Medical Stock Photo
495: John D. Cunningham/Visuals Unlimited
498: Custom Medical Stock Photo
501: Peter Menzel/Stock Boston
503: Wendy Neefus/ANIMALS ANIMALS
512: Philip A. Harrington/The Image Bank
524: Jim Cummins/FPG
531: C. Raines/Visuals Unlimited
533ALL: National Heart, Lung and Blood Institute/National Institutes of Health
545: Uniphoto
560: Visuals Unlimited
584: Douglas Struthers/Tony Stone Images
591: Will and Deni McIntyre/Photo Researchers
592: CNRI/SPL/Photo Researchers
607: Jessica Ehlers/Bruce Coleman Inc.
613: © 1992 by Michael W. Davidson, Institute of Molecular Biophysics, The Florida State University, Tallahassee, Florida. All Rights Reserved.
621: David Madison/Bruce Coleman Inc.
648: Fred Hossler/Visuals Unlimited

Index

Important Common Ions

1+		2+		3+		3−		2−		1−	
H^+	hydrogen ion	Mg^{2+}	magnesium ion	Al^{3+}	aluminum ion	N^{3-}	nitride ion	O^{2-}	oxide ion	F^-	fluoride ion
Li^+	lithium ion	Ca^{2+}	calcium ion			P^{3-}	phosphide ion	S^{2-}	sulfide ion	Cl^-	chloride ion
Na^+	sodium ion	Sr^{2+}	strontium ion							Br^-	bromide ion
K^+	potassium ion	Ba^{2+}	barium ion							I^-	iodide ion
Ag^+	silver ion	Zn^{2+}	zinc ion								

Important Ions with Variable Valence

1^+ or 2^+		2^+ or 3^+		1^+ or 3^+		2^+ or 4^+	
Cu^+	copper (I) ion	Fe^{2+}	iron (II) ion	Au^+	gold (I) ion	Sn^{2+}	tin (II) ion
Cu^{2+}	copper (II) ion	Fe^{3+}	iron (III) ion	Au^{3+}	gold (III) ion	Sn^{4+}	tin (IV) ion
						Pb^{2+}	lead (II) ion
						Pb^{4+}	lead (IV) ion

Important Polyatomic Ions

Cation		Anions					
1+		1−		2−		3−	
NH_4^+	ammonium ion	HCO_3^-	hydrogen carbonate (or bicarbonate) ion	CO_3^{2-}	carbonate ion	PO_4^{3-}	phosphate ion
		HSO_4^-	hydrogen sulfate (or bisulfate) ion	SO_4^{2-}	sulfate ion		
		HSO_3^-	hydrogen sulfite (or bisulfite) ion	SO_3^{2-}	sulfite ion		
		NO_3^-	nitrate ion				
		NO_2^-	nitrite ion				
		ClO_3^-	chlorate ion				
		ClO_2^-	chlorite ion				
		OH^-	hydroxide ion				

Some Useful Conversion Factors

Length
1 meter (m) = 100 centimeters (cm)
1 meter = 1000 millimeters (mm)
1 meter = 39.4 inches (in.)
1 inch = 2.54 cm

Volume
1 liter (L) = 1000 milliliters (mL) = 1000 cm^3
1 L = 1.06 quarts (qt)
1 qt = 946 mL

Mass
1 kilogram (kg) = 1000 grams (g)
1 kg = 2.20 pounds (lb)
1 lb = 454 g

Temperature
$°F = 1.8°C + 32$
$°C = \frac{(°F - 32)}{1.8}$
$K = °C + 273$

Gases
1 atm = 760 mm Hg
1 mole (STP) = 22.4 L

H_2O
density = 1.00 g/1 mL
melt: 80. cal/g
boil: 540 cal/g
SH = 1 cal/g°C

Energy
1 Cal = 1 kcal = 1000 cal
1 calorie (cal) = 4.18 joules(J)